T0136234

**THE UNIVERSITY OF ARIZONA SPACE SCIENCE SERIES**
Tom Gehrels, General Editor

**Planets, Stars and Nebulae, Studied with Photopolarimetry**
Tom Gehrels, editor, 1974, 1133 pages

**Jupiter**
Tom Gehrels, editor, 1976, 1254 pages

**Planetary Satellites**
Joseph A. Burns, editor, 1977, 598 pages

**Protostars and Planets**
Tom Gehrels, editor, 1978, 756 pages

**Asteroids**
Tom Gehrels, editor, 1979, 1181 pages

**Comets**
Laurel L. Wilkening, editor, 1982, 766 pages

**Satellites of Jupiter**
David Morrison, editor, 1982, 972 pages

**Venus**
D. M. Hunten, L. Colin, T. M. Donahue, and V. I. Moroz, editors, 1983, 1143 pages

**Saturn**
Tom Gehrels and Mildred S. Matthews, editors, 1984, 968 pages

**Planetary Rings**
Richard Greenberg and André Brahic, editors, 1984, 784 pages

**Protostars and Planets II**
David C. Black and Mildred S. Matthews, editors, 1985, 1293 pages

**Satellites**
Joseph A. Burns and Mildred S. Matthews, editors, 1986, 1021 pages

**The Galaxy and the Solar System**
Roman Smoluchowski, John N. Bahcall, and Mildred S. Matthews, editors, 1986, 483 pages

**Meteorites and the Early Solar System**
John F. Kerridge and Mildred S. Matthews, editors, 1988, 1269 pages

**Mercury**
Faith Vilas, Clark R. Chapman, and Mildred S. Matthews, editors, 1988, 794 pages

**Origin and Evolution of Planetary and Satellite Atmospheres**
S. K. Atreya, J. B. Pollack, and Mildred S. Matthews, editors, 1989, 1269 pages

**Asteroids II**
Richard P. Binzel, Tom Gehrels, and Mildred S. Matthews, editors, 1989, 1258 pages

**Uranus**
Jay T. Bergstralh, Ellis D. Miner, and Mildred S. Matthews, editors, 1991, 1076 pages

**The Sun in Time**
C. P. Sonett, M. S. Giampapa, and M. S. Matthews, editors, 1991, 990 pages

**Solar Interior and Atmosphere**
A. N. Cox, W. C. Livingston, and M. S. Matthews, editors, 1991, 1416 pages

**Mars**
Hugh H. Kieffer, Bruce M. Jakosky, Conway W. Snyder, and Mildred S. Matthews, 1992, 1498 pages

**Protostars and Planets III**
Eugene H. Levy and Jonathan I. Lunine, editors, 1993, 1596 pages

**Resources of Near-Earth Space**
John S. Lewis, Mildred S. Matthews, and Mary L. Guerrieri, editors, 1993, 977 pages

**Hazards Due to Comets and Asteroids**
Tom Gehrels, editor, 1994, 1300 pages

**Neptune and Triton**
Dale P. Cruikshank, editor, 1995, 1249 pages

**Cosmic Winds and the Heliosphere**
J. R. Jokipii, C. P. Sonett, and M. S. Giampapa, editors, 1997, 1013 pages

**Venus II—Geology, Geophysics, Atmosphere, and Solar Wind Environment**
S. W. Bougher, D. M. Hunten, and R. J. Phillips, editors, 1997, 1376 pages

**Pluto and Charon**
S. Alan Stern and David J. Tholen, editors, 1997, 728 pages

**Protostars and Planets IV**
Vincent Mannings, Alan P. Boss, and Sara S. Russell, editors, 2000, 1422 pages

**Origin of the Earth and Moon**
R. M. Canup and K. Righter, editors, 2000, 555 pages

# ORIGIN OF THE
# EARTH AND MOON

# ORIGIN OF THE EARTH AND MOON

**R. M. Canup**
**K. Righter**

Editors

With 69 collaborating authors

THE UNIVERSITY OF ARIZONA PRESS
Tucson

in collaboration with

LUNAR AND PLANETARY INSTITUTE
Houston

*About the front cover:*

The Earth-Moon system in its formative stages, some years to a century after the hypothetical collision that led to lunar formation. Earth is still surrounded by a remnant ring of debris inside the Roche limit. Outside the Roche limit, much of the material has aggregated into the proto-Moon, which is covered by a magma ocean and a patchy crust broken by impacts. Earth, reforming after the giant impact, also has regions of magma ocean and displays a ragged hemispheric fracture pattern that may be a residual of the Moon-forming impact (or a smaller, later impact?). Earth's surface is partly screened by clouds that are arranged not in familiar cyclonic storm patterns but in latitudinal belts, associated with Earth's rapid early rotation of 5–6 hours. In the background is the early Sun and a pattern of zodiacal light that is much stronger than today's zodiacal light, due to the high density of interplanetary dust and debris left over from the solar nebula, planetesimal collisions, and early cometary activity. Painting by William K. Hartmann.

*About the back cover:*

The Earth and Moon, as seen by the Near Earth Asteroid Rendezvous (NEAR) spacecraft in 1998. Credit: NASA and the National Space Science Data Center.

The University of Arizona Press
© 2000 The Arizona Board of Regents
First printing
All rights reserved
∞ This book is printed on acid-free, archival-quality paper.
Manufactured in the United States of America

Library of Congress Cataloging-in-Publication Data
Origin of the earth and moon / R. M. Canup, K. Righter, editors ; with 69 collaborating
authors.
p. cm. — (The University of Arizona space science series)
Includes bibliographical references and index.
ISBN 0-8165-2073-9 (acid-free)
1. Earth—Origin. 2. Moon—Origin. I. Canup, R. M. (Robin M.), 1968–  II. Righter, K.
(Kevin), 1965–  III. Series.
QB632.O76 2000
525—dc21
00-056386

# Contents

# Collaborating Authors

Abe Y., 413
Agnor C. B., 113
Amend J. P., 527
Bar-Nun A., 459
Borg L. E., 361
Canup R. M., 113, 145
Carlson R. W., 25
Cameron A. G. W., 133
Cassen P., 3
Dones L., 101, 493
Drake M., 413
Ertel W., 265
Floss C., 339
Gessmann C. K., 245
Grinspoon D., 493
Halliday A. N., 45
Hartmann W. K., 493
Hillgren V. J., 245
Holzheid A., 265
Hood L. L., 397
Humayun M., 3
Ida S., 145

Jacobsen S. B., 45
Jones J. H., 197
Koeberl C., 475
Kokubo E., 85, 145
Kortenkamp S., 85
Lee D.-C., 45
Li J., 245
Lissauer J. J., 101
Lugmair G. W., 25
Minarik W. G., 227
Mojzsis S., 475
Newsom H. E., 265
Nyquist L. E., 361
Okuchi T., 413
Ohtani E., 413
Ohtsuki K., 101
Owen T. C., 459
Ozima M., 63
Palme H., 197
Pepin R. O., 435
Podosek F. A., 63
Pollard D., 513

Porcelli D., 435
Pritchard M. E., 179
Righter K., 291, 413
Rushmer T., 227
Ryder G., 475, 493
Shearer C. K., 339
Shock E. L., 527
Snyder G. A., 361
Solomatov V. S., 323
Stevenson D. J., 179
Stewart G. R., 217
Taylor G. J., 227
Taylor L. A., 361
Touma J., 165
Walker R. J., 291
Walter M. J., 265
Ward W. R., 75
Warren P. H., 291
Weidenschilling S. J., 85
Williams D. M., 513
Zolotov M. Y., 527
Zuber M. T., 397

# Acknowledgment of Reviewers

*The editors gratefully acknowledge the following individuals for their
time and effort in reviewing chapters for this volume:*

Y. Abe
A. Anbar
A. Boss
R. Brett
R. W. Carlson
J. Chambers
L. Dones
M. J. Drake
G. Gaetani
S. J. Galer
E. Gaidos
R. Grieve
F. Guyot

A. Harris
E. Hauri
P. Hess
A. Holzheid
M. Humayun
S. Ida
J. H. Jones
J. Kasting
J. Lissauer
J. W. Morgan
S. Mojzsis
H. Newsom
L. Nyquist

U. Ott
T. Owen
M. Ozima
H. Palme
S. Peale
R. O. Pepin
R. Phillips
E. Pierazzo
S. Solomon
S. Sasaki
C. K. Shearer
J. Shervais
G. Snyder

F. Spera
D. Stevenson
G. Stewart
T. Swindle
S. R. Taylor
D. Trilling
M. Wadhwa
S. Weidenschilling
G. Wetherill
K. Zahnle

# *Preface*

The origin of the Earth-Moon system is one of the longstanding questions in planetary science. The "giant impact hypothesis," which has received increasingly widespread support over the past 15 years, implies that the origins of the Earth and Moon are fundamentally coupled by a collision late in Earth's formation history. This catastrophic event greatly affected the initial thermal, chemical, and dynamical state of the Earth-Moon system, leaving signatures still observable today. The subjects of the Earth and Moon have traditionally been reviewed separately; indeed, the most recent volume addressing both bodies is Ringwood's 1979 work. However, given what is now believed to be their common origin, it is clear that a discussion of the coupled formation and early evolution of both the Earth and Moon is overdue.

In December 1998, a conference entitled "Origin of the Earth and Moon" was held in Monterey, California. Organizers Michael J. Drake and Alex N. Halliday deserve much credit for bringing together people from many diverse fields to this highly interdisciplinary meeting, the first to jointly address both the Earth and Moon. The Monterey meeting was inspired by two very successful prior conferences: one in 1984 concerning the Moon (held in Kona, Hawai'i; see *Origin of the Moon,* 1986, edited by W. K. Hartmann, R. J. Phillips, and G. J. Taylor, Lunar and Planetary Institute, Houston) and another in 1988 concerning the Earth (held in Berkeley, California; see *Origin of the Earth,* 1990, edited by H. E. Newsom and J. H. Jones, Oxford University, New York). It was at the Kona meeting that the impact hypothesis (originally proposed nearly a decade earlier by Hartmann and Davis, 1975, and Cameron and Ward, 1976) first emerged as the clear frontrunner of the lunar origin theories, winning favor over the capture, coaccretion, and fission scenarios. At the Berkeley meeting, potential effects of the impact scenario on the Earth were considered, and in particular the question was raised of how to reconcile melting of the Earth by a giant impact with evidence that the Earth accreted cool and never melted to great depth.

The 1998 Monterey conference was lively and interactive, involving intense discussion of a decade of experimental, theoretical, and computational modeling related to the origin and early history of the Earth and Moon. An overall outcome of the Monterey meeting was a growing realization of the number (and complexity) of remaining outstanding issues concerning an impact-triggered origin of the Earth-Moon system. While the impact theory remains clearly favored, the original hypothesis has been somewhat revised. The scenario that has emerged involves an impact that occurred about 50 m.y. after the birth of the solar system, between an object with two to three times the mass of Mars and a proto-Earth that was only partially accreted. Such a collision is typical of those predicted by models of terrestrial accretion. The Moon accumulated quite rapidly (on a timescale of 1–100 years) from the impact-produced debris at an orbital distance of about 3–5 Earth radii. The energetics of accretion and core formation were high enough that both bodies maintained magma ocean systems — the Moon's was dry due to low-pressure and high-temperature accretion conditions, but the Earth's was likely volatile-rich ($H_2O$ and $CO_2$), as the great pressures at the base of the magma ocean and a thick early atmosphere would favor dissolution of volatiles into magma. Accretion onto the

Earth and Moon subsequent to lunar formation may have supplied additional siderophile-rich material to the Earth, while the largest later impacts onto the Moon left their signatures in the form of possibly temporally clustered impact basins. The later dynamical evolution of the Earth-Moon system can be reconciled with the Moon's current orbit, and reveals an important interplay between the Moon and the Earth's relatively stable obliquity.

Compositional constraints on this new scenario have emerged from both petrologic and isotopic studies. First, technical advances in mass spectrometry have allowed measurement of the isotopes of tungsten in a variety of terrestrial, lunar, and planetary materials, thus allowing application of the tungsten-hafnium short-lived chronometer to the topic of planetary accretion. Such measurements have provided new constraints on the timing of formation for the Moon and core formation in the Moon and Earth. In particular, they suggest that the Moon formed earlier than the Earth, 50 m.y. after the start of accretion. Second, an explosion of high-pressure experimentation in metal-silicate systems has forced a reevaluation of traditional core formation models for the Earth and Moon. Specifically, the "excess siderophile-element problem" of Earth's primitive upper mantle, long attributed to heterogeneous accretion or disequilibrium processes, can instead be alleviated by high-temperature and high-pressure metal-silicate equilibrium in the early Earth. The presence of terrestrial and lunar magma oceans (the latter long favored by petrologists) is consistent with a hot, early Earth-Moon system. Third, the discovery that comets have distinctly different deuterium-to-hydrogen ratios than terrestrial ocean water has forced consideration of the presence of water during accretion, instead of the previous assumption of dry accretion with later addition of cometary water.

New constraints have also arisen from dynamical modeling of the giant impact hypothesis. At the time of the Kona conference, only preliminary simulations of the impact event had been conducted. However, over the past decade vast increases in computational speed have led to greatly improved hydrodynamic models of large-impact events, with increasing resolution of the ejected protolunar material. The first generation of models describing how such material can accumulate to yield the Moon has revealed that accretion from a protolunar disk is quite inefficient, although the formation of a single moon appears likely. Together, the impact and disk-evolution models have placed significant new constraints on the type of impact that could produce the Moon. In particular, it has been found that an impact involving a total mass and angular momentum equal to that of the current Earth-Moon system (the originally envisioned impact scenario) does not yield a sufficiently massive moon. Instead, the successful models to date involve either an Earth with only about 60% of its current mass, or an impact with about twice the angular momentum of the Earth-Moon system.

Although the above-described scenario appears consistent with much of the data we have at hand, it is not without difficulty. There remains uncertainty about the specific way in which the Moon has attained its radiogenic tungsten relative to the Earth. Although much new work has emphasized the importance of high pressures and temperatures on metal-silicate equilibrium and core formation, very little data exist to evaluate alternative models for siderophile-element concentrations in the Earth's deep mantle. There is at this time very little understanding of the effect of water on phase relations, core compositions, and dynamics in the early Earth, and the history of the Earth's water may well involve both local and cometary sources. Simulations of potential moon-forming impacts have only spanned a small range of possible parameter space, and in particular have not yet investigated the effects of preimpact planetary spin. Dynamical models of lunar accretion that account for the thermal evolution of the protolunar disk have yet to be undertaken. And all the current impact models that yield a lunar-sized Moon require that the Earth-Moon system be significantly modified after the lunar-forming impact. We hope that these and the many other issues raised in this book help to shape and inspire fruitful research plans.

We wish to thank the many authors and reviewers who participated in this project, as well as the Lunar and Planetary Institute for their help both during the meeting in Monterey and during the production of this book. We are in particular indebted to Renée Dotson and Stephen Hokanson, whose patience and perseverance were central to this entire project. We also thank Christine Szuter and Anne Keyl of the University of Arizona Press. Special acknowledgements are due to Mike Drake and the editor-in-chief of the series, Tom Gehrels, for support and guidance in all stages of this project. We also wish to thank the NASA Origins of Solar Systems and Planetary Geology & Geophysics programs for their generous support of some of the publication costs for this volume.

*Robin Canup and Kevin Righter*
*May 2000*


*"Lady, by yonder blessed moon I vow*
*That tips with silver all these fruit-tree tops"*

"O swear not by the moon, inconstant moon,
That monthly changes in her circled orb,
Lest that thy love prove likewise variable."

*"What shall I swear by?"*

"Do not swear at all,
Or if thou wilt, swear by thy gracious self"

William Shakespeare
*Romeo and Juliet*; Act II, Scene 2

# Part I:
## Isotopic and Chemical Constraints on Accretion

# Processes Determining the Volatile Abundances of the Meteorites and Terrestrial Planets

**Munir Humayun**
*The University of Chicago*

**Patrick Cassen**
*NASA Ames Research Center*

The depletion of moderately volatile elements in chondrites is interpreted in terms of partial condensation from a cooling solar nebula. The patterns are evidence for an initially hot nebula (T ≈ 1400 K, condensation temperature of major elements) in which elements more volatile than Mg silicates and Fe-Ni alloy were present initially in the vapor. Models seeking to account for these depletions by localized heating of dust (e.g., during chondrule formation) produce K-isotopic fractionations that are not observed. It remains to be seen if the present model can account for the bulk compositions of the Earth and Moon, which are more depleted in moderately volatile elements than chondrites. It appears that the volatile-depletion processes were similar to those for chondrites (i.e., elemental fractionations are a function of condensation temperature, with $\delta^{41}K \approx 0$) but have proceeded to much greater extent.

Volatile depletion is a first-order chemical effect, occurring at scales of individual meteoritic components (e.g., chondrules) to individual planets, and even at the scale of the solar system. The terrestrial planets (rock + metal) are depleted relative to the outer planets in those constituents present as ice and gas. This chapter will focus on pre-giant-impact processes that controlled the moderately-volatile-element contents of the terrestrial planets, the Moon, and asteroids, particularly nebular processes. This discussion is generally applicable to all terrestrial planets, and provides a framework for later chapters on the chemical consequences of a giant impact on the Earth and Moon. Compositional data are readily available from terrestrial and lunar samples, with limited information from Mars and Venus. Compositional data on the various classes of meteorites provide the basis for a discussion of asteroidal evidence pertaining to volatile depletion. The depletions of the volatile elements appear to be thermodynamically controlled by processes occurring in the solar nebula, and provide evidence of nebular thermal history (*Cassen, 1996*). The objective here is to provide a genetic framework for the interpretation of volatile depletion in terms of the relevant nebular processes.

## 1. SOLAR NEBULA THERMAL STRUCTURE

It is reasonable to suppose that differences in the compositions of the planets reflect different conditions in the primitive solar nebula, the disk of gas and dust from which they all formed. Usually this idea is expressed in terms of variations with radial distance from the Sun, the most obvious discriminator. For instance, *Wasson* (1985) suggests that specific characteristics of primitive meteorites (such as oxidation state) reflect declining nebular temperature and pressure with distance. *Lewis* (1974), in a more quantitative way, postulated that planetary composition was determined by varying degrees of condensation of solids from a nebula in which the temperature declined smoothly from the silicate vaporization point within the orbit of Mercury to the ice condensation point beyond the orbit of Mars. Although one can hardly doubt the fundamental validity of the concept — the terrestrial planets are distinctly volatile poor compared to the outer planets and their satellites — such straightforward ideas do not necessarily fit seamlessly into other aspects of planet-building theories. Solid objects drifted, collided, and were gravitationally scattered; the nebula with which they exchanged mass and otherwise interacted was an evolving system. Nevertheless, temperature is one of the more obvious influences on chemical composition, so it is likely that the history of nebular temperature as a function of distance from the Sun will indeed turn out to be an essential aspect of a comprehensive theory. Thus we begin by outlining what might be learned about nebular thermal history from the theory of accretion disks and astronomical observations of disks around young, Sun-like stars.

The solar nebula was formed by the collapse of rotating interstellar matter (cold gas and dust). As the nebular disk formed, material in its inner regions moved radially inward (thereby contributing to the mass of the Sun), while material in its outer regions moved outward (*Lynden-Bell and Pringle,* 1974). Infall onto the nebular disk lasted for about $10^5$ yr (as inferred from observations of young stellar objects), but the process of mass redistribution continued for much longer, perhaps several times $10^6$–$10^7$ yr (the ages of the oldest revealed T Tauri stars). Mass redistribution in circumstellar disks releases gravitational energy, which heats the disks. It requires that angular momentum be transported in some irreversible way, either by gravitational torques, hydrodynamic (viscous) stresses, or magnetic fields. It is not known which among the several mechanisms that have been proposed were actually responsible for this evolution; therefore it is not yet possible to derive an evolutionary model from first principles. But it is possible to calculate the energy released within the disk as a function of distance from the Sun, as long as the mass flow rate toward the Sun is steady (or only slowly varying), a condition that is likely to prevail in the inner parts of the disk during much of the evolution. The energy released is radiated from the surfaces of the disk at an effective (photospheric) temperature given by

$$T_e^4 = \frac{3GM\dot{M}}{8\pi\sigma R^3} \qquad (1)$$

where $\dot{M}$ is the mass flow toward the Sun (accretion rate), M is the mass of the proto-Sun, R is the distance from the Sun, G is the gravitational constant, and $\sigma$ is the Stefan-Boltzmann constant. This formula was derived by *Lynden-Bell and Pringle* (1974) for a viscous accretion disk, but it does not require that viscosity be the agent of angular momentum transport (see the derivation by *Shu,* 1992).

The photospheric temperature $T_e$ is the temperature at the surface of the nebula. What is of interest for planetary materials is the temperature within the nebula, particularly at the midplane. These temperatures are related to $T_e$, but they depend on how opacity and energy dissipation are distributed with altitude (distance from the midplane) within the nebula. The formula for the midplane temperature $T_m$

$$T_m^4 = \frac{3\tau}{8} T_e^4 \qquad (2)$$

where $\tau$ is the optical depth to the nebular midplane, is valid as long as the nebula is optically thick ($\tau \gg 1$), convective transport is unimportant, and dissipation of energy is proportional to the local mass density, conditions that hold in many circumstances. Thus one derives

$$T_m^4 = \frac{9\tau GM\dot{M}}{64\pi\sigma R^3} \qquad (3)$$

Equation (3) is the basis for many models of nebular thermal structure (reviewed by *Boss,* 1998). The optical depth is frequently taken to be simply $\kappa\Sigma/2$, where $\kappa$ is the midplane Rosseland mean opacity (a function of $T_m$ itself; *Pollack et al.,* 1994; *Henning and Stognienko,* 1996) and $\Sigma(R)$ is the nebular surface density. *Cassen* (1994) derived expressions for $T_m$ that account for variations in $\kappa$ with altitude. To determine thermal evolution, the surface density and mass accretion rate should be derived from a dynamical model that accounts for angular momentum transport,

usually by some prescription for the viscosity (e.g., *Ruden and Pollack,* 1991; *Morfill and Wood,* 1989), or specified in a parameterized way (*Cassen,* 1994, 1996). Furthermore, because the dominant source of opacity is fine dust, the value of $\kappa$ evolved as dust settled and coagulated, which implies that a complete thermal history model must also be coupled to a model for coagulation. Such a synthesis demands considerably more knowledge of the physics of angular momentum transport and coagulation than we now have. Therefore, in what follows, we rely heavily on observations of T Tauri stars as analogs of the early solar system, and as guides to the temporal behavior of the factors in equation (3).

Before proceeding further, it is worth mentioning two factors that are not considered in equation (3). First, the only energy source was assumed to be that due to accretion within the nebula. External energy also heats the disk; illumination directly from the proto-Sun (*D'Alessio et al.,* 1998) or indirectly by sunlight is reprocessed by residual circumstellar material so as to shine back on the disk (e.g., *Natta,* 1993; *D'Alessio et al.,* 1997). While interstellar material still falls on the disk, shock heating at the surfaces also occurs. Based on numerical simulations of collapse, *Boss* (1993) concludes that compressional heating is also significant at that time. These sources enhance nebular temperatures above that predicted by equation (3). A second factor causes equation (3) to overestimate temperatures: the fact that the steady-state assumption is violated in the (outer) parts of the disk where material flows away from the Sun, rather than toward it (see the exact solution for a viscous disk in *Lynden-Bell and Pringle,* 1974). As one might suspect, internal energy release is more important than external heating in determining the midplane temperatures when accretion rates and optical depths are high. Beyond a few AU, however, external illumination often dominates, particularly in older disks (*D'Alessio et al.,* 1998). This subject is discussed further below.

We now turn to observational data to constrain the temporal behavior of the parameters appearing in equation (3). Mass accretion rates from disk to star for T Tauri stars have been derived by several groups (*Basri and Bertout,* 1989; *Valenti et al.,* 1993; *Hartigan et al.,* 1995; *Gullbring et al.,* 1998; *Hartmann et al.,* 1998). The method is based on an analysis of stellar spectra, in which the strengths of absorption and emission features are modeled by a "template" spectrum (adopted from an appropriate nonaccreting star), overlain by emission from the hot gas flowing from disk to star. To derive an accretion rate, the mass and radius of the star must also be modeled. Differences in assumptions and results from the analyses of several groups are discussed in *Gullbring et al.* (1998). In addition, *Hartmann et al.* (1998) exploit an empirical correlation (justified theoretically by *Calvet and Gullbring,* 1998) to derive mass accretion rates from infrared photometry. *Hartmann et al.* (1998) list $\dot{M}$ for 56 T Tauri stars, of which 44 lie between $10^{-9}$ and $10^{-7}$ $M_\odot$/ yr. These authors find a poorly constrained negative correlation with age, with $\dot{M}$ declining with time t as $t^{-\eta}$, 1.5 <

$\eta < 2.8$, in the age range (deduced from positions on the H-R diagram and comparison with theoretical evolutionary tracks) $3 \times 10^5$–$3 \times 10^6$ yr.

These mass accretion rates are considerably less than those that are believed to have occurred for younger, embedded sources. The techniques used to derive $\dot{M}$ cannot be applied to stars that are invisible at optical wavelengths, or even to those ("continuum") stars for which the accretion rate is sufficiently great so as to fill in the absorption lines used to identify spectral type. However, mass infall rates from the remaining stellar envelope onto embedded stars have been estimated, based on the amount of material required in a model envelope to reproduce the infrared spectrum of the object (*Whitney and Hartmann, 1993; Kenyon et al., 1993; Whitney et al., 1997*) and the microwave emission from envelope material (*Hayashi et al., 1993*). These values are typically several times $10^{-6}$ $M_{\odot}$/yr, consistent with the theoretically predicted time required to form a solar mass star (*Shu, 1977*), as long as the infalling material is fed at the same rate through the disk onto the star. In fact, the latter condition does not appear to be generally satisfied (*Kenyon et al., 1993*); rather it seems that the *average* accretion rate through the disk equals the infall rate, but the instantaneous accretion rate may be much greater yet, as in outburst, or smaller by an order of magnitude (see *Bell et al., 1999*). In either case, accretion rates greater than the typical T Tauri value of $10^{-8}$ $M_{\odot}$/yr, by at least a factor of 100, apparently occurred early in the evolution of the disks.

There are no comparable constraints on $\Sigma$. The main reason is that T Tauri disks, as expected for the solar nebula, are optically thick, even to microwave emission, within a few AU of the star. Disk masses have been estimated from the microwave emission of dust in the cool, outer regions, under assumptions regarding the microwave emission properties of the dust and an *ad hoc* surface density distribution (*Beckwith et al., 1990; Beckwith and Sargeant, 1993; Osterloh and Beckwith, 1995*). *Woolum and Cassen (1999)* used these masses and density distributions to calculate Rosseland mean optical depths in the terrestrial planet region, with the further assumption that fine dust exists there in the same relative proportion as it exists in the outer, optically thin regions. They then used the $\dot{M}$ referred to above to calculate midplane temperatures from equation (3), but modified to account for external illumination, as discussed next.

It was noticed by *Basri and Bertout (1989)* and others that T Tauri disks appeared to be more luminous than accretion disk theory predicted, for a given mass accretion rate derived from optical and ultraviolet data. Moreover, the observed spectral energy distribution in the infrared is usually flatter than the value predicted by accretion disk theory (*Beckwith et al., 1990*), indicating that $T_e$ declined less steeply than $R^{-3/4}$, as predicted by the theory. The most likely cause of these discrepancies is the interception of starlight by the disk and illumination by starlight that is reprocessed (absorbed and reemitted, or scattered) by residual circumstellar material (*Kenyon and Hartmann, 1987;*

*Calvet et al., 1994; Natta, 1993; Chiang and Goldreich, 1997; D'Alessio et al., 1998; Bell, 1999*), which heats the surface of the disk. Thus if one interprets the disk temperatures reported by *Beckwith et al. (1990)*, derived from fits to the spectral energy distributions, as $T_e$ due solely to disk accretion, one would derive substantially higher accretion rates than those derived by modeling the stellar accretion rates. *Woolum and Cassen (1999)* assumed that the contribution from external illumination, whatever its source, could be represented as an additional flux represented by $T_{ex}^4$, where the observed temperature (*Beckwith et al., 1990; Osterloh and Beckwith, 1995*) is given by

$$T_{obs}^4 = T_e^4 + T_{ex}^4 \qquad (4)$$

Using equations (3) and (4), *Woolum and Cassen (1999)* calculated $T_m$ for 26 T Tauri stars for which the required data was available. They derived temperatures at 1 AU mainly between 200 and 600 K, falling to 100–300 K at 2.5 AU. Thus it is likely that all but the most volatile substances had condensed in the terrestrial planet region by the time the solar nebula was a million years old. Their results also indicate that the importance of external illumination compared to internal heating increases with radial distance, becoming dominant beyond a few AU, as predicted by the models of *D'Alessio et al. (1998)*. A discussion of the ramifications of deviations from the model assumptions is given by *Woolum and Cassen (1999)*.

The much higher accretion rates inferred for younger objects, together with the likelihood that disks were dustier and more massive (both of which would promote higher optical depths), would result in hotter disks, with midplane temperatures increased by factors of $(100)^{1/4} \sim 3$ or greater. Equation (3), in agreement with detailed thermal structure models of disks at this early epoch (*Boss, 1993; Bell et al., 1997, 1999*), indicates that midplane temperatures sufficient to vaporize silicates would have existed under such conditions. It is not possible to define precisely the time period over which disks cool from these hot states, but several arguments suggest that accretion declined rapidly, probably in less than $10^5$ yr. At $5 \times 10^{-6}$ $M_{\odot}$/yr (a typical infall rate deduced for embedded objects), 0.5 $M_{\odot}$, the mass of a typical T Tauri star, is accumulated in $10^5$ yr. Some T Tauri stars with modest accretion rates have apparent ages as young as $10^5$ yr. In fact, it is likely that disk accretion rates settle down to the values cited above when infall diminishes to the point where the star is visible as an unobscured T Tauri star.

An examination of three relevant timescales also suggests that cooling was rapid. The first is the "evolution" time, which pertains to the diminishment of accretion rate. *Hartmann et al. (1998)* found that a value of the order $10^5$ yr was consistent with their results for accretion rates when compared with inferred stellar ages, although, as mentioned above, the correlation was not particularly good; shorter times may apply to very young stars. The second timescale is the intrinsic "cooling" time associated with the optical depth and thermal inertia of the nebula, i.e., the time in which the nebula would cool if all internal heating suddenly ceased. The cooling time is defined by

$$t_{cool} = \frac{3\Sigma c_v T_m \tau}{32\sigma T_m^4} \qquad (5)$$

For the specific heat $c_v = 9 \times 10^7$ and fiducial values of $\Sigma = 10^4$ g/cm$^2$ and $\tau = 2 \times 10^4$, one finds $t_{cool}(T_m = 1300$ K$) = 430$ yr and $t_{cool}(T_m = 200$ K$) = 10^5$ yr. These values indicate that the nebula readily adjusted to changes in accretion rate, and would have cooled rapidly if mass accretion ceased.

Finally, one can define a "coagulation" time, over which opacity is reduced by the conversion of fine dust to rocks. According to calculations reviewed by *Weidenschilling and Cuzzi* (1993), coagulation of fine dust could have reduced nebular opacity at 1 AU by an order of magnitude or more in $10^3$–$10^4$ yr (and longer by a factor of R$^{3/2}$ at greater radii). This result suggests that nebular cooling could have been driven by coagulation, as well as the reduction of accretion rate or surface density. But the observation that 1-m.y.-old T Tauri stars are still optically thick (*Strom et al.,* 1993) indicates that such a process is not the predominant factor in the long term. Either coagulation slows before the disks become optically thin or collisions among planetesimals supply a sufficient amount of fine dust to maintain the optical depth. [But it is worth remembering that there are also T Tauri stars without optically thick disks (*Walter,* 1986). Perhaps coagulation has proceeded to the point of transparency in these objects.]

One is led to the conclusion that the solar nebula was very likely hot enough at the midplane to vaporize all but the most refractory materials within a few astronomical units of the Sun, but only for a rather short interval during or soon after its formation, lasting ~$10^5$ yr. There is currently no reliable information available regarding the details of cooling as a function of time subsequent to the hot epoch, except for the conclusion that midplane temperatures at 1 AU had probably declined to less than 600 K within about a million years. [The midplane temperatures found by *Woolum and Cassen* (1999) are not correlated with inferred stellar age. Either the approximations used in the model are too crude to reveal a trend, or T Tauri stars evolve at a variety of rates, or both.] The evolution of nebular temperatures in terms of mass accretion rate is illustrated in Fig. 1 (adapted from *Bell et al.,* 1997), which shows a set of theoretical profiles more or less consistent with the ideas presented here. These profiles were calculated for steady state viscous disks, with viscous parameter $\alpha = 10^{-2}$ and opacity due to a full complement of dust in cosmic proportions; they do not include the effects of coagulation or external illumination. One should imagine a rapid decline from the highest accretion rates to values on the order of $10^{-8}$ M$_\odot$/yr after 1 m.y. Note that substantial radial gradients through the terrestrial planet region develop during cooling from the higher accretion rates, a condition possibly favorable for

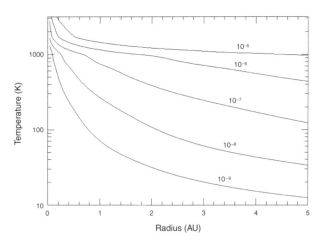

**Fig. 1.** Model nebular midplane temperatures in the terrestrial planet region for accretion rates of $10^{-5}$ to $10^{-9}$ M$_\odot$/yr. It is likely that the accretion rate in the solar nebula was originally at the upper end of this range and declined rapidly to lower values. The model assumes steady, viscous accretion, with viscous parameter $\alpha = 10^{-2}$ (*Bell et al.,* 1997). Figure supplied by K. R. Bell.

producing compositional gradients. The temporal decline, especially at the earliest times, was probably not monotonic. There is substantial evidence that brief (~100 years) but dramatic outbursts (FU Orionis events), caused by instability-induced enhancements of the disk accretion rate, occurred during the earliest phases of disk evolution (e.g., *Hartmann et al.,* 1993). Accretion rates of up to $10^{-4}$ M$_\odot$/yr caused these outbursts. Current theory (*Bell and Lin,* 1994) indicates that (1) only infall-fed disks with background accretion rates higher than about $10^{-7}$ M$_\odot$/yr experience the outburst instability, and (2) only the innermost disk (R $\leq$ 0.2 AU) is subject to the enhanced outburst accretion. Thus the effects of the outbursts at 1 AU and beyond were muted by the facts that the energy was released primarily very close to the star and the temperature at 1 AU was already high before the outburst occurred (see Fig. 1). Outbursts themselves may therefore have had little influence on the material that eventually coagulated to make the planets. For a more thorough discussion, see *Bell et al.* (1999).

We conclude this section with a few general remarks regarding the implications of the observations and conclusions presented above. First, it is notable that many T Tauri stars older than 1 m.y. are still active, in the sense that they are still accreting material from their disks. They cannot be regarded as completely "quiescent." Therefore, if a quiescent phase was required to initiate coagulation, one would expect the raw material to be cold and unfractionated. On the other hand, if the pervasive fractionation patterns observed in chondritic meteorites (*Grossman,* 1996) and planetary compositions reflect the memory of a hot nebula (*Wasson and Chou,* 1974; *Wasson,* 1985; *Palme and Boynton,* 1993; *Cassen,* 1996), coagulation (and the preservation of the products) must have begun very early. In this case, one would expect that planetesimal formation is already underway in the observed T Tauri disks. Early coagulation, however, presents a problem. Material preserved in bodies greater than several kilometers in radius would be melted by the decay of [26]Al (*Lee et al.,* 1977), a fate clearly avoided by primitive chondrites. The problem would not exist if the growth of many bodies in the asteroid belt to sizes greater than a few kilometers was delayed, perhaps by the gravitational action of an early-formed Jupiter; at this point the issue remains unresolved (*Kortenkamp and Wetherill,* 1998) (see *Woolum and Cassen,* 1999, for a further discussion).

## 2. METEORITIC AND PLANETARY VOLATILE-ELEMENT DEPLETION

The conservation of chemical species in the nebular system requires that for each species i

$$C_0^i = fC_{condensed}^i + (1-f)C_{vapor}^i \qquad (6)$$

where f is the mass fraction condensed, C is the chemical concentration of species i, and subscript 0 refers to the initial abundance, taken here to be the abundance in CI chondrites (using *Anders and Grevesse,* 1989, as the source for CI abundances). The fraction of an element condensed varies with condensation temperature, such that at chemical equilibrium f varies from 0 (above the condensation temperature) to 1 (below the condensation temperature) over a narrow temperature interval. A chemical fractionation results when condensed phases are removed from interaction with the gas phase before f = 1. This occurs when material is removed prior to the completion of condensation upon cooling, and upon evaporation of condensed material during heating. Since our samples of the solar nebula are condensed materials, a measure of volatile depletion is given by

$$F = \frac{C_c^i}{C_0^i} \qquad (7)$$

where F is the fraction of i remaining in the condensed phase. Because of the variations in major elements, C, $H_2O$ content, etc., concentrations are not directly comparable from meteorite to meteorite, a better measure of the depletion is given by reference to a refractory element, such that

$$F = \frac{\left(C^i / C^{ref}\right)_x}{\left(C^i / C^{ref}\right)_{CI}} \qquad (8)$$

where $C^{ref}$ is a major or refractory element, e.g., F for K is derived from the K/U or K/La ratio of a planet compared to that ratio in CI chondrites, or the K/Si ratio in a chondrite relative to that in CI chondrites (see *Humayun and Clayton,* 1995a). To record a chemical fractionation (e.g., volatile-element depletion) in meteoritic compositions, a physical separation of the two phases (e.g., gas and dust) must take place. Thus the elemental fractionations observed in meteorites are the chemical manifestations of physical processes operating in the nebula prior to, or during, the accretion of meteorite parent bodies and planets. Since volatile depletion is a first-order chemical effect, it can be expected to be a reflection of the first-order thermal structure of the solar nebula.

### 2.1. Elemental Fractionations in Nebular Materials

Relative to carbonaceous type 1 (CI) chondrites, all meteorite classes and the bulk compositions of the Moon and planets are depleted in volatile elements. Prior to any meaningful discussion of volatile-element depletion in nebular materials, a brief discussion of cosmic abundances is warranted. The selection of CI chondrites over other classes of chondrites is based on the similarity between the abundances of each class of elements in CI with those determined spectroscopically in the Sun (*Anders and Grevesse,* 1989), an attribute not shared by other meteorite classes. One consequence of this choice is that multistage processes are represented by the net fractionation induced as if this were a single-stage process. Since there is usually no known way of determining the magnitude of fractionation induced by each separate stage of a multistage process, volatile depletions are necessarily represented by a net fractionation relative to CI.

*Palme et al.* (1988) provide an elegant review of the abundance data and depletion factors for various chondrite classes, for planetary compositions, differentiated silicate (achondrites) and iron meteorites, and the interested reader is referred to their paper for further details. Bulk compositions of the major chondrite classes are given by *Wasson and Kallemeyn* (1988), and depletions as a function of condensation temperature are shown in Fig. 2. Bulk compositions of the Earth, Moon, and terrestrial planets can be found in *Taylor* (1979), *Morgan and Anders* (1980), *Anderson* (1983), *Wänke et al.* (1984), *Wänke and Dreibus* (1988), and *McDonough and Sun* (1995). The cosmochemical models of *Morgan and Anders* (1980) are based on combinations of components recognized in chondritic meteorites in proportions set to match the terrestrial K/U ratios, and several other geochemical inputs. The *Wänke and Dreibus* (1988) approach is to use two components, one of CI composition and the other a refractory material, with proportions of the two set by the K/U ratio. *Anderson* (1983) and *McDonough and Sun* (1995) calculate the bulk composition of the Earth's silicate mantle by estimating average compositions of natural materials, and estimates of their proportions in the Earth.

The estimated bulk compositions of the Earth by *Anderson* (1983) and by *McDonough and Sun* (1995) are compared in Fig. 3 with the cosmochemical model of *Morgan and Anders* (1980). It should be noted that the depletion in the Earth's mantle is larger than that for CO or CV chondrites (Fig. 2). It should also be noted that errors of 100 K in assignment of condensation temperature would not af-

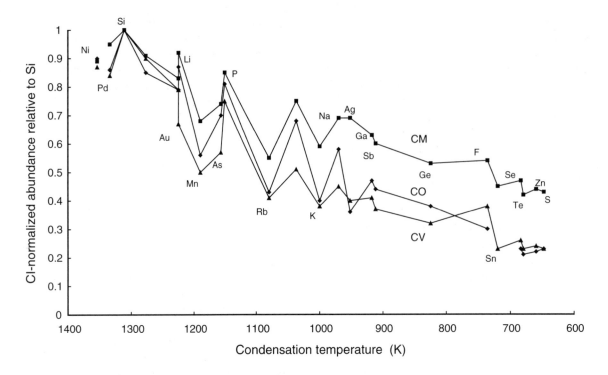

**Fig. 2.**  Abundances of moderately volatile elements in carbonaceous meteorites relative to Si, normalized by their relative abundances in CI meteorites, as a function of their 50% condensation temperatures at a nebular pressure of $10^{-4}$ atm. Squares denote values for CM meteorites, diamonds for CO, triangles for CV. The elements are, in order of increasing volatility, Ni, Pd, Si, Cr, Li, Au, Mn, As, P, Rb, Cu, K, Na, Ag, Ga, Sb, Ge, F, Sn, Se, Te, Zn, and S. Figure adapted from *Palme et al.* (1988) and *Cassen* (1996).

fect the patterns significantly. The correlation is observable despite the presence of weakly siderophile elements [strongly siderophile elements were not plotted but exhibit further depletions; cf. *Newsom* (1990) for a discussion]. The overall agreement between these estimates is sufficiently good to provide an approximate verification of the *Morgan and Anders* (1980) model. That model, following the approach of *Ganapathy and Anders* (1974), assumed that elemental abundances are uniform for particular classes of elements, e.g., moderately volatile elements all share an identical depletion factor (calculated from the K/U ratio for each planet). In detail, the abundance data on the lithophile elements K, Rb, Cs, Tl, and Pb in the terrestrial mantle are known precisely enough to recognize a distinct fractionation correlated with condensation temperature in Fig. 3. This aspect of volatile-element depletion is important for it allows a distinction to be made between two-component (*Wänke and Dreibus, 1988; Wänke et al., 1984*) or polycomponent (*Ganapathy and Anders,* 1974; *Morgan and Anders,* 1980) cosmochemical models and the models of *Cassen* (1996) involving condensation from a cooling solar nebula. Specifically, the cosmochemical models predict a step function between individual groups of volatile elements, while the Cassen model predicts a smooth decrease in depletion factors. Both aspects of these models can be seen in CM chondrites (Fig. 4) as recognized by *Wolf et al.* (1980), where the moderately volatile elements show a decrease in their depletion factors, while the highly volatile elements

exhibit a flat trend with condensation temperature. Evidence for such mixing has not been recognized in planetary compositions (Fig. 3), although the planetary compositions are more difficult to interpret as these must be filtered for the superposed effects of differentiation and core formation.

Volatile depletion is also exhibited by iron meteorites, the most depleted of which are the IVB iron meteorites where the moderately siderophile, volatile elements (Cu, Ga, Ge, As) are depleted relative to Ni and Ir when compared with other iron meteorite groups (*Kelly and Larimer,* 1977; *Rasmussen et al.,* 1984). The siderophile elements Pd, Pt, and Au are measurably depleted as well, relative to refractory Re, Ir, and Os (*Campbell and Humayun,* 1999), indicating that elements with condensation temperatures comparable to, and even higher than, those of Fe and Ni were depleted (i.e., Pd and Pt respectively; see *Palme and Wlotzka,* 1976, for condensation temperatures of these elements).

### 2.2.  Mechanisms of Volatile-Element Depletion

Two processes control the chemistry of volatiles during episodes of temperature variation: evaporation and condensation. These two processes reflect in turn first-order thermal structure of the solar nebula. In between these times of change, at stable temperatures, chemical equilibrium is maintained at the ambient temperature. The equilibrium condensation sequence of cooling gases has been developed

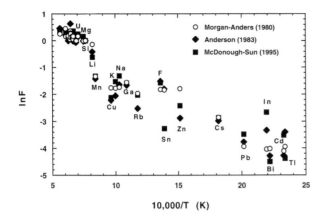

**Fig. 3.** Depletion of volatile elements in the Earth's mantle for lithophile and weakly siderophile elements, plotted as a function of 50% condensation temperature. Cosmochemical model of *Morgan and Anders* (1980) is compared with estimates by *Anderson* (1983) and *McDonough and Sun* (1995). Condensation temperatures modified from *Wasson* (1985) with Rb (850 K) and Cs (550 K) from *Wolf et al.* (1980). Elements plotted include Al, Ca, Sc, Ti, Sr, Y, Zr, Nb, Ba, REE, Hf, Ta, Th, and U, in addition to the labeled elements. Volatile siderophiles not plotted include P, S, Cr, Ge, Se, Ag, and Au. Refractory siderophiles are not plotted.

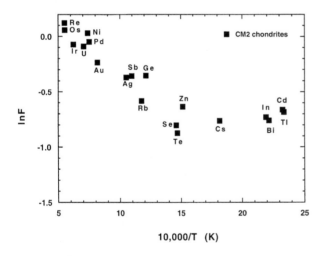

**Fig. 4.** Depletion of volatile elements in the CM2 carbonaceous chondrites as a function of condensation temperature showing that the most volatile elements require a two-component model (modified from *Wolf et al.*, 1980). Note that additional elements were plotted in the original figure of *Wolf et al.*, including C, N, and noble gases.

extensively for condensation of solids from nebular gas (*Larimer and Anders*, 1967; *Grossman*, 1972; *Grossman and Larimer*, 1974), and of liquids from nebular gases with enhanced dust/gas ratio (*Yoneda and Grossman*, 1995; *Ebel and Grossman*, 1999). Solid precursor material is first evaporated to form a "cosmic" composition gas from which condensates form upon cooling. During evaporation, sequestering of the evaporation residues will leave a signa-

ture of partial volatilization upon the composition of the solid material. The removal rate of the evaporation residues from the nebular gas must be high enough to avoid total evaporation (a process that leaves no chemical signature) during the heating phase, and to limit its reequilibration upon cooling. Thus evaporation residues can only be preserved under fairly specific nebular conditions: transient heating events that occur in a cool nebular disk ($T_{disk} \approx 300-600$ K). Evaporation residues may recondense volatiles on their surfaces, which can subsequently diffuse back into the bulk object. This has been proposed for chondrules (*Zanda et al.*, 1999; *Alexander*, 1995) and is a mechanism by which the large chemical and isotopic fractionation produced during evaporation can be reduced in magnitude on small spatial scales. If a net chemical change is to be brought about by volatilization, the recondensation process must be partial, preserving both chemical depletion and isotopic fractionation in bulk meteoritic material, the latter being detectable by sensitive mass spectrometric measurements. Thus to account for the volatile depletion patterns shown in Figs. 2 and 3, partial vaporization must operate far from the limit of chemical equilibrium. It is this limit in which kinetic isotope effects prevail, providing a powerful tool for distinguishing the conditions under which chemical fractionation occurred (*Humayun and Clayton*, 1995a).

### 2.3. Sites of Volatile-Element Depletion

*Palme et al.* (1988) observed that the absence "of volatile-rich meteorites argues against a local heat source for vaporization . . ." Despite that observation, many workers have argued that volatile-element depletions are the result of the repeated operation of local processes, particularly during chondrule formation events (*Wolf et al.*, 1980). Below, we review the possible sites for volatile-element-depletion processes.

*2.3.1. Presolar gas/grain separation.* Observations of the interstellar gas reveal relative abundance proportions (to solar abundances) that are depleted in the refractory elements (*Field*, 1974; *Hobbs et al.*, 1993). It can be inferred that these refractory constituents are present as dust grains. The question might be asked whether, during collapse of the protosolar cloud or earlier, clumping of interstellar dust grains to form refractory aggregates might contribute to the observed volatile depletion of chondritic meteorites. Such a fractionation would require the physical separation of grains from gas, and the preservation of this inhomogeneity within the nebula. *Cassen and Boss* (1988) discussed processes that might promote the concentration of grains during collapse, and concluded that large dust-to-gas concentrations were unlikely. It also seems doubtful that a memory of random presolar dust concentration fluctuations could explain the systematic differences found in chondritic meteorites, or would even be retained throughout the meteorite formation process.

*2.3.2. Grain growth and aggregation.* In those regions of the disk where temperatures were once sufficient to vaporize interstellar grains, condensation of new dust grains takes place. The refractory and major elements condensed into new phases under conditions approaching chemical equilibrium with the nebular gas, with compositions that can be determined from the equilibrium condensation sequence (*Grossman*, 1972; *Ebel and Grossman*, 1999). The major dust phases contribute most of the nebular opacity, so further cooling of the disk occurs when the grains are removed by aggregation (*Cassen*, 1996). Reductions in temperature also result from declining nebular surface density, as material accretes onto the Sun, and a declining accretion rate. The gradual removal of nebular gas, along with particles small enough to be entrained in the gas, would result in relative depletions of volatile elements in accumulated bodies. Also, if aggregation efficiently removed the smallest particles, the condensation of volatiles would be inhibited by the removal of substrates, since these elements are deposited through heterogeneous reactions with major-element substrates. Thus if surface area is limited by grain aggregation, there is a kinetic limitation to volatile-element condensation that depends on diffusion of the elements into grains (*Larimer*, 1967). Such processes are the most likely causes of the ubiquitous volatile-element depletions observed in chondrites.

*2.3.3. Chondrule formation.*    All chondrite classes, excluding CI, contain chondrules and exhibit depletions in volatile elements. It was natural to postulate that volatile loss during chondrule formation might explain the depletion of the moderately volatile elements, by allowing chondrules to be partial vaporization residues of CI composition material (*Ganapathy and Anders*, 1974). *Grossman and Olsen* (1974) showed that CM chondrites have about 52% "high-temperature materials," which included CAIs, chondrules, and isolated grains, the remainder comprising the fine-grained matrix. *Wolf et al.* (1980) observed that a two-component mixture between a matrix of CI composition and volatile-depleted "high-temperature fraction" could account for the elemental distributions in CM and CV chondrites, and perhaps in all chondrites. Thus there is circumstantial evidence that chondrule formation results in volatile depletion. Recent studies of individual chondrules have shown that chondrules have a wide range of volatile-element contents, and that chondrules are often surrounded by rims enriched in volatile elements and FeO (*Alexander*, 1995). Thus there is compelling evidence that volatiles were first lost from chondrules, and then subsequently recondensed on their surfaces. Is this a zero-sum process, or have there been net losses of the volatile elements during the chondrule-forming process? We will show below from K-isotopic data and bulk compositions of chondrites that any losses of volatile elements during chondrule formation must have been negligible, and that chondrites are not accumulations of partial volatilization residues. The formation of chondrules is a complex chemical process for which we find no role in determining the bulk compositions of asteroidal or planetary material, contrary to early postulations (*Larimer and Anders*, 1970; *Ganapathy and Anders*, 1974).

*2.3.4. Melting of planetesimals.*    Many workers (*Rasmussen et al.*, 1984; *Mittlefehldt*, 1987; *Wood*, 1998) have argued for volatile loss from differentiating planetesimals, despite cogent geochemical arguments to the contrary (*Jochum and Palme*, 1990; *Humayun and Clayton*, 1995a). Any mechanism of volatile loss on planetesimals must involve escape to space, and this may be conceived to occur in the case of basaltic volcanism (1) when volcanic gases are erupted and lost, and (2) by volatile loss from the surfaces of volcanic flows. The production of volcanic gases involves an equilibrium exchange between melt and vapor at depth, followed by a rapid release of gases into space. This process is relatively inefficient at losing alkalis, Pd, Au, and Mn, but may be important for highly volatile chalcophilic elements. Nonetheless, the chemical signatures produced involve both an equilibrium fractionation (between magmatic liquid and vapor) and the presence of flow-to-flow variations in the abundances of the volatiles (*Jochum and Palme*, 1990), neither of which are observed in the geochemical record.

*Rasmussen et al.* (1984) reported severe depletions of siderophile volatile elements in Group IVB irons including Cu, Ga, Ge, As $\approx 2.5 \times 10^{-2}$–$10^{-4} \times$ CI on a Ni-normalized basis. Also, Au is low, $4 \times 10^{-2}$, and Pd and Pt (evidenced by a higher than chondritic Ru/Pt ratio) are depleted (*Campbell and Humayun*, 1999). *Rasmussen et al.* (1984) propose that nebular fractionation cannot account for these depletions and, therefore, planetary outgassing of volatiles is required. It is hard to conceive of such severe volatile depletions during planetary outgassing. Even if an entire planetesimal were to be heated sufficiently hot to volatilize Pd, the siderophile-element budget will be rapidly sequestered in the core, out of contact with the surface where elements can be volatilized. Similarly, present models for nebular fractionations do not account for such severe depletions either. This discrepancy between observed depletions and possible mechanisms remains unresolved.

*2.3.5. Volatile loss from hot atmospheres during accretion or due to vaporizing collisions (hydrodynamic escape, impact erosion).*    Elemental fractionation can occur by mass-selective escape from a planetary atmosphere (e.g., *Hunten et al.*, 1987), but such fractionations usually affect only the most volatile species (H, C, N, noble gases). During accretion, a hot atmosphere containing moderately volatile elements might be sustained by large, vaporizing collisions, such as the one postulated to have formed the Moon. Of course, fractionations in such an atmosphere would be significant for a planetary object only if a substantial portion of the body was vaporized. It is possible that this was the case for the Moon (and perhaps some meteorite parent bodies, if involved in high-velocity collisions). *Abe et al.* (1998) have proposed that simultaneous hydrodynamic escape and condensation from a protolunar disk would produce volatile-element depletions. The details of such a model and its applicability to other objects have yet to be determined.

## 3. ISOTOPIC CONSTRAINTS ON VOLATILE-ELEMENT DEPLETION

The mechanism by which volatile elements are depleted from solar system materials can be reduced to a simple binary problem: partial condensation from a cooling gas, or partial evaporation from condensed dust or melt. The presence of a physical kinetic isotope effect during evaporation of nebular materials in those volatile elements that do not follow a simple two-component model then provides a measure of the extent of partial volatilization in the depletion of volatile elements. The class of elements to which this can be applied is that of the moderately volatile elements. The choice of elements requires that these (1) have more than one isotope, (2) have no geological fractionations of their isotopic compositions, and (3) have a well-determined depletion factor. These criteria are satisfied only by K and Rb. Potassium is the better of these choices, as it is both lighter and more abundant.

*Humayun and Clayton* (1995a) precisely determined the isotopic composition of potassium's stable isotopes ($^{41}K/^{39}K$ ratio) in planetary materials to estimate the amount of volatile loss that had occurred by partial evaporation. Their results (Fig. 5) indicated that evaporative losses of K had to be a negligible fraction of the total loss of K (and, by analogy, other moderately volatile elements) from planetary materials and meteorites. Thus local heating processes do not account for the depletion patterns in planetary materials, and a global explanation of these by partial condensation must be sought. The isotopic argument is detailed below, and a model of the condensation process is presented in the next section.

### 3.1. Stable Isotopes

Conservation of stable isotopes allows an analogous expression for isotopic fractionation [valid in the limit that $(1 + R_{vapor})/(1 + R_{condensed})$ is of order unity], as written for chemical fractionation above

$$R_0^i = F_{condensed}R_{condensed}^i + \left(1 - F_{condensed}\right)R_{vapor}^i \quad (9)$$

where R is an isotopic ratio, $R_0$ is the nebular value, which, for condensible elements, is best represented by CI chondrite values, and $F_{condensed}$ is the atomic fraction of i in the condensed phase. For refractory elements, $F_{condensed} = 1$ for all meteorites and bulk planets, so these should all have $R_0$ for the isotopic composition of refractory elements. This condition is not satisfied by moderately volatile elements, where $F_{condensed} < 1$. For these elements, $R_{condensed} = R_0$ is not an essential condition, and can only be fulfilled if $R_{condensed} = R_{vapor}$, i.e., no isotopic fractionation between gas and condensed phases existed. In principle, a physical kinetic isotope effect produced during evaporation of condensed materials can be used to place constraints on the nature of the physical mechanism by which these elements were de-

pleted. In practice, the stable isotope composition of K provides a critical test (*Humayun and Clayton*, 1995a). The generality of such a test has been challenged (e.g., *Esat*, 1996), so an extensive discussion of stable-isotope fractionation effects is provided below.

### 3.2. Equilibrium and Kinetic Isotope Effects in Chemical and Physical Processes

Isotopic mass fractionation, i.e., the separation by mass of the stable isotopes of an element, occurs during both chemical and physical processes. Three major types of isotopic mass fractionation can be recognized.

*Urey* (1947) first pointed out the importance of the differences in zero point energy ($\Delta ZPE$) of isotopically substituted molecules, resulting in an *equilibrium isotope effect*. At chemical equilibrium, there is a concentration of the heavier isotope on the stronger chemical bond between two molecules that can exchange atoms, e.g., O-isotope exchange between $CO_2$ and $H_2O$ molecules

$$\tfrac{1}{2}C^{16}O_2 + H_2{}^{18}O = \tfrac{1}{2}C^{18}O_2 + H_2{}^{16}O \quad (10)$$

where $^{18}O$ is concentrated by the $CO_2$ molecule by a factor of 1.040 at 298 K (*Urey*, 1947). The equilibrium constant for such an isotopic exchange reaction falls with increasing temperature, and equals 1.000 at infinite temperature (i.e., no isotopic preference for either molecule). This effectively occurs about 2000 K for O isotopes, where the isotopic differences between molecules with the largest free-energy differences become less than 0.1‰, the analytical detection limit. Covalently-bonded molecules show this effect most, hence producing important geochemical fractionations in the stable isotopes of H, B, C, N, O, Si, S, and Cl. The magnitude of the equilibrium isotope effect decreases with increasing mass of the element, roughly as $\Delta m/m$, and with decreasing molecular bond strengths (e.g., $H_2S > H_2Se > H_2Te$). The natural variations of Mg and K are less than analytical detection limits: 1–2‰ for $\delta^{26}Mg$, and 0.5‰ for $\delta^{41}K$ (*Humayun and Clayton*, 1995b), while that of $\delta^{44}Ca$ is about 2‰ (*Russell et al.*, 1978) in terrestrial materials. This makes ionically bonded elements ideal for the study of physical kinetic isotope effects, since the isotopic compositions of these elements are free of the equilibrium and chemical kinetic isotope effects that characterize covalently-bonded elements of comparable mass.

Kinetic isotope separation also occurs during chemical reactions as the result of differential rates of bond cleavage for isotopically substituted molecules, i.e., the rate constants of a chemical reaction are slightly different for each isotopic species, with those of the lighter species being faster than those of the heavier species. Such effects, termed *chemical kinetic isotope fractionation* here, occur widely in nature, e.g., C isotopes during photosynthesis, O isotopes during respiration and photosynthesis, S isotopes during microbial sulfate reduction, etc. Such effects are abundantly known from biologically mediated reactions, but are not exclusive to such processes. The chemical kinetic isotope fractionation arises from the same basic mechanism as the equilibrium isotope effect: The zero point energy of the heavy isotope substituted molecular bond is larger than that of the lighter isotope, making the heavy isotope substituted molecular bond stronger. The products of chemical reactions that have not proceeded to completion are isotopically lighter, while the residual reactants are isotopically heavier. Like the equilibrium isotope effect, the magnitude of the chemical kinetic isotope effect diminishes with increasing temperature, becoming negligible for high-temperature nebular processes.

Physical transport processes operating on gases or during gas-liquid/solid transitions such as diffusion, gravitational separation, and evaporation can separate isotopes by their differential velocity. This class of processes will be termed *physical kinetic isotope fractionation* here, to distinguish these processes from chemical kinetic isotope fractionation. The origin of isotope separation here arises from the equipartition of energy: The molecules of a substance having a Maxwellian distribution of velocities all have the same average kinetic energy. Thus for two isotopically substituted molecules of mass $(m_i)$ and velocity $(v_i)$

$$\tfrac{1}{2} m_1 v_1^2 = \tfrac{1}{2} m_2 v_2^2 = \tfrac{3}{2} kT \qquad (11)$$

from which it follows that

$$\frac{v_2}{v_1} = \sqrt{\frac{m_1}{m_2}} \qquad (12)$$

is the differential rate of any kinetic process that can separate isotopes by velocity, e.g., gaseous diffusion or evaporation. During gaseous diffusion from an orifice (e.g., Knudsen cell), or evaporation from a surface, the ratio of the fluxes leaving the orifice (surface) is derived from kinetic theory of gases to be

$$\frac{J_2}{J_1} = \frac{N_2 v_2}{N_1 v_1} = R \sqrt{\frac{m_1}{m_2}} \qquad (13)$$

where $R = N_2/N_1$ is the isotopic ratio in the condensed phase. Thus the vapor leaving the surface is isotopically distinct from the evaporating body's isotopic composition by a factor

$$\alpha = \sqrt{\frac{m_1}{m_2}}$$

where $\alpha$ is the vapor-solid or vapor-liquid isotope fractionation factor.

The physical kinetic isotope effect is temperature-independent in that the fractionation factor, $\alpha$, is a function only of the ratio of the masses of the evaporating species. The above discussion pertains to unidirectional transport away from the source. In bidirectional transport, no net isotope effect results when equal fluxes occur in both directions. For subequal fluxes, a net isotope effect results that can be written as $J_2/J_1 = \alpha R$, where $\alpha$ is the vapor-solid or vapor-liquid isotope fractionation factor, such that $\alpha \equiv R_{vapor}/R_c$, and

$$\sqrt{\frac{m_1}{m_2}} \leq \alpha \leq 1$$

and $J_i$ represents the net flux to or from the source of the ith isotopic species, and $R_{vapor}$ is the isotopic ratio of the instantaneous vapor leaving the surface.

The physical kinetic isotope effect is of importance in both natural and industrial (gaseous diffusion) isotope fractionation. Such effects are important in the large isotopic mass fractionation at global scales of Xe in the terrestrial atmosphere (fractionated by about 200‰), of N ($\delta^{15}N \approx 650‰$) (*Nier et al.*, 1976) and D/H ($\approx 4000–6000‰$) (*Owen et al.*, 1988; *Watson et al.*, 1994) in the martian atmosphere, and at local scales during evaporation of CAIs and other nebular materials (*Clayton et al.*, 1988; *Davis et al.*, 1990; *Wang et al.*, 1994). Physical kinetic isotope effects are responsible for the mass fractionation of O, Si, S, K, and Ca observed in the lunar soil (*Russell et al.*, 1977; *Garner et al.*, 1975; *Church et al.*, 1976; *Humayun and Clayton*, 1995b), where the accumulated effect of many micrometeorite impacts has resulted in a shift of the isotopic composition of K in bulk lunar soil to heavier values by 5–8‰. For ionically bonded elements that do not show measurable isotopic variation within a parent body, this isotopic effect provides a discriminant of the physical processes that generated the present chemical compositions of these objects.

### 3.3.  Isotopic Fractionation During Evaporation and Condensation

The chemical effects of the processes of evaporation and condensation are not always adequately distinguished in the literature. Therefore, here we will refer to *evaporation* as a process, beginning with initially condensed material, that removes elements selectively by volatility (i.e., in order of their condensation temperatures). If the process does not go to completion (*partial volatilization*), an evaporation residue remains, the chemical composition of which provides information on the physiochemical conditions of the thermal event leading to its production. We will refer to *condensation* as a process, beginning with initially gaseous (or vaporized) material, that results in elemental addition to a condensed phase forming upon cooling and nucleation from the vapor. When condensation from a vapor proceeds to completion, the chemical composition of the resulting condensate is identical to the initial composition of the vapor, and as such provides no information on the cooling path of the vapor. The products of *partial condensation* retain chemical information on the physiochemical conditions prevailing during vapor-liquid or vapor-solid separation involved in their genesis.

The moderately-volatile-element abundances in solar system materials imply that these are either the residues of partial evaporation or the products of partial condensation. Thus the presence of volatile-element depletion in meteoritic, lunar, and planetary materials provides information on the conditions of their genesis. The volatile depletion patterns cannot be uniquely ascribed to either partial evaporation or partial condensation based on the elemental abundances alone, since a set of conditions may always be found that duplicates the effects of the other process. On the other hand, the physical kinetic isotope effect produced during evaporation can allow the distinction of these two major processes (*Clayton et al.*, 1985; *Humayun and Clayton*, 1995). We will consider three cases of kinetic isotope fractionation during evaporation: at equilibrium (the null case), in vacuum, and during vapor exchange with the residue.

*3.3.1. Equilibrium.* The process of evaporation is a decrease in $F_{condensed}$ in equation (6) as molecules are transferred from the condensed phase into the vapor. In the case of equilibrium, where the net transfer to the vapor is zero, the thermally driven flux of molecules to the vapor must be balanced by a return flux of molecules from the vapor. For two isotopic species of an element, the return fluxes can be calculated from kinetic theory of gases to be

$$J_1 = \frac{P_{1,sat}}{\sqrt{2\pi m_1 kT}} \quad \text{and} \quad J_2 = \frac{P_{2,sat}}{\sqrt{2\pi m_2 kT}} \quad (14)$$

where $P_{sat}$ is the saturation vapor pressure of the species, m is the mass of the evaporating species, k is Boltzmann's constant, and T is the temperature in K. For isotopic equilibrium, the flux of isotope i from the gas to the condensed phase, $J_{i,gas}$, must equal the flux from the evaporating surface, $J_{i,c}$, such that

$$J_{i,gas} = J_{i,c} \quad (15)$$

In such a system there is *no* kinetic isotope effect. It is important to observe here that equilibrium isotope effects are then the only isotope mass fractionation effects remaining.

In such a case, with disequilibrium, the net isotopic effect is the sum of equilibrium and kinetic isotope effects. This distinction is relevant to the evaporation and condensation of meteoric water where both equilibrium and kinetic isotope effects play a role (*Craig et al.*, 1963). Equilibrium isotope effects decrease in magnitude with increasing temperature as $T^{-2}$, and with increasing mass, and are negligible for all but the covalently bonded light elements (e.g., O and Si, *Clayton et al.*, 1985) under conditions relevant in the solar nebula (i.e., temperatures of 1000–1500 K). Practically all the isotopic mass fractionation observed for elements like Mg, K, and Ca can be attributed to the kinetic isotope effect.

*3.3.2. Vacuum evaporation and Rayleigh fractionation.* Consider an element such as K, with no detectable equilibrium isotope effect, undergoing evaporation into vacuum.

For this case, the return flux from the vapor is zero. The ratio of the forward fluxes is

$$\frac{J_2}{J_1} = \frac{P_{2,sat}}{P_{1,sat}} \sqrt{\frac{m_1}{m_2}} = R_c \sqrt{\frac{m_1}{m_2}} \quad (16)$$

where $R_c$ is the isotopic ratio in the body. Thus the relative fluxes of isotopic species into the vapor is distinct from that of the evaporating body as a whole, with the result that there is a net loss of the lighter isotopic species to the vapor.

When the evaporating body is able to internally mix by convection or diffusion, the isotopic ratio at its surface remains identical to that of the body as a whole. For evaporation into vacuum, the ratio of the fluxes is

$$\frac{dN_2 \big/ dt}{dN_1 \big/ dt} = \frac{N_2 v_2}{N_1 v_1} = \frac{N_2}{N_1} \sqrt{\frac{m_1}{m_2}} \quad (17)$$

implying

$$d\ln N_2 = d\ln N_1 \sqrt{\frac{m_1}{m_2}} \quad (18)$$

Also, $R = N_2/N_1$, implying

$$d\ln R = d\ln N_2 - d\ln N_1 \quad (19)$$

Substituting yields

$$d\ln R = \left(\sqrt{\frac{m_1}{m_2}} - 1\right) d\ln N_1 \quad (20)$$

which can be integrated to yield

$$\ln\frac{R}{R_0} = \left(\sqrt{\frac{m_1}{m_2}} - 1\right)\ln\frac{N_1}{N_{1,0}} = \left(\sqrt{\frac{m_1}{m_2}} - 1\right)\ln F \quad (21)$$

where $F = N_1/N_{1,0}$ is the fraction of the evaporating species remaining in the body. We now arrive at the familiar Rayleigh fractionation equation

$$\frac{R}{R_0} = F^{\left(\sqrt{\frac{m_1}{m_2}} - 1\right)} = F^{(\alpha - 1)} \quad (22)$$

where

$$\alpha = \sqrt{\frac{m_1}{m_2}}$$

is the vapor-condensed phase fractionation factor. Isotope fractionation of Mg, Si, and O in vacuum evaporation experiments on forsterite and chondrite liquids reported by *Davis et al.* (1990) and *Wang et al.* (1994) closely followed a Rayleigh distillation law. Isotope fractionation of K in vacuum evaporation of chondrule compositions by *Wang et al.* (1999) similarly showed Rayleigh fractionation effects, although these results were less precise.

Figure 5a shows the exponential increase in isotopic fractionation with increasing depletion of K for the residue, for the instantaneous vapor coming off the surface, and for the cumulative vapor. Condensates from a vapor formed by partial evaporation are isotopically light. Thus the isotopic effects observed in materials derived by partial distillation and recondensation are isotopically heavy residues and isotopically light condensates. The removal of either constituent would result in a net isotopic fractionation, which is inconsistent with the K-isotopic data in Fig. 5c. Therefore it is not possible to produce a chemical fractionation of K by the removal of a distillation product and simultaneously produce isotopic properties consistent with the observed data. The process of Rayleigh fractionation is by no means limited to vacuum evaporation, as seen in Fig. 5b, but holds generally if two conditions are satisfied: the return flux from the surrounding environment is small compared to the forward flux, and the residue is isotopically well-mixed.

It has been argued that evaporation in the solar nebula may have occurred under "non-Rayleigh" conditions (*Young et al.*, 1998; *Wang et al.*, 1999), taken to imply the lack of internal mixing or diffusion-limited evaporation. However, no general class of planetary or meteoritic materials have been specifically identified as products of diffusion-limited evaporation; in particular, there is no compelling physical or chemical evidence that such a process had a role in controlling the volatile-element depletions of chondritic meteorites. Few materials in meteorites are large enough to undergo diffusion-limited evaporation, which requires dimensions larger than the characteristic diffusion length, $r_d = (Dt)^{1/2}$, where D is the diffusion coefficient and t is the duration of heating. Typical diffusion coefficients for K of $10^{-6}$ cm$^2$/s (silicate liquids; see *Hofmann*, 1980, for a review) and $10^{-10}$ cm$^2$/s (solids), and heating durations of $\approx 10^3$ s at T $\approx$ 1300 K yield $r_d \approx 300$ μm for liquids and $\approx 3$ μm for solids. Furthermore, diffusion-limited evaporation produces an outer rind depleted in volatile constituents, and an inner region free of the influence of evaporation. The chemical effect of such a diffusion-limited evaporation process is essentially identical to a two-component mixing model, evidence for which is not found in K and other elements with 800 K $\leq T_c \leq$ 1300 K. This is borne out by the experiments of *Wang et al.* (1999).

*3.3.3. Evaporation into a gas.* We develop the case where $J_{i,c} > J_{i,gas}$ and examine the effect of evaporation into a gas having a vapor pressure of the evaporating constituent i, $P_i$. A few simplifying assumptions are needed, including that the gas reservoir is large compared with the evaporating object, such that the isotopic composition of the gas remains unaffected by the evaporation of the object, and that the gas is well-mixed, which combine to give the isotopic composition of the gas, $R_{gas}$ = constant, which we will take to be equal to the isotopic composition of the bulk

system (i.e., no equilibrium or kinetic isotope effects involved in the production of this vapor). We will also regard the gas to be at a total pressure sufficiently low ($P_{total} \ll$ 1 atm) to be able to neglect the effect of collisions between evaporating molecules and molecules present in the gas (e.g., $H_2$, He), conditions that seem appropriate for the

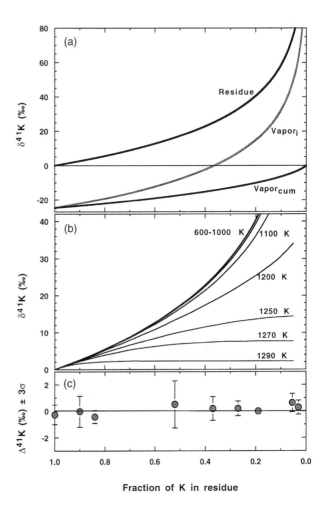

**Fig. 5.** **(a)** Calculated K isotopic compositions produced by evaporation of K into vacuum, or any rarified atmosphere where $P_{gas}/P_{sat} \leq 0.02$, with Rayleigh fractionation (solid curves), where $\delta^{41}K = 1000[(R/R_{std}) - 1]$. Curves are shown for the residue, the instantaneous vapor leaving the surface, and the cumulative vapor. **(b)** Calculated K-isotopic compositions produced by evaporation of K into an atmosphere having a finite vapor pressure, $P_{gas}$, determined by the temperature of the surrounding gas, $T_{gas}$. The evaporating object is held at $T_{surface}$ = 1300 K. Curves are shown for the residues only; $\delta^{41}K = 0$ for the surrounding gas. Note the significant deviation from Rayleigh fractionation occurring as $T_{surface} - 100 < T_{gas} < T_{surface}$. **(c)** Potassium-isotopic data for various analyzed materials normalized to terrestrial average (with $3\sigma_m$ error bars) from *Humayun and Clayton* (1995a), where $\Delta^{41}K = \delta^{41}K - \delta^{41}K_{Earth}$. Data points shown are (from right) Orgueil CI, mean EH/EL chondrites, mean H/L/LL chondrites, Murchison CM2, mean CV3 chondrites, SNC meteorites, Earth, Juvinas (EPB), mean lunar samples.

solar nebula. We will assume that the gas is warm enough ($T_{gas} \leq T_{condensation}$) to support a finite vapor pressure ($P_{gas}$) of the element under consideration (here K), giving rise to a flux, $J_{i,gas}$, which acts to damp the isotopic effect of evaporation. We examine the consequences below.

The relevant flux here is the differential flux of i into the vapor

$$J_i' = J_{i,c} - J_{i,gas} = \frac{P_{i,sat} - P_{i,gas}}{\sqrt{2\pi m_i kT}} \quad (23)$$

where $P_{i,sat}$ is the saturation vapor pressure (equilibrium vapor pressure). The ratio of the differential flux is given by

$$\frac{J_2'}{J_1'} = \left(\frac{P_{2,sat} - P_{2,gas}}{P_{1,sat} - P_{1,gas}}\right)\sqrt{\frac{m_1}{m_2}} \quad (24)$$

Since

$$J_i = \frac{1}{A}\frac{dN_i}{dt}$$

and

$$\frac{P_{1,gas}}{P_{1,sat}} \approx \frac{P_{gas}}{P_{sat}}$$

when $P_1$ is the dominant isotope, we then have

$$\frac{dN_2}{dN_1} = \frac{\left(R_c - R_{gas}\dfrac{P_{gas}}{P_{sat}}\right)}{\left(1 - \dfrac{P_{gas}}{P_{sat}}\right)}\sqrt{\frac{m_1}{m_2}} \quad (25)$$

where $R_c = N_2/N_1$, $R_{gas} =$ constant, and $P_{gas}/P_{sat}$ is fixed. This can be rewritten as

$$\frac{dN_2}{dN_1} = \frac{\sqrt{\dfrac{m_1}{m_2}}}{\left(1 - \dfrac{P_{gas}}{P_{sat}}\right)}\left(\frac{N_2}{N_1} - R_{gas}\frac{P_{gas}}{P_{sat}}\right) \quad (26)$$

We define

$$\beta \equiv \frac{\sqrt{\dfrac{m_1}{m_2}}}{\left(1 - \dfrac{P_{gas}}{P_{sat}}\right)} \quad (27)$$

yielding

$$\frac{dN_2}{dN_1} - \beta\frac{N_2}{N_1} = -R_{gas}\beta\frac{P_{gas}}{P_{sat}} \quad (28)$$

This is a linear ordinary differential equation with constants $\beta$, $R_{gas}$, and $P_{gas}/P_{sat}$, which has the solution

$$\frac{N_2}{N_1} = \left(\frac{N_{2,0}}{N_{1,0}} + \frac{R_{gas}\beta\dfrac{P_{gas}}{P_{sat}}}{1-\beta}\right)\left(\frac{N_1}{N_{1,0}}\right)^{(\beta-1)} - \left(\frac{R_{gas}\beta\dfrac{P_{gas}}{P_{sat}}}{1-\beta}\right) \quad (29)$$

For an infinite gas reservoir, $R_{gas} = R_0$, we can rewrite the above expression as

$$\frac{R}{R_0} = \left(1 + \frac{\beta\dfrac{P_{gas}}{P_{sat}}}{1-\beta}\right)F^{(\beta-1)} - \left(\frac{\beta\dfrac{P_{gas}}{P_{sat}}}{1-\beta}\right) \quad (30)$$

where $F = N_1/N_{1,0}$ as in equation (21) above. There are two limits: when $P_{gas}/P_{sat} = 0$, $\beta = \alpha$, and equation (30) converges to equation (22), the expression for Rayleigh fractionation; when $P_{gas}/P_{sat} = 1$, the solution is singular and is meaningful only if the differential flux vanishes (i.e., equilibrium is attained). Figure 5b shows the isotopic fractionation occurring in the residue for different values of $P_{gas}/P_{sat}$.

We next consider the partial pressure of K in the nebular gas and the value of $P_{gas}/P_{sat}$. If the depletion is to take place by devolatilization, some of the dust must experience a temperature ($T_{surface}$) higher than the ambient gas + dust mixture (at $T_{gas}$). The vapor pressure of K in the gas, $P_{gas}$, is set by the nebular temperature while the rate of loss of K from the heated object is determined by $P_{sat}$, with $T_{surface} > T_{gas}$

$$\left(\frac{P_{gas}}{P_{sat}}\right) = \exp\left(-\frac{\Delta E}{R}\left(\frac{1}{T_{gas}} - \frac{1}{T_{surface}}\right)\right) \quad (31)$$

where $\Delta E$ is the enthalpy of vaporization. Since Mg and Si are not depleted, the maximum temperatures at which K depletion could have occurred by devolatilization are $\approx 1300-1400$ K (for a nebular pressure of $10^{-4}$ atm), and the minimum temperatures required to evaporate K are on the order of $T_{cond} \approx 1000$ K (*Wasson*, 1985). Figure 6 shows values of $(1 - P_{gas}/P_{sat})$ as a function of $T_{gas}$, for $T_{surface}$ values of 1000–1300 K, appropriate for pressures of $10^{-4}$ atm. At higher nebular pressures, the condensation temperatures of all elements will scale upward. The vapor pressure data of *Gooding and Muenow* (1977) for K above chondrite melts give $\Delta E/R = 14,575$ K, and this value was used to calculate values of $\beta$ and $P_{gas}/P_{sat}$ as a function of the nebular gas temperature, $T_{gas}$ (Fig. 6). The isotopic effects produced in the residue by evaporation under these conditions with $T_{surface} = 1300$ K are shown in Fig. 5b. At $T_{gas} < T_{cond}$, the

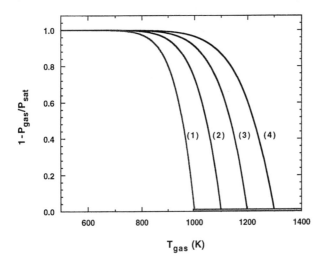

**Fig. 6.** The coefficient $(1 - P_{gas}/P_{sat})$ as a function of gas temperature, $T_{gas}$, for material of chondritic composition undergoing evaporation at surface temperatures of (1) 1000 K, (2) 1100 K, (3) 1200 K, and (4) 1300 K respectively. These temperatures were chosen to cover the range of possible conditions at $P_{total} = 10^{-4}$ atm from the condensation temperature of K to that of Si using condensation temperatures from *Wasson* (1985). The bar shows the range of conditions allowable by the data of *Humayun and Clayton* (1995a).

isotopic fractionation follows the Rayleigh law, as there is not enough K in the vapor to effect the residue. At higher values of $T_{gas}$, the relation between $\delta^{41}K$ and F falls below the Rayleigh curve. Even with $T_{gas}$ a mere 10 K below $T_{surface}$, a small but significant isotopic fractionation is evident. To achieve values of β and $P_{gas}/P_{sat}$ that are compatible with the precise K-isotopic data of *Humayun and Clayton* (1995a), $T_{gas} \approx T_{surface}$, implying nebular temperatures hot enough to place most of the moderately volatile elements in the gas. The presence of volatile elements in inner solar system materials then requires condensation from this gas. The volatile-element depletion in meteorites, the Earth, and Moon, cannot be produced by volatilization residues from a nebula with $T_{gas} < T_{cond}$ (K). Since any model of volatile depletion that fits K must also account for the depletion of all the moderately volatile elements, the nebular temperature at which the meteoritic and planetary volatile-element depletion was established must have been $T \leq 1400$ K, or equivalently the condensation temperatures of the major elements. Such a high nebular temperature is consistent with a thermostatically controlled nebula (*Morfill*, 1985) and with volatile-element abundances established during condensation by cooling of the disk (*Cassen*, 1996).

*3.3.4. Kinetic isotope effects during condensation.* It should also be obvious that condensation from a gas is a process that occurs only if $P_{gas}/P_{sat} \geq 1$, and hence with negligible kinetic isotope fractionation. Deviations from $P_{gas}/P_{sat} = 1$ are difficult to maintain during condensation, as a

significant oversaturation of the gas is not likely to persist with readily available nucleation sites (i.e., major silicate grains). The relationship between isotopic fractionation and $P_{gas}/P_{sat}$ indicates that undercooling of the condensates must have been less than 10 K to prevent measureable light-isotope enrichment in objects that have partially condensed K. This is consistent with global nebular cooling, but is not obviously satisfied by supercooling of K vapor on the surfaces of evaporation residues. Thus scenarios of the type envisaged by *Alexander* (1995), involving evaporation of volatiles from chondrules followed by recondensation of the vapor onto chondrule rims, require complete recondensation of the vapor. Such processes leave no chemical signature in the bulk composition of chondrites or planets (Figs. 2–4), the explanation of which is developed below in terms of nebular cooling.

## 4.  MODELS FOR FRACTIONATION DURING GLOBAL NEBULAR COOLING

The idea that the systematic depletions of moderately volatile elements found in chondritic meteorites might be caused by condensation in a nebula of diminishing mass was suggested by *Wasson and Chou* (1974) and further promoted by *Wai and Wasson* (1977) and *Wasson* (1985). Uncondensed elements would be removed with nebular gas if the nebula itself was in the process of being removed, so that volatile species would be underrepresented relative to more refractory species by the time the nebula was cool enough for all rock-forming elements to condense and accumulate. It was supposed that the nebula was lost either through a wind or by accretion onto the Sun.

Consider the moderately volatile depletion patterns for CM, CO, and CV carbonaceous chondrites shown in Fig. 2. The elements more volatile than Si have declining relative abundances with decreasing condensation temperature. If one attributes this trend to incomplete condensation, one must suppose that the nebula was at least hot enough to vaporize everything with condensation temperatures up to that of Si, at some time and place. If this were not true, there would be elements more volatile than Si (e.g., Cr, Li, Au) that would have condensed undepleted relative to Si. The fact that lithophile elements *more* refractory than Si tend to be present in carbonaceous chondrites with the same relative abundances (although enhanced as a group over Si) suggests that these elements condensed together at a temperature above the condensation temperature of Si, and were later incorporated into carbonaceous chondrites as a group. These patterns are, in fact, consistent with certain aspects of nebular thermal history theory. Silicate and metal grains, which evaporate or condense within about a hundred degrees of each other, provide most of the opacity in the nebula. Therefore, in a region where they are evaporated, the nebula tends to be optically thin, a condition that favors isothermality. *Morfill* (1985), *Boss* (1990, 1993), and *Cassen* (1994) noted that a nearly isothermal cavity, at or just above the silicate vaporization temperature, could be produced within a hot nebula, near the midplane, by such an effect. (In Fig. 1, the flattening of the temperature profile above about 1200 K is caused by the vaporization of dust.) Elements more refractory than Si could settle from the cool upper layers of the nebula into this cavity and accumulate in solar proportions; silicates and more volatile elements would remain gaseous at the midplane until the temperature had declined to their respective condensation values. If, in the meantime, nebular gas containing all uncondensed species was being removed, something like the volatility-correlated depletion pattern observed in the carbonaceous meteorites would be the expected result. The fact that elements more refractory than Si tend to be uniformly depleted in ordinary chondrites (in contrast to their overabundance in some carbonaceous chondrites, e.g., CV) indicates that, although these might have accumulated earlier than, or separate from, Mg-silicates, a simple sequential condensation and accumulation process is not the whole story.

The ubiquity and regularity of the depletion trend (particularly for the carbonaceous chondrites) implies a global process, which led *Cassen* (1996) to attempt to model the compositional trends produced by an evolving nebula. The goals were to determine if the condensation hypothesis was consistent with concepts of the evolution of protostellar disks derived from astrophysical considerations, and to examine its implications for other planet-building processes. The method used was as follows.

First, a set of evolutionary models of the nebula, starting at the time when addition of material from the protosolar cloud had just ceased, was quantitatively prescribed. This means that the nebular mass $M_d$, surface density distribution $\Sigma$, and accretion rate onto the Sun $\dot{M}$ were specified as functions of time and certain key parameters that were varied from model to model:

$$M_d = M_d(t)$$
$$\dot{M} = dM_d/dt$$
$$\Sigma = \Sigma(R,t)$$

Such formulae provide a complete description of the average motions of gas and entrained fine dust in the nebula. (In principle, one would like to derive these functions from a theory of dynamical evolution, but there is currently no theory that we regard to be more reliable than *ad hoc* postulates, judiciously constrained by astronomical observations.) To address meteoritic compositions, the evolutionary prescriptions must be augmented by some description of the coagulation process. In *Cassen* (1996), the results of coagulation theory (e.g., *Weidenschilling and Cuzzi,* 1993) were summarized by a simple expression for the rate at which rocky material coagulated. It was assumed that all material could be considered to exist either as gas, fine dust (the motion of which was coupled to the gas), or rocks sequestered in meteorite parent bodies (with motion decoupled from the gas). The surface density $\Sigma_i$ of elemental species i in fine dust was assumed to diminish due to coagulation to

rocks at a rate $F_i$ given by

$$F_i = \varepsilon\Omega\Sigma_i$$

wherever that species could condense at the midplane. Here $\Omega$ is the local orbital frequency. The efficiency $\varepsilon$ is assumed to be constant for all species, independent of kinetic effects and chemical affinities, as indicated by the moderately volatile element patterns in chondrites (Fig. 2). Thus $\Sigma_i$ is given by the mass conservation equation

$$\frac{\partial\Sigma_i}{\partial t} + \frac{1}{R}\frac{\partial(RV_R\Sigma_i)}{\partial R} = -F_i \tag{32}$$

where $\Sigma_i$ can be regarded as the surface density of element i coupled to the gas, if the righthand side is understood to be zero at times and radii for which the midplane temperature exceeds the condensation temperature of species i. The radial velocity $V_R$ is known from the evolutionary model. The total surface density of species i deposited and saved in rocky material is given by

$$\sigma_i = \int_{t_i}^{\infty} F_i dt = \int_{t_i}^{\infty} \varepsilon\Omega\Sigma_i dt \tag{33}$$

where $t_i$ is the time at which species i first condenses at the midplane. The abundances $R_i$ relative to Si in CI meteorites are then

$$R_i = \frac{\sigma_i/\sigma_{Si}}{f_{i0}/f_{Si0}}$$

where $f_i$ is the concentration of i defined by $\Sigma_i = f_i\Sigma$, and $f_{i0}$ is the initial concentration, taken to correspond to solar composition as indicated by CI meteorites.

Equation (3) was used to determine the midplane temperatures. The optical depth to the midplane $\tau$ was derived from the column density of silicates in the form of fine dust, $\tau = \Sigma_{si}\kappa/2$, using an average Rosseland mean opacity $\kappa$ based on the work of *Pollack et al.* (1994). Thus the abundances of each species in gas and fine-grained dust and rocky material can be calculated as functions of time and radial location.

To illustrate the general nature of the solutions, suppose that, at the radial location of interest, the surface density diminishes as $e^{-\alpha t}$, with a consequent decrease of the (inward) mass flux at the same rate. Suppose further that $f_i$ decreases exponentially as $e^{-\gamma(t-t_i)}$ due to coagulation, where $\gamma = \varepsilon\Omega$. Therefore

$$\sigma_i = \int_{t_i}^{\infty} \Sigma_0 e^{-\alpha t} f_{i0} e^{-\gamma(t-t_i)} dt$$
$$= \Sigma_0 f_{i0} e^{-\alpha t_i} \int_0^{\infty} e^{-\alpha s - \gamma s} ds \tag{34}$$

If we measure time from the moment that $T_m = T_{Si}$ at the radius of interest

$$\sigma_{Si} = \Sigma_0 f_{Si0} \int_0^\infty e^{-(\alpha+\gamma)t} dt \qquad (35)$$

and $R_i = e^{-\alpha t_i}$.

To find $t_i$, the equation for $T_m$ must be inverted, imposing the condition $T_m = T_i$ (the condensation temperature of species i). The midplane temperature is given by

$$T_m^4 = \frac{9\tau GM\dot{M}}{64\pi\sigma R^3} = \frac{9GM\dot{M}\Sigma\kappa}{128\pi\sigma R^3}$$
$$= T_{Si}^4 \frac{\dot{M}(t)}{\dot{M}(0)} \frac{\Sigma(t)}{\Sigma(0)} e^{-\gamma t} = T_{Si}^4 e^{-(2\alpha+\gamma)t} \qquad (36)$$

The three time-dependent factors in the fourth expression reflect cooling due to declining accretion rate and opacity, the latter caused by declining surface density and progressive coagulation of silicates. The solution for $t_i$ is found by setting $T_m = T_i$

$$t_i = \frac{4\ln(T_{Si}/T_i)}{(2\alpha+\gamma)}$$

Thus

$$R_i = (T_i/T_{Si})^{\frac{4}{\left(2+\frac{\gamma}{\alpha}\right)}} \qquad (37)$$

The relative abundances $R_i$ are plotted in Fig. 7 for different values of the ratio $\gamma/\alpha$. If $\gamma \gg \alpha$, coagulation is rapid once it has begun, cooling keeps pace with coagulation by the reduction in opacity, but all available rocky material accumulates before nebular mass has diminished; CI abundances are preserved and no fractionations are produced. On the other hand, for $\gamma$ comparable to or much less than $\alpha$, a fractionation pattern roughly like that seen in chondritic

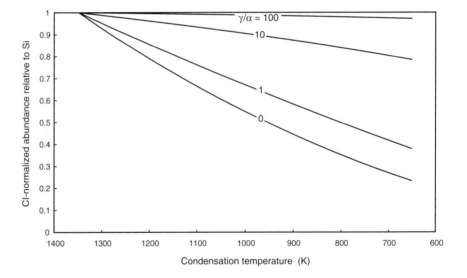

**Fig. 7.**  Relative abundances as a function of condensation temperature, predicted for a simple model in which nebular surface density and accretion rate are assumed to diminish as $e^{-\alpha t}$ while coagulation reduces the surface density of gas-entrained fine dust as $e^{-\gamma t}$.

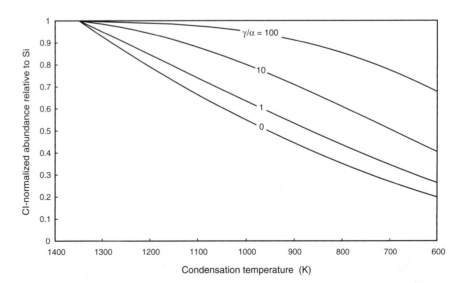

**Fig. 8.**  Relative abundances at 1 AU for the simple model in Fig. 7, but including the effects of an inward radial velocity whose value is inversely proportional to distance from the Sun. The local radial velocity is 100 cm/s.

meteorites results. Nebular removal induces cooling by a reduction in opacity so that coagulation proceeds to lower temperatures, with delays correlated with volatility.

In general, $f_i$ is more complicated than the simple exponential function of time used above. A realistic calculation must account for the fact that accumulation at one location affects abundances at other locations, as entrained dust moves in the nebula. This effect is contained in the advective term (second term on the righthand side) of equation (32). In terms of $f_i$ this equation is

$$\frac{\partial f_i}{\partial t} + V_R \frac{\partial f_i}{\partial R} = -\varepsilon \Omega f_i \qquad (38)$$

Consider now a nebula in which the surface density is uniform throughout the terrestrial planet region, but slowly decreasing with time, along with the mass accretion rate. For such a nebula, the radial velocity is inward and proportional to $1/R$, so we take $V_R = V_{Ra}(R_a/R)$, where $V_{Ra}$ is the velocity at radius $R_a$, taken, for instance, to be somewhere in the asteroid belt. Note also that $\Omega = \Omega_a(R/R_a)^{-3/2}$ where $\Omega_a$ is the orbital frequency at $R_a$. The solution of equation (38) can then be expressed as

$$f_i = f_{i0}e^{-y} \qquad (39)$$

where

$$y(R, t - t_i) = 2\gamma R^{1/2} \left\{ \left[ 1 + \frac{2(t - t_i)}{R^2} \right]^{1/4} - 1 \right\} \qquad (40)$$

$R$ and $t$ are measured in units of $R_a$ and the flow time $t_a = R_a/V_{Ra}$ respectively. The accumulation rate $\gamma$ is now defined by $\gamma = \varepsilon \Omega_a t_a$. (Again, it is understood that $\varepsilon = 0$ if $T_m > T_i$).

The rate of depletion of element i in fine-grained dust, taken to be the same as the rate of accumulation of i in meteorite parent bodies, is

$$F_i = (\gamma/t_a) R^{-3/2} \Sigma(t) f_{i0} e^{-y(R, t - t_i)} \qquad (41)$$

If we still assume that surface density and mass accretion each decline slowly as $e^{-\alpha t}$, $R_i$ is given by $e^{-\alpha(t_i - t_{Si})}$, where $t_i$ now satisfies

$$\ln \frac{T_{Si}^4}{T_i^4 R^3} = 2\gamma R^{1/2} \left[ \left( 1 + \frac{2t_i}{R^2} \right)^{1/4} - 1 \right] + 2\alpha t_i \qquad (42)$$

(If $T_m < T_i$ at $t = 0$, coagulation of that species begins immediately.) Relative abundances for this case are plotted in Fig. 8, for $R = 1$. The general nature of the solutions is similar to that of the previous example, but depletions of the most volatile species are somewhat enhanced because inward flowing gas has been depleted earlier by accumulation in the outer, cooler nebula.

Finding solutions for more realistic nebula models is complicated by several factors. Radial gas velocities are generally nonmonotonic functions of radius, with inward flow at small radii and outward flow at large radii. Midplane

condensation fronts usually move inward as the accretion rate and nebular opacity decrease with time, but may be overtaken by an inward gas flow, in which case they are actually seen as "evaporation" fronts by entrained dust. The entire coagulation history of a parcel of gas and dust affects subsequent coagulation at all other locations visited by that parcel. These factors are accounted for self-consistently in the models calculated by *Cassen* (1996), in which the nebula was assumed to have a mass and surface density that evolved according to

$$M_d(t) = M_{d1}[1 + t/t_e]^{-n}$$

$$\Sigma(R, t) = \Sigma_0(t) e^{-(R/R_0)^2}$$

The constants $M_{d1}$, $t_e$, and $n$ are parameters that characterize a given model, along with the coagulation efficiency $\varepsilon$ and the total angular momentum of the system; the values of $\Sigma_0$ and $R_0$ were constrained by mass and angular momentum conservation. Because the model predicts the total amount of rocky material deposited in parent bodies as a function of radial location, *Cassen* (1996) compared the predicted distribution with that of the solar nebula, as inferred from the masses of the rocky components of the planets (*Weidenschilling*, 1977; *Hayashi et al.*, 1985). The moderately volatile fractionation patterns in the asteroid belt were then calculated for those models with parameters for which the comparison was deemed favorable. It was found that most models that accounted for the planetary distribution of rocky mass also produced fractionation trends at distances between about 1.5 and 2.5 AU that matched well those found in CO and CV carbonaceous chondrites. These models were generally characterized by accretion times in the asteroid belt within an order of magnitude of $t_e$, in agreement with expectations suggested by the examples given above (see *Cassen*, 1996, for specific parameters and results). It was also a characteristic of the models that depletions at 1 AU were not substantially different from those predicted for the asteroid belt.

The precise shape of the CM pattern was not reproduced as well as those for COs and CVs, at least for the limited set of models explored. The chemical composition of CM chondrites is suitably described by a two-component model (*Wolf et al.*, 1980), where one component is volatile-depleted and the other is of CI composition (Fig. 4). The mixing of CI-composition material is largest at about 50% in CM, and smaller proportions are required for all other classes of chondrites. Thus for COs and CVs the mixing of CI-type material has comparatively less influence on the final composition, and better agreements with model calculations are observed in Figs. 7 and 8. Thus mixing of material processed under different nebular conditions is required to better account for the moderately-volatile-element fractionations in meteorites. The agreement between model calculations and meteoritic abundances is excellent in view of the necessity to mix materials processed at high temperatures in the nebula with unprocessed (or mildly

processed) presolar dust to account for the presence of ubiquitous presolar dust grains in all unmetamorphosed chondrite classes (*Anders,* 1988; *Huss,* 1990; *Huss and Lewis,*
1994). Additional constraints are provided by the presence
of interstellar materials with high D/H ratios (*Zinner,* 1988).
It should also be emphasized that this mixing is conceived
to involve material having a composition identical to that
of CI chondrites (e.g., nebular condensates, as for material
formed along the curve of $\gamma/\alpha = 100$ in Fig. 7, or interstellar dust that escaped nebular thermal processing above
$\approx 400$ K), and does not represent actual mixing of CI chondrites with other chondrites, which is not supported by O-
isotopic compositions of carbonaceous chondrites (*Clayton
and Mayeda,* 1999).

One can, of course, think of many factors not included
in the models that could have caused deviations from the
calculated abundances. Besides mixing, quantitative aspects
of the general trend of the depletions might be affected by
irregularities in the accretion history of the nebula, among
other things. The detailed deviations from monotonicity of
individual elements could reflect subtle aspects of chemistry, inaccuracies in the calculated condensation sequence or
the measured abundances, or the real effects of unsystem-
atic local processes.

Generally speaking, these calculations support the hypothesis that global fractionation patterns are the natural
result of the inherently coupled evolutions of condensing
and coagulating solids from a nebula of diminishing mass,
cooling from an initially hot state. They also indicate that
such primary fractionations were similar in magnitude,
within the radial distance in the nebula to which all silicate
materials were evaporated at the midplane. Further explo-
rations of the consequences of these processes should incorporate improvements to the model. For instance,
equation (3) is strictly valid only if the vertical thermal structure of the nebula adjusts rapidly to changes in heating rate
and opacity. It is possible to construct models that relax this
approximation, and that accommodate temperature-dependent opacities and the attendant variations in the vertical
opacity structure, as well as the pressure dependence of condensation temperatures. It also seems timely to couple
chemical equilibrium and thermal evolution calculations,
following the approach adopted herein to quantify the coagulation and dynamical evolution. Such models might reveal other interesting effects that might be manifested in the
meteoritic record, and that would provide quantitative tests
of theories of planet formation.

## 5. IMPLICATIONS FOR THE ORIGIN OF THE EARTH

Volatile depletion in the Earth can be explained as an
inheritance from the solar nebula, in which the precursors
to our planet's solid foundations were thermally processed.
Thus the major constituents of the Earth began to separate
from nebular gas when the temperature of the nebula was
on the order of the condensation temperature of major sili-

cates and iron metal ($\approx 1400$ K), a value attained only very
early in the evolution of the nebula. Models of nebular fractionation that account for moderately-volatile-element
depletions in chondritic meteorites produce depletions
within a factor of 2 of the observed terrestrial abundances.
An implication of this result is that the putative Moon-forming impact was not responsible for any major fractionations
beyond those that already existed in the proto-Earth, except
possibly for the most volatile species (H, C, N, noble gases).
Thus the depletions of elements like Mn and K are part of
a larger pattern of inner solar system depletion of volatiles
that extends to the constituents of our atmosphere, oceans,
and biosphere, and presumably to the constituents of the
other terrestrial planets (*Taylor,* 1992). The more severe
depletions of alkalis observed in eucrites and the Moon,
however, are not accounted for by any of the present models. Any additional processes of volatile loss that might be
invoked did not fractionate K isotopes (*Humayun and
Clayton,* 1995a). Because the Earth accumulated from planetesimals formed throughout a range of radii from the inner solar system (*Wetherill,* 1994), its composition probably
includes contributions from both more-depleted (e.g., eucrite) and less-depleted (e.g., CV/CO chondrite) material
than that of the bulk Earth.

***Acknowledgments.*** Frank Richter is thanked for discussions
on isotopic fractionation during evaporation. Reviews by Alan
Boss, Herbert Palme, Robert N. Clayton, Larry Grossman, and
Frank Richter helped improve the manuscript. Thanks to K. R. Bell
for supplying Fig. 1. Jennifer Lundal Humayun is thanked for her
assistance and support during this writing.

## REFERENCES

Abe Y., Zahnle K. J., and Hashimoto A. (1998) Elemental fractionation during rapid accretion of the Moon triggered by a giant
impact (abstract). In *Origin of the Earth and Moon*, p. 1. LPI
Contribution No. 957, Lunar and Planetary Institute, Houston.
Alexander C. M. O'D. (1995) Trace element contents of chondrule
rims and interchondrule matrix in ordinary chondrites. *Geo-
chim. Cosmochim. Acta, 59,* 3247–3266.
Anders E. (1988) Circumstellar material in meteorites: noble
gases, carbon and nitrogen. In *Meteorites and the Early Solar
System* (J. F. Kerridge and M. S. Matthews, eds.), pp. 927–955.
Univ. Arizona, Tucson.
Anders E. and Grevesse N. (1989) Abundances of the elements:
Meteoritic and solar. *Geochim. Cosmochim. Acta, 53,* 197–214.
Anderson D. L. (1983) Chemical composition of the mantle. *Proc.
Lunar Planet. Sci. Conf. 14th,* in *J. Geophys. Res., 88,* B41–
B52.
Basri G. and Bertout C. (1989) Accretion disks around T Tauri
stars. II. Balmer emission. *Astrophys. J., 341,* 340–358.
Beckwith S. V. W. and Sargent A. I. (1993) The occurrence and
properties of disks around young stars. In *Protostars and Plan-
ets III* (E. H. Levy and J. Lunine, eds.), pp. 521–541. Univ. of
Arizona, Tucson.
Beckwith S. V. W., Sargent A. I., Chini R. S., and Güsten R. (1990)
A survey for circumstellar disks around young stars. *Astron.
J., 99,* 924–945.

Bell K. R. (1999) Reprocessing in self luminous disks. *Astrophys. J.,* in press.

Bell K. R. and Lin D. N. C. (1994) Using FU Orionis outbursts to constrain self-regulated protostellar disk models. *Astrophys. J., 427,* 987–1004.

Bell K. R., Cassen P., Klahr H. H., and Henning Th. (1997) The structure and appearance of protostellar accretion disks: Limits on disk flaring. *Astrophys. J., 486,* 372–387.

Bell K. R., Cassen P., Wasson J. T., and Woolum D. S.(1999) The FU Orionis phenomenon and solar nebula material. In *Protostars and Planets IV* (V. Manning, A. P. Boss, and S. Russell, eds.). Univ. of Arizona, Tucson, in press.

Boss A. P. (1990) 3D solar nebula models: implications for Earth origin. In *Origin of the Earth* (H. E. Newsom and J. H. Jones, eds.), pp. 3–15. Oxford Univ., New York.

Boss A. P. (1993) Evolution of the solar nebula. II. Thermal structure during nebula formation. *Astrophys. J., 417,* 351–367.

Boss A. P. (1998) Temperatures in protoplanetary disks. *Annu. Rev. Earth Planet. Sci., 26,* 53–80.

Calvet N. and Gullbring E. (1998) The structure and emission of the accretion shock in T Tauri stars. *Astrophys. J., 509,* 802–818.

Calvet N., Hartmann L., Kenyon S. J., and Whitney B. A. (1994) Flat spectrum T Tauri stars: The case for infall. *Astrophys. J., 434,* 330–340.

Campbell A. J. and Humayun M. (1999) Microanalysis of platinum group elements in iron meteorites using laser ablation ICP-MS (abstract). In *Lunar and Planetary Science XXX,* Lunar and Planetary Institute, Houston.

Cassen P. (1994) Utilitarian models of the primitive solar nebula. *Icarus, 112,* 405–429.

Cassen P. (1996) Nebula models for the fractionation of moderately volatile elements. *Meteoritics & Planet. Sci., 31,* 793–806.

Cassen P. and Boss A. P. (1988) Protostellar collapse, dust grains, and solar system formation. In *Meteorites and the Early Solar System* (J. F. Kerridge and M. S. Matthews, eds.), pp. 304–328. Univ. of Arizona Press, Tucson.

Chiang E. and Goldreich P. (1997) Spectral energy distributions of T Tauri stars with passive circumstellar disks. *Astrophys. J., 490,* 368–370.

Church S. E., Tilton G. R., Wright J. E., and Lee-Hu C.-N. (1976) Volatile element depletion and $^{39}$K/$^{41}$K fractionation in lunar soils. *Proc. Lunar Sci. Conf. 7th,* pp. 423–439.

Clayton R. N. and Mayeda T. K. (1999) Oxygen isotope studies of carbonaceous chondrites. *Geochim. Cosmochim. Acta, 63,* 2089–2104.

Clayton R. N., Mayeda T. K., and Molini-Velsko C. A. (1985) Isotopic variations in solar system material: evaporation and condensation of silicates. In *Protostars and Planets II* (D. C. Black and M. S. Matthews, eds.), pp. 755–771. Univ. of Arizona, Tucson.

Clayton R. N., Hinton R. W., and Davis A. M. (1988) Isotopic variations in the rock-forming elements in meteorites. *Philos. Trans. R. Soc. Lond., A325,* 483–501.

Craig H., Gordon L. I., and Horibe Y. (1963) Isotopic exchange effects in the evaporation of water. 1. Low-temperature experimental results. *J. Geophys. Res., 68,* 5079–5087.

D'Alessio P., Calvet N., and Hartmann L. (1997) The structure and emission of accretion disks irradiated by infalling envelopes. *Astrophys. J., 474,* 397–406.

D'Alessio P., Cantó J., Calvet N., and Lizano S. (1998) Accretion disks around young objects. I. The detailed vertical structure. *Astrophys. J., 500,* 411–427.

Davis A. M., Hashimoto A., Clayton R. N., and Mayeda T. K. (1990) Isotope mass fractionation during evaporation of Mg$_2$SiO$_4$. *Nature, 347,* 655–658.

Ebel D. S. and Grossman L. (1999) Condensation in dust-enriched systems. *Geochim. Cosmochim. Acta,* in press.

Esat T. M. (1996) Comment on "Potassium isotope cosmochemistry: Genetic implications of volatile element depletion" by Munir Humayun and Robert N. Clayton. *Geochim. Cosmochim. Acta, 60,* 3755–3758.

Field G. B. (1974) The composition of interstellar dust. In *The Dusty Universe* (G. B. Field and A. G. W. Cameron, eds.), pp. 89–112. MacGraw-Hill, New York.

Ganapathy R. and Anders E. (1974) Bulk compositions of the Moon and Earth estimated from meteorites. *Proc. Lunar Sci. Conf. 5th,* pp. 1181–1206.

Garner E. L., Machlan L. A., and Barnes I. L. (1975) The isotopic composition of lithium, potassium, and rubidium in some Apollo 11, 12, 14, 15, and 16 samples. *Proc. Lunar Sci. Conf. 6th,* pp. 1845–1855.

Gooding J. L. and Muenow D. W. (1977) Experimental vaporization of the Holbrook chondrite. *Meteoritics, 12,* 401–408.

Grossman J. N. (1996) Chemical fractionations of chondrites: Signatures of events before chondrule formation. In *Chondrules and the Protoplanetary Disk* (R. Hewins, R. Jones, and E. Scott, eds.), pp. 243–253. Cambridge Univ., Cambridge.

Grossman L. (1972) Condensation in the primitive solar nebula. *Geochim. Cosmochim. Acta, 36,* 597–619.

Grossman L. and Larimer J. W. (1974) Early chemical history of the Solar System. *Rev. Geophys. Space Phys., 12,* 71–101.

Grossman L. and Olsen E. (1974) Origin of the high-temperature fraction of C2 chondrites. *Geochim. Cosmochim. Acta, 38,* 173–187.

Gullbring E., Hartmann L., Briceño C., and Calvet N. (1998) Disk accretion rates for T Tauri stars. *Astrophys. J., 492,* 323–341.

Hartigan P., Edwards S., and Ghandour L. (1995) Disk accretion and mass loss from young stars. *Astrophys. J., 452,* 736–768.

Hartmann L., Kenyon S., and Hartigan P. (1993) Young stars: episodic phenomena, activity, and variability. In *Protostars and Planets III* (E. H. Levy and J. Lunine, eds.), pp. 497–518. Univ. of Arizona, Tucson.

Hartmann L., Calvet N., Gullbring E., and D'Alessio P. (1998) Accretion and the evolution of T Tauri disks. *Astrophys. J., 495,* 385–400.

Hayashi C., Nakazawa K., and Nakagawa Y. (1985) Formation of the solar system. In *Protostars and Planets II* (D. C. Black and M. S. Matthews, eds.), pp. 1100–1153. Univ. of Arizona, Tucson.

Hayashi M., Ohashi N., and Miyama S. M. (1993) A dynamically accreting gas disk around HL Tauri. *Astrophys. J. Lett., 418,* L71–L76.

Henning Th. and Stognienko R. (1996) Dust opacities for protoplanetary accretion disks — Influence of dust aggregates. *Astron. & Astrophys., 311,* 291–303.

Hobbs L. M., Welty D. E., Morton D. C., Spitzer L., and York D. G. (1993) The interstellar abundances of tin and four other heavy elements. *Astrophys. J., 411,* 750–755.

Hofmann A. W. (1980) Diffusion in natural silicate melts: A critical review. In *Physics of Magmatic Processes* (R. B. Hargraves, ed.), pp. 385–417. Princeton Univ., Princeton, New Jersey.

Humayun M. and Clayton R. N. (1995a) Potassium isotope cosmochemistry: Genetic implications of volatile element depletion. *Geochim. Cosmochim. Acta, 59,* 2131–2148.

Humayun M. and Clayton R. N. (1995b) Precise determination of the isotopic composition of potassium: Application to terrestrial rocks and lunar soils. *Geochim. Cosmochim. Acta, 59,* 2115–2130.

Hunten D. M., Pepin R. O., and Walker J. C. G. (1987) Mass fractionation in hydrodynamic escape. *Icarus, 69,* 532–549.

Huss G. R. (1990) Ubiquitous interstellar diamond and SiC in primitive chondrites: abundances reflect metamorphism. *Nature, 347,* 159–162.

Huss G. R. and Lewis R. S. (1994) Noble gases in presolar diamonds I: Three distinct components and their implication for diamond origins. *Meteoritics, 29,* 791–810.

Jochum K. P. and Palme H. (1990) Alkali elements in eucrites and SNC meteorites: No evidence for volatility related losses during magma eruption or thermal metamorphism. *Meteoritics, 25,* 373–374.

Kelly W. R. and Larimer J. W. (1977) Chemical fractionations in meteorites. VIII. Iron meteorites and the cosmochemical history of the metal phase. *Geochim. Cosmochim Acta, 41,* 93–111.

Kenyon S. J. and Hartmann L. W. (1987) Spectral energy distributions of T Tauri stars: disk flaring and limits on accretion. *Astrophys. J., 323,* 714–733.

Kenyon S. J., Calvet N., and Hartmann L. (1993) The embedded young stars in the Taurus — Auriga molecular cloud. I. Models for spectral energy distributions. *Astrophys. J., 414,* 676–694.

Kortenkamp S. J. and Wetherill G. W. (1998) Terrestrial planet and asteroid formation in the presence of giant planets (abstract). In *Lunar and Planetary Science XXIX.* Lunar and Planetary Institute, Houston.

Larimer J. W. (1967) Chemical fractionations in meteorites — I. Condensation of the elements. *Geochim. Cosmochim Acta, 31,* 1215–1238.

Larimer J. W. and Anders E. (1967) Chemical fractionations in meteorites — II. Abundance patterns and their interpretation. *Geochim. Cosmochim. Acta, 31,* 1239–1270.

Larimer J. W. and Anders E. (1970) Chemical fractionations in meteorites — III. Major element fractionation in chondrites. *Geochim. Cosmochim. Acta, 34,* 367–387.

Lee T., Papanastassiou D. A., and Wasserburg G. J. (1977) $^{26}$Al in the early solar system: Fossil or fuel? *Astrophys. J. Lett., 211,* L107–L110.

Lewis J. S. (1974) The temperature gradient in the solar nebula. *Science, 186,* 440–443.

Lynden-Bell D. and Pringle J. E. (1974) The evolution of viscous discs and the origin of the nebular variables. *Mon. Not. R. Astron. Soc., 168,* 603–637.

McDonough W. F. and Sun S.-s. (1995) The composition of the Earth. *Chem. Geol., 120,* 223–253.

Mittlefehldt D. W. (1987) Volatile degassing of basaltic achondrite parent bodies: Evidence from alkali elements and phosphorus. *Geochim. Cosmochim. Acta, 51,* 267–278.

Morfill G. E. (1985) Physics and chemistry in the primitive solar nebula. In *Birth and Infancy of Stars* (R. Lucas and A. Omont, eds.), pp. 693–794. North-Holland.

Morfill G. E. and Wood J. A. (1989) Protoplanetary accretion disc models: the effects of several meteoritic, astronomical, and physical constraints. *Icarus, 82,* 225–243.

Morgan J. W. and Anders E. (1980) Chemical composition of Earth, Venus, and Mercury. *Proc. Natl. Acad. Sci. USA, 77,* 6973–6977.

Natta A. (1993) The temperature profile in T Tauri disks. *Astrophys. J., 412,* 761–770.

Newsom H. E. (1990) Accretion and core formation in the Earth: Evidence from siderophile elements. In *Origin of the Earth* (H. E. Newsom and J. H. Jones, eds.), pp. 273–288. Oxford Univ., New York.

Nier A. O., McElroy M. B., and Yung Y. L. (1976) Isotopic composition of the Martian atmosphere. *Science, 194,* 68–70.

Osterloh M. and Beckwith S. V. W. (1995) Millimeter-wave continuum measurements of young stars. *Astrophys. J., 439,* 288–302.

Owen T., Maillard J. P., de Bergh C., and Lutz B. L. (1988) Deuterium on Mars: The abundance of HDO and the value of D/H. *Science, 240,* 1767–1770.

Palme H. and Boynton W. V. (1993) Meteoritic constraints on conditions in the solar nebula. In *Protostars and Planets III* (E. H. Levy and J. I. Lunine, eds.), pp. 979–1004. Univ. of Arizona, Tucson.

Palme H. and Wlotzka F. (1976) A metal particle from a Ca, Al-rich inclusion from the meteorite Allende, and the condensation of refractory siderophile elements. *Earth Planet. Sci. Lett., 33,* 45–60.

Palme H., Larimer J. W., and Lipschutz M. E. (1988) Moderately volatile elements. In *Meteorites and the Early Solar System* (J. F. Kerridge and M. S. Matthews, eds.), pp. 436–461. Univ. of Arizona, Tucson.

Pollack J. B., Hollenbach D., Simonelli D., Beckwith S., Roush T., and Fong W. (1994) Optical properties of grains in molecular clouds and accretion disks. *Astrophys. J., 421,* 615–639.

Rasmussen K. L., Malvin D. J., Buchwald V. F., and Wasson J. T. (1984) Compositional trends and cooling rates of group IVB meteorites. *Geochim. Cosmochim. Acta, 30,* 805–813.

Ruden S. P. and Pollack J. B. (1991) The dynamical evolution of the protosolar nebula. *Astrophys. J., 375,* 740–760.

Russell W. A., Papanastassiou D. A., Tombrello T. A., and Epstein S. (1977) Ca isotope fractionation on the moon. *Proc. Lunar Sci. Conf. 8th,* pp. 3791–3805.

Russell W. A., Papanastassiou D. A., and Tombrello T. A. (1978) Ca isotope fractionation on the Earth and other solar system materials. *Geochim. Cosmochim. Acta, 42,* 1075–1090.

Shu F. H. (1977) Self-similar collapse of isothermal spheres and star formation. *Astrophys. J., 214,* 488–497.

Shu F. H. (1992) *The Physics of Astrophysics. Vol. II. Gas Dynamics.* University Science Books, Mill Valley. 90 pp.

Strom S. E., Edwards S., and Skrutskie M. F. (1993) Evolutionary timescales for circumstellar disks associated with intermediate and solar-type stars. In *Protostars and Planets III* (E. H. Levy and J. Lunine, eds.), pp. 837–881. Univ. of Arizona, Tucson.

Taylor S. R. (1979) Lunar and terrestrial potassium and uranium abundances: Implications for the fission hypothesis. *Proc. Lunar Planet. Sci. Conf. 10th,* pp. 2017–2030.

Taylor S. R. (1992) *Solar System Evolution: A New Perspective.* Cambridge Univ., New York. 307 pp.

Urey H. C. (1947) The thermodynamic properties of isotopic substances. *J. Chem. Soc.,* 562–581.

Valenti J. A., Basri G., and Johns C. M. (1993) T Tauri stars in blue. *Astron. J., 106,* 2024–2050.

Wai C. M. and Wasson J. T. (1977) Nebular condensation of moderately volatile elements and their abundances in ordinary chondrites. *Earth Planet. Sci. Lett., 36,* 1–13.

Walter F. (1986) X-ray sources in regions of star formation. I. The naked T Tauri stars. *Astrophys. J., 306,* 573–586.

Wang J., Davis A. M., Clayton R. N., and Mayeda T. K. (1994) Chemical and isotopic fractionation during the evaporation of the FeO-MgO-SiO$_2$-CaO-Al$_2$O$_3$-TiO$_2$-REE melt system (abstract). In *Lunar and Planetary Science XXV,* pp. 1457–1458. Lunar and Planetary Institute, Houston.

Wang J., Yu Y., Alexander C. M. O'D., and Hewins R. H. (1999) The influence of oxygen and hydrogen on the evaporation of K (abstract). In *Lunar and Planetary Science XXX.* Lunar and Planetary Institute, Houston.

Wänke H. and Dreibus G. (1988) Chemical composition and accretion history of terrestrial planets. *Philos. Trans. R. Soc. Lond., A325,* 545–557.

Wänke H., Dreibus G., and Jagoutz E. (1984) Mantle chemistry and accretion history of the Earth. In *Archaean Geochemistry* (A. Kröner et al., eds.), pp. 1–24. Springer-Verlag, Berlin.

Wasson J. T. (1985) *Meteorites, Their Record of Early Solar System History.* Freeman, New York. 155 pp.

Wasson J. T. and Chou C.-L. (1974) Fractionation of moderately volatile elements in ordinary chondrites. *Meteoritics, 9,* 69–84.

Wasson J. T. and Kallemeyn G. W. (1988) Compositions of chondrites. *Philos. Trans. R. Soc. London, A325,* 535–544.

Watson L. L., Hutcheon I. D., Epstein S., and Stolper E. M. (1994) Water on Mars: Clues from deuterium/hydrogen and water contents of hydrous phases in SNC meteorites. *Science, 265,* 85–90.

Weidenschilling S. J. (1977) The distribution of mass in the planetary system and solar nebula. *Astrophys. Space Sci., 51,* 153–158.

Weidenschilling S. J. and Cuzzi J. N. (1993) Formation of planetesimals in the solar nebula. In *Protostars and Planets III* (E. H. Levy and J. Lunine, eds.), pp. 1031–1060. Univ. of Arizona, Tucson.

Wetherill G. W. (1994) Provenance of the terrestrial planets. *Geochim. Cosmochim. Acta, 58,* 4513–4520.

Whitney B. A. and Hartmann L. (1993) Model scattering envelopes of young stellar objects. II. Infalling envelopes. *Astrophys. J., 402,* 605–622.

Whitney B. A., Kenyon S. J., and Gomez M. (1997) Near-infrared imaging polarimetry of embedded young stars in the Taurus-Auriga molecular cloud. *Astrophys. J., 485,* 703–784.

Wolf R., Richter G. R., Woodrow A. B., and Anders E. (1980) Chemical fractionations in meteorites — XI. C2 chondrites. *Geochim. Cosmochim. Acta, 44,* 711–717.

Wood J. A. (1998) The HED parent body: Thermal and Sr isotope evolution (abstract). In *Lunar and Planetary Science XXIX,* Abstract #1385. Lunar and Planetary Institute, Houston.

Woolum D. S. and Cassen P. (1999) Astronomical constraints on nebular temperatures: implications for planetesimal formation. *Meteoritics & Planet. Sci.,* in press.

Yoneda S. and Grossman L. (1995) Condensation of CaO-MgO-Al$_2$O$_3$-SiO$_2$ liquids from cosmic gases. *Geochim. Cosmochim. Acta, 59,* 3413–3444.

Young E. D., Nagahara H., Mysen B. O., and Audet D. M. (1998) Non-Rayleigh oxygen isotope fractionation by mineral evaporation: Theory and experiments in the system SiO$_2$. *Geochim. Cosmochim. Acta, 62,* 3109–3116.

Zanda B., Yu Y., Bourot-Denise M., Hewins R. H., and Connolly H. C. Jr. (1999) Sulfur behaviour in chondrule formation and metamorphism. *Geochim. Cosmochim. Acta,* submitted.

Zinner E. (1988) Interstellar cloud material in meteorites. In *Meteorites and the Early Solar System* (J. F. Kerridge and M. S. Matthews, eds.), pp. 956–983. Univ. of Arizona, Tucson.

# Timescales of Planetesimal Formation and Differentiation Based on Extinct and Extant Radioisotopes

**Richard W. Carlson**
*Carnegie Institution of Washington*

**Guenter W. Lugmair**
*Max Planck Institute for Chemistry and*
*University of California, San Diego*

At the time of their formation, many meteorites contained live, now extinct, radioisotopes with circa million-year half-lives. These extinct radioisotope systems show that condensation and accumulation of solids into bodies of sufficient size to undergo melting and internal differentiation occurred on a timescale of a few million years. For example, the oldest Pb-Pb age recorded for meteoritic material is the 4566 ± 8-Ma age for Allende high-temperature inclusions compared to a Mn-Cr age for differentiation of the howardite-eucrite-diogenite parent body of 4564.8 ± 0.9 Ma. This short timescale allows $^{26}Al$ to have been the primary heat source responsible for initial differentiation of planetesimals. Significantly younger ages for some meteorites, for example, the <4480-Ma cumulate eucrites, requires continued thermal processing of planetesimals over longer timescales. Nevertheless, the data suggest that the Earth and Moon accumulated from planetesimals that had already undergone internal differentiation.

## 1. INTRODUCTION

The question "How old is the Earth?" has been asked for as long as humankind has worried about its place in the universe. The answer "four and a half billion years" was discovered just over 40 years ago (*Patterson*, 1956). Since then, geochronology and cosmochronology have struggled to improve the precision of that estimate. Within the last 10–20 years, advances in analytical techniques have allowed exploitation of a number of radiometric systems, including those based on long-lived (half-lives of a billion years or more) radioactive isotopes and on isotopes that have decay half lives of a few million years or less (Table 1). Under ideal circumstances, some of these radiometric systems allow time resolution of a million years or less on events occurring 4.5 b.y. ago. The extinct radiometric systems, in particular, provide excellent chronological resolution during this early era of solar system evolution, but they cannot provide an absolute timescale. Fortunately, the relatively short half-life of the still extant isotope $^{235}U$ allows the U-Pb system to provide chronological resolution comparable to that of the extinct radiometric systems. Linking the timescale provided by U-Pb with that indicated by the extinct systems offers the potential for defining a high-precision chronological picture of the events that shaped the early solar system.

If solar nebula evolution was completed on the $10^5$-yr timescale suggested by star formation models (*Shu et al.,* 1993), then the million-year precision provided by extinct radiometric systems is inadequate to resolve events occurring during this process and would instead yield one "age"

for the solar system. However, if evolution of the protoplanetary disk is more prolonged (see *Podosek and Cassen,* 1994 for a review), to the tens-of-million-years timescale suggested by planetary accumulation models (*Wetherill,* 1990), modern radiometric chronology should be able to resolve various events involved in the planet formation process. The data becoming available for several radiometric systems are repeatedly showing "age" variations of several million years or more between different meteoritic materi-

TABLE 1.  Radioactive isotopes detected in early solar system materials.

| Parent | Daughter | Half-Life ($10^6$ yr) |
|---|---|---|
| $^{26}Al$ | $^{26}Mg$ | 0.73 |
| $^{40}K$ | $^{40}Ar$, $^{40}Ca$ | 1270 |
| $^{53}Mn$ | $^{53}Cr$ | 3.7 |
| $^{60}Fe$ | $^{60}Ni$ | 1.5 |
| $^{87}Rb$ | $^{87}Sr$ | 48,800 |
| $^{107}Pd$ | $^{107}Ag$ | 6.5 |
| $^{129}I$ | $^{129}Xe$ | 15.7 |
| $^{146}Sm$ | $^{142}Nd$ | 103 |
| $^{147}Sm$ | $^{143}Nd$ | 106,000 |
| $^{176}Lu$ | $^{176}Hf$ | 35,700 |
| $^{182}Hf$ | $^{182}W$ | 9 |
| $^{187}Re$ | $^{187}Os$ | 41,600 |
| $^{190}Pt$ | $^{186}Os$ | 450,000 |
| $^{232}Th$ | $^{208}Pb$ | 14,010 |
| $^{235}U$ | $^{207}Pb$ | 704 |
| $^{238}U$ | $^{206}Pb$ | 4469 |
| $^{244}Pu$ | Fission Xe | 80 |

als. Though alternate explanations exist, we believe that this range in apparent ages, for the most part, is real and that it reflects the timescale required to process the solid materials forming from the protoplanetary disk.

In this chapter, we review the chronological constraints available for the assembly and chemical differentiation of planetesimals in the early solar nebula. The data derive primarily from studies of meteorites and their constituent minerals. This information can be used to address the absolute time of formation of the first solids in the solar nebula and the time required to assemble material into planetesimals of sufficient size to experience thermal metamorphism, endogenous magmatism, and core formation.

## 2. STARTING CONDITIONS

Solar system evolution usually is defined as initiating with the collapse of a cold molecular cloud to form a centrally condensed star surrounded by a ring of dust and gas. Idealized "hot" solar nebula models (*Cameron,* 1962) then mix the interstellar flotsam and jetsam originally present in the molecular cloud by vaporizing the dust, stirring the protoplanetary nebula while it is in the gas phase, and then condensing minerals from a homogeneous gas as it cools (*Grossman,* 1972). The discovery that some primitive meteoritic materials have isotopic compositions distinct from the solar average (*Black and Pepin,* 1969; *Clayton et al.,* 1973; *Lee et al.,* 1978; *Lugmair et al.,* 1978; *McCulloch and Wasserburg,* 1978; *Niederer et al.,* 1980; *Niemeyer and Lugmair,* 1981; *Rotaru et al.,* 1992; *Podosek et al.,* 1997) and that discrete presolar grains are preserved in some meteorites (e.g., *Amari et al.,* 1994; *Bernatowicz and Zinner,* 1996) prove that this homogenization was incomplete. Intact presolar grains bearing the clear isotopic signature of distinct nucleosynthetic origins (*Zinner,* 1996) indicate that not all the material present in the protoplanetary disk reached temperatures sufficient for volatilization and mixing in the gas phase. The presence of more widespread isotopic anomalies (*Clayton et al.,* 1973; *Niederer et al.,* 1980; *Niemeyer and Lugmair,* 1981; *Rotaru et al.,* 1992; *Podosek et al.,* 1997) and the suggestion of an initial gradient in the $^{53}$Mn/$^{55}$Mn of the solar nebula (*Lugmair and Shukolyukov,* 1998) further indicates that complete chemical and isotopic homogeneity in the solar nebula was not achieved prior to condensation and planetesimal formation.

Initial chemical and isotopic heterogeneity in the disk and its implications for interpretation of the results from isotopic systems must be considered. In essence, one must distinguish ages that date events occurring in our solar system from those inherited from materials entering, but not mixing with, the presolar molecular cloud. This is particularly important when dealing with dating systems based on extinct radionuclides since the abundance of the parent isotope cannot be measured, but must be inferred from the variation in abundance of the daughter isotope. An "isochron" in a radiometric system based on an extinct nuclide provides not an age, but a measure of the abundance of the extinct

isotope when the analyzed materials were last in chemical and isotopic equilibrium. Variations in abundance of the extinct isotope between different samples could reflect variations in age of formation, but they also could indicate a variable abundance of this isotope in different regions of the solar nebula. The latter alternative becomes of particular concern in models where the collapse of the molecular cloud is triggered by a shock wave from a nearby supernova that also injects the cloud with newly produced elements (*Cameron and Truran,* 1977). Whether or not this newly introduced material has sufficient time to become homogeneously mixed into the solar nebula before planetesimal formation begins is a question awaiting resolution.

## 3. THE FIRST SOLIDS FORMED IN THE SOLAR NEBULA

The molecular cloud from which the solar nebula formed most likely consisted of a mixture of dust and gas (e.g., *Elmegreen,* 1985). The solids in this cloud presumably reflected the condensable products of stellar evolution and outflow occurring over the history of the galaxy with a diverse record of isotopic and chemical compositions determined by their nucleosynthetic paths. As mentioned above, the existence of discrete presolar grains in meteorites indicates that the solar nebula never reached complete homogeneity before planetesimal formation began. Nevertheless, the fact that isotopically anomalous material in meteorites was discovered only because of continuously improving analytical techniques within the last 35 years (*Reynolds and Turner,* 1964; *Black and Pepin,* 1969; *Clayton et al.,* 1973) indicates that a considerable degree of homogenization was achieved prior to planetesimal formation.

The first solids to form from a cooling hot nebula obviously would be those materials with the highest condensation temperatures (*Grossman,* 1972). A clear candidate for such materials are the calcium-aluminum-rich inclusions (CAIs) in primitive meteorites (*Grossman,* 1980). Calcium-aluminum-rich inclusions contain a variety of high-condensation temperature minerals such as corundum, perovskite, melilite, hibonite, and spinel. They also tend to give the oldest ages from both extant and extinct radiometric systems, which supports the concept that they are the first solids to condense following a high-temperature phase of the solar nebula. Most CAIs, however, are rich in $^{16}$O compared to other meteorites, Earth, and the Moon (*Clayton et al.,* 1973), and a small fraction of CAIs are isotopically anomalous in almost every element compared to normal solar system material (*Lee et al.,* 1978; *Lugmair et al.,* 1978; *McCulloch and Wasserburg,* 1978; *Loss et al.,* 1994). Thus, though the CAIs may be the best candidates for the first solids to form in the condensing solar nebula, they clearly sample a different nebular mix compared to the materials that make up Earth and the Moon. As will be discussed, this imposes some serious concerns in relating the extinct radiometric systems in CAIs with those of more common meteoritic and planetary materials.

Calcium-aluminum-rich inclusions extracted from the Allende carbonaceous chondrite provide the oldest Pb-Pb age of any material yet measured. Nine inclusions analyzed by *Chen and Wasserburg* (1981) define an excellent linear array on a plot of $^{207}Pb/^{206}Pb$ vs. $^{204}Pb/^{206}Pb$ (Fig. 1) that defines an age of 4566 ± 8 Ma [line fitting and 2σ error calculation performed with ISOPLOT (*Ludwig*, 1991)]. The inclusions show a very wide range of Pb isotopic composition with values both less and considerably more radiogenic than terrestrial Pb. Ages previously reported for these data (*Chen and Wasserburg*, 1981) are either single-stage model ages, calculated by subtracting an estimate of the solar system initial Pb isotopic composition from the measured isotopic compositions, or a Pb-Pb isochron fit to the six most radiogenic samples with the line forced to go through the estimate of solar system initial Pb. Although the data have not yet been published, more recent results discussed by *Göpel et al.* (1991) and *Allègre et al.* (1995) give a single-stage Pb model age for several CAIs of 4566 ± 2 Ma.

Of some concern with the Allende CAI Pb data is that the line defined by the inclusions passes substantially to the unradiogenic side of Canyon Diablo Pb (Fig. 1) (*Chen and Wasserburg*, 1981; *Tera and Carlson*, 1999). Canyon Diablo Pb is believed to represent the initial Pb isotopic composition of the solar nebula. Unlike the solar $^{87}Sr/^{86}Sr$, average solar Pb isotopic composition will not evolve rapidly because the solar $^{238}U/^{204}Pb$ ratio is quite low (~0.25) (*Anders and Grevasse*, 1989). For example, at this $^{238}U/^{204}Pb$ ratio, solar $^{206}Pb/^{204}Pb$ has changed from 9.307 (Canyon Diablo initial) to 9.564 today in comparison to an average modern terrestrial $^{206}Pb/^{204}Pb$ ratio of around 18. The initial Pb isotopic composition indicated by the Allende CAI data is $^{206}Pb/^{204}Pb$ = 9.068 and $^{207}Pb/^{204}Pb$ = 9.822 (*Tera and Carlson*, 1999) in comparison to Canyon Diablo Pb of 9.307 and 10.294 respectively (*Tatsumoto et al.*, 1973; *Göpel et*

*al.*, 1985). In contrast, chondrules extracted from Allende (*Chen and Tilton*, 1976) define an age of 4560 ± 67 Ma (the large uncertainty results primarily from a limited spread in Pb isotopic composition and the relatively unradiogenic composition of the chondrules) with an initial Pb isotopic composition consistent with Canyon Diablo Pb (Fig. 1).

There are three possible explanations for the discrepancy between Canyon Diablo Pb and the initial Pb isotopic composition indicated by the Allende CAIs: (1) Lead-lead isochrons that pass to the unradiogenic side of Canyon Diablo are not uncommon for meteoritic materials, but the deviation usually can be explained by contamination with terrestrial Pb (*Tera and Carlson*, 1999). However, since the Allende CAI data plot to both the radiogenic and unradiogenic side of terrestrial Pb, fortuitous control over the mixing proportions of indigenous meteoritic Pb and terrestrial contamination Pb would be required to maintain the excellent linearity of the inclusion data (*Tera and Carlson*, 1999). (2) The Allende CAIs include isotopically anomalous Pb compared to "normal" solar system materials. For example, the difference in $^{206}Pb/^{204}Pb$ and $^{207}Pb/^{204}Pb$ between the CAI initial Pb and Canyon Diablo is consistent with mass fractionation (S. J. G. Galer, personal communication, 1998), although the difference is very large (1.4% per AMU) for normal thermally induced mass fractionation. (3) The difference could reflect radiogenic ingrowth in the 6-m.y. age difference between Allende CAIs and chondrules. This would require a $^{238}U/^{204}Pb$ ratio of 125 (*Tera and Carlson*, 1999). Uranium/lead ratios this high are not uncommon in refractory meteoritic solids (*Lugmair and Galer*, 1992; *Göpel et al.*, 1994; *Tera et al.*, 1997), so the more radiogenic Pb in Canyon Diablo compared to CAIs conceivably could have been produced in the parent planetesimal before it segregated the core that would eventually produce the Canyon Diablo meteorite. However, Canyon Diablo Pb appears to be an appropriate estimate of the initial Pb isotopic composition of a wide variety of meteorite classes, many of which do not derive from parent bodies with high U/Pb. Consequently, appealing to a unique evolutionary history for the Canyon Diablo parent body may not be the correct explanation of the difference between Canyon Diablo Pb and the initial Pb isotopic composition indicated by CAIs. For the whole solar nebula to evolve with this high U/Pb ratio is unlikely. Any mass balance would eventually require the reincorporation of a large mass of unradiogenic Pb into the high U/Pb region/material in order to bring the nebular U/Pb ratio back to solar values.

Whatever the explanation, the data suggest that CAIs incorporated an initial Pb distinct from that of the average solar nebula. An important consequence is that single-stage model ages for CAIs calculated by subtracting Canyon Diablo Pb (*Göpel et al.*, 1991; *Allègre et al.*, 1995), instead of the initial Pb indicated by the CAI Pb-Pb isochron, from measured isotopic compositions will underestimate the true formation age by an amount that decreases as the sample becomes more radiogenic. The difference in initial Pb isotopic composition need not invalidate the Pb-Pb age ob-

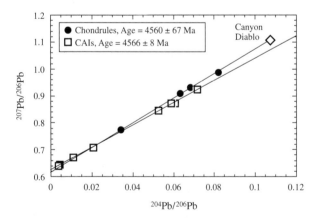

**Fig. 1.** Lead-lead systematics for Allende CAIs (*Chen and Wasserburg*, 1981) and chondrules (*Chen and Tilton*, 1976) showing the widespread Pb-isotopic composition of the CAIs. The line defined by the CAI data pass to the unradiogenic side of Canyon Diablo Pb, indicative of a nonsolar initial Pb-isotopic composition in CAIs.

tained for the Allende CAIs, as long as the range of U/Pb ratios observed in inclusions is a result of chemical processing during condensation/accumulation and they all formed with the same initial Pb isotopic composition. Variable incorporation of presolar material into CAIs, however, if sufficient to significantly modify the U/Pb ratio of individual inclusions, could introduce a "memory" of presolar events and cause the CAI Pb age to overestimate the beginning of condensation in the solar nebula.

In support of their antiquity, CAIs contain clear evidence for the presence of many now extinct nuclides, including $^{26}$Al (*Lee et al., 1977*), $^{53}$Mn (*Birck and Allègre, 1985*), $^{129}$I (*Wasserburg and Huneke, 1979*) and $^{146}$Sm (*Lugmair et al., 1983*). Aluminum-26 is a particularly important nuclide both because it can provide a very-high-resolution chronometer, and because it would serve as the main heat source for melting and initial differentiation of solar system objects if it were present in sufficient abundance (*Urey, 1955*). Whether or not $^{26}$Al was uniformly present in the early solar nebular, however, remains an issue of some discussion.

For those CAIs that contained live $^{26}$Al at the time of their formation, the majority have a $^{26}$Al/$^{27}$Al ratio of 3–6 × 10$^{-5}$ (*Lee et al., 1977; MacPherson et al., 1995*). Some CAIs, however, have no detectable $^{26}$Mg excess attributable to $^{26}$Al decay and provide an upper limit of $^{26}$Al/$^{27}$Al < 10$^{-7}$ (Fig. 2) (*Lee et al., 1976; Hutcheon, 1982*). Of the CAIs that lack evidence for live $^{26}$Al are those that also show distinct isotopic compositions for many elements, the so-called FUN (Fractionation and Unknown Nuclear) inclusions (*Lee et al., 1979*). The difference in $^{26}$Al/$^{27}$Al between CAIs translates to a >7-m.y. age difference between the "haves" and "have-nots" if it represents decay and not heterogeneity in initial $^{26}$Al abundance. The sense of the age difference, however, is unexpected. If the source of isotopically anomalous material in the FUN inclusions is recently introduced material into the solar nebula, one might expect it to be coupled with higher, not lower, $^{26}$Al abundance if the $^{26}$Al also was provided by outflow from a nearby star or supernova. One attractive alternative is that the FUN inclusions actually are among the earliest condensed solids in the solar nebula and sample the chemical and isotopic heterogeneity of the precollapse solar molecular cloud. In this case, FUN inclusions would not have contained the $^{26}$Al that may have been added at a later stage of solar nebula evolution. The non-FUN inclusions containing $^{26}$Al then would have formed by condensation and accretion in the solar nebula after molecular cloud collapse had caused vaporization and mixing of preexisting material with the newly introduced stellar outflow. A similar sequence of events can be invoked for "local" production of $^{26}$Al and other short-lived radionuclides in the X-wind (*Shu et al., 1993*) or by spallation (*Clayton and Jin, 1995*). Another alternative is that the CAIs simply did not sample average solar system material and that $^{26}$Al was not homogeneously distributed within the nebula (*Lee et al., 1977*). This alternative becomes less palatable as evidence for live $^{26}$Al is found in more "normal" meteoritic materials such as chondrules (*Hutcheon and Hutchison,*

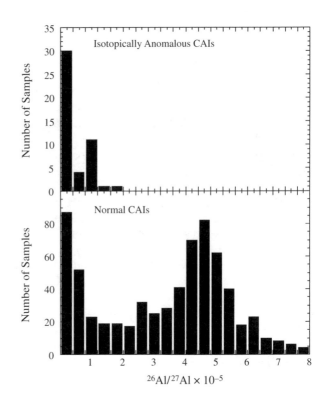

**Fig. 2.** Histogram of $^{26}$Al/$^{27}$Al for normal and isotopically anomalous (the so-called FUN and UN inclusions) from Allende. After *MacPherson et al.* (1995).

1989; *Russell et al., 1996*), chondrite feldspars (*Zinner and Göpel, 1992*) and perhaps even eucrites (*Srinivasan et al., 1998, 1999*).

Evidence for live $^{53}$Mn in Allende CAIs has been presented by *Birck and Allègre* (1985, 1988). A mineral isochron for one inclusion (BR-1B) indicates a $^{53}$Mn/$^{55}$Mn ratio of (3.1 ± 0.7) × 10$^{-5}$ [line fitting and 2σ error calculation performed with ISOPLOT (*Ludwig, 1991*)] at the time of closure of the Mn-Cr system in this inclusion. This value overlaps within uncertainty the $^{53}$Mn/$^{55}$Mn ratio of (4.53 ± 1.2) × 10$^{-5}$ derived by fitting the data for all the Allende CAIs reported by *Birck and Allègre* (1988). A problem with strict chronological interpretation of the Cr data for Allende CAIs, however, is that the Cr in these inclusions, and in several groups of primitive carbonaceous chondrites, displays anomalous abundances of $^{54}$Cr (Fig. 3) (*Birck and Allègre, 1988; Rotaru et al., 1992; Podosek et al., 1997*). This introduces the strong possibility that some of the observed variation in $^{53}$Cr may be due to a presolar component or is an analytical artifact resulting from variations in the value of the Cr-isotopic composition used to correct for instrumentally induced mass fractionation rather than directly attributable to *in situ* decay of $^{53}$Mn (*Birck and Allègre, 1988; Lugmair and Shukolyukov, 1998*).

In this sense, it is important to note that both the Pb-Pb and the bulk-inclusion Mn-Cr isochrons for Allende CAIs are, in essence, whole-rock isochrons that are taken to record the time when individual inclusions separated from a nebu-

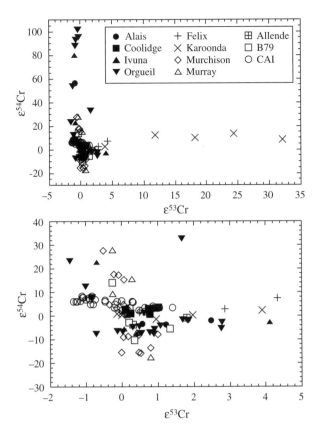

**Fig. 3.** Illustration of the magnitude of isotopically anomalous Cr in carbonaceous chondrites (*Rotaru et al.,* 1992). ε$^{53}$Cr and ε$^{54}$Cr are the relative deviation of the $^{53}$Cr/$^{52}$Cr and $^{54}$Cr/$^{52}$Cr, in parts per 10,000, compared to terrestrial Cr. The lower part of the figure is an expanded scale of the same data shown in the upper figure.

ration, as to cause reequilibration of the rim of the inclusion, but not its interior; or (3) the rim of the inclusion continued to interact with the solar nebula after its formation. Combined petrologic and isotopic studies of other individual refractory inclusions suggest that option 2 above is the correct explanation (*Podosek et al.,* 1991; *MacPherson and Davis,* 1993). In contrast, minerals from the rim and core of a type A CAI inclusion from Efremovka indicate both a higher $^{26}$Al/$^{27}$Al and initial $^{26}$Mg/$^{24}$Mg in the rim compared to the interior of the inclusion (*Fahey et al.,* 1987). This was used as evidence that the inclusion grew gradually from a solar nebula of varying isotopic composition and $^{26}$Al abundance (*Fahey et al.,* 1987) and reintroduces the issue of whether variations in the abundance of $^{26}$Al represent age differences between different samples or instead reflect heterogeneous distribution of the short-lived nuclide in the early solar nebula. It is possible that early $^{26}$Al-free material can become mixed with material containing freshly produced or injected $^{26}$Al (i.e., temporal heterogeneity). Chronological interpretation of such an assemblage would thus be meaningless.

## 4. EARLY IGNEOUS DIFFERENTIATION

As mentioned above, the Al-Mg systematics of CAIs can be interpreted as recording the first igneous events in the solar system, if these inclusions are in fact igneous assemblages. More clear-cut igneous products of the solar system are chondrules, igneous meteorites like the angrites and eucrites, and also the iron-rich meteorite classes that may represent the cores of planetesimals.

### 4.1. Chondrites

The bulk of chondritic material displays no evidence for live $^{26}$Al. Correlated Al and $^{26}$Mg excesses, however, have been found in chondrules from three ordinary chondrites: Semarkona (*Hutcheon and Hutchison,* 1989), Inman, and Chainpur (*Russell et al.,* 1996) at an abundance of $^{26}$Al/$^{27}$Al of ~8–9 × 10$^{-6}$, and in plagioclase in the H4 chondrite St. Marguerite (*Zinner and Göpel,* 1992) at an abundance of $^{26}$Al/$^{27}$Al = (2.0 ± 0.6) × 10$^{-7}$. Compared to a starting $^{26}$Al/$^{27}$Al ratio of 5 × 10$^{-5}$ in Allende CAIs, the data imply that the oldest chondrules are ~2 m.y. younger, and that the feldspars are 5.6 ± 0.4 m.y. younger than CAIs. It is important to note, however, that the majority of chondrules do not contain evidence for live $^{26}$Al at the time of their formation implying a chondrule formation/metamorphism interval in excess of 5–10 m.y.

Chondrules from the Chainpur and Bishunpur chondrites were found to contain variable abundances of $^{53}$Cr that correlate with their Mn/Cr ratios (*Nyquist et al.,* 1997b, 1999). The slope of these correlations correspond to $^{53}$Mn/$^{55}$Mn = (9.4 ± 1.7) × 10$^{-6}$ and (9.5 ± 3.1) × 10$^{-6}$ respectively (*Nyquist et al.,* 1999). Comparing these values to the 3.1 × 10$^{-5}$ $^{53}$Mn/$^{55}$Mn found for Allende CAI BR1-1B (*Birck and*

lar source with uniform Pb and Cr isotopic composition. The Cr and Pb systematics of the bulk inclusions would be disturbed if the inclusions later exchanged material with the solar nebula, but isochemical metamorphism and recrystallization within the individual inclusion would not alter them. In contrast, the Al-Mg and the BR-1B Mn-Cr correlations are measured for minerals extracted from individual inclusions. These correlations thus record the last equilibration of Al and Mg or Mn and Cr between the mineral phases of the inclusion. This last equilibration event could be the time of condensation of the individual phases from the nebula, the crystallization of the individual grains from a homogenous melt, or the time of a thermal metamorphism that allowed exchange of Mg and Cr between minerals. A good example of the latter process is illustrated in the data for one Allende CAI inclusion, 3529-26 (*Wasserburg,* 1985). Minerals separated from the outer edge of the inclusion indicate a $^{26}$Al/$^{27}$Al ratio of 2.3 × 10$^{-5}$ while minerals separated from the interior provide a higher value of 3.8 × 10$^{-5}$. These data suggest either that (1) the rim formed 550,000 years later than the interior; (2) the inclusion experienced transient reheating of sufficient magnitude, but limited du-

*Allègre,* 1988), and assuming homogeneous distribution of $^{53}Mn/^{55}Mn$ in the solar system, implies an age difference of ~6 m.y. between the chondrules and CAIs. Both the $^{26}Al$ and $^{53}Mn$ results for chondrules thus indicate that they began forming within 2–6 m.y. of CAI formation and that their formation and/or metamorphism continued past the point where $^{26}Al$ had decayed to undetectable levels, i.e., at least 5–10 m.y.

This evidence for rapid igneous processing of chondritic material is supported by the detection of live $^{53}Mn$ in Mn-rich minerals from unequilibrated enstatite chondrites (*Birck and Allègre,* 1988; *Wadhwa et al.,* 1997). These results give Mn-Cr correlations indicative of $^{53}Mn/^{55}Mn$ between $6 \times 10^{-6}$ and $5 \times 10^{-8}$. Again, assuming the Allende CAI $^{53}Mn/^{55}Mn$ of $3.1 \times 10^{-5}$ as the starting point and solar system homogeneity of $^{53}Mn/^{55}Mn$, this implies an age difference of 9–34 Ma between CAI formation and the last equilibration of Mn-Cr between the minerals of the unequilibrated enstatite chondrites. Results reported recently (*Shukolyukov and Lugmair,* 1999) for three enstatite chondrites (Abee, Indarch, Khairpur) show good correlations between $^{55}Mn/^{52}Cr$ and $^{53}Cr/^{52}Cr$ that give initial $^{53}Mn/^{55}Mn$ between $1.2 \times 10^{-6}$ and $3.0 \times 10^{-6}$. These values correspond to $\Delta T$ of 12–17 m.y. compared to the BR-1B CAI data.

While one must still face the issue of whether these variations in the initial abundance of short-lived nuclides reflect age differences, heterogeneous distributions of the radioactive nuclei, or nucleosynthetic variability in the isotopic composition of the daughter, the presence of live $^{26}Al$ and $^{53}Mn$ in "common" solar system materials such as the chondrites suggests that these short-lived nuclides were distributed at least throughout the inner part of the solar nebula and were not restricted to a special place where only CAIs formed.

Phosphates from the St. Severin LL6 ordinary chondrite provide unusually consistent Pb-Pb data (*Manhès et al.,* 1978; *Chen and Wasserburg,* 1981; *Göpel et al.,* 1991, 1994) that indicate an age of 4558 ± 6 Ma for closure of the U-Pb system in this meteorite. Phosphates from the H4 chondrite St. Marguerite provide a slightly older Pb-Pb age of 4563 ± 1 Ma with an initial Pb isotopic composition consistent with Canyon Diablo Pb (*Göpel et al.,* 1994). These findings are significant in that they provide an absolute chronometry for meteoritic material that agrees with the few-million-year age differences indicated by the extinct radiometric systems. For example, the age difference between St. Marguerite and CAIs is 3 ± 8 Ma from Pb-Pb and 5.6 ± 0.4 Ma from Al-Mg.

Uranium-lead data for phosphates, however, also clearly show that thermal processing of planetesimals did not stop on the million-year timescale but extended to many tens to hundreds of million years. For example, phosphates from several H5, H6, L5, and L6 chondrites define concordant U-Pb and Pb-Pb ages of 4518 ± 6.9 Ma and 4515 ± 6.4 Ma respectively, indicating closure of the U-Pb system some 50 m.y. after that in St. Marguerite phosphates (*Göpel et al.,* 1994; *Tera and Carlson,* 1999). Given the relatively low

diffusional closure temperature of phosphates, these relatively young ages most likely represent either shock metamorphic effects or perhaps an extended cooling interval for chondrites that may have been buried deep within a parent planetesimal.

## 4.2.  Primitive Achondrites

Two primitive achondrites, Acapulco and Brachina, show a very consistent picture in terms of their ages. Acapulco, with a chemical composition similar to H chondrites, yields a $^{53}Mn/^{55}Mn$ ratio of $(7.5 \pm 1.4) \times 10^{-7}$ and a calculated absolute "Mn-Cr" age of 4555.1 ± 1.2 Ma (*Zipfel et al.,* 1996), referenced to a 4558-Ma age for angrites. This is in good agreement with its Pb-Pb age of 4557 ± 2 Ma (*Göpel et al.,* 1992). For Brachina, the recent Mn-Cr study of *Wadhwa et al.* (1998) resulted in a very primitive $^{53}Mn/^{55}Mn$ ratio of $(3.8 \pm 0.4) \times 10^{-6}$. Since the Mn/Cr ratio in bulk Brachina, just as in Acapulco, is consistent with an ordinary chondritic Mn/Cr ratio, an absolute age can be calculated from the angrite Pb-Pb and Mn-Cr data. This yields the very old age of 4563.7 ± 0.9 Ma, attesting to a very early formation on or near an asteroidal surface.

## 4.3.  Angrites

Angrites are a rare and thinly populated meteorite class (only four samples are known) that consist predominantly of fassaitic clinopyroxene with variable proportions of olivine, plagioclase, and rare large whitlockite grains. Angrites figure prominently in efforts to reconstruct solar system chronology because they are unmetamorphosed examples of the crystallization products of melts on a planetesimal. For the following reasons, angrites may provide a better reference point than chondrites or CAIs for the relative chronologies provided by extinct radiometric systems: (1) Angrites provide concordant and very precise absolute U-Pb and Pb-Pb ages. (2) Angrites contained live $^{53}Mn$, $^{146}Sm$, and $^{244}Pu$ at the time of their crystallization. (3) Angrites clearly are the crystallization products of melts, whereas the petrogenesis of CAIs may be quite complex. (4) The degree of equilibration within chondrites is variable and its cause uncertain. (5) Angrites do not appear to contain the nucleosynthetic isotopic anomalies that interfere with the interpretation of the very small isotopic shifts in CAIs caused by the decay of extinct nuclides

Two angrites (Angra dos Reis and LEW 86010) have been studied extensively. They provide concordant Sm-Nd ages [LEW 86010: 4.53 ± 0.04 Ga (*Nyquist et al.,* 1994) and 4.55 ± 0.03 Ga (*Lugmair and Galer,* 1992); ADOR: 4.55 ± 0.04 Ga (*Lugmair and Marti,* 1977) and 4.56 ± 0.04 Ga (*Jacobsen and Wasserburg,* 1984)] and U-Pb ages (*Lugmair and Galer,* 1992). The very radiogenic Pb in the ADOR and LEW 86010 pyroxenes allows calculation of precise Pb model ages of 4557.8 ± 0.5 Ma that agree with the Pb-Pb isochron age of 4558.2 ± 3.4 Ma defined by LEW 86010 mineral data (*Lugmair and Galer,* 1992). Rubidium-strontium data for the two angrites fail to define a precise isochron because of the limited range, and very low, Rb/Sr characteristic of the angrites. Nevertheless, the initial Sr isotopic composition of the angrites is second only to Allende CAIs in recording the lowest values of any solar system material (*Lugmair and Galer,* 1992; *Nyquist et al.,* 1994).

In support of their antiquity, the angrites display evidence for several short-lived nuclides including $^{53}$Mn (*Nyquist et al.,* 1994; *Lugmair and Shukolyukov,* 1998), $^{146}$Sm (*Lugmair and Marti,* 1977; *Jacobsen and Wasserburg,* 1984; *Nyquist et al.,* 1994), and $^{244}$Pu (*Lugmair and Marti,* 1977; *Shukolyukov and Begemann,* 1996) (Fig. 4), but not $^{26}$Al with an upper limit of $^{26}$Al/$^{27}$Al of $2 \times 10^{-7}$ determined by Lugmair and Galer (1992). The $^{53}$Mn/$^{55}$Mn ratio of $(1.25-1.44) \times 10^{-6}$ determined for LEW 86010 (*Nyquist et al.,* 1994; *Lugmair and Shukolyukov,* 1998), at face value, corresponds to an age difference of 16–17 m.y. compared to the Allende CAI Mn-Cr data. This difference is at the high end of the $8 \pm 8$-m.y. difference suggested by Pb-Pb and may be an indication that at least some of the $^{53}$Cr excess in the Allende

CAIs is presolar in origin rather than from the *in situ* decay of $^{53}$Mn. The angrites also provide clear evidence for live $^{146}$Sm at an abundance of $^{146}$Sm/$^{144}$Sm of $0.0071 \pm 0.0017$ (*Lugmair and Galer,* 1992) to $0.0076 \pm 0.0009$ (*Nyquist et al.,* 1994) as measured for LEW 86010.

### 4.4. Eucrites

Compared to the angrites, eucrites are a much more abundant group of meteorites that seemingly offer many of the same chronometric strong points in that they are igneous rocks consisting predominantly of plagioclase and pyroxene that crystallized quickly from melts. In detail, however, radiometric data for the eucrites provide clear evidence that thermal processing of planetesimals did not stop within a few million years of molecular cloud collapse, but continued for as long as several hundred million years. This is shown both by relatively young ages for some eucrites and by disturbed isotope systematics in others. Whether these young ages reflect formation ages, thermal resetting through shock metamorphism, or slow cooling is a point of debate that will be discussed extensively below.

Several eucrites show evidence for the presence of live, now extinct, nuclides including $^{26}$Al, $^{53}$Mn, $^{60}$Fe (*Shukolyukov and Lugmair,* 1993), $^{146}$Sm, and $^{244}$Pu. Recently, *Srinivasan et al.* (1999) reported detection of $^{26}$Al in the Piplia Kalan eucrite at a level of $^{26}$Al/$^{27}$Al = $(7.5 \pm 0.9) \times 10^{-7}$. This report is of particular significance because it suggests that this eucrite formed 4–5 m.y. after CAIs, but more importantly indicates that $^{26}$Al was present in planetesimals that were sufficiently large to undergo differentiation within the solar nebula.

The short timescale for the beginning of eucrite formation suggested by the Piplia Kalan $^{26}$Al discovery is supported by the Mn-Cr results reported in *Lugmair and Shukolyukov* (1998). In this study, Chervony Kut gives the highest $^{53}$Mn/$^{55}$Mn of $(3.7 \pm 0.4) \times 10^{-6}$, corresponding to ages $5.8 \pm 0.8$ m.y. older than the angrites (Fig. 5). Using the Pb model age of 4557.8 Ma for LEW 86010 to translate the Mn-Cr results into absolute ages provides an age of $4563.6 \pm 0.9$ Ma for Chervony Kut; 2.4 m.y. younger, but overlapping within uncertainty the Allende CAI Pb-Pb age and also the age indicated by the $^{26}$Al results for Piplia Kalan. Other eucrites that show live $^{53}$Mn include Juvinas ($4.7 \pm 0.9$ m.y. older than the angrites), and Ibitira (0.9 +2 −4 m.y. younger than LEW 6010). The majority of eucrites that have been analyzed for their Mn-Cr isotope systematics, however, do not show the presence of live $^{53}$Mn at the time of their formation and/or metamorphism. These samples include the noncumulate eucrites Caldera, Pomozdino, EET 90020 (*Nyquist et al.,* 1997c), Padvarninkai (*Nyquist et al.,* 1996), and Y 792510 (*Nyquist et al.,* 1997a) and the cumulate eucrites EET 87520 and Moore County.

Another important finding from the Mn-Cr studies of eucrites and diogenites is that whole-rock data for 11 samples show a good positive correlation between $^{53}$Cr/$^{52}$Cr and $^{55}$Mn/$^{52}$Cr (Fig. 6) that provides $^{53}$Mn/$^{55}$Mn = $(4.7 \pm 0.5) \times$

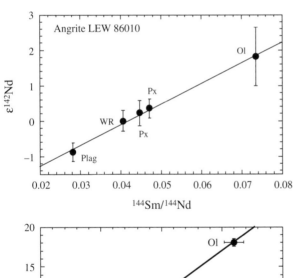
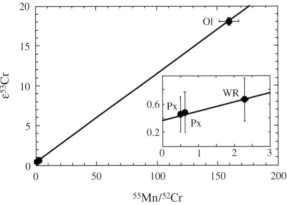

**Fig. 4.** Demonstration of the existence of the short-lived $^{53}$Mn (*Lugmair and Shukolyukov,* 1998) and $^{146}$Sm (*Lugmair and Galer,* 1992) at the time of formation of the angrite LEW 86010. $\varepsilon^{142}$Nd is the relative deviation, in parts per 10,000, of the measured $^{142}$Nd/$^{144}$Nd from a terrestrial standard. The inset in the lower figure shows the three points near the origin with expanded scale.

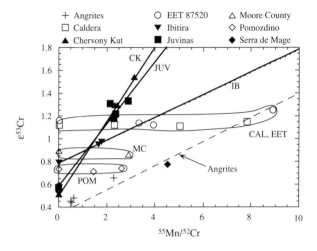

**Fig. 5.** Manganese-chromium systematics for the angrites and eucrites (*Lugmair and Shukolyukov, 1998*). Thick lines show the correlated variations in $^{53}Cr/^{52}Cr$ and $^{55}Mn/^{52}Cr$ found for the eucrites Chervony Kut (CK), Juvinas (JUV), and Ibitira (IB). Dotted line is the mineral isochron determined for the angrite LEW 86010. The horizontal outlined fields enclose data for eucrites that show no evidence for live $^{53}Mn$ at the time of their formation. These eucrites include Caldera (CAL), EET 87520 (EET), Moore County (MC), and Pomozdino (POM).

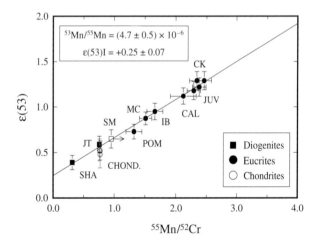

**Fig. 6.** Whole-rock Mn-Cr systematics of eucrites, diogenites, and chondrites (*Lugmair and Shukolyukov, 1998*). With the exception of EET 87520, the remaining eucrites and diogenites define a line that corresponds to a $^{53}Mn/^{55}Mn = (4.7 \pm 0.5) \times 10^{-6}$, which would correspond to an age $7.1 \pm 0.8$ m.y. older than the angrites.

$10^{-6}$ (*Lugmair and Shukolyukov, 1998*). Recent data for additional eucrites from the JSC laboratory show a similar correlation although with a lower slope and hence inferred $^{53}Mn/^{55}Mn = (2.9 \pm 0.9) \times 10^{-6}$ (*Nyquist et al., 1997c*). Presumably these whole-rock Mn-Cr "isochrons" reflect the time of major chemical differentiation on the howardite-

eucrite-diogenite (HED) parent planetesimal. The exact chronological significance of these correlations depends on whether the Mn/Cr measured for the individual samples was set when $^{53}Mn$ was still present, such as might be expected if the eucrites and diogenites are both products of a single magma ocean event, or whether the Mn/Cr was fractionated during a later partial melting event(s) that produced the eucrites. In either case, the correlation suggests that a planetesimal large enough (Vesta, r ~ 250 km) to differentiate into the source regions for the various diogenites and eucrites had formed and begun extensive melting of its interior by 4562–4565 Ma referenced to the 4558-Ma Pb-Pb age of the angrites. It is important to keep in mind, however, that this chronological comparison requires that $^{53}Mn$ was homogeneously distributed within the source regions of the meteorites under consideration. This was shown to be the case (*Lugmair and Shukolyukov, 1998*); the $^{53}Mn$-$^{53}Cr$ systematics of the HED parent body and of the angrites are consistent with a chondritic Mn/Cr ratio in their bulk parent bodies and their initial $^{53}Mn$ abundances were similar to that in ordinary chondrites.

Plutonium-xenon data (*Shukolyukov and Begemann, 1996*) indicate an ~100 m.y. range in ages for eucrites. Some eucrites have Pu-Xe ages similar to or slightly older than the angrites, but some samples provide Pu-Xe ages as much as 80–100 m.y. younger than the angrites. Qualitatively, the Pu-Xe and Mn-Cr age spectra for eucrites are similar. In detail, however, there are significant differences. For example, Chervony Kut, the oldest eucrite according to Mn-Cr systematics, gives a Pu-Xe age $20 \pm 19$ m.y. younger than the angrites. However, the Pu-Xe age of Pomozdino of $4559 \pm 15$ Ma is quite consistent with the upper limit of the Mn-Cr-derived age of $\leq 4554$ Ma. In general, the explanation for the indicated differences probably is the following: The "old" Pu-Xe ages (of, e.g., Ibitira, Pasamonte, Juvinas, Pomozdino) indicate almost complete retention of fission Xe and reflect the time of crystallization or recrystallization (e.g., Ibitira) of eucrites ~4550–4565 m.y. ago (this range also reflects the current experimental resolution afforded by the Pu-Xe technique). The "young" ages most likely indicate partial Xe loss due to secondary events (e.g., shock, brecciation) or slow cooling and do not bear direct chronological information unless the Xe loss during the second event was near total. This may be the case for the unbrecciated eucrite Caldera, whose Pu-Xe age probably dates the secondary recrystallization event (*Wadhwa and Lugmair, 1996*).

While the initial presence of short-lived isotopes in the eucrites indicates their formation on a few-million-year timescale, these old ages have not received strong support from the extant chronometers Sm-Nd and U-Pb. Several $^{147}Sm$-$^{143}Nd$ isochron ages greater than 4.55 Ga have been reported for eucrites (Table 2). These include Chervony Kut (4580 ± 30 Ma: *Wadhwa and Lugmair, 1995*), Juvinas (4560 ±80 Ma: *Lugmair, 1974*), Pasamonte (4580 ± 120 Ma: *Unruh et al., 1977*), and Pomozdino (4540 ± 120 Ma: *Karpenko et al., 1991*). In all cases, however, the

TABLE 2.  Radiometric ages for silicate meteorites.

| Sample | $^{26}$Al/$^{27}$Al ×10$^6$ | ΔT (Ma) | Ref. | $^{53}$Mn/$^{55}$Mn ×10$^6$ | ΔT (Ma) | Ref. | $^{146}$Sm/$^{144}$Nd | ΔT (Ma) | $^{147}$Sm (Ma) | Ref. | Pb-Pb (Ma) | Ref. | Pu-Xe ΔT (Ma) |
|---|---|---|---|---|---|---|---|---|---|---|---|---|---|
| CAIs | 42 (0,80) | ≡0 | [1] | 31±7 | +17 | [6] | | | | | 4566±8 | [27] | |
| *Chondrules* | | | | | | | | | | | | | |
| Allende | | | | | | | | | | | 4560±67 | [28] | |
| Semarkona | 7.7±2.1 | -1.8 | [2] | | | | | | | | | | |
| Inman | 9.4±6.3 | -1.6 | [3] | | | | | | | | | | |
| Chainpur | 7.9±2.7 | -1.8 | [3] | 9.4±1.7 | +10.8 | [7] | | | | | | | |
| Bishunpur | | | | 9.5±3.1 | +10.8 | [7] | | | | | | | |
| *Enstatite Chondrites* | | | | | | | | | | | | | |
| Mn-rich sulfides | | | | 0.05–0.74 | -17 to -2.8 | [8] | | | | | | | |
| Indarch | | | | 0.17–0.63 | -11 to -3.7 | [6] | | | | | | | |
| | | | | 2.8±0.2 | +4.3 | [9] | | | | | | | |
| Abee | | | | 3.0±0.6 | +4.7 | [9] | | | | | | | |
| Khairpur | | | | 1.22±0.07 | -0.1 | [9] | | | | | | | |
| *Ordinary Chondrites* | | | | | | | | | | | | | |
| Ste. Marguerite | 0.2±0.06 | -5.6 | [4] | | | | | | 4550±330 | [13] | 4563±1 | [29] | |
| St. Severin | | | | | | | | | | | 4558±6 | [29] | |
| Knyahinya (L5) | | | | | | | | | | | 4552±3 | [29] | |
| Homestead (L5) | | | | | | | | | | | 4552±3 | [29] | |
| L and LL | | | | | | | | | | | 4518±18 | [25] | |
| H | | | | | | | | | | | 4526±27 | [25] | |
| *Angrites* | | | | | | | | | | | | | |
| LEW 86010 | <0.2 | >-6 | | 1.25±0.07 | ≡0 | [10] | 0.0071±17 | ≡0 | 4553±34 | [14] | 4558±3 | [14] | -1±15 |
| | | | | 1.44±0.07 | +0.8 | [11] | 0.0076±9 | +10 | 4530±40 | [11] | | | |
| Angra dos Reis | | | | | | | 0.0047±23 | -61 | 4550±30 | [14] | 4558±3 | [14] | ≡0 |
| | | | | | | | 0.0118±32 | +75 | 4564±37 | [13] | | | |
| *Cumulate Eucrites* | | | | | | | | | | | | | |
| EET 87520 | | | | <0.25 | <-8.6 | [10] | 0.0069±4 | -4 | 4547±9 | [16] | 4420±20 | [16] | |
| Moama | | | | | | | 0.0041±13 | -82 | 4460±30 | [13] | 4426±94 | [25] | |
| Moore County | | | | <0.25 | <-8.6 | [10] | 0.0044±22 | -71 | 4456±25 | [25] | 4484±19 | [25] | |
| Serra de Mage | | | | 0.54±0.23 | -4.5 | [10] | | | 4410±20 | [26] | 4399±35 | [25] | -10 |

TABLE 2. (continued).

| Sample | $^{26}$Al/$^{27}$Al ×10$^6$ | ΔT (Ma) | Ref. | $^{53}$Mn/$^{55}$Mn ×10$^6$ | ΔT (Ma) | Ref. | $^{146}$Sm/$^{144}$Nd | ΔT (Ma) | $^{147}$Sm (Ma) | Ref. | Pb-Pb (Ma) | Ref. | Pu-Xe ΔT (Ma) |
|---|---|---|---|---|---|---|---|---|---|---|---|---|---|
| *Noncumulate Eucrites* | | | | | | | | | | | | | |
| Bereba | | | | | | | | | | | 4521 ± 0.4 | [25] | −60 ± 16 |
| Bouvante | | | | | | | | | | | 4510 ± 4 | [25] | −11 ± 15 |
| Cachari | | | | | | | | | | | 4453 ± 15 | [25] | 60 |
| Caldera | | | | <0.12 | <−12.5 | [12] | 0.0073 ± 11 | +4 | 4544 ± 19 | [12] | 4516 ± 3 | [30] | 45 |
| Chervony Kut | | | | 3.7 ± 0.4 | +5.8 | [10] | 0.0069 ± 15 | −4 | 4580 ± 30 | [15] | 4312 ± 2 | [30] | −20 ± 19 |
| EET 90020 | | | | | | | 0.0048 ± 20 | −68 | 4430 ± 30 | [17] | | | |
| Ibitira | | | | 1.06 ± 0.5 | −0.9 | [10] | 0.009 ± 1 | +35 | 4460 ± 20 | [18] | 4560 ± 3 | [31,32] | +23 ± 25 |
| Juvinas | | | | 3.0 ± 0.5 | +4.7 | [10] | | | 4560 ± 80 | [19] | 4539 ± 7 | [32] | −7 ± 15 |
| | | | | | | | | | | | 4321 ± 2 | [30] | |
| Nuevo Laredo | | | | | | | | | | | 4514 ± 15 | [25] | −51 ± 27 |
| Padvarninkai | 0.75 ± 0.09 | −4.2 | [5] | | | | 0.0050 ± 11 | −62 | 4510 ± 110 | [20] | | | |
| Pasamonte | | | | | | | | | 4580 ± 120 | [22] | 4530 ± 30 | [33] | +18 ± 19 |
| Piplia Kalan | | | | | | | | | | | | | |
| Pomozdino | | | | <0.6 | <−3.9 | [10] | | | 4540 ± 120 | [21] | | | +1 ± 15 |
| Sioux County | | | | | | | | | | | 4526 ± 10 | [34] | −59 ± 17 |
| Stannern | | | | | | | | | 4480 ± 70 | [23] | 4128 ± 16 | [25] | −98 ± 28 |
| Y 79510,62 | | | | | | | | | 4340 ± 60 | [24] | | | |
| Y 79510,65 | | | | | | | 0.0030 ± 10 | −128 | 4570 ± 90 | [24] | | | |

ΔT is the relative age calculated with respect to the sample indicated by ≡0. Positive ΔT implies an older age and negative ΔT implies a younger age than the reference sample. Data from or summarized by [1] *MacPherson et al.* (1995); [2] *Hutcheon and Hutchison* (1989); [3] *Russell et al.* (1996); [4] *Zinner and Göpel* (1992); [5] *Srinivasan et al.* (1999); [6] *Birck and Allègre* (1988); [7] *Nyquist et al.* (1999); [8] *Wadhwa et al.* (1997); [9] *Shukolyukov and Lugmair* (1999); [10] *Lugmair and Shukolyukov* (1998); [11] *Nyquist et al.* (1994); [12] *Wadhwa and Lugmair* (1996); [13] *Jacobsen and Wasserburg* (1984); [14] *Lugmair and Galer* (1992); [15] *Wadhwa and Lugmair* (1995); [16] *Lugmair et al.* (1991); [17] *Nyquist et al.* (1997c); [18] *Prinzhofer et al.* (1992); [19] *Lugmair* (1974); [20] *Nyquist et al.* (1996); [21] *Karpenko et al.* (1991); [22] *Unruh et al.* (1977); [23] *Lugmair and Scheinin* (1975); [24] *Nyquist et al.* (1997a); [25] *Lugmair et al.* (1977); [26] *Lugmair et al.* (1977); [27] *Chen and Wasserburg* (1981); [28] *Chen and Tilton* (1976); [29] *Göpel et al.* (1994); [30] *Galer and Lugmair* (1996); [31] *Chen and Wasserburg* (1985); [32] *Manhes et al.* (1987); [33] *Unruh et al.* (1977); [34] *Tatsumoto et al.* (1973). Pu-Xe data from *Shukolyukov and Begemann* (1996).

uncertainties of these ages are too large to resolve million-year time differences. In addition, disturbance of the Sm-Nd and U-Pb systematics is a common occurrence. Ibitira provides an excellent example of the problem. Ibitira is one of the few unbrecciated eucrites and contains vesicles suggestive of near surface eruption and quick cooling. Separated pyroxene and plagioclase from Ibitira define an excellent line on a Sm-Nd isochron diagram, yet provide a relatively young age of 4460 ± 20 Ma with an unusually high initial Nd-isotopic composition of $\varepsilon_{Nd}$ = 1.6 ± 0.8 (*Prinzhofer et al., 1992*). Similar Sm-Nd results have been reported for the eucrite EET 90020 (*Nyquist et al., 1997c*). The young Sm-Nd age obtained for Ibitira conflicts with its older Pb age and with the observation of live $^{53}$Mn (*Lugmair and Shukolyukov, 1998*) and $^{146}$Sm in this meteorite. The data for Ibitira indicate a $^{146}$Sm/$^{144}$Sm = 0.009 ± 0.001 at formation (*Prinzhofer et al., 1992*). This value is above that measured for the angrites ($^{146}$Sm/$^{144}$Sm = 0.0070 ± 0.0010) (*Lugmair and Galer, 1992; Nyquist et al., 1994*) in spite of the older $^{147}$Sm/$^{144}$Nd age of the angrites, and is an inconsistency indicative of disturbance of the Sm-Nd system in this meteorite (*Prinzhofer et al., 1992*). Elephant Moraine 90020, on the other hand, has a $^{146}$Sm/$^{144}$Sm = 0.0048 ± 0.0020, which is consistent with its "young" $^{147}$Sm/$^{144}$Nd age (*Nyquist et al., 1997c*).

Data for the unbrecciated eucrite Caldera provide another comparison. For this sample, *Wadhwa and Lugmair (1996)* produced a good $^{147}$Sm-$^{143}$Nd isochron, indicating an age of 4544 ± 19 Ma. The initial $^{146}$Sm/$^{144}$Sm determined for this eucrite is 0.0073 ± 0.0011. Both the $^{147}$Sm-$^{143}$Nd isochron age and the $^{146}$Sm/$^{144}$Sm of Caldera are within uncertainty of those determined for the angrite LEW 86010. However, the upper limit for the $^{53}$Mn/$^{55}$Mn in Caldera is 1.2 × 10$^{-7}$, at least an order of magnitude lower than found for the angrites and corresponding to an age difference of at least 12.5 m.y. This meteorite also has significantly younger Pu-Xe (*Shukolyukov and Begemann, 1996*) and Pb-Pb (*Galer and Lugmair, 1996*) ages of 4513–4516 Ma, suggesting that these systems have been strongly disturbed or reset in Caldera.

Uranium-lead ages potentially can provide sufficient chronological resolution for comparison with the extinct nuclide timescales, but disturbance of the U-Pb systems in eucrites tends to be the rule, rather than the exception. Uranium-lead ages greater than 4.55 Ga have been reported only for Ibitira (4560 ± 3: *Chen and Wasserburg, 1985; Manhès et al., 1987*), but these ages are whole-rock model ages and have not been substantiated by mineral Pb-Pb isochrons. Several noncumulate eucrites have Pb-Pb ages that range from 4510 to 4520 Ma (Table 2) (*Tera et al., 1989*). Juvinas gives a slightly older U-Pb age of 4539 ± 7 Ma (*Manhès et al., 1984*), but also shows evidence for disturbance of the U-Pb system and the addition of extraneous Pb. An attempt to remove this "extraneous" Pb by leaching resulted in a plagioclase-pyroxene tie line that provided a Pb-Pb age of 4321 ± 2 Ma (*Galer and Lugmair, 1996*). A similar Pb-Pb age of 4313 ± 2 Ma was obtained for Cher-

vony Kut (*Galer and Lugmair, 1996*) in spite of the Sm-Nd and Mn-Cr evidence indicating that this meteorite must be older than 4.55 Ga. Lead-lead ages for extensively shocked eucrites such as Cachari and Stannern range to much lower ages of 4453 ± 15 and 4128 ± 16 respectively (*Tera et al., 1989*).

Compared to the noncumulate eucrites, the cumulate eucrites appear to be much less affected by shock. They are not breccias, but are coarse-grained cumulate-textured rocks that do not contain veins of mafic glass as can be found in some of the noncumulate eucrites, e.g., Bereba, Cachari, and Nuevo Laredo (*Tera et al., 1997*). Cumulate eucrites, however, provide the youngest radiometric ages among the eucrites. Furthermore, their Sm-Nd and Pb-Pb ages show reasonable concordancy (Fig. 7). For example, Moore County gives a Pb-Pb age of 4484 ± 19 Ma, a $^{147}$Sm-$^{143}$Nd isochron of 4456 ± 25 Ma with chondritic initial Nd isotopic composition, and an initial $^{146}$Sm/$^{144}$Sm = 0.0044 ± 0.0022 (*Tera et al., 1997*). Compared to $^{146}$Sm/$^{144}$Sm = 0.007 as found for the angrite LEW 86010 (*Lugmair and Galer, 1992; Nyquist et al., 1994*), the $^{146}$Sm data for Moore County suggests that it is 69 m.y. younger than the angrites, in accord with its $^{147}$Sm-$^{143}$Nd and Pb-Pb ages. *Lugmair and Shukolyukov (1998)* examined the Mn-Cr systematics of Moore County and found no evidence for live $^{53}$Mn at the

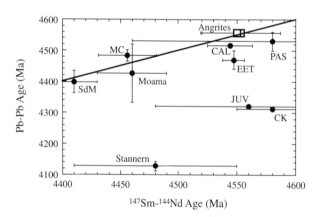

**Fig. 7.** Comparison of Pb-Pb and $^{147}$Sm-$^{143}$Nd ages for eucrites and angrites. The thick black line shows concordant ages. Many of the eucrites plot with concordant Pb-Pb and Sm-Nd ages, including the relatively young cumulate eucrites. Some samples fall off concordancy in the direction of anomalously young Pb-Pb ages, suggesting that the U-Pb system is more easily disturbed than the Sm-Nd system. After *Tera and Carlson (1999)*. Lead-lead data: Moore County (MC), Moama, Serra de Mage (SdM), and Stannern (*Tera et al., 1997*); Juvinas (JUV) and Chervony Kut (CK) (*Galer and Lugmair, 1996*). Samarium-neodymium data: Moore County (*Tera et al., 1997*); Serra de Mage (*Lugmair et al., 1977*); Moama (*Jacobsen and Wasserburg, 1984*); Juvinas (*Lugmair et al., 1976*); Chervony Kut (*Wadhwa and Lugmair, 1995*); Stannern (*Lugmair and Scheinin, 1975*). Lead-lead and Sm-Nd data from: angrites (*Lugmair and Galer, 1992*) and the eucrites Caldera (CAL) (*Wadhwa and Lugmair, 1996*), Pasamonte (*Unruh et al., 1977*), and EET 87520 (*Lugmair et al., 1991*).

time of its formation, which is consistent with its apparently young age. Another cumulate eucrite, Moama, gives similar results to Moore County with a Pb-Pb age of 4426 ± 94 Ma (*Tera et al.,* 1997), a $^{147}$Sm-$^{143}$Nd age of 4460 ± 30 Ma, and initial $^{146}$Sm/$^{144}$Sm = 0.0041 ± 0.0013 (*Jacobsen and Wasserburg,* 1984). Very old $^{147}$Sm-$^{143}$Nd ages have been reported for Moore County (4600 ± 40 Ma, *Nakamura et al.,* 1977) and Moama (4580 ± 50 Ma, *Hamet et al.,* 1978), but these are in disagreement with the newer results and should be viewed with caution.

A third cumulate eucrite, Serra de Mage, gives the youngest ages with Pb-Pb = 4399 ± 35 Ma (*Tera et al.,* 1997) and $^{147}$Sm-$^{143}$Nd = 4410 ± 20 Ma (*Lugmair et al.,* 1977). Complicating the picture for Serra de Mage, and by inference possibly for the other cumulate eucrites, *Lugmair and Shukolyukov* (1998) showed pyroxene and chromite separated from Serra de Mage to have different $^{53}$Cr/$^{52}$Cr ratios, suggesting that Serra de Mage formed with a $^{53}$Mn/$^{55}$Mn of (5.4 ± 2.3) × $10^{-7}$. This would suggest that Serra de Mage is only 5 m.y. younger than the angrites as opposed to the 150-m.y.-younger age indicated by the Sm-Nd and Pb-Pb data. An old age for Serra de Mage is also supported by a high $^{146}$Sm/$^{144}$Sm ~ 0.007, within error the same as angrite LEW 86010. The cause of this discrepancy is not clear and raises the troubling uncertainty of what event is being recorded by the Pb-Pb and $^{147}$Sm-$^{143}$Nd systems.

The possibility that the relatively young ages of the cumulate eucrites might be cooling ages rather than crystallization ages was investigated in a thermal model for the evolution of the presumed HED parent body, asteroid 4 Vesta (*Ghosh and McSween,* 1998). *Ghosh and McSween* (1998) model Vesta as being heated by $^{26}$Al decay that caused Fe-FeS melting and consequent core formation within 1.7 m.y. of accumulation followed by extensive melting of the silicate mantle within 2 m.y. Because of a combination of effective insulation from a ≈30-km-thick regolith and continued heating from short-lived radioactive species (other than $^{26}$Al), the temperature at 100 km depth within Vesta could remain above the melting point for over 100 m.y. (*Ghosh and McSween,* 1998). Thus, the ages of the cumulate eucrites could record the final stages of crystallization of an early "magma ocean" on the HED parent body. A problem with this interpretation, however, is that two mesosiderites (Horse Creek and Emery) show evidence for live $^{107}$Pd at the time of their formation (*Chen and Wasserburg,* 1996) and require closure of their Pd-Ag systems within at most a few million years, not a hundred million years. If the mesosiderites represent the core of the HED parent body, their old Pd-Ag ages suggest rapid core cooling and crystallization, but this is not predicted in the *Ghosh and McSween* (1998) model. Furthermore, recent studies of eucritic and diogenitic clasts in the Vaca Muerta mesosiderite (*Wadhwa et al.,* 1999) have shown that, although $^{53}$Mn was extant when the source magmas for these clasts were generated, the $^{53}$Mn/$^{55}$Mn ratio was already slightly lower than that in the HED parent body. Thus, the suggestion arises that these clasts originated on a parent body distinct from that of the HEDs. Although similar in composition, this parent body underwent global differentiation possibly ~1.7 m.y. after the HED parent body.

## 4.5.  Iron Meteorites

Segregation of metal from silicate can occur in a reducing nebular environment and at the planetary stage. The latter represents the single most significant chemical differentiation event to occur on a planet or planetesimal. Since Pb and W segregate with the metal, separation of core from mantle should be recorded in both the U-Pb and Hf-W isotopic systems of the mantles and crusts of differentiated planetesimals. The time of core crystallization, or at least the time when the core cooled below a diffusional blocking temperature, can be dated directly by application of the Re-Os and Pd-Ag radiometric systems to the iron meteorites.

Demonstrations of the existence of live $^{107}$Pd have been made for pallasites, mesosiderites, and most groups of irons (Table 3) (*Chen and Wasserburg,* 1996). In a detailed study of the group IVA iron Gibeon, *Chen and Wasserburg* (1990) produced an excellent correlation between $^{107}$Ag/$^{109}$Ag and Pd/Ag for different samples of the metal. The slope of the correlation corresponds to a $^{107}$Pd/$^{108}$Pd of (2.40 ± 0.05) × $10^{-5}$. Compared to this value, and recast as relative ages, most classes of iron meteorites indicate that closure of the Pd-Ag system occurred within 0–4 m.y. of Gibeon (*Chen and Wasserburg,* 1996) with the exception of Tlacotepec (IVB) and N'Goureyma, which give much younger closure ages (Table 3, Fig. 8). Pallasites and mesosiderites indicate distinctly younger closure times with mean Pd-Ag ages ~8–8.5 m.y. younger than Gibeon (*Chen and Wasserburg,* 1996). The Pd-Ag results thus suggest a formation interval <9 m.y. for most of the major iron-dominated meteorite groups. An important observation here is that "internal" Pd-Ag isochrons, such as that determined for Gibeon, record not the time of metal segregation, but the time when Ag and Pd diffusion stopped erasing the ingrowth of $^{107}$Ag between different pieces of metal separated by centimeters. Thus, the Ag-Pd age shows not only that the iron meteorites formed over a few-million-year interval, but that they cooled below the diffusional blocking temperature of Pd-Ag on this timescale. This implies either that the planetesimals from which the irons derive were quite small (≪100 km diameter) or that the parent planetesimal was disrupted by impact within a few million years of formation.

The very short formation interval suggested for iron meteorites from Pd-Ag is supported by recent data for the Hf-W isotopic system (Table 3, Fig. 8). Since W is a siderophile element while Hf is lithophile, metal segregation will result in a metal phase with a Hf/W ratio near zero and a silicate fraction with high Hf/W ratio. Iron meteorites are now known to be deficient in $^{182}$W compared to bulk chondritic material (*Harper et al.,* 1991; *Lee and Halliday,* 1995; *Harper and Jacobsen,* 1996). Twenty-three samples of iron meteorites from several different classes show an extremely

TABLE 3. Manganese-chromium, Ag-Pd, Hf-W, and Re-Os systematics of iron-rich meteorites.

| Sample | Group | $^{53}Mn/^{55}Mn$ $\times 10^{-6}$ | $\Delta T$ (Ma) | $^{107}Pd/^{108}Pd$ $\times 10^{-5}$ | $\Delta T$ (Ma) | $\varepsilon^{182}W$ | $\Delta T$ (Ma) | Re-Os |
|---|---|---|---|---|---|---|---|---|
| *Pallasites* | | | | | | | | |
| Eagle Station | | $2.3 \pm 0.3$ | +3.3 | <0.7 | <−10 | | | |
| Brenham | | | | $1.1 \pm 0.14$ | −5.2 | | | |
| Glorieta Mountain | | | | $0.91 \pm 0.22$ | −7 | | | |
| Springwater | | $14 \pm 4$ | +12.9 | | | | | |
| Omolon | | $1.29 \pm 0.19$ | +0.2 | | | | | |
| | | | | | | | | |
| *Mesosiderites* | | | | | | | | |
| Morristown | | | | $1.45 \pm 0.33$ | −2.6 | | | |
| Horse Creek | | | | $0.65 \pm 0.18$ | −1 | | | |
| Emery | | | | <14 | <+18.7 | | | |
| Bencubbin | | | | <0.53 | <−12 | | | |
| | | | | | | | | |
| *Irons* | | | | | | | | |
| IAB–IIICD | | | | | | $-3.3 \pm 0.2$ | −2.3 | $4529 \pm 23$ |
| Canyon Diablo | IA | | | <3.7 | <+6.2 | $-3.4 \pm 0.4$ | | |
| Toluca | IA | <0.05 | <−17 | <1.8 | <−0.6 | | | |
| Nantan | IIICD | | | <3.4 | <+5.4 | | | |
| Magura | | | | | | $-3.1 \pm 0.6$ | | |
| New Leipzig | | | | | | $-3.2 \pm 0.7$ | | |
| Shrewsbury | | | | | | $-3.4 \pm 0.9$ | | |
| Seelasgen | | | | | | $-3.6 \pm 0.5$ | | |
| | | | | | | | | |
| IIAB | | | | 1.83 | −0.4 | $-3.6 \pm 0.6$ | −1.3 | $4537 \pm 8$ |
| Tocopilla | IIA | | | $1.83 \pm 0.18$ | −0.4 | | | |
| Derrick Peak | IIB | | | $1.82 \pm 0.69$ | −0.5 | | | |
| Negrillos | IIA | | | | | $-3.1 \pm 0.5$ | | |
| Bennett County | IIA | | | | | $-4.6 \pm 0.9$ | | |
| Filomena | IIA | | | | | $-3.4 \pm 0.6$ | | |
| Lombard | IIA | | | | | $-4.3 \pm 0.3$ | | |
| Old Woman | | | | | | $-3.1 \pm 1.1$ | | |
| Navajo | | | | | | $-3.3 \pm 0.6$ | | |
| | | | | | | | | |
| IIIAB | IIIAB | | | $1.92 \pm 0.36$ | ≡0 | $-4.0 \pm 0.6$ | ≡0 | $4557 \pm 12$ |
| Cape York | IIIA | $22 \pm 10$ | +15.3 | $2.53 \pm 0.59$ | +2.6 | | | |
| El Sampal | IIIA | $0.8 \pm 0.1$ | −2.4 | $1.54 \pm 0.24$ | −2.1 | | | |
| Trenton | IIIA | | | $1.65 \pm 0.39$ | −1.4 | | | |
| Bear Creek | IIIB | $2.4 \pm 0.2$ | +3.5 | $1.81 \pm 0.38$ | −0.6 | | | |
| Mt. Edith | IIIB | $2.4 \pm 0.2$ | +3.5 | $2.11 \pm 0.36$ | +0.9 | | | |
| Chupaderos | IIIB | $2.4 \pm 0.2$ | +3.5 | $2.16 \pm 0.47$ | +1.1 | | | |
| Grant | IIIB | $1.0 \pm 0.4$ | −1.2 | $1.65 \pm 0.13$ | +1.4 | | | |
| Henbury | IIIA | | | | | $-4.1 \pm 0.6$ | | |
| Ssyromolotovo | IIIA | | | | | $-4.3 \pm 0.8$ | | |
| Susuman | IIIA | | | | | $-3.1 \pm 0.6$ | | |
| Mount Edith | | | | | | $-4.5 \pm 0.6$ | | |

limited range in W-isotopic composition with $^{182}W/^{184}W$ all within 3–5 parts in 10,000 lower than the bulk-chondritic value (*Horan et al.,* 1998). This narrow range in W-isotopic composition translates to a formation interval of <5 m.y. (*Horan et al.,* 1998) for the iron meteorites, in agreement with the 4-m.y. interval deduced from Pd-Ag. Even the sense of age difference between the different groups of iron meteorites, indicated by Hf-W and Pd-Ag, is the same.

Group IVA appear to be the oldest group followed by IIIAB, IVB, and IIAB, then IAB–IIICD as the youngest (Table 3). Curiously, the IVA group has the youngest Re-Os age (*Smoliar et al.,* 1996), but this group also shows disturbance in its Re-Os systematics that may be indicative of shock resetting (*Smoliar et al.,* 1996).

One problem with the Pd-Ag and Hf-W data for iron meteorites is relating their relative timescale to the absolute

TABLE 3.    (continued).

| Sample | Group | $^{53}$Mn/$^{55}$Mn × 10$^{-6}$ | ΔT (Ma) | $^{107}$Pd/$^{108}$Pd × 10$^{-5}$ | ΔT (Ma) | ε$^{182}$W | ΔT (Ma) | Re-Os |
|---|---|---|---|---|---|---|---|---|
| IVA | | | | 2.47 ± 0.32 | +2.3 | −4.2 ± 0.7 | +0.6 | 4456 ± 25 |
| Duchesne | IVA | | | 2.45 ± 0.02 | +2.3 | | | |
| Social Circle | IVA | | | 2.42 ± 0.04 | +2.2 | | | |
| Mart | IVA | | | 2.48 ± 0.06 | +2.4 | | | |
| Hill City | IVA | | | 2.63 ± 0.07 | +3.0 | | | |
| Bishop Canyon | IVA | | | 1.88 ± 0.05 | −0.2 | | | |
| Yanhuitlan | IVA | | | 3.05 ± 0.05 | +4.4 | | | |
| Ningbo | IVA | | | 2.40 ± 0.03 | +2.1 | | | |
| Gibeon | IVA | | | 2.46 ± 0.09 | +2.3 | | | |
| Jamestown | IVA | | | | | −3.5 ± 0.6 | | |
| Yanhiutlan | IVA | | | | | −4.4 ± 1.3 | | |
| Bushman Land | IVA | | | | | −3.8 ± 1.5 | | |
| Duel Hill | IVA | | | | | −5.1 ± 1.1 | | |
| IVB | | | | 1.33 ± 0.67 | −3.5 | −4.0 ± 0.4 | 0 | 4526 ± 27 |
| Tawallah Valley | IVB | | | 1.47 ± 0.02 | −2.5 | −4.0 ± 0.6 | | |
| Klondike | IVB | | | 1.65 ± 0.03 | −1.4 | | | |
| Hoba | IVB | | | 1.76 ± 0.12 | −0.8 | | | |
| Santa Clara | IVB | | | 1.2 ± 0.01 | −4.4 | | | |
| Warburton Range | IVB | | | 1.86 ± 0.08 | −0.3 | −4.0 ± 0.8 | | |
| Tlacotepec | IVB | | | 0.05 ± 0.02 | −35 | −4.5 ± 0.4 | | |
| Cape of Good Hope | IVB | | | | | −3.5 ± 0.9 | | |
| Tucson | AN | | | 1.72 ± 0.07 | −1.0 | | | |
| N'Goureyma | AN | | | 0.34 ± 0.06 | −16.3 | | | |
| Pinon | AN | | | 1.42 ± 0.04 | −2.8 | | | |
| Deep Spring | AN | | | 0.7 ± 0.04 | −9.5 | | | |
| Mundrabilla | AN | <0.15 | | <0.39 | <−15 | | | |

ΔT for the Mn-Cr data is calculated relative to a value of 1.25 × 10$^{-6}$ for $^{53}$Mn/$^{55}$Mn in angrites. ΔT for Pd-Ag and Hf-W are calculated relative to $^{107}$Pd/$^{108}$Pd = 1.92 × 10$^{-6}$ and ε$^{182}$W = −4, which are the average values determined for the group IIIAB samples. Manganese-chromium data from *Birck and Allègre* (1988); *Lugmair and Shukolyukov* (1998); *Hutcheon and Olsen* (1991). Palladium-silver data from *Chen and Wasserburg* (1996) and references therein. Hafnium-tungsten data from *Lee and Halliday* (1995); *Horan et al.* (1998). Rhenium-osmium data from *Smoliar et al.* (1996).

timescale available for other meteorite classes. Recent efforts have produced excellent Re-Os isochrons for individual groups of iron meteorites (Table 3) (*Shen et al.*, 1996; *Smoliar et al.*, 1996; *Birck and Allègre*, 1998), but these cannot be used to define an absolute scale since the decay constant of $^{187}$Re is not known with sufficient accuracy. Nevertheless, with the exception of the IVA group, the relative Re-Os age differences between the different groups of irons correlates with their Pd-Ag and Hf-W age differences, although the age range indicated by Re-Os is approximately a factor of 10 greater than for the extinct systems (Table 3).

Another opportunity exists, however, to cross-correlate the short-lived radiometric systems in irons with those in the silicate meteorites through combined Pd-Ag and Mn-Cr systematics. Pallasites show evidence for live $^{107}$Pd (*Chen and Wasserburg*, 1996) and $^{53}$Mn (*Birck and Allègre*, 1988; *Lugmair and Shukolyukov*, 1998). At this time, only the Eagle Station pallasite has been examined for both $^{107}$Pd and

$^{53}$Mn. This sample, unfortunately, had an undetectable $^{107}$Pd signal, probably as a result of its relatively low Pd/Ag, and provided only an upper limit of <7.2 × 10$^{-6}$ for $^{107}$Pd/$^{108}$Pd (*Chen and Wasserburg*, 1996). Eagle Station, however, did provide evidence for an initial $^{53}$Mn/$^{55}$Mn of (2.3 ± 0.3) × 10$^{-6}$ (*Birck and Allègre*, 1988). As discussed in *Birck and Allègre* (1988) and *Lugmair and Shukolyukov* (1998), these Mn-Cr results and the possible relationship of this meteorite to carbonaceous chondrites may raise doubts about the chronological meaning of these data, at least in the context of "normal" pallasites. Two other pallasites (Glorieta Mountain, Brenham) provided overlapping initial $^{107}$Pd/$^{108}$Pd of 0.91–1.1 × 10$^{-5}$ (*Chen and Wasserburg*, 1996) and one other pallasite (Omolon) indicates an initial $^{53}$Mn/$^{55}$Mn of (1.29 ± 0.19) × 10$^{-6}$ (*Lugmair and Shukolyukov*, 1998). The Omolon Mn-Cr results suggest pallasite formation essentially at the same time as angrites, whereas Eagle Station may have formed ~3 m.y. earlier, or 4561 Ma referenced to the 4558-

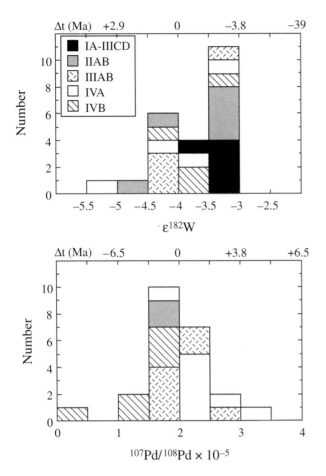

**Fig. 8.** Palladium-silver (*Chen and Wasserburg,* 1996) and Hf-W (*Horan et al.,* 1998) systematics for iron meteorites. Group IIIAB samples tend toward the lowest $\varepsilon^{182}W$ and highest $^{107}Pd/^{108}Pb$, suggesting that they may be among the oldest of the irons, but the group IVB irons have the lowest $\varepsilon^{182}W$, yet among the highest Pd-Ag. $\varepsilon^{182}W$ is the relative deviation in the sample's $^{182}W/^{180}W$, in parts per 10,000, compared to a terrestrial standard, which also overlaps the value determined for average chondritic material (*Lee and Halliday,* 1996).

Ma Pb-Pb age of the angrites. The upper limit for $^{107}Pd/^{108}Pd$ in Eagle Station requires its formation at least 2 m.y. after Glorieta Mountain, which apparently formed ~9 m.y. after Gibeon. If Eagle Station formed ~3 m.y. prior to the angrites and at least 11 m.y. after Gibeon, Gibeon would have to be at least 14 m.y. older than the angrites, or 4572 Ma referenced to the 4558-Ma age of the angrites.

Group IIIAB irons have Mn-rich phosphates that make them amenable for Mn-Cr study. *Hutcheon and Olsen* (1991) report detection of excess $^{53}Cr$ in six group IIIAB phosphates. Cape York provides a high $^{53}Mn/^{55}Mn$ ratio of $(22 \pm 10) \times 10^{-6}$ that would translate to an absolute age of 4573 Ma if referenced to the $^{53}Mn/^{55}Mn$ and 4558-Ma Pb-Pb age for angrites. This old age agrees with that inferred for Gibeon above, and the Pd-Ag ages of Cape York and Gibeon overlap within uncertainty.

Other group IIIABs, however, have $^{53}Mn/^{55}Mn$ ratios between 0.8 and $2.4 \times 10^{-6}$, which translate to ages between 11.8 and 17.7 m.y. younger than Cape York, yet these same IIIAB samples differ in Pd-Ag age by <4.5 m.y. The $^{53}Mn/^{55}Mn$ of IIIABs other than Cape York suggest that the IIIABs range from 3.5 m.y. older to 2.4 m.y. younger than the angrites, or from 4556 to 4562 Ma using the 4558-Ma Pb-Pb age of the angrites. According to Pd-Ag systematics, only the group IVA irons are older than the IIIABs with the Pd-Ag age difference implying that the IVAs formed at 4560 Ma. Resolution of the apparent conflict between the 4573-Ma Mn-Cr age for Cape York and the ca. 4560-Ma Mn-Cr ages for the other IIIABs awaits further study, but the widespread presence of the decay products of the short-lived isotopes $^{53}Mn$, $^{107}Pd$, and $^{182}Hf$ in irons show that iron segregation occurred quite early (<10–20 m.y. after CAI formation) in solar system history.

## 5.  HETEROGENEITY IN THE EARLY SOLAR NEBULA

As mentioned at the outset, the chronologies defined by extinct radiometric systems require the assumption that the apparent variations in the abundance of the extinct nuclide reflect decay and not heterogeneous distribution. The validity of this assumption can be evaluated using both an extinct ($^{53}Mn$) and extant ($^{87}Rb$) radiometric system.

### 5.1.  Rubidium-Strontium

The Rb/Sr of the solar nebula is quite high (*Anders and Grevasse,* 1989), which causes $^{87}Sr/^{86}Sr$ in the early solar nebula to increase at a rate where modern measurement precision can resolve a difference in initial $^{87}Sr/^{86}Sr$ equivalent to an ~2-m.y. age difference. Fortunately, several meteoritic materials, including CAIs, angrites, eucrites, and even the Moon, have sufficiently low Rb/Sr for their initial Sr-isotopic compositions to be determined with adequate accuracy to make chronological comparison possible (*Papanastassiou and Wasserburg,* 1969; *Podosek et al.,* 1991; *Lugmair and Galer,* 1992; *Smoliar,* 1993).

Allende CAIs have long been known to have the lowest $^{87}Sr/^{86}Sr$ of any meteoritic material (Table 4), in accord with their antiquity. The canonical starting point for Sr-isotope evolution in the nebula was defined by the Allende CAI (inclusion D7) with lowest $^{87}Sr/^{86}Sr$ (0.698880 ± 20: (*Gray et al.,* 1973) (all values discussed here are adjusted to a value of $^{87}Sr/^{86}Sr = 0.71025$ for the NBS 987 Sr standard). Most other inclusions, however, have a slightly higher $^{87}Sr/^{86}Sr$ averaging to 0.698920 ± 20 (*Gray et al.,* 1973; *Podosek et al.,* 1991). Although the difference between these values is marginal, and in fact has not been duplicated in repeat measurements of D7 (*Podosek et al.,* 1991), if taken as representing a prolonged formation interval for CAIs, this difference in $^{87}Sr/^{86}Sr$ would correspond to an age difference of about 3.1 m.y. using a solar $^{87}Rb/^{86}Sr$ ratio of 0.92, 4.56 G.y. ago. As for other isotopic systems in CAIs, one must consider the possibility of either nuclear anomalies or mass fractionation (*Patchett*, 1980a,b) that could influence their measured Sr-isotopic compositions.

Because of their exceptionally low Rb/Sr ratios, angrites also provide a precise measure of their initial Sr-isotopic compositions. Lewis Cliff 86010 and Angra dos Reis have initial $^{87}Sr/^{86}Sr$ equal to 0.698972 ± 8 and 0.698970 ± 18 respectively (*Lugmair and Galer*, 1992; *Nyquist et al.*, 1994). If compared to the average initial $^{87}Sr/^{86}Sr$ of Allende CAIs, this value for the initial Sr-isotopic composition of angrites implies an age difference of 3.9 m.y., about half the difference in Pb-Pb ages of these two materials. If compared to the lowest initial $^{87}Sr/^{86}Sr$ for CAIs, the initial Sr-isotopic composition of the angrites corresponds to an age difference of 7 m.y., which is within error of the Pb-Pb age difference. What must be kept in mind throughout this discussion is that ages calculated by comparison of Sr initial isotopic compositions are single-stage ages that assume growth of $^{87}Sr$ first in an environment with very high solar Rb/Sr followed by extremely limited growth of $^{87}Sr$ in the low Rb/Sr sample itself. If the sample experienced a more complex history, for example, formation by partial melting of a solid that had condensed from the solar nebula, the single-stage age will underestimate the true age of the sample if its source had a Rb/Sr less than solar.

In summarizing the data for initial Sr-isotope variation in eucrites, *Smoliar* (1993) suggested that the cumulate and noncumulate eucrites have slightly different initial Sr isotopic compositions of 0.698982 ± 14 and 0.699034 ± 14 respectively, in contrast to the previously defined eucrite initial of 0.69909 (BABI, *Papanastassiou and Wasserburg*, 1969). The lower of these two numbers overlaps the value determined for the angrites and supports the assertion of *Lugmair and Galer* (1992) that angrites and eucrites have overlapping initial Sr-isotopic compositions based on a comparison of the Moore County plagioclase with angrite initial Sr. The difference between the cumulate and noncumulate eucrite initials translates to a time difference of 4 m.y., which is similar to the age range for noncumulate eucrites indicated by Mn-Cr results (*Lugmair and Shukolyukov*, 1998). Of some surprise, however, is the suggestion that the cumulate source is older than the noncumulate source, in spite of the young ages obtained for the cumulates. One explanation would be that severe volatile depletion, and therefore alkali depletion, occurred throughout the accretionary history of the HED parent body and its precursor planetesimals. Once molten, fractionation further depleted Rb relative to Sr in the early formed (plagioclase-rich) source regions of the cumulate eucrites on the HED parent body, while a slightly higher Rb/Sr ratio was imprinted onto the source regions for the normal eucrites. Also, the Sr data for eucrites suggest that the HED parent body formed roughly at the same time as the angrite parent body, yet the whole-rock Mn-Cr systematics of eucrites and diogenites show that the HED parent body differentiated as much as 7 m.y. earlier than angrite crystallization. Again, the resolution of this apparent problem most likely lies in the tim-

ing and degree of Rb depletion in the precursor planetesimals of the actual parent bodies.

The Moon's crust has sufficiently low Rb/Sr to enable accurate estimation of the initial Sr-isotopic composition of the Moon. *Carlson and Lugmair* (1988) report an initial Sr isotopic composition for the 4.44-G.y.-old ferroan anorthosite 60025 of 0.699072 ± 20. This value is slightly higher, but within uncertainty, of the LUNI value of 0.699023 ± 60 (*Nyquist et al.*, 1974). A similar initial $^{87}Sr/^{86}Sr$ of 0.699056 was reported for the 4562 ± 68 Ma ferroan noritic anorthosite clast from 67016 by (*Alibert et al.*, 1994). Assuming a solar Rb/Sr, the initial Sr-isotopic composition of the ferroan anorthosites suggests that the materials that formed the Moon were isolated from the high Rb/Sr solar nebular only 8–12 m.y. after the CAIs.

This does not necessarily mean that the Moon formed before 4553 Ma. If one considers the giant impact model for lunar formation (*Hartmann and Davis*, 1975; *Stevenson*, 1987), some portion of the materials that now make up the Moon were present in the Earth before the giant impact. If during this residence in the proto-Earth, the future Moon materials were evolving $^{87}Sr$ with a Rb/Sr ratio equal to that to the bulk Earth ($^{87}Rb/^{86}Sr \approx 0.09$), Moon formation as late as the 4440-Ma age of 60025 can be accommodated (*Carlson and Lugmair*, 1988). Thus, the lunar Sr data do not necessarily contradict the assertion from Hf-W isotopic evidence that the Moon formed 50 ± 10 m.y. after the start of solar system evolution (*Lee et al.*, 1997). They do indicate, however, that if the Moon formed at least in part from the Earth, the Earth and its precursor planetesimals and, by inference, the putative impactor, had to acquire their low, decidedly nonsolar, Rb/Sr within <13 m.y. of CAI formation. To prolong Earth formation over more than 50 m.y. would require that Earth accreted *preferentially* from volatile-depleted planetesimals formed tens of millions of years earlier.

An important conclusion to be derived from this comparison of initial Sr-isotopic compositions is that the relative chronologies so derived do not deviate markedly from that provided by other radiometric systems. This also suggests that with present resolution, the Sr-isotopic composition of the solar nebula was fairly homogeneous.

## 5.2. Manganese-Chromium

In contrast to the apparent homogeneity suggested by Sr-isotopic data, Mn-Cr-isotope systematics indicate that there may have been a radial variation in the $^{53}Mn/^{55}Mn$ ratio at least in the inner solar nebula (*Lugmair and Shukolyukov*, 1998). This observation and its significance for the interpretation of data from extinct radiometric systems have been discussed at length by *Lugmair and Shukolyukov* (1998). In brief, these workers showed that samples from the Earth and Moon within error have the same $^{53}Cr/^{52}Cr$ ratios (defined as $\varepsilon^{53}Cr = 0$). Samples from Mars (as represented by the SNC meteorites) have slightly elevated $^{53}Cr$ abundances ($\varepsilon^{53}Cr \sim 0.22$) while angrites, ordinary chondrites, eucrites, the pallasite Omolon, and others have even higher bulk $^{53}Cr/$$^{52}Cr$ ratios ($\varepsilon^{53}Cr$ ~0.5). The cause of this radial variation in $^{53}Cr/^{52}Cr$ is unclear. The preferred possibility is that it reflects an initially heterogeneous distribution of live $^{53}Mn$ in the solar nebula. Another is that it reflects a fractionation of Mn/Cr with heliocentric distance in the solar nebula. This possibility is basically excluded by the newest Mn-Cr data on E chondrites (*Shukolyukov and Lugmair,* 1998, 1999), where their bulk Mn/Cr ratios are consistent with ordinary chondrites, but the bulk $^{53}Cr$ excess is only ~0.17 $\varepsilon$.

The ultimate cause of the observed variation in $^{53}Cr/^{52}Cr$ of solar system materials remains to be resolved. The mere observation reinforces the need to properly consider the assumptions involved in translating the relative ages obtained from extinct radiometric systems to the absolute timescales that are provided by systems like U-Pb. As has been mentioned above, the angrites, ordinary chondrites, eucrites, the pallasite Omolon, Acapulco, and Brachina have been shown to be consistent with an ordinary chondritic Mn/Cr ratio and a present-day relative $^{53}Cr$ excess of ~0.5 $\varepsilon$. Thus, their region of origin appears to have had homogeneously distributed $^{53}Mn$ and satisfies the conditions required for mapping the relative Mn-Cr ages onto an absolute timescale.

## 6. SUMMARY AND CONCLUSIONS

Overall, the data for extinct radiometric systems in meteorites indicates that planetesimal differentiation was well underway within 2–10 m.y. of the condensation of the first solids in the solar nebula. This extends to the iron meteorites, showing that core-mantle-crust differentiation on planetesimals occurred within this same short timeframe. Consequently, the probability is high that larger objects, such as Earth, were formed by the accumulation of already differentiated planetesimals. Given the short time involved in this initial differentiation, decay of $^{26}Al$ could have provided the main energy source for this first stage of planetesimal differentiation.

In detail, there are some discrepancies between the timescales indicated by different extinct chronometers and between the extinct and extant chronometers. Some of these problems are most likely related to postformation metamorphic disturbance of the isotopic systems in the studied meteorites. Avoiding, or better understanding, the effects of such disturbance will be aided by more studies that apply many different chronometers to the same samples. Perhaps the most intriguing problem facing solar system chronology is defining to high precision the absolute time of the start of condensation in the solar nebula. Lead-lead data for angrites and chondrites clearly indicate that chemical processing of solids in the solar nebula had begun by 4558–4563 Ma. Older ages are found only for CAIs, the expected products of early condensation, but CAI ages suffer uncertainty because of the clear observation of nucleosynthetic and/or mass-fractionation-induced isotope heterogeneity in CAIs. The $\Delta T$ indicated by Al-Mg systematics of Allende CAIs and St. Marguerite feldspars suggests a CAI age of

4569 Ma referenced to the 4563-Ma Pb-Pb age of St. Marguerite phosphates. The Mn-Cr $\Delta T$ between Allende CAIs and angrites suggests a CAI age of 4575 Ma referenced to the 4558-Ma Pb-Pb age of angrite LEW 86010. Both ages are at the high end of, but within uncertainty, of the 4566 ± 8-Ma Pb-Pb isochron measured for Allende CAIs and suggest the start of solar system condensation is presently known to a precision of only ±4–5 m.y.

***Acknowledgments.*** We thank A. Shukolyukov for his constructive comments on the original manuscript. Fouad Tera is gratefully acknowledged for numerous highly stimulating plumbology discussions. Reviews by Steve Galer, Larry Nyquist, and Meenakshi Wadhwa contributed to substantial clarification of the arguments presented here and a significant, and beneficial, expansion of the reference list.

## REFERENCES

Alibert C., Norman M. D., and McCulloch M. T. (1994) An ancient Sm-Nd age for a ferroan noritic anorthosite clast from lunar breccia 67016. *Geochim. Cosmochim. Acta, 58,* 2921–2926.

Allègre C. J., Manhès G., and Göpel C. (1995) The age of the Earth. *Geochim. Cosmochim. Acta, 59,* 1445–1456.

Amari S., Lewis R. S., and Anders E. (1994) Interstellar grains in meteorites: I. Isolation of SiC, graphite, and diamond; size distributions of SiC and graphite. *Geochim. Cosmochim. Acta, 58,* 459–470.

Anders E. and Grevasse N. (1989) Abundances of the elements: Meteoritic and solar. *Geochim. Cosmochim. Acta, 53,* 197–214.

Bernatowicz T. J. and Zinner E. (1996) *Astrophysical Implications of the Laboratory Study of Presolar Material, Vol. 402,* p. 750. AIP Conf. Proc., New York.

Birck J.-L. and Allègre C. J. (1985) Evidence for the presence of $^{53}Mn$ in the early solar system. *Geophys. Res. Lett., 12,* 745–748.

Birck J.-L. and Allègre C. J. (1988) Manganese-chromium isotope systematics and development of the early solar system. *Nature, 331,* 579–584.

Birck J. L. and Allègre C. J. (1998) Rhenium-187–osmium-187 in iron meteorites and the strange origin of the Kodaikanal meteorite. *Meteoritics & Planet. Sci., 33,* 647–654.

Black D. C. and Pepin R. O. (1969) Trapped neon in meteorites — II. *Earth Planet. Sci. Lett., 6,* 395–405.

Cameron A. G. W. (1962) The formation of the sun and planets. *Icarus, 1,* 13–69.

Cameron A. G. W. and Truran J. W. (1977) The supernova trigger for formation of the solar system. *Icarus, 30,* 447–461.

Carlson R. W. and Lugmair G. W. (1988) The age of ferroan anorthosite 60025: oldest crust on a young Moon? *Earth Planet. Sci. Lett., 90,* 119–130.

Chen J. H. and Tilton G. R. (1976) Isotopic lead investigations on the Allende carbonaceous chondrite. *Geochim. Cosmochim. Acta, 40,* 635–643.

Chen J. H. and Wasserburg G. J. (1981) The isotopic composition of uranium and lead in Allende inclusions and meteorite phosphates. *Earth Planet. Sci. Lett., 52,* 1–15.

Chen J. H. and Wasserburg G. J. (1985) U-Th-Pb isotopic studies on meteorite ALAH81005 and Ibitira (abstract). In *Lunar and*

*Planetary Science XVI*, pp. 119–120. Lunar and Planetary Institute, Houston.

Chen J. H. and Wasserburg G. J. (1990) The isotopic composition of Ag in meteorites and the presence of [107]Pd in protoplanets. *Geochim. Cosmochim. Acta, 47*, 1725–1737.

Chen J. H. and Wasserburg G. J. (1996) Live [107]Pd in the early solar system and implications for planetary evolution. In *Earth Processes: Reading the Isotope Code* (A. Basu and S. R. Hart, eds.), pp. 1–20. AGU Monograph 95, Washington, DC.

Clayton D. D. and Jin L. (1995) Gamma rays, cosmic rays, and extinct radioactivity in molecular clouds. *Astrophys. J., 451*, 681–699.

Clayton R. N., Grossman L., and Mayeda T. K. (1973) A component of primitive nuclear composition in carbonaceous meteorites. *Science, 182*, 485–488.

Elmegreen B. G. (1985) Molecular clouds and star formation: an overview. In *Protostars and Planets II* (D. C. Black and M. S. Matthews, eds.), pp. 33–58. Univ. of Arizona, Tucson.

Fahey A. J., Zinner E. K., Crozaz G., and Kornacki A. S. (1987) Microdistributions of Mg isotopes and REE abundances in a Type A calcium-aluminum-rich inclusion from Efremovka. *Geochim. Cosmochim. Acta, 51*, 3215–3229.

Galer S. J. G. and Lugmair G. W. (1996) Lead isotope systematics of noncumulate eucrites. *Meteoritics & Planet. Sci., 31*, A47–A48.

Ghosh A. and McSween H. Y. Jr. (1998) A thermal model for the differentiation of Asteroid 4 Vesta, based on radiogenic heating. *Icarus, 134*, 187–206.

Göpel C., Manhès G., and Allègre C. J. (1985) U-Pb systematics of iron meteorites: uniformity of primordial Pb. *Geochim. Cosmochim. Acta, 49*, 1681–1695.

Göpel C., Manhès G., and Allègre C. J. (1991) Constraints on the time of accretion and thermal evolution of chondrite parent body. *Meteoritics, 26*, 338.

Göpel C., Manhès G., and Allègre C. J. (1992) U-Pb study of the Acapulco meteorite. *Meteoritics, 27*, 226.

Göpel C., Manhès G., and Allègre C. J. (1994) U-Pb systematics of phosphates from equilibrated ordinary chondrites. *Earth Planet. Sci. Lett., 121*, 153–171.

Gray C. M., Papanastassiou D. A., and Wasserburg G. J. (1973) The identification of early condensates from the solar nebula. *Icarus, 20*, 213–239.

Grossman L. (1972) Condensation in the primitive solar nebula. *Geochim. Cosmochim. Acta, 36*, 597–619.

Grossman L. (1980) Refractory inclusions in the Allende meteorite. *Annu. Rev. Earth Planet. Sci., 8*, 559–608.

Hamet J., Nakamura N., Unruh D. M., and Tatsumoto M. (1978) Origin and history of the cumulate eucrite Moama as inferred from REE abundances, Sm-Nd and U-Pb systematics. *Proc. Lunar Planet. Sci. Conf. 9th*, pp. 1115–1136.

Harper C. L. and Jacobsen S. B. (1996) Evidence for [182]Hf in the early solar system and constraints on the timescale of terrestrial core formation. *Geochim. Cosmochim. Acta, 60*, 1131–1153.

Harper C. L., Volkening J., Heumann K. G., Shih C.-Y., and Wiesmann H. (1991) [182]Hf-[182]W: New cosmochronometric constraints on terrestrial accretion, core formation, the astrophysical site of the r-process, and the origin of the solar system (abstract). In *Lunar and Planetary Science XXII*, pp. 515–516. Lunar and Planetary Institute, Houston.

Hartmann W. K. and Davis D. R. (1975) Satellite-sized planetesimals and lunar origin. *Icarus, 24*, 504.

Horan M. F., Smoliar M. I., and Walker R. J. (1998) [182]W and [187]Re-[187]Os systematics of iron meteorites: chronology for melting, differentiation, and crystallization in asteroids. *Geochim. Cosmochim. Acta, 62*, 545–554.

Hutcheon I. D. (1982) Ion probe magnesium isotopic measurements of Allende inclusions. In *Nuclear and Chemical Dating Techniques: Interpreting the Environment Record, Vol. 176* (L. A. Curie, ed.), pp. 95–128. American Chemical Society Symposium Series.

Hutcheon I. D. and Hutchison R. (1989) Evidence from the Semarkona ordinary chondrite for [26]Al heating of small planets. *Nature, 337*, 238–241.

Hutcheon I. D. and Olsen E. J. (1991) Cr isotopic composition of differentiated meteorites: A search for [53]Mn (abstract). In *Lunar and Planetary Science XXII*, pp. 605–606. Lunar and Planetary Institute, Houston.

Jacobsen S. B. and Wasserburg G. J. (1984) Sm-Nd evolution of chondrites and achondrites, II. *Earth Planet. Sci. Lett., 67*, 137–150.

Karpenko S. F., Smoliar M. I., Petaev M. I., and Shukolyukov Y. (1991) Rb-Sr and Sm-Nd systematics in Pomozdino meteorite. *16th Symp. Antarct. Meteorites*, pp. 178–179.

Lee D.-C. and Halliday A. N. (1995) Hafnium-tungsten chronometry and the timing of terrestrial core formation. *Nature, 378*, 771–774.

Lee D.-C. and Halliday A. N. (1996) Hf-W isotopic evidence for rapid accretion and differentiation in the early solar system. *Science, 274*, 1876–1879.

Lee D.-C., Halliday A. N., Snyder G. A., and Taylor L. A. (1997) Age and origin of the Moon. *Science, 278*, 1098–1103.

Lee T., Papanastassiou D. A., and Wasserburg G. J. (1976) Demonstration of [26]Mg excess in Allende and evidence for [26]Al. *Geophys. Res. Lett., 3*, 109–112.

Lee T., Papanastassiou D. A., and Wasserburg G. J. (1977) Aluminum-26 in the early solar system: fossil or fuel? *Astrophys. J. Lett., 211*, L107–L110.

Lee T., Papanastassiou D. A., and Wasserburg G. J. (1978) Calcium isotopic anomalies in the Allende meteorite. *Astrophys. J. Lett., 220*, L21–L25.

Lee T., Russell W. A., and Wasserburg G. J. (1979) Calcium isotopic anomalies and the lack of aluminum-26 in an unusual Allende inclusion. *Astrophys. J. Lett., 228*, L93–L98.

Loss R. D., Lugmair G. W., Davis A. M., and MacPherson G. J. (1994) Isotopically distinct reservoirs in the solar nebula: isotopic anomalies in Vigarano meteorite inclusions. *Astrophys. J. Lett., 436*, L193–L196.

Ludwig K. R. (1991) *ISOPLOT: A Plotting and Regression Program for Radiogenic-Isotope Data*. U.S. Geol. Survey Open-File Report 91-445. 39 pp.

Lugmair G. W. (1974) Sm-Nd ages: A new dating method. *Meteoritics, 9*, 369.

Lugmair G. W. and Galer S. J. G. (1992) Age and isotopic relationships among angrites Lewis Cliff 86010 and Angra dos Reis. *Geochim. Cosmochim. Acta, 56*, 1673–1694.

Lugmair G. W. and Marti K. (1977) Sm-Nd-Pu timepieces in the Angra dos Reis meteorite. *Earth Planet. Sci. Lett., 35*, 273–284.

Lugmair G. W. and Scheinin N. B. (1975) Sm-Nd systematics of the Stannern meteorite. *Meteoritics, 10*, 447–448.

Lugmair G. W. and Shukolyukov A. (1998) Early solar system timescales according to [53]Mn-[53]Cr systematics. *Geochim. Cosmochim. Acta, 62*, 2863–2886.

Lugmair G. W., Marti K., Kurtz J. P., and Scheinin N. B. (1976) History and genesis of lunar troctolite 76535 or: How old is old? *Proc. Lunar Sci. Conf. 7th*, pp. 2009–2033.

Lugmair G. W., Scheinin N. B., and Carlson R. W. (1977) Sm-Nd systematics of the Serra de Mage eucrite. *Meteoritics, 10*, 300–301.

Lugmair G. W., Marti K., and Scheinin N. B. (1978) Incomplete mixing of products from R-, P-, and S-process nucleosynthesis: Sm-Nd systematics in Allende inclusion EK 1-04-1 (abstract). In *Lunar and Planetary Science IX*, pp. 672–674. Lunar and Planetary Institute, Houston.

Lugmair G. W., Shimamura T., Lewis R. S., and Anders E. (1983) Samarium-146 in the early solar system: evidence from neodymium in the Allende meteorite. *Science, 222*, 1015–1018.

Lugmair G. W., Galer S. J. G., and Carlson R. W. (1991) Isotope systematics of cumulate eucrite EET-87520. *Meteoritics, 26*, 368.

MacPherson G. J. and Davis A. M. (1993) A petrologic and ion microprobe study of a Vigarano Type B refractory inclusion: evolution by multiple stages of alteration and melting. *Geochim. Cosmochim. Acta, 57*, 231–243.

MacPherson G. J., Davis A. M., and Zinner E. K. (1995) The distribution of aluminum-26 in the early Solar System — A reappraisal. *Meteoritics, 30*, 365–386.

Manhès G., Allègre C. J., and Provost A. (1978) Comparative U-Th-Pb and Rb-Sr study of the Saint Severin amphoterite: consequences for early solar system chronology. *Earth Planet. Sci. Lett., 39*, 14–24.

Manhès G., Göpel C., and Allègre C. J. (1984) U-Th-Pb systematics of the eucrite "Juvinas" Precise age determination and evidence for exotic lead. *Geochim. Cosmochim. Acta, 48*, 2247–2264.

Manhès G., Göpel C., and Allègre C. J. (1987) High resolution chronology of the early solar system based on lead isotopes. *Meteoritics, 22*, 453–454.

McCulloch M. T. and Wasserburg G. J. (1978) Barium and neodymium isotopic anomalies in the Allende meteorite. *Astrophys. J. Lett., 220*, L15–L19.

Nakamura N., Unruh D. M., Gensho R., and Tatsumoto M. (1977) Evolution history of lunar mare basalt, Apollo 15 samples revisited. *Proc. Lunar Planet. Sci. Conf. 8th*, pp. 712–713.

Niederer F. R., Papanastassiou D. A., and Wasserburg G. J. (1980) Endemic isotopic anomalies in titanium. *Astrophys. J. Lett., 240*, L73–L77.

Niemeyer S. and Lugmair G. W. (1981) Ubiquitous isotopic anomalies in Ti from normal Allende inclusions. *Earth Planet. Sci. Lett., 53*, 211–225.

Nyquist L. E., Bansal B. M., Wiesmann H., and Jahn B.-M. (1974) Taurus-Littrow chronology: some constraints on early lunar crustal development. *Proc. Lunar Sci. Conf. 5th*, pp. 1515–1539.

Nyquist L. E., Bansal B., Wiesmann H., and Shih C.-Y. (1994) Neodymium, strontium, and chromium isotopic studies of the LEW86010 and Angra dos Reis meteorites and chronology of the angrite parent body. *Meteoritics, 29*, 872–885.

Nyquist L. E., Bogard D., Wiesmann H., Shih C.-Y., Yamaguchi A., and Takeda H. (1996) Early history of the Padvarninkai eucrite. *Meteoritics & Planet. Sci., 31*, A101–A102.

Nyquist L., Bogard D., Takeda H., Bansal B., Wiesmann H., and Shih C.-Y. (1997a) Crystallization, recrystallization, and impact-metamorphic ages of eucrites Y792510 and Y791186. *Geochim. Cosmochim. Acta, 61*, 2119–2138.

Nyquist L., Lindstrom D., Shih C.-Y., Wiesmann H., Mittlefehldt

D., Wentworth S., and Martinez R. (1997b) Mn-Cr isotopic systematics of chondrules from the Bishunpur and Chainpur meteorites (abstract). In *Lunar and Planetary Science XXVIII*, pp. 1033–1034. Lunar and Planetary Institute, Houston.

Nyquist L. E., Wiesmann H., Reese Y., Shih C.-Y., and Borg L. E. (1997c) Samarium-neodymium age and maganese-chromite systematics of eucrite Elephant Moraine 90020. *Meteoritics & Planet. Sci., 32*, A101–A102.

Nyquist L. E., Shih C.-Y., Wiesmann H., Reese Y., Ulyanov A. A., and Takeda H. (1999) Towards a Mn-Cr timescale for the early solar system (abstract). In *Lunar and Planetary Science XXX*, Abstract #1604. Lunar and Planetary Institute, Houston (CD-ROM).

Papanastassiou D. A. and Wasserburg G. J. (1969) Initial strontium isotopic abundances and the resolution of small time differences in the formation of planetary objects. *Earth Planet. Sci. Lett., 5*, 361–376.

Patchett P. J. (1980a) Sr isotopic fractionation in Allende chondrules: a reflection of solar nebular processes. *Earth Planet. Sci. Lett., 50*, 181–188.

Patchett P. J. (1980b) Sr isotopic fractionation in Ca-Al inclusions from the Allende meteorite. *Nature, 283*, 438–441.

Patterson C. (1956) Age of meteorites and the Earth. *Geochim. Cosmochim. Acta, 10*, 230.

Podosek F. A. and Cassen P. (1994) Theoretical, observational, and isotopic estimates of the lifetime of the solar nebula. *Meteoritics, 29*, 6–25.

Podosek F. A., Zinner E. K., MacPherson G. J., Lundberg L. L., Brannon J. C., and Fahey A. J. (1991) Correlated study of initial $^{87}Sr/^{86}Sr$ and Al/Mg isotopic systematics and petrologic properties in a suite of refractory inclusions from the Allende meteorite. *Geochim. Cosmochim. Acta, 55*, 1083–1110.

Podosek F. A., Ott U., Brannon J. C., Neal R. C., Bernatowicz T. J., Swan P., and Mahan S. E. (1997) Thoroughly anomalous chromium in Orgueil. *Meteoritics & Planet. Sci., 32*, 617–627.

Prinzhofer A., Papanastassiou D. A., and Wasserburg G. J. (1992) Samarium-neodymium evolution of meteorites. *Geochim. Cosmochim. Acta, 56*, 797–815.

Reynolds J. H. and Turner G. J. (1964) Rare gases in the chondrite Renazzo. *J. Geophys. Res., 69*, 3263–3281.

Rotaru M., Birck J. L., and Allègre C. J. (1992) Clues to early solar system history from chromium isotopes in carbonaceous chondrites. *Nature, 358*, 465–470.

Russell S. S., Srinivasan G., Huss G. R., Wasserburg G. J., and MacPherson G. J. (1996) Evidence for widespread $^{26}Al$ in the solar nebula and constraints for nebula timescales. *Science, 273*, 757–762.

Shen J. J., Papanastassiou D. A., and Wasserburg G. J. (1996) Precise Re-Os determinations and systematics of iron meteorites. *Geochim. Cosmochim. Acta, 60*, 2887–2900.

Shu F. H., Najita J., Galli D., Ostriker E., and Lizano S. (1993) The collapse of clouds and the formation and evolution of stars and disks. In *Protostars and Planets III* (E. H. Levy and J. Lunine, eds.), pp. 3–45. Univ. of Arizona, Tucson.

Shukolyukov A. and Begemann F. (1996) Pu-Xe dating of eucrites. *Geochim. Cosmochim. Acta, 60*, 2453–2471.

Shukolyukov A. and Lugmair G. W. (1993) $^{60}Fe$ in eucrites. *Earth Planet. Sci. Lett., 119*, 159–166.

Shukolyukov A. and Lugmair G. W. (1998) The $^{53}Mn$-$^{53}Cr$ isotope system in the Indarch EH4 chondrite: further argument for $^{53}Mn$ heterogeneity in the early solar system (abstract). In *Lunar and Planetary Science XXIX*, pp. 1208–1209. Lunar and

Planetary Institute, Houston.

Shukolyukov A. and Lugmair G. W. (1999) The $^{53}$Mn-$^{53}$Cr isotope systematics of the enstatite chondrites (abstract). In *Lunar and Planetary Science XXX*, Abstract #1093. Lunar and Planetary Institute, Houston (CD-ROM).

Smoliar M. I. (1993) A survey of Rb-Sr systematics of eucrites. *Meteoritics, 28*, 105–113.

Smoliar M. I., Walker R. J., and Morgan J. W. (1996) Re-Os ages of group IIA, IIIA, IVA, and IVB iron meteorites. *Science, 271*, 1099–1102.

Srinivasan G., Goswami J. N., and Bhandari N. (1998) Search for extinct aluminum-26 in the Piplia Kalan eucrite. *Meteoritics & Planet. Sci., 33*, A148–A149.

Srinivasan G., Goswami J. N., and Bhandari N. (1999) $^{26}$Al in eucrite Piplia Kalan: plausible heat source and formation chronology. *Science, 284*, 1348–1350.

Stevenson D. J. (1987) Origin of the Moon — the collision hypothesis. *Annu. Rev. Earth Planet. Sci., 15*, 271–315.

Tatsumoto M., Knight R. J., and Allègre C. J. (1973) Time differences in the formation of meteorites as determined from the ratio of lead-207 to lead-206. *Science, 180*, 1279–1283.

Tera F. and Carlson R. W. (1999) Assessment of the Pb-Pb and U-Pb chronometry of the early Solar System. *Geochim. Cosmochim. Acta*, in press.

Tera F., Carlson R. W., and Boctor N. Z. (1989) Contrasting Pb-Pb ages of the cumulate and non-cumulate eucrites (abstract). In *Lunar and Planetary Science XX*, pp. 1111–1112. Lunar and Planetary Institute, Houston.

Tera F., Carlson R. W., and Boctor N. Z. (1997) Radiometric ages of basaltic achondrites and their relation to the early history of the solar system. *Geochim. Cosmochim. Acta, 61*, 1713–1731.

Unruh D. M., Nakamura N., and Tatsumoto M. (1977) History of the Pasamonte achondrite: relative susceptability of the Sm-Nd, Rb-Sr, and U-Pb systems to metamorphic events. *Earth Planet. Sci. Lett., 37*, 1–12.

Urey H. C. (1955) The cosmic abundances of potassium, uranium, and thorium and the heat balances of the Earth, the Moon and Mars. *Proc. Natl. Acad. Sci. U.S., 41*, 127–144.

Wadhwa M. and Lugmair G. W. (1995) Sm-Nd systematics of the eucrite Chervony Kut (abstract). In *Lunar and Planetary Science XXVI*, pp. 1453–1454. Lunar and Planetary Institute, Houston.

Wadhwa M. and Lugmair G. W. (1996) Age of the eucrite "Caldera" from convergence of long-lived and short-lived chronometers. *Geochim. Cosmochim. Acta, 60*, 4889–4893.

Wadhwa M., Zinner E. K., and Crozaz G. (1997) Manganese-chromium systematics in sulfides of unequilibrated enstatite chondrites. *Meteoritics & Planet. Sci., 32*, 281–292.

Wadhwa M., Shukolyukov A., and Lugmair G. W. (1998) $^{53}$Mn-$^{53}$Cr systematics in Brachina: a record of one of the earliest phases of igneous activity on an asteroid? (abstract). In *Lunar and Planetary Science XXIX*, pp. 1480–1480.Lunar and Planetary Institute, Houston.

Wadhwa M., Shukolyukov A., Davis A. M., and Lugmair G. W. (1999) Origin of silicate clasts in mesosiderites: trace element microdistributions and Mn-Cr systematics tell the tale (abstract). In *Lunar and Planetary Science XXX*, Abstract #1707. Lunar and Planetary Institute, Houston (CD-ROM).

Wasserburg G. J. (1985) Short-lived nuclei in the early solar system. In *Protostars and Planets II* (D. C. Black and M. S. Matthews, eds.), pp. 703–737. Univ. of Arizona, Tucson.

Wasserburg G. J. and Huneke J. C. (1979) I-Xe dating of I bearing phases in Allende (abstract). In *Lunar and Planetary Science X*, pp. 1307–1309. Lunar and Planetary Institute, Houston.

Wetherill G. W. (1990) Formation of the Earth. *Annu. Rev. Earth Planet. Sci., 18*, 205–256.

Zinner E. (1996) Presolar material in meteorites: an overview. In *Astrophysical Implications of the Laboratory Study of Presolar Materials, Vol. 402* (T. J. Bernatowicz and E. K. Zinner, eds.), pp. 3–26. AIP Conf. Proc., New York.

Zinner E. and Göpel C. (1992) Evidence for $^{26}$Al in feldspars from the H4 chondrite Ste. Marguerite. *Meteoritics, 27*, 311–312.

Zipfel J., Shukolyukov A., and Lugmair G. W. (1996) Manganese-chromium systematics in the Acapulco meteorite. *Meteoritics & Planet. Sci., 31*, A160.

# Tungsten Isotopes, the Timing of Metal-Silicate Fractionation, and the Origin of the Earth and Moon

**A. N. Halliday**
*Eidgenössische Technische Hochschule Zürich*

**D-C. Lee**
*University of Michigan*

**Stein B. Jacobsen**
*Harvard University*

The chondritic W-isotopic composition of the silicate Earth is inconsistent with early and rapid core formation as in certain heterogeneous accretion models or the timescales of formation of asteroidal cores. Protracted homogeneous accretion and continuous core formation at decreasing rates provide plausible explanations. The Hf-W data for lunar samples can, with various caveats, be reconciled with a major Moon-forming impact ~25–65 m.y. after the start of the solar system. The recent suggestion that the proto-Earth to impactor-mass ratio was more like 7:3 is easily reconciled with W-isotopic data. However, the W data can also be satisfactorily modeled with the "traditional" near-Earth-sized proto-Earth and Mars-sized impactor depending on the amount and timing of postimpact accretion.

## 1. ACCRETION MODELS

Accretion theory has undergone dramatic developments over the past 40 years, and a critical part of this development has revolved around the mechanisms of metal-silicate fractionation. The new Hf-W chronometer (*Harper et al.,* 1991; *Lee and Halliday,* 1995b, 1996, 1997, 1998, 2000; *Harper and Jacobsen,* 1996a; *Jacobsen and Harper,* 1996; *Lee et al.,* 1997; *Horan et al.,* 1998) is ideal for determining the timing of core formation. It can therefore be used to address the timescales for accretion and provide tests of accretionary models (*Halliday et al.,* 1996; *Kramers,* 1998; *Halliday and Lee,* 1999; *Halliday,* 2000).

The chemical condensation sequence modeled thermodynamically for a nebula gas cooling from 2000 K has long been considered a starting point for understanding the basic chemistry of the material accreting in the inner solar system (*Grossman and Larimer,* 1974). However, the application of condensation theory to the striking variations in the densities and compositions of the terrestrial planets, and how metal and silicate form in distinct reservoirs has been problematic. In some early models (*Ringwood,* 1966) it was considered that the Fe metal in the Earth's core formed by reduction of Fe in silicates and oxides, requiring a huge CO atmosphere. Large atmospheres have at various times been considered a fundamental feature of the early Earth. Based on noble gas evidence, *Harper and Jacobsen* (1996b) suggested that Fe was reduced to form the core during a stage with such a massive early $H_2$-He atmosphere. Such models serve to illustrate the difficulty in explaining large amounts of Fe metal at the center of an Earth with an oxidized

mantle. *Okuchi et al.* (1998) have presented the alternative theory that the dissociation of water and the extraction of large amounts of H into the core resulted in the oxidation of iron in the Earth's mantle. However, at one time the only likely alternative explanation appeared to be that the silicate portion of the Earth accreted separately from the metal. *Eucken* (1944) first proposed such a heterogeneous accretion model in which early-condensed metal formed a core to the Earth around which silicate accreted after condensation at lower temperatures. In this context, the silicate-depleted/iron-enriched nature of Mercury makes sense as a body that accreted in an area of the solar nebula that was too hot to condense the same proportion of silicate as is found in the Earth (*Grossman,* 1972). Conversely, the low density of Mars could partly reflect collection of an excess of silicate in cooler reaches of the inner solar nebula. So the concept of heliocentric "feeding zones" for accretion fitted this nicely. The discovery that Fe metal condenses at a lower temperature than some refractory silicates (*Larimer and Anders,* 1970) changed things somewhat; nevertheless, a series of models involving progressive heterogeneous accretion at successively lower condensation temperatures was developed for Earth (e.g., *Turekian and Clark,* 1969). These heterogeneous accretion models produced a zoned Earth with an early metallic core surrounded by silicate, without the need for massive reduction of iron on Earth itself.

The timescales that were proposed for accretion and core formation in all these early models were typically short. Those that assumed that most accretion was accomplished as a gaseous protoplanet that was subsequently stripped of its volatiles (*Cameron, 1978*) produced short accretionary timescales of <$10^6$ yr. Clearly, if Earth's core formed by reduction of iron in a large atmosphere, this process must have been early unless the nebula lasted longer than currently considered likely (<10 m.y.). So the heterogeneous accretion models of *Turekian and Clark* (1969) and others require fast accretion and core formation if these processes reflect condensation in the nebula. *Hanks and Anderson* (1969) also proposed extremely short timescales for accretion and core formation (<$10^6$ yr).

However, *Safronov* (1954) argued that the formation of the planets would be dominated by collisions between planetesimals and raised the possibility that accretion was highly protracted. *Wetherill* (1986) showed that runaway planetary embryos would collide and cause further growth of planets. Whereas models that are based on accretion with runaway growth as the final step complete the process within $10^6$ yr (*Lin and Papaloizou, 1985*), the last collisional stages of accretion extend major growth over periods of up to $10^8$ yr. *Wetherill* (1986) predicted that Earth would accrete to about half its present mass in roughly 10 m.y., but that growth would continue as a result of major collisions until ~100 m.y. after the start of the solar system.

This late collisional growth process explains the low density of the Moon. Clearly, some refractory-enriched but iron-depleted bodies can be made or captured in the same vicinity as Earth. This is hard to explain simply by coaccretion. *Wetherill* (1994a) went on to challenge the whole notion of feeding zones by showing that although on average the terrestrial planets would sample material from different heliocentric orbits, the source of materials would be extremely broad for each planet. The idea that heliocentric orbits dictated the Fe/Si of a planet therefore became more difficult to accommodate. Eventually, even the strongly Si-depleted nature of Mercury was explained by a giant impact (*Benz et al., 1987*). In fact, such impacts have been recognized as fundamental features of the growth of all planets.

So the view of how the early solar system accreted has changed from an orderly, condensation-related, geometric sequence of composition with localized feeding zones, to a violent, stochastic, collision-dominated history with a broad range of sources. Our present understanding is far from complete. We do not understand how accretion started and by what mechanism sticking is accomplished in the hot inner solar system (*Boss, 1990*). We do not know how or when the jovian planets accreted (*Wetherill, 1994b*). Nor do we understand how their accretion affected accretion of the terrestrial planets. However, we can test some of the above theories for the terrestrial planets by using short-lived nuclides.

## 2.  SHORT-LIVED NUCLIDES

The atomic abundances of the daughters of extinct nuclides provide the most powerful method for elucidating the history of chemical fractionation and mass transfer relevant

to the early development of the Moon, Earth, and Mars (cf. *Jacobsen and Harper, 1996*). In general, the shortest lived of these are mainly used for studying chondrites and nebular timescales. Nuclides with intermediate half-lives are ideal for studying differentiated meteorites. The longest half-life short-lived nuclides, such as $^{182}$Hf (half-life = 9 m.y.) are the only ones suitable for studying the later stages of accretion of terrestrial planets.

Studies of $^{107}$Pd ($T_{1/2}$ ~6.5 m.y.) in iron meteorites provided evidence that the earliest planetary differentiation processes occurred within a few $^{107}$Pd half-lives of the formation of our solar system (*Chen and Wasserburg, 1996*). Studies of $^{53}$Mn-$^{53}$Cr (half-life = 3.7 m.y.) have refined the timescales for the parent bodies of stony meteorites and allowed precise time differences to be determined between eucrites and a group of very rare achondrites called angrites. The Pb-Pb age of the angrites has been determined at 4.5575 ± 0.0005 Ga using a select group of pyroxenes (*Lugmair and Galer, 1992*). An internal Mn-Cr isochron for the angrites allows the initial abundance of $^{53}$Mn at this time to be ascertained, allowing one to convert the Mn-Cr data for eucrites into an absolute age. This indicates that the eucrite parent body may have formed as early as the U-Pb ages of CAI in Allende (4.566 ± 0.002 Ga) the canonical start of the solar system (*Göpel et al., 1991, 1994; Lugmair and Shukolyukov, 1998*). The eucrite parent body is widely thought to be asteroid 4 Vesta. Hence, we are at the point of having Mn-Cr and Pd-Ag evidence that sizable planetesimals were already formed and differentiated at a time currently indistinguishable from the U-Pb ages of the objects thought to be the oldest formed within the solar system. For a review of this subject see the chapter by *Carlson and Lugmair* (2000).

## 3.  HAFNIUM-TUNGSTEN SYSTEMATICS

The Hf-W-isotopic system has long been recognized as having great potential for determining the age of Earth's core (e.g., *Harper et al., 1991*). Hafnium-182 decays to $^{182}$W with a half-life of 9 m.y., a period of time similar to that commonly considered relevant to accretion (Fig. 1). Both parent and daughter element are in known chondritic proportions (~1:1) in Earth because they are highly refractory. Hafnium is lithophile ("rock-loving") and is partitioned strongly into the silicate Earth during metal segregation. Tungsten, on the other hand, is moderately siderophile ("metal-loving") and should partition preferentially into a metallic phase (*Rammensee and Wänke, 1977; Newsom et al., 1996*). So the chondritic Hf/W ratio of Earth should be internally fractionated by core formation (Fig. 2). If this takes place during the lifetime of $^{182}$Hf, an excess in the atomic abundance of $^{182}$W should develop in the silicate Earth as a consequence of enhanced Hf/W. Conversely, the W-isotopic compositions of early metals should be deficient in $^{182}$W relative to chondritic atomic abundances because of the isolation of W before $^{182}$Hf decayed. *Lee and Halliday* (1995b, 1996, 1997, 2000) showed that the abundance

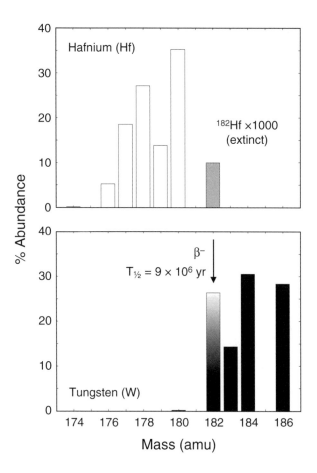

**Fig. 1.** Isotopic abundances of Hf and W. Also shown is the amount of ¹⁸²Hf at the start of the solar system, magnified by a factor of 1000.

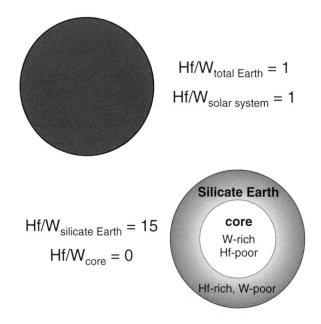

$$Hf/W_{total\ Earth} = 1$$
$$Hf/W_{solar\ system} = 1$$

**Silicate Earth**

**core**
W-rich
Hf-poor

Hf-rich, W-poor

$$Hf/W_{silicate\ Earth} = 15$$
$$Hf/W_{core} = 0$$

**Fig. 2.** Schematic showing the fractionation of Hf from W in the early solar system.

of ¹⁸²Hf at the start of the solar system was relatively high (¹⁸²Hf/¹⁸⁰Hf >10⁻⁴), rendering this chronometer particularly accurate for a range of objects and timescales.

Despite its considerable potential, Hf-W was unusable because nobody had developed a suitable technique for measuring W-isotopic compositions on small samples. We knew the isotopic composition of the plentiful W as found in the silicate Earth and used in chisels and drill bits. But nobody was able to measure the isotopic compositions of the trace W in meteorites to draw comparisons with the W-isotopic composition of the silicate Earth. The first ionization potential for W is too high to permit efficient ionization using a thermal source. Over the past decade the development of negative ion thermal ionization mass spectrometry (NTIMS) permitted the first precise measurements of W at high sensitivity (*Völkening et al.*, 1991; *Harper et al.*, 1991; *Harper and Jacobsen*, 1996a). Over the same timeframe, multiple collector inductively coupled plasma mass spectrometry (*Walder and Freedman*, 1992) was developed and this soon led to a relatively straightforward method for measuring W-isotopic compositions (*Halliday et al.*, 1995; *Lee and Halliday*, 1995a). Both of these techniques appear to work well (*Halliday and Lee*, 1999). *Harper et al.* (1991) were the first to publish preliminary findings that indicated that ¹⁸²Hf may have been live in the early solar system. They found a hint of a deficiency in the atomic abundance of ¹⁸²W in the iron meteorite Toluca and later confirmed this with a more precise set of NTIMS measurements (*Harper and Jacobsen*, 1996a; *Jacobsen and Harper*, 1996). As shown in Fig. 3, W-isotopic measurements on many other early metals have demonstrated that this deficiency is very common (*Lee and Halliday*, 1995b, 1996, 1998, 2000; *Horan et al.*, 1998). Precise W-isotopic data are now available for a wide range of early solar system objects including iron meteorites (*Lee and Halliday*, 1995b, 1996; *Harper and Jacobsen*, 1996a; *Horan et al.*, 1998), ordinary chondrites (*Lee and Halliday*, 1996, 2000), enstatite chondrites (*Lee and Halliday*, 1998), carbonaceous chondrites (*Lee and Halliday*, 1995b, 1996), eucrites (*Lee and Halliday*, 1997), martian meteorites (*Lee and Halliday*, 1997), and lunar samples (*Lee et al.*, 1997).

The W-isotopic composition of a sample or a reservoir j is conveniently expressed as ε deviations from average chondrites (CHUR)

$$\varepsilon_{W,j} = \left( \frac{\left( ^{182}W/^{183}W \right)_j}{\left( ^{182}W/^{183}W \right)_{CHUR}} - 1 \right) \times 10^4$$

In practice, both ¹⁸²W/¹⁸³W and ¹⁸²W/¹⁸⁴W have been reported and used, but the deployment of ε_W notation means that this is of no consequence. Also note that all the ε_W data of *Lee and Halliday* (1995b, 1996, 1997, 2000) and *Lee et al.* (1997) quote W-isotopic compositions relative to the ¹⁸²W/¹⁸⁴W of the present-day bulk silicate Earth, because this is based on a very precisely determined laboratory stan-

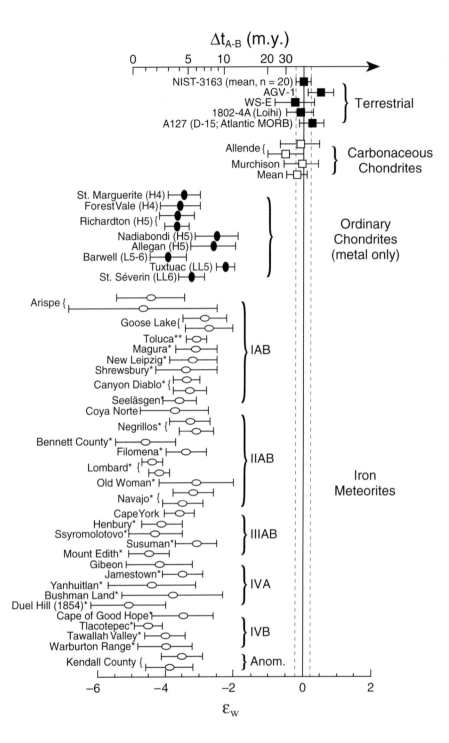

**Fig. 3.** Tungsten-isotopic compositions for terrestrial samples, whole-rock carbonaceous chondrites, and early segregated metals in ordinary chondrites and iron meteorites, expressed in ε units as deviations from the terrestrial value (see text). The W-isotopic model age for the time differences between separation of metal from a primitive chondritic reservoir is shown at the top of the figure. Data from *Lee and Halliday* (1995b, 1996), *Jacobsen and Harper* (1996), *Lee et al.* (1997), and *Horan et al.* (1998). From *Halliday and Lee* (1999), with permission.

dard isotopic composition. The values determined for carbonaceous chondrites are currently unresolvable from that of the bulk silicate Earth and agree to within <70 ppm. All data are reported as $\varepsilon_W$ values relative to the (terrestrial) laboratory standard in Figs. 3 and 4. The modeling presented

in this and other papers is actually aimed at addressing the issue of why the carbonaceous chondrites and silicate Earth have such similar composition and therefore use $\varepsilon_W$ as a theoretical deviation from a chondritic value, as in the above equation, but this is also of no consequence. In these first

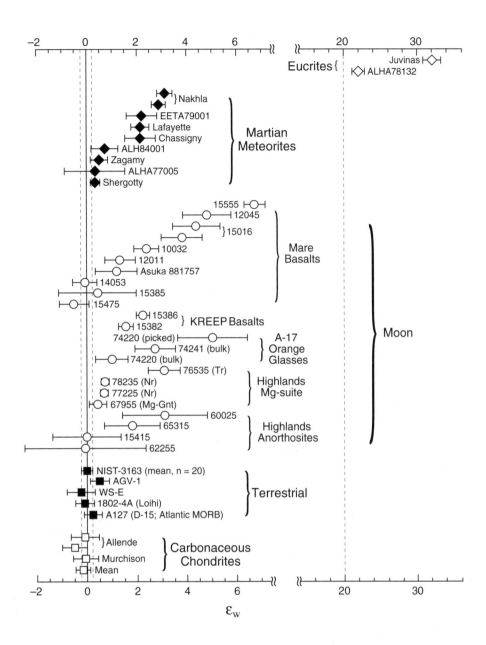

**Fig. 4.**  Tungsten-isotopic compositions for terrestrial samples, whole-rock carbonaceous chondrites, lunar samples, martian meteorites, and eucrites expressed in ε units as deviations from the terrestrial value (see text). Data from *Lee and Halliday* (1996, 1997) and *Lee et al.* (1997). From *Halliday and Lee* (1999), with permission.

few years, several of the Hf-W-isotopic results have been surprising and have raised important new constraints on the accretion history of the inner solar system.

## 4. CHONDRITIC TUNGSTEN-ISOTOPIC COMPOSITION OF EARTH

With a Hf/W ratio of 10 to 40 for the silicate Earth, there should be about a percent-level excess in $^{182}W/^{184}W$ (or $^{182}W/^{183}W$) if Earth's core formed at the same time as many iron meteorites. In fact, the W in the silicate Earth (the same

W we use in everyday life) is identical in isotopic composition to that found in carbonaceous chondrites, to within <70 ppm (Fig. 4). This lack of a resolvable difference between the silicate Earth and chondrites (which may be called the solar system standard for average Hf-W systematics) is a fundamental observation that places important constraints on the early history of Earth, discussed below. Note that the W-isotopic composition of the bulk silicate Earth (BSE) is, by definition, zero ($\varepsilon_W = 0$), because this is the composition of laboratory standards to which all other data are compared. So, somewhat confusingly, the big issue to explain is why the chondritic W-isotopic composition is also zero. Put in nontechnical terms, how does the BSE have chondritic W if its Hf/W ratio is so strongly nonchondritic?

An important test of the Hf-W system is to demonstrate that early bodies with high Hf/W do indeed have radiogenic W and that the systematics of Hf/W isochrons are consistent with decay of former [182]Hf. The Hf/W ratio of the BSE is not so different from that of eucrites, which we know, from Mn-Cr data (*Lugmair and Shukolyukov, 1998*) form early. Highly radiogenic W is found in the eucrites (Fig. 4) consistent with the early differentiation age, and high Hf/W ratios (>10) for the meteorites and their calculated parent body (probably close to 20) (*Lee and Halliday, 1997*). So these data provide confirmation of the basic Hf-W theory. Now internal isochrons have also been obtained for ordinary chondrites (*Lee and Halliday, 2000*), establishing without doubt that the effect is the result of formerly live [182]Hf with a [182]Hf/[180]Hf of >10$^{-4}$.

The obvious explanation for the difference between eucrites and the silicate Earth is that the fractionation of Hf from W took place relatively late in the early history of Earth. Such a fractionation would occur as a consequence of core formation. So either Earth's core formed late or the accretion of Earth was protracted. The W-isotopic data are clearly inconsistent with early and rapid accretion and core formation for Earth. For example, the major classes of heterogeneous accretion models in which the various metal/silicate ratios among the terrestrial planets have been explained as a consequence of differing ambient temperatures during condensation from a hot nebula gas, involve early condensed metal that is added almost directly to the core (*Eucken, 1944; Turekian and Clark, 1969; Clark et al., 1972*). If the earliest accreted material represented a high-temperature reduced metal-rich condensate that quickly formed Earth's core, it would leave the inner solar system fractionated with respect to Hf/W and should result in radiogenic W in the silicate Earth. It is clear that this cannot have happened, otherwise the silicate Earth would have W that is vastly radiogenic rather than chondritic. No matter how much early segregation of metal from silicate takes place in the early solar system, it must remix and reequilibrate during the early stages of the formation of Earth or else there would be a resolvable residual [182]W excess in the silicate portion.

If the W in the silicate Earth is largely added under oxidizing conditions after the major phase of terrestrial core formation has ceased, one might expect a chondritic W-isotopic composition for the silicate Earth (*Newsom, 1996; Shearer and Newsom, 1998*). However, as pointed out by *Lee and Halliday* (1995b) and *Harper and Jacobsen* (1996a), this is not the case for the following reason (Fig. 5): We know that the Hf/W ratio of the total Earth is chondritic and the Hf/W ratio of the silicate Earth is 10 to 40 (*Newsom et al., 1996*). Most adopt a value of ~15 (*Halliday and Lee, 1999*). Let us assume that terrestrial core formation was early but that 99.9% of the W in the silicate Earth was added after the main stage of core formation. If this is the case,

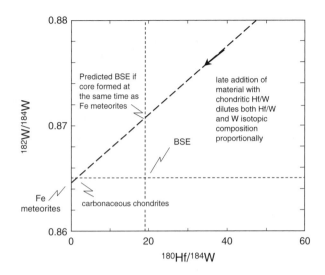

**Fig. 5.** The addition of chondritic material to the silicate Earth after early core formation cannot in and of itself explain the W-isotopic composition of the silicate Earth, because the W-isotopic effect will be diluted in proportion to the decrease in Hf/W (see text for discussion). From *Halliday and Lee* (1999), with permission.

the Hf/W ratio of the silicate Earth prior to core formation must have been 15,000. It will therefore have generated exceedingly radiogenic W. The effects on the W-isotopic composition and Hf/W ratio, of diluting with chondritic W, are no different from the effects of having lower ratios in the first place (Fig. 5). The W-isotopic composition and Hf/W will be lowered proportionally.

It seems inescapable that a silicate Earth with high Hf/W but chondritic W mandates core formation >50 m.y. after the start of the solar system, unless accretion was slow. Some early physical models (e.g., *Solomon, 1979*) have considered that core formation on Earth and Mars developed only after tens to hundreds of millions of years, following the late buildup of accretional energy and melting. The W-isotopic data are consistent with such models. Although such dynamic models are no longer considered realistic, Pb-isotopic data for the silicate Earth also are consistent with core formation taking place tens of millions of years after the start of the solar system (*Allègre et al., 1995; Galer and Goldstein, 1996; Halliday et al., 1996*).

However, both the W- and Pb-isotopic data can be also modeled with gradual accretion and continuous core formation (*Halliday et al., 1996; Harper and Jacobsen, 1996a; Jacobsen and Harper, 1996; Halliday and Lee, 1999; Halliday, 2000*). Most recent physical models (*Hanks and Anderson, 1969; Sasaki and Nakasawa, 1986; Stevenson, 1981; Shaw, 1978*) regard core formation as an integral part of the density-driven differentiation of a planet heated rapidly by accretional energy. Melting temperatures may build up on timescales of <20 m.y., particularly as there is now strong evidence that the accreting inner solar system was hotter than previously considered as a consequence of frictional heating during collapse of the solar nebula (*Boss,* 1990). If the core formed as the Earth accreted, the fractional (not absolute) growth of the planet with time becomes the dominant parameter dictating the magnitude of the W-isotopic effect.

## 5. PROTRACTED ACCRETION

Isotope data for inner solar system objects lend support to the argument that protracted accretion is a good explanation for the chondritic W of the silicate Earth. Iron meteorites generally come from small (20–500 km) parent bodies (*Wasson,* 1985). They have the least radiogenic W yet measured (*Lee and Halliday,* 1995b, 1996; *Harper and Jacobsen,* 1996a; *Horan et al.,* 1998). A corresponding silicate reservoir of the parent body would be expected to have $\varepsilon_W$ ~50–100 if it had a Hf/W ratio similar to eucrites, the Moon, or the silicate Earth. Eucrites are highly radiogenic ($\varepsilon_W$ ~30) (*Lee and Halliday,* 1997) and are thought to be derived from the asteroid 4 Vesta (525 km diameter). Martian meteorites range from chondritic to slightly radiogenic ($\varepsilon_W$ = 0–3) (*Lee and Halliday,* 1997). Earth, the largest of these bodies, has $\varepsilon_W$ = 0, similar to carbonaceous chondrites (*Lee and Halliday,* 1995b, 1996). So with these scant data there is a pattern that the smaller the differentiated body, the greater the excess $^{182}$W in the silicate portion. This is not expected if core formation does not occur until a critical stage of planetary development. Indeed, both U-Pb and Hf-W isotopic data for martian meteorites provide confirmation that Mars segregated its core rapidly (*Chen and Wasserburg,* 1986; *Lee and Halliday,* 1997). Palladium-silver and Hf-W data demonstrate that the small asteroidal cores represented in iron meteorites formed fast (*Chen and Wasserburg,* 1986; *Lee and Halliday,* 1996). Finally, Mn-Cr and Hf-W data show that the asteroidal eucrite parent body was formed and differentiated in ~10 m.y. Clearly, core formation, however it occurs, can take place extremely early. The W-isotopic effects are as expected if larger planets simply take longer to accrete than smaller planets, which in turn take longer to accrete than planetesimals the size of asteroids. So protracted accretion with continuous core formation seems a more likely explanation for chondritic W in the silicate Earth than delayed core formation. The magnitude of the W-isotopic effects depends on the type of transport mechanism and isotopic equilibration in the silicate Earth (*Harper and Jacobsen,* 1996a).

In models for continuous reservoir evolution (such as core formation) it is useful to characterize the timescale by the concept of a mean age, $\langle T \rangle$ (*Jacobsen and Wasserburg,* 1979). For early processes we often use the mean time of formation, $\langle t \rangle$, measured from $T_0$ (4566 m.y. = the age of the solar system) instead of the mean age. Thus, the mean time of accretion is $\langle t \rangle = T_0 - \langle T \rangle$. For example, if $\langle T \rangle =$ 4546 m.y., then $\langle t \rangle$ = 20 m.y.

A particularly simple model that gives a good first-order approximation to the *Wetherill* (1986) accretion model is a model with an exponentially decaying rate of accretion.

In this case the mass of the Earth as a function of time is (equation (18) of *Jacobsen and Harper,* 1996)

$$M(t) = M(\infty)(1 - e^{-\alpha t})$$

where $M(\infty)$ is the mass at the end of the accretion process and $\alpha$ is the time constant for accretion. The cumulative fractional mass of the Earth at time t relative to the present day is $X(t) = M(t)/M(\infty)$. This simple approximation to the accretion process has been used in a number of recent models (*Halliday et al.,* 1996; *Harper and Jacobsen,* 1996a; *Jacobsen and Harper,* 1996; *Halliday and Lee,* 1999; *Halliday,* 2000). For this model the mean age of accretion is (equation (19) of *Jacobsen and Harper,* 1996)

$$\langle T \rangle = \frac{t}{1 - e^{-\alpha t}} - \frac{1}{\alpha}$$

When looking at the system after accretion this can be simplified to $\alpha^{-1} \sim (T_0 - \langle T \rangle) = \langle t \rangle$. This means that the mean time of accretion, $\langle t \rangle$, is the inverse of the time constant ($\alpha$) for accretion and corresponds to the time taken to achieve ~63% growth (*Harper and Jacobsen,* 1996a). This is identical to the "accretionary mean life" used by *Halliday et al.* (1996), *Halliday and Lee* (1999), and *Halliday* (2000). However, to calculate the mean time of accretion as a function of time, one needs to use the full equation above.

The Hf/W ratio of the silicate portion of a planet is also a critical parameter. The Hf/W ratio in a reservoir j is sometimes given in terms of the Hf/W fractionation factor relative to a chondritic reservoir [with $(Hf/W)_{CHUR}$ ~1]

$$f_j^{Hf/W} = \frac{(Hf/W)_j}{(Hf/W)_{CHUR}} - 1$$

In all the models considered here, a Hf/W ratio of ~15 or $f^{Hf/W}$ ~14 is used for the silicate Earth. This is the most widely agreed upon value (*Halliday and Lee,* 1999) and is at the lower, more conservative end of the range of 10 to 40 considered permissible by *Newsom et al.* (1996). However the uncertainties on this figure need to be considered carefully.

The W-isotopic evolution of the silicate Earth can be modeled with continuous segregation of a core in present-day proportions. In these models the Earth grows from material with chondritic Hf-W compositions, and the calculated $\varepsilon_W$-isotopic effects can either be expressed relative to the present-day W-isotopic of carbonaceous chondrites, which equals the silicate Earth (*Halliday et al.,* 1996; *Halliday and Lee,* 1999; *Halliday,* 2000) or they can be given relative to the $^{182}$W/$^{183}$W in an evolving chondritic reservoir

$$\varepsilon_{W,j}(t) = \left( \frac{\left(^{182}W/^{183}W\right)_j}{\left(^{182}W/^{183}W\right)_{CHUR}} - 1 \right) \times 10^4$$

As discussed above and by *Harper and Jacobsen* (1996a) the $\varepsilon_W$ evolution depends on the type of transport mechanism assumed to form the core. The simplest case is that of a primitive differentiation model where each parcel added to the Earth is immediately segregated into metal and silicate, with the metal added to the core without further equilibration. The solution to the $\varepsilon_W$ evolution in the silicate Earth is in this case given by equation (43) of *Jacobsen and Harper* (1996)

$$\varepsilon_W(t) = \lambda_{182} Q_W \left(\frac{^{182}Hf}{^{180}Hf}\right)_i \frac{f^{Hf/W}}{M(T_0)} \int_0^t M(\xi) e^{-\lambda_{182}\xi} d\xi$$

Here M is the mass of the Earth as a function of time, $\lambda_{182} = 0.077$ m.y.$^{-1}$ is the decay constant of $^{182}$Hf, $Q_W = 1.55 \times 10^4$ (*Jacobsen and Harper*, 1996) and ($^{182}$Hf/$^{180}$Hf) = $2.4 \times 10^{-4}$ (*Lee et al.*, 1997).

In an alternative scenario, *Harper and Jacobsen* (1996a), *Halliday et al.* (1996), *Halliday and Lee* (1999), *Jacobsen and Yin* (1998), *Jacobsen* (1999), and *Halliday* (2000) consider the case in which the metal completely equilibrates with the surrounding mantle before being added to the core. This is most likely to occur in the incremental growth of small mass fractions of material to Earth and in a magma ocean. Therefore, this is sometimes referred to as the magma ocean differentiation model. A similar scenario for W equilibration is shown in Fig. 6. The $\varepsilon_W$ evolution in this case was calculated using equation (12) of *Harper and Jacobsen* (1996a).

The average age of the core is the same as the average age of the Earth in both of the above cases. The W-isotopic effect in such a growing and differentiating planet is independent of the absolute rate of growth (mass per unit time).

It is a function of the relative rate of growth (fractional mass of a planet per unit time, $M(t)/M(T_0)$) and how this evolves. So small and large planets alike would have the same W-isotopic effect in their silicate portions if their fractional growth curves were the same. The shorter the value of $\langle t \rangle$, the greater will be the excess of $^{182}$W in the silicate Earth because a greater proportion of the planet will have formed and differentiated at an earlier stage.

A simple comparison of $\varepsilon_W$ effects for the two models are shown in Fig. 7, assuming exponential accretion and $\langle t \rangle = 25$ m.y. It is clear that the two models predict very different $\varepsilon_W$ for the silicate Earth. The primitive differentiation model shows much higher $\varepsilon_W$ effects than the magma ocean model. They both initially increase toward a maximum value (~20 m.y. for the magma ocean model and 40 m.y. for the primitive differentiation model) and then decrease with increasing time toward the final value at the end of accretion. Only in the magma ocean model is the effect reduced to a very low value at the end of accretion. So the absence of a W-isotopic effect in the silicate Earth is difficult to explain unless there was thorough isotopic equilibration of W added to the Earth during accretion prior to metal segregation into the core.

A variety of accretionary growth curves are shown in Fig. 8a. The fractional mass of the Earth is shown as a function of time using a simple exponentially decreasing rate of accretion. Growth curves are shown for various values for the mean time of accretion, $\langle t \rangle$, ranging from 10 to 100 m.y. Also shown are the model results for the growth of the Earth of *Wetherill* (1986). The fastest exponential growth curves are similar to the curve of *Wetherill* (1986), and this corresponds to the slowest of the various dynamic models discussed above. This model is similar to the simple

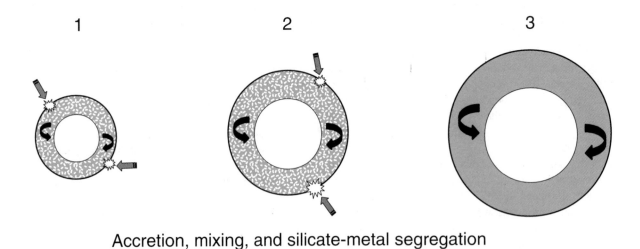

Accretion, mixing, and silicate-metal segregation

**Fig. 6.** Cartoon illustrating the assumptions in a possible continuous core formation model. Earth is assumed to accrete from material that is, on average, chondritic in Hf/W- and W-isotopic composition. As it does so the newly accreted material mixes with the W in the silicate Earth before segregating new core material. In this model the average age of Earth is the same as the average age of the core.

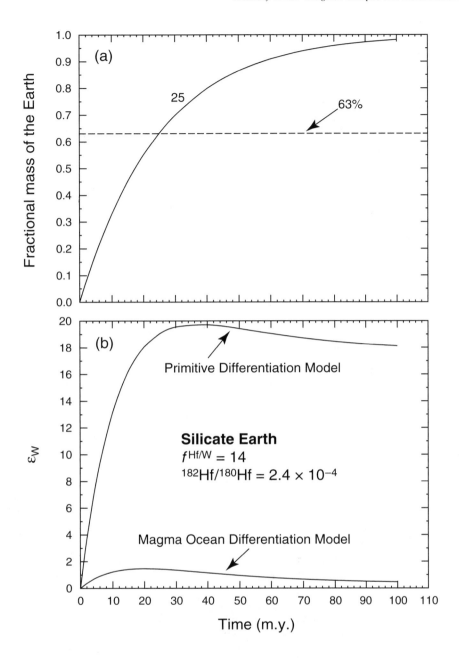

**Fig. 7.** (a) The fractional mass of the Earth as a function of time using an exponentially decreasing continuous growth and a mean time of accretion of 25 m.y. (b) The W-isotopic effects in the silicate Earth calculated with continuous core formation for the accretion curve shown in (a). The result for both a primitive differentiation model as well as a magma ocean differentiation model is shown.

exponential growth curves, but shows significant deviations with increasing time and overall corresponds to a mean time of accretion of 16 m.y.

The W-isotopic effects ($\varepsilon_W$) in the silicate Earth were calculated for continuous core formation both with the exponential accretion model as well as the *Wetherill* (1986) accretion model (Fig. 8b). For the *Wetherill* model, a smooth curve was fitted through the calculated points shown in Fig. 8a. Such a smooth curve misses the total mass of the Earth by about 2.5% at 100 m.y. after the start of the solar system. To reach the final mass this was all added at 100 m.y. to see what the effect of such a relatively small

"giant" impact would be at 100 m.y. after the start of the solar system (*Jacobsen,* 1999, 2000). For exactly exponential accretion, the mean time of accretion must be at least ~25 m.y. or longer to fit the constraint of $\varepsilon_W$ <0.7 for the present silicate Earth (*Halliday et al.,* 1996; *Halliday and Lee,* 1999; *Halliday,* 2000). However, a model with a small deviation from the exponential accretion, such as the *Wetherill* (1986) scenario will also fit the constraint for a substantially smaller mean time of accretion of 16 m.y. (*Jacobsen,* 1999, 2000), provided there are percent levels of accretion after 100 m.y. The likelihood that bodies greater than the size of the Moon accreted to the Earth at such a

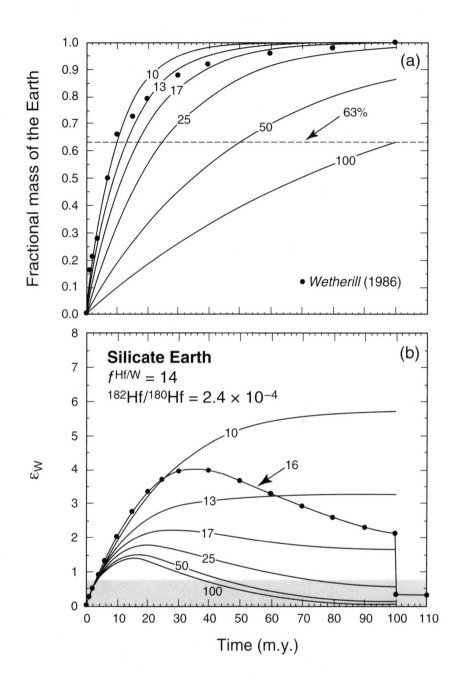

**Fig. 8.** **(a)** The fractional mass of the Earth as a function of time using: (1) A simple exponentially decreasing rate of accretion. Growth curves are shown for various values for the mean time of accretion, $\langle t \rangle$, ranging from 10 to 100 m.y. (calculated using equations (18) and (19) of *Jacobsen and Harper,* 1996). For this exponential model, the mean time shown is the time that is required to build ~63% of Earth's mass. (2) The model results for the growth of the Earth of *Wetherill* (1986). This model is similar to the simple exponential growth curves, but shows significant deviations with increasing time and corresponds to a mean time of accretion of 16 m.y. **(b)** The W-isotopic effects ($\varepsilon_W$) in the silicate Earth were calculated for continuous core formation both with the exponential accretion model as well as the *Wetherill* (1986) accretion model. For the *Wetherill* model, a smooth curve was fitted through the calculated points shown in **(a)**. Such a smooth curve misses the total mass of Earth by about 2.5% at 100 m.y. To reach the final mass this was all added at 100 m.y. to see what the effect of such a relatively small "giant" impact would be at 100 m.y. (*Jacobsen,* 1999, 2000). The $\varepsilon_W$ evolution was in both cases calculated using equation (12) of *Harper and Jacobsen* (1996). The shaded bar indicates the uncertainty in the difference between the present-day W-isotopic compositions of the bulk silicate Earth and carbonaceous chondrites of $\varepsilon_W$ <0.7 (*Lee and Halliday,* 1996). Thus, for exactly exponential accretion the mean time of accretion must be at least ~25 m.y. or longer to fit the constraint of $\varepsilon_W$ <0.7. However, a model with a small deviation from the exponential accretion, such as the *Wetherill* (1986) scenario will also fit the constraint for a substantially smaller mean time of accretion of 16 m.y. (*Jacobsen,* 1999, 2000).

late stage is unclear. What is clear is that early rapid accretion and later slower accretion provides a general mechanism by which the silicate Earth could have developed W that was more radiogenic than chondritic in its early history, with elimination of the $^{182}$W excess by the addition of chondritic material in the later stages when $^{182}$Hf was almost extinct. This general feature is a highly robust result that has been independently deduced by at least three groups of investigators (*Halliday et al.*, 1996; *Harper and Jacobsen*, 1996a; *Kramers*, 1998; *Jacobsen and Yin*, 1998; *Halliday and Lee*, 1999; *Halliday*, 2000).

It has also been argued that to provide better and more comprehensive constraints on the age of the core, it is necessary to include Pb-isotopic constraints (*Halliday et al.*, 1996; *Halliday and Lee*, 1999). *Kramers and Tolstihkin* (1997) provide an integrated study of W- and Pb-isotopic modeling together with simulations of the formation of the siderophile element abundances in the silicate Earth. They incorporate a small delay to the onset of core formation and a considerable amount of late accretion. *Halliday* (2000) shows with W and Pb isotopes that the mean time for exponentially decreasing accretion of the Earth must lie in the range of 25–40 m.y. However, there are problems in using Pb isotopes to constrain the timing of core formation (see *Jacobsen and Harper*, 1996; *Harper and Jacobsen*, 1996a; *Halliday et al.*, 1996).

As with the single-stage core-formation age calculation, the conclusion that accretion was protracted is independent of changes in oxidation state and postcore addition of any chondritic veneer (*Jacobsen and Harper*, 1996; *Halliday and Lee*, 1999). The silicate liquid/metallic liquid partition coefficients for W depend greatly on oxidation state (*Rammenssee and Wänke*, 1977). If conditions were highly oxidizing on the early Earth, W might not fractionate greatly from Hf in the bulk silicate Earth and radiogenic W would not be formed. At some later time, the terrestrial mantle clearly became sufficiently reducing that W partitioned strongly into the core, resulting in the high Hf/W of the bulk silicate Earth. As far as Hf-W is concerned, this is the equivalent effect to that of simply delaying core formation and does not change the long-term effects (*Halliday et al.*, 1996). Most accretionary models have considered decreasing the metal/silicate partition coefficients during the later stages of core formation as Earth becomes more oxidizing (*Newsom*, 1990, 1996; *O'Neill*, 1991; *Shearer and Newsom*, 1998). The effects of this on continuous core formation models are essentially the same as slowing down the rate of core formation relative to accretion. The W-isotopic effects are still distinctive no matter by how much the partition coefficients change and at what stage, provided one sets the final Hf/W of the silicate Earth to be in the range 10–40, as found today (*Halliday and Lee*, 1999). None of the heterogeneous accretion or late veneer models, in and of themselves, provides an explanation for the chondritic W-isotopic composition of the silicate Earth at the accretion rates suggested by *Wetherill* (1986).

## 6. AGE OF THE MOON

Tungsten isotopes provide important evidence for how the Moon formed. However, the Moon can also provide evidence for how Earth may have accreted (*Halliday et al.*, 1996). Tungsten isotopes can be used to study the age of the Moon and the development of the material from which it formed. The prevalent view is that the Moon represents the silicate-rich portion of a planet that formed in a collision between Earth and an impactor. As such, its bulk W-isotopic composition provides us with a sample of what early solar-system silicate reservoirs were like during the late stages of accretion. Therefore, we can use its composition to test some of the above accretionary scenarios (*Halliday et al.*, 1996; *Halliday*, 2000). Here we first discuss the W-isotopic evidence for the Moon's exact age.

Previous isotopic ages for the Moon have mainly focused on precise Nd- and Pb-isotopic constraints (*Tera et al.*, 1973; *Wasserburg et al.*, 1977; *Hanan and Tilton*, 1987; *Carlson and Lugmair*, 1988; *Shih et al.*, 1993; *Alibert et al.*, 1994). As with Pb isotopes, the use of W isotopes is complicated because the radiogenic effects measured in lunar samples could be partly inherited from its parent planet. Assuming the Moon formed when there was still sufficient live $^{182}$Hf around to generate radiogenic $^{182}$W within its interior, how do we determine an age? As the Moon is silicate rich, it has a high Hf/W ratio. Most workers consider the Hf/W of the bulk Moon to be at least as high as that of the bulk silicate Earth (*Palme*, 1999). The Moon appears to have developed magmas at an early magma ocean stage, from which a crust and core formed, leaving a residual lunar mantle. The crust and core should both have low Hf/W. As a consequence, the residual lunar mantle, which is rich in cumulates from the magma ocean, has even higher Hf/W than the bulk Moon. This is evident from the fact that the Hf/W ratios of all lunar basalts are higher than that of the bulk silicate Earth, even though W is more incompatible than Hf during melting (*Lee et al.*, 1997). Furthermore, the formation of ilmenite-rich cumulate layers in the magma ocean may have created reservoirs with particularly high Hf/W ratios in parts of the lunar mantle.

The timing of all these fractionation processes will affect the resultant W-isotopic compositions. If the Moon formed sufficiently late, its W-isotopic composition would only reflect that of the silicate-rich portion of the planet from which it was derived, because the effects of decay within the Moon itself would be negligible. If the Moon formed early but differentiated late, its W-isotopic composition would be uniform, reflecting decay within the bulk Moon prior to differentiation, superimposed on the isotopic composition of silicate portion of the parent planet from which it was derived. If the Moon both formed and differentiated sufficiently early, one might expect to find a range of W-isotopic compositions in lunar samples, reflecting components of isotopic evolution on the parent planet and in the lunar mantle.

Most dynamic simulations of the giant impact derive the Moon from the silicate portion of the impactor with little admixed material from Earth or the core of the impactor. So the W-isotopic composition of the Moon would reflect the W-isotopic composition of the impactor at the time of the impact plus any effects from decay within the Moon itself.

Independent constraints on the age of the Moon leave considerable scope for what one might expect to find in terms of W-isotopic compositions (*Wasserburg et al.,* 1977; *Halliday et al.,* 1996; *Carlson and Lugmair,* 2000). There are considerable data to support a late origin with an age >4.42 Ga. The most compelling evidence comes from early ferroan anorthosites such as 60025 that have relatively low first-stage μ values and define an age of ~4.5 Ga. *Carlson and Lugmair* (1988) reviewed all the most precise and concordant data and concluded that the Moon had to have formed in the time interval 4.44–4.51 Ga. This is consistent with the estimate of 4.47 ± 0.02 Ga of *Tera et al.* (1973).

Lunar samples yield a range of $\varepsilon_W$ values from chondritic values of ~0 to fairly radiogenic values of 6.5 (*Lee et al.,* 1997) (Fig. 4). There are three aspects of the W data that require explanation. The first is the preservation of W-isotopic heterogeneity in the Moon. The second is the lower limit on the W-isotopic composition, which is chondritic and is probably an initial ratio. The third is how to generate W-isotopic compositions that range up to 6.5 but have an average of ~2 for the samples shown in Fig. 4.

The explanation for the first of these issues depends on how the Moon formed. *Lee et al.* (1997) considered the possibility that the variability in the lunar W-isotopic data reflected incomplete mixing of heterogeneous debris from parent planet(s). However, they dismissed this idea because it is unlikely that the energetic conditions of a giant impact followed by an early lunar magma ocean environment would permit the survival of such inherited W-isotopic heterogeneity. If one accepts the giant impact theory, it is more likely that the W-isotopic heterogeneity was generated within the Moon itself by radioactive decay.

It follows from this that the chondritic lower limit on the range of W-isotopic compositions is an initial ratio from the time of formation of the Moon. This initial isotopic composition can be explained in either of two ways. The first explanation is that the Moon formed no earlier than ~50 m.y. after the start of the solar system from a parent planet that was accreting slowly. By this time, the W-isotopic composition of the silicate-rich portion of such a body could have decreased to chondritic values (Fig. 8). This is only possible if the Moon forms no earlier than roughly 50 m.y. after the start of the solar system.

A second mechanism for generating a chondritic initial W-isotopic composition in the Moon would be a major impact on the parent planet. Giant impacts effectively reduce the $\varepsilon_W$ of the silicate Earth to ~0 at any stage during accretion (*Jacobsen,* 1999, 2000; *Halliday and Lee,* 1999). The same would be true in the accretion history of the impactor. Major impacts are thought to have been relatively

"common" occurrences at this stage of accretion. So, if the parent planet of the Moon (usually thought to be the impactor) happened to suffer another impact just prior to that which produced the Moon itself, the initial W-isotopic composition of the Moon would be chondritic.

Although this could happen at any time, with both of these models there is a major problem explaining how chondritic isotopic compositions are subsequently preserved within the high Hf/W Moon unless the giant impact that produced this body took place at a relatively late stage (≥50 m.y. after the start of the solar system). Tungsten-isotopic evolution curves ($\varepsilon_W$) for a bulk Moon starting out with $\varepsilon_W$ ~0 are shown in Fig. 9. These $\varepsilon_W$ curves for the bulk Moon are shown for various times of origin of the Moon ranging from 0 to 60 m.y. after the start of the solar system and assuming the bulk Moon and the Earth's mantle have identical $f^{Hf/W}$ values of 14. The curves were calculated using equation (3) of *Harper and Jacobsen* (1996a) with $T_{cf}$ as the time of formation of the Moon. The shaded bar indicates the range of present-day W-isotopic effects ($\varepsilon_W$) observed in lunar rocks (Fig. 4). It can be seen that, prior to ~60 m.y. after the start of the solar system, it is impossible to maintain a chondritic W-isotopic composition in lunar rocks, given the Hf/W of the bulk Moon. One explanation that would be consistent with an earlier age would be that these lunar samples are sampling a low Hf/W reservoir that formed early in lunar history. The earlier the age of the Moon, the earlier this reservoir needs to form and the more isolated it needs to be. Given the common occurrence of near chondritic W-isotopic compositions among lunar basalts that were formed by melting high Hf/W lunar mantle, this explanation is difficult to support. Another possible explanation for chondritic W in some lunar samples may be contamination of lunar magma sources by chondritic W from late impactors on the Moon (*Lee et al.,* 1997).

The third significant observation is the radiogenic W-isotopic composition of the Moon. This places more specific constraints on the timing of the giant impact. As shown in Fig. 9, prior to 25 m.y. after the start of the solar system the average W-isotopic composition of the Moon would be more radiogenic than that of the W-isotopic composition of any normal lunar samples. So the Moon cannot have formed prior to ~25 m.y. after the start of the solar system. If one assumes that the radiogenic W in the Moon was formed by decay within the lunar mantle, the constraints may be more severe. The Hf/W ratios of normal mare basalts and basaltic glasses range between 20 and 70 (*Lee et al.,* 1997). Because W is significantly more incompatible than Hf during mantle melting, these represent minima for their source regions. The Hf/W ratio of the lunar mantle (Moon minus crust and core) has been estimated at $f^{Hf/W}$ ~30 (*Lee et al.,* 1997), about twice that of the bulk silicate Earth. Using this Hf/W ratio for the lunar mantle would result in an *average* present-day W-isotopic composition of $\varepsilon_W$ >10 if the Moon formed at 30 m.y. after the start of the solar system (*Halliday,* 2000). This is well outside the range of all published W-isotopic compositions for lunar samples (Fig. 4). The

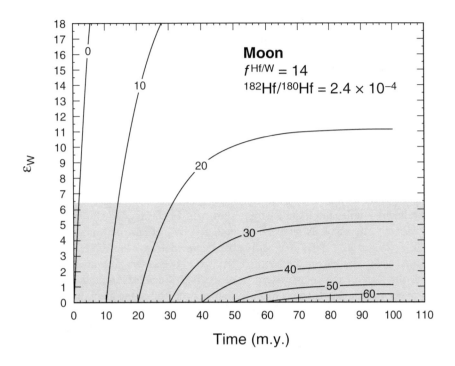

**Fig. 9.**   Calculated W-isotopic effects ($\varepsilon_W$) in the Moon assuming it starts out with $\varepsilon_W$ ~0. Giant impacts effectively reduce the $\varepsilon_W$ of the silicate Earth to about 0 at any stage during accretion (*Jacobsen,* 1999, 2000; *Halliday,* 2000). Closed system W-isotopic growth curves are shown for the bulk Moon for various times of origin of the Moon ranging from 0 to 60 m.y. after the start of the solar system and assuming the bulk Moon and the Earth's mantle have identical $f^{Hf/W}$ values of 14. The curves were calculated using equation (3) of *Harper and Jacobsen* (1996) with $T_{cf}$ as the time of formation of the Moon. The shaded bar indicates the range of observed W-isotopic effects (present-day $\varepsilon_W$) in lunar rocks (Fig. 4). As shown, prior to about 50 m.y. after the start of the solar system, it is hard to generate the near-chondritic W-isotopic compositions found in many lunar rocks. Subsequent to 20–30 m.y. after the start of the solar system, it is hard to generate the high $\varepsilon_W$ values in the Moon with $f^{Hf/W}$ values of 14. The Hf/W ratio of the lunar mantle (Moon minus crust and core) has been estimated at $f^{Hf/W}$ ~30 (*Lee et al.,* 1997), about twice that of the bulk silicate Earth. If reservoirs with such high Hf/W ratios formed immediately at the time of formation of the Moon, it is possible to have the Moon form fairly late (i.e., at ~50 m.y. after the start of the solar system). However, if such high Hf/W ratios are due to the formation of cumulate layers in the lunar magma ocean at a late time (~100 m.y.) then the Moon must have formed prior to ~30 m.y. after the start of the solar system.

interpretation of $\varepsilon_W$ values in lunar samples depends both on the age of the Moon as well as what is considered to be reasonable $f^{Hf/W}$ values for lunar magma sources. For example, to obtain the upper limit for $\varepsilon_W$ in the Moon of 6.5 we need $f^{Hf/W}$ ~177 if the age of the Moon is 60 m.y. after the start of the solar system; for 40 m.y. we obtain $f^{Hf/W}$ ~38, and for 25 m.y. we obtain $f^{Hf/W}$ ~12. It is clear that the timing and magnitude of Hf/W fractionation is the critical issue. Thus, subsequent to ~25 m.y. after the start of the solar system it is hard to generate the observed range of $\varepsilon_W$ values in the Moon if all lunar magma sources have $f^{Hf/W}$ similar to Earth's mantle. If reservoirs with very high Hf/W ratios ($f^{Hf/W}$ ~180) formed immediately at the time of formation of the Moon, then it is possible to have the Moon form fairly late (i.e., ~60 m.y. after the start of the solar system). However, if such high Hf/W ratios are due to the formation of cumulate layers in the lunar magma ocean at a late time (~100 m.y. after the start of the solar system), then the Moon must have formed prior to ~30 m.y. after the

start of the solar system. It then becomes problematic explaining the chondritic W-isotopic compositions.

*Lee et al.* (1997) favored an alternative approach that also assumes that the radiogenic W was generated from a chondritic initial W-isotopic composition, affected to varying degrees by late growth of $^{182}$W as a result of fractionation of Hf/W to extreme values in the lunar interior. They used both a fossil isochron approach as well as Hf-W model ages to estimate an age of 4.52–4.50 Ga for the Moon, 55 ± 10 m.y. after the start of the solar system, lending support to late-stage giant impact models of lunar origin.

Two other new sets of constraints have recently been acquired that are relevant to the issue of the Hf-W age of the Moon. First, some of the lunar samples contain an important component of cosmogenic W (*Leya et al.,* 2000). So it is unwise at this stage to overinterpret the data for individual samples. Some samples (e.g., 15555) still clearly have a strong radiogenic effect and the broad Hf-W constraints on the age of the Moon appear robust. Nonetheless, caution is needed at this stage. Second, if the effects are radiogenic, one expects to see some correlation between the W-isotopic data and the source region Hf/W. This has now been found in the form of very high Hf/W (50–150) and very radiogenic W ($\varepsilon_W$ = 5–11) in high-Ti mare basalts (D-C. Lee, unpublished data), and these unpublished data favor the ~50-m.y. radiogenic interpretation.

In summary, we are faced with a paradox that we cannot resolve without a better understanding of the timing and processes of chemical fractionation in lunar magma sources. The problem is that it is easiest to explain lunar samples with $\varepsilon_W$ close to zero by having the Moon form late, while the high-$\varepsilon_W$ values in some lunar samples are more easily explained by having the Moon form early, as shown in Fig. 9. However, as argued by *Lee et al.* (1997) it is still possible to explain the high-$\varepsilon_W$ values by having the Moon form late, at ~55 m.y. after the start of the solar system. While the W-isotopic data do not yield a unique age, a plausible range for the time of formation of the Moon would be 25–65 m.y. after the start of the solar system. This corresponds to an age of 4.54–4.50 Ga, just overlapping with the estimate of *Carlson and Lugmair* (1988) at 4.51–4.44 Ga, but a little earlier than that of *Tera et al.* (1973) at 4.47 ± 0.02 Ga.

## 7.  ORIGIN OF THE MOON

The age of the Moon determined above provides support for the giant impact theory. Models such as the co-accretion or capture hypotheses provide no explanation for why the Moon should be young. The fission hypothesis is so difficult dynamically speaking that it is seldom considered viable these days. In the giant impact theory, it is essential that the colliding planets have reached a certain mass and therefore a late age for the Moon is to be expected. However, the exact masses under consideration have been the subject of considerable uncertainty and here too the Hf-W isotopic system can be brought to bear on the various models.

In the first giant impact simulations, the Earth was considered to have almost completely formed by the time of the impact. The impactor was generally considered to be about the size of Mars (*Cameron and Benz*, 1991). One model calculation of $\varepsilon_W$ for this giant impact simulation predicts a present-day W-isotopic composition that is more radiogenic ($\varepsilon_W$ >1) than chondritic (*Halliday*, 2000), assuming the Moon formed at ~50 m.y. after the start of the solar system. However, *Jacobsen* (1999) has shown by considering a broader range of accretion functions that the "traditional" giant impact theory can, on its own, explain the W-isotopic data.

*Cameron and Canup* (1998) have recently revised the giant impact theory and now propose that Earth was only half formed at the time the Moon was created. They reason that the impactor may have been significantly larger than Mars and the critical parameter in their model is the mass

of the proto-Earth relative to that of the impactor, which needs to be in the ratio 7:3. Mass-accretion curves are shown in Fig. 10a for 10%, 20%, and 30% growth of Earth subsequent to the addition from a giant impact at 50 m.y. after the start of the solar system. Note that the rate of accretion prior to the impact would have been more than 5× slower than in the dynamic simulation of *Wetherill* (1986), if the Earth was only half formed by the time of a giant impact at 50 m.y. after the start of the solar system.

The W-isotopic effects ($\varepsilon_W$) in the bulk silicate that follow from the revised giant impact theory of *Cameron and Canup* (1998) using the accretion curves in Fig. 9a are shown in Fig. 10b. The $\varepsilon_W$ curves were calculated using equation (12) of *Harper and Jacobsen* (1996). The shaded bar again indicates the 0.7-$\varepsilon_W$ unit uncertainty in the difference between the present-day W-isotopic compositions of the bulk silicate Earth and carbonaceous chondrites. As shown, even before the giant impact at 50 m.y. after the start of the solar system, the $\varepsilon_W$ is reduced to below 0.7 and the giant impact effectively reduces $\varepsilon_W$ to ~0 at 50 m.y. Following the giant impact, $\varepsilon_W$ rises to at most 0.6 for the 10% case and decays to ~0.4 by 100 m.y. after the start of the solar system. The 20% and 30% cases generate even smaller postimpact $\varepsilon_W$ values in the silicate Earth. No matter how radiogenic the W prior to the impact, the W-isotopic composition of the silicate Earth will always be reduced to chondritic by the impact and does not get much higher subsequently (*Halliday*, 2000) in the revised giant impact scenario. This model for the Moon can accommodate the chondritic W-isotopic composition of the Earth with <10% of post-giant impact accretion. However, as shown in Fig. 8, a much smaller impact (2.5%) at the end of accretion (~100 m.y. after the start of the solar system) is sufficient to have the same effect of reducing the $\varepsilon_W$ in the silicate Earth to ~0. Giant impacts always eliminate the excess $^{182}W$ in the silicate Earth. Prior to ~50 m.y. after the start of the solar system the W then becomes radiogenic again. Only subsequent to ~50 m.y. will there be insignificant recovery of the $^{182}W$ excess in the silicate Earth (*Jacobsen*, 1999, 2000).

## 8.  CONCLUSIONS

With the development of the Hf-W system we can now place some first-order constraints on how fast Earth accreted and what the mechanisms of core formation were. First, the formation of a metal core at an early stage from a selective high-temperature condensate, as in many heterogeneous accretion models, should yield very radiogenic W in the silicate Earth, which is not observed. The Hf-W data for iron meteorites and eucrites provide powerful evidence of early and rapid segregation of metal in planetesimals. However, W-isotopic data require a more protracted history for Earth. Either Earth's core formed late, or accretion and core formation were both gradual over the first 100 m.y. of Earth's history. The W-isotopic data for Earth require a mean time

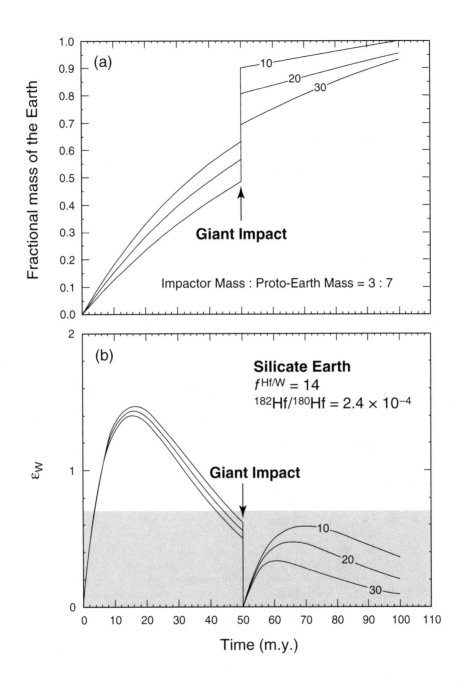

**Fig. 10.** Tungsten-isotopic effects ($\varepsilon_W$) of slow accretion followed by the addition of an impactor with a proto-Earth to impactor mass ratio of 7:3, corresponding to the revised giant impact theory of *Cameron and Canup* (1998). **(a)** Curves are shown for 10%, 20%, and 30% growth of the Earth subsequent to the addition from the giant impact at 50 m.y. after the start of the solar system. **(b)** The W-isotopic effects ($\varepsilon_W$) in the bulk silicate Earth that follow from the accretion curves in **(a)** were calculated using equation (12) of *Harper and Jacobsen* (1996). The shaded bar again indicates the uncertainty in the difference between the W-isotopic compositions of the bulk silicate Earth and carbonaceous chondrites (*Lee and Halliday*, 1996). As shown, even before the giant impact at 50 m.y. after the start of the solar system the $\varepsilon_W$ is reduced to below 0.7 and the giant impact effectively reduces $\varepsilon_W$ to ~0 at 50 m.y. Following the giant impact $\varepsilon_W$ rises to at most 0.6 for the 10% case and decays to ~0.4 by 100 m.y. The 20% and 30% cases generate even smaller postimpact $\varepsilon_W$ values in the silicate Earth. As shown in Fig. 7 a much smaller impact (2.5%) at the end of accretion is sufficient to have the same effect of reducing the $\varepsilon_W$ in the silicate Earth to ~0. Giant impacts always eliminate the excess [182]W in the silicate Earth, but only subsequent to ~50 m.y. after the start of the solar system will there be insignificant recovery of the [182]W excess in the silicate Earth to leave it with chondritic W (*Halliday and Lee*, 1999, *Jacobsen*, 1999, 2000; *Halliday*, 2000).

of accretion that could be as short as 16 m.y., depending on the exact late-stage history. But the whole process probably took ~100 m.y., consistent with the model of *Wetherill* (1986). The later stages of accretion would likely involve the catastrophic effects of giant impactor(s), leading to the formation of the Moon. Hf-W isotopic data for lunar samples provide extremely useful constraints on the time of formation of the Moon. An age of ~50 m.y. after the start of the solar system is consistent with all the data. An earlier age (back to ~25 m.y. after solar system formation) is possible, but this requires more in the way of special circumstances to explain both the chondritic and the radiogenic W-isotopic compositions of lunar samples. Better constraints are likely to come from future refinements of both Hf-W isotopic data and Hf-W partitioning data and a better understanding of Hf-W isotopic systematics in lunar samples. The more recent giant impact models in which the proto-Earth to impactor mass ratio was about 7:3 readily yield chondritic W in the silicate Earth. However, the original giant impact scenario with a Mars-sized impactor at the end of accretion can also be reconciled with the Hf-W data, depending on the exact age of the Moon and the amount and timing of postimpact accretion.

*Acknowledgments.* Part of this work was funded by NASA and NSF grants to ANH and SBJ.

## REFERENCES

Alibert C., Norman M. D., and McCulloch M. T. (1994) An ancient Sm-Nd age for a ferroan noritic anorthosite clast from lunar breccia 67016. *Geochim. Cosmochim. Acta, 58,* 2921–2926.

Allègre C. J., Manhès G., and Göpel C. (1995) The age of the Earth. *Geochim. Cosmochim. Acta, 59,* 1445–1456.

Benz W., Cameron A. G. W., and Slattery W. L. (1987) Collisional stripping of Mercury's mantle. *Icarus, 74,* 516–528.

Boss A. P. (1990) 3D Solar nebula models: implications for Earth origin. In *Origin of the Earth* (H. E. Newsom and J. H. Jones, eds.), pp. 3–15. Oxford Univ., New York.

Carlson R. W. and Lugmair G. W. (1988) The age of ferroan anorthosite 60025: oldest crust on a young Moon? *Earth Planet. Sci. Lett., 90,* 119–130.

Cameron A. G. W. (1978) Physics of the primitive solar accretion disk. *Moon and Planets, 18,* 5–40.

Cameron A. G. W. and Benz W. (1991) Origin of the Moon and the single impact hypothesis IV. *Icarus, 92,* 204–216.

Cameron A. G. W. and Canup R. M. (1998) The giant impact occurred during Earth accretion (abstract). In *Lunar and Planetary Science XXIX,* pp. 1062–1063. Lunar and Planetary Institute, Houston.

Carlson R. W. and Lugmair G. W. (2000) Timescales of planetesimal formation and differentiation based on extinct and extant radioisotopes. In *Origin of the Earth and Moon* (R. M. Canup and K. Righter, eds.), this volume. Univ. of Arizona, Tucson.

Chen J. H. and Wasserburg G. J. (1986) Formation ages and evolution of Shergotty and its parent planet from U-Th-Pb systematics. *Geochim. Cosmochim. Acta, 50,* 955–968.

Chen J. H. and Wasserburg G. J. (1996) Live [107]Pd in the early solar system and implications for planetary evolution. In *Earth Processes: Reading the Isotope Code* (A. Basu and S. Hart,

eds.), pp. 1–20. American Geophysical Union, Washington, DC.

Clark S. P. Jr., Turekian K. K., and Grossman L. (1972) Model for the early history of the Earth. In *The Nature of the Solid Earth* (E. C. Robertson, ed.), pp. 3–18. McGraw-Hill, New York.

Eucken A. (1944) Physikalisch-Chemische Betrachtungen über die früheste Entwicklungsgeschichte der Erde. *Nachr. Akad. Wiss. Göttingen,* Math-Phys. Kl., Heft 1, 1–25.

Galer S. J. G. and Goldstein S. L. (1996) Influence of accretion on lead in the Earth. In *Isotopic Studies of Crust-Mantle Evolution* (A. Basu and S. R. Hart, eds.), pp. 75–98. American Geophysical Union, Washington, DC.

Göpel C., Manhès G., and Allègre C. J. (1991) Constraints on the time of accretion and thermal evolution of chondrite parent bodies by precise U-Pb dating of phosphates. *Meteoritics, 26,* 73.

Göpel C., Manhès G., and Allègre C. J. (1994) U-Pb systematics of phosphates from equilibrated ordinary chondrites. *Earth Planet. Sci. Lett., 121,* 153–171.

Grossman L. (1972) Condensation in the primitive solar nebula. *Geochim. Cosmochim. Acta, 36,* 597–619.

Grossman L. and Larimer J. W. (1974) Early chemical history of the solar system. *Rev. Geophys. Space Phys., 12,* 71–101.

Halliday A. N. (2000) Terrestrial accretion rates and the origin of the Moon. *Earth Planet. Sci. Lett., 176,* 17–30.

Halliday A. N. and Lee D-C. (1999) Tungsten isotopes and the early development of the Earth and Moon. *Geochim. Cosmochim. Acta, 63,* 4157–4179.

Halliday A. N., Lee D-C., Christensen J. N., Walder A. J., Freedman P. A., Jones C. E., Hall C. M., Yi W., and Teagle D. (1995) Recent developments in inductively coupled plasma magnetic sector multiple collector mass spectrometry. *Intl. J. Mass Spec. Ion Proc., 146/147,* 21–33.

Halliday A. N., Rehkämper M., Lee D-C., and Yi W. (1996) Early evolution of the Earth and Moon: new constraints from Hf-W isotope geochemistry. *Earth Planet. Sci. Lett., 142,* 75–89.

Hanan B. B. and Tilton G. R. (1987) 60025: Relict of primitive lunar crust? *Earth Planet. Sci. Lett., 84,* 15–21.

Hanks T. C. and Anderson D. L. (1969) The early thermal history of the Earth. *Phys. Earth Planet. Inter., 2,* 19–29.

Harper C. L. and Jacobsen S. B. (1996a) Evidence for [182]Hf in the early solar system and constraints on the timescale of terrestrial core formation. *Geochim. Cosmochim. Acta, 60,* 1131–1153.

Harper C. L. and Jacobsen S. B. (1996b) Noble gases and Earth's accretion. *Science, 273,* 1814–1818.

Harper C. L., Völkening J., Heumann K. G., Shih C.-Y., and Wiesmann H. (1991) [182]Hf-[182]W: new cosmochronometric constraints on terrestrial accretion, core formation, the astrophysical site of the r-process, and the origin of the solar system (abstract). In *Lunar and Planetary Science XXII,* pp. 515–516. Lunar and Planetary Institute, Houston.

Horan M. F., Smoliar M. I., and Walker R. J. (1998) [182]W and [187]Re-[187]Os systematics of iron meteorites: chronology for melting, differentiation, and crystallization in asteroids. *Geochim. Cosmochim. Acta, 62,* 545–554.

Jacobsen S. B. (1999) Accretion and core formation models based on extinct radionuclides (abstract). In *Lunar and Planetary Science XXX,* Abstract #1978. Lunar and Planetary Institute, Houston (CD-ROM).

Jacobsen S. B. (2000) Accretion and core formation models: constraints based on extinct and long-lived radionuclides. *Geochemistry, Geophysics, Geosystems (G³),* submitted.

Jacobsen S. B. and Harper C. L. Jr. (1996) Accretion and early differentiation history of the Earth based on extinct radionuclides. In *Earth Processes: Reading the Isotope Code* (A. Basu and S. Hart, eds.), pp. 47–74. American Geophysical Union, Washington, DC.

Jacobsen S. B. and Wasserburg G. J. (1979) The mean age of mantle and crust reservoirs. *J. Geophys. Res., 84,* 7411–7427.

Jacobsen S. B. and Yin Q. (1998) W isotope variations and the time of formation of asteroidal cores and the Earth's core (abstract). In *Lunar and Planetary Science XXIX,* pp. 1852–1853. Lunar and Planetary Institute, Houston.

Kramers J. D. (1998) Reconciling siderophile element data in the Earth and Moon, W isotopes and the upper lunar age limit in a simple model of homogeneous accretion. *Chem. Geol., 145,* 461–478.

Kramers J. D. and Tolstikhin I. N. (1997) Two terrestrial lead isotope paradoxes, forward transport modelling, core formation and the history of the continental crust. *Chem. Geol., 139,* 75–110.

Larimer J. W. and Anders E. (1970) Chemical fractionations in meteorites III: major element fractionation in chondrites. *Geochim. Cosmochim. Acta, 34,* 367–387.

Lee D.-C. and Halliday A. N. (1995a) Precise determinations of the isotopic compositions and atomic weights of molybdenum, tellurium, tin and tungsten using ICP magnetic sector multiple collector mass spectrometry. *Intl. J. Mass Spec. Ion Proc., 146/147,* 35–46.

Lee D.-C. and Halliday A. N. (1995b) Hafnium-tungsten chronometry and the timing of terrestrial core formation. *Nature, 378,* 771–774.

Lee D.-C. and Halliday A. N. (1996) Hf-W isotopic evidence for rapid accretion and differentiation in the early solar system. *Science, 274,* 1876–1879.

Lee D.-C. and Halliday A. N. (1997) Core formation on Mars and differentiated asteroids. *Nature, 388,* 854–857.

Lee D.-C. and Halliday A. N. (1998) Tungsten isotopes, the initial $^{182}Hf/^{180}Hf$ of the solar system and the origin of enstatite chondrites. *Mineral. Mag., 62A,* 868–869.

Lee D.-C. and Halliday A. N. (2000) Hf-W internal isochrons for ordinary chondrites and the initial $^{182}Hf/^{180}Hf$ of the solar system. *Chem. Geol.,* in press.

Lee D.-C., Halliday A. N., Snyder G. A., and Taylor L. A. (1997) Age and origin of the Moon. *Science, 278,* 1098–1103.

Leya I., Wieler R., and Halliday A. N. (2000) Cosmic-ray production of tungsten isotopes in lunar samples and meteorites and its implications for Hf-W cosmochemistry. *Earth Planet. Sci. Lett.,* in press.

Lin D. N. C. and Papaloizou J. (1985) On the dynamical origin of the solar system. In *Protostars and Planets II* (D. C. Black and M. S. Matthews, eds.), pp. 981–1072. Univ. of Arizona, Tucson.

Lugmair G. W. and Galer S. J. G. (1992) Age and isotopic relationships between the angrites Lewis Cliff 86010 and Angra dos Reis. *Geochim. Cosmochim. Acta, 56,* 1673–1694.

Lugmair G. W. and Shukolyukov A. (1998) Early solar system timescales according to $^{53}Mn$-$^{53}Cr$ systematics. *Geochim. Cosmochim. Acta, 62,* 2863–2886.

Newsom H. E. (1990) Accretion and core formation in the Earth: evidence from siderophile elements. In *Origin of the Earth* (H. E. Newsom and J. H. Jones, eds.), pp. 273–288. Oxford Univ., New York.

Newsom H. E. (1996) W/Hf fractionation in chondrites and the Earth: constraints on timing of core formation (abstract). In *Lunar and Planetary Science XXVII,* pp. 957–958. Lunar and Planetary Institute, Houston.

Newsom H. E., Sims K. W. W., Noll P. D. Jr., Jaeger W. L., Maehr S. A., and Bessera T. B. (1996) The depletion of W in the bulk silicate Earth. *Geochm. Cosmochim. Acta, 60,* 1155–1169.

Okuchi T., Abe Y., and Iwamori H. (1998) Hydrogen in the core: evidence for the state of the lost protoatmosphere (abstract). In *Origin of the Earth and Moon,* p. 29. LPI Contribution No. 957, Lunar and Planetary Institute, Houston.

O'Neill H. St. C. (1991) The origin of the Moon and the early history of the Earth: a chemical model; Part 1: The Moon. *Geochim. Cosmochim. Acta, 55,* 1135–1158.

Palme H. (1999) The lunar Hf/W ratio and the significance of $^{182}W/^{184}W$ ratios in lunar samples (abstract). In *Lunar and Planetary Science XXX,* Abstract #1763. Lunar and Planetary Institute, Houston (CD-ROM).

Rammensee W. and Wänke H. (1977) On the partition coefficient of tungsten between metal and silicate and its bearing on the origin of the Moon. *Proc. Lunar Sci. Conf. 8th,* pp. 399–409.

Ringwood A. E. (1966) The chemical composition and origin of the Earth. In *Advances in Earth Sciences* (P. M. Hurley, ed.), pp. 287–356. MIT, Cambridge.

Safronov V. S. (1954) On the growth of planets in the protoplanetary cloud. *Astron. Zh., 31,* 499–510.

Sasaki S. and Nakazawa K. (1986) Metal-silicate fractionation in the growing Earth: energy source for the terrestrial magma ocean. *J. Geophys. Res., 91,* 9231–9238.

Shaw G. H. (1978) Effects of core formation. *Phys. Earth Planet. Inter., 16,* 361–369.

Shearer C. K. and Newsom H. E. (1998) W-Hf isotope abundances and the early origin and evolution of the Moon and Earth (abstract). In *Lunar and Planetary Science XXIX,* pp. 1759–1760. Lunar and Planetary Institute, Houston.

Shih C.-Y., Nyquist L. E., Dasch E. J., Bogard D. D., Bansal B. M., and Wiesmann H. (1993) Age of pristine noritic clasts from lunar breccias 15445 and 15455. *Geochim. Cosmochim. Acta, 57,* 915–931.

Solomon S. C. (1979) Formation, history and energetics of cores in the terrestrial planets. *Earth Planet. Sci. Lett., 19,* 168–182.

Stevenson D. J. (1981) Models of the Earth's core. *Science, 214,* 611–619.

Tera F., Papanastassiou D. A., and Wasserburg G. J. (1973) A lunar cataclysm at ~3.95 AE and the structure of the lunar crust (abstract). In *Lunar Science IV,* pp. 723–725. Lunar Science Institute, Houston.

Turekian K. K. and Clark S. P. Jr. (1969) Inhomogeneous accumulation of the Earth from the primitive solar nebula. *Earth Planet. Sci. Lett., 6,* 346–348.

Völkening J., Köppe M., and Heumann K. G. (1991) Tungsten isotope ratio determinations by negative thermal ionization mass spectrometry. *Intl. J. Mass Spec. Ion Proc., 107,* 361–368.

Walder A. J. and Freedman P. A. (1992) Isotopic ratio measurement using a double focusing magnetic sector mass analyzer with an inductively coupled plasma as an ion source. *J. Anal. Atomic Spect., 7,* 571–575.

Wasserburg G. J., Papanastassiou D. A., Tera F., and Huneke J. C. (1977) Outline of a lunar chronology. *Philos. Trans. R. Soc. Lond., A285,* 7–22.

Wasson J. T. (1985) *Meteorites: Their Record of Early Solar-System History.* W. H. Freeman and Company, New York. 251 pp.

Wetherill G. W. (1986) Accumulation of the terrestrial planets and implications concerning lunar origin. In *Origin of the Moon*

(W. K. Hartmann et al., eds.), pp. 519–550. Lunar and Planetary Institute, Houston.

Wetherill G. W. (1994a) Provenance of the terrestrial planets. *Geochim. Cosmochim. Acta, 58,* 4513–4520.

Wetherill G. W. (1994b) Possible consequences of absences of "Jupiters" in planetary systems. *Astrophys. Space Sci., 212,* 23–32.

# The Xenon Age of the Earth

**Frank A. Podosek**
*Washington University*

**Minoru Ozima**
*Tokyo University*

Relative to plausible primordial composition, the Earth's atmospheric Xe contains excesses of $^{129}$Xe and of the heaviest isotopes, conveniently represented by $^{136}$Xe, which plausibly reflect radiogenic contributions from the short-lived radionuclides $^{129}$I (by $\beta$ decay) and $^{244}$Pu (by spontaneous fission) respectively. The corresponding abundances of these parent radionuclides are low compared to those prevailing in the early solar system, as inferred from analyses of meteorites. Although there are a number of uncertainties involved in translating the observations into time, it is hard to avoid the conclusion that the Earth did not "form," in the sense of becoming closed to Xe loss, until about 100 m.y. after formation of the solar system as a whole. This Xe formation interval is in rough agreement with comparable conclusions reached from other isotopic systems, but in detail it is not easy to say just what this interval means in terms of Earth for-mation because it remains unclear how chemical events such as Xe closure on a planetary scale may be related to physical events such as accretion of mass. An interesting but speculative possibility is that the Xe formation interval dates loss of atmospheric gases in a giant (moon-forming) impact.

## 1. INTRODUCTION

The issue of how old the Earth is has presumably always been a part of human culture, and many ways to address this issue, scientific and otherwise, have been advanced. In the past century, radioactive decay has become widely recognized as not only the best and most precise but also essentially the only truly robust chronometer available for dating events on a geological timescale, and during this century the problem of the Earth's age has become progressively more narrowly circumscribed through radiometric geochronology. In the past few generations it has also become recognized that it makes little sense to inquire about either the timing or mechanism of the formation of the Earth as an isolated problem; instead the origin of the Earth must be seen as part of a grander process, the origin of the solar system.

Somewhat ironically, considering that we live on the Earth, it turns out that dating the origin of the solar system as a whole is easier than dating the origin of the Earth as a planet. The age of the solar system is some 4.57 Ga, and, absent some bizarre chain of analytical errors or some serious misconception about how stars and planetary systems form, it is hard to imagine that this number could be wrong by more than about 0.01 Ga. There are many ways to estimate the age of the solar system, but without rehearsing the list or the history the rationale behind this statement can be explained fairly succintly: 4.57 Ga is the absolute age, rather precisely determined by the $^{207}$Pb-$^{206}$Pb method (e.g., *Allègre et al.,* 1995), of certain refractory-element rich objects, commonly called CAIs (an acronym for Ca-Al-rich

inclusions), in undifferentiated meteorites. There is no substantial evidence that any other objects formed in the solar system are older than CAIs; to the contrary, there is persuasive evidence that no object formed in the solar system *could* be significantly older than CAIs. This last conclusion follows from the observation that when CAIs formed they included some natural radionuclides with very short lifetimes, notably $^{26}$Al (e.g., *MacPherson et al.,* 1995), which has a halflife of 0.74 m.y. The $^{26}$Al (and other comparably short-lived radionuclides) was perhaps synthesized concomitantly with formation of the Sun itself (e.g., *Lee et al.,* 1998), but is more widely held to have been synthesized in the interior of another star (or stars), then ejected into the interstellar medium from which the solar system formed (e.g., *Cameron et al.,* 1997). In either case, for plausible models the interval between synthesis of $^{26}$Al and its incorporation in CAIs was no more than 1–2 m.y., whence the solar system as a whole could not be more than a few million years older than CAIs.

Since the Earth cannot be older than the solar system, 4.57 Ga is an upper limit to its age. The difficulty in trying to assess the actual age of the Earth is that there are no known materials analogous to CAIs, i.e., no terrestrial rocks or minerals whose formation can be inferred to be essentially coeval with the formation of the planet. The oldest known terrestrial minerals formed about 4.2 G.y. ago (*Froude et al.,* 1983), so this is the highest available direct lower limit to the age of the Earth.

There are, however, a handful of indirect ways to estimate the age of the Earth (e.g., *Podosek,* 1998, 1999), indirect in that they are more complicated than simply dating a rock. These are model ages based on various isotopic systems, collectively and rather compellingly indicating that the Earth formed between 4.4 and 4.5 Ga. It is not easy to pin down an exact formation date, however, because the models tend to be complicated and involve assumptions difficult to verify and/or parameters difficult to measure. One such approach, the most venerable, centers around models for the evolution of Pb isotopic compositions in the Earth's mantle (e.g., *Dalrymple*, 1991; *Allègre et al.*, 1995) and typically deals in absolute ages, i.e., time intervals measured backward from today. Other approaches involve evolution of $^{87}Sr/^{86}Sr$ in preaccretionary Earth materials, as recorded in the Earth's *initial* planetary $^{87}Sr/^{86}Sr$ (e.g., *McCulloch*, 1994), or the accumulation of daughters of short-lived radionuclides, now extinct but present in the early solar system; these include the Xe daughters of $^{129}I$ and $^{244}Pu$, and $^{182}W$, the daughter of $^{182}Hf$ (e.g., *Halliday and Lee*, 1999). All these approaches yield relative ages: Chronological information emerges as an interval, measured forward from the time of origin of the solar system as a whole. There is also at least one useful approach that does not involve isotopes: theoretical dynamic modeling of how long it should take the Earth's precursor planetesimals to accrete and make a planet (e.g., *Wetherill*, 1986); this approach too yields a relative age.

Because Earth must have formed early in solar system history in a time window narrow compared to absolute age, and because most of the relevant chronological information is expressed as age relative to the time of formation of the solar system, it is customary to discuss the time of the Earth's origin in terms of a *formation interval*, the interval between formation of the solar system and formation of the Earth. Thus, the absolute age of the oldest known terrestrial minerals restricts the formation interval of the Earth to no more than about 400 m.y. In contrast, the various indirect approaches cited in the previous paragraph all suggest a shorter interval, about 100 m.y. in round numbers, spanning a range of roughly 50 to perhaps 150 m.y. It should be noted that while this interval is small compared to that imposed by the oldest known minerals and even smaller compared to the total age of the solar system, it is still a long time. In particular, estimates of timescales for major changes in the nature of the solar nebula, estimated from theoretical calculations and astronomical observations, and isotope-based estimates of the intervals needed to accrete and melt at least some terrestrial-type (i.e., rock or rock plus metal) planetary bodies, are considerably shorter, a few to several but not more than about 10 m.y. (e.g., *Podosek and Cassen*, 1994; *Lugmair and Galer*, 1992; *Lugmair and Shukolyukov*, 1998). Viewed from this perspective, it took a long time to make the Earth.

The focus of this paper is a critical review and evaluation of one of the indirect methods for estimating the Earth's formation interval, based on accumulation of radiogenic isotopes of Xe. This will also involve some comparison with the other approaches, and also consideration of current views about the origin of the Earth's Moon. Before proceeding to the details, it is well to note that the various isotopic approaches to the Earth's formation interval are based on different chemical systems and are not necessarily dating the same events, so it is not really clear whether they are in agreement or disagreement. Most importantly, it must be appreciated that the term "formation," which is used rather glibly above, involves unavoidable ambiguities. When applied to the relatively simple problem of radiometric dating of a rock, formation is usually understood, explicitly or implicitly, to designate isotopic closure (e.g., *Dodson*, 1973), a somewhat complicated but at least reasonably straightforward concept. When applied to a planet such as the Earth, most investigators would probably interpret formation in terms of accretion of some threshold level of mass, not thermal or chemical isotopic closure events or processes. There are considerable uncertainties involved in attempting to relate the chronological information carried by isotopic systems to the physical events and processes involved in building a planet.

## 2. XENOLOGY

A reasonable case can be made that of all the elements in the periodic table, the one that carries the most information about both solar nebular and planetary processes, in its elemental abundances and especially in its isotopic compositions, is Xe (e.g., *Reynolds*, 1963; *Pepin and Phinney*, 1978; *Ozima and Podosek*, 1983; *Pepin*, 1991; *Porcelli and Pepin*, 2000). The converse of this proposition is that it is also the most difficult to explain in quantitative detail. Xenon isotopic compositions are commonly substantially different from one meteorite sample to another, and within individual undifferentiated meteorites there are distinct reservoirs of different origins. Particularly relevant in present context, the isotopic composition of terrestrial (atmospheric) Xe is distinct from that of Xe known in any other (extraterrestrial) samples of solar system materials (see Fig. 1). Some aspects of terrestrial Xe isotopic structure are thought to be reasonably well understood, but others continue to challenge quantitative understanding despite four decades of scrutiny. A general review of this situation is beyond the scope of this chapter, but some aspects are relevant in the subsequent discussion. In broad terms, it is commonly held that terrestrial atmospheric Xe is a strongly fractionated (more than 3% per amu) variant of a primordial solar nebula component, to which the radiogenic components (from $^{129}I$ and actinides) described below have been added. The primordial component, called U-Xe by *Pepin and Phinney* (1978), is also present in meteorites, but typically with still other superposed components that complicate exact determination of its composition.

### 2.1. Radiogenic Xenon-129 from Iodine-129

Most meteorites exhibit variable relative abundance of $^{129}Xe$, from one to another and between different internal hosts, as revealed by examination of Xe released in step-

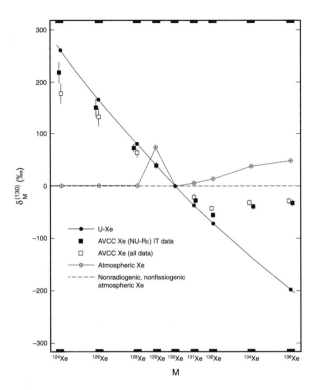

**Fig. 1.** Display of Xe isotopic ratios, normalized to $^{130}$Xe, shown as permil deviations from a reference composition. "U-xenon" (solid circles connected by solid lines), an end-member composition calculated from meteorite data, is interpreted as a primordial composition for terrestrial Xe. The reference composition for this display (horizontal broken line at $\delta = 0$) is mass-fractionated U-Xe. Terrestrial atmospheric Xe composition (Earth symbols, connected by dotted lines) is interpreted as mass-fractionated U-Xe plus superposed radiogenic contributions: $^{129}$Xe from $\beta$ decay of $^{129}$I and all the isotopes, most prominently $^{136}$Xe, which lack more neutron-rich stable isobars, from spontaneous fission, primarily of $^{244}$Pu. The squares show various meteorite data. From *Pepin and Phinney* (1978).

wise heating. In some cases the variations are minor, in others they are not subtle in the least. Commonly, at least some of the $^{129}$Xe variation can be shown to correlate with the element I, and there is no reasonable doubt that many meteorites contain substantial contributions of radiogenic $^{129}$Xe produced *in situ* by $\beta$ decay of $^{129}$I (e.g., *Reynolds*, 1960; *Jeffery and Reynolds*, 1961; *Swindle and Podosek*, 1988). Iodine-129 (halflife about 16 m.y.) is one of a handful of radioisotopes commonly termed "short-lived radionuclides," the first one discovered. These radionuclides have lifetimes long enough to have survived the interval between nucleosynthesis and the formation of solids in the early solar system but short enough that they are now effectively extinct. Because of their relatively short lifetimes they are particularly useful in studying the chronology of solar system formation and early evolution (e.g., *Podosek and Nichols*, 1997). A representative abundance in the early solar

system is commonly cited as $^{129}$I/$^{127}$I $\approx 1 \times 10^{-4}$ (e.g., *Hohenberg et al.*, 1967; *Swindle and Podosek*, 1988). In cosmic (solar) composition, $^{127}$I/$^{129}$Xe $\approx 0.7$ (*Anders and Grevesse*, 1989), so the decay of $^{129}$I adds but little to the inventory of $^{129}$Xe. In the materials of which the terrestrial planets are made, however, the noble gas Xe is very strongly depleted relative to cosmic composition; in particular, the Xe/I ratio is not uncommonly a few to several orders of magnitude lower than the cosmic value, so that in comparison with primordial $^{129}$Xe the radiogenic component of $^{129}$Xe from decay of $^{129}$I is often prominent, sometimes overwhelming.

Given the near ubiquity of observable radiogenic $^{129}$Xe in meteorites, it is not unexpected that the Earth also exhibits excess $^{129}$Xe produced from decay of the $^{129}$I incorporated at the time of its formation (Fig. 1). As discussed below, the most noteworthy aspect of the Earth's radiogenic $^{129}$Xe is not that there is any at all, but rather that there is not more of it. The inventory of radiogenic $^{129}$Xe, compared to the inventory of I, determines the $^{129}$I/$^{127}$I ratio at the time of formation; comparison with the ratio obtaining at the time of formation of the solar system then yields the Earth's formation interval (*Wetherill*, 1975; *Bernatowicz and Podosek*, 1978). Formally, this I-Xe age is analogous to a whole-rock gas retention age based on the more familiar K-Ar or U-Th-He parent-daughter systems, except that, because $^{129}$I is now extinct, only a relative (to meteorites) age is obtained, not an absolute age, i.e., the calculation leads directly to a formation interval. Because of the short lifetime of $^{129}$I the age result is relatively precise despite substantial uncertainties in input parameters and assumptions. As described below, there are some interesting twists to this

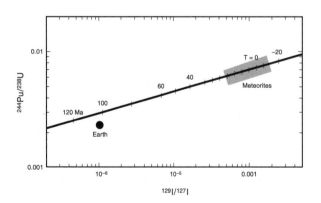

**Fig. 2.** Display of the relationship of $^{129}$I/$^{127}$I (abscissa) and $^{244}$Pu/$^{238}$U (ordinate) in meteorites (shaded area) and the Earth (labeled solid point; see text for description of calculation). The locus of variations generated solely by radioactive decay is a straight line; the line shown originates in the meteorite field (see text), and is labeled in million-year intervals. This figure illustrates the Earth's depletion in $^{244}$Pu, by a factor of a few, and in $^{129}$I, by more than 2 orders of magnitude, relative to meteorites, suggesting a formation interval on the order of 100 m.y. From *Ozima and Podosek* (1999).

problem, but it is hard to stray very far from the conclusion that at its formation the Earth's $^{129}I/^{127}I$ ratio was about 2 orders of magnitude lower than the characteristic ratio for meteorites (Fig. 2). This figure translates to a formation interval of about 100 m.y.

The most easily quantifiable part of the Earth's inventory of radiogenic $^{129}Xe$ is that in air. Although the overall issue of the isotopic composition of atmospheric Xe is complex, identification of the contribution of radiogenic $^{129}Xe$ is fairly straightforward; at least estimates have not changed significantly over more than two decades. We will adopt the estimate that 6.8% (*Pepin and Phinney*, 1978; *Pepin*, 1991; see Fig. 1) of atmospheric $^{129}Xe$ (i.e., $2.76 \times 10^{11}$ mol) is radiogenic. In the approximation that the Earth is extensively degassed of primordial and early-formed radiogenic gases, this is also the total terrestrial inventory. Actually, the solid Earth is not completely degassed, since gas concentrations in mantle samples are not nil, and some mantle samples even exhibit excess radiogenic $^{129}Xe$ relative to air (e.g., *Ozima*, 1994). The total quantities in the mantle are estimated to be rather small, however. *Allègre et al.* (1995), following *Allègre and Staudacher* (1994), for example, estimate that the Earth is 85% degassed, which is not a significant distinction on the scale of interest here.

The amount of the element I to be associated with radiogenic Xe is a much greater source of uncertainty, since the geochemical behavior of I is not very well understood. Recent estimates lean heavily on the work of *Deruelle et al.* (1992). *Allègre et al.* (1995) use 10–12 ppb, and *Zhang* (1998) uses 10 ppb, both for bulk silicate Earth, for example, severalfold higher than the value used by *Wetherill* (1975).

*Ozima and Podosek* (1999) adopt a slightly different approach, ascribing generation of the atmosphere (including the radiogenic components) to degassing of some unspecified part of the mantle but assuming that the corresponding I is all in the crust (including the oceans), citing crustal I of 1.55 ppm, again from *Deruelle et al.* (1992).

None of these subtleties matter very much. *Ozima and Podosek* (1999), for example, infer that the Earth's initial $^{129}I/^{127}I$ was $8.1 \times 10^{-7}$, more than 2 orders of magnitude (a factor of 123) lower than the commonly cited "representative" meteorite value, which corresponds to a formation interval of about 111 m.y. (Fig. 2). For comparison, the figure reported by *Allègre et al.* (1995) is 108 m.y., that by *Zhang* (1998) is 109 m.y. The near identity of these numbers is coincidental, but they illustrate the point that the calculated I-Xe formation interval does not depend sensitively on just how the calculation is done. These numbers are also not qualitatively different from considerably less recent estimates, e.g., by *Wetherill* (1975), *Bernatowicz and Podosek* (1978), or *Pepin and Phinney* (1978).

It is difficult to assign formal errors to such calculations because they involve assumptions or geochemical parameters, such as global or crustal I concentration, that do not come with formal error limits. It is probably not unreasonable to think in terms of a factor-of-2 level uncertainty in

initial $^{129}I/^{127}I$. It might also be noted in passing that the highest $^{129}I/^{127}I$ ratios reported in meteorites are closer to $1.5 \times 10^{-4}$ than to $1.0 \times 10^{-4}$ (e.g., *Swindle and Podosek*, 1988). Even so, factor-of-2 uncertainty, or perhaps a bit more, in $^{129}I$ abundance translates to an age uncertainty of only about 20 Ma. It is difficult, without very different parameters or assumptions, to move away from the qualitative conclusion that formation of the Earth did not occur until about 100 m.y. after formation of the solar system.

## 2.2.  Radiogenic Xenon-136 from Plutonium-244

The other short-lived radionuclide of interest in present context is $^{244}Pu$ (halflife 81 m.y.), coincidentally the second to be discovered (*Rowe and Kuroda*, 1965). Plutonium-244 decays primarily by α emission, but its prior presence cannot be detected through this channel because it leads, through very short-lived intermediaries, to monoisotopic $^{232}Th$. There is, however, a weak (0.125%) branch for spontaneous fission, which is detectable by fission tracks (linear trails of lattice disorder caused by recoiling fission fragments) or by production of isotopes of Xe and Kr (again, evident in isotopic structures only because the noble gases are so scarce in inner solar system planetary materials). In practice, the best way for quantitative evaluation of $^{244}Pu$ abundance is usually through fission Xe. Fission produces several isotopes of Xe, all those that do not have isobars on the neutron-rich side of β stability; when fission Xe is added to primordial Xe the biggest fractional effect occurs at $^{136}Xe$, which is therefore commonly used as a proxy for the entire fission component. With some liberty, a fission component is subsumed as a "radiogenic" component.

Terrestrial atmospheric Xe also contains an evident contribution of radiogenic Xe from fission (Fig. 1). Some of this must be due to $^{238}U$, the only other naturally occurring nuclide significantly subject to spontaneous fission, but only very little, because the fission branching ratio for $^{238}U$ is much smaller than for $^{244}Pu$. Direct support for identification of $^{244}Pu$ as the source of the heavy-isotope Xe excesses in air (Fig. 1) also follows from observation that the pattern of heavy-isotope excesses matches that of $^{244}Pu$ spontaneous fission satisfactorily well (*Pepin and Phinney*, 1978; *Pepin*, 1991) but is distinct from that of $^{238}U$. The situation for $^{244}Pu$ — a short-lived radionuclide that generates a Xe daughter that accumulates in the atmosphere — is qualitatively similar to that of $^{129}I$, and a Xe formation interval can be computed for the $^{244}Pu$-$^{136}Xe$ system just as for the $^{129}I$-$^{129}Xe$ system. In practice, however, there are important differences in detail.

One such difference is that it is harder to identify just how much radiogenic $^{136}Xe$ there is in air. This is essentially a matter of uncertainty in estimating the underlying nonradiogenic component (Fig. 1), harder than in the case of $^{129}Xe$ because the meteorite data on which the U-Xe composition (Fig. 1) is based contain substantial heavy-isotope components that must be subtracted away, and also because the fractionation required to fit atmospheric Xe composition is based on the lightest isotopes, and uncertainty in extrapolation to the heaviest isotopes is nontrivial. To illustrate the scale, many investigators adopt *Pepin and Phinney's* (1978) estimate that 4.6% of atmospheric $^{136}$Xe is radiogenic, but *Igarashi* (1995) defends an alternative calculation of the nonradiogenic composition that leads to only 2.8% radiogenic $^{136}$Xe.

Another difference arises in estimating a representative early solar system value for $^{244}$Pu abundance. The element Pu has no stable isotopes, nor any longer-lived than $^{244}$Pu. The abundance of $^{244}$Pu must thus be indexed to an isotope of a different element. The usual choice is $^{238}$U. Plutonium and U are both refractory lithophile elements, and for undifferentiated meteorites (chondrites) and the bulk silicate Earth it is probably safe to assume that Pu/U is not fractionated from cosmic composition, and thus that differences in $^{244}$Pu/$^{238}$U are due to radioactive decay, i.e., time. But the early solar system value for this ratio is not precisely established. Most investigators adopt $^{244}$Pu/$^{238}$U = 0.0068 in the early solar system, the chondrite value of *Hudson et al.* (1989). But this is a difficult measurement, and the recommended value is based on a single meteorite. Measurement of $^{244}$Pu abundance is considerably easier in some differentiated meteorites, and on the basis of apparent coherence of Pu and Nd within such meteorites *Marti et al.* (1977) recommend a substantially lower value, $^{244}$Pu/$^{238}$U ≈ 0.004, for the early solar system.

Issues relating to how much of the Earth's radiogenic $^{136}$Xe is in the atmosphere are similar to those described above for $^{129}$Xe. Using the same atmosphere-crust reckoning as for $^{129}$Xe-$^{129}$I, crustal U of 0.91 ppm (from *Taylor and McLennan,* 1985), and the lower (*Igarashi,* 1995) estimate of radiogenic $^{136}$Xe in air, *Ozima and Podosek* (1999) calculate an initial $^{244}$Pu/$^{238}$U for the Earth of 0.0025, which is a factor of 2.7 lower than the *Hudson et al.* (1989) meteoritic value. This shortfall factor is much smaller than that for $^{129}$I, but because of the longer halflife of $^{244}$Pu the corresponding formation intervals are quite consistent (Fig. 2); indeed, considering the uncertainties, the degree of concordancy appearing in Fig. 2 must be viewed as fortuitous.

Overall, the level of uncertainty that should be attached to initial $^{244}$Pu/$^{238}$U shortfall is no better than that for initial $^{129}$I/$^{127}$I, at least a factor of 2 or somewhat worse. Because of the longer lifetime of $^{244}$Pu, this translates to a much larger age (formation interval) uncertainty than in the case of $^{129}$I. In practice, then, the $^{244}$Pu-$^{136}$Xe system, although a very interesting story in its own right, is useful primarily as a consistency check in estimation of the Earth's formation interval, but does not independently provide a formation interval with a precision competitive with that of the $^{129}$I-$^{129}$Xe system.

It has been noted that because both daughters of interest in this discussion are isotopes of the same element it is possible to calculate a combined I-Pu-Xe age (formation interval) based on the isotopic ratio of radiogenic $^{129}$Xe to radiogenic $^{136}$Xe and the "parent" ratio I/U (cf. *Wetherill,* 1975; *Pepin and Phinney,* 1978; *Bernatowicz and Podosek,*

1978; *Zhang,* 1998). This is analogous to a Pb-Pb model age, except that the parents must be related as an elemental (I/U) rather than an isotopic ($^{235}$U/$^{238}$U) ratio. A combined I-Pu-Xe age is an attractive idea in principle, but in practice it is not very useful for constraining the age of the Earth: It is neither much easier nor more precise to estimate the ratio of the radiogenic daughters than each one separately, and there is no particular geochemical or cosmochemical coherence between I and U that would make it easier to estimate their ratio than their individual inventories. The I-Pu-Xe approach is more useful in the cases of the Moon and Mars (discussed below), however, in which arguments can be based on ratios independently of inventories.

## 2.3. The Relevance of "Missing Xenon"?

It has long been recognized that the pattern of primordial gas *elemental* ratios is rather similar from one meteorite to another even though the overall absolute gas concentrations vary over several orders of magnitude. This pattern, one of progressively greater depletion of lighter gases, came to be called the *planetary* pattern because of its resemblance to the noble gas abundance pattern in the terrestrial atmosphere. Relative to the meteorite abundance pattern, however, air is conspicuously depleted in Xe, by about an order of magnitude. This observation has led to the recurrent suggestion that there really should be, and perhaps once was, about an order of magnitude more Xe in air than there actually is, i.e., that most of the Earth's atmospheric Xe is somehow "missing" (see *Ozima and Podosek,* 1983, for a review), perhaps because Xe was selectively retained in the interior of the Earth and not degassed into the atmosphere, perhaps because it was selectively removed from the atmosphere, etc. One possible mechanism for selective removal is tectonic subduction. Another possibility, one that seems to have attracted the most attention, is incorporation of Xe into sediments, but this seems to be quantitatively unsubstantiated (e.g., *Bernatowicz et al.,* 1984). The issue has never been satisfactorily resolved: The hypothesized missing Xe has not been found, and it is unclear whether the analogy with meteorites is strong enough to justify belief that there is any. Recently, *Ozima and Podosek* (1999) have argued that comparison of atmospheric noble gas elemental abundances with the solar abundance pattern suggests that there should indeed be about an order of magnitude more Xe in air than there actually is, i.e., that there really is a "missing Xe" problem even though all attempts to find it have failed.

This possibility is relevant to the issue at hand because if most of the primordial Xe that should be in the atmosphere is actually missing, it is likely that some of the radiogenic Xe is missing as well. As pointed out by *Bernatowicz and Podosek* (1978), and quantitatively modeled by *Ozima and Podosek* (1999), allowance for missing radiogenic Xe decreases the formation interval calculated as above. As further noted by *Ozima and Podosek* (1999), it is also important to consider just *when* the hypothesized missing Xe becomes missing. If, for example, it is supposed that the missing Xe is removed essentially contemporaneously with or shortly (compared to the lifetime of [129]I) after Earth formation, the age calculation is unaffected, because no radiogenic Xe has yet been produced so none of it is missing. At the other extreme, if the missing Xe is removed much later (compared to the lifetime of [244]Pu), e.g., more or less continuously over geologic history, so that radiogenic [136]Xe is removed with nearly the same efficiency as radiogenic [129]Xe, then the degree of removal cannot be very large because the deficiency of [136]Xe, compared to meteorites, is no more than a factor of 2–3. The maximum effect on the Xe age calculation occurs when the Xe becomes missing on a characteristic timescale long compared to the lifetime of [129]I but not long compared to the lifetime of [244]Pu, i.e., a few to several tens of millions of years after Earth formation. In such case it is possible to imagine that the Earth has about an order of magnitude more radiogenic [129]Xe than is now present in air without violating plausible inventory considerations for radiogenic [136]Xe, and the nominal formation interval is correspondingly reduced by a few halflives of [129]I, i.e., by about 50 m.y.

Even if the hypothesis of missing atmospheric Xe is correct, it affects the Xe age calculation only if the unknown mechanism for hiding the Xe acts promptly but not too promptly after formation of the Earth, assuming also that radiogenic Xe produced to that time is already in the reservoir from which the Xe is removed, presumably air. It is thus probably best to view the relevance of missing Xe as broadening the range of Xe formation interval consistent with available geochemical data. For a nominal 110-m.y. interval as calculated above, for example, allowance for missing Xe without changing any of the other parameters permits a shorter formation interval, down to about 60 m.y.

## 2.4. Radiogenic Xenon on the Moon?

Since the origins of the Earth and Moon are likely closely linked, it is pertinent to inquire whether radiogenic Xe isotopes are evident on the Moon. The short answer to this question is yes, they are. There are several lunar samples whose absolute ages are reliably inferred to be greater than 4.4 Ga, old enough that at least [244]Pu was not effectively extinct at the time of rock formation. It might thus be expected that in favorable cases the oldest lunar rocks would exhibit evidence for *in situ* decay of [244]Pu, and indeed such observations have been made (e.g., *Caffee et al.*, 1981). The broader question, however, is whether the Moon, like the Earth, provides evidence for its planetary inventory of [129]I and [244]Pu established prior to the formation of its oldest rocks. The answer to this question is somewhat more ambiguous.

The Moon does not have a permanent atmosphere: Gases emanating from lunar solids enter a transient atmosphere that is quickly removed by photoionization and/or solar wind interactions (e.g., *Hodges et al.*, 1974). It is thus

not possible to estimate initial radionuclide inventories by daughter accumulation in the atmosphere, as is done for the Earth. There is, however, the so-called "excess fission Xe" effect: Highland samples, mostly soil breccias, are not uncommonly found to contain surface-correlated [244]Pu-fission Xe, often accompanied by [129]Xe attributed to [129]I, in quantities well beyond *in situ* production (e.g., *Drozd et al.*, 1972, 1976; *Swindle et al.*, 1986a). This effect is generally interpreted in terms of incorporation into regolith samples of radiogenic Xe degassed from the lunar interior; either by adsorption as the gas percolates through the regolith or by electromagnetic implantation from the transient atmosphere. This effect cannot support inventory-based age calculations but with some assumptions chronological constraints can be obtained from the ratio of excess [129]Xe to excess [136]Xe (i.e., I-Pu-Xe dating, as discussed above). The situation has been reviewed by *Swindle et al.* (1986a), who calculate a model lunar formation interval of 63 ± 42 m.y. This result is broadly consistent with absolute ages of the oldest reliably-dated lunar rocks and also with the various estimates of the Earth's formation interval, including that based on Xe. While this result nominally favors the shorter end of the range usually considered for the Earth, it is also consistent with the benchmark 100 m.y., and in view of the assumptions involved in this calculation (*Swindle et al.*, 1986a) it would seem difficult to argue that a formation interval longer by perhaps a few tens of millions of years is precluded.

## 2.5. Radiogenic Xenon on Mars?

While the focus of this review is the Earth-Moon system, Mars provides an interesting comparison. There can be little doubt that the atmosphere of Mars contains radiogenic [129]Xe: The most prominent feature of the isotopic spectrum of martian atmospheric Xe is a relative abundance of [129]Xe that is higher (than in terrestrial atmospheric Xe) by more than a factor of 2. This feature is quite evident in the Viking lander data (*Owen et al.*, 1977) as well as in the more precise laboratory analysis of EET 79001 (*Swindle et al.*, 1986b), a meteorite widely held to come from Mars and to incorporate martian atmospheric gases.

Although the fractional isotopic excess of radiogenic [129]Xe in the martian atmosphere is large, the total quantity of radiogenic [129]Xe, per gram of planet, is relatively small, just over half the terrestrial value. Moreover, the planetary abundance of I in the model of *Dreibus and Wänke* (1987) is about 25 ppm, more than twice the terrestrial value. The inferred [129]I/[127]I ratio is thus severalfold lower than for Earth. In itself, however, this is not a strong constraint on "formation" of Mars (see discussion below) because Mars is generally considered to have experienced substantial and late (on the scale of the lifetime of [129]I) exospheric loss of atmospheric gases (e.g., *Pepin*, 1991, 1994).

*Swindle and Jones* (1997) point out that if nonradiogenic Xe in the atmosphere of Mars is interpreted as fractionated U-Xe, as for the Earth, there is no significant heavy-isotope contribution attributable to actinide fission. They argue that this is unreasonable, especially for apparent late formation (because of the difference in lifetimes, the ratio of $^{244}$Pu-derived $^{136}$Xe to $^{129}$I-derived $^{129}$Xe should be higher for later times of establishment of planetary inventories). They thus advocate a different interpretation of nonradiogenic Xe composition, based on solar wind Xe rather than U-Xe, that would imply the presence of actinide (presumably $^{244}$Pu) fission in martian atmospheric Xe. Even with favorable assumptions, however, *Swindle and Jones* (1997) conclude that the ratio of radiogenic $^{129}$Xe to radiogenic $^{136}$Xe in the martian atmosphere is at least an order of magnitude higher than in the terrestrial atmosphere, in one-stage atmospheric degassing models corresponding to a formation interval of 53 m.y. It is noteworthy that relatively early formation of Mars, particularly in comparison with Earth, is also suggested by excesses of $^{182}$W (ascribed to 9-m.y. $^{182}$Hf) in some martian rocks (SNC meteorites) but not in terrestrial rocks (*Halliday and Lee,* 1999), indicating that Mars differentiated metal from silicate while $^{182}$Hf was still significantly extant.

## 2.6. Radical Heterogeneity of Iodine-129 and/or Plutonium-244?

A much-discussed aspect of the lore of short-lived radionuclides (e.g., *Podosek and Nichols,* 1997) is that they must have been added to the mix that constitutes the solar system relatively recently, in comparison with stable nuclides, before solar system formation. This follows simply from their short lifetimes: If they were not added recently, they would not have survived to appear in the solar system. Because of late addition they may be more susceptible to heterogeneous distribution than stable or long-lived nuclides. It is thus germane to consider the possibility that $^{129}$I and/or $^{244}$Pu were heterogeneously distributed in the early solar system, i.e., the possibility that the low values inferred for the Earth (or the Moon) are actually a reflection of isotopic heterogeneity rather than radioactive decay over a long interval.

The only two short-lived radionuclides for which any direct evidence suggesting major heterogeneity has been advanced are $^{26}$Al (*Esat et al.,* 1978; *Lee et al.,* 1979) and $^{53}$Mn (*Lugmair and Shukolyukov,* 1998) (halflives 0.74 m.y. and 3.7 m.y. respectively). Possible heterogeneity in other short-lived radionuclides, such as $^{129}$I and $^{244}$Pu, is suspected only because of "guilt by association" reasoning. Suspicion of heterogeneous distribution of $^{129}$I, in particular, has often been voiced in connection with chronological interpretations of meteorite abundances that do not correlate well with other meteorite characteristics, but even so no specific scheme of $^{129}$I distribution and corresponding chronological reinterpretation has been advanced.

On the other hand, there are significant arguments that the distributions of $^{129}$I and $^{244}$Pu should not be radically heterogeneous, or at least should be much less susceptible to heterogeneity than those of much shorter-lived species

such as $^{26}$Al and $^{53}$Mn. The early solar system abundance of $^{129}$I is about an order of magnitude *less* than that which would follow from more-or-less continuous addition of fresh nucleosynthetic products into the local interstellar medium. This observation gave rise to the concept of a free-decay interval $\Delta \approx 10^8$ yr, the interval between the last addition of $^{129}$I to the local interstellar medium and the formation of the solar system from that medium (e.g., *Wasserburg et al.,* 1969; *Podosek and Swindle,* 1988). The early solar system abundance of $^{244}$Pu is readily accommodated within this scenario, and a comparable free-decay interval for actinides is independently suggested by the absence of detectable $^{247}$Cm (*Chen and Wasserburg,* 1981). Prior to solar system formation the time of last significant addition of $^{129}$I and $^{244}$Pu is thus inferred to be much greater than the time of last addition of species such as $^{26}$Al and $^{53}$Mn, and plausibly long enough to permit substantial mixing into presolar materials (e.g., *Clayton,* 1983). We thus conclude that there is no substantive basis for believing that the low abundances of $^{129}$I and $^{244}$Pu in the Earth reflects radical isotopic heterogeneity rather than radioactive decay.

## 3. THE SIGNIFICANCE OF A PLANETARY XENON RETENTION AGE

Having considered above the parameters and assumptions involved in calculating a Xe-retention formation interval for the Earth, we must now confront the issue of what this means for the planet Earth, i.e., how to infer chronological constraints on important aspects of Earth history. This is not a straightforward task.

When applied to the traditional geochemical problem of dating a (meteorite) rock, formation defined by Xe retention, in common with other isotopic systems, is interpreted in terms of isotopic closure, typically downward passage through an effective closure temperature as described by *Dodson* (1973). The actual value of this closure temperature, which can in some cases be estimated from laboratory stepwise degassing results, is likely variable, but is on the order of a few to several hundred degrees, definitely subigneous but higher than most meteorites have experienced throughout most of solar system history. The actual physical event involved is evidently variable as well (e.g., *Swindle and Podosek,* 1988). It was once thought that I-Xe formation represented condensation of meteoritic solids from nebular gas, and perhaps this is approximately so for some samples, but it now appears more reasonable that isotopic closure represents relaxation of parent body metamorphic conditions for undifferentiated meteorites or igneous differentiation for differentiated meteorites. In a few cases closure may represent low-temperature parent-body aqueous alteration or impact-generated shock.

None of this seems particularly relevant to the problem of planetary formation, in which all materials have experienced igneous temperatures during planetary evolution and in which closure is effected not by slow diffusion in cool materials but by accumulation in a gravitationally bound atmosphere. We could imagine forming the Earth by accreting known types of meteoritic materials [excluding lunar and SNC (martian) meteorites] in such a way that Xe was retained in the atmosphere even if the incoming rocks or planetesimals were themselves vaporized by the impact. If that were the case the calculated Xe formation interval of the Earth would likely be less than 10 m.y. Evidently, it did not happen that way. Moreover, Xe closure in such a scenario would be something that happened prior to accretion; the accretion itself could have happened at any later time.

We might also imagine forming the Earth by accreting materials like known meteorites except that they did not form until some 60–110 m.y. later. The terrestrial data do not preclude this possibility. Nevertheless, no such materials are recognized among known meteorites. It can also be argued that the major processes responsible for the properties of known meteorites — nebular fractionation, accretion into parent-body planetesimals of size a few hundred kilometers or less, parent-body heating by short-lived radionuclides (mainly $^{26}$Al) or enhanced solar activity — should have happened on a timescale not much more (and probably less) than about 10 m.y. It thus seems most likely that when pre-Earth materials had reached the stage of aggregation comparable to that of presently known meteorites they should have had Xe formation intervals comparable to or less than those of known meteorites, i.e., about 10 m.y. or less. They must have lost most of their radiogenic Xe later, something that known meteorites did not do, and it seems reasonable that this loss was related to a history also not experienced by known meteorites, namely continued accretion into larger bodies.

One possibility is that preaccretionary materials lost volatiles, specifically Xe, during protracted (10–100 m.y.) evolution of planetesimals larger than present asteroids but still small compared to present planets. There must have been some such history between planetesimals and planets, but speculations in this regard can be written on a nearly clean slate; there are no relevant samples, and models for this aspect of planet building are mostly concerned with mass distribution, gravitational interactions, and collisions, without much heed to what conditions were like on these bodies.

It may be suggested, for example, that increasingly violent collisions among progressively larger bodies led to efficient devolatilization by impact heating/melting/vaporization. In the extreme, it could be supposed that bodies accreting onto proto-Earth lost their volatiles when they finally impacted the growing Earth. It is far from clear whether this is a quantitatively viable scenario. Even if impactors are totally vaporized, it is not clear whether volatiles would escape from the gravitational field of a proto-Earth that is a significant fraction of its present mass. It should also be noted explicitly that even if volatiles are lost on impact, the necessary boundary condition — a long Xe formation interval — will not be achieved unless Xe is lost preferentially to I (as well as U and Pu), a nontrivial

condition. If impact degassing is indeed responsible, then the Xe formation interval must represent some effective mean time of mass accretion — rather longer than anticipated on the basis of dynamical models (e.g., *Wetherill,* 1986).

It is obviously tempting to try to relate the Xe formation interval to the giant impact scenario for formation of the Moon (e.g., *Cameron and Benz,* 1991; *Ida et al.,* 1997; *Cameron and Canup,* 1998; *Halliday and Drake,* 1999). Both the lunar and terrestrial Xe isotopic data would be consistent with this hypothesis, and this would allow an earlier accretion of mass more in accord with dynamical models (e.g., *Wetherill,* 1986). The Xe data would then likely date the impact at 60–110 Ma. This suggestion must presently be regarded as rather speculative, however, since available dynamical models for impact and formation of the Moon do not make quantitative predictions for the chemical events on which isotopic chronometers are based. It is unclear, for example, how much of the Moon is derived from the impactor rather than the target, or whether any previously existing structure of core-mantle-crust-atmosphere in the target would be rehomogenized or (isotopically) reequilibrated in the impact. With particular regard to the present context, if a giant impact is to be invoked to account for the long formation interval inferred from Xe the most likely mechanism would seem to be loss of an early atmosphere already containing essentially all the radiogenic Xe produced up to that point. It is not clear whether or not this would happen. It should also be noted that loss of atmospheric Xe in an impact is not itself sufficient; it must be further supposed that the corresponding I, which would presumably be resident in the crust, was *not* lost at the same time.

## 4.  CONCLUSIONS

Xenon data indicate that the Earth did not "form" until some 110 m.y. after formation of the solar system as a whole. This statement reflects the inference that the Earth's initial $^{129}$I/$^{127}$I ratio, calculated from the atmospheric inventory of radiogenic $^{129}$Xe (the daughter of $^{129}$I) and the corresponding I inventory, was more than 2 orders of magnitude lower than the value prevailing in the early solar system. Uncertainty in the $^{129}$I/$^{127}$I ratio is substantial, at least a factor of 2, but because the halflife of $^{129}$I (16 m.y.) is so short the corresponding uncertainty in formation interval is modest, perhaps 20–30 m.y. A qualitatively similar Xe formation interval can be calculated from radiogenic $^{136}$Xe, the daughter (by spontaneous fission) of $^{244}$Pu; this result is consistent with that based on $^{129}$I, but because the halflife of $^{244}$Pu (81 m.y.) is so much larger the Pu-Xe system, on its own, is much less useful in this regard. If allowance is made for the possibility that most of the Xe, including radiogenic Xe, that should be in the atmosphere has somehow been removed or hidden, the I-Xe and Pu-Xe formation interval could be reduced to perhaps 60 m.y. This is still a long time in view of the pace of nebular evolution, planetesimal formation, and dynamical models for construction of the Earth.

This is a strong constraint on models for the formation of the Earth and the Earth-Moon system. It has been available for decades, but seems not to have attracted the attention it deserves, perhaps because it is a chemical constraint that is difficult to relate to the physical and dynamic events usually thought to define "formation" of a planet. The same is true of other isotopic systems that suggest Earth formation intervals comparable to that obtained from Xe. It should be noted that the Xe formation interval, as presently understood, is *not* the age of some internal Earth process, i.e., it is not the age of the atmosphere, or of core-mantle-crust differentiation; it is a constraint on establishment of global chemical inventories. A straightforward interpretation of this formation interval is that the Earth did not accrete the major part of its mass until later than 60–110 m.y. after formation of the solar system. Alternatively, major mass accretion might have occurred earlier if it is postulated that the chemical fractionation reflects some later event, such as the giant impact often suggested to be responsible for formation of the Moon; it remains unclear, however, whether the necessary chemical processes are consistent with the giant impact scenario.

*Acknowledgments.*   We appreciate helpful conversations with R. O. Pepin. This work was supported by NASA grant NAG5-7149 and by the McDonnell Center for the Space Sciences.

# REFERENCES

Allègre C. J. and Staudacher T. (1994) Rare gases systematics and mantle structure. *Mineral. Mag., 58,* 14.

Allègre C. J., Manhes G., and Göpel C. (1995) The age of the Earth. *Geochim. Cosmochim. Acta, 59,* 1445–1456.

Anders E. and Grevesse N. (1989) Abundances of the elements: Meteoritic and solar. *Geochim. Cosmochim. Acta, 53,* 197–214.

Bernatowicz T. J. and Podosek F. A. (1978) Nuclear components in the atmosphere. In *Terrestrial Noble Gases* (E. C. Alexander Jr. and M. Ozima, eds.), pp. 99–135. Japan Sci. Soc. Press, Tokyo.

Bernatowicz T. J., Podosek F. A., Honda M., and Kramer F. E. (1984) The atmospheric inventory of xenon and noble gases in shales: The plastic bag experiment. *J. Geophys. Res., 89,* 4597–4611.

Caffee M., Hohenberg C. M., and Hudson G. B. (1981) Troctolite 76535: A study in the preservation of early isotopic records. *Proc. Lunar Planet. Sci. 12B,* pp. 99–115.

Cameron A. G. W. and Benz W. (1991) The origin of the moon and the single impact hypothesis IV. *Icarus, 92,* 204–216.

Cameron A. G. W. and Canup R. M. (1998) The giant impact occurred during Earth accretion (abstract). In *Lunar and Planetary Science XXIX,* Abstract #1062. Lunar and Planetary Institute, Houston.

Cameron A. G. W., Vanhala H., and Höflich P. (1997) Some aspects of triggered star formation. In *Astrophysical Implications of the Laboratory Study of Presolar Materials* (T. J. Bernatowicz and E. K. Zinner, eds.), pp. 665–693. AIP Conf. Proc. 402.

Chen J. H. and Wasserburg G. J. (1981) The isotopic composition of uranium and lead in Allende inclusions and meteoritic phosphates. *Earth Planet. Sci. Lett., 52,* 1–15.

Clayton D. D. (1983) Extinct radioactivities: A three-phase mixing model. *Astrophys. J., 268,* 381–384.

Dalrymple B. G. (1991) *The Age of the Earth.* Stanford Univ., Stanford. 474 pp.

Deruelle B., Dreibus G., and Jambon A. (1992) Iodine abundances in oceanic basalts: Implications for Earth dynamics. *Earth Planet. Sci. Lett., 108,* 217–227.

Dodson M. H. (1973) Closure temperature in cooling geochronological and petrological systems. *Contrib. Mineral. Petrol., 40,* 259–274.

Dreibus G. and Wänke H. (1987) Volatiles on Earth and Mars: A comparison. *Icarus, 71,* 225–240.

Drozd R. J., Hohenberg C. M., and Ragan D. (1972) Fission xenon from extinct $^{244}$Pu in 14301. *Earth Planet. Sci. Lett., 15,* 338–346.

Drozd R. J., Kennedy B. M., Morgan C. J., Podosek F. A., and Taylor G. J. (1976) The excess fission xenon problem in lunar samples. *Proc. Lunar Sci. Conf. 7th,* pp. 599–623.

Esat T., Lee T., Papanastassiou D. A., and Wasserburg G. J. (1978) Search for $^{26}$Al effects in the Allende FUN inclusion C1. *Geophys. Res. Lett., 5,* 807–810.

Froude D. O., Ireland T. R., Kinney P. D., Williams I. S., Compston W., Williams I. R., and Myers J. S. (1983) Ion microprobe identification of 4100 to 4200 Myr-old terrestrial zircons. *Nature, 304,* 616–618.

Halliday A. N. and Drake M. J. (1999) Colliding theories. *Science, 283,* 1861–1863.

Halliday A. N. and Lee D.-C. (1999) Tungsten isotopes and the early development of the Earth and moon. *Geochim. Cosmochim. Acta,* in press.

Hodges R. R., Hoffman J. H., and Johnson F. S. (1974) The lunar atmosphere. *Icarus, 21,* 415–426.

Hohenberg C. M., Podosek F. A., and Reynolds J. H. (1967) Xenon-iodine dating: Sharp isochronism in chondrites. *Science, 156,* 233–236.

Hudson G. B., Kennedy B. M., Podosek F. A., and Hohenberg C. M. (1989) The early solar system abundance of $^{244}$Pu as inferred from the St. Severin chondrite. *Proc. Lunar Planet. Sci. Conf. 19th,* pp. 547–557.

Ida S., Canup R. M., and Stuart G. R. (1997) Lunar accretion from an impact-generated disk. *Nature, 389,* 353–357.

Igarashi G. (1995) Primitive xenon in the Earth. In *Volatiles in the Earth and Solar System* (K. A. Farley, ed.), pp. 70–80. AIP Conf. Proc. 341.

Jeffery P. M. and Reynolds J. H. (1961) Origin of excess Xe$^{129}$ in stone meteorites. *J. Geophys. Res., 66,* 3582–3583.

Lee T., Russell W. A., and Wasserburg. G. J. (1979) Calcium isotopic anomalies and the lack of aluminum-26 in an unusual Allende inclusion. *Astrophys. J. Lett., 228,* L93–L98.

Lee T., Shu F. H., Shang H., Glasgold A. E., and Rehm K. (1998) Protostellar cosmic rays and extinct radioactivities in meteorites. *Astrophys. J., 506,* 898–912.

Lugmair G. W. and Galer S. J. G. (1992) Age and isotopic relationships among the angrites Lewis Cliff 86010 and Angra dos Reis. *Geochim. Cosmochim. Acta, 56,* 1673–1694.

Lugmair G. W. and Shukolyukov A. (1998) Early solar system timescales according to $^{53}$Mn-$^{53}$Cr systematics. *Geochim. Cosmochim. Acta, 62,* 2863–2886.

MacPherson G. J., Davis A. M., and Zinner E. K. (1995) The distribution of aluminum-26 in the early solar system — A reappraisal. *Meteoritics, 30,* 365–386.

Marti K., Lugmair G. W., and Scheinin N. B. (1977) Sm-Nd-Pu systematics in the early solar system (abstract). In *Lunar and Planetary Science VIII,* pp. 619–621. Lunar and Planetary Institute, Houston.

McCulloch M. T. (1994) Primitive $^{87}Sr/^{86}Sr$ from an Archean barite and conjecture on the Earth's age and origin. *Earth Planet. Sci. Lett., 126,* 1–13.

Owen T., Biemann K., Rushneck D. R., Biller J. E., Howarth D. W., and Lafleur A. L. (1977) The composition of the atmosphere at the surface of Mars. *J. Geophys. Res., 82,* 4635–4639.

Ozima M. (1994) Noble gas state in the mantle. *Rev. Geophys., 32,* 405–426.

Ozima M. and Podosek F. A. (1983) *Noble Gas Geochemistry.* Cambridge Univ., Cambridge. 367 pp.

Ozima M. and Podosek F. A. (1999) Formation age of the Earth from $^{129}I/^{127}I$ and $^{244}Pu/^{238}U$ systematics and the missing Xe. *J. Geophys. Res.,* in press.

Pepin R. O. (1991) On the origin and early evolution of terrestrial planet atmospheres and meteoritic volatiles. *Icarus, 92,* 2–79.

Pepin R. O. (1994) Evolution of the martian atmosphere. *Icarus, 111,* 289–304.

Pepin R. O. and Phinney D. (1978) Components of xenon in the solar system. Unpublished manuscript, University of Minnesota.

Podosek F. A. (1998) The age of the Earth and moon. In *Origin of the Earth and Moon,* pp. 34–35. LPI Contribution No. 957, Lunar and Planetary Institute, Houston.

Podosek F. A. (1999) A couple of uncertain age. *Science, 283,* 1863–1864.

Podosek F. A. and Cassen P. (1994) Theoretical, observational and isotopic estimates of the lifetime of the solar nebula. *Meteoritics, 29,* 6–25.

Podosek F. A. and Nichols R. H. Jr. (1997) Short-lived radionuclides in the solar nebula. In *Astrophysical Implications of the Laboratory Study of Presolar Materials* (T. J. Bernatowicz and E. K. Zinner, eds.), pp. 617–647. AIP Conf. Proc. 402.

Podosek F. A. and Swindle T. D. (1988) Nucleocosmochronology. In *Meteorites and the Early Solar System* (J. F. Kerridge and M. S. Matthews, eds.), pp. 1114–1126. Univ. of Arizona, Tucson.

Porcelli D. and Pepin R. O. (2000) Rare gas constraints on early Earth history. In *Origin of the Earth and Moon* (R. M. Canup and K. Righter, eds.), this volume. Univ. of Arizona, Tucson.

Reynolds J. H. (1960) Determination of the age of the elements. *Phys. Rev. Lett., 4,* 8–10.

Reynolds J. H. (1963) Xenology. *J. Geophys. Res., 68,* 2939–2956.

Rowe M. W. and Kuroda P. K. (1965) Fissiogenic xenon from the Pasamonte meteorite. *J. Geophys. Res., 70,* 709–714.

Swindle T. D. and Jones J. H. (1997) The xenon isotopic composition of the primordial martian atmosphere: Contributions from solar and fission components. *J. Geophys. Res., 102,* 1671–1678.

Swindle T. D. and Podosek F. A. (1988) Iodine-xenon dating. In *Meteorites and the Early Solar System* (J. F. Kerridge and M. S. Matthews, eds.), pp. 1127–1146. Univ. of Arizona, Tucson.

Swindle T. D., Caffee M. W., Hohenberg C. M., and Taylor S. R. (1986a) I-Pu-Xe dating and the relative ages of the Earth and moon. In *Origin of the Moon* (W. K. Hartmann, R. J. Phillips, and G. J. Taylor, eds.), pp. 331–357. Lunar and Planetary Institute, Houston.

Swindle T. D., Caffee M. W., and Hohenberg C. M. (1986b) Xenon and other noble gases in shergottites. *Geochim. Cosmochim. Acta, 50,* 1001–1019.

Taylor S. R. and McLennan S. M. (1985) *The Continental Crust: Its Composition and Evolution.* Blackwell, Oxford. 312 pp.

Wasserburg G. J., Schramm D. N., and Huneke J. C. (1969) Nuclear chronologies for the galaxy. *Astrophys. J. Lett., 157,* L91–L96.

Wetherill G. W. (1975) Radiometric chronology of the early solar system. *Annu. Rev. Nucl. Sci., 25,* 283–328.

Wetherill G. W. (1986) Accumulation of the terrestrial planets and implications concerning lunar origin. In *Origin of the Moon* (W. K. Hartmann, R. J. Phillips, and G. J. Taylor, eds.), pp. 519–550. Lunar and Planetary Institute, Houston.

Zhang Y. (1998) The young age of Earth. *Geochim. Cosmochim. Acta, 62,* 3185–3189.

*Part II:*

*Dynamics of Terrestrial*
*Planet Formation*

# On Planetesimal Formation:
# The Role of Collective Particle Behavior

## William R. Ward
### *Southwest Research Institute*

The case for and against collective particle behavior as a means of planetesimal formation in the early solar system is reviewed. It is argued that recent models of the excitation of planetesimal dispersion velocities by turbulence at the gas-particle interface are sensitive to poorly known properties of the boundary layer and do not preclude gravitational instability of the planetesimal disk. The uncertain efficacy of binary accretion through the problematic centimeter to kilometer size range provides motivation to explore conditions that would permit some form of instability to operate. Of particular interest are axisymmetric modes destabilized by drag effects that can manifest themselves at Q values considerably greater than unity. The timescale for these modes is determined by gas drag interaction with the remnant nebula, and their wavelengths can be much longer than the length scale of the largest turbulent eddies.

## 1. INTRODUCTION

The exact nature of planetesimal accumulation in the early solar nebula has long been a puzzle. The starting conditions for this process are grains that have either condensed from the cooling nebula or possibly survived from presolar material. Planetesimals must somehow progressively form from smaller bodies, and it is likely that most of their growth history is dominated by binary accretion, i.e., accretion that occurs as a result of particle-particle collisions. Solid particles mixed throughout a cooling circumstellar gas disk are envisioned to settle toward the midplane of the nebula, forming a relatively thin layer of debris of surface density $\sigma$. Gravity alone can result in binary accretion if collisions are sufficiently inelastic; this generally requires that the specific impact energy be not much greater than $v_{esc}^2$, where

$$v_{esc} \equiv \left( \frac{2GM}{R} \right)^{1/2} \sim 7.5 \times 10^{-4}$$
$$\left( \frac{\rho_p}{g/cm^3} \right)^{1/2} \left( \frac{R}{cm} \right) cm/s \qquad (1)$$

is the escape velocity of a target of mass M, radius R and density $\rho_p$. On the other hand, there is a minimum dispersion velocity $c_{crit}$ for a Keplerian sheet of material to be stable against its own self-gravity. This is the velocity at which the Toomre stability parameter, $Q \equiv c\kappa/\pi G\sigma$, is of order unity (*Toomre*, 1969; see section 2), where $\kappa$ is the disk's epicycle frequency; in a Keplerian disk, $\kappa$ is simply the orbital frequency $\Omega(r) = (GM_\star/r^3)^{1/2}$. For a planetesimal disk with a power-law radial surface density [$\sigma = \sigma_o(r/AU)^{-k}$] orbiting a solar mass star [$M_\star = M_\odot$], this critical velocity reads

$$c_{crit} = \frac{\pi G\sigma}{\kappa} \sim 1 \times \left( \frac{\sigma_o}{g/cm^2} \right) \left( \frac{r}{AU} \right)^{3/2-k} cm/s \qquad (2)$$

With our solar system, the mass of the terrestrial planets implies a minimum surface density at 1 AU of $\sigma_o \sim 10$ g/cm²; this value could have been a factor of several higher if accretion was inefficient. It is often useful to define the so-called normalized disk mass, $\mu_d \equiv \pi G\sigma/r\Omega^2 = \pi\sigma r^2/M_\odot$ as a proxy for the surface density, in which case $c_{crit} \sim \mu_d r\Omega$.

The critical velocity does not depend explicitly on particle size; the ratio of impact to gravitational binding energy for particles comprising a marginally stable disk is of order

$$\mathcal{E} \equiv \left( \frac{c_{crit}}{v_{esc}} \right)^2 \sim 2 \times 10^6 \left( \frac{\sigma_o}{\rho_p R} \right)^2 \left( \frac{\rho_p}{g \, cm^{-3}} \right) \left( \frac{r}{AU} \right)^{3-2k} \qquad (3)$$

Thus, for a $\sigma_o \sim 10$ g/cm² disk composed of 2 g/cm³ particles at 1 AU, $\mathcal{E} \sim 10^8$ (R/cm)$^{-2}$. Dissipation of this energy excess would require a coefficient of restitution $\varepsilon < v_{esc}/c_{crit} \approx 10^{-4}$(R/cm); if the dispersion velocity exceeds $c_{crit}$, $\varepsilon$ would need to be still smaller. This stringent requirement renders it unlikely that gravity alone can account for planetesimal growth through the centimeter to kilometer size range via binary accretion, and that some nongravitational force(s) would have to be invoked.

*Weidenschilling* (1984, 1988, 1997) has constructed numerical models of planetesimal growth in which accretion is assumed to proceed via nongravitational forces. The outcomes of particle collisions are modeled by supposing that the total mass excavated from the target is proportional to the impact energy, i.e.,

$$\Delta m \approx C_{ex} \left( mv^2/2 \right) \qquad (4)$$

where m is the mass of the projectile. Consequently, there is a threshold velocity $v_* = (2/C_{ex})^{1/2}$ for which $\Delta m \sim m$. Experimental data for impacts of competent particles into unconsolidated powders from *Hartmann* (1984) were used to fix $v_*$ at ~10³ cm/s, which in turn implies a proportionality constant of $C_{ex} \sim 2 \times 10^{-6}$ g/erg. The difference, $m - \Delta m$, is taken to be the change in the target mass, $\Delta m_t$. The ratio

$\Delta m_t/m = 1 - (v/v_*)^2$ is the accretion efficiency, so that there is net accretion for impact velocities less than $v_*$, but net erosion for larger velocities. Under this prescription, impact velocities much in excess of $c_{crit}$ can still lead to accretion. We will not review this model in detail here, but refer the reader to recent expositions by *Weidenschilling* (1997), *Weidenschilling and Cuzzi* (1993), and *Weidenschilling et al.* (1989). Here we simply note that there is a pivotal assumption whose validity remains unclear, viz., that the entire projectile mass is added (i.e., "sticks") to the target before the excavated mass is removed. Indeed, fracturing the target does not seem especially conducive to bond formation. Nonetheless, it appears plausible that nongravitational forces could account for the accumulation of particles to at least centimeters in size, while for objects on the order of 0.10 km or larger, gravity can become the dominant mechanism for adhesion. However, between these sizes growth remains problematic and is the subject of continuing research.

## 2.    GRAVITATIONAL INSTABILITY

A possible growth strategy through this problematic size range was suggested by *Safronov* (1969) and by *Goldreich and Ward* (1973; hereafter *GW*) that exploits collective behavior of particles in a disk. If $Q \leq 1$, the disk is unstable to perturbations of the form $e^{i(kr + \omega t)}$ for a range of wavenumbers, k. In a collisionless stellar disk, the overall effect of the instabilities may be to heat the disk until $Q \sim 1$ and $c \sim c_{crit}$. However, in the case of a planetesimal disk where physical collisions are commonplace, the disk may not be able to recover stability because the dispersion velocity is subject to damping. Thus, although individual binary collisions may not result in aggregation, their inelastic nature could promote a breakup of the disk into gravitationally bound associations of particles, some of which might survive and eventually coalesce.

Following its introduction, the instability mechanism became quite popular as a means to bridge the problematic size range of planetesimal growth. Further calculations supported the viability of the process in the presence of a quiescent gas phase (e.g., *Coradini et al.*, 1981; *Sekiya*, 1983). However, a possible difficulty was recognized by *Weidenschilling* (1980), who cautioned that gas disk turbulence might keep c (and thus Q) too high for instability to set in. Before we examine this issue, it is useful to review the stability criterion.

### 2.1.    The Critical Q

The dispersion relationship for axisymmetric perturbations of the form $e^{i(kr + \omega t)}$ in a two-dimensional particulate disk reads (e.g., *GW*)

$$\omega^2 = \kappa^2 - 2\pi G\sigma|k| + c^2 k^2 \equiv f(k)\kappa^2 \qquad (5)$$

If the disk's half-thickness, H, is taken into account, the second term is to be multiplied by a dilution factor $\mathcal{F} = (1 + kH)^{-1}$ (*Genkin and Safronov*, 1975). The frequency, $\omega$, is real only if $f(k) \geq 0$, which can limit the allowable wavenumbers for stable oscillations. If $f < 0$, $\omega$ is imaginary and the perturbation varies as $e^{ikx}e^{nt}$ with

$$n \equiv i\omega = \pm\kappa\sqrt{-f}$$

so that there are both decaying and growing solutions.

Introducing the scale length $h \equiv c/\kappa$, the function $f(k)$ can be rewritten in terms of h and Q as

$$f = 1 - (2\mathcal{F}/Q)|kh| + (kh)^2 \qquad (6)$$

which goes through a minimum at some wavenumber $k_*$ as illustrated in Fig. 1a for the case $\mathcal{F} = 1$. Figure 1b displays the real and imaginary parts of $\omega$ for the two solutions of equation (1) as functions of f. If $f_{min} \equiv f(k_*) \geq 0$, the disk is stable at all wavenumbers. [Note that this is not strictly true if there is any friction in the system (e.g., *Spiegel*, 1972; *Lynden-Bell and Pringle*, 1974; *Ward*, 1974; also section 4 below).] For a two-dimensional disk, $k_*h = Q^{-1}$ and $f_{min} = 1 - (k_*h)^2 = 1 - Q^{-2}$, implying that $Q > 1$ disks are stable. If the dilution factor is included, but we ignore the effect of the disk gravity on its thickness and set $H \sim h$, we get $k_*h(1 + k_*h)^2 = Q^{-1}$ and $f_{min} = 1 - (k_*h)^2[1 + 2(k_*h)]$ for which $Q_{crit} = 0.553$ and $k_*h = 0.658$. In this case, the stability criterion can also be expressed in terms of the average spatial density $\rho = \sigma/2h$ of solids as $\rho_{crit}/\rho_* = 2/3Q_{crit} = 1.21$, where

$$\rho_* \equiv 3M_\odot/4\pi r^3 = 1.42 \times 10^{-7} \left(\frac{r}{AU}\right)^{-3} g/cm^3 \qquad (7)$$

is used as a convenient reference density (e.g., *Safronov*, 1991; *Cuzzi et al.*, 1993). To include the effect of the disk's own gravity on its thickness, we need the additional relationship

$$h/H - H/h \sim 2/Q \qquad (8)$$

which shows that the half-thickness varies from $\sim$h for $Q \gg 1$, wherein the thickness is maintained by the vertical component of the primary's gravity, to $\sim Qh/2 = c^2/2\pi G\sigma$ for $Q \ll 1$, wherein the thickness is due to the disk's self-gravity. In this situation, we find that $k_*h(1 + k_*H)^2 = Q^{-1}$ and $f_{min} = 1 - (k_*h)^2 [1 - 2(k_*H)]$. The critical stability parameter turns out to be $Q_{crit} = 0.767$ with $k_*h = 0.804$, and

$$H/h = \left[\sqrt{Q^2 + 1} - 1\right]/Q = F_Q/Q = 0.339$$

This stability criterion can again be expressed in terms of the average spatial density, $\rho = \sigma/2H$, of material as $\rho_{crit}/\rho_* = (2/3)F_Q^{-1} \approx 2.56$. The higher value does not mean, however, that instability is more difficult to achieve if disk gravity in the vertical direction is included. It is quite the contrary, as indicated by the higher value of $Q_{crit}$. This is why Q is a better stability marker than $\rho_{crit}$. However, given the large uncertainties in the problem, it is usually sufficient to consider $Q \sim 1$ as the stability threshold.

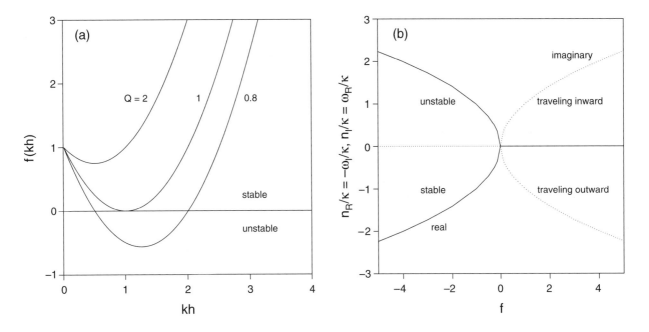

**Fig. 1.** (a) The dispersion relationship, f(kh), for a two-dimensional disk for various values of the Toomre stability parameter, Q. Dynamic instability occurs for f < 0. The minimum of the curve occurs at $k_*h = 1/Q$; its value at that point is $f_{min} = 1 - 1/Q^2$. (b) The real (solid curve) and imaginary (dotted curve) parts of $n = i\omega$ are plotted separately as functions of f. There are two modes whose curves are symmetrical around neutral stability axis, n = 0. For f ≥ 0, these are traveling waves that propagate in both directions, but for f ≤ 0 they are stationary with one mode growing and the other decaying.

## 2.2.  Turbulence

*Goldreich and Ward* (1973) were the first to point out that a turbulent boundary layer of high Reynolds number would develop between a thin disk of particles orbiting at nearly Keplerian velocity and a more slowly moving gas disk partially supported by a pressure gradient (*Whipple*, 1972). The estimated turbulent stress on the disk was used by *GW* to assess its lifetime against orbital decay. Denoting the effective turbulent viscosity, $\nu_t$, the thickness of the Ekman layer is $\delta \sim (\nu_t/\Omega)^{1/2}$, while the viscosity is $\nu_t \sim \delta\Delta v/\mathfrak{R}_e^*$, with $\mathfrak{R}_e^*$ being the so-called critical Reynolds number at which turbulence sets in. The velocity difference between the particle disk and the bulk of the gas disk with sound speed $c_g$ is $\Delta v = \eta r\Omega$, where a value $\eta \sim (c_g/r\Omega)^2 \sim$ few × $10^{-3}$ is typical. Solving self-consistently gives (*GW*)

$$\delta \sim \Delta v/\Omega\mathfrak{R}_e^* \; ; \; \nu_t \sim (\Delta v)^2/\Omega\mathfrak{R}_e^{*2} \qquad (9)$$

*Weidenschilling* (1980, 1988) pointed out that boundary layer turbulence might also stir up the particle layer enough to stabilize it. The velocity $V_t$ and wavenumber $k_t$ of the largest turbulent eddies satisfy the relationships (e.g., *Dubrulle*, 1995)

$$V_t k_t \sim \Omega \; ; \; V_t/k_t \sim \nu_t \qquad (10)$$

Thus, the eddy velocity is roughly $V_t \sim \Delta v/\mathfrak{R}_e^* \sim (\eta/\mathfrak{R}_e^*)r\Omega$.

This turbulence excites random motions among the particles through drag interaction. Resulting particle velocities are size dependent through the so-called stopping time, $t_s$, which increases with some power, q, of the radius of the particles depending on the appropriate drag law, i.e., q = 1 for Epstein's law and q = 2 for Stokes' law (e.g., *Weidenschilling*, 1977). To reasonably good approximation we can combine these into a single expression

$$\Omega t_s \approx \frac{\rho_p R}{\sigma_g}\left(1 + \frac{R}{\lambda}\right) \qquad (11)$$

which is valid up to $R/\lambda \sim O(10)$, where $\lambda \approx (r/AU)^{11/4}$ cm is the mean free path of gas molecules in a minimum mass nebula (*Cuzzi et al.*, 1993). Note that $\Omega t_s$ equals unity for $R_1 \sim (\lambda/2)[(1 + 4\sigma_g/\lambda\rho_p)^{1/2} - 1]$.

For a gas disk surface density of 1500 g/cm² at 1 AU, $R_1 \sim 20$ cm, for $\rho_p \sim 3$ g/cm³ particles. In this case the second (Stokes') term in equation (11) dominates and $\Omega t_s \sim (R/R_1)^2$. For $\sigma_g = 50$ g/cm² at 10 AU, $\rho_p = 1$ g/cm³ particles have $R_1 \sim 50$ cm, and $\Omega t_s \sim R/R_1$, with the first (Epstein's) term dominating in equation (11). The maximum particle velocities can be approximated by $c \sim V_t/Sc$ where $Sc \equiv 1 + St$ is called the Schmidt number (e.g., *Weidenschilling and Cuzzi*, 1993). The quantity $St = \Omega_e t_s$ is the Stokes number, where $1/\Omega_e$ is the eddy turnover time and is usually assumed to be $\sim 1/\Omega$. (It is sometimes convenient to use the reciprocal of the stopping time, $\alpha \equiv t_s^{-1}$, in which case, $St \sim \Omega/\alpha$.)

A disk comprised of particles with dispersion velocity, c, has a $Q = \kappa c/\pi G \sigma \sim (\eta/\mu_d \mathfrak{R}_e^*) Sc^{-1}$, so stability requires

$$Sc \leq (\eta/\mu_d \mathfrak{R}_e^*) \qquad (12)$$

Since Sc cannot be smaller than unity, stability cannot be established if $\eta/\mu_d \leq \mathfrak{R}_e^*$. This is because in this case, even the maximum eddy velocity does not exceed $c_{crit}$. *Goldreich and Ward* (1973) used $\mathfrak{R}_e^* \sim 500$ for the critical Reynolds number, which would make the RHS of equation (12) of order unity for $\eta \sim 0.005$, and $\mu_d \approx 10^{-5}$, which are close to the conditions expected for much of a minimum mass nebula. If $\mathfrak{R}_e^*$ is significantly smaller, stability can be maintained for particles smaller than a limiting size given by $(R/R_1)^q \leq \eta/\mu_d \mathfrak{R}_e^*$, where the appropriate value of q is to be used.

### 2.3.  CDC Model

*Cuzzi et al.* (1993, hereafter *CDC*) developed a two-fluid numerical model of a turbulent disk. The *CDC* model employs state-of-the-art representation of turbulent motions in the computations. Detailed descriptions of gas and particle density profiles as a function of height from the midplane are obtained, which illuminate the behavior of the two populations as shown in Fig. 2. The authors conclude that instability is precluded because particles settle to a subdisk with a spatial density significantly less than $\rho_*$ given in equation (7). However, although many aspects of the *CDC* turbulence model are more sophisticated than previous models, the turbulence strength is still set by selecting an effective critical Reynolds number. The authors prefer $45 < \mathfrak{R}_e^* < 180$ (recall that a lower $\mathfrak{R}_e^*$ implies more vigorous turbulence), and make an effort to justify their choice by quoting experimental results. Nevertheless, there is still much uncertainty in extrapolating from laboratory to solar system scales. [Indeed, in more recent work exploring the possibility of particle damping of turbulence the authors revise their estimate downward to $\mathfrak{R}_e^* = 25$ (*Dobrovolskis et al.,* 1999).]

To illustrate the sensitivity of the results to this parameter, one can obtain an order-of-magnitude estimate of the thickness of the disk by equating the terminal settling velocity, $w_{term} \sim z\Omega^2/\alpha$, to the upward diffusion velocity, $w_{diff} \sim (\nu_t/S_c)|\rho^{-1}\partial\rho/\partial z|$ (*CDC,* equation (62)), and solving for the height at which they are equal. Setting $\partial/\partial z \approx 1/z$ one finds

$$h \approx \left(\frac{\alpha\nu_t}{Sc\Omega^2}\right)^{1/2} \sim \frac{\Delta v}{\Omega\mathfrak{R}_e^*}\left(\frac{\alpha}{Sc\Omega}\right)^{1/2} \qquad (13)$$

This expression corresponds to a

$$Q = 2\rho_*/3\rho \sim \left(\eta/\mu_d\mathfrak{R}_e^*\right)/\sqrt{Sc\Omega t_s}$$

which appears similar to the estimate in the last section; however, there are some subtle but important differences. The *CDC* model does not adopt equation (10); it is argued that the velocity of the largest eddy is of order

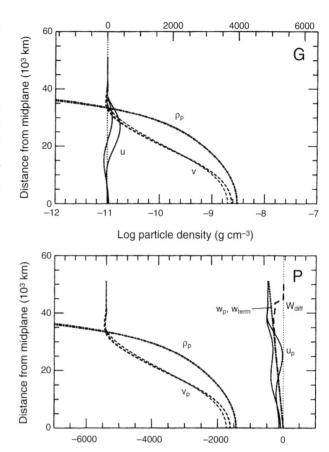

**Fig. 2.**  Results of the *CDC* model for 10-cm-radius particles at 1 AU. The radial, longitudinal, and vertical velocities of the particles are $u_p$, $v_p$, and $w_p$ respectively. Particle and gas radial velocities are the solid curves; particle and gas azimuthal (relative) velocities are the short dashed curves. Particle velocities are shown relative to Keplerian velocity, and gas velocities are shown relative to pressure-supported circular motion. The heavy dotted and dashed curves are the particle terminal velocity $w_F$ and diffusion velocity $w_{diff}$.

$$V_t \sim \Delta v/\sqrt{\mathfrak{R}_e^*}$$

so that $V_t k_t \sim \Omega\mathfrak{R}_e^* \equiv \Omega_e$, implying a much shorter eddy turnover time $\sim 1/\Omega_e$. This changes the Stokes parameter to $St = \Omega_e t_s = \mathfrak{R}_e^*\Omega t_s$ and introduces another $\mathfrak{R}_e^{*-1/2}$ dependency into Q when $\Omega t_s \gg 1/\mathfrak{R}_e^*$. Accordingly, stability requires

$$\Omega t_s < (\eta/\mu_d\mathfrak{R}_e^{*3/2}) \qquad (14)$$

Although smaller critical Reynolds numbers are adopted in *CDC*, this is largely offset by the increased sensitivity to $\mathfrak{R}_e^*$, so that the upper limit for stability is still R ~ few tens of centimeters.

The numerical results of *CDC* indicate that for 10-cm particles at 1 AU with $\mathfrak{R}_e^* = 180$ the midplane density is $\rho \sim$

$3.2 \times 10^{-9}$ g/cm³. Because this result contains no disk self-gravity, it corresponds to $Q = (2/3)\rho_*/\rho \sim 65$. For 60-cm particles, they employ $\mathfrak{R}_e^* = 55$ and the density increases by a factor of ~14 to $\rho = 4.5 \times 10^{-8}$, and $Q < 5$. There is an increase in St by a factor of ~36 because the appropriate drag law is Stokes at the 1-AU environment. Note that using the scaling argument of equation (13), had $\mathfrak{R}_e^* = 180$ been used as in the R = 10-cm case, Q would be very close to unity.

The *CDC* paper also presents results for 20-cm and 60-cm particle disks at 10 AU, although it is unclear what value of the Reynolds number is used in these simulations. The respective densities are $\rho \sim 1 \times 10^{-10}$ g/cm², $3 \times 10^{-10}$ g/cm³, while the reference density is $\rho_* = 1.4 \times 10^{-10}$ g/cm³. This implies Q values of {0.93, 0.31}. In this nebula environment, the drag law is Epstein's, accounting for the factor of 3 density difference for the two values of R.

Unfortunately, the Reynolds number is not varied much in the *CDC* computations. It would be very useful if the threshold values of $\mathfrak{R}_e^*$ for marginal stability for various particle sizes and orbital distances could be determined.

## 2.4. Density Stratification

If instead particles are much smaller than a centimeter, the predicted thickness of the layer may become so great that the vertical shear in the velocity field, dv/dz, is insufficient to overcome the stabilizing influence of density stratification, dρ/dz. *Sekiya* (1998) has developed an analytic prescription for the vertical structure of the planetesimal layer based on the Richardson criterion (e.g., *Chandrasekar*, 1961). Assuming small particles (i.e., ≤1 mm), the particle layer is assumed to be maintained near the onset of turbulence by a local vertical density gradient that keeps the Richardson number

$$Ri \equiv \frac{-g(\partial\rho/\partial z)}{(\rho_g + \rho)(\partial v/\partial z)^2} \quad (15)$$

at its critical value of 1/4, where g is the vertical gravity, and $\rho$, $\rho_g$ are the spatial densities of solids and gas respectively. This leads to a relationship between the surface density of solid material, $\sigma$, and the ratio $\chi \equiv \rho/\rho_g$ of the spatial densities of solids and gas in the midplane of the disk, *viz.*, $\sigma = \eta r \rho_g D(q,\chi)$, where $q \equiv 4\pi G \rho_g/\Omega^2$, $X \equiv q + 1/(1 + \chi)$, and

$$D(q,\chi) \equiv (1 + q)\ln\left[\frac{1 + q + \sqrt{(1+q)^2 - X^2}}{X}\right] \quad (16)$$
$$- \sqrt{(1+q)^2 - X^2}$$

The value of D ranges from $D(q, 0) = 0$ to $D(q, \infty) \approx \ln(2/q)$ for $q \ll 1$, so that the surface densities greater than $\sigma_{crit} \sim \eta r \rho_g \ln(M_\star/2\pi\rho_g r^3)$ will have "infinite" midplane densities. For the values employed above at r = 1 AU, $\rho_g \sim 1 \times 10^{-9}$ g/

cm³, $\eta \sim 0.005$, one finds $\sigma_{crit} \sim 340$ g/cm², while at r = 10 AU, $\rho_g \sim 2 \times 10^{-12}$ g/cm³, and $\sigma_{crit} \sim 5.8$ g/cm². However, these values would decrease if the gas density diminishes, and *Sekiya* (1998) suggests that instabilities might have been triggered during the dissipation of the gaseous disk.

## 3. GAS-INDUCED DIFFERENTIAL VELOCITIES

*Weidenschilling* (1995) recognized that even with its low Reynolds numbers, the *CDC* model still implies instability could develop in a disk of meter-sized particles, especially in the outer parts of the disk. However, he raised a second issue, *viz.*, whether the spread of orbital decay velocities due to size dependent gas drag could itself inhibit the growth of unstable modes, even when $c < c_{crit}$. Again, the bulk of the gas disk is partially supported by a radial pressure gradient, and consequently rotates at a rate $\Omega_{gas} \sim (1 - \eta)\Omega$ slightly slower than Keplerian (*Whipple*, 1972). The much-denser particles do not enjoy this pressure support, and in the particle layer the gas is partially dragged with the solids depending on the particle sizes and the degree of mass loading (see Fig. 2 of *CDC*; also *Nakagawa et al.*, 1986). Accordingly, $\eta \le 10^{-4}$ will be adopted as a reasonable estimate near the midplane. Drag results in a radial drift rate (e.g., *Goldreich and Ward*, 1973; *Weidenschilling*, 1977)

$$\dot{r} \approx -2\eta r\Omega\left(\frac{\Omega t_s}{1 + (\Omega t_s)^2}\right) \quad (17)$$

*Weidenschilling* (1995) states that "The drag induced velocities are size dependent; they would not produce any net velocity dispersion in a system of identical particles. In any realistic size distribution produced by coagulation, particles will have a range of radial velocities comparable to the mean value." Using this rule of thumb, he argued that instability will be suppressed as long as $\dot{r}(\bar{R}) > c_{crit}$, with the diversity of decay rates corresponding directly with the quantity c in the dispersion relationship (equation (5)). The ratio of the lead term of equation (17) to $c_{crit}$ can be written $2\eta/\mu_d \sim O(10^2)$, implying that $\dot{r} \sim c_{crit}$ if $\Omega t_s \sim O(10^{\pm2})$.

To test this assertion more rigorously, a drag acceleration term, $\mathbf{F}_{drag}/m = -\alpha(\mathbf{v} - r\Omega_g\hat{\mathbf{e}}_\theta)$, can be added to the hydrodynamic equation of motion, whose first-order perturbation equations take the form

$$\frac{\partial u}{\partial t} - 2\Omega v = -\frac{\partial\phi}{\partial r} - \alpha u \ ;$$
$$\frac{\partial v}{\partial t} + 2Bu = -\alpha(v + r\Delta\Omega) \quad (18)$$

where u, v are the first-order radial and azimuthal velocity perturbations; $\Delta\Omega \equiv \Omega - \Omega_g = \eta\Omega$, $B = \Omega/4$ is the vorticity; $\alpha$ is the reciprocal of stopping time; and $\phi$ is the disk's self-gravity potential. The disk is assumed to be dynamically cold (c = 0) in order to rely only on a possible stabilizing effect of the drag-induced velocities. Setting $\partial/\partial t = \partial/\partial r \to 0$ yields the familiar secular solution to equation (18)

$$u_{sec} = -2\eta r\Omega\left(\frac{\alpha\Omega}{\kappa^2 + \alpha^2}\right) ;$$

$$v_{sec} = -\eta r\Omega\left(\frac{\alpha^2}{\kappa^2 + \alpha^2}\right) \tag{19}$$

The first expression is identical to equation (17) and describes a radial drift rate that peaks at $St = 1$; the azimuthal velocity approaches the limiting value $-\eta r\Omega$ as $\alpha \to \infty$, implying a particle co-moving with the gas.

### 3.1. Unimodal Disk

Consider first a unimodal disk. Since equation (18) is linear, solutions due to $\partial\phi/\partial r$ will not couple to $u_{sec}$, $v_{sec}$. Finding their interaction requires keeping some higher-order cross terms, the most important of which are due to the contribution of $u_{sec}$ to the convective derivative (G. Stewart, personal communication). Perturbation velocities can be written as a combination of the secular drag terms and oscillatory terms, $\{u, v\} = \{u_{sec}, v_{sec}\} + \{u_k, v_k\}e^{i(kr + \omega t)}$, where the oscillatory terms obey

$$i\omega u_k + iku_{sec}u_k - 2\Omega v_k = -\partial\phi/\partial r - \alpha u_k ;$$

$$i\omega v_k + iku_{sec}v_k + 2Bu_k = -\alpha v_k \tag{20}$$

These can be solved together with the linearized continuity and Poisson's equations

$$i\omega\sigma + iku_{sec}\sigma + ik\sigma_o u_k = 0 ;$$

$$\partial\phi/\partial r = -2\pi iG\sigma\,\text{sgn}(k) \tag{21}$$

where $\sigma_o$, $\sigma$ are the unperturbed and the first-order surface densities. This leads to the dispersion relationship

$$(i\omega' + \alpha)\left[\kappa^2 - 2\pi G\sigma_o|k| + i\omega'(i\omega' + \alpha)\right] - \alpha\kappa^2 = 0 \tag{22}$$

where $\omega' \equiv \omega + ku_{sec}$. The oscillatory term behaves as $e^{ik(r - u_{sec}t)}e^{i\omega't}$. As $\alpha \to 0$, equation (22) splits into a neutrally stable mode, $\omega' = 0$, and solutions given by equation (5) with $c = 0$. If $\alpha \ll \kappa$, the usual unstable branch $f < 0$ has

$$n \approx \kappa\sqrt{-f} - \frac{\alpha}{2}\left(1 + \frac{1}{f}\right) \tag{23}$$

so that the growth rate of the unstable branch is decreased for $f < -1$, but only modestly. (Note that for $f > -1$ the rate is increased.)

### 3.2. Bimodal Disk

Now consider a second population of particles of surface density $\Sigma$, perturbation velocities $U$, $V$, self-gravity potential $\Phi$, and its own size-dependent drag parameter, $A$. The relevant equations for axisymmetric modes in this population are

$$i\tilde{\omega}U_k - 2\Omega V_k = -\partial(\phi + \Phi)/\partial r - AU_k ;$$

$$i\tilde{\omega}V_k + 2BU_k = -AV_k \tag{24}$$

$$i\tilde{\omega}\Sigma + ik\Sigma_o U_k = 0 ;$$

$$\partial\Phi/\partial r = -2\pi iG\Sigma\,\text{sgn}(k) \tag{25}$$

where $\tilde{\omega} \equiv \omega + kU_{sec}$ with $U_{sec}$ denoting the orbital decay rate of the $\Sigma$ population. Note that the radial equation couples to the first population through the combined gravitational potential. In a like manner, the combined potential $\phi + \Phi$ is to be used in the radial equation (20). The dispersion relationship for the full set of equations is then

$$i\omega'i\tilde{\omega}\left[\left(i\omega' + \alpha\right)^2 + \kappa^2\right]\left[\left(i\tilde{\omega} + A\right)^2 + \kappa^2\right] -$$
$$2\pi G\sigma_o|k|i\tilde{\omega}\left(i\omega' + \alpha\right)\left[\left(i\tilde{\omega} + A\right)^2 + \kappa^2\right] \tag{26}$$
$$-2\pi G\Sigma_o|k|i\omega'\left(i\tilde{\omega} + A\right)\left[\left(i\omega' + \alpha\right)^2 + \kappa^2\right] = 0$$

This is a six-degree polynomial implying six possible modes of behavior, a full exploration of which is beyond the scope of this chapter. However, if we assume a weak drag case where $A \ll \tilde{\omega}$, $\alpha \ll \omega'$, we can approximate equation (26) with

$$\left(\kappa^2 - \omega'^2\right)\left(\kappa^2 - \tilde{\omega}^2\right) - 2\pi G\sigma_o|k|\left(\kappa^2 - \tilde{\omega}^2\right) -$$
$$2\pi G\Sigma_o|k|\left(\kappa^2 - \omega'^2\right) = 0 \tag{27}$$

In the limits $\{\sigma_o \to 0; \Sigma_o \to 0\}$ the usual stability relationships are recovered

$$\{\omega'_o ; \tilde{\omega}_o\} = \pm\sqrt{\kappa^2 - 2\pi G\{\sigma_o; \Sigma_o\}|k|}$$

However, unstable wavelengths are stationary with respect to the moving frame of each population. Thus, seen in a frame moving at the average velocity $\bar{u} = (u_{sec} + U_{sec})/2$ the unstable maxima move in opposite directions at the rate $u_{sec} - \bar{u} = \bar{u} - U_{sec} = (u_{sec} - U_{sec})/2 \equiv \Delta u$. This is the essence of Weidenschilling's mechanism; if $\Delta u$ is too large the populations will decouple and not benefit from their combined gravity.

Equation (27) is a quartic equation that can be solved algebraically, albeit rather tediously. However, it is simple to demonstrate the behavior for the special case $\Sigma_o = \sigma_o = \sigma_T/2$ by defining $\bar{\omega} \equiv (\omega' + \tilde{\omega})/2$, for which equation (27) reduces to

$$\bar{\omega}^4 - 2\left[\kappa^2 - \pi G\sigma_T|k| + (k\Delta u)^2\right]\bar{\omega}^2 +$$
$$\left(\kappa^2 - (k\Delta u)^2\right)\left[\kappa^2 - 2\pi G\sigma_T|k| - (k\Delta u)^2\right] = 0 \tag{28}$$

which is quadratic in $\bar{\omega}^2$. The solutions read

$$\bar{\omega}^2 = \kappa^2 - \pi G\sigma_T |k| + k^2(\Delta u)^2 \pm$$
$$\pi G\sigma_T |k| \sqrt{1 + \left[ \kappa^2 - \pi G\sigma_T |k| \right] (2\Delta u / \pi G\sigma_T)^2} \quad (29)$$

If $\Delta u \ll \pi G\sigma_T/\kappa = c_{crit}$, the lower root is $\bar{\omega}^2 \approx \kappa^2 - 2\pi G\sigma_T |k| + k^2(\Delta u)^2$ and $\Delta u$ seems to play a role similar to a dispersion velocity as suggested by *Weidenschilling* (1995). However, low values of $\Delta u$ are insufficient to stabilize the disk. If we raise $\Delta u \gg c_{crit}$, $\bar{\omega}^2$ will have an imaginary part if the expression under the radical goes negative. This behavior can be traced to instabilities in the individual populations, which we assumed were cold. Concentrating on wavenumbers small enough to avoid these, we can write $\bar{\omega}^2 \approx (\kappa - |k\Delta u|)^2 - \pi G\sigma_T |k|$. For any $\Delta u$, $\bar{\omega}^2$ becomes negative for a range of $k$ near $|k| \approx \kappa/|\Delta u|$, with a corresponding growth rate of order

$$n_{inst} \approx \left| \frac{c_{crit}}{\Delta u} \right|^{1/2} \Omega \quad (30)$$

Thus the relative velocity $\Delta u$ does not by itself ensure stability, although the range of unstable wavelengths is restricted.

On the other hand, we should mention that there is another way in which drag-induced differential velocities could enter the stability picture. Collisions between subpopulations could be a source of $c$ by randomizing a portion of the directed velocity field. If collisions are very infrequent the random component will be damped by drag and the situation will not change much. If collisions are too frequent, i.e., with a frequency $\omega_c \gg \alpha$, size-dependent velocity differences will not have time to develop completely. However, for comparable collision vs. stopping timescales, a nonnegligible dispersion velocity may well be sustainable, and this important issue should also be further investigated. Nevertheless, even if a convincing case can eventually be made that fast dynamic modes are suppressed, there are other evolutionary routes for the disk involving collective behavior to which we now turn.

## 4. DESTABILIZED NEUTRAL MODES

In his Kuiper prize lecture, *Safronov* (1991) argued that friction due to gas drag destabilizes the disk by modifying the dispersion relation to $n(n + \alpha) = -\kappa^2 f(k)$. Solving for

$$n = \left[ \pm\sqrt{\alpha^2 - 4\kappa^2 f} - \alpha \right]/2$$

growth still requires $f < 0$. However, this is not the strongest case that can be made for a destabilizing influence of gas drag. If gas is present, it not only can have a stirring effect, but introduces a source of friction that alters the stability of neutral modes (*Spiegel*, 1972). During their initial study, *GW* derived equation (22) (see *Ward*, 1974), which

can be rearranged to read

$$n(n + \alpha) - \frac{\alpha\kappa^2}{n + \alpha} = -\kappa^2 f(k) \quad (31)$$

This relationship was later independently found by *Coradini et al.* (1981). Figure 3 plots the real ($n_R = -\omega_I$) and imaginary ($n_I = \omega_R$) parts of n as a function of f for the case $\alpha/\Omega = 1$, superimposed on the $\alpha/\Omega = 0$ case of Fig. 1b. Compared to Safronov's expression, equation (31) contains an additional term that forces the LHS to approach $-\kappa^2$ instead of vanishing as $n \to 0$. Consequently, the upper branch is always constrained to pass through the point $(f, n) = (1, 0)$ for any value of $\alpha$. Note that the range of traveling waves is extended to lower f but have associated real parts of n that damp them. However, there is now a stationary mode that damps above $f = 1$, but grows for smaller values. The additional range $0 < f < 1$ destabilized by friction encompasses wavenumbers approaching zero ($0 < k < k_{stable}$), i.e., long wavelengths are rendered unstable. Expanding equation (31) about $f = 1$ to first order in n yields (*Ward*, 1974)

$$n \approx \kappa \left( \frac{\alpha\kappa}{\alpha^2 + \kappa^2} \right)(1 - f) =$$
$$\Omega \left( \frac{\Omega t_s}{1 + (\Omega t_s)^2} \right) kh(2\mathcal{F}/Q - kh) \quad (32)$$

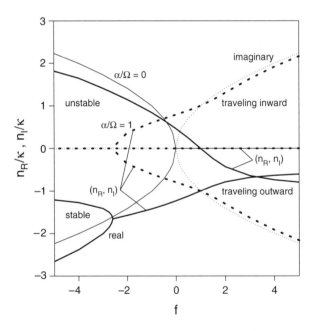

**Fig. 3.** Real and imaginary parts of $n = i\omega$ for $\alpha/\Omega = 1$ superimposed on gas free $\alpha/\Omega = 0$ case of Fig. 1b. Inward and outward traveling waves extend to negative values of f, but are damped. At still more negative f these modes are stationary ($n_I = 0$) and damp ($n_R < 0$) at two different rates. In addition, there is a third, nontraveling mode that damps for $f > 1$, but grows ($n_R > 0$) for $f < 1$. The unstable range thus includes wavenumbers to zero for which $f \to 1$.

Thus n approaches zero for both high and low values of $\Omega t_s$, with the fastest growth occurring for $t_s \sim 1/\Omega$, i.e., when the stopping time is comparable to the orbital period. Indeed, this the same functional dependence on $\Omega t_s$ exhibited by the orbital decay of particles (equation (17)). Modes have time to grow if $nr/\dot{r} \gg 1$, or $kh(2\mathcal{F}/Q - kh) \gg 2\eta$. The fastest growing mode is still $k_*$ for which $f(k)$ is a minimum, and for a two-dimensional disk has $nr/\dot{r} \approx (2\eta Q^2)^{-1}$. Substantial growth is possible for

$$Q < 1/\sqrt{2\eta} \sim \text{few} \times 10$$

## 5.    DRAG INSTABILITY

Recently, another form of instability has been described that does not rely on self-gravity but is instead generated by the very drag stresses it has been argued prevent gravitational instability (*Goodman and Pindor*, 1999; *Ward and Weidenschilling*, 2000). The stress across the gas-particle interface due to boundary layer turbulence has a maximum strength of order (*GW*)

$$S \sim \rho_g \nu(dv_g/dz) \approx \rho_g(\Delta v)^2/\mathfrak{R}_e^* \quad (33)$$

Assuming collisions in the particle layer are frequent enough for it to respond as a unit, the orbital decay rate is $u_{sec} \sim 4S/\sigma\Omega$, where the stress is assumed to be applied to both sides of the sheet. If an annulus becomes overdense, its decay rate is less than adjacent "upstream" regions and additional material will overtake it. This increases the surface density and further slows the decay rate of the annulus. A simple demonstration of this feedback behavior can be made by replacing the drag acceleration used to derive equation (18) by $\sim 2S\hat{e}_\theta/\sigma$. Together with the continuity equation, this yields a dispersion relationship of the form (*Ward and Weidenschilling*, 2000)

$$\hat{\omega}^3 - \kappa^2 f\hat{\omega} + ku_{sec}/\Omega = 0 \quad (34)$$

where $\hat{\omega} \equiv \omega'/\Omega$, and we have set $\kappa = \Omega$. This relationship was first given by *Goodman and Pindor* (1999) in the nongravitating, cold-disk limit $f \to 1$, and is a reduced cubic that has three real roots if $f \geq 3\,|ku_{sec}/2\Omega|^{2/3}$. This condition defines a modified stability variable (*Ward and Weidenschilling*, 2000)

$$f_* \equiv 1 - 3\,|ku_{sec}/2\Omega|^{2/3} -$$
$$2\pi G\sigma|k|/\Omega^2 + (kc/\Omega)^2 \quad (35)$$

which is plotted in Fig. 4 for a $Q = 1$, marginally stable disk, i.e., $c = c_{crit}$. Secular velocities and wavenumbers have been normalized by $\hat{u} \equiv u_{sec}/c_{crit}$ and $\hat{k} \equiv kc_{crit}/\Omega$ respectively. A widening range of wavenumbers becomes unstable with increasing $\hat{u}$. A new critical $Q_*$ can be defined by setting equation (35) and its derivative $df_*/dk$ equal to zero. This results in

$$Q_* = \frac{1}{\hat{k}_*}\sqrt{(1 + \hat{k}_*)/2} \quad \text{where}$$
$$\hat{k}_* + 2^{1/3}\,|\hat{u}|^{2/3}\,\hat{k}_*^{2/3} = 1 \quad (36)$$

$\hat{k}_*$ being the wavenumber for which $f_*$ is a minimum. These values are shown in Fig. 5 as a function of $\hat{u}$. As $\hat{u} \to 0$, $\hat{k}_* \to 1$ and $Q_* \to 1$ as required, but as $\hat{u} \gg 1$, $\hat{k}_* \to 1/\sqrt{2}\hat{u}$ and $Q_* \to 1/\sqrt{2}\hat{u}\hat{k}_* \to \hat{u} \gg 1$. This implies that the particle dispersion velocity $c$ must exceed the secular decay velocity to insure stability. Recalling from section 2.2 that $c \approx V_t/Sc \sim \Delta v/\mathfrak{R}_e^*Sc$ and using equation (33) to write $u_{sec} \sim 4\rho_g(\Delta v)^2/\sigma\Omega\mathfrak{R}_e^*$, the ratio

$$Q/\hat{u} = c/u_{sec} \approx \frac{1}{4\,Sc}\left(\frac{\sigma}{\eta r\rho_g}\right) \quad (37)$$

For the values used by the *CDC* model at 10 AU, i.e., $\sigma \sim 0.6$ g/cm$^2$, $\rho_g \sim 2 \times 10^{-12}$ g/cm$^3$, $\eta \sim 0.005$, we find $Q/\hat{u} \sim 0.1$ Sc, implying instability. This criterion is valid provided $\hat{u} \ll \Omega^2/2\pi G\rho_g$ at 10 AU. (A somewhat different criterion applies at 1 AU; see *Ward and Weidenschilling*, 2000.)

When $f_* < 0$, the solutions to equation (34) read

$$\hat{\omega} = |w|^{1/3}\,e^{i(\varphi,\varphi \pm 2\pi)/3} + \frac{f}{3}|w|^{-1/3}\,e^{-i(\varphi,\varphi \pm 2\pi)/3} \quad (38)$$

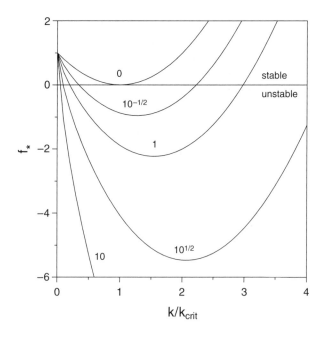

**Fig. 4.** Plot of drag-modified stability variable $f_*$ vs. k for several values of $u_{sec}$. The wavenumber is normalized to $k_{crit} \equiv \Omega/c_{crit}$ and the orbital decay velocity is normalized to $c_{crit} \equiv \pi G\sigma/\Omega$. The disk is marginally stable in the absence of drag, i.e., $c = c_{crit}$. Instability sets in when $f_* < 0$. As $\hat{u}$ increases, a widening range of wavenumbers is unstable.

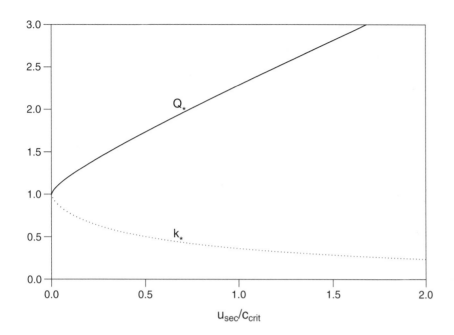

**Fig. 5.** Critical Toomre stability number, $Q_*$, and wavenumber $\hat{k}_*$ for a disk subject to a boundary layer drag that induces secular orbit decay at a rate $u_{sec} = \hat{u}c_{crit}$.

where

$$w = \left[ \left( ku_{sec}/2\Omega \right)^2 - \left( f/3 \right)^3 \right]^{1/2} - ku_{sec}/2\Omega \; ;$$
$$\varphi = \left[ 1 - \mathrm{sgn}(w) \right]\pi/2 \tag{39}$$

and the growth rate of the unstable mode is given by the imaginary part

$$n = -\hat{\omega}_I\Omega = -\Omega\left[ |w|^{1/3} - (f/3)|w|^{-1/3} \right]$$
$$\sin(\varphi, \varphi \pm 2\pi)/3 \tag{40}$$

If $f_* \ll 0$, $(f/3)^3 \ll (\hat{k}\hat{u}/2)^2$, and the growth rate of the unstable mode is

$$n \approx \frac{\sqrt{3}}{2}\left|\hat{k}\hat{u}\right|^{1/3}\left( 1 - \frac{f/3}{\left|\hat{k}\hat{u}\right|^{2/3}} \right)\Omega \tag{41}$$

Growth is faster than the disk's characteristic decay rate $u_{sec}/r = \hat{u}c_{crit}/r = (\hat{u}/rk_{crit})\Omega$ for $rk \geq (u_{sec}/r\Omega)^2$, which is satisfied throughout most of the unstable range.

## 6. DISCUSSION

It has been over a quarter century since gravitational instability in the planetesimal disk was suggested as a mechanism to account for planetesimal growth through the centimeter to kilometer size range. The essence of the mechanism is that collective behavior may promote accumulation even if individual binary collisions do not result in immediate accretion. The motivation behind the proposal was the apparent difficulty particles have in adhering to one another in this size range as alluded to in section 1. If this is not a problem, then whether or not the disk is collectively unstable becomes an ancillary question of much less significance; accretion could proceed quickly in either case. In this context, it is important to realize that disproving the instability hypothesis does not itself solve the accretion problem.

Even so, Weidenschilling's long-standing caution that turbulence generated at the particle-gas layer interface could excite particle dispersion velocities and possibly stabilize the planetesimal disk is an appropriate one. However, the case for stability is not especially robust either. Whether the disk is predicted to be stable or not depends on quantities such as the Reynolds number, surface density, particle sizes, etc., that are known only to an order of magnitude, if that. Reasonable values concerning disk conditions can lead to either conclusion, especially in the outer parts of the disk. Size-dependent orbital decay velocities due to drag may also work to stabilize the disk, but more work is needed on this problem to settle the issue. If gravitational instability is avoided by having dispersion velocities in excess of $c_{crit} \sim \pi G\sigma/\Omega \sim \mu_d r\Omega$, a better case must be made that the requisite adhesion can indeed occur at such velocities among particles with sizes extending to a fraction of a kilometer in radius. There are nongravitational sticking mechanisms, but these tend to work in a small size range. Further experimental and theoretical work is needed on this important topic.

Even if fast gravitational modes are suppressed, collective particle behavior could manifest itself through frictionally destabilized neutral modes (*Ward*, 1974; *Coradini et al.*, 1981) or the newly described drag instability (*Goodman and*

*Pindor,* 1999; *Ward and Weidenschilling,* 2000). These can persist at Q values well in excess of unity. The axisymmetric modes take the form of annuli of increasing density that grow on a timescales given by equations (32) and (41). And, as the local surface density increases, the Toomre stability number will decrease for a given dispersion velocity. An increase in σ by a factor Q may eventually allow local fast modes to be triggered. It would be most worthwhile to develop a nonlinear treatment of these types of instabilities to see whether disk fragmentation can be induced.

Indeed, the arguments presented here are based on linear stability analysis and cannot adequately describe the behavior into the nonlinear regime. *GW* attempted to sketch out traits of the nonlinear behavior via dimensional arguments in the case of an optically thick disk, $\tau \sim \sigma/\rho_p R \gg 1$, for which the particle collision frequency, $\omega_c \sim \Omega\tau$, is high. This included an estimate of a fragment scale with low enough internal angular momentum to contract to solid density if particle dispersion velocities were collisionally damped during the collapse, i.e., the so-called first generation objects. [*CDC* incorrectly refers to this one particular aspect as the Goldreich-Ward instability (GWI), but a more common usage of GWI is to refer in general to the notion of gravitational instability in a planetesimal disk.] However, since gas turbulence can excite small-particle dispersion velocities, a sufficiently thin debris disk may require particles some ten of centimeters in size, resulting in an optical depth below unity. This invalidates the conditions assumed by *GW,* and leaves the nonlinear behavior as largely undetermined. On the other hand, this problem may now be amenable to modern computational methods, and should be a high-priority undertaking by numerical modelers. Given the advantages of collective mechanism, it seems prudent to keep them under active consideration, and to explore conditions that might permit their operation in some form.

***Acknowledgments.*** This work was supported by funds from NASA's Origins of Solar Systems and Planetary Geology and Geophysics Programs. The author thanks G. Stewart, S. J. Weidenschilling, R. M. Canup, and H. Levison for several constructive comments that have helped clarify the presentation.

# REFERENCES

Chandrasekar S. (1961) *Hydrodynamic and Hydromagnetic Stability.* Oxford Univ., Oxford.

Coradini A., Federico C., and Magni G. (1981) Formation of planetesimals in an evolving protoplanetary disk. *Astron. Astrophys., 98,* 173–185.

Cuzzi J. N., Dobrovolskis A. R., and Champney J. M. (1993) Particle-gas dynamics in the mid-plane of the protoplanetary nebula. *Icarus, 106,* 102–134.

Dobrovolskis A. R., Dacles-Mariani J. S., and Cuzzi J. N. (1999) Production and damping of turbulence by particles in the solar nebula. *J. Geophys. Res.,* in press.

Dubrulle B., Morfill G., and Sterzik M. (1995) The dust subdisk in the protoplanetary nebula. *Icarus, 114,* 237–246.

Genkin I. L. and Safronov V. S. (1975) Instability of rotating gravitating systems with radial perturbations. *Astron. Zh., 52,* 306–315.

Goldreich P. and Ward W. R. (1973) The formation of planetesimals. *Astrophys. J., 183,* 1051–1061.

Goodman J. and Pindor B. (1999) Secular instability and planetesimal formation in the dust layer. *Icarus,* submitted.

Lynden-Bell D. and Pringle J. B. (1974) The evolution of viscous discs and the origin of the nebular variables. *Mon. Not. R. Astron. Soc., 168,* 603–637.

Nakagawa Y., Sekiya M., and Hayashi C. (1986) Settling and growth of dust particles in a laminar phase of a low-mass solar nebula. *Icarus, 67,* 375–390.

Safronov V. S. (1969) *Evolution of the Protoplanetary Cloud and the Formation of the Earth and Planets.* Nauka, Moscow. (Translated in English as NASA TTF-677.)

Safronov V. S. (1991) Kuiper Prize lecture: Some problems in the formation of the planets. *Icarus, 94,* 260–271.

Sekiya M. (1983) Gravitational instabilities in a dust-gas layer and formation of planetesimals in the solar nebula. *Prog. Theor. Phys., 69,* 1116–1130.

Sekiya M. (1998) Quasi-equilibrium density distributions of small dust aggregations in the solar nebula. *Icarus, 133,* 298–309.

Spiegel E. (1972) Some fluid dynamical problems in cosmogony. In *Proc. Symposium on the Origin of the Solar System,* Nice, April 3–7, 1972. Edition du Centre National de la Recherche Scientifique.

Toomre A. (1969) On gravitational instability of a disk of stars. *Astrophys. J., 139,* 1217–1238.

Ward W. R. (1974) The formation of the solar system. In *Frontiers in Astrophysics* (E. H. Avrett, ed.), pp. 1–40. Harvard Univ., Cambridge.

Ward W. R. and Weidenschilling S. J. (2000) Drag instability in the planetesimal disk. *Astrophys. J. Lett.,* submitted.

Weidenschilling S. J. (1977) Aerodynamics of solid bodies in the solar nebula. *Mon. Not. R. Astron. Soc., 108,* 57–70.

Weidenschilling S. J. (1980) Dust to planetesimals. *Icarus, 44,* 172–189.

Weidenschilling S. J. (1984) Evolution of grains in a turbulent solar nebula. *Icarus, 60,* 555–567.

Weidenschilling S. J. (1988) Formation processes and timescale for meteorite parent bodies. In *Meteorites and the Early Solar System* (J. F. Kerridge and M. S. Matthews, eds.). Univ. of Arizona, Tucson.

Weidenschilling S. J. (1995) Can gravitational instability form planetesimals? *Icarus, 116,* 433–435.

Weidenschilling S. J. (1997) The origin of comets in the solar nebula: A unified model. *Icarus, 127,* 290–306.

Weidenschilling S. J. and Cuzzi J. N. (1993) Formation of planetesimals in the solar nebula. In *Protostars and Planets III* (E. H. Levy and J. L. Lunine, eds.). Univ. of Arizona, Tucson.

Weidenschilling S. J., Donn B., and Meakin P. (1989) Physics of planetesimal formation. In *The Formation and Evolution of Planetary Systems* (H. A. Weaver and L. Danly, eds.). Cambridge Univ.

Whipple F. L. (1972) On certain aerodynamic processes for asteroids and comets. In *From Plasma to Planet: Proceedings of the Nobel Symposium 21* (A. Elvius, ed.). Wiley, New York.

# Formation of Planetary Embryos

**S. J. Kortenkamp**
*Carnegie Institution of Washington*

**E. Kokubo**
*University of Tokyo*

**S. J. Weidenschilling**
*Planetary Science Institute*

We review simulations of the embryo formation stage in the standard model of terrestrial planet formation. This stage is distinguished by the emergence of a few dozen $10^{26}$–$10^{27}$-g (Mercury- to Mars-sized) planetary embryos from a swarm of $10^{12}$–$10^{18}$-g (kilometer-sized) planetesimals distributed between 0.5 and 1.5 AU. Near 1 AU, in a narrow accumulation zone of semimajor axis width $\Delta a \approx 0.05$ AU, formation of a planetary embryo is characterized by runaway growth of the largest body in the zone on a timescale of order $10^5$ yr. This is a robust finding corroborated by several independent research groups using distinct numerical methods: gas dynamic statistical simulations, direct N-body orbital integrations, and hybrid statistical simulations refined using results from N-body integrations. The fundamental mechanism that facilitates runaway growth is the exchange of random orbital kinetic energy during gravitational encounters between large and small bodies in the planetesimal swarm. This tendency toward energy equipartition, dubbed "dynamical friction," lowers the relative velocities of the larger bodies with respect to each other, enhancing their mutual gravitational cross-sections and thereby increasing the rate at which they accumulate one another.

## 1. INTRODUCTION

Most conventional models of planet formation begin with a protoplanetary disk of gas and dust orbiting a central protostar. Direct observational confirmation of this critical initial state is found in the stunning crop of "proplyds" (**pro**to**pl**anetar**y d**isks) observed in the star-forming regions of the Orion nebula (*O'Dell et al.*, 1993; *McCaughren and O'Dell,* 1996) and the Taurus molecular cloud (*Burrows et al.,* 1996; see also brief reviews by *Beckwith and Sargent,* 1996, and *O'Dell and Beckwith,* 1997). More mature circumstellar disks may betray the presence of embedded planetary systems by displaying structure similar to that found in the tenuous debris disk in our own solar system (the zodiacal cloud). These planetary signatures include warps, central offsets, dust bands, resonant rings, and dust clouds (*Dermott et al.,* 1999). Recent observations have already revealed some of these features, such as a warp in the disk around β Pictoris (*Burrows et al.,* 1995; *Mouillet et al.,* 1997) and the asymmetric double-lobed features in the disks around HR 4796A (*Jayawardhana et al.,* 1998; *Koerner et al.,* 1998) and Fomalhaut (*Holland et al.,* 1998). Follow-up observations of the HR 4796A disk (*Telesco et al.,* 2000) combined with modeling (*Wyatt et al.,* 1999) suggest that the brightness asymmetry is an indicator of an unseen planet. Submillimeter images of a dust ring around the star ε Eridani (*Greaves et al.,* 1998) have also revealed substructure within the ring that could be due to perturbations from a Neptune-sized planet (*Liou and Zook,* 1999).

About 30–40 orders of magnitude in mass evolution separate the dust particles in young proplyds from the planets responsible for the distortions of more adolescent disks. Over the past 30 years the collective efforts of numerous workers have led to the emergence of a standard model describing how this mass evolution proceeds. The development of this model can be traced through reviews by *Wetherill* (1980, 1990, 2001), *Lissauer* (1993), and *Lissauer and Stewart* (1993). Descriptions of the standard model usually involve three stages of growth: (1) collisional coagulation of dust particles into $10^{12}$–$10^{18}$-g planetesimals, (2) gravitational accumulation of planetesimals into $10^{26}$–$10^{27}$-g planetary embryos, and (3) giant impacts between embryos that result in full-sized $10^{27}$–$10^{28}$-g terrestrial planets. The somewhat arbitrary transitions between these three stages clearly involve considerable overlap in the mass distribution of bodies in the disk. Nonetheless, each stage is typically modeled as an independent process. This chapter reviews the second stage, formation of planetary embryos. In the interest of providing proper context we offer below a synopsis of all three stages.

### 1.1. Initial Planetesimal Formation

The most widely accepted theory for the earliest stage of planet formation in very young protoplanetary disks involves the collision and coagulation of microscopic grains into macroscopic aggregates (*Weidenschilling and Cuzzi,* 1993; *Weidenschilling,* 1997). The motion of these tiny particles is coupled to the gaseous nebula and collisions are facilitated by turbulence and size-dependent gas drag, which results in differential settling toward the disk midplane and differential orbital decay toward the protostar. The crucial range for these growing aggregates is centimeter- to meter-sized, at which point the particles are most strongly affected by nebular gas and quickly removed by orbital decay. If approximately meter-sized bodies can form, then presumably growth can continue to kilometer-sized planetesimals. The timescale for growth from micrometer-sized grains to meter-sized bodies at 1 AU is thought to be on the order of $10^4$ yr (*Weidenschilling and Cuzzi*, 1993).

An alternative to the collisional coagulation theory is the idea that a thin dense central particle layer may suffer gravitational instabilities, fragmenting into clumps that collapse to form kilometer-sized planetesimals (*Goldreich and Ward*, 1973). However, it is generally believed that even in the most quiescent protoplanetary disks, turbulence would be too high to allow a central particle layer to reach the critical density required for the instabilities to develop (*Weidenschilling and Cuzzi*, 1993; *Weidenschilling*, 1997). Despite this problem the theory remains attractive (see *Ward*, 2000) because it is not strongly dependent on poorly understood particle properties or sticking mechanisms.

Once planetesimals reach kilometer-sized (by whichever method you subscribe to) they are effectively decoupled from the gaseous nebula. At this point they enter the evolutionary stage where their own mutual gravitational interactions become an important factor in their further accumulation.

## 1.2. Gravitational Accumulation of Planetesimals

One Earth-mass of planetesimals ranging from $10^{12}$ to $10^{18}$ g (kilometer-sized) would constitute a swarm of $10^{10}$–$10^{16}$ bodies. Based on our current understanding this swarm would require ~$10^8$ yr to produce a mature terrestrial planet. Direct N-body numerical integration of the complete evolution of such a swarm, running on the fastest of our current generation of computers, would require far longer than a typical research career and probably exceed the longevity of most advanced civilizations. Thus, until the inevitable advent of "infinitely" fast, massively parallel quantum computers (*Milburn*, 1998), direct N-body integrations will be intractable except — as we shall see — for small regions in the swarm and over a very limited range of masses and evolution times. To study mass evolution over the full range of planetesimal sizes, statistical methods have been developed that are akin to the kinetic theory of gases. In these models nebular gas drag acting on a swarm of kilometer-sized planetesimals leads to nearly co-planar, nearly circular concentric orbits. Mutual gravitational interactions among the planetesimals perturb their eccentricities and inclinations to small nonnegligible values ($10^{-4}$–$10^{-5}$). Assuming the orbits are randomly oriented in their longitudes of perihelion and node, the eccentricity (e) and inclination (i) distributions of the swarm can be characterized by the root mean square velocity with respect to a circular zero-inclination reference orbit, or

$$v_{RMS} \propto \sqrt{\tfrac{5}{8}e^2 + \tfrac{1}{2}i^2}$$

These RMS velocities are analogous to the random thermal velocities of molecules of a gas confined to a container of known temperature. Therefore, these statistical gas dynamic models of planetesimal growth are sometimes referred to as "particle-in-a-box" calculations.

One of the most advanced statistical models in use today is that of *Wetherill and Stewart* (1989, 1993). In the 10 years since it was first introduced the model has grown to incorporate many of the physical processes important to planetesimal growth, including fragmentation (the proverbial kitchen sink). [See *Stewart and Ida* (2000) and *Wetherill and Inaba* (1999) for up-to-date descriptions of this model.] As the model evolved to become more physically realistic and explore larger areas of parameter space there emerged one general characteristic that remained essentially unchanged, that of the results. Near 1 AU, in a narrow zone of semimajor axis, a swarm of kilometer-sized planetesimals evolves in such a way that a single massive body emerges from the distribution and grows to become a Mercury- to Mars-sized planetary embryo on a timescale of order $10^5$ yr. The phrase "runaway growth" is used to describe this mode of planetesimal growth where a single dominant body grows much more rapidly than (or "runs away from") the smaller bodies in the swarm.

Typically a runaway planetary embryo is $10^2$–$10^3$ times more massive than the second largest body in the accumulation zone. This result presents somewhat of a problem for particle-in-a box techniques because they involve an implicit assumption of local uniformity of the planetesimal swarm. While this condition occurs during the earliest stage of accretion, as the swarm evolves it transitions to a nonuniform distribution due to the formation of a small number of large bodies. In other words, the very phenomenon of interest — planet formation — begins to render the method less valid. Therefore, in order to model this intermediate stage of planetary growth, a hybrid model was developed (*Spaute et al.*, 1991; *Weidenschilling et al.*, 1997) that includes aspects from the purely statistical technique as well as from N-body numerical integration. The hybrid model uses multiple zones of semimajor axis spanning the range from 0.5 to 1.5 AU and treats the larger planetesimals as discrete bodies with individual values of semimajor axis, eccentricity, and inclination. The smaller bodies in each semimajor axis zone are treated as a uniform distribution across the zone. This multizone model does not include fragmentation, instead assuming perfect accretion of colliding bodies. These simulations confirmed the runaway mode of planetesimal growth, producing a few dozen Mercury- to Mars-sized planetary embryos separated by about 0.05 AU. The timescale for growth was also found to be of order $10^5$ yr near 1 AU, confirming the results from the purely statistical simulations.

Despite the intractable nature of full scale N-body simulations, on a limited scale they have proven to be complementary to the statistical simulations. In N-body simulations, the gravitational interactions of bodies are directly calculated and the orbits of the bodies are numerically integrated. As the change in the spatial distribution of bodies is automatically calculated, this approach is suitable for studying the later stage of planetesimal accumulation when the spatial uniformity of the planetesimal distribution begins to break down. On the other hand, the high calculation cost ($\propto N^2$) restricts the number of bodies that can be treated in a timely fashion. The recent development of specialized software (e.g., *Makino and Aarseth, 1992; Kokubo et al., 1998*) and hardware (*Sugimoto et al., 1990; Makino et al., 1993, 1997; Makino and Taiji, 1998*) for N-body integrations has made it possible to treat ~$10^3$ bodies in simulations lasting for a few times $10^5$ yr. Two important physical processes barred by the limitation on N are fragmentation and the dynamical effect of a swarm of countless small bodies.

Results of planetesimal accumulation simulations by N-body (e.g., *Kokubo and Ida, 1995, 1996, 1998, 2000*) and statistical methods have been directly compared (*Wetherill et al., 1996; Inaba et al., 1999a,b*). In general, the smaller swarm of larger planetesimals used in N-body simulations evolves via runaway growth to produce Mercury- to Mars-sized embryos on a timescale of order $10^5$ yr near 1 AU. The final masses, orbital separations, and the timescale for growth are in good agreement with the results from the hybrid multizone statistical simulations.

## 1.3. Giant Impacts

In the third and final stage of the standard model of terrestrial planet formation the isolated embryos perturb each other into crossing orbits. Occasionally two embryos collide and merge into a larger body. Such "giant impacts" continue until a few remaining planets are isolated from each other on stable orbits. The timescale for this stage is of order $10^8$ yr. Various elements of this stage, including potential Moon-forming events, are reviewed by *Lissauer et al.* (2000) and *Canup and Agnor* (2000).

## 1.4. Brief Outline of Chapter

In the next two sections of this chapter we describe in general terms the methods and results from statistical and N-body simulations of embryo formation. Following these sections is a discussion of how the mechanism of giant planet formation may play a role in embryo formation. We conclude with a summary and brief discussion.

## 2. STATISTICAL CALCULATIONS

In the solar nebula, as a planetesimal of radius r sweeps up smaller bodies in the swarm, its mass m grows according to the relation

$$\frac{dm}{dt} = \underbrace{\overbrace{\pi r^2 F_g}^{\text{collision cross section}} \bar{V}_{rel} \rho}_{\text{volume swept out per unit time}} \tag{1}$$

where t is time, $\rho$ is the spatial mass density of small bodies in the nebula, $\bar{V}_{rel}$ is the average relative velocity between the growing planetesimal and the smaller bodies, and $F_g$ is the gravitational enhancement of the geometric cross-section. Following *Öpik* (1951), the two-body approximation of this enhancement factor is

$$F_g = 1 + V_e^2 / \bar{V}_{rel}^2$$

where $V_e$ is the surface escape velocity of the growing planetesimal. *Wetherill and Cox* (1985) have shown numerically that even for extremely low encounter velocities the two-body approximation is still valid — albeit in a statistical sense — but approaches an upper limit of $F_g \approx 3000$.

From the simplified expression for dm/dt we see that the rate of growth is controlled by the relative velocities. At the same time, the relative velocities are determined by the gravitationally interacting bodies, i.e., by the mass distribution of bodies in the swarm. This coupled evolution of the mass and velocity distributions in the swarm is the key to understanding how planetesimals grow into planetary embryos.

Investigation of the mass-velocity coupling (e.g., *Stewart and Kaula, 1980; Hornung et al., 1985*) has led to the recognition of several mechanisms that can affect the velocity evolution of the swarm. The relations, in a form applicable to statistical simulations (random velocities with respect to a circular orbit), were given by *Stewart and Wetherill* (1988). The acceleration of a body of mass and random velocity ($m_1$, $v_1$) interacting with a swarm of field bodies of mass and random velocity ($m_2$, $v_2$) is given by

A. Viscous stirring caused by gravitational encounters:

$$\frac{dv_1}{dt} \propto (m_1 + m_2)v_1^2 + (m_2 v_2^2 - m_1 v_1^2) \tag{2}$$

B. Viscous stirring caused by inelastic collisions:

$$\frac{dv_1}{dt} \propto m_2(v_1^2 - v_2^2) + 2m_1 v_1^2 \tag{3}$$

C. Energy damping caused by inelastic collisions:

$$\frac{dv_1}{dt} \propto -\left[m_2(v_1^2 - v_2^2) + 2m_1 v_1^2\right] \tag{4}$$

D. Energy equipartition caused by dynamical friction:

$$\frac{dv_1}{dt} \propto m_2 v_2^2 - m_1 v_1^2 \tag{5}$$

E. Energy damping caused by gas drag:

$$\frac{dv_1}{dt} \propto -v_1^2 \qquad (6)$$

Equation (5) and the second term of equation (2) were first derived by *Stewart and Kaula* (1980) and represent an equipartition of kinetic energy. These terms are both potential damping mechanisms. As an example, consider the relation for dynamical friction in equation (5). If $v_1 \approx v_2$ and $m_1 > m_2$ then $dv_1/dt < 0$ and the velocity of the more massive body will be damped. A similar expression for the smaller swarm bodies [simply reverse the subscripts in equation (5)] indicates that $dv_2/dt > 0$, pumping up the velocities of the smaller bodies. Thus, equipartition of random kinetic energy will tend to increase the random velocities of the smaller bodies while decreasing the random velocities of the larger bodies.

Early work using the statistical method was done by *Safronov* (1969). His derivations of the velocity evolution found the mutual gravitational perturbations in the swarm to be positive-definite ($dv/dt > 0$, no energy equipartition). Safronov included gas drag but not damping due to collisions. His calculations were extended by *Nakagawa et al.* (1983), who also used positive-definite gravitational stirring. Nakagawa et al. differed from Safronov by including collisional damping but ignoring gas drag. Both Safronov and Nakagawa et al. found that a swarm of planetesimals initially of equal or nearly equal mass would evolve such that the largest bodies were always of nearly equal mass and that

these largest bodies contained most of the mass of the swarm. Such characteristics describe what is termed "orderly growth." *Wetherill and Stewart* (1989) reproduced the Safronov and Nakagawa et al. models for the purpose of comparison (see Fig. 1). The mass distribution of the initial planetesimal swarm in Fig. 1 is given by

$$\frac{dn}{dm} \propto e^{-m/m_d} \qquad (7)$$

where $m_d = 3 \times 10^{18}$ g. A total mass of $6.3 \times 10^{26}$ g was distributed in the semimajor axis range 0.99–1.01 AU. In both the Safronov and Nakagawa cases after $\sim 10^6$ yr there are $\sim 100$ bodies in the mass range $10^{24}$ to $10^{25}$ g.

In the orderly growth found by Safronov and Nakagawa et al. the effects of equipartition of energy were not included. These terms (equation (5) and second term of equation (2)) influence the velocity evolution of the swarm by increasing the random velocities of the smaller bodies and decreasing the random velocities of the larger bodies. *Wetherill and Stewart* (1989) demonstrated that including these terms led to an alternative form of planetesimal growth termed "runaway growth." Runaway growth is characterized by the emergence of a single dominant body that grows much more rapidly than the remaining smaller bodies in the swarm. Eventually this single largest body detaches from the continuous mass distribution and ultimately accretes most of the mass in the accumulation zone, with the remaining mass contained in much smaller bodies. An example of runaway growth is shown in Fig. 2. In this case the initial

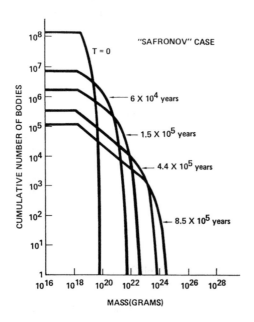 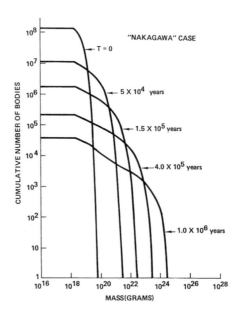

**Fig. 1.** Evolution of the mass distribution for conditions simulating the models of *Safronov* (1969) and *Nakagawa et al.* (1983). The planetesimal swarm is spread between 0.99 and 1.01 AU with an initial exponential mass distribution given by equation (7). Both cases use only positive-definite gravitational stirring [only first term in equation (2)] and ignore dynamical friction [equation (5)]. The Safronov case includes collisional damping but not gas drag, while the Nakagawa case includes gas drag but has no collisional damping. In both cases, after $10^6$ yr most of the mass of the swarm is contained in $\sim 100$ bodies between $\sim 10^{24}$ and $10^{25}$ g. Plots are similar to those in *Wetherill and Stewart* (1989).

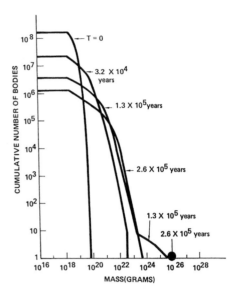

**Fig. 2.** The effect of adding equipartition of energy [equation (5) and second term in equation (2)] to the model. Identical initial distribution to that used in Fig. 1. By $1.3 \times 10^5$ yr a bulge at the high-mass end of the distribution indicates the transition into runaway growth. By $2.6 \times 10^5$ yr the largest body in the bulge has swept up the others, leading to a bimodal mass distribution, with a single body of order $10^{26}$ g (Mercury-sized) and a swarm of bodies $<10^{24}$ g. Plot is from *Wetherill and Stewart* (1989).

conditions are identical to those used in the Safronov and Nakagawa cases shown in Fig. 1. After $3 \times 10^4$ yr the velocities of the largest bodies have fallen an order of magnitude below the velocities of bodies at the median mass. As a result of this tendency toward equipartition of energy, the relative velocities of the largest bodies with respect to each other are well below their mutual escape velocities. This leads to high mutual gravitational cross-sections and enhanced growth rates of the larger bodies compared to the rest of the swarm. By $1.3 \times 10^5$ yr this produces a bulge of "multiple runaways" in the mass distribution. By $2.6 \times 10^5$ yr the bodies in the bulge have coalesced into a single body $\sim 10^{26}$ g (Mercury- to Mars-sized) while the remaining swarm is composed of bodies below $\sim 10^{24}$ g. Growth of this runaway body would continue until all the bodies in the 0.02 AU-wide accumulation zone are accreted, resulting in a final embryo mass $\sim 6 \times 10^{26}$ g.

Similar characteristics of rapid runaway growth had been reported earlier by *Greenberg et al.* (1978), despite apparently different expressions for gravitational stirring and dynamical friction (see discussion and comparison in *Kolvoord and Greenberg*, 1992).

### 2.1. Including Other Processes

There are a number of important processes that were not included in the work of *Wetherill and Stewart* (1989), but were studied in a subsequent paper (*Wetherill and Stewart*,

1993). These include (1) the effects of collisional fragmentation, (2) velocity perturbations in the low relative velocity regime where solar gravity becomes important, and (3) independent evolution of eccentricity (e) and inclination (i) rather than assuming the fixed ratio e/i = 2.

The effects of fragmentation are modeled in two ways. The first is cratering, where during a collision a certain amount of material is ejected and escapes from the mutual gravitational field of the two colliding bodies. The second is catastrophic fragmentation, where the amount of material ejected during a collision is greater than 50% of the combined mass of the two colliding bodies. Details of the cratering and catastrophic disruption criteria and the size distribution of the fragments can be found in the appendix of *Wetherill and Stewart* (1993). In the model, fragments larger than $\sim 1$ m in radius ($\sim 10^7$ g) are added to the planetesimal swarm and explicitly modeled the same as all other sized bodies. Fragments smaller than about 1 m are assumed to be lost via gas drag and removed from the swarm.

The tendency toward equipartition of energy would lower relative velocities into the range where solar gravity becomes an important factor. *Wetherill and Stewart* (1989) included the effect of this on the gravitational cross-sections using work subsequently published by *Greenzweig and Lissauer* (1990). By 1993 the effects of solar gravity on the velocity evolution of the swarm had been studied as well (*Ida*, 1990; *Ida and Makino*, 1992a), and these effects were included in the 1993 paper.

Earlier work on the velocity evolution of a planetesimal swarm (*Stewart and Wetherill*, 1988) assumed that eccentricity was always equal to twice the inclination (in radians). This relation does not hold in the low-velocity regime (*Ida*, 1990). In their 1993 modeling, Wetherill and Stewart allowed eccentricity (e) and inclination (i) to evolve independently. They found that during most of the growth the relative velocities are high enough that the constant ratio e/i = 2 is appropriate. It becomes inappropriate during interactions between small collision fragments (strongly damped by gas drag) and the larger runaway bodies as well as mutual interactions between the runaway bodies. During these low-velocity encounters modifications to the eccentricity and inclination evolution were made following the work of *Ida* (1990). These modifications are important during a $\sim 10^4$-yr period in the early stages of evolution when runaway growth is being initiated. In this critical interval the low-velocity effects lead to a ratio e/i $\gg$ 2 for the pair of bodies involved in an encounter.

Figure 3 shows the results from these extended calculations. A total mass of about $4 \times 10^{27}$ g was distributed in an initial swarm of equal mass ($4.8 \times 10^{18}$ g) bodies centered on 1 AU in a zone of width $\Delta a = 0.17$ AU. After $10^3$ yr the swarm resembles the initial exponential distributions used in the earlier calculations shown in Fig. 2. At this stage the largest bodies are already 30 times more massive than the original bodies. Within $7 \times 10^3$ yr equipartition of energy results in faster growth rates for the largest bodies because of their lower mutual encounter velocities. By this

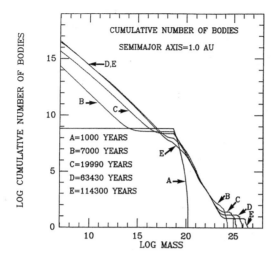

**Fig. 3.** Effects of including fragmentation, low-velocity modifications, and independent evolution of eccentricity and inclination. The initial distribution of equal mass bodies quickly evolves to the exponential distribution used in Figs. 1 and 2. A low-mass fragmentation tail develops and these small bodies are strongly damped by gas drag and swept up by the emerging runaway bodies in the high-mass bulge of the distribution. Plot is similar to that in *Wetherill and Stewart* (1993).

time a bulge in the distribution appears at the high mass end. Such a bulge is the graphical signature of runaway growth. The rate at which the larger bodies in the bulge merge with one another is increasing as a result of their low relative velocities and low mutual inclinations (recall i evolves independently of e). The swarm has also developed a low-mass fragmentation tail below the initial mass of $4.8 \times 10^{18}$ g. By $2 \times 10^4$ yr the runaway bulge contains about 10 bodies above $10^{25}$ g. Merger of these and other runaway bodies in the bulge continues and their further growth is becoming increasingly dependent on capture of the small fragments. These small bodies are significantly damped by gas drag to very low random velocities. Accretion of these low-velocity fragments prolongs the rapid runaway phase of growth. By ~$10^5$ yr one-third of the original mass is contained in a few runaway bodies of mass ~$10^{26}$ g (Mercury-to Mars-sized).

### 2.2. Hybrid Multizone Simulations

The final outcome of the runaway growth process is difficult to determine from the single accumulation zone models of *Wetherill and Stewart* (1989, 1993). These statistical gas dynamic models rely on the assumption of uniformity in the planetesimal swarm. The number density, size distribution, orbital eccentricities and inclinations, etc., do not vary over the volume of space in which the bodies interact

by collisions and gravitational stirring. This uniformity begins to break down with the emergence of a small number of runaway bodies. In order to avoid this problem a hybrid model was developed (*Spaute et al.,* 1991; *Weidenschilling et al.,* 1997). This model divides the planetesimal swarm into a series of zones in semimajor axis. The numerous small bodies are treated as a continuum represented by a series of logarithmic mass bins, each with a distribution of eccentricity and inclination. Bodies with masses greater than a specified threshold are treated as discrete, with individual values of semimajor axis, eccentricity, and inclination. The algorithms describing the mutual interactions of the larger discrete bodies in different zones have been tested and refined using direct N-body numerical integration (*Ida and Makino,* 1993; *Kokubo and Ida,* 1995), hence the "hybrid" nature of the model. Collisional and gravitational interactions are allowed among the continuum bins and discrete bodies, including those in different zones. The spatial resolution of the model provides information on the orbital spacing of the large bodies and variation of the swarm's parameters with heliocentric distance. Unlike the purely statistical models of Wetherill and Stewart, fragmentation is not yet included in the hybrid multizone model; collisions are assumed to result in perfect accretion.

Details of the model and tests of its features are given by *Weidenschilling et al.* (1997); here we describe only the results from the largest simulation. This assumed a swarm containing two Earth-masses in the range of semimajor axis, a, from 0.5 to 1.5 AU, with surface density proportional to $a^{-3/2}$. Spatial resolution was 0.01 AU (100 zones). The initial state consisted of planetesimals 15 km in diameter. Accretion was most rapid at the inner edge of the swarm, due to the higher surface density and shorter orbital period. Runaway growth of large bodies occurred as a wave propagating outward, reaching masses of order $10^{26}$ g on a typical timescale of $10^5$ orbital periods. These bodies became detached from the size distribution, with a lack of objects in the range $10^{24}$–$10^{26}$ g. The orbital spacing and low eccentricities of the large bodies rarely allowed them to collide; most of their growth was by accreting smaller bodies from the continuum. After discrete bodies reached masses ~$10^{26}$ g, their growth was much slower due to stirring of the continuum bodies to high eccentricities, with the consequent decrease in gravitational cross-section. The simulation was halted after a model time of $10^6$ yr, at which time the 21 largest bodies had a median mass of $5 \times 10^{26}$ g and contained nearly 90% of the mass of the swarm (Fig. 4). A notable feature is that their masses are nearly uniform, independent of heliocentric distance. The orbital separations are approximately 10 times the Hill radius. This value is in qualitative agreement with results obtained by *Kokubo and Ida* (1998); using orbital integrations they found "oligarchic" growth of planetary embryos to similar masses and separations (see section 3). Quantitative comparison is uncertain, as their simulation covered a much narrower region and shorter timescale.

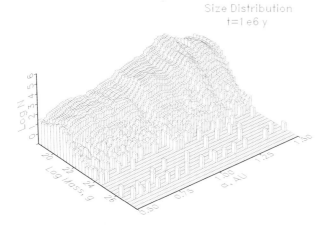

**Fig. 4.** The distribution of masses in the zones of semimajor axes for the multizone simulation of *Weidenschilling et al.* (1997) at model time $10^6$ yrs. Runaway growth has produced a gap in the mass distribution. The largest bodies have typical masses $\sim 5 \times 10^{26}$ g (Mars-sized) and orbital spacings $\sim 0.05$ AU.

## 3.  N-BODY SIMULATIONS

Important elements of the formation of planetary embryos can also be described using direct N-body orbital integrations. The dynamics of a planetesimal swarm are dominated by the central force, with individual orbits deviating from pure Keplerian motion due to secondary perturbing forces. The equation of motion of planetesimals is given by

$$\frac{d\mathbf{v}_i}{dt} = \underbrace{-GM_\odot \frac{\mathbf{x}_i}{|\mathbf{x}_i|^3}}_{\text{central force}} - \underbrace{\sum_{\substack{j=1 \\ j \neq i}}^{N} Gm_j \frac{\mathbf{x}_i - \mathbf{x}_j}{|\mathbf{x}_i - \mathbf{x}_j|^3}}_{\text{gravitational interactions}} + \underbrace{\mathbf{f}_{\text{gas}}}_{\substack{\text{gas} \\ \text{drag}}} + \underbrace{\mathbf{f}_{\text{col}}}_{\substack{\text{collisional} \\ \text{interactions}}} \quad (8)$$

where t is time, m is planetesimal mass, $\mathbf{x}$ and $\mathbf{v}$ are the planetesimal position and velocity vectors, G is the gravitational constant, and $M_\odot$ is the solar mass. The four terms on the right of equation (8) represent accelerations due to the central force of solar gravity, mutual gravitational interactions of planetesimals, gas drag from the solar nebula, and collisions between planetesimals (accretion). Equation (8) in the N-body simulations plays a role analogous to equations (2)–(6) in the statistical simulations.

The gas component of the solar nebula orbited the proto-Sun with a velocity slightly slower than the Keplerian circular velocity $\mathbf{v}_c$ due to the pressure gradient in the nebula (*Adachi et al., 1976*). The expression for gas velocity takes the form $\mathbf{v}_{\text{gas}} = (1-2\eta)^{1/2}\mathbf{v}_c$, where $\eta$ is a small quantity depending on the radial pressure gradient given by $\eta = 0.0019a^{1/2}$. For simplicity, no turbulent motions are assumed in the nebula. The gas drag force (per unit mass) on a plan-

etesimal with mass m and radius r takes the form

$$\mathbf{f}_{\text{gas}} = -\frac{1}{2m}C_D\pi r^2\rho|\mathbf{u}|\mathbf{u} \quad (9)$$

where $C_D$ is dimensionless drag coefficient, $\rho$ is the gas density, and $\mathbf{u} = \mathbf{v} - \mathbf{v}_{\text{gas}}$ is the relative velocity between the planetesimal and the gas. *Kokubo and Ida* (2000) adopted $C_D = 2$ and the internal density of planetesimals $\rho_p = 2$ g cm$^{-3}$, which are within a range suitable for planetesimals. The velocity change by collision is an impulsive force that reduces the relative velocity of colliding bodies and changes its direction.

In N-body simulations, the orbits of planetesimals are calculated by numerically integrating equation (8). For the numerical integration, *Kokubo and Ida* (2000) used the predictor-corrector type Hermite scheme (*Makino and Aarseth, 1992; Kokubo et al., 1998*) with the hierarchical time-step (*Makino, 1991*). The most time-consuming part of N-body simulations is the calculation of the mutual gravitational interactions, with the number of operations increasing in proportion to the square of the number of bodies. *Ida and Makino* (1992a,b, 1993) and *Kokubo and Ida* (1995, 1996, 1998, 2000) calculated the force by directly summing up interactions of all planetesimal pairs on the special-purpose computer for N-body simulation: GRAPE/HARP (*Sugimoto et al., 1990; Makino et al., 1993, 1997; Makino and Taiji, 1998*). This software/hardware combination allows simulations limited to N $\sim 10^3$ to evolve for $\sim 10^5$ yr in a reasonable amount of CPU time.

In N-body simulations, a collision is assumed to take place when the close encounter distance between two bodies is smaller than the sum of their physical radii. Because of the limitations on the size of N, *Kokubo and Ida* (1996, 1998, 2000) assumed perfect accretion, with collisions producing a single body with the combined mass of the two colliding bodies (no fragmentation). In the assumption of perfect accretion, the position and velocity of the center of mass of the colliding bodies are conserved. The lack of collisional fragmentation does not change the mode of planetesimal growth (still runaway), although it does affect the duration and final embryo mass reached in the runaway mode (*Wetherill and Stewart, 1993*; see also the earlier discussion pertaining to Fig. 3).

In the most recent model (*Kokubo and Ida, 2000*) planetesimals are initially distributed in a ring around semimajor axis a = 1 AU with a surface density of solids $\Sigma = 10$ g cm$^{-2}$. The density of the gas component of the nebula is $\rho = 2 \times 10^{-9}$ g cm$^{-3}$. This model is 50% more massive than the minimum-mass solar nebula model (*Hayashi, 1981*). Initially there are 3000 equal mass planetesimals ($\sim 10^{23}$ g). At this stage, gravitational stirring and gas drag are the dominant perturbations on the planetesimal dynamics. The surface density of the planetesimal swarm is kept constant by replacing any planetesimals scattered out of the ring. The width of the planetesimal swarm is $\Delta a = 0.021$ AU.

The initial eccentricities (e) and inclinations (i) of the planetesimals are given by the Rayleigh distribution (*Ida and Makino,* 1992a) with the equilibrium dispersion $\langle e^2 \rangle^{1/2} = 2\langle i^2 \rangle^{1/2} = 0.004$ (*Ohtsuki et al.,* 1993). Figure 5 shows four snapshots of the system on the (a,e) plane for t = 0, 0.5, 1 and $2 \times 10^5$ yr. In $2 \times 10^5$ yr, the number of bodies has decreased to 1322 and the mass of the largest body (filled circle) reaches about 200 times the initial mass while the mean mass has become only about twice as large. This runaway body keeps growing and then isolates from the continuous power-law mass distribution. In this stage, the runaway body predominantly grows in its feeding zone as a sink of the mass flow from the continuous power-law mass distribution.

The evolution of the mass distribution is shown in Fig. 6 where $n_c$ is the cumulative number of bodies. The mass distribution relaxes to a distribution that is well approximated by a power-law distribution. The largest body at t = $2 \times 10^5$ yr is shown by a dot that is separated from the continuous mass distribution. The mass range $10^{23}$ g $\leq$ m $\leq$ $10^{24}$ g, which contains most of the system mass, can be approximated by $dn_c/d_m \propto n \propto m^\alpha$, where n is the number of bodies in a linear mass bin. The power indexes calculated by using the least-square-fit method are $\alpha = -2.6$ for t = $10^5$ yr and $\alpha = -2.2$ for t = $2 \times 10^5$ yr. These values are consistent with the results of the gas-free calculation by *Kokubo and Ida* (1996) and references therein. The power index gradually decreases with time since there is no supply of small bodies in the simulation. A power index smaller than –2 is a characteristic of runaway growth, which means most of the total system mass exists in small bodies. *Kokubo and Ida* (2000) defined runaway growth as the mode that produces a power-law mass distribution with the index $\alpha < -2$.

The evolution of the distributions of the RMS eccentricity and the RMS inclination is plotted in Fig. 7. Let us focus on the mass range $10^{23}$ g $\leq$ m $\leq$ $10^{24}$ g. The values for mass larger than this range are not statistically significant since each of the larger mass bins often has only one body. First, the distributions tend to relax to a decreasing function of mass through dynamical friction among bodies (t = $0.5 \times 10^5$ and $10^5$ yr) (*Ida and Makino,* 1992b). Second, the distributions tend to flatten (t = $2 \times 10^5$ yr). This is because as a runway body grows, the system is mainly excited by the runaway body (*Ida and Makino,* 1993). In this late runaway stage, the eccentricity and inclination of planetesimals are scaled by the Hill radius of the runaway body, which is

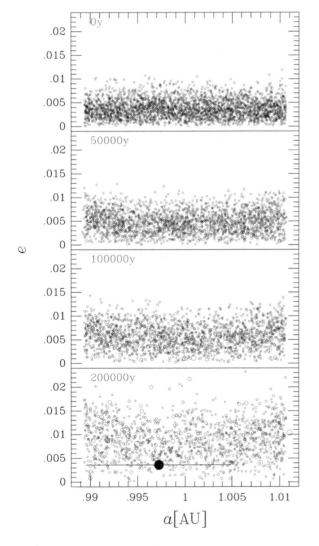

**Fig. 5.** Snapshots of a planetesimal system on the (a,e) plane. The circles represent planetesimals and their radii are proportional to the radii of planetesimals. The system initially consists of 3000 equal-mass ($10^{23}$ g) planetesimals. The numbers of planetesimals are 2215 (t = 50,000 yr), 1787 (t = 100,000 yr), and 1322 (t = 200,000 yr). In the t = 200,000-yr panel, the filled circles represent an emerging runaway embryo and lines from the center of the embryo to both sides have the length of 5 $r_H$.

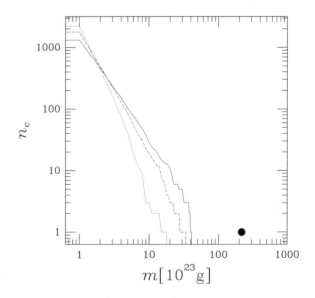

**Fig. 6.** For the same simulation as shown in Fig. 5 the cumulative number of bodies is plotted against mass at t = 50,000 yr (dotted curve), t = 100,000 yr (dashed curve), and t = 200,000 yr (solid curve). A runaway body at t = 200,000 yr is shown by a dot.

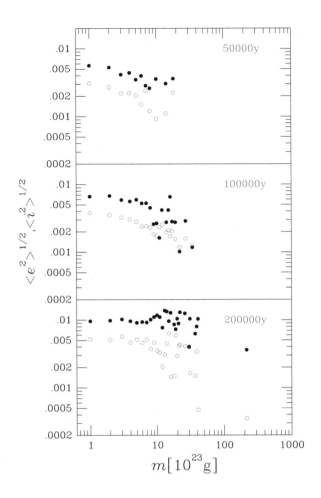

**Fig. 7.** For the same simulation as shown in Fig. 5 time evolution of the RMS eccentricities (filled circles) and inclinations (open circles) for each mass bin.

almost independent of the mass of planetesimals. In addition, as the gas drag is stronger for small bodies with high eccentricity and inclination, the eccentricity and the inclination of small bodies are more damped by the gas drag than those of large bodies, which also leads to the flat distributions of the eccentricity and the inclination. Note that the eccentricity and the inclination of the largest body is always kept small due to dynamical friction from smaller bodies. These small eccentricities and inclinations facilitate runaway growth (*Wetherill and Stewart,* 1989, 1993; *Ohtsuki and Ida,* 1990; *Kokubo and Ida,* 1996).

The relation between the mass and the random velocity (eccentricity and inclination) distributions is derived by a simple analytical argument when the random velocity distribution is approximated by a power-law distribution. Let us assume that the mass distribution is proportional to $m^{\alpha}$ and the random velocity distribution proportional to $m^{\beta}$. The stationary power-law state of the mass distribution is realized when the mass flux becomes independent of mass (see *Makino et al.,* 1998)

$$nm\frac{dm}{dt} = \text{constant} \qquad (10)$$

When the mass distribution is considered, the growth rate of a planetesimal is given by

$$\frac{dm}{dt} = \int_0^m \pi \left(r+r'\right)^2 F_g V_{rel} \frac{n_s(m')m'}{\max(H,H')}dm' \qquad (11)$$

where $n_s$ is the surface number density of planetesimals and H is the scale height of the planetesimal system. When gravitational focusing is effective, the growth rate reduces to

$$\frac{dm}{dt} \propto m^{\alpha - 2\beta + 10/3} \qquad (12)$$

where we used $H \propto V_{ran}$ and $V_{rel}^2 \simeq V_{ran}^2 + V_{ran}'^2$, $V_{ran}$ being the random velocity of planetesimals. From equation (10), for the stationary state, the relation between the mass and velocity distributions is given by

$$\alpha = \beta - \frac{13}{6} \qquad (13)$$

This relation gives $\alpha = -8/3$ for $\beta = -1/2$ (energy equipartition). In the present calculation, $\beta \simeq -1/2$ at $10^5$ yr and the corresponding approximated power index agrees well with the above estimate. This analytical argument assumes a supply of planetesimals at the low-mass end and the removal of planetesimals at the high-mass end. In the actual simulation, this assumption does not hold. Therefore, as the number of bodies decreases, the distribution gradually becomes more gentle due to the lack of the supply of planetesimals at the low-mass end. When $\alpha = -8/3$, the growth rate becomes $dm/dt \propto m^{5/3}$, which means that the growth is driven by collisions with similar-sized bodies. In other words, the growth is hierarchical.

In the stationary state, the relative growth rate is given as

$$\frac{1}{m}\frac{dm}{dt} \propto m^{-\alpha - 2} \qquad (14)$$

This shows that the definition of runaway growth, $\alpha < -2$, is equivalent to the "classical" definition that the relative growth rate increases with mass. That is, for $m_1 > m_2$ the mass ratio $m_1/m_2$ increases with time. It should be noted that from the above argument the sufficient condition for runaway growth, taking into account the mass distribution of planetesimals, is $\beta < 1/6$.

### 3.1.   Oligarchic Growth

Like the multizone statistical simulations, N-body simulations have been used over wide semimajor axis regions where multiple runaway embryos grow and interact with each other. *Kokubo and Ida* (1998) found that while most planetesimals remained small, the runaway embryos grew at a similar rate and kept their orbital separation greater than about five Hill radii [$r_H = (2M/3M_\odot)^{1/3}a$, where M is the embryo mass]. This so-called "oligarchic growth" is a result of the self-limiting nature of runaway growth and orbital repulsion among growing embryos. Through oligarchic growth a bimodal embryo-planetesimal distribution formed, similar to that found using the hybrid multizone statistical method.

The formation of similar-sized embryos is explained by the slowdown of runaway growth (*Lissauer,* 1987; *Ida and Makino,* 1993). Once the runaway body becomes isolated from the continuous power-law mass distribution, the system can be approximated by a bimodal distribution: an embryo and the planetesimal swarm. In this stage, the relative growth rate is given as

$$\frac{1}{M}\frac{dM}{dt} \propto \Sigma M^{1/3} v_m^{-2} \qquad (15)$$

where $\Sigma$ is the surface density of the solid material and $v_m$ is the velocity dispersion of planetesimals. In the early runaway stage, $\Sigma$ and $v_m$ are independent of M (*Ida and Makino,* 1993), so that $(1/M)dM/dt \propto M^{1/3}$, which leads to runaway growth. When the mass of an embryo exceeds the critical value ~50–100 m, where m is the typical mass of planetesimals, the embryo mainly excites the velocity dispersion of neighboring planetesimals and the velocity dispersion of its neighborhood comes to depend on the embryo's escape velocity (and thus its mass M), so that $v_m \propto M^{1/3}$ (*Ida and Makino,* 1993). In this case, the relative growth rate becomes

$$\frac{1}{M}\frac{dM}{dt} \propto \Sigma M^{-1/3} \propto M^\gamma \qquad (16)$$

The power index $\gamma$ could be smaller than –1/3, since the local $\Sigma$ decreases through accretion of planetesimals by the embryo as M increases (*Lissauer,* 1987). This means that when embryos exceed the critical mass, the growth mode among embryos is orderly. In other words, the mass ratios between neighboring embryos tend toward unity, so that the embryos grow while keeping similar masses. Note that in this stage, the mass ratio of an embryo to its neighboring planetesimals continues to increase with a relative growth rate that is $(M/m)^{(1/3)}$ larger than that of the planetesimals when $v_m \propto M^{1/3}$. Thus, the bimodal embryo-planetesimal distribution is preserved.

While the embryos grow, orbital repulsion keeps their orbital separations wider than about 5 $r_H$ (*Kokubo and Ida,* 1995). The orbital repulsion is a coupling effect of scattering between large bodies and dynamical friction from small bodies. Scattering between two embryos on nearly circular orbits increases their eccentricities and orbital separation. After scattering, dynamical friction reduces the eccentricities while the orbital separation remains nearly constant.

Thus, the orbital separation of two embryos increases while they keep nearly circular orbits. If the separation is less than 5 $r_H$, relatively strong scattering occurs and the separation expands rapidly. As embryos grow, their orbital separation normalized by the Hill radius becomes small, since $r_H \propto M^{1/3}$. This implies that they repeat the orbital repulsion while growing. Consequently, the orbital separation of embryos remains larger than about 5 $r_H$. The typical orbital separation is about 10 $r_H$, which only weakly depends on the mass and the semimajor axis of embryos and the surface density (*Kokubo and Ida,* 1998).

An example N-body simulation demonstrates the principle of oligarchic growth where the effect of gas drag has been taken into account and realistically sized planetesimals have been used (*Kokubo and Ida,* 2000). The planetesimal system initially consists of 4000 bodies. The initial mass distribution is given by a power-law mass distribution with power-law index $\alpha = -2.5$ for the mass range $2 \times 10^{23}$ g $\leq$ m $\leq 4 \times 10^{24}$ g. The width of the planetesimal ring is $\Delta a = 0.092$ AU. The total mass of the ring is $1.3 \times 10^{27}$ g, which is consistent with the surface density $\Sigma = 10$ g cm$^{-2}$. The initial dispersion of the eccentricity (e) and the inclination (i) of bodies are given by the Rayleigh distribution with $\langle e^2 \rangle^{1/2} = 2\langle i^2 \rangle^{1/2} = 0.002$.

Figure 8 shows five snapshots of the system on the (a, e) plane for t = 1, 2, 3, 4 and $5 \times 10^5$ yr. In $5 \times 10^5$ yr, the number of bodies decreases to 1257. The filled circles in Fig. 8 represent emerging embryos with masses larger than $2 \times 10^{25}$ g, or >100 times the minimum mass of the system. The horizontal bars on each embryo represent 5 $r_H$. The five stages in the evolution of the system are summarized as follows: *($10^5$ yr)* — Four embryos are formed through runaway growth and their separation of about 5 $r_H$ is kept by orbital repulsion between the neighboring embryos. *(2 × $10^5$ yr)* — Seven embryos have emerged with masses of the same order. As they grow, their Hill radii increase and it becomes difficult to keep the orbital separation by orbital repulsion since there is no room to expand the orbits. The embryos with high eccentricity (due to recent scattering) sometimes collide with other embryos. *(3 × $10^5$ yr)* — Two large embryos and three smaller ones dominate the system. *(4 × $10^5$ yr)* — A large embryo formed by the collision of two of the three smaller ones. *(5 × $10^5$ yr)* — The final state shows three large embryos with masses $1$–$2 \times 10^{26}$ g on nearly circular noninclined orbits with the orbital separation of 5–10 $r_H$. The three large embryos contain 41% of the total system mass and 47% is in all five embryos.

Embryos may also migrate radially due to the tidal interaction with the solar nebula (see *Ward,* 1986, 1997). This effect is potentially important but has not been included in any of the simulations described above. *Tanaka and Ida* (1999) recently showed that fast inward migration of embryos accelerates their growth since unperturbed planetesimals with the low random velocity are encountered as an embryo migrates. Slow inward migration slows embryo growth because a gap is opened as planetesimals are shepherded ahead of the embryo. The full extent and speed of migra-

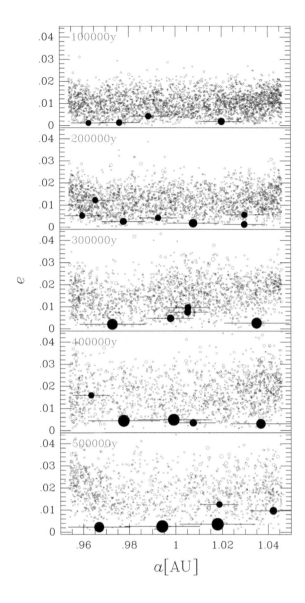

**Fig. 8.** Snapshots of a planetesimal system on the (a,e) plane. The circles represent planetesimals and their radii are proportional to the radii of planetesimals. The system initially consists of 4000 planetesimals whose total mass is $1.3 \times 10^{27}$ g. The initial power-law mass distribution has an index $\alpha = -2.5$ within the mass range $2 \times 10^{23}$ g $\leq$ m $\leq 4 \times 10^{24}$ g. The numbers of planetesimals are 2712 (t = 100,000 yr), 2200 (t = 200,000 yr), 1784 (t = 300,000 yr), 1488 (t = 400,000 yr), and 1257 (t = 500,000 yr). The filled circles represent protoplanets with mass larger than $2 \times 10^{25}$ g and lines from the center of the protoplanet to both sides have a length of $5\ r_H$.

tion, as well as its effects on the growth of an ensemble of migrating embryos, are still uncertain.

It is also difficult to include collisional fragmentation in N-body simulation because collisional fragmentation increases the number of planetesimals by orders of magnitude. *Beaugé and Aarseth* (1990) and *Alexander and Agnor* (1998) implemented it to some degree by adopting a low-

mass cutoff that does not cause a large increase of planetesimals. In the final stage, collisional fragmentation must play an important role. Particularly, the accretion of small fragmented bodies whose velocity dispersion is small due to gas drag can prolong the runaway phase of embryo growth (*Wetherill and Stewart,* 1993).

The implementation of the above effects are subjects of future work. As hardware and software for N-body simulations develop, a larger number of bodies can be handled, allowing the simulations to become more global and more realistic. *Richardson et al.* (1997, 2000) have been pursuing rather ambitious plans to model $10^6$ bodies over a timescale of $10^6$ yr. It should be noted that the most important role of N-body simulation is to explore a nonuniform nonlinear system and to clarify the physics behind basic processes.

## 4. EFFECTS OF GIANT PLANETS

Embryo formation in the inner solar system is not necessarily isolated from the process of giant planet formation taking place in the outer solar system. There are currently two competing mechanisms of giant planet formation — core accretion and disk instability (see detailed comparison by *Boss,* 1998a). In the core accretion model (*Pollack et al.,* 1996) protoplanetary cores form by collisional accumulation of planetesimals. These cores eventually reach a critical mass of about 10 Earth-masses, at which point they can begin accreting massive gaseous mantles in a phase of growth requiring $10^6$–$10^7$ yr to reach Jupiter-mass. As described in the earlier sections, by this time accretion in the inner solar system is well underway, with a small population of Mercury- to Mars-sized planetary embryos distributed throughout the terrestrial planet region. Models pertaining to the final stage of terrestrial planet formation (e.g., *Chambers and Wetherill,* 1998) typically permit these embryos to evolve for 5–10 m.y. before introducing Jupiter and Saturn, assuming that they formed via core accretion.

In the disk instability model (*Boss,* 1997, 1998b) giant gaseous protoplanets form from gravitational instabilities in the cool outer regions of the primitive solar nebula. Formation timescales for this process are of order $10^2$ yr, with the giant protoplanets forming toward the end of the accretion of the central star. This mechanism is so fast that the giant planets could have attained their large masses well before the stage of disk evolution during which planetesimal formation is conventionally considered to occur. If this was the case then the relative velocities of small planetesimals would be dominated by secular perturbations from the giant planets rather than from their own mutual perturbations. The dependence of planetesimal growth models on the formation mechanism of giant planets is currently being investigated by at least two groups.

*Thébault and Brahic* (1999) and *Thébault et al.* (1999) studied planetesimal evolution in the asteroidal region of a gas-free disk subject to Jupiter perturbations. They found that the high velocities reached in regions of mean-motion resonances could diffuse inward via planetesimal collisions. In some of their models the high velocities in this collisional diffusion wave propagated in as far as the orbit of Mars before being damped by collisional dissipation. Thébault's group estimated that these high velocities could have limited growth in the asteroid belt to bodies less than 500 km. They suggested that the lack of larger bodies in the belt could be a consequence of an early formed Jupiter. However, their models did not include fragmentation and it is likely that including it would considerably limit the propagation of the collisional diffusion waves.

*Kortenkamp and Wetherill* (2000; *Wetherill and Kortenkamp,* 1999) studied planetesimal evolution subject to Jupiter and Saturn perturbations and gas drag. Under both mechanisms of giant planet formation, gravitational interactions between the massive protoplanets and the even more massive gaseous disk cause the protoplanets to migrate inward toward the central star (*Ward,* 1997), presumably giv-

ing rise to the small orbits of some of the known extrasolar giant planets (*Trilling et al.,* 1998). To allow for some migration, Kortenkamp and Wetherill placed Jupiter and Saturn at somewhat greater distances ($a_J = 6.2$ and $a_S = 10.5$ AU) while using the present values of all other orbital elements. For a given planetesimal density, gas drag causes the orbits of smaller bodies to decay faster than those of larger bodies. Over a wide range of planetesimal sizes, as the orbits of smaller bodies decay past the orbits of larger bodies the orbits of the two populations are not in phase, that is, they do not evolve on coplanar concentric orbits. Figure 9 shows the size-dependent inclinations and longitudes of ascending node for planetesimals on orbits near 1 AU. A similar size-dependency exists for the eccentricities and arguments of pericenter. Figure 10 shows an example of the mean encounter velocities with respect to an arbitrary $10^{17}$ g reference body. The size-dependent phasing of the evolving orbital elements leads to a pronounced dip in encounter velocities between bodies of similar mass. This dip allows bodies to grow despite the fact that Jupiter and Saturn are perturbing them to relatively high eccentricities and inclinations. When the mass distribution has evolved to the point where the largest bodies are $10^{24}$–$10^{26}$ g, the dip becomes less significant because stochastic perturbations between the largest bodies begin to dominate the size-dependent phased perturbations from Jupiter and Saturn. However, as the distribution enters this regime, dynamical friction may prevent the encounter velocities of the largest bodies from growing, thus preserving a size-dependent dip in encounter velocities. Calculations using the gas dynamic model of

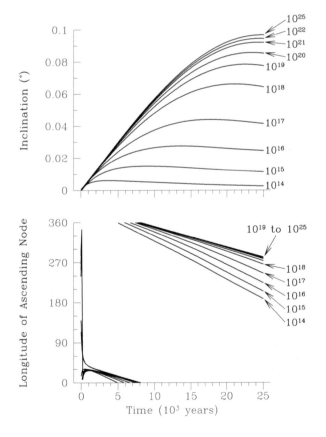

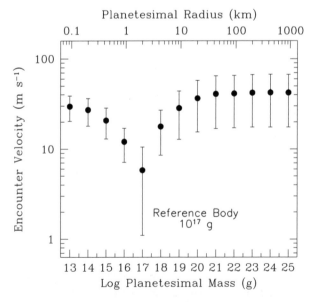

**Fig. 9.** Size-dependent secular variations of inclination and longitude of ascending node for planetesimals of various mass near 1 AU subject to Jupiter and Saturn perturbations and gas drag. Inclinations are referred to the invariable plane of the Jupiter-Saturn system. The planetesimal mass in grams is indicated to the right of each plot. A similar size-dependency exists for the eccentricity and argument of pericenter.

**Fig. 10.** Planetesimal encounter velocities near 1 AU with respect to $10^{17}$-g reference bodies. The size-dependent phasing of orbital elements shown in Fig. 9 leads to a dip in encounter velocities between similar-sized bodies. Velocities shown are mean values (±1 σ) over a 25,000-yr period.

*Wetherill and Stewart* (1993) suggest that under these conditions runaway growth at 1 AU would be significantly delayed but not necessarily prevented. On the other hand, it is not clear whether runaway growth can occur in the asteroid belt on the $10^6$–$10^7$-yr timescale for which nebular gas is likely to have been present.

While the jury is certainly still out on the topic, at this point it appears that the standard model of terrestrial planet formation is robust enough to accommodate the early formation of Jupiter and Saturn. This scenario may also provide a mechanism for limiting growth in the asteroid region, either by a collisional diffusion wave or size-dependent secular perturbations or some combination of each.

## 5. SUMMARY AND DISCUSSION

Statistical simulations treat the random velocities of planetesimals (with respect to a circular orbit) as analogous to the random thermal velocities of molecules in the kinetic theory of gases. This approach can model the evolution of a huge number of bodies over a wide range of masses (about 10–12 orders of magnitude are used in current modeling). Important physical effects such as dynamical friction, gas drag, and fragmentation can be easily incorporated into the simulations. The fundamental assumption of statistical simulations is that all the properties of the planetesimal swarm are uniformly distributed throughout the volume of space being studied. This assumption begins to break down for the relatively few large bodies produced by the runaway growth process. N-body simulations avoid the assumption of uniformity but are limited in the number of bodies that can be modeled. This limitation affects the extent to which dynamical friction can be studied and usually precludes fragmentation from being included at all.

Despite the shortcomings of each approach, comparison of the results show remarkable agreement. The statistical simulations showed that on a timescale of order $10^5$ yr, in a swarm of planetesimals distributed over a narrow accumulation zone near 1 AU, a single dominant runaway body emerges (an embryo) with a mass ~$10^{26}$ g (*Wetherill and Stewart*, 1989, 1993). N-body simulations confirm runaway growth (*Kokubo and Ida*, 1996), with the planetesimal distribution relaxing into a bimodal system of isolated runaway bodies and a continuous power-law mass distribution with $dn_c/dm \propto m^\alpha$, where $n_c$ is the cumulative number of bodies and $\alpha \simeq -2.5$. The eccentricity and inclination of runaway bodies are kept small due to dynamical friction from smaller planetesimals. Over a wider range of semimajor axis that encompasses multiple accumulation zones several such embryos form. Following the merger of some runaway bodies and the subsequent orbital isolation of those remaining, the growth mode shifts to oligarchic growth (*Kokubo and Ida*, 1998, 2000) where larger embryos tend to grow more slowly than smaller ones, while the growth of embryos is still faster than that of planetesimals. The orbital separation of the embryos is kept greater than about 5 Hill radii through orbital repulsion.

Runaway growth is fostered by the tendency toward equipartition of random kinetic energy (dynamical friction) between the large and small planetesimals in the swarm (*Wetherill and Stewart*, 1989). This effect lowers the mutual encounter velocities of the larger bodies and allows them to merge and grow at a rate that is faster than the growth of smaller bodies. Runaway growth is accelerated by any damping mechanism that lowers the relative velocities of any bodies with respect to the larger bodies. Two such mechanisms are inelastic collisions and gas drag. Gas drag has significant influence on the small collision fragments in the swarm. Accretion of these low velocity fragments extends the duration of runaway growth and increases the final mass reached by embryos during the runaway phase of growth (*Wetherill and Stewart*, 1993).

Runaway growth is inhibited by mechanisms that increase the relative velocities of any bodies with respect to the larger bodies. One such mechanism, considered in the previous sections, is perturbations by the growing embryos themselves, which excite neighboring planetesimals. Another mechanism is distant perturbations by massive bodies farther out in the protoplanetary disk. Examples of possible perturbers include brown dwarf or stellar companions or early-formed giant planets like Jupiter and Saturn (*Kortenkamp and Wetherill*, 2000). The relative encounter velocities of planetesimals at 1 AU in the presence of early-formed Jupiter and Saturn and the gas of the protoplanetary disk are considerably lower than they would be if it was not for the direct effects of gas drag and the tendency for secular perturbations of the orbital elements to remain in phase when the semimajor axes are similar. At the same time, the presence of gas reduces greatly the tendency for the perturbations to remain in phase, except for bodies very nearly equal in mass. For bodies of other masses, relative velocities approaching 100 m s$^{-1}$ are found for bodies near 1 AU and several times larger near 2.6 AU. These high relative encounter velocities are a handicap to growth of $10^{26}$-g embryos near 1 AU and could prevent their growth entirely in the asteroid belt.

The results of N-body simulations are also important as basic data for modeling the evolution of a planetesimal system by statistical simulations. Using N-body results, *Stewart and Ida* (2000) recently refined the analytical expressions (equations (2)–(6)) for the velocity evolution of a planetesimal swarm. Earlier, *Weidenschilling et al.* (1997) used N-body results from *Ida and Makino* (1993) and *Kokubo and Ida* (1995) to refine the hybrid multizone accretion model. Results from this hybrid model show that approximately 20 Mars-sized runaway embryos form between 0.5 and 1.5 AU on a timescale of order $10^5$ orbital periods (~$10^5$ yr near 1 AU). These results are among the initial conditions used by *Agnor et al.* (1999; see also *Canup and Agnor*, 2000) for N-body integrations of accretion in the late giant impact stage. They found that such a system was unstable, with perturbations leading to crossing orbits and collisions until a few large terrestrial-sized planets remained. Similar outcomes were found by *Chambers and*

*Wetherill* (1998) for different initial conditions. It appears that the final number, masses, and spacings of planets are stochastic, and not very dependent on the intermediate embryo formation stage of accretion. There is one discrepancy, however: In the multizone simulation, the orbits of the large bodies had low eccentricities and remained stable for times well in excess of $10^5$ yr. In the continuation by orbital integrations, mutual perturbations led to crossing orbits on a timescale of only $10^4$ yr. It is possible that the algorithms used in the multizone simulation underestimate distant perturbations by bodies in noncrossing orbits, artificially stabilizing the system of embryos. However, that simulation included nebular gas drag and dynamical friction, which coupled the large bodies to the residual population of small planetesimals. These comprised only about 10% of the mass of the swarm, but damping by gas drag allowed them to dissipate energy without limit. This damping was absent from the N-body integrations that showed rapid transition to chaos. Preliminary tests of N-body integrations with added damping suggest that a relatively low-mass background swarm of small planetesimals could stabilize such a system (Agnor, personal communication, 1999). Thus, the onset of the final stage of accretion with crossing orbits and giant impacts may have depended on the sweeping up of the smaller planetesimals and/or the removal of gas from the inner region of the solar nebula.

*Acknowledgments.*    We thank George Wetherill, Shigeru Ida, and two anonymous reviewers for their suggestions and comments concerning the manuscript.

# REFERENCES

Adachi I., Hayashi C., and Nakazawa K. (1976) The gas drag effect on the elliptic motion of a solid body in the primordial solar nebula. *Progr. Theor. Phys., 56,* 1756–1771.

Agnor C. B., Canup R. M., and Levison H. F. (1999) On the character and consequences of large impacts in the late stage of terrestrial planet formation. *Icarus, 142,* 219–237.

Alexander S. G. and Agnor C. B. (1998) N-body simulations of late stage planetary formation with a simple fragmentation model. *Icarus, 132,* 113–124.

Beaugé C. and Aarseth S. J. (1990) N-body simulation of planetary formation. *Mon. Not. R. Astron. Soc., 245,* 30–39.

Beckwith S. V. W. and Sargent A. I. (1996) Circumstellar disks and the search for neighbouring planetary systems. *Nature, 383,* 139–144.

Boss A. P. (1997) Giant planet formation by gravitational instability. *Science, 276,* 1836–1839.

Boss A. P. (1998a) Formation of extrasolar giant planets: Core accretion or disk instability? *Earth, Moon, Planets, 81,* 19–26.

Boss A. P. (1998b) Evolution of the solar nebula IV: Giant gaseous protoplanet formation. *Astrophys. J., 503,* 923–937.

Burrows C. J., Krist J. E., Stapelfeldt K. R., and WFPC2 Investigation Definition Team (1995) HST observations of the Beta Pictoris circumstellar disk (abstract). *Bull. Amer. Astron. Soc., 27,* 1329.

Burrows C. J., Stapelfeldt K. R., Watson A. M., Krist J. E., Ballester G. E., Clarke J. T., Crisp D., Gallagher J. S. III,

Griffiths R. E., Hester J. J., Hoessel J. G., Holtzman J. A., Mould J. R., Scowen P. A., Trauger J. T., and Westphal J. A. (1996) Hubble Space Telescope observations of the disk and jet of HH 30. *Astrophys. J., 473,* 437–451.

Canup R. M. and Agnor C. B. (2000) Accretion of the terrestrial planets and the Earth-Moon system. In *Origin of the Earth and Moon* (R. M. Canup and K. Righter, eds.), this volume. Univ. of Arizona, Tucson.

Chambers J. E. and Wetherill G. W. (1998) Making the terrestrial planets: N-body integrations of planetary embryos in three dimensions. *Icarus, 136,* 304–327.

Dermott S. F., Grogan K., Holmes E. K., and Wyatt M. C. (1999) Signatures of planets in circumstellar disks. In *Modern Astrometry and Astrodynamics Honouring Heinrich Eichhorn* (R. Dvorak, H. F. Haupt, and K. Wodnar, eds.), pp. 189–199. Austrian Academy of Sciences, Vienna.

Goldreich P. and Ward W. (1973) The formation of planetesimals. *Astrophys. J., 183,* 1051–1061.

Greaves J. S., Holland W. S., Moriarty-Schieven G., Jenness T., Dent W. R. F., Zuckerman B., McCarthy C., Webb R. A., Butner H. M., Gear W. K., and Walker H. J. (1998) A dust ring around ε Eridani: Analog to the young solar system. *Astrophys. J. Lett., 506,* 133–137.

Greenberg R., Wacker J., Chapman C. R., and Hartmann W. K. (1978) Planetesimals to planets: Numerical simulation of collisional evolution. *Icarus, 35,* 1–26.

Greenzweig Y. and Lissauer J. J. (1990) Accretion rates of protoplanets. *Icarus, 87,* 40–77.

Hayashi C. (1981) Structure of the solar nebula, growth and decay of magnetic fields and effects of magnetic and turbulent viscosities on the nebula. *Progr. Theor. Phys. Suppl., 70,* 35–53.

Holland W. S., Greaves J. S., Zuckerman B., Webb R. A., McCarthy C., Coulson I. M., Walther D. M., Dent W. R. F., Gear W. K., and Robson I. (1998) Submillimetre images of dusty debris around nearby stars. *Nature, 392,* 788–790.

Hornung P., Pellat R., and Barge P. (1985) Thermal velocity equilibrium in the protoplanetary cloud. *Icarus, 64,* 295–307.

Ida S. (1990) Stirring and dynamic friction rates of planetesimals in the solar gravitational field. *Icarus, 88,* 129–145.

Ida S. and Makino J. (1992a) N-body simulation of gravitational interaction between planetesimals and a protoplanet I. Velocity distribution of planetesimals. *Icarus, 96,* 107–120.

Ida S. and Makino J. (1992b) N-body simulation of gravitational interaction between planetesimals and a protoplanet II. Dynamical friction. *Icarus, 98,* 28–37.

Ida S. and Makino J. (1993) Scattering of planetesimals by a protoplanet: Slowing down of runaway growth. *Icarus, 106,* 210–227.

Inaba S., Tanaka H., Ohtsuki K., and Nakazawa K. (1999a) High-accuracy statistical simulation of planetary accretion: I. Test of the accuracy by comparison with the solution to the stochastic coagulation equation. *Earth Planets Space, 51,* 205–217.

Inaba S., Tanaka H., Nakazawa K., Wetherill G. W., and Kokubo E. (1999b) High-accuracy statistical simulation of planetary accretion: II. Comparison with N-body simulation. *Icarus,* submitted.

Jayawardhana R., Fisher S., Hartmann L., Telesco C., Piña R., and Fazio G. (1998) A disk of dust surrounding the young star HR 4796A. *Astrophys. J. Lett., 503,* 79–82.

Koerner D. W., Ressler M. E., Werner M. W., and Backman D. E. (1998) Mid-infrared imaging of a circumstellar disk around HR

4796A: Mapping the debris of planetary formation. *Astrophys. J. Lett., 503,* 83–86.

Kokubo E. and Ida S. (1995) Orbital evolution of protoplanets embedded in a swarm of planetesimals. *Icarus, 114,* 247–257.

Kokubo E. and Ida S. (1996) On runaway growth of planetesimals. *Icarus, 123,* 180–191.

Kokubo E. and Ida S. (1998) Oligarchic growth of protoplanets. *Icarus, 131,* 171–178.

Kokubo E. and Ida S. (2000) Formation of protoplanets from planetesimals in the solar nebula. *Icarus, 143,* 15–27.

Kokubo E., Yoshinaga K., and Makino J. (1998) On a time-symmetric Hermite integrator for planetary N-body simulation. *Mon. Not. R. Astron. Soc., 297,* 1067–1072.

Kolvoord R. A. and Greenberg R. (1992) A critical reanalysis of planetary accretion models. *Icarus, 98,* 2–19.

Kortenkamp S. J. and Wetherill G. W. (2000) Terrestrial planet and asteroid formation in the presence of giant planets I. Relative velocities of planetesimals subject to Jupiter and Saturn perturbations. *Icarus, 143,* 60–73.

Liou J. C. and Zook H. A. (1999) Signatures of giant planets on the solar system Kuiper Belt dust disk and implications for extrasolar planet in Epsilon Eridani (abstract). In *Lunar and Planetary Science XXX,* Abstract #1698. Lunar and Planetary Institute, Houston (CD-ROM).

Lissauer J. J. (1987) Timescales for planetary accretion and the structure of the protoplanetary disk. *Icarus, 69,* 249–265.

Lissauer J. J. (1993) Planet formation. *Annu. Rev. Astron. Astrophys., 31,* 129–174.

Lissauer J. J. and Stewart G. R. (1993) Growth of planets from planetesimals. In *Protostars and Planets III* (E. H. Levy and J. I. Lunine, eds.), pp. 1061–1088. Univ. of Arizona, Tucson.

Lissauer J. J., Dones L., and Ohtsuki K. (2000) Origin and evolution of terrestrial planet rotation. In *Origin of the Earth and Moon* (R. M. Canup and K. Righter, eds.), this volume. Univ. of Arizona, Tucson.

Makino J. (1991) A modified Aarseth code for GRAPE and Vector processors. *Publ. Astron. Soc. Japan, 43,* 859–876.

Makino J. and Aarseth S. J. (1992) On a Hermite integrator with Ahmad-Cohen scheme for gravitational many-body problems. *Publ. Astron. Soc. Japan, 44,* 141–151.

Makino J. and Taiji M. (1998) *Scientific Simulations with Special-Purpose Computers — The GRAPE Systems.* Wiley, Chichester.

Makino J., Kokubo E., and Taiji M. (1993) HARP: A special-purpose computer for the N-body problem. *Publ. Astron. Soc. Japan, 45,* 349–360.

Makino J., Taiji M., Ebisuzaki T., and Sugimoto D. (1997) GRAPE-4: A massively-parallel special-purpose computer for collisional N-body simulations. *Astrophys. J., 480,* 432–446.

Makino J., Fukushige T., Funato Y., and Kokubo E. (1998) On the mass distribution of planetesimals in the early runaway stage. *New Astronomy, 3,* 411–417.

McCaughrean M. J. and O'Dell C. R. (1996) Direct imaging of circumstellar disks in the Orion nebula. *Astron. J., 111,* 1977–1986.

Milburn G. J. (1998) *The Feynman Processor: Quantum Entanglement and the Computing Revolution.* Perseus, Reading.

Mouillet D., Larwood J. D., Papaloizou J. C. B., and Lagrange A. M. (1997) A planet on an inclined orbit as an explanation of the warp in the Beta Pictoris disc. *Mon. Not. R. Astron. Soc., 292,* 896–904.

Nakagawa Y., Hayashi C., and Nakazawa K. (1983) Accumulation of planetesimals in the solar nebula. *Icarus, 54,* 361–376.

O'Dell C. R. and Beckwith S. V. W. (1997) Young stars and their surroundings. *Science, 276,* 1355–1359.

O'Dell C. R., Wen Z., and Hu X. (1993) Discovery of new objects in the Orion nebula on HST images: Shocks, compact sources, and protoplanetary disks. *Astrophys. J., 410,* 696–700.

Ohtsuki K. and Ida S. (1990) Runaway planetary growth with collision rate in the solar gravitational field. *Icarus, 85,* 499–511.

Ohtsuki K., Ida S., Nakagawa Y., and Nakazawa K. (1993) Planetary accretion in the solar gravitational field. In *Protostars and Planets III* (E. H. Levy and J. I. Lunine, eds.), pp. 1061–1088. Univ. of Arizona, Tucson.

Öpik E. J. (1951) Collision probabilities with the planets and the distribution of interplanetary matter. *Proc. Royal Irish Acad., 54-A,* 165–199.

Pollack J. B., Hubickyj O., and Greenzweig Y. (1996) Formation of the giant planets by concurrent accretion of solids and gas. *Icarus, 124,* 62–85.

Richardson D. C., Quinn T., and Lake G. (1997) Direct simulation of planet formation with a million planetesimals (abstract). *Bull. Amer. Astron. Soc., 29,* 1027–1028.

Richardson D. C., Quinn T., Stadel J., and Lake G. (2000) Direct large-scale N-body simulations of planetesimal dynamics. *Icarus, 143,* 45–59.

Safronov V. S. (1969) *Evolution of the Protoplanetary Cloud and Formation of the Earth and Planets.* Nauka, Moscow. (Translation, 1972, NASA TT F-677.)

Spaute D., Weidenschilling S., Davis D. R., and Marzari F. (1991) Accretional evolution of a planetesimal swarm: I. A new simulation. *Icarus, 92,* 147–164.

Stewart G. R. and Ida S. (2000) Velocity evolution of planetesimals: Unified analytical formulae and comparison with N-body simulations. *Icarus, 143,* 28–44.

Stewart G. R. and Kaula W. M. (1980) A gravitational kinetic theory for planetesimals. *Icarus, 44,* 154–171.

Stewart G. R. and Wetherill G. W. (1988) Evolution of planetesimal velocities. *Icarus, 74,* 542–553.

Sugimoto D., Chikada Y., Makino J., Ito T., Ebisuzaki T., and Umemura M. (1990) A special-purpose computer for gravitational many-body problems. *Nature, 345,* 33–35.

Tanaka H. and Ida S. (1999) Growth of migrating protoplanet. *Icarus, 139,* 350–366.

Telesco C. M., Fisher R. S., Piña R. K., Knacke R. F., Dermott S. F., Wyatt M. C., Grogan K., Holmes E. K., Ghez A. M., Prato L., Hartmann L. W., and Jayawardhana R. (2000) Deep 10 and 18 μm imaging of the HR 4796A circumstellar disk: Transient dust particles and tentative evidence for a brightness asymmetry. *Astrophys. J., 530,* 329–341.

Thébault P. and Brahic A. (1999) Dynamical influence of a proto-Jupiter on a disc of colliding planetesimals. *Planet. Space Sci., 47,* 233–243.

Thébault P., Brahic A., Perrot C., and Charnoz S. (1999) Inhibition of planetesimal accretion by a proto-Jupiter. *Icarus,* submitted.

Trilling D. E., Benz W., Guillot T., Lunine J. I., Hubbard W. B., and Burrows A. (1998) Orbital evolution and migration of giant planets: Modeling extrasolar planets. *Astrophys. J., 500,* 428–439.

Ward W. R. (1986) Density waves in the solar nebula: Differential Lindblad torque. *Icarus, 67,* 164–180.

Ward W. R. (1997) Survival of planetary systems. *Astrophys. J. Lett., 482,* 211–214.

Ward W. R. (2000) On planetesimal formation. In *Origin of the Earth and Moon* (R. M. Canup and K. Righter, eds.), this volume. Univ. of Arizona, Tucson.

Weidenschilling S. J. (1997) The origin of comets in the solar nebula: A unified model. *Icarus, 127,* 290–306.

Weidenschilling S. J. and Cuzzi J. N. (1993) Formation of planetesimals in the solar nebula. In *Protostars and Planets III* (E. H. Levy and J. I. Lunine, eds.), pp. 1031–1060. Univ. of Arizona, Tucson.

Weidenschilling S. J., Spaute D., Davis D. R., Marzari F., and Ohtsuki K. (1997) Accretional evolution of a planetesimal swarm II. The terrestrial zone. *Icarus, 128,* 429–455.

Wetherill G. W. (1980) Formation of the terrestrial planets. *Annu. Rev. Astron. Astrophys., 18,* 77–113.

Wetherill G. W. (1990) Formation of the Earth. *Annu. Rev. Earth Planet. Sci, 18,* 205–256.

Wetherill G. W. (2001) Planet formation. *Annu. Rev. Earth Planet. Sci., 29,* in preparation.

Wetherill G. W. and Cox L. P. (1985) The range of validity of the two-body approximation in models of terrestrial planet accumulation II: Gravitational cross-sections and runaway accretion. *Icarus, 63,* 290–303.

Wetherill G. W. and Inaba S. (1999) Planetary accumulation with a continuous supply of planetesimals. In *From Dust to Terrestrial Planets* (R. Kallenbach, ed.), in press. Kluwer, Dordrecht.

Wetherill G. W. and Kortenkamp S. J. (1999) Asteroid belt formation with an early formed Jupiter and Saturn (abstract). In *Lunar and Planetary Science XXX,* Abstract #1767. Lunar and Planetary Institute, Houston (CD-ROM).

Wetherill G. W. and Stewart G. R. (1989) Accumulation of a swarm of small planetesimals. *Icarus, 77,* 330–357.

Wetherill G. W. and Stewart G. R. (1993) Formation of planetary embryos: Effects of fragmentation, low relative velocity, and independent variation of eccentricity and inclination. *Icarus, 106,* 190–209.

Wetherill G. W., Kokubo E., Ida S., and Chambers J. (1996) Comparison of numerical integration and "gas dynamic" modeling of runaway planetesimal growth (abstract). In *Lunar and Planetary Science XXVII,* pp. 1425–1426. Lunar and Planetary Institute, Houston.

Wyatt M. C., Dermott S. F., Telesco C. M., Fisher R. S., Grogan K., Holmes E. K., and Piña R. K. (1999) How observations of circumstellar disk asymmetries can reveal hidden planets: Pericentre glow and its application to the HR 4796 disk. *Astrophys. J., 527,* 918–944.

# Origin and Evolution of Terrestrial Planet Rotation

**Jack J. Lissauer**
*NASA Ames Research Center*

**Luke Dones**
*Southwest Research Institute*

**Keiji Ohtsuki**
*Yamagata University and University of Colorado*

Rotation is one of the most fundamental of planetary properties. A planet's rotation state is determined by a combination of processes that occur during and after its accretion. We review the current understanding of the accumulation of spin angular momentum by a growing terrestrial planet and the evolution of planetary spin subsequent to the accretionary epoch. Considerable progress toward understanding the origin of planetary rotation has been made over the past decade. Calculations have clearly shown that large impacts are likely to play a major role in determining planetary spin; these same impacts can also eject a disk of debris into orbit about a planet. It is not yet known whether or not terrestrial planets also possess a significant "systematic" component of spin produced by a biased distribution in the incoming directions of accumulated small planetesimals.

## 1. INTRODUCTION

The origin of planetary rotation is one of the most fundamental questions of cosmogony, and in most models of the growth of the planets it is intimately tied to the formation of planetary satellite systems (e.g., *Laplace*, 1796; *Lissauer*, 1993; *Lissauer et al.*, 1995). Questions concerning the origin of planetary rotation have proven to be extremely difficult to answer (cf. *Safronov*, 1969). Even the data provide a very complex set of clues: Six of the nine planets in our solar system have an obliquity (the angle between the rotational angular momentum and orbital angular momentum vectors) of less than 30°, a distribution that would have only a $10^{-5}$ probability of occurring randomly (*Lissauer and Kary*, 1991). The spin periods of six of the nine planets, including the five most massive ones, lie within a factor of 2.5 of each other. However, the observed obliquities of Venus, Uranus, Pluto, and the asteroids, together with calculations of rotational evolution, suggest that the initial spin axes may have been random, at least for the planets that accreted most of their angular momentum from solid planetesimals (*Dones and Tremaine*, 1993a).

The maximum rate at which a planet can spin and still retain loosely bound material at its equator can be calculated by setting the centrifugal force equal to the gravitational force. For a spherical planet of density ρ, this yields a minimum rotational period

$$T_{min} = \sqrt{\frac{3\pi}{G\rho}} \qquad (1a)$$

where G is the gravitational constant. Measuring density in g cm$^{-3}$, equation (1a) becomes

$$T_{min} \text{ (hours)} = 3.3\rho^{-1/2} \qquad (1b)$$

Note that the minimum rotation period for retaining material does not depend on the size of the body. Real planets are not rigid spheres, but rather more elastic bodies that bulge at the equator in response to the centrifugal force induced by rotation. Small asteroids and comets are more rigid, but usually spin about their axis of greatest moment of inertia (which is the lowest energy state for a given angular momentum). Both of these types of nonsphericity imply that material at the equator is farther from the rotation axis; thus gravity is diminished and centrifugal force increased, so equation (1) underestimates the actual minimum rotation period of planets. (But note that equation (1) does not apply to small, cohesive bodies, which can rotate faster.) Some asteroids rotate near breakup speed (*Pravec et al.*, 1998), but the giant planets' spin periods are at least twice that given by equation (1), and terrestrial planets spin at less than one-tenth breakup. However, if the orbital angular momentum of the Moon were added to the Earth's spin, it would rotate at approximately half the rate required for breakup.

An alternative and well-defined characteristic timescale is the orbital period of the planet. Most planets rotate many times per orbital period. In this chapter, we shall denote planetary spin rates by the number of siderial rotation periods per orbit, $|\Re|$, where positive values of $\Re$ signify prograde (direct) rotation (obliquity less than 90°) and negative values denote retrograde spin (obliquity greater than 90°).

Planets obtain rotational angular momentum from the relative motions at collision of the material from which they accrete. The angular momentum, L, acquired in an individual collision can be in any direction. However, the symmetry of the system about the plane of the planet's orbit, $(z, \dot{z}) \rightarrow (-z, -\dot{z})$, implies that there is no systematic preference for positive or negative $L_x$ or $L_y$. An ordered component to $L_z$ is possible, producing a net planetary spin angular momentum either in the same direction as the orbital angular momentum (prograde rotation) or in the opposite direction (retrograde rotation). Additionally, the stochastic nature of planetary accretion from planetesimals allows for a random component to the spin angular momentum of a planet (*Safronov*, 1966).

As planetary growth progresses, random velocities of planetesimals increase as a result of scattering by increasingly massive bodies in the disk (*Safronov*, 1969). A more massive planet has a deeper gravitational potential well for the planetesimal to fall into, and the moment arm of a growing planet increases as its radius. These three factors imply that a planet is likely to accumulate the bulk of its rotational angular momentum during the latter stages of accretion, regardless of whether the systematic or random component dominates. Simulations that use scaled parameters to model variations in conditions as a planet grows are thus likely to yield approximately correct results, provided the parameters applicable toward the end of the accretionary epoch are used.

Planetary rotation is closely related to the issue of the origin of the Moon. At present, the orbital angular momentum of the Moon is about 5× as large as the spin angular momentum of the Earth. The Earth and the Moon have tidally interacted with each other since the formation of the Moon, and these interactions have decelerated the spin of the Earth while they have expanded the Moon's orbit. Therefore models of the formation of the Earth-Moon system need to account for the present total angular momentum of the system, but the initial angular momenta of the components are not well constrained. If the present Earth-Moon angular momentum was provided by a single large impact, the impactor must have been about as large as Mars (e.g., *Hartmann and Davis*, 1975; *Cameron and Ward*, 1976). On the other hand, in order for fission to have occurred, the spin angular momentum of the proto-Earth must have been at least 3.4× as large as the present total angular momentum of the Earth-Moon system (e.g., *Durisen and Gingold*, 1986); a substantial amount of excess angular momentum would thus have had to have been removed from the system subsequent to fission.

Considerable progress toward understanding the origin of the spins of terrestrial planets has been made over the past decade. In this chapter, we review current models of the accumulation of rotational angular momentum by growing solid planets. The systematic component of rotation that a planet obtains from a uniform surface density disk of small planetesimals is reviewed in section 2. The rotation of planets that accumulate from small planetesimals in nonuniform disks is discussed in section 3. The random component of planetary spin imparted by large impactors is reviewed in

section 4. Rotation results from recent numerical simulations of the final stages of planetary growth are summarized in section 5. Section 6 lists mechanisms that can alter planetary rotation subsequent to the accretionary epoch. We conclude in section 7 with a summary of both the well-understood and the controversial aspects of the origin of the rotational angular momentum of terrestrial planets.

Questions concerning the origin of planetary rotation have historically played a significant role in the development of planetary cosmogonies. Until almost 1900, cosmogonists believed that planets that accumulated from material in Keplerian orbits would have retrograde rotation. It was reasoned that if two bodies in Keplerian orbits stuck together, the excess speed of the inner body would lead to retrograde rotation of the resultant assemblage. As the rotation directions of most planets were known to be prograde, various "solutions" to this dilemma were proposed. *Laplace* (1796) postulated that the rings from which he believed the planets to have formed each rotated with constant angular velocity (due to viscosity), and thus the linear velocity of material within a given planet-forming ring increased outward. *Chamberlin* (1897) was apparently the first to realize that the arguments for the accumulation of planets from particles on Keplerian orbits producing retrograde spins depended on the assumption that the orbits were initially circular, and that accretion of bodies on eccentric orbits could produce prograde rotation. *Safronov* (1966) proposed that the obliquities of the planets were produced by stochastic effects of large impacts by individual "planetesimals" having up to 5–10% of the planet's total mass. Excellent historical accounts of the development of models for the origin of planetary rotation are presented by *Brush* (1978, 1981), who concentrates on the work done prior to 1920, and *Safronov* (1969), who emphasizes work during the first seven decades of the twentieth century. A more succinct historical review is given in the final appendix of *Lissauer and Kary* (1991; henceforth *LK91*) and updated in the final appendix of *Lissauer et al.* (1997; henceforth *LBGK97*) and again in the appendix of this chapter.

## 2.  SYSTEMATIC COMPONENT IN A UNIFORM DISK

The problem of calculating the rotation rate of a planet that grows by accumulating many small planetesimals from a uniform surface density disk appears at first glance to be quite simple. However, this appearance is deceptive, and intuition can be a poor guide in this problem. The Keplerian motion of planetesimals in a disk suggests that the part of the planet closest to the Sun should move the fastest, which implies retrograde rotation. In contrast, viewed as a fluid, a Keplerian disk has positive vorticity, suggesting that planets should rotate in the prograde direction. The problem was finally solved in the 1990s through a combination of analytic calculations for nongravitating planets and numerical three-body experiments for massive planets. We review the results of these calculations in this section.

## 2.1. Analytic Calculations

The rotation rate of a nongravitating homogeneous spherical planet of radius $R_p$ orbiting with semimajor axis, a, that accretes small planetesimals from a uniform disk can be computed analytically in various regimes. Consider a planet on a circular orbit in a two-dimensional disk of planetesimals that have eccentricities, e, that satisfy

$$R_p/a \ll e \ll 1 \qquad (2)$$

The planet's rotation rate is given by

$$\Re = \left[1 - \frac{1}{4}\frac{\mathbf{K}\left(\sqrt{3/2}\right)}{\mathbf{E}\left(\sqrt{3/2}\right)} + \frac{1}{12\pi\mathbf{E}\left(\sqrt{3/2}\right)}\right] \approx 0.57673 \quad (3)$$

where $\mathbf{E}(k)$ and $\mathbf{K}(k)$ are the complete elliptic integral of the second and first kinds respectively, with $\mathbf{E}(\sqrt{3/2}) \approx$ 1.211 and $\mathbf{K}(\sqrt{3/2}) \approx 2.157$ (*LK91; Dones and Tremaine,* 1993b). The same result is obtained for the rotation of a nongravitating planet on an eccentric orbit within a disk of planetesimals on either circular or eccentric orbits (*LBGK97*). If the orbits of the planet and the planetesimals are circular and coplanar, then

$$\Re = 3/8 \qquad (4)$$

(*LK91*).

The rotation rate of a nongravitating planet embedded in a disk of planetesimals with eccentricities satisfying the assumptions in equation (2) and inclinations, i, satisfying the similar criteria

$$R_p/a \ll \sin i \ll 1 \qquad (5)$$

is given by

$$\Re = \frac{3}{4} - \frac{3\left(3I^2+1\right)}{16\left(I^2+1\right)}\frac{\mathbf{K}\left(\frac{\sqrt{3}}{2\sqrt{1+I^2}}\right)}{\mathbf{E}\left(\frac{\sqrt{3}}{2\sqrt{1+I^2}}\right)} + \frac{1}{16\pi\mathbf{E}\left(\frac{\sqrt{3}}{2\sqrt{1+I^2}}\right)} \quad (6)$$

where

$$I \equiv \frac{\sin i}{e} \qquad (7)$$

[*LBGK97*; see Fig. 3 of that paper for a plot of $\Re(I)$]. Two interesting limiting values from equation (7) are

$$\Re(I \to \infty) = \frac{3}{16} + \frac{1}{8\pi^2} \approx 0.200165 \qquad (8a)$$

and

$$\Re(I \to 0) = \frac{3}{4} - \frac{3}{16}\frac{\mathbf{K}\left(\frac{\sqrt{3}}{2}\right)}{\mathbf{E}\left(\frac{\sqrt{3}}{2}\right)} + \frac{1}{16\pi\mathbf{E}\left(\frac{\sqrt{3}}{2}\right)} \approx 0.43255 \quad (8b)$$

Note that this latter value is three-fourths the two-dimensional rotation rate, because the limit is taken subject to equation (5).

The above results demonstrate that a nongravitating planet that accumulated from very small planetesimals in a uniform disk would rotate very slowly (less than one rotation per orbit) in the prograde direction.

## 2.2. Numerical Experiments

Calculation of the accretion of rotational angular momentum by gravitating planets is a more complex problem that has been studied numerically. Individual planetesimals can impart a large amount of specific spin angular momentum; however, a high degree of cancellation occurs within uniform surface density disks of small planetesimals. Thus, simulations must include a very large number of test particles in order to obtain statistically robust results. Fortunately, this problem scales well with the radius of planet's Hill sphere

$$h = \left(\frac{M_p}{3M_\star}\right)^{\frac{1}{3}} a \qquad (9)$$

provided the radius of the planet as well as preencounter eccentricities, inclinations, and separations in semimajor axis, $\Delta a$, are scaled appropriately

$$r_H \equiv \frac{R_p}{h} \qquad (10a)$$

$$e_H \equiv e\frac{a}{h} \qquad (10b)$$

$$i_H \equiv \sin i \frac{a}{h} \qquad (10c)$$

$$b_H \equiv \frac{\Delta a}{h} \qquad (10d)$$

This scaling means that calculations need only be performed for one value of the planet's mass. Moreover, while h becomes larger as the planet grows, the value of $r_H$ does not depend on the planet's mass provided its density and distance from the Sun remain unchanged. Eccentricities and inclinations are likely to increase as the planet grows, plausibly at a rate that allows $e_H$ and $i_H$ to remain constant, and if that is the case, then the planet's feeding zone scales in an analogous manner. Thus, the systematic component of spin acquired by a given planet may be well represented by the results of a single calculation for fixed distributions of $e_H$ and $i_H$.

Several groups have independently performed extensive three-body integrations to determine the rotation of planets accreting from small planetesimals of various eccentricities and inclinations in uniform surface density disks. If the planet and planetesimals are all on circular orbits prior to

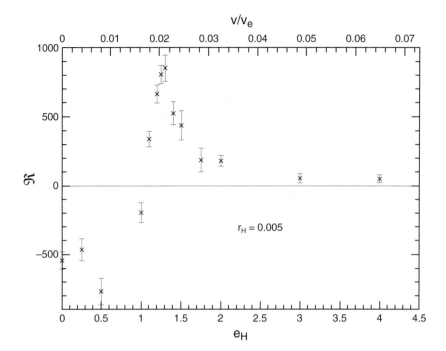

**Fig. 1.** The rotation rate, $\mathfrak{R}$, of a planet of radius $r_p = 0.005$ that accreted from a uniform disk of small planetesimals, all of which had the same eccentricity prior to encounter, is shown as a function of that eccentricity. Plot from *LK91*.

approaching one another, the planet accumulates enough spin angular momentum to rotate at roughly the rates of Earth and Mars, but in the retrograde direction (*Giuli*, 1968). There is a very small range of eccentricities that produce prograde rotation at a comparable rate (Fig. 1; cf. *LK91*; *Greenberg et al.*, 1997; *Ida and Nakazawa*, 1990), but for a realistic distribution of planetesimal eccentricities, a planet accreting from a uniform surface density disk of small bodies rotates much slower than does Earth or Mars (*LK91*; *LBGK97*; *Dones and Tremaine*, 1993a; *Ohtsuki and Ida*, 1998). Planetesimal inclination typically reduces the magnitude of planetary rotation; however, when the eccentricity of the planet and planetesimals are nearly equal, the inclusion of inclination can turn retrograde rotation into prograde rotation by reducing the accretion cross-section advantage of planetesimals with low relative eccentricities (*LBGK97*).

## 3. SYSTEMATIC COMPONENT IN A NONUNIFORM DISK

When the mass of a protoplanet becomes much larger than a typical mass of neighboring planetesimals as a result of runaway growth (*Wetherill and Stewart*, 1989, 1993), gravitational perturbations by the protoplanet are more important than mutual interactions between planetesimals. The strong gravitational perturbations produce a gap in the semimajor axis distribution of planetesimals centered on the orbit of the protoplanet, as demonstrated using N-body simulations by *Ida and Makino* (1993). If the effect of gas drag on planetesimals is also taken into account, the gap becomes clearer (*Tanaka and Ida*, 1997), and only sufficiently small planetesimals whose orbits decay rapidly can enter the protoplanet's feeding zone (*Weidenschilling and Davis*,

1985; *Kary et al.*, 1993; *Kary and Lissauer*, 1995). In this case, the protoplanet may well accumulate material preferentially from the edges of its accretion zone.

Numerical simulations by *Giuli* (1968) demonstrated that planetesimals accreted from the outer extremities of a planet's feeding zone provide, on average, a substantial amount of positive rotational angular momentum. This has been confirmed by later more detailed simulations (*Tanikawa et al.*, 1989; *Ida and Nakazawa*, 1990; *LK91*; *LBGK97*; *Greenberg et al.*, 1997; see Fig. 2). These results imply that a planet accreting material primarily from the edges of a gap centered on its orbit can acquire rapid prograde rotation.

If planetesimal trajectories are perturbed exclusively by a single protoplanet on a circular orbit, the value of their Jacobi integral

$$E_J = \frac{1}{2}\left(e_H^2 + i_H^2\right) - \frac{3}{8}b_H^2 + \frac{9}{2} \qquad (11)$$

is conserved. *Ohtsuki and Ida* (1998) assume that the combined effects of the protoplanet and the disk remove objects with high $E_J$, thereby producing a "gap" in the phase-space distribution of the Jacobi integral for the surviving planetesimals. The Jacobi integral is strongly correlated with semimajor axis, especially when planetesimal eccentricities and inclinations are small (equation (11)), but because of the conservation of $E_J$ in the three-body problem under the Hill's approximations (*Hill*, 1878), such gaps have a sounder theoretical basis than those defined in terms of $b_H$ only. Substantial positive angular momentum is accreted from planetesimals with small $E_J$ (Fig. 3). The net effect of such a gap can lead to spin angular momentum comparable to that of the Earth/Moon system, Mars, Uranus, and Neptune,

$e_H = 2.0$, $i_H = 0.0$, $r_H = 0.1$

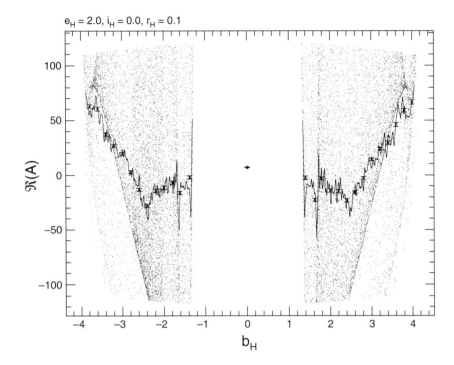

**Fig. 2.** The induced rotation rate, $\Re$, of a planet of radius $r_H = 0.01$ on a circular orbit within a flat disk of planetesimals with eccentricities $e_H = 2$ is shown as a function of initial separation in semimajor axis between the planetesimal and the planet, $b_H$. The points represent individual planetesimal trajectories. The curves are triangularly weighted running averages of 100–200 consecutive points. The crosses represent regions of width 0.2–0.4 in $b_H$; the error bars give statistical uncertainties. Plot from *LK91*; see *Kary* (1993) for similar plots using other parameters.

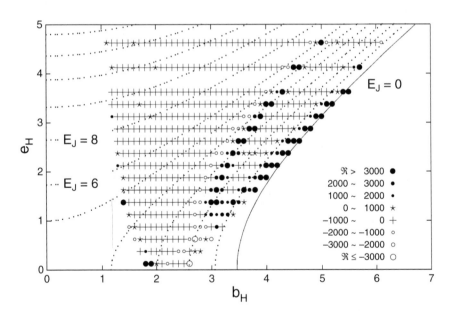

**Fig. 3.** The average spin angular momentum (measured in spins per orbit) brought by planetesimals to a planet with a radius of $r_H = 0.005$, as a function of planetesimals' initial $b_H$ and $e_H$, with $\langle e_H^2 \rangle^{1/2} = 2$ and $i_H = 0$ (after *Ohtsuki and Ida*, 1998). The contour of the Jacobi integral $E_J$ is also plotted by the dashed curves, with the interval of 1 for $0 \le E_J \le 6$, and 2 for $E_J \ge 8$.

if the assumed maximum value for $E_J$ is as small as unity (*Ohtsuki and Ida*, 1998).

Since formation of such gaps is more distinct in the presence of solar nebula gas (*Tanaka and Ida*, 1997), effects of the gas on the planetesimals' orbital evolution need to be directly included for more accurate evaluation of spin accretion. Gas drag causes orbital decay of planetesimals and protoplanets, and protoplanets may also suffer radial migration due to tidal interaction with the solar nebula (*Goldreich and Tremaine*, 1980; *Ward*, 1986, 1997; *Tanaka and Ida*, 1999). If planetesimals' orbits decay slowly as a result of gas drag, they enter the planet's accretion zone at large $b_H$

(Fig. 4; cf. *Kary et al.*, 1993). As spin angular momentum accumulation from planetesimals is roughly symmetrical in $b_H$, planets that migrate inward relative to planetesimals accrete spin angular momentum in an analogous manner (H. Tanaka and K. Ohtsuki, personal communication, 1999). These numerical simulations directly incorporate the effect of orbital decay, and they show that, for certain drag rates and preencounter eccentricities, a planet at the location of the present Earth could have acquired rapid prograde rotation. However, in most of the cases investigated so far by the simulations that include orbital decay (*Kary et al.*, 1993; *Kary and Lissauer*, 1995; H. Tanaka and K. Ohtsuki, per-

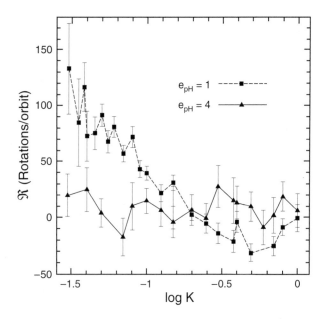

**Fig. 4.** The rotation rate (measured in spins per orbit) of a $10^{-6} M_\star$ ($M_\star$ is the mass of a central star) planet with a radius of $r_H = 0.05$ that accretes most of its spin angular momentum from migrating planetesimals with drag parameter K [see *Kary and Lissauer* (1995) for the definition of K; K is inversely proportional to the radius of a planetesimal, and the orbits of particles with large K decay more rapidly], as a function of the logarithm of K, for two different planetary eccentricities (after *Kary and Lissauer,* 1995).

sonal communication, 1999), the rotation rate has been found not to be large enough to account for the angular momentum of the Earth/Moon system. If this is the case, the random component of planetary spin imparted by large impactors must provide most of the spin angular momentum of the terrestrial planets (section 4).

Formation of gaps and resultant rapid prograde rotation would be prevented or weakened by several effects. First, rapid migration of planetesimals or a planet does not allow gaps to fully develop. If drag is slow, then the planet "sees" a gap, and accumulates positive spin angular momentum, whereas if drag is rapid, then little net rotational angular momentum is received (Fig. 4; cf. *Kary et al.,* 1993). Similarly, a migrating planet acquires rapid prograde rotation only if its migration is sufficiently slow (H. Tanaka and K. Ohtsuki, personal communication, 1999). For planetesimals larger than meter-sized, orbits of smaller planetesimals decay more rapidly than those of larger ones (*Adachi et al.,* 1976; *Weidenschilling,* 1977). In a realistic swarm of planetesimals with a size distribution, particles with various sizes but smaller than a certain critical size would enter the planet's feeding zone. These particles would provide both prograde and retrograde rotation for a planet depending on their rates of orbital decay. Thus a swarm of planetesimals with a broad size distribution could not produce very rapid prograde rotation. Second, large orbital eccentricity of a planet is another factor that prevents acquisition of rapid

prograde rotation. A planet can acquire rapid prograde rotation due to the effect of nonuniform planetesimal distribution if it is on a circular orbit (*LK91*) or an orbit with small eccentricity (i.e., the planet's radial excursions due to its eccentricity must be smaller than the radius of its Hill sphere), but it cannot if the planet follows a more eccentric orbit (*Kary and Lissauer,* 1995; *LBGK97*; Fig. 4). Finally, size and separation between protoplanets would also affect planetary spins. If two massive protoplanets are formed with a small radial separation, planetesimals scattered gravitationally by one protoplanet would be distributed almost uniformly within the feeding zone of the other (*Tanaka and Ida,* 1997). In this case, the net rotation acquired by these protoplanets would be nearly zero or retrograde, as expected for spin accretion in a uniform disk (section 2).

## 4. THE RANDOM COMPONENT OF PLANETARY SPIN

As discussed in section 1, accretion of small bodies produces a nonzero spin angular momentum oriented along the z direction, that is, perpendicular to the planet's orbital plane. However, the spin angular momenta of the planets also have components within their orbital planes, so that the obliquities are nonzero. Many authors have proposed that large impacts may give rise to the obliquities of the terrestrial planets (*Safronov,* 1966; *Hartmann and Vail,* 1986, *Lissauer and Safronov,* 1991). In this view, most of the angular momenta of the terrestrial planets still arises from the accretion of small bodies (sections 2 and 3). However, it appears equally plausible that most or all of the spin angular momenta of the terrestrial planets resulted from one or a few large impacts (*Dones and Tremaine,* 1993a).

The magnitude of the spin angular momentum, **L**, imparted to a planet of radius $R_p$ by an impactor of mass $m_1$ with impact velocity $v_i$ is of order

$$L = \frac{1}{2} m_1 R_p v_i \qquad (12)$$

where the factor 1/2 assumes that the impactor undergoes strong gravitational focusing, so that the mean impact parameter $\langle p \rangle = R_p/2$ ($\langle p \rangle = R_p/\sqrt{2}$ in the opposite limit of a nongravitating planet). If we assume that the planet's spin results from a single impact, the resulting spin rate is

$$\omega = L/I \qquad (13)$$

where

$$I = \alpha M_p R_p^2 \qquad (14)$$

$M_p$ is the planet's mass, and $\alpha$ is a constant that equals 0.4 for a uniform sphere and has a value between 0.33 and 0.37 for the terrestrial planets. In the strong focusing limit, the impact velocity is of order the planet's escape velocity

$$v_e = \sqrt{\frac{2GM_p}{R_p}} \qquad (15)$$

We then find that the typical spin rate

$$\omega \propto \frac{m_1}{M_p} \sqrt{\frac{GM_p}{R_p^3}} \qquad (16)$$

or equivalently, the typical spin period

$$T \equiv \frac{2\pi}{\omega} \propto (G\rho)^{-1/2} \frac{M_p}{m_1} \qquad (17)$$

*Dones and Tremaine* (1993a,b) and *Tremaine and Dones* (1993) investigated the possibility that large impacts provided most of the spin angular momentum of the terrestrial planets, assuming that the terrestrial planets and their impactors belonged to the same distribution. Specifically, the number of objects N of mass greater than m was assumed to follow $N(m) \propto m^{-\gamma}$, with $\gamma \leq 1$. The condition $\gamma \leq 1$ implies that most of the mass and angular momentum are accreted in a few giant impacts. For $0.5 \leq \gamma < 1$, *Dones and Tremaine* (1993b) and *Tremaine and Dones* (1993) find a root-mean-square spin rate

$$\omega_{rms} = 2 \frac{m_{max}}{M_p} \sqrt{\frac{GM_p}{R_p^3}} \qquad (18)$$

or, equivalently, a spin period

$$P = 1.5 \text{ hours} \frac{M_p}{m_{max}} \rho^{-1/2} \qquad (19)$$

(cf. equation (1b)), where $\rho$ is measured in g cm$^{-3}$, and the mean mass of the largest impactor

$$\left\langle \frac{m_{max}}{M_p} \right\rangle \sim 0.2 \qquad (20)$$

Predicted spin periods for Earth and Mars from equations (19) and (20) are of order 3–4 hr. While the Earth's primordial spin period may have been of this order (see *Touma*, 2000), Mars' primordial period was probably comparable to its present value, 24.6 h. However, the predicted distribution of spin periods has a tail at long periods, reflecting the wide statistical range possible in the mass of the largest impactor. Giant impacts thus provide a plausible explanation for the primordial spin periods of both Earth and Mars.

These results on the random component of spin are broadly similar to those of *Lissauer and Safronov* (1991). Lissauer and Safronov focus mostly on scenarios in which only part of the planet's spin angular momentum arises from large impacts, with the rest arising from small-body accretion, whereas here we have assumed that all of the spin comes from large impacts. Also, Lissauer and Safronov assume a truncated power-law mass distribution, so that $m_{max}$ and $\gamma$ are independent variables, whereas in the above discussion, $\gamma$ is the only independent variable. Lissauer and Safronov's scaling relation (spin rate $\omega \propto m_{max}^{1/2}$) is there-fore different from equation (18). However, for typical values of $m_{max}/M_p \sim 0.1$–0.2, the spin rates predicted by the two models are comparable unless $\gamma$ is very close to 1.

The maximum spin rate for a spherical planet (equation (1)) implies minimum spin periods of less than 2 h for Earth and Mars. A single grazing impact into a homogeneous planet can produce a spin exceeding this rate if $m_{max} \geq 0.28 M_p$, suggesting that large impacts can result in incomplete accretion of the impactor. "Incomplete accretion" is, of course, the whole point of the Moon-forming impact hypothesis! In a more realistic model, some ejecta will be placed in orbit in somewhat smaller impacts by tidal torques and/or hydrodynamic effects. *Cameron and Benz* (1991; see also *Cameron*, 2000) have argued that large amounts of mass and angular momentum were lost in the impact that formed the Moon. An improved model of the spin imparted by giant impacts will have to consider fragmentation and loss of ejecta, as in models of the rotation of asteroids (*Farinella et al.*, 1992).

Large impacts should produce a broad distribution of obliquities, with both prograde and retrograde spin, although in some cases the theoretical distribution is not isotropic (for an isotropic distribution, the probability, $\mathcal{P}$, of an obliquity, $\varepsilon$, is $\mathcal{P}(\varepsilon) = 1/2 \sin \varepsilon$). The small prograde obliquities of Earth and Mars have sometimes been used to argue against giant impacts as the source of these planets' spin. However, the obliquities of both planets may have evolved substantially since the epoch of planetary accretion, so the present obliquities do not provide a strong constraint (see section 6).

## 5. N-BODY SIMULATIONS OF PLANETARY ACCRETION

Computational power has recently advanced to the point that the late stages of growth of terrestrial planets can be simulated using direct N-body integrations of (initially) several dozen large planetesimals. Such simulations can also follow the rotational angular momentum that the growing planets accumulate through accretional impacts.

*Agnor et al.* (1999) and J. Chambers (personal communication, 1999) have found that collisions during the late stages of terrestrial planet accumulation do not have a strong bias in the direction of the rotational angular momentum that they provide the planet (Fig. 5). Thus substantial cancellation of the contributions of different impacts occurs, and planetary rotation axes are (at least approximately) randomly distributed in space. The impacts provide large amounts of spin angular momentum, so despite their quasirandom directions, they cause the growing planets to rotate very rapidly, at nearly the breakup rate.

*Rivera et al.* (1998) have modeled the dynamics leading to the giant impact that is believed to have formed the Earth-Moon system. They begin with the current system of planets from Mercury to Neptune, except that the Earth-Moon system is replaced by two bodies on heliocentric orbits between Venus and Mars that have total mass and orbital

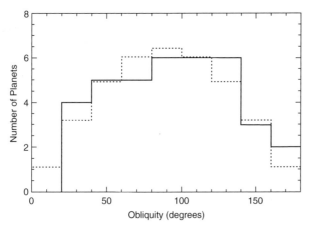

**Fig. 5.** Distribution of the obliquities for the "final planets" that experienced at least one large impact in the numerical simulations of the late stages of planetary growth performed by *Agnor et al.* (1999). The dotted line indicates the theoretical obliquity distribution for randomly oriented spin axes.

angular momentum equal to the Earth and Moon. Integrations are stopped when the first collision occurs. The spin of the body resulting from this merger appears to be randomly oriented and is usually rapid, although only occasionally is it within a factor of 1.5 of breakup. Impacts between two bodies of comparable mass typically produce the most rapidly rotating planets.

Various approximations were made to simplify the above-mentioned calculations. Fragmentation was neglected, and small bodies and gas were omitted from the simulations. Moreover, possibly as a result of these approximations, the "final" planets resulting from the calculations have more eccentric and inclined orbits than do the terrestrial planets in our solar system (cf. *Chambers and Wetherill*, 1998). Thus, although the accretion simulations add strong support to the hypothesis that the random component dominates the spin of terrestrial planets, the calculations run to date should not be viewed as definitive.

## 6. EVOLUTION OF PLANETARY SPINS

A number of physical processes can change the spins of planets after their formation (*Harris and Ward*, 1982). Indeed, the spin states of all four of the terrestrial planets have evolved substantially since their formation. We list below the relevant processes, describe the evolution of the spin states of the Earth-Moon system and Mars, and briefly comment upon the spins of Mercury and Venus. The rotation of asteroids can be substantially modified by impacts that occur at velocities far greater than the escape speeds of the bodies (*Davis et al.*, 1989; *Love and Ahrens*, 1997; *Asphaug and Scheeres*, 1999); as this process is not believed to be important for the major terrestrial planets, we do not discuss it here.

### 6.1. Spin-Orbit Coupling

The Sun exerts a torque on the equatorial bulge of a planet that makes the planet's spin axis precess like a top. (For the Earth, the Moon also exerts a torque that is twice as large as the solar torque.) The orbits of the planets also precess as a result of their mutual gravitational interactions. Thus a planet's obliquity can change because of changes in the orientation of either its spin axis or its orbit pole. If the spin axis precesses rapidly compared with the orbital precession, the spin axis follows the instantaneous orbit pole (i.e., vector normal to the orbit), and the obliquity remains nearly constant. Mercury, Venus, and Earth currently are in this regime, with Earth's obliquity remaining in the range of 22.0°–24.6° over the last 18 m.y. (*Laskar et al.*, 1993a). On the other hand, if the spin axis precesses slowly, it follows the time-averaged orbit pole, and the obliquity varies in proportion to the planet's inclination to the invariable plane of the solar system. The giant planets exhibit this type of variation.

The model of spin/orbit coupling described above assumes that the planetary orbits vary with time in a regular fashion. In the simplest "secular" theory, which treats a system of N planets as coupled harmonic oscillators, the orbital inclinations of the planets can be described as the sum of N–1 sinusoidal terms with periods ranging from roughly 50,000 yr to 2 m.y. (*Harris and Ward*, 1982). If the precession period of a planet's spin axis is close to one of these secular periods, a near-resonance results, and large obliquity variations can occur. It has been known since the work of *Ward* (1973) that a near-resonance causes Mars's obliquity to vary from 15° to 35° over timescales of 1 m.y.

The actual variations in Mars' obliquity are likely to be even larger for two reasons, involving changes in either the secular periods or the planetary spin precession period that may have brought the two periods closer to resonance. First, we now believe that the orbits of the planets are chaotic. That is, the orbits of the planets (and consequently, their obliquities) cannot be predicted 100 m.y. into the past or future, because gravitational interactions among the planets cause the uncertainties in the orbits to grow exponentially with time. [Said chaos probably does not result in drastic changes in the sizes and shapes of the planetary orbits over the age of the solar system (*Laskar*, 1997), but this result has not been rigorously shown.] Because of chaos, the planetary orbits vary with a large number of frequencies; in fact, the planetary orbits cannot be well-represented over long timescales as a sum of sine waves. Thus there are many more opportunities for near-resonances than in the simple secular theory; in general, obliquity variations are larger when the full complexity of the planetary orbits is considered. Over long timescales, only statistical statements can be made about the likely variations of Mars's obliquity. *Touma and Wisdom* (1993) infer a range of 11°–49° during the last 80 m.y., and *Laskar and Robutel* (1993) argue that obliquities can range from 0° to 60° further in the past.

Second, obliquity variations can be amplified by changes in the period of precession of a planet's spin axis. This period is inversely proportional to the planet's departure from a spherical shape (i.e., its oblateness). Prior to the formation of the Tharsis Ridge of volcanos on Mars, Mars' spin precession period would have been closer to resonance with one of the secular periods (*Ward et al.,* 1979; *Bills,* 1990; *Ward,* 1992). *D. Williams et al.* (1998) propose a controversial theory that a feedback between the Earth's obliquity and its oblateness could have reduced the Earth's obliquity by tens of degrees in less than 100 m.y. some 500 m.y. ago, if the continents were situated so as to promote the formation of large polar ice sheets. *D. Williams and Pollard* (2000) discuss this scenario in detail.

In contrast with Mars, the Earth at present experiences very small obliquity variations because its spin axis has a precession period of 26,000 yr, shorter than any of the secular periods. The Moon is largely responsible for causing this rapid precession (*Laskar et al.,* 1993b). Without the Moon, the Earth's precession period would be ~81,000 yr, and its obliquity oscillations would be like those of Mars (*Ward,* 1982, 1992). However, in the long term, as the Moon continues to evolve outward from the Earth due to tides, a resonance will occur, resulting in large obliquity variations (*Ward,* 1982, 1992; *Tomasella et al.,* 1996; *Néron de Surgy and Laskar,* 1997; see below).

### 6.2. Tides

Whereas spin/orbit coupling changes obliquities without changing spin rates, tidal friction changes both spin rates and obliquities. "Solid-body" tides in the planets driven by the Sun (and the Moon, in the case of the Earth) dissipate energy in the planets, and transfer angular momentum between rotational and orbital motion (*Goldreich,* 1966). For a planet whose rotation period is shorter than the orbital period of its "satellite" (the Sun or a moon), and whose spin is prograde, tidal friction slows the planet's rotation, and causes the satellite's orbit to spiral outward. Thus, in the past, the Earth spun faster, and the Moon was closer to the Earth. These changes have been measured from the Apollo lunar laser ranging experiment (*Dickey et al.,* 1994), astronomical records, and bands in fossils and sedimentary layers (*G. E. Williams,* 1993; *Sonett et al.,* 1996; *Sonett and Chan,* 1998) for almost 1 G.y. into the past.

Tidal friction causes a planet's obliquity to evolve toward a specific prograde value for any initial rotation rate; this value is close to zero for a slow rotator and close to 90° for a fast rotator (*Goldreich and Peale,* 1970). Since the Earth rotates rapidly, its obliquity is increasing slowly over geologic time, and may have been only 10° some 4 G.y. ago (*Goldreich,* 1966; *Touma and Wisdom,* 1994). [*G. E. Williams* (1993) discusses possible evidence that the Earth's obliquity was higher until ~0.5 G.y. ago; *D. Williams et al.* (1998; also *D. Williams and Pollard,* 2000) provide a possible mechanism for a recent rapid decrease of Earth's obliquity.]

However, Earth's spin state at very early times is uncertain (*Touma and Wisdom,* 1994, 1998): When the Moon was much closer to Earth, tidal interactions were complex. Before the Earth acquired the Moon, it may have undergone large, chaotic obliquity variations, as likely occurred for the other terrestrial planets (*Laskar and Robutel,* 1993). The usual end result of tidal evolution is synchronous rotation such as that of the Moon and the Pluto/Charon system, in which the spin period equals the orbital period. In the distant future, the Earth's rotation may reach the synchronous state, but this state is unstable due to solar tides.

*Néron de Surgy and Laskar* (1997) simulated 500 possible scenarios for the evolution of the Earth's spin for the next 5 G.y. The large number of simulations is necessary because the spin evolves chaotically, and because of uncertainties in the future value of the Earth's tidal coefficient and the amount of core-mantle coupling within the Earth. In their standard model (i.e., present-day tidal coefficient, no core-mantle coupling), the Earth's spin state begins to undergo chaotic variations 1.5 G.y. in the future, and in a typical case, the obliquity sometimes reaches values larger than 80°. This evolution is qualitatively like that predicted by *Ward* (1982), but the obliquity reaches even larger values than Ward predicted. Néron de Surgy and Laskar also point out that the scenario by *G. E. Williams* (1993) for rapid decrease of the Earth's obliquity after 630 Ma implies an implausibly large despinning of the Earth.

Mercury and Venus have been despun by solar tides. Atmospheric tides have also affected Venus's spin (*Dobrovolskis,* 1980). Possible rotation histories of Mercury are discussed by *Peale* (1988), while *Yoder* (1997) and *Néron de Surgy and Laskar* (1999) treat Venus. Precise measurements by spacecraft might be able to constrain the primordial rotation states of both planets (*Peale,* 1989).

## 7. CONCLUSIONS

Terrestrial planets accumulate rotational angular momentum from the relative motions of colliding and accreting planetesimals. The angular momentum provided by individual impacts can be in any direction. The net rotation of a planet may be produced by a systematic directional bias in the distribution of accreted spin angular momentum and/or from the stochastic effects of accretion of large impactors. The low obliquities of six of the eight largest planets in the solar system suggest the possible importance of an ordered component of spin angular momentum accumulated during the growth of planetary bodies. However, planetary rotation can be altered subsequent to accretion, so spins may have been more randomly oriented early in the solar system's history. Moreover, calculations of terrestrial planet growth from planetesimals yield planets whose rotational angular momenta are dominated by one or a few large impacts, and are quasirandomly directed. Thus, despite considerable advances over the past decade, our understanding of the process of planetary growth is not yet sufficient to reach definitive conclusions regarding the accumulation of planetary spin angular momentum. We believe that when the obliquities of many extrasolar terrestrial planets becomes known, a wide variety of values will be represented. However, we do not wish to speculate what biases such a distribution might contain (apart from a likely concentration of low obliquities among strongly tidally slowed airless planets).

## APPENDIX

*LK91* presented analytic and numerical calculations of the origin of the systematic component of the rotation of a planet accreting while on a circular orbit; *LBGK97* extended these calculations to planets on eccentric orbits. The final appendix of *LK91* briefly summarizes the history of studies of the origin of planetary rotation. The final appendix of *LBGK97* discusses studies published from 1991 to 1997 as well as some earlier work neglected by the previous summary. We update these historical accounts in this appendix.

*Greenberg et al.* (1997) performed "backward" numerical integrations (i.e., outward from a planet's Hill sphere) for planetesimals orbiting in the planet's equator plane. Their goal was to map out how the angular momentum imparted by the planetesimals the planet varies with the planetesimals' semimajor axis and eccentricity. Their results are consistent with *LK91* and with *Dones and Tremaine* (1993a).

*Ohtsuki and Ida* (1998) presented detailed numerical simulations of the accumulation of planetary rotational angular momentum by accretion of planetesimals in a nonuniform disk. Their results are reviewed in section 3 of this chapter. Several groups (including researchers at the Southwest Research Institute in Boulder, the University of Washington, and NASA Ames Research Center) are following the spin of growing planets as part of N-body simulations of the late stages of terrestrial planet growth; this work is discussed in section 5.

*Néron de Surgy and Laskar* (1999) have shown that planetary perturbations and thermal atmospheric tides may have produced chaotic variations in the obliquity of Venus so large to have reversed the direction of the planet's spin. Thus Venus's current retrograde rotation does not necessarily imply that it was spinning backward at the end of the accretionary epoch. *Touma and Wisdom* (1998) have shown that passage through a resonance with the Sun may have altered the angular momentum of the Earth-Moon system when the Moon was less than 10 planetary radii from Earth; this study is reviewed in *Touma* (2000).

***Acknowledgments.*** We thank Anthony Dobrovolskis and Eugenio Rivera for valuable comments on the manuscript. This work was supported by the NASA Planetary Geology and Geophysics program through RTOPs 344-30-50-01 and 344-30-51-04.

## REFERENCES

Adachi I., Hayashi C., and Nakazawa K. (1976) The gas drag effect on the elliptical motion of a solid body in the primordial solar nebula. *Prog. Theor. Phys., 56,* 1756–1771.

Agnor C. B., Canup R. M., and Levison H. F. (1999) On the character and consequences of large impacts in the late stage of terrestrial planet formation. *Icarus, 142,* 219–237.

Asphaug E. and Scheeres D. J. (1999). Deconstructing Castalia: Evaluating a post-impact state. *Icarus, 139,* 383–386.

Bills B. G. (1990) The rigid body obliquity history of Mars. *J. Geophys. Res., 95,* 14137–14153.

Brush S. G. (1978) The Chamberlin-Moulton cosmogony. *J. Hist. Astron., 9,* 1–41, 77–104.

Brush S. G. (1981) From bump to clump: Theories of the origin of the Solar System 1900–1960. In *Space Science Comes of Age: Perspectives in the History of the Space Sciences,* pp. 78–100. Smithsonian Institution, Washington, DC.

Cameron A. G. W. (2000) Higher-resolution simulations of the giant impact. In *Origin of the Earth and Moon* (R. M. Canup and K. Righter, eds.), this volume. Univ. of Arizona, Tucson.

Cameron A. G. W. and Benz W. (1991) The origin of the Moon and the single impact hypothesis IV. *Icarus, 92,* 204–216.

Cameron A. G. W. and Ward W. R. (1976) The origin of the Moon (abstract). In *Lunar Science VII,* p. 120. The Lunar Science Institute, Houston.

Chamberlin T. C. (1897). A group of hypotheses bearing on climate change. *J. Geol., 5,* 653–683.

Chambers J. E. and Wetherill G. W. (1998) Making the terrestrial planets: N-body integrations of planetary embryos in three dimensions. *Icarus, 136,* 304–327.

Davis D. E., Weidenschilling S. J., Farinella P., Paolicchi P., and Binzel R. P. (1989) Asteroid collisional history: Effects on sizes and spins. In *Asteroids II* (R. P. Binzel, T. Gehrels, and M. S. Matthews, eds.), pp. 805–826. Univ. of Arizona, Tucson.

Dickey J. O., Bender P. L., Faller J. E., Newhall X X, Ricklefs R. L., Ries J. G., Shelus P. J., Veillet C., Whipple A. L., Wiant J. R., Williams J. G., and Yoder C. F. (1994) Lunar laser ranging: A continuing legacy of the Apollo program. *Science, 265,* 482–490.

Dobrovolskis A. R. (1980) Atmospheric tides and the rotation of Venus. II. Spin evolution. *Icarus, 41,* 19–35.

Dones L. and Tremaine S. (1993a) On the origin of planetary spins. *Icarus, 103,* 67–92.

Dones L. and Tremaine S. (1993b) Why does the Earth spin forward? *Science, 259,* 350–354.

Durisen R. H. and Gingold R. A. (1986) Numerical simulations of fission. In *Origin of the Moon* (W. K. Hartmann, R. J. Phillips, and G. J. Taylor, eds.), pp. 487–498. Lunar and Planetary Institute, Houston.

Farinella P., Davis D. R., Paolicchi P., Cellino A., and Zappalà V. (1992) Asteroid collisional evolution: An integrated model for the evolution of asteroid rotation rates. *Astron. Astrophys., 253,* 604–614.

Giuli R. T. (1968) On the rotation of the Earth produced by gravitational accretion of particles. *Icarus, 8,* 301–323.

Goldreich P. (1966) History of the lunar orbit. *Rev. Geophys., 4,* 411–439.

Goldreich P. and Peale S. J. (1970) The obliquity of Venus. *Astron. J., 75,* 273–283.

Goldreich P. and Tremaine S. (1980) Disk-satellite interactions. *Astrophys. J., 241,* 425–441.

Greenberg R., Fischer M., Valsecchi G. B., and Carusi A. (1997) Sources of planetary rotation: Mapping planetesimals' contributions to angular momentum. *Icarus, 129,* 384–400.

Harris A. W. and Ward W. R. (1982) Dynamical constraints on the formation and evolution of planetary bodies. *Annu. Rev.*

*Earth Planet. Sci., 10,* 61–108.

Hartmann W. K. and Davis D. R. (1975) Satellite-sized planetesimals and lunar origin. *Icarus, 24,* 504–515.

Hartmann W. K. and Vail S. M. (1986) Giant impactors: Plausible sizes and populations. In *Origin of the Moon* (W. K. Hartmann, R. J. Phillips, and G. J. Taylor, eds.), pp. 551–566. Lunar and Planetary Institute, Houston.

Hill G. W. (1878) Researches in the lunar theory. *Am. J. Math., 1,* 5–26, 129–147, 245–260.

Ida S. and Makino J. (1993) Scattering of planetesimals by a protoplanet: Slowing down of runaway growth. *Icarus, 106,* 210–227.

Ida S. and Nakazawa K. (1990) Did rotation of the protoplanets originate from the successive collisions of planetesimals? *Icarus, 86,* 561–573.

Kary D. M. (1993) Planetesimal dynamics in the vicinity of a growing protoplanet. Ph.D. thesis, State University of New York at Stony Brook.

Kary D. M. and Lissauer J. J. (1995) Nebular gas drag and planetary accretion. II. Planet on an eccentric orbit. *Icarus, 117,* 1–24.

Kary D. M., Lissauer J. J., and Greenzweig Y. (1993) Nebular gas drag and planetary accretion. *Icarus, 106,* 288–307.

Laplace P. S. (1796) *Exposition du Système du Monde.* Cercle-Social, Paris, Vol. II. (English translation by Pond J. (1809), Richard Phillips, London.)

Laskar J. (1997) Large scale chaos and the spacing of the inner planets. *Astron. Astrophys., 317,* L75–L78.

Laskar J. and Robutel P. (1993) The chaotic obliquity of the planets. *Nature, 361,* 608–612.

Laskar J., Joutel F., and Boudin F. (1993a) Orbital, precessional, and insolation quantities for the Earth from –20 Myr to +10 Myr. *Astron. Astrophys., 270,* 522–533.

Laskar J., Joutel F., and Robutel P. (1993b) Stabilization of the earth's obliquity by the moon. *Nature, 361,* 615–617.

Lissauer J. J. (1993) Planet formation. *Annu. Rev. Astron. Astrophys., 31,* 129–174.

Lissauer J. J. and Kary D. M. (1991) The origin of the systematic component of planetary rotation. I. Planet on a circular orbit. *Icarus, 94,* 126–159.

Lissauer J. J. and Safronov V. S. (1991) The random component of planetary rotation. *Icarus, 93,* 288–297.

Lissauer J. J., Pollack J. B., Wetherill G. W., and Stevenson D. J. (1995) Formation of the Neptune system. In *Neptune and Triton* (D. P. Cruikshank, ed.), pp. 37–108. Univ. of Arizona, Tucson.

Lissauer J. J., Berman A. F., Greenzweig Y., and Kary D. M. (1997) Accretion of mass and spin angular momentum by a planet on an eccentric orbit. *Icarus, 127,* 65–92.

Love S. G. and Ahrens T. J. (1997) Origin of asteroid rotation rates in catastrophic impacts. *Nature, 386,* 154–156.

Néron de Surgy O. and Laskar J. (1997) On the long term evolution of the spin of the Earth. *Astron. Astrophys., 318,* 975–989.

Néron de Surgy O. and Laskar J. (1999) On the past evolution of the spin of Venus. *Astron. Astrophys.,* in press.

Ohtsuki K. and Ida S. (1998) Planetary rotation by accretion of planetesimals with nonuniform spatial distribution formed by the planet's gravitational perturbation. *Icarus, 131,* 393–420.

Peale S. J. (1988) The rotational dynamics of Mercury and the state of its core. In *Mercury* (F. Vilas, C. R. Chapman, and M. S. Matthews, eds.), pp. 461–493. Univ. of Arizona, Tucson.

Peale S. J. (1989) Some unsolved problems in evolutionary dynamics in the solar system. *Cel. Mech. Dyn. Astron., 46,* 253–275.

Pravec P., Wolf M., and Sarounova L. (1998) Lightcurves of 28 near-Earth asteroids. *Icarus, 136,* 124–153.

Rivera E., Lissauer J. J., Duncan M. J., and Levison H. F. (1998) The dynamical evolution of the Earth-Moon progenitors: II. Results and interpretation (abstract). In *Origin of the Earth and Moon,* p. 38. LPI Contribution No. 957, Lunar and Planetary Institute, Houston.

Safronov V. S. (1966) Sizes of the largest bodies falling onto the planets during their formation. *Sov. Astron., 9,* 987–991.

Safronov V. S. (1969) *Evolution of the Protoplanetary Cloud and Formation of the Earth and Planets.* Nauka, Moscow (in Russian). (English translation NASA TTF–677, 1972.)

Sonett C. P. and Chan M. A. (1998) Neoproterozoic Earth-Moon dynamics: Rework of the 900 Ma Big Cottonwood Canyon tidal laminae. *Geophys. Res. Lett., 25,* 539–542.

Sonett C. P., Kvale E. P., Zakharian A., Chan M. A., and Demko T. M. (1996) Late Proterozoic and Paleozoic tides, retreat of the Moon, and rotation of the Earth. *Science, 273,* 100–104. (Corrections in *Science, 273,* 1325, and *Science, 274,* 1065–1069.)

Tanaka H. and Ida S. (1997) Distribution of planetesimals around a protoplanet in the nebula gas. II. Numerical simulations. *Icarus, 125,* 302–316.

Tanaka H. and Ida S. (1999) Growth of a migrating protoplanet. *Icarus, 139,* 350–366.

Tanikawa K., Manabe S., and Broucke R. (1989) On the origin of the spin angular momentum by accretion of planetesimals: Property of collision orbits. *Icarus, 79,* 208–222.

Tomasella L., Marzari F., and Vanzani V. (1996). Evolution of the Earth obliquity after the tidal expansion of the Moon orbit. *Planet. Space Sci., 44,* 427–430.

Touma J. (2000) The phase space adventure of Earth and Moon. In *Origin of the Earth and Moon* (R. M. Canup and K. Righter, eds.), this volume. Univ. of Arizona, Tucson.

Touma J. and Wisdom J. (1993) The chaotic obliquity of Mars. *Science, 259,* 1294–1297.

Touma J. and Wisdom J. (1994) Evolution of the Earth-Moon system. *Astron. J., 108,* 1943–1961.

Touma J. and Wisdom J. (1998) Resonances in the early evolution of the Earth-Moon system. *Astron. J., 115,* 1653–1663.

Tremaine S. and Dones L. (1993) On the statistical distribution of massive impactors. *Icarus, 106,* 335–341.

Ward W. R. (1973) Large-scale variations in the obliquity of Mars. *Science, 181,* 260–262.

Ward W. R. (1982) Comments on the long-term stability of the Earth's obliquity. *Icarus, 50,* 444–448.

Ward W. R. (1986) Density waves in the solar nebula: Differential Lindblad torque. *Icarus, 67,* 164–180.

Ward W. R. (1992) Long-term orbital and spin dynamics of Mars. In *Mars* (H. Kieffer, B. Jakosky, and C. Snyder, eds.), pp. 298–320. Univ. of Arizona, Tucson.

Ward W. R. (1997) Protoplanet migration by nebula tides. *Icarus, 126,* 261–281.

Ward W. R., Burns J. A., and Toon O. B. (1979) Past obliquity oscillations of Mars — The role of the Tharsis uplift. *J. Geophys. Res., 84,* 243–259.

Weidenschilling S. J. (1977) Aerodynamics of solid bodies in the solar nebula. *Mon. Not. R. Astron. Soc., 180,* 57–70.

Weidenschilling S. J. and Davis D. R. (1985) Orbital resonances in the solar nebula: Implications for planetary accretion. *Icarus, 62,* 16–29.

Wetherill G. W. and Stewart G. R. (1989) Accumulation of a swarm of small planetesimals. *Icarus, 77,* 330–367.

Wetherill G. W. and Stewart G. R. (1993) Formation of planetary embryos: Effects of fragmentation, low relative velocity, and independent variation of eccentricity and inclination. *Icarus, 106,* 190–204.

Williams D. M. and Pollard D. (2000) Earth-Moon interactions: Implications for terrestrial climate and life. In *Origin of the Earth and Moon* (R. M. Canup and K. Righter, eds.), this volume. Univ. of Arizona, Tucson.

Williams D. M., Kasting J. F., and Frakes L. A. (1998) Low-latitude glaciation and rapid changes in the Earth's obliquity explained by obliquity-oblateness feedback. *Nature, 396,* 453–455.

Williams G. E. (1993) History of the Earth's obliquity. *Earth Sci. Rev., 34,* 1–45.

Yoder C. F. (1997) Venusian spin dynamics. In *Venus II: Geology, Geophysics, Atmosphere, and Solar Wind Environment* (S. W. Bougher, D. M. Hunten, and R. J. Phillips, eds.), pp. 1087–1124. Univ. of Arizona, Tucson.

# Accretion of the Terrestrial Planets and the Earth-Moon System

**Robin M. Canup**
*Southwest Research Institute*

**Craig B. Agnor**
*University of Colorado*

Current models for the formation of the terrestrial planets suggest that the final stage of planetary accretion is characterized by collisions between tens to hundreds of lunar to Mars-sized planetary embryos. In this view, large impacts are an inevitable outcome as a system of embryos destabilizes to yield the final few planets. One such impact is believed to be responsible for the origin of the Moon. Improvements in numerical methods have recently allowed for the first direct orbit integrations of the final stage of accretion, which is believed to persist for ~$10^8$ yr. The planetary systems produced by these simulations bear a general resemblence to the terrestrial planets, but on average differ from our system in the final number of planets (fewer), their orbital spacings (wider) and their eccentricities and inclinations (larger). The discrepancy between these predictions and the nearly circular orbits of both Earth and Venus is significant, and is likely a result of the approximations inherent to the late-stage accretion simulations performed to date. Results from these works further highlight the important role of stochastic impact events in determining final planetary characteristics. In particular, impacts capable of supplying the angular momentum of the Earth-Moon system are predicted to be common.

## 1. INTRODUCTION

In the planetesimal hypothesis, the growth of terrestrial planets is the result of the process of collisional accumulation from initially small particles in the protoplanetary disk. The accretion process is typically described in terms of three stages of growth, which are distinguished by our basic understanding of the relevant physical processes involved in forming solid bodies in a particular size range. The first stage involves the formation of kilometer-sized planetesimals from an initial protoplanetary disk of gas and dust. By the end of this stage of growth (discussed in chapter by *Ward, 2000*), planetesimals have reached sizes large enough so that their dynamical evolution is determined primarily by gravitational interactions with the central star and with other planetesimals, rather than by surface, electromagnetic, or sticking forces. The middle stage consists of the accumulation of a swarm of kilometer-sized planetesimals into lunar- to Mars-sized planetary embryos (see chapter by *Kortenkamp et al., 2000*). Numerous works have demonstrated that in this stage, dynamical friction acts to reduce encounter velocities with the largest bodies, facilitating the "runaway" growth of ~$10^{25}$–$10^{27}$ g ($M_\oplus = 5.98 \times 10^{27}$ g) planetary embryos in as little as $10^5$ yr (e.g., *Greenberg et al., 1978; Wetherill and Stewart, 1993*). The last stage then consists of the formation of the final few planets via the collision and merger of tens to hundreds of planetary embryos. Evolution during this period is thought to be driven by distant interactions between the embryos, and requires a few times ~$10^8$ yr (e.g., *Wetherill, 1992*).

In one of the first modern works to examine terrestrial planet formation, *Safronov* (1969) proposed that planets accreted in radially confined feeding zones, in a relatively quiescent manner through the accumulation of small bodies. Developments in the past two decades suggest that the generally localized, runaway stage persists only until bodies grow to the size of the Moon or Mars, leaving many planetary embryos throughout the terrestrial region. Such a system of embryos is dynamically unstable on timescales that are short (~$10^6$ yr) compared to the age of the solar system, and highly energetic collisions between embryos then occur to yield the four terrestrial planets. In this scenario, the characteristics of the final planets are determined mainly by the specifics of the last few large impacts that each planet experiences. The stochastic nature of the final stage thus yields an inherent degree of uncertainty for the outcome of accretion in any given system. Indeed, a wide range of possible planetary architectures is found to arise from even nearly identical initial conditions, suggestive of the great variety of terrestrial planet systems that might exist in extrasolar systems.

The physical and dynamical environment in which the accretion of the terrestrial planets took place is directly relevant to models of lunar formation. In the giant impact scenario, the Moon forms as a result of a single impact with Earth late in its formation history. While works to date have generally considered the various stages of planet accretion and the formation of the Moon separately, a more holistic approach may be required. In particular, models of the proposed lunar-forming impact and the accretion of the Moon have yet to identify a single impact that can yield the final masses of the Earth and Moon, together with the current system angular momentum (*Cameron and Canup*, 1998; *Cameron*, 2000). However, recent terrestrial accretion studies suggest that impacts subsequent to the lunar-forming event may have contributed significantly to the final mass and/or angular momentum of the Earth-Moon system, offering a possible resolution to this dilemma (*Agnor et al.*, 1999).

In this chapter, we review recent simulations of late-stage accretion and discuss the successes and weaknesses of these models (section 2). We then address issues especially relevant to the formation of the Moon via giant impact, including late-stage impact statistics and the potential role of multiple impacts in affecting planetary spin angular momenta (section 3). A brief discussion of open issues is included in section 4.

## 2. ACCRETION OF THE TERRESTRIAL PLANETS

The environment in which the final accretion of terrestrial-type planets takes place is dependent upon the outcome of the preceeding runaway growth stage. Midstage accretion models utilizing both statistical treatments that model the entire terrestrial region, and N-body simulations of runaway growth within a local radial zone of the disk, yield qualitatively similar results: embryos with masses ~0.01–0.1$M_\oplus$, occupying nearly circular, low-inclination orbits after $10^5$–$10^6$ yr. The embryos have typical orbital spacings of $\Delta a \sim 10\,R_{Hill}$, where $R_{Hill}$ is the mutual Hill radius given by

$$R_{Hill} \equiv \frac{a_1 + a_2}{2}\left(\frac{m_1 + m_2}{3M_\odot}\right)^{1/3} \qquad (1)$$

where $a_1$, $a_2$, $m_1$, and $m_2$ are the semimajor axes and masses of adjacent embryos, and $M_\odot$ is the mass of the Sun. A system of two bodies on initially circular orbits will be stable against mutual collision so long as $\Delta a > 3.5\,R_{Hill}$ (e.g., *Gladman*, 1993). However, multiplanet systems (or planets with nonzero initial eccentricities) require larger separations for stability. *Chambers et al.* (1996) performed numerical integrations of like-sized planets intially on circular orbits and found a separation of $\Delta a \approx 8$–$10\,R_{Hill}$ provided stability for ~$10^6$–$10^7$ yr, in fair agreement with the predicted spacings from the midstage accretion simulations (e.g., *Weidenschilling et al.*, 1997). *Ito and Tanikawa* (1999) conducted numerical stability analyses of planetary embryos, including initial eccentricities and inclinations, a range of embryo masses, and Jupiter and Saturn. They found much shorter stability times of ~$10^4$–$10^5$ yr for a system of N = 14 embryos with $\Delta a \approx 8$–$10\,R_{Hill}$ separations and initial values of $\langle e \rangle^{1/2} = 2\langle i \rangle^{1/2} = 0.005$. The latter timescales are close to those found by the late-stage N-body simulations (described below) that begin with similar initial eccentricities. For comparison, the current terrestrial planets have mutual separations ranging from about 26 to 40 $R_{Hill}$.

The boundary between the middle and late stages is generally believed to be representative of a dynamical transition: from a stage in which growth is dominated by collisions with local material, to one in which distant interactions among the embryos lead to collisions on much longer timescales. Recent simulations that model embryo formation in the full terrestrial zone (0.5–1.5 AU) find that 90% of the system mass is contained in a few tens of embryos after about a million years, with the remaining ~10% of the mass contained in a swarm of much smaller planetesimals (*Weidenschilling et al.*, 1997). These simulations have included effects due to gas drag and a parameterization of distant perturbations between embryos, but to date have not included collisional fragmentation. At present, it is not clear to what extent the planetesimal swarm persists or is regenerated via collisional erosion throughout the late stage. It is often assumed that all the small material in the disk would be rapidly swept up by the embryos, since the timescale for embryo formation ($10^5$–$10^6$ yr) is much shorter than the timescale for the accumulation of the final planets (~2 × $10^8$ yr). In this case, the dynamics of the final stage are governed solely by gravitational interactions among and collisions between the large embryos. Interactions among embryos initially on nearly circular orbits lead to eccentricity growth, and then to orbit crossing; once orbital isolation is overcome, the embryos collide and merge. The accumulation of embryos into larger bodies then proceeds until the secular orbital oscillations (primarily the eccentricities) of the remaining bodies are insufficient to allow bodies to encounter each other, and a few planets remain on well-separated orbits. Until recently, simulations of this final stage were limited to statistical treatments due to the large number of orbital times involved. However, numerical techniques now exist that allow for direct integration of systems of N ~ 10–100 embryos for ~$10^8$ yr. Results from simulations using both methods are reviewed in the next section.

### 2.1. Late-Stage Simulations

Late-stage terrestrial accretion has been modeled using two basic (and complementary) techniques. Monte Carlo simulations follow the orbital evolution of embryos in a statistical manner based on two-body scattering events (e.g., *Wetherill*, 1985), while N-body orbital integrations directly track the trajectories of each embryo at all times (*Chambers and Wetherill*, 1998, hereafter *CW98*; *Agnor et al.*, 1999, hereafter *ACL99*). Under comparable sets of assumptions, both methods produce similar configurations of final planets.

*Wetherill* was the first to model the dynamical evolution of systems of planetary embryos throughout the terrestrial region. His Monte Carlo scheme, with its approximate treatment of dynamical interactions, is computationally fast and has been used to generate large numbers of planetary systems for a range of starting conditions and assumptions (e.g., *Wetherill*, 1985, 1992, 1994, 1996). In this method, the probability of each body experiencing a close encounter with any other body is determined using an adaptation of the *Opik* (1951) formalism; the outcome of the encounter is then a function of a randomly selected distance of closest approach. *Wetherill* (1992) presented results of 434 Monte Carlo simulations of the evolution of a few hundred embryos throughout the terrestrial and asteroid belt region. Fragmentation was assumed to occur when the impact energy exceeded a chosen threshold, at which point the total mass involved in a collision was divided into a small number (i.e., 4) of equal-sized pieces. Bodies with masses smaller than $8 \times 10^{25}$ g (or about the mass of the Moon) were removed from the simulation, in order to limit the total number of objects to a computationally manageable level. Effects due to perturbations from and resonances with Jupiter and Saturn were included in the form of simple parameterizations. The planetary systems resulting from these simulations broadly resembled the current planets in number and size, with some discrepancies. In particular, Earth-like planets with large eccentricities were common outcomes; of the final planets with masses greater than $3.5 \times 10^{27}$ g ($0.58$ $M_{\oplus}$), ~40% had eccentricities greater than 0.05 and ~75% had eccentricities larger that of Earth or Venus.

Direct integration methods explicitly track the gravitational interactions among all bodies in a simulation. However, this increase in dynamical accuracy comes at the cost of computational speed, which limits the number of simulations and bodies in a simulation that can be considered. Direct integrations of the late stage have recently become possible due to developments in symplectic integration techniques, which afford system energy and angular momentum conservation for the requisite ~$10^8$ orbits (e.g., *Wisdom and Holman*, 1991; *Duncan et al.*, 1998; *Chambers*, 1998b).

*CW98* performed 27 late-stage integrations that each began with a system of roughly 25–50 embryos on circular orbits distributed throughout the terrestrial region (0.5–1.8 AU). All collisions were assumed to result in complete and inelastic merger. Figure 1 shows the evolution of a system of planetary embryos in mass and semimajor axis space for 100 m.y. (*CW98*, their Fig. 1). The first encounter between embryos often occurs in the inner part of the disk, and results in a scattering of the two bodies involved such that their neighbors then become subject to close encounters. This leads to what is described as a "wave" of close encounters that travels outward through the terrestrial region. Interactions among the embryos, including both resonances and scattering events, drive the accretion process until the final planets are so well-separated and few in number that their mutual perturbations are insufficient to allow for further collisions. Note that in the last frame of Fig. 1 there are still crossing orbits, and so accretion is this case is likely not yet complete.

*CW98* also investigated the effect of Jupiter and Saturn on embryo accretion in the 0.55–4.0-AU region by adding the giant planets (with their current masses and orbits) to a subset of their simulations after $10^7$ yr (see Fig. 2). The extent to which the giant planets were present and at their current locations during the late stage is uncertain. Requir-

ing that Jupiter and Saturn formed before the nebula dispersed implies a formation time of less than ~$10^7$ yr, suggesting that they were most likely present during most of the final accretion period. However, these planets may have migrated significantly during and subsequent to their formation. In general, perturbations from outer giant planets act to decrease the stability of a system of protoplanets (e.g., *Ito and Tanikawa*, 1999), as both mean motion and secular resonances drive large-amplitude eccentricity oscillations. Of particular importance are the $\nu_5$ and the $\nu_6$ secular resonances, which occur where the apsidal precession rate of orbiting material is near to that of Jupiter or Saturn respectively (or at about 0.6 AU for the $\nu_5$ and 2.1 AU for the $\nu_6$, assuming current planetary orbital elements and no nebular gas disk).

The *CW98* simulations indicate that planetary embryos with orbital radii larger than ~1.2 AU are typically scattered into the $\nu_6$ or other, mean-motion resonances where they become dynamically coupled to the outer planets, leading to either collisions with embryos in the terrestrial region or ejection from the solar system following close encounters with Jupiter. In their current positions, Jupiter and Saturn thus appear to prevent the accumulation of embryos into terrestrial-sized planets for a > 1.2 AU. Given that the mass of Mars is only $\approx 0.1$ $M_{\oplus}$, it is possible that Mars itself is a leftover planetary embryo formed via runaway growth (*Wetherill*, 1992; *CW98*). In the inner disk, the $\nu_5$ resonance can lead to embryo removal via collisions with the Sun. In simulations that include Jupiter and Saturn, at least 15% of the total initial embryo mass is typically lost through collisions with the Sun and hyperbolic ejection events (*CW98*).

*ACL99* performed direct integrations of the late stage that began with either 50 (m ~ 0.04 $M_{\oplus}$) or 22 (m ~ 0.10 $M_{\oplus}$) planetary embryos distributed between 0.5 and 1.5 AU with small eccentricities and inclinations (e ~ 0.01 and i ~ 0.05°). In the latter case, the initial embryos were those produced by the *Weidenschilling et al.* (1997) full terrestrial zone simulation of the midstage (see chapter by *Kortenkamp et al.*, 2000). Giant planets were not included. Figure 3 shows the evolution of the mass distribution of embryos from one of the *ACL99* simulations using the *Weidenschilling et al.* (1997) initial conditions. The general pattern of the dynamics is similar in the simulations of both *CW98* and *ACL99*, despite their use of different initial distributions of embryos and different numerical integration methods.

Figure 4 shows all the final planets from the 37 simulations performed in both *CW98* and *ACL99*. Note that the *ACL99* simulations that began with the Weidenschilling et al. initial conditions considered a smaller total planetary embryo mass (~1.8 $M_{\oplus}$) than the *CW98* runs, and thus yield somewhat smaller final planets. The most massive planets tend to form near or interior to 1 AU, where the surface density and the collision frequency are the highest. All simulations of the late stage, whether direct integrations or Monte Carlo simulations, appear to produce planets with distributions similar to Fig. 4 when considering a total mass ~2 $M_{\oplus}$ in the terrestrial region (see, e.g., *Wetherill*, 1992).

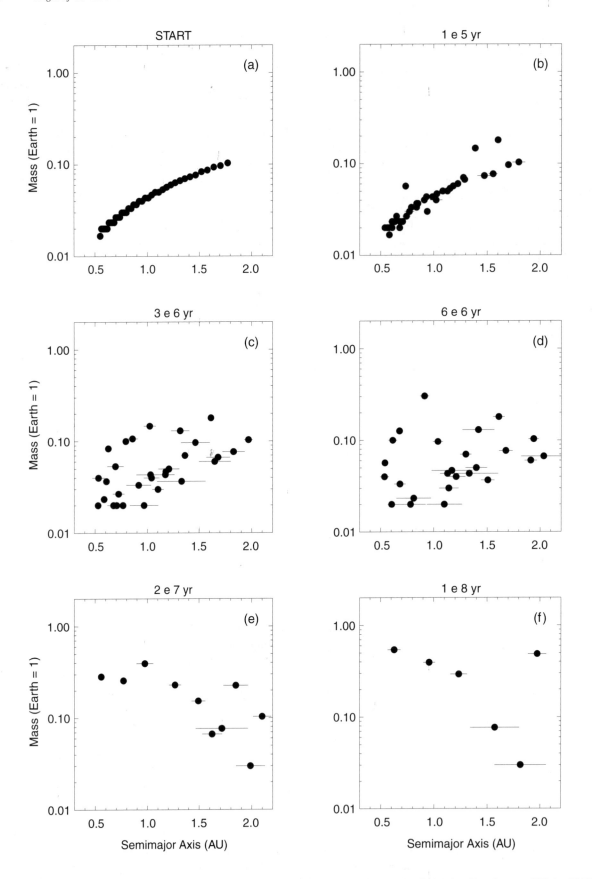

**Fig. 1.** Masses and semimajor axes of the surviving objects at six different times from a simulation by *Chambers and Wetherill* (1998) that began with embryos in the terrestrial region and did not include Jupiter and Saturn (their model A). The horizontal line through each symbol connects the perihelion and aphelion of the embryo's orbit. Mutual interactions between embryos perturb them into crossing orbits and drive the accretion process until the final bodies' orbits are well separated.

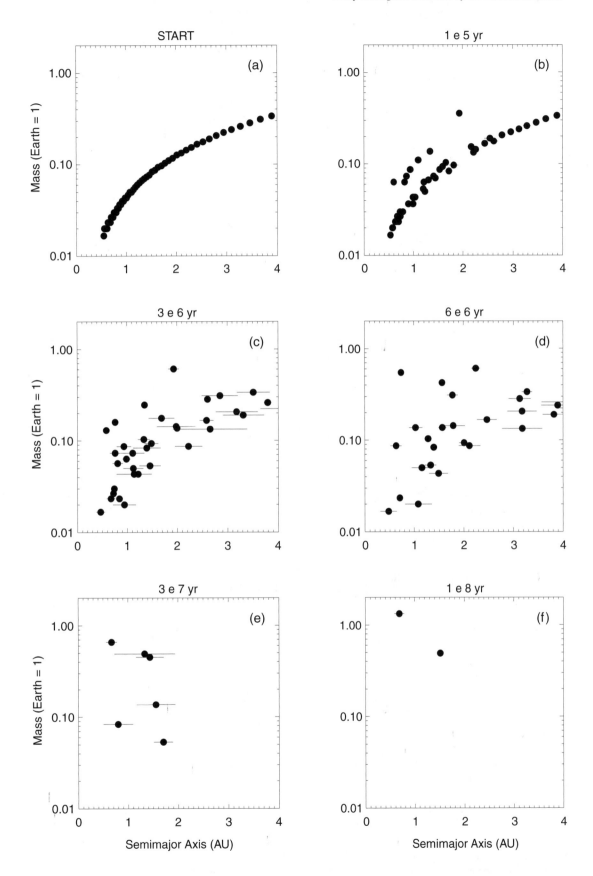

**Fig. 2.** Same as Fig. 1, except this simulation included embryos in the present-day asteroid belt and Jupiter and Saturn (from *Chambers and Wetherill,* 1998, an example of their model C). In this case, the asteroid belt is cleared of embryos and the terrestrial region has a smaller number of final planets than in Fig. 1.

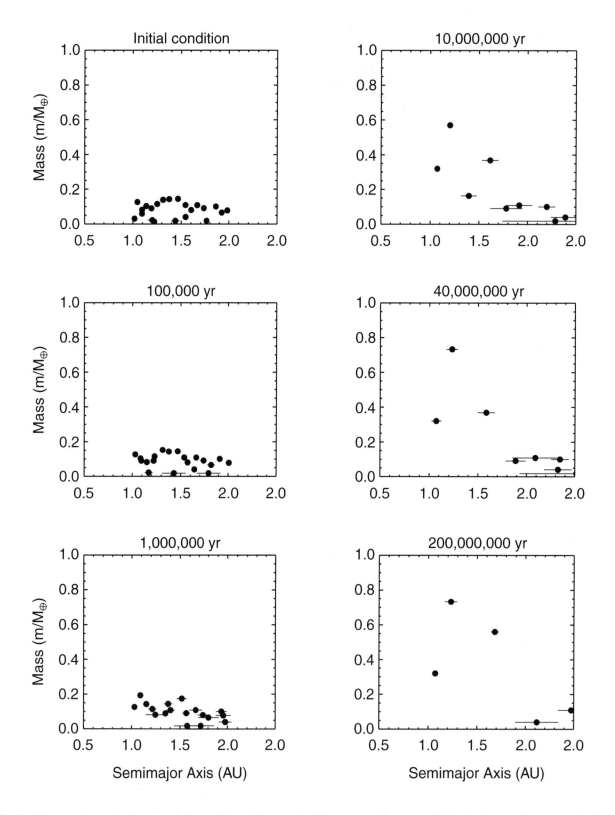

**Fig. 3.**    Masses and semimajor axes of the surviving objects at six different times from one of the simulations of *Agnor et al.* (1999, their Fig. 3). This simulation used as its starting condition the 22 largest embryos from the multizone embryo formation calculation of *Weidenschilling et al.* (1997). Despite the different integration techniques and initial conditions, the evolution of the system is broadly similar to those shown in Figs. 1 and 2.

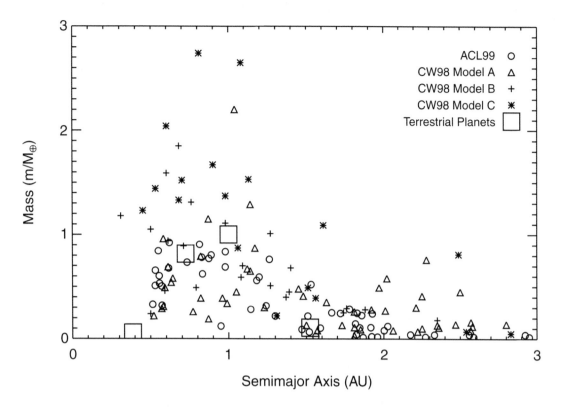

**Fig. 4.** Masses and semimajor axes of all of the final planets produced by the 27 simulations of *Chambers and Wetherill* (1998) and the 10 simulations of *Agnor et al.* (1999). Initial conditions for models A and C are shown in Figs. 1a and 2a. Note that model B of *Chambers and Wetherill* (1998) had an identical distribution of initial embryos to that in model A, except that in model B Jupiter and Saturn were included in the simulation. The values for the terrestrial planets are included for comparison.

Figure 5 shows the time-averaged eccentricities and inclinations of the final planets with masses greater than 0.5 $M_\oplus$ from *CW98* and *ACL99*. The average final eccentricity of the largest body from each simulation done by *ACL99* (who did not include Jupiter and Saturn) was 0.08, with time-averaged eccentricities all greater than 0.05. *CW98* report comparable eccentricities for simulations performed with similar initial conditions. Even larger eccentricities resulted for runs that included Jupiter and Saturn; in this case the average eccentricity of the most massive final planet in each system was 0.18.

The large eccentricities of the planets produced by simulations also yields larger angular momentum deficits for the systems formed. The angular momentum deficit, or AMD, is given by

$$AMD = \sum_{k=1}^{N} m_k n_k a_k^2 \left[ 1 - \sqrt{1 - e_k^2} \cos i_k \right] \qquad (2)$$

where N is the total number of planets, $n_k$ is the planet's mean-motion about the Sun, and $m_k$ is the planet's mass. In the 10 simulations of *ACL99*, the average value of the AMD of the final planets formed was between 4.5 and 17 times larger than that of the terrestrial planets, despite an initial system angular momentum that exceeded the terres-

trial system by no more than 5%. Larger AMD values require final planets more widely spaced (and therefore typically fewer in number) than their terrestrial counterparts for system stability. In the *ACL99* simulations that yielded two adjacent planets with masses greater than 0.5 $M_\oplus$ (80% of their runs), the two massive planets had an average spacing of $\langle \Delta a \rangle = 44$ $R_{Hill}$ (with individual values ranging from 33 to 55 $R_{Hill}$), in comparison to the $\Delta a = 26.25$ $R_{Hill}$ separation between Earth and Venus.

While the results obtained with various models of the late stage are thus in fairly close agreement, they continue to yield systems that are different in character than the current planets. An important question is why such discrepancies persist, even as modeling methods of embryo interactions have improved. The answer is likely that computational and model limitations continue to restrict simulations to simplified scenarios that neglect potentially influential processes. For example, current simulations are not capable of handling the large numbers of smaller planetesimals present during the postrunaway evolution of planetary embryos, and perhaps during the final stage as well. Dynamical interactions with such a small body population might yield lower planetary eccentricities, although this has yet to be convincingly demonstrated. This issue is discussed in more detail in section 4. However, assuming the basic con-

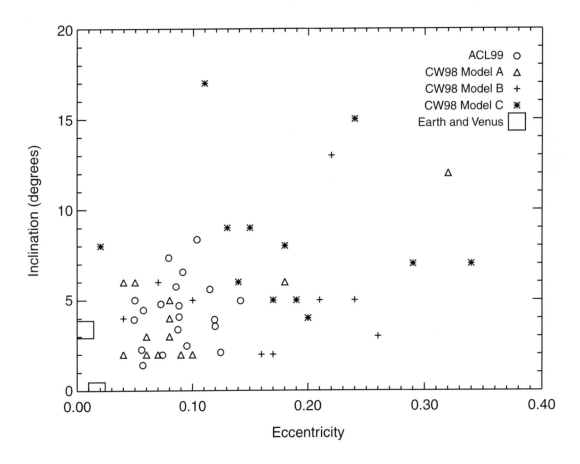

**Fig. 5.** The time-averaged inclinations and eccentricities of the final planets larger than 0.50 M⊕ are shown. Note the general absence of planets with eccentricities of 0.03 or less. The current values for Earth and Venus are shown for comparison (squares). Time-averaged values are e ~ 0.03, and I ~ 2° for Earth and Venus.

clusion that runaway growth yields embryos significantly smaller than either Earth or Venus is correct, the final accretion stage must have been characterized by large impacts between embryos. In section 3, we review the implications of such impacts for the final stages of growth of a terrestrial planet.

## 2.2.  Radial Mixing During Late-Stage Accretion

The Earth, Moon, Mars, and meteorites originating in the asteroid belt display isotopic commonalities and differences that were generally thought to reflect their formation from distinct radial reservoirs of material in the protoplanetary disk. For example, the O-isotopic signatures of terrestrial and lunar material fall on the same fractionation line, which is distinct from that of meteorites believed to have originated in the asteroid belt or Mars. The standard explanation is that the Earth and Moon formed from material contained in the same radial zone. This conceptually agreed well with earlier views of planet formation, which employed the concept of planetary growth from a local "feeding zone" (e.g., *Safronov,* 1969; *Lewis,* 1972). Recent studies suggest that the concept of feeding zones may be relevant to the

midstages of planet formation, when eccentricities and inclinations are low.

However, the evolution of a swarm of embryos into a few planets yields a large degree of mixing throughout the terrestrial zone. *Wetherill* (1994) studied the initial location of the embryos that comprised the final planets that formed in his Monte Carlo simulations. In general, all the embryos initially in the region between 0.5 and 2.5 AU became radially mixed, and the final planets were comprised of material that had its origins throughout this region. However, the average provenance of a planet was found to be somewhat correlated to its semimajor axis, i.e., planets in the outer terrestrial region accreted more material from larger heliocentric distances than the inner planets. This general result is also evident in the recent N-body simulations, as shown in Fig. 6. *CW98* find that final planets with a < 1 AU are comprised of material that originated primarily in the 0.5–1.5-AU region (~75% of their final mass), but also contain material from further out in the protoplanetary disk (~25% from material originating between 2.5 and 3.5 AU). It is thus consistent with the current accretion models that terrestrial planets may "remember" the initial compositional zoning of the protoplanetary disk to a limited degree. Whether

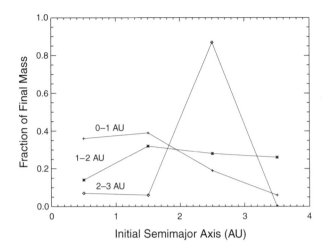

**Fig. 6.** The fraction of the planet mass obtained from different regions of the protoplanetary disk are shown for the final planets of the simulations presented in *Chambers and Wetherill* (1998, their Fig. 21). The three curves are for planets with final semimajor axes of 0 < a < 1 AU, 1 < a < 2 AU, and 2 < a < 3 AU. While planets form from some material near their end orbital radius, radial mixing of material causes planets to acquire a significant fraction of their mass from more distant regions.

such predictions can be reconciled with geochemical constraints that appear to favor compositional zoning is an important and open question.

## 3. LATE-STAGE IMPACT EVENTS

It is during the final stage that most of the end planetary characteristics — e.g., mass, spacing, orbital elements, spin angular momentum, presence of impact-generated satellites, etc. — are determined. For example, growing planets acquire their spins and obliquities as they accrete material, with the relative motions of the bodies at impact determining the contributions of each accreted body to the spin angular momentum of the final planet. As a protoplanet grows, both the average impact velocity and average moment arm of colliding bodies increase, so that the rotation state of a planet is determined primarily during its final accretion (see chapter by *Lissauer et al.*, 2000).

Large impact events could be either accretionary or erosive. The fate of debris generated during a planet-scale impact event would depend upon the specifics of the impact, e.g., impact angle, velocity relative to escape velocity, and mass ratio of impactor to target. The two-body escape velocity is

$$v_{esc} \equiv \sqrt{2G\,(m_1 + m_2)/(R_1 + R_2)} \qquad (3)$$

where R is radius. Collisions with $v_{imp} \gg v_{esc}$ would likely be primarily erosive or even disruptive; such an impact has been invoked to account for the loss of portions of Mercury's early mantle, resulting in the planet's current anom-

alously high density (e.g., *Benz et al.*, 1988). Oblique impacts can result in the ejection of material into bound orbit around the target protoplanet, which can then rapidly accrete to form satellites. Satellite formation via this mechanism has been proposed for the origin of the Earth-Moon system (*Hartmann and Davis*, 1975; *Cameron and Ward*, 1976), the Pluto-Charon binary (e.g., *Stern et al.*, 1997, and references therein), and for asteroid-satellite systems (e.g., *Durda*, 1996).

Determination of impact outcome as a function of specific impact energy, material strength, and target size has been the focus of a great body of experimental and theoretical research. General scaling laws that predict quantities such as crater size and total ejected mass exist for impacts that span the size range from centimeter-sized particles up to target radii of about ~200 km (e.g., *Housen and Holsapple*, 1990; *Ryan and Melosh*, 1998; *Benz and Asphaug*, 1999). To date, numerical simulations of planet-scale impacts have focused primarily on the particular impact believed responsible for the Earth-Moon system (*Benz et al.*, 1986, 1987, 1989; *Cameron and Benz*, 1991). In general, an approximately Mars-sized body with an impact angular momentum of at least the current angular momentum of the Earth-Moon system appears to be required to yield a lunar-sized moon. However, general scaling relationships for planet-scale impacts are just starting to be developed (e.g., *Canup et al.*, 2000, and discussion in section 4). Given the lack of such relationships, late-stage accretion simulations have modeled collisional outcomes through the use of simple extrapolations from scaling laws derived for much smaller impacts (in the case of Wetherill's Monte Carlo simulations), or by simply assuming that every impact results in complete accretion (in the case of the N-body simulations).

### 3.1. Occurrence and Timing of Large Impacts

The Monte Carlo simulations of *Wetherill* (e.g., 1985, 1986, 1992) were among the first to examine the predicted timing and size of the largest impactors to collide with Earth during its accretion (see also *Hartmann and Vail*, 1986). *Wetherill* (1985) performed three-dimensional simulations that each began with 500 bodies whose masses ranged from about $6 \times 10^{24}$ to $1 \times 10^{27}$ g, and with orbital distances between 0.7 and 1.1 AU. In every case giant impacts (i.e., impacts by objects with at least the mass of Mars) occurred, typically ~20 m.y. after the start of a simulation. In 1992, Wetherill performed similar calculations that began with an initial distribution of embryos that spanned the entire terrestrial region and the asteroid belt (0.4–3.8 AU). Similar numbers of giant impacts occurred, but at somewhat later times of $5 \times 10^7$–$10^8$ yr. In both studies, about one impact onto a large planet by a body at least as massive as Mars occurred for each simulation that contained a final planet with mass similar to that of Venus or Earth.

*ACL99* characterized the collisions in their N-body simulations in terms of impactor mass, velocity, and impact an-

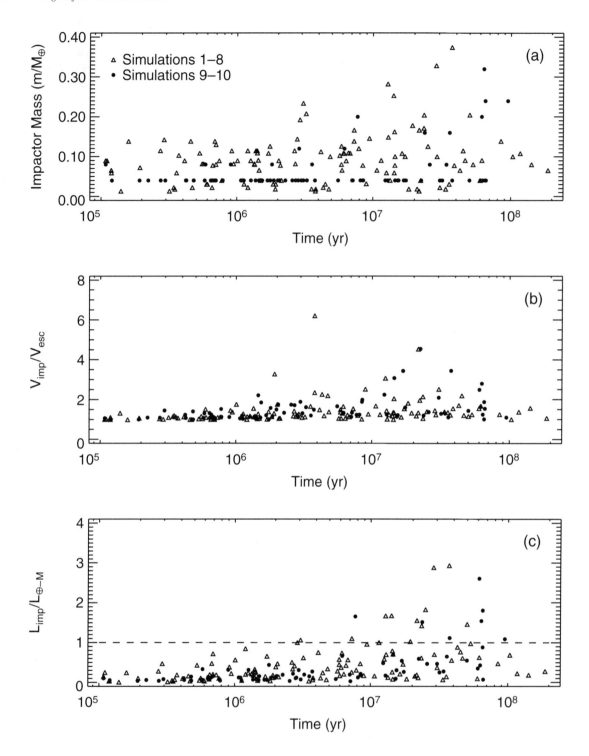

**Fig. 7.** The impactor mass, impact velocity, and the angular momentum of the collisional encounter are shown as a function of the time at which each impact occurred. All collisions from the 10 simulations of *Agnor et al.* (1999, their Fig. 8) are shown. The angular momentum of the Earth-Moon system ($L_{\oplus-M}$) is shown by the dashed line. Simulations 1–8 utilized the *Weidenschilling et al.* (1997) embryos as their starting conditions.

gular momentum as a function of time (see Fig. 7). The impactor was defined to be the less massive member of a colliding pair. Figure 8 gives a breakdown of the impact velocity and collision angular momentum as a function of impactor mass from the same simulations. The initial em-

bryos considered in *ACL99* had masses of 0.04–0.15 $M_{\oplus}$. As a simulation proceeds, embryos collide and merge, and a spectrum of embryo masses develops. A weak degree of dynamical friction (see discussion in chapter by *Kortenkamp et al.,* 2000) causes the smaller bodies in the disk to obtain

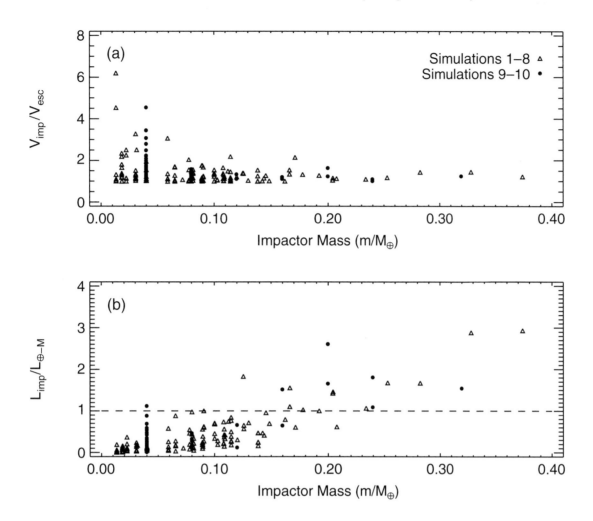

**Fig. 8.** The impact velocity and angular momentum of the collisional encounter are shown as a function of the impactor mass for the same collisions displayed in Fig. 7 (*Agnor et al.*, 1999, their Fig. 9). Note that some small, high-velocity impactors are capable of delivering more than 1 $L_{\oplus-M}$ to the system.

larger eccentricities and inclinations (and thus higher impact velocities) than their more massive counterparts. After a few million years (at which time the embryo masses span about a factor of five), some impacts occur at several times the escape velocity (see Fig. 7b).

These results generally resemble those of *Wetherill* (1992) in terms of the timing of the largest impacts. The *ACL99* final planets with m > 0.5 $M_{\oplus}$ experienced their largest impact at an average time of 29 m.y. (with individual values ranging from 1.4 to 95 m.y.); the last impact onto these planets occured at an average time of 46 m.y. (with individual values ranging from 3.1 to 108 m.y.). These timings are in relatively good agreement with isotopic constraints on the age of the Earth and Moon. The Hf-W chronometer places the time of core formation in the Earth and Moon at ~50 m.y. (e.g., *Lee et al.*, 1997; see chapter by *Halliday et al.*, 2000), while the formation interval for Earth based on the terrestrial excess of the heaviest isotopes of Xe is ~100 m.y. (see chapter by *Podosek and Ozima*, 2000).

In an attempt to identify impacts that could result in the formation of a lunar-sized satellite, *ACL99* described those collisions with an encounter angular momentum equal to or exceeding the current angular momentum of the Earth-Moon system ($L_{\oplus-M}$) as "potential moon-forming impacts." This classification reflects the very rudimentary knowledge of the impact dynamics required to form an impact-generated satellite, and is somewhat different than the "giant impacts" of *Wetherill* (1985, 1992), which were classified by the impactor mass. In the 10 *ACL99* accretion simulations, there were a total of 20 potential moon-forming impacts (see Figs. 7b and 8a). In addition, 25% of the final planets larger than 0.50 $M_{\oplus}$ experienced more than one such impact.

In the *ACL99* simulations that followed the evolution of the 22 embryos from *Weidenschilling et al.* (1997), the number of moon-forming collisions that occurred in each simulation ranged from 0 to 3, with an average of about 1.6 per simulation. In the simulations that began with 50 initial embryos (and a slightly larger total embryo mass of 2 $M_\oplus$), there were an average of 3.5 moon-forming collisions per simulation. The impact velocities of the moon-forming collisions were typically somewhat larger than the the two-body escape velocity. With the exception of one (0.04 $M_\oplus$) impactor that had an impact velocity of 3.44 $v_{esc}$, the $L_{imp} > L_{\oplus-M}$ collisions occurred with velocities in the range 1.00– 1.63 $v_{esc}$, with an average of 1.20 $v_{esc}$ (see Fig. 8a). High-resolution simulations of the moon-forming impact performed to date (e.g., *Cameron*, 1997; *Cameron*, 2000) assume $v_{imp} = v_{esc}$, and so have considered only the lower limit of possible specific impact energies.

### 3.2. Following the Growth of a Planet

The spin angular momentum of an embryo will evolve due to each of the large impacts it experiences. As a basic model for this process, *ACL99* assumed that for each collision, the spin angular momentum of the merged body was just the sum of the spin angular momenta of the two colliding bodies and the orbital angular momentum of the two bodies about their center of mass. The rotational periods of the growing planets were then calculated by assuming that all bodies were spheres of uniform density. In general, the accretion of material and angular momentum was likely less than 100% efficient during collisions between planetary embryos, and the assumption of inelastic mergers to model collisions will tend to overestimate both of these quantities. In addition, no account was made for the precession of spin axes that will result for oblate planets, or for the evolution of planetary obliquities due to spin-orbit coupling (see chapter by *Williams and Pollard*, 2000).

Figure 9 shows the evolution of the mass, the magnitude of the spin angular momentum (in units of $L_{\oplus-M}$), obliquity, and rotation period of a typical Earth-like planet (defined to be a planet with $m \geq 0.50\ M_\oplus$). In this particular case, each collision results in a net increase in the magnitude of the spin angular momentum of the planet. Impacts reorient the direction of the spin angular momentum vector, and in general, the spins and obliquities of the growing planets during the late stage are quite large (see chapter by *Lissauer et al.*, 2000). For collisions with $v_{imp} = v_{esc}$, the angular momentum of the impact, $L_{imp}$, scaled by the critical angular momentum for rotational stability, $L_{crit}$, is just

$$\frac{L_{imp}}{L_{crit}} = \frac{\sqrt{2}}{K} c_i \left(1 - c_i\right) \sqrt{c_i^{1/3} + \left(1 - c_i\right)^{1/3}} \sin \xi \quad (4)$$

with

$$L_{crit} \equiv M_T^{5/3} K \sqrt{G} \left[3/(4\pi\rho)\right]^{1/6} \quad (5)$$

where $M_T$ and $R_T$ is the total colliding mass and the radius of the combined body, $\rho$ and $K$ are the density and gyration constants of the colliding bodies, $c_i$ is the fraction of the total mass contained in the smaller of the colliding bodies, $K = I/M_T R_T^2$ where $I$ is the moment of inertia, and $\xi$ is the impact angle. The angular momentum imparted from a single collision can approach or exceed $L_{crit}$ for large $c_i$. For

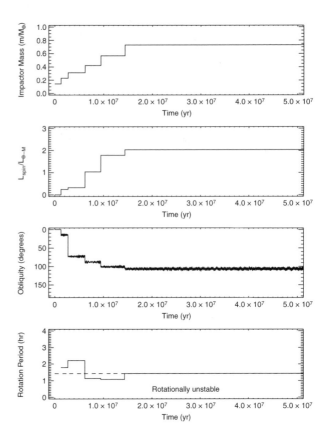

**Fig. 9.** The mass, magnitude of the spin angular momentum in units of the Earth-Moon system, obliquity, and rotation period of a typical planet (the 0.73 $M_\oplus$ final planet shown in Fig. 3) are plotted as a function of time (*Agnor et al.*, 1999, their Fig. 10). The dotted line of the rotation period vs. time graph indicates the rotational stability limit. The small oscillations in the obliquity are due to the motion of the body's orbit; the spin axis of the body was assumed to remain fixed in the inertial frame between collisions.

example, with like-sized bodies (and assuming a gyration constant equal to that of Earth, or $K = 0.335$), $L_{imp}/L_{crit} = 1.33 \sin \xi$. Indeed, the *ACL99* simulations find planetary rotation rates that often exceed the rotational stability limit. Clearly the assumption of merger in such cases is invalid, as the excess angular momentum would instead be carried away in ejected debris or in debris placed into circumplanetary orbit.

Even small embryo collisions ($m_{imp} \approx 0.04\ M_\oplus$) can make angular momentum contributions to a planet that are significant in comparison to $L_{\oplus-M}$. In Fig. 10 the mass and the magnitude of the spin angular momentum of a 0.80 $M_\oplus$ planet from one of the *ACL99* simulations are shown as a function of time. This planet experiences two collisions with initial embryos (at $t \sim 32 \times 10^6$ and $\sim 50 \times 10^6$ yr), which both act to reduce the magnitude of the planet's spin angular momentum, slowing its spin rate. The second of these collisions occurred with $v_{imp} = 1.4\ v_{esc}$, and had a collisional angular momentum of 0.55 $L_{\oplus-M}$. The largest impact to strike this planet was also the last impact the planet experienced, which was the case for about one-half the Earth-like

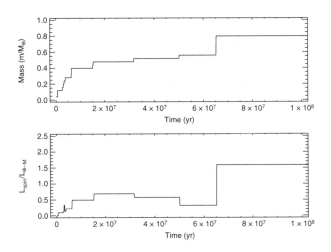

**Fig. 10.** The mass and magnitude of the spin angular momentum in units of the Earth-Moon system are shown as a function of time for a planet from a simulation that began with smaller but more closely spaced initial embryos (*Agnor et al.,* 1999, their Fig. 13). Small impacts (~0.04 $M_\oplus$), such as those occurring at $3.2 \times 10^6$ yr and $5.0 \times 10^7$ yr, can make significant contributions to the angular momentum accretion of the planets.

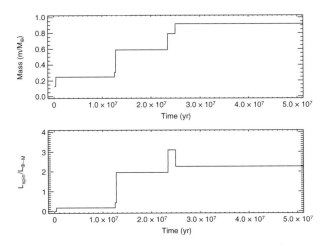

**Fig. 11.** The planet mass and spin angular momentum as a function of time for a planet that experienced a net reduction in its spin angular momentum due to cancellation between contributions from multiple impacts (*Agnor et al.,* 1999, their Fig. 11).

planets in the *ACL99* simulations. During the growth of the planet shown in Fig. 11, the largest impact (at t = 13 × $10^6$ yr) was followed by two smaller impacts, the last of which resulted in a net decrease of 0.85 $L_{\oplus-M}$ in the magnitude of the spin angular momentum. Approximately 40% of the Earth-like planets produced by the *ACL99* simulations experienced collisions that decreased the magnitude of the spin angular momentum of the planet by 0.10 $L_{\oplus-M}$ or more.

For the Earth-like planets produced in *ACL99,* the largest and second largest impactor contributed an average of 30%

and 19% of the final planet mass and contributed an average of 1.44 $L_{\oplus-M}$ and 0.67 $L_{\oplus-M}$ respectively. For nearly 50% of these planets, the last impactor was also the largest. However, for an equal number of final planets, the last impactor was less massive than both the largest and second-largest impactor. In these cases, the last impact contributed an average of only 8.3% of the final planet mass, but contributed an average angular momentum of 0.76 $L_{\oplus-M}$ with values ranging from 0.10 to 1.83 $L_{\oplus-M}$. Furthermore, these planets accreted an average of 31% (range 7–62%) of their final planet mass after experiencing their first $L_{imp} \geq L_{\oplus-M}$ impact. These statistics were derived from a limited number of simulations; however, they suggest that the spin angular momentum of a terrestrial planet is likely the result of more than one large impact, and that moon-forming collisions can occur before planetary growth is completed.

### 3.3. Implications for Lunar Origin

Two fundamental constraints on models of the formation of the Earth-Moon system are the masses of the bodies and the current system angular momentum. Models of lunar formation have generally made the assumption that the Moon-forming impact was the last impact Earth experienced. However, recently this constraint has been relaxed in an order to yield a lunar-mass Moon as described below.

SPH ("smoothed-particle hydrodynamics") simulations of the lunar-forming impact have typically considered collisions with impactor to target mass ratios of 3:7 and 2:8, no initial spin angular momentum of the impactor or target, and $v_{imp} = v_{esc}$ (*Cameron,* 1997; see also chapter by *Cameron,* 2000). These studies have investigated a variety of impact angular momenta and system masses (0.5–1.0 $M_\oplus$). Recent N-body simulations of the accumulation of the Moon from an impact-generated disk indicate that a disk mass of at least two lunar masses is required to form the Moon (e.g., *Ida et al.,* 1997; see chapter by *Kokubo et al.,* 2000). The only impacts that SPH simulations have shown to be capable of producing a circumterrestrial disk massive enough to form the Moon are those with a total mass of 1.0 $M_\oplus$ and an impact angular momentum greater than 2.0 $L_{\oplus-M}$, or those with a total mass of ~0.6 $M_\oplus$ and 1.0 $L_{\oplus-M}$ (*Cameron and Canup,* 1998; *Canup et al.,* 2000). Each of these scenarios is unable to simultaneously account for both the mass and angular momentum of the Earth-Moon system with a single impact. The first case, with $L_{imp} \sim 2\ L_{\oplus-M}$, requires that the Earth-Moon system somehow rid itself of an angular momentum excess ~1 $L_{\oplus-M}$ after the moon-forming event. As solar tides could remove only a small fraction of this, this scenario would seem to require that Earth experienced another impact to reset Earth's spin. In the second case, Earth acquires another ~0.3–0.4 $M_\oplus$ of material after the Moon has formed. This scenario also requires that Earth experienced later impacts.

The results of the recent late-stage simulations show that several large impacts typically affect a planet's end state, implying that Moon-forming impacts with total masses and angular momenta different from that of the Earth-Moon system may be plausible. The ACL99 simulations find that planets occasionally experienced glancing collisions at several times the escape velocity with smaller (~0.05 $M_\oplus$) impactors that were capable of delivering up to 1.1 $L_{\oplus-M}$. At present it isn't known whether or not this type of collision would be efficient at producing a circumplanetary disk. If this type of collision occurred after the Moon had already formed, it could conceivably reset the spin angular momentum of Earth, possibly offering a resolution to the high angular momentum impact scenario described above. The late-stage simulations also predict that embryos have rapid spin rates throughout the final accumulation stage. In particular, the final planets of the *ACL99* simulations that experienced $L_{imp} \geq L_{\oplus-M}$ impacts had an average angular momentum of ~0.91 $L_{\oplus-M}$ (an average rotational period of 2.4 hr) immediately prior to these collisions. While the future refinements of planet formation models (e.g., inclusion of the effects of small bodies during the final accretion stage) may modify this estimate, it is significant enough to suggest that the spin of Earth prior to the Moon-forming event was likely not negligible, and that Moon-forming collisions in which the impactor and/or the target are spinning should be studied. Perhaps, e.g., an Earth with a prograde spin prior to the lunar-forming impact would result in higher yields of bound orbiting debris for a given impact angular momentum. Finally, the late-stage simulations indicate that large impacts occur with $v_{imp} > v_{esc}$, and so such collisions should also be explored in the context of the lunar-forming event. Higher specific impact energies ($\propto v_{imp}^2$) might act to increase the yield of ejected material for a given impact angular momentum ($\propto v_{imp}$), and potentially help to mitigate the difficulty to date in forming a massive protolunar disk with an $L_{imp} \sim L_{\oplus-M}$ impact.

## 3.4. Implications for the Frequency of Impact-triggered Satellites

To some extent, the early perception that large impact events were rare coincided well with the fact that among the current terrestrial planets, only Earth has a sizeable, impact-generated moon. However, the accretion simulations performed over the past decade depict a terrestrial planet formation process whose final stages are dominated by stochastic impact events. In fact, it seems likely that terrestrial planets may have experienced multiple giant impacts in their histories, and that many such impacts may have led to the formation of satellites (e.g., *Canup et al., 2000*).

If accretion models now suggest that most terrestrial planets undergo such impact events, is it not a contradiction that, e.g., Venus does not currently possess a satellite? Not necessarily. First, while giant impacts appear to have been commonplace in the inner solar system, there are still only certain impact orientations that have been shown to yield significant amounts of debris in circumplanetary orbit. So while oblique, relatively low-velocity collisions might generate a circumplanetary disk, head-on collisions,

near misses, or collisions with $v_{imp} \gg v_{esc}$ might generate debris that reimpacts the planet or escapes entirely to heliocentric orbit. Large impacts may be common, but satellite-generating impacts will be less so.

Second, it is also plausible that other terrestrial planets once had impact-generated satellites that simply did not persist against continuing bombardment during the late stage or against orbital evolution due to tidal interaction. A simple criterion for satellite stability is that the orbit initially expands rather than contracts as it tidally evolves, or $a > a_{co}$, where a is the initial semimajor axis of the satellite, and $a_{co}$ is the co-rotation radius

$$a_{co} \equiv \left( \frac{GM_p}{\omega^2} \right)^{1/3} \qquad (6)$$

where $M_p$ and $\omega$ are the mass and angular velocity of the planet. Since an impact-generated satellite cannot conceivably accrete within the Roche radius (see chapter by *Kokubo et al., 2000*), $a \geq a_{Roche}$, with

$$a_{Roche} \approx 2.5 \left( \frac{\rho_p}{\rho_s} \right)^{1/3} R_p \qquad (7)$$

where $\rho_s$ is the satellite density, and $\rho_p$ and $R_p$ are the density and radius of the planet. The requirement that $a_{Roche} > a_{co}$ then yields a constraint on the initial spin rate of the planet

$$\omega_o \geq \left[ \frac{4/3\,\pi G}{(2.5)^3} \rho_s \right]^{1/2} \qquad (8)$$

For a satellite density equal to that of the Moon, a planet's day must be shorter than about 7 hr for co-rotation to fall within the Roche limit. Slower rotators would quickly lose their impact-generated moons to inward tidal decay on very short timescales (e.g., approximately years for an Earth-sized planet and a lunar-mass satellite with $a = 3\ R_\oplus$). The long-term survival of a satellite that initially forms outside the co-rotation radius is not guaranteed, and will depend on the interplay of solar and satellite tides, which both act to slow the planet's rotation and cause $a_{co}$ to evolve outward (*Burns, 1973*; *Ward and Reid, 1973*). If $a_{co}$ overtakes a, the eventual fate of the satellite will again be a collision with the planet (although in this case the timescale for the orbital decay can be much longer).

## 4.  OPEN ISSUES

Models of the final stage of terrestrial accretion are generally successful, producing a few planets from a system of planetary embryos in a timeframe consistent with isotopic constraints for the age of the Earth and Moon. Some disparities between current predictions and the terrestrial planets persist, in particular with regard to accounting for the very low eccentricities of Earth and Venus. However, such differences are likely the result of simplified assumptions employed by the late-stage models, which to date have all ignored the production and dynamical influence of small material. From studies of the runaway growth phase, it is well known that the general role of a background population of small bodies is to decrease the relative velocities (and therefore eccentricities and inclinations) of the largest bodies via dynamical friction. A key open question is then how the presence of a planetesimal swarm will alter the dynamical environment in which embryos accrete to form planets. Whether the influence of small bodies in the late stage could be sufficient to yield Earth-like planets with nearly circular orbits remains uncertain, and addressing this question will require an improved understanding of both the generation of debris during planet-scale collisions, and the persistence of a background population as it co-evolves with a system of embryos.

The outcomes of planet-scale collisions are still poorly understood, and the type of hydrodynamic simulations utilized to model the lunar-forming impact could be extended to a broader range of collisional parameter space in order to better generalize impact outcomes. Accounting for the amount of ejected debris, its velocity, and its angular momentum will affect estimates of both the potential background population of small material and planetary spin rates. Current scaling relationships derived for meter- to 100-kilometer-sized targets relate $Q_D^*$, the critical specific energy required to disperse one-half the target mass, to target size (e.g., *Benz and Asphaug*, 1999). Given these relationships for $Q_D^*$, and assuming a linear relationship between the specific impact energy, Q, and the amount of material ejected (as implied in *Love and Ahrens*, 1996), ACL99 estimated the fraction of material that would be dispersed for each of the collisions in their simulations. This fraction ranged from ~1% to 20% of the total mass involved in a collision (their Fig. 15). SPH simulations of the lunar-forming impact (*Cameron*, 2000; *Canup et al.*, 2000) suggest that the amount of ejected mass will also depend upon target-to-impactor mass ratio and impact angular momentum, in addition to the specific impact energy.

Given that some nonneglible fraction of material is ejected during collisions between planetary embryos, the question is then how long this material will persist as a background population during the final stage. Much of the debris ejected in an impact might rapidly reaccumulate onto its parent body during subsequent orbital passes, unless it is dynamically perturbed by other embryos or the giant planets. If an impact was in fact responsible for the loss of much of Mercury's mantle, we have at least one example case in which the great majority of the ejected material must not have reaccreted onto the source planet. Assuming immediate reaccretion is avoided, the timescales for sweep-up of the remaining debris could be comparable to (or longer than) the time between embryo collisions. A simple fragmentation model, which assumed that a few percent of the mass

involved in each collision was ejected as fragments, was included in two-dimensional N-body simulations of the accretion of embryos in the terrestrial region by *Alexander and Agnor* (1998). Interestingly, that work found that the ejected fragments quickly contained up to 50% of the total system mass. Another possible contribution to a late-stage background population could be leftover remants from the earlier runaway stage. The *Weidenschilling et al.* (1997) simulation did not include fragmentation, and its prediction that 10% of the system mass remained in small bodies after a million years should therefore be viewed as a lower limit. Recent orbit integrations of test particles placed throughout the current terrestrial region (see Fig. 5 in chapter by *Hartmann et al.*, 2000) find that near 50% of the particles remain in the 0.7–1.3-AU region after 100 m.y.

Modeling the possible interactions between a system of embryos and a background population of small material for timescales ~$10^8$ yr presents many computational and algorithmic challenges for the next generation of planet formation simulations. An important step in this regard will involve an improved understanding of the transition from the runaway growth period to the late stage. While the embryos in the *Weidenschilling et al.* (1997) simulations (which contained about 90% of the system mass) remained in low-eccentricity, stable orbits for times longer than $10^5$ yr, the evolution of the same embryos using direct integrations in ACL99 (but ignoring the 10% of mass contained in smaller material) destabilized on times ~$10^4$ yr. Determining the source of these differences will be closely related to efforts to determine the role of small material in late-stage accretion.

While important processes relevant to the final accumulation of the terrestrial planets are thus not yet completely understood, the basic predictions of existing models are in fairly good agreement with the current terrestrial planets. Additional attention to issues neglected in simulations performed to date may help to resolve the outstanding question of forming low-eccentricity planets. But in general, a stochastic phase dominated by large impacts appears an inevitable consequence of the evolution of tens to hundreds of planetary embryos into a final few planets. Barring a major change in our understanding of the outcome of the preceeding runaway growth stage, this conclusion appears robust, and will likely persist even as late-stage accretion models are further refined.

*Acknowledgments.* The authors wish to acknowledge helpful reviews by G. Wetherill and J. Chambers, and support from NASA's Origins of Solar Systems and Graduate Student Researchers programs.

## REFERENCES

Agnor C. B., Canup R. M., and Levison H. (1999) On the character and consequences of large impacts in the late stage of terrestrial planet formation. *Icarus, 142,* 219–237.

Alexander S. G. and Agnor C. B. (1998) N-body simulations of late stage planetary formation with a simple fragmentation model. *Icarus, 132,* 113–124.

Benz W. and Asphaug E. (1999) Catastrophic disruptions revisited. *Icarus,* in press.

Benz W., Cameron A. G. W., and Melosh H. J. (1989). The origin of the Moon and the single-impact hypothesis III. *Icarus, 81,* 113–131.

Benz W., Slattery W. L., and Cameron A. G. W. (1986) The origin of the Moon and the single-impact hypothesis I. *Icarus, 66,* 515–535.

Benz, W., Slattery,W. L., and Cameron, A. G. W. (1987). The origin of the Moon and the single-impact hypothesis II. *Icarus, 71,* 30–45.

Benz W., Slattery W. L., and Cameron A. G. W. (1988) Collisional stripping of Mercury's mantle. *Icarus, 74,* 516–528.

Burns J. A. (1973) Where are the satellites of the inner planets? *Nature, 242,* 23–25.

Cameron A. G. W. (1997) The origin of the Moon and the single impact hypothesis V. *Icarus, 126,* 126–137.

Cameron A. G. W. (2000) Higher-resolution simulations of the giant impact. In *Origin of the Earth and Moon* (R. M. Canup and K. Righter, eds.), this volume. Univ. of Arizona, Tucson.

Cameron A. G. W. and Benz W. (1991) The origin of the Moon and the single impact hypothesis IV. *Icarus, 92,* 165–168.

Cameron A. G. W. and Ward W. R. (1976) The origin of the Moon (abstract). In *Lunar Science VII,* pp. 120–122. Lunar Science Institute, Houston.

Cameron A. G. W. and Canup R. M. (1998) The giant impact occurred during Earth accretion (abstract). In *Lunar and Planetary Science XXIX,* Abstract #1062. Lunar and Planetary Institute, Houston (CD-ROM).

Canup R. M., Ward W. R., and Cameron A. G. W. (2000) A scaling relationship for satellite-forming impacts. *Icarus,* submitted.

Chambers J. E. (1998b) A hybrid symplectic integrator that permits close encounters between massive bodies. *Mon. Not. R. Astron. Soc., 304,* 793–799.

Chambers J. E. and Wetherill G. W. (1998) Making the terrestrial planets: N-body integrations of planetary embryos in three dimensions. *Icarus, 136,* 304–327.

Chambers J. E., Wetherill G. W., and Boss A. (1996) The stability of multi-planet systems. *Icarus, 119,* 261–268.

Duncan M., Levison H. F., and Lee M. H. (1998) A multiple timestep symplectic algorithm for integrating close encounters. *Astron. J., 116,* 2067–2077.

Durda D. D. (1996). The formation of asteroidal satellites in catastrophic collisions. *Icarus, 120,* 212–219.

Gladman B. (1993) Dynamics of systems of two close planets. *Icarus, 106,* 247–263.

Greenberg R. J., Wacker J., Chapman C. R., and Hartmann W. K. (1978) Planetesimals to planets: Numerical simulations of collisional evolution. *Icarus, 35,* 1–26.

Halliday A. N., Lee D.-C., and Jacobsen S. B. (2000) Tungsten isotopes, the timing of metal-silicate fractionation, and the origin of the Earth and Moon. In *Origin of the Earth and Moon* (R. M. Canup and K. Righter, eds.), this volume. Univ. of Arizona, Tucson.

Hartmann W. K. and Davis D. R. (1975) Satellite-sized planetesimals and lunar origin. *Icarus, 24,* 504–515.

Hartmann W. K. and Vail S. R. (1986) Giant impactors: Plausible sizes and populations. In *Origin of the Moon* (W. K. Hartmann, R. J. Phillips, and G. J. Taylor, eds.), pp. 551–566. Lunar and Planetary Institute, Houston.

Hartmann W. K., Ryder G., Dones L., and Grinspoon D. (2000) The time-dependent intense bombardment of the primordial Earth/Moon system. In *Origin of the Earth and Moon* (R. M. Canup and K. Righter, eds.), this volume. Univ. of Arizona, Tucson.

Housen K. R. and Holsapple K. A. (1990) On the fragmentation of asteroids and planetary satellites. *Icarus, 84,* 226–253.

Ida S., Canup R. M., and Stewart G. R. (1997) Lunar formation from an impact-generated disk. *Nature, 389,* 353–357.

Ito T. and Tanikawa K. (1999) Stability and instability of the terrestrial protoplanet system and their possible roles in the final stage of planet formation. *Icarus, 139,* 336–349.

Kokubo E., Canup R. M., and Ida S. (2000) Lunar accretion from an impact-generated disk. In *Origin of the Earth and Moon* (R. M. Canup and K. Righter, eds.), this volume. Univ. of Arizona, Tucson.

Kortenkamp S. J., Kokubo E., and Weidenschilling S. J. (2000) Formation of planetary embryos. In *Origin of the Earth and Moon* (R. M. Canup and K. Righter, eds.), this volume. Univ. of Arizona, Tucson.

Lee D-C., Halliday A. N., Snyder G. A., and Taylor L. A. (1997) Age and origin of the Moon. *Science, 278,* 1098–1103.

Lewis J. S. (1972) Metal/silicate fractionation in the solar system. *Earth Planet. Sci. Lett., 15,* 286–290.

Lissauer J. J., Dones L., and Ohtsuki K. (2000) Origin and evolution of terrestrial planet rotation. In *Origin of the Earth and Moon* (R. M. Canup and K. Righter, eds.), this volume. Univ. of Arizona, Tucson.

Love S. G. and Ahrens T. J. (1996) Catastrophic impacts on gravity dominated asteroids. *Icarus, 124,* 141–155.

Opik E. J. (1951) Collision probabilities with the planets and the distribution of interplanetary matter. *Proc. R. Irish Acad., A54,* 165–199.

Podosek F. A. and Ozima M. (2000) The xenon age of the Earth. In *Origin of the Earth and Moon* (R. M. Canup and K. Righter, eds.), this volume. Univ. of Arizona, Tucson.

Ryan E. V. and Melosh H. J. (1998) Impact fragmentation: From the laboratory to asteroids. *Icarus, 133,* 1–24.

Safronov V. S. (1969) *Evolution of the Protoplanetary Cloud and Formation of the Earth and Planets.* Nauka, Moscow (Translated in 1972 as NASA TT F-677).

Stern A. S., McKinnon W. B., and Lunine J. I. (1997) On the origin of Pluto, Charon, and the Pluto-Charon binary. In *Pluto and Charon* (S. A. Stern and D. J. Tholen, eds.), pp. 605-664. Univ. of Arizona, Tucson.

Ward W. R. (2000) On planetesimal formation. In *Origin of the Earth and Moon* (R. M. Canup and K. Righter, eds.), this volume. Univ. of Arizona, Tucson.

Ward W. R. and Reid M. J. (1973) Solar tidal friction and satellite loss. *Mon. Not. R. Astron. Soc., 164,* 21–32.

Weidenschilling S. J., Spaute D., Davis D. R., Marzari F., and Ohtsuki K. (1997) Accretional evolution of a planetesimal swarm. *Icarus, 128,* 429–455.

Wetherill G. W. (1985) Occurrence of giant impacts during the growth of the terrestrial planets. *Science, 228,* 877–879.

Wetherill G. W. (1986) Accumulation of the terrestrial planets and implications concerning lunar origin. In *Origin of the Moon* (W. K. Hartmann, R. J. Phillips, and G. J. Taylor, eds.), pp. 519–550. Lunar and Planetary Institute, Houston.

Wetherill G. W. (1992) An alternative model for the formation of the asteroid belt. *Icarus, 100,* 307–325.

Wetherill G. W. (1994) The provenance of the terrestrial planets. *Geochim. Cosmochim. Acta, 58,* 4513–4520.

Wetherill G. W. (1996) The formation and habitability of extrasolar planets. *Icarus, 119,* 219–238.

Wetherill G. W. and Stewart G. R. (1993) Formation of planetary embryos: Effects of fragmentation, low relative velocity, and independent variation of eccentricity and inclination. *Icarus, 106,* 190–209.

Williams D. M. and Pollard D. (2000) Earth-Moon interactions: Implications for terrestrial climate and life. In *Origin of the Earth and Moon* (R. M. Canup and K. Righter, eds.), this volume. Univ. of Arizona, Tucson.

Wisdom J. and Holman M. (1991) Symplectic maps for the N-body problem. *Astron. J., 102,* 1528–1538.

*Part III:*

*Impact-triggered Formation
of the Earth-Moon System*

# Higher-Resolution Simulations of the Giant Impact

## A. G. W. Cameron

*Harvard-Smithsonian Center for Astrophysics*

This study reports the results of recent simulations (until early 1999) of the giant impact that is a candidate to have formed the Moon, using smooth particle hydrodynamics (SPH). The early part of the work consisted of a survey of much of parameter space (but only for a 7:3 mass ratio of proto-Earth/impactor) for the giant impact using low-resolution calculations (10,000 particles). It was found that to produce a reasonable amount of material in orbit at the end of the giant impact (about 2 lunar masses, or $M_{\mathbb{C}}$), and with an collisional angular momentum just a little larger than that in the Earth-Moon system now, the optimum total colliding mass should be about two-thirds of an Earth mass ($M_{\oplus}$). The later part of the work reports the detailed simulations of three cases of much higher resolution (100,000 particles) in the vicinity of the impact optimum parameters. The giant impact is actually a dynamically complicated series of two collisions, in which the impactor is destroyed only in the second collision, with the Fe core draining into the proto-Earth and much of the mantle drawn out into a trailing spiral arm. This arm breaks up into a large number of gravitationally bound clumps in orbit near the Roche limit. The subsequent evolution of this cloud of material is dominated by the effects of tidal deformation and stripping. The simulations are described through the period in which particles near the Roche limit are dispersed around their orbits, but before a major amount of accumulation into a larger body has taken place.

## 1. INTRODUCTION

In ancient times (more than a quarter century ago), "lunar scientists" were in a considerable quandary concerning the origin of the Moon. For a long time there had been three competing hypotheses about this problem: that the Moon had resulted in a rotationally induced fission of the Earth, that the Moon and Earth had been formed together in a common orbit, and that the Moon had been formed elsewhere and had been captured by Earth. Indeed, part of the scientific motivation for Project Apollo, a motivation strongly urged by Harold Urey among others, had been to determine which of these three hypotheses was the correct one. But the scientific results from Apollo, including those from *in situ* instruments on the surface of the Moon as well as those from lunar sample analyses, had failed to persuade more than a handful of lunar scientists to change their opinions about the preferred hypothesis, and many of them considered all the hypotheses to be largely irrelevant speculation.

The giant impact hypothesis was introduced in the middle 1970s to provide a fresh approach to the problem, but initially the scientific rationale was flawed. *Hartmann and Davis* (1975) had suggested that a large collision with the proto-Earth during its accumulation would throw up a lot of material from the surface of Earth, from which the Moon could possibly be formed, but the present author pointed out to them (see a footnote in their paper) that the requirement that the Earth-Moon system should have a large angular momentum placed a lower limit on the mass of the impactor (at least as large as the mass of Mars). In turn, *Cameron and Ward* (1976) elaborated on this issue, but their hypothesis was that the collision with a large body would have generated an immense cloud of gas, whose expansion would rather inefficiently inject the seed material for lunar accumulation into Earth orbit. In later work it turned out that the actual mechanism consisted of gravitational torques on collisional material resulting from its nonspherical mass distribution.

After a decade in which it seemed that the above ideas had been quietly ignored, the Kona conference on the origin of the moon was held in 1984, and the organizers were surprised and pleased to find that many papers elaborating on the giant impact hypothesis were to be presented. A popular description of the event was that a bandwagon of support for the giant impact hypothesis had formed and that most of the participants in the conference hastened to jump on board, after the detailed reviews of the three classical theories at the meeting had convinced these participants that the classical ideas were unsatisfactory. But an important outcome of the conference was that it attracted new people to take a fresh look at the giant impact hypothesis and to start making simulations of it on supercomputers. Early reports on such computations made it into the conference proceedings (*Hartmann et al.,* 1986).

According to the original form of the giant impact hypothesis, toward the end of the planetary accumulation process, the proto-Earth collided with a planetary body having a substantial fraction of the proto-Earth mass. The collision took place with a fairly large impact parameter, so that the combined angular momentum of the two bodies was approximately that now in the Earth-Moon system (for convenience we call this unit of angular momentum 1 lem, equal to $3.5 \times 10^{41}$ cgs units). The collision would leave much debris in orbit, forming a disk (*Cameron and Ward,*

1976; *Cameron, 1986*). The dissipation of this disk was then thought to spread matter radially so that the Moon could collect together gravitationally beyond the Roche lobe (*Ward and Cameron, 1978; Thompson and Stevenson, 1983, 1988*). The scenario has been reviewed by *Stevenson* (1987).

The simulations of the giant impact were of necessity numerical hydrodynamic calculations, and two types of hydrodynamic code became involved. Conventional hydrodynamic codes divide space into discrete cells and calculate the passage of matter through the cells. H. J. Melosh (along with M. E. Kipp of the Sandia National Laboratory) carried out a number of simulations of the giant impact, but the computational requirements to carry out a simulation in three dimensions were so severe that they had to do the problem in two dimensions, studying collisions between cylinders, and with necessarily very rough approximations to the proper gravitational effects (*Melosh and Sonett, 1986; Kipp and Melosh, 1986*). Such hydrodynamical calculations work very well with low-density gases that would be generated in a giant impact, and Melosh has emphasized the effects of jets in producing these. However, when *Melosh and Kipp* (1989) first made a few three-dimensional simulations, they had to use several million cells and even then they were unable to treat self-gravity accurately.

Also, starting in 1986, a series of studies of this giant impact in the early history of Earth was carried out utilizing smooth particle hydrodynamics (SPH). (The name used in the literature is usually "smoothed," but I prefer "smooth" because the particles used are born smooth and have never been "smoothed.") In this method the domain of the computation is not divided into spatial cells, as in ordinary hydrodynamics, but rather the matter in the domain of computation is represented by smooth overlapping spheres. For computational purposes a sphere has a radial density distribution that is bell-shaped, with highest density at the center and a sharp outer edge. The particles carry individual internal energies, but such properties as density and pressure are collectively determined by the overlap between the density distributions of the particles; these are only meaningful when typically a few tens of particles contribute to the overlap. The method is intrinsically three-dimensional, but in the early years of its use for giant impact calculations it was not feasible to use very large numbers of particles because of computer limitations. For a long time simulations were done with just 3008 particles, chosen to be a multiple of 64 to optimize the use of Cray XMP supercomputers, but now I am routinely using 100,000 particles. Our SPH studies have been reported in a series of papers by the author, with W. Benz and others [Moon1 (*Benz et al.,* 1986); Moon2 (*Benz et al.,* 1987); Moon3 (*Benz et al.,* 1989); Moon4 (*Cameron and Benz,* 1991); Moon5 (*Cameron,* 1997)].

In the more recent papers mentioned above, all particles in any one of the colliding objects were given the same mass (but they could vary in composition), and their smoothing lengths were adjusted to overlap with a few tens of their neighbors. The objects were composed of Fe in their cores and had dunite mantles, using the ANalytical Equation of State (ANEOS) software developed at the Sandia National Laboratory (this software is described in Moon3). The equation of state of dunite was developed by H. J. Melosh, who also recently found that the equation of state for Fe obtained from Sandia National Laboratory underestimates the vapor phase at the highest temperatures, but this is expected to have only a minimal effect on the range of conditions found in the present calculations. Each object contained 5000 particles, so that 10,000 particles were involved in each computation. The materials did not exhibit rigidity, so that they always acted like fluids; thus the starting temperatures of all particles used in the studies described here were set to 2000 K in order to minimize evaporation from the colliding bodies due solely to high initial temperatures.

Subsequent to the publication of Moon5, several works modeled the accretionary dynamics of a centrally condensed impact-generated protolunar disk (*Canup and Esposito,* 1996; *Ida et al.,* 1997). These works demonstrated that the incorporation of initial disk material into a moon or moons is inefficient, due mainly to the material's proximity to the Roche limit. In particular, *Ida et al.* (1997) concluded that, for a wide range of initial disk parameters, only between 15% and 40% of the initial disk material is incorporated into one or two moons, with the remainder scattered onto Earth.

In the series of papers Moon1–5, it was always assumed that the sum of the two masses involved in the giant impact would be 1 $M_\oplus$. The most recent paper (Moon5) showed that enough material could be placed in orbit by such a collision to form the Moon for mass ratios of proto-Earth to impactor ranging from 5:5 to 8:2, but only if the colliding masses had about twice the angular momentum now in the Earth-Moon system. Owing to the difficulty of extracting such a large amount of angular momentum from the collision products, it was suggested that the giant impact required to form the Moon would have occurred when the accumulation of the proto-Earth was very incomplete, perhaps about 0.5 $M_\oplus$.

The first stage of the present investigation was designed to find an optimum total mass in the collision that would place approximately 2 $M_{\mathbb{C}}$ of material in orbit when the angular momentum in the collision was in the vicinity of 1 lem, or perhaps a little more, to allow for some extraction of angular momentum from the rapidly rotating proto-Earth by solar tides. This was a rough criterion for being able to form the Moon, allowing for inefficiencies (*Ida et al.,* 1997). In this stage the calculations utilized 10,000 particles in the SPH code; the main concern was to survey a lot of parameter space at low resolution when the principal objective was to determine the amount of material left in orbit for different collision parameters. Later calculations used higher resolutions, primarily 100,000 particles.

The general plan of the exploration was as follows. The first step was to determine the yield of material in orbit as a function of collisional angular momentum for high and low colliding total masses (1.0 and 0.5 $M_\oplus$). The second step was to determine the yields in orbit at 1.0 and 1.25 lem for total colliding masses in the interval between these limits. This produced an optimum mass of 0.65 $M_\oplus$. The third step was to determine the yields at 0.65 total colliding masses within the interval 1.0–1.25 lem. In order to produce good yields of material in orbit following a giant impact, a mass ratio of 7:3 was used throughout for the proto-Earth and the impactor, for all values of the collisional mass. The impactor was always taken to have been at rest at infinity prior to the calculation, since early in this series it was found that the main effect of increasing this velocity is to increase slightly the amount of escaping mass.

## 2. DEPENDENCE OF ORBITING YIELDS UPON ANGULAR MOMENTUM

The general characteristics of giant impacts upon the proto-Earth have been established in the papers previously cited. For lower values of the angular momentum the lower mass body (the impactor) strikes the proto-Earth and is destroyed, being drawn out into a long filament, primarily by shearing. The part of the impactor closest to the proto-Earth is directly decelerated in the collision, while the part on the opposite side is slowed to a lesser extent by propagated shocks and by gravity. Most of the mass then has sub-orbital velocity and falls out of orbit onto the proto-Earth. The material remaining in orbit is thus derived primarily from the farside of the impactor. The Fe core of the impactor is normally part of the material that falls out, but because of its higher density it normally falls right through the mantle of the proto-Earth and becomes combined with the Fe in the proto-Earth core.

For higher angular momentum collisions the impact is less destructive, and the bulk of the impactor continues onward in an elongated elliptical orbit. Once clear of the Roche lobe of the proto-Earth, it pulls itself together under its own self-gravitational attraction. Then it comes around to impact the proto-Earth for a second time. This time the collision is destructive to the impactor, most of which falls out of orbit into the proto-Earth as described above. However, considerably more material does remain in orbit, initially forming a long spiral arm of debris, usually with a slightly more massive blob of material at the outer end. Gravitational torques exerted along the spiral arm are generally effective in transferring angular momentum to the outer parts of the arm, raising the apogee of that material. This is the situation that will be found below for all the cases in which a giant impact produces a satisfactory amount of material in orbit for the formation of the Moon.

For a total collisional mass of 1 $M_\oplus$ the yield of material in orbit is very small for giant impacts having smaller angular momenta, near 1 lem. With increasing angular momentum, the yield progressively increases until the impactor nearly misses the proto-Earth. Then increasing amounts of mass that might have wound up in orbit actually escape from the gravitational binding of the proto-Earth.

This is shown in Fig. 1. For very small values of the angular momentum it is interesting to note that the amount of material that escapes from the proto-Earth after the giant impact exceeds the amount of material that is left in orbit. The reason for this is that when the surfaces of the impac-

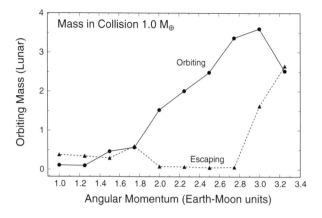

**Fig. 1.** The amount of mass in lunar mass units left in orbit and escaping from a proto-Earth of 1 $M_\oplus$ following a giant impact, as a function of the angular momentum in the collision, in Earth-Moon units (lem).

tor and proto-Earth meet at the beginning of the collision, they form a rather small angle and are nearly perpendicular to the direction of the collision, so that material is compressed to a high pressure. This sets up a large negative pressure gradient away from the initial impact point, and material is squirted out sideways at fairly high velocity.

The amount of material left in orbit then increases with the angular momentum $J_{coll}$, while the amount of material escaping remains small. Since the formation of the Moon from this orbital debris is inefficient, our working rule of thumb has been that forming the Moon requires about 2 $M_{\mathbb{C}}$ of material to be left in orbit, not necessarily inside the Roche lobe. Figure 1 shows that for a collisional mass of 1 $M_\oplus$, the collision must have an angular momentum of at least 2.2 lem to be able to form the Moon according to this criterion. For collisions beyond 2.8 lem the amount of material left in orbit decreases and the amount of material escaping from the proto-Earth starts rapidly increasing. Beyond 3.25 lem the impactor misses the proto-Earth so there is no giant impact. The maximum amount of mass left in orbit is only 3.6 $M_{\mathbb{C}}$, or less than 5% of 1 $M_\oplus$.

Following the giant impact the bulk of the angular momentum involved in the collision is left as rotation of the proto-Earth. Some of this rotational angular momentum can then be transferred to orbital angular momentum of the proto-Earth around the Sun; much of the remainder goes into the orbital angular momentum of the Moon once it is formed. However, solar tides now are weaker than lunar tides, and this would certainly have been even more true in the early history of the Earth-Moon system (*Munk and MacDonald*, 1975). However, for scenarios in which the giant impact occurred when Earth was only partly accumulated, it is clear that large secondary collisions during the subsequent accumulation could modify the spin of Earth (and the tilt of the equatorial plane relative to the plane of the lunar orbit may in part be due to such collisions). As long as such secondary collisions involve much smaller masses than in the giant impact, their effects are likely to

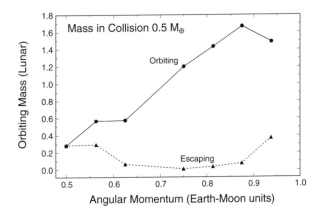

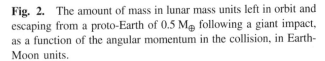

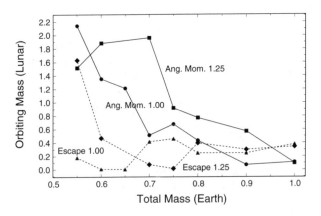

**Fig. 2.** The amount of mass in lunar mass units left in orbit and escaping from a proto-Earth of 0.5 $M_\oplus$ following a giant impact, as a function of the angular momentum in the collision, in Earth-Moon units.

**Fig. 3.** The amount of mass in lunar mass units left in orbit and escaping from a proto-Earth of varying mass following a giant impact, for fixed angular momenta in the collision of 1.00 and 1.25 Earth-Moon units.

be largely self-cancelling, and the early rotational angular momentum of Earth was probably minimally affected. At this stage in the investigation it was assumed that the giant impact had an angular momentum in the interval 1.0–1.25 lem.

In any case, it was clear that a total collisional mass of 1 $M_\oplus$ was an upper limit to the mass that could give an optimum set of parameters for the giant impact. Therefore the next step was to establish a lower limit for the optimum mass, and to determine whether the yield pattern for material left in orbit following the impact would resemble that shown in Fig. 1. Therefore the runs were carried out whose results are presented in Fig. 2. Here the total collisional mass was taken to be 0.5 $M_\oplus$, and again the angular momentum in the collision was varied. It was necessary to investigate collisions with angular momenta less than 1 lem, since for 1 lem the impactor missed the proto-Earth. However, the pattern displayed in Fig. 2 is similar to that in Fig. 1, so it was concluded that it was safe to generalize these results.

The most important guideline indicated by these results is that the best yields of material left in orbit after the giant impact come from those cases in which the angular momentum approaches the maximum possible value, beyond which no collision will occur. However, it should also be noted in Fig. 2 that the maximum yield in orbit of any of the runs is less than the 2 $M_{\mathbb{C}}$ that appears to be desirable for accumulating the Moon. Thus a total collisional mass of 0.5 $M_\oplus$ is a lower limit to the optimum total collisional mass for the giant impact.

## 3. DEPENDENCE OF ORBITING YIELDS UPON COLLISIONAL MASS

The next step was to compute simulations of giant impacts for total colliding masses distributed between the upper and lower limits established for the total colliding mass,

and to do so for the two values of the collisional angular momentum discussed above. The results of these computations are shown in Fig. 3.

It may be seen that for a collisional angular momentum of 1 lem the only case with more than 2 $M_{\mathbb{C}}$ left in orbit is for a total colliding mass of 0.55 $M_\oplus$; already at 0.6 $M_\oplus$ the residual orbiting mass has been reduced to <1.4 $M_{\mathbb{C}}$. The question then arises as to whether the 0.55 $M_\oplus$ case is a viable candidate for an acceptable giant impact. It turns out that it is not. There is a great deal of Fe included within the residual orbiting debris. If this material should be collected together into a single orbiting body, then that body would have far too much Fe to be a good representation of the Moon.

In Fig. 3 the curve representing a collisional angular momentum of 1.25 lem is much more promising. This shows that approximately 2 $M_{\mathbb{C}}$ of residual orbiting material are produced from 0.6 to 0.7 $M_\oplus$ of total colliding material. This indicates that the optimum conditions for the giant impact should be sought at about 0.65 $M_\oplus$ of colliding material. At this mass there is no significant amount of orbiting Fe produced, whereas for 0.6 $M_\oplus$ more Fe was left in orbit than was desirable, although considerably less than in the 0.55 $M_\oplus$ case. For other choices of the ratio of the masses of the impactor and proto-Earth this optimum collisional mass would be modified slightly, but that range of parameter space is at present inadequately explored.

Since there was a rather substantial increase in the amount of residual orbiting mass at 0.65 $M_\oplus$ in Fig. 3, the next step was to compute simulations of giant impacts at this total colliding mass in the angular momentum range 1.0 to 1.25 lem. Some results in this interval are presented in Fig. 4. It may be seen that almost 2 $M_{\mathbb{C}}$ are left in orbit at angular momenta of both 1.1 and 1.2 lem. This angular momentum range was then chosen as the most promising parameter region for further investigation.

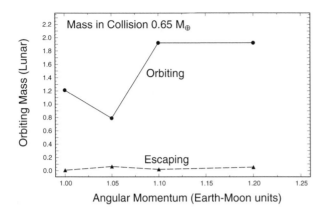

**Fig. 4.** The amount of mass in lunar mass units left in orbit and escaping from a proto-Earth of 0.65 $M_\oplus$ following a giant impact, for angular momenta in the collision in the interval 1.00–1.25 Earth-Moon units.

## 4. RUNS WITH INTERMEDIATE RESOLUTION

It may be noted that the orbiting yields exhibited in the first four figures do not form very smooth curves. This is because the future trajectories of the particles in a complicated dynamical system such as this are highly sensitive to initial conditions, and it only takes a slight displacement of a particle interacting with many nearby attracting bodies to magnify that displacement. The main constraint on the evolution of the system is the necessity for a continuing conservation of angular momentum. But in itself this is not definitive, since angular momentum can be conserved by having more mass in orbit at a smaller distance, or less mass in orbit at a larger distance. Thus small changes in the initial conditions may lead to modest changes in the orbiting mass.

You might think that increasing the resolution of the system by increasing the number of particles in the simulation would improve the accuracy of determination of the amount of mass in orbit, but I have found that this is not the case for the range of resolutions that I have used here. For some cases near the optimum initial conditions given above (where I have used 10,000, 20,000, and 100,000 particles in the simulation), I have found that increasing the resolution for the same set of initial conditions may either increase or decrease the amount of mass in orbit, but only modestly.

An extreme example of the governance of angular momentum in the outcome of a simulation is given by the results of an experiment made in response to many queries we have received as to whether the impact would be greatly modified if the proto-Earth had a substantial initial spin. The experiment involved 20,000 particles and the masses of the proto-Earth and impactor were as usual in the ratio 7:3. The impactor was spun on an axis perpendicular to the plane of

the collision with an angular momentum of 0.4 lem, which was considerably more than it was likely to have acquired from accumulating impacts of small planetesimals. The impact parameter of the giant impact was then adjusted so that the total angular momentum in the system was 1.1 lem. The amount of material placed in orbit was less than but within about 20% of the ordinary case with no initial spin and an angular momentum of 1.1 lem, and it resembled the situation where the orbiting mass had had some time to dissipate with some loss of mass from orbit.

## 5. RUNS WITH HIGHER RESOLUTION

Three runs employing 100,000 particles, AS04 (1.10 lem), AS05 (1.15 lem), and AS06 (1.20 lem), have simulated giant impacts and their aftermaths; the present paper reports on these simulations for roughly two days of real time. The general characteristics of these runs are very similar. In the following I shall show a series of images of AS04 in four color pages, Plates 1–4. In order to show the similarities among the simulations, at times corresponding to every second image in the early part of a run, and to every third image in the later part of the run, a compressed series of images are shown in Plates 5 and 6 for AS05 and in Plates 7 and 8 for AS06 that cover the same elapsed time period. The bottom half of the final image for each of the runs shows an enlargement of the scene at the final time, together with the same image rotated 90° to show a view in the equatorial plane of the proto-Earth. The proto-Earth mass in these runs is 0.455 $M_\oplus$ and the impactor mass is 0.195 $M_\oplus$, for a total colliding mass of 0.65 $M_\oplus$, in the region predicted to be optimum for accumulating the Moon.

In the SPH code the time steps are variable, and are adjusted so that the estimated errors in the positions and velocities of the particles, summed over all of them, do not exceed a predetermined criterion of accuracy. If they do, the time step is reduced and repeated. When the elapsed time exceeds an "archival time," an archival dump of the quantities in the computation is made. For the higher-resolution runs the interval between such dumps is approximately 2 minutes. In these three runs the computation was started some 11 hr before the midpoint in the giant impact (at about dump 60) and in the present paper I report on the results through dump 1500, 51 hr after the start of the computation and 40 hr after the midpoint in the first collision in the giant impact. The SPH technique has the unique property that the particles preserve their identity through the entire computation, and therefore the entire archival dump collection forms a database that can be examined to investigate details of the simulation in ways that might not have been foreseen ahead of time.

For the AS04 run (1.1 lem) Plate 1 shows images of the giant impact at intervals of 10 archival dumps (starting at 60, 11 hr from the beginning of the run, and thereafter every 20 minutes). Plate 2 (starting at dump 180) shows images at intervals of 20 dumps (40 minutes) to dump 300, and thereafter at intervals of 50 dumps (200 minutes) to dump

550. Plate 3 continues at intervals of 50 dumps to dump 1150. Plate 4 continues at intervals of 50 dumps through the first six images, to dump 1450, and in the bottom half of the sheet an image of dump 1500 is shown scaled up by a factor of 2 in each dimension for added resolution; the left two-thirds of the width shows the usual pole-on view centered on the protoplanet, and the righthand strip shows the same dump rotated through 90° so that the view is in the plane of the equator.

The initial models consisted of Fe cores surrounded by dunite mantles with ANEOS equations of state. The equation of state does not exhibit the property of rigidity, so a consequence is that the material acts like a fluid within SPH no matter how low the temperature is. Thus, a conservative approach was taken by setting the initial temperature throughout the colliding bodies at 2000 K, which inhibits evaporation of the rocky materials due only to a mild heating, as had been done for the lower-resolution cases. The viscosity is very low in SPH compared to other hydrodynamic codes. A paper showing the magnitude of viscous effects in this code is that of *Benz et al.* (1990). The SPH code itself is initially used with a damping term to let the particles in the bodies separately settle into a quiescent hydrostatic equilibrium. They are then launched from a separation of about 10 proto-Earth radii onto a collisional trajectory with the chosen collisional angular momentum.

In the color images of Plates 1–8, the internal energy of the particles is indicated to be within one of four energy ranges by the use of four different colors each for rock (dunite) and Fe; this representation turns out to be much more useful for diagnostic purposes than using many colors in a pseudocontinuous distribution. In making these plots, the Fe particles were plotted after the rock ones so that their images appeared superposed on the rock ones in order to see the behavior of Fe within the planet without obscuration. For the dunite rock used to represent the mantle material the four colors used, in order of increasing internal energy, are dark red, light red or pink, brownish-yellow, and white. For Fe used to represent the core material the colors used were dark blue, light blue, dark green, and light green in ascending order of internal energy.

The color images for the AS04 run begin with Plate 1, the first panel (of 12) showing the state of the system about halfway through the first collision between the proto-Earth and the impactor. The view looks down on the plane of the collision. Since the interval between images is approximately 20 minutes of real time, each row of three images is about 1 hr. Both bodies are significantly deformed in the collision, although the greater deformation is in the impactor. The purpose of starting the calculation several hours before the impact is to allow the impactor to become deformed as it enters into the gravitational field of the target, which is a noticeable effect. The Fe core of the impactor is denser than the mantle, so that following the collision it pushes forward to the leading edge of the deformed impactor as it undergoes a gravity swing around the proto-Earth. Some material from the impactor falls onto the proto-Earth,

and in panels 3 through 7 some Fe core material can be seen penetrating the proto-Earth mantle and accreting to its core. The double-armed configuration of panels 5 to 7 is clearly a demonstration that the conservation of angular momentum can impose some unusual dynamic effects on the post-collision material. In the final panels of the figure the impactor pulls itself together into an approximate spherical shape, and the material in the space between the two bodies is mostly lost through fallback onto one or the other of them. The impactor is now in a bound orbit around the proto-Earth, and in the last two panels of the figure the impactor has passed through apogee and swings back for its second collision with the proto-Earth, becoming again strongly deformed as it begins the second collision in the last panel.

The series of images continues in Plate 2. The first panel shows that the impactor has moved into the proto-Earth. Note the rock bulge at the top of that panel. This shows an expanding arc of material driven by a large pressure gradient that developed where the surfaces of the two bodies met at a small angle. The trajectory of the impactor has now brought it closer to the proto-Earth, so that this second collision is more violent and leads to a stronger deformation in the impactor. The impactor material forms an initially short but fat trailing arm on the proto-Earth, and in panels 2 through 7 the Fe from the impactor drains into the proto-Earth, wrapping itself around the proto-Earth core and gradually merging with it. The rock spiral arm continues to elongate and to become thinner through panel 8 in the figure.

It is important to note that this spiral arm is not a continuous fluid. I asked W. Benz, who has been studying problems involving structural rigidity using SPH, to estimate a characteristic scale length along the spiral arm where fracturing could be expected, and his estimate was about 0.1 km, very small compared to the size of the smooth particles employed in this calculation. Thus what we are really seeing along the spiral arm is a rubble pile held together by self-gravity but elongated by the dynamics of the collision. The importance of the shape of the spiral arm is that this configuration is very efficient at transporting angular momentum outward via gravitational torques.

The spiral arm continues to lengthen, but beyond panel 7 it now exhibits several instabilities. In panel 7 we see that a fracture has developed close to a sharp bend in the arm, also fairly close to the protoplanet. The material between the bend and the break pulls itself together into a small cluster. The protoplanet has a marked elongation. With this we can see the protoplanet rotate; it has a period of approximately 2.5 hr. This forms a rotating quadrupole that can contribute to the gravitational transfer of angular momentum outward. Meanwhile, the spiral arm sags inward toward one of the elongated ends of the protoplanet and collides with it, making a closed loop. Half a revolution of the protoplanet later, the spiral arm collides with the other elongated end and forms another closed loop. But at this point the rest of the spiral arm is fragmenting into many smaller clusters that are spreading themselves around the protoplanet in the general vicinity of the Roche limit. Throughout the panels the outer end of the arm has been thickening and growing into a sizable cluster. The transferal of angular momentum through the spiral pattern to this cluster causes it to recede toward a large apogee and to leave the panels toward the right.

The image sequence continues in the lower panels of Plate 2. What we see being established in these panels and in those throughout Plate 3 is a pattern of behavior that has not previously been noted in discussions of the evolution of the giant impact debris in its orbits around the protoplanet. The spiral arm discussed in the two previous figures has deposited most of the material remaining external to the protoplanet into orbits in the vicinity of the Roche distance and slightly beyond. This material is extensively clustered into rubble piles held together by self-gravity. The gravitational interactions among these clusters frequently perturb their orbits so that their perigees fall well within the Roche lobes of the protoplanet. Such clusters then undergo tidal shearing, which draws them out into long curving arcs in the images; the innermost ends of the arcs are those at a smaller distance from the protoplanet surface, since the approximately Keplerian orbital velocity is greater at these shorter distances. Depending on the angle at which the clusters descended into the Roche lobe, all, some, or none of the cluster material may be accreted onto the protoplanet. However, since the mutual orbital perturbations among the clusters are rather weak, the clusters usually descend only a relatively short distance into the Roche lobe, and usually there will be no accretion of material onto the protoplanet. The usual end result will therefore be the disruption of the cluster rubble pile, partially or entirely, with the individual particles placed on independent orbits. These orbits usually have apogees beyond the Roche distance, and since most of the closer gravitational encounters with other particles will occur near these apogees, the perigees of the orbits will usually be raised out of the Roche lobe. These individual particles can then be reaccreted onto existing clusters, or they may collide with one another to generate new clusters. The rate of accretion onto the protoplanet is small compared to what it would be if the clusters were not disrupted (which is what has usually been assumed in previous treatments of this problem). Throughout the panels of Plate 3 many striking examples may be seen of the tidal shearing of these clusters into long arcs.

This general pattern of behavior continues for an extended time. Many examples of long arcs produced by tidal disruption may be seen throughout Plate 3 and the first half of Plate 4. However, one general trend can be seen through the sequence of panels. There is a gradual depletion in the number of large clusters orbiting the protoplanet. These clusters have been engaged in a process of mutual destruction, but this is far from complete and is proceeding much more slowly in the final panel than in the first one. Also, the cloud of small individual SPH particles, mostly near the Roche limit and outward for about another factor 2 in radius, gradually thins out as the major clumps of particles grow. The degree to which the clusters of particles grow within less than two days after the giant impact is quite remarkable. During this period of time there is a rather low density of these particles within the Roche lobe of the protoplanet. Occasionally, as may be seen in some of the panels, a cloud of particles invades this region, clearly due to collective gravitational interactions, and then leaves it again. Such collective effects need a more extended study.

Detailed large-scale images of the protoplanet at the time of dump 1500 are shown in the bottom half of Plate 4, showing the view from on top of a pole in the left two-thirds of the page and the view in the plane of the equator in the right-hand strip in the final one-third of the page width. Recall that the image of the Fe is plotted on top of the image of the rock. There are a few blobs of impactor Fe that have not completely settled down on top of the protoplanet core, but are close to it. Within the core itself there are patches of light blue and of dark and light green, indicating inhomogeneities in the temperature distribution within the core (the impactor Fe has had the greater amount of shock heating and will form most of the light green material). The core has not had time to become completely stirred. The rock mantle is mostly light red, but yellow and white particles are sprinkled throughout it (these are the hotter ones), indicating a more complete stirring. The simulation does not allow heat diffusion within the proto-Earth since the elapsed time is too short for this to be of major importance. It may also be seen that there is a large amount of rotational flattening of the proto-Earth; the equatorial axis is 40% larger than the polar axis. The small amount of impactor Fe not completely accumulated onto the core is in the equatorial plane. It may also be seen that the external orbiting rock forms a fairly thin distribution in the equatorial plane. Because of this rotational flattening, the external potential in the equatorial plane has significant departures from an inverse square law dependence, and the orbital motions of particles are non-Keplerian. Near the Roche lobe the difference of the potential is several percent of the value it would have if the protoplanetary mass was spherically distributed.

At dump 1500 AS04 has 1.37 $M_{\mathbb{C}}$ in orbit. The largest cluster has 1037 particles with a total mass of 0.389 $M_{\mathbb{C}}$. The next three largest clusters have in the vicinity of 330 particles with masses of 0.149, 0.134, and 0.134 $M_{\mathbb{C}}$, and together these four clusters have 0.543 $M_{\mathbb{C}}$. The largest cluster is not shown in the lower part of Plate 4, as it is the cluster formed at the end of the spiral arm that had vanished after panel 9 of Plate 2, and is at a distance of $1.2 \times 10^{10}$ cm from the center of the proto-Earth. There are 16 other clusters with more than 10 particles, and 45 with fewer than 10 particles, mostly just pairs of particles. How this would sort itself out is an interesting game of celestial billiards that would surely be quite different for a small change in initial conditions, but the accumulation achieved in less than two days has taken place much more quickly than many people would estimate. And we have seen the operation of a number of physical processes that are not included in simulations of planetary accumulation that use simple N-body codes with permanent merging of material upon collision. These include the huge rotational flattening of the proto-Earth, which causes Keplerian orbits to rotate around the proto-Earth, the tidal stripping that largely prevents material from colliding with the proto-Earth, and the frequent gain and loss of particles that I have found to take place on individual clusters of particles.

Plates 5 and 6 show the highlights of the run AS05, which has a collisional angular momentum of 1.15 lem. The images in Plate 5 may be compared to their counterparts in Plates 1 and 2, since these images correspond to the times for AS04 for every second image in Plates 1 and 2. The images in the first half of Plate 6 correspond to the times of every third image in Plate 3 and the top half of Plate 4. The enlarged images in the bottom half of Plate 6 correspond to dump 1500, just as do those in the bottom half of Plate 4. It may be seen that the giant impact simulation in AS05 closely parallels that of AS04 as far as the general features are concerned.

Plates 7 and 8 were prepared for the same dump numbers as Plates 5 and 6, but for the AS06 run (1.20 lem). Again there is a close correspondence between the images at the same dump numbers in all three runs. However, a larger amount of material was placed into orbit in run AS06, perhaps as a result of the higher collisional angular momentum. As a result, the clusters of particles in orbit near the Roche limit for AS06 appear to be more massive than those for AS05. AS05 has 5 clusters with roughly 300 particles (2 of which are not in the image) with a summed mass of $0.60\ M_C$, and a total orbiting mass of $1.28\ M_C$. AS06 has five large clusters with 1007, 565, 541, 471, and 394 particles, summing to $1.17\ M_C$, with a total orbiting mass of $1.79\ M_C$, thus confirming the visual impression in Plate 8. The 1007 particle cluster is not seen in Plate 8, being at a radial distance of $2.3 \times 10^{10}$ cm from the protoplanet in dump 1500. The larger orbiting mass in AS06 is also likely to be a function of the larger collisional angular momentum in that case. Of the orbiting particles, 4.3% are Fe; these would all be derived from the impactor, and therefore the fraction of orbiting mass that is Fe would be somewhat smaller than this, close to 3%. This would be close to the fractional mass of the lunar Fe core, so that if this mass were (inefficiently) collected into the Moon in proportion to this fraction, then there would not be much room to add representative terrestrial material to the Moon during the final stages of accumulation of Earth. Thus AS06 must be close to the upper limit of what is acceptable as a candidate to form the Earth-Moon system from the point of view of composition, angular momentum, and orbiting mass.

As the system evolves after the giant impact the equatorial radius in run AS05 settles down to be just over $7 \times 10^8$ cm. Figure 5 shows an analysis of the material in orbit during this time. The total number of SPH particles above $10^9$ cm is plotted as a function of time at the top of the figure. This starts at 11 hr into the run, which corresponds to panel 10 in Plate 5, where a significant spiral arm has developed but before it starts breaking up into clusters of

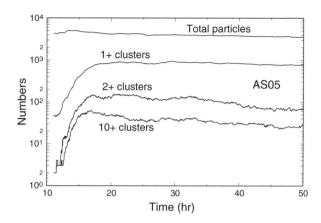

**Fig. 5.** The amount of material in orbit for case AS05, plotted as a function of elapsed time since the beginning of the simulation. The top line is the total number of particles, whether in clusters or not. "1+ Clusters" refers to the total number of clusters, including individual isolated particles. "2+ Clusters" refers to the total number of clusters containing at least two particles. "10+ Clusters'" refers to the total number of clusters containing at least 10 particles. This is typical of the other runs as well.

particles. The total number of particles slightly exceeds 5000 near the beginning of Fig. 5, but at the end it is only slightly less than 3600, indicating the slow loss of orbiting mass in the simulation. Technically the spiral arm forms a large cluster at the beginning, together with some individual particles placed into orbit earlier during the collisions. The second curve downward from the top is the number of clusters present in orbit, including isolated individual particles. This curve climbs near the beginning as the spiral arm fragments, but thereafter it stays remarkably constant, with small fluctuations as some of the particles dip below $10^9$ cm and then rise above that level, which often happens somewhat collectively as may be seen in Plates 3, 4, 6, and 8. The total number of real clusters, containing two or more particles, is the third curve from the top. It also rises steeply at the beginning, but then it goes through a broad peak from 15 hr to about 33 hr, after which it has a significant decline, indicating the erosion of the clusters due to tidal forces. This curve fluctuates much more than the preceding two curves, which must be attributed primarily to fluctuations in the number of clustered pairs of particles. The fourth curve shows the number of clusters having 10 or more particles. The shape of this is quite similar to the preceding one, but the fluctuations have a smaller amplitude, and in this case they must be attributed to gains and losses of particles from the larger clusters, which causes the number of clusters at the threshold of 10 particles to vary. The general message of Fig. 5 is that there is only a slow loss of mass from orbit, but the clusters of particles erode somewhat more quickly. Similar plots for AS04 and AS06 exhibit the same trends.

Figure 6 shows the mass external to the proto-Earth for the AS04, AS05, and AS06 runs. AS05 has significantly

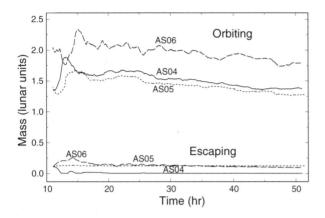

**Fig. 6.** The amount of mass in orbit for the three cases, AS04, AS05, and AS06, as a function of elapsed time since the beginning of their simulations. Also shown are the amounts of mass on escape trajectories.

more mass escaping from the Earth-Moon system, which lowers the orbiting mass slightly compared to AS04, but otherwise the two runs are very similar, again showing the slow rate of accretion of mass from orbit onto the proto-Earth. AS06 again has a considerable amount of mass escaping, but the amount of mass in orbit is considerably greater than for the other two cases. How this mass will be divided between proto-Earth accretion and lunar accretion is not determined.

Figures 7 and 8 show the temperatures of the particles within the proto-Earth and in orbit at dump 1500 for AS04, Fig. 7 for particles originally in the proto-Earth, and Fig. 8 for particles originally in the impactor.

These two figures display an odd behavior within the proto-Earth that requires some interpretation. In the central region of the proto-Earth some of the temperatures of both the Fe and the dunite have decreased to almost 0°. This temperature decrement smoothly goes away in the dunite mantle as the surface is approached, and the temperature approaches 2000 K, the value which both the proto-Earth and the impactor had initially in their interiors. The equation of state used for both the Fe and the dunite was ANEOS, developed at the Sandia National Laboratory. The ANEOS parameters for Fe were developed at Sandia, and those for dunite were developed by H. J. Melosh (see Moon3). Like most equations of state used to describe the behavior of solids, the internal binding energy of the solid is represented through a negative internal pressure. The giant impact stirs the Fe in the proto-Earth core and the dunite in the mantle, and Fe and dunite are added to each, so that individual particles may end up at higher or lower pressures than their initial values, and the thermodynamics is untrustworthy. Where the Fe and the dunite have been shock heated, the temperature is raised to the vicinity of as much as 7000 K for dunite and 9000 K for Fe. Melosh has found that the Sandia treatment of Fe at high temperatures is unreliable as it develops a vapor phase, so the highest values of the Fe temperature in these two figures are also unreliable. They

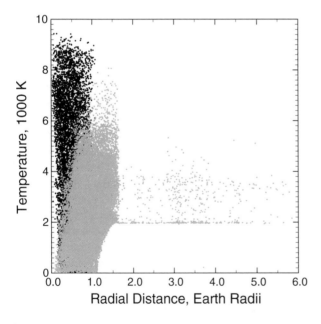

**Fig. 7.** The temperature distribution of the particles in run AS04 that were originally in the proto-Earth, for dump 1500 some 40 hr after the giant impact. See the text for a discussion of the peculiar low temperatures in the deeper parts of the proto-Earth. Iron particles are plotted in black and rock (dunite) particles are plotted in gray.

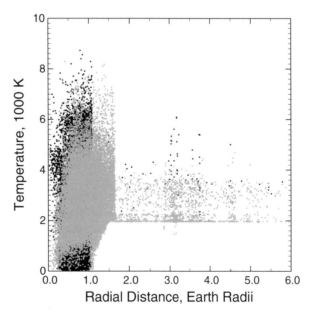

**Fig. 8.** The temperature distribution of the particles in run AS04 that were originally in the impactor, for dump 1500 some 40 hr after the giant impact. See the text for a discussion of the peculiar low temperatures in the deeper parts of the proto-Earth. Iron particles are plotted in black and rock (dunite) particles are plotted in gray.

should, however, be of the right order of magnitude, and not as bad as in simulations I have done with a total mass in the giant impact of 1 $M_\oplus$, since with only 0.65 $M_\oplus$ in the giant impact here, the energy released is significantly reduced and somewhat lower temperatures will be produced. These uncertainties should also have a negligible effect on the shape of the postimpact proto-Earth.

Figures 7 and 8 also show the temperatures of the orbiting particles in dump 1500 for AS04 out to 6 present Earth radii. It may be seen, as has been known for many years about the giant impact, that there are many fewer orbiting particles originating in the proto-Earth (Fig. 7) than in the impactor (Fig. 8). Only a relatively few particles have temperatures above 4000 K, and only a few in Fig. 8 have temperatures that dip a little below 2000 K, and these are in a couple of the larger clusters that have formed.

## 6.  THE ACCUMULATION OF A LUNAR BODY

The calculations presented above show that the accumulation of the Moon is a more complicated process than originally envisaged by *Ida et al.* (1997) and in subsequent work by these authors and others. The original assumptions made about the accumulation were that the giant impact would produce a disk of finely divided material smoothly distributed about the proto-Earth. Whenever particles in the disk collided with one another, they were generally expected to stick together, beginning a process of accumulation that would produce a hierarchy of sizes of small bodies. The larger bodies would then sweep up the smaller ones or else perturb them so that they would collide with the proto-Earth. Eventually one or only a small number of accumulated particles would survive in orbit, and it was hoped that the major survivor would have a mass comparable to that of the Moon.

The images in Plates 1–8 cover a time period of just over two days of real time for all three of the higher-resolution simulations. The general expectation has been that an accumulation time of months to a few years is likely to be required to assemble the Moon, a long time compared to any feasible SPH simulation. Therefore the orbital integrations that have been done used other methods. In the work of *Ida et al.* (1997), discussed above, orbits of particles exterior to the proto-Earth were followed using an N-body integrator that has the ability to detect and deal with collisions between orbiting bodies.

The first problem to be encountered here is that the high degree of rotational flattening of the proto-Earth causes departures of the gravitational potential from Keplerian (which would consider all the mass to be concentrated at a central point and which is usually assumed in N-body integrations) by several percent for distances comparable to the Roche limit.

The second problem with this approach has to do with the assumptions involving collisions between particles. Such collisions were assumed to be either pure mergers or to be inelastic bounces. Neither of these assumptions is consistent with a rubble pile model of an accumulated cluster. As remarked above in connection with the breakup of the spiral arm formed after the second giant impact collision, the strength of materials in a rocky assemblage fails to prevent fracture at distances exceeding about 0.1 km, so larger bodies assumed to be merged are nevertheless going to be broken up again by tidal distortion and tidal shredding. If the rock material is hot enough to form a magma, it is even less resistant to these deformation processes. Also, when a body collides with a rubble pile, the collision will be cushioned by the many loose constituents of the rubble pile, which will share the energy of the impact and will not present the rigid surface needed for a significant rebound. If the incoming body is itself a rubble pile, this aspect will be even more prevalent.

Thus a fully satisfactory postimpact simulation scheme has yet to be developed. Nevertheless, from the work that has been done we can make some conjectures about the accumulation of the Moon, even though we cannot give a good estimate of its final mass in any one of these cases at the present time. The first point to note is that as long as the major bodies in orbit near the Roche limit are several in number and comparable in mass (but very small compared to the proto-Earth), the mutual perturbations among them tend to lead to partial shredding, so that the mass in these bodies slowly diminishes, as may be seen in the image sequences presented in this chapter. But if one of these bodies should become much more massive than the others in orbit, then its orbit would be much less easily perturbed into a shredding situation. Meanwhile, it is likely to accumulate the smaller bodies that run into it. So the key to lunar accumulation appears to be the establishment of a dominant orbiting mass near the Roche limit.

The simulations shown here, covering a period of some 40 hr after the giant impact, depict an astonishingly rapid growth of sizable clusters of particles in orbit near the Roche limit, and a larger cluster formed at the end of the postimpact spiral arm that is placed into an orbit with a large apogee. The size of this latter cluster and the size distribution of the clusters orbiting near the Roche limit illustrate the individual stochastic pathways that the accumulation of the the orbiting debris can take. One might think that this could make the actual accumulation of the Moon nonpredictable, but with most of the orbiting material located in the radial region beyond, but not too far beyond, the Roche limit, the conservation of angular momentum will prevent too large a variation in this orbiting mass, so that the many different accumulation pathways are likely to converge on much the same mass for the body that is produced. This is particularly true because of the very slow accumulation of orbiting mass on the proto-Earth, which is a consequence of gravitational perturbations among the orbiting clusters of particles that lead mostly to some of the smaller ones being placed onto paths that dip only mildly into the Roche lobe, where moderate or extensive tidal stripping will take place among the rubble piles, and most of the material will be returned to the region just beyond the Roche limit and approximately recircularized. The clusters that are placed on paths with a large apogee will return to a much lower apogee where they will gravitationally interact with the other clusters, leading to major modifications in their orbits.

There are small amounts of Fe in the orbiting material in two of the cases studied here, but these amounts are highly variable and appear to be statistical accidents of the postimpact dynamics. Late in the runs AS04 has 1% orbiting Fe, and AS05 has 0.14% orbiting Fe. As we have seen, AS06 has significantly more orbiting Fe, but not enough alone to prove an embarrassment for lunar composition. However, these are merely lower limits on the amount of Fe likely to end up in the Moon. Following the giant impact, the proto-Earth must still accumulate about one-third of its mass. At that time the escape velocity from the Roche distance of the proto-Earth is about 7 km/s, which is the minimum collisional velocity of an incoming planetesimal. When the Moon is formed it will start in orbit near the Roche lobe and gradually recede from there due to tidal torques. There will, of course, be many collisions of the Moon with incoming planetesimals; collisions that overtake the Moon in orbit will have a reduced velocity, while those that meet the Moon head-on will be more energetic. The cratering events that are more energetic are likely to eject from the Moon more mass than possessed by the incoming planetesimal, while the less-energetic events may lead to a net accumulation of material. It is not clear whether there will be a net gain or loss of material to or from the Moon in these events. But it is clear that there will be an exchange of material; even if there is a net loss of material the Moon will retain some material from the incoming planetesimals. This retained material will include an Fe component. If the planetesimal had a normal inner planet composition, then nearly one-third of it is likely to be Fe. Thus the Moon is likely to acqire a few percent of its interior mass in the form of Fe, much of which is likely to wind up in the core, while the remainder may form mantle silicates.

The giant impact is almost certainly the largest collision that the proto-Earth had with an incoming planetesimal, since larger impactors are unlikely to be found at much earlier times, and the low-resolution survey reported earlier in this paper indicates that a collision with the angular momentum of the present Earth-Moon system (1 lem) onto a now more massive proto-Earth is unlikely to inject much orbiting mass at a later time but may very well tilt the spin axis by unacceptable amounts. However, the subsequent accumulation of material must also have involved some substantial collisions with quite massive planetesimals, and these will have led to some modifications to the angular momentum of the proto-Earth. The fact that the equatorial plane of Earth differs from that of the lunar orbit gives some indication of the violence of these subsequent collisions, but one cannot deduce any detailed history from a single data value of this sort. *Ward and Canup* (2000) have suggested that this difference in the two planes could be produced by resonant interactions between the Moon (after formation)

with a residual disk of debris within the Roche lobe. However, inspection of the color images accompanying this chapter fails to support the hypothesis that such a residual disk existed.

The time of formation of the Moon has recently been determined from measurements of the $^{182}$Hf–$^{182}$W ratio in lunar rocks (*Lee et al.,* 1997). It happened approximately 50 m.y. after the formation of the primitive solar nebula. That date appears nicely consistent with the scenario outlined here. It is often estimated that Earth took about 100 m.y. to accumulate, although that cannot be a very precise number because the rate of Earth accumulation probably declined roughly exponentially, so its ending is ill-defined. However, we do have the picture that when Earth was half accumulated the giant impact occurred and the Moon was in orbit very shortly thereafter.

## 7. DISCUSSION

The giant impact theory for the formation of the Moon is still very inadequately explored, although some additional insights do emerge from the present work. This work started with a low-resolution survey of possible giant impact parameters. That survey was incomplete since it only dealt with an assumed 7:3 mass ratio of proto-Earth to impactor, which was chosen as a plausible guess at a promising mass ratio to place significant amounts of mass in proto-Earth orbit. The survey needs to be extended to other mass ratios to determine their effect upon the apparent optimum parameter set.

Perhaps the most important conclusion reached here is that the actual process of accumulating the Moon in orbit around the proto-Earth is more complicated than had been thought and that the important effects of tidal stripping must be properly taken into account in further developments of the theory. A further complication is that N-body orbital integrations that have been done to simulate the lunar accretion process are at least partially invalidated by the rotational flattening of the proto-Earth, which causes significant departures from a Keplerian potential in the vicininty of the Roche lobe, and by tidal stripping processes. These are interesting challenges for the refinement of the giant impact theory. Since giant impacts are thought to have played important roles in the general process of accumulation of the terrestrial planets and possibly also some aspects of the accumulation of the giant planets as well, the resolution of these problems will probably have some broader applications.

## REFERENCES

Benz W., Slattery W. L., and Cameron A. G. W. (Moon1) (1986) The origin of the Moon and the single impact hypothesis I. *Icarus, 66,* 515–535.
Benz W., Slattery W. L., and Cameron A. G. W. (Moon2) (1987) The origin of the Moon and the single impact hypothesis II. *Icarus, 71,* 30–45.

Benz W., Cameron A. G. W., and Melosh H. J. (Moon3) (1989) The origin of the Moon and the single impact hypothesis III. *Icarus, 81,* 113–131.

Benz W., Bowers R. L., Cameron A. G. W., and Press W. H. (1990) Dynamic mass exchange in doubly degenerate binaries. I. *Astrophys. J., 348,* 647.

Cameron A. G. W. (1986) The impact theory for origin of the Moon. In *Origin of the Moon* (W. K. Hartmann, R. J. Phillips, and G. J. Taylor, eds.), pp. 609–616. Lunar and Planetary Institute, Houston.

Cameron A. G. W. (Moon5) (1997) The origin of the Moon and the single impact hypothesis V. *Icarus, 126,* 126–137.

Cameron A. G. W. and Benz W. (Moon4) (1991) The origin of the Moon and the single impact hypothesis IV. *Icarus, 92,* 204–216.

Cameron A. G. W. and Ward W. R. (1976) The origin of the Moon (abstract). In *Lunar Science VII,* pp. 120–122. Lunar Science Institute, Houston.

Canup R. M. and Esposito L. W. (1996) Accretion of the Moon from an impact-generated disk. *Icarus, 119,* 427–446.

Hartmann W. K. and Davis D. R. (1975) Satellite-sized planetesimals and lunar origin. *Icarus, 24,* 504–515.

Hartmann W. K., Phillips R. J., and Taylor G. J., eds. (1986) *Origin of the Moon.* Lunar and Planetary Institute, Houston. 781 pp.

Ida S., Canup R. M., and Stewart G. R. (1997) Lunar formation from an impact-generated disk. *Nature, 389,* 353–357.

Kipp M. E. and Melosh H. J. (1986) Short note: A preliminary study of colliding planets. In *Origin of the Moon* (W. K. Hartmann, R. J. Phillips, and G. J. Taylor, eds.), pp. 643–647. Lunar and Planetary Institute, Houston.

Lee D.-C., Halliday A. N., Snyder G. A., and Taylor L. A. (1997) Age and origin of the Moon (abstract). *Meteoritics & Planet. Sci., 32,* A78.

Melosh H. J. and Kipp M. E. (1989) Giant impact theory of the Moon's origin: First 3-D hydrocode results (abstract). In *Lunar and Planetary Science XX,* pp. 685–686. Lunar and Planetary Institute, Houston.

Melosh H. J. and Sonett C. P. (1986) When worlds collide: Jetted vapor plumes and the Moon's origin. In *Origin of the Moon* (W. K. Hartmann, R. J. Phillips, and G. J. Taylor, eds.), pp. 621–642. Lunar and Planetary Institute, Houston.

Munk W. H. and MacDonald G. J. F. (1975) *The Rotation of the Earth.* Cambridge Univ., Cambridge. 323 pp.

Stevenson D. J. (1987) Origin of the Moon: the collision hypothesis. *Annu. Rev. Earth Planet. Sci., 15,* 271–315.

Thompson A. C. and Stevenson D. J. (1983) Two-phase gravitational instabilities in thin disks with application to the origin of the Moon (abstract). In *Lunar and Planetary Science XIV,* pp. 787–788. Lunar and Planetary Institute, Houston.

Thompson A. C. and Stevenson D. J. (1988) Gravitational instability in two-phase disks and the origin of the Moon. *Astrophys. J., 333,* 452–481.

Ward W. R. and Cameron A. G. W. (1978) Disk evolution within the Roche limit (abstract). In *Lunar and Planetary Science IX,* pp. 1205–1207. Lunar and Planetary Institute, Houston.

Ward W. R. and Canup R. M. (2000) Origin of the Moon's orbital inclination from resonant disk interactions. *Nature, 403,* 741–743.

# Lunar Accretion from an Impact-Generated Disk

## E. Kokubo
*National Astronomical Observatory*

## R. M. Canup
*Southwest Research Institute*

## S. Ida
*Tokyo Institute of Technology*

We review current models for the accumulation of the Moon from an impact-generated debris disk. Such a disk is dynamically distinguished by its substantial mass relative to the Earth and a very centrally condensed radial profile, with a mean orbital radius near the classical Roche limit. In the inner protolunar disk, accretion is inhibited by tidal forces. Typically, a single large moon accretes just outside the Roche limit, at a distance of about 3.5–4.0 × the Earth's radius. A simple relationship between the fraction of the disk mass that is incorporated into the final moon and the initial disk angular momentum has been determined from simulations spanning a wide range of initial conditions, collisional parameterizations, and numerical resolutions. Predicted accretion yields range from 10% to 55%, with most of the remaining material scattered onto the Earth. Recent N-body simulations show the formation of transient gravitational instabilities in the inner disk, leading to rapid disk-spreading rates. These results may, however, be affected by current models' neglect of the thermal state of the disk material. Analyses of the orbital evolution of material due to tidal interaction with the Earth suggest that remnants of the initial accretion phase will likely be accumulated by either the largest moon or the Earth, leaving a single moon in most cases.

## 1. INTRODUCTION

Hydrodynamic simulations of potential lunar-forming impacts (*Kipp and Melosh*, 1986, 1987; *Benz et al.*, 1986, 1987, 1989; *Cameron and Benz*, 1991; *Cameron*, 1997; *Cameron and Canup*, 1998; see also chapter by *Cameron*, 2000) demonstrate the plausibility of the basic impact hypothesis. These simulations predict the ejection of roughly a lunar-mass worth of material into orbit following an off-center impact by an object with a mass close to that of Mars. The resulting debris cloud is centrally condensed, with a mean orbital radius of 2–3 $R_\oplus$, or at about the Roche limit for silicate density materials ($a_R = 2.9\ R_\oplus$), where $R_\oplus$ is the radius of the Earth. The predicted initial state of material ejected into this protolunar cloud is dependent on the simulation specifics, with temperatures ranging from those of a vaporous cloud to those of a mixture of solid and molten material.

Initial modeling of the evolution of an impact-generated disk focused on how such a disk might viscously spread and become subject to collapse due to gravitational instability (*Ward and Cameron*, 1977, hereafter *WC77*; *Thompson and Stevenson*, 1988). Another question was why a disk of material roughly centered on the Roche limit should yield a single large moon, while similarly located systems around the outer planets consisted of rings and multiple small satellites. The first model of lunar accumulation from an impact-generated protolunar disk utilized a statistical model

of accretion that included tidal inhibition of accretion in the region surrounding the Roche limit (*Canup and Esposito*, 1996). Canup and Esposito found that systems of multiple small moons appeared to be probable outcomes. They suggested that the easiest way to form the Moon was to begin with a lunar-mass of material exterior to $a_R$, where $a_R$ is the Roche limit radius, and that the most favorable impacts appeared to be those with about twice the angular momentum of the Earth-Moon system.

Accretion simulations utilizing direct N-body orbit integrations with ~$10^3$ initial ~100-km-sized bodies (*Ida et al.*, 1997, hereafter *ICS97*) revealed disk-wide scattering among the moonlets. These interactions cleared the inner protolunar disk, leaving a single large moon at 3–4 $R_\oplus$ in two-thirds of cases, and two large moons in one-third of cases. Scattering onto the Earth resulted in significant mass loss from the disk and net accretion yields below 50%. Thus an initial disk mass of at least 2 $M_C$ (where $M_C$ is lunar mass) appeared required to yield a lunar-sized moon.

Recently, *Kokubo et al.* (2000, hereafter *KIM00*) have performed similar simulations using N = 10,000 particles. While the accretion yields found by *KIM00* are similar to those in *ICS97*, the *KIM00* simulations resolve the development of spatial structure in the disk that was only vaguely observable in the *ICS97* runs. This structure is found to be the dominant mechanism for angular momentum and mass transfer in the N-body simulations. About 10% of the *KIM00* runs produced two large moons outside the Roche limit.

Modeling of the tidal evolution of multiple bodies in terrestrial orbit (*Canup et al.,* 1999) suggests that the two moon states will destabilize (through either mutual collision or collision of one of the moons with the Earth), and that an inner massive moon will likely sweep up smaller outer debris as it tidally evolves outward.

While simulations of both lunar accretion and the long-term evolution of bodies in Earth orbit appear to naturally predict an end state of a single moon, forming a lunar-mass moon remains problematic. Comparisons between predictions of impact simulations and results of the lunar accretion simulations are currently underway. However, to date a single impact has yet to be identified that can simultaneously account for the masses of the Earth and the Moon, as well as the current system angular momentum (*Cameron and Canup,* 1998; *Canup et al.,* 2000; see also chapter by *Cameron,* 2000). The only impacts thus far simulated that produce sufficient amounts of ejecta involve either impacts with angular momenta of about twice that of the Earth-Moon system, or an impact with a reduced-mass Earth that is only about 70% accreted after the moon-forming impact (see chapter by *Cameron,* 2000).

In this chapter, we review the accretion models, which describe the evolution of a particulate protolunar disk composed of a distribution of solid bodies. We note that this likely may not correspond well to the earliest state of the protolunar disk (a state that the impact simulations do not yet unambiguously constrain), or to the physical state of the disk material as it collisionally evolves. Below we first outline basic timescale arguments for the postimpact evolution of the disk; for additional discussion, see *Thompson and Stevenson* (1988), and reviews by *Stevenson* (1987) and *Pritchard and Stevenson* (2000).

Timescales for cooling from a lunar mass, optically thick protolunar disk radiating as a blackbody are ~10–100 yr. By far the fastest process in the disk is collisions between ejected bodies, with the characteristic time between collisions given by $t_{col} \sim 1/(\tau\Omega) \sim 4 \times 10^3 (a/3\ R_\oplus)^{3/2}(1/\tau)$ s, where $\tau$ is optical depth and $\Omega \equiv (GM_\oplus/a^3)^{1/2}$ is the orbital frequency at a semimajor axis a. For $\tau \sim 1$ at $a = 3\ R_\oplus$, $t_{col}$ is about 1 hr. Collisions damp relative energies, causing the disk to flatten, and exchange angular momentum, causing the disk to spread. The timescale for disk spreading is $t_{spread} \sim a^2/\nu$, where $\nu$ is viscosity; $t_{spread}$ is much longer than the collision time in most disks. The standard kinematic viscosity is a function of the velocity dispersion v and $t_{col}$: $\nu \sim v^2 t_{col}$. For a 2 $M_{\mathbb{C}}$, uniform surface density disk composed of mass m particles extending to $a \sim 3\ R_\oplus$, $t_{spread} \sim 6 \times 10^{20}/m$ yr, where m is in grams and we have assumed $v \sim v_{esc}$, where $v_{esc}$ is the surface escape velocity of the disk particles. However, disk-spreading times can be much shorter than this estimate for a massive disk subject to gravitational instability (*WC77*). Instability-induced clumps are not stable within the Roche limit, but lead to enhanced collision rates that in turn yield a much larger effective viscosity, $\nu_{eff} \propto G^2\Sigma^2/\Omega^3$ where $\Sigma$ is the disk surface density (*WC77*; see also *Lin and Pringle,* 1987). This is the same functional form

for viscosity that was later found by *KIM00* using a somewhat different physical argument (see section 3, equation (31)). N-body simulations of the protolunar disk have confirmed the rapid timescale for disk spreading predicted by Ward and Cameron, with $t_{spread} \sim$ months. However, such rapid rates may be physically unrealistic when the thermal and radiative properties of the disk are taken into account. *Thompson and Stevenson* (1988) recognized that the rate of spreading in the protolunar disk may be fundamentally regulated by the ability of the disk to radiate the gravitational binding energy liberated as the disk spreads. In this case, $t_{spread}$ is on the order of the disk cooling time, or 10–100 yr, vastly longer than that predicted using the viscosity derived in *WC77*.

When material in the disk has cooled and solidified, mutual collisions will result in fragmentation for high impact velocities, and in accretion if relative velocities are $\leq v_{esc}$. However, in a massive protolunar disk, the rate of collision may be so high as to remelt or even revaporize disk material during the accretion process (see discussion in chapter by *Pritchard and Stevenson,* 2000). Once massive bodies are present, they will experience orbital evolution due to tidal interaction with both the disk and the Earth. Using the current terrestrial tidal dissipation factor, the orbital evolution time for a lunar mass body with $a = 3\ R_\oplus$ due to terrestrial tides is on the order of years to decades, longer than nominal accretion times implied by N-body simulations.

It thus should be recognized that the protolunar disk may evolve significantly prior to the point at which its state could be aptly described by a particulate distribution of solid bodies, which is the assumption inherent to all the simulations described in this chapter. Indeed, the nature of the protolunar disk viscosity and its associated spreading time will be functions of the initial thermal state of the ejected material, as well as the subsequent evolution of the disk's energy budget as it spreads and accretes. To date, models have not included such processes, and it is thus unclear whether current simulations offer an adequate description of the disk viscosity and temporal evolution. This is particularly true in the inner disk, where material is most likely to have been significantly heated. The need for further investigation of these issues has been highlighted by a new theory for the origin of the Moon's orbital inclination (*Ward and Canup,* 2000). This theory relies on a resonant interaction between the newly formed Moon and an inner remnant disk, the effectiveness of which is dependent upon the viscosity, lifetime, and mass of the inner disk.

However, the main finding of the accretion simulations to date — a relationship between the mass of the moon and the initial angular momentum of the disk — should be relatively independent of the exact physical nature of the disk material. Assuming that the disk angular momentum is provided by the original impact event, the final size of the moon that can accrete just outside the Roche limit is constrained by a simple conservation of angular momentum argument, regardless of the mode of angular momentum transport in the disk.

In this chapter, we describe the general results of the accretion simulations, which are relatively insensitive to choice of initial disk conditions over the parameter space explored to date. In section 2, we outline a parameterization for modeling the tidal inhibition of accretion near the Roche limit. Section 3 describes the results of N-body simulations of the protolunar disk, and section 4 discusses the long-term evolution and stability of material in terrestrial orbit. Section 5 offers a summary and discusses areas for future research.

## 2.  MODELING ACCRETION NEAR THE ROCHE LIMIT

Traditionally, simulations of accretion in a circumsolar protoplanetary disk (e.g., *Greenberg et al.,* 1978; *Nakagawa et al.,* 1983; *Spaute et al.,* 1991; *Wetherill and Stewart,* 1993; *Weidenschilling et al.,* 1997) have utilized two-body approximations to describe interactions between orbiting bodies. For example, a standard accretion criterion is that the rebound velocity following a collision must be less than the two-body escape velocity of the colliding pair. Such approaches are valid if the physical size of an orbiting object is much smaller than its Hill sphere, as is the case for orbits well outside the classical Roche limit. For an impact-generated disk, $\langle a \rangle \sim a_R$, and a three-body treatment is required to account for tidal inhibition of accretion near and within the Roche limit. In this section, we review developments in tidal accretion models, focusing on the model utilized in lunar accretion simulations.

### 2.1.  The Hill Three-Body Formalism

The Hill approximation describes the motion of two bodies orbiting a much more massive central body using a rotating coordinate system. The Hill coordinate system is defined so that the x axis points radially outward, the y axis is tangent to a circular orbit, and the z axis is normal to the orbital plane. The angular velocity of the coordinate system is just the Keplerian orbital frequency, $\Omega = (GM_c/a_0^3)^{1/2}$, where $a_0$ is the reference orbital radius and $M_c$ is the mass of the central body. For a complete derivation, see *Nakazawa and Ida* (1988).

Hill's equations are often written in nondimensionalized form, with time scaled by $\Omega^{-1}$ and length scaled by the product $(ha_0)$, where h is the reduced Hill radius. For a pair of orbiting bodies with masses $m_1$ and $m_2$, $h \equiv [(m_1 + m_2)/(3 M_c)]^{1/3}$. The linearized equations of relative motion in nondimensionalized Hill units are

$$\ddot{x} = \quad 2\dot{y} + 3x - \frac{3x}{r^3}$$
$$\ddot{y} = -2\dot{x} \qquad\quad - \frac{3y}{r^3} \qquad (1)$$
$$\ddot{z} = \qquad\qquad - z - \frac{3z}{r^3}$$

where x, y, and z are the relative coordinates in the rotat-

ing frame and $r = (x^2 + y^2 + z^2)^{1/2}$. The $(2\dot{y})$ and $(-2\dot{x})$ terms represent Coriolis forces, and those proportional to $1/r^3$ are the mutual gravity terms. The tidal terms are $(3x)$ and $(-z)$: The tidal acceleration is positive in the radial direction (acting to increase the separation of the orbiting bodies), negative in the vertical direction (acting to decrease the relative separation of the orbiting bodies), and has no component in the azimuthal direction.

A constant of the motion described by equation (1) is the Jacobi energy

$$E_J = \frac{1}{2}\left(\dot{x}^2 + \dot{y}^2 + \dot{z}^2\right) + U(x,y,z) \qquad (2)$$

where $U(x,y,z)$ is the Hill potential

$$U(x,y,z) = -\frac{3}{2}x^2 + \frac{1}{2}z^2 - \frac{3}{r} + \frac{9}{2} \qquad (3)$$

The first two terms of U are the tidal potential, and the third is the mutual gravity between the orbiting objects. The constant 9/2 has been added so that U vanishes at $(x,y,z) = (\pm 1,0,0)$, i.e., the $L_1$ and $L_2$ Lagrangian points. Figure 1 shows the Hill potential in the $z = 0$ plane. The $U = 0$ surface defines the Hill "sphere," which is actually lemon-shaped with a half-width (in Hill units) of unity in the radial direction, 2/3 in the azimuthal direction, and ≈0.64 in the vertical direction. The Hill sphere is roughly the region of space within which the gravity of an orbiting object dominates the motion of nearby particles.

For orbits near the Roche limit, the physical size of an orbiting body becomes comparable to the size of its Hill sphere. The Hill radius of an isolated orbiting body of mass

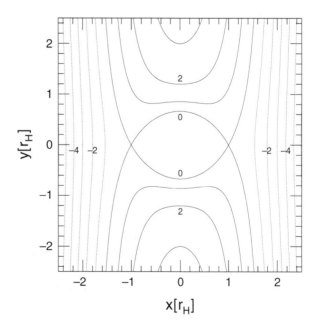

**Fig. 1.**   The Hill potential for the $z = 0$ plane; units are in Hill units (time scaled by $\Omega^{-1}$ and length scaled by $r_H = ha_0$).

m is just $r_H = [m/(3M_c)]^{1/3}a$. The ratio between the physical radius of a body and its Hill radius is

$$\frac{r}{r_H} = 3^{1/3} \left( \frac{\rho}{\rho_c} \right)^{-1/3} \frac{R_c}{a} \tag{4}$$

where $R_c$ and $\rho_c$ are the radius and density of the central body, and $\rho$ is the density of the orbiting body. This ratio can also be expressed in terms of the classical Roche limit for a fluid, strengthless body

$$\frac{r}{r_H} \approx 0.6 \left( \frac{a_R}{a} \right) \tag{5}$$

where

$$a_R \equiv 2.456 \left( \frac{\rho}{\rho_c} \right)^{-1/3} R_c \tag{6}$$

For $a \gg a_R$ (and so $r \ll r_H$), encounters between objects are well described by two-body approximations that ignore the gravity of the central body. However, if $a \sim a_R$ (or if relative velocities are comparable or less than the quantity $ha_0\Omega$, known as the Hill velocity), a three-body approach is required to describe interactions and collisional outcomes.

## 2.2. Tidal Accretion Criteria

Early models for accretion near the Roche limit were developed to describe collisions in planetary rings. *Weidenschilling et al.* (1984) incorporated tidal effects into their ring evolution model by allowing for disruption of bodies when a size-dependent tidal stress exceeded an assumed internal strength. However, it was later argued (*Longaretti*, 1989) that gravity must be the dominant mechanism for accretion of bodies larger than a few centimeters. *Longaretti* (1989) derived a tidal accretion condition by determining the equilibrium between the tidal force and the mutual gravitational force between a pair of orbiting particles aligned radially with respect to the central planet. This defined a critical mass ratio for a pair of colliding bodies below which gravitational accretion could occur, which was equivalent to the requirement that the sum of the radii of the colliding bodies must be less than $(ha_0)$ for accretion. Henceforth we will refer to the quantity $(ha_0)$ as the mutual Hill radius.

A somewhat different set of conditions results if a tidal model is developed using an energy rather than a force approach. *Ohtsuki* (1993) observed that since the Hill potential $U = 0$ surface is closed, two objects cannot escape their mutual Hill sphere if their postcollision relative energy, $E_J^{'}$, is negative. From this $E_J^{'} < 0$ condition, Ohtsuki derived a maximum coefficient of restitution that would allow for accretion assuming $r \ll r_H$ and neglecting the tidal terms in equation (3). For the case of $r \sim r_H$, *Ohtsuki* (1993) performed numerical integrations of collisions, concluding that the probability for accretion dropped rapidly for $r \geq 0.7\,r_H$. *Canup and Esposito* (1995, hereafter *CE95*) expanded on

Ohtsuki's approach, including the tidal terms in U and deriving capture criteria for the $r \sim r_H$ and $a \sim a_R$ case. These criteria include (1) a critical mass ratio for accretion [of the same form as that found by *Longaretti* (1989) but somewhat more restrictive] and (2) a critical coefficient of restitution. These constraints define a "Roche zone": the region surrounding the classical Roche limit where tidally modified accretion occurs (*CE95*). Below we review the derivation of these basic tidal accretion criteria, which can be easily incorporated into statistical or direct integration accretion simulations.

As two bodies collide, $E_J$ is just

$$E_J = \frac{1}{2} v_{imp}^2 - \frac{3}{2} x_p^2 + \frac{1}{2} z_p^2 - \frac{3}{r_p} + \frac{9}{2} \tag{7}$$

where $v_{imp}$ is the scaled impact velocity, and $x_p$, $y_p$, and $z_p$ are the coordinates of the impact point, such that $x_p^2 + y_p^2 + z_p^2 = r_p^2$. Here $r_p$ is

$$r_p \equiv \frac{r_1 + r_2}{ha_0} \tag{8}$$

which can also be expressed as

$$r_p = \frac{R_c}{a_0} \left( \frac{\rho}{3\,\rho_c} \right)^{-1/3} \frac{1 + \mu^{1/3}}{(1+\mu)^{1/3}} \approx 0.6 \left( \frac{a_R}{a_0} \right) \frac{1 + \mu^{1/3}}{(1+\mu)^{1/3}} \tag{9}$$

where $\mu$ is the mass ratio of the colliding objects, with $0 < \mu \leq 1$. The postcollision energy is (*Ohtsuki*, 1993)

$$E_J^{'} = \frac{1}{2} \varepsilon^2 v_{imp}^2 - \frac{3}{2} x_p^2 + \frac{1}{2} z_p^2 - \frac{3}{r_p} + \frac{9}{2} \tag{10}$$

Here $\varepsilon$ is an effective coefficient of restitution given by

$$\varepsilon = \left( \frac{\varepsilon_n^2 v_n^2 + \varepsilon_t^2 v_t^2}{v_n^2 + v_t^2} \right)^{1/2} \tag{11}$$

where the $v_n$ and $v_t$ are the normal and tangential components of the relative impact velocity and $\varepsilon_n$ and $\varepsilon_t$ are the normal and tangential coefficients of restitution. When the velocity and orientation of an impact are both known (e.g., in an N-body simulation), the $E_J^{'} < 0$ test for accretion for a given collision can utilize equation (10) directly. Below we derive accretion criteria averaged over all impact orientations.

If two bodies collide with random orientation, averaging equation (10) over all impact orientations gives the angle-averaged rebound energy (*CE95*)

$$E_J^{'} = \frac{1}{2} \varepsilon^2 v_{imp}^2 - \frac{3}{r_p} - \frac{1}{3} r_p^2 + \frac{9}{2} \tag{12}$$

The specific choice of impact orientation affects the coefficient of the $r_p^2$ term; here we have assumed random impact orientation. The $r_p^2$ coefficient is 3/2 in the case of impacts occurring in the radial direction, –1/2 for impacts in the vertical direction, and 0 for impacts oriented in an azimuthal direction. The above equation can also be written as

$$E_J^{\cdot} = \frac{1}{2}\left(\varepsilon^2 v_{imp}^2 - v_{esc,3B}^2\right) \tag{13}$$

where $v_{esc,3B}$ is an angle-averaged three-body escape velocity (*CE95*)

$$v_{esc,3B} \equiv \sqrt{6/r_p + 2r_p^2/3 - 9} \tag{14}$$

The scaled two-body escape velocity is just $\sqrt{6/r_p}$.

A necessary condition for $E_J^{\cdot} < 0$ is that the term $v_{esc,3B}^2$ is positive, because the term $\varepsilon^2 v_{imp}^2$ is always positive. This condition requires that

$$r_p \leq 0.691 \tag{15}$$

Note that $v_{esc,3B}^2$ is also positive for $r_p > 3.3$, but in this case the bodies are well outside their Hill sphere and cannot remain gravitationally bound. The physical meaning of equation (15) is that $(r_1 + r_2)$ must be less than the angle-averaged Hill radius. For $E_J^{\cdot}$ to be negative, $v_{esc,3B}^2$ must also be larger than $\varepsilon^2 v_{imp}^2$. The latter condition yields a critical coefficient of restitution

$$\varepsilon < \varepsilon_{cr,3B} \equiv \frac{v_{esc,3B}}{v_{imp}} \tag{16}$$

Thus when the physical size of colliding bodies exceeds about 70% of their mutual Hill radius they will not on average remain gravitationally bound, even if collisions are completely inelastic. This differs from the two-body approximation, in which completely inelastic collisions always result in accretion. Equation (15) is also a more stringent requirement than that obtained by *Longaretti* (1989) using a force approach to model escape in the radial direction, which yields an $r_p < 1$ criterion. This is because escape from the Hill sphere is also possible azimuthally and vertically, and the Hill "sphere" is actually narrower in these directions.

Equations (9) and (15) define a critical mass ratio for accretion for a completely inelastic ($\varepsilon = 0$) collision as a function of orbital location and particle density (*CE95*)

$$\frac{(1 + \mu_{cr})^{1/3}}{1 + \mu_{cr}^{1/3}} = \left(\frac{1}{0.691}\right)3^{1/3}\frac{R_c}{a_0}\left(\frac{\rho}{\rho_c}\right)^{-1/3} \tag{17}$$

where $\mu_{cr}$ is the maximum mass ratio that two bodies can have in order to remain gravitationally bound after a completely inelastic collision. Figure 2 is a plot of the critical mass ratio for accretion as a function of orbital radius, with the classical Roche limit shown for comparison. The critical accretion curve in Fig. 2 was derived assuming random impact orientation; a choice of a specific impact orientation (e.g., a radial impact) shifts the curves along the x-axis but does not change their form (see, e.g., Fig. 4). Note that the

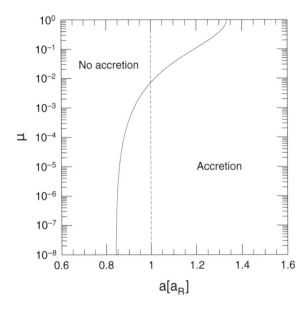

**Fig. 2.** The angle-averaged critical mass ratio for accretion as a function of scaled orbital radius and particle density for completely inelastic collisions ($\varepsilon = 0$). Here $a_R$ is the classical Roche limit (dashed line), which is a function of the ratio of particle to planetary density (see text for details). Accretion is precluded by tidal forces on average for impacts occurring with random impact orientation to the left of the mass ratio curve.

accretion of like-sized ($\mu = 1$) bodies is tidally inhibited within a region that extends beyond the classical Roche limit, while pairs with low mass ratios can accrete interior to $a_R$.

### 2.3. Character of Tidal Accretion

The three-body accretion criteria discussed above define three basic regimes of accretional growth surrounding the Roche radius. For orbits interior to about 0.85 $a_R$, bodies overflow the mean width of their Hill sphere and tidal effects on average preclude accretion. Accretion in the range of 0.85 $a_R \leq a \leq 1.4$ $a_R$ (defined by *CE95* to be the "Roche zone") is mass ratio dependent: collisions between bodies with a mass ratio less than $\mu_{cr}$ may result in accretion if rebound energies are low enough. Bodies with mass ratios larger than $\mu_{cr}$ cannot on average remain gravitationally bound, even for completely inelastic collisions. For orbits exterior to about 1.4 $a_R$, accretion is possible between objects of all sizes.

The parameterizations of tidal effects discussed here are simplistic, and represent only a first-order approach to the problem. Potentially important physical effects have yet to be included, or really even investigated. For example, implicit in the derivation has been the assumption that colliding bodies are spherical and will bounce in a manner similar to billiard balls, rebounding in an intact state with a given coefficient of restitution. The outcome of a collision near the Roche limit might be quite different if the collision energy were high enough to fragment a significant amount of the mass of one or both of the colliding bodies. In this case, the pulverized fragments might more easily be able to avoid physically protruding from the Hill sphere. Another assumption involves the use of the standard Hill approximation to describe the local potential of two bodies in contact that are similar in size with $r_p \sim 1$. In this case, the distribution of mass is far from spherically symmetric, and would likely cause a distortion of the local potential and the shape of the Hill "sphere" from the standard Hill approximation. Orbiting bodies would also likely be rotating before and after collisions, the effects of which were not included in *Ohtsuki* (1993) or *CE95*.

Finally, the criteria described here have been derived using a three-body approach, and any local clumping of material in the region of a collision could collectively influence collisional outcomes in a manner different than that predicted here. For example, local simulations of collisional evolution within Saturn's rings suggest that the formation of gravitational wakes fosters the buildup of temporary aggregates (*Salo*, 1992; see also discussion in *CE95*); such clumping is also observed in the lunar accretion simulations. However, direct N-body integrations that treat all collisions as inelastic rebounds and explicitly model mutual interactions show the growth of aggregates in a similar manner as that predicted by the tidal accretion criteria reviewed above (see section 3).

The tidal accretion criteria do a credible job of accounting for the gross characteristics of the planetary ring systems around the outer planets (e.g., *Longaretti*, 1989; *CE95*; *Canup and Esposito*, 1997). The location of the Roche zone, the variation of the inhibition of accretion with orbital radius, and the mass-ratio dependence of tidal accretion appear to coincide qualitatively well with a fundamental observed transition in all the outer satellite systems — from inner rings, to coexisting rings and moons, to outer isolated moons. This basic agreement is important, as the current ring and inner satellite systems are a direct observable relevant to a dynamical state through which the protolunar disk may have evolved.

## 3. N-BODY SIMULATION OF LUNAR ACCRETION

In this section, we focus on lunar accretion as modeled using direct N-body orbital integrations. In the evolution of a protolunar disk, global effects such as radial migration of lunar material, interaction of formed moons with the disk, and collective effects such as the formation of particle aggregates are potentially important. The merit of N-body simulation is that all gravitational interactions are explicitly accounted for, while the main disadvantage of this technique is its high computational cost. *ICS97* were the first to use N-body simulation to investigate the evolution of a particulate protolunar disk and lunar formation. Inspired by the work of *ICS97*, *KIM00* performed higher-resolution N-body simulations of a protolunar disk and investigated the evolution of the spatial structure of the disk in detail. The main

result of the simulations by both *ICS97* and *KIM00* is that, in most cases, a single large moon is formed just outside the Roche limit on a timescale of a month, with a nearly circular orbit close to the equatorial plane of the initial disk. The mass of the moon is linearly dependent on the initial disk angular momentum.

In this section, we focus on three points: (1) why a single moon is the typical outcome of the disk evolution, (2) what determines the timescale of lunar accretion from a particulate disk, and (3) the relation between the mass of the accreted moon and the initial disk condition.

### 3.1. Method of Calculation

*3.1.1. Orbital integration.* In an N-body simulation, the orbits of particles are calculated by numerically integrating the equation of motion

$$\frac{d\mathbf{v}_i}{dt} = -GM_\oplus \frac{\mathbf{x}_i}{|\mathbf{x}_i|^3} - \sum_{j \neq i}^{N} Gm_j \frac{\mathbf{x}_i - \mathbf{x}_j}{|\mathbf{x}_i - \mathbf{x}_j|^3} \quad (18)$$

where m, **x**, and **v** are the mass, position, and velocity of disk particles respectively, and G is the gravitational constant. The first term of the righthand side of equation (18) represents the Earth's gravity, and the second is the mutual gravitational interaction of disk particles. *ICS97* and *KIM00* did not include the force due to tidal bulges raised on the Earth by orbiting bodies, since the timescale of orbital evolution due to terrestrial tides is much longer than the accretional timescale (*CE96*). The $J_2$ component of Earth's gravity ($\simeq 10^{-3}$ at present; $\simeq 10^{-2}$ at the time of formation) was also not included. For numerical integration, both *ICS97* and *KIM00* used the predictor-corrector type Hermite scheme (*Makino and Aarseth*, 1992; *Kokubo et al.*, 1998).

The most expensive part of N-body simulation is the calculation of the mutual gravitational force, whose cost increases in proportion to the square of the number of particles. However, the recent development of software and hardware for N-body simulation has made it possible to consider more than $10^4$ particles in a protolunar disk simulation, compared to the $10^3$ particles utilized in *ICS97*. In the *KIM00* simulations, mutual gravitational forces were calculated by directly summing up interactions of all pairs of particles on the special-purpose computer, HARP-3/GRAPE-4 (*Makino et al.*, 1993, 1997, *Makino and Taiji*, 1998).

*3.1.2. Collision and accretion.* Collisions between particles play an important role in the evolution of a protolunar disk. In an N-body simulation, a collision occurs when the distance between two particles equals the sum of their radii. It is assumed that two colliding particles rebound with a relative rebound velocity **v'**, which is determined by the relative impact velocity **v** and the coefficients of restitution

$$\begin{aligned} \mathbf{v}'_n &= -\varepsilon_n \mathbf{v}_n \\ \mathbf{v}'_t &= \varepsilon_t \mathbf{v}_t \end{aligned} \quad (19)$$

The velocity of each particle after the collision is then calculated based on conservation of momentum. *KIM00* and

*ICS97* performed simulations with two values for the normal coefficient of restitution, $\varepsilon_n = 0.1$ and $0.01$; the tangential component was fixed at $\varepsilon_t = 1$ for simplicity. For these values, the effective coefficient of restitution given by equation (11) is $\varepsilon \simeq 0.7$.

In N-body accretion simulations, the initial particles in the disk are assumed to have infinite strength, so that they remain intact even at arbitrarily close distances to the Earth. This prevents the total number of bodies in a simulation from rapidly growing to a computationally unmanigable quantity. However, this assumption means that the simulated size distribution of material interior to the Roche limit likely contains bodies that are larger than those that in reality could have accreted there; for example, in the high-resolution simulations of *KIM00*, the smallest initial particles have m ~ $10^{-5}$ M$_{\mathbb{C}}$, or a diameter of about 60 km.

Given an initial distribution of disk particles, conditions for when collisions will result in accretion must be then specified. The necessary and sufficient conditions for gravitational binding between two orbiting particles are that (1) the Jacobi energy of the two bodies (equation (10)) after the collision is negative, and (2) the centers of mass of both colliding bodies are within their mutual Hill sphere. Because the Hill potential is nonaxisymmetric, both conditions depend on the angle of impact, as discussed in section 2. The *ICS97* simulations used the angle-averaged *CE95* accretion criteria, assuming that any collision with $E_{J'} < 0$ and $(r_1 + r_2) < 0.7 \, r_H$ resulted in a merger. The merged spherical body was assigned a total mass equal to that of the colliding bodies, and its position and velocity were set equal to those of the center of mass of the collision. In a merging event, some fraction of the orbital angular momentum of two colliding bodies would in reality be transferred into the spin of the merged body. However, the spin angular momentum obtained by many merging events is generally much smaller than the orbital angular momentum, and so the total orbital angular momentum of a system during a simulation is very nearly conserved. Merged bodies were assumed to have infinite strength, so that a merged body that accreted outside the Roche limit remained intact even if it later strayed within $a_R$.

The *KIM00* simulations expanded on *ICS97* by considering three different formalisms for collisional outcomes. The first, called the "partial accretion model," was identical to that utilized in *ICS97*. In the second model, the "total accretion model," the condition for merger was relaxed so that accretion was assumed for collisions after which $E_{J'} < 0$ and $(r_1 + r_2) < r_H$. In a final set of simulations, *KIM00* did not allow for any mergers, and instead simply allowed particles to bounce inelastically. In this "rubble pile model," gravitationally bound aggregates of particles form outside the Roche limit, and are tidally disrupted when they stray too close to Earth. In both the rubble pile and sometimes in the partial accretion model, gravitationally bound particles can remain in contact with one another even though they are not formally merged.

While the total and partial accretion models both assume that mergers create bodies of infinite strength, the rubble pile model assumes that merged aggregates have no strength and are held together only by their self-gravity. In the case of a strengthless, deformable fluid body, the classical Roche limit defines the minimum distance for an object to remain gravitationally bound, which from equation (5) implies r ≤ 0.6 $r_H$ for stability. Physical reality would fall somewhere in between the infinite strength and strengthless approximations.

*KIM00* found that the results of the total and partial accretion models and the rubble pile model are quantitatively similar over relatively short dynamical timescales. Over longer dynamical timescales, the rubble pile model differs slightly because, in this case, the moon loses some mass during collisions with other aggregates and through tidal stripping. The results of both the total and partial accretion models are essentially the same over long dynamical timescales.

*3.1.3. Initial conditions.* *ICS97* and *KIM00* started their simulations of lunar accretion assuming a solid particle disk. As the initial properties of an impact-generated disk are uncertain, and because the disk may significantly evolve before it cools and may be able to be treated as a particulate distribution, both *ICS97* and *KIM00* modeled the protolunar disk using a wide array of initial conditions. The initial mass distribution of disk particles was modeled by a power-law mass distribution

$$n(m)dm \propto m^{-\alpha}dm \tag{20}$$

where n(m) is the number of particles of mass m. The density of disk particles is $\rho = 3.3$ g cm$^{-3}$ (the bulk lunar density) and the density of the Earth is $\rho_{\oplus} = 5.5$ g cm$^{-3}$. Disk particles are assumed to be spheres. The initial disk is axisymmetric, with a power-law surface density distribution given by

$$\Sigma(a)da \propto a^{-\beta}da \tag{21}$$

where a is the distance from the Earth, with inner and outer cutoffs, $a_{in}$ and $a_{out}$. The assumption of disk axisymmetry should be valid because a nonaxisymmetric disk becomes axisymmetric due to Keplerian shear on a timescale of several Kepler times (approximately days). The initial eccentricities and inclinations of particles are assumed to be Rayleigh distributed. The ratio of the RMS eccentricity to the RMS inclination was fixed as $\langle e^2 \rangle^{1/2}/\langle i^2 \rangle^{1/2} = 2$. In general, the initial distributions of eccentricities and inclinations do not affect the disk evolution since they relax with a timescale on the order of the Kepler period due to collisional damping.

*KIM00* studied the evolution of two initial disk masses, 2 M$_{\mathbb{C}}$ and 4 M$_{\mathbb{C}}$. They also varied the power index of the surface density distribution ($\beta = 1,3,5$) and the outer cutoff of the disk ($a_{out} = 0.5,1,1.5,2 \, a_R$), which is equivalent to changing the initial specific angular momentum of the disk, $j_{disk}$. The $j_{disk}$ values were varied over the range

$$0.62\sqrt{GM_{\oplus}a_R} \leq j_{disk} \leq 1.0\sqrt{GM_{\oplus}a_R}$$

The effects of the power index of the mass distribution and

the initial velocity dispersion on the result were also tested. The power-law exponent of the mass distribution was chosen to be $\alpha = 0.5, 1.5, 2.5, \infty$, with a dynamic range in mass of $m_{max}/m_{min} = 1000$ for the $\alpha \neq \infty$ cases ($\alpha = \infty$ corresponds to an equal-mass case).

## 3.2.  Evolution of a Protolunar Disk

*KIM00* performed 60 simulations with the total and partial accretion models, and $10^4$ initial particles. They followed the evolution of the disk for 1000 $T_K$, where $T_K$ is the Kepler period at the distance of the Roche limit and $T_K \simeq 7$ hr. The disk evolution is qualitatively similar for all the simulations in *KIM00* and *ICS97*, although the initial disk conditions, the accretion model utilized, and the number of initial disk particles were varied.

In Figs. 3–5, we show an example of the evolution of a $10^4$-particle disk as simulated by *KIM00*. The initial disk has a mass $M_{disk} = 4\,M_{\mathbb{C}}$ with $\alpha = 1.5$, $\beta = 3$, $a_{in} = R_{\oplus}$, $a_{out} = a_R$, and $\langle e^2 \rangle^{1/2} = 0.3$. The coefficients of restitution were assumed to be $\varepsilon_n = 0.1$ and $\varepsilon_t = 1$ and the total accretion model was adopted.

Figure 3 shows snapshots of the protolunar disk in the R–z plane for t = 0, 10, 20, 100, 1000 $T_K$. The protolunar disk first flattens through collisional damping and then expands radially. A single large moon forms around $R \simeq 1.4\,a_R$ on a nearly noninclined circular orbit on a timescale of ~100 $T_K$. These are universal characteristics of the accreted moon in all the simulations by *KIM00* and most of the *ICS97* simulations, and appear to be nearly independent of initial disk conditions.

The orbital radius of the location of each merging collision and the mass ratio of the accreted particles are plotted in Fig. 4. The angle-averaged *CE95* critical mass-ratio is shown as well as the total accretion model condition, $r_p < r_H$. As the total accretion model was adopted here, accretion was possible to the right of the curve $r_p < r_H$. In t = 0–10 $T_K$, disk particles spread outward and start to accrete with one another if the accretion conditions are satisfied. In the total accretion model, accretion becomes possible for $m_2/m_1 = 10^{-3}$ beyond $a \simeq 0.65\,a_R$. The minimum mass ratio of $10^{-3}$ at this stage reflects the initial mass dynamic range. The accretion location spreads outward as the disk expands, and accretion between particles with small mass difference occurs in t = 10–20 $T_K$. The rapid formation of the moon occurs in t = 20–100 $T_K$. In this stage, the formation of relatively large moonlets and collisions among them make a single large moon. The accretion between particles that differ greatly in mass around $R = 1.3\,a_R$ indicates the accretion of disk particles by the growing moon. The moon gradually migrates outward due to interaction with the disk while still accreting some material in t = 100–1000 $T_K$, although the growth rate is low.

The mass of the largest moon, M, and the mass fallen to the Earth, $M_{fall}$, are plotted vs. time in Fig. 5. The mass of material that escapes from the gravitational field of the Earth is usually smaller than 5% of the initial mass of the disk.

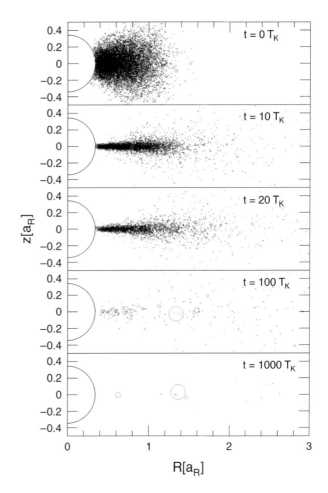

**Fig. 3.**  Snapshots of the protolunar disk on the R–z plane at t = 0, 10, 20, 100, and 1000 $T_K$. The semicircle centered at the coordinate origin represents the Earth. Disk particles are shown as circles whose size is proportional to the physical size of disk particles.

The mass of the moon at t = 1000 $T_K$ is 0.85 $M_{\mathbb{C}}$, while 3.1 $M_{\mathbb{C}}$ of the initial disk mass has fallen to the Earth. The fraction of the disk mass incorporated into the moon varies with the initial disk conditions.

The evolution of the disk is divided into two stages, namely, the rapid growth and slow growth stages as seen in Figs. 3 and 5. The duration of the rapid growth stage is ~100 $T_K$, or about 1 month. In this stage, the redistribution of disk mass through angular momentum transfer supplies material for accretion outside the Roche limit: Most of the disk mass falls to the Earth while some of the mass is transported outward (see Fig. 5). The formation of the moon is almost completed in this stage. The slow growth stage after ~100 $T_K$ is the "cleaning up stage," where the moon sweeps up and scatters away the residual disk mass.

*3.2.1.  Formation of a single moon.*  In order to see why a single large moon is a typical outcome of the disk evolution, we examine the rapid growth stage in detail. The evolution of the spatial structure of the disk in the rapid growth

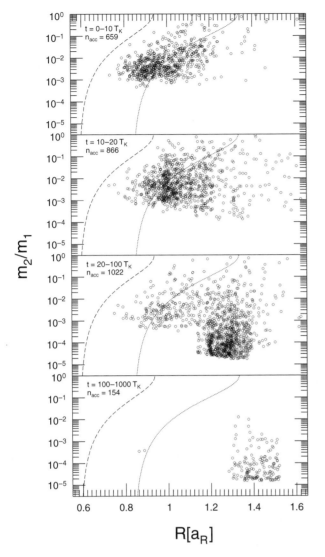

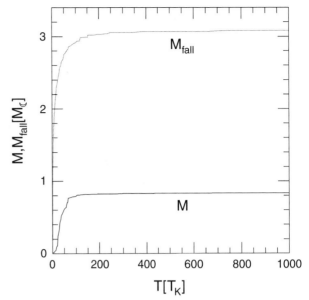

**Fig. 5.** The moon mass M (solid curve) and the mass fallen to the Earth $M_{fall}$ (dotted curve) as a function of time.

**Fig. 4.** The mass ratio $m_2/m_1$ of two accreted particles vs. the orbital radius of the location of the accreting event for the time periods t = 0–10, 10–20, 20–100, and 100–1000 $T_K$. The total number accreting events in each period is $n_{acc}$. The angle-averaged tidal accretion condition (dotted curve) and the total accretion model condition $r_p \leq r_H$ (dashed curve) are also shown.

stage is most easily seen by the rubble pile model, since in this model gravitationally bound particles are not merged but form particle aggregates. Snapshots of the disk in the x–y plane are shown for t = 0, 1, 5, 10, 20, and 40 $T_K$ in Fig. 6. The initial condition of the disk here is the same as that shown above except here an equal-mass initial distribution was considered. Figure 7 shows snapshots of the radial profile of the surface density. The surface density drops to zero at the surface of the Earth ($R_\oplus = 0.34\ a_R$). At t = 40 $T_K$ of Fig. 6, a large bound aggregate with a mass of about one-half the present Moon is formed at R ≃ 1.3 $a_R$, which is consistent with the result of the total and partial accretion models.

Before examining the disk evolution in further detail, we briefly discuss the stability of a differentially rotating disk. Disk stability has been studied extensively in the context of galactic and circumstellar disks (see, e.g., *Binney and Tremaine,* 1987). In a disk, self-gravity tends to produce density contrasts, while the random motion of constituent particles and the tidal force (shear) smooth it. It is convenient to introduce Toomre's Q parameter (*Toomre,* 1964)

$$Q \equiv \frac{v_R \Omega}{\pi G \Sigma} \qquad (22)$$

where $v_R$ is the radial velocity dispersion of disk particles and $\Omega$ is the angular velocity of the disk. When Q > 1, that is, when the effect of the tidal force or the random motion overwhelms that of the self-gravity of the disk, the disk is gravitationally stable and density contrasts do not grow in the disk. In fact, a particulate protolunar disk is marginally stable, but instability still plays an important role in the disk evolution. In terms of $a_R$, Q is given by

$$Q \sim 0.1 \left(\frac{\rho_{disk}}{\rho}\right)^{-1} \left(\frac{a}{a_R}\right)^{-3} \qquad (23)$$

where $\rho_{disk}$ is the spatial density of disk material and $\rho$ is the internal density of the disk particles (*Ida et al.,* 2000).

The evolution of an initially compact disk in the rapid growth stage is described below:

1. Contraction of the disk. The initial disk is dynamically "hot" (high relative velocities) and Toomre's Q value is much larger than unity (t = 0 $T_K$ of Fig. 6), so that the

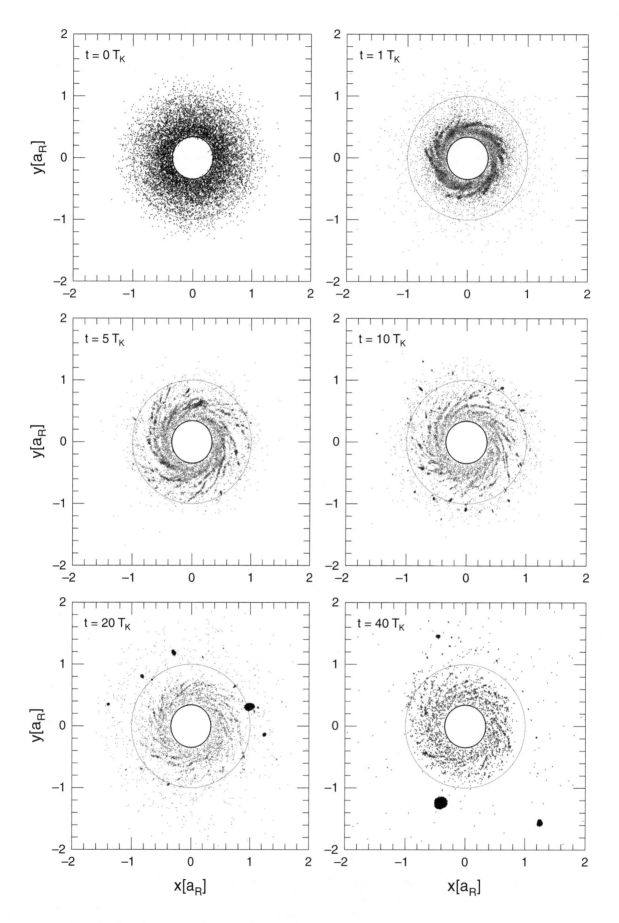

**Fig. 6.**  Snapshots of the circumterrestrial disk on the x–y plane at t = 0, 1, 5, 10, 20, and 40 T$_K$.

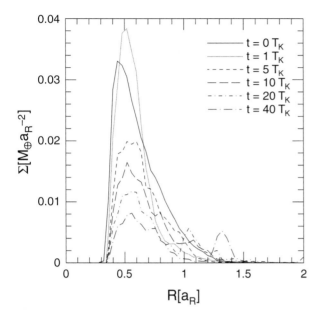

**Fig. 7.** The surface density $\Sigma$ is plotted as a function of the distance from the Earth R for t = 0, 1, 5, 10, 20, and 40 $T_K$.

disk is gravitationally stable. First the disk shrinks radially and vertically because the velocity dispersion decreases through collisional damping. The timescale of this contraction is of the order of $T_K$ since the initial optical depth of the particulate disk in the simulation is of order of unity and the coefficient of restitution for particles is less than unity.

2. Formation of clumps. As the velocity dispersion decreases, Q decreases. At this time particle clumps grow near a ≃ 0.5 $a_R$ where Q has its minimum value Q ~ 1 (t = 1 $T_K$).

3. Formation of spiral arms. The clumps are soon destroyed by the tidal force because they are within the Roche limit. They become elongated due to Keplerian shear, which results in the formation of spiral arm-like structures (t = 5 $T_K$). The radial wavelength of the spirals, as well as the size of the clumps, is roughly consistent with the critical wavelength with Q = 1 expected from linear stability analysis (e.g., *Toomre*, 1964)

$$\lambda \sim \lambda_c \equiv \frac{2\pi^2 G\Sigma}{\Omega^2} \sim 2\pi a \frac{M_{disk}}{M_\oplus} \quad (24)$$

For a = 0.5 $a_R$ and $\Sigma$ = 0.03 $M_\oplus a_R^{-2}$, $\lambda_c$ ~ 0.1 $a_R$, which is consistent with the results in Fig. 6. Since the pitch angle of the spiral arms is moderate, the number of spiral arms is estimated by

$$n_s \simeq \frac{2\pi a}{\lambda_c} \sim \frac{M_\oplus}{M_{disk}} \quad (25)$$

For $M_{disk}$ = 4 $M_C$, $n_s$ ~ 20, which is consistent with results of N-body simulations. The spiral arms are transient material waves, not pattern waves, and are not gravitationally

bound inside the Roche limit. The spiral arms are sheared out as they wind up, and then the cycle is repeated, as gravitational instability leads to the formation of clumps that are elongated to form spiral arms again.

4. Mass transfer by spiral arms. Particles are transferred outside the Roche limit through the gravitational torque exerted by the spiral arms, in compensation for the inward evolution of many particles to Earth. The surface density inside the Roche limit decreases with time due to mass transfer to Earth and beyond the Roche limit. On the other hand, since the simulated disk was initially entirely within the Roche limit, the surface density outside $a_R$ increases with time as material spreads outward. While the mass (and angular momentum) is effectively transferred by spiral waves, Q is kept around 2 (e.g., *Salo*, 1995; *Daisaka and Ida*, 1999).

5. Collapse of aggregates. When a tip of a spiral arm extends beyond the Roche limit, it collapses into a small aggregate (t = 10 $T_K$). Outside the Roche limit, particles in contact in the aggregate are gravitationally bound. The mass of the aggregate is approximately given by

$$m \simeq \pi\Sigma\left(\frac{\lambda_c}{2}\right)^2 \sim \pi^2\left(\frac{M_{disk}}{M_\oplus}\right)^2 M_{disk} \quad (26)$$

The formation of aggregates outside the Roche limit in the rubble pile model corresponds to the formation of moonlets in the accretion models.

6. Formation of a lunar seed. By sweeping up the small aggregates, a large aggregate quickly grows on a timescale of 10 $T_K$ (t = 20 $T_K$). A single large aggregate (the lunar seed) is formed at a ≃ 1.1 $a_R$, with a nearly noninclined and circular orbit.

7. Growth of the lunar seed. The lunar seed stays just outside the Roche limit and continues to sweep up particles spreading beyond the Roche limit. As the lunar seed grows, it moves gradually outward due to interaction with the inner disk. The peak of the surface density at t = 40 $T_K$ (R ≃ 1.3 $a_R$) corresponds to the lunar seed.

As a massive and compact particulate disk evolves in the manner described above, a single large moon forms inevitably. The relations (24), (25), and (26) also hold in simulations with somewhat different values of $M_{disk}$. Note that if the initial disk is radially extended past the Roche limit, accretion is immediately possible in the outer disk, which may result in a temporary multiple moon system (*Canup et al.*, 1999).

*3.2.2. Timescale of lunar accretion.* KIM00 showed that the timescale of the rapid growth stage is of the order of 100 $T_K$ (approximately 1 month), relatively independent of the initial conditions they simulated. Assuming that the initial disk is contained primarily within the Roche limit, the moon forms from material spreading beyond $a_R$, so that the predicted timescale of lunar formation is almost equivalent to the timescale of the mass and angular momentum transfer due to the gravitational torque by the spiral arms. The angular momentum flux through a right circular cylinder centered on the disk axis is given by

$$F_g = \frac{1}{4\pi G} \int_0^{2\pi} R d\theta \int_{-\infty}^{\infty} dz \frac{\partial \Phi}{\partial R} \frac{\partial \Phi}{\partial \theta} \qquad (27)$$

where $\Phi$ is the disk potential (*Lynden-Bell and Kalnajs, 1972*). For a disk with a spiral pattern whose potential can be represented by

$$\Phi_s(R, \theta) = -\frac{2\pi G}{|k|} \Sigma_s(R) e^{i[n_s \theta + f(R)]} \qquad (28)$$

where k is a radial wavenumber of the spiral pattern in the tight-winding approximation, $\Sigma_s(R)$ gives the amplitude of the spiral pattern, and f(R) is the shape function of the spiral pattern (see, e.g., *Binney and Tremaine, 1987*), the angular momentum flux is given by (*Lynden-Bell and Kalnajs, 1972*)

$$F_g = \frac{\pi^2 n_s G R \Sigma_s^2}{k^2} \qquad (29)$$

Substituting the critical wavenumber $k_c = 2\pi/\lambda_c = \Omega^2/(\pi G \Sigma)$ for k and $\Sigma$ for $\Sigma_s$ and using $n_s = kR \tan i$, we obtain

$$F_g = \frac{\pi^3 G^2 R^2 \Sigma^3 \tan i}{\Omega^2} \qquad (30)$$

where i is the pitch angle of the spiral arms. The effective viscosity for the angular momentum flux due to the gravitational torque exerted by the spiral arms, defined as $v_g = F_g/(3\pi R^2 \Sigma \Omega)$ (*Lynden-Bell and Pringle, 1974*), is thus

$$v_g = \frac{\pi^2 \tan i}{3} \frac{G^2 \Sigma^2}{\Omega^3} \qquad (31)$$

Using this effective viscosity, the timescale of the angular momentum transfer by the spiral arms is estimated as

$$T_g \equiv \frac{(\Delta R)^2}{v_g}$$
$$\sim 10^2 \left(\frac{\Sigma}{0.01 \, M_\oplus \, a_R^{-2}}\right)^{-2} \left(\frac{\Delta R}{0.5 \, a_R}\right)^2 \left(\frac{a}{a_R}\right)^{-9/2} T_K \qquad (32)$$

where $\Delta R$ is the radial shift of material due to angular momentum transfer and we have used $\tan i \simeq 1$. This timescale agrees well with the results of the N-body simulations by *KIM00* and *ICS97*. The functional form of $T_g$ shows that the timescale of angular momentum transfer, in other words, the timescale of lunar accretion, depends on not the individual mass of disk particles but rather on the surface density of the disk. *WC77* obtained almost the same viscosity and timescale by considering the energy dissipation in the clumps formed by gravitational instability.

The spiral structure is not always clear in the disk since it is often destroyed by gravitational scattering by large moonlets inside the Roche limit. However, the mass trans-

fer rate hardly changes. This is because for the mass transfer, the important point is not an exact spiral structure but a nonaxisymmetric structure. Detailed investigation of the angular momentum transfer in a particulate protolunar disk (T. Takeda, personal communication) shows that the gravitational torque exerted by the spiral arms is the dominant driver for angular momentum transfer near the Roche limit as long as the initial number of disk particles is larger than a few thousand for the disks modeled here.

For a compact disk (i.e., one initially within the Roche limit), the results of the rubble pile model show that the lunar seed is formed not by gradual pairwise collision of disk particles but collective particle processes: formation of clumps by gravitational instability, angular momentum transfer due to the gravitational torque due to the spiral arm-like structures, and collapse and collision of particle aggregates. The size of the clumps and the spiral arms are in this case determined by the critical wavelength $\lambda_c$ of the disk, which is a function of the surface density. Mass transfer is driven by the gravitational torque by the spiral arms, whose timescale depends on the surface density. Overall, the N-body simulations show that it is the surface density of the disk, rather than the properties of the individual particles, that governs the evolution of the disk. However, these interpretations are dependent on the assumption that the protolunar disk can be adequately modeled with $10^3$–$10^4$ particles and that the thermal evolution of the disk material can be neglected.

*3.2.3. Dynamical characteristics of the moon.* In this section, we consider the relationship between the dynamical characteristics of the accreted moon and the initial protolunar disk that was investigated by both *ICS97* and *KIM00*. However, the results that a single large moon is formed at $R \simeq 1.3 \, a_R$ and that a linear relationship exists between the mass of the moon and the initial disk angular momentum are essentially the same in all the simulations.

The orbital elements of the moon for all of the *KIM00* simulations are shown in Fig. 8. The semimajor axis of the moon in all cases is between $a_R$ and $1.7 \, a_R$, determined mainly by the formation location of the lunar seed and the subsequent interaction with the disk. The lunar seed forms just outside the Roche limit and it is pushed outward from its birthplace somewhat by recoil from the inner disk (shepherding). The eccentricity and inclination of the moon are small due to dynamical friction and collisional damping; in most cases, they are <0.1. These values are almost independent of the detailed initial conditions of the disk, and are similar to the results in *ICS97*. The resultant semimajor axis of the moon is small compared with the present lunar semimajor axis. On a longer timescale, the moon migrates outward by the tidal interaction with the Earth, presumably sweeping up outer residual mass (see section 4). Material inside the co-rotation radius ($\simeq 2.3 \, R_\oplus$ for an initial 5-hr terrestrial day) will tidally evolve inward and fall to the Earth.

In the majority of cases, the largest moon that accretes is much more massive than any other remaining body. However, in about one-third of the *ICS97* simulations a "two-moon" system was formed, defined to be one in which the

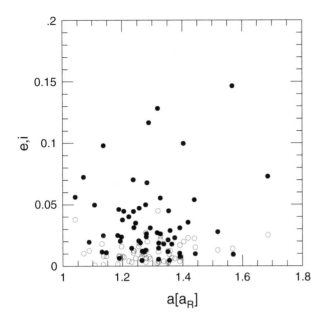

**Fig. 8.** The eccentricity (filled circles) and inclination (open circles) of the moon is plotted vs. the semimajor axis of the moon for all the runs.

*ICS97* explained the relationship between the moon mass M and the specific angular momentum of the protolunar disk, $j_{disk}$, by using a conservation of mass and angular momentum argument. From conservation of mass, we have

$$M_{disk} = M + M_{fall} + M_{esc} \qquad (33)$$

where $M_{esc}$ is the total mass of material that escapes. Conservation of angular momentum gives

$$M_{disk}j_{disk} = Mj + M_{fall}j_{fall} + M_{esc}j_{esc} \qquad (34)$$

where j, $j_{fall}$, and $j_{esc}$ are the mean specific angular momenta of the final moon, the mass that impacts the Earth, and the escaping mass, respectively, which are given by

$$j = \sqrt{GM_{\oplus}(1 - e^2)a}$$

$$j_{fall} = \sqrt{GM_{\oplus}(1 + e_{fall})q_{fall}}$$

$$j_{esc} = \sqrt{GM_{\oplus}(1 + e_{esc})q_{esc}}$$

where a, e, $q_{fall}$, $e_{fall}$, $q_{esc}$, and $e_{esc}$ are the mean semimajor

mass of the second largest body exceeded 20% of the mass of the largest moon. In the *ICS97* two-moon cases, many of the second largest moonlets had orbital radii within the Roche limit, as they had been scattered inward subsequent to their formation. When estimating the number of two-moon systems formed from their simulations, *KIM00* ignored any moonlets inside $a_R$, assuming that in reality such bodies would be tidally disrupted. Given this assumption, *KIM00* found that only about 10% of their simulations yielded two-moon systems, and that in most of these cases the second largest moon was on a horseshoe orbit (i.e., in a 1:1 resonance) with the largest moon. A horseshoe moonlet is the survivor of the rapid moon formation stage when moonlets are formed and collide with one another. As collisions in this stage are stochastic, a moonlet can sometimes survive by being on a horseshoe orbit with the most massive moonlet.

In Fig. 9, the mass of the accreted moon, M, scaled by the initial disk mass is plotted vs. the initial specific angular momentum of the disk, $j_{disk}$, for all the *KIM00* simulations. For cases in which the moon had a companion on a horseshoe orbit, the sum of the moon and the horseshoe companion is plotted.

The results of *KIM00* and *ICS97* (their Fig. 5) show that $M/M_{disk}$ increases linearly with $j_{disk}$. This is because in a small $j_{disk}$ disk, (i.e., in a more compact disk), a greater amount of mass must fall to the Earth in order for some mass to spread beyond the Roche limit, yielding a smaller final moon. The fraction of material escaping from the Earth also increases with $j_{disk}$, although this fraction is usually less than 5% of the disk mass. The overall yield of incorporation of disk material into a moon(s) ranges from 10% to 55%.

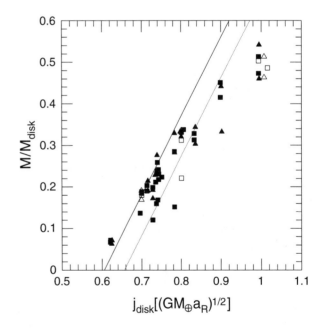

**Fig. 9.** The fraction of the initial disk mass incorporated into the moon, $M/M_{disk}$, is plotted against the initial specific angular momentum of the disk, $j_{disk}$, for all the *KIM00* runs. The triangles correspond to runs with an initial disk mass of $M_{disk} = 2\,M_{\mathbb{C}}$ and the squares to runs with $M_{disk} = 4\,M_{\mathbb{C}}$. The filled triangles and squares are for those runs assuming a coefficient of restitution $\varepsilon_n = 0.1$, and the open ones for those assuming $\varepsilon_n = 0.01$. The circles indicate runs that ended with two moons, defined to be those where the second largest moonlet has more than 20% of the mass of the largest moon. In these cases, the second-largest moon is on a horseshoe orbit with the largest moon, and the sum of the mass of the moon and the horseshoe orbiter is plotted. The theoretical estimate is also shown for $M_{esc} = 0$ (solid line) and $M_{esc} = 0.05\,M_{disk}$ (dotted line).

axis and eccentricity of the moon, and the mean perigee distance and eccentricity of the Earth impactors and the escaping material respectively. This conservation argument assumes that the accretion disk is flat ($\langle i^2 \rangle^{1/2} \simeq 0$) and that all material left in Earth orbit has been accreted into a single moon.

From equations (33) and (34), we obtain

$$M = \frac{(j_{disk} - j_{fall})M_{disk} + (j_{fall} - j_{esc})M_{esc}}{j - j_{fall}} \quad (35)$$

The mean values of each orbital element obtained by *KIM00* are $\bar{a} = 1.3\ a_R$, $\bar{e} = 0.04$, $\bar{q}_{fall} = 0.3\ a_R \simeq R_e$, $\bar{e}_{fall} = 0.2$, $\bar{q}_{esc} = 1.3\ a_R$, and $\bar{e}_{esc} = 1.1$. The relation $\bar{q}_{esc} \simeq \bar{a}$ reflects the fact that mass is ejected mainly due to gravitational scattering by the moon. Substituting these mean values into equation (35) yields

$$\frac{M}{M_{disk}} \simeq 1.9 \frac{j_{disk}}{\sqrt{GM_\oplus a_R}} - 1.1 - 1.9 \frac{M_{esc}}{M_{disk}} \quad (36)$$

This estimate is also shown in Fig. 9. The results of the high-resolution N-body simulations (*KIM00*) agree somewhat better than those of *ICS97* with the above analytic estimate, as *KIM00* found a somewhat larger moon mass.

Since $M_{esc}$ is always much smaller than $M_{disk}$, we can neglect the $M_{esc}$ terms in equation (35). In this case M is a function of $j$, $j_{fall}$, $j_{disk}$, and $M_{disk}$. However, $j$ and $j_{fall}$ are not free parameters but always have almost the same values since $j$ is determined by the fact that the moon forms just outside the Roche limit and $j_{fall}$ by the fact that remaining particles collide with the Earth. Then, the distribution of the disk mass to the moon and the Earth impactors is determined by the conservation of angular momentum. As the mass of the escapers is small compared with the disk mass, we can predict the mass of the moon from equation (36) when the mass and the angular momentum of the disk are given. *KIM00* confirmed that equation (36) holds for disks with masses in the range of $M_{disk} = 0.2$–$8\ M_\mathbb{C}$.

The results of the N-body simulation deviate a little from the analytical estimate at low (~0.6) and high (~1.0) $j_{disk}$. At the low end, the mass of the moon predicted by the simulations is larger than the analytical estimate because the semi-major axis of these moons and the specific angular momentum of the escaping material are smaller in these cases than the mean values used in equation (35). As the moons in the low $j_{disk}$ cases tend to be smaller in general, they suffer less gravitational recoil from the disk and move outward by a smaller distance, yielding a smaller moon semimajor axis than the mean value. For the high $j_{disk}$ cases, the analytical estimate of the lunar mass is larger than that obtained by the N-body simulations. At the end of these simulations, there are still about 1000 particles exterior to the moon, so that accretion is not yet complete. In fact, the sum of the mass of the moon and the mass of the particles bound to the Earth exterior to the moon (which would likely be the final moon mass), is on average ~15% larger than the lunar

mass at $t = 1000\ T_K$, and more consistent with the analytical estimate.

In summary, as a consequence of the evolution of a particulate protolunar disk, a single large moon on a nearly noninclined circular orbit is formed just outside the Roche limit. This result hardly depends on the initial condition of the particulate disk, as long as

$$0.62\sqrt{GM_\oplus a_R} \le j_{disk} \le 1.0\sqrt{GM_\oplus a_R}$$

$M_{disk} = 0.2$–$8\ M_\mathbb{C}$, and $\varepsilon_n = 0.01$–$0.1$, which may include the plausible conditions for the impact-generated disk. The moon is always formed around $a \simeq 1.3\ a_R$. In this case the mass of the moon is predicted simply by conservation of angular momentum from the initial disk. The accretion yields (the fraction of disk material incorporated into the moon) range from 10% to 55%.

## 4. EVOLUTION OF CIRCUMTERRESTRIAL MATERIAL

Two-thirds of the *ICS97* simulations produced a single large moon together with smaller bodies in exterior orbits; one-third yielded systems with two large moons. The great majority of the *KIM00* simulations yield the former case, while 10% yield two moons. While most accretion is complete after about a year, the final sweepup of material will occur over a longer time. Any bodies that remain on stable, noncolliding orbits after the initial accretion phase eventually must either collide with the Earth or be swept up by the moon as it orbitally evolves outward due to tidal interaction with the Earth. To date it has been assumed that the accretional stage of growth can be accurately modeled without including the effects of tidal evolution, as in general the tidal timescales are much longer than accretion times.

In this section, we address the question of whether or not a single moon will result from the likely end configurations of accretion in an impact-generated protolunar disk. For a complete discussion, see *Canup et al.* (1999; hereafter *CLS99*). Here we describe the basic tidal evolution process, and then discuss circumstances whereby moonlets and debris could become captured in mean-motion resonances as they tidally evolve. Such resonances are common among the satellites of the gas giant planets, and help to stabilize multiple moon systems in those cases. However, a terrestrial satellite system differs from the outer satellite systems in several key respects that predispose the terrestrial system to a single moon state.

### 4.1. Tidal Evolution of Moonlets

Exterior to synchronous orbit (the distance at which the orbital frequency equals the angular rotation rate of the Earth, $\simeq 2.3\ R_\oplus$ for a 5-hr terrestrial day), tides raised on Earth by an orbiting satellite lead to a transfer of angular momentum from Earth's rotation to the satellite's orbit, causing an increase in the orbital radius of the satellite.

Conversely, satellites within $a_{sync}$ lose angular momentum and evolve inward due to terrestrial tides. A simple model for the rate of evolution of orbital radius due to this process can be used to estimate when two moons that are initially orbitally separated will evolve into orbits that are close enough to be unstable (*Canup and Esposito,* 1996). Once mutual collisions are possible between objects with $a > a_R$, the material involved will likely eventually accrete into a single body. Here we consider the system evolution until this occurs (for a description of the later tidal evolution of the Moon, see chapter by *Touma,* 2000).

The rate of evolution of orbital radius due to terrestrial tides is approximately given by

$$\left.\frac{da}{dt}\right|_{\oplus} \simeq 3\,k_2\sqrt{\frac{G}{M_{\oplus}}}R_{\oplus}^5 m a^{-11/2}\sin(2\delta) \qquad (37)$$

where $k_2$ is the Earth's second order Love number, m and a are the mass and orbital radius of the orbiting body, and $\delta$ is the tidal lag angle (e.g., *Burns,* 1986). For a constant lag angle, equation (37) can be integrated to yield the orbital position as a function of time

$$a(t) = \left(\frac{13}{2}Kmt + a_0^{13/2}\right)^{2/13} \qquad (38)$$

where

$$K \equiv 3\,k_2\sin(2\delta)\frac{R_{\oplus}^{13/2}}{M_{\oplus}}\sqrt{\frac{GM_{\oplus}}{R_{\oplus}^3}} \qquad (39)$$

For sufficiently large t, $a(t) \propto (mt)^{2/13}$, and so the most massive moonlet will have the largest a value.

Consider two moonlets 1 and 2 with masses $m_1$ and $m_2$ and semimajor axes $a_1$ and $a_2$ (with $a_1 < a_2$). The evolution of the ratio $(a_1/a_2)$ as two moonlets tidally evolve is important, because mean-motion resonances (which each occur at some characteristic $(a_1/a_2)$ value) affect the system stability. A mean-motion resonance occurs when the ratio of the orbital motions of the two bodies is nearly a ratio of integers, e.g., for the (p + q):p resonance, $\Omega_1/\Omega_2 \simeq (p + q)/p$ where p and q are integers, and q is the order of the resonance. When two moonlets evolve through a resonance, the outcome is dependent upon whether $d(a_1/a_2)/dt$ is positive (typically referred to as the "converging" case) or negative (the "diverging" case).

From equation (37)

$$\frac{d}{dt}\left(\frac{a_1}{a_2}\right) = \frac{K}{a_2^2}\left(m_1 a_2 a_1^{-11/2} - m_2 a_1 a_2^{-11/2}\right) \qquad (40)$$

The ratio $(a_1/a_2)$ asymptotes to $(m_1/m_2)^{2/13}$, the value at which $d(a_1/a_2)/dt = 0$, (see *Canup and Esposito,* 1996). As two moonlets evolve to this asymptotic value, equation (40) implies three possible evolution paths: (1) $m_1/m_2 > 1$: moonlet 1 overtakes moonlet 2; (2) $(a_1/a_2)^{13/2} < m_1/m_2 < 1$: moonlet 1 does not overtake moonlet 2 and $(a_1/a_2)$ increases to the

asymptotic value since $d(a_1/a_2)/dt > 0$; (3) $m_1/m_2 < (a_1/a_2)^{13/2}$: moonlet 1 does not overtake moonlet 2 and $(a_1/a_2)$ decreases to the asymptotic value since $d(a_1/a_2)/dt < 0$.

In both (1) and (2), capture into resonance is possible, while capture is precluded in case (3). Before discussing further the effects of mean-motion resonances in the next section, we need to comment briefly on the evolution of satellite eccentricities due to tidal interaction.

From *Kaula* (1964) and *Goldreich and Soter* (1966), the rate of change of eccentricity is

$$\left.\frac{de}{dt}\right|_{tot} \simeq \frac{19e}{8a}\left.\frac{da}{dt}\right|_{\oplus}\left[\text{sgn}(\sigma) - \frac{28}{19}A\right] \qquad (41)$$

where the first and second terms are due to tides raised on Earth and the satellite, respectively, $da/dt|_{\oplus}$ is given in equation (37), $\sigma = (2\omega - 3\Omega)$, $\omega$ is the angular rotation rate of the Earth, and A is defined as

$$A = \frac{k_2^* \sin(2\delta^*)}{k_2 \sin(2\delta)}\left(\frac{m}{M_{\oplus}}\right)^{-2}\left(\frac{R^*}{R_{\oplus}}\right)^5 \qquad (42)$$

the ratio of satellite-to-planet effects used in *Mignard* (1980, 1981; see also *Kaula,* 1964; *Burns,* 1986), where the starred quantities are those of the satellite. The lag angle $\delta$ is related to the tidal dissipation factor, Q, by $Q \sim 1/\sin(2\delta)$. For the current Earth-Moon system, $(k_2/Q)_{\mathbb{C}} \simeq 0.0011$, $(k_2/Q)_{\oplus} \simeq 0.021$, and so $A \sim 0.5$ (*Burns,* 1986; *Dickey et al.,* 1994); thus currently de/dt from equation (41) is positive. However, we know that $Q_{\oplus}$ has varied over the Moon's history, as its current value implies that the Moon achieved its present position after only about 2 b.y. (see *Burns,* 1986). Given this uncertainty, a range of A values from 0 to 20 is plausible during the Moon's evolutionary history, the latter representing a case where only solid body tides contribute to terrestrial dissipation.

## 4.2.  Mean-Motion Resonances between Moonlets

As moonlets orbitally evolve due to tides they will pass through mutual mean-motion resonances. The evolution of the system during passage through or capture into an isolated resonance can be described by means of the adiabatic theorem (see, e.g., *Dermott et al.,* 1988). Capture into resonance is only possible for converging orbits with $d(a_1/a_2)/dt > 0$; for co-planar orbits with $d(a_1/a_2)/dt < 0$, passage through resonance results only in a jump in eccentricity and not permanent capture (see *Dermott et al.,* 1988; *Peale,* 1986).

Figure 10 is a plot of the asymptotic $(a_1/a_2)$ value [$= (m_1/m_2)^{2/13}$] due to tidal evolution as a function of moonlet mass ratio; also shown are the locations of first- and second-order mean-motion resonances. Below the solid curve orbits are tidally converging ($d(a_1/a_2)/dt > 0$), and capture into resonance is possible, depending on factors such as moonlet eccentricity and the rate of orbital evolution. Also shown (dashed line) is the critical $(a_1/a_2)$ ratio for two-body sta-

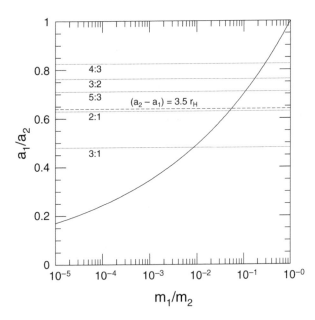

**Fig. 10.** The solid curve is the asymptotic value of $(a_1/a_2)$ due to tidal interaction with the Earth as a function of moonlet mass ratio. Above and to the left of the curve, $(a_1/a_2)$ decreases as moonlets tidally evolve; below and to the right, $(a_1/a_2)$ increases due to tides. Also shown are the positions of first- and second-order mean-motion resonances (dotted lines). The dashed horizontal line is the $(a_1/a_2)$ separation required for two-body stability with $(m_1 + m_2) = M_\mathbb{C}$. The only first- or second-order resonance that is well outside the $3.5\,r_H$ stability separation in this case is the 3:1 (from *CLS99*).

bility [i.e., $(a_2 - a_1) \leq 3.5\,r_H$] (*Gladman*, 1993) for $(m_1 + m_2) = M_\mathbb{C}$. Above this ratio, resonances are not isolated, orbits are chaotic, and mutual collisions can occur. Because of the large mass ratio of the Moon to the Earth, the only low-order resonance that lies well outside the two-body stability separation for bodies totaling a lunar mass is the 3:1. Thus there are a limited number of low-order resonances in which two Earth-orbiting moonlets whose total mass is a lunar mass could become captured.

*CLS99* conducted a stability study for mean-motion eccentricity resonances in the protolunar disk using both analytical techniques and numerical simulations. In general, resonances lead to eccentricity growth, and so long-lived capture in a resonance requires that this growth is offset by eccentricity damping due to tidal evolution. From equation (41), it is seen that the effectiveness of the latter process is a function of A. For resonances where capture was possible, *CLS99* calculated equilibrium eccentricities due to the combined effects of the resonance and terrestrial and satellite tides, as a function of A. This approach, coupled with N-body integrations including the acceleration on orbiting bodies due to terrestrial tides, showed that the typical end states predicted by the *ICS97* simulations were unstable, and would likely yield a single moon in each case. Their findings are most easily summarized in terms of the initial relative positions and masses of the moonlets.

*4.2.1. $m_1 > m_2$ case.* In this case, orbits converge due to tides and it would appear likely that the inner moonlet would overtake and accrete the outer moonlet. However, capture into resonance can occur, which could prevent mutual collision. The equilibrium eccentricity for the outer body in an exterior mean-motion eccentricity resonance due to the combined effect of the resonance, and satellite and terrestrial tides is, in general, less than unity only for A ≥ 20. Thus for $0 \leq A \leq 20$, exterior eccentricity resonances are unstable, and a massive inner moonlet will likely sweep up smaller outer moonlets as it tidally evolves outward.

*4.2.2. $m_1 \sim m_2$ case.* Here capture into resonance can occur, and stable equilibrium values of moonlet eccentricities in resonance are achieved for plausibly high rates of satellite dissipation (*CLS99*). However, in this case resonances destabilize as the relative importance of satellite to planetary tides approaches its current value of A ~ 1.

For the two-moon cases found in *ICS97* (which fall into this category), a more immediate issue for determining stability is the proximity of the inner moon to synchronous orbit, interior to which terrestrial tides lead to a decrease in orbital radius. *CLS99* found that the inner moon in all of the *ICS97* two-moon cases evolved inward and collided with the Earth in times as short as a year (assuming a terrestrial day of 5 hr).

*4.2.3. $m_1 \ll m_2$ case.* In this case, the asymptotic value of $(a_1/a_2)$ achieved as bodies tidally evolve is smaller than that needed for instability [$(a_1/a_2) > 0.64$ for instability with two moonlets totaling a lunar mass]. The initial value of $(a_1/a_2)$ would be greater than ~0.4–0.5 for potentially long-lived $m_1 \ll m_2$ pairs, assuming an outer moonlet with a ~ 1–1.5 $a_R$ and an inner particle just outside the co-rotation radius (2.3 $R_\oplus$). In these cases, orbits tidally diverge [$(a_1/a_2)$ decreases] and capture into resonance is precluded. The larger exterior body would tidally evolve outward and leave smaller inner bodies behind, potentially likely yielding a stable, multiple moon system. However, simulations do not predict that this configuration should persist after accretion from a protolunar disk, since perturbations by a moon that forms with close to a lunar mass appear to cause inner debris to collide with the Earth.

*4.2.4. $a_1 \simeq a_2$ case.* Some recent accretion simulations predict the formation of moon pairs occupying horseshoe orbits (*KIM00*, see also section 3). The 1:1 resonance represents an interesting case, as tidal torques will cause the libration amplitude to decrease, increasing the stability of the resonance with time (e.g., *Yoder et al.*, 1983; *Peale*, 1986; *Fleming and Hamilton*, 2000). Depending on the libration amplitude, the coorbital configuration could be destabilized through physical collisions with exterior objects encountered as the system tidally evolved outward. Another possibility is that scattering events with nearby objects could sufficiently increase eccentricities to allow for close encounters between the coorbitals.

Thus several factors appear to predispose a terrestrial system to a single moon state. First is the rapid rate of orbital evolution of satellites due to tidal interaction with the Earth.

Even for solid-body tidal Q values (Q ~ 100s), a moon that forms close to the Earth evolves out to 20 $R_\oplus$ (a typical outer limit for an impact-generated debris cloud; *Cameron and Benz,* 1991) in only $10^7$–$10^8$ yr. Second, terrestrial Q values are within an order-of-magnitude of likely tidal Q values for orbiting satellites. This means that the plausible range of "A values" — the relative role of satellite to planetary tides in affecting satellite eccentricity evolution extends only to A ~ 20, with a current value of A ~ 0.5–1. For a satellite orbiting a gaseous planet, A ~ 1000, and satellite orbits are circularized by satellite tides. In a terrestrial system, planetary tides act to increase satellite eccentricities, destabilizing resonances and increasing mutual collisions. The large mass-ratio of the Moon to the Earth, coupled with lunar formation from a centrally condensed disk appears to insure that small inner disk material inside the Roche limit is effectively perturbed onto the Earth (*ICS97, KIM00*). However, an open question remains as to whether moonlet pairs that form in horseshoe orbits could remain stable over long times, and this issue merits investigation.

## 5. CONCLUSIONS

We have reviewed lunar accretion from a particulate protolunar disk that might result from a giant impact event. The typical radial extent of such a disk is believed to be on the order of the Roche limit, $a_R$. In the accretion process, terrestrial tidal forces and collective effects such as the development of spiral arms thus can play an important role in the protolunar disk case. The Earth's tidal force partially inhibits accretion of particles in the Roche zone (0.85 $a_R \le$ a $\le$ 1.4 $a_R$), where accretion is dependent on the mass ratio of colliding bodies.

N-body integrations have been utilized to simulate the evolution of particulate protolunar disks, and have revealed that accretion in most such disks results in the formation of a single large moon. The moon is forms with a $\simeq$ 1.3 $a_R$ on a nearly noninclined, circular orbit. The evolution of a particulate protolunar disk consists of two basic stages. The first stage is a rapid growth stage, where material transferred outside the Roche limit as the disk spreads (together with material initially outside the Roche limit) self-gravitationally collapses and subsequently accretes to form a moon. The timescale for this stage is on the order of a month. Rapid angular momentum transfer by transient instabilities in the disk leads to the short (approximately 1 month) disk-spreading times, and orbital periods of only several hours yield a comparably short accretion time. These results hardly depend on the assumed initial condition of the disk, as long as the disk mass is on the order of 1 $M_{\mathbb{C}}$ and it is assumed to be well represented by a particulate distribution. The second stage, in which the moon accretes material spreading outward from the inner disk, persists for about 1 yr. The moon masses predicted by the N-body simulations coincide well with analytical estimates based on conservation of angular momentum of the disk. The efficiency of incorporation of disk material into a moon is 10–55%, and the yield

increases linearly with the initial specific angular momentum of the disk. Recent simulations (with N = 10,000 particles) tend to predict a a slightly larger moon mass and a higher probability of a single moon than previous simulations (with N ~ 1000). In cases of initial disks that radially extend beyond the Roche limit, multiple moons may result from the initial accretion phase. Simulations of the long-term evolution of multiple moons in terrestrial orbit, or of an inner moon with smaller exterior debris, find that all such systems destabilize as they tidally evolve, yielding a single moon in most cases.

The obtained ($j_{disk}$ vs. $M/M_{disk}$) relationship tells us that in order to form a moon with a present lunar mass, we need

$$j_{disk} \simeq 0.9\sqrt{GM_\oplus a_R}$$

for $M_{disk} = 2\ M_{\mathbb{C}}$ and

$$j_{disk} \simeq 0.7\sqrt{GM_\oplus a_R}$$

for $M_{disk} = 4\ M_{\mathbb{C}}$ respectively. Thus, in order to form the present-sized moon from a light disk, the disk must be extended, while a compact disk may also yield a lunar-sized moon if it is very massive. This relationship thus provides an important constraint on the type of disk that must be created by a giant impact to yield the Moon. Simulations of the impact event to date suggest that to obtain the required disk we may need an impact with angular momentum significantly larger than the present Earth-Moon angular momentum, or an impact with a reduced-mass Earth (e.g., *Cameron and Canup,* 1998; *Canup et al.,* 2000; see also chapters by *Cameron,* 2000, and *Canup and Agnor,* 2000).

An important factor that is not considered in previous, purely dynamical models is the thermal evolution of disk material. As a first step in investigating the evolution of a protolunar disk, the disk was assumed to be a particulate distribution. It is, however, believed that an entirely particulate distribution is not the most probable state for the protolunar disk when it condenses from the silicate vapor or liquid droplet cloud produced by the giant impact. Furthermore, the accretion timescale predicted by the N-body simulations is so short that it would be difficult for particles to cool by radiation in the course of accretion (e.g., *Thompson and Stevenson,* 1988). Indeed, a significant fraction of the accreted moon might be remelted or even reevaporated during accretion.

A coexistence of vapor/liquid and solid phases would be likely in the disk. Disk spreading and accretion might then instead proceed on the cooling timescale of the entire disk (~10–100 yr), much longer than the accretion time predicted by the N-body simulations of a particulate disk. However, a single large moon with mass predicted by equation (36) would likely still be the end result, as this is a basic consequence of conservation of angular momentum as we discuss below.

For example, consider lunar accretion from a disk composed of vapor, liquid, and solid phases. The outer disk would likely be cooler than the inner disk (due to a greater surface area for radiative cooling and lower disk surface densities), so that the disk might consist of vapor/liquid components at small radii and of solid particles at large radii. As the outer particulate disk becomes dynamically cold through collisional damping, instabilities develop and rapidly transfer angular momentum in the outer disk. For this situation, we can estimate the mass of the accreted moon formed from only the outer particulate portion of the disk. Applying conservation of mass and angular momentum to the outer particulate disk, we can obtain a similar result as equation (36), where in this case we use the radius of the inner edge of the outer particulate disk for $q_{fall}$. Because the angular momentum transfer rate in the inner vapor/liquid disk would be significantly smaller than that in the outer particulate disk where spiral structure is prominent, it would be valid to consider mass and angular momentum transfer only within the outer disk. As the disk cools, the particulate region then extends inward. Finally when the disk becomes an entirely particulate disk, we may have the same relation as equation (36) as long as the accreted moon is located just outside of the Roche limit. This is because the lunar accretion is controlled by the mass and angular momentum conservations of the disk that are independent of the phase of lunar material. (If the inner vapor/liquid disk diffuses well beyond the Roche limit, moons might accrete well outside the Roche limit and the characteristics of the accreted moons might change.)

While the mass of the final moon may not be overly sensitive to thermal considerations, the specific properties of the Moon's orbit could be. Recently, *Ward and Canup* (2000) have shown that a single resonant interaction between a lunar-sized moon formed outside the Roche limit and an inner disk can increase the moon's orbital inclination from an initially low value (on the order of 1°) to values as high as 15°. This may offer a natural explanation for the origin of the Moon's initial inclination, which is known to have been ~10° from back integrations of the Moon's current orbit (see chapter by *Touma*, 2000). However, the *Ward and Canup* (2000) mechanism is effective only if an inner disk with at least 25% of a lunar mass persists for decades to centuries after the Moon accretes. Examination of these and other issues will require a new generation of protolunar disk models, including a detailed investigation of the evolution of a multiphase protolunar disk.

## REFERENCES

Benz W., Slattery W. L., and Cameron A. G. W. (1986) The origin of the moon and the single impact hypothesis I. *Icarus, 66*, 515–535.

Benz W., Slattery W. L., and Cameron A. G. W. (1987) The origin of the moon and the single impact hypothesis II. *Icarus, 71*, 30–45.

Benz W., Cameron A. G. W., and Melosh H. J. (1989) The origin of the moon and the single impact hypothesis III. *Icarus, 81*, 113–131.

Binney J. and Tremaine S. (1987) *Galactic Dynamics.* Princeton Univ., Princeton.

Burns J. A. (1986) The evolution of satellite orbits. In *Satellites* (J. A. Burns and M. S. Matthews, eds.), pp. 117–158. Univ. of Arizona, Tucson.

Cameron A. G. W. (1997) The origin of the moon and the single impact hypothesis V. *Icarus, 126*, 126–137.

Cameron A. G. W. (2000) Higher-resolution simulations of the giant impact. In *Origin of the Earth and Moon* (R. M. Canup and K. Righter, eds.), this volume. Univ. of Arizona, Tucson.

Cameron A. G. W. and Benz W. (1991) The origin of the moon and the single impact hypothesis IV. *Icarus, 92*, 204–216.

Cameron A. G. W. and Canup R. M. (1998) The giant impact occurred during Earth accretion (abstract). In *Lunar and Planetary Science XXIX*, Abstract #1062. Lunar and Planetary Institute, Houston (CD-ROM).

Canup R. M. and Agnor C. B. (2000) Accretion of the terrestrial planets and the Earth-Moon system. In *Origin of the Earth and Moon* (R. M. Canup and K. Righter, eds.), this volume. Univ. of Arizona, Tucson.

Canup R. M. and Esposito L. W. (1995) Accretion in the Roche zone: Co-existence of rings and ringmoons. *Icarus, 113*, 331–352.

Canup R. M. and Esposito L. W. (1996) Accretion of the Moon from an impact-generated disk. *Icarus, 119*, 427–446.

Canup R. M. and Esposito L. W. (1997) Evolution of the G-ring and the population of macroscopic ring particles. *Icarus, 126*, 28–41.

Canup R. M., Levison H. F., and Stewart G. R. (1999) Evolution of a terrestrial multiple-moon system. *Astroph. J., 117*, 603–620.

Canup R. M., Ward W. R., and Cameron A. G. W. (2000) A scaling relationship for satellite-forming impacts. *Icarus*, submitted.

Daisaka H. and Ida S. (1999) Spatial structure and coherent motion in dense planetary rings induced by self-gravitational instability. *Earth Planet Space, 51*, 1195–1213.

Dermott S. F., Malhotra R., and Murray C. D. (1988) Dynamics of the uranian and saturnian satellite systems: A chaotic route to melting Miranda? *Icarus, 76*, 295–334.

Dickey J. O., Bender P. L., Faller J. E., Newhall X X, Ricklefs R. L., Ries J. G., Shelus P. J., Veillet C., Whipple A. L., Wiant J. R., Williams J. G., and Yoder C. F. (1994) Lunar laser ranging: A continuing legacy of the Apollo program. *Science, 265*, 482–490.

Fleming H. J. and Hamilton D. P. (2000) On the origin of the Trojan asteroids: Effects of Jupiter's mass accretion and radial migration. *Icarus*, submitted.

Gladman B. (1993) Dynamics of systems of two close planets. *Icarus, 106*, 247–263.

Goldreich P. and Soter S. (1966) Q in the solar system. *Icarus, 5*, 375–389.

Greenberg R., Wacker J. F., Hartmann W. K., and Chapman C. R. (1978) Planetesimals to planets: Numerical simulation of collisional evolution. *Icarus, 35*, 1–26.

Ida S., Canup R. M., and Stewart G. R. (1997) Lunar accretion from an impact-generated disk. *Nature, 389*, 353–357.

Ida S., Kokubo E., and Takeda T. (2000) N-body simulations of moon accretion. In *Collisional Processes in the Solar System* (H. Rickman and M. Marov, eds.). Kluwer, in press.

Kaula W. M. (1964) Tidal dissipation by solid friction and the resulting orbital evolution. *Rev. Geophys., 2*, 661–685.

Kipp M. E. and Melosh H. J. (1986) Short note: A preliminary numerical study of colliding planets. In *Origin of the Moon* (W. K. Hartmann, R. J. Phillips, and G. J. Taylor, eds.), pp. 643–

648. Lunar and Planetary Institute, Houston.

Kipp M. E. and Melosh H. J. (1987) A numerical study of the giant impact origin of the Moon: The first half hour (abstract). In *Lunar and Planetary Science XVIII*, pp. 491–492. Lunar and Planetary Institute, Houston.

Kokubo E., Yoshinaga K., and Makino J. (1998) On a time-symmetric Hermite integrator for planetary N-body simulation. *Mon. Not. R. Astron. Soc., 297*, 1067–1072.

Kokubo E., Makino J., and Ida S. (2000) Evolution of a circumterrestrial disk and formation of a single moon. *Icarus*, submitted.

Lin D. N. C. and Pringle J. E. (1987) A viscosity prescription for a self-gravitating accretion disc. *Mon. Not. R. Astron. Soc., 255*, 607–613.

Longaretti P. (1989) Saturn's main ring particle size distribution: An analytic approach. *Icarus, 81*, 51–73.

Lynden-Bell D. and Kalnajs A. J. (1972) On the generating mechanism of spiral structure. *Mon Not. R. Astron. Soc., 157*, 1–30.

Lynden-Bell D. and Pringle J. E. (1974) The evolution of viscous discs and the origin of the nebular variable. *Mon. Not. R. Astron. Soc., 168*, 603–637.

Makino J. and Aarseth S. J. (1992) On a Hermite integrator with Ahmad-Cohen scheme for gravitational many-body problems. *Publ. Astron. Soc. Jpn., 44*, 141–151.

Makino J., and Taiji M. (1998) *Scientific Simulations with Special-Purpose Computers — The GRAPE Systems*. Wiley and Sons, Chichester.

Makino J., Kokubo E., and Taiji M. (1993) HARP: A special-purpose computer for N-body problem. *Publ. Astron. Soc. Jpn., 45*, 349–360.

Makino J., Taiji M., Ebisuzaki T., and Sugimoto D. (1997) GRAPE-4: A massively-parallel special-purpose computer for collisional N-body simulations. *Astrophys. J., 480*, 432–446.

Mignard F. (1980) The evolution of the lunar orbit revisited, II. *Moon and Planets, 23*, 185–201.

Mignard F. (1981) The lunar orbit revisited, III. *Moon and Planets, 24*, 189–207.

Nakagawa Y., Hayashi C., and Nakazawa K. (1983) Accumulation of planetesimals in the solar nebula. *Icarus, 54*, 361–376.

Nakazawa K. and Ida S. (1988) Hill's approximation in the three-body problem. *Prog. Theor. Physics Supp., 96*, 167–174.

Ohtsuki K. (1993) Capture probability of colliding planetesimals: Dynamical constraints on the accretion of planets, satellites and ring particles. *Icarus, 106*, 228–246.

Peale S. J. (1986) Orbital resonances, unusual configurations and exotic rotation states among planetary satellites. In *Satellites* (J. A. Burns and M. S. Matthews, eds.), pp. 159–223. Univ. of Arizona, Tucson.

Pritchard M. E. and Stevenson D. J. (2000) Thermal aspects of a lunar origin by giant impact. In *Origin of the Earth and Moon* (R. M. Canup and K. Righter, eds.), this volume. Univ. of Arizona, Tucson.

Salo H. (1992) Gravitational wakes in Saturn's rings. *Nature, 359*, 619–621.

Salo H. (1995) Simulations of dense planetary rings III. *Icarus, 117*, 287–312.

Spaute D., Weidenschilling S. J., Davis D. R., and Marzari F. (1991) Accretional evolution of a planetesimal swarm: I. A new simulation. *Icarus, 92*, 147–164.

Stevenson D. J. (1987) Origin of the Moon. *Annu. Rev. Earth Planet. Sci., 15*, 271–315.

Thompson C. and D. J. Stevenson (1988) Gravitational instability in two-phase disks and the origin of the Moon. *Astrophys. J., 333*, 452–481.

Toomre A. (1964) On the gravitational stability of a disk of stars. *Astrophys. J., 139*, 1217–1238.

Touma J. (2000) The phase space adventure of the Earth and Moon. In *Origin of the Earth and Moon* (R. M. Canup and K. Righter, eds.), this volume. Univ. of Arizona, Tucson.

Ward W. R. and Cameron A. G. W. (1977) Disk evolution within the Roche limit (abstract). In *Lunar and Planetary Science IX*, pp. 1205–1207. Lunar and Planetary Institute, Houston.

Ward W. R. and Canup R. M. (2000) Origin of the Moon's orbital inclination through resonant disk interactions. *Nature, 403*, 741–743.

Weidenschilling S. J., Chapman C. R., Davis D. R., and Greenberg R. (1984) Ring particles: collisional interactions and physical nature. In *Planetary Rings* (R. Greenberg and A. Brahic, eds.), pp. 367–415. Univ. of Arizona, Tucson.

Weidenschilling S. J., Spaute D., Davis D. R., Marzari F., and Ohtsuki K. (1997) Accretional evolution of a planetesimal swarm: 2. The terrestrial zone. *Icarus, 128*, 429–455.

Wetherill G. W. and Stewart G. R. (1993) Formation of planetary embryos: Effects of fragmentation, low relative velocity, and independent variation of eccentricity and inclination. *Icarus, 106*, 190–209.

Yoder C. F., Colombo G., Synnott S. P., and Yoder K. A. (1983) Theory of motion of Saturn's coorbiting satellites. *Icarus, 53*, 431–443.

# The Phase Space Adventure of the Earth and Moon

## Jihad Touma
*American University of Beirut*

The Sun and Moon raise earthly tides that dissipate energy and transfer angular momentum from Earth's rotation to the Moon's orbit. This interplay has been going on for a very long time, and will continue for a while longer, forcing the Earth and Moon to sample exciting resonant islands in their phase-space. We will travel back in time, and watch with consternation, as the young Moon is captured into a devious network of resonances — courtesy of the Earth and Sun — that mercilessly distort its orbit, pump the eccentricity (thus heating the Moon's body, perhaps melting it), and push the inclination to values that can explain the current orbital configuration. Looking ahead, we witness how the Moon — vengeful, patient Moon — steers Earth to a spin-orbit trap that threatens to disrupt its obliquity and the climate with it. Ultimately, Earth might have the final word in this saga, when tides reverse their action, and the Moon finds itself spiraling into the final showdown.

## 1. IN THE BEGINNING

The story I am about to recount is a strand, rather involved, in the larger narrative that explores dynamical constraints on the origin of the Moon. That narrative was initiated primarily by *Darwin* (1880), who, reversing the action of dissipative tides, concluded that Moon originated very close to Earth, and suggested that it formed as a result of the fission of a fast-spinning Earth. Many years later, following numerous dynamical studies (see *Boss and Peale*, 1986, and *Burns,* 1986, for a review), as well as improved knowledge of lunar geology and geochemistry, a strong candidate for the origin of Moon emerged that appeared in embarrassing contradiction with its current orbital state. The candidate is of course the giant impact scenario, which naturally accounts for the substantial angular momentum of the Earth-Moon system, solves most of the geochemical constraints, is a likely outcome of planetary formation process, and appears to work in increasingly sophisticated simulation of lunar accretion from a circumterrestrial disk of ejecta (*Ida et al.,* 1997).

The contradiction was first pointed out by *Goldreich* (1966), who, following Darwin's work, along with computer-aided simulations of a multiaveraged theory, concluded that a degree of lunar inclination to the ecliptic in the present leads to 2° of inclination to Earth's equator, when the Moon was close to Earth. This result challenges any formation scenario in which the Moon starts out in Earth's equator (co-accretion was current when Goldreich was writing), particularly the giant impact scenario. One way out, of course, is to imagine a second impact that would push Earth's equator away from the lunar orbit (see *Stevenson,* 1987). Such an impact seems likely, following both statistical arguments as well as direct N-body simulations of the later stages of terrestrial planet formation. Another very recent (and to this author, more compelling) suggestion is that of *Ward and Canup* (1999), who propose a mechanism involving resonant interactions between the Moon and disk material within

the Roche radius to excite the lunar inclination. The process sketched in their paper appears quite robust and generates the desired inclination for a range of plausible conditions, but it awaits actualization in more realistic impact and accretion simulations. Together with Jack Wisdom, I have preferred to work with as few assumptions as possible to see whether dynamics could offer a way out of the paradoxes that dynamics devised. Essentially, we chose to play the game that Darwin defined, the game that *Gerstenkorn* (1955), *MacDonald* (1964), *Kaula* (1964), *Goldreich* (1966), and *Mignard* (1981) played over the years, a game that works with the simplest model of tidal interactions, and as complete a representation of the dynamics as possible, to first work back in time and see what we can learn. When we first decided to follow this adventure (*Touma and Wisdom,* 1994b, hereafter referred to as *TW1*), we learned that, with the exception of resonances missed because of artificially accelerated tides, Goldreich's observations withstood the increased sophistication in modeling that was made possible by dramatic improvements in computational speed. Later, more realistic simulations into the past (Touma and Wisdom, unpublished data) revealed strong orbital resonances. This outcome necessitated a different approach. Passage through resonance is not time reversible. Capture into resonance and resulting amplification in one direction is ruled out in the other. So instead of starting in the present and working backward, we (*Touma and Wisdom,* 1998, hereafter referred to as *TW2*) started in plausible equatorial configurations, and studied the outcome of forward tidal evolution. In the process, we opened a can of worms that, provided some conditions are met, suggests a solution to the inclination problem, a possible source of heat large enough to cause substantial melting of the Moon, and an occasion to reopen the discussion of the thermal evolution of the Moon when coupled with a complete modeling of its orbit.

The story I am about to recount is the story of this can, opened. The gory details can be found in *TW1* and *TW2* and references therein. Here I have collected significant results, numerical and analytic, as I put together an account of the lunar orbit from the edge of the Roche radius in Earth equator, through a turbulent web of resonances, to the fortunately quiet present, on to the ultimate spiraling back into Earth. My account will naturally highlight a consistent scenario, but it will draw attention, at appropriate junctures, to uncertainties about, and variations on, the conditions that make it possible. First though, I pause to give the reader a bird's-eye view of the dynamics.

## 2.  SNEAK PREVIEW

The trio Sun, Earth, and Moon dances to rhythms set by mutual gravitational interactions, with strong perturbations from Earth's bulge and the rest of the planets. Earth orbit evolves secularly as a result of slow forcing by the planets. The present lunar orbit is notoriously difficult to analyze. The main effects result from forcing by the Sun's and Earth's bulges, on an ellipse osculating around the Earth-Moon barycenter. Primary among these changes is the regression of the line of nodes with a period of 18 years, and a progression of the lunar perigee with a period of 9 years. The evolution of this last period with time will play an important role in our story. Earth rotation is coupled to the Sun and the Moon via a substantial bulge. The resulting torques force the precession of the equinox. Furthermore, secular variations in the ecliptic excite small-amplitude variations of the obliquity.

Had energy been conserved, we would not have much of a story to tell. However, Earth is not rigid and the gravitational field is not uniform. So differential forces develop that deform Earth. Had the response been ideal, the Moon and the Sun would raise bulges that follow them on their separate courses relative to Earth. However, friction (mostly in the shallow seas) delays the response that we model by a delayed bulge. In the case of Earth, which spins faster than either the Moon or the Sun "revolve" around it, the bulge leads the perturbing body. Such a bulge will experience retarding torques from the tide-raising body, which attempts to slow Earth down. The day is currently lengthening at the rate of 0.0016 s/century. An equal and opposite torque acts on the orbit of the tide-raising body, making it expand, attempting to increase its eccentricity, and affecting inclinations in nontrivial ways that we explore later. The orbits of both the Earth and Moon are affected by this transfer of angular momentum, with the effect being stronger and faster on the Moon than on Earth. From lunar laser ranging, we know that the Moon is receding at the rate of $3.82 \pm 0.07$ cm/yr (*Dickey et al.*, 1994). This process will continue until the tide-raising bodies are evolving synchronously with the spin of Earth, a possibility we contemplate later in the tale.

Tides are also raised by Earth on the Moon. Such tides are quite effective early on in the evolution of the Moon, braking its rotation to a probable state of synchronous rotation with its orbit, a state that the Moon is currently in. Besides synchronization, these tides damp the eccentricity of the Moon, countering the effect of Earth tides. This they

do because of the periodic flexing of the Moon as it travels from perigee to apogee around Earth. This behavior dissipates energy while conserving angular momentum. It damps the eccentricity and decreases the semimajor axis. Currently the Moon's eccentricity is increasing because Earth tides dominate the effect of the Moon's tides on its orbit.

The conservative dynamics governs the evolution of three planes: the plane of the osculating ecliptic, the plane of the osculating lunar orbit, and the nodding-precessing plane of Earth's equator. In and between these planes are motions that are widely separated in frequency: the earthly day, the lunar month, the solar year, and various precession frequencies ranging from a few years to $10^5$ yr for the ecliptic. As a result of dissipation of energy and transfer of angular momentum from one component of the dynamical system to another, actions (semimajor axis, eccentricity, inclination, obliquity) are evolving secularly and the frequencies set up by the conservative interactions are changing appreciably over the age of the system. In that sense, the dissipative tides are "powering" the adventure of the Earth and Moon into regions of phase space that could have been much more exciting than anything we experience in the present; regions in which resonances might have existed between the frequencies; resonances that could amplify the actions to levels all but erased by the subsequent action of tides.

To explore these dynamics, we can follow one of two complementary routes. We can reverse the action of the tides and work backward from the present to find out what, if any, constraints the present orbital configuration sets on the place and conditions of origin of the Moon. Along this path, we learn that the current rate of tidal of evolution of the Moon is inconsistent with origin 4.5 b.y. ago, the current rate taking the Moon to Earth's surface in less than 2 b.y. This is the infamous timescale problem. The traditional way out argues that the current tidal Q of Earth is smaller than its value in the past, the decrease resulting from a resonant configuration brought about by the drifting continents. I will not have much to add to this argument, and choose to bury questions related to the timescale in the evolution of the semimajor axis of the lunar orbit, which will serve as our dependent variable. We also run, with Goldreich, into the aforementioned inclination impass, followed by indications (unpublished calculations) of strong orbital resonances that are missed in Goldreich's treatment. The irreversibility of passage through resonance forced us to study an alternative route, namely the forward evolution of the Moon from a variety of initial conditions (orbits, spins, physical properties) soon after formation in Earth equator. Hence, we watched a web of resonances open up ahead of the lunar orbit. The first, the evection, couples the precession of the lunar perigee (faster then since the Moon was closer and Earth's bulge bigger) to the yearly orbital motion of Earth and can pump the lunar eccentricity to values as large as 0.5. A large postevection eccentricity enhances another resonance, the eviction, that couples the eccentricity to the inclination, and can force the Moon out of Earth's equator by a maximum of 3°, not enough to solve the inclination problem. Assumptions about initial interior properties of the Moon, and its subsequent heating as its eccentricity is pumped by the evection, suggested a scenario according to which the Moon, after a first forward passage through evection, regresses and is captured into evection, thanks to enhanced tidal dissipation in a hotter Moon. Capture into evection can push the mutual inclination to as much as the 12° needed around 6 $R_e$ to solve the inclination problem. Subsequently, the Moon finds its way out of evection, and follows a natural course to its present orbit. In the process, it suffers a dose of tidal heating large enough to cause substantial melting. This scenario relies critically on an adiabatic tidal evolution, an initially cold Moon, and a sharp increase in lunar dissipation resulting from a phase of large orbital eccentricity. These special conditions, though conceivable within current uncertainties about the initial geophysical and lunar interior properties, made us wonder about plausible alternatives that, for the most part, did not offer a consistent way out of the equator to the present orbit. I will start by describing the trip back, partly to review history, partly to introduce the numerical algorithms that we worked with. Then, I will flesh out the forward route from a remote origin in a giant impact to a more peaceful present orbit, a few degrees out off the ecliptic, with a surface that holds clues to a much hotter past. I will close with a brief description of the ultimate evolution of the Earth-Moon system, followed by a discussion of serious uncertainties and promising developments.

## 3. BACK TO THE PAST

Past studies of lunar histories, including Goldreich's, make strong approximation in the dynamics of the system to reduce the problem to manageable size. We wondered whether the inclination problem could not be resolved by relaxing some of those assumptions, relying on faster machines and algorithms (*TW1; TW2*).

On the way, we reexamined various tidal models in the context of a Hamiltonian reformulation of the multiply averaged theory presented by *Goldreich* (1966). Goldreich ignores planetary perturbations, and assumes that the orbit of the Earth-Moon barycenter about the Sun is a fixed circular orbit. The orbit of the Moon, whose finite size is ignored, is likewise taken to be a circular orbit, but is inclined relative to the ecliptic. The Earth is assumed to be axisymmetric, and its gravitational potential limited to the second-order moment. The disturbing potential of the Sun on the Earth-Moon system is truncated after second-order terms in the ratio of the Earth-Moon distance to the distance to the Sun. The orbital period effects are first removed by analytical averaging, motivated by the fact that the orbital timescales (month and year) are shorter than the node regression timescale (18 yr). Then motion on the nodal regression timescale is removed by numerical averaging, motivated by the fact that the nodal regression timescale is much shorter than the timescale associated with tidal evolution. The assumption of a fixed circular Earth-Moon orbit with respect to the

Sun drastically reduces the number of degrees of freedom. The Earth-Moon relative vector has 3° of freedom. The Earth orientation has 3° of freedom. Axisymmetry plus principal axis rotation reduces to 1 the number of degrees of freedom in the orientation of the Earth. Assumption of a circular lunar orbit and averaging over orbital periods leaves 1° of freedom in the orbital motion, which is removed by further averaging over the precession timescale. Thus, one ends up with only 1° of freedom, that associated with the obliquity and precession of the equinox of the Earth, which is coupled to the orbital inclination through the integrals of motion guaranteed by averaging. Tidal interactions induce changes in the semimajor axis and the inclination of the lunar orbit, and affect the orientation of the Earth.

Our Hamiltonian formulation of the "restricted" dynamics (*TW1*) gave a thorough independent check on Goldreich's calculations. It confirmed his main conclusions, particularly the inclination problem. Furthermore, it revealed the essential indifference of the dynamics to any particular member of the class of tidal models that we considered.

We then proceeded to eliminate most of the constraints, numerically exploring the available phase space for interesting dynamics, then developing the analytical theories appropriate for these dynamics.

### 3.1. Putting Machines to Work

Anxious to relax the assumptions that were introduced earlier, we incorporate *Mignard's* (1981) model of tides in Lie-Poisson algorithms for the integration of the solar system coupled to Earth's rotation.

Mignard assumes a tidal bulge delayed by a constant $\delta t$, then Taylor-expands the delayed dynamic variables, assuming a small $\delta t$. The tide-raising potential can be expanded in terms of Legendre polynomials, out of which we keep only the second-order terms

$$U_0 = \frac{\mu_p}{2r_p^5}\left[3\left(\mathbf{r}\times\mathbf{r}_p\right)^2 - r^2 r_p^2\right] \qquad (1)$$

where $\mu_p$ is the gravitational constant times the mass of the tide-raising body, $\mathbf{r}_p$ is the position vector of the tide-raising body, and $\mathbf{r}$ is the point where the potential is calculated. The tide-raising potential produces an elastic deformation of Earth that leads to an additional potential at position $\mathbf{r}$ given by

$$U(\mathbf{r}) = k_2\frac{\mu_p R_e^5}{2r_p^5 r^5}\left[3\left(\mathbf{r}\times\mathbf{r}_p\right)^2 - r^2 r_p^2\right] \qquad (2)$$

where $k_2$ is the potential Love number of the Earth and $R_e$ is the radius of Earth. Dissipation causes a delay of the response by a time $\delta t$ that is small compared to the rotational and orbital periods in the problem. Thus the tidal bulge potential at time t, evaluated at position $\mathbf{r}(t)$, is equal to the potential produced by an ideal response of Earth with the perturbing body at $\mathbf{r}_d(t) = \mathbf{r}_p\,(t - \delta t) + \delta t[\boldsymbol{\omega}(t) \times \mathbf{r}_p\,(t - \delta t)]$.

We expand $\mathbf{r}_d$ for small $\delta t$

$$\mathbf{r}_d(t) = \mathbf{r}_p(t) - \delta t[\mathbf{v}_p(t) + \boldsymbol{\omega}(t) \times \mathbf{r}_p(t)] \tag{3}$$

where $\mathbf{v}_p(t)$ is the velocity of the perturbing body at time t. Then, we carry out the following steps: (1) replace $\mathbf{r}_p$ in U by the approximate expression for $\mathbf{r}_d$; (2) expand the potential up to terms of order $\delta t$; (3) find the additional force at $\mathbf{r}$ due to a perturber at $\mathbf{r}_p$ by taking the gradient of the potential energy with respect to $\mathbf{r}$; (4) equate $\mathbf{r}$ to $\mathbf{r}_p$ in the expression of the force to find the force experienced by the perturbing body (the Moon or the Sun) due to the deformations it produces on the Earth.

We recover the force acting on the Moon due to a delayed tidal bulge on the Earth

$$\mathbf{F}_{tides} = -\frac{3k_2 GM_m^2 R_e^5}{r_{em}^{10}}$$
$$\left\{ r_{em}^2 \mathbf{r}_{em} + \delta t \left[ 2\mathbf{r}_{em}(\mathbf{r}_{em} \times \mathbf{v}_{em}) \right. \right. \tag{4}$$
$$\left. \left. + r_{em}^2 (\mathbf{r}_{em} \times \boldsymbol{\omega} + \mathbf{v}_{em}) \right] \right\}$$

The tidal torque acting on the lunar orbit is given by

$$\mathbf{T} = \mathbf{r}_{em} \times \mathbf{F} = -\frac{3k_2 GM_m^2 R_e^5 \delta t}{r_{em}^8}$$
$$\left[ (\mathbf{r}_{em} \times \boldsymbol{\omega}) \mathbf{r}_{em} - r_{em}^2 \boldsymbol{\omega} + \mathbf{r}_{em} \times \mathbf{v}_{em} \right] \tag{5}$$

An equal and opposite torque acts on the Earth figure. The geocentric tidal acceleration of the Moon takes the form

$$\ddot{\mathbf{r}}_{tides} = -\frac{3k_2 \mu_m \left(1 + \frac{\mu_m}{\mu_e}\right) R_e^5}{r_{em}^{10}}$$
$$\left\{ r_{em}^2 \mathbf{r}_{em} + \delta t \left[ 2\mathbf{r}_{em}(\mathbf{r}_{em} \times \mathbf{v}_{em}) \right. \right. \tag{6}$$
$$\left. \left. + r_{em}^2 (\mathbf{r}_{em} \times \boldsymbol{\omega} + \mathbf{v}_{em}) \right] \right\}$$

where $\mu_e = GM_e$ and $\mu_m = GM_m$. Sun-raised tides admit an equally straightforward representation. Expressions for the cross-terms (torques on lunar orbit resulting from Sun-raised tides) can be derived by taking the derivative of the tidal potential with respect to the position vector of the perturbed body (the Moon) and keeping the position vector of the perturber (the Sun) in the equations.

We had derived Lie-Poisson integrators for rigid-body dynamics in the solar system [*Wisdom and Holman* (1991) together with *Touma and Wisdom* (1994a)], which follow the free rigid-body dynamics, the Keplerian motion, point-point interactions, and rigid-body-point interactions. The integrators preserve the symplectic structure and the total angular momentum of the system. Dissipation will affect the symplectic structure, but the total angular momentum should remain constant. The motion is governed by the differential equation

$$\frac{d}{dt}\mathbf{x} = \mathbf{v}_H + \mathbf{v}_D \tag{7}$$

where $\mathbf{v}_H$ denotes the Hamiltonian vector field and $\mathbf{v}_D$ denotes the dissipative vector field. We approximate the solution with a first-order expansion

$$\mathbf{x}(t) = \exp(\mathbf{v}_H + \mathbf{v}_D t)\mathbf{x}(0) =$$
$$\exp(\mathbf{v}_H t)\exp(\mathbf{v}_D t)\mathbf{x}(0) + O(t^2) \tag{8}$$

Of course, higher-order approximations of the dynamics can be obtained in the usual fashion. The first order splitting corresponds to the intuitive idea of a dissipative kick to the conservative dynamics.

The Hamiltonian contribution is followed with our Lie-Poisson integrator. The dissipative kicks, given by equations (5) and (6), act at a fixed radius, and depend linearly on the momenta. The integral curves are given by the exponential of a constant matrix. We approximate this action by keeping the first two terms in the Taylor expansion of the solution. Explicitly, over a time-step $\Delta t$, the tide raised by the Moon on Earth kicks the spin angular momentum of Earth

$$\Delta \mathbf{L}_{Earth} = \mathbf{T}\Delta t \tag{9}$$

and the geocentric velocity of the Moon by

$$\Delta \mathbf{v}_{em} + \ddot{\mathbf{r}}_{tides}\Delta t \tag{10}$$

where $\mathbf{T}$ and $\ddot{\mathbf{r}}_{tides}$ are the torque and geocentric acceleration vectors, given by equations (5) and (6). These vectors are evaluated at the current geocentric position and velocity vectors of the Moon and the current angular velocity vector of Earth. Kicks due to solar tides and cross tidal interactions can be similarly applied.

Tides change the spin rate of Earth, and change $J_2$ in the process. We account for this change by updating $J_2$ after each cycle of the full integrator. We apply Hamiltonian and dissipative kicks as if the tensor of inertia were constant in time, then use the resulting spin angular velocity of Earth to update $J_2$. We neglected off-diagonal contributions to the tensor of inertia, mostly because, in our simulations, Earth did not develop a substantial wobble.

Our first calculations followed a Moon that interacts gravitationally with Earth, its bulge, and the rest of the solar system. We kept only the $J_2$ term in the potential of solid Earth. Earthly tides were raised by both the Sun and Moon. We ignored dissipation in the Moon. Now, despite algorithmic improvements, and despite the gain in computer speed, a full history of the Earth-Moon system with the current dissipation rates would have taken about a year to complete. So, we opted for judicious efficiency over slow realism. During the slow phase of the evolution, 60–30 $R_e$, we increased the dissipation rate in Earth to 4000× its current value, and set it back to 100× its current value before 30 $R_e$ where the dissipation is much more efficient. This tidally enhanced calculation did not show any fundamental departures from the Goldreich picture. The average behavior was exactly that obtained in Goldreich's theory, with the mutual inclination continuing to pose a problem.

## 4. FORWARD TO THE PRESENT

The fast tide calculations were undertaken with foreknowledge that by accelerating the evolution some resonance effects might be missed. We subsequently carried out additional simulations with more realistic rates of tidal evolution (100× current value between 60 and 30 $R_e$, and 10× the current value before), and found that the system does indeed pass through strong resonances when the Moon is close to Earth. As noted earlier, the encounter with resonances forced us to consider an alternative approach to the problem. Rather than starting at the present and working backward, we studied the forward evolution from a variety of initial, equatorial configurations, very close to Earth. Thus, we learned that the dynamical evolution of the Earth-Moon system is much more complicated than the averaged secular behavior led one to believe.

Let us take a look at a sample evolution of the Earth-Moon system through these resonances. Starting with the Moon initially in the equator plane, with an initial eccentricity of 0.01, and a semimajor axis of 3.5 $R_e$, we marched forward with the help of the full numerical model described in *TW1*. The initial obliquity of the Earth was 10°. The initial rotation period of the Earth was 5.0 hr. The initial rate of the semimajor axis evolution was set at 1 km/yr.

In Fig. 1, we show the eccentricity e of the lunar orbit as a function of its semimajor axis a. The system encounters its first strong resonance, around 4.6 $R_e$, when the period of precession of the pericenter of the Moon is near one year. This resonance, known as the "evection," was already discussed by *Kaula and Yoder* (1976), who discounted its importance because their attention was then focused on coaccretion formation scenarios in which the Moon is initially outside the resonance. While the system is captured into the evection, the eccentricity of the Moon increases to values above 0.5 where the system escapes from resonance. In the postevection evolution the eccentricity continues to increase. Had we allowed tides on the Moon, it would have left the evection wtih a smaller but still substantial eccentricity; the postevection evolution of the eccentricity would have been decided by the magnitude of Moon tides raised by Earth as compared to Earth tides raised by the Moon.

In Fig. 2 we show the mutual inclination ε of the lunar orbit to the Earth's equator vs. the semimajor axis a of the Moon. The dominant feature is an excitation of the mutual inclination to about 2.5° near a = 6 $R_e$. This excitation occurred as the system passed through a resonance between the argument of the evection resonance and the node of the lunar orbit on Earth's equator. We call this resonance the "evection."

The rest of my story is concerned with a detailed description of these two resonances, and the consequences of the passage of physically plausible Moons through one, then the other.

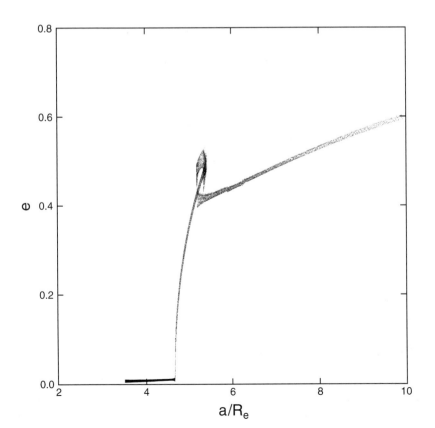

**Fig. 1.** A plot of the eccentricity vs. the semimajor axis of the lunar orbit illustrates the very strong effect the evection resonance may have on the evolution of the lunar orbit.

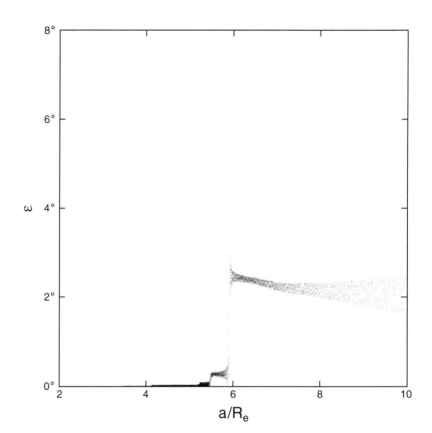

**Fig. 2.** A plot of the Moon's inclination to Earth's equator vs. the semimajor axis of the lunar orbit shows the Moon's eviction from the equator at about 6 $R_e$.

## 5. A HAMILTONIAN FOR THE DYNAMIC DUO

We make a series of approximations that will take us from the full model of the numerical simulations to a simple Hamiltonian that governs the dynamics of the evection and eviction resonances.

We ignore planetary perturbations of the Earth-Moon system. We take Earth's spin axis to be fixed in space with constant obliquity I and fixed longitude of the ascending node of the equator $\Omega_e$. Thus, the relevant kinetic energy is that of point masses Sun, Earth, and Moon. Considering only the quadrupole contribution from Earth's bulge, the potential energy of the remaining system has the form

$$V = -\frac{Gm_0m_1}{r_{01}}\left[1 - \frac{J_2R_e^2}{r_{01}^2}P_2(\cos\theta_{01})\right]$$
$$-\frac{Gm_0m_2}{r_{02}}\left[1 - \frac{J_2R_e^2}{r_{02}^2}P_2(\cos\theta_{02})\right] - \frac{Gm_1m_2}{r_{12}} \quad (11)$$

where the indexes 0, 1, and 2 refer to the Earth, Moon, and Sun respectively. The Hamiltonian is the usual sum of kinetic and potential energies.

Next, we take the Hamiltonian along a sequence of coordinate changes and approximations:

1. We express kinetic and potential energy in terms of the Jacobi coordinates. These coordinates separate the center of mass motion from the relative motion, and preserve the form of the kinetic energy as the sum over planets of squares of momenta divided by mass factors. An appropriate choice for the problem at hand is to position the Moon relative to Earth ($r_1'$), then the Sun relative to the Earth-Moon barycenter ($r_2'$). One recovers the conjugate momenta from a generating function that executes the desired coordinate transformation. The inverse distance $r_{12}$ is expanded in the small ratio $r_1'/r_2'$, and only the principal terms are kept in the expanded potential.

2. We divide the resulting Hamiltonian into the Keplerian terms and perturbations, omit terms that govern the potential interaction of the Earth and Sun, and express the Keplerian motion of the Moon around Earth in terms of Delaunay variables (see *Brouwer and Clemence*, 1961, for details).

We end up with

$$H = -\frac{m_1'\mu^2}{2L_1^2} + \frac{Gm_0m_1}{r_1'}\frac{J_2R_e^2}{(r_1')^2}P_2(\cos\theta_1')$$
$$-\frac{Gm_1m_2}{r_2'}\frac{m_0}{\eta_1}\frac{(r_1')^2}{(r_2')^2}P_2\left(\frac{x_1'\times x_2'}{r_1'r_2'}\right) \quad (12)$$

The first term represents the gravitational interaction of the Earth-Moon pair, the second term represents the effect of Earth's oblateness, and the last term is the leading term in the interaction of the Moon and the Sun. Here $\mathbf{x}_1'$ and $\mathbf{x}_2'$ are the Jacobi vector positions of the Moon and the Sun, and $r_1'$ and $r_2'$ are the magnitudes of these vectors. The Jacobi masses $m_i'$ are given by $m_i' = m_i \eta_{i-1}/\eta_i$, with $\eta_i = \Sigma_{j=0}^{j=i} m_j$. We also have $\mu = Gm_0 m_1$. The Delaunay momentum conjugate to the mean anomaly $l_1$ of the lunar orbit is

$$L_1 = \sqrt{m_1 \mu a_1}$$

where $a_1$ is the semimajor axis of the lunar orbit. The Delaunay momentum

$$G_1 = L_1\sqrt{1 - e_1^2}$$

is conjugate to the argument of pericenter $g_1 = \omega_1$, where $e_1$ is the eccentricity of the lunar orbit, and the Delaunay momentum $H_1 = G_1 \cos i_1$ is conjugate to the longitude of the ascending node $h_1 = \Omega_1$, where $i_1$ is the inclination of the lunar orbit. The angle between the symmetry axis of the Earth and the vector from the Earth to the Moon is denoted by $\theta_1'$.

We take the orbit of the Earth-Moon barycenter about the Sun to be circular with semimajor axis $a_2$ and mean motion $n_2$. Note that our assumptions thus far have left us with the dynamics of an osculating lunar orbit around Earth, perturbed by the Sun on a fixed circular, and Earth's bulge whose orientation is also fixed in space.

Let us finally express the various vectors in an equatorial reference frame with the ascending node of the equator on the Earth-Sun orbit plane as the reference longitude. The components of vector $\mathbf{x}_2'$ to the Sun are $[\cos(n_2 t - \Omega_e),$ $\cos I \sin(n_2 t - \Omega_e), -\sin I \sin(n_2 t - \Omega_e)]$; the vector $\mathbf{x}_1'$ from the Earth to the Moon has components $R_z(\Omega)R_x(i)R_z(\omega)(r_1 \cos\theta, r_1 \sin\theta, 0)$, where $r_1$ is the distance from the Earth to the Moon, $\theta$ is the true anomaly, $\omega$ is the argument of pericenter, $\varepsilon$ is the inclination of the lunar orbit to the equator, and $\Omega$ is the ascending node of the lunar orbit relative to the equinox. The functions $R_x$ and $R_z$ generate active righthand rotations about the x and z axis respectively. We express the Hamiltonian as a Poisson series, and pick out terms that correspond to the resonances of interest

$$H_{res} = -\frac{m_1' \mu_2}{2L_1^2} + \frac{Gm_0 m_1}{r_1'} \frac{J_2 R_e^2}{(r_1')^2}$$

$$\left(\frac{3}{4}\sin^2\varepsilon(1 - \cos 2(\theta + \omega)) - \frac{1}{2}\right)$$

$$-\frac{Gm_1 m_2}{a_2} \frac{m_0}{\eta_1} \frac{(r_1')^2}{(a_2)^2}\left[\frac{3}{4}\cos^4\frac{\varepsilon}{2}\cos^4\frac{I}{2}\cos 2 \right. \tag{13}$$

$$(n_2 t - \Omega_e - \theta - \omega - \Omega) - \frac{3}{4}\cos^2\frac{\varepsilon}{2}\cos^2$$

$$\left. \frac{I}{2}\sin\varepsilon\sin I \cos(2(n_2 t - \Omega_e - \theta - \omega - \Omega) + \Omega)\right]$$

We have omitted other resonance terms that are higher order in mutual inclination and obliquity and that may be more important for slower rates of tidal evolution.

We average over the lunar orbital period and make a canonical transformation to the resonance variable $\sigma = n_2 t - \Omega_e - \omega - \Omega$ with the generating function

$$F_2 = (l_1 + g_1 + h_1)L' + (n_2 t - \Omega_e - g_1 - h_1)\Sigma - h_1\Psi \tag{14}$$

to get

$$\bar{H}_{res}' = n_2 \Sigma + n_1 J_2\left(\frac{R_e}{a_1}\right)^2$$

$$L'\left(\frac{3}{4}\sin^2\varepsilon - \frac{1}{2}\right)\frac{1}{(1-e^2)^{3/2}}$$

$$-\frac{15}{8}e^2 n_2 \frac{n_2}{n_1}L'\cos^4\frac{\varepsilon}{2}\cos^4\frac{I}{2}\cos(2\sigma) \tag{15}$$

$$+\frac{15}{8}e^2 n_2 \frac{n_2}{n_1}L'\cos^2\frac{\varepsilon}{2}\cos^2\frac{I}{2}$$

$$\sin\varepsilon\sin I\cos(2\sigma - \psi)$$

Note that we have dropped the constant term that controls the mean longitude. The mean longitude $\lambda = l_1 + g_1 + h_1$ is conjugate to the momentum $L' = L_1$. The momentum $\Sigma = L_1 - G_1$ is conjugate to the resonance angle $\sigma$, which is the difference between the longitude of the Sun and the longitude of the pericenter. The momentum $\Psi = G_1 - H_1$ is conjugate to $\psi = -h_1 = -\Omega$. We also have

$$e^2 = \frac{\Sigma}{L'}\left(2 - \frac{\Sigma}{L'}\right) \tag{16}$$

and that

$$\sin^2\left(\frac{\varepsilon}{2}\right) = \frac{\Psi}{2(L' - \Sigma)} \tag{17}$$

The evection resonance is associated with the term with argument $2\sigma$, and the eviction resonance is associated with the term with argument $2\sigma - \psi$. The term associated with the eviction resonance is closely related to the evection resonance term, but has a stronger effect on the inclination of the Moon's orbit than on the eccentricity.

## 6. THE EVECTION

I refer the interested reader to *TW2* for additional details on the evection. Here, I describe some of the analytic results and summarize the qualitative conclusions.

A Hamiltonian governing dynamics around the evection is simply obtained by ignoring the eviction terms, and dropping the mutual inclination and obliquity dependent terms

$$H_e = n_2 \Sigma - \frac{1}{2}n_1 J_2\left(\frac{R_e}{a_1}\right)^2$$

$$L'\frac{1}{(1-e^2)^{3/2}} - \frac{n_2^2}{n_1}L'e^2\frac{15}{8}\cos 2\sigma \tag{18}$$

We can take this Hamiltonian to the familiar form for second-order resonances by further examining it under the assumption of small eccentricity

$$H_e \approx \left( n_2 - \frac{3}{2} J_2 \frac{R_e^2}{a_1^2} n_1 \right) \Sigma - 3n_1 J_2$$

$$\frac{R_e^2}{a_1^2} \frac{1}{L'} \Sigma^2 - \frac{n_2^2}{n_1} \frac{15}{4} \Sigma \cos 2\sigma \qquad (19)$$

where we have kept nonresonant terms up to order $\Sigma^2$ and resonant terms to order $\Sigma$. We introduce rectangular coordinates via the canonical transformation

$$\xi = \sqrt{2\Sigma} \cos \sigma$$

and

$$\eta = \sqrt{2\Sigma} \cos \sigma$$

where $\xi$ is the momentum conjugate to $\eta$, and end up with the truncated evection Hamiltonian

$$H_e \approx \delta \left( \frac{\xi^2 + \eta^2}{2} \right)$$

$$-\alpha \left( \frac{\xi^2 + \eta^2}{2} \right) - \beta \left( \frac{\xi^2 - \eta^2}{2} \right) \qquad (20)$$

with

$$\delta = n_2 - \frac{3}{2} J_2 \frac{R_e^2}{a_1^2} n_1 \qquad (21)$$

$$\alpha = 3n_1 J_2 \frac{R_e^2}{a_1^2} \frac{1}{L'} \qquad (22)$$

$$\beta = \frac{n_2^2}{n_1} \frac{15}{4} \qquad (23)$$

With the help of this Hamiltonian, we can ask a number of questions regarding the behavior of the Moon in and around the evection:

1. *Location of resonance.* The parameter $\delta$ provides a measure of the distance to the resonance. The oblateness of the Earth depends on its rotation rate: We assume $J_2 = J_{20} (\omega/\omega_0)_2$, where $J_{20}$ is the value of $J_2$ for the rotation rate $\omega_0$. The present value of $J_2$ is about $J_{20} = 0.001083$. For a rotation rate of 5.2 hr, we find $J_2 \approx 0.023$. Thus the resonance occurs near $a_1 \approx 4.64 \, R_e$, which is indeed about where we find it.

2. *Equilibria.* The fixed points have either $\xi = 0$ or $\eta = 0$. For $\xi = 0$ the fixed points have $\eta = 0$ or

$$\eta = \sqrt{(\delta + \beta)/\alpha}$$

For $\eta = 0$ the fixed points have $\xi = 0$ or

$$\xi = \sqrt{(\delta - \beta)/\alpha}$$

The fixed point at the origin always exists. The fixed points away from the origin only exist if the quantity under the square root is positive. For $\delta \geq -\beta$, there is a pair of symmetrically placed fixed points with $\xi = 0$. For $\delta \geq \beta$ there are, in addition, symmetrically placed fixed points with $\eta = 0$. The origin is an unstable fixed point for $-\beta < \delta < \beta$, the fixed points on the $\eta$ axis with $\xi = 0$ are stable, and the fixed points on the $\xi$ axis with $\eta = 0$ are unstable.

3. *Eccentricity evolution.* The eccentricity of the stable fixed point of the truncated averaged evection resonance Hamiltonian compares well with the eccentricity of the numerical evolution. At high eccentricity, the agreement is acceptable, but naturally the full resonance model does better.

4. *Effect of obliquity.* The actual nonzero obliquity has very little effect.

5. *Adiabatic capture.* In the adiabatic limit, the calculation of capture probabilities during passage through a second-order resonance is straightforward (*Yoder,* 1979; *Henrard,* 1982; *Borderies and Goldreich,* 1984). For increasing $\delta$, capture is certain if the resonance is encountered with an eccentricity that is smaller than a critical value, 0. 0773 as the Moon is evolving away from Earth. If the resonance is encountered with larger eccentricity the system may be captured or not depending on the phases.

6. *Nonadiabatic capture.* The early tidal evolution of Earth-Moon system is plagued with uncertainties. Rates of separation might be too fast for the adiabatic assumption. In the limit of very fast evolution the resonance is hardly felt. In the intermediate limit, the situation is more complicated and we rely on numerical simulations for insight. The probabilities for a tidal evolution rate of 1 km/yr are close to those predicted by the adiabatic theory. For small eccentricity, capture is not certain even for this rate of evolution. For a faster evolution of 10 km/yr, the capture probabilities at small eccentricity are significantly smaller.

7. *Passage without capture.* For eccentricities that are smaller than the critical value, and in the case of nonadiabatic rates of evolution, the Moon is no longer captured in the evection, but its eccentricity is still modified. Small initial eccentricities are typically increased; initial eccentricities closer to the critical eccentricity are typically decreased. After a nonadiabatic passage through the evection resonance the system does not have a characteristic eccentricity. This is unfortunate because the subsequent evolution of the eccentricity depends sensitively on the initial eccentricity. The rate of growth of eccentricity is proportional to eccentricity, so the initial seed for the eccentricity dramatically affects the subsequent evolution. Even with dissipation in the Moon many of these evolutionary tracks develop large eccentricity, which can only be damped by enhanced tidal dissipation during some phase of the subsequent evolution.

8. *Capture and escape from the evection resonance.* Once captured in the evection, the system tracks the stable fixed point of the model Hamiltonian. At an eccentricity near 0.5 the system escapes from the resonance for the following reason: As the eccentricity is pumped by the evection resonance, it reaches a value at which the angular motion of the

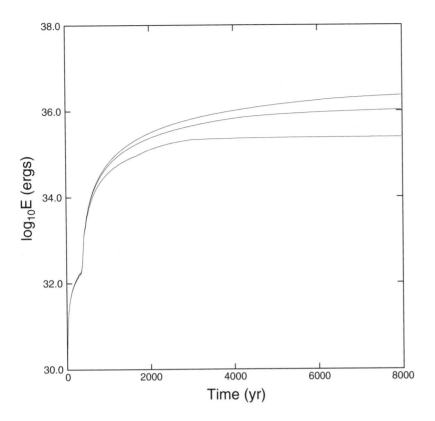

**Fig. 3.** The accumulated total energy deposited in the Moon for A = 1, A = 3, and A = 10.

Moon at perigee increases past the angular rate of rotation of Earth, a point at which the semimajor axis ceases to evolve. This state is unstable to further tidal evolution. The eccentricity continues to evolve and the amplitude of the libration increases until the system leaves the resonance. Capture in the evection resonance is always temporary.

9. *Tidal dissipation in the Moon.*   Dissipation in the Moon can strongly affect the evolution of its eccentricity (*Goldreich*, 1963). In particular, the eccentricity at which the system escapes the evection resonance depends on how much energy dissipation there is in the Moon. We model the effect of lunar tidal dissipation with the help of average *Mignard* (1981) tides for zero lunar obliquity. The parameter

$$A = \frac{k_2'}{k_2} \frac{\delta t'}{\delta t} \left( \frac{m_0}{m_1} \right)^2 \left( \frac{R_m}{R_e} \right)^3 \qquad (24)$$

measures the relative rates of energy dissipation in the Earth and Moon. Here $k_2$ and $k_2'$ are the potential Love numbers of the Earth and Moon, $m_0$ and $m_1$ are the masses of the Earth and Moon, $\delta t$ and $\delta t'$ are the tidal time delay for the Earth and Moon, and $R_e$ and $R_m$ are the radii of the Earth and Moon. The current value of A is around 1.1. Its value when the Earth-Moon system was near the evection resonance is unknown. With dissipation in the Moon, there is a critical value of the eccentricity for each semimajor axis above which the semimajor axis decreases rather than increases. At the critical eccentricity the rate of change of the semimajor axis is zero. This eccentricity approximates the eccentricity of the lunar orbit after the system escapes from

the evection resonance. The time evolution of the eccentricity during the evection resonance passage determines the extent of tidal heating in the Moon. The evolution of the eccentricity depends on the relative rates of tidal dissipation in the Earth and Moon. As the dissipation in the Moon increases, the peak eccentricity decreases. The Moon is strongly heated if the system is captured in the evection resonance. In Fig. 3 the cumulative energy deposited is plotted vs. time. For the rate of energy deposition we use the expression of *Yoder and Peale* (1981)

$$\frac{dE}{dt} = \frac{42}{19} \frac{\pi e^2 \rho^2 R^7 n^5}{\mu Q} \qquad (25)$$

where e is the orbital eccentricity, $\rho$ is the mean lunar density, R is the lunar radius, n is the mean motion, $\mu$ is the effective rigidity, and Q is the effective tidal Q for the Moon. The relationship between an A parameter and the effective lunar Q involves too many uncertain parameters. So as a point of reference we adopt, following *Peale and Cassen* (1978), $\mu = 6.5 \times 10^{11}$ dynes/cm and Q = 100 and estimate that the energy deposited in the Moon during this epoch is of order $10^{36}$ ergs. Using a specific heat of $10^7$ ergs/g K we find the temperature of the Moon as a whole rises by more than 1000 K, surely sufficient to melt substantial portions of the Moon. Strong tidal heating during evection passage provides an abundant energy source for the creation of the lunar magma ocean, and for interior heating that is later manifested in mare volcanism. However, an issue that must be addressed is whether such tidal heating can be ruled out on the basis of the radius constraint of *Solomon and Chaiken* (1976). They found that the observed paucity of extensional and compressional features on the Moon implies that the radius of the Moon has changed by less than ±1 km since the end of the heavy bombardment, and that this could only be accomplished with a magma ocean if the interior of the Moon was initially cool. Essentially the same conclusions were drawn by *Solomon and Longhi* (1977), *Cassen et al.* (1979), *Kirk and Stevenson* (1989), and others, who have considered different aspects of the thermal evolution of the Moon. The difficulty is that tidal heating is strongest in the center of the Moon (*Peale and Cassen*, 1978). So if tidal heating were responsible for the magma ocean, the interior would initially have been hot. Perhaps it is time to revisit the lunar thermal histories, with a possible passage through the evection resonance in mind.

10. *Evection and lunar formation.* The evection could affect the evolution of the protolunar disk through resonance-disk interactions. The evection could provide a holding point for runaway moonlets that get captured into the resonance, perhaps mitigating the problem of the runaway disk material seen in the simulations of *Ida et al.* (1997). The evection resonance could even be the point of formation of the Moon.

## 7.  THE EVICTION

Following capture in the evection, the Moon leaves with a large eccentricity that brings into play mixed eccentricity-inclination resonances. The strongest is the eviction that excites the mutual inclination. It is associated with the last term in the average resonance Hamiltonian. Neglecting the evection resonance, and assuming the eccentricity is fixed at its postevection value, the approximate eviction Hamiltonian is then

$$
\begin{aligned}
H_1 = & n_2\Sigma + n_1 J_2 \left(\frac{R_e}{a_1}\right)^2 \\
& L'\left(\frac{3}{4}\sin^2\varepsilon - \frac{1}{2}\right)\frac{1}{(1-e^2)^{3/2}} \\
& + \frac{n_2^2}{n_1}L'e^2\frac{15}{8}\cos^2 I/2\sin I\cos^2 \\
& \qquad \varepsilon/2\sin\varepsilon\cos(\psi-2\sigma)
\end{aligned}
\tag{26}
$$

We can put the Hamiltonian in the standard form for a first-order resonance by first carrying a canonical transformation with generating function

$$
F_2 = (\psi - 2\sigma)R + \sigma S
\tag{27}
$$

then expanding the eviction resonance Hamiltonian in powers of S/L' and powers of R/L', and keeping nonresonance terms to order $R^2$ and the lowest-order resonant terms

$$
H_i = AR^2 + BR + \sqrt{2R}\cos r
\tag{28}
$$

with

$$
A = -\frac{99}{4}\frac{n_1 J_2}{L'}\left(\frac{R_e}{a_1}\right)^2\left[1 + 5\frac{S}{L'} + 15\left(\frac{S}{L'}\right)^2 + \ldots\right]
$$

$$
B = -2n_2 + \frac{9}{2}n_1 J_2\left(\frac{R_e}{a_1}\right)^2
$$

$$
\left[1 + 4\frac{S}{L'} + 10\left(\frac{S}{L'}\right)^2 + 20\left(\frac{S}{L'}\right)^3 + \ldots\right]
$$

$$
C = \frac{15}{8}S\frac{1}{\sqrt{L'}}\frac{n_2^2}{n_1}\cos^2\frac{I}{2}\sin I
\tag{29}
$$

The coefficient B is a measure of the distance from the resonance. We change coordinates again to

$$
p = \sqrt{2R}\cos r
$$

and

$$
q = \sqrt{2R}\sin r
$$

to put the Hamiltonian in the canonical form

$$
H_i = A\left(\frac{p^2 + q^2}{2}\right)^2 + B\left(\frac{p^2 + q^2}{2}\right) + Cp
\tag{30}
$$

The Moon leaves the eviction with B initially positive, and a small mutual inclination [a correspondingly small area on the (p,q) eviction phase plane]; A is negative all during this phase. The system admits three fixed points: two stable and one unstable. Initially, the Moon is close to the stable fixed point closest to the origin. As the Moon evolves outward, B decreases while remaining positive, until it reaches the bifurcation value $B = B_c = -3(C^2 A/4)^{1/3}$, at which the fixed point that the Moon is near disappears in a collision with the unstable fixed point. Following the bifurcation, the system now encircles the sole remaining fixed point, with a finite area on the phase space. The area of the phase plane at the point of bifurcation is $6\pi(C^2/4A^2)^{1/3}$. Far from the resonance where the mutual inclination is constant, this area is $\pi(p^2 + q^2) = 2\pi R = 4\pi(L' - S)\sin^2\varepsilon/2$. Thus, the mutual inclination after passage through the eviction resonance satisfies

$$
\sin^2\frac{\varepsilon}{2} = \frac{3}{L' - S}\left(\frac{C^2}{4A^{2'}}\right)^{1/3}
\tag{31}
$$

leading to the approximate expression for the posteviction mutual inclination

$$
\sin\varepsilon \approx 2\sqrt{3}\left(\frac{5}{264}\right)^{1/3}\left(\frac{n_2^2}{n_1^2}\frac{a_1^2}{J_2 R_e^2}e^2\sin I\right)^{1/3}
$$
$$
\left(1 - \frac{7e^2}{12} + \ldots\right)
\tag{32}
$$

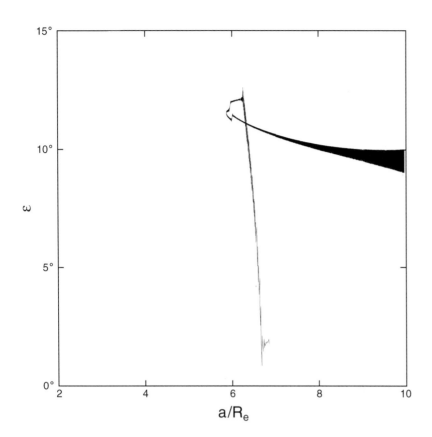

**Fig. 4.** The evolution of the mutual inclination if the system is captured by the eviction during an interval of enhanced lunar dissipation. The system escapes with mutual inclination of about 12°, just what is needed to resolve the mutual inclination problem.

With the parameters of the run displayed in Fig. 2, we estimate a postevection mutual inclination of about 3.1°, a little higher than the value observed in the numerical experiment. In fact, the tidal rates in the experiment are too fast for the adiabatic approximation and we found that slower rates lead to excitations that are in closer agreement with adiabatic estimates.

Unfortunately then, the encounter of the eviction resonance is in the wrong direction for capture, and the amplitude of the excitation is not sufficient to solve the mutual inclination problem. Is there a way out to the present?

## 8.  BACK INTO THE EVICTION

The way out is back at first, and the way back is found by a Moon that becomes increasingly dissipative.

If the Moon's A is initially large, it traverses the eviction with a relatively small excitation of the eccentricity, which cuts down on the already small eviction excitation. The eccentricity continues to decrease to the present value, and the A parameter could, perhaps as a result of cooling, decrease to the observed value near 1. But the mutual inclination would not match the current value.

For a small initial A, capture in eviction excites the eccentricity to a large value that continues to grow after escape from eviction. Tidal dissipation in the Moon must increase to damp out the large eccentricity. It seems likely that the high dissipation interval coincides with the eviction resonance passage itself or soon thereafter. The large

postevection eccentricities enhance the tidal dissipation in the Moon, lowering its effective Q and increasing A. A sudden increase in A resulting from passage through eviction alters the subsequent evolution dramatically. Strong dissipation in the Moon not only damps the eccentricity of the lunar orbit, but can cause the semimajor axis of the Moon to go through a period of regression: tides in the Moon dissipate energy but conserve angular momentum; the orbital angular momentum involves the product $a(1 - e^2)$; conservation of angular momentum causes a to decrease as e decreases. If the dissipation in the Moon is strong enough, this tendency of lunar tides to decrease the semimajor axis can win over the tendency of Earth tides to increase the lunar semimajor axis. The passage through the eviction resonance is followed closely by the passage through the eviction resonance. If this interval of enhanced tidal dissipation occurs after the passage through eviction then the regression of the semimajor axis can carry the system once more through the eviction resonance. This time, though, the passage through the resonance is in the direction for which capture is possible. Indeed, if the eccentricity has not decayed too much, capture into the eviction is almost certain.

We have made quite a few simulations to explore this process. We have taken a typical A = 0 postevection-eviction system state, and evolved the system with a variety of A parameters and rates of tidal evolution. We investigated A = 1, 3, 5, 8, and 10, and Earth tidal dissipation rates that, if acting alone, would result in rate of semimajor axis evolution of $da/dt|_E$ = 0.1 km/yr, 0.2 km/yr, 0.3 km/yr, and

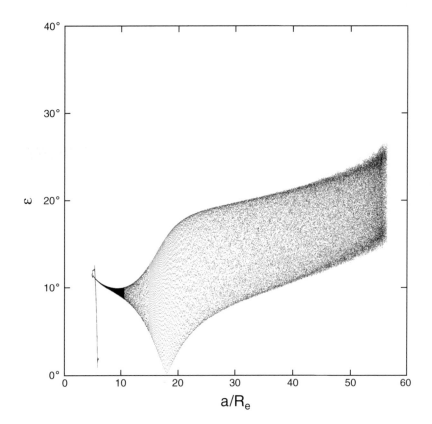

**Fig. 5.** The evolution of the Earth-Moon system from an initial equatorial lunar orbit, through the evection and eviction resonances, to the present time is consistent with the current inclination of the lunar orbit.

0.4 km/yr. We found that runs with A > 1 and A ~ da/dt|$_E$ ≤ 2 km/yr are captured by the eviction. Capture in the eviction during this phase requires somewhat slower evolution than is required for capture into the evection. The evolution of the mutual inclination is shown in Fig. 4. For this simulation A = 10 and da/dt|$_E$ = 0.2 km/yr.

For those runs that are captured the subsequent evolution is remarkably insensitive to the parameter values. In every case the mutual obliquity shoots up to values in the range of 9° to 13°, before the system escapes.

The mechanism of escape from the eviction is simple, and can be understood in terms of the eviction action, which is the area on the p–q eviction phase plane (see section 7). After capture the eviction action has a value near the area of the separatrix at the point of bifurcation, which forced the initial excitation in the forward passage through eviction. We observe in the simulations that the action grows as the evolution in the eviction proceeds. We also note that the allowed area in the libration region is shrinking because the strength of the eviction resonance is proportional to the square of the eccentricity, which is decreasing. So eventually the system must escape the eviction.

After escape from the eviction, the mutual inclination is left in the range of 9° to 13°. The eccentricity continues to decrease because of the enhanced A. After the eccentricity declines sufficiently, the semimajor axis resumes its outward journey. The eviction resonance is passed a third time, but this time the eccentricity is small enough that there is no substantial excitation.

The eccentricity can evolve to the present eccentricity if the A parameter is reduced to the present value (presumably by cooling) in an appropriate way.

The excitation of the mutual inclination is in just the right range so as to solve the mutual inclination problem. If we evolve the system onward from this point to the present the current configuration of the system, including the inclination of the Moon, can be recovered. Figure 5 shows the evolution of the mutual inclination. In this simulation the final obliquity of the Earth is a couple of degrees too small. A small increase of the initial obliquity would remove this discrepancy without substantially changing the overall evolution.

## 9. BACK TO THE FUTURE

Now that we made an original equatorial orbit consistent with the present orbit, we hand the reigns over to the Moon as it guides us into the stormy future(s)that awaits the system.

*Goldreich* (1966) noted that as the Moon evolves outward, and Earth spins down, there will come a point (around 75 R$_e$), when the day is equal to the month, i.e. , when the Moon revolves synchronously with the spinning Earth. The likeliest outcome is for the synchronous state to be maintained by nonaxisymmetric contributions of Earth's figure. The Sun's tidal action on the locked couple will extract angular momentum, forcing Earth to spin faster and the Moon to spiral inward as it maintains the geosynchronous

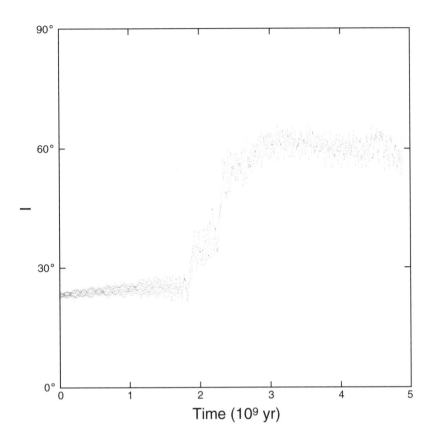

**Fig. 6.** A likely path to disaster: the obliquity of Earth over the next 5 b.y.

state. Another possibility is for Earth not to get locked in this state, and for the day to get longer than the lunar month as the Sun continues to slow Earth. Then, the bulge raised by the Moon on Earth will lag behind the Moon. The action of the lunar tides reverses, and continues to dominate the solar tides as it forces Earth to spin up and the Moon to spiral in. Another possibility is for the system to get locked in a higher-order resonance, in which case solar tides would again force the couple to get tighter as it remains locked in resonance. In all cases it seems that the ultimate fate of the Moon is to spiral back in toward an Earth that spins up.

On the way to the synchronous state, Earth experiences a dramatic turn of events. It was known to *Ward* (1982) that as the Moon recedes, Earth's precession frequency decreases to a point (around 63 $R_e$, or a couple of billion years in the future, assuming current tidal rates) when it enters in resonance with a cluster of orbital frequencies. This cluster will temporarily capture the precession of the spin axis of Earth in a chaotic secular spin orbit resonance, analogous to the ones examined by *Laskar and Robutel* (1993) and *Touma and Wisdom* (1993). *Neron de Surgy and Laskar* (1997) used an orbit-averaged model of a tidally evolving lunar orbit, coupled to Laskar's secular representation of Earth orbit, to study the evolution of Earth obliquity. They find, with current tidal rates, that Earth obliquity, which is naturally chaotic during this phase, can get excited to values larger than 81°, with a maximum of 89.5°. On the other hand, I simulated the future evolution of Earth spin in a full integration, coupling a tidally evolving Earth-Moon system to a chaoti-

cally evolving solar system. I set the dissipation rate at 100× its current value. Around 1.9 b.y. in the future, the obliquity of the Earth, shown in Fig. 6, begins a climb that takes it from an average of 29° to an average of 60° in a period of 0.5 b.y. Keep in mind that the enhanced tidal rate reduces the impact of the resonance.

A few billion years after this disaster, the Sun threatens both the Earth and Moon as it enters its expansive red giant phase.

## 10. A CAN OF WORMS

It is a serious problem indeed that under an apparent contradiction buries a can of worms. The inclination problem is apparently such a problem. We set out to solve it with dynamics alone, initially questioning the validity of the averaged secular theory. We unraveled a network of resonances, which though exciting in itself, did not offer any direct way out of an equatorial past to the present. Instead, its properties suggested an admittedly circuitous route that forced us to couple dynamics to the more uncertain terrain of the geophysical properties of the Earth and Moon, a few million years after formation. It was demanding enough to map out the interesting dynamics, let alone to attempt a serious exploration of the coupling between dynamics and geophysics. Instead, we traced a plausible scenario, relying in its details on hard-earned experience with other satellites (Io in particular), a scenario that solves the inclination problem and provides a source of heat for the lunar magma ocean. For that to happen, the Moon starts out cold, weakly dissipative, and is changed by passage through evection into a strongly dissipative body. Compelling as it is, this scenario raises many questions, as it answers two longstanding ones. Whether the Moon started out cool or not is surely dependent on its formation process, the speed of accretion, and the size distribution of the material out of which it forms. We will surely learn more about this in the future. A more serious concern involves the heat deposition in the central region of the Moon, which is apprently ruled out by the radius constraint of *Solomon and Chaiken* (1976). Perhaps a revisiting of the geophysics will offer a way out of this dilemma. This is quite an exciting situation to be in, one that opens the conversation on the early geophysical properties of the Earth and Moon to crucial input from strongly disruptive dynamics [that of *Ward and Canup* (1999) only adds to the excitment by introducing the remnant protolunar disk as a significant perturber of the lunar orbit]. Another related question involves the rotational dynamics of the Moon, when dragged through these resonances. Assuming as we did, that the synchronous rotation state is established very soon after formation, how is it affected by the large orbital eccentricity of the Moon during evection? How do the Cassini tracks [described by *Ward* (1975)] look like along the orbits we described? Finally, we point out that other resonances could be encountered on evolutionary tracks that we have not fully explored. In particular, a Moon that is allowed to develop a large eccentricity (around 0.6) at a distance of 28 $R_e$ encounters a secular inclination-eccentricity resonance that excites the mutual inclination. Clearly, there is much left to be told about the wild history of this most familiar of systems.

## REFERENCES

Boss A. P. and Peale S. J. (1986) Dynamical constraints on the origin of the Moon. In *Origin of the Moon* (W. K. Hartmann, R. J. Phillips, and G. J. Taylor, eds.), pp. 59–101. Lunar and Planetary Institute, Houston.

Borderies N. and Goldreich P. (1984) A simple derivation of capture probabilities for the J + 1 : J and J + 2 : J orbit-orbit resonance problems. *Celestial Mech., 32*, 127–136.

Brouwer D. and Clemence G. M. (1961) *Methods of Celestial Mechanics.* Academic, New York. 598 pp.

Burns J. A. (1986) The evolution of satellite orbits. In *Satellites* (J. A. Burns and M. S. Matthews, eds.), pp. 117–158. Univ. of Arizona, Tucson.

Cassen P., Reynolds R. T., Graziani F., Summers A., McNellis J., and Blalock L. (1979) Convection and lunar thermal history. *Phys. Earth Planet. Inter., 19*, 183.

Darwin G. H. (1880) On the secular change in the elements of the orbit of a satellite revolving about a tidally distorted planet. *Philos. Trans. Roy. Soc. London, 171*, 713.

Dickey J. O., Bender P. L., Faller J. E., Newhall X X, Ricklefs R. L., Ries J. G., Shelus P. J., Ceillet C., Whipple A. L., Wiant

J. R., Williams J. G., and Yoder C. F. (1994) Lunar laser ranging: A continuing legacy of the Apollo program. *Science, 265*, 482–490.

Gerstenkorn H. (1955) Uber Gezeitenreibung beim Zweikorperproblem. *Z. Astrophys., 26*, 245.

Goldreich P. (1963) On the eccentricity of satellite orbits in the solar system. *Mon. Not. Roy. Astron. Soc., 126*, 256.

Goldreich P. (1966) History of the lunar orbit. *Rev. Geophys., 4*, 411–439.

Henrard J. (1982) Capture into resonance: An extension of the use of adiabatic invariants. *Celestial Mech., 27*, 3–22.

Ida S., Canup R. M., and Stewart G. R. (1997) Lunar accretion from an impact-generated disk. *Nature, 389*, 353–357.

Kaula W. M. (1964) Tidal dissipation by solid friction and the resulting orbital evolution. *Rev. Geophys., 2*, 661–685.

Kaula W. M. and Yoder C. (1976) Lunar orbit evolution and tidal heating of the Moon (abstract). In *Lunar Science VII*, pp. 440–442. Lunar Science Institute, Houston.

Kirk R. L. and Stevenson D. J. (1989) The competition between thermal contraction and differentiation in the stress history of the Moon. *J. Geophys. Res., 94*, 12133–12144.

Laskar J. and Robutel P. (1993) The chaotic obliquity of the planets. *Nature, 361*, 608–612.

MacDonald G. J. F. (1964) Tidal friction. *Rev. Geophys., 2*, 467–541.

Mignard F. (1981) The lunar orbit revisited, III. *Moon and Planets, 24*, 189–207.

Neron de Surgy O. and Laskar J. (1997) On the long term evolution of the spin of the Earth. *Astron. Astrophys., 318*, 975–989.

Peale S. J. and Cassen P. (1978) Contribution of tidal dissipation to lunar thermal history. *Icarus, 36*, 245–269.

Solomon S. C. and Chaiken J. (1976) Thermal expansion and thermal stress in the Moon and terrestrial planets: Clues to early thermal history. *Proc. Lunar. Sci. Conf. 7th*, pp. 3229–3243.

Solomon S. C. and Longhi J. (1977) Magma oceanography: 1. Thermal evolution. *Proc. Lunar. Sci. Conf. 8th*, pp. 583–599.

Stevenson D. J. (1987) Origin of the Moon — The collision hypothesis. *Annu. Rev. Earth Planet. Sci., 15*, 271–315.

Touma J. and Wisdom J. (1993) The chaotic obliquity of Mars. *Science, 259*, 1294–1297.

Touma J. and Wisdom J. (1994a) Lie-Poisson integrators for rigid body dynamics in the solar system. *Astron. J., 107*, 1189–1202.

Touma J. and Wisdom J. (1994b) Evolution of the Earth-Moon system. *Astron. J., 108*, 1943–1961.

Touma J. and Wisdom J. (1998) Resonances in the early evolution of the Earth-Moon system. *Astron. J., 115*, 1653–1663.

Yoder C. (1979) Diagrammatic theory of transition of pendulum like systems. *Celestial Mech., 19*, 3–29.

Yoder C. and Peale S. J. (1981) The tides of Io. *Icarus, 47*, 1–35.

Ward W. R. (1975) Past orientation of the lunar spin axis. *Science, 189*, 377–379.

Ward W. R. (1982) Comments on the long-term stability of the Earth's obliquity. *Icarus, 50*, 444–448.

Ward W. R. and Canup R. M. (1999) Origin of the Moon's orbital inclination from resonant disk interactions. *Nature, 403*, 741–743.

Wisdom J. and Holman M. (1991) Symplectic maps for the n-body problem. *Astron. J., 102*, 1528–1538.

# Thermal Aspects of a Lunar Origin by Giant Impact

## M. E. Pritchard and D. J. Stevenson

*California Institute of Technology*

We develop and analyze the arguments that giant impact scenarios of lunar origin appear to require an initially hot Moon (near or above the solidus for most of the mass). Pre-giant-impact heating, impact heating, disk evolution heating and cooling, and the accretion of the Moon are all taken into account. However, the current poor understanding of the dynamical continuum disk phase that is expected to intervene between the endpoint of standard giant impact simulations (Cameron et al.) and the beginning of lunatesimal aggregation scenarios (Canup and coworkers), together with other uncertainties, prevent precise conclusions for initial lunar temperatures. Isotopic considerations (especially potassium) do not argue against extensive devolatilization during lunar formation. We also assess the geological and geochemical evidence suggesting that a substantial fraction of the initial Moon was cold (~1000 K or less). These arguments are difficult to quantify with high certainty for several reasons, especially the uncertain mechanical properties of the near-surface materials. We conclude that the current uncertainties prevent a firm conclusion about possible conflict between the giant impact scenario and lunar history. However, the bulk of the evidence and modeling suggests hot (near solidus or above) initial conditions for nearly all the lunar mass.

## 1.  INTRODUCTION

The principal argument favoring a giant impact origin of the Moon is dynamical: The angular momentum of the Earth-Moon system can be explained as the outcome of the oblique impact of a body of Mars mass or greater on proto-Earth. Alternative explanations of lunar origin such as fission or coaccretion have great difficulty explaining this angular momentum budget, and the capture hypothesis, while not rigorously excluded, requires very special and unreasonable assumptions.

Although the giant impact origin hypothesis owes some of its support to the unsatisfactory features of the alternatives, it also has a naturalness almost approaching inevitability: Giant impacts appear to be a natural consequence of planetary accumulation (*Wetherill,* 1990), the emplacement in Earth orbit of material "splashed out" from the impact is highly likely (see *Cameron,* 2000), and the accumulation into one Moon appears to be dynamically plausible (*Ida et al.,* 1997; *Canup et al.,* 1999). However, a giant impact origin has very striking implications for the thermal state of early Earth and the material from which the Moon formed, as well as the timescales for lunar accumulation. In this chapter, we will assume that the Moon did indeed arise from a giant impact on Earth, and that this occurred in the context of a "standard" picture of terrestrial planet formation in which nearly all Earth's mass accumulated over a total period of around 100 m.y. We will, however, allow ourselves the freedom of considering a variety of timings for this giant impact, so that we do not limit ourselves to the assumption that the Moon-forming event was the last, large impact on Earth. Our goal here is to assess whether the Moon we know is compatible with the Moon that might form following a giant impact. This assessment will be ul-timately somewhat unsatisfying because it will highlight the uncertainties that still exist in our understanding of the dynamics (both before and subsequent to lunar formation) and our understanding of the thermal state of the Moon.

One might imagine that there is no memory of origin in the Moon we now see. After all, we cannot easily decide by looking at a block of ice whether the water molecules that comprise it were once in the form of steam. However, the Moon has several attributes that might serve as a record of how it came to be. One is its relative inactivity, compared to Earth, which allows the possibility that the lunar surface has a cumulative memory of the thermal history, in particular the net thermal contraction or expansion arising from a change in mean temperature over geologic time. Another is the chemical makeup of the Moon, particularly the absence of volatiles and metallic iron. A third attribute is the layering the Moon possesses and the chronology of that layering, particularly the early formation of the lunar highland crust, and the timing of mare genesis. A fourth and very important attribute is the isotopic data that may constrain the timing and energetics of earliest events, the primordial layering of the Moon, and the extent to which the Moon is subsequently well mixed. Last but not least, the geophysical attributes of the Moon (gravity and topography, evidence for or against mantle convection, explanation of lateral variability) provide a constraint on the history of the internal thermal state.

The first of these lunar attributes, the possibility of "reading" the surface for net strain, has figured prominently in discussions of lunar history, and seems to pose the severest puzzle for the giant impact hypothesis because it argues for an early Moon that was cold on the inside (*Solomon,* 1986). It accordingly receives considerable attention in this chapter. If the Moon were perfectly fluid, no memory of average thermal expansion or contraction could exist (the net straining of the fluid would not be apparent in any observable property — that is the definition of a fluid). If the Moon were perfectly elastic, then one could imagine measuring stress in the surface rocks and thence deducing net strain, though there would be no manifest evidence of this in large-scale photographs of the surface. In fact, we possess no such stress measurements, but more importantly, the outer part of the Moon is neither liquid nor an elastic solid, but a real, complex material that faults under sufficient stress. One could expect to see surface features (scarps) were the Moon's external radius (and hence surface area) to decrease sufficiently. This "record" might only be expressed well for the period after the late terminal bombardment (the decline of impact flux to the current levels, which occurred around 3.8 Ga), but that is already sufficient, according to the standard story, to suggest severe constraints on the change in mean lunar temperature over the past 3.5 G.y., constraints that might not be met by a Moon that started as hot as the giant impact story would suggest.

The other lunar attributes, especially the volatile depletion and rapid formation of lunar highland crust (suggestive of a lunar magma ocean epoch), are often considered to be supportive of the giant impact hypothesis. But as we will discuss, the issues posed by the isotopic and geophysical data are not so readily compatible with the high temperatures of lunar formation from the giant impact.

There are many issues that arise from the giant impact scenario for lunar formation, but our emphasis here is on the *thermal budget* and the implications of that thermal budget for the formation process and, especially, the Moon that results. In essence, this budget involves three aspects: gravitational energy, radiative loss, and radioactivity. Gravitational energy release is enormous, radiative loss is potentially enormous, and radioactive heating is modest by comparison (but has occurred much more recently and is hence disproportionately important in understanding the Moon as we now see it). To appreciate the issues, it is useful to think through all the contributions that could arise as we follow a parcel of material that ends up being part of the Moon. Along with each contribution, we list the chapter section in which it is discussed: (1) the preimpact state of the target (proto-Earth) and projectile (section 2); (2) the change in thermodynamic state of the material arising from the impact and the emplacement of material in Earth orbit (section 2); (3) the extent to which the disk material cools radiatively as the disk evolves, and heats by gravitational energy release as the disk spreads and coagulates (section 3); (4) the heating arising from the energy of formation of the Moon from dispersed material or small proto-Moons, and the radiative cooling that can occur during this accretion (section 4); (5) the cooling of a lunar magma ocean in the early postaccretion phase of the Moon and the affects of subsequent impacts (section 5); (6) radioactive heating and the linking of early thermal history to conventional thermal histories based on the processes of conduction, melt migration, and subsolidus convection (section 6).

Section 6 will also attempt to answer the fundamental question of the extent to which we can link the Moon we see to the Moon that might arise from a giant impact, and section 7 identifies the unresolved issues that arise in this attempt.

## 2.   THE THERMAL AND DYNAMICAL STATE AFTER A GIANT IMPACT

In the currently favored picture of terrestrial planet formation, some planetesimals form very quickly (perhaps while $^{26}$Al is still alive) and runaway accretion occurs on a timescale $\sim 10^5$ yr, leading to bodies of the order of 1 $M_{\mathbb{C}}$ (lunar mass) as the building blocks for making Earth. These bodies are already internally hot (by $^{26}$Al heating or accretion) and have likely formed cores. Certainly, the ages of some iron meteorites dictates the formation of cores in many planetesimals, and it seems likely that these will merge to larger cores in the gentle runaway accretion phase.

In the spirit of Safronov's pioneering ideas, and consistent with more recent simulations, the subsequent accumulation into planets typically involves collisions with relative velocities "at infinity" that are only a fraction of the escape velocity from the largest body in the swarm (*Wetherill,* 1990). Accordingly, the dominant energy release from an impact is around GM/R per unit mass of projectile, where G is Newton's gravitational constant, M is the accreting body's mass, and R is its radius. Consider, first, the thermal state of the proto-Earth and giant projectile prior to the giant impact. If we balance $GM^2/R$ by $MC_p\Delta T$, where $C_p$ is the specific heat ($\sim 8 \times 10^6$ erg/gK) and $\Delta T$ is the average temperature rise due to gravitational accretion, then $\Delta T \sim 40,000f^{2/3}$ K, where f is the mass in Earth units, and the same mean density as Earth ($\sim 5.5$ g/cm$^3$) is assumed. [Note: 0.8 $GM^2/R$ is actually a better approximation than the standard uniform density result $3 GM^2/5R$ because of gravitational self-compressions and because the iron core will be present, even at this stage, and additional core formation is expected to be contemporaneous with accretion (*Stevenson,* 1989).] Of course, this ignores melting and vaporization. Latent heat of melting is comparatively small for rocks (comparable or less than the sensible heat required to heat the rock from room temperature to melting, for example). However, vaporization requires a very large amount of energy. Indeed, GM/RL ~1 for $M_{\oplus}$ (Earth mass), where L is the latent heat of vaporization for rock, so only a small fraction of mass is likely to be vaporized. Evidently, the energy of accretion is much more than enough to completely melt everything (even after allowing for the high melting point of materials subjected to pressures similar to those encountered near Earth's center).

This does not automatically mean that the bodies are hot, because radiative losses can be prodigious. If the surface of the body radiates at a constant temperature $T_e$ for time $\tau$, then we can try equating $GM^2/R = 4\pi R^2\sigma(T_e^4 - T_0^4)\tau$ (where $\sigma$ is the Stefan-Boltzman constant and $T_0$ is a background temperature due to the Sun or a solar nebula). In the limit $T_e^4 \gg T_0^4$, this yields an "equilibrium" temperature $T_e \sim$ 450 f$^{3/4}$ $(10^7 \text{ yr}/\tau)^{1/4}$. At first sight, this might seem to suggest that the material can be efficiently cooled, but that is a fallacy, at least for f > 0.05. The reason is that this heat must be supplied from inside the body to the surface, since the delivery of this energy was mainly through the impact of substantial-sized bodies (rather than, say, dust particles). This implies a high-temperature gradient (typically sufficient to reach the melting point only tens to hundreds of meters below surface on average) and the material must be at least partially molten at all depths of energy deposition (up to hundreds of kilometers) in order for the material to be sufficiently fluid to carry the heat outward at the required rate. Interestingly, lava lakes on Hawai'i typically show effective radiating temperatures of around 500 K (because of a thin skin overlying vigorously stirred magma), and many regions on Io (which has predominantly silicate volcanism) show similar temperatures [with only small patches showing much higher values (*McEwen et al.,* 1999)]. The heat flows of interest here are even larger than the current heat flow on Io by some 2 orders of magnitude. To summarize, substantial-sized bodies (and a mass of order 1 $M_\sigma$ or even 0.5 $M_\sigma$ qualifies) must be extensively partially molten in order to lose heat efficiently, and would be even more molten if they failed to lose heat efficiently! The cooling time is on the order of the accretion time (100 m.y.) or longer. Cooling of material to subsolidus conditions is inefficient and most of the mantle lies between solidus and liquidus (*Sasaki and Nakazawa,* 1986; *Abe,* 1993). In the detailed calculations of *Solomatov and Stevenson* (1993), the deeper mantle stays "frozen" near the solidus, despite the steep rise of solidus with pressure, because of this inefficiency of subsolidus convection and because the adiabat (including the entropy effects of melting) lies between solidus and liquidus over an extended pressure range. *The preimpact proto-Earth and giant projectile were likely at or above the solidus throughout* (except, perhaps, for a surface layer of negligible thickness).

We assume that as a consequence of the impact, a mass of at most of order 2 $M_C$ is placed in near-Earth orbit, and that this mass is small compared to the projectile mass (consistent with the simulations). Consequently, the gravitational and orbital kinetic energy of this material is at most ~0.01 GM$^2$/R larger than if no orbital injection had occurred. (We are assuming that negligible mass is ejected "to infinity". It should be noted that if ejected mass leaves at very high velocity compared to escape, it could be important in the total energy budget, but the simulations do not provide any evidence of this kind of jetting.) We also assume that the oblique impact provides an angular momentum about equal to that of the Earth-Moon system. This means that it must provide ~I$\Omega^2$/2 energy of Earth rotation, where I is Earth's moment of inertia and $\Omega$ corresponds to a length of day of around 5 h. This is also ~0.01 GM$^2$/R. Provided the impactor mass is much larger than ~0.02 M, these kinetic energy requirements are small and most of the

impact energy will go into heating the projectile and target. From the numbers mentioned previously, it is evident that an incoming mass of order 1 $M_\sigma$ has sufficient energy to melt all of Earth, particularly as only the latent heat of melting is required. As discussed by *Tonks and Melosh* (1993) and *Pierazzo and Melosh* (1999), and foreshadowed in part in *Stevenson* (1987), sufficiency does not imply inevitability since the entropy production in the shock heating process is highly nonuniform, with the "far" hemisphere receiving much less heating than nearby material. The material that finds its way into orbit is very likely not the most severely heated (e.g., material on the "farside" of the projectile from the impact site). Nonetheless, Tonks and Melosh tend to understate the consequences of a giant impact melting by assuming cold starting conditions. It should also be stressed that material that is already at the solidus at depth will actually melt more upon pressure release (e.g., into Earth orbit) because specific entropy increases along the melting curve (i.e., the adiabat is less steep than the melting curve). Of course, vaporization entails a much higher entropy cost, and can easily accommodate the energy budget once the temperature exceeds ~2000 K. There has not yet been a simulation that can predict with high confidence the thermal outcome (the free-energy state of the material placed in orbit), but it would seem likely that it is mostly molten and at least partially vaporized, simply because there is no other way to balance the energy budget.

In a few of the simulations discussed by Cameron, an intact Moon may form promptly (i.e., on a timescale on the order of 1 day). If this body has a sufficiently high periapse (thus avoiding tidal disruption), it will survive and evolve outward in its orbit. In this limiting scenario, there is negligible time available to radiate away the energy produced prior to and during the giant impact, and it seems unavoidable that the initial Moon was completely molten. We will concentrate instead on "slower" scenarios, those involving a circumterrestrial disk, but even these scenarios do not necessarily avoid a completely molten initial Moon.

## 3. COOLING IN A CIRCUMTERRESTRIAL DISK

In the immediate aftermath of a giant impact, the distribution of material splashed out into Earth orbit is highly three-dimensional (bars, blobs, and spiral density waves). On a very short dynamical timescale (a few Keplerian orbits) it is possible (but not certain) that this will smear out into an azimuthally uniform disk that contains perhaps 2 $M_C$ and sufficient angular momentum that it significantly extends beyond (or can partially evolve beyond) the Roche limit. *Thompson and Stevenson* (1988) show that the action of Earth's noncentral gravity (i.e., the precession due to $J_2$) will assure the realignment of angular momentum vectors of Earth spin and disk orbital motion on a timescale of days. On these dynamical timescales, cooling is negligible. Although extensive melt seems likely in this disk, it is less clear whether the liquid-vapor ("foamy") state discussed at length by Thompson and Stevenson occurs or is important. They invoked this as a source of patch instabilities (and thus turbulence and viscosity) for the disk, thereby providing a means for disk spreading on a timescale of years to centuries. Developments in disk evolution over the past decade (e.g., for the solar nebula) suggest that turbulent viscosity models for disk spreading may often fail (*Balbus and Hawley,* 1998) or at least be less efficient than other methods of angular momentum redistribution. Earth may possess a dynamo at this time, providing seeds for a magnetic field mechanism of angular momentum transfer (*Balbus and Hawley,* 1998), or there may be wave mechanisms available. This problem remains unsolved and may well require a major computational effort for progress.

One viewpoint would be to assume no significant orbital ("viscous") evolution of the disk prior to the formation of a planetesimal swarm. This is the implicit assumption of the work of Canup and collaborators (*Ida et al.,* 1997; *Kokubo et al.,* 2000), since they use the output of Cameron's simulations as input for their calculation (cf. *Cameron and Canup,* 1998). It is important to realize that this is a major assumption, with high probability of being incorrect, but there is no obvious better choice at present. The significance of this approach is that it "works" in the sense that they succeed in making a single Moon from the assumed starting configuration of planetesimals. There is another problem, however, in that the timescale of accumulation from planetesimals to a single, dominant Moon (with very little left over in smaller bodies) is very short indeed, essentially negligible relative to cooling times. This may be self-inconsistent if the accumulation process releases enough heat to cause melting and thus invalidate the planetesimal approximation embodied in their approach. In any event, the timescale of lunar formation (defined, say, as the time from the giant impact to the accumulation of ~90% of the mass) is simply not known because the disk evolution is not known. If we write the disk-spreading timescale

$$\tau \sim \frac{\Delta R_{disk}^{2}}{\upsilon}$$

where $\Delta R_{disk}$ is the radial extent of the disk and $\upsilon$ is the effective viscosity, then

$$\alpha \equiv \frac{\upsilon}{C_s H_{disk}} \sim 10^{-4}$$

for $\tau \sim 10^3$ yr. Here, $C_s$ is the sound speed, and $\alpha$ is the parameter often used to characterize disks (e.g., the solar nebula). The mechanisms commonly discussed for the solar nebula [turbulence, magnetic field instabilities, waves (cf. *Adams and Lin,* 1993)] are also of relevance here and all yield values for $\alpha$ larger than this. Accordingly, spreading timescales in excess of thousands of years seem dynamically unreasonable and timescales as short as years to decades are certainly conceivable. We need also to remember

that the Moon thus formed may be less massive than the Moon we now see, with part of lunar accretion delayed by up to tens of millions of years by the arrival of more Sun-orbiting material.

While the exact physical state of the circumterrestrial disk is not known, we can estimate the maximum amount of cooling possible within the disk, and thus determine a lower bound for the initial temperature of the material that would accrete to form the Moon. In this simple model we imagine that there is a lunar mass of material distributed uniformly in a torus around the Earth

$$M_{moon} = \rho H_{disk} \pi \left( R_o^2 - R_i^2 \right) \qquad (1)$$

where $\rho = 3.3$ g/cm$^3$ is the mean density of the lunar material, $H_{disk}$ is the thickness of the disk, and $R_i$ and $R_o$ are the inner and outer radii of the torus respectively. In order to estimate a minimum cooling time, we assume that the disk is fluidlike (i.e., if it is particulate, then the particles in this disk are much smaller than the thickness of the disk) and convecting vigorously so that the entire disk and the interiors of the particles have the same temperature. If $R_o$ is 4 $R_\oplus$ and $R_i$ is 2 $R_\oplus$ then the disk thickness would be 10 km, so the particles must be much smaller than this. Of course, one could in principle envisage this as a particulate swarm with intervening space (thus allowing the disk to be thicker), but only so long as the thermal equilibration among the particles is very fast. We further assume that the top and bottom of this disk are radiating heat away, causing the disk temperature to decrease according to the differential equation

$$\frac{\partial}{\partial t} \left[ \rho C_p H_{disk} T_d \right] = -2\sigma \left( T_d^4 - T_a^4 \right) \qquad (2)$$

where $T_d$ is the temperature of the isothermal disk, $T_a$ is the ambient temperature, and other variables are as previously defined. The exact ambient temperature that the disk sees will depend on the surface temperature of the Earth (perhaps of order 1000–2000 K) and the exact physical state of the disk (e.g., the density of particles as a function of radius as well as distance above the equatorial plane). A simple estimate of the radiative equilibrium of a particle in orbit between 2 and 5 $R_\oplus$ (where $R_\oplus$ is one Earth radius) with the above surface temperatures of the Earth could easily be around 300–600 K, with even higher values likely at the disk interior and immediately following the giant impact. We account for phase changes in our model disk by keeping the disk temperature at the solidus until the entire disk has radiated enough heat to overcome the latent heat of freezing and solidify.

Figure 1 shows some results of our disk-cooling model. This plot shows that it takes our idealized disk of order 10 yr [as suggested in dynamical models (e.g., *Ida et al.,* 1997)] to lose memory of the high temperature of the giant impact event. Thus, if the Moon accretes on timescales of less than 10 yr after the giant impact, the material that forms the Moon will be unavoidably hot. Implicitly assumed here is that the ejecta from the giant impact is of order 1000–2000 K, but as we discussed above, this also appears inescapable.

It should be emphasized that our assumptions have been made to estimate the maximum cooling of the disk. Larger particles within a disk that is convecting less vigorously will not have a uniform temperature and thus the interiors of these bodies and the center of the disk will cool much more slowly. In addition, we have neglected energy produced in the disk as a result of either accretional energy or viscous dissipation. We now make some rough estimates of the energy derived from dissipation within the disk and leave a discussion of accretional heating to a later section. We will estimate the amount of energy dissipated as equivalent to the gravitational energy associated with tidal evolution of the disk. As an example, if the process of accretion is only 50% efficient (e.g., *Ida et al.,* 1997) we start with 2 $M_{\mathbb{C}}$ of material in orbit at $R_i$. If 1 $M_{\mathbb{C}}$ evolves to 2 $R_i$ the other half migrates to

$$\left[2-\sqrt{2}\right]^2 R_i \approx 0.34\, R_i$$

to conserve angular momentum, and so the gravitational energy release is

$$\frac{GM_eM_l}{R_i}\left(\frac{1}{2}+\frac{1}{0.34}-2\right)$$

We assume that this energy is dissipated throughout the entire disk region from

$$\left[2-\sqrt{2}\right]^2 R_i$$

to $R_i$ and over the timescale necessary for the material to migrate this distance. The timescale is an adjustable parameter because its exact value is unknown due to uncertainties in the size distribution of the particles, and how these objects interact with each other and the Earth near the Roche limit. *Thompson and Stevenson* (1988) estimated that possible disk evolution times might be 100–1000 yr, but dynamical simulations have disk-evolution times much less than this, perhaps ~1–10 yr or less.

The evolution of the disk cooling governed by equation (2) with additional energy generated from dissipation is shown in Fig. 2. Several plausible values of the time necessary for the disk to evolve from 2 $R_\oplus$ to 4 $R_\oplus$ are shown. We see that the energy generated within the disk can have a significant effect, allowing disk temperature to remain much higher than the ambient temperature even with our optimistic assumptions about cooling. If disk evolution times are really of order a few years (*Ida et al.,* 1997) these calculations show that any significant cooling from the high temperatures immediately following the giant impact is unlikely. As shown in Fig. 2, if the disk accretional efficiency is about 90% instead of the 50%, the energy gener-

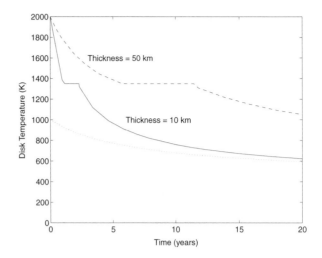

**Fig. 1.** Maximum cooling of a circumterrestrial disk. The flat portions of the plots are an artifact of the cooling process where the temperature remains at the solidus until the entire disk material is solidified. Depending on the thickness of the disk, the disk can cool to the ambient temperature on decadal timescales or slightly longer (still less than 50 yr or so). Values used in the calculation are: latent heat of freezing $4 \times 10^9$ erg/g, solidus of 1350 K, $C_p = 12 \times 10^6$ erg/gK, and the ambient temperature = 500 K.

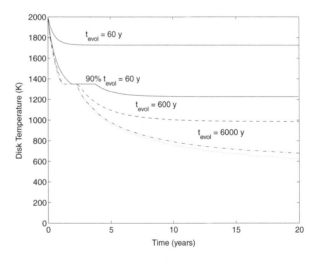

**Fig. 2.** Cooling in a circumterrestrial disk with energy from viscous dissipation. Viscous dissipation is estimated as being the gravitational energy of moving 1 $M_{\mathbb{C}}$ from 2 $R_\oplus$ to 4 $R_\oplus$ while an equal mass moves inward to conserve angular momentum. This energy release is assumed to be spread equally over the area that the two lunar masses move and during the time period as shown on the graph. The line labeled 90% shows how the disk temperature is reduced if the efficiency of disk material accreting into the Moon is 90% instead of the 50% assumed in the other plots, with a 60-yr timescale for disk evolution used. The dotted line corresponds to no viscous heating and is the same as the middle line in Fig. 1. Disk thickness is 10 km for all calculations and other parameters are the same as in Fig. 1.

ated is about a factor of 4 less for a given evolution time, but the temperature is still well above ambient conditions. A higher efficiency may be possible in a disk with spiral density waves (W. R. Ward, personal communication, 1998).

In the above calculation we have approximated the disk as particles that are much smaller than the disk thickness (objects ≪10 km) and assumed infinitely efficient convection from the interior to the disk surfaces. In reality, a continuous disk of molten rock will possess upper and lower thermal boundary layers, within which conduction from the interior must support radiation from the surface. Moreover, this disk presumably fragments into lunatesimals. These bodies cannot freely radiate if they are smaller than ~10 km (since the disk is then thicker than the individual bodies and they will shadow each other) but will lose heat less efficiently than a continuous disk if they are much larger, because of reduced area to volume ratio. In the following model, we take the best case (maximal cooling rate) in which the material can be thought of as either a continuous disk or a closely packed monolayer of lunatesimals. Think of a hockey-puck-shaped element of this disk with unit surface area top and bottom and thickness H. Suppose that at some instant in time, these two unit surface areas of this hockey puck consist of material at the solidus, $T_m$. These surfaces then solidify and cool to the temperature $T_s$. During the time it takes to cool, the surface temperature declines from $T_m$ to $T_s$, but most of this decline occurs quickly and for the rest of the time, the temperature is close to $T_s$ and the heat flow declines approximately as the inverse square root of time (the usual conductive thickening of a boundary layer). The average heat flow from the hockey puck during this is then ~$4\sigma(T_s^4 - T_a^4)$. (The factor of 4 comes from a factor of 2 due to integrating heat flux over the time interval, assuming $T_a \ll T_s$, and another factor of 2 arising from the top and bottom surfaces.) Eventually, the surface will founder because it is more dense (a conventional convective instability) or, more probably, due to a surface disruption caused by impact-generated waves or other disturbances in the disk. The average temperature of this foundering material will be ~$\frac{1}{2}(T_m + T_s)$.

For the purpose of this model, we assume that the material will sink to the center of the lunatesimal or disk. We can test the viability of this assumption by comparing the timescale of thermal diffusion to the time it takes the surface material to sink assuming it falls according to Stokes velocity. The critical condition is

$$\frac{\frac{1}{2}H}{v} = \frac{\frac{1}{2}H}{\frac{2}{9}\frac{g\Delta\rho\delta^2}{\eta}} \leq \frac{\delta^2}{\kappa} \qquad (3)$$

where $\kappa$ is the thermal diffusivity = $10^{-2}$ cm²/s, $\delta$ is the size of the thermal boundary layer ~$2\sqrt{\kappa\tau}$, $\tau$ is the timescale, g is the gravitational acceleration within the lunatesimal or disk ~$4\pi\rho GH$ (but declining to zero at the center), $\eta$ is the viscosity of the magma-rich interior, and $\Delta\rho$ is the den-

sity difference between the sinking slab and the interior. Using the expressions above, we can solve for time

$$\tau^2 \geq \frac{27\eta}{\kappa\Delta\rho 2^4 4\pi G\rho} \qquad (4)$$

Taking $\eta \sim 100$ p, $\Delta\rho \sim 0.1$ g/cm³ gives a timescale of order $10^5$ s, which is much shorter than the evolution timescales that we are interested in. This calculation is an idealization, especially since it uses a fully fluid viscosity, whereas the actual viscosity will increase as the interior crystallizes. In a more realistic model, the sinking time will increase as the interior temperature begins to drop, allowing for further cooling the interior (which will further increase sinking time, and so on), thus reducing the efficiency of our heat removal mechanism.

Once the surface slab has begun to sink, a new batch of the interior at the solidus will be exposed at the surface, where it radiates, initially at a temperature $T_m$. This layer cools to a surface temperature $T_s$, founders and sinks, exposing a new layer, and so on until all the mass has been cycled through. The equation governing this model is given by

$$4\sigma\left(T_s^4 - T_a^4\right) t_{cool} = \frac{1}{2}\left(T_m - T_s\right)\rho C_p H + \rho LH \qquad (5)$$

where the lefthand side gives the energy radiated away from the body and $t_{cool}$ is an independent variable (the total time needed to cycle all the mass through the surface layer). The first term on the righthand side refers to the energy needed to cool the sinking material, and the second term is the energy used in freezing the material, where L is the latent heat. This mechanism would allow for maximum cooling of the object since all the interior is cycled to the surface and able to freeze there. In addition, we have neglected accretional heating and growth, because both of these would reduce the ability of the lunatesimals to cool as efficiently as our model.

Figure 3 shows the mean temperature that can be achieved by this idealized mechanism for lunatesimals of different sizes. Results for H ~ 0.1 km are shown, but remember that this is unrealistic for lunar formation (except at the lower density periphery of the disk) because the disk thickness is much larger. These calculations show that even on a decadal timescale, it is difficult for objects larger than 10 km to have a mean temperature less than 1000 K. One of our assumptions — that all the lunatesimals were at the solidus — might be considered extreme. For example, if the giant impactor struck the Earth obliquely, some of the ejecta might remain internally undisturbed (*Pierazzo and Melosh*, 1999). However, even if material were not significantly heated by the impact event, as was discussed in section 2, the outer layers of both the Earth and the impactor were initially hot, so it is difficult to imagine that ejecta would be too far from the solidus.

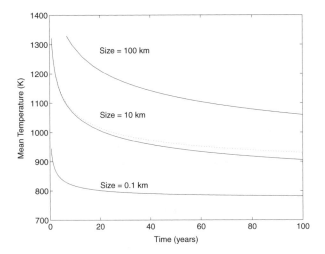

**Fig. 3.** Mean internal temperature of lunatesimals using idealized cooling mechanism calculated using equation (5) for different-sized objects. The ambient temperature used in all calculations was 200 K except for the dotted line, which shows the calculation for the 10-km-sized object with an ambient temperature of 400 K for comparison. Unlike Fig. 1 and Fig. 2, the horizontal coordinate is total cooling time, not instantaneous elapsed time.

We turn now to a brief consideration of chemical and isotopic effects. There are two issues here: (1) Is the disk chemically and isotopically homogeneous (i.e., well mixed)? (2) Is there significant loss and fractionation by hydrodynamic outflow? Neither question turns out to have a clear-cut answer, but here are the relevant considerations. First, it is not assured that the initial condition (e.g., the distribution of material one day after the impact) is homogeneous, because the orbital placement of material may be related to its depth in the (possibly heterogeneous) projectile or target. [Note that it is not correct to assume that a projectile that is extensively partially molten has a homogeneous mantle! Some aspects of this are addressed by *Solomatov and Stevenson* (1993).] Cameron's simulations exhibit substantial mixing, however, presumably a form of shear- (Kelvin-Helmholtz) or pressure-gradient (Rayleigh-Taylor)-induced turbulence, so homogenization might occur. Second, the disk evolution may involve an "eddy diffusivity" that is much larger than the effective viscosity (the latter being defined by the timescale to redistribute angular momentum in the disk). This may be true of convective motions (*Balbus and Hawley*, 1998) and may well arise from the instabilities analyzed by *Thompson and Stevenson* (1988). Since the angular momentum redistribution timescale is presumably of order the lunar formation timescale or less, it is likely that passive tracers will mix efficiently. Most constituents are not "passive," however, because they partition strongly between vapor and liquid phase and have strongly temperature-dependent vapor pressures. If the disk were all one temperature, then this would not matter (evapo-

ration and condensation would be in balance at each location). More probably there are substantial differences in disk surface temperature with radius (with lower temperatures presumed to be at the outer edge). This could lead to a "cold finger" phenomenon whereby there is a net flux of volatile constituents from hot to cold regions. Unlike a conventional laboratory situation, this redistribution is strongly inhibited by the angular momentum constraints (gas cannot flow radially except by changing its angular momentum) and is thus limited by the strength of the eddy diffusive mechanism [cf. similar situations discussed by *Morfill et al.* (1985), *Stevenson and Lunine* (1988), and *Stevenson* (1990)]. If the total mass in the vapor phase is small compared to the disk mass then this will not be important at least for major elements.

Minor and especially volatile elements are of particular interest. In the case of water (or, more precisely, H derived from dissociation of water) it seems likely that efficient escape is possible (*Thompson and Stevenson,* 1988). Even major-element fractionation could be conceived, as argued by *Abe et al.* (1999). Here, we confine ourselves to the issue of K isotopes. *Humayun and Clayton* (1995) report an undetectably small difference between the K-isotopic ratio ($^{41}$K/$^{39}$K) among samples that have large differences in volatile depletion (with the lunar samples exhibiting net K depletion of a factor of about 50 relative to CI meteorites). Is this compatible with an evaporating disk? First, it is important to remember that only net loss matters: Evaporation following recondensation cannot possibly have any effect in a well-mixed, closed system. However, the depletion of volatiles exhibited by the Moon is startling compared to Earth (around a factor of 5–10), so we need to pose this question: Can we devolatilize the Moon-forming material by evaporation and escape, without violating the isotopic constraint? Significant K-isotopic fractionation only arises kinetically (the equilibrium fractionation at relevant high temperatures is negligible). Isotopic fractionation can then arise in two ways: at the evaporating surface of the disk, and in the exosphere (the outgoing, low-density, supersonic gas flow). The former process is considered by Humayun and Clayton; for this they derive a Rayleigh distillation equation showing large fractionation can result from large volatile loss. The latter process (mass fractionation in the outgoing flow) is the basis for Xe-isotopic fractionation models (*Pepin*, 1997) and many models of isotopic fractionation in planetary atmospheres. Typically this process is effective under relatively gentle outflow conditions (mass loss on million-year timescales or more), unlike the highly energetic conditions we consider here for lunar formation. When the hydrodynamic outflow is sufficiently vigorous (far from Jeans' escape conditions), the fractionation is small. We focus here on the former mechanism of fractionation at the disk/atmosphere interface (precisely those conditions that most concerned Humayun and Clayton).

Consider a unit area of disk that is undergoing net evaporation. Near the disk surface, this causes a gentle upward flow of high-density gas that gradually evolves into a supersonic low-density flow at and above the exobase. (This is just like a planetary atmosphere, with many atmospheric scale heights between the surface and the exobase.) For isotopic species i, assume that the number of atoms in the magma reservoir (per unit disk area) is $N_{ir}$. The net flux (atoms or molecules per unit area per unit time) from the melt into the vapor phase is

$$J_i = \beta(P_{ie} - P_i)/c_i \quad m_i = -dN_{ir}/dt = n_i v_w \quad (6)$$

where $\beta$ is a universal constant of order unity ($\beta = 1/\sqrt{2\pi}$ in ideal kinetic theory), $P_{ie}$ is the equilibrium vapor pressure (depends on temperature T and magma composition), $P_i$ is the actual partial pressure in the atmosphere overlying the magma, $c_i$ is the sound speed, $m_i$ is the relevant atomic or molecular mass, $n_i$ is the number density in the gas (just above the magma), and $v_w$ is the net vertical flow (the "wind") and applies to all species. By the ideal gas law, $P_i = n_i kT$, where k is Boltzman's constant so it follows that $P_{ie} = n_i kT(1 + v_w/\beta c_i)$. We assume negligible equilibrium partitioning, a well-mixed disk, and Henry's Law, which means that $P_{ie}$ is proportional to $N_{ir}$. It follows that

$$\frac{N_{1r}}{N_{2r}} = \frac{P_{1e}}{P_{2e}} = \frac{n_1\left(1 + \dfrac{v_w}{\beta c_1}\right)}{n_2\left(1 + \dfrac{v_w}{\beta c_2}\right)} \quad (7)$$

where "1" and "2" refer to two isotopic species. However, $J_1/J_2 = dN_{1r}/dN_{2r} = n_1/n_2$, from which we conclude that

$$\frac{dN_{1r}}{dN_{2r}} = \frac{N_{1r}\left(1 + \dfrac{v_w}{\beta c_2}\right)}{N_{2r}\left(1 + \dfrac{v_w}{\beta c_1}\right)} \quad (8)$$

This reduces to the Rayleigh distillation result quoted by Humayun and Clayton in the limit of very high wind speed (direct evaporation into vacuum). Using their notation, we get $\delta^{41}K = 24.7 (v_w/\beta c)\ln(1/F)$, in the limit that the surface Mach number $M = v_w/c \ll 1$ (where c is the mean sound speed); this compares with their result (plotted in their Fig. 5) of $\delta^{41}K = 24.7\ln(1/F)$, where F is the fraction of K remaining. This shows that if $M \ll 1$, fractionation is small. In essence, M is a measure the closeness of the liquid-vapor interface to equilibrium.

If M is near unity, then the disk can evaporate completely on a timescale of order the mass per unit area ($\rho H$) divided by the mass evaporation rate (P/c). This timescale $\tau \sim \rho Hc/P \sim (10^5 s)(1 \text{ bar}/P)$, which is short compared to the expected timescale over which the disk evaporates significantly, since major species can be expected to have vapor pressures of order 1 bar or more. If we consider a minor species (or rare isotope), then the mass we must lose and the vapor pressure are reduced by the same factor and the timescale re-

mains the same. In other words, we do not need high wind speed to eliminate the volatiles from the disk if we have a timescale of order years at our disposal! Consequently, isotopic fractionation is small. Here is an illustration: Suppose we wish to lose 1% of the disk mass. If we evaporate for a timescale $\tau$ then the required "wind" is given by $0.01 \rho H_{disk} \sim v_w P\tau/c^2$, which for $P \sim 1$ mbar ($10^3$ dynes/cm$^2$) and $\tau \sim 30$ yr ($\sim 10^9$ s) implies $v_w \sim 30$ cm/s and $\delta^{41}K \sim 0.1\%$ (compatible with observations). Here is one instance where the giant impact scenario of lunar formation could be said to be "slow"!

## 4. ACCRETIONAL HEATING OF THE MOON

As shown in the previous sections, even under the most optimistic conditions, it is unlikely that a circumterrestrial disk can cool significantly on timescales of order a decade following the giant impact. Because of our poor understanding of processes that occur in the circumterrestrial disk, we cannot rule out the possibility that accretion could be delayed by 100–1000 yr following the giant impact (*Thompson and Stevenson,* 1988). Such a delay might allow for cooling of some disk material. Perhaps accretion with a short timescale, of order a decade (*Ida et al.,* 1997), occurred after these longer disk-evolution times such that accreting lunatesimals began cold. In the following section, we present a simple model of accretional heating to show that for reasonable initial lunatesimal temperatures, the initial lunar thermal state is unavoidably hot.

Our model for the interior temperature of the Moon during accretion will balance the accretional energy with energy lost due to radiation, heating of lunar material, and phase changes. We write this as

$$\frac{\partial}{\partial t}\left\{\int_0^M C_p\left[T(m) - T_i\right]dm\right\} = $$
$$\left(\frac{v_{rel}^2}{2} + \frac{GM}{R}\right)\frac{\partial M}{\partial t} - 4\pi R^2\sigma\left(T_s^4 - T_a^4\right) \quad (9)$$

where $\partial M/\partial t$ is the accretion rate, R is the radius of the proto-Moon, and $v_{rel}$ is the relative velocity between the proto-Moon and the lunatesimals (before acceleration due to infall in the proto-Moon gravity). T(m) is the temperature of material at the radius of the sphere containing mass m (less than the total mass at a given instant in time M), $T_i$ is the temperature of material before accretion, $T_a$ is the ambient temperature in the disk, $T_s$ is the surface temperature, and other variables are as defined previously. Our model will assume that the temperature changes only in the outermost layers (i.e., we neglect internal transport processes because they will be negligible over the radius of the Moon during the timescales that we are interested in), so that we can write the lefthand side of the above expression as

$$C_p \left[ T(m) - T_i \right] \frac{\partial M}{\partial t}$$

We divide equation (9) into two separate processes to be modeled: (1) heating material in the outer layer to the surface radiating temperature $T_s$, and (2) heating from $T_s$ to $T(M)$, with the latter being thought of as the "internal temperature," but still occurring in a thin layer immediately below the surface layer, which has temperature $T_s$. In order to account for the fact that energy from a large impact is deposited both in the surface layer and beneath the surface, *Kaula* (1979) uses a parameter h, which is the fraction of heat dumped below the surface. This parameter has a value between 0 (no heat below the surface) and 1 (all heat below the surface) and is poorly known, although plausible values could be between 0.1 and 0.5 (*Ransford and Kaula*, 1980; *Davies*, 1985). For the purpose of our model, we will assume that all the heat deposited below the surface is used in the process of heating the material from $T_s$ to $T(M)$. [This is a reasonable assumption for the short accretion times that we are considering, since this layer will be several kilometers thick, which is the depth interval over which the impact energy is deposited (*Kaula*, 1979).] Mathematically, this expression can be written as

$$T(M) - T_s = \frac{h \left( \dfrac{v_{rel}^2}{2} + \dfrac{GM}{R} \right)}{C_p} \qquad (10)$$

The second process of heating the from $T_i$ to $T_s$ is thus governed by

$$4\pi R^2 \sigma \left( T_s^4 - T_o^4 \right) =$$

$$\left[ (1-h) \left( \frac{v_{rel}^2}{2} + \frac{GM}{R} \right) - C_p \left( T_s - T_i \right) \right] \frac{\partial M}{\partial t} \qquad (11)$$

As the Moon accretes, this formulation will allow us to determine $T(M)$ as a function of depth, and since the processes of internal heat transport are negligible over the time intervals of interest, this profile will be the same once accretion is complete. Our model assumes that mass is added uniformly over the surface, and that impactors are small compared to the proto-Moon. It is now recognized that such models underestimate the heating of planetary embryos because thermal evolution is strongly affected by the existence of giant impacts, which are not uniformly distributed (*Melosh*, 1990, and references therein). It is unclear whether "giant" impacts are relevant to lunar formation (where "giant" now means impacts on the proto-Moon by bodies that are $\geq 0.1 \, M_{\mathcal{C}}$). However, as we show below, it is of more importance that over the timescale of interest (decades), the accretional energy so overwhelms the heat loss mechanisms that the exact details of h or giant impacts will be relatively unimportant.

Figure 4 shows results from our model for different accretional conditions, and compares them to the basalt solidus and the initial temperature profile from the geologic constraint (*Solomon and Chaiken*, 1976; *Kirk and Stevenson*, 1989), which is discussed in more detail in section 6. For clarity, we only show temperature profiles for lunatesimals with initial temperatures of 550 K, because the shape of profiles with hotter initial temperatures are similar, but displaced vertically such that the y intercept is their initial temperature. In addition, an initial temperature of 550 K was chosen because it is the warmest temperature profile that can satisfy the geologic constraint for these accretionary timescales. (By this we mean that the thermal energy in the Moon beneath the magma ocean for this accretion model is equal to the energy in the models that satisfy the geologic constraint.) We have found that the temperature profile is nearly identical for all values of h when the total accretional timescale is between 1 and 10 yr. This is expected since the accretional heat overwhelms all heat loss mechanisms during this short interval. As shown in Fig. 4, for accretion times of 100 yr, the temperature profile does depend on h, with h = 0.1 yielding cooling temperatures since most energy is deposited at the surface, where it can be radiated away. The $\partial M / \partial t$ used in the calculations in Fig. 4 was constant (an equal amount of mass was deposited in each time step). We also tested the effect of an accretion rate that decreases exponentially in time and found that it had only a minor effect (cooling of the outer 100 km or so) for timescales of order 10 yr. For longer timescales, how the accretion rate is specified is more important, but we believe that our model is more unreliable at these longer time periods since the role of individual giant impactors becomes more important to the heating of the outer layers. Although we believe the general trend and temperatures of our calculations, the wiggles of the temperature profiles shown in Fig. 4 are artifacts of how we parameterized phase changes, and should not be taken too literally. Our model shows the temperature distribution above the solidus at large radius immediately following accretion. This temperature profile is only a transitory event, because convection at these high temperatures and low viscosities is very efficient, such that the temperature will be reduced to the solidus in a short amount of time.

Our calculations show that for the short times of accretion that are indicated by dynamical calculations, significant heating is unavoidable and is independent of exactly how the heat is distributed (i.e., exactly what the value of h is). In addition, we have found that in order to satisfy the geologic constraint with accretion on these timescales, the mean initial temperature of the accreting material must be of order 550 K. Figure 3 shows that it is difficult to reach such a low mean initial temperature for reasonably sized bodies, even if accretion is delayed by 100–1000 yr. Our calculated accretional temperature profiles are probably lower bounds over the specified time interval since we have used rather conservative estimates for the eccentricities and

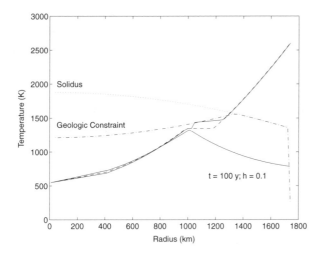

**Fig. 4.** Temperature profile within the Moon for different accretional conditions over a short time interval. All plots start with the initial radius of the proto-Moon of 10 km, accreting matter with a temperature of 550 K, and an ambient temperature of 300 K. The relative velocity for all planetesimals was calculated using

$$v_{rel} = \left[\tfrac{5}{8}e^2 + 0.5\sin^2(i)\right]v_{Kep}$$

(*Kenyon and Luu,* 1998), where e and i are the eccentricity and inclination of the accreting material respectively, and $v_{Kep}$ is the Keplerian velocity. Results are shown using an orbital radius of 3 $R_{\oplus}$, e = 0.1, and i = 5°. Using extreme values (e = 0.3, i = 25°) results in temperature profiles that are uniformly ~400 K warmer at all depths. The line labeled geologic constraint refers to the initial temperature profile that is consistent with no net change in lunar radius since 3.5 Ga (*Kirk and Stevenson,* 1989). Solidus refers to the basalt solidus that increases as a function of depth (*Ringwood and Essene,* 1970). The dashed line and the upper solid line refer to the temperature profiles when the accretion time is 1–10 yr. The difference between the two lines is that the dashed line has an h = 0.1 and the solid line has h = 0.9, showing that the temperature distribution is only slightly affected by this parameter for this timescale. The solid line labeled "t = 100 y; h = 0.1," refers to an accretion timescale of 100 yr, with 10% of the accretional energy deposited into the interior. For a timescale of 100 yr with h = 0.9, the temperature profile is nearly identical to the profile for accretion times of 1–10 yr.

inclinations appropriate for objects in the disk (see caption of Fig. 4 for details). Our models have assumed that heat loss is primarily radiative, which is an idealization. However, it is unclear how wrong this idealization is since radiative heat loss is likely to dominate at the surface layer, and might be fairly efficient during the process of fast accretion, which will mix the upper layer and throw ejecta far above the body. One possible means of having cooler accretion than our model calculations would be to have a few proto-Moons collect the high eccentricity/inclination debris when they are small and then to coalesce into a final object at low relative velocity once their orbits have evolved closer together. More sophisticated models incorporating

the effects of the giant collisions involved in this scenario will be required to determine how much cooler this model would be.

## 5. EARLIEST STAGES OF LUNAR EVOLUTION

In the simplest and standard approach to lunar history, we can think of the lunar thermal evolution problem as being defined by the initial condition (temperature and composition as a function of radius) and internal heating due to radioactivity, together with the relevant thermodynamics and physics of heat transport. Here, we comment briefly on two other factors: tidal heating and later impacts.

*Touma and Wisdom* (1998) have shown how to reconcile the current lunar orbit with tidal theory and the initially equatorial orbit implied by giant impact scenarios that involve disk formation. However, the passage through the evection resonance leads to large orbital eccentricity, the dampening of which implies a potentially large amount of tidal heating. They suggest that this guarantees a molten initial Moon (irrespective of arguments presented here for initial thermal state). It can easily be shown that passage through this resonance will still occur even if the Moon is not then at its final mass, though possibly with a somewhat smaller net tidal energy dissipation.

The effect of tidal heating is much harder to assess than radioactive or impact heating because it may be highly nonuniform in space and time. In the standard approach to this problem (*Peale and Cassen,* 1978), first applied to the Moon and subsequently to Io, melting may proceed from the inside outward because the maximum deformation is occurring deep down. However, this assumes that the "local" Q (a measure of anelasticity) is uniform throughout the body. In the case of a body that possesses a crust, a magma ocean, and a "slushy" interface between magma ocean and deeper cold material, it is possible that decoupling will occur, with most of the tidal heating occurring near the surface. This is also the region that has greatest ability in eliminating heat. The early tidal evolution is extremely rapid (the orbital radius can double in around 10,000 yr) and the heating events described by *Touma and Wisdom* (1998) are probably encountered when the Moon has no stable, conductive crust but a vigorously convecting magma ocean that eliminates heat quickly. From the perspective of current lunar evolution, it does not matter if the lunar magma ocean portion of the Moon receives a factor of 2 larger energy input (i.e., if tides inject an energy equal to that provided by accretion) if this occurs when this heat can be quickly eliminated (i.e., millions of years but not hundreds of millions of years). It is thus far from clear that tidal energy input is important for lunar thermal history.

Impacts during accretion certainly contribute importantly to the initial temperature of the Moon. However, impacts spread out over a long time after accumulation of most of the mass are relatively unimportant and may even help cool the Moon. To appreciate the latter point, consider the case where the Moon is developing a thick crust/thermal lithosphere (perhaps because of crystal flotation leading to the lunar highlands). To make matters concrete, imagine a region of the surface of order $10^4$ km$^2$ in area, which receives an impactor of radius 10 km every $\tau$ years. Because impacting bodies disrupt a region far larger in extent than the bolide radius, one could imagine that the entire region in question is "reset" (i.e., the cold lithosphere of thickness $\sim \sqrt{\kappa \tau}$ is replaced by hot material from depth). This enhances the heat loss by allowing a volume of order $10^4 \sqrt{\kappa \tau}$ km$^3$ to cool efficiently (before the next impact comes along). In this example, the energy input due to the impact is of order the bolide mass times specific energy of impact (most of which is radiated away immediately after impact) and the enhanced energy output is much larger because the resurfaced volume is much larger than the bolide volume, provided $\tau > 10^4$ yr. This would apply for all but very high impact fluxes. The exception is very large infrequent impact events that may bury substantial energy at great depth.

## 6. IMPLICATIONS OF INITIAL THERMAL STATE FOR LUNAR HISTORY

The initial lunar thermal state has important implications for subsequent thermal evolution. In this section, we will review whether certain geological or geochemical observations can be used to constrain the initial thermal state. In particular, we will investigate: the extent of differentiation and the size of the lunar magma ocean, the timing and composition of mare volcanism, and the geologic observation that there are no global-scale faults, interpreted to mean that there has been no net lunar radius change since ~3.5 Ga. Unfortunately, as we outline below, the first two considerations are rather nonunique and can be satisfied by a large number of accretion models. The last constraint has been previously used to argue for a cold lunar interior initially and is difficult to reconcile with thermal models that have accretion timescales of order 10 yr (see Fig. 4). But, as we will show below, this constraint can be significantly relaxed, such that a hot lunar interior is allowed.

The Apollo missions revealed the presence of a global layer of anorthosite, which along with other evidence, has been used to argue for the existence of a lunar magma ocean [a review of these arguments is provided by *Warren* (1985)]. Before the advent of the giant impact hypothesis, a major question in lunar science was the energy source for the melting of a lunar magma ocean. As we have outlined above, there is no difficulty in melting a considerable part of the Moon if lunar accretion is completed in a short amount of time immediately following the giant impact event. Knowledge of the original depth of the magma ocean would be a useful constraint on the initial thermal state of the Moon. However, there is no consensus on the exact depth of the magma ocean and it plausibly lies between 250 and 1000 km (*Warren*, 1985). A large number of accretional models are thus allowed, with many different possible accretional timescales and temperatures for accreting ma-

terial (see Fig. 4), such that there is currently little hope of using this constraint on the initial thermal state.

Two observations regarding the composition of mare basalts have been used to support the premise that the lunar mantle has been poorly mixed subsequent to 4.4 Ga: (1) heterogeneity of the mare source compared to the MORB source on Earth (*Turcotte and Kellogg,* 1986), and (2) isotopic closure of the source region very early in lunar history (*Brett,* 1977). From these observations it has been inferred that convection in the Moon has been weak or nonexistent, and thus that the Moon's interior was initially cold, since a hot initial state would lead to vigorous convection and mixing. Using this line of reasoning, *Turcotte and Kellogg* (1986) concluded that a cold initial state as implied by the geochemical data is inconsistent with the giant impact hypothesis. However, as *Turcotte and Kellogg* (1986) themselves observe, compositional gradients are likely to form in the mantle during the process of differentiation, which would inhibit or weaken convection even if the Moon is hot. It is not obvious that weak convection would be inconsistent with observations cited above, because convection in this scenario would be a poor mixer, and the exact size and depth of the mare source and the heterogeneities within it is unknown. Thus, several thermal history models, including some with "hot" interiors consistent with our accretionary models for the giant impact, might still be allowed.

Another possible constraint on the initial thermal state is the timing of mare volcanism, which occurred mostly in the first billion years or so. However, as *Cassen et al.* (1979) observed, most thermal models with an internal temperature >500 K and a magma ocean have temperature profiles near the basalt solidus at the appropriate depths and times for mare volcanism, and are consistent with this constraint. These thermal models calculate only the mean temperature at a given depth and still require local thermal perturbations to explain the remelting of the mare basalt source. Although there is no consensus on the cause of the remelting of the source region, it is possible that it occurred due to internal processes that are only indirectly related to the initial thermal state [e.g., models by *Hess and Parmentier* (1995) and *Wieczorek and Phillips* (1999)].

Binder and co-workers have advocated an initially molten Moon, but their arguments are unfortunately nonunique [e.g., mass balance of trace elements does not require melting of the entire Moon (*Binder,* 1986)] or suffer from a lack of data — the origin and occurrence of young, highland thrust faults (*Binder,* 1982; *Binder and Gunga,* 1985) and shallow moonquakes (*Binder and Oberst,* 1985) are poorly constrained (a summary of these points is made in *Solomon,* 1986).

Because the surface of the Moon has been relatively unmodified for the past 3.5 G.y. or so, several workers have used the lack of global-scale thrust faults on the surface of the Moon to conclude that global stresses within the lithosphere have been below the strength of lunar materials. Elastic stresses generated in the lunar lithosphere can be related to the strain caused by a net change in lunar radius through (*Solomon*, 1986)

$$\frac{\Delta R}{R} = \frac{1-\upsilon}{E}\sigma_t \qquad (12)$$

where $\Delta R/R$ is the fractional radius change of the Moon, $\sigma_t$ is the horizontal stress in the lithosphere, $\upsilon$ is Poisson's ratio, and E is Young's modulus, assumed horizontally uniform and applicable to a near-surface spherical shell. Using an E of $10^{12}$ dynes/cm$^2$, $\upsilon$ of 0.25, and a compressive yield strength of 1 kbar, the net lunar radius change must have been no more than ~1 km during the past 3.5 G.y. in order to explain the lack of global faults on the surface (*Mac-Donald*, 1960; *Solomon and Chaiken*, 1976). Thermal history models indicate that in order to offset cooling and contraction of the outer lunar layers that were part of a magma ocean and maintain a nearly zero net radius change, the lunar interior would have to have begun cold, and heated and expanded through time (*Solomon*, 1977, 1986). This stringent constraint on the initial thermal condition is relaxed only slightly by allowing for the expansion that occurs from phase changes of $Al_2O_3$ when the mare lavas reach the surface from depth (*Kirk and Stevenson*, 1989). In Fig. 4, we see that in order to satisfy the geologic constraint and dynamical models for accretion on decadal timescales, the accreting lunatesimals would need unreasonably cold initial temperatures. Therefore, a cold initial lunar interior consistent with the geologic constraint is incompatible with current dynamical models of accretion of the Moon. For comparison, we calculate that an initially molten Moon would have a maximum radius change of no more than 10 km in the past 3.5 G.y. using the volume effect of *Kirk and Stevenson* (1989). In the following section, we introduce three considerations that could relieve this conclusion and allow the hot initial lunar interior to be consistent with the geologic constraint. These considerations are (1) a better model of the stress-strain behavior of the lunar lithosphere, (2) nonmonotonic models of lunar thermal history, and (3) testing whether faults formed by lunar contraction would be visible through the regolith.

Relating the strain generated in the elastic lithosphere by thermal expansion and contraction within the Moon to the stress necessary for the initiation of faulting requires knowledge of the elastic moduli of the lithosphere. The value of E cited above is appropriate for the deep lithosphere, but data collected by seismometers placed on the Moon by the Apollo astronauts indicates that the outer 20 km of the lithosphere have low seismic velocities and strongly scatter seismic waves. This data has been interpreted to mean that these layers are porous and extensively fractured and broken (e.g., *Toksöz et al.*, 1974). A material that is pervasively fractured can undergo much more compressive strain before failure than an uncracked sample. This extra strain is believed to be accommodated by closing of cracks and pores and sliding along crack boundaries (see, e.g., *Li et al.*, 1998; *Lehner*

*and Kachanov*, 1995; and references therein). Mathematically, this means that the effective Young's modulus of the material is less than the Young's modulus of the sample calculated using the standard equation (e.g., *Golombek*, 1985)

$$V_{shear} = \sqrt{\frac{E}{2\rho(1+\upsilon)}}$$

or the Young's modulus of the undamaged sample. It is important to note the distinction between the effective Young's modulus and seismically determined Young's modulus — the effective modulus is relevant for the stress-strain behavior of the material over geologic timescales (i.e., global contraction), while the seismic modulus is a measure of the elastic properties over a much shorter time interval. The seismic modulus "now" (at the time of Apollo) may also be larger than the seismic modulus billions of years ago because of the intervening contraction and crack closure (e.g., *O'Connell and Budiansky*, 1974). The low effective Young's modulus of the cracked outer lunar lithosphere allows a greater strain of the material before failure, and consequently a greater change in lunar radius, or a hotter initial thermal state.

*Binder and Gunga* (1985) used a rough calculation based on the porosity of the lunar regolith to estimate that the closing of cracks and pores could accommodate perhaps 2–8 km of radius change before the lithosphere would break on visible faults. In order to more robustly quantify how large this effect might be, we use an empirical relationship between the effective Young's modulus of an impact-shocked gabbro sample and the damage parameter of the sample (*He and Ahrens*, 1994). The damage parameter is defined as

$$D_o = 1 - \left(\frac{C_p}{C_{po}}\right)^2$$

where $C_p$ is the compressional velocity of the damaged sample and $C_{po}$ is the compressional velocity of the intact sample. We have fit their data with

$$E = 1200 \times \exp(-7.27 D_0) \text{ [kbar]} \qquad (13)$$

because this provides a better fit than their linear relation, and is more realistic since their result predicts the unphysical result that the effective Young's modulus goes to zero for a finite, achievable $D_0 < 1$. Using these seismic velocities for the Moon (*Toksöz et al.*, 1974; *Khan et al.*, 1999) and equation (13), we have plotted in Fig. 5 the effective Young's modulus as a function of depth within the lunar crust, and compared this modulus to previous estimates of the Young's modulus.

From Fig. 5, we see that the effective Young's modulus in the outer layers might be a factor of order 10–20 lower than previous estimates, allowing for a significant radius change without causing the formation of faults. However, there are at least two reasons that we are not certain whether

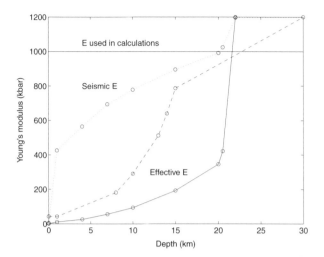

**Fig. 5.** Young's modulus as a function of depth in the lunar lithosphere. The line labeled "E used in calculations" shows the Young's modulus used in previous work in equation (12). "Seismic E" refers to the Young's modulus calculated using the standard equation

$$V_{shear} = \sqrt{\frac{E}{2\rho(1 + \upsilon)}}$$

where the shear wave velocities were taken from *Toksöz et al.* (1974), following the assumption that $\upsilon = 0.25$. *Golombek* (1985) used this method with seismic velocities from the shallow lunar lithosphere (depths of 1–2 km) to calculate elastic moduli 1–2 orders of magnitude lower than those used in previous calculations. The two curves "Effective E" are taken from the experiments of *He and Ahrens* (1994) and equation (13) appropriate for seismic velocities as determined by *Toksöz et al.* (1974) (solid line) and *Khan et al.* (1999) (dashed line). It is important to use the correct value of E (for highly fractured rock and relevant for deformation over geologic as opposed to seismic timescales) in order to properly relate global compression to lithospheric stress and thus determine when faults would form.

such a large strain could be accommodated. Because the outer layers are significantly cracked, the strength of this material is also reduced. *He and Ahrens* (1994) describe that their heavily impact damaged sample is able to be strained a factor of 3× more than the undamaged sample before failing, although a slightly larger value could be allowed for the Moon since the damage parameter of the outer lunar layers is less than any samples they tested. Our second reservation about the values plotted in Fig. 5 is that equation (13) was derived for laboratory samples under terrestrial conditions and high strain rates. Using Apollo measurements of seismic velocities to determine the damage parameter (defined under laboratory conditions) might not be exactly correct, but because it is well accepted that the outer lithosphere is cracked by impacts, the effect of crack closure in reducing Young's modulus is still likely to be important. Another difficulty in extrapolating the experimental results to lunar conditions is the likely presence of volatiles in the laboratory samples, while the real lunar lithosphere is al-

most completely devoid of volatiles. Exactly what this effect will be is not certain, but usually volatiles reduce the strength of a sample, and so the real lunar lithosphere might accommodate more strain before failure. A final difference between the laboratory experiments and lunar conditions is the strain rate. Superimposed upon the low strain rate of global contraction are higher strain rate deformations associated with impacts and adjustments of lateral thermal heterogeneity. It is unknown what effect these transitory events would have upon the stress-strain behavior of the outer lithosphere, but it is plausible that these events would fracture or "shake up" the lithosphere, allowing for some stress adjustment, and thus a larger strain accommodation. In the final analysis, it appears that cracks and pores within the lunar lithosphere can allow a great deal of lithospheric contraction without forming faults, perhaps corresponding to upward of 5 km of radius change, although further investigation is required.

If the outer layers of the Moon have elastic moduli as low as those suggested in Fig. 5, there are at least two observations that must be explained. First, a lithosphere with the lower Young's modulus must be able to account for the extensional and contractional features found around mare basins associated with flexure of the lithosphere. Our (as yet unpublished) calculations indicate that flexural models with the lowest E shown in Fig. 5 is consistent with both the existence and type of faulting observed, within the uncertainty in the model and parameters used. Second, Mercury exhibits signs of global contraction (*Strom et al.,* 1975 — although poorly constrained in magnitude, see *Binder and Lange,* 1980) and its outer layers are believed to have similar mechanical properties to the lunar lithosphere. However, global contraction on Mercury would have involved solidification of at least part of its massive core, and therefore radius change might have been several times larger than on the Moon, because the volumetric change in Fe solidification is much greater than thermal contraction in silicate materials. Thus, the larger radius change (strain) at Mercury could have eventually accumulated to the point where failure would occur even with the low effective moduli in the pervasively fractured outer lithosphere. The details of how this failure would form faults in the surface need to be further explored in order to explain the recent geologic observations that mercurian radius change was <1 km (*Watters et al.,* 1998).

The thermal models that have been used to constrain zero net radius change — both convective (*Cassen et al.,* 1979) and conductive with melting (e.g., *Solomon and Chaiken,* 1976; *Kirk and Stevenson,* 1989) — have relied upon the assumption that the mean lunar temperature has decreased monotonically. If the temperature decrease is not monotonic, the Moon will not cool as fast, and thus the Moon can start with a hotter initial condition and still have a close to zero radius change. A possible thermal history in which this could occur would be if the surface boundary condition of mantle convection changes, for example, during solidification of a magma ocean (the idea for this was suggested by

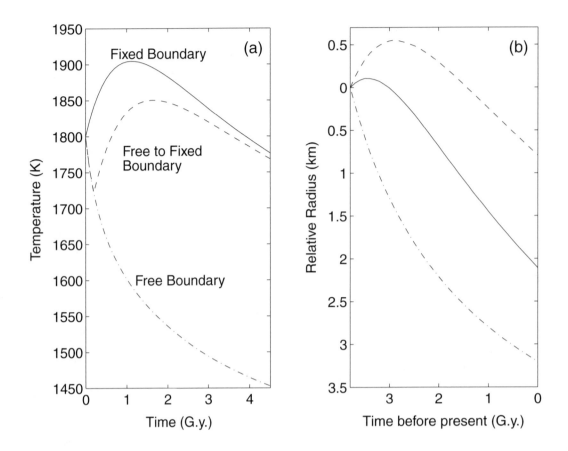

**Fig. 6.** (a) Mantle temperature from parameterized convection of the Moon using different boundary conditions. The solid line shows the mantle temperature from a model with a fixed surface boundary condition, while the dot-dashed line shows the temperature from a model with the free boundary condition. The dashed line shows results from a model with a free boundary condition until 4.4 Ga (approximate solidification time of the magma ocean) and a fixed boundary subsequently, showing how this combination can result in very little temperature change (and thus radius change) during lunar history. **(b)** Estimates of radius change for the three thermal models shown on left. The radius change was calculated by using a thermal model similar to that of *Schubert and Spohn* (1990), which entails placing a conductive lid (where the temperatures were calculated using finite differences) on top of an interior where the temperature and heat flux were calculated using parameterized convection. Radius change is shown for only the last 3.8 G.y., the end of the heavy bombardment, and the initial radius is defined to be zero at that time. The model with the free and fixed surface boundary conditions result in more than 2 km of radius decrease, while the model with the transition from free to fixed boundary condition has much less radius change.

N. Sleep, personal communication, 1998). The idea is that before the magma ocean solidifies, the surface boundary condition is free, but once the ocean solidifies, the boundary condition becomes rigid so that heat is removed less efficiently. (It is also possible to have a thermal history model that has a nonmonotonic temperature decrease with a fixed surface boundary condition for certain initial conditions, as shown in Fig. 6.) We can observe this effect by making a simple parameterized model of heat removal during convection

$$MC_p \frac{\partial T}{\partial t} = H(t) - Q(t) \qquad (14)$$

where the lefthand side is the secular change in mantle heat content, the first term on the righthand side is the radioactive heating, and the second term on the right is the convective heat flux out of the mantle. We adjust the radioactive heat flux with time appropriate for the radioactive decay constants and energies for the four main radioactive isotopes, and concentrations U = 20 ppb, K/U = 2000, Th/U = 3.6 (*Kirk and Stevenson*, 1989). The convective heat flow with the free boundary condition is given by

$$Q = \frac{k\Delta T Nu}{d}$$

where Nu is the Nusselt number and is equal to $0.27 * Ra^{0.319}$

(e.g., *Kiefer and Hager,* 1991), k is the thermal conductivity, and $\Delta T$ is the temperature drop between the mantle and bottom of the magma ocean (defined as 1500 K). With the fixed boundary condition, the heat flux is the same with the Nusselt number defined as (*Grasset and Parmentier,* 1998 — the numbers are slightly different due to different formulations of Ra) $0.33*Ra^{0.294}$, where the appropriate $\Delta T$ is defined as $2.23\,RT_m^2/A$, where A is the activation energy $\sim 300 \times 10^{10}$ erg/mol, and R is the gas constant. We calculated the Rayleigh number, $Ra = g\alpha\rho\Delta Td^3/\eta\kappa$, using g = 163 cm/s², $\alpha = 3 \times 10^{-5}$ 1/K, $\rho = 3.3$ g/cm³, $\eta$ using the equation for diffusion creep from *Karato and Wu* (1993) with a grain size of 0.2 mm, the size of the convecting layer, d, is the lunar radius ($\sim$1740 km), and other values are the same as defined previously. We do not necessarily believe that mantle convection has dominated the release of heat on the Moon (*Pritchard and Stevenson,* 1999), but have done the following calculations to show how changing surface boundary conditions affect the radius change and thermal evolution of a convecting Moon.

The results of our model are shown in Fig. 6. As expected, we see that the mantle temperature from convection with the free boundary condition is much lower than with the fixed boundary condition. When the free boundary condition is changed instantaneously to the fixed condition as the magma ocean crystallizes, we see a brief period of temperature increase such that the mantle temperature at the present is similar to the temperature when the ocean solidified. This model should not be taken too literally; in particular, it omits the large upward migration of heat sources through time. This may be why the models typically predict supersolidus conditions, whereas we believe that the lunar interior is mostly subsolidus today. The model nevertheless serves as an example of how changing boundary conditions can effect how the interior cools. We have included a conductive lithosphere into our model of parameterized convection in analogy to the approach taken by *Schubert and Spohn* (1990) to estimate the radius change since the end of the heavy bombardment for the thermal history models shown in Fig. 6. While the lunar radius contracted more than 2 km in the thermal history models with the fixed and free surface boundaries, the model with the instantaneous switch from free to fixed boundaries has a much smaller radius change. This last model predicts that there should be a slight expansion in the lunar radius since the end of the heavy bombardment, although there is no indication of such global expansion in the geologic record (*Golombek and McGill,* 1983; *Golombek et al.,* 1992; *Golombek and Banerdt,* 1993). It should first be noted that there are many adjustable and unconstrained parameters in the parameterized convection model that can be used to minimize this expansion to match the observations. But, alternatively, the geologic evidence might not be providing a completely accurate picture of global expansion. Perhaps the low cohesion and elastic modulus of the outer layers of the Moon require that global expansion is relieved on multiple small faults that would be difficult to observe instead of a few

larger faults. Such a mechanism has been proposed by *Weisberg and Hager* (1999) for global contraction, and is discussed below. The geologic observation that all grabens on the Moon appear to be older than 3.6 Ga (*Lucchitta and Watkins,* 1978) was used by *Solomon and Head* (1979) as a constraint on the change of the global stress from extension to compression. Unfortunately, the cessation of graben formation at 3.6 Ga is a rather nonunique constraint because local effects (flexure, magmatic pressures, etc.) could mask out the signal from global stresses.

Another example of a thermal history model that has a nonmonotonic decrease in temperature that might effect lunar radius change is given by *Parmentier and Hess* (1998). This model has an ilmenite-cumulate layer that forms in the final stages of the magma ocean enriched in radioactive elements that sinks to the bottom of the lunar mantle where it warms the interior. However, it is unclear at this time how successful this model will be in reducing lunar radius change since this part of the mantle is volumetrically unimportant.

A third means of relaxing the geological constraint on the initial thermal condition has been made by *Weisberg and Hager* (1999). They have used a finite-element model with physical properties appropriate for the outer few kilometers of the lunar lithosphere (low cohesion and elastic moduli). Under these conditions, lunar contraction can be accommodated on numerous small faults instead of fewer large ones that would be observable. While this model is promising, more work is required to ensure that it can explain the formation of compressional and extensional faults associated with mascon loading and explain the occurrence of faults related to global compression on Mercury. The model of Weisberg and Hager and the mechanisms proposed here (especially regarding the low effective E in the outer lithosphere) are independent of each other but are not mutually exclusive. In fact, they might be complementary in reducing the predicted surface expression of global contraction — more compression can be accommodated before failure in a lithosphere with low effective E (as suggested here), and once failure occurs, the surface expression is subdued because of the low cohesion in the upper lunar lithosphere (*Weisberg and Hager,* 1999).

A final alternative is that lunar radius contraction might be accommodated by folding of the lunar lithosphere instead of faulting. The idea here is that under certain circumstances, the energetically preferred mode of deformation could be buckling of the lithosphere instead of straining it. However, such a situation does not appear to exist on the Moon, since our analysis reveals that the power in the topographic spectrum [as determined by *Clementine* (*Lemoine et al.,* 1997)] at appropriate wavelengths is insufficient by at least an order of magnitude to account for significant global contraction from folding.

In summary, we conclude that there is no strong geochemical or geological evidence regarding the exact initial thermal state of the Moon. It was previously believed that the geologic constraint placed a strict limit on the initial temperature of the lunar interior. A consideration of the mechanical properties of the lunar lithosphere relevant for deformation over geologic as opposed to seismic timescales and alternative thermal history models indicates that a hotter initial condition is allowed, but it is unclear whether the Moon could be initially molten. Although neither line of evidence favors the existence of a very hot lunar interior, because of the uncertainties involved in these calculations we should be hesitant to use these constraints to rule out the giant impact hypothesis for the origin of the Moon.

## 7.  OUTSTANDING PROBLEMS

Giant impact scenarios imply a hot initial Moon. The geologic evidence is most easily understood if we assume that the deep interior of the initial Moon is not at or above the solidus. Although this suggests a possible conflict between the Moon we know and the Moon that would arise from a giant impact, we have seen that there are sufficient uncertainties in our understanding of the formation, subsequent history, and current state to preclude firm conclusions. A clearer picture would emerge if we could answer the two central questions:

1. *How much time elapsed between the giant impact and the "planetesimal phase" of lunar accretion? What happens during this period?*  A period of order 100 yr may need to elapse before the planetesimal simulations have relevance, since this is the likely delay before solid bodies can form. During this period, the spatial distribution of the disk of partially molten material will have changed in some as yet unknown ways. Chemical changes may also occur during this period. The amount of cooling before lunar accretion has important consequences for the thermal state of the deep interior of the initial Moon.

2. *Is there a "memory" of lunar initial thermal state in the lunar rock record and geological/geophysical properties (beyond that implicit in the standard lunar magma ocean scenario)?*  The absence of contraction scarps has long been invoked as a memory of thermal history, but this has always been suspect because it seemed to require special pleading (a balance of cooling the outer region and heating the deeper region). As our discussion shows, it is difficult to state the implications of this absence with quantitative precision. It seems likely that the argument is merely insufficiently precise, at least for the Moon, to infer a useful constraint. Geochemical arguments are also suspect although more work needs to be done on the expected extent of mixing within a convective or differentiating Moon.

*Acknowledgments.*  We thank S. Solomon and Y. Abe for critical reviews and acknowledge N. Sleep, J. Melosh, O. Weisberg, W. Kiefer, A. Freed, J. Eiler, S. Solomatov, M. Golombek, and R. Canup for useful discussions. M.E.P. was partly supported from the NASA Planetary Geology and Geophysics Program.

## REFERENCES

Abe Y. (1993) Physical state of very early Earth. *Lithos, 30,* 223–235.

Abe Y., Zahnle K. J., and Hashimoto A. (1999) Elemental fractionation during rapid accretion of the Moon triggered by a giant impact. *Geochim. Cosmochim. Acta,* submitted.

Adams F. C. and Lin D. N. C. (1993) Transport processes and the evolution of disks. In *Protostars and Planets III* (E. H. Levy and J. I. Lunine, eds.), pp. 721–748. Univ. of Arizona, Tucson.

Balbus S. A. and Hawley J. F. (1998) Instability, turbulence, and enhanced transport in accretion disks. *Rev. Modern Phys., 70,* 1–53.

Binder A. B. (1982) Post-Imbrium global tectonism: Evidence for an initially totally molten moon. *Moon and Planets, 26,* 117–133.

Binder A. B. (1986) The initial thermal state of the Moon. In *Origin of the Moon* (W. K. Hartmann, R. J. Phillips, and G. J. Taylor, eds.), pp. 425–433. Lunar and Planetary Institute, Houston.

Binder A. and Gunga H.-C. (1985) Young thrust-fault scarps in the highlands: Evidence for an initially totally molten Moon. *Icarus, 63,* 421–441.

Binder A. B. and Lange M. A. (1980) On the thermal history of a moon of fission origin. *The Moon, 17,* 29–45.

Binder A. B. and Oberst J. (1985) High stress shallow moonquakes: Evidence for an initially totally molten moon. *Earth Planet. Sci. Lett., 74,* 149–154.

Brett R. (1977) The case against early melting of the bulk of the Moon. *Geochim. Cosmochim. Acta, 41,* 443–445.

Cameron A. G. W. (2000) Higher-resolution simulations of the giant impact. In *Origin of the Earth and Moon* (R. M. Canup and K. Righter, eds.), this volume. Univ. of Arizona, Tucson.

Cameron A. G. W. and Canup R. M. (1998) The giant impact and the formation of the Moon. In *Origin of the Earth and Moon,* p. 3. LPI Contribution No. 957, Lunar and Planetary Institute, Houston.

Canup R. M., Levison H. F., and Stewart G. R. (1999) Evolution of a terrestrial multiple-moon system. *Astron. J., 117,* 603–620.

Cassen P., Reynolds R. T., Graziani F., Summers A., McNellis J., and Blalock J. (1979) Convection and lunar thermal history. *Phys. Earth Planet. Inter., 19,* 183–196.

Davies G. F. (1985) Heat deposition and retention in a solid body growing by impacts. *Icarus, 63,* 45–68.

Golombek M. P. (1985) Fault type predictions from stress distributions on planetary surfaces: Importance of fault initiation depth. *J. Geophys. Res., 90,* 3065–3074.

Golombek M. P. and Banerdt W. B. (1993) Importance of expansion and contraction in the formation of tectonic features on the Moon (abstract). In *Lunar and Planetary Science XXIV,* pp. 545–546. Lunar and Planetary Institute, Houston.

Golombek M. P. and McGill G. E. (1983) Grabens, basin tectonics, and the maximum total expansion of the Moon. *J. Geophys. Res., 88,* 3563–3578.

Golombek M. P., Banerdt W. B., and Franklin B. J. (1992) Limits on the expansion and contraction of the Moon (abstract). In *Lunar and Planetary Science XXIII,* pp. 425–426. Lunar and Planetary Institute, Houston.

Grasset O. and Parmentier E. M. (1998) Thermal convection in a volumetrically heated, infinite Prantl number fluid with strongly temperature-dependent viscosity: Implications for planetary thermal evolution. *J. Geophys. Res., 103,* 18171–18181.

He H. and Ahrens T. J. (1994) Mechanical properties of shock-damaged rocks. *Intl. J. Rock Mech. Min. Sci. Geomech. Abstr., 31,* 525–533.

Hess P. C. and Parmentier E. M. (1995) A model for the thermal and chemical evolution of the Moon's interior: implications for the onset of mare volcanism. *Earth Planet. Sci. Lett., 134*, 501–514.

Humayun M. and Clayton R. N. (1995) Potassium isotope cosmochemistry: Genetic implications of volatile element depletion. *Geochim. Cosmochim. Acta, 59*, 2131–2148.

Ida S., Canup R. M., and Stewart G. R. (1997) Lunar accretion from an impact generated disk. *Nature, 389*, 353–357.

Karato S.-I. and Wu P. (1993) Rheology of the upper mantle: A synthesis. *Science, 260*, 771–778.

Kaula W. M. (1979) Thermal evolution of Earth and Moon growing by planetesimal impacts. *J. Geophys. Res., 84*, 999–1008.

Kenyon S. J. and Luu J. X. (1998) Accretion in the early Kuiper belt I. Coagulation and velocity evolution. *Astron. J., 115*, 2136–2160.

Khan A., Mosegaard K., and Rasmussen L. (1999) A reassessment of the Apollo lunar seismic data and the lunar interior (abstract). In *Lunar and Planetary Science XXX*, Abstract #1259. Lunar and Planetary Institute, Houston (CD-ROM).

Kiefer W. S. and Hager B. H. (1991) Geoid anomalies and dynamic topography from convection in cylindrical geometry: Applications to mantle plumes on Earth and Venus. *Geophys. J. Intl., 108*, 198–214.

Kirk R. L. and Stevenson D. J. (1989) The competition between thermal contraction and differentiation in the stress history of the moon. *J. Geophys. Res., 94*, 12133–12144.

Kokubo E., Canup R. M., and Ida S. (2000) Lunar accretion from an impact-generated disk. In *Origin of the Earth and Moon* (R. M. Canup and K. Righter, eds.), this volume. Univ. of Arizona, Tucson.

Lehner F. K. and Kachanov M. (1995) On the stress-strain relations for cracked elastic materials in compression. In *Mechanics of Jointed and Faulted Rock* (H.-P. Rossmanith, ed.), pp. 49–61. Balkema, Rotterdam.

Lemoine F. G. R., Smith D. E., Zuber M. T., Neumann G. A., and Rowlands D. D. (1997) A 70th degree lunar gravity model (GLGM-2) from *Clementine* and other tracking data. *J. Geophys. Res., 102*, 16339–16359.

Li C., Prikryl R., and Nordlund E. (1998) The stress-strain behavior of rock material related to fracture under compression. *Eng. Geol., 49*, 293–302.

Lucchitta B. K. and Watkins J. A. (1978) Age of graben systems on the Moon. *Proc. Lunar Planet. Sci. Conf. 9th*, pp. 3459–3472.

MacDonald G. J. F. (1960) Stress history of the Moon. *Planet. Space Sci., 2*, 249–255.

McEwen A. S., Lopes-Gautier R., Keszthelyi L., and Kieffer S. W. (1999) Extreme volcanism on Jupiter's moon Io. In *Environmental Effects of Volcanic Eruptions: From Deep Oceans to Deep Space* (T. Gregg and J. Zimbleman, eds.), in press.

Melosh H. J. (1989) Giant impacts and the thermal state of the early Earth. In *Origin of the Earth* (H. E. Newsom and J. H. Jones, eds.), pp. 69–83. Oxford Univ., Oxford.

Morfill G. E., Tscharnuter W., and Volk H. J. (1985) Dynamical and chemical evolution of the protoplanetary nebula. In *Protostars and Planets II* (D. C. Black and M. S. Matthews, eds.), pp. 493–542. Univ. of Arizona, Tucson.

O'Connell R. J. and Budiansky B. (1974) Seismic velocities in dry and saturated cracked solids. *J. Geophys. Res., 79*, 5412–5426.

Parmentier E. M. and Hess P. C. (1998) On the possible role of stratification in the evolution of the Moon (abstract). In *Lunar and Planetary Science XXIX*, Abstract #1182. Lunar and Planetary Institute, Houston (CD-ROM).

Peale S. J. and Cassen P. (1978) Contribution of tidal dissipation to lunar thermal history. *Icarus, 36*, 245–269.

Pepin R. O. (1997) Evolution of earth's noble gases: Consequences of assuming hydrodynamic loss driven by giant impact. *Icarus, 126*, 148–156.

Pierazzo E. and Melosh H. J. (1999) Hydrocode modeling of oblique impacts: The fate of the projectile. *Meteoritics & Planet. Sci.*, submitted.

Pritchard M. E. and Stevenson D. J. (1999) How has the Moon released its internal heat? In *Lunar and Planetary Science XXX*, Abstract #1981. Lunar and Planetary Institute, Houston (CD-ROM).

Ransford G. A. and Kaula W. M. (1980) Heating of the Moon by heterogeneous accretion. *J. Geophys. Res., 85*, 6615–6627.

Ringwood A. E. and Essene E. (1970) Petrogenesis of lunar basalts and the internal constitution of the Moon. *Science, 167*, 607–608.

Sasaki S. and Nakazawa K. (1986) Metal-silicate fractionation in the growing Earth: Energy source for the terrestrial magma ocean. *J. Geophys. Res., 91*, 9231–9238.

Schubert G. and Spohn T. (1990) Thermal history of Mars and the sulfur content of its core. *J. Geophys. Res., 95*, 14095–14104.

Solomatov V. and Stevenson D. J. (1993) Suspension in convective layers and style of differentiation in a terrestrial magma ocean. *J. Geophys. Res., 98*, 5375–5390.

Solomon S. C. (1977) The relationship between crustal tectonics and internal evolution in the Moon and Mercury. *Phys. Earth Planet. Inter., 15*, 135–145.

Solomon S. C. (1986) On the early thermal state of the Moon. In *Origin of the Moon* (W. K. Hartmann, R. J. Phillips, and G. J. Taylor, eds.) pp. 435–452. Lunar and Planetary Institute, Houston.

Solomon S. C. and Chaiken J. (1976) Thermal expansion and thermal stress in the Moon and terrestrial planets: Clues to early thermal history. *Proc. Lunar Sci. Conf. 7th*, pp. 3229–3243.

Solomon S. C. and Head J. W. (1979) Vertical movement in the mare basins: Relation to mare emplacement, basin tectonics, and lunar thermal history. *J. Geophys. Res., 80*, 1667–1682.

Stevenson D. J. (1987) Origin of the Moon: The collision hypothesis. *Annu. Rev. Earth Planet. Sci., 15*, 271–315.

Stevenson D. J. (1989) Fluid dynamics of core formation. In *Origin of the Earth* (H. E. Newsom and J. H. Jones, eds.), pp. 231–249. Oxford Univ., Oxford.

Stevenson D. J. (1990) Chemical heterogeneity and imperfect mixing in the solar nebula. *Astrophys. J., 348*, 730–737.

Stevenson D. J. and Lunine J. I. (1988) Rapid formation of Jupiter by diffusive redistribution of water vapor in the solar nebula. *Icarus, 75*, 146–155.

Strom R. G., Trask N. J., and Guest J. E. (1975) Tectonism and volcanism on Mercury. *J. Geophys. Res., 80*, 2478–2507.

Thompson C. and Stevenson D. J. (1988) Gravitational instability of two-phase disks and the origin of the Moon. *Astrophys. J., 333*, 452–481.

Toksöz M. N., Dainty A. M., Solomon S. C., and Anderson K. R. (1974) Structure of the Moon. *Rev. Geophys. Space Phys., 12*, 539–567.

Tonks W. B. and Melosh H. J. (1993) Magma ocean formation due to giant impacts. *J. Geophys. Res., 98*, 5319–5333.

Touma J. and Wisdom J. (1998) Resonances in the early evolution of the earth-moon system. *Astron. J., 115*, 1653–1663.

Turcotte D. L. and Kellogg L. H. (1986) Implications of isotope data for the origin of the Moon. In *Origin of the Moon* (W. K. Hartmann, R. J. Phillips, and G. J. Taylor, eds.), pp. 311–329. Lunar and Planetary Institute, Houston.

Warren P. H. (1985) The magma ocean concept and lunar evolution. *Annu. Rev. Earth Planet. Sci., 13,* 201–249.

Watters T. R., Robinson M. R., and Cook A. C. (1998) Topography of lobate scarps on Mercury: New constraints on the planet's contraction. *Geology, 26,* 991–994.

Weisberg O. and Hager B. H. (1999) Global lunar contraction with subdued surface topography. *J. Geophys. Res.,* submitted.

Wetherill G. W. (1990) Formation of the Earth. *Annu. Rev. Earth Planet. Sci., 18,* 205–256.

Wieczorek M. A. and Phillips R. J. (1999) The "Procellarum KREEP terrane": Implications for mare volcanism and lunar evolution. *J. Geophys. Res.,* in press.

# Geochemical Constraints on the
# Origin of the Earth and Moon

**John H. Jones**
*NASA Johnson Space Center*


**Herbert Palme**
*Universität zu Köln*

---

As originally conceived, the giant impact theory for the origin of the Moon made several predictions: (1) the Moon's complement of metal resides in Earth's core; (2) similarities in some siderophile-element abundances between the Earth and Moon exist because the Moon inherited them from Earth; (3) low abundances of volatiles in the Moon are the result of thermal processing of terrestrial materials; and (4) Earth passed through a magma ocean stage as a result of the giant impact. The geochemical and geophysical evidence now suggests that (1) the Moon has a small core, so prior removal of siderophiles into Earth's core is not required; (2) the similarity in W abundances between the Earth and Moon is coincidental; (3) alkalis in the Moon were not simply inherited from Earth; and (4) there is evidence from terrestrial mantle lherzolites that Earth may never have had a magma ocean. These observations are sufficient to cast doubt on the giant impact hypothesis.

## 1. INTRODUCTION

A possible relationship between Earth and the Moon, based on the chemical compositions of both planets, was extensively discussed at the Kona conference on the Origin of the Moon in 1984. The chemical results, as presented in several articles in *Origin of the Moon* (1986, Lunar and Planetary Institute, Houston) may be broadly summarized as follows:

1. The lower density of the Moon requires a lower bulk Fe content than that of Earth. However, the higher FeO content of lunar basalts implies that the silicate portion of the Moon has a higher FeO content than that of Earth.

2. Estimates for the refractory element content of the Moon range from 2–4× CI, compared to 2.75× CI for Earth's mantle. Relative abundances of refractory elements in both bodies appear to be chondritic. In addition, the abundances of Cr and V are similar in the mantles of The Earth and Moon, but different from CI chondrites.

3. The abundances of siderophile elements in the lunar mantle decrease with increasing siderophility, i.e., in the sequence Fe, W, Co, Ga, P, Ni, Mo, Re. This is strong circumstantial evidence for the presence of a small lunar core.

4. Moderately volatile elements are depleted in the Moon relative to Earth. Sodium, K, and Zn are all depleted in the Moon. A notable exception is Mn, which is higher in the Moon than in the Earth's mantle.

5. Highly volatile elements (e.g., Cd, Tl) are depleted in the Moon by about a factor of 50, relative to the Earth, with the exception of Cs. The reason for the comparatively high Cs abundance in the Moon is unknown.

6. The most abundant element in the Earth and Moon is O. And, within analytical error, the Earth and Moon are identical in terms of their O isotopic compositions.

Taken as a group, these chemical characteristics were believed most compatible with a giant impact origin for the Moon, whereby a Mars-sized planetesimal struck Earth and ejected protolunar material into Earth orbit. Although some specific points were difficult to reconcile with this collisional ejection hypothesis, this model was favored by most researchers. In part this may have been because the model offered a high degree of flexibility in explaining the composition of the Moon, as reflected in the wide range of estimates for the contribution of the projectile to the bulk Moon. For example, *Ringwood* (1990) believed that terrestrial material dominated the Moon and estimated, on the basis of siderophile elements, that there had been less than ~1% chondritic material added to the Moon. *MacFarlane* (1989), in comparing the compositions of the Earth and Moon to various chondrite groups, concluded that Earth probably contributed no more than 50% of the protolunar material. *Newsom and Taylor* (1989) believed that the Moon's composition was dominated by the impactor. And *O'Neill* (1991) calculated that Earth contributed 70% of the present lunar mass. And, of course, this multiplicity of views concerning the relative contributions of the target and impactor complicates any discussion of lunar origin.

Since the Kona conference in 1984, several new analytical techniques have been developed or refined, and application of these new methods are providing new insights into the question of the lunar origin:

1. The precise determination of the Cr isotopic composition of lunar samples confirmed the isotopic similarity of lunar and terrestrial material, as shown by *Lugmair and*

*Shukolyukov* (1998), who showed that the $^{53}Cr/^{52}Cr$ is variable for inner solar system materials, but identical in the Earth and Moon.

2. The analysis of the W isotopic composition of meteorites, lunar, and terrestrial rocks must be considered as the most important geochemical parameter applied to problems of lunar origin that has been made since the Kona conference. The implications of these new $^{182}W/^{184}W$ data (*Lee et al.,* 1997) will be discussed in a separate section.

3. The isotopic composition of K has been precisely measured, and it was found that terrestrial and lunar samples are indistinguishable in their $^{41}K/^{39}K$ ratios (*Humayun and Clayton,* 1995).

4. In the last 10 years it has become increasingly clear that metal/silicate partition coefficients may strongly depend on pressure and temperature. It has been suggested that the high concentrations of Ni and Co in the mantle of Earth are the result of silicate equilibration with metal at the base of a terrestrial magma ocean, created by the impact of a Mars-sized object during the formation of the Moon (*Li and Agee,* 1996; *Righter et al.,* 1997).

5. The Os isotopic composition of xenoliths of the terrestrial mantle have been measured. Osmium isotopes in mantle samples approach those of chondrites as the bulk compositions of the xenoliths themselves approach estimates of the bulk silicate Earth (*Meisel et al.,* 1996). Therefore, there is a correlation between siderophile elements (Re/Os) and lithophile elements such as Al and REE. The most fertile xenoliths have Os isotopic compositions that are closest to chondritic.

Here we will examine both old and new geochemical observations in order to evaluate the giant impact hypothesis. First we will review the most relevant older data, and then we will discuss several different geochemical topics in more depth.

## 2.  BULK CHEMISTRY AND PHYSICAL PROPERTIES

It is generally agreed that the Moon and Earth differ in substantive ways. The Moon is more reduced than Earth, but the Moon has a lower Mg# [molar Mg/(Mg + Fe) × 100] and higher FeO than Earth's mantle. The Moon's uncompressed density is lower, and its moment of inertia is significantly higher. We shall discuss these noncontroversial aspects of the Earth and Moon before turning to those properties that are less agreed upon. Often we will compare chemical abundances in the Earth and Moon to those of chondrites, and in these discussions we will use the data compiled by *Newsom* (1995). Unless stated otherwise, we will not normalize elemental abundances to elements such as Mg or Si. Thus, normalizations to CI chondrites have not been corrected for volatile loss.

We will often refer to the bulk silicate Earth (BSE), distinguishing between the bulk Earth (including the core) and its silicate portion. Strictly speaking, this composition is well known only for Earth's upper mantle. However, we know of no substantive evidence that the lower mantle has a different composition than the upper mantle. And current models of mantle convection favor good communication between the upper and lower mantle (e.g., *Davies,* 1988). Hereafter, we assume that the bulk compositions of the upper and lower mantle are indistinguishable.

### 2.1.  Oxygen Isotopes

The O isotopic compositions of different meteorite groups are usually different. Thus, O isotopes have become the "industry standard" of the meteoritics community for determining the provenance of rocky materials within the solar system (e.g., *Clayton,* 1993), even if the geographic location of that provenance is unknown. The Earth and Moon have essentially identical O isotopic signatures, indicating that they were derived from isotopically similar materials. We note, however, that enstatite chondrites, the most reduced solar system materials, and CI chondrites, the most oxidized meteorites, fall on or near the same fractionation line as the Earth and Moon (e.g., *Clayton,* 1993). Thus, it is unclear whether the O isotopic signature of the Earth and Moon is generally indicative of all rocky materials within 1 AU of the Sun or whether it is a local property of the Earth-Moon system. For this reason alone, it would be important to know the O isotopic compositions of Venus and Mercury. Unfortunately, the prospects for such measurements in the foreseeable future are poor.

### 2.2.  Redox State

*2.2.1. Earth.* The overall redox state of the Earth is not known precisely, but this particular aspect of the Earth is not a very useful parameter. This is because different portions of the Earth record different $f_{O_2}$, and thus appear to be in disequilibrium. Earth's surface and atmosphere are highly oxidizing, as indicated by the presence of water and oceans. Earth's mantle is more reduced, having an $f_{O_2}$ near that of the QFM (quartz-fayalite-magnetite) oxygen buffer (e.g., *Ringwood,* 1979). And *Christie et al.* (1986) have argued that, in detail, Earth's upper mantle is somewhat more reducing than QFM and that QFM-1 (one $\log_{10}$ unit below QFM) is a better approximation. In detail, a fairly wide variation in inferred $f_{O_2}$ can be found within the mantle-derived sample suite, but we prefer to use estimates based on basalts, such as mid-ocean ridge basalts (MORB), that sample large mantle volumes.

The Earth's core, of course, must be considerably below QFM and even below IW (iron-wüstite). In general Fe metal is only stable at conditions below IW, since the activity of FeO in bodies of chondritic composition is below unity. However, this observation is tempered by the presence of Ni, which can persist in the metallic form to redox states near or exceeding QFM. Thus, at redox states of IW-1 or below, Fe-rich metal is possible, but at redox states above IW-1 metal becomes increasingly Ni-rich.

It is of interest, however, to calculate what $f_{O_2}$ would be necessary to account for the current depletion in FeO from chondritic, assuming that the mantle and core were once in equilibrium. Assuming ideality and metal/silicate equilibration at low pressure, this calculation implies that the core formed from the mantle at around IW-2.5 (*Jones and Drake,* 1986; *Holzheid and Palme,* 1996). This is quite reducing and, if the core and current mantle were never in equilibrium, could be a conservative upper limit to the true $f_{O_2}$ that pertained during core formation. This latter possibility has been explicitly incorporated into the bulk compositional models of *Wänke* (1981), who postulated that, during core formation, there was very little FeO in the silicate portion of Earth. In the Wänke model, nearly all FeO is accreted and added to the mantle after the main epoch of core formation is over. Evidence for this scenario comes from the high abundances of siderophile elements such as W and Mo in Earth's mantle (*Wänke,* 1981). We will discuss this model and alternative models of core formation in more detail below.

To summarize, the Earth's different physical domains have different redox states and are apparently in disequilibrium. These domains appear to have maintained approximately the same redox conditions over most of Earth's history. The oldest sediments (about 3.9 Ga) yield evidence of liquid water on Earth's surface, and the oldest basalts imply a redox state near QFM [3.8 Ga (*Delano,* 1993); see also *Canil* (1999)]. These Archean basalts were already low in FeO compared to lunar mare basalts, implying that core formation was essentially complete by 4 Ga. Some have argued (in order to explain "future" Pb isotopic compositions) that core formation has continued over Earth's entire history (e.g., *Oversby and Ringwood,* 1971). However, there is no actual evidence that core formation has occurred since ~4 Ga (*Newsom et al.,* 1986). Conversely, it has been suggested that some plume basalts may contain trace amounts of core materials entrained at the core-mantle boundary (*Walker et al.,* 1995). Still, the dominant situation appears to be disequilibrium between reservoirs that have communicated poorly since the time of Earth's formation.

*2.2.2. Moon.* Comparatively, the redox state of the Moon appears quite simple. Lunar basalts are very nearly saturated in Fe metal. Iron-nickel metal is a common late-stage crystallization product of lunar basalts, and lunar basaltic liquids do not gain or lose significant FeO when they are contained in high-purity Fe metal capsules (*Walker et al.,* 1977). In addition, lunar basalts and eucrites have similar FeO concentrations. Taken together, these observations imply that the lunar $f_{O_2}$ is similar to the eucrite parent body, which has been experimentally determined to be IW-1 (*Stolper,* 1977).

Another point of view can also be taken. Since the lunar mantle contains nearly 500 ppm Ni and correspondingly high Co contents, *Seifert et al.* (1988) concluded that the lunar core is Ni-rich, indicating a slightly higher $f_{O_2}$ than defined by pure FeO-Fe equilibrium (~IW + 0.5). These two sets of observations and calculations probably

bracket the true redox state of the Moon. And we, the authors, do not agree as to which is most likely correct. However, the differences in these estimates are not large, making it likely that the redox state of the Moon is between IW and IW-1.

### 2.3. Mg#, CaO/Al₂O₃, and Total Iron

*2.3.1. Earth.* The Mg# of the Earth is well constrained from mantle xenoliths to be about 89 (*O'Neill and Palme,* 1998; *BVSP,* 1981; *Taylor,* 1982). And the FeO content used above to calculate the $f_{O_2}$ of core-mantle equilibration (~IW-2.5) was ~8 wt%. If the Earth's core contains ~85 wt% Fe [assuming ~10 wt% of a light element (e.g., *Jeanloz,* 1990) and ~5 wt% Ni], Earth's total Fe is calculated to be ~31 wt%. This translates into a chondrite-normalized concentration of ~1.7× CI, compared to 1.55× CI for Mg, assuming a Mg-free core. This calculation assumes that the bulk Earth should have an approximately chondritic Fe/Ni ratio.

It is assumed that the Ca/Al ratio of Earth is chondritic (e.g., *Palme and Nickel,* 1985). However, Ca/Al in primitive spinel lherzolites from the mantle are often super-chondritic (*Palme and Nickel,* 1985). This is probably because these rocks have lost small amounts of low-degree (<5%) partial melts, which typically have low Ca/Al ratios (*Baker et al.,* 1995). The best estimates for CaO and Al₂O₃ in the BSE are 3.8 and 4.7 wt% respectively (*O'Neill and Palme,* 1998).

*2.3.2. Moon.* The Mg# of the Moon is not well known. Because we do not have lunar materials that escaped the differentiation of the lunar magma ocean, the composition of the bulk Moon must be reconstructed, and this is difficult. Still, most models of lunar composition converge on an Mg# of 80–85 (e.g., *Taylor,* 1982; *Ringwood et al.,* 1987; *Jones and Delano,* 1989). This translates into a bulk lunar FeO content of about 13 wt%, significantly higher than that of Earth's mantle. *Jones and Hood* (1990) and *Mueller et al.* (1988) found that this amount of FeO was necessary to simultaneously reconcile the lunar geophysical properties of density and moment of inertia. In models that postulated that the Moon had the same chemical composition as the Earth's upper mantle, the Moon was not dense enough. Another means of increasing the Moon's density (i.e., increasing the size of a hypothetical iron metal core; see below) causes the moment of inertia to be too small. Consequently, it has become generally accepted that the Moon has 1.5–1.6× more FeO than Earth's mantle. However, the FeO/MgO of the lunar mantle is unlikely to be uniform. The existence of the lunar Mg suite, which contains lithologies with Mg# ≥ 90, argues that there must be a lunar reservoir having an Mg# similar to that of Earth (e.g., *Warren,* 1986). Therefore, it seems to us that deriving total FeO (and Mg#) from bulk geophysical properties is the least model-dependent approach.

Similarly, the total Fe content of the Moon is also uncertain. However, a maximum Fe content may be calculated by taking 13 wt% FeO in the silicate Moon and assuming a pure Fe core that constitutes 5 wt% of the Moon. This translates into a total Fe content of 14.6 wt% or ~0.8× CI, at least a factor of 2 lower than that for the Earth. If the lunar core is Ni rich (~45% Ni; *Seifert et al.,* 1988), then cores larger than 1 wt% result in superchondritic Ni/Fe ratios in the Moon. Regardless, it seems inescapable that the bulk Fe (bulk FeO) contents of the Earth and Moon are different.

The Ca/Al ratio of the Moon is also presumed to be chondritic. *Jones and Delano* (1989) calculated a wide variety of models for the Moon's bulk composition and found that bulk CaO and $Al_2O_3$ concentrations were fairly insensitive to the exact model. Consequently, we adopt the *Jones and Delano* (1989) values of lunar CaO and $Al_2O_3$ as 3.0 and 3.7 wt% respectively.

## 2.4.  Geophysical Properties

*2.4.1.  Earth.*    The moment of inertia of the Earth about its polar axis, $C/(MR^2)$, is 0.3305, which should be compared to the value of 0.4 expected for a homogeneous sphere (*Verhoogen et al.,* 1970). This smaller value is because of the mass concentrated in the Earth's core. Similarly, the calculated average density of the Earth (5.517 g/cm³), which is too high for rock alone, indicates that Earth must contain a dense metallic component. These inferences are confirmed by seismographic observations that also indicate that there is a large metallic core at the center of Earth (e.g., *Macelwane,* 1951).

*2.4.2.  Moon.*    In contrast to the Earth, the Moon's moment of inertia of 0.3905 is very close to what would be expected for a homogeneous sphere (*Hood,* 1986; *Hood and Jones,* 1987). Further, the Moon's average density of 3.344 is consistent with that expected for ultramafic rocks whose mineralogy is dominated by olivine with an Mg# of ~90.

Therefore, based on these observations, it is reasonable to expect that the Moon either has no core or has only a very small core. Initially, it was believed that there was no core present in the Moon; and this observation was the driver behind models that postulated that the Moon had been substantially derived from the silicate portion of Earth (e.g., *Ringwood,* 1979). Effectively, in this scenario, the Moon's complement of metal now resides in the Earth's core.

However, further investigation indicates that the Moon probably does have a small core ~200–500 km in radius (*Hood and Jones,* 1987; *Mueller et al.,* 1988; *Jones and Hood,* 1990). This calculation is sensitive to many assumptions, particularly the bulk silicate composition, the assumed differentiation history of the Moon, and the current thermal state of the lunar interior. However, permuting these various parameters results in a series of models, most of which require a small metallic core. And as long as this core is 300–450 km, its size would also be consistent with electromagnetic sounding data that place size limits on the lunar core (*Hood et al.,* 1999).

In summary, we reiterate that a small lunar core is consistent with measurements of lunar density and moment of inertia and is actually helpful in modeling these parameters in a self-consistent way (*Hood and Jones,* 1987). There are also good geochemical reasons for invoking a small lunar core that we will discuss in more detail below.

## 2.5.  Summary

The Earth and Moon have similar O-isotopic compositions and therefore are probably closely related in terms of the materials that comprise them. However, there are also significant chemical differences, particularly in Mg# and redox state. If the Moon formed by a giant impact upon Earth, there must have been significant amounts of impactor material incorporated into the Moon to account for these chemical differences. In an extreme version of this model, the composition of the Moon could be dominated by the composition of the impactor (e.g., *McFarlane,* 1989). In the latter case, any chemical similarities between the Earth and Moon would be largely coincidental.

## 3.  VANADIUM, CHROMIUM, AND MANGANESE

There are indeed apparent chemical similarities between the Earth and Moon. The elements V, Cr, and Mn were considered crucial by *Wänke and Dreibus* (1986) in this regard. All three elements are depleted in the Earth's mantle relative to chondrites. And *Wänke and Dreibus* (1986) assumed that V, Cr, and Mn all partitioned into Earth's core early (prior to the formation of the Moon) under very reducing conditions. Subsequently, small amounts of oxidized material were added to Earth's mantle, so that it acquired its current redox state. Thus, *Wänke and Dreibus* (1986) considered the depletions of Cr, Mn, and V to be a uniquely terrestrial signature. Analyses of meteorites considered to be of martian origin (the so called "SNC" meteorites) and samples from the eucrite parent body (hereafter, EPB; probably asteroid 4 Vesta) indicated an absence of this kind of fractionation.

*Ringwood et al.* (1991) bolstered this view by providing experimental evidence that V, Cr, and Mn would not partition into the core of a Mars-sized planet. Therefore, these authors postulated that the depletions of V, Cr, and Mn in the Moon could not be due to depletions inherited from the impactor (in a giant impact scenario) but must reflect their depletion in Earth during core formation at very high pressures (~1000 kbar) and very reducing conditions. The abundances of V, Cr, and Mn in the Moon may therefore provide clues to its origin.

But there is little evidence that Mn partitions into Fe-Ni metal, even at reducing conditions (*Rammensee et al.,* 1983). *O'Neill and Palme* (1998) noted that the ratio of the two similarly volatile elements, Mn and Na, is chondritic in the Earth's mantle. Additional loss of Mn into the core is therefore cosmochemically unlikely. The low concentrations of Mn and Na in the Earth are the result of a general depletion in moderately volatile elements. Such depletions are not unusual. Carbonaceous chondrites are increasingly depleted in Mn and Na going from type 1 to type 3, reflecting nebular conditions during the formation of solid matter. Manganese depletion in the Earth is almost certainly not the result of an indigenous terrestrial process.

The situation is different for V and Cr. The depletion of these two elements in the Earth cannot be ascribed to volatility-related depletions. Chromium is similar in volatility to Mg; and V is nominally considered even less volatile. But, although V is generally considered a refractory element, it is less enriched in Ca,Al-rich inclusions from the Allende meteorite than other more refractory elements such as Al and REE (*Wänke et al.,* 1974a). This leads to a bulk depletion of V in CV chondrites, such that the Al/V ratio in CV chondrites is 182 (*Wasson and Kallemeyn,* 1988), whereas the CI ratio is 159 (*Palme and Beer,* 1993). On the other hand, the BSE has an Al/V ratio of 291 (*O'Neill and Palme,* 1998), a value clearly larger than observed in bulk meteorites. This would argue against volatility as the cause of the terrestrial V depletion. But *Shaffer et al.* (1991) showed that the volatilization rate of V was sensitive to $f_{O_2}$ and therefore could be a function of the local dust/gas ratio during condensation. Alternatively, some V may reside in Earth's core, as suggested by the experiments of *Rammensee et al.* (1983) and *Ringwood et al.* (1991). Consequently, interpretation of the Earth's V depletion is not straightforward.

The most model-independent element of the three appears to be Cr. Chromium is the most fractionated, as seen in the Mg/Cr ratio, which is 87 in the mantle of Earth and 36 in CI chondrites. Chromium in the Earth's mantle may indeed be possibly ascribed to early fractionation into the Earth's core, as suggested by several authors. Consequently, we will concentrate the remainder of our discussion on Cr.

First, it is significant that the Cr isotopic compositions of the Earth and Moon are the same (*Lugmair and Shukolyukov,* 1998). As in the case of O, the Cr isotopic signature of the Earth and Moon indicates that both bodies formed from similar materials. And in the context of a giant impact, it appears that large contributions (more than approximately two-thirds) from an isotopically dissimilar impactor are not allowed (*Lugmair and Shukolyukov,* 1998). Of course, this estimate depends in detail on the exact magnitude of the impactor's Cr isotopic anomaly.

Chromium is a compatible element and that makes estimating its abundance in the bulk Moon problematic. A lunar Cr abundance of 2200 ppm (*Ringwood et al.,* 1987) was primarily derived from the Cr/Fe correlation among mare basalts (*Seifert and Ringwood,* 1987; *O'Neill,* 1991). However, mare basalts may be derived from rather differentiated source regions, as demonstrated by their ubiquitous Eu anomalies (i.e., denoting plagioclase fractionation). Thus, the Moon may have a large reservoir of cumulate opx and olivine that has not been sampled. In calculating the lunar Cr content, the amount of Cr in this reservoir is very important but not well known.

We have recalculated the Cr content of the Moon in two different ways that complement earlier calculations. The first makes use of the good Cr vs. V correlation in lunar samples

(*Seifert and Ringwood,* 1987; *Haskin and Warren,* 1991) and the assumption that V is present at 2.2× CI in the Moon (124 ppm), along with other refractory elements. This calculation yields 2500 ppm Cr in the bulk silicate Moon and a lunar Cr/V ratio of 20. If V is truly depleted relative to other refractory lithophile elements, then bulk lunar Cr will be correspondingly lower. For comparison the BSE Cr/V ratio is 30 and the Cr content of the BSE is 2540 ppm (e.g., *O'Neill and Palme,* 1998).

A second method is to assume that the Apollo 15 green glass (an olivine-normative composition with a fairly flat REE pattern and a small, negative Eu anomaly) was only in equilibrium with olivine at the time of its eruption and use an appropriate olivine/liquid partition coefficient for Cr to arrive at the Cr content of the lunar mantle. In this calculation we assume that Cr in the crust is negligible and that green glass A (*Delano,* 1986a) was produced by 20% partial melting. For a melt of Apollo 15 green A composition (*Delano,* 1986a) and a Cr speciation dominated by $Cr^{2+}$, the $^{ol/liq}D_{Cr}$ is ~0.5 (*Hanson and Jones,* 1998). This translates into a Cr content of 2000 ppm for the bulk silicate Moon.

These calculations are clearly oversimplified. But several different approaches, with very different assumptions, converge on Cr abundances of 2000–2500 ppm; and the 2500 ppm value may be an upper limit if V is actually depleted compared to other refractory elements. Here, we will adopt the *Ringwood et al.* (1987) value of 2200 ppm that is intermediate to our two calculations.

(As an aside, we return to the possibility of a "hidden" reservoir that is enriched in Cr and Mg and that is possibly the source of the lunar highland Mg suite. However, our adopted value for bulk lunar Cr of 2200 ppm is not very different from the Cr content of olivine-rich Mg-suite rocks. For example, the dunite 72415 has a Cr content of 2300–2500 ppm. Consequently, it is not obvious to us that ignoring the Mg-suite reservoir in our calculation is a serious problem.)

For a bulk lunar MgO content of 37% and our adopted Cr concentration, the Mg/Cr ratio of the Moon is ~100 and similar to that of Earth's mantle (~86). Within the error of the calculation, the Mg/Cr ratios of the mantles of the Earth and Moon are the same. Thus, although Mg/Cr shows little variation in chondrites, Cr appears to be significantly and similarly depleted in both the Earth and Moon. This may be the best evidence for a terrestrial origin of the Moon. However, as we shall see below, at least one other element, W, has similar depletions in the Earth and Moon, but that this is a coincidence.

## 4. ALKALI METALS AND THEIR ISOTOPES

### 4.1. Alkali Metals

The alkalis in the Moon are depleted relative to Earth, although the exact amounts of these depletions are subject to controversy.

*4.1.1. Earth.* *Dreibus et al.* (1976) presented Li-Zr systematics for terrestrial basalts, indicating that the Li/Zr ratio of the Earth is ~0.1. If refractory lithophiles such as Zr exist at 2.75× CI in the Earth (*Jagoutz et al.,* 1979), the Li concentration is 1 ppm. Both *Taylor and McLennan* (1985) and *Ringwood* (1991) give a Na concentration of ~2500 ppm for the bulk silicate Earth (BSE). The K content of the Earth is constrained from both $^{40}$Ar systematics and from correlations with nonvolatile, incompatible elements from igneous rocks. *McDonough et al.* (1992) calculate a K content for the BSE of 240 ± 30 ppm. This number is based on a terrestrial K/U ratio of $1.3 \times 10^4$ and a BSE U concentration of 21 ppb. In addition, from correlations of K and Rb, a BSE Rb concentration of 0.64 ppm can be derived. *Jones and Drake* (1993), while accepting the *McDonough et al.* (1992) calculation of K and Rb, were not convinced that the BSE Rb/Cs of 28 estimated by these authors was correct. This was because *McDonough et al.* (1992) chose to invoke an unsampled mantle reservoir possessing chemical properties that they found convenient. *Jones and Drake* (1993) used mass balance arguments and the Rb/Cs ratios of continental crust and oceanic mantle to calculate a BSE Rb/Cs ratio of 40, which translates into a BSE Cs concentration of 16 ppb. These calculated abundances for Li, Na, K, Rb, and Cs correspond to depletions relative to CI chondrites of 0.72, 0.50, 0.43, 0.27, and 0.086 respectively. Thus, the alkalis are somewhat depleted in the Earth, and these depletions appear to increase with increasing volatility (*Kreutzberger et al.,* 1986).

*4.1.2. Moon.* The *Dreibus et al.* (1976) Li-Zr correlations for lunar rocks translate into a lunar Li concentration of 0.35 ppm, assuming refractory lithophile elements in the Moon exist at ~2.2× CI (*Ringwood et al.,* 1987; *Jones and Delano,* 1989). Similarly, a Rb concentration of 0.086 ppm and a K concentration of 36 ppm for the Moon are derived from the lunar Ba/Rb and K/Ba ratios of 60 and 7 respectively (*BVSP,* 1981; *Wänke,* 1981). The Rb/Cs ratio of 21 for mare basalts and KREEP yields a lunar Cs concentration of 4.1 ppb. And taking Taylor's Na/K ratio of 7.2 for the bulk Moon (*Taylor,* 1980) translates into a bulk Na concentration 260 ppm. This calculation for Na is difficult since Na may have acted somewhat compatibly during early lunar petrogenesis. For example, anorthositic highland samples may contain nearly as much Na as KREEP. And using the *Taylor* (1980) Na value in this calculation results in a Na/K ratio that is slightly subchondritic. Therefore, we prefer to use the Na estimate of *Ringwood et al.* (1987) of 450 ppm and will do so in subsequent discussions. These calculated abundances for Li, Na, K, Rb, and Cs correspond to depletions relative to CI chondrites of 0.24, 0.090, 0.066, 0.038, and 0.021 respectively.

*4.1.3. Discussion.* Earth. Figure 1 shows terrestrial alkali depletions plotted vs. the square root of their atomic masses. This linear correlation is considerably better than that based on 50% condensation temperatures (e.g., *Wasson,* 1985) or the boiling temperatures for the pure elements and is suggestive of fractionation by a diffusive transport process or Jeans escape (i.e., fractionations that depend on

$M^{-0.5}$). Conceptually, this idea has attractive aspects. If depletions in "volatile" elements were actually due to a transport process, as opposed to simple volatility, a long-standing contradiction might be explained: If a moderately volatile element is modestly depleted, a highly volatile element ought not to have condensed at all. In fact, what is observed in chondrites is that the highly volatile elements are more depleted than moderately volatile elements, but not to the extent thermodynamics would predict.

However, if the systematics of Fig. 1 are due to a nebular process, then alkali abundances of chondrites should have the same functional form. In fact, this is not observed. Although there are aspects of the alkali patterns for Earth and for carbonaceous chondrites that are similar, imagination would be required to equate them. The strongest similarity, in terms of the slope of the depletion trend, is between the Earth and one of the experiments of *Kreutzberger et al.* (1986), where a suite of alkalis were volatilized from a Di$_{75}$-An$_{25}$ silicate liquid.

The *Kreutzberger et al.* (1986) experiment cannot be indicative of a transport process *per se.* The charges themselves are homogeneous with respect to the alkali elements (M. J. Drake, personal communication). The $M^{0.5}$ parameterization therefore reflects some process at the liquid-gas interface (or, possibly, it is merely a convenient parameterization). Consequently, it is tempting to interpret the terrestrial alkali pattern as due to partial volatilization of dust or

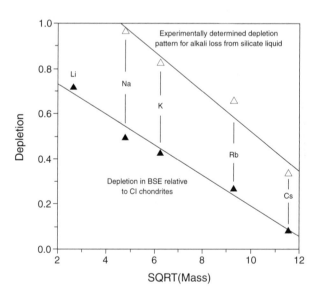

**Fig. 1.** Alkali depletions in the Earth vs. the square root of their atomic weight (solid symbols). It is generally believed that alkali depletions in the Earth are due to volatility. However, condensation temperatures are not available for all alkalis. Because of this, we have used the square root of atomic weight as a parameterization of alkali depletions. Shown for comparison are depletions produced in a 1-bar volatility experiment (open symbols; *Kreutzberger et al.,* 1986). Both patterns have similar slopes and could possibly reflect the operation of similar processes; but the exact interpretation of the terrestrial depletion pattern is still uncertain (see text).

early nebular condensates. But this hypothesis, too, has its difficulties. The data of *Humayun and Clayton* (1995) constrain the depletion process to be one that does not fractionate K isotopes (see below); for most volatile-loss scenarios, this is difficult (although not impossible).

Alternatively, there could be a fractionation that is indigenous to Earth that would produce the observed relationship between abundance and mass. Metallization of K, Rb, and Cs has been suggested as a possible mechanism for incorporating these elements into the core (e.g., *Bukowinski*, 1976). However, in this case, the inclusion of Na and Li by the parameterization would have to be accidental.

We currently have no unambiguous explanation for the systematics of Fig. 1. We tentatively prefer some partial volatilization scenario. Conceivably, there are nebular processes that could totally volatilize all the K from some dust grains but leave other grains largely unaffected. Alternatively, if the devolatilized dust and gas were to maintain chemical equilibrium, isotopic fractionation might be avoided (*Humayun and Clayton*, 1995). But in this case, we are again faced with the thermodynamic problem of how a suite of elements with rather different condensation temperatures can be depleted to similar degrees.

Moon. Figure 2 shows the lunar alkali pattern normalized both to the BSE (top) and to H chondrites (bottom). Lunar alkalis are depleted and fractionated relative to those of Earth. However, because Cs is enriched compared to what would be expected on the basis of volatility alone (*Kreutzberger et al.*, 1986; *Wulf et al.*, 1995), other factors must

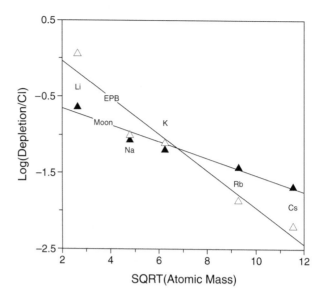

**Fig. 3.** Log depletion vs. square root of atomic weight for alkalis in the Moon (solid symbols) and the eucrite parent body (open symbols). Note the difference in functional form from Fig. 1. While alkalis in the Moon and EPB are not identical, they are broadly similar. Because we have no reason to believe that the EPB suffered a giant impact, we question whether the Moon did either.

have played a role in establishing lunar alkali abundances. *O'Neill* (1991) suggested that addition of 4% H-chondrite material, in addition to volatile loss from terrestrial parental materials, could explain the observed lunar alkali abundance pattern. Figure 2 shows that, indeed, the lunar Rb and Cs abundances are in H-chondritic relative proportions (i.e., flat pattern). And, if this H-chondrite model is correct, approximately two-thirds of the lunar K budget came from that source as well. But, again, lunar alkalis cannot be directly related to terrestrial alkalis in any simple way.

Figure 3 compares the alkali depletions for the Moon and the EPB. The EPB relative depletions were calculated by taking the alkali/Ba analyses of *Tera et al.* (1970) and assuming that all these elements are incompatible during silicate partial melting. Absolute depletions were calculated assuming that Ba in the EPB is 2× CI. Note the change of scale from Fig. 1, which is linear, whereas the ordinate of Fig. 3 is a log scale. The square root of atomic weight parameterization is not as good for these bodies as for Earth, but the main point of Fig. 3 is that the alkali depletions in the Moon and EPB are more similar than they are different. Currently we have no reason to suspect that an event as energetic as the hypothetical giant impact ever affected the EPB. Consequently, the similarity of elemental depletion patterns between the Moon and the EPB does not seem to require a giant impact origin for the Moon.

Finally, we note that some of the experimental alkali depletions produced by *Kreutzberger et al.* (1986) appear similar to the Moon and the EPB. For short experiments, like that shown in Fig. 1, depletion is best plotted vs. $M^{0.5}$. For long experiments with greater alkali loss, the best pa-

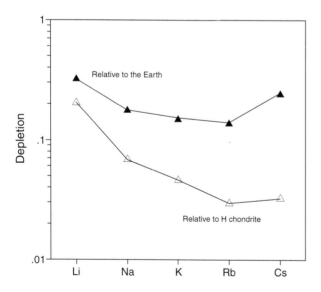

**Fig. 2.** Lunar alkali abundances normalized to those of the Earth and to an H chondrite. The Rb/Cs ratios of the Earth and Moon make derivation of the Moon from the Earth problematic. If the Moon were derived from the Earth, but depleted in alkalis by a high-temperature event, the most volatile element, Cs, should be the most depleted and this is not observed. But as noted by *O'Neill* (1991), lunar Rb and Cs abundances may be best explained by minor addition of chondritic material to the proto-Moon. The lunar Rb/Cs ratio is very similar to that of an H chondrite.

rameterization is log(depletion) vs. $M^{0.5}$. The shorter experiments appear "Earth-like," whereas the longer experiments appear "EPB-like." Thus, it is possible that a single loss mechanism, operating to differing degrees, can explain the observed variation in alkali depletion patterns for the terrestrial planets.

### 4.2.  Constraints from Potassium Isotopes

The foregoing discussion has mostly neglected the important observation that the K isotopic composition of solar system materials is the same everywhere (*Humayun and Clayton,* 1995). The implication of this observation is that the K depletions in the rocky bodies of the inner solar system must have acquired these depletions in a manner that does not fractionate isotopes. This rules out fractional evaporation, diffusion, and Jeans escape as viable depletion mechanisms. It also appears to rule out the hypothetical transport process that was alluded to above as a means of explaining the systematics of Fig. 1.

The slope of the line in Fig. 1 corresponds to a ~4‰ per mass unit fractionation, if all alkalis acted as isotopes of a single hypothetical element (‰, "per mil," is a measurement in parts per thousand). This implies, based on Fig. 1, that the terrestrial $^{41}K/^{39}K$ ratio should be 8‰ lighter than CI, which is far outside the analytical errors of the *Humayun and Clayton* (1995) analyses (~1‰), and fractionations of this magnitude are not observed.

The lack of a difference between the K isotopic compositions of the Earth and Moon also implies that, if the Moon was derived from terrestrial materials, then the observed depletion in lunar K cannot be due to a process that fractionates isotopes. Possibly, though, the addition of unfractionated H-chondrite material (*O'Neill,* 1991) after depletion/fractionation masks isotopic changes. To a certain extent, we can evaluate this possibility. Even if two-thirds of the Moon's K was added after the hypothetical depletion/fractionation process (Fig. 2), it should be possible to detect the fractionations associated with a giant impact if the K isotopic fractionation from this process was greater than 6‰. For comparison, in simple Rayleigh fractionation, it is expected that a 6.5× depletion (the observed depletion) would cause a ~60‰ effect (*Humayun and Clayton,* 1995). And if one-third of the Moon's K were made of this material, the effect would be observable. If two-thirds of lunar K were added as a chondritic veneer, the original depletion would have actually been larger, with concomitant increases in isotopic fractionation. Again, the K isotopic compositions of the Earth and Moon are identical within analytical uncertainty.

### 4.3.  Summary

In many respects, lunar alkalis are similar to those of other small, volatile-depleted bodies of the solar system, such as the EPB. There is no evidence from this suite of elements that the Moon was primarily derived from terrestrial materials. In particular, the K isotopes of lunar rocks are unfractionated relative to Earth and CI chondrites, whereas Rayleigh fractionation following a giant impact should have resulted in isotopically heavy K in the Moon relative to Earth. We conclude that there is no evidence of a giant impact in a suite of elements that is expected to be highly sensitive to volatile element fractionations. There is no reason why conclusions regarding alkali elements, in particular the highly volatile element Cs, should not be extended to other elements with similar volatility (Br, Pb, Tl, etc.). These elements are depleted in lunar basalts relative to terrestrial basalts by nearly a constant factor (0.026; *Wolf and Anders,* 1980).

## 5.  THE LUNAR HAFNIUM/TUNGSTEN RATIO AND THE SIGNIFICANCE OF $^{182}W/^{184}W$ RATIOS IN LUNAR SAMPLES

The work of *Lee and Halliday* (1995, 1996) has demonstrated the usefulness of the $^{182}Hf$-$^{182}W$ system for dating the separation of metal from silicates in planetary and protoplanetary environments. The decay of $^{182}Hf$ ($t_{1/2}$ = 9 m.y.) to $^{182}W$ produces a $^{182}W$ excess in systems with high Hf/W ratios. Because Hf is lithophile and W is siderophile, metal that separated from silicates early in solar system history has lower than chondritic $^{182}W/^{184}W$ ratios. And the silicates that are complementary to this metal have high Hf/W ratios and have correspondingly higher $^{182}W/^{184}W$ ratios. These excesses or deficiencies in $^{182}W$ are commonly expressed as $\varepsilon$ units, which are deviations from the chondritic value expressed as parts in $10^4$. For example, iron meteorites and chondritic metals, with Hf/W ratios near zero, typically have $\varepsilon(^{182}W)$ values of –3 to –4, whereas eucrites, with Hf/W ratios ~20× CI, have $\varepsilon(^{182}W)$ values of +30 to +40.

### 5.1.  Terrestrial and Lunar Tungsten Isotopes

The $^{182}W/^{184}W$ ratio in terrestrial samples is indistinguishable from bulk chondrites [$\varepsilon(^{182}W)$ = 0 (*Lee and Halliday,* 1995)]. Because W and Hf are both refractory elements, we do not expect them to fractionate significantly during nebular condensation. However, W and Hf are observed to be fractionated by a factor of ~20 in the BSE, presumably by terrestrial core formation. Consequently, if the bulk Earth has chondritic Hf/W and the BSE has chondritic $^{182}W/^{184}W$ but superchondritic Hf/W, then separation of W from Hf by core formation must have occurred late, more than 50 m.y. after the formation of chondrites (*Lee and Halliday,* 1995), when all $^{182}Hf$ had decayed.

In contrast to Earth, there is a clear $^{182}W/^{184}W$ excess in most of the lunar samples analyzed by *Lee et al.* (1997). As discussed in detail by *Lee et al.* (1997), the most reasonable explanation for these $^{182}W$ excesses is that the Moon formed and differentiated before all $^{182}Hf$ had decayed. However, Wieler and coworkers (personal communication, 1999) have recently presented calculations indicating that,

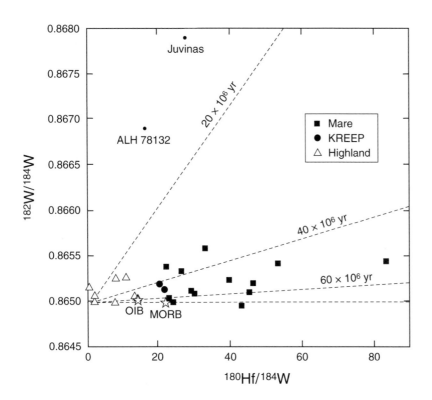

**Fig. 4.** Tungsten-182/tungsten-184 vs. $^{180}Hf/^{184}W$ for lunar samples. Also shown are data for two eucrites (ALH 78132 and Juvinas) and two terrestrial basalts (OIB and MORB; stars). Dashed lines are isochrons with the time of formation dating from that of chondrites. Most mantle-derived lunar samples show a positive (radiogenic) $^{182}W$ anomaly. It is difficult for the Moon to have inherited its W from the Earth, whose W is not distinguishable from chondritic.

because the cosmic-ray-exposure ages of lunar rocks can be large, $^{182}W$-isotopic analyses of lunar samples must be corrected for neutron capture on $^{181}Ta$. But even after appropriate corrections are applied to the data of *Lee et al.* (1997), $^{182}W$ excesses remain, although the sizes of the anomalies are reduced. Consequently, we will discuss the *Lee et al.* (1997) data as they were published. In Fig. 4 the lunar samples analyzed by Lee et al. are plotted on a $^{182}W/^{184}W$ vs. $^{180}Hf/^{184}W$ diagram. Lines with constant $^{182}Hf/^{180}Hf$ ratios at 20, 40, and 60 m.y. after chondrite formation are indicated, assuming a $^{182}Hf/^{180}Hf$ ratio of 2.61 × $10^{-4}$ at the beginning of the solar system. Data for two eucrites, Juvinas and ALH 78132, and two terrestrial basalts, MORB and ocean island basalts (OIB), are shown for comparison (*Lee and Halliday,* 1997).

In models where the Moon formed from Earth's mantle after terrestrial core formation (e.g., *Ringwood,* 1979), one should not expect excess $^{182}W$ in lunar samples. Lunar samples should lie along a horizontal line with terrestrial basalts (Fig. 4). If, however, the formation of the Moon resulted in a significantly higher lunar Hf/W ratio, it is possible that a small $^{182}W/^{184}W$ anomaly that cannot be resolved in terrestrial rocks could be transformed into an analytically resolvable anomaly in lunar rocks. This is essentially the interpretation of *Lee et al.* (1997).

Let us explore this scenario in more detail. In Fig. 4 the terrestrial OIB sample is slightly above the average terrestrial ratio of 0.86500, but with overlapping error bars. The mare basalt plotting on the same line (60 × 10⁶ yr) has a clearly resolvable anomaly. Thus, one way to produce the observed lunar anomalies from young (i.e., ~60 m.y. after

chondrites), terrestrial mantle material would be to drastically increase the lunar Hf/W ratio during the formation of the Moon. Later, after all $^{182}Hf$ has decayed, lunar samples would all plot along the 60-m.y. line. Subsequently, the Hf/W ratios must then be shifted back to the presently observed values without changing the $^{182}W/^{184}W$ ratio. But because Hf and W both act incompatibly during silicate differentiation, it is hard to make large changes in the Hf/W ratio. Consequently, even though the scenario we have just given is mathematically possible, it appears physically implausible. In effect, this scenario requires a "hidden reservoir" for either W or Hf in the Moon.

We can discuss this possibility more quantitatively. The basic prerequisites for understanding the W-isotopic evolution of the Moon and its significance for a terrestrial origin are (1) knowledge of the average W/Hf ratio of the Earth and Moon and (2) the extent of early Hf/W fractionation in the Moon.

### 5.2. Lunar and Terrestrial Hafnium/Tungsten Reservoirs

*Wänke et al.* (1974b) found a correlation of La with W in a large number of lunar samples of different origin, and the lunar La/W ratio was about 20× CI. *Rammensee and Wänke* (1977) experimentally determined the metal/silicate liquid partition coefficient of W and concluded that, in order to achieve the observed W depletion, the Moon would require a much more massive metal core than was compatible with lunar density. From this it was inferred that the observed depletion of W in the Moon was achieved during core formation in the Earth, supporting a terrestrial origin for the Moon (*Rammensee and Wänke, 1977*).

The constancy of La/W ratios in lunar samples suggests a similar behavior of the two elements during global lunar differentiation. Crystallization of the lunar magma ocean resulted in the formation of the anorthositic highland crust, the residual mafic lunar interior, and a small residual melt layer. This residual melt possessed high concentrations of incompatible elements that did not partition into the mafic cumulate layer or the anorthositic crust and was enriched in K, REE, and P, leading to the acronym KREEP for this component (*Hubbard et al., 1971*). After the formation of the Moon, impacts tapped the KREEP reservoir and distributed KREEP-rich material on the surface of the Moon, in particular at the Apollo 14 and (in part) 15 sites (*Warren, 1985,* and references therein). The relative abundances of the incompatible elements in all KREEP-containing samples is surprisingly uniform (*Palme and Wänke, 1975; Warren and Wasson, 1979*), strongly indicating that the abundance sequence of these incompatible elements reflects a Moon-wide differentiation event.

In Fig. 5 the KREEP-rich soil sample 14163 represents the KREEP component. Elements in the upper part of Fig. 5 are arranged in order of decreasing CI-normalized abundance. The sequence thus reflects the degree of incompatibility during crystallization of the lunar magma ocean.

Mare basalts formed by remelting of the mafic cumulate layer. This layer must have had an incompatible-element pattern complementary to KREEP. Elements such as U or Ba are so incompatible that they can hardly be accommodated in the mafic cumulate layer; elements more compatible with mafic silicates, such as Yb and Sc, have higher

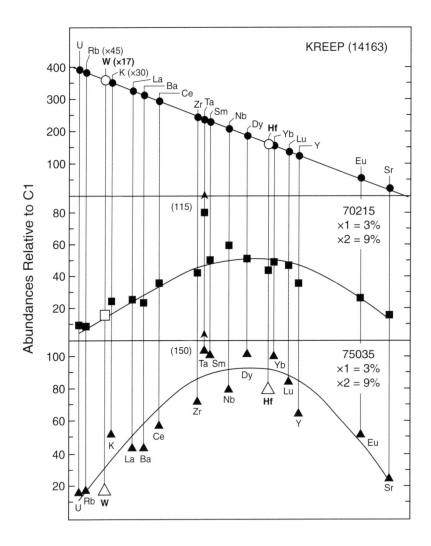

**Fig. 5.** Trace-element abundances (relative to CI chondrites) for lunar samples. Two complementary reservoirs for Hf and W exist on the Moon, KREEP and the cumulates from the lunar magma ocean. For KREEP the elements are ordered in increasing incompatibility and are believed to be the result of 97% fractional crystallization of the lunar magma ocean. The mare basalts 70215 and 75035 can be modeled by 9% partial melting of the cumulates. Tungsten is more incompatible than Hf during lunar petrogenesis and if lunar [182]W anomalies are generated from terrestrial material by producing regions of extreme Hf/W ratios, KREEP should have no [182]W anomaly. KREEP does have a positive [182]W anomaly, making derivation of the Moon from terrestrial material unlikely.

concentrations in mare basalts and lower concentrations in KREEP. In a second event, partial melting of the cumulate layer produced the characteristic patterns of mare basalts, as shown for two Apollo 17 basalts in Fig. 5 [see *Wänke et al.* (1974b) and *Palme and Wänke* (1975) for details].

The complementary relationship between KREEP and mare basalts allows us to "quantify" the incompatibility of W. The abundance of W in Apollo 14 soil 14163 and in the two mare basalts can be fitted into this scheme by placing W between La and U in Fig. 5. Thus, in the lunar environment W is more incompatible than La and less incompatible than U. The positions of the moderately volatile elements Rb and K in Fig. 5 are determined in the same way. From Fig. 5 it is also clear that the Hf/W ratio of the bulk Moon is bracketed by KREEP (from below) and mare basalts (from above).

*Palme and Rammensee* (1981) noticed a good correlation of W and U in lunar rocks with systematically higher ratios of U/W in KREEP-dominated rocks and lower U/W ratios in mare basalts, as expected from Fig. 5. An average lunar U/W ratio of 1.93 was calculated. This leads to a W depletion factor, $(W/U)_{Moon}/(W/U)_{CI}$, of about 22×. (The depletion factor of 17× indicated in Fig. 5 is based on less-accurate La/W ratios and older CI normalizing values.) From this U/W ratio and the CI abundances of refractory elements, an average lunar Hf/W of 25.2 is calculated, corresponding to a $^{180}Hf/^{184}W$ ratio of 29.7.

A similar picture is obtained for Earth. In Fig. 6 we have plotted the average incompatible element contents of continental crust normalized to the BSE. Terrestrially, W is apparently the most incompatible element, with the largest crust/mantle ratio of 69, using the data of *Newsom et al.* (1996) with 1.1 ppm for BCC (bulk continental crust) and

16 ppb for the BSE, significantly above Rb with a corresponding factor of 53 (BSE, *McDonough and Sun,* 1995; BCC, *Taylor and McLennan,* 1985). The high extraction efficiency of W into the crust is compensated by low W in the depleted mantle as sampled by MORB. The similarity of Figs. 5 and 6 reflects the strongly incompatible behavior of W in the Earth and Moon. The depletion of W in the Moon (22×) is within error the same as that of the terrestrial mantle (21×). But although this similarity in Hf/W has been used to support a common origin for the Earth and Moon, the presence of a W-isotopic anomaly on the Moon and its absence on Earth cannot be reconciled with a completely terrestrial origin for the lunar W.

In summary, there is no evidence for a "hidden reservoir" for either W or Hf in the Moon. The basaltic samples analyzed by *Lee et al.* (1997) fit comfortably into the scheme outlined above. They are either representative of the depleted mantle or the residual KREEPy liquid from the crystallization of the lunar magma ocean. We argue, therefore, that the geochemistry of W and Hf in the Moon is well understood.

The logic seems inescapable. If the Moon formed from the terrestrial mantle late (~50 m.y. after chondrites), the Hf/W ratio of the Moon must increase so that any remaining $^{182}Hf$ contributes significantly to the lunar W budget. There are only reasonable two ways to do this: (1) extraction of W to the lunar core or (2) extreme silicate differentiation. The first scenario is ruled out since the Hf/W ratios of the Earth and Moon are the same; and the second scenario makes a simple prediction that is not fulfilled. One of the complementary lunar reservoirs, either KREEP or the mafic cumulates, should have no W-isotopic anomaly. In this case, the prediction is that the low Hf/W reservoir (KREEP) should have no anomaly. In fact, KREEP-rich basalts do have a W isotopic anomaly of at least 2ε (*Lee et al.,* 1997). We see no way to make the Moon solely from the terrestrial mantle.

If one wishes to reconcile the W-isotopic evidence of the Earth and Moon with the giant impact hypothesis for the formation of the Moon, more complicated models are required. There are two general kinds of such models:

*5.2.1. Early formation of the Moon.* The Moon could have formed by a giant impact earlier — for example, 35–40 m.y. after chondrite formation, as indicated in Fig. 4. This, however, requires that the core of Earth formed before the giant impact. But core formation in the Earth earlier than ~50–60 m.y. after chondrites would produce a $^{182}W/^{184}W$ anomaly in the mantle that is presently not observed. Later incoming material and additional core formation, after the formation of the Moon, must then have diluted the $^{182}W$ excess on Earth (but not on the Moon!) to an extent that would not allow detection. If, for example, the Moon had formed from Earth 40 m.y. after the formation of iron meteorites, the Earth and Moon could have had a $^{182}W/^{184}W$ ratio of 0.86520 (about 2ε units above the chondritic value, at a Hf/W ratio of 20) to explain the presently observed lunar anomalies. To reduce the $^{182}W/^{184}W$ ratio to

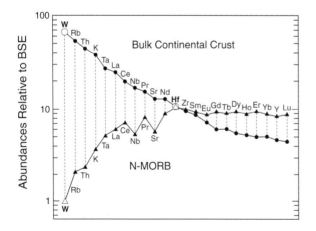

**Fig. 6.** Trace-element abundances in the Earth's continental crust and oceanic upper mantle, normalized to the bulk silicate earth (BSE). Though produced by a different process than on the Moon (*Hofmann,* 1988), the Earth also has complementary reservoirs, with the continental crust resembling KREEP and the MORB mantle resembling the lunar cumulate mantle.

the present value in the Earth would require the later removal of at least 50% of the W present at that time into Earth's core and the addition of the same amount of W with chondritic $^{182}W/^{184}W$.

This model may have physical difficulties. The later addition of accretional components to the Earth required by the model occurred (according to present models of Earth accretion) by impacts of ~$10^{26}$ g bodies, so-called embryos (*Wetherill*, 1994). If these embryos had differentiated into core and mantle with correspondingly elevated $^{182}W/^{184}W$ ratios in their silicates (*Taylor and Norman*, 1990), then the iron cores would have to completely disaggregate and reequilibrate with the silicates during impact in order to avoid any $^{182}W$ excess in Earth. But in many of the simulations of the impact process (*Benz and Cameron*, 1990; *Benz et al.*, 1989; *Cameron and Canup*, 1998) there is little interaction between the impactor core and the target mantle. Thus, for this general class of models to work, the late-accreting impactors would probably have to be of a different size than is currently envisioned.

However, regardless of the exact physical means by which W is added to Earth after the Moon's separation, highly improbable coincidences are required. If W and Hf in the Moon today are representative of the terrestrial mantle prior to the Moon's formation, then terrestrial radiogenic W must have been sequestered (to the core?) and nonradiogenic W must have been added to Earth's mantle. These additions and subtractions must be done in such a way that the Hf/W ratio remains constant. Therefore, the W added must exactly counterbalance the W that was removed. Whether this delicate balance can be achieved is problematic. It seems more likely to us that the similarity of Hf/W between the Earth and Moon is coincidental.

*5.2.2. Contribution of the impactor.* Of course, there may have been contributions from the impactor to the Moon. In fact, in some models the impactor dominates the bulk lunar composition (e.g., *Newsom and Taylor*, 1989), although it is not clear to us that this type of model is testable. An undifferentiated projectile would not provide isotopically unusual W. But if, as discussed above, the projectile were differentiated into a core and mantle, the mantle should have a significantly enhanced $^{182}W/^{184}W$ ratio. The core of the impactor with the complementary low $^{182}W/^{184}W$ is required to separate from the impactor mantle in order to preserve the $^{182}W$ anomaly of that mantle and import it into the Moon. Presumably the ultimate fate of that core is the core of Earth (e.g., *Benz et al.*, 1989). Earth's mantle should also receive radiogenic W from the impactor mantle, although presumably less than the Moon. Therefore, this model's attractiveness may possibly be enhanced if the impactor core is allowed to equilibrate or partially equilibrate with the terrestrial mantle. If the impact occurred 50 m.y. after chondrites (or later), then $^{182}Hf$ would be essentially all decayed, and we cannot expect lunar differentiation to have had any influence on lunar W isotopic composition. We would expect a uniform, elevated $^{182}W/^{184}W$

ratio in the Moon. But this scenario seems unlikely given the observed variability in lunar $^{182}W/^{184}W$ measurements.

Note the dichotomy within a single scenario: The impactor core is not allowed to equilibrate with protolunar material but may be required to equilibrate with Earth's mantle. To prevent the former it would be convenient if the impactor core remained intact. To facilitate the latter it would be convenient if the core were dispersed. Thus, the fate of the core of a differentiated impactor is a very important question that needs to be addressed by detailed physical modeling.

Of course, it may be that, despite the similarities between the Earth and Moon that we have discussed, the origins of the Earth and Moon are not strongly coupled. In models of a separate origin of the Moon (indistinguishable from models where the entire Moon comes from the impactor), $^{182}W/^{184}W$ ratio of the Moon would have to be below that of the mare basalts and above that of KREEP, something around 0.86537, which would, with a $^{180}Hf/^{184}W$ ratio of 30, correspond to core formation in the Moon (or in the impactor) at ~40 m.y. after the formation of chondrites.

We note as an aside that it is possible that the W-isotopic anomaly of the Moon is significantly larger than estimates based on the currently analyzed samples. *Jones* (1998) presented evidence that mixing has occurred between a radiogenic lunar mantle and a nonradiogenic crust. His preferred age for the Moon (~34 m.y. after chondrites) was derived from the sample with the largest $^{182}W$ anomaly (15555). But if 15555 itself contains a nonradiogenic W "contaminant," the Moon could be older still. More lunar rocks need to be analyzed to discover the most radiogenic W isotopic signature. However, it is not out of the question that the Moon is significantly older than we currently believe.

### 5.3. Summary

In summary, the W-isotopic compositions of the Earth and Moon do not support a terrestrial origin of the Moon, although it is possible to construct models that are compatible with W isotopes and the giant impact hypothesis. However, these models are complex and alternately appear to require either additions of undifferentiated or differentiated projectiles that either equilibrate with some reservoirs but not others. Models where the impact occurs early (~35 m.y. after chondrites) only work if later-accreting material removes the signature of radiogenic W from Earth's mantle. Models where the impact occurs late (>50 m.y. after chondrites) are best accommodated by an impactor that differentiated early (e.g., *Taylor and Norman*, 1990). These late impact models work best if the impactor and target (Earth) equilibrate their W isotopes. However, we do not favor late impacts and inherited anomalies because of the observed variation in $^{182}W/^{184}W$ within the lunar sample suite. We agree with *Lee et al.* (1997) that the overall variation in lunar

$^{182}$W/$^{184}$W is best explained by decay of $^{182}$Hf within the Moon.

## 6. IRON, COBALT, AND NICKEL AND THE NATURE OF THE LUNAR CORE

The major element whose abundance is best known in both the Earth and the Moon is FeO. Based on analyses of lunar rocks and on geophysical data, it seems well established that the FeO content of the silicate fraction of the Moon is about 50% higher than the FeO content of Earth's mantle (see above discussion). Cosmochemically iron is associated with Ni and Co. The three elements Fe, Co, and Ni will therefore be discussed together. Estimates for the bulk silicate Earth and the bulk silicate Moon are summarized in Table 1.

According to Table 1 the silicate part of the Moon has about 50% more FeO than Earth's mantle, the Co content of the lunar mantle is similar to that of Earth's mantle, and the Ni content of the lunar mantle is a factor of 4 below that of Earth's mantle. As the metal/silicate partition coefficients increase in the sequence from Fe through Co to Ni, it is clear that the abundances of these elements (relative to chondrites) in the lunar mantle decrease with increasing siderophility.

If the Moon is predominantly made of material with the composition of the present terrestrial upper mantle, then an addition of ~5 wt% FeO from the impactor is required to satisfy the FeO content of the lunar mantle — even more if a lunar core is considered (Table 1). For example, with a devolatilized CI projectile (27.3% Fe; *O'Neill*, 1991), this leads to about 18% of impactor component in the Moon's silicate composition.

$$\text{Fe}_{\text{Earth-mantle}} \times 0.82 + \text{Fe}_{\text{impactor}}\, 0.18 = \text{Fe}_{\text{silicate.Moon}} \quad (1)$$

If the impactor was undifferentiated, as the above calculation implies, it follows that at least 2900 ppm Ni and 136 ppm Co must have been added to the Moon in the same event. These Ni and Co additions are not strongly depen-

dent on the type of impactor, as long as the projectile is of broadly chondritic composition. The impactor contribution of Ni and Co significantly exceeds the presently observed concentration in the Moon (Table 1).

This extra Ni and Co could have been extracted into a lunar core. To accommodate this Ni and Co requires an ~1 wt% lunar core with some 40% Ni (*O'Neill*, 1991). Higher impactor contributions or lower FeO in the proto-Earth mantle imply larger cores. Using this CI-addition model, we can calculate the composition of the lunar core and calculate core/mantle "partition coefficients" of 6.2, 111, and 851 for Fe, Co, and Ni respectively. Experimentally determined metal/silicate liquid partition coefficients (calculated relative to a partition coefficient of 6.2 for Fe) give 215 for Co and 2533 for Ni respectively (*Holzheid et al.*, 1997). These values are considerably larger than we just calculated. Thus, for a given Fe distribution between lunar core and mantle, there is too much Ni and Co in the lunar mantle. A larger core with less Ni and more Fe would make the fit worse, as this implies higher core/mantle ratios for Fe and correspondingly higher effective metal/silicate partition coefficients for Ni and Co. Thus, as we saw in our discussion of W isotopes (above), making the Moon by adding undifferentiated chondritic material to terrestrial mantle material is difficult.

To avoid the problem of excess Ni and Co it has been suggested that the impactor was differentiated into core and mantle and that the core of the impactor mixed with that of the Earth without reequilibrating with silicates (*Taylor and Norman*, 1990). The mantle of the impactor would then have to have significantly higher FeO contents than the Earth's mantle in order to produce the elevated FeO of the Moon. A larger fraction of differentiated impactor material is required to produce the Moon's FeO content compared to the addition of chondritic matter, because (by definition) some Fe has been used to make a core. But, as discussed above, an impactor that is differentiated into core and mantle may have a substantial $^{182}$W/$^{180}$W excess in its mantle (see section on W isotopes). If the impactor core does not reequilibrate with silicate, a significant W anomaly may be imposed on Earth's mantle.

Another possible solution to the excess Ni and Co problem would be to have lunar core formation occur, not from a totally molten Moon, but from a partially molten Moon. *Jones and Delano* (1989) used this concept to explain the high Mg# source for the lunar Mg-suite lithologies. Again, taking the data from Table 1 at face value, we can perform the same mass-balance calculation but with three reservoirs (metal, olivine, silicate liquid) rather than two. If we take the metal/silicate liquid partition coefficients given above and arbitrarily assume 20% silicate partial melting during core formation, we can calculate what values of $D_{Co}$(olivine/silicate liquid) and $D_{Ni}$(olivine/silicate liquid) would be necessary to explain the "excess" Co and Ni. For this calculation, we require that 20% silicate liquid and 80% olivine must account for the lunar silicate NiO and CoO and

TABLE 1.   Iron, Ni, and Co abundances in the Earth and Moon.

| | Fe % | Co ppm | Ni ppm | Fe/Ni (oxides) | Ni/Co (oxides) |
|---|---|---|---|---|---|
| Earth mantle | 6.35 | 102 | 1860 | 34 | 18 |
| Lunar mantle | 10.1 | 90 | 470 | 205 | 5.2 |
| Bulk Moon* | 10.1 | 159 | 4430 | | |
| Earth/Moon | 0.65 | 1.17 | 4.23 | | |
| CI | 18.23 | 506 | 10770 | 16.9 | 21.3 |

Sources: Earth: *O'Neill and Palme* (1998); Moon: Fe, *Jones and Delano* (1989); Ni and Co, *Delano* (1986b); CI: *Palme and Beer* (1993).

\* 82% Earth mantle + 18% CI-volatile free impactor (27.3% Fe).

that the addition of 1 wt% metal to this silicate mixture constitutes bulk lunar Ni and Co. The values of D(olivine/silicate liquid) that we calculate are ~3–4 for both elements. We do not expect that $D_{Ni}$(olivine/silicate liquid) should equal $D_{Co}$(olivine/silicate liquid), but considering the oversimplified nature of the calculation, we consider core formation from a partially molten Moon a viable solution to the Ni-Co problem.

In summary, the higher FeO content of the lunar mantle requires addition of Fe from an impactor if the Moon is primarily made of Earth mantle material. Because of constraints from W isotopes, the additional FeO cannot easily come from an undifferentiated projectile. However, addition of ~18% CI-like material to Earth's mantle would satisfy FeO mass balance. If the FeO content of the impactor mantle resembled that of Mars or the EPB (~19 wt% Fe), we would need about 30% impactor material to make the Moon from Earth's mantle. The present-day abundances of Ni and Co are better explained if core formation occurred in a partially, rather than a totally molten Moon.

### 6.1.  Other Siderophile Elements

As mentioned above, the abundances of Fe, Co, and Ni in the silicate Moon decrease with increasing siderophility. This trend extends to the more siderophile elements, such as Mo, Ge, and Re as shown by *Newsom* (1984). There is a clear trend: the greater its siderophility, the greater an element's depletion in the lunar mantle. Thus, the pattern of siderophile elements in the Moon appears to require the presence of a lunar core. Assuming that the Moon dominantly formed from terrestrial materials, the extent of the contamination of the Moon with an impactor is unclear and depends strongly on the poorly known size of the actual lunar core. A larger impactor contribution to the proto-Moon would require a larger core, if the impactor were chondritic. However, as we have argued above, this model is acceptable for the Moon but may have undesired consequences for terrestrial W isotopes.

Calculations by *Jones and Hood* (1990) show that it is difficult to model the lunar siderophile-element pattern by separation of metal from a lunar bulk composition having the siderophile-element abundances of the terrestrial upper mantle. Trivial amounts of metal will deplete terrestrial siderophiles to lunar abundances and no commonality between different siderophile-element models was found (i.e., each siderophile element required a different model). In contrast, great commonality (i.e, intersection of model solutions) was found if a modified chondritic source was assumed (modified in the sense that phosphorus was somewhat volatile depleted). *Jones and Hood* (1990) found intersections between different siderophile-element solutions when the degree of silicate partial melting was ~20% and the size of the metallic core was 3–5 wt%. This solution is in general agreement with that found for Ni and Co above. And although the *Jones and Hood* (1990) core size was larger than the 1 wt% assumed above, *Jones and Hood*

(1990) assumed more reducing conditions than *O'Neill* (1991), with commensurately less Ni in the metal phase, requiring a larger core.

### 6.2.  Summary

In summary, the lunar siderophile-trace-element pattern appears to require core formation in the Moon. The best fits to the siderophile-element data are models that assume core formation from an essentially chondritic, partially molten Moon. Starting compositions for the Moon that resemble the present-day upper mantle do not fit the siderophile data well. The core sizes (3–5 wt%) required to deplete approximately chondritic levels of siderophiles in the silicate portion of such a Moon (to the levels that are currently inferred) are compatible with core sizes estimated from the most recent geophysical data (1–3 wt%). These conclusions from other siderophiles are in general agreement with those from W isotopes, provided core formation on the Moon occurred ~35 m.y. after chondrites. The only problem for this scenario is the lack of a linkage with the Earth. We reiterate that W-isotopic systematics imply that terrestrial core formation appears to have occurred considerably later than on the Moon. If the giant impact formed the Moon and triggered core formation in the Earth, then this should not be the case.

## 7.  EVIDENCE FROM TERRESTRIAL MANTLE XENOLITHS

The hypothesized giant impact would have been the most important thermal event Earth ever experienced. The kinetic energy gained during such an impact is sufficient to melt most of the Earth and to vaporize the rest (*Melosh*, 1990). But impacts do not distribute their energy uniformly and the physical state of the Earth after the giant impact is uncertain. If for no other reason, the exact geometry of the impact will likely play an important role. Grazing impacts will impart less energy to the Earth than a dead-center collision. Even so, the giant impact is widely believed to have produced a terrestrial magma ocean, and it is possible that this large degree of partial melting facilitated the separation of Earth's core (*Li and Agee*, 1996; *Righter et al.*, 1997). If core formation did indeed occur immediately following the giant impact, the Earth would almost certainly have been left in a totally molten state, since the core-formation event would have also released large amounts of potential energy (e.g., *Jones and Drake*, 1986).

In addition, most models of accretion in the early solar system favor the growth of a relatively small number of embryotic planetesimals (e.g., *Wetherill*, 1994). As these planetesimals accrete into planets, they do so violently, because of their size. All these considerations, taken as a whole, argue that Earth began life very hot and perhaps mostly molten.

In an important paper, *Jagoutz et al.* (1979) noted that several fertile mantle xenoliths (spinel lherzolites) from around the world had major-element abundances that ap-

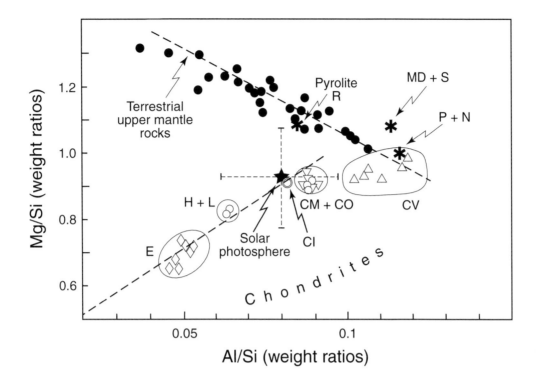

**Fig. 7.** Magnesium/silicon vs. Al/Si for mantle peridotites and chondrites. As peridotites (solid circles) become more fertile (decreasing Mg/Si and increasing Al/Si), they approach a trend defined by chondrites (open symbols) that was produced by nebular processes. In turn, our best estimates for the composition of the BSE [Pyrolite R = *Ringwood* (1979); MD + S = *McDonough and Sun* (1995); P + N = *Palme and Nickel* (1985)] fall near the intersection of these two trends, as do some of the mantle peridotites. This suggests that some portions of the terrestrial mantle have survived without much igneous processing. In turn, this sheds doubt on whether the Earth ever passed through a magma ocean stage.

proached a trend defined by chondritic meteorites (Fig. 7). They reasoned that these xenoliths were nearly unprocessed samples of the BSE and used them to reconstruct the BSE composition. *Jones and Drake* (1986), in their review of terrestrial core formation, were impressed by the *Jagoutz et al.* (1979) observation. Consequently, *Jones and Drake* (1986) deliberately kept the degree of silicate partial melting in their core formation models small, fearing that large degrees of silicate melting would lead to differentiation, eradicating the primitive sample suite identified by *Jagoutz et al.* (1979).

*Ringwood* (1990) stated his opposition to the giant impact hypothesis by noting that, if the Earth had ever been totally molten, it would tend to crystallize in such a way that it would be stratified and stable against convective mixing. *Tonks and Melosh* (1990) countered that any terrestrial magma ocean would have convected turbulently and crystallizing solids would remain suspended and not settle, as in a normal magma chamber. Thus, a magma ocean need not imply differentiation.

*Jones* (1996) reiterated Ringwood's concern in a different way. He noted that, even if the magma ocean turbulently convected and crystals remained suspended, crystallization would eventually proceed to a point where the mantle was

mostly solid and could not continue to convect in a turbulent manner. At this point or at some later time, the residual liquid may not remain in the mantle but may erupt as basaltic magma, depleting the mantle of its basaltic components. Thus, an important question is whether the "primitive" samples identified by *Jagoutz et al.* (1979) could have survived a terrestrial magma ocean. Or can mantle mixing effectively undo the differentiation we expect the magma ocean to have done?

One means of answering this question is to assess the degree to which fertile mantle lherzolites approach chondritic compositions, and one of the most interesting studies utilized Os isotopes. Rhenium-187 decays to $^{187}$Os with a $t_{1/2}$ of 50 G.y. *Meisel et al.* (1996) measured the Os-isotopic compositions of a suite of mantle xenoliths possessing varying degrees of fertility. Figure 8 shows the correlation between $^{187}$Os/$^{188}$Os and two different indexes of fertility, Al content and Lu content. As Al and Lu approach the values we expect for the BSE, the Os-isotopic composition approaches chondritic. Thus, for the most fertile lherzolites, the time-integrated Re/Os ratio of these mantle samples has remained within ±3% of chondritic over the lifetime of the solar system.

It seems unlikely to us that differentiation in a magma

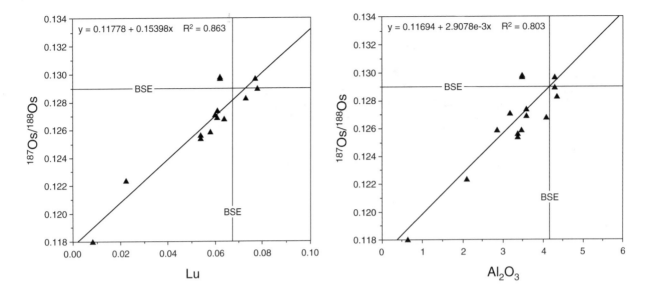

**Fig. 8.** Osmium-187/osmium-188 vs. Lu and $Al_2O_3$ in mantle peridotites. As mantle peridotites become more fertile and their Lu and Al abundances approach that of the BSE, their $^{187}Os/^{188}Os$ ratio approaches chondritic. For the most fertile peridotites, this means that their time-integrated Re/Os ratio has remained within ~±3% of chondritic. The Re-Os system has apparently suffered minimal igneous processing over the age of the Earth. Again, this sheds doubt on whether the Earth ever experienced a magma ocean.

ocean and subsequent homogenization (perhaps by mantle convective processes over the age of the Earth) could have conspired to produce such an Os isotopic outcome. During basalt production Os behaves as a compatible element and remains in the residue, whereas Re behaves moderately incompatibly and is concentrated in the basaltic liquid. (Thus, the Os isotopic composition of the continental crust is quite radiogenic because the crustal Re/Os ratio is high.) Considering the ease with which Re and Os can fractionate during igneous processes, it seems more reasonable to us that the Re and Os in fertile lherzolites have never been much fractionated. We therefore assume that they were added as a late veneer (along with other highly siderophile elements) after core formation had ceased and were then mixed into Earth's mantle in the solid state.

In addition, another aspect of the *Meisel et al.* (1996) study appears to have gone underappreciated. The good correlations between the xenoliths' Os isotopic compositions and their Lu and Al concentrations appear to imply that the veneer was mixed into a mantle that was approximately chondritic and homogeneous (except for its previously removed metal component). Otherwise, if the late veneer had been mixed into a heterogeneous, nonchondritic mantle, there is no reason for the correlations seen in Fig. 8 unless later remixing were complete.

However, there is evidence that mixing of the veneer itself into the mantle was not as complete as was originally thought by *Jones and Drake* (1986). *Spettel et al.* (1991) have analyzed several suites of spinel lherzolites; and Ni and Ir analyses of these samples are shown in Fig. 9. Nickel/Ir ratios vary from sample to sample (which may reflect our ability to acquire a representative sample) and, more impor-

tantly, from locality to locality. Sample localities are designated by either solid or open symbols depending on the average Ni/Ir ratio of that locality. With few exceptions, individual samples from a given locality either have an Ir content of 2–3.5 ppb and a Ni/Ir ratio ≥ 600 or have an Ir content of 3.5–6 ppb and a Ni/Ir ratio ≤ 600. Almost all variation in Ni/Ir in Fig. 9 is due to variability in Ir and is much larger than analytical uncertainty (~10% relative). Because Ir acts compatibly during normal igneous processes, we take this variation to reflect the completeness of mixing of the late veneer (*Spettel et al.*, 1991). Although more work on this issue is needed, we take the *Meisel et al.* (1996) and *Spettel et al.* (1991) data to mean that the early terrestrial mantle was largely undifferentiated at the time the veneer was added. Mixing was apparently incomplete for Ir and, consequently, significant mantle heterogeneities that were present at the time the veneer was added are unlikely to have been totally erased by subsequent mixing.

But if the early mantle did differentiate, then it had remixed and was homogeneous by the time the Ir, Re, and Os were added. So when did addition of this late veneer occur? This is a difficult question to answer. Because Re and Os in fertile lherzolites have not obviously fractionated from their chondritic ratio, there is no event to date. However, we believe it unlikely that the veneer was added after the late heavy bombardment recorded on the Moon (~3.8–3.9 Ga). Remixing of Earth's mantle would have to have been complete by that time.

However, we see no evidence of large, early differentiation. We believe that, except for the removal of metal to the core, large regions of the mantle of the early Earth were left

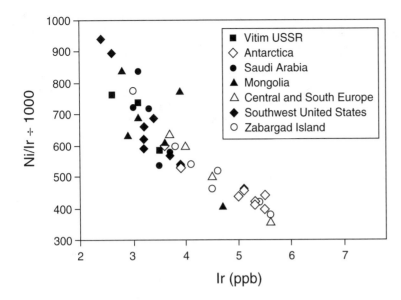

**Fig. 9.** Nickel/iridium vs. Ir for spinel lherzolites from several different localities (*Spettel et al.,* 1991). These localities are given in the legend. A given locality has been assigned either a solid symbol or an open symbol based on the average Ni/Ir ratio of that locality. Variations in Ni/Ir are almost entirely associated with variation in Ir. Individual samples do not span the entire range of the data and, consequently, the solid and open symbols are highly segregated. We interpret these data to reflect original heterogeneity from the mixing of the late veneer into the Earth's mantle (see text).

relatively unprocessed and that some remnants of this pristine mantle "survive" today. We use "survive" in quotations because even the most pristine mantle lherzolites have had some removal of highly incompatible elements, presumably by extraction of silicate liquid (e.g., *Hirschmann et al.,* 1998). These extractions have sometimes been minor enough that major-element abundances and the Re-Os system have not been much affected. This is presumably because Re and Os either reside compatibly in silicate (or oxide), as Os is presumed to, or reside in minor sulfides that are not very soluble in low-degree silicate partial melts.

Thus, if the Earth differentiated and was then remixed (to the hand-specimen scale), this remixing is required to be so complete that we cannot detect it. We see no clear evidence that either differentiation or remixing occurred. Clearly, if this general conclusion is correct, it is difficult for Earth to have passed through a magma ocean stage. And if the Earth did not have a magma ocean, we believe it is reasonable to question the giant impact hypothesis as well. As alluded to above, if some mantle samples have survived unprocessed and undifferentiated, we may have to not only rethink the giant impact hypothesis but the general problem of accretion in the inner solar system.

## 8. CONCLUSIONS

We have reviewed several important aspects of the chemical composition of the Earth and Moon. In summary: (1) It appears that the bulk compositions of the Earth and Moon are different. (2) The depletions of volatile alkali elements in the Moon cannot simply be the residues that remain following the partial volatilization of terrestrial materials. (3) Taken at face value, $^{182}Hf$-$^{182}W$ systematics would indicate that the Moon is older than Earth. Or, stating it more precisely, core formation on the Moon must have predated core formation on Earth. (4) The presence of a

significant lunar core (1–3 wt%) has obviated the need for the Moon's metal component to reside in the Earth, even though the Moon must be depleted in Fe relative to chondrites. (5) It is not evident from our sampling of the terrestrial mantle that the Earth ever passed through a magma ocean stage.

Although none of these observations actually disproves the giant impact hypothesis, we find it disquieting that the obvious consequences expected of a giant impact are not observed to be fulfilled. In its original conception, the giant impact predicted that (1) the Moon's depletion in volatile elements occurred as a result of the impact (*Wänke and Dreibus,* 1986); (2) the Moon, being derivative from the Earth, should be younger; and (3) the thermal consequences of the impact would result in a terrestrial magma ocean (*Melosh,* 1990; *Tonks and Melosh,* 1990). Other simple predictions, such as that the Moon should have no core and that the Mg# of the Moon should match that of the Earth, were known or suspected to be unfulfilled, even at the time the giant impact hypothesis was proposed. However, it was believed at the time that contributions from the impactor might resolve these discrepancies. Contributions from the impactor are notoriously difficult to disprove. In their extreme limit, some impactor-contribution models may not even be testable. However, as we have shown above, an analysis that includes both consequences for Earth, as well as for the Moon, may preclude some models or render them unlikely.

Our current best estimate is that the Moon formed from material having nearly chondritic abundances of siderophile elements at ~35 m.y. after chondrites. We also favor a partial magma ocean for the Moon, rather than complete melting. The main problem with our overall scenario is that it does not accommodate most versions of the giant impact hypothesis very well. It may, however, be compatible with the circumterrestrial disk model of *Weidenschilling et al.*

(1986). In the Weidenschilling model, silicate is preferentially trapped in a circumterrestrial disk compared to iron metal. Metallic cores or pieces of metallic cores, being stronger and less easily disrupted than silicate, preferentially pass through the disk. Thus, when the disk finally accretes into a moon, that moon is depleted in metal.

# REFERENCES

Baker M. B., Hirschmann M. M., Ghiorso M. S., and Stolper E. M. (1995) Compositions of near-solidus peridotite melts from experiments and thermodynamic calculations. *Nature, 375,* 308–311.

Basaltic Volcanism Study Project (1981) *Basaltic Volcanism on the Terrestrial Planets.* Pergamon, New York. 1286 pp.

Benz W. and Cameron A. G. W. (1990) Terrestrial effects of the giant impact. In *Origin of the Earth* (H. E. Newsom and J. H. Jones, eds.), pp. 61–67. Oxford Univ., New York.

Benz W., Cameron A. G. W., and Melosh H. J. (1989) The origin of the Moon and the single impact hypothesis. *Icarus, 81,* 113–131.

Bukowinski M. S. T. (1976) The effect of pressure on the physics and chemistry of potassium. *Geophys. Res., Lett., 3,* 491–494.

Cameron A. G. W. and Canup R. M. (1998) The giant impact and the formation of the Moon (abstract). In *Origin of the Earth and Moon,* pp. 3–4. LPI Contribution No. 957, Lunar and Planetary Institute, Houston.

Canil D. (1999) Vanadium partitioning between orthopyroxene, spinel and silicate melt and the redox states of mantle source regions for primary magmas. *Geochim. Cosmochim. Acta, 63,* 557–572.

Christie D. M., Carmichael I. S. E., and Langmuir C. H. (1986) Oxidation states of mid-ocean ridge basalt glass. *Earth Planet. Sci. Lett., 79,* 397–411.

Clayton R. N. (1993) Oxygen isotopes in meteorites. *Annu. Rev. Earth Planet. Sci., 21,* 115–149.

Davies, G. F. (1988) Ocean bathymetry and mantle convection 1. Large-scale flow and hot spots. *J. Geophys. Res., 93,* 10467–10480.

Delano J. W. (1986a) Pristine lunar glasses: Criteria, data, and implications. *Proc. Lunar Planet. Sci. Conf. 16th,* in *J. Geophys. Res., 91,* D201–D213.

Delano J. W. (1986b) Abundances of cobalt, nickel, and volatiles in the silicate portion of the Moon. In *Origin of the Moon* (W. K. Hartmann, R. J. Phillips, and G. J. Taylor, eds.), pp. 231–248. Lunar and Planetary Institute, Houston.

Delano J. W. (1993) Oxidation state of the Earth's upper mantle during the last 3800 million years: implications for the origin of life (abstract). In *Lunar and Planetary Science XXIV,* pp. 395–396. Lunar and Planetary Institute, Houston.

Dreibus G., Spettel B., and Wänke H. (1976) Lithium as a correlated element, its condensation behavior, and its use to estimate the bulk composition of the moon and the eucrite parent body. *Proc. Lunar Sci. Conf. 7th,* pp. 3383–3396.

Hanson B. Z. and Jones J. H. (1998) The systematics of $Cr^{3+}$ and $Cr^{2+}$ partitioning between olivine and liquid in the presence of spinel. *Amer. Mineral., 83,* 669–684.

Haskin L. and Warren P. (1991) Lunar chemistry. In *Lunar Sourcebook: A User's Guide to the Moon* (G. H. Heiken, D. T. Vaniman, and B. M. French, eds.), pp.357–474. Cambridge Univ., New York.

Hirschmann M. M., Ghiorso M. S., Wasylenki L. E., Asimov P. D., and Stolper E. M. (1998) Calculation of peridotite partial melting from thermodynamic models of minerals and melts. I. Review of methods and comparison with experiments. *J. Petrol., 39,* 1091–1115.

Hofmann A. W. (1988) Chemical differentiation of the Earth: The relationship between mantle, continental crust, and oceanic crust. *Earth. Planet. Sci. Lett., 40,* 297–314.

Holzheid A. and Palme H. (1996) The influence of FeO on the solubilities of cobalt and nickel in silicate melts. *Geochim. Cosmochim. Acta, 60,* 1181–1193.

Holzheid A., Palme H., and Chakraborty S. (1997) The activities of NiO, CoO and FeO in silicate melts. *Chem. Geol., 139,* 21–38.

Hood L. L. (1986) Geophysical constraints on the lunar interior. In *Origin of the Moon* (W. K. Hartmann, R. J. Phillips, and G. J. Taylor, eds.), pp. 361–410. Lunar and Planetary Institute, Houston.

Hood L. L. and Jones J. H.(1987) Geophysical constraints on lunar bulk composition and structure: A reassessment. *Proc. Lunar Planet. Sci. Conf. 17th,* in *J. Geophys. Res., 92,* E396–E410.

Hood L. L., Lin R. P., Mitchell D. L., Acuna M., and Binder A. (1999) Initial measurement of the lunar induced magnetic moment in the geomagnetic tail using Lunar Prospector data (bstract). In *Lunar and Planetary Science XXX,* Abstract #1402. Lunar and Planetary Institute, Houston (CD-ROM).

Hubbard N. J., Meyer C. Jr., Gast P. W., and Wiesmann H. (1971) The composition of derivation of Apollo 12 soils. *Earth Planet. Sci. Lett., 10,* 341–350.

Humayun M. and Clayton R. N. (1995) Potassium isotope geochemistry: Genetic implications of volatile element depletion. *Geochim. Cosmochim. Acta, 59,* 2131–2148.

Jagoutz E., Palme H., Baddenhausen H., Blum K. Cendales M., Dreibus G., Spettel B., Lorenz V., and Wänke H. (1979) The abundances of major, minor and trace elements in the earth's mantle as derived from primitive ultramafic nodules. *Proc. Lunar Planet. Sci. Conf. 10th,* pp. 1141–1175.

Jeanloz R. (1990) The nature of the Earth's core. *Annu. Rev. Earth Planet. Sci., 18,* 357–386.

Jones J. H. (1996) Chondrite models for the composition of the Earth's mantle and core. *Philos. Trans. Roy. Soc. Lond., A354,* 1481–1494.

Jones J. H. (1998) Tungsten isotopic anomalies in lunar samples and the early differentiation of the Moon (abstract). In *Lunar and Planetary Science XXIX,* Abstract #1814. Lunar and Planetary Institute, Houston (CD-ROM).

Jones J. H. and Delano J. W. (1989) A three component model for the bulk composition of the Moon. *Geochim. Cosmochim. Acta, 53,* 513–527.

Jones J. H. and Drake M. J. (1986) Geochemical constraints on core formation in the Earth. *Nature, 322,* 211–228.

Jones J. H. and Drake M. J. (1993) Rubidium and cesium in the Earth and the Moon. *Geochim. Cosmochim. Acta, 57,* 3785–3792.

Jones J. H. and Hood L. L. (1990) Does the Moon have the same chemical composition as the Earth's upper mantle? In *Origin of the Earth* (H. E. Newsom and J. H. Jones, eds.), pp. 85–98. Oxford Univ., New York.

Kreutzberger M. E., Drake M. J., and Jones J. H.(1986) Origin of the Earth's Moon: Constraints from alkali volatile trace elements. *Geochim. Cosmochim. Acta, 50,* 91–98.

Lee D.-C. and Halliday A. N. (1995) Hafnium-tungsten chronometry and the timing of terrestrial core formation. *Nature, 378,* 771–774.

Lee D. C. and Halliday A. N. (1996) Hf-W isotopic evidence for rapid accretion and differentiation in the early solar system. *Science, 274,* 1876–1879.

Lee D. C. and Halliday A. N. (1997) Core formation on Mars and differentiated asteroids. *Nature, 388,* 854–857.

Lee D. C., Halliday A. N., Snyder G. A., and Taylor L. A. (1997) Age and origin of the Moon. *Science, 278,* 1098–1103.

Li, J. and Agee C. B. (1996) Geochemistry of mantle-core differentiation at high pressure. *Nature, 381,* 686–689.

Lugmair G. W. and Shukolyukov A. (1998) Early solar system timescales according to $^{53}$Mn-$^{53}$Cr isotope systematics. *Geochim. Cosmochim. Acta, 62,* 2863–2886.

Macelwane J. B. (1951) Evidence on the interior of the Earth derived from seismic sources. In *Internal Constitution of the Earth* (B. Gutenberg, ed.), pp. 227–304. Dover, New York.

McDonough W. F. and Sun S.-s. (1995) The composition of the Earth. *Chem. Geol., 120,* 223–253.

McDonough W. F., Sun S.-s., Ringwood A. E., Jagoutz E., and Hofmann A. W. (1992) Postassium, rubidium, and cesium in the Earth and Moon and the evolution of the mantle of the Earth. *Geochim. Cosmochim. Acta, 56,* 1001–1012.

McFarlane E. A. (1989) Formation of the Moon in a giant impact: Composition of the impactor. *Proc. Lunar Planet. Sci. Conf. 19th,* pp. 593–605.

Meisel T., Walker R. J., and Morgan J. W. (1996) The osmium isotopic content of the Earth's primitive upper mantle. *Nature, 383,* 517–520.

Melosh H. J. (1990) Giant impacts and the thermal state of the early Earth. In *Origin of the Earth* (H. E. Newsom and J. H. Jones, eds.), pp. 69–83. Oxford Univ., New York.

Mueller S., Taylor G. J., and Phillips R. (1988) Lunar composition: A geophysical and petrological synthesis. *J. Geophys. Res., 93,* 6338–6352.

Newsom H. (1984) Constraints on the origin of the Moon from the abundance of molybdenum and other siderophile elements. In *Origin of the Moon* (W. K. Hartmann, R. J. Phillips, and G. J. Taylor, eds.), pp. 203–229. Lunar and Planetary Institute, Houston.

Newsom H. E. (1995) Composition of the solar system, planets, meteorites, and major terrestrial reservoirs. In *Global Earth Physics, A Handbook of Physical Constants. AGU Reference Shelf 1,* pp. 159–189. American Geophysical Union, Washington, DC.

Newsom H. E. and Taylor S. R. (1989) Geochemical implications of the formation of the Moon by a single giant impact. *Nature, 338,* 29–34.

Newsom H. E., White W. M., Jochum K. P., and Hofmann A. W. (1986) Siderophile and chalcophile element abundances in oceanic basalts, lead isotope evolution and growth of the Earth's core. *Earth Planet. Sci. Lett., 80,* 299–313.

Newsom H. E., Sims K. W. W., Noll P. D., Jaeger W. L., Maehr S. A., and Beserra T. B. (1996) The depletion of tungsten in the bulk silicate earth: constraints on core formation. *Geochim. Cosmochim. Acta, 60,* 1155–1169.

O'Neill H. St. C. (1991) The origin of the Moon and the early history of the Earth — A chemical model, Part 1: The Moon. *Geochim. Cosmochim. Acta, 55,* 1135–1157.

O'Neill H. St. C. and Palme H. (1998) Composition of the silicate Earth: Implications for accretion and core formation. In *The Earth's Mantle, Composition, Structure and Evolution* (I. Jackson, ed.), pp. 3–126. Cambridge Univ., London.

Oversby V. M. and Ringwood A. E. (1971) Time of formation of the Earth's core. *Nature, 234,* 463–465.

Palme H. and Beer H. (1993) Abundances of the elements in the solar system. In *Instruments; Methods; Solar System* (H. H. Voigt, ed.), pp. 196–221. Landolt-Börnstein, Group VI: *Astronomy and Astrophysics, Vol. 3a.* Springer-Verlag, Berlin.

Palme H. and Nickel K. (1985) Ca/Al ratio and composition of the Earth's upper mantle. *Geochim. Cosmochim. Acta, 49,* 2123–2132.

Palme H. and Rammensee W. (1981) The significance of W in planetary differentiation processes: Evidence from new data on eucrites. *Proc. Lunar Planet. Sci. 12B,* pp. 949–964.

Palme H. and Wänke H. (1975) A unified trace element model for the evolution of the lunar crust and mantle. *Proc. Lunar Sci. Conf. 6th,* pp. 1179–1202.

Rammensee W. and Wänke H. (1977) On the partition coefficient of tungsten between metal and silicate and its bearing on the origin of the Moon. *Proc. Lunar Sci. Conf. 8th,* pp. 399–409.

Rammensee W., Palme H., and Wänke H. (1983) Experimental investigation of metal-silicate partitioning of some lithophile elements (Ta, Mn, V, Cr) (abstract). In *Lunar and Planetary Science XIV,* pp. 628–629. Lunar and Planetary Institute, Houston.

Righter K., Drake M. J., and Yaxley G. (1997) Prediction of siderophile element metal-silicate partition coefficients to 20 GPa and 2800°C: the effects of pressure, temperature, oxygen fugacity, and silicate and metallic melt compositions. *Phys. Earth Planet. Inter., 100,* 115–134.

Ringwood A. E.(1979) *Origin of the Earth and Moon.* Springer-Verlag, New York. 295 pp.

Ringwood A. E. (1990) Earliest history of the Earth-Moon system. In *Origin of the Earth* (H. E. Newsom and J. H. Jones, eds.), pp. 101–134. Oxford Univ., New York.

Ringwood A. E. (1991) Phase transformations and their bearing on the constitution and dynamics of the mantle. *Geochim. Cosmochim. Acta, 55,* 2083–2110.

Ringwood A. E., Seifert S., and Wänke H. (1987) A komatiite component in Apollo 16 highland breccias: Implications for the nickel-cobalt systematics and bulk composition of the Moon. *Earth Planet. Sci. Lett., 81,* 105–117.

Ringwood A. E., Kato T., Hibberson W., and Ware N. (1991) Partitioning of Cr, V, and Mn between mantles and cores of differentiated planetesimals: implications for giant impact hypothesis of lunar origin. *Icarus, 89,* 122–128.

Seifert S. and Ringwood A. E. (1987) The lunar geochemistry of chromium and vanadium. *Earth, Moon, Planets, 40,* 45–70.

Seifert S., O'Neill H. St. C., and Brey G. (1988) The partitioning of Fe, Ni, and Co between olivine, metal, and basaltic liquid: an experimental and thermodynamic investigation, with application to the composition of the lunar core. *Geochim. Cosmochim. Acta, 52,* 603–616.

Shaffer E. E., Jurewicz A. J. G., and Jones J. H. (1991) Experimental studies of the volatility of V and Mn (abstract). In *Lunar and Planetary Science XXII,* pp. 1221–1222. Lunar and Planetary Institute, Houston.

Spettel B., Palme H., Ionov D. A., and Kogarko L. N. (1991) Variations in the iridium content of the upper mantle of the Earth (abstract). In *Lunar and Planetary Science XXII,* pp. 1301–1302. Lunar and Planetary Institute, Houston.

Stolper E. M. (1977) Experimental petrology of the eucritic meteorites. *Geochim. Cosmochim. Acta, 41,* 587–611.

Taylor S. R.(1980) Refractory and moderately volatile element abundances in the earth, moon and meteorites. *Proc. Lunar Planet. Sci. Conf. 11th,* pp. 333–348.

Taylor S. R. (1982) *Planetary Science: A Lunar Perspective.* Lunar and Planetary Institute, Houston. 481 pp.

Taylor S. R. and McLennan S. M. (1985) *The Continental Crust: Its Composition and Evolution.* Blackwell, Oxford. 312 pp.

Taylor S. R. and Norman M. D. (1990) Accretion of differentiated planetesimals to the Earth. In *Origin of the Earth* (H. E. Newsom and J. H. Jones, eds.), pp. 69–83. Oxford Univ., New York.

Tera F., Eugster O., Burnett D. S., and Wasserburg G. J. (1970) Comparative study of Li, Na, Rb, Cs, Ca, Sr, and Ba abundances in achondrites and in Apollo 11 lunar samples. *Proc. Apollo 11 Lunar Sci. Conf.,* pp. 1637–1657.

Tonks W. B. and Melosh H. J. (1990) The physics of crystal settling and suspension in a turbulent magma ocean. In *Origin of the Earth* (H. E. Newsom and J. H. Jones, eds.), pp. 151–174. Oxford Univ., New York.

Verhoogen J., Turner F. J., Weiss L. E., Wahrhaftig C. and Fyfe W. S. (1970) *The Earth,* pp. 610–611. Holt, Rinehart and Winston, New York.

Walker D., Longhi J., Lasaga A. C., Stolper E. M., Grove T. L., and Hays J. F. (1977) Slowly cooled microgabbros 15555 and 15565. *Proc. Lunar Sci. Conf. 8th,* pp. 1521–1547.

Walker R. J., Morgan J. W., and Horan M. F. (1995) [187]Os enrichment in some mantle plume sources: Evidence for core-mantle interaction? *Science, 269,* 819–822.

Wänke H. (1981) Constitution of terrestrial planets. *Philos. Trans. Roy. Soc. Lond., A303,* 287–302.

Wänke H. and Dreibus G. (1986) Geochemical evidence for the formation of the Moon by impact induced fission of the proto-Earth. In *Origin of the Moon* (W. K. Hartmann, R. J. Phillips, and G. J. Taylor, eds.), pp. 649–672. Lunar and Planetary Institute, Houston.

Wänke H., Baddenhausen H., Palme H. and Spettel B. (1974a) On the chemistry of the Allende inclusions and their origin as high temperature condensates. *Earth Planet. Sci. Lett., 23,* 1–7.

Wänke H., Palme H., Baddenhausen H. Dreibus G., Jagoutz E., Kruse H., Spettel B., Teschke F., and Thacker R. (1974b) Chemistry of Apollo 16 and 17 samples: Bulk composition, late stage accumulation and early differentiation of the moon. *Proc. Lunar Sci. Conf. 5th,* pp. 1307–1335.

Warren P. H. (1985) The magma ocean concept and the lunar evolution. *Annu. Rev. Earth Planet. Sci., 13,* 210–204.

Warren P. H. (1986) The bulk-Moon MgO/FeO ratio: A highlands perspective. In *Origin of the Moon* (W. K. Hartmann, R. J. Phillips, and G. J. Taylor, eds.), pp. 279–310. Lunar and Planetary Institute, Houston.

Warren P. H. and Wasson J. T. (1979) The origin of KREEP. *Rev. Geophys. Space Phys., 17,* 73–88.

Wasson J. T. (1985) *Meteorites: Their Record of Early Solar-System History.* W. H. Freeman, San Francisco. 267 pp.

Wasson J. T. and Kallemeyn G. W. (1988) Composition of chondrites. *Philos. Trans. Roy. Soc. Lond., A325,* 535–544.

Weidenschilling S. J., Greenberg R., Chapman C. R., Herbert F., Davis D. R., Drake M. J., Jones J. H., and Hartmann W. K. (1986) Origin of the Moon from a circumterrestrial disk. In *Origin of the Moon* (W. K. Hartmann, R. J. Phillips, and G. J. Taylor, eds.), pp. 731–762. Lunar and Planetary Institute, Houston.

Wetherill G. W. (1994) Provenance of the terrestrial planets. *Geochim. Cosmochim. Acta, 58,* 4513–4520.

Wolf R. and Anders E. (1980) Moon and Earth: Compositional differences inferred from siderophiles, volatiles and alkalis in basalts. *Geochim. Comochim. Acta., 44,* 2111–2124.

Wulf A. V., Palme H., and Jochum K. P. (1995) Fractionation of volatile elements in the early solar system: evidence from heating experiments on primitive meteorites. *Planet. Space Sci., 43,* 451–468.

# Outstanding Questions for the Giant Impact Hypothesis

## Glen R. Stewart

*University of Colorado*

Recent efforts to fill in the details of the giant impact hypothesis (GIH) for the origin of the Moon have revealed difficulties in reproducing some of the observed physical, thermal, and geochemical properties of the Earth-Moon system. Many of these problems may arise from the overly simplistic nature of the original statement of the GIH as well as from shortcomings in current models of terrestrial planet formation. Plausible modifications of the GIH include: (1) the Moon was formed when Earth was only two-thirds of its present mass; (2) the Moon accreted slowly from the outer edge of a protolunar disk while the inner disk remained hot and molten; (3) resonant interactions with the remaining protolunar disk material increased the inclination of the Moon's orbit as it receded from Earth; and (4) impacts on the Earth and Moon subsequent to the giant impact modified the chemical composition of each body. New lines of research required to develop and test these ideas are identified.

## 1. INTRODUCTION

The most attractive attribute of the giant impact scenario for the formation of the Moon is the dynamical plausibility of large impacts during the formation of the terrestrial planets. Statistical simulations of the early stages of planetesimal accretion in the terrestrial zone appear to yield a population of protoplanets with masses ranging from 2% to 20% of an Earth mass (*Wetherill and Stewart*, 1993; *Weidenschilling et al.*, 1997) and recent N-body simulations apparently confirm this result (*Kokubo and Ida*, 1998). It seems to be inevitable that the merger of several hundred of these protoplanets into four terrestrial planets must have required numerous giant impacts between massive protoplanets. Monte Carlo simulations of the final stages of terrestrial planet accumulation (*Chambers and Wetherill*, 1998; *Agnor et al.*, 1999) from a population of lunar-mass protoplanets allowed *Wetherill* (1985, 1986) to estimate the number and mass distribution of giant impacts. Recently, more accurate N-body simulations of the final stages of accumulation have allowed investigators to tabulate the relative timing of giant impacts as well as the relative velocities and impact parameters characterizing the collisions (*Agnor et al.*, 1999). This body of work shows that giant impacts between planet-sized objects were probably a ubiquitous feature of terrestrial planet accumulation.

At the same time it is important to state the limitations of our understanding of planetary accretion. Simulations of the final stages of accretion have so far completely neglected the effects of collisional fragmentation. A strong hint that collisional fragmentation was important during the final stages of accretion can be found in *Agnor et al.'s* (1999) result that many impacts have such large impact parameters that their assumption of perfect accretion leads to excessively high planetary spin rates. In reality, collisions with large impact parameters and large relative velocities likely produce net erosion rather than perfect accretion, which tends to broaden the size distribution of planetesimals. The formation of an Fe-rich Mercury by collisional stripping of its silicate mantle may be a vivid example of this process (*Benz et al.*, 1988). Statistical simulations of the early runaway stage of planetary accretion that include collisional fragmentation find that a broad size distribution of planetesimals is maintained even as runaway growth of protoplanets proceeds (*Wetherill and Stewart*, 1993). The production of additional small planetesimals by collisional disruption of some protoplanets would tend to blur the idealized transition between the "early runaway stage" of protoplanet formation and the "slow final merger of protoplanets" into four terrestrial planets. It will be necessary to model the collisional outcomes of a wide variety of giant impacts in order to model the evolution of the planetesimal size distribution throughout the entire accumulation of the terrestrial planets. Much work remains to be done before we can predict the statistics and timing of giant impacts with real confidence.

Even if giant impacts are a natural characteristic of planetary accretion, it still must be shown that such an impact can actually produce an Earth-Moon system with the observed physical properties. These physical properties are conveniently divided into three categories: (1) dynamical properties, including the masses of the Earth and Moon as well as the angular momentum of the Earth-Moon system; (2) thermal properties, such as the presence or absence of a magma ocean on Earth and whether or not the Moon was ever completely molten; and (3) geochemical properties, such as the abundance of metals and siderophile elements in the Moon. All these properties are discussed in various chapters of this book. My purpose in this chapter is not to review all these other chapters, but rather to highlight some of the outstanding difficulties that remain with the giant impact scenario for the formation of the Moon and to suggest future lines of inquiry that could possibly resolve some of these difficulties.

## 2.  POSSIBLE PROBLEMS WITH THE GIANT IMPACT HYPOTHESIS

1. The preservation of mantle xenoliths that strongly resemble chondritic meteorites in their major-element abundances has been used to argue that a magma ocean never existed on Earth (see chapter by *Jones and Palme,* 2000). This is a controversial argument because the depth of the magma ocean and the degree of differentiation of material brought in by subsequent impacts is vigorously debated in the literature (see chapter by *Solomatov,* 2000). Available petrologic data has not settled this question because the relevant chemical partition coefficients are poorly constrained (*Presnall et al.,* 1998).

*Jones and Palme's* (2000) argument is also less compelling if one does not insist that the giant impact happen at the very end of Earth's formation. In fact, the size distribution of impactors is likely to be much broader than suggested by recent N-body simulations. If collisional disruption of some portion of the initial population of protoplanets is allowed, then the final assembly of Earth after the giant impact may be less dominated by Mars-sized impactors, allowing the magma ocean to solidify before a substantial fraction of Earth is accreted. Large impactors are more effective at maintaining a magma ocean than more numerous small impactors because they deposit their accretion energy at greater depths in Earth's mantle (*Kaula,* 1979). It is clearly important to determine the frequency and size distribution of impactors required to sustain a magma ocean on the accreting Earth (see chapter by *Pritchard and Stevenson,* 2000).

2. Hafnium-182–tungsten-182 dating of lunar and terrestrial rocks suggest that the Moon formed ~50 m.y. before Earth was completely formed (*Lee et al.,* 1997). Recent N-body simulations yield impact statistics that are not incompatible with the giant impact occurring this early. A possible problem with this picture is that subsequent impacts on Earth could lead to an excess of [182]W on Earth (see chapter by *Jones and Palme,* 2000). An open question that needs more work is whether or not the metallic cores of all impactors are neatly segregated into Earth's core without mixing in the mantle, as suggested by the giant impact simulations (see chapter by *Cameron,* 2000). The limited resolution of current impact simulations may well underestimate the degree of mixing of impactor metals in Earth's mantle if the relevant two-fluid instabilities occur at wavelengths that are small compared to the particle size in an SPH simulation. In any case, collisional disruption of some protoplanets will tend to mechanically mix the cores and mantles in the population of impactors, leading to a more chemically homogenous source of new material that is added to Earth's mantle.

3. Lunar accretion from a protolunar disk is very inefficient, since only about half the initial disk material is accreted into the Moon. The remaining material is pushed back onto Earth by the Moon (*Ida et al.,* 1997; see chapter by *Kokubo et al.,* 2000). To get 2 M$_{\mathbb{C}}$ (lunar mass) of material into Earth orbit probably requires a giant impact with substantially more angular momentum than currently exists in the Earth-Moon system. *Cameron and Canup* (1998) suggest that this problem is alleviated by allowing the giant impact to occur when Earth is less than two-thirds its present mass (see chapter by *Cameron,* 2000). An early-formed Moon has the additional advantage that subsequent large impacts on Earth could reduce the angular momentum of the Earth-Moon system and thereby widen the window of possible impact parameters for the Moon-forming event (*Agnor et al.,* 1999).

4. Lunar accretion from a protolunar disk of solid particles is very rapid, leading to a Moon that is initially completely molten (*Ida et al.,* 1997). This result apparently contradicts the observed lack of evidence for substantial contraction or expansion in the Moon's crust that would seem to require an initially cold lunar interior (*Solomon,* 1984; see chapter by *Pritchard and Stevenson,* 2000). Some petrologic models are also at odds with a completely molten Moon (see chapter by *Snyder et al.,* 2000). A possible solution to the problem of rapid hot accretion is that the rate of accretion is governed by the timescale to condense solids in the outer disk and not by the rapid sweepup of solids by a growing Moon. A complete understanding of this process will require detailed models of the dynamical and thermal evolution of the protolunar disk because radial spreading of material outward beyond Earth's Roche radius is necessarily accompanied by angular momentum and energy transport throughout the disk.

5. Lunar accretion from a protolunar disk would tend to produce a Moon in an orbit that lies in the plane of Earth's equator. Attempts to reconstruct the evolution of the Moon's orbit from its current dynamical state find that the initial inclination of the orbit relative to Earth's equator was about 10° (*Goldreich,* 1966; *Touma and Wisdom,* 1994). An attractive solution to this problem is that resonant interactions between the Moon and the remaining disk material can plausibly pump up the Moon's orbital inclination (*Ward and Canup,* 2000). In addition, as the Moon's orbit migrates outward to 6 Earth radii, the Moon may have been temporarily captured into a couple of resonances associated with solar gravitational perturbations, which could also increase the eccentricity and inclination of the Moon's orbit (*Touma and Wisdom,* 1998; see chapter by *Touma,* 2000). A major uncertainty with this model is the rate of tidal dissipation of energy in the Moon's interior that is associated with major increases in the Moon's orbital eccentricity. Strong tidal heating of the Moon exacerbates the problem of how to avoid a completely molten lunar interior.

## 3.  NEW ASPECTS OF THE GIANT IMPACT THEORY

In this section I outline three lines of inquiry that are motivated by the problems listed in the preceding section.

### 3.1.  Early Evolution of the Protolunar Disk

In Cameron's highest-resolution smoothed particle hydrodynamics (SPH) simulations of the giant impact, the protolunar disk is formed by the tidal disruption and shearing out of a "spiral arm" of material (see chapter by *Cameron, 2000*). After some 30 hr of dynamical evolution, the protolunar disk continues to exhibit significant nonaxisymmetric structure. In addition, several moonlets on eccentric orbits sweep back and forth across the Roche radius of the proto-Earth and continue to shed particles into the disk and onto the proto-Earth as the calculation proceeds. This is clearly not the kind of smooth axisymmetric disk that theorists are used to modeling with fluid dynamical simulations! Is the protolunar disk really this peculiar, or are we being misled by the limitations of the SPH simulations?

The persistent graininess of the disk after 10 rotation periods can partly be explained by insufficient numerical resolution in the disk. Inside the Roche radius, molten silicate bodies should eventually be fragmented by tidal forces down to a size that is stabilized by the surface tension of liquid silicate (300 ergs cm$^{-2}$). A rough estimate of the stable size of liquid drops can be obtained in the following manner. Assume the tidal force deforms a droplet into a prolate spheroid of the form

$$r(\theta, \varphi) = \frac{r_o \left(1 - \varepsilon^2\right)^{1/6}}{\sqrt{1 - \varepsilon^2 \cos^2 \theta}}$$

where $r_o$ is the radius of the sphere with the same volume and $\varepsilon$ is the ellipticity. Incompressible prolate spheroids become unstable when $\varepsilon$ exceeds 0.88 (*Chandrasekhar, 1987*). When surface tension (proportional to the surface area) is balanced against the tidal force (proportional to the quadrupole moments), the drop radius that has this ellipticity is

$$r_o \approx 37 \left( \frac{a}{R_E} \right) cm$$

where a is the distance from the proto-Earth and $R_E$ is the radius of the proto-Earth. More accurate calculations that include the drop's self-gravity and rotation do not change this estimate significantly near Earth's surface, but can substantially increase the maximum drop radius near the Roche radius of Earth.

If the $3.2 \times 10^{22}$ g SPH particles could be subdivided into submeter-sized droplets of liquid silicate, the simulation would most likely produce a much smoother disk after 10 rotation periods because the collision rate between droplets would exceed the orbital rotation rate by several orders of magnitude. In particular, the orbital eccentricities of these droplets would be damped to near zero in all but the most rarefied regions of the disk. Does this mean that one can transform Cameron's disks into smooth fluid disks by some kind of averaging procedure? Not necessarily! A serious limitation of Cameron's SPH simulations is that they cannot avoid angular momentum transport due to numerical viscosity effects. In SPH simulations, the effective kinematic viscosity is proportional to the smoothing length of the particles (*Nelson et al., 1998; Lombardi et al., 1999*). Since the

smoothing length is continually adjusted to allow particles to interact with their nearest neighbors, the effective viscosity is tied to the local mean spacing between particles and thus is proportional to $n(\mathbf{r})^{-1/3}$, where $n(\mathbf{r})$ is the local number density of particles in the simulation. Low-density regions where the spacing between particles is large will be more strongly affected by this numerical viscosity. Even in Cameron's highest-resolution simulations containing 100,000 particles, the smoothing length in the protolunar disk exceeds several hundred kilometers. The large effective viscosity in these simulations appears to deplete the prolunar disk of material in a few tens of orbits (see chapter by *Cameron, 2000*). In reality, submeter-sized droplets of molten silicate would have mean free paths that are also measured in meters. The details of the radial surface density profile obtained in these simulations are therefore likely to be modified in future higher-resolution simulations.

As the protolunar material begins to form a symmetric disk, it may be subject to global gravitational instabilities. Initial attempts to model the protolunar disk using a fluid dynamical simulation have produced the interesting result that nonaxisymmetric modes in the interior of Earth participate in forming spiral density waves in the disk (*Pickett et al., 1998*). These fluid dynamical calculations assume an inviscid Earth, which may actually be appropriate if the outer layer of Earth is a deep magma ocean after the giant impact. Tidal interactions between Earth and nonaxisymmetric modes in the protolunar disk may have unexpected consequences for the radial evolution of the protolunar disk (*Ward, 1998*).

The wavelength $\lambda$ of gravitational instabilities in the disk is given by

$$\frac{\lambda}{a} = 2 \frac{M_E}{a^2 \sigma} \left( \frac{H}{a} \right)^2$$

where $M_E$ is the proto-Earth mass, a is the orbit radius, $\sigma$ is the surface density of the protolunar disk, and H is the disk scale height. If the protolunar disk consists entirely of rock vapor, the scale height would be on the order of 10% of the orbit radius and one would find an instability scale that is comparable to the radius of the disk. The large latent heat required to vaporize silicate makes this kind of global instability seem unlikely. If the bulk of the disk mass is in the liquid state, then the scale height was determined by the criteria for marginal gravitational instability

$$\frac{H}{a} = \frac{\pi a^2 \sigma}{M_E}$$

which implies a wavelength for instabilities that is substantially smaller than the radius of the disk. The N-body simulations of a protolunar disk consisting of inelastic particles shown in Fig. 7 of the chapter by *Kokubo et al.* (2000) provide a good illustration of how gravitational instabilities develop in an unstable disk with a small scale height. The spreading rate in this simulation is roughly consistent with the effective disk viscosity estimated by *Ward and Cameron*

(1978). In reality, the disk scale height must have been maintained at a larger value than the above estimate because a marginally unstable disk liberates energy faster than it can radiate the energy to space (*Thompson and Stevenson*, 1988; see chapter by *Pritchard and Stevenson, 2000*). Standard accretion disk theory dictates that the inner regions of the disk will remain hot due to the release of gravitational potential energy as the disk tries to spread radially. Thus the outer regions of the disk will cool first and be subject to gravitational instabilities while the inner disk remains hot and stable. It is entirely conceivable that the inner disk will remain molten throughout the accretion of the Moon from condensed material in the outer region of the disk.

Further progress on this subject will require improved models of the rheology of the postimpact Earth as well as the dynamical and thermal evolution of the disk. The first-order question that remains to be answered is how rapidly and how efficiently disk material can be pushed out beyond Earth's Roche radius where it can then be accreted into a moon.

### 3.2.  How Rapid was Lunar Accretion?

N-body simulations of the accretion of the Moon from a disk of particles yields the impressive result that the Moon is formed on a timescale of months (*Ida et al., 1997*). In retrospect, this result is an artifact of the assumed initial condition that the entire protolunar disk consists of solid material. Realistically, the disk initially consists of liquid rock and possibly some rock vapor. Since the cooling time of the disk greatly exceeds the accretion time of condensed material, the actual rate of lunar accretion is likely to be governed by the rate-limiting step, namely, the rate of supply of solid rock in the Roche zone of the disk [the Roche zone extends from 0.8 to 1.35× the classical Roche radius; see chapter by *Kokubo et al.* (2000) for a discussion of the Roche zone]. Liquid rock is not a suitable building material because liquid structures cannot withstand the tidal stresses inside the Roche zone. Experimental data shows that the yield stress of silicate rock is an exponentially decreasing function of temperature in the melting range, so a rather sharp transition is likely.

The cooling time for a uniform density disk is estimated to be about 10 yr (see chapter by *Pritchard and Stevenson, 2000*). Since the surface density of the disk will decrease rapidly at its outer edge, the cooling time in the outer edge is likely to be shorter than the time required to cool the bulk of the disk. Viscous spreading of the disk will liberate additional gravitational energy, which will tend to keep the inner disk hot on a longer timescale. In this scenario, the rate of lunar accretion will be determined by the rate of supply of condensable solids in the outer edge of the protolunar disk. If the initial disk contains too much material above Earth's Roche radius, then lunar accretion will be so rapid that a completely molten Moon is inevitable. Vigorous gravitational instabilities in the disk that rapidly push a lunar mass of material outside the Roche radius would also lead to a molten Moon. Perhaps the only way

to form a cold Moon is to have a relatively stable, low-viscosity disk that gradually pushes material outward on a timescale that is governed by the cooling time of the bulk of the disk.

An interesting consequence of slow accretion on the outer edge of a mostly fluid disk is the possibility of satellite-disk interactions that push small moonlets outward away from the edge of the disk. The timescale for radial migration a distance $\Delta a$ (small compared to a) due to satellite-disk interactions is (*Goldreich and Tremaine, 1982*)

$$t = \frac{M_\oplus^2}{1.68\Omega a^2 \sigma m}\left|\frac{\Delta a}{a}\right|^3$$

where $\Omega$ is the orbit frequency, a is the orbit radius, $\sigma$ is the surface density of the protolunar disk, and m is the mass of the moonlet. For a rough estimate, I assume $\sigma = 3 \times 10^7$ g/cm$^2$, $1/\Omega = 4.2 \times 10^3$ s, and a = 3 R$_\oplus$. The timescale for radial migration becomes

$$t = (0.35\text{yr})\left(\frac{M_\mathbb{C}}{m}\right)\left|\frac{\Delta a}{a}\right|^3$$

This estimate suggests that even moonlets as small as 1% of 1 M$_\mathbb{C}$ could be pushed away from the outer edge of the disk on a time scale that is short compared to the 10-yr cooling time. On a 100-yr timescale tidal evolution due to Earth tides can also move moonlets outward. Detailed simulations could determine if each moonlet is rapidly accreted by its predecessor or if a moonlet belt is formed that requires an even longer timescale to merge into a single large moon. Dynamical simulations of multiple moons subject to both tidal evolution and their mutual gravitational interactions suggest that tens of years are required to allow collisions between these moons (*Canup et al., 1999*; see chapter by *Kokubo et al., 2000*). Future simulations should also include gravitational interactions with the remaining protolunar disk (*Canup and Ward, 2000*). In any case, this scenario implies a more leisurely rate of lunar accretion that may avoid the completely molten Moon indicated by the results of *Ida et al.* (1997).

### 3.3.  Early Formation of the Moon

The amount of angular momentum required by the giant impact to successfully inject 2 M$_\mathbb{C}$ of material into Earth orbit substantially exceeds the current angular momentum budget of the Earth-Moon system unless Earth is reduced to about two-thirds its current mass (*Cameron and Canup, 1998*). The obvious solution to this dilemma is to form the Moon early before Earth is fully formed. There are a couple of additional theoretical results that lend support to this idea.

First, simulations of the final stages of terrestrial planet formation yield statistics of giant impacts that favor multiple giant impacts distributed over the 100-m.y. timescale of planetary accretion (*Wetherill, 1985, 1986; Agnor et al., 1999*; see chapter by *Canup and Agnor, 2000*). Not all giant impacts will have sufficient angular momentum to produce a Moon, but they are all capable of producing significant changes to the spin state of the proto-Earth. Indeed, the impact statistics of *Agnor et al.* (1999) suggest that the angular momentum of Earth could easily be changed after the Moon-forming event by a subsequent impact, which mitigates the necessity of exactly reproducing the present angular momentum budget with the same event that formed the Moon.

Second, an early-formed Moon may have a more complicated accretion history compared to a late-formed Moon because it will inevitably be subjected to an intense impact flux as the proto-Earth accretes the remaining approximately one-third of its mass. To assess the effects of large impacts on an early-formed Moon, I will make a rough estimate of the number of impacts on the Moon after its formation. The Moon is initially formed in low-Earth orbit where it would have essentially the same gravitational enhancement of its collision cross section as Earth, leading to a total impact rate equal to Earth's reduced by the ratio of the Moon and Earth's geometric cross sections (the current ratio of geometric cross sections is 1:13). However, tidal evolution of the Moon's orbit from 3 Earth radii out to 20 Earth radii is rapid on the 100-m.y. timescale of planetary accretion, which reduces Earth's gravitational enhancement of the lunar cross section to a few tens of percent. I will therefore neglect the influence of Earth on the Moon's impact rate. Assuming that the approach velocity of planetesimals is likely to be about 6 km/s (about half the surface escape velocity of the present Earth), the ratio of impact fluxes on the Moon and Earth was probably less than 1:30 once the Moon's orbit evolved out beyond 20 Earth radii.

Next I assume that the remaining one-third Earth mass of planetesimals impacting Earth is distributed in a power-law size distribution of the form $dN/dm = C/m^q$ with $m_s < m < m_b$. The total number of planetesimals in this size distribution is

$$N = \frac{2-q}{1-q} \frac{M_\oplus}{3} \frac{m_b^{1-q} - m_s^{1-q}}{m_b^{2-q} - m_s^{2-q}}$$

For q values less than 2, most of the mass will be contained in the largest bodies, which is often the case in simulations of planetary accretion. If we set $m_b = 0.1\ M_\oplus$ and $m_s = 0.001\ M_\oplus$, the total number of planetesimals ranges between 33 and 72 as q ranges between 1.5 and 2.0. Since 1 in 30 of these bodies will likely hit the Moon, I find that the Moon suffers at least one impact by a body greater than one-tenth the mass of the Moon. If the largest mass in our distribution, $m_b$ is reduced, the number of large impacts on the Moon increases. The number of large impacts on the Moon can be reduced if I assume that most of the impactors are contained in objects substantially smaller than the Moon, but this assumption would reduce the probability of our initial hypothesis that a giant impact on Earth formed the Moon. Note that one should not take the size distribution of impactors found in published N-body simulations of planet formation too seriously, because most of these simulations neglect collisional fragmentation, which would tend to increase the number of smaller planetesimals by disrupting larger objects. It would be erroneous, for example, to believe that there are no impactors smaller than the Moon simply because recent N-body simulations assume that all the initial planetesimals are large!

The tentative conclusion from the above discussion is that an early-formed Moon is likely to suffer one or more "giant impacts" after its formation. Can these impacts collisionally disrupt the Moon? Will the Moon gain or lose mass in such impacts? Although the relevant simulation of a giant impact on the Moon remains to be done, the SPH simulations of *Love and Ahrens* (1996) can be extrapolated to estimate the energy density, Q*, required to disrupt and disperse one-half the mass of the Moon. I find that Q* is roughly between $1 \times 10^7$ and $2 \times 10^7$ J/kg. The mass of an impactor required to deliver this much energy to the Moon is

$$m = \frac{2Q^*}{v^2} M_{\mathbb{C}}$$

where the impact velocity v is assumed to be equal to 6 km/s. I find that the impactor mass required to disrupt the Moon is in the range of $0.0068\ M_\oplus$ and $0.0136\ M_\oplus$. Since this mass is comparable to the mass of the Moon itself, it appears unlikely that an early-formed Moon will be collisionally disrupted. It not so unlikely that the Moon will lose perhaps 10% of its mass in a large impact. I base this conclusion on the simulation result that the fraction of mass eroded in a large impact is proportional to the impact energy (*Love and Ahrens,* 1996). The reader is warned that Love and Ahrens' proportionality may not apply to cases where the velocity of the impactor does not greatly exceed the escape velocity of the target.

Orbital integrations of lunar ejecta show that the vast majority of impact ejecta escapes from Earth orbit before it has a chance to reimpact the Moon (*Gladman et al.,* 1995). This occurs because the velocity perturbations caused by close encounters with the Moon substantially exceed the escape velocity from Earth at 60 Earth radii. When the Moon was only at 20 Earth radii, the escape velocity from Earth is comparable to the velocity perturbations caused by the Moon, so some ejecta will surely reimpact the Moon. This argument points toward net erosion of the Moon, but cannot rule out the possibility that the Moon receives significant amounts of siderophile-rich material by large impacts after its formation, because giant impacts tend to mix some impactor material with the target regardless of the final mass balance. Recent measurements of the size of the lunar core suggest that siderophile elements will be efficiently sequestered in the lunar core, so these latter additions of material may not pose a real problem for an early-formed Moon (see chapter by *Jones and Palme,* 2000).

## 4.  CONCLUSIONS

Many of the perceived problems with the giant impact hypothesis for the formation of the Moon may well originate from our primitive understanding of how the terrestrial planets formed. Idealized models of planetary accretion make for entertaining computer simulations and facilitate the communication of important concepts across broad fields of expertise, but they can also mislead the unwary into thinking that our level of understanding is deeper than it really is. Collisions between planetesimals of all sizes are inherently messy processes that are difficult to characterize with simple models and a small number of parameters. The computational methods used to model these collisions, such as SPH simulations, are imperfect tools that could clearly benefit from additional refinements. The incorporation of collisional disruptions into the final stages of planetary accretion is a challenging problem for the future that will require innovative simulations that possibly combine N-body simulations with statistical methods. Nevertheless, our current understanding of planetary accretion clearly predicts that giant impacts do occur, and the full consequences of this for the formation of the Moon have only begun to be explored.

***Acknowledgments.***    The author thanks D. J. Stevenson, S. Ida, and K. Righter for their constructive reviews. This work was supported by NASA's Planetary Geology and Geophysics Program.

## REFERENCES

Agnor C. B., Canup R. M., and Levison H. F. (1999) On the character and consequences of large impacts in the late stage of terrestrial planet formation. *Icarus, 142,* 219–237.

Benz W., Slattery W. L., and Cameron A. G. W. (1988) Collisional stripping of Mercury's mantle. *Icarus, 74,* 516–528.

Cameron A. G. W. (2000) Higher-resolution simulations of the giant impact. In *Origin of the Earth and Moon* (R. M. Canup and K. Righter, eds.), this volume. Univ. of Arizona, Tucson.

Cameron A. G. W. and Canup R. M. (1998) The giant impact occurred during Earth accretion (abstract). In *Lunar and Planetary Science XXIX,* Abstract #1062. Lunar and Planetary Institute, Houston (CD-ROM).

Canup R. M. and Agnor C. B. (2000) Accretion of the terrestrial planets and the Earth-Moon system. In *Origin of the Earth and Moon* (R. M. Canup and K. Righter, eds.), this volume. Univ. of Arizona, Tucson.

Canup R. M. and Ward W. R. (2000) A hybrid fluid/N-body model for lunar accretion (abstract). In *Lunar and Planetary Science XXXI,* Abstract #1916. Lunar and Planetary Institute, Houston (CD-ROM).

Canup R. M., Levison H. F., and Stewart G. R. (1999) Evolution of a terrestrial multiple moon system. *Astron. J., 117,* 603–620.

Chambers J. E. and Wetherill G. W. (1998) Making the terrestrial planets: N-body integrations of planetary embryos in three dimensions. *Icarus, 136,* 304–327.

Chandrasekhar S. (1987) *Ellipsoidal Figures of Equilibrium.* Dover, New York.

Gladman B. J., Burns J. A., Duncan M. J., and Levison H. F. (1995) The dynamical evolution of lunar impact ejecta. *Icarus, 118,* 302–321.

Goldreich P. (1966) History of the lunar orbit. *Rev. Geophys., 4,* 411–439.

Goldreich P. and Tremaine S. (1982) The dynamics of planetary rings. *Annu. Rev. Astron. Astrophys., 20,* 249–283.

Ida S., Canup R. M., and Stewart G. R. (1997) Lunar accretion from an impact-generated disk. *Nature, 389,* 353–357.

Jones J. H. and Palme H. (2000) Geochemical constraints on the origin of the Earth and Moon. In *Origin of the Earth and Moon* (R. M. Canup and K. Righter, eds.), this volume. Univ. of Arizona, Tucson.

Kaula W. M. (1979) Thermal evolution of the Earth and moon growing by planetesimal impacts. *J. Geophys. Res., 84,* 999–1008.

Kokubo E. and Ida S. (1998) Oligarthic growth of protoplanets. *Icarus, 131,* 171–178.

Kokubo E., Canup R. M., and Ida S. (2000) Lunar accretion from an impact-generated disk. In *Origin of the Earth and Moon* (R. M. Canup and K. Righter, eds.), this volume. Univ. of Arizona, Tucson.

Lee D. C., Halliday A. N., Snyder G. A., and Taylor L. A. (1997) Age and origin of the Moon. *Science, 278,* 1098–1103.

Lombardi J. C., Sills A., Rasio F. A., and Shapiro S. L. (1999) Tests of spurious transport in smoothed particle hydrodynamics. *J. Comput. Phys., 152,* 687–735.

Love S. G. and Ahrens T. J. (1996) Catastrophic impacts on gravity dominated asteroids. *Icarus, 124,* 141–155.

Nelson A. F., Benz W., Adams F. C., and Arnett D. (1998) Dynamics of circumstellar disks. *Astrophys. J., 502,* 342–371.

Pickett B. K., Durisen R. H., and Stewart G. R. (1998) Three-dimensional hydrodynamic simulations of the post-impact protoearth (abstract). In *Origin of the Earth and Moon,* pp. 32–33. LPI Contribution No. 957, Lunar and Planetary Institute, Houston.

Presnall D. C., Weng Y.-H., Milholland C. S., and Walker M. J. (1998) Liquidus phase relations in the system $Mg-MgSiO_3$ at pressures up to 25 GPa — constraints on crystallization of a Hadean mantle. *Phys. Earth Planet. Inter., 170,* 83–95.

Pritchard M. E. and Stevenson D. J. (2000) Thermal aspects of a lunar origin by giant impact. In *Origin of the Earth and Moon* (R. M. Canup and K. Righter, eds.), this volume. Univ. of Arizona, Tucson.

Snyder G. A., Borg L. E., Nyquist L. E., and Taylor L. A. (2000) Chronology and isotopic constraints on lunar evolution. In *Origin of the Earth and Moon* (R. M. Canup and K. Righter, eds.), this volume. Univ. of Arizona, Tucson.

Solomatov V. S. (2000) Fluid dynamics of a terrestrial magma ocean. In *Origin of the Earth and Moon* (R. M. Canup and K. Righter, eds.), this volume. Univ. of Arizona, Tucson.

Solomon S. C. (1984) On the early thermal state of the Moon. In *Origin of the Moon* (W. K. Hartmann, R. J. Phillips, and G. J. Taylor, eds.), pp. 435–452. Lunar and Planetary Institute, Houston.

Thompson A. C. and Stevenson D. J. (1988) Gravitational instability in two-phase disks and the origin of the Moon. *Astrophys. J., 333,* 452–481.

Touma J. (2000) The phase space adventure of the Earth and Moon. In *Origin of the Earth and Moon* (R. M. Canup and K. Righter, eds.), this volume. Univ. of Arizona, Tucson.

Touma J. and Wisdom J. (1994) Evolution of the Earth-Moon system. *Astron. J., 108,* 1943–1961.

Touma J. and Wisdom J. (1998) Resonances in the early evolution of the Earth-Moon system. *Astron. J., 115,* 1653–1663.

Ward W. R. (1998) Earth interactions with an impact-generated disk (abstract). In *Origin of the Earth and Moon,* p. 52. LPI Contribution No. 957, Lunar and Planetary Institute, Houston.

Ward W. R. and Cameron A. G. W. (1978) Disk evolution within the Roche limit. *Proc. Lunar Planet. Sci. Conf. 9th*, pp. 1205–1207.

Ward W. R. and Canup R. M. (2000) Origin of the Moon's orbital inclination from resonant disk interactions. *Nature, 403,* 741–743.

Weidenschilling S. J., Spaute D., Davis D., Marzari F., and Ohtsuki K. (1997) Accretional evolution of a planetesimal swarm: 2. The terrestrial zone. *Icarus, 128,* 429–455.

Wetherill G. W. (1985) Occurance of giant impacts during the growth of the terrestrial planets. *Science, 228,* 877–879.

Wetherill G. W. (1986) Accumulation of the terrestrial planets and implications concerning lunar origin. In *Origin of the Moon* (W. K. Hartman, R. J. Phillips, and G. J. Taylor, eds.), pp. 519–550. Lunar and Planetary Institute, Houston.

Wetherill G. W. and Stewart G. R. (1993) Formation of planetary embryos: Effects of fragmentation, low relative velocity, and independent variation of eccentricity and inclination. *Icarus, 106,* 190–209.

*Part IV:*

*Differentiation of the
Earth and Moon*

# Physical Processes of Core Formation

## T. Rushmer
*University of Vermont*

## W. G. Minarik
*Carnegie Institution of Washington and University of Maryland, College Park*

## G. J. Taylor
*University of Hawai'i*

Separation of metal from silicate was one of the earliest processes active during the accretion of terrestrial solar system bodies, contemporaneous with the earliest recorded silicate melting, hydrothermal metamorphism, and large-impact brecciation. Experiments at planetary pressures and temperatures on the microstructure of metal-sulfide melt in a nondeforming silicate matrix have borne out the insights from metallurgy that metal-rich sulfide melts are unable to migrate through a static, solid silicate matrix at any pressure within the upper mantle or transition zone. Several experiments have suggested that metallic-melt percolation may be possible in the lower mantle, but these pressures are only obtained among the terrestrial bodies within the mantles of Earth and Venus. Recent reconnaissance experiments on deforming systems have shown that deformation profoundly alters the static microstructure, both at high differential stress and at high strain and more moderate stress. These processes are certainly a part of the differentiation process at various times on all the terrestrial bodies known or suspected to have cores. Future, ongoing studies will determine the range of pressures, temperatures, deformation rates, melt fraction, and compositions under which core-forming melts can segregate from silicate mantle material.

## 1. INTRODUCTION

The formation of the Earth's metallic core was one of the most significant events in Earth's developmental history. As knowledge of the core-forming event has important implications for not only the thermal state of the early Earth and the subsequent compositions of the core and mantle, but also for the nature of accretion processes in general, scientists have been intensely interested in determining the core formation process. Successful models of core formation must provide both an explanation for the composition of the mantle [e.g., the excess siderophile-element problem (*Ringwood,* 1966)] and a physically realistic mechanism by which metal separates from its silicate matrix. This is partly in response to the fact that most core-formation models [e.g., inefficient core formation (*Jones and Drake,* 1986), equilibrium between Fe-S-O liquid and mantle (*Brett,* 1984), or heterogeneous accretion (*Wänke et al.,* 1984)] have been based on geochemical arguments and need to provide an attractive means for metal-liquid segregation that will also explain the geochemical data. Indeed, one of the most frequently debated questions is whether the core grew by percolation of metallic liquid through a solid silicate mantle, or by separation of immiscible metal and silicate melts within a planetary-scale magma ocean. Fortunately, significant progress has been made toward experimentally determining metal-silicate siderophile-element partitioning at

extreme pressure-temperature conditions, and these results may be able to satisfy some of the equilibrium constraints that could not be reconciled with earlier partition coefficients determined at 1 bar (*Li and Agee,* 1996; *Righter and Drake,* 1997b; *Walter et al.,* 2000). In fact, the current early Earth models have, for the most part, been able to explain the geochemical observations, although no one model seems yet to be able to adequately account for all observations (*Jones and Drake,* 1986).

Clearly, researchers must continue to work toward reconciling geochemical constraints with the results from experimental studies on liquid metal distribution to arrive at a viable model for core formation. As will be described in this chapter, there are now several studies carried out on metal-silicate systems to determine the mobility of core-forming melts in a solid silicate under both static and dynamic conditions (*Minarik et al.,* 1996; *Minarik and Ryerson,* 1996; *Shannon and Agee,* 1996; *Ballhaus and Ellis,* 1996; *Gaetani and Grove,* 1999; *Bruhn et al.,* 1998; *Rushmer and Jones,* 1998a,b). The static experiments show that there is an overall lack of mobility for core-forming melts at low melt fractions and this implies that heterogeneous accretion models that require core growth by accumulation of small amounts of metal-sulfide melts are improbable. Trapping of metallic melts in the mantle helps to assuage the problem of the apparent disequilibrium between core and mantle, so the inefficient core formation model (*Jones and Drake,* 1986) is not radically changed by these results. Most significantly, however, the results have suggested that extensive melting of silicates is necessary to segregate the core. This in turn has promoted research at very high temperatures and pressures as mentioned above, conditions equivalent to that at the base of a hypothetical magma ocean. It has been suggested that segregation of metal through a magma ocean by raining may result in polybaric/polythermal geochemical equilibration, but the consequences of this on the mantle composition has not been determined (*Walter,* 1998). Even given a molten upper mantle, there still may be a solid deeper mantle through which core-forming metals must descend (*Li and Agee,* 1996; *Righter and Drake,* 1997a,b). This pushes the problem of metal-liquid segregation from a solid silicate matrix deeper into the mantle.

One mechanism that has been recently discussed that would overcome the high dihedral angle problem of molten metals is the effect of deformation on the mobility of core forming melts (*Rushmer and Jones,* 1998a). The interfacial energy that controls how liquid metal is distributed within the silicate matrix is only applicable in static or very slowly deforming environments. As soon as any pressure differences exist in the system, interfacial energy no longer solely controls the distribution of melt. Instead, melt accumulates along grain edges and faces and, if present, in microcracks and shear zones, all of which can form an interconnected permeable network even if the melt-solid assemblage has a high interfacial energy. As it is quite likely that larger planetesimals were convecting during heating and differentiation, it is probable that some amount of deformation accompanied core formation. The implications of active deformation during core formation for the different models for physically separating metal from a silicate matrix are significant. A magma ocean may not be required for physical segregation of the core if the silicate matrix is undergoing deformation. This possibility has been suggested before based on geochemical observations that are inconsistent with whole-scale mantle melting (*Ringwood,* 1990; *Jones,* 1996). For example, there are fertile mantle xenoliths that have some trace-element ratios that are close to chondritic. This is either fortuitous or this material has remained unmodified (unmelted?) over most of Earth's history (*Jones,* 1996). There is also no trace-element or isotopic evidence for fractionation of a perovskite-rich lower mantle from the upper mantle (*Walter,* 1998).

Regardless of which core formation model is correct, physical models for metal segregation need to be extended from high silicate melt fractions to pressure-temperature conditions near or below the silicate solidus. At present there is more data available on static metal-silicate distribution than on deformation results, but enough is known to suggest that the presence or absence of a silicate melt may be an important variable under both conditions (*Rushmer and Jones,* 1998b, 1999). Deformation without the presence of silicate melt will also likely allow for migration of molten metal/sulfide, but the nature of the pathways will depend on the dynamic environment (e.g., the amount of strain and magnitude of stress). The mechanism by which melt is able to interconnect at the grain boundary may also have a significant effect on Fe-S–silicate equilibria. If metal/sulfide can segregate both with and without extensive silicate melting (but with the aid of deformation), then it is important to explore the ramifications of such a process on the potential origin(s) of the Earth's core. One approach is to formulate physical regimes of core formation and then to explore the geochemical implications within those regimes. With this method, specific questions can be addressed: What is the fate of late-accreted metal-bearing materials? How might metal/sulfide–silicate separation occur in large convecting bodies vs. in asteroids? What is the nature of the metal/silicate differentiation in the planetesimals that accreted to form the planets and the Moon-forming impactor? Do they equilibrate with the Earth's mantle? And under which physical regime is this possible? Under which regime is it not possible?

Considering the above, an important focus now is to integrate the recent experimental advances made in geochemistry with physical models of core formation. This will lead to a new generation of viable core formation models that can answer questions regarding how core-forming melts have been able to physically separate from the silicate mantle, descend to form the core, and still fit geochemical constraints. To start, theoretical models (e.g., *Stevenson,* 1990; *Taylor,* 1992) need to be considered with the experimental studies. This chapter will present both theoretical considerations and experimental results on distribution of core-forming melts, under static and dynamic conditions, and then discuss the main implications of the results in regard to core forming processes.

## 2.  THEORETICAL CONSIDERATIONS OF METAL SEGREGATION

*Stevenson* (1990) clearly and thoroughly discussed the fluid dynamics of core formation, and *Taylor* (1992) discussed the specific case of core formation in asteroids. We will not repeat all the issues raised in those papers here. Instead, we emphasize the key physical principles of metal migration in planets and which experiments are needed to quantify them. Experiments are essential because the physics of metal segregation is complicated by numerous factors, such as differences in the composition of the metallic phase (particularly the S and O contents), with paired variations in S and O fugacity, textural disequilibrium, anisotropic interfacial tension, and differential stresses. Several experimental studies have been carried out during the past decade and they are reviewed in the next two sections. In this section we first review the phase equilibria of the Fe-S system and its relation to silicate phase equilibria. We then discuss the process of metal segregation by porous flow, a process dominated by the interfacial energy between molten metal and surrounding silicates. This leads to an examination of the role of deformation in metal segregation, which changes the role of interfacial energy in melt distribution. Finally, we discuss sinking of molten metal from molten silicate. *Stevenson* (1990) investigated the case of solid metal settling through solid silicate and found that it is far too slow a process to have formed a planetary core. It also did not happen in asteroidal-sized bodies unless a process can be found to make metallic masses in the 1–10-m range (*Taylor,* 1992). In this paper, we do not consider the problem of solids settling through solids.

## 2.1.  Core Formation Timing

An important constraint on core formation is the time at which segregation processes began to occur because this helps determine the size of the core-forming body. Isotopic systems have shown that differentiated meteorites melted and cooled very quickly after the first solar system materials formed (taken to be CAI inclusions), perhaps powered by the decay of $^{26}$Al (0.7-m.y. half-life; see *Srinivasan et al.,* 1999). Short-lived, extinct, radioactive systems give the highest resolution of these earliest events, and the most powerful for timing core formation is the $^{182}$Hf-$^{182}$W parent-daughter pair (half-life of 9 m.y.). Both elements are thought to be present in chondritic abundances within the Earth, but W is strongly partitioned into metal, whereas Hf partitions into coexisting silicate. If metal-silicate separation (core formation) occurs before $^{182}$Hf has completely decayed, then excess $^{182}$W will be present in the mantle. Most differentiated bodies (including the Moon and Mars) record very early metal-silicate separation, within several half-lives of the oldest solar-system solids. Iron meteorites all formed within 5 m.y. of each other (*Horan et al.,* 1998). This suggests that after the first 5–10 m.y. of accretion, all colliding planetesimals larger than several hundred kilometers diameter will be differentiated bodies with cores. The exception is Earth, whose upper mantle has a chondritic W-isotopic abundance, indicating equilibration between mantle and core after $^{182}$Hf had decayed completely, more than 50–100 m.y. after CAI formation (*Lee et al.,* 1997). This can be explained in one of three ways: core formation in the Earth much later than any other solar system body, late reequilibration of mantle and core after $^{182}$Hf had decayed, or continuous and very efficient separation of metal from silicate throughout accretion, except for the last 1% of material (the late veneer), which imparted chondritic W abundances on the mantle.

These scenarios all imply different pressure regimes for metal-silicate segregation. If core formation took place continuously during accretion of the Earth, high-pressure phases may have been relatively unimportant. On the other hand, if core melting occurred late (after more than 50% of the Earth was accreted), then high-pressure phase assemblages would have played an important role. Furthermore, late accretion may have allowed differentiation to take place in small planetesimals before they accreted to the growing Earth. Because the details of accretion rate vs. the rate and timing of core formation are not fully understood, we dis-

cuss the physics of melt segregation under a wide range of conditions.

## 2.2.  Phase Equilibria

Phase equilibria (Figs. 1a,b) for silicates and Fe-FeS assemblages can help define the physical setting in which metal segregation took place. Figure 1a combines experimental data on a carbonaceous chondrite (*Agee,* 1997) and a mantle peridotite (*Herzberg and Zhang,* 1996) with metal alloy melting curves (*Boehler,* 1992, 1993, 1996; *Hirayama et al.,* 1993; *Fei et al.,* 1997; *Pike et al.,* 1999). For Fe-FeS, curves for the melting of pure Fe (curve 1), the melting of Fe-FeS (curve 2), and the Fe-FeS eutectic (curve 3) are shown. At pressures below about 5 GPa, corresponding to the interiors of asteroids and the Moon (reasonable sizes for most of the planetesimals that accreted to the growing Earth), the liquidus of metal-sulfide assemblages is above the silicate solidus. The eutectic in the Fe-FeS system forms well below the silicate solidus (Figs. 1a,b), but iron meteorites (almost certainly the cores of asteroids) do not have eutectic compositions (*Taylor,* 1992); metal-sulfide melts of H chondrites correspond approximately to melting temperatures between curves 1 and 2 in Fig. 1a. Because iron meteorites appear to have crystallized from totally molten cores, metal must have segregated in the molten state, and therefore must have been accompanied by a considerable amount of silicate melting (*Taylor,* 1992). Although pressures are higher in the Moon, a similar argument can be made, although the thermally required amount of silicate melting would be less. This implies that metal segregated at least from a silicate mush, possibly from a magma ocean.

Conditions are very different in the present-day Earth and other terrestrial planets. At higher pressures (>5 GPa), the Fe-FeS liquidus is below the silicate solidus at all pressures, except for a region near the olivine-spinel transition, and even there only a pure Fe melt is above the silicate solidus. Thus, in a fully formed Earth, molten metal would have to segregate from a solid silicate rock unless it was heated substantially above the metallic liquidus. Furthermore, the nature of the solid silicate changes with pressure: Olivine dominates up to 17 GPa, spinel and majorite between 17 and 24 GPa, and perovskite above 24 GPa. The interfacial energies between metallic melts and silicate solids might be quite different in these distinct silicate assemblages, leading to changes in the mechanism of metal segregation with pressure. This was a prime motivation for experiments at high pressure (see below).

Some experiments in the Fe-FeS system indicate that S-rich melts (more S than the eutectic melts) might segregate by porous flow more readily than anion-poor metallic melts. If this is an important effect, its significance varies with pressure because the composition of the eutectic varies with pressure (*Fei et al.,* 1997; *Urakawa et al.,* 1987). At 1-bar pressure, the eutectic contains 44 mol% S, but that decreases with increasing pressure and is about 33 mol% at 30 GPa (Fig. 1b). The composition remains roughly constant to

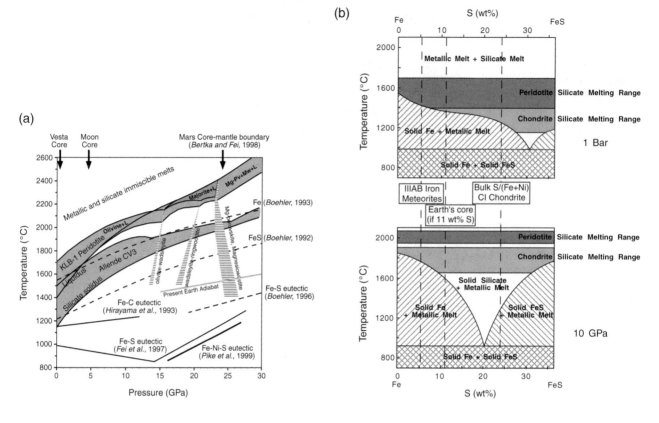

**Fig. 1.** (a) Experimentally determined melting temperatures as a function of pressure for a carbonaceous chondrite (Allende CV; *Agee,* 1997) and a fertile peridotite (KLB-1; *Herzberg and Zhang,* 1996). Also plotted are the melting temperatures of the Fe-Ni-S eutectic, Fe-S eutectic and Fe-C eutectic compositions, stoichiometric FeS, and Fe (sources given on figure). Major silicate phase boundaries are also noted, as are the interior pressures of three solar system bodies with cores: Vesta, the Earth's Moon, and Mars. *Boehler's* (1992, 1993, 1996) data (dashed lines) were determined using a laser-heated diamond anvil cell; other data were determined using the multianvil apparatus. *Boehler's* data extend to 60 GPa, but he has experiments at only three pressures below 40 GPa, leaving the slope and position of the curve relatively unconstrained. (b) Fe-FeS melt compositions (curved lines) as a function of temperature and S content at two pressures: 1 bar and 10 GPa from *Fei et al.* (1997). The composition of the eutectic melt shifts to more anion-poor compositions with pressure. Also plotted are the silicate solidus and liquidus temperatures (horizontal lines) of Allende chondrite (*Agee,* 1997) and KLB peridotite (*Herzberg and Zhang,* 1996). The compositions of chondrites, Earth's core, and iron meteorites are also indicated.

45 GPa, and then becomes more S-rich again, rising to 42 mol% at 135 GPa (the pressure at the core-mantle boundary) (*Williams and Jeanloz,* 1990). Thus, if the wetting properties of Fe-FeS melts in a silicate matrix vary with S content, they are more likely to be important in the uppermost mantle and in small bodies such as asteroids and deep inside the mantle.

## 2.3. Metal Segregation by Porous Flow

One potential mechanism for metal segregation is that the metal melts and forms an interconnected network, like silicate melts do, and flows downward to the center of the planet. Whether this happens or not depends on the dihedral angle, θ (Fig. 2). The dihedral angle of a liquid is the angle formed by the liquid in contact with two solid grains, in this case solid silicate or oxide grains (*Smith,* 1964; *von Bargen and Waff,* 1986). If the dihedral angle is <60°, the

melt will fill channels between the solid grains and form an interconnected network, even at small melt fractions. On the other hand, if θ is >60°, the melt is confined to pockets at grain corners and cannot flow readily. Thus, many experiments have been done to measure θ under a variety of conditions.

The value of the dihedral angle reflects the interfacial energies between phases in contact (e.g., *Smith,* 1964; *von Bargen and Waff,* 1986; *Stevenson,* 1990). If the interfacial energies are independent of grain orientation and the pressure is hydrostatic, $\theta = 2 \cos^{-1}(\gamma_{ss}/2\gamma_{sl})$, where $\gamma_{ss}$ is the interfacial surface energy between solid grains (in our case, solid silicate grains) and $\gamma_{sl}$ is the interfacial energy between solid and liquid (in our case, solid silicate and metallic liquid). For example, if $\gamma_{ss}$ is 900 ergs/cm² and $\gamma_{sl}$ is 1000 ergs/cm² (values typical of metal-silicate systems; *Stevenson,* 1990), then θ is 126°, far greater than 60°. However, $\gamma_{sl}$ can be greatly affected by other constituents such as S and O (see below), temperature (e.g., *Murr,* 1975), and pressure (see next section).

Even when θ >60°, at high melt fractions some melt can migrate because the melt completely fills the grain edges. *von Bargen and Waff* (1986) computed how much melt would be stranded in a solid-liquid system as a function of θ and the melt fraction. On a plot of melt fraction vs. θ, *von Bargen and Waff* (1986) define a connection boundary, the transition between connected and isolated melt pockets, and a pinch-off boundary above which melt is stranded. These boundaries diverge with increasing θ, which means that the greater the dihedral angle, the higher the melt fraction needed to allow connectivity. They show that the critical melt fraction below which pinch-off takes place is equal to 0.009 (θ – 60)$^{0.5}$ (applies to θ between 60° and 100°). For θ = 100°, the pinch-off value is 0.057 (5.7 vol%), so even if some metal were able to migrate toward the core, a substantial amount would be left behind. Such a large amount of metal is completely incompatible with siderophile-element contents of the mantle. Additionally, melts with large dihedral angles tend to form larger, heterogeneously distributed melt pockets surrounded by multiple grain boundaries (termed "negative diffusion" by *Stevenson,* 1990), raising the melt fraction needed for sustained melt migration.

If metallic melt is able to connect, its rate of migration is quite rapid, as can be calculated from Darcy's Law

$$v = (k/\eta)\ dP/dh \qquad (1)$$

where v is the velocity of the melt relative to the solid matrix, k is the permeability, η is the viscosity of the melt, and dP/dh is the hydrostatic pressure gradient. The latter is

equal to (Δρ)g, where Δρ is the difference in density between the melt and the solid and g is the gravitational acceleration. Permeability has several formulations. *Turcotte and Schubert* (1982) suggest

$$k = a^2\Phi/24\pi \qquad (2)$$

where a is the mean grain radius and Φ is the melt fraction. Using a grain radius of 10$^{-3}$ (1 mm), Φ of 0.1 (10% melt), Δρ of 3500 kg m$^{-3}$, g of 10 m s$^{-2}$ (large planetesimals), and a viscosity of 0.005 Pa s, the migration velocity is 9 × 10$^{-3}$ m s$^{-1}$, or 290 km yr$^{-1}$. Thus, any metallic melt that can form an interconnected network ought to readily and quickly sink to form a core. The central issue is the extent to which the metal connects.

Experiments (see below) indicate that θ is >60° for anion-poor metallic melts. However, there is some indication that melts rich in the anions S and O might have θ <60° and, hence, flow readily (*Gaetani and Grove,* 1999), about as rapidly as indicated above. This would happen because γ$_{sl}$ is affected by S and O (*Ip and Torguri,* 1993). These elements change the electronic configuration of the metallic melt, affecting its interaction with solid silicates. The extent of the effect is not known. If the effect of S and/or O is strong, however, removal of the eutectic melt would leave a residue richer in Fe, possibly almost pure Fe (with some Ni), which would be stranded. Thus, although under some circumstances a S- or O-rich metallic melt might migrate toward the core, it would probably not lead to complete segregation.

Some investigators have suggested that increasing pressure may lead to a lower dihedral angle between solid silicate and metallic melts (e.g., *Urakawa et al.,* 1987). This could happen either because of the direct effect of pressure on interfacial energies, or because of the change in mineralogy of the silicates or the composition of the liquid. Both of these factors can, in principle, change γ$_{ss}$ and γ$_{sl}$, hence θ. If correct, metals could segregate rapidly once some critical pressure was reached, but less efficiently at lower pressures. Consequently, the effect of pressure on the wetting properties of metallic melts has been studied experimentally, as described in section 3.

### 2.4. Metal Segregation in a Dynamic Environment

The relationship between interfacial energies and dihedral angle discussed above apply to cases of hydrostatic equilibrium. However, as soon as there is any differential stress, interfacial energy does not solely control the distribution of melt. Instead of metallic melts tending to concentrate in isolated pockets, the melt may accumulate along grain edges and faces, or in microcracks and shear zones. This could allow the metallic melt to form an interconnected network and migrate toward the planet's center. Melt migration enhanced by differential stresses has been documented for partially molten crustal rocks (*Rushmer,* 1996; *Rutter and Neumann,* 1995). As it is quite likely that the

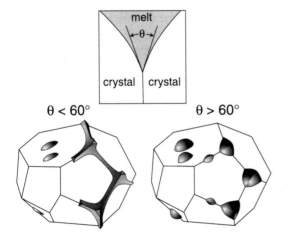

**Fig. 2.** The definition of the dihedral angle θ and a depiction of the two endmember microstructures for static systems: wetting (melt forms an interconnected network along grain edges) and nonwetting (melt forms isolated pockets at grain corners).

mantle was convecting during core formation, some amount of deformation must have taken place.

If a deforming mantle containing molten metal and solid silicate contained a network of veins filled with molten metal, the melts could segregate rapidly. We can illustrate this by calculating the flow velocity by using Darcy's Law, but permeability is difficult to estimate. Using preliminary results from experiments (below), we can estimate permeability from $k = Nb^3/12$, where N is the number of fractures per meter and b is the width of the fractures (*Snow*, 1969; see also *Rutter and Neumann*, 1995). Experiments suggest that N is about $10^4$ channels per meter and b is $10^{-5}$ m. This yields a permeability of about $10^{-12}$ m$^2$. For comparison, *Connolly et al.* (1997) estimated a permeability of $10^{-13}$ to $10^{-15}$ m$^2$ from the distribution of melt-induced fractures formed during dehydration melting of a muscovite-bearing metaquartzite. Using a density difference between liquid and solid of 3500 kg m$^{-3}$ and the melt viscosity of 0.005 Pa s, Darcy's Law suggests that the melt moves at a velocity of $7 \times 10^{-6}$ m s$^{-1}$, or about 220 m yr$^{-1}$. Although there are enormous uncertainties in this calculation, it shows that melt segregation is likely to be rapid if a network of veins forms.

The concept of vein formation as the driving force for metal segregation was applied to asteroidal-sized bodies by *Keil and Wilson* (1993). They showed that even without the presence of volatiles, melting of the metal-sulfide assemblage causes the local pressure to rise and exceed the tensile strength of the surrounding rock, forming veins. [This process was demonstrated by heating samples of the type 3 chondrite Krymka at 1173 K in vacuum (*McSween et al.,* 1978); the sample displayed numerous veins throughout the silicate matrix. A sample heated at 1273 K formed an interconnected network of veins throughout the chondrite matrix.] With even small amounts of volatiles present, veins form readily on bodies smaller than about 250 km in radius, and, surprisingly, the metallic melts may contain enough exsolved gas bubbles to rise instead of sink. Keil and Wilson call on this mechanism to explain the low S contents of most iron meteorite groups.

Such fracture mechanisms would seem unlikely inside the Earth, especially in the hot mantle, where rocks are ductile. However, recent experiments, discussed below, show that fracture-forming pathways for melt migration can occur deep in the Earth, even in previously ductile rock. Fracture of the partially molten rock, even at great depth, can take place if melt pressure (equal to pore-fluid pressure) is able to build up. This occurs when the rate of melt production is rapid and melt cannot escape the system quickly. With the increase of melt pressure, the effective normal stress (imposed deformation) on the rock is reduced by an amount equal to the melt pressure (law of effective stress). If melt pressure continues to build, eventually the rock can no longer support the imposed differential stress and will fracture. This process of microcracking in the presence of a melt phase is called melt-enhanced embrittlement or melt-induced microcracking (*Davidson et al.,* 1994). Once the rock has fractured and the melt pressure lowered by melt leaving the system, the deformation will switch back to

ductile and the rock will flow again. This mechanism is therefore likely to be cyclical and repeat several times during a single melting event. It may also change siderophile-element patterns in the mantle as the cycling might approximate fractional melting, rather than the batch melting usually assumed.

## 2.5. Segregation of Molten Metal Through Molten Silicate

The role of giant impacts is central to theories for the formation of the terrestrial planets. Such impacts are expected to produce magma oceans, although there is a lack of definitive geochemical evidence for a terrestrial magma ocean. Giant impacts on the Earth will lead to the production of an emulsion of droplets of liquid iron in a silicate liquid (*Stevenson,* 1990), and core formation can take place readily in this environment. There are numerous complexities in the process of the metal sinking in from an emulsion (see *Stevenson,* 1990, for details). The initial distribution of droplet sizes or the evolution of the droplet size spectrum in a turbulent convection system are not known. Droplets may become larger, hence sink easier, by Ostwald coarsening. The vigorous convection in the magma ocean may prevent droplets from sinking readily. Nevertheless, *Stevenson* (1990) concludes that metal droplets would sink to the base of a 700- or 800-km-deep magma ocean in only 10–100 yr.

## 2.6. Summary of Theoretical Considerations

In spite of the possibility that many of the cores of the terrestrial planets formed as the result of magma-ocean-producing giant impacts, all the processes discussed above may have operated at different times in different places as the Earth formed, and in the planetesimals from which it accreted. Segregation of molten metal from solid silicate is especially relevant for understanding differentiation in the initial stages of planetary accretion, as objects grew to asteroidal sizes. It is also relevant for the later stages of accretion (perhaps during addition of the hypothetical late veneer) and in the Moon. Static experiments need to explore the variations in dihedral angle with pressure, temperature, and the concentrations of nonmetallic alloying light elements such as S and O. The role of metal segregation during deformation also needs to be studied experimentally. Such experiments are described in the next two sections.

## 3.   EXPERIMENTAL STUDIES

### 3.1.   Static Experiments on Molten Metal/Silicate Systems

Spurred on by the reviews of the physical mechanisms of core formation (*Stevenson,* 1990; *Taylor,* 1992) in the Earth and other terrestrial bodies, five separate laboratories have independently undertaken studies of the microstructure of Fe-FeS (± Ni ± O ± Co) melts in contact with silicates. The studies have in part also been inspired by *Urakawa et al.* (1987), who suggested that the eutectic metallic melts become significantly more O-rich with pressure, and that this O content may serve to reduce the interfacial energy of Fe-rich melts against solid silicate sufficiently to allow wetting of the grain edges.

Each lab approached the problem differently. The study of *Minarik et al.* (1996) used a synthetic Fe-Ni-S melt (starting with Urakawa's eutectic composition at 10 GPa) in contact with olivine and oxide at a range of pressures from 3.5 to 11 GPa, 1500°C, in graphite or diamond capsules (see Fig. 3 for representative textures). *Ballhaus and Ellis* (1996) systematically studied a range of synthetic Fe-Ni-Co-S melt compositions in contact with olivine at 2 GPa, and several different temperatures, in graphite capsules. *Shannon and Agee* (1996, 1998), in contrast, used an ordinary chondrite (Homestead L5) as a starting material, producing a Fe-rich melt at pressures between 2 and 25 GPa. Temperatures increased with pressure in order to remain above the metal solidus and below the silicate solidus. The melt was in contact with a number of silicate and oxide phases, including olivine, pyroxene, garnet, ferripericlase, wadsleyite, ring-

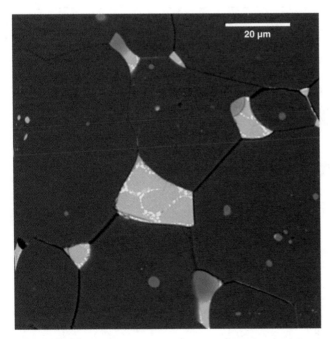

**Fig. 3.** A backscattered electron image of a quenched Fe-Ni-S melt in an olivine and ferropericlase matrix. Olivine is dark gray and ferropericlase is light gray in color. The experiment was performed in a piston-cylinder device under isostatic pressure conditions of 3.5 GPa and 1500°C. The homogenous melt has unmixed during the several-second quench into Fe- and sulfide-rich domains. Note the near 120° equilibrium microstructure of the olivine three-grain boundaries, and the high apparent melt-olivine dihedral angels. The melt generally occupies the corners formed by four grain intersections. Because olivine-melt grain edges are energetically unfavorable, during grain growth some melt pockets have coalesced into larger melt pockets surrounded by multiple grains, further reducing the area of melt-silicate contact.

woodite, and silicate perovskite. *Gaetani and Grove* (1999) performed one atmosphere, controlled O- and S-fugacity experiments on a range of Fe-S-O (+ minor Ni, Mn) melts at 1350°C in contact with olivine in olivine capsules (but in a C-bearing atmosphere). *Herpfer* (1992) and *Herpfer and Larimer* (1993) performed 1-GPa experiments at 1300°C on a range of Fe-S melts in contact with olivine in olivine capsules.

These studies therefore differ in pressure and temperature, melt composition, solid assemblage (olivine ± oxide vs. natural), and control of (or knowledge of) the volatile fugacities. The results are generally consistent among these experiments, despite their differences. Anion-poor Fe-Ni-S-O melts do not wet silicates (i.e., the dihedral angles are much greater than the critical 60°), and hence are unable to percolate at low melt fractions. This will be true in nondeforming systems and systems deforming at rates slower than the rate of recrystallization of grain boundaries. The exceptions to this statement seem to be anion-rich melts [(S + O)/(Fe + Ni) > 1] and metal-sulfide melts in contact with silicate perovskite at pressures greater than 25 GPa (how much greater is not yet known). Figure 4 is a summary of the static experiments to date on olivine and ordinary chondrite matrixes. The wetting ($\theta$ <60°) melts of *Gaetani and Grove* (1999) occurred only at the highest O-to-S fugacity ratios but unknown S and O fugacities (conditions not unreasonable for today's upper mantle), and their results are supported by the high S metallic liquids (but unknown S and O fugacities) of *Minarik et al.* (1996), *Ballhaus and Ellis* (1996), and *Herpfer* (1992). The continued presence of sulfide in the Earth's upper mantle, and the lack of isotopic and geochemical evidence (see *Allègre et al.*, 1995; *Galer and Goldstein,* 1996; *Shirey and Walker,* 1998) for continuous core formation over Earth's history, suggest that these high O- and S-fugacity conditions are not common and have been attained only sporadically, if at all. The choice of alloying elements is based on both abundance and availability (see *Hillgren et al.,* 2000). Metal in iron meteorites is alloyed primarily with S, O (as described above), and C. Other alloying elements either are not present in sufficient abundance to change the physical properties (P, etc.), require implausible O fugacities (silicon), or may only allow for very high pressures (K).

There have been several reconnaissance diamond anvil experiments examining the microstructure of metallic melts plus silicates at lower mantle pressures (e.g., *Knittle and Jeanloz,* 1991; *Goarant et al.,* 1992). These experiments are difficult to interpret due to the small number of grain boundaries and the large temperature, pressure, and O-fugacity gradients. They seem to indicate partial wetting of silicates by metallic melts at core-mantle boundary pressures (135 GPa), consistent with the perovskite-bearing experiments of *Shannon and Agee* (1998).

The use of measured metallic-melt/solid silicate dihedral angles to understand rock microstructure is only applicable to systems in which the metallic melt is the only melt present. The presence of a second, silicate melt complicates understanding of the microstructure and transport proper-

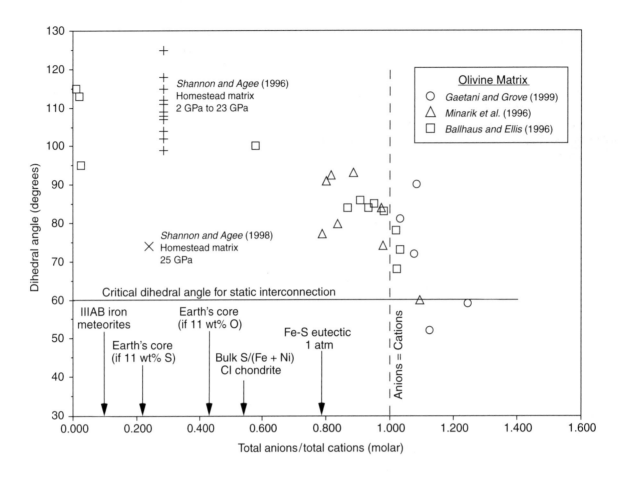

**Fig. 4.** A summary of the measured olivine-metallic melt dihedral angles as a function of the anion (S + O) to cation (Fe + Ni + Co + Mn + Cr) ratio of the melt. Carbon contents are ignored, and O contents have not been measured for many experiments. The experiments cover a range of temperatures (1100°–2100°C) and pressures (1–14 GPa). *Herpfer* (1992) did not publish liquid compositions, and so this data is not plotted. Also plotted are the dihedral angles determined for Homestead L5 ordinary chondrite matrix in contact with metallic melt; the matrix phases depend on pressure and include olivine, pyroxene, oxide, garnet, spinel, and octahedrally coordinated silicates. The composition of the low-pressure Fe-S eutectic and bulk CI chondrite are noted, as are estimates for the Earth's core and IIIAB meteorites (which are linked to main-group pallasites, the HED meteorites, and the asteroid 4 Vesta) (*Mittlefehldt et al.*, 1998). Only metallic melts with high S and O contents have dihedral angles near the critical 60° for wetting. Dihedral angles are insensitive to pressure and temperature and silicate mineralogy, with the exception of silicate perovskite. Dihedral angles are generally known to within ±5°, and composition errors can be quite large, but have been omitted for clarity.

ties of the system. Silicate melts have lower interfacial energies with respect to silicate and oxide minerals, and will preferentially wet the grain-boundary channels. The metallic melts will then be relegated to the centers of melt pockets at grain corners, isolated from the grain boundaries by the silicate melts. Movement of this mixture of immiscible melts can be impeded at low total melt fractions due to the energy of curvature of the melt-melt interface. Flow of two liquid mixtures has been investigated extensively in shallow aqueous systems (e.g., *Alder and Brenner,* 1988) and theoretically for $H_2O$-$CO_2$ immiscible mixtures in the mid-crust (*Bailey,* 1994). Early-forming melts in primitive terrestrial planetesimals could conceivably be a mixture of metallic melt, silicate melt, and vapor (in the H-O-C-S system). Depending on the bulk density of this fluid mixture, these melts might rise instead of sink, carrying the entrained

metal-sulfide melt with them. As these melts form interconnected channels at low melt fraction (because of the silicate melt), or may cause fracturing on melting, they could separate independent of the metal phase, and without simultaneous core formation. This type of scenario may help explain the loss of silicate melt and S from primitive chondrites that have not lost metal or highly siderophile elements (*Wilson and Keil,* 1991). In addition, microstructures in deforming systems can also allow interconnected melts at lower S contents, as will be explored in the next section.

### 3.2. Experiments on Metal Segregation in Dynamic Environments

As mentioned in the theoretical section, we need to understand the rheology of partially molten systems to consider metallic melt segregation aided by deformation processes. Recent experimental studies have substantially improved our understanding of the rheology of partially molten silicate rocks, and we find that melt distribution is a function of not only the interfacial energies of minerals and melts, but is also determined by state of stress, the active deformation mechanisms, and even the deformation history (e.g., *Rutter and Neumann,* 1995; *Hirth and Kohlstedt,* 1995a,b; *Daines,* 1997). For crustal rocks (as an analog to metallic melts in silicates) the experimental stress conditions combined with the generation of a high melt pore pressure increases the possibility for melt-enhanced fracturing, which leads to cataclastic flow and eventually viscous flow at high melt volumes. Figures 5a,b summarize experimental rock deformation results under these conditions and illustrate the different rheologies for crustal assemblages. There is a shift from fracture- and mineral-dominated microstructures to melt-dominated structures at high melt fractions. However, the experiments suggest that melt segregation is not controlled by this rheological transition as had been previously believed, but more by the ability of the melt to interconnect. This is possible at low melt fractions, when deformation in the presence of melt enhances fracture and forms flow paths for melt migration. This transition between fracture and viscous flow will also likely occur at different melt fractions for different assemblages depending on pressure,

which helps determine the melting reaction, melt composition (which influences melt viscosity), rate of melt production (development of pore pressure), and deformation rate. In general, the deformation experiments show that the presence of a melt phase can induce fracture and cataclastic deformation in previously ductile rock at low melt fractions and allow for efficient melt segregation. As a consequence, silicate melt segregation has been found to be possible under a much wider range of melt fractions than previously believed. Can this process of interconnecting melt through fracture be applicable in silicate-metal assemblages? If the metallic melt phase behaves similarly to the silicate melt phase in partially molten deformation studies on crustal rocks, then melt-enhanced embrittlement may be an important mechanism by which metallic melt segregates and would allow for migration at low degrees of melting. Conversely, there are enough differences between the physical properties of molten metal and molten silicate (e.g., viscosity of the melt phase, in particular) that they may not be comparable. Experiments under deformation are underway to assess the behavior of molten metal under deformation (*Rushmer and Jones,* 1998b, 1999; *Bruhn et al.,* 1998).

*3.2.1. Results on a natural chondrite.* Experiments are currently determining the deformation behavior of a partially molten chondrite in order to establish the metallic melt distribution under dynamic conditions. The main question to

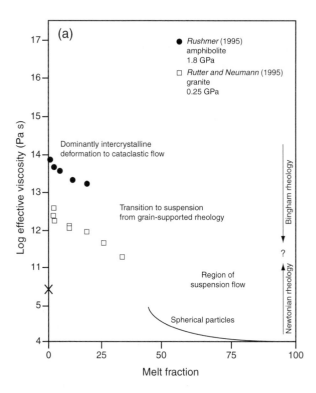

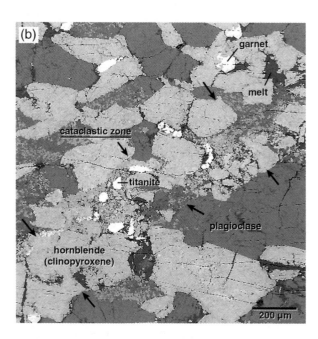

**Fig. 5.** **(a)** Log effective viscosity (shear stress/strain rate) vs. melt fraction for experiments on partially molten granite (*Rutter and Neumann,* 1995) and amphibolite (*Rushmer,* 1995). Grain-supported rheology dominates between 0 and ~30 vol% melt (there are variations depending on the type of melting reactions) up to the transition zone to suspension, where crystals can no longer interact. This transition zone is the "rheological critical melt fraction." Also included is the theoretical relationship for the viscosity of suspended spheres (*Roscoe,* 1952). **(b)** Backscatter image of cataclastic behavior observed in the amphibolite at ~15 vol% melt (*Rushmer,* 1995).

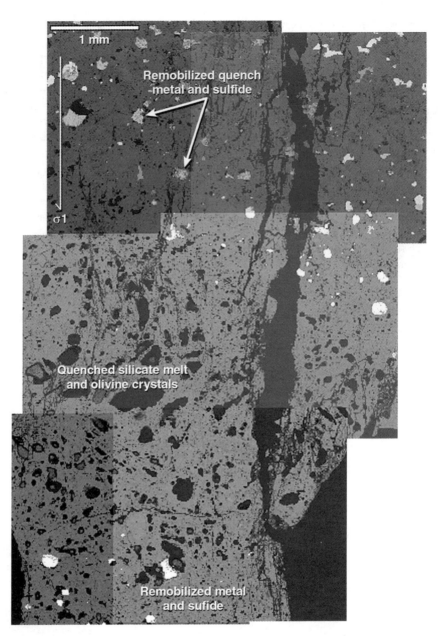

**Fig. 6.** Experiment KM-2 is the hottest of the experiments on the Kernouve H6 chondrite. Backscatter electron images form a mosaic from the mid to lower portion of the experimental charge where extensive silicate melting (>50%) has occurred. Note that all metal and sulfide is no longer present in this area. Analyses of the metal and sulfide show the presence of an Fe-P phase. Sigma 1 arrow shows main compression direction.

be addressed is whether or not the molten metallic portion of the meteorite can be mobilized during deformation. Cores were made from a slab of the ordinary H6 chondrite Kernouve (supplied by the Smithsonian Institution). The chondrite contains 20–25% Fe-Ni metal and sulfide (FeS), and its matrix is composed of olivine, orthopyroxene, plagioclase, clinopyroxene, chromite, and chlorapatite. Deformation experiments (performed at confining pressures of 1.0 GPa at $10^{-5}$/s strain rate with temperatures ranging between ~950°C and >1050°C) produced variable amounts of silicate melt; from significant (>50 vol%) to silicate melt free. In the hottest runs where significant silicate melt is present at the hot spot, complete mobilization of metal and sulfide is observed (Fig. 6). Iron-nickel metal and sulfide are observed as globular forms with clear quench textures.

In these experiments separation of metal from silicate through a silicate melt appears quite possible (*Stevenson,* 1990), but the thermal gradient in the capsule may also be a driving force (see also *Schmitz et al.,* 1998). In the moderate- to low-temperature experiments, some mobilization of metal is also observed, but is not a function of the presence of the silicate melt as observed in the hottest experiments. Here flattening of Fe-Ni in lenses perpendicular to the main compression direction is observed (Fig. 7). In the low-temperature runs, deformation textures in the silicate-melt-free areas appear cataclastic (textural evidence of a melt-enhanced embrittlement process), and the preliminary results suggest that FeS appears to be more mobile than Fe-Ni metal, and perhaps inducing fracture (Fig. 8). Some of the cataclastic texture is also associated with narrow reac-

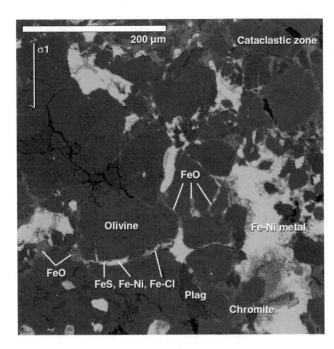

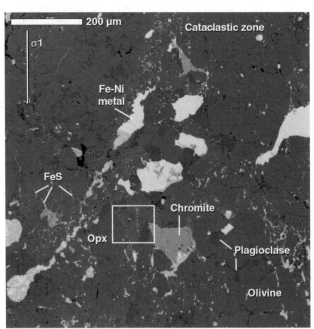

**Fig. 7.** Experiment KM-8 contains no silicate melt and extensive cataclasis is seen in the most deformed central portion of the sample. Ferrous oxide is seen in fractures, but is unclear at this point as to how it is mobilized; FeS links up flattened lenses of Fe-Ni metal and is also found in fractures. Iron chloride compounds are found in this and other samples and there are still questions on its formation. Silicate phases present in this image are olivine and plagioclase. Sigma 1 arrow shows main compression direction.

tion/diffusion zones in the olivine grains. Olivines in these reaction/diffusion zones are more Fe-rich and contain micrometer-sized FeS (Fig. 9). One interpretation is that they record evidence of FeS movement. Iron-nickel metal is also found lining cracks in some grains. Dissemination of sulfide is observed on the very fine scale in the cataclastic zones, suggesting ease of mobility of FeS and its potential ability to induce fracture.

The preliminary results suggest different mechanisms are activated under varying conditions for mobilization of metallic/sulfide melts. In the high-temperature samples rich in silicate melt, Fe metal and sulfide have been removed from the melted portion of the sample, are intimately mixed, and are spheroid-shaped (with some evidence of metal liquid immisicibility involving a Fe-P phase). In the lower-temperature runs, Fe-metal textures are either thick lenses (unmobilized?) or in thin stringers. Preferential movement of FeS appears to be possible. Some textures in the lowest temperature runs (no silicate melt present) suggest a cataclastic response to deformation forming interconnected fracture systems. Here sulfide is molten and appears quite mobile, whereas Fe metal remains solid but still deformed. The effect of the low viscosity of the FeS melt [~5 × $10^{-3}$ Pa s (*Iida and Guthrie,* 1988)] on the behavior of the partially molten chondrite during deformation has yet to be established concretely. However, the brittle texture in the

**Fig. 8.** Experiment KM-9, which contains silicate melt in the lower portion of the sample, but none at the top portion where this backscatter image is taken. In the lower portion, again metal and sulfide (with the Fe-P phase) are not present in the silicate melt rich areas. In this upper portion, no silicate melt is present and cataclasis is observed. Ferrous sulfide is seen finely disseminated along fractures and within the matrix; Fe-Ni metal is also observed in cracks. The zone of most intense deformation, which is seen by the orientation of metal and sulfide, is probably a small fault zone. Movement appears to be down on the left relative to the right. Focused deformation also appears in conjugate form, observed by another zone of oriented metal and quench sulfide. Orientations are at a moderate angle to the main compression direction. Sulfide shows good quench textures. Chromite, olivine, plagioclase, and orthopyroxene are all present in this image. The box shows the close-up area of Fig. 9.

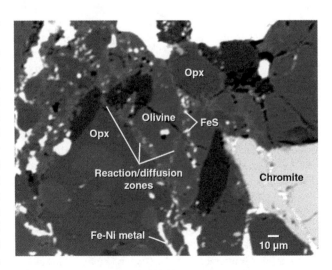

**Fig. 9.** Close-up of reaction/diffusion zones in olivine. Olivine in these areas are more Fe-rich and contain micrometer-sized FeS. One possibility is that the reaction/diffusion zones record FeS movement within the matrix during deformation.

deformation zones in the chondritic samples suggest it is almost a breccia, with fine dissemination of quench FeS melt throughout the zones. In the partially molten silicate systems where the melt is more viscous [$10^4$–$10^6$ Pa s (*Baker,* 1998) for hydrous granitic melt], cataclasis is on a larger scale and quenched melt is not as finely distributed within the matrix. These textural differences can be seen when Fig. 5b and Fig. 8 are compared. In general, though, cataclasis produced by melt-enhanced embrittlement is the deformation response to applied stress in both systems. Further work needs to be done to determine the extent of mobility of molten Fe metal and its distribution as a function of deformation.

### 3.2.2. *Results on olivine-Fe/Ni-sulfide aggregates.*

*Bruhn et al.* (1998) performed a different type of deformation experiment. They used hot-pressed aggregates of fine-grained olivine with 7% Fe/Ni metal and FeS. The experiments are similar in set-up to those performed on mantle rock types (peridotites) in the presence of small amounts of silicate melt (*Hirth and Kohlstedt,* 1995a,b; *Bai et al.,* 1997; *Daines,* 1997). In contrast to the experiments described above, these workers investigating mantle materials have opted for high-strain, low-stress experimental conditions with low melt fractions (up to 5 vol%), which are more applicable to current mantle rheology. Experimental results have shown that deformation under these low-stress conditions can result in the development of a lattice preferred orientation in the olivine crystals and a fabric defined by elongation and flattening of grains. If crystalline anisotropy is strong, the lattice-preferred orientation will also control melt distribution so it too possesses a preferred orientation. The deformation fabric can also orient melt channels into parallelism. Higher stress experiments can result in melt orientation responding to the stress field instead of lattice-preferred orientations (*Daines,* 1997), and will eventually lead to melt-enhanced fracture if melt pressure reaches the point where it reduces the effective confining pressure.

For the experiments on olivine + 7% Fe/Ni metal and sulfide, the conditions are set so that the differential stresses range between 40 and 100 MPa and strain is between 200% and 300% (in simple shear). Under these conditions the olivine matrix deforms by dislocation creep. Temperature and pressure conditions are 1250°C and 300–400 MPa respectively, at which the Fe/Ni-sulfide is molten. Deformation had a pronounced effect on the metal/sulfide melt topology. *Bruhn et al.* (1998) observe that many of the melt pockets were elongated and oriented antithetically to the direction of shear at angle of ~20° to the shear plane. This microstructure and the orientations of the metal and sulfide pockets are similar to those observed in experimentally sheared olivine + MORB samples, in which the basaltic melt forms a highly interconnected network (*Zimmerman et al.,* 1999).

Under both experimental conditions, high strain and high stress, preliminary results suggest that molten core-forming metals may be mobile even if the silicate matrix is solid.

However, the pressure and temperature conditions are different enough between the two sets of experiments that much more work will need to be done to determine the range of deformation behavior and subsequent metal/sulfide distribution within a given pressure-temperature regime (the high-strain experiments are all at low pressure, 300–400 MPa, and high temperature, 1200°C. The experiments on the natural chondrite are at higher pressures (1.0 GPa) but lower temperatures; between 950°C and >1050°C). Whether or not the molten metal and sulfide will escape along fractures or elongated melt channels will depend on both the dynamic environment and the pressure-temperature conditions, which control if silicate melt is present and the nature of the deformation mechanisms active in the silicate matrix. Even with these kinds of data, in order to apply them to early Earth, some major assumptions about the thermal regime, the convection rate, and the fate of impacts of different-sized bodies need to be made. However, these experimental results allow us to begin to integrate the recent advances made in geochemistry with different physical regimes of core formation. It is then important to explore the ramifications of such processes on the mantle we sample today. This will lead to a new generation of core formation models that can be tested against geochemical constraints.

## 4. IMPLICATIONS

Evidence from meteorites indicate that metallic cores formed in many asteroids, including about 80 distinct groups (some with only one or two members) of iron meteorites, pallasites, and the meteorites from the HED parent body (howardites, eucrites, and diogenites). Isotopic data indicate that these differentiation events took place early, essentially contemporaneous with a few million years after the formation of CAIs, the oldest materials in the solar system (*Lugmair and Shukolyukov,* 1998; *Carlson and Lugmair,* 2000). Thus, even small bodies differentiated, and they did so before (or simultaneously with) the formation of the terrestrial planets. This suggests that the planetesimals accreting to the growing Earth may have already differentiated into cores and mantles (*Taylor and Norman,* 1990). In fact, if differentiation occurred in small bodies, it may also have occurred in the larger bodies that accreted to the Earth. Thus, in a sense, formation of the Earth's core began before the planet formed, and there are a variety of settings in which fractionation of metal from silicate took place. These settings vary in pressure, amount of silicate melting, and strain rate.

We consider three different major regimes (Fig. 10) in which metal and sulfide segregation may have occurred, and discuss general geochemical implications of metal segregation in each: (1) A magma ocean (if present), where metal sulfide can readily move, but may be subjected to continuously changing pressure and temperatures. Under conditions of low stress and strain, as in planetesimals up to roughly lunar size, metallic melts do not form an interconnected network and so segregation appears to require the presence of considerable amounts of melt, consistent with a magma ocean. (2) Metal/sulfide segregation in conditions where the matrix is solid, but elongated melt pools become interconnected (when strain is high and stress is low). This regime can be visualized where convection is mostly under solid-state conditions with only metallic melt present. If convection is active, this will impart a directed stress that can be either significant (producing transient fractures?) or moderate. Under moderate to low applied stress conditions, orientation of melt pools may occur with time. Low-stress conditions may also lead to formation of an interconnected network if the metallic assemblage is rich in O or S. (3) Segregation when fracturing occurs (high stress) and preferential movement of metal and sulfide compounds is possible. This regime is probably the most important during impact of large bodies where brecciation can occur and metallic melt may be easily distributed in fractures, and in actively convecting bodies with high metallic melt pore pressure. It is quite likely that different mechanisms operated at different times and places during core formation, leading to complicated, but interesting, geochemical signatures.

## 4.1. Regime 1: Segregation in a Magma Ocean

Regime 1 is defined here as metal sulfide segregation in the presence of a high silicate melt fraction (e.g., >40%, a magma ocean model) either in a large body (perhaps after a large impact that produced an emulsified silicate-FeS melt) or in smaller ones (≤Moon-sized) where static melting conditions may occur (due to the decay of $^{26}$Al or $^{60}$Fe). The complete segregation of metal sulfide is possible in the presence of significant silicate melt. This is observed both experimentally and is feasible theoretically. Under these conditions, experimental static and deformation observations show that metal sulfide is completely swept out of zones containing a high silicate melt fraction. If a magma ocean were present on early Earth, then core formation could be a swift and extremely efficient process. However, there are

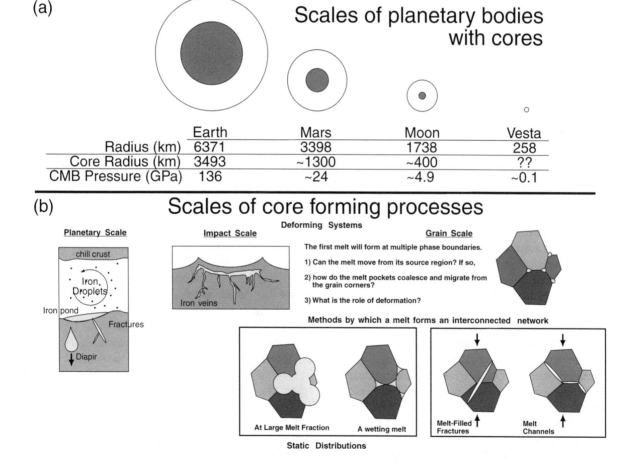

| | Earth | Mars | Moon | Vesta |
|---|---|---|---|---|
| Radius (km) | 6371 | 3398 | 1738 | 258 |
| Core Radius (km) | 3493 | ~1300 | ~400 | ?? |
| CMB Pressure (GPa) | 136 | ~24 | ~4.9 | ~0.1 |

**Fig. 10.** (a) A true scale comparison of the sizes and core conditions for the terrestrial bodies known to have cores. No or minimum convection is considered in bodies Vesta-sized and smaller. In these bodies regime 1 for metal-silicate separation may be dominant. In larger-sized bodies, Mars-sized and larger, regimes 2 and 3 may become important as active convection likely accompanied core formation. (b) A comparison of the scales and mechanisms considered for the separation of core metal from mantle silicate. Here regimes 1, 2, and 3 are shown and note that more than one regime may occur during the formation of a single planetary core, even in the presence of a hypothetical magma ocean. See text for further discussion.

interesting complications. As *Walter* (1998) points out, as the metal sulfide melt sinks, it will continuously equilibrate at increasing pressures. The silicate magma would be convecting vigorously, hence would be well mixed. Thus, the siderophile-element signature in the mantle would reflect equilibration of silicate and metallic melts at a range of pressures. Because siderophile-element partition coefficients change with pressure (e.g., *Righter and Drake*, 1997b), this polybaric equilibration would affect the resulting concentrations of siderophile elements, possibly accounting for the excess of moderately siderophile elements in the primitive upper mantle.

Another complication is that we do not know the size distribution of the metallic masses sinking through the silicate magma ocean. If undifferentiated material dominates the accretion process, then the metal would surely be finely disseminated after impacting the Earth. On the other hand, the fate of a large core in an accreting planetesimal is not clear. *Stevenson* (1990) argues that shear stresses during impact produce an emulsion, a fine mixture of metal and silicate, that would equilibrate readily. However, it is possible that accreting cores larger than a few tens or hundreds of kilometers remain relatively intact, falling rapidly through the mantle to a growing core. In this case, the metal may not equilibrate with the silicate portion of the Earth. The accreting silicates would reflect a range of pressures lower than one might calculate for the average of the Earth's mantle.

## 4.2.  Regime 2: High Strain, Low Stress (Late Veneer)

Most workers now believe that a late veneer is required to explain primitive upper mantle highly-siderophile-element (HSE) concentrations, and this leads us to the process of metal and sulfide separation from either metal-bearing impactors or mantle without large-scale mantle melting. Regime 2 is considered here where conditions in a convecting but solid-silicate early Earth are such that strain is high and stress is quite low, somewhat similar to rheological conditions in the current mantle. Here, oriented melt pockets of molten metal and sulfide might provide a network within a solid silicate matrix. How the high grain-boundary energies of the molten metal phases and the silicate matrix might affect distribution is not well known, but there is a possibility that under very low stresses they can influence some distribution behavior. Also, a potentially important control on the distribution of melt is the nature of the deforming materials. If the relative strength of the matrix is low, then the plastic flow of the matrix can accommodate internal fluid pressures and fracture will not occur. Molten metal/sulfide widely distributed on a grain scale along melt channels or in pockets might easily reach equilibration with the silicate matrix. This may apply to conditions where core-bearing planetesimals are impacting the early Earth and equilibration of metal and sulfide with at least some portion of the silicate mantle is possible. How-

ever, if the impactor is large and imparts enough stress, then segregation may fall into another regime, where fracturing of the matrix provides interconnected pathways for segregation.

## 4.3.  Regime 3: High Stresses

Regime 3 is under fracturing conditions, where stresses are high and/or where the strength of the matrix is such that it cannot support internal melt pore pressure. Under such conditions, melt-induced embrittlement can occur. Here preferential movement along grain boundaries and through grains in cataclastic zones of mainly FeS appears to be a possibility from the results of the high-stress experiments. For larger-scale linkage of the fractures, focused shear zones might be a common texture. At lower temperatures, lenses of Fe metal connected by FeS could form localized sites of deformation and migration. The separation of metal and sulfide into distinct phases, rather than a metal sulfide melt, changes the geochemical conditions during equilibration. This process would tend to isolate parts of the early mantle, preventing large-scale equilibration. It might also require a longer time for core formation.

If the early Earth was solid or only partially molten (e.g., silicate fractions ≪20%), then fracturing or melt channels may be the main mechanism by which at least a portion of the core was formed. This mechanism is supported by the lack of geochemical evidence for substantial silicate melting in the early Earth described earlier. If shear zones dominated the pathways for metal segregation, metal-silicate disequilibrium may be common. Determining what the specific resulting geochemical signature might look like is beyond the scope of this paper, but certainly bears investigation.

## 4.4.  Multiple Regimes

Core formation may have involved all these regimes, and may have taken place continuously from the time planetesimals were large enough to melt. On small bodies, substantial silicate melting, perhaps driven by short-lived radionuclides, would have accompanied metal-silicate segregation. These objects accreted to form larger bodies, perhaps along with undifferentiated planetesimals. The cores of the accreting planetesimals may have sunk to the core of the larger body or perhaps stranded in the larger body's upper mantle, perhaps as finely disseminated grains if the cores were disrupted on impact or in fracture systems. Once planetesimals had grown to much greater than lunar size, impacts may have been energetic enough to cause the formation of magma oceans on the growing Earth. This could have happened many times. In these cases, metal sinking through silicate melt to the core would equilibrate over a range of pressures with silicates if the bodies were small enough, although if the bodies remained in large masses disequilibration could also occur with metal segregation focused along fracture pathways. Smaller objects may have also accreted to the Earth, adding metal to the upper mantle. Metal might partially segregate if stress conditions were correct, leading to an inefficient process of core formation and oxidation of remaining metal, leaving the record we now call the late veneer.

The scale at which different physical regimes of core formation are most dominant also needs to be considered. These include convection rate and therefore deformation, which may be high in larger bodies (greater than Mars-sized) or nonexistent in asteroid-sized and smaller bodies (<100 km). Scale will also affect the extent of geochemical equilibrium between silicate and metal. Distribution of core-forming materials at a microscopic scale, either statically or during deformation, will strongly increase chances for chemical equilibrium. However, macroscopically distributed fracture zones may channellize melt, leaving large portions of the silicate mantle unaffected and out of equilibrium with metal/sulfide. Geochemical evolution will also depend on the nature and size of the impactor, in particular the degree of differentiation (*Taylor and Norman*, 1990).

## 5.  CONCLUSIONS

As we continue to explore early Earth evolution, it is clear that different physical processes active during core formation can lead to distinct geochemical signatures. Therefore, it is important now and in the future to address core-forming processes under a variety of different conditions, such as pressure, temperature, and stress. These variables will be controlled by the size of the body and, ultimately, help to produce the observable geochemistry we see in the terrestrial planets and in the Moon today. From the theoretical and experimental considerations discussed in this chapter, we can now come to several conclusions that can also serve as guideposts for future research:

1. Under static conditions, metallic melts will not segregate because interfacial energies between molten metal and solid silicate are too high. Segregation under static conditions requires either high contents of O or S in the metallic melt, or a high percentage of silicate melt (>~40%).

2. Deformation may lead to metal segregation by forming either interconnected melt channels or fracture networks. The efficiency and the effect on silicate geochemistry of the process, however, require further experiments to quantify.

3. It is possible that core formation began in the planetesimals that accreted to form the Earth, and continued in the growing Earth. The process is thus likely to reflect variations in the extent to which metal-silicate reached equilibration, creating challenges for realistic geochemical modeling. However, better estimates of the timing of core formation are helping to constrain the rates of segregation processes.

4. Core formation in terrestrial planets most likely did not take place by one mechanism only. It involved metal sinking through molten silicates potentially reequilibrating on the way down, metal segregating from solid silicate with the help of deformation producing a range of equilibrium

conditions, and possibly highly O- or S-rich metallic melts segregating from solid silicate. All these processes will leave a specific geochemical signature in the silicate mantle.

## 6.  FUTURE WORK

There are many questions left unanswered, and even though we have made significant progress in the past decade (particularly experimentally), we regard the following as particularly fruitful directions for future research:

1. Integrating physical process with silicate mantle geochemistry. Theoretically, this can involve calculations on the critical core-bearing impactor size for more detailed constraints on mantle-core equilibration during bombardment. Experimentally, addressing the nature and extent of the equilibration in the metal/silicate system during active deformation is important, as is determining the rate at which equilibration can be achieved at various pressures, temperatures, and stress conditions (from static to fracture regimes), or whether it can be achieved at all. In addition, we have very little partitioning data for solid-silicate/metal for the siderophile elements. Until we have more, we cannot fully test the geochemical ramifications of the separation of metal from solid silicate. The role of FeS in core formation also seems to be a particularly intriguing problem as, physically, it appears to be quite mobile in both oxidizing and deforming environments.

2. Modeling of solid silicate/metallic melt rheology in a variety of geologic settings and scale so that the conditions under which transitions between static, pervasive ductile flow and fracture regimes can be estimated. This in turn can help determine the dominant physical process of core formation as a function of body size.

3. Combined knowledge in these areas will provide a much firmer base on which to build future core-forming hypotheses.

***Acknowledgments.***  We would all like to thank the editors (Robin Canup and Kevin Righter) for the effort made in putting together this book. The authors also appreciate the thoughtful and helpful reviews by Kevin Righter, Glenn Gaetani, and Horton Newsom. TR thankfully acknowledges funding from the NASA Cosmochemistry Program (NAG 5-7247). Support for WGM during this study was provided by the Carnegie Institution of Washington, and preliminary work was done with the support of Rick Ryerson and the Lawrence Livermore National Laboratory. GJT is funded by NAG 5-4212. TR also thanks John Jones and Glenn Gaetani, who have continued to provide particularly helpful suggestions and discussions in regard to the deformation experiments on the Kernouve sample. This is HIGP Publication No. 1075 and SOEST Publication No. 4881.

## REFERENCES

Agee C. (1997) Melting temperatures of the Allende meteorite: Implications for a Hadean magma ocean. *Phys. Earth Planet. Inter., 100,* 41–47.

Alder P. M. and Brenner H. (1988) Multiphase flow in porous

media. *Annu. Rev. Fluid Mechanics, 20,* 35–59.

Allègre C. J., Manhès G., and Göpel C. (1995) The age of the Earth. *Geochim. Cosmochim. Acta, 59,* 1445–1456.

Bai Q., Jin Z.-M., and Harry Green H. II (1997) Experimental investigation of the rheology of partially molten peridotite at upper mantle pressures and temperatures. In *Deformation-enhanced Fluid Transport in the Earth's Crust and Mantle* (M. B. Holness, ed.), pp. 40–61. Chapman and Hall, London.

Bailey R. C. (1994) Fluid trapping in mid-crustal reservoirs by $H_2O$-$CO_2$ mixtures. *Nature, 371,* 238–240.

Baker D. R. (1998) Granitic melt viscosity and dike formation. *J. Structural Geol., 20,* 1395–1404.

Ballhaus C. and Ellis D. J. (1996) Mobility of core melts during Earth's accretion. *Earth Planet. Sci. Lett., 143,* 137–145.

Bertka C. M. and Fei Y. W. (1998) Density profile of an SNC model Martian interior and the moment-of-inertia factor of Mars. *Earth Planet. Sci. Lett., 157,* 79–88.

Boehler R. (1992) Melting of the Fe-FeO and the Fe-FeS systems at high pressure — constraints on core temperatures. *Earth Planet. Sci. Lett., 111,* 217–227.

Boehler R. (1993) Temperatures in the Earth's core from melting-point measurements of iron at high static pressures. *Nature, 363,* 534–536.

Boehler R. (1996) Fe-FeS eutectic temperatures to 620 kbar. *Phys. Earth Planet. Inter., 96,* 181–186.

Brett R. (1984) Chemical equilibration of the Earth's core and upper mantle. *Geochim. Cosmochim. Acta, 48,* 1183–1188.

Bruhn D., Groebner N., and Kohlstedt D. L. (1998) Interconnectivity of metal melts in experimentally sheared mantle rocks: Implications for formation of Earth's core. *Eos Trans. AGU, 79,* F1020.

Carlson R. W. and Lugmair G. W. (2000) Timescales of planetesimal formation and differentiation based on extinct and extant radioisotopes. In *Origin of the Earth and Moon* (R. Canup and K. Righter, eds.), this volume. Univ. of Arizona, Tucson.

Connolly J. A. D., Holness M. B., Rubie D., and Rushmer T. (1997) Reaction-induced microcracking: An experimental investigation of a mechanism for enhancing anatectic melt extraction. *Geology, 25,* 591–594.

Daines M. J. (1997) Melt distribution in partially molten peridotites: implications for permeability and melt migration in the upper mantle. In *Deformation-enhanced Fluid Transport in the Earth's Crust and Mantle* (M. B. Holness, ed.), pp. 62–81. Chapman and Hall, London.

Davidson C., Schmid S. M., and Hollister L. S. (1994) Role of melt during deformation in the deep crust. *Terra Nova, 6,* 133–142.

Fei Y., Bertka C., and Finger L. (1997) High-pressure iron sulfur compound, $Fe_3S_2$, and melting relations in the Fe-FeS system. *Science, 275,* 1621–1623.

Gaetani G. A. and Grove T. L. (1999) Wetting of mantle olivine by core-forming melts: The influence of variable $fO_2/fS_2$ conditions. *Earth Planet. Sci. Lett.,* in press.

Galer S. J. G. and Goldstein S. L. (1996) Influence of accretion on lead in the Earth. In *Earth Processes: Reading the Isotopic Code* (A. Basu and S. Hart, eds.), pp. 75–98. AGU, Washington, DC.

Goarant F., Guyot F., Peyronneau J., and Poirier J.-P. (1992) High-pressure and high-temperature reactions between silicates and liquid iron alloys, in the diamond anvil cell, studied by analytical electron microscopy. *J. Geophys. Res., 97,* 4477–4487.

Herpfer M. A. (1992) Solid state diffusion and melt microstruc-

tures in metal-silicate systems. Ph.D. dissertation, Arizona State University, Tempe, Arizona.

Herpfer M. A. and Larimer J. W. (1993) Core formation: An experimental study of metallic melt-silicate segregation. *Meteoritics, 28,* 362.

Herzberg C. and Zhang J. Z. (1996) Melting experiments on anhydrous peridotite KLB-1 — compositions of magmas in the upper mantle and transition zone. *J. Geophys. Res., 101,* 8271–8295.

Hillgren V. J., Gessmann C. K., and Li J. (2000) An experimental perspective on the light element in the Earth's core. In *Origin of the Earth and Moon* (R. Canup and K. Righter, eds.), this volume. Univ. of Arizona, Tucson.

Hirayama Y., Fujii T., and Kurita K. (1993) The melting relation of the system, iron and carbon at high pressure and its bearing on the early stage of the earth. *Geophys. Res. Lett., 20,* 2095–2098.

Hirth G. and Kohlstedt D. L. (1995a) Experimental constraints on the dynamics of the partially molten upper mantle — deformation in the diffusion creep regime. *J. Geophys. Res., 100,* 1981–2001.

Hirth G. and Kohlstedt D. L. (1995b) Experimental constraints on the dynamics of the partially molten upper mantle 2. Deformation in the dislocation creep regime. *J. Geophys. Res., 100,* 15441–15449.

Horan M. F., Smoliar M. I., and Walker R. J. (1998) W-182 and Re-187-Os-187 systematics of iron meteorites: Chronology for melting, differentiation, and crystallization in asteroids. *Geochim. Cosmochim. Acta, 62,* 545–554.

Iida T. and Guthrie R. I. L. (1988) *The Physical Properties of Liquid Metals.* Oxford Univ., New York. 288 pp.

Ip S. W. and Toguri J. M. (1993) Surface and interfacial tension of the Ni-Fe-S, Ni-Cu-S, and fayalite slag systems. *Metall. Trans., 24B,* 657–668.

Jones J. H. (1996) Chondrite models for the composition of the earth's mantle and core. *Philos. Trans. Roy. Soc. London, A354,* 1481–1494.

Jones J. H. and Drake M. J. (1986) Geochemical constraints on core formation in the Earth. *Nature, 322,* 221–228.

Keil K. and Wilson L. (1993) Explosive volcanism and the compositions of cores of differentiated asteroids. *Earth Planet. Sci. Lett., 117,* 111–124.

Knittle E. and Jeanloz R. (1991) Earth's core-mantle boundary: Results of experiments at high pressures and temperatures. *Science, 251,* 1438–1443.

Lee D. C., Halliday A. N., Snyder G. A., and Taylor L. A. (1997) Age and origin of the moon. *Science, 278,* 1098–1103.

Li J. and Agee C. B. (1996) Geochemistry of mantle-core differentiation at high pressure. *Nature, 381,* 686–689.

Lugmair G. W. and Shukolyukov A. (1998) Early solar system timescales according to Mn-53-Cr-53 systematics. *Geochim. Cosmochim. Acta, 62,* 2863–2886.

McSween H. Y. Jr., Taylor L. A., and Lipschutz M. E. (1978) Metamorphic effects in experimentally heated Krymka (L3) chondrite. *Proc. Lunar Planet Sci. Conf. 9th,* pp. 1437–1447.

Minarik W. G. and Ryerson F. J. (1996) Metal-sulfide melt noninterconnectivity in silicates, even at high pressure, high-temperature, and high melt fractions (abstract). In *Lunar and Planetary Science XXVII,* pp. 881–882. Lunar and Planetary Institute, Houston.

Minarik W. G., Ryerson F. J., and Watson E. B. (1996) Textural entrapment of core-forming melts. *Science, 272,* 530–533.

Mittlefehldt D. W., Jones J. H., Palme H., Jarosewich E., and Grady M. M. (1998) Sioux County — A study in why sleeping dogs are best left alone (abstract). In *Lunar and Planetary Science XXX*, Abstract #1220. Lunar and Planetary Institute, Houston.

Murr L. E. (1975) *Interfacial Phenomena in Metals and Alloys.* Addison-Wesley, Reading, Massachusetts. 376 pp.

Pike W. A., Bertka C. M., and Fei Y. (1999) Melting temperatures in the Fe-Ni-S system at high pressures: Implications for the state of the martian core (abstract). In *Lunar and Planetary Science XXX*, Abstract #1489. Lunar and Planetary Institute, Houston.

Righter K. and Drake M. (1997a) A magma ocean on Vesta: Core formation and petrogenesis of eucrites and diogenites. *Meteoritics & Planet. Sci., 32*, 929–944.

Righter K. and Drake M. (1997b) Metal-silicate equilibrium in a homogeneously accreting earth: New results for Re. *Earth Planet. Sci. Lett., 146*, 541–553.

Ringwood A. E. (1966) Chemical evolution of the terrestrial planets. *Geochim. Cosmoschim. Acta, 30*, 40–104.

Ringwood A. E. (1990) Earliest history of the Earth-Moon system. In *Origin of the Earth* (H. E. Newsom and J. H. Jones, eds.), pp. 101–134. Oxford Univ., New York

Roscoe R. (1952) The viscosity of suspensions of rigid spheres. *British J. Appl. Phys., 3*, 267–269.

Rushmer T. (1995) An experimental deformation study of partially molten amphibolite — application to low-melt fraction segregation. *J. Geophys. Res., 100*, 15681–15695.

Rushmer T. (1996) Melt segregation in the lower crust: How have experiments helped us? In *Third Hutton Symposium on The Origin of Granites and Related Rocks* (M. Brown et al., eds.), pp. 73–83. GSA, Denver, Colorado.

Rushmer T. and Jones J. H. (1998a) Core formation under dynamic conditions: An experimental approach (abstract). In *Lunar and Planetary Science XXIX*, Abstract #1274. Lunar and Planetary Institute, Houston.

Rushmer T. and Jones J. H. (1998b) Core formation under dynamic conditions: An experimental approach (abstract). In *Origin of the Earth and Moon*, p. 39. LPI Contribution #957. Lunar and Planetary Institute, Houston.

Rushmer T. and Jones J. H. (1999) Core formation under dynamic conditions: Initial results on silicate-metal distribution under differential stress (abstract). In *Lunar and Planetary Science XXX*, Abstract #1369. Lunar and Planetary Institute, Houston.

Rutter E. H. and Neumann D. H. K. (1995) Experimental deformation of partially molten westerly granite under fluid-absent conditions, with implications for the extraction of granitic magmas. *J. Geophys. Res., 100*, 15697–15715.

Schmitz M. D., Holzheid A., and Grove T. L. (1998) Core formation in the earth: Insights from textural equilibria of coexisting silicate melt, solid silicate, and Fe-S-O melt (abstract). In *Lunar and Planetary Science XXIX*, Abstract #1026. Lunar and Planetary Institute, Houston.

Shannon M. C. and Agee C. B. (1996) High pressure constraints on percolative core formation. *Geophys. Res. Lett., 23*, 2717–2720.

Shannon M. C. and Agee C. B. (1998) Percolation of core melts at lower mantle conditions. *Science, 280*, 1059–1061.

Shirey S. B. and Walker R. J. (1998) The Re-Os isotope system in and high-temperature geochemistry. *Annu. Rev. Earth Planet. Sci., 26*, 423–500.

Smith C. S. (1964) Some elementary principles of polycrystalline microstructure. *Metall. Rev., 9*, 1–48.

Snow D. T. (1969) Anisotropic permeability of fractured media. *Water Resour. Res., 5*, 1272–1289.

Srinivasan G., Goswami J. N., and Bhandari N. (1999) Al-26 in eucrite Piplia Kalan: plausible heat source and formation chronology. *Science, 284*, 1348–1350.

Stevenson D. J. (1990) Fluid dynamics of core formation. In *Origin of the Earth* (H. E. Newsom and J. H. Jones, eds.), pp. 231–249. Oxford Univ., New York.

Taylor G. J. (1992) Core formation in asteroids. *J. Geophys. Res., 97*, 14717–14726.

Taylor S. R. and Norman M. D. (1990) Accretion of differentiated planetesimals to the Earth. In *Origin of the Earth* (H. E. Newsom and J. H. Jones, eds.), pp. 29–44. Oxford Univ., New York.

Turcotte D. L. and Schubert G. (1982) *Geodynamics Applications of Continuum Physics to Geological Problems.* Wiley, New York. 450 pp.

Urakawa S., Kato M., and Kumazawa M. (1987) Experimental study on the phase relations in the system Fe-Ni-O-S up to 15 GPa. In *High Pressure Research in Mineral Physics* (M. H. Manghnani and Y. Syono, eds.), pp. 95–111. TERRAPUB/AGU, Tokyo/Washington, DC.

von Bargen N. and Waff H. S. (1986) Permeabilities, interfacial areas and curvatures of partially molten systems: Results of numerical computations of equilibrium microstructures. *J. Geophys. Res., 91*, 9261–9276.

Walter M. J. (1998) Segregation of Earth's core in Hadean magma ocean (abstract). In *Origin of the Earth and Moon*, p. 51. LPI Contribution #957. Lunar and Planetary Institute, Houston.

Walter M. J., Newsom H. E., Ertel W., and Holzheid A. (2000) Siderophile elements in the Earth and Moon: Metal/silicate partitioning and implications for core formation. In *Origin of the Earth and Moon* (R. Canup and K. Righter, eds.), this volume. Univ. of Arizona, Tucson.

Wänke H., Dreibus G., and Jagoutz E. (1984) Mantle chemistry and accretion history of the Earth. In *Archean Geochemistry* (A. Kroner et al., eds.), pp. 1–24. Springer-Verlag, New York.

Williams Q. and Jeanloz R. (1990) Melting relations in the iron-sulfur system at ultra-high pressures: Implications for the thermal state of the Earth. *J. Geophys. Res., 95*, 19299–19310.

Wilson L. and Keil K. (1991). Consequences of explosive eruptions on small solar system bodies: the case of the missing basalts on the aubrite parent body. *Earth Planet. Sci. Lett., 104*, 505–512.

Zimmerman M. E., Zhang S., Kohlstedt D. L., and Karato S. (1999) Melt distribution in mantle rocks deformed in shear. *Geophys. Res. Lett., 26*, 1505–1508.

# An Experimental Perspective on the Light Element in Earth's Core

**Valerie J. Hillgren**
*Max Planck Institut für Chemie*

**Christine K. Gessmann**
*University of Bristol*

**Jie Li**
*Geophysical Laboratory*

Since the recognition that the outer core is less dense than pure Fe at the same conditions, there has been a healthy debate over what the alloying element might be. The elements most often proposed are Si, O, S, C, and H. In recent years, there has been a considerable amount of experimental work directed at the possibilities of these various elements as the light element. The combination of experimental results with theoretical and cosmochemical considerations shows that it is unlikely that any one of these elements can account for the density deficit on its own. Based on the results of experimental studies combinations of Si with O and Si with S can probably be ruled out. The experimental work shows that S with C, S with O, and Si with C may be promising combinations. Thus, future research should focus on these and other potential ternary systems.

## 1. INTRODUCTION

As soon as it was recognized that the outer liquid portion of Earth's metallic core is ~10% less dense than pure Fe (*Birch*, 1952), the debate began over which element or elements is responsible for the density deficit. In order to be a viable candidate for either a sole or primary light element, an element must meet three relatively obvious constraints. The first is that it must be lighter than Fe, and how much lighter will determine how much of the element is required. Second, it must alloy readily with Fe. Third, the element must be cosmochemically abundant enough that Earth could reasonably contain enough to account for the amount observed in the mantle plus the amount presumed to be in the core. An additional constraint arises from the fact that the solid inner core contains considerably less light elements, perhaps even none (*Jephcoat and Olson,* 1987; and reviews by *Poirier,* 1994; and *Stevenson,* 1981). Therefore, the light element must partition favorably into the liquid phase at core pressures and temperatures. These constraints with an emphasis on the first three obvious ones have resulted in S, O, Si, C, and H being the preferred candidates with each element having its proponents.

The earlier debates had to rely on low-pressure experimental data, limited shock data, and theory on the high-pressure behavior of Fe alloys. In other words, there were no experimental data at high pressures on the properties of these potential alloys with Fe. Because of advances in high-pressure experimental petrology over the last 15 years, there has been a surge in experimental studies aimed at determin-ing the light element. Although the problem is far from solved, these new high-pressure experimental data have tightened the constraints on the identity and abundance of the light element(s).

It should be pointed out that the core likely contains ~5 wt% Ni (e.g., *McDonough and Sun,* 1995). This Ni is often overlooked in discussions of the light element because it has essentially no effect on the density of Fe (*Mao et al.,* 1990). However, it could potentially have important effects on the chemistry of the light element, and where these have been observed, it will be noted. Throughout the discussion, the reader should keep in mind that although the core is often referred to as an alloy of Fe and light element(s) and much of the experimental work pertains to Fe-light-element pairs, the core should be more properly thought of as an FeNi-light-element alloy.

It is the goal of this chapter to review the experimental work on the light element, emphasizing the more recent high-pressure work, in conjunction with theoretical and cosmochemical arguments, and then to assess what the sum of this information tells us about the light element(s) in the core. To do this we will examine each of the five most favored elements in turn and then consider combinations of these elements. However, before delving into the specific arguments and data associated with each element, it is worthwhile to look at why the identity of the light element is so critical to Earth science, particularly because many of the reasons for its importance also provide additional constraints on the physical and chemical properties of the light element.

## 2.  WHY THE IDENTITY OF THE LIGHT ELEMENT IS IMPORTANT

### 2.1.  Composition of Earth

As will be seen in the sections on each element, several to >10 wt% of an element can be required for it to be the sole or primary light element, and this amount of light element can have significant effects on the bulk composition of Earth. The exact composition of Earth in turn has implications for accretion scenarios as it will dictate what materials are required to make up Earth. This is, of course, closely linked to the constraint that the element must be cosmochemically abundant enough, but the constraint is tightened here in that not only must the element be abundant enough but it should be present in Earth in a cosmochemically reasonable proportion. In other words, the resulting bulk composition of Earth should grossly resemble chondritic meteorites, the presumed building blocks of Earth. In fact the opposite approach — looking at the composition of meteorites and deducing the composition of the core — is often taken. Cosmochemical abundance and the composition of Earth are particularly relevant to S and Si as light elements and are discussed in more detail in the sections dealing with those elements. For each candidate light element the amount necessary is discussed in the relevant section as well as whether this amount is plausible.

### 2.2.  Metal-Silicate Partitioning of Siderophile Elements

The amount of an element left in the mantle after core formation depends on how it partitions between metal and silicate. The abundance in the mantle of elements generally expected to have partitioned into metal during core formation (i.e., siderophile elements) are often used to glean something about the early history of Earth and core-formation scenarios. Redox conditions have important and well-studied effects on the partitioning of siderophile elements (see review by *Walter et al.,* 2000). Thus the redox conditions necessary to incorporate the light element should also be able to produce the correct observed abundances of siderophile elements or should not preclude the observed abundances if other reasonable processes are taken into account. "Reasonable" is of course a relative term, and any extra process would have to stand up to scrutiny. The composition of the core forming metal may also affect siderophile-element partitioning (see review by *Walter et al.,* 2000). Thus, the composition of the core and the redox conditions necessary to incorporate the light element into the core must be reconcilable with the observed abundances of the siderophile elements in the mantle. Conversely, identifying the light element in the core may be key in explaining the siderophile-element abundance pattern in the mantle.

### 2.3.  Core-Mantle Interaction

Although there is controversy over whether the experimental work shows unequivocally that mantle silicates would react with Fe-metal at core-mantle boundary pressures and form D" (*Goarant et al.,* 1992; *Hillgren and Boehler,* 1997, 1998a; *Knittle and Jeanloz,* 1989, 1991a), it is clear that the nature and extent of any possible reactions will be influenced by the exact composition of the core. The product of any possible reaction between the mantle and core must have seismic properties that are compatible with those observed for D". Thus, though any number of alloys may react readily with mantle silicates, if the properties of the reaction products are incompatible with D", then a particular light element may be ruled out. If a reaction is not responsible for the entirety of D", then the reaction product's seismic properties must be consistent with those observed right at the core-mantle boundary.

The light element plays yet another role in possible core-mantle reactions. Any material added to the mantle during such a reaction will bear a chemical signature of the core. The light element may effect the partitioning of trace elements between the inner and outer core, and thus the chemical signature imprinted on the mantle through the addition of core material. Imposing such a chemical signature onto the mantle has been discussed primarily in regards to the highly siderophile elements (i.e., Pt, Re, Os, etc.) (see review by *Righter et al.,* 2000).

### 2.4.  Compositional Convection in the Outer Core

Convection in the outer core must be maintained in order to sustain the geodynamo. However, because metallic fluids are so thermally conducting the core could become isothermal. Convection can be maintained, however, through compositional convection. Traditionally, compositional convection is thought to be driven from the bottom by the concentration of the light element into the outer core as the inner core freezes, leaving a less dense light-element-enriched liquid at the base of the outer core which then rises (*Braginsky,* 1964; *Stevenson et al.,* 1983). An alternative method to drive compositional convection is from the top through concentration of the light element into the mantle during a core-mantle reaction, leaving a denser liquid at the top of the core which then sinks (*Guyot et al.,* 1997). A combination of both processes could also be responsible for compositional convection. Driving compositional convection requires that the light element be either depleted enough at the top of the outer core or enriched enough at the base, as a slight enhancement or depletion of the concentration of the light element may not be able to do so. How much an element must be depleted or enriched to drive compositional convection will depend on the individual element because each element affects the density of Fe to differing extents. Because top-driven compositional convection is produced through a core-mantle reaction, it has the same constraint placed on it: The reaction product must be reconcilable with either D" or the materials directly overlying the core. Bottom-driven compositional convection, like the observed density difference between the inner and outer core, necessitates that the light element partitions favorably into the liquid phase. The concentration of the light element into the liquid outer core places constraints on the phase relations in the Fe–light-element system. If the Fe–light-element system is a simple binary eutectic, the bulk composition of the system must lie to the Fe side of the eutectic. If this system is a continuous solid solution, then the light-element end member must have the lower melting temperature. A third possibility is that the system is a solid solution with an azeotrope (either a temperature minimum or maximum on the solidus and liquidus curves). If the azeotrope is a minimum, then the core composition must fall to the Fe-rich side of the azeotrope, and if the system contains an azeotropic maximum, then the core composition must fall to the light-element side of it.

## 2.5. Temperature in the Core and at the Core-Mantle Boundary

Knowledge of the thermal structure of Earth is necessary to understand convection in Earth and the evolution of Earth. The temperature at the base of the mantle, the temperature in the core, and the resulting temperature jump across the core-mantle boundary are key components of the thermal structure. The temperature jump across the core-mantle boundary is particularly interesting in light of the recent observations of an ultra-low-velocity zone just at the core-mantle boundary, which is interpreted to be a partial melt of the mantle (*Garnero et al.,* 1993; *Zerr et al.,* 1998). Details of how to estimate the thermal structure and the data used can be found in *Boehler* (1996b). Briefly, the process involves calculating an adiabatic temperature gradient from the 670-km discontinuity [the pressure-temperature conditions at this depth are constrained by the transition of $\gamma$-(Mg, Fe)$_2$SiO$_4$ to magnesiowüstite and Mg, Si-perovskite] down to the core-mantle boundary. The inner-outer core boundary is at the freezing point of the core material, thus calculating an adiabatic temperature gradient upward from that point would give the temperature at the core-mantle boundary on the core side. However, only an upper limit can be placed on this temperature using melting data on Fe because the light element and how much it depresses the melting point of pure Fe at core pressures is unknown. There is even controversy over this upper limit because the melting temperature of Fe has to be extrapolated from ~220 GPa to 330 GPa, the pressure at the boundary of the inner and outer core (e.g., *Boehler,* 1996b, and *Anderson and Duba,* 1997). Thus, knowing the identity and abundance of the light element in the core and how much it depresses the melting point of Fe would allow us to determine a more precise thermal structure for Earth.

## 3. SILICON

Because Si is one of the most abundant elements in Earth, it was proposed immediately that it may at least partly account for the density deficit of the core (*Birch,* 1952). The earliest studies on Si as a light element focused on the amount of Si necessary to account for the density deficit

and the maximum amount of Si that could be dissolved into liquid metal during core-formation processes. Most of the earlier work suggested that the amount of Si in the core was on the order of 20 wt% (e.g., *Balchan and Cowan,* 1966; *Knopoff and MacDonald,* 1960; *MacDonald and Knopoff,* 1958; *Ringwood,* 1959; *Stewart,* 1973), an amount that according to simple-density models would be sufficient to account for the density deficit solely by Si [for example, *Poirier* (1994) estimates 18 wt% Si would meet the density constraints]. On the other hand, various workers taking cosmochemical arguments and constraints such as mantle composition into account (discussed in more detail below) have suggested the core should contain between 5 wt% and 14 wt% Si (*O'Neill,* 1991; *Allègre et al.,* 1995; *Wänke,* 1981; *Wänke and Dreibus,* 1997). If the simple density estimates are correct, then these lower values would likely require additional light components to be present in the core. However, based on calculations of the thermal pressure in the core, *Sherman* (1997) argues that amounts of Si in the core as low as 7.3 wt% are consistent with limits placed on core temperature through the melting of Fe, and the maximum possible Si content in the core is 10.7 wt%. These estimates are much lower than those placed by simple-density calculations and well within the range of Si contents derived from cosmochemical considerations.

*MacDonald and Knopoff* (1958) and *Ringwood* (1959) first pointed out that the Mg/Si ratio of Earth's upper mantle was elevated relative to that of chondritic meteorites, the likely parental material of the planets. If the lower mantle has the same composition as the upper mantle, then the Si depletion in the mantle could be explained if the missing Si resides in the core. Later, however, *Ringwood* (1977) favored O instead of Si as the major light component in the core (see the section on O) and suggested that perhaps the chondrites were enriched in Si rather than the mantle being depleted (*Ringwood,* 1989). Nevertheless, as was noted above, there are still many proponents of Si as the light element based on the apparent Si depletion in the upper mantle (e.g., *Allègre et al.,* 1995; *Wänke,* 1981; *Wänke and Dreibus,* 1997; *Javoy,* 1995). Their various estimates of the Si content of the core are arrived at by comparing element trends in Earth to those in various classes of meteorites and determining those that most closely match Earth.

On the other hand, *Murthy and Hall* (1970, 1972) argued that to incorporate Si in the core, it has to be reduced from silicates, requiring very high temperatures, thus also volatilizing more-volatile elements such as S. Sulfur, however, is still present in Earth. In answer to this problem, *Wänke* (1981) suggested a two-step accretion model, in which he also explains the abundances of siderophile elements in Earth's mantle that do not seem to show a signature of metal segregation, let alone a signature of segregation under the reducing conditions required to incorporate large amounts of Si into the core. The model proposes that during the first 80% or so of accretion, the material forming Earth was very reduced and stripped of volatile elements. During this time a Si-rich core was formed. The conditions are so reduced that essentially all Fe and siderophile elements are stripped from the mantle into the core. During the last 20% or so of accretion, the material is much more oxidized, and this material adds volatile elements and Fe as well as other siderophile elements to the mantle. A small amount of metal may still be segregated to the core at this time. Thus, the core is out of equilibrium with the mantle and will potentially react with it.

Another objection to Si (and O) in the core was presented by *Ringwood and Hibberson* (1990). Their experimental results show that within a sequence of oxides $TiO_2$ is more soluble in liquid metal than $SiO_2$. Therefore, segregating significant amounts of Si into the core would also require Ti to be depleted in the mantle, which is not the case. However, *Poirier* (1994) pointed out that "this argument holds only if one assumes that Si in the core comes from dissolution of the mantle at high pressure." The Ti argument has recently been weakened by contradicting experimental results: *Gessmann et al.* (1995), *Gessmann and Rubie* (1998a), and *O'Neill et al.* (1998) show that the Ti content in liquid metal is always significantly lower than that of Si over a broad range of pressure and temperature conditions. These studies indicate that in the presence of mantle minerals and/or silicate melt Ti is far less "metal loving" than Si. Thus, Si in the core does not necessarily imply that Ti should be depleted in the mantle.

In summary, the various methods of estimating possible Si contents of the core give widely varying values. Therefore, the goal of recent experimental work has been to determine under what conditions it is possible to incorporate significant amounts of Si into liquid metal in the presence of silicate melt and/or mantle minerals and to determine whether those conditions would impress chemical signatures upon the mantle that are or are not observed.

### 3.1. Experimental Work

Some of the early experimental work on Si was performed by *Balchan and Cowan* (1966), *Knittle and Jeanloz* (1989, 1991a), *Ringwood and Hibberson* (1991), and *Goarant et al.* (1992). All these studies showed that Si could be dissolved in liquid Fe alloys at various experimental conditions. *Balchan and Cowan* (1966) performed shock-wave experiments on Fe-Si alloys, and their results were consistent with 14–20 wt% Si in the core. *Knittle and Jeanloz* (1989, 1991a) reported that in diamond anvil cell experiments (25–70 GPa, 2700°–3000°C) metal and perovskite reacted to form stishovite and a Si and O containing Fe-alloy. Dissolution of Si in liquid metal was also reported by *Goarant et al.* (1992) in diamond anvil cell experiments at 70 GPa performed with an Fe plus forsterite assemblage. *Ringwood and Hibberson* (1991) showed that $SiO_2$ and other oxides dissolve (although in small amounts) in liquid Fe metal above 16 GPa and 1900°C. However, they argued against Si in the core due to higher solubility of oxides of Ti, which was discussed in the previous section.

Significant progress in determining the solubility of Si in liquid metal experimentally at high pressure has been achieved by recent studies: *Ito et al.* (1995), *Kilburn and Wood* (1997), *Li* (1998), *Agee and Li* (1998), *Gessmann and Rubie* (1998a,b,c), *O'Neill et al.* (1998), *Hillgren and Boehler* (1998a,b). These studies cover a broad range of conditions and employed a variety of techniques. They are discussed in some more detail below, but two particularly important contributions of these studies are worth noting. The first is the systematic determination of Si solubility in liquid metal as a function of single variables [e.g., pressure, temperature, oxygen fugacity ($f_{O_2}$), and composition], thus allowing the influence of various variables to be deciphered and resolved (e.g., *Gessmann and Rubie*, 1998a,b; *Kilburn and Wood*, 1997). The second is the investigation of Si solubility in liquid metal up to 100 GPa, providing a significantly extended database (*Hillgren and Boehler*, 1998a,b).

*Ito et al.* (1995) performed experiments between Fe and enstatite or olivine in a multianvil apparatus from 10 to 26 GPa, up to 2500°C, and an estimated $f_{O_2}$ slightly below the iron-wüstite (IW) buffer curve. The resulting amounts of Si in the Fe-metal do not exceed 2 wt%, which is slightly higher than the findings of *Ringwood and Hibberson* (1991) at lower pressures and temperatures. These results have been corroborated by *Li* (1998), *Agee and Li* (1998), *Gessmann and Rubie* (1998a), and *O'Neill et al* (1998), all of whom consistently find very limited Si solubility in liquid metal (S-bearing and S-free) in equilibrium with silicate melt and/or mantle minerals. Over a broad range of pressure and temperature conditions, these workers generally find <1.4 wt% Si in the metal. It should be noted that these studies have been performed over limited ranges of $f_{O_2}$, and in the cases of *Li* (1998) and *O'Neill et al.* (1998) almost constant $f_{O_2}$ (around IW-2). Only *Gessmann and Rubie* (1998a) varied the redox conditions slightly (IW-1.7–IW-3.3) and found that the Si contents in the Fe-Ni metal alloy increased with decreasing $f_{O_2}$.

The importance of the variable $f_{O_2}$ on the solubility of light components in liquid metal has been clearly demonstrated by *Kilburn and Wood* (1997). They determined Si solubility (along with that of S, see below) in liquid metal as a function of $f_{O_2}$ in a piston cylinder at constant pressure and temperature. They showed that, despite the moderate pressure and temperature conditions of their experiments, significant and increasing amounts of Si dissolve in liquid Fe metal with decreasing $f_{O_2}$. They found up to 24.4 wt% Si in the quenched metal at extremely reducing conditions (IW-6.6). This indicates that Si may contribute significantly to the light-element budget in the core, if metal segregation during the early history of Earth is thought to take place in a very reducing environment as in heterogeneous accretion scenarios.

*Gessmann and Rubie* (1998b,c) extended their earlier study to a wider range of redox conditions and pressures, finding higher Si contents in the quenched liquid Fe-Ni metal (e.g., 5.5 wt% at IW-4, 9 GPa, 2000°C or 3.5 wt% at IW-3, 18 GPa, 2200°C) than reported in earlier work. They were able to discern pressure and temperature dependencies for the solubility of Si in liquid metal. At a given $f_{O_2}$, both pressure (at constant T) and temperature (at constant P) enhance the amount of Si dissolved in liquid metal. These trends along with some of the results of *Kilburn and Wood* (1997) are illustrated in Fig. 1. Extrapolating the observed trends, *Gessmann and Rubie* (1998c) suggest that at pressures and temperatures exceeding the experimental conditions, significant amounts of Si are likely to be dissolved in liquid metal even at moderate $f_{O_2}$, i.e., close to the IW buffer curve. Such redox conditions are relevant and plausible for core formation scenarios, and *Gessmann and Rubie* (1998b,c) therefore suggest that Si may be one of the major constituents of the light-element budget in Earth's core.

*Hillgren and Boehler* (1998a,b) performed diamond anvil cell experiments investigating Si solubility in liquid metals for various compositions. In experiments with liquid Fe and San Carlos olivine (which converts to perovskite and magnesiowüstite at the run pressures and temperatures), they do not find significant amounts of Si (<0.2 wt%) in quenched liquid Fe metal, even at pressures up to 100 GPa (Fig. 2). In experiments using iron-silicides containing 9 wt% and 17 wt% Si as starting materials, the liquid metal contained significant amounts of Si (up to 11 wt% at 56 GPa, 2800 K). However, they found the FeO content of the silicate coexisting with the Si-rich liquid metal to be much lower than presently observed in the mantle, and therefore argue that Si is not likely a major component of the core. Recently, they have used their cumulative results to argue against Si as the light element in the core (*Hillgren and Boehler,* 1999). Their argument is threefold. First, a core in equilibrium with

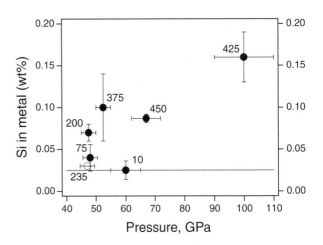

**Fig. 2.** The data of *Hillgren and Boehler* (1998a) showing Si contents in Fe metal vs. pressure for Fe-metal in contact with San Carlos olivine that is converted to magnesiowüstite and perovskite at run pressures and temperatures. The numbers next to each data point are an indication of the experimental temperature: They are the number of degrees above the melting point of Fe. The open symbol is an experiment that also contained a small amount of S. The Si content increases with increasing temperature and pressure, but even at 100 GPa less than 0.2 wt% Si is dissolved in the metal (*Hillgren and Boehler,* 1998a).

the mantle cannot contain enough Si to account for the density deficit. Second, a Si-rich core formed under extremely reducing conditions would react with the mantle, forming a layer richer in $MgSiO_3$-perovskite than the surrounding mantle and would therefore be less dense than the surrounding mantle which is not consistent with interpretations of D" (*Lay et al.,* 1998). Finally, in some of their most reducing experiments, they found Ca in the metal, and they argue that this may indicate that under extremely reducing conditions, refractory lithophile elements could be fractionated from one another which is not observed in Earth (*Jagoutz et al.,* 1979).

The experimental results and conclusions of *Gessmann and Rubie* (1998b,c) and *Hillgren and Boehler* (1998a,b, 1999) appear to be at odds with one another. However, thermodynamic calculations suggest that the seemingly conflicting results may simply be a natural consequence of the phase changes in silicates with pressure. *Gessmann et al.* (1999) recently performed thermodynamic modeling of the experimental results of *Gessmann and Rubie* (1998b,c) and *Kilburn and Wood* (1997) at an $f_{O_2}$ set to IW-2. They compared the obtained solution models to thermodynamic data and extrapolated them to core pressures and variable temperatures. These preliminary modeling results indicate that although at pressures below 25 GPa significant Si can be dissolved in liquid metal, at pressures within the perovskite stability field (above 25 GPa), Si solubility in liquid metal decreases with increasing pressure. These results are in agreement with earlier work by *Guyot et al.* (1997), who performed pressure-volume-temperature measurements of

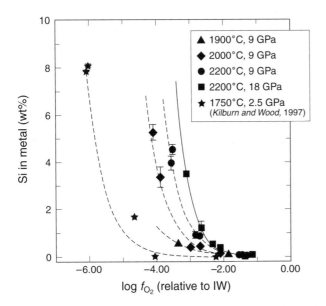

**Fig. 1.** The experimental data of *Gessmann and Rubie* (1998b) and *Kilburn and Wood* (1997) showing Si contents in quenched Fe-Ni liquid metal as a function of $f_{O_2}$. Regression lines are shown for sets of experiments performed at constant pressure and temperature. For a given value of the $f_{O_2}$, an increase in P (at constant T) and also increasing T (at constant P) both increase Si solubility (*Gessmann and Rubie,* 1998b).

ε-FeSi. They included their experimental results in a thermodynamic model of reactions between Fe and $SiO_2$. Their results indicate a maximum solubility of Si in metal between 8 and 15 GPa depending on temperature and a decrease in the solubility of Si once the stability field of stishovite is reached. Based on different experimental data sets and thermodynamic approaches, both studies (*Guyot et al.,* 1997; *Gessmann et al.,* 1999) arrive at similar conclusions: The incorporation of significant amounts of Si in metal is plausible at medium pressures (i.e., below 25 GPa), but at much higher pressures, such as those at the core-mantle boundary, Si present in liquid metal may exsolve from the core, thus arguing against Si as a major light constituent in Earth's core.

In summary, the experimental work indicates that Si can be incorporated into liquid Fe in large amounts under certain pressure, temperature, and $f_{O_2}$ conditions. However, the results of thermodynamic modeling (based on experimental results from <25 GPa) indicate that Si could exsolve from the liquid metal at core-mantle boundary pressures. This result agrees with the low-Si solubility observed in diamond anvil cell experiments at pressures >25 GPa and moderate $f_{O_2}$. The consequences of incorporating large amounts of Si into the core and then exsolving it still warrant further investigation, though, both experimentally and theoretically in order to fully evaluate heterogeneous accretion, potential core-mantle interaction, ant the possibility for top-driven compositional convection.

## 4. OXYGEN

Oxygen was proposed in the early 1970s as the light element in the core (e.g., *Bullen,* 1973; *Dubrovskiy and Pan'kov,* 1972), but serious discussion about it really began after *Ringwood* (1977) suggested that the miscibility gap observed between liquid Fe and liquid FeO at 1 atm closes at elevated pressures. He also suggested that FeO becomes metallic at higher pressures, an idea that remains controversial today (e.g., *Knittle and Jeanloz,* 1991b, and *Sherman,* 1995). Ringwood estimated that approximately 10 wt% of O in the core would be sufficient to account for the density deficit, and *Poirier* (1994) calculated that 9 wt% O would need to be present in the core to meet density constraints.

A potential drawback to O as a light element is that incorporating O into metal is a high-pressure process, and much of core formation may have occurred under near-surface, low-pressure conditions in a growing Earth. During the early stages of accretion and core formation, Earth was much smaller with a considerably lower central pressure. Later, as Earth approached its current size, metal was segregated from silicates near the surface of Earth as material was added to the outside of the planet. However, the core-mantle boundary currently sits at 1.3 Mbar pressure, so there is no reason to presume that although initial core formation processes were low pressure, the core did not continually reequilibrate (or attempt to reequilibrate even if full equilibrium was never attained) as its pressure regime in-

creased. High-pressure reequilibration is rarely considered in core formation models, and although it is debatable whether high-pressure reequilibration would effect upper mantle geochemistry, it may be a very important process for determining the light element. Two examples of this have already been given in this chapter: Si exsolves from metal at lower mantle pressures, and high pressures are necessary to incorporate O into metal.

*O'Neill et al.* (1998) and *O'Neill and Palme* (1998) have raised an interesting objection to O as the light element: They argue that O is too volatile. Oxygen may not condense out of the solar nebula in a form that could be incorporated into the core (i.e., as FeO) until after S has begun to condense, and as will be seen below, it is argued that S may be too volatile for enough to have been accreted to Earth to account for the density deficit in the core. Thus if S is too volatile to account for the light element budget, then O is even more so.

A potentially more serious drawback to O as the light element comes from the melting and phase relations in the Fe-FeO system. At one atmosphere pressure the system Fe-FeO is a eutectic system with less than 0.2 wt% O in the eutectic composition. A core with 8–10 wt% O would lie to the FeO side of the eutectic, and thus the phase crystallizing out to form the inner core would be FeO. This would concentrate the light element into the inner core, which is contrary to observation and would inhibit compositional convection. *Kato and Ringwood* (1989) and *Ringwood and Hibberson* (1990) have shown that at 16 GPa the eutectic composition does shift toward FeO, but even at 16 GPa the eutectic composition only contains between 2 wt% and 3 wt% O. Thus, for O to be the sole light element in the core, the eutectic would need to shift considerably more toward FeO with increasing pressure. A similar problem arises if the Fe-FeO system becomes a continuous solid solution at very high pressures. The melting temperature of FeO at high pressure is higher than that of Fe (*Boehler,* 1992, 1993). Therefore the phase crystallizing out to form the inner core would be enriched in FeO over the outer core, once again concentrating the light element into the inner core and inhibiting compositional convection. It is probably more likely that a eutectic system would evolve into a solid solution with an azeotropic minimum replacing the eutectic. In this case the bulk composition of the core must fall to the Fe side of the azeotrope so that the first solid is not enriched in O over the liquid. Thus, if O is to be the sole light element in the core, then the system must either remain a eutectic or evolve to a system with an azeotropic minimum replacing the eutectic, and the eutectic composition must shift even further toward FeO than already observed.

Although there are several potential pitfalls for O as the sole light element in the core, it has still been pursued experimentally because even if it is not the sole light element, perhaps it is still a major contributor. For example, in their model of the composition of Earth, *Allègre et al.* (1995) felt that the amount of Si and S that they surmised should be in the core was not enough to account for the entire density deficit of the core. Therefore, they treated O as essentially a "filler" element and suggested that the core contains 4.1 wt% O. The work on the Fe-FeO phase diagram and solubility of O in Fe (i.e., *Kato and Ringwood, 1989; Ohtani et al., 1984; Ringwood and Hibberson, 1990; Ringwood and Hibberson, 1991*) certainly shows that some O could be dissolved in the core. The more recent experimental work has not concentrated on the topology of the phase diagram, rather on the amount of O that can be incorporated into Fe when it is in equilibrium with mantle minerals.

## 4.1. Experimental Work

As already stated, the work of *Ohtani et al.* (1984), *Kato and Ringwood* (1989), and *Ringwood and Hibberson* (1990, 1991) indicated that O can certainly be dissolved in the metal of Earth's core. Diamond anvil cell experiments by *Knittle and Jeanloz* (1991b) and *Goarant et al.* (1992), to 70 GPa and 130 GPa respectively, demonstrated that reactions between liquid Fe and lower mantle minerals could also dissolve FeO into the metal. However, the diamond cell work was only qualitative and did not measure the actual solubility of O in Fe-metal.

Recently, additional studies in systems relevant to core-mantle chemistry have been carried out using both the multianvil apparatus and the diamond anvil cell (*Agee and Li,* 1998; *Gessmann and Rubie,* 1998a,b,c; *Hillgren and Boehler,* 1998a,b; *Ito et al.,* 1995; *O'Neill et al.,* 1998). These studies have quantitatively measured the amount of O that dissolves in the metal over a range of pressure and temperature conditions.

The detected O content in the quenched liquid metal does not greatly exceed ~1 wt% in any of these recent systematic studies using core-mantle analogs as experimental starting materials. This result holds irrespective of pressure, temperature, and $f_{O_2}$ of the experiments. In fact, the amount of O found in the metal is extremely low (0.1–0.4 wt%) in the majority of the experiments performed so far. In systematic multianvil experiments, *Gessmann and Rubie* (1998b,c) observed the most O dissolved in liquid metal: ~1.2 wt% at 9 GPa, 2200°C, and $f_{O_2}$ ~1.5 log bar units below the iron-wüstite buffer. At the most extreme conditions studied to date, 100 GPa and 3300 K, *Hillgren and Boehler* (1998a,b) found only 1 wt% O in liquid Fe in equilibrium with magnesiowüstite and Mg,Si-perovskite. In their experiments temperature and pressure were increased in tandem so they were unable to discern specific temperature and pressure trends. *Gessmann and Rubie* (1998b,c), on the other hand, systematically varied temperature, pressure, and $f_{O_2}$ and found that the O content in liquid Fe-Ni metal increases with increasing $f_{O_2}$, increasing temperature, and decreasing pressure. The latter dependency is also reported by *O'Neill et al.* (1998). The effect of temperature on the O solubility is in agreement with earlier investigations (*Kato and Ringwood,* 1989; *Ohtani et al.,* 1984), while the determined pressure dependency contradicts the results and predictions of

*Ohtani et al.* (1984), *Kato and Ringwood* (1989), and *Ringwood and Hibberson* (1990). Thus there is some discrepancy among the experimental results, but even so the generally low O contents found in metals argue strongly against O being a major component in the core. However, the experimental results do not rule out contributions of O to the light-element budget in the outer core on the order of 1–2 wt%.

## 5. SULFUR

Sulfur as a light element was popularized by *Murthy and Hall* (1970, 1972) who noted that S was depleted in the crust and mantle by several orders of magnitude relative to other volatile species such as halogens, water, and rare gases. Depletion of S of this same magnitude is not observed in meteorites, precluding preterrestrial fractionation as the cause. Neither could it be due to volatile loss during accretion, as elements that are more volatile than S were less depleted. Therefore the most likely explanation is that S is hidden in the core.

In fact, in addition to the cosmochemical argument given above, S is a particularly strong candidate for the principle light element in the core on several other grounds. First, S is moderately to highly siderophile during iron-silicate interaction at pressures near Earth's surface as well as in its deep interior (see below), therefore most of the S present in Earth was probably sequestered to the core during core-mantle differentiation.

The phase and melting relationships in the Fe-FeS binary are also favorable to S being the light element. At ambient pressure, Fe and S can form a series of compounds, and Fe and pyrrhotite ($Fe_{0.9}S$) form a eutectic binary system on the Fe-rich side with Fe and FeS liquids completely miscible (Fig. 3). Calculations based on electronic band structure predict that the Fe-FeS system should become a solid solution above 100 GPa (*Boness and Brown,* 1990; *Sherman,* 1991, 1995), thus enhancing alloying between Fe and S. Experimental studies on the melting of Fe-FeS mixtures at high pressures show that the melting temperature of Fe-FeS mixtures is lower than that of Fe (see below). Thus, if the system is a continuous solid solution, then the solid crystallizing out to form the inner core would be depleted in S relative to the liquid outer core, which is in accordance with the observed density difference between the outer and inner core and would also help drive compositional convection.

Most estimates of the amount of S required to account for the density deficit in Earth's outer core relative to pure liquid Fe are on the order of 10 wt% [9–12 wt% (*Ahrens,* 1979), 10 ± 4 wt% (*Brown et al.,* 1984), or 11 ± 2 wt% (*Ahrens and Jeanloz,* 1987)]. This estimate assumes ideal mixing between the end members Fe and pyrrhotite ($Fe_{0.9}S$) or pyrite ($FeS_2$). The densities of these phases were measured at pressures and temperatures near the core-mantle boundary by shock-wave experiments. The uncertainties are associated with those in the equation of state for solid phases,

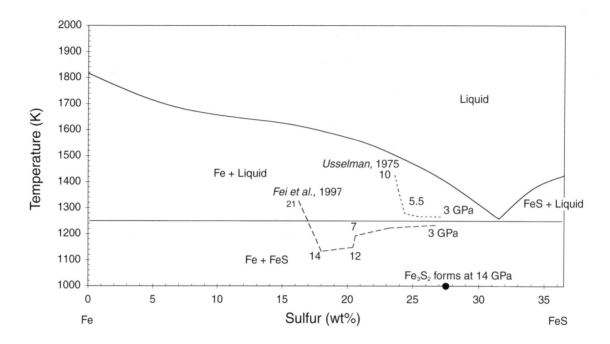

**Fig. 3.**  Simplified phase relations in Fe-S binary system at 1 bar (after *Raghavan,* 1988). The dashed and dotted lines illustrate the change of eutectic composition with pressure as measured by *Usselman* (1975) and *Fei et al.* (1997). The large dot marks the composition of Fe$_3$S$_2$, which forms at 14 GPa (*Fei et al.,* 1995).

differences between solid and liquid phases, geotherms, and different seismic models. *Sherman* (1997) considering non-ideal mixing in the Fe-FeS binary and the thermal pressure in the core calculates that the core could contain as little as 5 wt% S. It is these estimates of the amounts of S required that are the biggest stumbling block to S as the sole light element: The lowest estimate of 5 wt% just overlaps with the upper limits estimated by the cosmochemists, which are discussed below.

To determine how much S may be in the core, cosmochemists use mass balance between the core and mantle and an estimate of the bulk Earth S content. The bulk Earth S content is estimated by assuming that the mantle is homogeneous and that the S abundance in Earth is similar to the abundance of a lithophile element (i.e., an element not expected to have segregated to the core) of comparable volatility. Any S missing from the mantle relative to the comparison element is assumed to be in the core. Following this method, *Ganapathy and Anders* (1974) estimated the S content of the core to be ~5 wt%, assuming that the S to K ratio of Earth is the same as the solar system S/K. Later, *Morgan and Anders* (1980) pointed out that the geochemical processes involving S and K are poorly understood, therefore 5 wt% is not necessarily an upper limit for S content in the core. However, *Kargel and Lewis* (1993) have estimated that the core can only contain between 1.8 wt% and 4.1 wt% S by how far S falls off a volatility trend they have constructed for Earth and taking into account uncertainties in the S condensation temperature. Recently *Dreibus and Palme* (1996) have whittled down the potential amount of

S in the core further by estimating the bulk Earth S content by comparing it to Zn, and arguing the core cannot have more than 2 wt% S. Therefore it appears that more S is required to account for the density deficit than may actually be available in Earth.

### 5.1.  Experimental Work

Because there is little doubt that if S is present, it will alloy with Fe and segregate to the core, the focus of experimental work on S has been quite different from the work on O and Si, which has concentrated on the conditions that are required to incorporate these elements into the core in sufficient quantity to account for the density deficit. Rather the work on S has focused on the density of Fe-S compounds at high pressure and thus how much S is required to account for the density deficit. In addition, it has also concentrated on what could be loosely termed the consequences for Earth of a S-rich core: the abundance of siderophile elements and S in the mantle, melting relations in the Fe-FeS binary that have consequences for the core temperature and inner core composition, and the role S may play in aiding core formation.

### 5.2.  Density

The amount of S needed to account for the density deficit in the core can be best estimated from a direct comparison between density of the outer core and that of liquid Fe with various amounts of S under the corresponding pressure and temperature conditions. However, such data are not yet available. Instead, the density of endmember Fe-S compounds have been determined by shock wave experiments to outer core pressure and temperature ranges (*Ahrens, 1979; Ahrens and Jeanloz, 1987; Brown et al., 1984; Jephcoat, 1985*). These densities are then used to model probable S contents in the core as discussed above. Meanwhile, in order to remove some of the uncertainties related to shock measurements, initial efforts have been undertaken to make static measurements of Fe-S compounds at high pressure and high temperature (*Fei et al., 1995; Jephcoat, 1985*). The data on density vs. pressure are summarized in Fig. 4.

### 5.3. Melting Relations of the Iron-rich Portion of the Iron-Sulfur Binary System

Melting relations in the Fe-S system have many implications for the composition and thermal state of the core. If S is the light element in the core, the melting temperature of the Fe-S alloy constrains the temperature at the inner-outer-core boundary and gives the lower limit for the temperature within the outer core. If Fe-S forms a eutectic binary system within the pressure range of the core, then the composition of the core must fall on the Fe-rich side of the eutectic, in order to match the observation that the inner core is much denser than the outer core. Moreover, the eutectic temperature gives a lower limit of the temperature at the inner-outer-core boundary, while the eutectic composition gives the upper limit of S content in the outer core.

If the system becomes a continuous solid solution, the melting temperature of the S-rich end member must be lower than that of Fe as outlined above for O, and the width of the phase loop will determine the partitioning of S between the inner and outer core.

For these reasons, the composition and temperature of the eutectic point in the Fe-rich portion of Fe-S system have been studied extensively over the past decades. Experiments with the piston-cylinder and belt apparatus up to 10 GPa showed that with increasing pressure the eutectic temperature increases and the eutectic composition becomes more Fe-rich (Fig. 3) (*Brett, 1969; Ryzhenko and Kennedy, 1973; Usselman, 1975*). *Williams and Jeanloz* (1990) and *Boehler* (1992) measured the melting of Fe-S mixtures with fixed composition, particularly end members such as troilite (FeS) and pyrite ($FeS_2$), in the diamond anvil cell. Measurements were carried out at pressures up to 120 GPa. Although there is disagreement in the exact melting temperatures, both studies found the melting temperature of sulfides are lower than that of pure Fe by at least several hundreds of degrees. Later, *Boehler* (1996a) measured the eutectic temperature between Fe and FeS up to 62 GPa. He found that the eutectic temperature increases with pressure, and the magnitude of eutectic depression becomes smaller at high pressure, which supports the possibility of solid-solution between Fe and sulfide at core pressures. Experimental data on melting temperature vs. pressure are summarized in Fig. 5.

Recently, a new high-pressure Fe-S compound $Fe_3S_2$ was synthesized in experiments in the multianvil apparatus (*Fei*

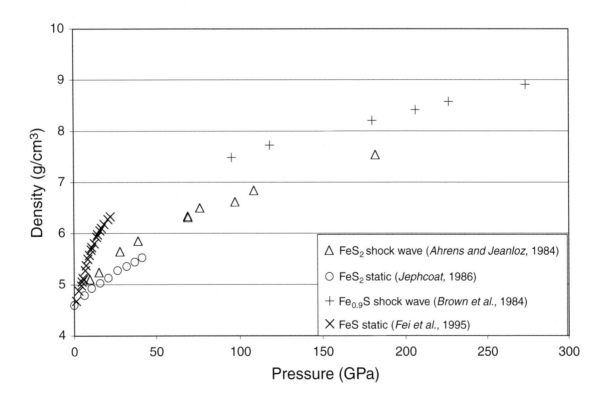

**Fig. 4.** A summary of the experimentally determined density vs. pressure data for compounds in the Fe-S binary system. The source of the data and the type (static vs. shock wave) is indicated in the legend.

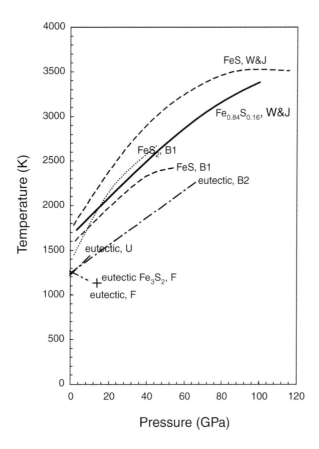

**Fig. 5.** Melting of Fe-S alloys with S contents ranging from ~10 wt% ($Fe_{0.84}S_{0.16}$) to ~53 wt% ($FeS_2$). In the order of increasing S content, the melting curves are plotted as solid, dashed, dash-dotted, and dotted lines. Eutectic refers to that between Fe and FeS, except for the cross, which represents the eutectic between Fe and $Fe_3S_2$. U = *Usselman* (1975); B1 = *Boehler* (1992); B2 = *Boehler* (1996a); F = *Fei et al.* (1997); W&J = *William and Jeanloz* (1990).

*et al.,* 1997). The formation of this new intermediate compound changes the melting relations in the Fe-S binary system. At 14 GPa, the eutectic temperature in Fe-$Fe_3S_2$ system is hundreds of degrees lower than that extrapolated from data on Fe-FeS system at lower pressure (Fig. 3). This implies that the lower limit of the outer core temperature may be even lower than previously thought. Furthermore, the maximum amount of S possible in the core as constrained by phase relations in the Fe-S system would be lower, as $Fe_3S_2$ has a lower S content than any of the stable Fe-S compounds at ambient pressure.

### 5.4. Partitioning of Sulfur Between Metal and Silicate

It is generally accepted that S is moderately siderophile to highly siderophile. However, *Jones and Drake* (1986) raised the concern that equilibrium between a S-rich core and mantle silicates may leave far more S in the mantle than presently observed. More recent studies on the behavior of S, though, show that the affinity of S for Fe is enhanced by

increasing pressure and weakened by increasing temperature (e.g., *Li and Agee,* 1996; *Ohtani et al.,* 1997; *Walker et al.,* 1993; *Wendlandt,* 1982). Thus, it is possible that the combined effect of pressure and temperature on the partitioning of S will resolve this problem.

### 5.5.  Role of Sulfur in Core Formation

Mantle abundances of moderately and highly siderophile elements are much higher than expected from equilibrium partitioning between iron and silicate measured at ambient pressure. In other words, these elements' metal-silicate partition coefficients are too large to be consistent with near-surface core-mantle equilibrium. *Brett* (1984) noticed that partition coefficients between S-bearing Fe-rich liquid and silicate are much smaller than those between Fe and silicate and suggested that the apparently high abundances of siderophile elements may reflect equilibrium between a S-rich core and mantle.

*Li and Agee* (1996) studied partitioning of Ni and Co, two moderately siderophile elements, in a S-bearing system up to pressures of 20 GPa. Comparison between their results and those in a S-free system by *Thibault and Walter* (1995) showed that S had little effect on the partitioning of Ni and Co, especially at high pressures. Later, *Jana and Walker* (1997b) did a systematic study of the effect of S on partitioning of a large suite of siderophile elements. They found that the effect can be large and intricate for some elements while negligible for others. In general, Ni, Co, and Au became slightly less siderophile in the presence of S-bearing metallic liquids, while W, Ge, and P became much less siderophile. Iridium, Pt, and Re avoid S-bearing metallic liquids but do not really become less siderophile because they form refractory alloys that would still likely segregate to form the core. Thus, the segregation of a S-bearing metallic liquid can potentially reduce the Fe-loving nature of many elements to explain the excess paradox but it does not work for all elements, especially some highly siderophile elements.

Sulfur may play still another role during core segregation. *Murr* (1975) discovered that small amounts of S can cause large decreases in the surface tension of liquid Fe, therefore facilitating the settling of core-forming alloy. Experimental studies at higher pressures (*Shannon and Agee,* 1996; *Minarik et al.,* 1996) with variable S content confirm that S reduces the dihedral angle between Fe-rich alloy and solid silicate matrix, which could expedite core segregation by allowing the core-forming metal to produce an interconnected network through solid silicate grain boundaries and then percolate to the center of Earth. This phenomenon is discussed in more detail by *Rushmer et al.* (2000).

### 6.  CARBON

Carbon is often mentioned as a potential light element but rarely championed by anyone. It would require 6.6 wt% C in the core to completely account for the density deficit [calculated following the method described in *Poirier* (1994), and taking parameters for liquid C from *Wood* (1993)], and such large amounts seem precluded by the very high volatility of C. The 50% condensation temperature of C in the solar nebula is on the order of 500 K, as compared to ~1500 K for an Fe-rich metal phase (*Grossman and Larimer*, 1974). Less than 3 ppm C is expected to be present in Fe condensed from the solar gas, and even this small amount of C may be partially lost during high-temperature processes during or after accretion. Therefore, although the abundance of C in Earth's mantle is comparable to S (*McDonough and Sun*, 1995), the bulk Earth abundance of C is believed to be lower than S by more than an order of magnitude (*Kargel and Lewis*, 1993). Consequently, the amount of C available for the core should be well below 1 wt%, far too little to explain the observed density deficit.

On the other hand, C alloys easily with Fe. At ambient pressure, several weight percent C can easily dissolve in liquid Fe at moderately high temperature, and C becomes even more soluble in Fe at higher temperatures (discussed more below). In addition, Fe and C form cohenite ($Fe_3C$), and cohenite is one of the most common meteoritic minerals (*Wasson*, 1985). However, the total amount of C present in iron meteorites is not more than about a half weight percent. The low concentration of C in iron meteorites would seem to indicate that it is unlikely that C is one of the major light elements in the core, if core formation was mostly a near-surface, low-pressure process.

### 6.1. Experimental Work

Because of industrial interests (namely the production of steel), there are numerous experimental studies on the Fe-C system at ambient pressure. However, the study by *Wood*

(1993), a combination of experimental and theoretical work, appears to be the only study directed at C as a component of the core. At ambient pressure, Fe and graphite form a eutectic system. The eutectic composition contains 4.2 wt% C, and the eutectic temperature is about 400° below the melting temperature of Fe, and $Fe_3C$ is stable below the solidus (Fig. 6). Increasing pressure enhances C solubility on the C-rich side of the eutectic point, shifts the eutectic composition toward Fe, and expands the stability field of $Fe_3C$ (*Wood*, 1993). At 9 GPa, a new compound $Fe_7C_3$ becomes stable (*Shterenberg et al.*, 1975).

*Wood* (1993) argues that at pressures greater than 0.01 GPa, C is no longer volatile, and thus C accreted to Earth would remain to potentially dissolve as much as 2–4 wt% in the core. He uses experimental results and theoretical calculations to predict the topology of the Fe-C phase diagram at core pressures. He finds that if the $Fe_3C$ stability field continues to grow with pressure, then under core conditions, the presence of even a small amount of C (as low as 0.3 wt%) would lead to solidification of $Fe_3C$ to form the solid inner core. He estimates that the density of $Fe_3C$ under the relevant conditions is a good match to the density of the inner core. No experimental work has yet been done to test the validity of such predictions. Crystallization of an $Fe_3C$ inner core would require that the core contain a second light element because concentration of C in the inner core would not be consistent with the inner-outer core density difference and would inhibit compositional convection. *Wood* (1993) argues that the second light element would likely be S, and this scenario is discussed in the section on element combinations.

If C is present in the core, it is important to understand what if any effects it may have on siderophile-element partitioning: Perhaps C can be ruled out as a major light com-

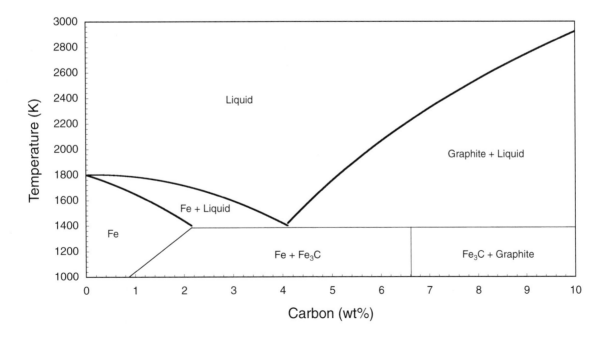

**Fig. 6.** A simplified version of the Fe-rich portion of the Fe-C phase diagram at 1 bar (after *Wood*, 1993).

ponent based on the abundance of siderophile elements in the mantle, or perhaps it holds the key to the excess-siderophile-element problem. Another important reason to understand the effects of C on siderophile-element partitioning is that in many studies on element partitioning at high temperatures, graphite capsules were used as a sample container, as it has a very high melting temperature (e.g., *Thibault and Walter,* 1995). Usually the C contents in run products were not reported. In most cases, C was probably not measured due to analytical difficulties. Nevertheless, a small amount of C was probably present, as sometimes indicated by low totals in the analyses of Fe-rich alloy. Recently, *Jana and Walker* (1997a) studied the effect of C on the distribution of a number of siderophile elements between metal and silicate. For Ni and Co, they found a small decrease in siderophility with increasing C content of the metal, and for Mo and W they found a small increase in siderophility. Germanium and P became significantly less siderophile. They find that this combination of effects is not consistent with the observed abundances of siderophile elements in the mantle and simple equilibrium segregation of a C-saturated Fe-liquid near the surface of Earth.

In sum, if C is retained in Earth during accretion, then several weight percent of C could have easily segregated to the core. Calculations then suggest that the inner core is $Fe_3C$, and thus C must have a companion light element. The presence of C in the core is also not the magic bullet that explains the abundance of siderophile elements in the mantle.

## 7.  HYDROGEN

From the standpoint of lightness and abundance, H is an excellent candidate for the light element. However, it has received little attention primarily because it takes high pressures to alloy H with Fe. One early advocate was *Stevenson* (1977), who argued that H should alloy with Fe at modest pressures (a few kilobars) and a core consisting of FeH could account for the density deficit (which amounts to ~1 wt% H in the core). He further argued that Earth need contain only 10% of a low-temperature component (i.e., a component with a full range of volatiles such as carbonaceous chondrites) to supply the necessary H. Later, however, he argued that H was not a good candidate for the light element in the core because the bulk of the metal that segregated to form the core equilibrated at low pressures as it was added to the outside of the planet, and high pressures are required to partition H into Fe (*Stevenson,* 1981).

### 7.1.  Experimental Work

Interest in H as a light element has been maintained through a combination of theoretical work on the formation and stability of Fe-hydrides and experimental work on reactions between hydrous silicates and oxides and Fe-metal (e.g., *Antonov et al.,* 1982; *Fukai,* 1984, 1992; *Fukai and Suzuki,* 1986; *Suzuki et al.,* 1984). *Suzuki et al.* (1984) stud-

ied mixtures of $MgSiO_3$, $Mg_3Si_4O_{10}(OH)_2$, MgO, $SiO_2 \times 0.636H_2O$, and Fe in a tetrahedral-anvil apparatus at 5 GPa and temperatures ranging from 1000° to 1200°C. Their final phase assemblages contained Fe-bearing olivine and pyroxene and the Fe appeared to have been molten at run conditions that were 500°C lower than the melting point of Fe. Therefore *Suzuki et al.* (1984) deduced that the water and OH in the silicates and oxides reacted with the Fe to produce FeO, which was incorporated into the silicates, and $FeH_x$, which melted. Based on this experimental work and a model Earth composed of 10% of a low-temperature CI chondrite component and 90% of a high-temperature enstatite chondrite component, *Fukai and Suzuki* (1986) determined that the core composition should be $FeH_{0.41}C_{0.05}$ $O_{0.13}S_{0.03}$, and they also calculated that this would be consistent with the observed density deficit in the core.

Because $FeH_x$ decomposes upon decompression and the H then escapes, *Suzuki et al.* (1984) were not able to analyze the produced $FeH_x$ and thus prove conclusively that it formed. *Badding et al.* (1991, 1992) formed solid $FeH_{0.94}$ in the diamond cell and detected it *in situ* by X-ray. They found that at room temperature $FeH_{0.94}$ formed at 3.5 GPa, and they studied the crystal structure and volume-pressure relations to 62 GPa. Based on their data they estimated that the core composition could range between $FeH_{0.3}$ and $FeH_{0.9}$. *Yagi and Hishinuma* (1995) also studied the formation of $FeH_x$ *in situ* with X-rays, but they studied the reactions between $Mg(OH)_2$, $SiO_2$, and Fe in a cubic-anvil apparatus between 2.2 and 4.9 GPa and up to 1350°C. They observe the formation of $FeH_x$ even at their lowest pressures when the temperature of the experiment reaches the decomposition temperature of the $Mg(OH)_2$ and conclude that $FeH_x$ could form at very shallow depths in Earth. They estimate the composition of the Fe-hydride in their experiments was $FeH_{0.3-0.4}$.

In order to really determine how much H might be incorporated into the core, it is necessary not only to know that solid $FeH_x$ can form, but to study the partitioning of H into liquid Fe in systems with compositional relevance to Earth. The difficulty of doing such experiments, of course, is that the H is not retained upon quenching and X-ray diffraction cannot be used on a liquid to determine its composition so the liquid cannot be studied *in situ*. Recently *Okuchi* (1997) and *Okuchi and Takahashi* (1998) have developed a technique to estimate the amount of H that was incorporated into liquid Fe at high pressures and temperatures. They studied reactions between $Mg(OH)_2$, $SiO_2 \times xH_2O$, and Fe in a multianvil apparatus at 7.5 GPa and temperatures between 1100° and 1500°C. After temperature quenching an experiment, they rapidly released the pressure as opposed to the normal slow decompression used in multianvil experiments. Although the $FeH_x$ still decomposed and the H escaped, a quench texture of exsolved bubbles was produced in the Fe. The volume fraction of the bubbles was converted into H concentrations. Figure 7 summarizes the results of *Okuchi* (1997) and *Okuchi and Takahashi* (1998): They found that at 7.5 GPa the molar portion of H

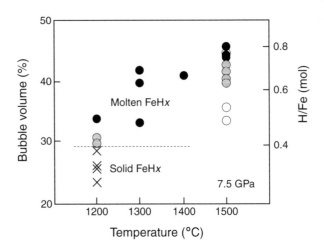

**Fig. 7.** The experimental data of *Okuchi* (1997) and *Okuchi and Takahashi* (1998) showing the H/Fe ratio of Fe metal vs. temperature at 7.5 GPa. The crosses represent solid $FeH_x$, and the circles molten $FeH_x$. The shading of the circles represents different starting fractions of water: open, the lowest; hatched, intermediate; solid, the most (*Okuchi*, 1997).

in the liquid metal increases from 0.41 at 1200°C to 0.69 at 1500°C. Their runs at 1200°C contained both solid and liquid $FeH_x$, and they observed that the H partitions favorably into the liquid (the liquid-solid partition coefficient is 1.2). They used their results to calculate the equilibrium constant and enthalpy change for the Fe-water reaction. Using these thermodynamic parameters and the same compositional model of Earth as *Fukai and Suzuki* (1986), they calculate that if the core segregated from a hydrous magma ocean, the H to Fe ratio in the liquid outer core could be as high as 0.41, and the H to Fe ratio of the inner core could then be as high as 0.34. The ratio of 0.41 for the outer core is at the lower end of what *Badding et al.* (1991, 1992) calculate as a possible core composition if H is the only light element. If one follows the density calculations of *Fukai and Suzuki* (1986) this ratio is also less than 0.6, which would be required to account for the total density deficit in the core. *Okuchi* (1997) argues that the ratio of 0.34 in the inner core could produce a density deficit in the inner core of about 4.5%, which is within the range of possible density deficits in the inner core.

In summary, the recent experimental work shows that H possesses many qualities that make it a good candidate for the light element in the core. $FeH_x$ can easily be formed at relatively shallow depths in Earth. Hydrogen would also partition favorably into the outer core, helping to drive compositional convection and meeting the constraint that the density deficit of the inner core be less than that of the outer core. However, whether enough H can be incorporated into the core to account for the entire density deficit depends on compositional models of Earth, accretion scenarios (i.e., Was there a hydrous magma ocean and how much water did it

contain? Did most of the metal equilibrate near the surface of the planet as it was added?) and theoretical models of the density of $FeH_x$

## 8. ELEMENT COMBINATIONS

Perhaps an overriding result of the combined experimental and theoretical work is that it is difficult for any one element to account for the entire density deficit (see Table 1). Incorporation of large amounts of Si into the core may have undesirable effects on mantle chemistry. The experimental work to date indicates that it is unlikely that more than about 1 wt% O could be dissolved into the core. Sulfur would be readily partitioned into the core, but it appears that it is not abundant enough in Earth to be the sole light element. The phase relationships in the Fe-C system at core pressures would require at least one other light element. Finally, for enough H to be incorporated into the core requires that Earth accretes enough water and the presence of a hydrous magma ocean. So, is the light element really a mixture consisting of a little bit of each of many elements and little hope of sorting it out? The answer to this question is likely no because the great strength of the experimental work to date is that it gives some very clear evidence about allowable and inadmissible combinations of elements that are discussed below and summarized in Table 2.

The following arguments for element combinations assume that the core is a single liquid and not a mixture of immiscible liquids. It is difficult to conceive of two immiscible liquids of differing densities remaining well mixed: They will certainly segregate with the lighter liquid rising and the denser liquid sinking, forming a stratified outer core. There is some evidence for a thin (<100 km) stably stratified layer at the top of the core (e.g., *Braginsky*, 1999, and references therein). If this layer does exist, it is so thin that it should have little effect on estimates of the density deficit in the core, which would be based on the bulk seismic properties of the outer core. However, the layer might have implications for the chemistry of core-mantle reactions and, in fact, might itself be the product of a core-mantle reaction.

*O'Neill et al.* (1998) and *O'Neill and Palme* (1998) pointed out that it was well known within the steel-making industry that at atmospheric pressure both Si and O could not be simultaneously dissolved in Fe-metal, and the experimental results of *O'Neill et al.* (1998) from higher pressures were consistent with the mutual exclusiveness of Si and O. *Gessmann and Rubie* (1998a,b,c) studied the combined solubilities of Si and O under a wide range of pressure and temperature conditions and found clear opposing trends for the solubility of O and Si as a function of $f_{O_2}$ (Fig. 8). Thus the low pressure observation holds at higher pressures and temperatures. This relationship, if it holds at core pressures, suggests that if there is any amount of Si in the core, there can be no significant contribution to the density deficit from O.

TABLE 1.  Summary of the candidate light elements.

| Element | Density Models | Ref. | Cosmochemical Availability | Ref. | Experimental Limits |
|---|---|---|---|---|---|
| Si | 18 wt% <br> 7.3–10.7 wt% | [11] <br> [14] | 5 wt% <br> 7.3 wt% <br> 12.5 wt% <br> 14 wt% | [8] <br> [2] <br> [16] <br> [18] | Partitions into Fe at the pressures of the deep upper mantle but not at lower mantle pressures; may remove too much Fe from mantle; depletion of refractory lithophile elements? |
| O | 9–10 wt% | [11] <br> [12] | Too volatile? | [9] <br> [10] | 1–2 wt% |
| S | ~10 wt% <br><br> as little as 5 wt% | [1] <br> [4] <br> [3] <br> [14] | 5 wt% <br> 1.8–4.1 wt% <br> <2 wt% | [6] <br> [7] <br> [5] | During core formation would readily partition into the metal phase |
| C | 6.6 wt% | [15] | 2–4 wt% | [17] | During core formation would readily partition into the metal phase; would be concentrated into the inner core |
| H | 1 wt% | [13] | ??? | | During core formation would readily partition into the metal phase |

Shows amounts required to account for the density deficit, amounts available in Earth based on cosmochemical constraints, and limits placed on each element through synthesis of all the experimental results. A more detailed discussion of the experimental limits with references can be found in the sections on the individual elements. References: [1] *Ahrens* (1979); [2] *Allègre et al.* (1995); [3] *Ahrens and Jeanloz* (1987); [4] *Brown et al.* (1984); [5] *Dreibus and Palme* (1996); [6] *Ganapathy and Anders* (1974); [7] *Kargel and Lewis* (1993); [8] *O'Neill* (1991); [9] *O'Neill et al.* (1998); [10] *O'Neill and Palme* (1998); [11] *Poirier* (1994); [12] *Ringwood* (1977); [13] *Stevenson* (1977); [14] *Sherman* (1997); [15] this work; [16] *Wänke* (1981); [17] *Wood* (1993); [18] *Wänke and Dreibus* (1997).

TABLE 2.  Summary of likelihood of proposed element combinations discussed in text.

| Element | Proposed Companion | Probability |
|---|---|---|
| Si | O | unlikely |
| | S | unlikely |
| | C | possibility |
| S | C | possibility |
| | O | possibility |
| H | Any | unknown |

Oxygen is not the only element that is incompatible with Si in a metallic liquid. *Poirier* (1994) pointed out that there was a miscibility gap in the liquid in the Fe-Si-S system at atmospheric pressure that would restrict plausible core compositions containing both Si and S to fall outside the miscibility gap. However, *Kilburn and Wood* (1997) performed metal-silicate partitioning experiments for Si and S on compositions within the region where the liquids are miscible, and they found that the metal-silicate partition coefficients of Si and S have opposite dependencies with respect to $f_{O_2}$: At low $f_{O_2}$, Si would be incorporated into the metal, and all the S would partition into the silicate liquid. At higher $f_{O_2}$ the opposite would happen. Thus, even in the range of compositions where the liquids are technically miscible,

they will not both partition into the metal simultaneously in a system containing both metal and silicate liquid. *Li* (1998) found similar results, and *Hillgren and Boehler* (1998a) also reported a decreased solubility of Si in liquid metal if a small amount of S is present at 47 GPa (see also Fig. 2). Thus, the incompatibility of S and Si appears to hold at even higher pressures. A potential way out of this dilemma is to incorporate Si and S during different stages of core formation in a heterogeneous accretion scenario. Silicon would be segregated to the core during the initial reducing phase of core formation, and S would be added during the later oxidizing phase. However, the core would likely try to reequilibrate with the new more oxidizing environment, and it is not clear then that Si would remain in the core. It would need to be directly addressed experimentally whether such a core could be in equilibrium with mantle minerals at core-mantle boundary pressures, and if not, whether the core-mantle reaction could leave both S and Si in the core and/or produce reasonable reaction products. The same set of arguments could be applied to the possibility of Si and O as companion elements if a field of miscibility forms in the Fe-Si-O system at high pressures. However, to date there is little evidence that this might occur (e.g., *O'Neill et al.* 1998).

The 1-bar phase relations in the Fe-Si-C ternary system suggest that Si and C are potential companion elements. *Schurman and Kramer* (1969) derived an empirical formula based on experimental results for the amount of C that can

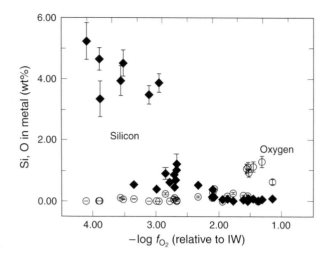

**Fig. 8.** The experimental data of *Gessmann and Rubie* (1998a) showing the Si (closed diamonds) and O (open circles) contents of quenched Fe-Ni-liquid metal as a function of $f_{O_2}$. The data cover a P,T range of 5–23 GPa and 1800°–2400°C. The solubilities of Si and O show opposite trends as a function of redox conditions.

be dissolved into Fe-Si liquids that is dependant on temperature and Si content (see also *Schmid*, 1981; *Raghavan*, 1992). According to this formula, a liquid at 1 bar and 1536°C (the melting point of Fe) with 5 wt% Si could contain 3.7 wt% C. An increase in temperature of 200°C would increase the amount of C by ~0.5 wt%, while an increase in the Si content by 1 wt% would decrease the amount of C by ~0.3 wt% (this number slowly decreases with increasing Si content because of the changing activity coefficients of both C and Si in the liquid). Detailed density models would be required to determine if amounts of this order would sufficiently lower the density of liquid Fe, but because both amounts are approximately half of what is required of either element on its own (depending on the model), it appears that combinations of Si and C could readily account for the density deficit. The amounts required are also within the cosmochemical limits (Table 1). The system certainly warrants further study, particularly whether such liquids could be equilibrated with mantle minerals.

*Wang et al.* (1991) showed that there is a miscibility gap in the Fe-S-C system at ambient pressure, which implies that S and C could be mutually exclusive as light elements. *Wood* (1993) studied the Fe-S-C system up to 5 GPa and temperatures up to 2400 K in a piston cylinder apparatus and observed a gradual closing of miscibility gap with increasing pressure and temperature. Therefore it is probable that at core pressures and temperatures the miscibility gap becomes small enough that most of the Fe-S-C liquid composition relevant to the core falls into the one stable liquid field. In fact, the miscibility gap would not need to shrink much for possible core compositions, such as about 2 wt% S (the lower end of the cosmochemical upper limits) and about 5 wt% C (a little less than the amount of C necessary to completely account for the density deficit) or 4 wt% S (the

cosmochemical upper limit) with 2–3 wt% C to lower the density to be compatible with the observed density deficit. Thus, if C is indeed not extremely volatile during the accretion of Earth and is therefore retained, S and C would be compatible light elements.

Another potential companion element with S is O. There is a miscibility gap in the Fe-S-O ternary system at low pressures, but *Urakawa et al.* (1987) found that the gap is narrowed with pressure and with the addition of Ni. They predict that the gap could disappear by 25 GPa. At 47 GPa, *Hillgren and Boehler* (1998a,b) found that the solubility of O in the metal increased from 0.3 wt% to 1 wt% when a small amount of S was present. Although more study is needed, O and S certainly show promise as companion light elements.

Because not much work has been done regarding H as a light element, there is not really any experimental information about permissible combinations with H. However, the work of *Okuchi* (1997) and *Okuchi and Takahashi* (1998) does hint that combinations of H with O or Si may not be likely. They report only a qualitative EDS analysis of their Fe grain matrix and find only peaks for Fe and the C coat. This indicates that at least under their experimental conditions significant amounts of O and/or Si are not likely to dissolve in Fe-metal in combination with H. However, we cannot draw a firm conclusion on this until further experimental work is performed investigating a larger range of variables.

Thus, the experimental work shows that light-element combinations of Si with O or S are unlikely. Phase relations in the 1-bar Fe-Si-C ternary suggest that Si and C are a combination worth further study. It is not known what elements could be likely companions with H, although the experimental results to date suggest that either Si or O is unlikely. The experimental results also show that a combination of S and C is possible, and a combination of S and O appears promising. In both of these cases, it remains to be seen if the preferential sequestering of either C or O into the inner core would still allow enough concentration of light element into the outer core to drive compositional convection. Also, the inner core compound must have a density that matches the observed inner core density, and by corollary, the density difference between the inner core compound and the proposed outer core composition must be consistent with that observed. *Wood* (1993) argues that the density of $Fe_3C$ does match the inner core density. However, he did not consider the possibility of the Fe-FeS system becoming a solid solution at core pressures, which would also effect the phase relationships and the composition of the first solid in the Fe-C-S ternary.

## 9. CONCLUSIONS

The enigma of the identity of the light element or elements is far from solved. However, experimental studies and theoretical considerations have provided much insight to the problem, and it seems one of the most fundamental insights is that the core likely contains at least two light elements, a conclusion reached by many authors (e.g., *Poirier*, 1994; *Stevenson*, 1981). The recent experimental work has also provided us some clues about possible combinations of light elements. Combinations of Si with S or O appear unlikely, particularly if the core is not grossly out of equilibrium with the mantle, and combinations of Si with C and S with C or O merit further detailed study. Certainly, systems involving H warrant more study for possible companion elements.

In regard to any proposed element combination for the light element, there are many experimental avenues to be explored. The combination of high-pressure melting and phase relations in candidate ternary systems, the high-pressure partitioning of the light elements between the liquid and solid metallic phases, and studies of the densities of solidus phases at core pressures gives the information necessary to assess whether a proposed element combination would produce an inner core with a reasonable density, and thus provide the necessary density contrast between the inner and outer core, as well as concentrating enough light element into the outer core to drive compositional convection. It will also be very important to study the partitioning of the proposed element combinations between the metallic and silicate phases as illustrated by the example of Si and S where their opposite metal-silicate partitioning behavior rules them out as companion elements even though they could form a single miscible liquid alloy with Fe. Also, further experimental study can reveal whether a proposed core composition is out of equilibrium with the mantle and if the resulting reaction products can be reconciled with D″. Finally, partitioning studies of siderophile and other elements between proposed core compositions and silicates could help constrain core composition and the conditions under which the core formed. In summary, probable Fe and FeNi ternary core compositions clearly present us with a vast field of research, and it may not be all that unlikely that the next review written on this subject will be recommending the quaternary systems on which research should focus.

*Acknowledgments.* The authors thank A. Holzheid, F. Guyot, and K. Righter for providing thought-provoking reviews that greatly improved the chapter.

## REFERENCES

Agee C. B. and Li J. (1998) Experimental constraints on the light elements in the core (abstract). *Mineral. Mag., 62A,* 19–20.

Ahrens T. J. (1979) Equations of state of iron sulfide and constraints on the sulfur content of the Earth. *J. Geophys. Res., 84,* 985–998.

Ahrens T. J. and Jeanloz R. (1987) Pyrite: shock compression, isentropic release, and composition of the Earth's core. *J. Geophys. Res., 92,* 10363–10375.

Allègre C. J., Poirier J.-P., Humler E., and Hofmann A. W. (1995) The chemical composition of the Earth. *Earth Planet. Sci. Lett., 134,* 515–526.

Anderson O. L. and Duba A. (1997) The experimental melting curve of iron revisited. *J. Geophys. Res., 102,* 22659–22669.

Antonov V. E., Belash I. T., and Ponyatovsky E. G. (1982) T-P phase diagram of the Fe-H system at temperatures to 450°C and pressures to 6.7 GPa. *Script. Metall., 16,* 203–208.

Badding J. V., Hemley R. J., and Mao H. K. (1991) High-pressure chemistry of hydrogen in metals: In situ study of iron hydride. *Science, 253,* 421–424.

Badding J. V., Mao H. K., and Hemley R. J. (1992) High-pressure crystal structure and equation of state of iron hydride: implications for the Earth's core. In *High Pressure Research: Application to Earth and Planetary Sciences* (Y. Syono and M. H. Manghnani, eds.), pp. 363–371. Terra Scientific, Tokyo.

Balchan A. S. and Cowan G. R. (1966) Shock compression of two iron-silicon alloys to 2.7 Megabars. *J. Geophys. Res., 71,* 3577–3588.

Birch F. (1952) Elasticity and constitution of the Earth's interior. *J. Geophys. Res., 69,* 227–286.

Boehler R. (1992) Melting of the Fe-FeO and the Fe-FeS systems at high pressure: constraints on core temperatures. *Earth Planet. Sci. Lett., 111,* 217–227.

Boehler R. (1993) Temperatures in the Earth's core from melting-point measurements of iron at high static pressures. *Nature, 363,* 534–536.

Boehler R. (1996a) Fe-FeS eutectic temperatures to 620 kbar. *Phys. Earth Planet. Inter., 96,* 181–186.

Boehler R. (1996b) Melting temperature of the Earth's mantle and core: Earth's thermal structure. *Annu. Rev. Earth Planet. Sci., 24,* 15–40.

Boness D. A. and Brown J. M. (1990) The electronic band structures of iron, sulfur, and oxygen at high pressures and the Earth's core. *J. Geophys. Res., 95,* 21721–21730.

Braginsky S. I. (1964) Kinematic models of the Earth's hydromagnetic dynamo. *Geomagn. Aeron. Engl. Transl., 4,* 572–583.

Braginsky S. I. (1999) Dynamics of the stably stratified ocean at the top of the core. *Phys. Earth Planet. Inter., 111,* 21–34.

Brett R. (1969) Melting relations in the Fe-rich portion of the system Fe-FeS at 30 kbar pressure. *Earth Planet. Sci. Lett., 6,* 479–482.

Brett R. (1984) Chemical equilibrium of the Earth's core and upper mantle. *Geochim. Cosmochim. Acta, 48,* 1183–1188.

Brown J. M., Ahrens T. J., and Shampine D. L. (1984) Hugoniot data for pyrrhotite and the Earth's core. *J. Geophys. Res., 89,* 6041–6048.

Bullen K. E. (1973) Cores of terrestrial planets. *Nature, 243,* 68–70.

Dreibus G. and Palme H. (1996) Cosmochemical constraints on the sulfur content in the Earth's core. *Geochim. Cosmochim. Acta, 60,* 1125–1130.

Dubrovskiy V. A. and Pan'kov V. L. (1972) On the composition of the Earth's core. *Acad. Sci. USSR Phys. Solid Earth (Transl.), 7,* 452–455.

Fei Y., Prewitt C. T., Mao H.-K., and Bertka C. M. (1995) Structure and density of FeS at high pressure and high temperature and the internal structure of Mars. *Science, 268,* 1892–1894.

Fei Y., Bertka C. M., and Finger L. W. (1997) High-pressure iron-sulfur compound Fe$_3$S$_2$, and melting relations in the system Fe-FeS. *Science, 275,* 1621–1623.

Fukai Y. (1984) The iron-water reaction and the evolution of the Earth. *Nature, 308,* 174–175.

Fukai Y. (1992) Some properties of the Fe-H system at high pressures and temperatures, and their implications for the Earth's core. In *High-Pressure Research: Application to Earth and Planetary Sciences* (Y. Syono and M. H. Manghnani, eds.), pp. 373–385. Terra Scientific, Tokyo.

Fukai Y. and Suzuki T. (1986) Iron-water reaction under high pressure and its implication in the evolution of the Earth. *J. Geophys. Res., 91,* 9222–9230.

Ganapathy R. and Anders E. (1974) Bulk compositions for the Moon and Earth, estimated from meteorites. In *Chemical and Isotope Analyses; Organic Chemistry* (W. A. Gose, ed.), pp. 1181–1206. Pergamon, New York.

Garnero E. J., Grand S. P., and Helmberger D. V. (1993) Low p-wave velocity at the base of the mantle. *Geophys. Res. Lett., 20,* 1843–1846.

Gessmann C. K. and Rubie D. C. (1998a) The effect of temperature on the partitioning of nickel, cobalt, manganese, chromium, and vanadium at 9 GPa and constraints on formation of the Earth's core. *Geochim. Cosmochim. Acta, 62,* 867–882.

Gessmann C. K. and Rubie D. C. (1998b) Metal-oxide equilibria at high pressures and temperatures: Are Si and O the light elements in the core (abstract). *Mineral. Mag., 62A,* 517–518.

Gessmann C. K. and Rubie D. C. (1998c) Si and O solubilities in liquid metal as a function of pressure, temperature and oxygen fugacity (abstract). In *Origin of the Earth and Moon,* p. 10. LPI Contribution No. 957, Lunar and Planetary Institute, Houston.

Gessmann C. K., Rubie D. C., and O'Neill H. S. C. (1995) Partitioning of siderophile elements between liquid metal, silicate melt and magnesiowüstite: Dependence of P, T, $f_{O_2}$ and implications for the formation of the Earth's core (abstract). *Eos Trans. AGU, 76,* F663.

Gessmann C. K., Kilburn M., Wood B. J., and Rubie D. C.(1999) Solubility of Si and O in the Earth's core (abstract). *IUGG XXII, Vol. A, A.25,* JSS02/W/18-A1.

Goarant F., Guyot F., Peyronneau J., and Porier J.-P. (1992) High-pressure and high-temperature reactions between silicates and liquid iron and its implications in the evolution of the Earth. *J. Geophys. Res., 97,* 4477–4487.

Grossman L. and Larimer J. W. (1974) Early chemical history of the solar system. *Rev. Geophys. Space Phys., 12,* 71–101.

Guyot F., Zhang J., Martinez I., Matas J., Ricard Y., and Javoy M. (1997) P-V-T measurements of iron silicide (ε-FeSi) — Implications for silicate-metal interactions in the early Earth. *Eur. J. Mineral., 9,* 277–285.

Hillgren V. J. and Boehler R. (1997) Diamond anvil cell study of interactions between Fe-metal and mantle silicates up to 1 Mbar and 3300 K. *Eos Trans. AGU, 78,* F757.

Hillgren V. J. and Boehler R. (1998a) High pressure reactions between metals and silicates: Implications for the light element in the core and core-mantle interactions (abstract). *Mineral. Mag., 62A,* 624–625.

Hillgren V. J. and Boehler R. (1998b) The light element in the core and core-mantle interactions (abstract). In *Origin of the Earth and Moon,* p. 10. LPI Contribution No. 957, Lunar and Planetary Institute, Houston.

Hillgren V. J. and Boehler R. (1999) A case against Si as the light element in the Earth's core (abstract). In *Lunar and Planetary Science XXX.* Lunar and Planetary Institute, Houston (CD-ROM).

Ito E., Morooka K., Ujike O., and Katsura T. (1995) Reactions between molten iron and silicate melts at high pressure: Implications for the chemical evolution of the Earth's core. *J. Geophys. Res., 100,* 5901–5910.

Jagoutz E., Palme H., Baddenhausen H., Blum K., Cendales M., Dreibus G., Spettel B., Lorenz V., and Wänke H. (1979) The abundances of major, minor and trace elements in the Earth's mantle as derived from primitive ultramafic nodules. *Proc. Lunar Planet. Sci. Conf. 10th,* pp. 2031–2050.

Jana D. and Walker D. (1997a) The impact of carbon on element distribution during core formation. *Geochim. Cosmochim. Acta, 61,* 2759–2763.

Jana D. and Walker D. (1997b) The influence of sulfur on partitioning of siderophile elements. *Geochim. Cosmochim. Acta, 61,* 5255–5277.

Javoy M. (1995) The integral enstatite chondrite model of the Earth. *Geophys. Res. Lett., 22,* 2219–2222.

Jephcoat A. (1985) *Hydrostatic Compression Studies on Iron and Pyrite to High Pressures: The Composition of the Earth's Core and the Equation of State of Solid Argon.* John Hopkins, Baltimore, Maryland. 214 pp.

Jephcoat A. and Olson P. (1987) Is the inner core of the Earth pure iron? *Nature, 325,* 332–335.

Jones J. H. and Drake M. J. (1986) Geochemical constraints on core formation in the Earth. *Nature, 322,* 221–228.

Kargel J. S. and Lewis J. S. (1993) The composition and early evolution of Earth. *Icarus, 105,* 1–25.

Kato T. and Ringwood A. E. (1989) Melting relationships in the system Fe-FeO at high pressures: implications for the composition of the Earth's core. *Phys. Chem. Minerals, 16,* 524–538.

Kilburn M. R. and Wood B. J. (1997) Metal-silicate partitioning and the incompatibility of S and Si during core formation. *Earth Planet. Sci. Lett., 152,* 139–148.

Knittle E. and Jeanloz R. (1989) Simulating the core-mantle boundary: An experimental study of high-pressure reactions between silicates and liquid iron. *Geophys. Res. Lett., 16,* 609–612.

Knittle E. and Jeanloz R. (1991a) Earth's core-mantle boundary: results of experiments at high pressures and temperatures. *Science, 251,* 1438–1443.

Knittle E. and Jeanloz R. (1991b) The high pressure phase diagram of $Fe_{0.94}O$: A possible constituent of the Earth's core. *J. Geophys. Res., 96,* 16169–16180.

Knopoff L. and MacDonald G. J. F. (1960) An equation of state for the core of the Earth. *Geophys. J. Roy. Astron. Soc., 3,* 68–77.

Lay T., Williams Q., and Garnero E. J. (1998) The core-mantle boundary layer and deep Earth dynamics. *Nature, 392,* 461–468.

Li J. (1998) Element partitioning constraints on core-mantle differentiation. In *Earth and Planetary Sciences,* p. 105. Harvard University, Cambridge, Massachusetts.

Li J. and Agee C. B. (1996) Geochemistry of mantle-core differentiation at high pressure. *Nature, 381,* 686–689.

MacDonald G. J. F. and Knopoff L. (1958) On the chemical composition of the outer core. *Geophys. J. Roy. Astron. Soc., 1,* 284–297.

McDonough W. F. and Sun S.-s. (1995) The composition of the Earth. *Chem. Geol., 120,* 223–253.

Minarik W. G., Ryerson F. J., and Watson E. B. (1996) Textural entrapment of core-forming melts. *Science, 272,* 530–533.

Mao H. K., Wu Y., Chen L. C., Shu J. F., and Jephcoat A. P. (1990) Static compression of iron to 300 GPa and $Fe_{0.8}Ni_{0.2}$ alloy to 260 GPa: Implications for composition of the core. *J. Geophys. Res., 95,* 21737–21742.

Morgan J. W. and Anders E. (1980) Chemical composition of Earth, Venus and Mercury. *Proc. Natl. Acad. Sci., 77,* 6973–6977.

Murr L. E. (1975) *Interfacial Phenomena in Metals and Alloys.* Addison-Wesley. 376 pp.

Murthy V. R. and Hall H. T. (1970) The chemical composition of the Earth's core: Possibility of sulfur in the core. *Phys. Earth Planet. Inter., 2,* 276–282.

Murthy V. R. and Hall H. T. (1972) The origin and chemical composition of the Earth's core. *Phys. Earth Planet. Inter., 6,* 123–130.

O'Neill H. S. C. (1991) The origin of the Moon and the early history of the Earth — A chemical model. Part 2: The Earth. *Geochim. Cosmochim. Acta, 55,* 1159–1172.

O'Neill H. S. C. and Palme H. (1998) Composition of the silicate Earth: Implications for accretion and core formation. In *The Earth's Mantle, Composition, Structure, and Evolution* (I. Jackson, ed.), pp. 3–126. Cambridge Univ., Cambridge.

O'Neill H. S. C., Canil D., and Rubie D. C. (1998) Oxide-metal equilibrium to 2500°C and 25 GPa: Implications for core formation and the light component in the Earth's core. *J. Geophys. Res., 103,* 12237–12260.

Ohtani E., Ringwood A. E., and Hibberson W. (1984) Composition of the core II: Effect of high pressure on solubility of FeO in molten iron. *Earth Planet. Sci. Lett., 71,* 94–103.

Ohtani E., Yurimoto H., and Seto S. (1997) Element partitioning between metallic liquid, silicate liquid and lower-mantle minerals: Implications for core formation of the Earth. *Phys. Earth Planet. Inter., 100,* 97–114.

Okuchi T. (1997) Hydrogen partitioning into molten iron at high pressure: implications for Earth's core. *Science, 278,* 1781–1784.

Okuchi T. and Takahashi E. (1998) Hydrogen in molten iron at high pressure: The first measurement. In *Properties of Earth and Planetary Materials at High Pressure and Temperature.* American Geophysical Union, Washington, DC.

Poirier J.-P. (1994) Light elements in the Earth's outer core: A critical review. *Phys. Earth Planet. Inter., 85,* 319–337.

Raghavan V. (1988) *Phase Diagrams of Ternary Iron Alloys. Part 2: Ternary Systems Containing Iron and Sulphur.* Indian Institute of Metals, Calcutta.

Raghavan V. (1992) *Phase Diagrams of Ternary Iron Alloys. Part 6A.* Indian Institute of Metals, Calcutta.

Righter K., Walker R. J., and Warren P. H. (2000) Significance of highly siderophile elements and osmium isotopes in the lunar and terrestrial mantles. In *Origin of the Earth and Moon* (R. M. Canup and K. Righter, eds.), this volume. Univ. of Arizona, Tucson.

Ringwood A. E. (1959) On the chemical evolution and density of planets. *Geochim. Cosmochim. Acta, 15,* 257–283.

Ringwood A. E. (1977) Composition of the core and implications for the origin of the Earth. *Geochem. J., 11,* 111–135.

Ringwood A. E. (1989) Significance of the terrestrial Mg/Si ratio. *Earth. Planet. Sci. Lett., 95,* 1–7.

Ringwood A. E. and Hibberson W. (1990) The system Fe-FeO revisited. *Phys. Chem. Minerals, 17,* 313–319.

Ringwood A. E. and Hibberson W. (1991) Solubilities of mantle oxides in molten iron at high pressures and temperatures: Implications for the compositions of the Earth's core. *Earth Planet. Sci. Lett., 102,* 235–251.

Rushmer T., Minarik W., and Taylor G. J. (2000) Physical processes of core formation. In *Origin of the Earth and Moon* (R. M. Canup and K. Righter, eds.), this volume. Univ. of Arizona, Tucson.

Ryzhenko B. and Kennedy G. C. (1973) The effect of pressure on the eutectic in the system Fe-FeS. *Am. J. Sci., 273,* 803–810.

Schmid R. (1981) Thermodynamics of Fe-C-Si melts. *Calphad, 5,* 255–266.

Schurman E. and Kramer D. (1969) Influence of temperature and the equivalent effects of alloying elements on carbon solubility in iron-rich carbon-saturated ternary and multicomponent melts (in German). *Geissereiforsch., 21,* 29–42.

Shannon M. C. and Agee C. B. (1996) High pressure constraints on percolative core formation. *Geophys. Res. Lett., 23,* 2717–2720.

Sherman D. M. (1991) Chemical bonding in the outer core: high-pressure electronic structures of oxygen and sulfur in metallic iron. *J. Geophys. Res., 96,* 18029–18036.

Sherman D. M. (1995) Stability of possible Fe-FeS and Fe-FeO alloy phases at high pressure and the composition of the Earth's core. *Earth Planet. Sci. Lett., 132,* 87–98.

Sherman D. M. (1997) The composition of the Earth's core: constraints on S and Si vs. temperature. *Earth Planet. Sci. Lett., 153,* 149–155.

Shterenberg L. E., Slesarev V. N., Korsunskaya I. A., and Kamenetskaya D. S. (1975) The experimental study of the interaction between the melt carbides and diamond in the iron-carbon system at high pressures. *High Temp.-High Pressures, 7,* 517–522.

Stevenson D. J. (1977) Hydrogen in the Earth's core. *Nature, 268,* 130–131.

Stevenson D. J. (1981) Models of the Earth's core. *Science, 214,* 611–619.

Stevenson D. J., Spohn T., and Schubert G. (1983) Magnetism and thermal evolution of the terrestrial planets. *Icarus, 54,* 466–489.

Stewart R. M. (1973) Composition and temperature of the outer core. *J. Geophys. Res., 78,* 2586–2597.

Suzuki T., Akimoto S., and Fukai Y. (1984) The system iron-enstatite-water at high pressures and temperatures — Formation of iron hydride and some geophysical implications. *Phys. Earth Planet. Inter., 36,* 135–144.

Thibault Y. and Walter M. J. (1995) The influence of pressure and temperature on the metal-silicate partition coefficients of nickel and cobalt in a model C1 chondrite and implications for metal segregation in a deep magma ocean. *Geochim. Cosmochim. Acta, 59,* 991–1002.

Urakawa S., Kato M., and Kumazawa M. (1987) Experimental study on the phase relationships in the system Fe-Ni-O-S up to 15 GPa. In *High Pressure Research in Mineral Physics* (M. H. Manghnani and Y. Syono, eds.), pp. 95–111. Terra Scientific, Tokyo.

Usselman T. M. (1975) Experimental approach to the state of the core: part 1. the liquidus relations of the Fe-rich portion of the Fe-Ni-S system from 30 to 100 kb. *Am. J. Sci., 275,* 278–290.

Walker D., Norby L., and Jones J. H. (1993) Superheating effects on metal/silicate partitioning of siderophile elements. *Science, 262,* 1858–1861.

Walter M. J., Newsom H. E., Ertel W., and Holzheid A. (2000) Siderophile elements in the Earth and Moon: Metal/silicate partitioning and implications for core formation. In *Origin of the Earth and Moon* (R. M. Canup and K. Righter, eds.), this volume. Univ. of Arizona, Tucson.

Wang C., Hirama J., Nagasaka T., and Ban-Ya S. (1991) Phase equilibria of liquid Fe-S-C ternary system. *ISIJ International, 31,* 1292–1299.

Wänke H. (1981) Constitution of terrestrial planets. *Philos. Trans.*

*R. Soc. Lond., A303,* 287–302.

Wänke H. and Dreibus G. (1997) New evidence for Si as the major light element in the Earth's core (abstract). In *Lunar and Planetary Science XXVIII.* Lunar and Planetary Institute, Houston.

Wasson J. T. (1985) *Meteorites — Their Record of Early Solar-System History.* Freeman, New York. 267 pp.

Wendlandt R. F. (1982) Sulfide saturation of basaltic and andesitic melts at high pressures and temperatures. *Am. Mineral., 67,* 877–885.

Williams Q. and Jeanloz R. (1990) Melting relations in the iron-sulfur system at ultra-high pressures: implications for the thermal state of the Earth. *J. Geophys. Res., 95,* 19299–19310.

Wood B. J. (1993) Carbon in the core. *Earth Planet. Sci. Lett., 117,* 593–607.

Yagi T. and Hishinuma T. (1995) Iron hydride formed by the reaction of iron, silicate, and water: Implications for the light element of the Earth's core. *Geophys. Res. Lett., 22,* 1933–1936.

Zerr A., Diegeler A., and Boehler R. (1998) Solidus of the Earth's deep mantle. *Science, 281,* 243–245.

# Siderophile Elements in the Earth and Moon: Metal/Silicate Partitioning and Implications for Core Formation

**M. J. Walter**
*Institute for Study of the Earth's Interior*

**H. E. Newsom**
*University of New Mexico*

**W. Ertel**
*University of Arizona*

**A. Holzheid**
*Massachusetts Institute of Technology*

We estimate the depletion in siderophile elements due to core formation in the silicate mantles of the Earth and Moon. Uncertainties in the depletions are assessed; these arise from limited analytical data, corrections for element volatility, and uncertainties in the initial abundances when assuming that protoplanets accreted from chondritic materials. We review the experimentally determined partitioning behavior of siderophile elements between Fe-rich metal and silicate melt, and consider the results with respect to core-formation models. Low-temperature and low-pressure single-stage metal/silicate equilibration cannot explain siderophile-element depletions in the Earth and Moon. More complex multistage equilibrium models (e.g., heterogeneous accretion) can generally account for element depletions, but the number of free parameters makes such models relatively nonunique. Single-stage equilibrium core segregation at high temperature and pressure in a deep magma ocean can apparently account for the depletions of several moderately siderophile elements in Earth's mantle, although more partitioning data at high pressures are needed to confirm and extend this result. Highly siderophile elemental abundances are probably derived by the addition of chondritic material after core formation. Uncertainties in the bulk siderophile content of the Moon presently preclude unique models of core formation.

## 1. INTRODUCTION

The Earth and Moon contain Fe-rich metallic cores that segregated from mantle silicate very early in their 4.5-b.y. history (e.g., *Halliday et al.*, 1996). One consequence of the formation of a metallic core in a planetary body is that the bulk of the "metal-loving" or siderophile elements (see Table 1) are strongly partitioned into the metallic core. This is exemplified in both the Earth and Moon by siderophile-element depletions in mantle rocks relative to primitive solar system abundances (see reviews by *Jones and Drake*, 1986; *Newsom*, 1990). However, trace amounts of siderophile elements are retained in mantle silicates and record the chemical history of metal segregation, and if metal segregation was an equilibrium process, the relative depletions of siderophile elements give important clues for deducing the conditions of core formation.

The manner in which the siderophile elements partition between metal and silicate depends strongly on intensive variables such as pressure ($P$), temperature ($T$), composition ($X$), and oxygen fugacity ($f_{O_2}$). Therefore, in order to

chemically reconstruct the history of core formation, one must accurately know the partitioning behavior of siderophile elements between metal and silicate over a wide range of conditions, as well as the degree of depletion of siderophile elements in the silicate mantles of the Earth and Moon.

An assumption implicit in many early core-formation models was that metal segregation was contemporaneous with accretion, and that metal and silicate equilibrated at near-surface, low-temperature, and low-pressure conditions (e.g., *Stevenson*, 1981; *Jones and Drake*, 1986). However, relative to the depletions expected on the basis of experimentally determined low-temperature and low-pressure metal/silicate partition coefficients, many siderophile elements are significantly overabundant in the mantles of the Earth and Moon (e.g., *Ringwood*, 1966, 1979); by up to 2 orders of magnitude in the case of some moderately siderophile elements (MSE), and perhaps by as much as 5–10 orders of magnitude for the highly siderophile elements (HSE) (see Figs. 1 and 2). Further, the nearly chondritic (i.e., undifferentiated primitive solar system values) Ni/Co and interelement HSE abundance ratios in Earth's mantle are

TABLE 1.    General chemical classification of the siderophile elements.

| Volatility* | Slightly Siderophile[†§] | Moderately Siderophile[§] | Highly Siderophile |
|---|---|---|---|
| Refractory | V | Mo, W | Re, Os, Ir, Pt, Ru, Rh, |
| Transitional | Cr | Fe, Co, Ni | Pd |
| Moderately Volatile | Mn | P, Cu, Ga, Ge, As, Ag, Sb | Au |
| Volatile | — | Sn, Tl, Bi | — |

*Refers to 50% condensation temperatures at $10^{-4}$ atm where refractory ≥1400 K, transitional = 1250–1400 K, moderately volatile = 800–1250 K, volatile ≤800 K.

[†] The elements, V, Cr, and Mn are nominally lithophile, but may develop slightly siderophile tendencies at some conditions relevant to core formation.

[§] Here, we define elements with metal/silicate partition coefficients >1 but <10 as slightly siderophile, <10,000 as moderately siderophile, and ≥10,000 as highly siderophile. This classification is generalized and the designation of an element may change depending on the conditions of partitioning (e.g., $f_{O_2}$, T, P, etc.).

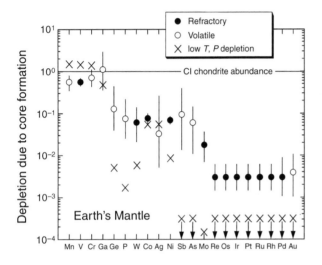

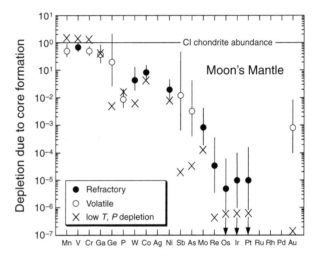

**Fig. 1.**   The depletion of siderophile elements in the Earth's mantle due to core formation (solid circles, see Table 2), normalized to Fe, except for V, normalized to Al, and Cr, normalized to Mg (see Table 2, this study). Note that the relative depletions (identical within 10%) of the highly siderophile elements are much better known than their absolute depletions. The low *T-P* model depletions (shown as symbol ×) for the slightly and moderately siderophile elements (Mn through Mo) are made assuming 30% metal (67:33 ratio of solid metal to S-bearing liquid metal) segregates from 70% silicate (90:10 ratio of solid to liquid silicate), and using partition coefficients at 1 atm, ~1250°C, and log $f_{O_2}$ = –12.4 [data from Table 2 of *Newsom* (1990) and references therein]. Low *T-P* model depletions for the highly siderophile elements (Re-Au) are made on the basis of the metal/silicate partition coefficients given in Table 7, assuming 30% metal segregates from 70% silicate.

**Fig. 2.**   The depletion of siderophile elements in the Moon due to core formation (solid circles, see Table 2) normalized to refractory elements, except for V, normalized to Al, and Cr, normalized to Mg. Model low *T-P* depletions (symbol ×) are calculated as in Fig. 1.

clearly inconsistent with low-temperature, low-pressure metal/silicate equilibration (e.g., *Jones and Drake,* 1986; *Newsom,* 1990).

The failure of low-temperature, low-pressure metal/silicate equilibrium core-formation models to explain the depletion of siderophile elements in Earth's mantle inspired the invention of more complex models involving equilibration

at low *T* and *P*, such as (1) sulfide-metal/silicate equilibrium in the Fe-S-O system (*Brett,* 1984), (2) inefficient core formation wherein small amounts of metal or sulfide remain behind in the mantle after core segregation (e.g., *Arculus and Delano,* 1981; *Jones and Drake,* 1986), (3) heterogeneous accretion (e.g., *Wänke,* 1981; *Ringwood,* 1984; *Newsom,* 1990; *O'Neill,* 1991b), and (4) a combination of preterrestrial low-temperature, low-pressure equilibration and terrestrial high-temperature, high-pressure equilibration (*Murthy and Karato,* 1997). Among these, heterogeneous accretion models have been especially popular in the last decades, but evaluating whether such models are realistic is difficult because the large number of adjustable parameters and hypothetical accretion components make the models relatively nonunique. Each of these types of models has strengths and weaknesses, and, in our opinion, no model involving substantial low-temperature, low-pressure metal/silicate equilibration has provided an entirely satisfactory account of siderophile-element depletions in the mantle.

Modern accretion theory has provided the impetus for a shift away from low-temperature, low-pressure models, and significant attention is now being given to the consequences of metal segregation at high temperature and pressure. In recent years it has become increasingly accepted that a large fraction of Earth's mass accreted by impacting of planetesimals (ranging from tens to hundreds of kilometers in diameter), with a "giant impact" of a Mars-sized body possibly responsible for the formation of the Moon (e.g., *Wetherill,* 1985, 1990). Dynamic models have shown that the kinetic energy imparted to the proto-Earth by large impactors, as well as the potential energy released upon metal segregation, is sufficient to melt large portions of the silicate mantle, creating a deep magma ocean (e.g., *Benz and Cameron,* 1990; *Melosh,* 1990; *Tonks and Melosh,* 1993). A magma ocean early in the Moon's history is also predicted on the basis of anorthosites, plagioclase-rich rocks that presumably originated as floatation cumulates, which make up a substantial fraction of the Moon's crust (see *Taylor,* 1992). On the basis of modern theory, metal segregation in a magma ocean would occur as small liquid-metal particles, like a metallic rain, which would remain in equilibrium with molten silicate over a range of high temperature and pressure (*Stevenson,* 1990). In light of these considerations it has become apparent that equilibrium metal segregation in a deep magma ocean is a valid hypothesis, and that the intensive variables of $T$ and $P$ might be critically important in our understanding of core formation (e.g., *Murthy,* 1991). This inspired a furious experimental effort to measure siderophile-element partition coefficients at high $T$ and $P$ over the last decade, and the results have provided new insights into core formation.

Here we give an account of the current "state of the subject" regarding core formation models, and provide a compilation of information we hope will prove useful to both the initiated and uninitiated reader. First, we give estimates and uncertainties for the depletion of siderophile elements due to core formation in the mantles of the Earth and Moon. Second, we provide a review of the experimentally determined partitioning behavior of siderophile elements between metal alloys and silicate melt. And finally, we focus on the modern concept of core formation based on metal segregation in a deep magma ocean.

## 2. SIDEROPHILE-ELEMENT ABUNDANCES IN THE EARTH AND MOON

The use of siderophile elements as probes of planetary accretion and core formation depends on a quantitative knowledge of the extent of their depletion in planetary silicate mantles resulting from metal segregation. Determining the depletion factors due to core formation requires knowledge of the initial abundances of siderophile elements in the planet, the siderophile-element abundances after core formation in the silicate portion of the planet (primitive mantle), and the amount of depletion relative to CI chondrites due to volatility. The uncertainties in determining the initial abundances are discussed first.

### 2.1. Bulk Planetary Siderophile-Element Abundances Inferred from Chondrites

Bulk planetary abundances of siderophile elements are usually based on the abundance of siderophile elements in chondritic meteorites, especially the CI carbonaceous chondrites (e.g., *McDonough and Sun,* 1995). The CI chondrites probably represent the most primitive solar system material, and ordinary chondritic meteorites have refractory-lithophile-element ratios that are similar to CI values. Similarly, refractory lithophile elements appear to be present in the bulk Earth in near-chondritic relative proportions (e.g., *McDonough and Sun,* 1995), and it is assumed that the same is true for refractory siderophile elements. Indeed, this assumption is implicit in our estimation of bulk planetary siderophile-element proportions, and we use CI chondrite abundances as a "bulk planet" normalization factor. However, because the Earth and Moon may have accreted from a variety of materials of chondritic affinity, we use the abundance variation of siderophile elements among the different chondrite groups to assign an uncertainty on the chondrite normalization factors.

*2.1.1. Refractory siderophile elements.* Absolute abundances of siderophile elements in ordinary chondrites are fractionated relative to CI chondrites, so they are usually normalized to either highly refractory elements or the amount of Fe present, as shown in Figs. 3a,b. The normalization of siderophile elements to the abundance of refractory elements arises naturally from the use of refractory lithophile elements as normalizing elements (e.g., W/Th or Mo/Ce) to correct for igneous fractionation on parent bodies. However, as shown in Fig. 3a, the siderophile elements in chondrites do not correlate very well with refractory elements. The resulting uncertainty ranges from ≈30% for Cr and W to close to a factor of 2 for the other refractory siderophile elements. The abundances of most refractory siderophile elements in chondrites are best correlated with the abundance of Fe (Fig. 3b), suggesting a strong control in the early solar nebula by metal/silicate fractionation. Therefore, normalization of siderophile elements in ordinary chondrites to their bulk Fe content and to the siderophile-element/Fe content of CI chondrites provides the tightest constraint on the initial abundances of siderophile elements in planetary materials, and we use this approach in our normalization. The uncertainties in normalized initial or "bulk" planetary abundances arising from the variations in chondritic proportions range from ≈15% for Ni and Co to ≈30% for other siderophile elements such as Mo, Ru, Pd, Re, Os, Ir, and Pt.

*2.1.2. Volatile siderophile elements.* A source of considerable uncertainty for some siderophile elements is the amount of depletion due to element volatility (see Table 1).

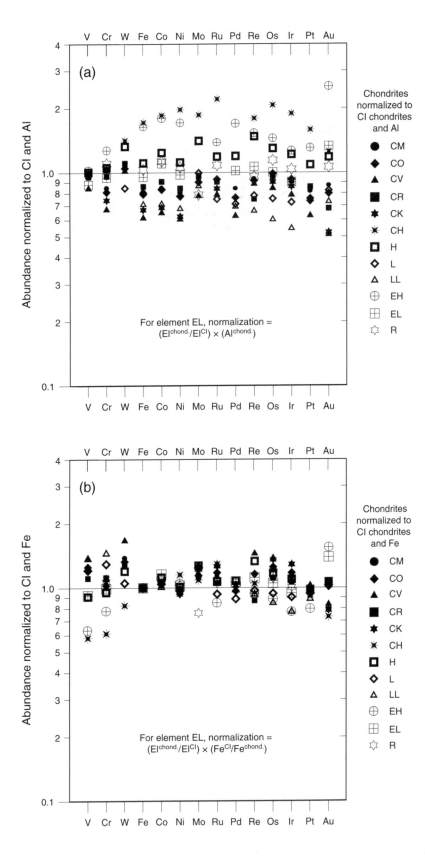

**Fig. 3.** The abundance in chondrite groups of nonvolatile siderophile elements, the moderately volatile element Au, and the nominally lithophile elements V and Cr relative to mean CI chondrites and **(a)** Al and **(b)** Fe. Data compiled in *Newsom* (1995) and *Kargel and Lewis* (1993). Note that data for all elements are not available for all chondrite groups.

The origin and timing of volatile-element depletion in solar system materials is controversial. However, the absence of any substantial fractionation of K isotopes in chondrites, chondrules, CAIs, achondrites, Earth, and the Moon is compelling evidence that volatile depletion was related to an early thermal event in the nebula and not to accretionary processes (*Humayun and Clayton,* 1995; *Humayun and Cassen,* 2000). The extent of loss of volatile siderophile elements in a planet can be estimated from the depletion of similarly volatile lithophile elements, the assumption being that the depletions of all volatile elements are similar as a function of condensation temperature. Figure 4 shows vola-

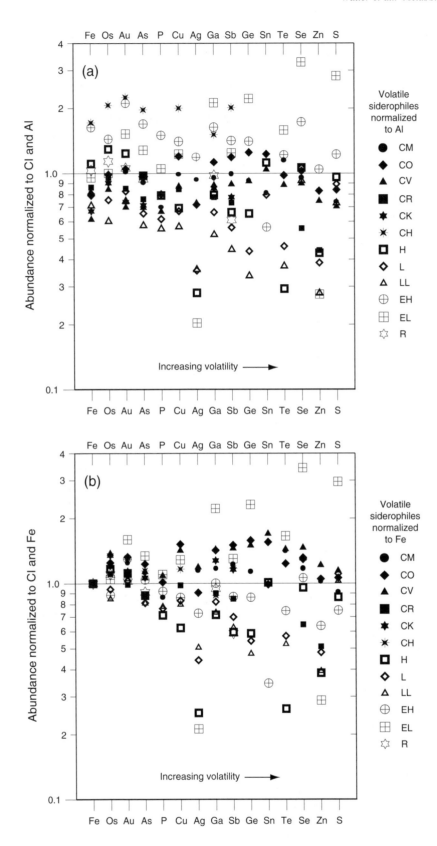

**Fig. 4.** The volatility-corrected abundance in chondrites of volatile siderophile and lithophile elements, normalized to CI chondrites and **(a)** Al and **(b)** Fe. The abundances of the volatile siderophile elements have been corrected for volatile depletion based on the depletion of volatile lithophile elements (see text). The nonvolatile elements Os and Fe are shown for reference. Note that data for all elements are not available for all chondrite groups.

tile-siderophile-element abundances for ordinary chondrites, after correcting for volatile depletion in this way, compared to CI abundances with normalization to Al (Fig. 4a) and Fe (Fig. 4b). Volatile siderophile elements are highly variable and are generally depleted in most ordinary chondrite groups relative to CI chondrites. The volatile siderophile

elements are very poorly correlated to refractory elements in chondrites (Fig. 4a), resulting in uncertainties greater than a factor of 2 in initial abundances if the initial Fe content of a planet is not known, as for the Moon. Like the refractory siderophile elements, the abundances of the volatile siderophile elements in chondrites are better correlated to their Fe content (Fig. 4b), and we use this normalization in our calculations. Uncertainties range from ±30% for Au, As, and P, and up to a factor of 2 for more volatile elements. Thus, uncertainties in the chondrite abundances increase greatly with increasing volatility.

A complicating feature for volatile siderophile elements is that the Earth and Moon are apparently depleted in volatile lithophile elements relative to all chondrite groups (e.g., *Taylor,* 1992; *McDonough and Sun,* 1995). Thus, either there was excess volatile loss during accretion of these bodies or during the formation of the Moon, which apparently can be discounted by the lack of K-isotopic fractionations (*Humayun and Clayton,* 1995), or the accretionary materials were of a nonchondritic provenance (at least for volatile elements). The corrections and associated uncertainties for volatile loss were made by visually defining an envelope around a plot of the depletions of volatile lithophile elements in the Earth and Moon relative to CI abundances and refractory elements as a function of their volatility (*Newsom,* 1990). The extent of the envelope at a given volatility for each volatile siderophile element determines the magnitude and uncertainty in the volatility correction. Unfortunately, because the abundances of the volatile lithophile elements in the Earth and Moon often have large uncertainties, this volatile depletion correction adds greatly to the uncertainty of the depletion due to core formation for these elements. The implication for volatile siderophile elements by using this procedure is that the depletion ascribed to core formation by normalizing to chondrites is a maximum. The overall large uncertainties for the volatile siderophile elements in determining the depletion due to core formation means that they cannot presently provide tight constraints on core formation models.

## 2.2. Siderophile-Element Abundances in the Silicate Mantles of Earth and Moon

*2.2.1. Earth.* Abundances of the siderophile elements in Earth's silicate mantle have been estimated on the basis of literature data for abundances in peridotites (e.g., oceanic, xenoliths, alpine), komatiites, and basalts from the upper mantle (made on the basis of compilations in *Newsom,* 1990; *McDonough and Sun,* 1995; and *Newsom et al.,* 1996, and references therein). Uncertainties in establishing mantle abundances arise from analytical uncertainty, corrections for igneous fractionation, and the assumption that the upper mantle is representative of the bulk mantle. Formal uncertainties have been estimated for the first two of these, but cannot be constrained for the third.

Several advances have been made in determining the depletions of siderophile elements in the last few years. For the MSE, a study by *Newsom et al.* (1996) provided improved constraints on the depletion of W in the mantle, adding more confidence to models for the timing of core formation based on the evolution of the Hf-W-isotopic system. For the HSE, the discovery of the chondritic Re-Os-isotopic evolution in the mantle confirms the original chon-

dritic relative abundance pattern in the primitive mantle within about ±6% (*Meisel et al.,* 1996; *Righter et al.,* 2000). High-precision measurements of Ir/Os ratios (e.g., *Humayun et al.,* 1997) also directly confirm the chondritic relative abundances of these elements in mantle peridotites. Recent high-precision analytical data suggest that subtleties in the HSE abundance patterns in the Earth and Moon may reflect a common source in the late influx of material (*Morgan et al.,* 1999). Alternatively, the data could suggest internal processes such as fractionation in fluid-rich environments (*Rehkämper et al.,* 1997), or mixing with outer-core materials (*Snow and Schmidt,* 1998).

*2.2.2. Moon.* Abundances of siderophile elements in the Moon's mantle are from *Newsom and Bessera* (1990) and *Newsom and Runcorn* (1991), as well as *Heiken et al.* (1991, Chapter 8). The available data for siderophile elements from lunar samples, excluding soils, was plotted against refractory lithophile elements of similar geochemical behavior to obtain their depletion and uncertainty relative to chondritic ratios. Unfortunately, little progress has been made estimating siderophile-element abundances in the source regions of the picritic glasses from the Moon due to analytical difficulties, and possible contamination from meteoritic material in the case of highly siderophile elements.

*2.2.3. Summary.* Given in Table 2, and shown in Figs. 1 and 2, are our best estimates with attendant uncertainties of mantle depletions of siderophile elements in the Earth and Moon due to core formation. To ascertain these values we divided the siderophile-element to Fe ratios in the model primitive mantles of the Earth and Moon by the siderophile-element to Fe ratios in CI chondrite (i.e., assumed bulk Earth). There are three principle sources of uncertainty in our model calculated depletions due to core formation, and these arise from estimates of (1) primitive mantle composition (e.g., corrections for igneous fractionation), (2) bulk planet composition (i.e., variations in chondritic siderophile/Fe ratios as in Figs. 3 and 4), and (3) volatile-element depletion. Such uncertainties should be propagated into core formation models of the Earth and Moon. Further, as discussed below (see section 5), there is considerable ambiguity in the origin of the Moon, so that CI chondrite may not be a suitable normalization factor.

## 3. FUNDAMENTALS OF SIDEROPHILE-ELEMENT PARTITIONING

The fundamental operators of geochemical models for core formation are siderophile-element partition coefficients. The partition coefficient is a scalar value that describes how a given element is apportioned between two coexisting phases — that is, the concentration ratio of the element of interest in the two phases. In the case of core formation models, the two phases of primary interest are metal and silicate, either as solids or liquids. In this section, we provide a brief review of the fundamentals of partitioning, as well as a discussion of how partition coefficients are

TABLE 2.  Depletions attributable to core formation of refractory and volatile
siderophile elements in the primitive mantles of the Earth and Moon.

| Element | Earth Primitive Mantle Depletions | | | Moon Primitive Mantle Depletions | | |
|---|---|---|---|---|---|---|
|  | Median | High | Low | Median | High | Low |
| Mn* | 0.55 | 0.8 | 0.33 | 0.5 | 0.95 | 0.31 |
| V | 0.56 | 0.68 | 0.45 | 0.68 | 0.95 | 0.49 |
| Cr* | 0.7 | 1.1 | 0.42 | 0.51 | 0.77 | 0.34 |
| Ga* | 1.1 | 2.9 | 0.34 | 0.38 | 0.84 | 0.17 |
| Ge* | 0.13 | 0.42 | 0.04 | 0.2 | 2 | 0.024 |
| As* | 0.06 | 0.15 | 0.011 | 0.0032 | 0.04 | 0.0004 |
| Sb* | 0.093 | 0.38 | 0.013 | 0.012 | 0.46 | 0.00064 |
| Ag* | 0.033 | 0.26 | 0.005 | — | — | — |
| P* | 0.073 | 0.21 | 0.025 | 0.0087 | 0.02 | 0.0039 |
| Co | 0.076 | 0.092 | 0.062 | 0.083 | 0.16 | 0.043 |
| Ni | 0.069 | 0.084 | 0.056 | 0.02 | 0.046 | 0.009 |
| W | 0.06 | 0.14 | 0.021 | 0.045 | 0.12 | 0.019 |
| Mo | 0.018 | 0.036 | 0.007 | 0.00083 | 0.004 | 0.00019 |
| Au* | 0.0039 | 0.011 | 0.0011 | 0.0008 | 0.008 | 0.0001 |
| Ru | 0.003 | 0.006 | 0.0015 | — | — | — |
| Pd | 0.003 | 0.0084 | 0.0011 | — | — | — |
| Re | 0.003 | 0.006 | 0.0015 | 3.3E-05 | 0.00034 | 3.3E-06 |
| Os | 0.003 | 0.006 | 0.0015 | 5E-06 | 6E-05 | 4.2E-07 |
| Ir | 0.003 | 0.006 | 0.0015 | 0.00001 | 0.0001 | 1.0E-6 |
| Pt | 0.003 | 0.006 | 0.0015 | 0.00001 | 0.00016 | 6.50E-07 |

\* Corrections have been made for the depletion due to volatility of these siderophile elements.
The associated uncertainties reflect our knowledge of the applicable initial chondritic abundances, the primitive mantle abundances, and corrections for volatile loss for certain elements.

obtained experimentally, concentrating on partitioning of siderophile elements between Fe-rich metal (solid and liquid)
and silicate melt.

## 3.1.  Theoretical

*3.1.1.  Metal/silicate partition coefficients.*  The partitioning of siderophile elements between Fe metal and silicate is usually expressed as the concentration ratio of the
element in the metal to that in the silicate phase. Here, we
adopt the terminology of *Beattie et al.* (1993). Throughout
the text, $^{met/sil}D_M$ (or simply $^{met/sil}D$) refers to the partition
coefficient of element M between Fe-rich metal (solid or
liquid) and silicate melt, calculated on a weight fraction
basis

$$^{met/sil}D_M = {}^{met}C_M / {}^{sil}C_M \qquad (1)$$

where $^{met}C_M$ and $^{sil}C_M$ are the concentrations of element M
in the metal and silicate respectively. Partition coefficients
can be calculated on the basis of either mol or weight fractions, and the choice is essentially arbitrary, but since the
raw data of phase analyses are usually in weight percent
(wt%), the convention in the literature is typically to report
partition coefficients as weight ratios.

In general, trace amounts of siderophile elements in
metal and silicate phases are difficult to measure with high
precision and accuracy. For this reason, it is common to

make partitioning experiments with siderophile elements as
major components in the metal phase, or to determine the
solubility of a pure metal in silicate melt and calculate a
partition coefficient at infinite dilution in Fe-rich metal.
However, partition coefficients may depend on the composition of the silicate and metallic phases. Thus, direct comparison of partitioning results among experimental studies
in which a variety of metal alloy compositions were used
requires recalculation of the experimentally determined
solubility of the siderophile elements in the silicate melt to
reflect equilibrium with pure metal by using the activity
coefficients of the siderophile elements in the metal alloy;
metal/silicate partition coefficients calculated in this way are
designated as $^{met/sil}D_M{}^{sol}$. These partition coefficients are the
inverse of the solubilities and are calculated as

$$^{met/sil}D_M{}^{sol} = 1/C_M{}^{sol} = {}^{met}\gamma_M \times$$
$$[{}^{met}C_M/{}^{sil}C_M] = {}^{met}\gamma_M \times {}^{met/sil}D_M \qquad (2)$$

where $C_M{}^{sol}$ is the solubility of element M in the silicate
and $^{met}\gamma_M$ is the activity coefficient of element M in the Fe-
metal alloy (for details see *Borisov et al.,* 1994; *Holzheid
and Palme,* 1996).

*3.1.2.  Two-component distribution coefficients.*  Factor
$^{met/sil}D_M$ depends strongly on the redox state of the system,
which in turn depends on temperature. For this reason,
*O'Neill* (1992) suggested expressing partitioning of siderophile elements between Fe-metal and FeO-containing sili-

cate as an exchange reaction

$$x/2 \ ^{met}Fe + \ ^{sil}MO_{x/2} = x/2 \ ^{sil}FeO + \ ^{met}M \quad (3)$$

where the corresponding two-component metal/silicate distribution coefficient, $K_D$, of reaction (2) is

$$^{met/sil}K_D^{M/Fe} = (^{met}X_M/^{sil}X_{MOx/2})/ \\ (^{met}X_{Fe}/^{sil}X_{FeO})^{x/2} = \ ^{met/sil}D_{M*}/^{met/sil}D_{Fe*} \quad (4)$$

Here, x refers to the formal valence of the siderophile element, $^{met}X_{Fe}$ and $^{met}X_M$ are mole fractions of Fe and M in the metal phase, $^{sil}X_{FeO}$ and $^{sil}X_{MOx/2}$ are mole fractions of FeO and $MO_{x/2}$ in the silicate phase, and $^{met/sil}D_{M*}$ refers to mol fraction partition coefficients ($K_D$ can be calculated using either mol or weight fractions). This exchange distribution coefficient is independent of $f_{O_2}$, and is often insensitive to temperature (*Takahashi and Irvine*, 1981). Metal/silicate partitioning expressed by an exchange coefficient allows direct comparison of partitioning data determined in various studies without considering absolute values of $f_{O_2}$. This is particularly important for high-pressure experiments where absolute $f_{O_2}$ cannot be measured *in situ* during the experiment. However, when using this equation it is implicit that the valence state of the siderophile element in the silicate is known.

*3.1.3. Valence state.* The solubility and partitioning of siderophile elements in silicate melt are strongly $f_{O_2}$ dependent, with solubility decreasing and $^{met/sil}D_M$ increasing with decreasing $f_{O_2}$. If a siderophile element dissolves in silicate melt as a metal oxide, the formal valence state of the ionic species can be calculated using the $f_{O_2}$ dependence of the experimentally determined solubility or partitioning.

The equilibrium distribution of an element between coexisting metal and silicate phases can be expressed as

$$M + \ ^{x}/_4 \ O_2 = MO_{x/2} \quad (5)$$

where M is the element of interest in the metal phase, $MO_{x/2}$ is the dissolved metal oxide in the silicate, and x is the valence of the element in the metal oxide. The equilibrium constant, K, of reaction (5) is

$$K_M = a_{MOx/2}/[a_M \times (f_{O_2})^{x/4}] \quad (6)$$

where $a_{MOx/2}$ and $a_M$, respectively, are the activities of the metal oxide in the silicate and the element in the metal phase at a defined $f_{O_2}$. Substituting the activities of the metal oxide and metal in equation (6) by the products of their mole fractions, X, and activity coefficients, $\gamma$, yields

$$K_M = (X_{MOx/2} \times \gamma_{MOx/2})/[(X_M \times \gamma_M) \times (f_{O_2})^{x/4}] \quad (7)$$

Rearranging equation (7) yields a linear relationship between the mole fraction of the oxide species in the silicate and $f_{O_2}$

$$logX_{MOx/2} = \ ^{x}/_4 \times log \ f_{O_2} + log \ K_M + \\ log(X_M) - log \ (\gamma_M/\gamma_{MOx/2}) \quad (8)$$

Because the ratio of the mol fractions, $X_M/X_{MOx/2}$, is the metal silicate partition coefficient, $^{met/sil}D_{M*}$, equation (8) can also be written as

$$log \ ^{met/sil}D_{M*} = -^{x}/_4 \times log \ f_{O_2} - \\ log \ K_M + log \ (\gamma_M/\gamma_{MOx/2}) \quad (9)$$

At constant $K_M$, and if the activity coefficient ratio, ($\gamma_M/\gamma_{MOx/2}$), is constant (i.e., Henry's law behavior), a linear correlation between solubility or $^{met/sil}D_{M*}$ and $f_{O_2}$ is produced. In terms of solubility, the general equation is

$$logX_{MOx/2} = x/4 \times log \ f_{O_2} + constant \quad (10)$$

In terms of the partition coefficient, the general equation is

$$log \ ^{met/sil}D_{M*} = -x/4 \times log \ f_{O_2} + constant \quad (11)$$

The slope of the correlation is a function of the valence state of the metal oxide (x = 4 × slope), whereas the intercept depends on the equilibrium constant of reaction (5) and the activity coefficient ratio, ($\gamma_M/\gamma_{MOx/2}$).

## 3.2. Experimental and Analytical Techniques

In the last decade a tremendous effort was made for determining $^{met/sil}D_M$ for siderophile elements over a wide range of conditions. Critical to this effort was the availability of experimental apparatus that allowed precise and accurate control of intensive parameters, as well as sophisticated analytical techniques for accurate measurement of trace amounts of siderophile elements in silicate run products. Here, we briefly review the common experimental and analytical techniques used in solubility and partitioning studies.

*3.2.1. Experimental.* A variety of experimental apparatus, from 1-atm gas-mixing furnaces to ultrahigh-pressure devices, allows experimental coverage over a $P$-$T$-$f_{O_2}$ space ranging from crustal conditions to the core-mantle boundary. In Table 3 the general attributes of the four most commonly used experimental apparatus are provided.

Experiments at 1 atm are made in gas-mixing furnaces, which provide very precise control of temperature with minimal temperature gradients, precise regulation of $f_{O_2}$, and permit *in situ* sampling of silicate melts throughout the duration of the experiment. Two common methods are used for solubility experiments at 1 atm. In the "wire loop" method, a bead of silicate melt is hung on a wire loop made of the metal of interest. The sample (or several samples at once) is suspended and equilibrated in the hot zone of the furnace at the temperature and $f_{O_2}$ of interest (see *Borisov et al.*, 1994, and *Holzheid et al.*, 1994, for details). In a recent method developed by *Dingwell et al.* (1994), silicate starting mixtures are placed within a crucible made of the

TABLE 3. Attributes of common experimental apparatus used in siderophile-element solubility and partitioning studies.

| | Gas-Mixing Furnace | Piston Cylinder | Multianvil | Diamond Anvil |
|---|---|---|---|---|
| Pressure | $10^{-4}$ GPa | ~0.5 to 5 GPa ± 2% | ~0.3 to 50 GPa ± 5% | ~1 to >300 GPa ± 10% |
| Temperature | <2000 ± 2 K<br>no $T$ gradient | <2600 ± 10 K<br><5 K/mm $T$ gradient | <3000 ± 20 to 200 K<br>20–200 K/mm $T$ gradient | up to >4000 ± 50 to 500 K<br>5–25 K/μm $T$ gradient |
| Oxygen Fugacity | pure $O_2$ to IW-3<br>precisely controlled gas mixing<br>in situ $f_{O_2}$ measurement | $f_{O_2}$ imparted by capsule, sample, or external buffer<br>$f_{O_2}$ calculated after experiment | $f_{O_2}$ imparted by capsule, sample, or external buffer<br>$f_{O_2}$ calculated after experiment<br>may have large $f_{O_2}$ gradients | $f_{O_2}$ imparted by capsule, sample, or external buffer<br>$f_{O_2}$ calculated after experiment<br>may have large $f_{O_2}$ gradients |
| Sample | wire loops: ~10 mm³<br>up to 10 samples<br>crucibles: ~35 cm³<br>multiple sampling<br>stirring rods | noble metal, ceramic, or graphite containers<br>up to 1 cm³ | noble metal, ceramic, or graphite containers<br><1 to 50 mm³ | MgO, Al$_2$O$_3$, KBr, NaCl, Ar, Ne pressure media<br>~10$^{-5}$ mm³ |

Values given are meant to be approximate and can vary considerably depending on experimental technique. See *Holloway and Wood* (1988) for a detailed review of experimental apparatus and techniques.

metal of interest, and the crucible is placed within a 1-atm furnace at high temperature and at fixed $f_{O_2}$. A rod, made of the element of interest and attached to a motor-driven spindle, is then placed within the molten sample and the sample is continuously stirred to promote equilibration. The melt is sampled periodically (hours to days) and equilibrium is determined when a steady state concentration of the metal in the silicate melt is attained. The $f_{O_2}$ is then changed and the procedure is repeated.

Large-volume solid-media apparatus, including piston-cylinder and multianvil devices, can be used within a range of ~0.5–50 GPa (pressures above ~25 GPa can be generated in multianvil devices using sintered diamond anvils) and at temperatures up to ~3000 K with good control of *P* and *T*, except at the most extreme conditions. However, $f_{O_2}$ is difficult to measure or regulate. Diamond anvil cells permit exploration at ultrahigh pressures (to >1 Mbar), and laser-heating techniques permit very high temperatures (>4000 K), but very small sample size and comparatively large uncertainties in *T* and *P* have limited the utility of this device to date. However, the future of this device for partitioning experiments seems promising as shown in a recent study by *Tschauner et al.* (1999).

For experiments made in high-*P* solid-media apparatus, it is imperative to continue the effort to reduce experimental uncertainties. Of particular importance are (1) reduction of the sometimes very large *T* uncertainties at *P* greater than ~20 GPa, (2) improvements in the ability to regulate and measure $f_{O_2}$, (3) the need to find new capsule materials that do not react with silicate and metallic melts at high *T*, and (4) adherence to the accepted use of reversal-type and/or time series experiments to establish equilibrium.

*3.2.2. Analytical.* The solubilities of siderophile elements in silicates are typically very low (ppm–ppt levels), so special effort is needed for their precise and accurate measurement. Table 4 gives the general attributes of the analytical techniques typically used to measure siderophile elements abundances in rocks and experimental samples. Analytical techniques can be characterized as either bulk sample or small volume (spot) techniques. In general, bulk techniques offer low detection limits (ppm–ppb) but require large amounts of sample (>0.5–1 g) and afford no spatial information. Thus, information on sample heterogeneity is lost and there is a risk of contamination by micrometer- or submicrometer-sized metallic particles in the analysis of silicates. In contrast, spot techniques permit measurement of element concentrations while preserving spatial information, the resolution of which depends on the size of the laser, ion, or electron beam. In particular, laser-ablation ICP-MS is becoming the analytical tool of choice for measuring siderophile elements in both natural and experimental samples because of good spatial resolution, good sample imaging, and low detection limits.

*3.2.3. The highly-siderophile-element nugget problem.* In some solubility experiments at 1 atm, it has been observed that with decreasing $f_{O_2}$, HSE solubilities in silicate melt do not decrease beyond an element-specific concentration. This was interpreted as either dissolved zero-valent HSE or contamination by very small "nuggets" of metal (*Borisov et al.*, 1994; *Borisov and Palme*, 1995; *O'Neill et al.*, 1995). In these studies, samples were analyzed in bulk, so questions regarding small-scale heterogeneity could not be resolved. In a recent Pt-solubility study, a small volume analytical method (LA-ICP-MS) was used, and large concentration spikes in the analytical signal were observed, up to several orders of magnitude above the normal signal. This was interpreted as contamination by a heterogeneous distribution of metal nuggets within the glass matrix *(Ertel et al.*, 1999). These nuggets have not been visually identified and may have diameters on the order of nanometers; even if particles of this size were isolated, they are presently impossible to analyze. Such nuggets are apparently formed exclusively by HSE and at concentration levels below ~100 ppm; they are formed in a single reduction process (decrease of $f_{O_2}$), and their redissolution is kinetically slow. Some speculations regarding their origin are (1) kinetic artifacts due to short run durations (*O'Neill et al.*, 1995), (2) byproducts formed by preexisting contamination (*Borisov and Palme*, 1997), and (3) a mechanically and thermodynamically stable phase in the melt (i.e., zero-valent HSE).

A possible mechanism for the origin of HSE nuggets is that a homogeneous distribution of HSE tends to precipitate in the melt during reduction rather than reprecipitating on the metal surface of the wire, capsule, spindle, or crucible used in the experiment. As reduction proceeds, the nuggets form progressively larger, elemental conglomerates. A nugget formed in this way would then have the same chemical potential as the other metal surfaces in equilibrium with the melt. Because of their small size, surface tension and surface-interaction forces with the melt might overcome density differences between the metal and silicate melt precluding physical separation. As a result, the nugget remains stable in the melt and further reduction facilitates growth by absorbing more neutral noble metal atoms. The nugget effect is presently a serious problem for accurate determinations of HSE solubility in silicate melts at reducing condition (e.g., approximately below the QFM buffer).

## 4. EXPERIMENTALLY DETERMINED PARTITION COEFFICIENTS

### 4.1. Slightly Siderophile Elements

*4.1.1. Manganese, vanadium, and chromium.* These elements are often considered lithophile elements, but their depletion in both the Earth and Moon relative to CI chondrite (see Figs. 1 and 2) has led to significant speculation as to their possible siderophile behavior, especially at elevated T and P (e.g., *Ringwood*, 1966; *Drake et al.*, 1989; *Ringwood et al.*, 1990). However, metal/silicate partitioning experiments show that, while these elements tend to become less lithophile with increase in *T* and *P*, they remain lithophile at least up to 20 GPa and 2500°C at $f_{O_2}$ appropriate for metal segregation (*Drake et al.*, 1989; *Walker et al.*,

TABLE 4. Attributes of common analytical techniques used in siderophile-element solubility and partitioning studies.

| | Bulk Sample | | | Small Volume (spot) | | |
|---|---|---|---|---|---|---|
| | INAA | ICP-AES | Diss-ICP-MS | EMP (~2-μm spot) | SIMS (~10-μm spot) | LA-ICP-MS (~50-μm spot) |
| Sample Size | >0.5–1 g | >0.5–1 g | >0.5–1 g | ~50 μm³ | ~80 μm³ | ~2000 μm³ |
| Detection Limit | ppm | ppm | sub ppb | 50–500 ppm | ppb | ppb |
| Imaging | none | none | none | excellent | poor | good |
| Notes | The values for detection limit and sample size for bulk techniques are interdependent. Detection limit is also element specific, and will vary depending on other factors such as counting time and neutron flux. | | | excellent for analysis of metals<br><br>detection limit too high for many siderophile elements in silicates | poor ionization and isobaric interference limits use for many elements | low detection limit, good resolution and imaging |

INAA = instrumental neutron activation analysis; ICP-AES = inductively coupled plasma atomic emission spectroscopy; Diss-ICP-MS = dissolution inductively coupled plasma mass spectroscopy; EMP = electron microprobe; SIMS = secondary ion mass spectroscopy; LA-ICP-MS = laser ablation inductively coupled plasma mass spectroscopy. Values given are general and can vary considerably depending on sample and analytical technique. See *Potts* (1987), *Thompson and Walsh* (1989), *Reed* (1996), and *Sylvester and Eggins* (1997) for more detailed reviews of these techniques.

TABLE 5.   Summary of experimentally determined valence states of
siderophile elements dissolved in S-free silicate melts at 1 atm.

| Element | $T$ range (°C) | log $f_{O_2}$ range | Valence State | References |
|---|---|---|---|---|
| Fe | 1300–1600 | −8 to −13 | 2+ (FeO) | [1,2] |
| Co | 1300–1600 | −1 to −13 | 2+ (CoO) | [2,6,7] |
| Ni | 1300–1600 | −1 to −13 | 2+ (NiO) | [2,6–9] |
| Ga | 1260–1600 | −8 to −13 | 3+ ($Ga_2O_3$) | [10] |
| Ge | 1260–1600 | −8 to −13 | 2+, 4+ (GeO, $GeO_2$) | [1,10] |
| P | 1190–1600 | −8 to −14 | 4+, 5+ ($P_2O_4$, $P_2O_5$) | [1,3] |
| Mo | 1300–1450 | −10 to −15 | 4+, 6+ ($MoO_2$, $MoO_3$) | [1,7] |
| Sn | 1260 | −8 to −17 | 2+, 4+ (SnO, $SnO_2$) | [10] |
| W | 1300–1600 | −8 to −15 | 4+, 6+ ($WO_2$, $WO_3$) | [1,4,5] |
| Re | 1150–1400 | −6 to −11 | 1+, 2+ ($Re_2O$, ReO) | [11–13] |
| Os | 1400 | −8 to −12 | 3+ ($Os_2O_3$) | [14] |
| Ir | 1300–1480 | 0 to −12 | 1+, 2+ ($Ir_2O$, IrO) | [11,15] |
| Pt | 1300–1550 | 0 to −8 | 2+ (PtO) | [16,17] |
| Ru | 1400 | −1 to −3 | 3+ ($Ru_2O_3$) | [18] |
| Rh | 1300 | 0 to −6 | 2+ (RhO) | [17] |
| Pd | 1350–1470 | 0 to −10 | 0, 1+, 2+ (Pd, $Pd_2O$, PdO) | [19] |
| Au | 1300–1480 | 0 to −8 | 1+ ($Au_2O$) | [20] |

The references given here are not meant to be exhaustive, and the valence states given here reflect our assessment of the most convincing results, and are only valid within the specified ranges in $f_{O_2}$ and $T$ investigated. [1] *Schmitt et al.* (1989), [2] *Holzheid and Palme* (1996), [3] *Newsom and Drake* (1983), [4] *Rammensee and Wänke* (1977), [5] *Ertel et al.* (1996), [6] *Capobianco and Amelin* (1994), [7] *Holzheid et al.* (1994), [8] *Colson et al.* (1991), [9] *Dingwell et al.* (1994), [10] *Capobianco et al.* (1999) [11] *O'Neill et al.* (1995), [12] *Righter and Drake* (1997), [13] *Ertel* (1996), [14] *Borisov and Walker* (1998), [15] *Borisov and Palme* (1995), [16] *Borisov and Palme* (1997), [17] *Ertel et al.* (1999), [18] *Borisov and Nachtweyh* (1998), [19] *Borisov et al.* (1994), [20] *Borisov and Palme* (1996).

1993; *Hillgren et al.*, 1994; *Ohtani et al.*, 1997). Thus, the depletions of these elements in the Earth and Moon are apparently inherited features, and probably reflect depletion due to volatility (*Drake et al.*, 1989).

## 4.2.  Moderately Siderophile Elements (MSE)

The solubility and partitioning behavior of many MSE between metal and silicate melt have been studied extensively at 1 atm, and increasingly at high $T$ and $P$ as well. Here we present a summary of a large body of experimental data with emphasis placed on the effects of $f_{O_2}$, $T$, $P$, and $X$.

*4.2.1. Iron, cobalt, and nickel.* On the basis of both 1-atm and high-$P$ experiments, Fe, Co, and Ni have the common feature that they dissolve in silicate melts in a divalent oxidation state over a wide range of $f_{O_2}$ (Table 5, Fig. 5). Activity coefficients for FeO, NiO, and CoO in silicate melts, calculated using the intercept-γ-relationship of equation (8), have been found to be greater than 1 and independent of $f_{O_2}$ and $T$. This reflects the tendency of the metal oxide to be concentrated in the metal relative to the silicate melt. Furthermore, variable FeO and MgO do not affect the activity coefficients of these elements in the silicate melt (e.g., *Holzheid et al.*, 1997; *Ertel et al.*, 1997).

The solubilities of Co and Ni in silicate melt decrease with increase in $T$ at constant absolute $f_{O_2}$ (*Capobianco and*

*Amelin*, 1994; *Dingwell et al.*, 1994; *Holzheid et al.*, 1994). However, an increase in $T$ causes calculated $^{met/sil}D_M^{sol}$ for both Co and Ni to decrease at constant $\Delta f_{O_2}$ relative to the IW buffer — i.e., absolute $f_{O_2}$ increases with increase in $T$ (*Capobianco and Amelin*, 1994; *Holzheid et al.*, 1994). A decrease in $^{met/sil}D$ for both elements with increase in $T$ at constant $\Delta f_{O_2}$ (i.e., a decrease in $K_D$) has also been observed in experiments at high $T$ and $P$ (*Thibault and Walter*, 1995).

The effect of $P$ on Co and Ni partitioning has been studied up to 26 GPa. Figure 6 shows a compilation of data from high-$P$ experiments over a limited range of high $T$ (2000°–2600°C) showing the dependence of $K_D$ on $P$. The silicate melts in these experiments are somewhat variable in composition but are generally peridotitic, and the liquid metals are all Fe-rich but contain variable amounts of S, C, and siderophile elements. Even though all these variables have some effect on partitioning (e.g., see *Jana and Walker*, 1997a,b,c, and below), the data indicate clearly that $K_D$ for both elements decreases systematically as $P$ increases, at least up to 26 GPa (*Walker et al.*, 1993; *Hillgren et al.*, 1994; *Thibault and Walter*, 1995; *Li and Agee*, 1996; *Ohtani et al.*, 1997; *Righter et al.*, 1997; and *Ito et al.*, 1998). An increase in the solubility of Co and Ni in silicate melt with increase in $P$ was predicted on the basis of the observed preference of these cations for octahedral sites with increase in $P$ (*Keppler and Rubie*, 1993). A significant observation

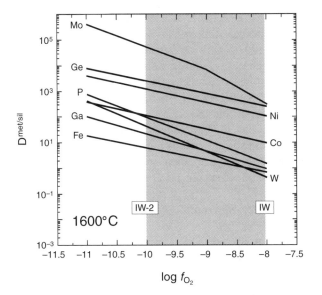

**Fig. 5.** A summary of $^{met/sil}D$ as a function of $f_{O_2}$ for moderately siderophile elements dissolved in silicate melts at 1 atm and 1600°C. The trends are not idealized but are based directly on experimental data, and their slopes give the valence state of the metal cation according to equation (8) in the text. At these conditions, Fe, Ni, and Co are observed to dissolve as divalent species, Ga dissolves as a trivalent species, Ge dissolves as either divalent or tetravalent species, W and P dissolve primarily as tetravalent species, and Mo dissolves as both tetravalent and hexavalent species. The data sources are as given in Table 5.

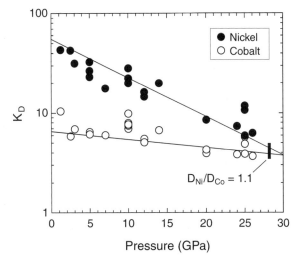

**Fig. 6.** $^{met/sil}K_D$ as a function of pressure for Ni and Co. The data cover a range in temperature from 2000°–2600°C. The silicate melts are generally peridotitic, and the Fe-rich liquid metals have variable S and C contents (data from *Walker et al.*, 1993; *Hillgren et al.*, 1994; *Thibault and Walter*, 1995; *Li and Agee*, 1996; *Ohtani et al.*, 1997; *Ito et al.*, 1998). The curves are fitted to the data of *Li and Agee* (1996), which are at a constant temperature of 2000°C, and relatively constant liquid-metal (S-bearing) and silicate-melt compositions. A ratio of $D_{Ni}/D_{Co}$ of about 1.1 is required to account for the Ni/Co ratio of Earth's mantle, and occurs at a pressure of about 28 GPa, implying high temperature and pressure core segregation.

with respect to core-formation models, as will be discussed below, is that the combined effects of $T$ and $P$ on partitioning of Co and Ni conspire to produce conditions at which the $^{met/sil}D_{Ni}/^{met/sil}D_{Co}$ ratio reaches an appropriate value to account for their relative depletions in Earth's mantle, and at $^{met/sil}D$ values that are remarkably close to those needed to account for their absolute abundances (*Li and Agee*, 1996; *Righter et al.*, 1997).

The value of $^{met/sil}D$ for Co and Ni is apparently little influenced by variable FeO and MgO in the silicate melt or by the Fe content of the metal phase (*Holzheid and Palme*, 1996; *Holzheid et al.*, 1997). However, the overall degree of melt polymerization does have some effect on partitioning. For example, *Ertel et al.* (1997) found that variable $Na_2SiO_3$ increases $^{met/sil}D$ for Ni, although progressive addition of $Na_2SiO_3$ to the silicate melt eventually reverses the trend due to progressive depolymerization of the silicate melt structure at Na/Al-ratios > 1. This is generally consistent with the results of *Jana and Walker* (1997a) who found that $^{met/sil}D$ for Fe, Co, and Ni decreased modestly as the silicate melt became progressively depolymerized.

*4.2.2. Phosphorus, molybdenum, and tungsten.* These three elements dissolve in silicate melts in high valence states over a wide range of $f_{O_2}$, $T$, and $P$ (Table 5, Fig. 5). Phosphorus apparently dissolves in either a penta- or tetravalent oxidation state, whereas W and Mo dissolve in either hexa- or tetravalent oxidation states (*Schmitt et al.*, 1989; *Holzheid et al.*, 1994; *Walter and Thibault*, 1995; *Ertel et al.*, 1996). The data for Mo suggest a relatively abrupt valence change from $Mo^{6+}$ to $Mo^{4+}$ at an $f_{O_2}$ slightly below the IW-1 buffer at 1 atm (*Holzheid et al.*, 1994).

With increase in $T$ at constant absolute $f_{O_2}$, the solubilities of both W and Mo in silicate melt decrease at 1 atm (*Ertel et al.* 1996; *Holzheid et al.*, 1994). At constant $\Delta f_{O_2}$ relative to the IW buffer, *Ertel et al.* (1996) found that, with increase in $T$, $^{met/sil}D$ for W decreases, whereas *Schmitt et al.* (1989) found that it increases; in the case of Mo, $^{met/sil}D$ decreases (*Holzheid et al.*, 1994), although for P it may increase (*Schmitt et al.*, 1989). In higher $T$ and $P$ partitioning experiments, *Walter and Thibault* (1995) could not resolve either a $T$ or $P$ effect on $^{met/sil}D$ for W, but found a decrease in $^{met/sil}D$ for Mo with both increasing $T$ and $P$ at constant $\Delta f_{O_2}$. The data of *Ohtani et al.* (1997) indicate that in combination, the effect of increasing $T$ and $P$ is to decrease $^{met/sil}K_D$ for all three elements. *Righter and Drake* (1999), on the basis of a parameterization of data from several labs and over a wide range of conditions (see section 5 for further details), predict that an increase in $T$ causes a decrease in $^{met/sil}D$ for Mo and P but an increase for W, whereas an increase in $P$ causes an increase in $^{met/sil}D$ for W and P, but a decrease for Mo at constant $\Delta f_{O_2}$. Further experimentation is needed to more precisely isolate the effects of $T$ and $P$ for these three elements.

The value of $^{met/sil}D$ for these high-valence cations shows a remarkable dependence on silicate melt composition, which is probably linked to the structure of the silicate melt (Fig. 7). All three elements become distinctly less sidero-

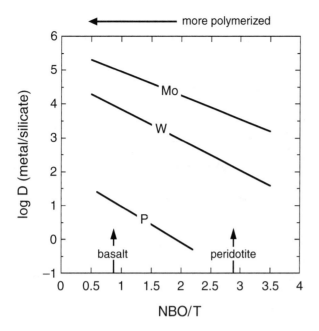

**Fig. 7.** met/silD as a function of degree of polymerization of the silicate melt as expressed by the factor nbo/t (number of nonbridging oxygens to tetrahedrally coordinated cations). This diagram illustrates that the partitioning of the high valence cations P, W, and Mo is highly dependent on melt composition. For example, in changing from a basaltic to a peridotitic melt, met/silD changes by up to about 2 orders of magnitude for these elements. The trends are based on the data of *Walter and Thibault* (1995) at 1 and 2 GPa, 2200–2300 K, at IW-1.9, and from *Jana and Walker* (1997a) at 1 GPa, 1500°C, and about IW-1.3.

phile as the melt becomes more depolymerized. Indeed, met/silD can decrease by up to 2 orders of magnitude between a relatively polymerized basaltic melt and a relatively depolymerized peridotitic melt. This behavior has important implications for core formation because if metal segregation occurs at low silicate melt fractions at relatively low *P*, the melt is basaltic, whereas in a magma ocean the melt would be peridotitic (*Walter and Thibault*, 1995; *Hillgren et al.*, 1996; *Jana and Walker*, 1997a).

*4.2.3. Copper, gallium, germanium, arsenic, silver, antimony, and tin.* Relatively fewer partitioning data are available for these elements, probably resulting from their more limited use in constraining core formation models because of uncertainties in their abundance due to volatility. In experiments at 1 atm, *Schmitt et al.* (1989) and *Capobianco et al.* (1999) have observed that Ga dissolves in silicate melt as a trivalent species. *Schmitt et al.* (1989) observed that at 1 atm, Ge dissolves in silicate melt in a divalent oxidation state and that there is a modest positive *T* effect on met/silD. However, the recent 1 atm data of *Capobianco et al.* (1999) show a tetravalent oxidation state for Ge in Fe-Ni alloys at 1260°C. The experiments of *Walker et al.* (1993) show a decrease in met/silD for Ge with simultaneous increase in *T* and *P*, and if met/silD continues to have

a positive *T* dependence at high *P*, this may indicate a negative *P* dependence. Both *Hillgren et al.* (1996) and *Jana and Walker* (1997a) have shown that met/silD for Ge is moderately dependent on silicate melt composition, decreasing in more depolymerized melts. *Jones and Drake* (1986) determined met/silD for Ag at 1 atm and 1270°C and under reducing conditions, but the effects of intensive parameters have not been isolated. *Capobianco et al.* (1999) observed that Sn dissolves in silicate melt as a tetravalent species at 1260°C and at $f_{O_2}$ greater than about the IW buffer, whereas it dissolves as a divalent species at lower $f_{O_2}$.

### 4.3. Highly Siderophile Elements

As a group, the HSE are present in Earth's mantle in near-chondritic relative proportions (Fig. 1; see *Righter et al.*, 2000). From the point of view of HSE chemistry, this is surprising considering the large interelemental differences in their (1) oxide reduction potentials, (2) Henry's Law activity coefficients in Fe-rich metal, (3) metal/sulfide-liquid partitioning behavior, and (4) partitioning behavior between basaltic melt and solid silicates and oxides (e.g., olivine, spinel) (*Barnes et al.*, 1985; *Capobianco and Drake*, 1990; *Capobianco et al.*, 1991, 1992, 1994; *Fleet and Stone*, 1991;

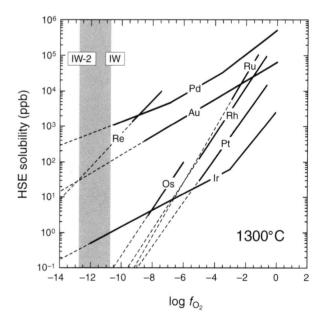

**Fig. 8.** A summary of the observed dependence of HSE solubility on $f_{O_2}$ in haplobasaltic silicate melt at 1 atm and 1300°C (recalculated for solubilities in equilibrium with pure metal). Palladium, Au, and Ir are dissolved in silicate melt predominantly as monovalent species, whereas Pt, Rh, and Re are dominantly divalent, and Os and Ru primarily exist as trivalent species. Although some elements undergo changes in oxidation state at high $f_{O_2}$, in general the oxidation states of HSE in silicate melts are unexpectedly low relative to known low-temperature HSE oxide chemistry (*O'Neill et al.*, 1995). Data sources are as given in Table 5.

TABLE 6. General variation in $^{met/sil}D$ of siderophile elements as a function of $T$, $P$, and silicate melt and metal alloy composition.

| Element | $(dD/dT)_{P, X, \Delta f_{O_2}}$ | $(dD/dP)_{T, X, \Delta f_{O_2}}$ | $(dD/dX)_{T, P, \Delta f_{O_2}}$ | References |
|---------|-----------|-----------|-----------|------------|
| Co | negative | negative | minor | [2–7,10,13–22] |
| Ni | negative | negative | minor | [2–7,10,13–24] |
| Ga | negative | unknown | unknown | [2] |
| Ge | positive | negative | moderate | [2–5] |
| P | negative | positive | large | [1–7] |
| Mo | negative | negative | large | [4–6,11,16] |
| W | negative | positive | large | [2,4,5–13] |
| Re | negative | positive | modest | [25–28] |
| Os | unknown | unknown | unknown | [29] |
| Ir | none | unknown | unknown | [26,30–31] |
| Pt | negative | none | minor | [32–35] |
| Ru | negative | unknown | moderate | [36–39] |
| Rh | unknown | unknown | unknown | [35–37] |
| Pd | negative | none | minor | [32,36,40] |
| Au | negative | unknown | unknown | [41,42] |

The references given here are meant to provide the reader with an extensive coverage of the experimental database. In some cases there are apparent discrepancies among different studies, and we leave it to the reader to ferret these out. The dependencies of $^{met/sil}D$ given here are general and reflect our assessment of the most convincing results, and are only valid within the specified ranges in $\Delta f_{O_2}$ (i.e., relative to the IW buffer), $T$, $P$, and $X$ investigated. Further studies are required for many of the siderophile elements in order to establish the behavior of $^{met/sil}D$ over the wide range of conditions relevant to core formation. [1] *Newsom and Drake* (1983), [2] *Schmitt et al.* (1989), [3] *Walker et al.* (1993), [4] *Hillgren et al.* (1996), [5] *Jana and Walker* (1997a,b,c), [6] *Righter et al.* (1997; 1999), [7] *Ohtani et al.* (1997), [8] *Rammensee and Wänke* (1977), [9] *Newsom and Drake* (1982), [10] *Hillgren et al.* (1994), [11] *Walter and Thibault* (1995), [12] *Ertel et al.* (1996), [13] *Gaetani and Grove* (1997), [14] *Seiffert et al.* (1988), [15] *Capobianco and Amelin* (1994), [16] *Holzheid et al.* (1994), [17] *Thibault and Walter* (1995), [18] *Holzheid et al.* (1995), [19] *Agee et al.* (1995), [20] *Holzheid and Palme* (1996), [21] *Li and Agee* (1996), [22] *Ito et al.* (1998), [23] *Dingwell et al.* (1994), [24] *Ertel et al.* (1997), [25] *Kimura et al.* (1974), [26] *O'Neill et al.* (1995), [27] *Ertel* (1996), [28] *Righter and Drake* (1997), [29] *Borisov and Palme* (1998), [30] *Amosse and Allibert* (1993), [31] *Borisov and Palme* (1995), [32] *Azif et al.* (1996), [33] *Borisov and Palme* (1997), [34] *Holzheid et al.* (1998), [35] *Ertel et al.* (1999), [36] *Capobianco and Drake* (1990), [37] *Capobianco et al.* (1994), [38] *Capobianco and Hervig* (1997), [39] *Borisov and Nachtweyh* (1998), [40] *Borisov et al.* (1994), [41] *Capobianco* (1990), [42] *Borisov and Palme* (1996).

*Fleet et al.*, 1991a,b; *O'Neill et al.*, 1995). Thus, the expectation is that geological processes, and specifically core formation, would produce significant interelemental fractionations of the HSE. Overall, the partitioning behavior of the HSE over the range of conditions relevant for core formation is poorly known. Here we review the current state of knowledge of HSE solubility in silicate melts and metal/silicate partitioning at 1 atm.

*4.3.1. Rhenium.* *O'Neill et al.* (1995) presented preliminary results for Re solubility in silicate melt of composition Di-An + 10 wt% CaO. They observed an unusual monovalent oxidation state for dissolved Re at relatively low $f_{O_2}$ (below $10^{-6}$ atm) where nugget effects tend to level out solubilities to a constant value. *Ertel* (1996) also performed Re solubility experiments in An-Di eutectic melt over 6 log units of $f_{O_2}$, and observed a divalent oxidation state for Re

(Table 5, Fig. 8). *Righter and Drake* (1997) investigated Re solubility in haplobasaltic melt in the system CaO-MgO-$Al_2O_3$-$SiO_2$-$TiO_2$ between 1150°C and 1350°C at 1 atm using a NiRe alloy. They also observed that Re dissolved in a divalent oxidation state, but found higher Re solubilities than *Ertel* (1996) when calculated for a pure Re phase; this may be related to differences in melt composition.

On the basis of a parameterization of their own experiments and other existing partitioning data for Re, *Righter and Drake* (1997) predict that $^{met/sil}D$ decreases with increase in $T$, increases with increase in $P$, and shows a modest dependence on silicate melt composition (Table 6). However, many more data are needed to test the robustness of these predictions.

*4.3.2. Osmium.* Experimentally Os is difficult to handle, as it is very brittle and has the highest melting point of all

HSE (3320°C at 1 atm). In addition, its oxidation produces the extremely toxic $OsO_4$ even at room temperature. The initial results of *Borisov and Palme* (1998) on the solubility of Os in An-Di eutectic melt at 1400°C show a large scatter (determined by INAA), due possibly to nugget formation, and indicated a monovalent oxidation state. Improvements in experimental technique (*Borisov and Walker*, 1998) resulted in new data showing an $f_{O_2}$-solubility dependence indicating that Os dissolves as a trivalent species in silicate melt between $10^{-6}$ to $10^{-8}$ atm at 1400°C (Table 5, Fig. 7). The effects of *T*, *P*, and *X* on Os solubility and partitioning are unknown.

*4.3.3. Iridium.* Iridium-solubility experiments in haplobasaltic melts were performed by *Amosse and Allibert* (1993), *Borisov and Palme* (1995), and *O'Neill et al.* (1995). A positive correlation between Ir solubility and $f_{O_2}$ indicating that Ir is dissolved in a monovalent oxidation state was observed in the high-$f_{O_2}$ regime (above the QFM buffer) by *O'Neill et al.* (1995), and over a wide range in $f_{O_2}$ in the study by *Borisov and Palme* (1995; Table 5, Fig. 8). Conversely, *Amosse and Allibert* (1993) observed a negative correlation over a limited range of $f_{O_2}$ ($\sim 10^{-4}$ to $10^{-7}$ atm), and ascribed this odd behavior to zero-valent Ir dissolved in the silicate melt, which is passivated by an oxide layer on Ir metal. However, this interesting result has not been repeated in other studies. At the highest $f_{O_2}$ (air and pure oxygen), Ir may change its oxidation state from monovalent to trivalent (*Borisov and Palme*, 1995). At low $f_{O_2}$, Ir solubility was observed to become independent of $f_{O_2}$ in the study by *O'Neill et al.* (1995), which could be interpreted as zero-valent Ir. However, the inhomogeneous distribution of Ir (*Borisov and Palme*, 1995) and the slow kinetics observed at rather reducing conditions in stirred crucible experiments (*O'Neill et al.*, 1995) show strong evidence that Ir forms nuggets at such low $f_{O_2}$, making the interpretation of zero-valent Ir an artifact. A temperature dependence of Ir solubility was not observed over the relatively small temperature range investigated by *Borisov and Palme* (1995), and the effects of *P* and *X* on Ir solubility are unknown.

*4.3.4. Platinum.* Several groups have systematically investigated the solubility of Pt in haplobasaltic melt at 1 atm (e.g., *Amosse and Allibert*, 1993; *Azif et al.*, 1996; *Borisov and Palme*, 1997; *Ertel et al.*, 1999). Similar to what had been found with Ir, *Amosse and Allibert* (1993) found a negative correlation between solubility and $f_{O_2}$, and ascribed this behavior to a zero-valent Pt and a passivation reaction. Again, this result has not been repeated in the other studies that concentrated on the dependence of solubility on $f_{O_2}$ and temperature, applying either wire-loop or stirred-crucible techniques. In these studies it was observed that Pt dissolved in a divalent state in silicate melt at ppb to ppm levels (Table 5, Fig. 8). At the highest $f_{O_2}$ studied, the existence of some tetravalent Pt could not be excluded, whereas at lower $f_{O_2}$, solubilities are strongly influenced by nugget formation. On the basis of the studies referred to in Table 6, Pt partitioning shows small to negligible dependence on *T*, *P*, and *X*, although much more work is needed to confirm this prediction.

*4.3.5. Ruthenium.* *Capobianco and Hervig* (1997) reported Ru solubilities in silicate melt in equilibrium with $RuO_2$. Their results, however, showed Ru solubilities that were about 1 order of magnitude higher than solubilities of Ru in Di-An eutectic melt equilibrated with pure Ru metal under oxidizing conditions (*Borisov and Nachtweyh*, 1998). In the latter study, Ru was observed to dissolve in silicate melt in a trivalent oxidation state (Table 5, Fig. 8). A slight decrease of Ru solubility with increasing temperature from 1350° to 1450°C was observed; between 1450° and 1500°C no change was found. Analogous discontinuities in solubility with temperature have also been observed for Au (*Borisov and Palme*, 1995) and Pt (*Borisov and Palme*, 1997). The effect of *P* on Ru solubility is unknown, although *Capobianco and Hervig* (1997) reports substantial silicate melt composition dependence.

*4.3.6. Rhodium.* Only a few Rh solubility studies have been made to date (*Capobianco and Drake*, 1990; *Capobianco et al.*, 1994; *Ertel et al.*, 1999). *Ertel et al.* (1999) have made a systematic investigation of the solubility dependence on $f_{O_2}$ in haplobasaltic melt using the stirred crucible method. They found that Rh dissolves in a divalent oxidation state at $f_{O_2}$ greater than $10^{-6}$ atm at 1300°C, although some evidence was found for an increased oxidation state ($Rh^{3+}$ and/or $Rh^{4+}$) at the most oxidized conditions (Table 5, Fig. 8). *Ertel et al.* (1999) found that Rh solubility became independent of $f_{O_2}$ below a value of $\sim 10^{-6}$ atm, and this is interpreted as being due to the nugget effect as found for other HSE. The effects of *T*, *P*, and *X* on Rh solubility are unknown.

*4.3.7. Palladium.* There have been a number of studies on the solubility of Pd in basaltic silicate melts and, generally, Pd dissolves at ppb to ppm levels, typically the highest of all HSE (*Capobianco and Drake*, 1990; *Capobianco et al.*, 1991, 1992; *Borisov et al.*, 1994; *Azif et al.*, 1996). In all these studies it was found that Pd dissolved predominantly in a monovalent oxidation state (Table 5). However, the $f_{O_2}$ dependence of Pd is separated into three regimes: Above an $f_{O_2}$ of $10^{-3}$ atm a change toward a higher oxidation state or a partial $Pd^{2+}$-complex component is evident from the data of *Borisov et al.* (1994), but a pure divalent oxidation state was not observed (Table 5, Fig. 8). Below an $f_{O_2}$ of $10^{-7}$, there is evidence for a decrease in oxidation state indicating the presence of some zero-valent Pd. *Capobianco et al.* (1991, 1992) reported similar results. *Borisov et al.* (1994) observed a slight increase in Pd solubility with increasing temperature, although only by $\sim 40\%$ over a temperature range of 100°C. *Holzheid et al.* (1998) reported no change in $^{met/sil}D$ for Pd at *P* up to 20 GPa at 2000°C, and *Capobianco and Hervig* (1997) reported a minor effect of melt composition on Pd solubility.

*4.3.8. Gold.* *Borisov and Palme* (1996) presented experiments for Au solubility in a silicate melt composition close to the 1-atm Di-An eutectic using a $Au_{45}Pd_{55}$-loop technique over a wide range of $f_{O_2}$ ($10^{-9}$ to 0.21 atm). They observed an $f_{O_2}$-solubility dependence indicating that Au dissolved in a monovalent oxidation state (Table 5, Fig. 8). The Au solubilities reported by *Borisov and Palme* (1996) were about 30% higher than reported by *Capobianco* (1990) for glasses in equilibrium with PdAu alloys in air, and a difference in silicate melt compositions was proposed as a possible reason for the discrepancy. *Borisov and Palme* (1996) observed an increase in Au solubility with increase in temperature, although stronger temperature dependence was observed below 1350°C than at higher *T*. The effects of *P* and *X* on Au solubility in silicate melts are unknown.

## 4.4. The Effect of Light Elements on Siderophile-Element Partitioning

Earth's outer core is most probably a molten alloy composed of Fe and about 5% Ni (based on a chondritic Fe/Ni ratio of about 18). However, on the basis of seismic data it has been known for many years that the outer core is about 10% less dense than pure Fe (the presence of Ni having a relatively minor effect on bulk density), whereas the inner core may be pure metallic Fe (e.g., *Birch,* 1952; *Jephcoat and Olson,* 1987). Thus, the outer core presumably contains one or more lighter elements, the most commonly proposed of which are H, C, O, Si, S, and K (e.g., see review by *Poirier,* 1994). Recently, there has been considerable experimental work at high *T* and *P* for determining the solubilities of these elements in Fe over a range of conditions in order to predict which element(s) can best explain the density deficit. While certain element combinations can probably be eliminated, a wide range of possibilities still exist and no consensus has emerged as of yet (see *Hillgren et al.,* 2000).

With respect to siderophile-element partitioning, a knowledge of which element(s) alloyed with metal during core segregation is important because partitioning can be sensitive to the alloy and silicate melt composition. For example, *Brett* (1984) and *Agee et al.* (1995) suggested that if the core segregated as an Fe-rich sulfide liquid, $^{\text{sulfide/sil}}D$ could be considerably lower than $^{\text{met/sil}}D$ for many siderophile elements. While the sulfide liquids referred to in those studies have too much S to be directly applicable to Earth's core (~25 wt% compared to a predicted value of $11 \pm 2$ wt% needed to account for the outer core density deficit, e.g., *Ahrens and Jeanloz,* 1987), S may have been an important component during metal segregation and it can have a dramatic effect on siderophile-element partitioning.

There is now enough experimental data to predict quantitatively the effects that variable amounts of S and C have on partitioning for several of the MSE. For example, *Righter and Drake* (1999) fitted an empirical expression, which includes terms for S and C, to a large number of experimental data for partitioning of P, W, Co, Ni, and Mo between molten Fe and silicate melt (see section 5.1 for details of the parameterization method). On the basis of their parameterizations, C causes a decrease in $^{\text{met/sil}}D$ for Ni, Co, P, and Mo, but an increase for W. Sulfur causes an increase in $^{\text{met/sil}}D$ for Ni, Co, and Mo, but a decrease in $^{\text{met/sil}}D$ for P and W. Therefore, the effects of S and C are opposite for all elements except P. Righter and Drake found that the effects of S and C on $^{\text{met/sil}}D$ can be large for W and Mo, and

the amounts of S and C dissolved in metal are important factors for modeling the depletions of W and Mo due to core formation.

Sulfur also has a large effect on the partitioning of HSE among Fe-rich metal, sulfide melt, and silicate melt. *Fleet and Stone* (1991) and *Fleet et al.* (1991a) showed experimentally that heavier HSE such as Os, Ir, and Pt are concentrated in metal relative to sulfide, whereas lighter HSE like Pt are concentrated in sulfide. Further, *Lauer and Jones* (1998) working in the Fe-Ni-S system showed that Os is more siderophile than Pt. HSE partitioning between sulfide melt and silicate melt is also sensitive to sulfide composition, although $^{\text{sulfide/sil}}D$ is similar for the HSE over a range of conditions. The presence of S in silicate melt tends to reduce $^{\text{met/sil}}D$ for the HSE, although distinctly different partitioning behavior has been observed for Pt and Pd between Fe-metal alloy and S-bearing silicate melts (*Fleet et al.,* 1991b). Thus, if Fe-rich metal alloy segregated from a S-bearing silicate melt, fractionations among the HSE would be expected. Clearly, more experimental data on the effects of light elements on partitioning are needed for all the siderophile elements.

## 4.5. Partitioning Between Metal and Solid Silicates

Metal segregation in planetary bodies may occur over a wide range of temperatures, and the coexisting phases among which siderophile elements are partitioned may be solid or liquid. Thus far, we have been considering partitioning between Fe-rich metal (solid or liquid) and liquid silicate. However, the participation of solid silicates during metal segregation would be expected in a variety of circumstances. For example, metal segregation in an accreting planetesimal might occur at subsolidus or near-solidus conditions. Alternatively, metal segregation in a magma ocean might involve near-liquidus phases, or perhaps equilibration with solids at the base of the magma ocean. For this reason, knowledge of partitioning involving solid mantle phases over the wide array of conditions possible for core formation is desirable.

It is conceivable that metal segregation involved at least four different kinds of phases including solid metal alloy (SM), liquid metal alloy (LM), solid silicate or oxide (SS) and liquid silicate (LS); for example, the 1-atm depletion model shown in Fig. 1 was calculated assuming the presence of all four kinds of phases (see also *Jones and Drake,* 1986; *Newsom,* 1990). Thus, six kinds of partition coefficients are possible ($^{\text{SM/LM}}D$, $^{\text{SM/SS}}D$, $^{\text{SM/LS}}D$, $^{\text{LM/SS}}D$, $^{\text{LM/LS}}D$, and $^{\text{SS/LS}}D$), with a minimum of three required to derive all six. Presently, there are much fewer data for metal/solid silicate partitioning than for metal/liquid silicate, and this is a fertile ground for future experimentation.

Partitioning of Ni and Co between Fe-rich metal/sulfide and olivine has been well studied at 1 atm (e.g., *Seifert et al.,* 1988; *Ehlers et al.,* 1992; *Gaetani and Grove,* 1997), and these elements both partition strongly into metal and sulfide, but are compatible in olivine relative to silicate melt. Thus, if olivine and silicate melt were both present during metal segregation, the bulk $^{met/sil}D$ for Ni and Co would be lower than if only silicate melt were present. The partitioning behavior of a number of siderophile elements between Fe-rich metal and magnesiowüstite has been determined at high *P* (*Ohtani and Yurimoto,* 1996; *Ito et al.,* 1998; *Gessmann and Rubie,* 1998; *O'Neill et al.,* 1998). These studies show that V, Cr, Co, and Ni partition more strongly into magnesiowüstite than into silicate melt in the presence of Fe, but Mo, W, and Re prefer silicate melt to magnesiowüstite. Furthermore, *O'Neill et al.* (1998) found that at constant $\Delta f_{O_2}$, Ni and Co become less siderophile with increase in *P*, but Cr becomes more siderophile. In a recent diamond anvil study *Tschauner et al.* (1999) found that $^{met/sil}D_{Co}/$ $^{met/sil}D_{Ni}$ (metal/Mg-silicate perovskite) is approximately 1 at lower mantle pressures for Ni and Co, and absolute $^{met/sil}D$ decreases for both elements with increase in pressure, reaching unity at ~80 GPa.

## 5. CORE FORMATION MODELS

### 5.1. Earth

Presently the Earth's core must be in gross disequilibrium with the upper mantle and probably with the bulk of the lower mantle, since the upper mantle is relatively oxidized at well above the equilibrium Fe-FeO buffer (e.g., *O'Neill and Wall,* 1987), and exists at a much lower *T* and *P* than the core. If core formation were a disequilibrium process, then there is little hope for deducing the conditions of metal segregation from siderophile-element depletions in the mantle. But if equilibrium predominated as the core segregated, then observed siderophile-element depletions can place powerful constraints on core formation models.

Early models for Earth accretion supposed a gradual accretion of material, but these are generally discounted on dynamic grounds (see review in *Taylor,* 1992). Earth accretion is now considered to have been a stochastic process, wherein the bulk of its mass was accreted by "catastrophic" impacts by numerous planetesimals (tens to hundreds of kilometers in diameter), which became progressively larger as accretion proceeded (e.g., *Wetherill,* 1990). Accretion may have been homogeneous, with planetesimals having essentially constant bulk chemical composition, or it may have been heterogeneous with bodies of variable composition. Planetesimals may have contained disseminated metal, or they may have previously differentiated into mantle and core (see *Carlson and Lugmair,* 2000). If equilibrium between metal and silicate and metal segregation were essentially contemporaneous with each accretionary event (*Stevenson,* 1981), then the depletion of siderophile elements in Earth's mantle would record the aggregate effect of all the impact and metal segregation events. For homogeneous accretion, the amalgamation of multistage metal segregation could now appear chemically as a "single" equilibrium event. In contrast, heterogeneous accretion would record a more complex multistage chemical equilibrium. For this reason,

core-formation models are often linked conceptually with accretion models.

With the addition of each new planetesimal to the growing Earth, metal segregation may have occurred at relatively low *T* and *P* in the absence of high-degree melting of silicates, or it may have occurred at much higher *T* and *P*, perhaps in a deep magma ocean. The extent of melting due to impacts depends largely on impactor size, velocity, and angle of incidence, and each event was probably unique. If an impactor had a metallic core, the core material may have rapidly been sequestered to the Earth's core with little reequilibration (e.g., *Benz and Cameron,* 1990; *Melosh,* 1990). The presently most acceptable dynamic models for the Earth-Moon system have the Moon forming as a byproduct of a giant impact between the proto-Earth and a Mars-sized planetesimal. In a recent model, the impact occurs when the proto-Earth had accreted only about 50% of its current mass, either homogeneously or heterogeneously, and the post-impact Earth would have about 60–70% of its present mass (*Cameron and Canup,* 1998). This relatively low initial mass is needed to avoid ending up with too much angular momentum in the Earth-Moon system, and implies that substantial accretion of material, perhaps by additional giant impacts, occurred on the Earth after the formation of the Moon. The Moon-forming giant impact would have spewed a large portion of both the impactor and proto-Earth's mantle into space, and caused extensive silicate melting. Some of the silicate from the impactor remained in orbit and accreted into the Moon, whereas much of the silicate material, and perhaps nearly all the metal from the impactor, returned to Earth. Mixing of this metal with the magma ocean on Earth, perhaps accompanied by metal excavated from Earth's protocore during impact, may have resulted in large-scale reequilibration between metal and silicate at high *T* and *P* in a deep magma ocean, a truly singular equilibrium event. However, the addition of the final 30–40% of Earth material must also have played a role in the siderophile-element budget of Earth. For our purposes, we naively assume that anything is possible, and proceed to consider the consequences for the siderophile-element depletion in Earth's mantle.

As shown in Fig. 1, single-stage low-temperature, low-pressure metal/silicate equilibrium cannot account for the observed siderophile-element depletion in the mantle, and we will consider a high-temperature, high-pressure magma-ocean single-stage equilibrium model below. In heterogeneous accretion models, core formation has been considered as a series of equilibrium metal segregation events, with progressively changing intensive parameters, most notably composition and $f_{O_2}$. For example, *Newsom* (1990) presented a typical model, reminiscent of the two-component model of *Wänke et al.* (1984). In this model the first 93% of Earth's mass accretes from reduced materials, with metal segregation during this stage stripping essentially all the siderophile elements into the core according to low *T-P* partitioning behavior. In the next accretion stage nearly all of Earth's remaining mass is added as more oxidized material, reinjecting siderophile elements into the mantle. Metal segregation is inefficient during this stage due to the higher $f_{O_2}$, and only a small amount segregates to the core. After this stage the mantle inventory of MSE is established, whereas the HSE are still depleted and fractionated relative to present abundances. Lastly a "late veneer" of oxidized material adds chondritic proportions of siderophile elements to the mantle, affecting the abundances of MSE only slightly, but establishing the HSE mantle inventory. In this specific model, there are no "giant" impacts, although in general heterogeneous accretion can accommodate "giant" impacts and magma oceans (see *O'Neill,* 1991a,b). Heterogeneous accretion models are variably successful in accounting for siderophile-element depletions in the mantle; all models encounter some difficulties in simultaneously reproducing the abundance of all siderophile elements, and most notably Ga. We leave it to the reader to compare and contrast among the available versions (see for example *Wänke,* 1981; *Ringwood,* 1984; *Sun,* 1984; *Wänke and Dreibus,* 1988; *Newsom,* 1990; *O'Neill,* 1991b). However, we note that if core formation was a heterogeneous multistage process, then the vastness of the parameter space created when considering that the intensive variables of $f_{O_2}$, $T$, $P$, and $X$ are free to vary during each stage makes geochemical solutions to mantle siderophile-element abundances relatively nonunique, and other types of evidence must be sought to constrain core formation models.

Before considering a magma ocean model, it is worth noting that there is no independent evidence that a magma ocean ever existed on Earth. Unlike the Moon, there is no thick cumulate crust, and if one ever existed, the tectonic engine of Earth would have reworked it long ago. The absence of any significant chemical fractionation in samples from the upper mantle, as might be expected if the magma ocean underwent crystallization differentiation, has been used as an argument against a magma ocean (e.g., *Jones and Drake,* 1986; *Kato et al.,* 1988). However, this lack of fractionation does not preclude a magma ocean either, as theoretical models indicate that in a turbulent magma ocean equilibrium crystallization might be favored over fractional crystallization (*Tonks and Melosh,* 1990; *Solomatov and Stevenson,* 1993), such that significant chemical fractionation might be impeded. Here, we acknowledge that direct evidence of a magma ocean on Earth is lacking, but consider a deep magma ocean to be a likely consequence if accretion included giant impacts as predicted in modern theory.

In order to develop a magma-ocean core-segregation model using siderophile elements, it is necessary to describe $^{met/sil}D$ as a function of all intensive variables. Even though considerable experimental progress has been made in mapping out the extensive parameter space, especially for some of the MSEs, the data are still too few for most elements to permit a comprehensive test of the magma ocean model. *Righter et al.* (1997) and *Righter and Drake* (1997, 1999) have adopted the approach of least-squares fitting of experimental partitioning data to an empirically derived linear

equation. For example, in one of their most recent formulations, *Righter and Drake* (1999) use the equation

$$\ln {}^{met/sil}D_M = a\ln f_{O_2} + b/T + cP/T + d(nbo/t) + e\ln(1 - X_s) + f\ln(1 - X_c) + g \qquad (12)$$

where a–g are regression coefficients, nbo/t is a melt polymerization parameter (a proxy for composition-oxide), and $X_s$ and $X_c$ are the mol fractions of S and C in the metal, respectively. Figure 9 shows the predicted siderophile-element depletions in Earth's mantle for a deduced set of conditions based on their parameterizations for P, W, Co, Ni, and Mo. The model fits the observed data within uncertainty for all elements, and on this basis *Righter and Drake* (1999) predict metal/silicate equilibration at about 27 ± 0.6 GPa and 2250 ± 300 K, a result similar to the prediction of *Li and Agee* (1996) that was based only on Co and Ni.

For a simple high-temperature, high-pressure equilibrium model to satisfy the depletions of these five siderophile elements within uncertainty is an interesting result, considering that the model must simultaneously account for both absolute and relative depletions (e.g., Co and Ni absolute depletions and the near-chondritic Co/Ni ratio). There are two other features of the model that are particularly noteworthy. First, the model requires about 5 wt% each of C and S dissolved in the core-forming metal, enough to account for the density deficit of the outer core. Second, up to several weight percent $H_2O$ is required to be dissolved in the silicate melt, lowering the solidus $T$ of the mantle enough to produce a magma ocean; the model $T$ of 2250 K is below the anhydrous peridotite solidus.

Figure 10 illustrates a conceptual model that has emerged to account for metal/silicate equilibration in a magma ocean

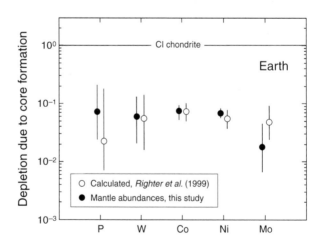

**Fig. 9.** Depletion due to core formation in Earth for P, W, Co, Ni, and Mo (solid circles) relative to the calculated depletion for equilibrium segregation in a hydrous peridotite magma ocean at 27 GPa and 2250 K (open circles) on the basis of the parameterizations of *Righter and Drake* (1999).

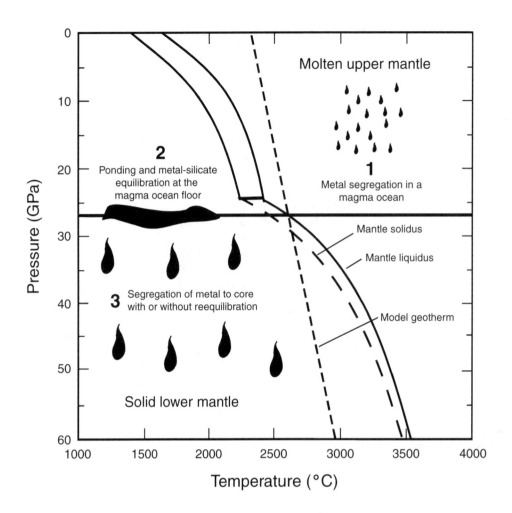

**Fig. 10.** Pressure-temperature diagram showing a conceptual model of metal/silicate equilibrium and segregation in a deep magma ocean. The solidus and liquidus are for pyrolitic mantle and are based on experimental data of *Zhang and Herzberg* (1994) at <25 GPa and *Zerr et al.* (1998) at >25 GPa. The cusp along the solidus at the upper-mantle/lower-mantle boundary is placed at about 25 GPa and corresponds to the subsolidus transition from a ringwoodite-rich ($\gamma$ phase) upper mantle to a Mg-perovskite-rich lower mantle. For the model magma ocean geotherm shown, the entire upper mantle would be molten, a "magma ocean," whereas the lower mantle would be solid. Such a situation might arise as a consequence of a "giant impact" with a Mars-sized body. Equilibrium metal segregation from the upper mantle occurs by precipitation of small liquid metal globules, a "metallic rain" over a wide range in $T$ and $P$ (see *Stevenson,* 1990), and the metal ponds at the solid magma ocean floor. The metal finally equilibrates at this depth giving the upper mantle its present siderophile elements signature (see Fig. 9), before subsequent gravitational instability permits segregation and sinking of large metal diapirs through the lower mantle with little reequilibration. Alternatively, the magma ocean floor may exist at a greater depth and the upper mantle siderophile-element signature may reflect an accumulated average of polybaric metal segregation in the upper mantle (see *Li and Agee,* 1996; *Righter et al.,* 1997; *Righter and Drake,* 1999).

at high $T$ and $P$. For a peridotite mantle, a pressure of 27 GPa is generally coincident with the transition from an upper mantle dominated by olivine and its polymorphs to a lower mantle dominated by silicate perovskite (e.g., *Ito and Takahashi,* 1987). Experimental studies in laser-heated diamond anvil cells show that the $dT/dP$ slope of the melting curve for peridotite in the lower mantle is relatively steep (e.g., *Zerr et al.,* 1998), such that a cusp occurs along the peridotite solidus at ~25 GPa, corresponding to the subsolidus dissociation of ringwoodite ($\gamma$-spinel) to form Mg-perovskite + magnesiowüstite. Depending on the magma ocean geotherm and the amount of $H_2O$ in the mantle, this could

result in an entirely or partially molten upper mantle and a solid lower mantle, forming a floor to the magma ocean. Metal segregated in a molten upper mantle may have ponded and reached final equilibration at a magma ocean floor at ~27 GPa, or perhaps metal continuously equilibrated with silicate melt over a range of depths averaging to ~27 GPa within a deeper magma ocean.

How robust is the chemical evidence for metal segregation in a magma ocean? There is a sufficiently large experimental database available for Co and Ni to adequately cover metal segregation at upper mantle conditions, and the uncertainties in their mantle abundances are the lowest of all siderophile elements. Because $^{met/sil}D$ for Ni and Co are particularly sensitive to *P*, these elements provide the most leverage on this variable. However, partitioning data for P, W, and Mo are much fewer than for Co and Ni, and are especially sparse at *P* > 10 GPa. Consequently, the formal errors in the parameterization of *Righter and Drake* (1999) are larger for these elements, and partitioning at very high P is not well constrained. When combined with the relatively larger uncertainties in mantle abundances for P, W, and Mo, there is a significant flexibility in producing a "match" between model and observation. For example, the predicted W/Mo ratio of ~1.2 in *Righter and Drake's* model is somewhat at odds with the observed mantle ratio of ~3.5, but the combined uncertainties permit reconciliation of the model with the data. Indeed, *Righter and Drake* (1999) note that at the 95% confidence level, their model uncertainties do not require a hydrous magma ocean. Thus, while the match for Ni and Co in the equilibrium deep magma ocean model is compelling, other siderophile elements cannot yet place tight constraints on the model. We emphasize that there is a pressing need for more experimental data, especially at *P* > 10 GPa and over a range of experimental conditions, for all siderophile elements to more fully test the magma ocean model.

*5.1.1. The highly siderophile elements.* There are too few data for the HSE at high *T* and *P* to test the magma ocean core-segregation model using these elements. Table 7 shows calculated HSE $^{met/sil}D$ at low *T* and *P* conditions and at $f_{O_2}$ relevant to metal segregation, and Fig. 1 shows calculated mantle abundances for equilibrium core segregation on the basis of these partition coefficients. These results show that extreme differences in partitioning behavior must be overcome at higher *T* and *P* to account for the mantle abundances and near-chondritic proportions of these elements, a feature requiring that all HSE had essentially identical partition coefficients. Calculated $^{met/sil}D$ for the HSE at 1 atm vary over 9 orders of magnitude and in the order

$$Pt > Rh > Ru > Ir > Os > Au > Pd > Re$$

Given their vastly different partitioning behavior at low *T* and *P*, it would be incredible if there exists any set of conditions in which all HSE partition coefficients converged to a single value. Preliminary data for the effect of increased *T* and *P* on partitioning do not indicate convergence of HSE $^{met/sil}D$ (*Holzheid et al.*, 1998). These considerations strongly reinforce the argument that, subsequent to core segregation, HSE abundances in upper mantle rocks were fixed by the addition of a late veneer of oxidized, HSE-enriched material with chondritic relative abundance ratios (e.g., *Kimura et al.*, 1974). Thus, even if core segregation at high *T* and *P* can be accounted for in a single equilibrium event, heterogeneous accretion in some form may still be required to account for the HSE.

## 5.2. The Moon

Figure 2 shows the depletion of siderophile elements in the Moon's mantle and, like Earth, the siderophile-element depletions cannot be attributed to low *T* and *P* core formation. However, it is intrinsically more difficult to model core formation in the Moon because of the uncertainty surrounding the Moon's origin and, consequently, its bulk composition. It has been previously postulated that Moon's bulk composition may be approximated by that of Earth's mantle (e.g., *Ringwood,* 1990; *O'Neill,* 1991a), in which case calculating depletion relative to CI chondrite, as in Fig. 2, is clearly inappropriate. Using the proposed bulk Moon composition of *O'Neill* (1991a), *Righter and Drake* (1996) used a parameterization similar to that described above to model core formation in the Moon assuming a core comprising 5% by mass of the Moon, and found that metal segregation at ~3.5 GPa and 1925°C can account for the depletion of P, W, Co, Ni, and Mo. Like Earth, this model is consistent with metal segregation in a magma ocean. Unlike Earth, there is direct evidence for a magma ocean in the form of a thick crust of anorthosite cumulates.

As discussed above, a presently popular model is that the Moon originated as a byproduct of a "giant" impact between a Mars-sized body and the proto-Earth, perhaps after Earth had accreted about 50% of its present mass (*Cameron and Canup,* 1998). This model is generally consistent with stochastic Earth accretion models (*Wetherill,* 1985) and can account for the angular momentum of the Earth-Moon system. Simulations show that the majority of material remaining in orbit after such an impact would be nearly free of metal and derived primarily from the impactor mantle (*Benz and Cameron,* 1990), with the remainder from Earth's mantle, so that the Moon would be primarily composed of impactor material. The model of *Cameron and Canup* (1998) also indicates that very little material would need to be added to the Moon after the impact. These results are generally consistent with the recent results of Lunar Prospector, which from the moment of inertia (*Konopliv et al.,* 1998) and from magnetic measurements constrain the size of Moon's core to be relatively small (1–3 wt%, see *Hood*

TABLE 7. Calculated $^{met/sil}D$ of the HSE at 1 atm.

| Element | Relative $f_{O_2}$ | *T* (°C) | $D^{met/sil}$ | References |
|---------|------------|----------|-----------|------------|
| Re | IW-1 | 1400 | >8 × 10^6 | [1–3] |
| Os | IW-1 | 1400 | 6 × 10^10 | [4] |
| Ir | IW-2 | 1300 | >10^12 | [5] |
| Pt | IW-2 | 1300 | 7 × 10^15 | [6] |
| Ru | IW-2 | 1300 | 2 × 10^12 | [7] |
| Rh | IW-2 | 1300 | 5 × 10^12 | [6] |
| Pd | IW-2 | 1350 | 2 × 10^7 | [8] |
| Au | IW-2 | 1350 | 3 × 10^7 | [9] |

Partition coefficients at the conditions listed are calculated on the basis of extrapolation of the solubility data from the following references: [1] *O'Neill et al.* (1995); [2] *Ertel* (1996); [3] *Righter and Drake* (1997); [4] *Borisov and Palme* (1998); [5] *Borisov and Palme* (1995); [6] *Ertel et al.* (1999); [7] *Borisov and Nachtweyh* (1998); [8] *Borisov and Palme* (1994b); [9] *Borisov and Palme* (1996).

*and Zuber*, 2000), making it virtually impossible to construct the Moon solely from material with the composition of the Earth's mantle (see also *Jones and Hood*, 1990). The presence of a large fraction of impactor material in the Moon invalidates using the siderophile-element depletion pattern of Earth's mantle as a possible initial state for core formation in the Moon. Some combination of impactor (chondrite?) and Earth's mantle may be appropriate, although a compounding problem is that siderophile-element abundances in Earth's mantle at the time of the giant impact are not known. The provenance of the Moon-forming impactor is likely to be similar to that of Earth based on their nearly identical O-isotopic compositions, although the extremely depleted volatile element and enriched refractory element content in the Moon relative to Earth suggests that the impactor was a unique object (*Taylor*, 1992). Thus, as there is considerable uncertainty on estimates of bulk Moon composition, models for core formation in the Moon remain ambiguous.

## 6. FUTURE WORK

Improvements in our understanding of core formation will come from a continued multidisciplinary approach. Further efforts in modeling the dynamics and thermal physics of accretion by large impacts, and the consequences for the thermal and compositional history of the impactor and proto-Earth, will continue to help constrain chemical models. Tighter constraints on the bulk chemical compositions of Earth and especially the Moon are needed, most notably for volatile elements. Improvements in precision and accuracy, both analytically and experimentally, will make our understanding of planetary compositions and partitioning behavior increasingly precise. The experimental effort to determine partitioning behavior of siderophile elements has been tremendous, but much more work is needed to determine the effects of all intensive variables on partitioning behavior. To test the magma ocean model more critically, it is of immediate importance to determine $^{met/sil}D$ for the MSE at >10 GPa, and for the HSE at high $T$ and $P$. Further, in order to develop more complex and, hopefully, realistic models of core segregation, it is desirable to know the partitioning behavior over a wide range of conditions among all phases that may have played a role in core formation, such as olivine and its polymorphs in the upper mantle and perovskite and magnesiowüstite in the lower mantle.

*Acknowledgments.*    The authors thank R. Brett, M. Drake, and K. Righter for thoughtful and constructive reviews that improved this manuscript. Special thanks go to K. Righter and R. Canup for their patience, help, and encouragement as editors.

## REFERENCES

Agee C. B., Li J., Shannon M. C., and Circone S. (1995) Pressure-temperature phase diagram for the Allende meteorite. *J. Geophys. Res., 100,* 17225–17740.

Ahrens T. J. and Jeanloz R. (1987) Pyrite: shock compression, isentropic release, and the composition of Earth's core. *J. Geo-*

*phys. Res., 92,* 10363–10375.

Amosse J. and Allibert M. (1993) Partitioning of iridium and platinum between metals and silicate melts: evidence for passivation of the metals depending on $f_{O_2}$. *Geochim. Cosmochim. Acta, 57,* 2395–2398.

Arculus R. J. and Delano J. W. (1981). Siderophile element abundances in the upper mantle: evidence for a sulfide signature and equilibrium with the core. *Geochim. Comochim. Acta, 45,* 1331–1343.

Azif E., Pichavant M., and Auge T. (1996) Solubility of Pt and Pd in Dio-An melts: methodology and effects of $f_{O_2}$ and addition of iron. *Terra Nova Abstract Suppl. 1,* VIth International EMPG Symposium.

Barnes S., Naldrett A. J., and Gorton M. P. (1985) The origin of the fractionations of platinum-group elements in terrestrial magmas. *Chem. Geol., 81,* 45–53.

Beattie P., Drake M., Jones J., Leeman W., Longhi J., MCKay G., Nielson R., Palme H., Shaw D., Takahashi E., and Watson B. (1993) Terminology for trace-element partitioning. *Geochim. Cosmochim. Acta, 57,* 1605–1606.

Benz W. and Cameron G. W. (1990) Terrestrial effects of the giant impact. In *Origin of the Earth* (H. E. Newsom and J. H. Jones, eds.), pp. 61–68. Oxford Univ., New York.

Birch F. (1952) Elasticity and constitution of the Earth's interior. *J. Geophys. Res., 69,* 227–286.

Borisov A. and Nachtweyh K. (1998) Ruthenium solubility in silicate melts: experimental results at oxidizing conditions. In *Lunar and Planetary Science XXIX,* Abstract #1320. Lunar and Planetary Institute, Houston (CD-ROM).

Borisov A. and Palme H. (1995) The solubility of iridium in silicate melts: new data from experiments with Ir10Pt90 alloys. *Geochim. Cosmochim. Acta, 59,* 481–485.

Borisov A. and Palme H. (1996) Experimental determination of the solubility of gold in silicate melts. *Mineral. Petrol., 56,* 297–312.

Borisov A. and Palme H. (1997) Experimental determination of the solubility of platinum in silicate melts. *Geochim. Cosmochim. Acta, 61,* 4349–4357.

Borisov A. and Palme H. (1998) Experimental determination of osmium-metal/silicate partition coefficients. *N. Jb. Miner. Abh., 172,* 347–356.

Borisov A. and Walker R. J. (1998) Osmium solubility in silicate melts: new efforts and new results (abstract). In *Origin of the Earth and Moon,* pp. 2–3. LPI Contribution No. 957, Lunar and Planetary Institute, Houston.

Borisov A., Palme H., and Spettel B. (1994) Solubility of palladium in silicate melts: implications for core formation in the Earth. *Geochim. Cosmochim. Acta, 58,* 705–716.

Brett R. (1984) Chemical equilibration of the Earth's core and upper mantle. *Geochim. Cosmochim. Acta, 48,* 1183–1188.

Cameron A. G. W. and Canup R. M. (1998) The giant impact occurred during Earth accretion. In *Lunar and Planetary Science XXIX,* Abstract #1062. Lunar and Planetary Institute, Houston (CD-ROM).

Capobianco C. J. (1990) A method for the extraction of thermodynamic properties of alloys which are surprisingly soluble in silicate melts at high temperature. In *Lunar and Planetary Science XXI,* pp. 164–165. Lunar and Planetary Institute, Houston.

Capobianco C. J. and Amelin A. A. (1994) Metal/silicate partitioning of nickel and cobalt: the influence of temperature and oxygen fugacity. *Geochim. Cosmochim. Acta, 58,* 125–140.

Capobianco C. J. and Drake M. J. (1990) Partitioning of ruthenium, rhodium, and palladium between spinel and silicate melt and implications for platinum-group element fractionation trends. *Geochim. Cosmochim. Acta, 54,* 869–874.

Capobianco C. J. and Hervig R. L. (1996) Solubility of Ru and Pd in silicate melts: the effect of melt composition. In *Lunar and Planetary Science XXVII,* pp. 197–198. Lunar and Planetary Institute, Houston.

Capobianco C. J., Drake M. J., and Rogers P. S. Z. (1991) Crystal/melt partitioning of Ru, Rh, and Pd for silicate and oxide basaltic liquidus phases. In *Lunar and Planetary Science XXII,* pp. 179–180. Lunar and Planetary Institute, Houston.

Capobianco C. J., Hervig R. L., and Amelin A. A. (1992) Effect of oxygen fugacity and melt composition on PGE silicate melt solubilities. *Eos Trans. AGU, 73,* 344.

Capobianco C. J., Hervig R. L., and Drake M. J. (1994) Experiments on crystal/liquid partitioning of Ru, Rh, and Pd for magnetite and hematite solutions crystallized from silicate melt. *Chem. Geol., 113,* 23–43.

Capobianco C. J., Drake M. J., and de'Aro J. (1999) Siderophile geochemistry of Ga, Ge and Sn: cationic oxidation states in silicate melts and the effect of composition in iron-nickel alloys. *Geochim. Cosmochim. Acta,* in press.

Carlson R. W. and Lugmair G. W. (2000) Timescales of planetesimal formation and differentiation based on extinct and extant radioisotopes. In *Origin of the Earth and Moon* (R. M. Canup and K. Righter, eds.), this volume. Univ. of Arizona, Tucson.

Colson R. O., Haskin L. A., and Keedy C. R. (1991) Reinterpretation of reducing potential measurements done by linear sweep voltammetry in silicate melts. *Geochim. Cosmochim. Acta, 55,* 2831–2838.

Dingwell D. B., O'Neill H. St. C., Ertel W., and Spettel B. (1994) The solubility and oxidation state of Ni in silicate melt at low oxygen fugacity: results using a mechanically assisted equilibrium technique. *Geochim. Cosmochim. Acta, 58,* 1967–1974.

Drake M. J., Newsom H. E., and Capobianco C. (1989) V, Cr, and Mn in the Earth, Moon, EPB, and SPB and the origin of the Moon: Experimental studies. *Geochim. Cosmochim. Acta, 53,* 2101–2111.

Ehlers K., Grove T. L., Sisson T. W., Recca S. I., and Zervas D. A. (1992) The effect of oxygen fugacity on the partitioning of nickel and cobalt between olivine, silicate melt, and metal. *Geochim. Cosmochim. Acta, 56,* 3733–3743.

Ertel W. (1996) Bestimmung des loeslichkeitsverhaltens und der metall-silikat-vertei-lungskoeffizienten der siderophilen elemente (Ni, W, Ir, Pt, Rh und Re) in einer haplobasaltischen schmelze bei hohen temperaturen. Ph.D. dissertation, Universitaet Bayreuth.

Ertel W., O'Neill H. St. C., Dingwell D. B., and Spettel B. (1996) Solubility of tungsten in a haplobasaltic melt as a function of temperature and oxygen fugacity. *Geochim. Cosmochim. Acta, 60,* 1171–1180.

Ertel W., Dingwell D. B., and O'Neill H. St. C. (1997) Composition dependence of the activity of Ni in silicate melts. *Geochim. Cosmochim. Acta, 61,* 4707–4721.

Ertel W., O'Neill H. St. C., Sylvester P. J., and Dingwell D. B. (1999) Solubilities of Pt and Rh in a haplobasaltic silicate melt at 1300°C. *Geochim. Cosmochim. Acta,* in press.

Fleet M. E. and Stone W. E. (1991) Partitioning of platinum-group elements in the Fe-Ni-S system and their fractionation in nature. *Geochim. Cosmochim. Acta, 55,* 245–253.

Fleet M. E., Stone W. E., and Crocket J. H. (1991a) Partitioning of palladium, iridium, and platinum between sulfide liquid and basalt melt: effects of melt composition, concentration, and oxygen fugacity. *Geochim. Cosmochim. Acta, 55,* 2545–2554.

Fleet M. E., Tronnes R. G., and Stone W. E. (1991b) Partitioning of platinum group elements in the Fe-O-S system to 11 GPa

and their fractionation in the mantle and meteorites. *J. Geophys. Res., 96,* 21949–21958.

Gaetani G. and Grove T. L. (1997) Partitioning of moderately siderophile elements among olivine, silicate melt, and sulfide melt: constraints on core formation in the Earth and Mars. *Geochim. Cosmochim. Acta, 61,* 1829–1846.

Gessmann C. K. and Rubie D. C. (1998) The effect of temperature on the partitioning of nickel, cobalt, manganese, chromium, and vanadium at 9 GPa and constraints on formation of Earth's core. *Geochim. Cosmochim. Acta, 62,* 867–882.

Halliday A. N., Rehkämper M., Lee D.-C., and Yi W. (1996) Early evolution of the Earth and Moon: new constraints from Hf-W isotope geochemistry. *Earth Planet. Sci. Lett., 142,* 75–89.

Heiken G., Vaniman D., and French B. M., eds. (1991) *Lunar Sourcebook: A User's Guide to the Moon.* Cambridge, New York. 736 pp.

Hillgren V. J., Drake M. J., and Rubie D. C. (1994) High-pressure and high-temperature experiments on core-mantle segregation in the accreting earth. *Science, 264,* 1442–1445.

Hillgren V. J., Drake M. J., and Rubie D. C. (1996) High pressure and high temperature metal-silicate partitioning of siderophile elements: the importance of silicate liquid composition. *Geochim. Cosmochim. Acta, 60,* 2257–2263.

Hillgren V. J., Gessmann C. K., and Li J. (2000) An experimental perspective on the light element in Earth's core. In *Origin of the Earth and Moon* (R. M. Canup and K. Righter, eds.), this volume. Univ. of Arizona, Tucson.

Holloway J. and Wood B. J. (1988) *Simulating the Earth: Experimental Geochemistry.* Unwin Hyman, London. 196 pp.

Holzheid A. and Palme H. (1996) The influence of FeO on the solubilities of Co and Ni in silicate melts. *Geochim. Cosmochim. Acta, 60,* 1181–1193.

Holzheid A., Borisov A., and Palme H. (1994) The effect of oxygen fugacity and temperature on solubilities of nickel, cobalt, and molybdenum in silicate melts. *Geochim. Cosmochim. Acta, 58,* 1975–1981.

Holzheid A., Rubie D. C., O'Neill H. St. C., and Palme H. (1995) The influence of pressure on the metal-silicate partition coefficients of Ni and Co. *Suppl. Eur. J. Mineral., 7,* 110.

Holzheid A., Borisov A., and Palme H (1997) The activities of NiO, CoO, and FeO in silicate melts. *Chem. Geol., 139,* 21–38.

Holzheid A., Sylvester P., Palme H., Borisov A., and Rubie D. C. (1998) Solubilities of Pt, Ir and Pd in silicate melts at high pressures. In *Lunar and Planetary Science XXIX,* Abstract #1296. Lunar and Planetary Institute, Houston (CD-ROM).

Hood L. L. and Zuber M. T. (2000) Recent refinements in geophysical constraints on lunar origin and evolution. In *Origin of the Earth and Moon* (R. M. Canup and K. Righter, eds.), this volume. Univ. of Arizona, Tucson.

Humayun M. and Cassen P. (2000) Processes determining the volatile abundances of the meteorites and terrestrial planets. In *Origin of the Earth and Moon* (R. M. Canup and K. Righter, eds.), this volume. Univ. of Arizona, Tucson.

Humayun M. and Clayton R. N. (1995) Potassium isotope cosmochemistry: genetic implications of volatile element depletion. *Geochim. Cosmochim. Acta, 59,* 2121–2148.

Humayun M., Brandon A. D., Dick H. J. B., and Shirey S. B. (1997) Ir/Os constraints on terrestrial accretion and core formation. In *Lunar and Planetary Science XXVII,* pp. 613–614. Lunar and Planetary Institute, Houston.

Ito E. and Takahashi E. (1987) Ultra high-pressure phase transformations and the constitution of the deep mantle. In *High Pressure Research in Mineral Physics* (M. H. Manghnani and

Y. Syono, eds.), pp. 221–229. American Geophysical Union, Washington, DC.

Ito E., Katsura T., and Suzuki T. (1998) Metal/silicate partitioning of Mn, Co, and Ni at high-pressures and high-temperatures and implications for core formation in a deep magma ocean. In *Properties of Earth and Planetary Materials at High Pressure and Temperature* (M. H. Manghnani and T. Yagi, eds.), pp. 215–225. American Geophysical Union, Washington, DC.

Jana D. and Walker D. (1997a) The influence of silicate melt composition on distribution of siderophile elements among metal and silicate liquids. *Earth Planet. Sci. Lett., 150,* 463–472.

Jana D. and Walker D. (1997b) The impact of carbon on element distribution during core formation. *Geochim. Cosmochim. Acta, 61,* 2759–2763.

Jana D. and Walker D. (1997c) The influence of sulfur on partitioning of siderophile elements. *Geochim. Cosmochim. Acta, 61,* 5255–5277.

Jephcoat A. and Olson P. (1987) Is the inner core of the Earth pure iron? *Nature, 325,* 332–335.

Jones J. H. and Drake M. J. (1986) Geochemical constraints on core formation in the Earth. *Nature, 322,* 221–228.

Jones J. H. and Hood L. L. (1990) Does the Moon have the same chemical composition as the Earth's upper mantle? In *Origin of the Earth* (H. E. Newsom and J. H. Jones, eds.), pp. 85–98. Oxford Univ., New York.

Kargel J. S. and Lewis J. S. (1993) The composition and early evolution of the Earth. *Icarus, 105,* 1–25.

Kato T., Ringwood A. E., and Irifune T. (1988) Experimental determination of element partitioning between silicate perovskites, garnets and liquids: constraints on early differentiation of the mantle. *Earth Planet. Sci. Lett., 89,* 123–145.

Keppler H. and Rubie D. C. (1993) Pressure-induced coordination changes of transition-metal ions in silicate melts. *Nature, 364,* 54–56.

Kimura K., Lewis R. S., and Anders E. (1974) Distribution of gold and rhenium between nickel-iron and silicate melts: implications for the abundance of siderophile elements on the Earth and Moon. *Geochim. Cosmochim. Acta, 38,* 683–701.

Konopliv A. S., Binder A. B., Hood L. L., Kucinskas A. B., Sjogren W. L., and Williams J. G. (1998) Improved gravity-field of the Moon from Lunar Prospector. *Science, 281,* 1476–1480.

Lauer H. V. Jr. and Jones J. H. (1998) Partitioning of Pt and Os between solid and liquid metal in the iron-nickel-sulfur system. In *Lunar and Planetary Science XXIX,* Abstract #1796. Lunar and Planetary Institute, Houston (CD-ROM).

Li J. and Agee C. (1996) Geochemistry of mantle-core differentiation at high pressure. *Nature, 381,* 686–689.

McDonough W. F. and Sun S. S. (1995) The composition of the Earth. *Chem. Geol., 120,* 223–253.

Meisel T., Walker R. J., and Morgan J. W. (1996) The osmium isotopic composition of the Earth's primitive upper mantle. *Nature, 383,* 517–520.

Melosh H. J. (1990) Giant impacts and the thermal state of the early Earth. In *Origin of the Earth* (H. E. Newsom and J. H. Jones, eds.), pp. 69–83. Oxford Univ., New York.

Morgan J. W., Walker R. J., and Brandon A. D. (1999) Siderophile elements in the Earth's upper mantle and the Moon's ancient impact breccias: manifestations of the same late influx? In *Lunar and Planetary Science XXX,* Abstract #1207. Lunar and Planetary Institute, Houston (CD-ROM).

Murthy V. R. (1991) Early differentiation of the Earth and the problem of mantle siderophile elements: a new approach. *Science, 253,* 303–306.

Murthy V. R. and Karato S. (1997) Core formation and chemical equilibrium in the Earth-II: chemical consequences for the mantle and core. *Phys. Earth Planet. Inter., 100,* 81–95.

Newsom H. E. (1990) Accretion and core formation in the Earth: evidence from siderophile elements. In *Origin of the Earth* (H. E. Newsom and J. H. Jones, eds.), pp. 273–288. Oxford Univ., New York.

Newsom H. E. (1995) Composition of the solar system, planets, meteorites, and major terrestrial reservoirs. In *Global Earth Physics: A Handbook of Physical Constants* (T. J. Ahrens, ed.), pp. 159–189. American Geophysical Union, Washington, DC.

Newsom H. E. and Beserra T. B. (1990) Geochemical constraints on the origin of the Moon. In *Lunar and Planetary Science XXI,* pp. 875–876. Lunar and Planetary Institute, Houston.

Newsom H. and Drake M. J. (1982) The metal content of the eucrite parent body: constraints from the partitioning behavior of tungsten. *Geochim. Cosmochim. Acta, 46,* 2483–2489.

Newsom H. and Drake M. J. (1983) Experimental investigations of the partitioning of phosphorous between metal and silicate phases: implications for the Earth, Moon and eucrite parent body. *Geochim. Cosmochim. Acta, 47,* 93–100.

Newsom H. E. and Runcorn S. K. (1991) New constraints on the size of the lunar core and the origin of the Moon. In *Lunar and Planetary Science XXII,* pp. 973–974. Lunar and Planetary Institute, Houston.

Newsom H. E., Sims K. W. W., Noll P. D. Jr., Jaeger W. L, Maehr S. A., and Beserra T. B. (1996) The abundance of W in the bulk silicate Earth: constraints on core formation. *Geochim Cosmochim. Acta, 60,* 1155–1169.

Ohtani E. and Yurimoto H. (1996) Element partitioning between metallic liquid, magnesiowüstite, and silicate liquid at 20 GPa and 2500°C: a secondary ion mass spectrometric study. *Geophys. Res. Lett., 23,* 1993–1996.

Ohtani E., Yurimoto H., and Seto S. (1997) Element partitioning between metallic liquid, silicate liquid, and lower-mantle minerals: implications for core formation of the Earth. *Phys. Earth Planet. Inter., 100,* 97–114.

O'Neill H. St. C. (1991a) The origin of the Moon and the early history of the Earth: a chemical model. Part 1: the Moon. *Geochim. Cosmochim. Acta, 55,* 1135–1157.

O'Neill H. St. C. (1991b) The origin of the Moon and the early history of the Earth: a chemical model. Part 2: the Earth. *Geochim. Cosmochim. Acta, 55,* 1159–1172.

O'Neill H. St. C. (1992) Siderophile elements and the Earth's formation. *Science, 257,* 1282–1284.

O'Neill H. St. C. and Wall V. J. (1987) The olivine-orthopyroxene-spinel oxygen barometer, the nickel precipitation curve, and the oxygen fugacity of the Earth's upper mantle. *J. Petrol., 28,* 1169–1191.

O'Neill H. St. C., Dingwell D. B., Borisov A., Spettel B., and Palme H. (1995) Experimental petrochemistry of some highly siderophile elements at high temperatures, and some implications for core formation and the mantle's early history. *Chem. Geol., 120,* 255–273.

O'Neill H. St. C., Canil D., and Rubie D. C. (1998) Oxide-metal equilibria to 2500°C and 25 GPa: Implications for core formation and the light component in the Earth's core. *J. Geophys. Res., 103,* 12239–12260.

Poirier J-P. (1994) Light elements in the Earth's outer core: a critical review. *Phys. Earth Planet. Inter., 85,* 319–337.

Potts P. J. (1987) *A Handbook of Silicate Rock Analysis.* Blackie and Sons, London. 621 pp.

Rammensee W. and Wänke H. (1977) On the partition coefficient of tungsten between metal and silicate and its bearing on the origin of the Moon. *Proc. Lunar Sci. Conf. 8th,* pp. 399–409.

Reed S. J. B. (1996) *Electron Microprobe Analysis and Scanning Electron Microscopy in Geology.* Cambridge Univ., New York. 201 pp.

Rehkämper M., Halliday A. N., Barfod D., Fitton J. G., and Dawson J. B. (1997) Platinum-group element abundance patterns in different mantle environments. *Science, 278,* 1595–1598.

Righter K. and Drake M. (1996) Core formation in Earth's Moon, Mars and Vesta. *Icarus, 124,* 513–529.

Righter K. and Drake M. (1997) Metal-silicate equilibrium in a homogeneously accreting Earth: new results for Re. *Earth. Planet. Sci. Lett., 146,* 541–553.

Righter K. and Drake M. (1999) Effect of water on metal-silicate partitioning of siderophile elements: a high pressure and temperature terrestrial magma ocean and core formation. *Earth Planet. Sci. Lett.,* in press.

Righter K., Drake M., and Yaxley G. (1997) Prediction of siderophile element metal-silicate partition coefficients to 20 GPa and 2800°C: the effects of pressure, temperature, oxygen fugacity, and silicate and metallic melt compositions. *Phys. Earth Planet. Inter., 100,* 115–134.

Righter K., Walker R. J., and Warren P. H. (2000) Significance of highly siderophile elements and osmium isotopes in the lunar and terrestrial mantles. In *Origin of the Earth and Moon* (R. M. Canup and K. Righter, eds.), this volume. Univ. of Arizona, Tucson.

Ringwood A. E. (1966) Chemical evolution of the terrestrial planets. *Geochim. Cosmochim. Acta, 30,* 41–104.

Ringwood A. E. (1979) *Origin of the Earth and Moon.* Springer, New York. 250 pp.

Ringwood A. E. (1984) The Earth's core: its composition, formation and bearing upon the origin of the Earth. *Proc. Roy. Soc. Lond., A395,* 1–46.

Ringwood A. E. (1990) Earliest history of the Earth-Moon system. In *Origin of the Earth* (H. E. Newsom and J. H. Jones, eds.), pp. 101–134. Oxford Univ., New York.

Ringwood A. E., Kato T., Hibberson W., and Ware N. (1990) High pressure geochemistry of Cr, V, and Mn and implications for the origin of the Moon. *Nature, 347,* 174–176.

Schmitt W., Palme H., and Wänke H. (1989) Experimental determination of metal/silicate partition coefficients for P, Co, Ni, Cu, Ga, Ge, Mo, and W and some implications for the early evolution of the Earth. *Geochim. Cosmochim. Acta, 53,* 173–185.

Seifert S., O'Neill H. S. C., and Brey G. (1988) The partitioning of Fe, Ni and Co between olivine, metal, and basaltic liquid: an experimental and thermodynamic investigation, with application to the composition of the lunar core. *Geochim. Cosmochim. Acta, 52,* 603–616.

Snow J. E. and Schmidt G. (1998) Constraints on Earth accretion deduced from noble metals in the oceanic mantle. *Nature, 391,* 166–169.

Solomatov V. S. and Stevenson D. J. (1993) Suspension in convective layers and style of differentiation of a terrestrial magma ocean. *J. Geophys. Res., 98,* 5375–5390.

Stevenson D. J. (1981) Models of the Earth's core. *Science, 214,* 611–619.

Stevenson D. J. (1990) Fluid dynamics of core formation. In *Origin of the Earth* (H. E. Newsom and J. H. Jones, eds.), pp. 231–250. Oxford Univ., New York.

Sun S-S. (1984) Geochemical characteristics of Archean ultramafic and mafic volcanic rocks: implications for mantle composition and evolution. In *Archaean Geochemistry: The Origin and Evolution of the Archaean Continental Crust* (A. Kröner et al., eds.), pp. 25–46. Springer-Verlag, Berlin.

Sylvester P. J. and Eggins S. M. (1997) Analysis of Re, Au, Pd, Pt and Rh in NIST glass certified reference materials and natural basalt glasses by laser ablation ICP-MS. *Geostandards Newsletter, 21,* 215–229.

Takahashi E. and Irvine T. N. (1981) Stoichiometric control of crystal/liquid single-component partition coefficients. *Geochim. Cosmochim. Acta, 45,* 1181–1185.

Taylor S. R. (1992) *Solar System Evolution: A New Perspective.* Cambridge Univ., New York. 307 pp.

Thibault Y. and Walter M. J. (1995) The influence of pressure and temperature on the metal-silicate partition coefficients of Ni and Co in a model C1 chondrite and implications for metal segregation in a deep magma ocean. *Geochim. Cosmochim. Acta, 59,* 991–1002.

Thompson M. and Walsh N. J. (1989) *Handbook of Inductively Coupled Plasma Spectrometry,* 2nd edition. Chapman and Hall, London. 316 pp.

Tonks W. B. and Melosh H. J. (1990) The physics of crystal settling and suspension in a turbulent magma ocean. In *Origin of the Earth* (H. E. Newsom and J. H. Jones, eds.), pp. 151–174. Oxford Univ., New York.

Tonks W. B. and Melosh H. J. (1993) Magma ocean formation due to giant impacts. *J. Geophys. Res., 98,* 5319–5333.

Tschauner O., Zerr A., Specht S., Rocholl A., Boehler R., and Palme H. (1999) Partitioning of nickel and cobalt between silicate perovskite and metal at pressures up to 80 GPa. *Nature, 398,* 604–607.

Walker D., Norby L., and Jones J. H. (1993) Superheating effects on metal-silicate partitioning of siderophile elements. *Science, 262,* 1858–1861.

Walter M. J. and Thibault Y. (1995) Partitioning of tungsten and molybdenum between metallic liquid and silicate melt. *Science, 270,* 1186–1189.

Wänke H. (1981) Constitution of terrestrial planets. *Philos. Trans. Roy. Soc. Lond., A303,* 287–302.

Wänke H. and Dreibus G. (1988) Chemical composition and accretion history of terrestrial planets. *Philos. Trans. Roy. Soc. Lond., A325,* 545–557.

Wänke H., Dreibus G., and Jagoutz E. (1984) Mantle chemistry and accretion history of the Earth. In *Archaean Geochemistry: The Origin and Evolution of the Archaean Continental Crust* (A. Kröner et al., eds.), pp. 1–24. Springer-Verlag, Berlin.

Wetherill G. W. (1985) Occurrence of giant impacts during the growth of the terrestrial planets. *Science, 228,* 877–879.

Wetherill G. W. (1990) Formation of the Earth. *Annu. Rev. Earth Planet. Sci., 18,* 205–256.

Zerr A., Diegeler A., and Boehler R. (1998) Solidus of Earth's deep mantle. *Science, 281,* 243–246.

Zhang J. and Herzberg C. (1994) Melting experiments on anhydrous peridotite KLB-1 from 5 to 22.5 GPa. *J. Geophys. Res., 99,* 17729–17742.

# Significance of Highly Siderophile Elements and Osmium Isotopes in the Lunar and Terrestrial Mantles

**Kevin Righter**
*University of Arizona*

**Richard J. Walker**
*University of Maryland*

**Paul H. Warren**
*University of California, Los Angeles*

Analytical and experimental advances in the past decade have increased our understanding of the highly siderophile elements (HSE) and the $^{187}$Re-$^{187}$Os and $^{190}$Pt-$^{186}$Os isotopic systems in terrestrial, meteoritic, and planetary materials. Analyses of a broad range of natural materials and experiments on natural analogs have demonstrated that HSE reside primarily in metal alloys, sulfides, and oxides. These elements will be strongly partitioned into a metallic core during planetary differentiation, leaving low concentrations ($\ll$1 μg/g) in silicate mantle. The terrestrial upper mantle contains HSE in ~0.008× chondritic abundances; these high, near-chondritic relative levels are commonly cited in support of accretion of chondritic material to the Earth after the metallic core had formed. Determining the HSE contents of the lunar mantle is a challenge due to the lack of any true mantle samples. Data for the most MgO-rich of the low-Ti basalts suggest that the lunar mantle has lower HSE concentrations than the terrestrial upper mantle, but only by a relatively small factor, ~5. The Ni/Ir ratio is nearly the same in the lunar mantle as it is in its terrestrial counterpart. Osmium-isotopic data indicate that the terrestrial primitive upper mantle has time-integrated Re/Os and Pt/Os ratios within ±6% and ±25% respectively of chondritic values. Some terrestrial mantle reservoirs that feed plume systems contain suprachondritic $^{187}$Os/$^{188}$Os and $^{186}$Os/$^{188}$Os that could reflect primordial mantle heterogeneities that formed in early Earth history, either as trapped residual liquid from a magma ocean or via preserved stratification within the mantle. Alternately, these heterogeneities could reflect early crustal recycling or present-day outer core-lower mantle interaction. Future experimental and analytical efforts must focus on collecting data that can be used to resolve such issues as (1) What was the timing, composition, and areal extent of the late veneer on the Earth and Moon? (2) Have the HSE been fractionated between the outer core, inner core, lower mantle and upper mantle? (3) What are the HSE contents of the lunar mantle?

## 1. INTRODUCTION

The highly siderophile elements (HSE) include the platinum group elements (PGE: Pt, Pd, Rh, Ru, Os, Ir), Au, and Re. Aside from great historical interest in these noble metals and the ore bodies in which they are found, this group of elements holds important information bearing on the origin of the Earth, Moon, terrestrial planets, and chondritic materials. These elements are so named because of their tendency to alloy with Fe metal (siderophile "iron-loving") when present, or at least to have metal-silicate partition coefficients greater than 10,000 (e.g., *Jones and Drake*, 1986). Because of this tendency, most of the chondritic budget of these elements in the terrestrial planets has been concentrated into the metallic core, leaving very low concentrations in the mantle and crust. As a result, the concentrations of these eight elements have been used to assess the role of metallic and chondritic material in various planetary accretion and differentiation processes (e.g., *Kimura et al.,*

1974). A good example is the model for terrestrial upper mantle HSE concentrations, which holds that the chondritic relative proportions of HSE were produced by late addition of a small amount (0.8% of the Earth's mass) of chondritic material, after core formation stripped the mantle of HSE (e.g., *Kimura et al.,* 1974; *Chou et al.,* 1983).

Three recent analytical and experimental developments have heightened interest in these elements. First, because of their very low concentrations in natural materials, there have been few analytical techniques sufficiently sensitive to allow precise and accurate determinations in non-ore materials. The primary technique has been radiochemical neutron activation analysis (RNAA; e.g., UCLA, Chicago, USGS, Mainz groups). In the past decade, however, new techniques such as inductively-coupled plasma mass spectrometry (ICP-MS; e.g., *Halliday et al.,* 1998; *Pattou et al.,* 1996) have become more widely available and thus have significantly increased the number of bulk-rock HSE data published. Second, the development of negative thermal ionization mass spectrometry (NTIMS) techniques (*Creaser et al.,* 1991; *Volkening et al.,* 1991) has permitted the $^{187}$Re-$^{187}$Os isotopic system to mature considerably in the last decade (cf. review in *Shirey and Walker,* 1998). Also, the analytical barriers preventing use of the $^{190}$Pt-$^{186}$Os system have been partly overcome (*Walker et al.,* 1997a; *Brandon et al.,* 1998a). These two isotopic systems offer new perspectives on age-old geo- and cosmochemical problems. And third, high-pressure experimentation has become more widespread, and thus allowed access to P-T space corresponding to the deep upper mantle and top of the lower mantle (e.g., *Mao and Hemley,* 1998). Because of the wealth of new analytical and experimental data, traditional models for terrestrial accretion, mantle differentiation, and crustal recycling have been challenged. The intent of this paper is to review these recent developments, discuss the new kinds of models being considered as a result of the new data, and to emphasize what future efforts and approaches may be most fruitful.

## 2.  PARTITIONING BEHAVIOR AND HOST PHASES FOR HIGHLY SIDEROPHILE ELEMENTS

Although it is well known that the distribution of many of the HSE in planetary crusts, mantles, and cores is determined by sulfide and metal, there is an increasing awareness of important roles for oxides and silicates. Insights provided by both natural and experimental systems are discussed below.

### 2.1.  Constraints from Natural Systems

That metal can dominate the distribution of HSE in natural systems is demonstrated by HSE concentrations in certain iron meteorites. For example, there are µg/g levels of the HSE in the least-evolved magmatic iron meteorites (Table 1). Liquid-crystal fractionation can lead to major reductions in the evolved melt concentrations, however, as evidenced by relatively low HSE concentrations in some highly evolved iron meteorites (*Pernicka and Wasson,* 1987; *Morgan et al.,* 1995). Metal will even concentrate HSE over that of sulfide (e.g., *Shen et al.,* 1996), normally the major HSE host phase in terrestrial samples.

There are a wide variety of HSE sulfides and arsenides, such as laurite — $(Ru,Os)S_2$, bowieite — $(Rh, Ir)_2S_3$, vysotskite — $(Pd,Ni)S$, braggite — $(Pt,Pd,Ni)S$, sperrylite — $PtAs_2$, and Hollingsworthite — $(Rh,Pt,Ru)AsS$ (*Daltry and Wilson,* 1997), and $ReS_2$ (*Korzhinsky et al.,* 1994). These are often found in ore bodies, but are much rarer in common rocks and probably not the most common host phases in most of these rocks. In addition, many of these phases are not stable at high temperatures and thus may be secondary phases formed upon cooling from higher magmatic temperatures. More common host phases for HSE can be monosulfide solid solutions (mss), pyrite, chalcopy-

rite, bornite, and molybdenite (e.g., *Freydier et al.,* 1997; *McCandless and Ruiz,* 1993; *Stein et al.,* 1998a,b). Slightly more common are PGE alloys and intermetallic compounds such as rutheniridosmine — OsIrRu, and $Ir_3Fe$, $Pt_2FeNi$, $Pt_3Fe$, and $Pt_2FeCu$, which have been found in ophiolites and layered intrusions (*Hattori and Hart,* 1991, *Hart and Kinloch,* 1989; *Daltry and Wilson,* 1997). Laurite can also be a dominant phase for some HSE in the ultramafic portions of ophiolites (*Walker et al.,* 1996).

In sulfide-absent systems, HSE can be concentrated into other phases such as oxides: chromite, magnetite, ferrite, spinel, and hematite-ilmenite (ss). For instance, chromite contained within the Bushveld and Stillwater layered mafic intrusions have Ir, Os, and Ru concentrations that range to higher than 100, 100, and 600 ng/g respectively (*Gijbels et al.,* 1974; *Lambert et al.,* 1989; *Marcantonio et al.,* 1993). Relatively high concentrations of some HSE have also been reported in chromites and spinels from the ultramafic portions of ophiolites and various other ultramafic systems including komatiites (*Agiorgitis and Wolf,* 1978; *Parry,* 1984; *Page,* 1984; *Razin and Khomenko,* 1969; *Walker et al.,* 1997b, 1999a). Chromite from the Archean Fiskenaesset anorthosite complex (West Greenland) is also enriched in Re and Ir (*Morgan et al.,* 1976). The compatibility of Re in magnetite was demonstrated in a basalt to rhyolite differentiation suite of samples from the Galapagos (*Righter et al.,* 1998). This compatibility is also supported by measurements of Re in a series of magnetite separates from Columbia River basalts (*Chesley and Ruiz,* 1998). In comparison, the Suwalki anorthosite contains both sulfide and titanomagnetite, and the Re and Os is hosted mainly by the former (*Stein et al.,* 1998c), reinforcing the idea that sulfide will dominate the HSE distribution when present. On the other hand, there may be important crystal chemical controls on the affinity of various HSE in oxides (see next section), but a systematic study has not yet been undertaken. It should also be noted that some of the HSE in oxides may reside in HSE alloys or sulfides, such as laurite (*Stockman and Hlava,* 1984; *Auge,* 1988; *Prichard and Tarkian,* 1988); such inclusions are mainly in the rims of oxides, but can also occur interstitial to the oxides in association with silicates. But *Peach and Mathez* (1996) emphasize that Ir must be contained in chromite because fractionation of Ir from Pd cannot be explained in scenarios where Ir is hosted solely by an Os-Ir alloy.

The role of silicates in the distribution of HSE is largely unknown, but there are important silicate phases involved in fractionating these elements. Olivine may be an important host for Os, Ir, and Ru based on the analysis of mineral separates from a fertile peridotite (*Hart and Ravizza,* 1996) and whole-rock komatiite trends (*Brügmann et al.,* 1987). Mineral separate analyses from komatiites and picrites, however, have revealed that Re and Os are, at least in some instances, incompatible in olivine (*Walker et al.,* 1997b, 1999a). This suggests that Os, Ru, and Ir may be compatible in Cr-rich spinel inclusions in olivine, rather than olivine, in the komatiite suites studied by *Brügmann et al.*

TABLE 1.  Concentrations of HSE measured in mineral concentrates (in ng/g, except * = µg/g).

| Mineral | Re | Au | Pd | Pt | Rh | Ru | Os | Ir | Refs. |
|---|---|---|---|---|---|---|---|---|---|
| *Metal and Phosphide* | | | | | | | | | |
| Iron meteorites | 0.034–5.0* | 0.25–7.0* | 1.5–20* | 0.5–30* | 0.14–2.5* | 0.16–36* | 0.14–70* | 0.01–50* | [3,5,6,7,18,22,27,28,29,30] |
| Schreibersite | 1.89–29.92 | — | — | — | — | — | 5.65–76.8 | — | [28] |
| *Sulfides* | | | | | | | | | |
| Molybdenite | 2.39–4200* | — | — | — | — | — | 25.2–3536 | — | [12,14,31,32] |
| Chalcopyrite | 0.085–0.116 | 0.72–15 | 5–41 | 0.46–5.6 | 0.6–0.7 | — | 0.082–0.874 | 1.4–5.8 | [8,11] |
| Pyrrhotite | 70–350 | 0.15–2.86 | 3.7–7.3 | 0.4–9.0 | 53–138 | 114–206 | 1–91 | 53–105 | [11,33] |
| Pyrite | 0.09–99.99 | — | — | — | — | — | 0.016–1.98 | — | [8,31] |
| Bornite | 0.06–0.18 | 0.09 | — | 1.24 | — | — | 0.042–0.087 | — | [8,11] |
| MORB sulfide | 2.4–468 | 10.5–13.0* | 27.8–32.7* | 29.3* | — | 140 | 0.25–996 | 520–706 | [20,26] |
| MSS — peridotite | — | 0.46–2.38* | — | — | — | 5–32* | 3.5–9.5* | 3.2–9.22* | [10,17] |
| MSS in diamond (eclogitic) | 52–370 | — | — | — | — | — | 4.7–122 | — | [2,21] |
| MSS in diamond (peridotitic) | 0.3–2.55* | — | <12–50* | <55* | <7–169* | <7–1313* | 5.968–6.097* | <73* | [2,21] |
| Troilite | 0.51,1.07 | — | — | — | — | — | 1.26,1.77 | — | [28] |
| *Oxides* | | | | | | | | | |
| Chromite | 0.28–1.95 | 0.5–5.0 | — | — | — | — | 10–100 | 3–96 | [13,19] |
| Stillwater | 0.1–7.44 | — | 50–500 | 50–500 | — | 370–680 | 19–310 | 51.5–114.4 | [9,15] |
| Bushveld | — | — | — | — | — | 6.3–715 | 1.5–1065 | 2.7–555 | [1] |
| Greek ultramafic | — | — | — | — | — | 80–200 | — | 200–900 | [23] |
| Siberia/Ural | — | — | — | — | — | — | — | — | |
| Magnetite | 1.3–125 | — | — | — | — | — | 0.5–169 | — | [4,24,34] |
| Spinel | 3.65–7.43 | 0.4–1.2 | 0.5–8.6 | — | — | — | 0.036, 9.8–37.2 | 1.9–3.4 | [16,35] |
| *Silicates* | | | | | | | | | |
| Orthopyroxene | — | — | — | — | — | 2.6–235 | 0.055–22.6 | 0.4–33.7 | [9,10] |
| Plagioclase | — | — | — | — | — | 1.7–58.9 | 0.39–9.0 | 0.55–10.1 | [9] |
| Garnet | 12.6 | 0.03–0.1 | 0.3–0.7 | — | — | — | — | 0.03–0.7 | [16,25] |
| Olivine | 0.120–0.442 | 0.03–0.3 | 0.07–1.7 | — | — | — | 0.036, 0.184–0.212 | 0.07–0.9 | [10,16,35] |

References:   [1] *Agiorgitis and Wolf (1978)*; [2] *Bulanova et al. (1996)*; [3] *Chen and Wasserburg (1996)*; [4] *Chesley and Ruiz (1998)*; [5] *Choi et al. (1995)*; [6] *Cobb (1967)*; [7] *Crocket (1972)*; [8] *Freydier et al. (1997)*; [9] *Gijbels et al. (1974)*; [10] *Hart and Ravizza (1996)*; [11] *Li et al. (1993)*; [12] *Markey et al. (1998)*; [13] *Marcantono et al. (1993)*; [14] *McCandless et al. (1993)*; [15] *McCandless et al. (1999)*; [16] *Mitchell and Keays (1981)*; [17] *Morgan and Baedecker (1983)*; [18] *Nichiporuk and Brown (1965)*; [19] *Parry (1984)*; [20] *Peach et al. (1990)*; [21] *Pearson et al. (1998)*; [22] *Pernicka and Wasson (1987)*; [23] *Razin and Khomenko (1969)*; [24] *Righter et al. (1998)*; [25] *Roy-Barman et al. (1996)*; [26] *Roy-Barman et al. (1998)*; [27] *Scott and Wasson (1976)*; [28] *Shen et al. (1996)*; [29] *Smales et al. (1967)*; [30] *Smoliar et al. (1996)*; [31] *Stein et al. (1998a)*; [32] *Stein et al. (1998b)*; [33] *Walker et al. (1991)*; [34] *Walker et al. (1997a)*; [35] *Walker et al. (1999a)*.

(1987). Garnet-rich separates from a garnet clinopyroxenite have elevated Re concentrations relative to the whole rock, suggesting garnet is a host phase for Re (*Roy-Barman et al.*, 1996), but garnet-rich separates from garnet lherzolites do not have unusually high concentrations of Au, Pd, or Ir (*Mitchell and Keays*, 1981). Such measurements have not been made for the other HSE, so the question of whether garnet is a host phase for Pt, Ru, Rh, or Os remains open. Very low concentrations of Re and Os have been reported in glimmerites or biotite-bearing xenoliths (*Handler et al.*, 1997), indicating that biotite may not be a significant host phase for these HSE. *Peucker-Ehrenbrink and Blum* (1998) also report very low (<60 pg/g) Re and Os concentrations in biotite from a gneiss. Very little is known about the compatibility of Pt or Rh in silicates due to the paucity of data for these two elements relative to Re, Au, Pd, Ir, Os, and Ru. Although feldspars and amphiboles are not likely to be significant host phases for HSE, there are too few data for any of the HSE to make a quantitative evaluation.

Fluids can be important agents for redistributing HSE in a variety of environments. $H_2S$-rich volcanic gases can precipitate Mo or Re sulfides at elevated temperatures (500°–600°C), indicating that volcanic gases can contain significant amounts of Re (*Bernard et al.*, 1990). The similarity of Au differentiation trends to those of Cu in arc magmas (*Stanton*, 1993) have led some to conclude that Au distribution (like Cu) can be controlled by a vapor phase in systems that are vapor saturated, such as volatile-laden arc magmas. Elevated Au, Ir, Os, and Re concentrations have also been measured in volcanic fumes at Mauna Loa and Kilauea (*Finnegan et al.*, 1990). *Wood* (1987) showed that although Os and Ru oxides have small volatilities (±1 pg/g) at typical magmatic $f_{O_2}$, Pd and Ru have comparatively large volatilities (500 pg/g and 20 pg/g respectively) as chloride species. His thermodynamic calculations show that the conditions that enhance HSE volatilities are high temperatures, high $f_{O_2}$ and $f_{HCl}$, and low $f_{H_2O}$ (*Wood*, 1987). These results are of particular interest with respect to the Re-Os isotopic system; it has been argued that Os may be mobile in a fluid phase in the subarc mantle (*Brandon et al.*, 1996).

## 2.2.  Experimental Studies

Because core-mantle segregation is potentially the most important control of HSE distribution in terrestrial planets, study of the partitioning of HSE between metal and silicate liquid has received much attention (see *Walter et al.*, 2000). These experimental studies have demonstrated that bulk distribution coefficients (D = weight ratio of an element between two phases) for the HSE are dependent upon $f_{O_2}$, temperature, pressure, melt composition, and metallic liquid composition. Despite this recognition, there remain many open questions about HSE metal-silicate partitioning, and many of these parameters have not been quantified for every element. For instance, there remains uncertainty about the valences of some of the HSE at high temperature and

pressure; some experimental evidence suggests that many of these elements are stable at lower than expected valences such as 1+ or 2+ (*Borisov and Palme*, 1996, 1997). There have been many experiments completed on S-bearing systems (for good reason because of the common occurrence of S-bearing PGE ore bodies), but the effects of other alloying light elements, such as O, P, and C, are virtually unknown, yet relevant to planetary core formation.

Sulfide liquids are common HSE hosts in magmatic ore deposits (e.g., *Naldrett et al.*, 1979; *Li et al.*, 1993; *Walker et al.*, 1991) and thus sulfide-silicate melt partitioning has been studied experimentally by several different groups (*Fleet et al.*, 1991a; *Peach et al.*, 1994, *Stone et al.*, 1990, *Crocket et al.*, 1997). After numerous studies involving multiple elements, it has become clear that sulfide liquid/ silicate liquid D is very large for most of the HSE. The range of $10^3$–$10^5$ can be attributed to differences in experimental conditions (e.g., $f_{O_2}$ and temperature) as well as analytical techniques, but within a given set of experiments the values of the partition coefficients are very similar. Despite this first-order observation, there are several lingering questions regarding these experiments. First, why is the D for such a geochemically diverse set of elements similar for elements that are commonly fractionated in natural samples? This may be due to the contrasting behavior of the HSE in S-bearing systems, as compared to S-free. That is, metal/sulfide D is very large (100–1000) for those elements (Ir, Os, Ru) that have large metal/silicate D, whereas metal/sulfide D is small (<100) for those elements (Au, Pd, Rh) that have smaller metal/silicate D (*Fleet et al.*, 1991b; *Fleet and Stone*, 1991; *Lauer and Jones*, 1998). As a result, the effect of sulfur will be to equalize the D. Second, why is there sometimes scatter in the values of D, despite constant experimental conditions? This may be due to the presence of micronuggets of HSE-rich alloys in the silicate melts, interfering with analysis of the silicate phase (see discussion in *Walter et al.*, 2000).

That oxides will be important phases in controlling the distribution of HSE in rocks is borne out by the observations in natural systems described above. Thermochemical properties of several HSE oxides, $RuO_2$, $IrO_2$, PdO, and $ReO_2$ (*O'Neill and Nell*, 1997; *Nell and O'Neill*, 1997a), have been the focus of experimental studies; a complete understanding of simple HSE oxide systems is critical to our understanding of their distribution in nature. *Capobianco* (1993) and *Nell and O'Neill* (1997b) demonstrated the stability of $MgRh_2O_4$ spinel at oxidized conditions. Certain Pt-bearing oxides, $NiPt_2O_4$ and $ZnPt_2O_4$ (*Muller and Roy*, 1965), and Ru-bearing oxides (*Dulac*, 1969; *Krutzsch and Kemmler-Sack*, 1983) are also stable at high O pressures. *Capobianco and Drake* (1990) and *Capobianco et al.* (1994) measured very large (D = 100–800) partition coefficients for Ru and Rh between titanomagnetite-melt and ilmenite/hematite-melt, and found evidence for the stability of $Rh^{3+}$. In addition, *Capobianco* (1998) found that considerable amounts of Ru dissolve in rhombohedral (ilmenite/ hematite ss) oxides at 1 bar and 1200°–1400°C; at mantle

TABLE 2.  Experimental partition coefficients for HSE.

| System | Re | Au | Pd | Pt | Rh | Ru | Os | Ir | Refs. |
|---|---|---|---|---|---|---|---|---|---|
| *Metal/silicate*\* | | | | | | | | | |
| *Sulfides* | | | | | | | | | |
| Sulfide liquid/silicate liquid (×10³) | — | 1–3 | 1.8–120 | 0.9–12 | — | 4.4–46 | 3.7–430 | 3.2–200 | [4,5,6,10, 12,15] |
| Sulfide liquid/FeS  (Fe–O–S) | — | — | 22–60 | 26–50 | 3–20 | 0.8–4 | 1–2 | 15 | [7,8] |
| Sulfide liquid/FeS  (Fe–Ni–Cu–S) | — | 11.4 | 2.3–7.5 | 3.9–5.8 | 0.26–0.43 | 0.19–0.31 | 0.19–0.27 | 0.17–0.47 | [9] |
| *Oxide/silicate liquid* | | | | | | | | | |
| Magnetite | — | — | 0.4–1.2 | — | 110–430 | 110–8000 | — | — | [3] |
| Spinel | — | — | <0.02 | 3 | 78–90 | 25–28 | — | 25 | [1,2,11] |
| Chromite | — | — | — | — | — | — | — | — | |
| Hematite/ilmenite (ss) | — | — | 0.7 | — | 310 | 380 | — | — | [3] |
| *Silicate/silicate liquid* | | | | | | | | | |
| Olivine | <0.01 | <1 | <0.2–0.4 | 0.22 | 1.8–12 | 0.2–0.8 | >1 | >7 | [2,11,13] |
| Diopside | 0.035 | — | <0.3 | 1.4–2.0 | 3 | 1.9 | 0.075 | — | [2,11,13,16] |
| Orthopyroxene | 0.18 | — | — | — | — | — | — | — | [14] |
| Garnet | 2–5 | — | — | — | — | — | — | — | [14] |
| Anorthite | — | — | <0.2 | — | <0.4 | <0.3 | — | — | [2] |

\* See review of this topic by *Walter et al.* (2000).

References:  [1] *Capobianco and Drake* (1990); [2] *Capobianco et al.* (1991); [3] *Capobianco et al.* (1994); [4] *Crocket et al.* (1992); [5] *Crocket et al.* (1997); [6] *Fleet et al.* (1991a); [7] *Fleet and Stone* (1991); [8] *Fleet et al.* (1991b); [9] *Fleet et al.* (1993); [10] *Fleet et al.* (1996); [11] *Malvin et al.* (1986); [12] *Peach et al.* (1994); [13] *Righter et al.* (1995); [14] *Righter and Hauri* (1998); [15] *Stone et al.* (1990); [16] *Watson et al.* (1987).

$f_{O_2}$, Ru may be present in 1–10 μg/g in natural ilmenites and thus easily detectable by a number of the newer microanalytical techniques. The ability of ilmenite to host HSE such as Ru is also relevant to models for lunar magma genesis, because ilmenite is thought to be a component of some parts of the lunar mantle (e.g., *Snyder et al.*, 2000).

Experimental studies of HSE partitioning in silicates are few compared to other siderophile elements such as Ni (e.g., *Beattie et al.*, 1991; *Jones*, 1995). Despite the identification of olivine as a potential host for Os, there have been no experimental measurements of a partition coefficient for Os between olivine and silicate melt. It should be noted that the report of Os, Ir, and Pt compatibility in olivine reported by *Malvin et al.* (1986) is perhaps not valid due to the presence of many interfering metallic particles in crystals from their runs (C. J. Capobianco, personal communication, 1999). It has been determined at high $f_{O_2}$ (NNO buffer and higher), however, that Rh is compatible in forsterite, Ru is mildly incompatible (*Capobianco et al.*, 1991), and Pd and Re are incompatible (*Capobianco et al.*, 1991; *Righter et al.*, 1995) (Table 2). In fact, *Capobianco et al.* (1991) found Rh and Ru partitioning to be temperature dependent, much the same as Ni and Co (e.g., *Leeman and Lindstrom*, 1978). Such a temperature dependence may account for the discrepancy among reported olivine-liquid D reported by *Hart and Ravizza* (1996) and *Walker et al.* (1999a) — the lower values in komatiitic systems may simply reflect a temperature dependence that favors higher values in basaltic systems. This should be one focus of future work.

There is a dearth of data for HSE partitioning between pyroxene and silicate melt. *Righter and Hauri* (1998) measured D(Re) orthopyroxene/melt = 0.18 in a basaltic system, at the FMQ buffer. Clinopyroxene has been examined for Re and Os partitioning by *Watson et al.* (1987) and *Righter et al.* (1995) and both are incompatible (D ~0.05) at oxidized conditions. *Capobianco et al.* (1991) measured mild incompatibility of Ru, Rh, and Pd between diopside and $CaO$-$MgO$-$Al_2O_3$-$SiO_2$ melt at high $f_{O_2}$. There have been no other experimental studies of pyroxene/melt for the other HSE. Rhenium is compatible in garnet with a D ~3, and this has been important to models of crustal recycling (high Re/Os oceanic crust will be transformed into eclogite), mantle melting (OIB and MORB generation), and understanding the long-term evolution of Re/Os in the mantle. Although garnet is clearly an important phase for Re, almost nothing is known of the behavior of all other HSE in garnet-bearing systems. D (Ru, Rh, and Pd) anorthite /liquid has been determined by *Capobianco et al.* (1991); the role of feldspar is unknown for the other HSE, but insight from natural systems suggests that they are incompatible (*Gijbels et al.*, 1974).

Finally, the potential of various deep mantle silicate phases (majorite-rich garnet, perovskite, aluminous pyroxene) as hosts for HSE remains an open question; such information will be important in evaluating models of upper and lower mantle evolution and exchange, especially in light of Re-Pt-Os isotopic systematics.

## 3.  ABUNDANCES OF HIGHLY SIDEROPHILE ELEMENTS IN THE TERRESTRIAL MANTLE

Because we have no samples of the lunar mantle, a solid understanding of the controls on HSE distribution in terrestrial mantle samples is critical to any understanding or interpretation of lunar samples. These terrestrial suites will be used as a baseline for comparison to the lunar data. The abundances of HSE in the terrestrial mantle have been heavily discussed in the past literature (e.g., *Morgan*, 1986) and are currently under renewed scrutiny. Here we review only a few of the basic facts and will rely on the published record for further details.

The most common samples of Earth's mantle are peridotite xenoliths (xeno = foreign; lith = rock; chunks of rock found in volcanic deposits formed during eruptions) and peridotite-bearing ultramafic massifs. Careful mineralogical, petrological, and geochemical studies have identified certain xenolith and massif peridotites as primitive, meaning that they represent material that is likely to be most similar to the original mantle composition at ~4.4 Ga. Such compositional similarities are illustrated by the elements

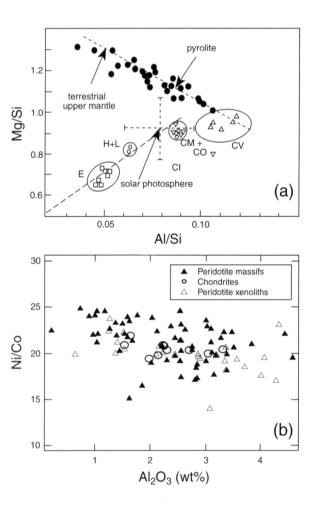

**Fig. 1.** Primitive upper mantle **(a)** Mg/Si vs. Al/Si and **(b)** Ni/Co vs. $Al_2O_3$ (after a figure from *Jagoutz et al.*, 1979).

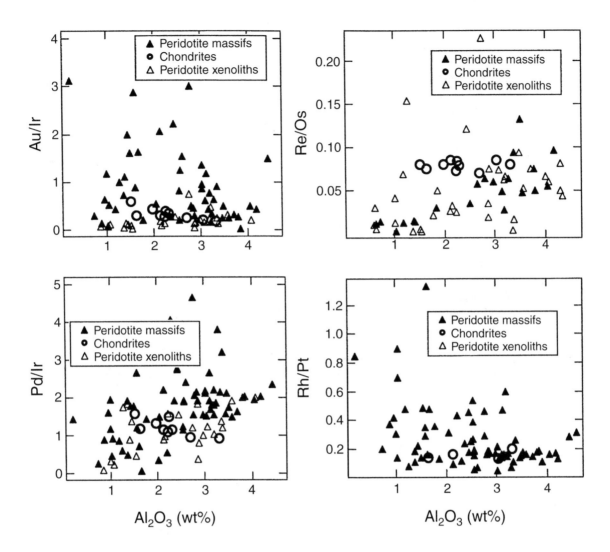

**Fig. 2.** Terrestrial upper mantle HSE (Re/Os, Au/Pt, Pd/Ir, Rh/Pt). Data for massif peridotites from *Gueddari et al.* (1996), *Pattou et al.* (1996) and *Garuti et al.* (1997); xenolith data from *Morgan* (1986), *Meisel et al.* (1996), *Brueckner et al.* (1995), *Rehkämper et al.* (1997); and chondrite data is from compilation of *Newsom* (1995).

Mg, Al, and Si; the chondritic meteorite trend is approached by the mantle peridotite trend, with the most fertile (primitive) peridotites lying close to the intersection (Fig. 1a). These same peridotite samples also have chondritic ratios of many refractory lithophile elements, and globally similar abundances of moderately (e.g., Ni/Co; Fig. 1b) and highly siderophile elements. In particular, *Morgan et al.* (1981) and *Morgan* (1986) determined that primitive upper mantle samples have near-chondritic relative HSE abundances, ~0.008× CI (i.e., 0.008× concentration by weight in CI chondrites). Sometimes this depletion is also normalized to a refractory element and is presented as a value of 0.003× CI. We refer the reader to the chapter by *Walter et al.* (2000) or *Newsom* (1990) for further details of how terrestrial and lunar mantle siderophile-element depletions are calculated. Although more recent studies of primitive peridotite suites have identified some deviations in the near-chondritic ratios of HSE (e.g., *Pattou et al.*, 1996), there is

clearly a convergence of many HSE ratios (e.g., Pd/Pt, Au/Pt, Rh/Pt) to chondritic values (Fig. 2). Several suites are characterized by Pd/Ir and Ru/Ir ratios demonstrably higher than chondritic; whether this is a primary feature of the primitive mantle or due to a later secondary process is still debated and uncertain (e.g., *Rehkämper et al.*, 1997; *Snow and Schmitt*, 1998).

The concentrations of HSE in terrestrial basaltic and ultramafic rocks has also been reviewed extensively by *Barnes et al.* (1985), *Crocket* (1979), and *Mathez and Peach* (1990) and only general features will be discussed here. New data since these reviews have added to our understanding of HSE in basaltic rocks, especially Re and Os in OIB and MORB from Os-isotopic studies. However, there is still much to learn about other HSE in these kinds of rocks. The concentrations of HSE in a mantle or crustal melt will be controlled largely by sulfide (MSS), chromite (or other cubic oxides), and silicate phases exhibiting compatibility such as

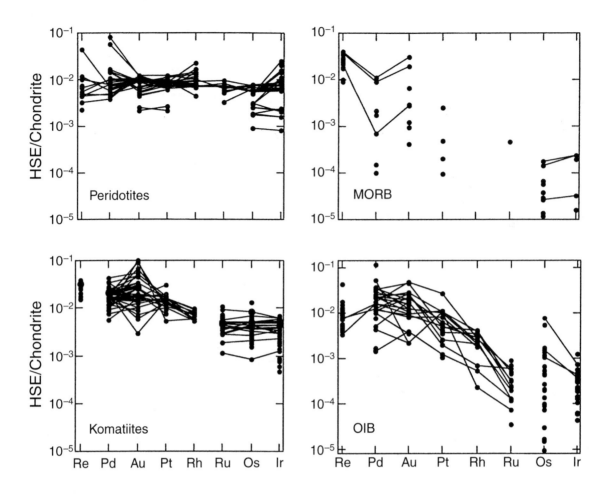

**Fig. 3.** HSE concentrations of terrestrial peridotite, komatiite, and ocean island and mid-ocean ridge basalt, all normalized to HSE concentrations of CI chondrites (from compilation of *Newsom*, 1995). Peridotite data are from *Morgan et al.* (1981), *Morgan* (1986), *Meisel et al.* (1996), *Pattou et al.* (1996), *Rehkämper et al.* (1997), *Garuti et al.* (1997), *Gueddari et al.* (1996), and *Mitchell and Keays* (1981). Komatiite data are from *Brügmann et al.* (1987), *Barnes and Picard* (1993), *Foster et al.* (1996), *Rehkämper et al.* (1999), *Walker et al.* (1999a), and *Crocket and MacRae* (1986). OIB data are from *Fryer and Greenough* (1992), *Rehkämper et al.* (1999), *Widom and Shirey* (1996), *Pegram and Allègre* (1992), *Hauri and Hart* (1993), *Reisberg et al.* (1993), *Martin et al.* (1994). MORB data are from *Hertogen et al.* (1980), *Rehkämper et al.* (1999), *Schiano et al.* (1997), *Ravizza and Pyle* (1997), *Roy-Barman and Allègre* (1994), and *Roy-Barman et al.* (1998).

olivine, orthopyroxene, or garnet. In general, large-degree partial melts (>20–25%) are sulfide-undersaturated and thus the mantle budget of HSEs has been released into the melt (no residual sulfide left in the source). Small-degree partial melts, such as MORB and OIB, are sulfide-saturated and thus the mantle budget of the HSE still resides in the residual sulfide (e.g., *Mathez and Peach*, 1990). These "primary" fractionations due to sulfide can be overprinted with "secondary" fractionation due to chromite (spinel) or silicates. In general, komatiites (sulfide-undersaturated) exhibit the most regular behavior, showing an overall incompatibility of Au, Re, Pd, and Pt, and compatibility of Rh, Ru, Ir, and Os (Fig. 3). Concentrations measured in five or six different komatiite suites are very similar (Fig. 3). The concentrations in OIB and MORB, however, show a much larger range for each element. The range of concentrations in the elements Rh, Ru, Ir, and Os is most likely due to the

combined effects of silicate, oxide, and sulfide fractionation, whereas the scatter in Au, Re, Pt, and Pd is due mainly to oxide and sulfide. With more analyses of a wider range of basaltic rock types, the variation of certain elements within rock types will be better understood.

## 4. ABUNDANCES OF HIGHLY SIDEROPHILE ELEMENTS IN THE LUNAR MANTLE

### 4.1. Lunar Mare Basalts: Imperfect Analogs of Terrestrial Mafic Igneous Rocks

In this section, we estimate the HSE composition of the lunar mantle for purposes of comparison with Earth's mantle. This is a difficult and controversial task. Peridotite of manifest lunar mantle origin has not been sampled (even as a xenolith), so estimates must be derived entirely from data for crustal rocks. Most of the Moon's crust consists of ancient anorthositic cumulates that are of limited value in this connection. Accessible samples of this ancient material tend to be polymict breccias, with HSE compositions contaminated by meteoritic impact debris. Despite some similarities in Re and Ir concentrations in lunar and terrestrial anorthosites (*Morgan et al.,* 1976), the general behavior of the HSE in anorthositic cumulates is poorly understood.

Fortunately, the Apollo lunar collection includes many samples of the distinctive type of mafic extrusive (distinguishable from other lunar rocks by high FeO, $TiO_2$, and Ca/Al, and typically also young age) termed mare basalt. A compositionally very similar (although usually more MgO-rich) set of mare pyroclastic materials, found as spherules concentrated in portions of the lunar regolith, augment the collection of mare basalts.

The term mare basalt is potentially misleading. Most such rocks are far more melanocratic (MgO + FeO rich) than terrestrial basalt. Their typical contents of feldspar and $Al_2O_3$ (generally 8–10 wt%) are barely over half the levels typical of terrestrial basalt. These rocks are arguably more analogous to relatively low-MgO komatiite (*Echeverría,* 1982) than to basalt. Also, the mantle setting for melting and melt/mantle segregation is fundamentally different for the Moon vs. the Earth. Lacking plate tectonics, the Moon convects only below its single, thick lithospheric plate. Throughout the era of significantly sampled mare volcanism (i.e., post-3.9 Ga), convection was confined to well below the surface (probably >100 km deep). Even if convection spanned all the remaining lunar volume, the maximum pressure range was slightly less than 4 GPa. In contrast, the Earth's mantle convects from within a short distance of the surface (at spreading centers), down to at least 670 km, i.e., 24.0 GPa. Dispersion in isotopic ratios such as $\varepsilon_{Nd}$ and $\varepsilon_{Hf}$ (*Nyquist and Shih,* 1992) confirm that homogenization by convective stirring was far less thorough within the post-magma-ocean lunar mantle compared to its terrestrial counterpart. "Layer-cake" stratigraphy engendered by magma ocean fractional crystallization must have been reduced by convective stirring in both mantles, but far greater heterogeneity survived in the case of the Moon. Thus it seems likely that lunar mantle melting was comparatively less sensitive to pressure release and other effects of convective tectonics, and more sensitive to local compositional quirks [e.g., high Th, U, and K, leading to locally higher T; high $Al_2O_3$ and/or low *mg* ratio (molar Mg/Mg + Fe), leading to lower melting T]. More obviously, because pressure and $f_{O_2}$ are vastly higher in the Earth, phase equilibria during mantle melting are shifted and diversified. Possible implications of these complexities for HSE and associated compatible elements are discussed below.

The bulk lithophile compositions of the two mantles are probably similar but not identical. A common interpretation is that the FeO concentration is higher, and MgO lower, in the Moon than in the Earth. For example, *Wänke* (1981) and *Taylor* (1982) inferred that the MgO concentration is only 0.8× as great in the bulk Moon (31–32 wt%) as in Earth's primitive mantle (38–40 wt%). However, such interpretations rely on the dubious (see above) representativeness of mare basalts. Many occurrences of extremely high *mg* among highland rocks argue for Moon-Earth similarity of *mg* (*Warren,* 1986; *Seifert and Ringwood,* 1988). *Taylor* (1982) also estimated that Cr (4200 μg/g) and V (150 μg/g) are considerably higher in the bulk Moon than in Earth's mantle (3000 and 84 μg/g). In general, however, literature estimates imply that these elements are at approximately the same concentration in the two mantles, with preferred values of about 3000 μg/g for Cr and 80 μg/g for V (*Wänke,* 1981; *Seifert and Ringwood,* 1988; *Drake et al.,* 1989; *O'Neill,* 1991a).

## 4.2. Limitations and Difficulties of the Lunar Database

The HSE are difficult to determine at the low concentrations typical of basaltic rocks. Data for HSE in lunar rocks obtained to date have come almost exclusively from the highly sensitive technique of radiochemical neutron activation analysis (RNAA). For mare basalts, reliable data have come mainly from just two RNAA laboratories. For one of these labs, that of E. Anders at the University of Chicago, the data were compiled by *Wolf et al.* (1979). Reliable mare basalt HSE data from the UCLA lab of J. Wasson (and in later years P. Warren) were published by *Chou et al.* (1975), *Wasson et al.* (1976), and *Warren et al.* (1986, 1987). Many early data, even from these otherwise reputable labs, have been discredited (*Wolf et al.,* 1979) as suspiciously high in comparison to later analyses of the same or closely similar samples. The most frequent cause of spurious data is probably laboratory contamination, a problem enhanced for lunar basalts by low HSE concentrations (even by basalt standards) and by the small sample masses available for analysis. The potential for contamination is particularly severe for small lithic fragments, whether found as clasts within polymict breccias or as small rocklets in the regolith (in which case the contamination might originate from adhering materials and/or laboratory substances). We follow the recommendations of *Wolf et al.* (1979) regarding reliability of older Chicago data, and for UCLA data, we choose to disregard HSE results from analyses begun prior to late 1974. Of course, purely analytical problems may also lead to spurious data.

Yet another serious question with HSE data for lunar basalts is how representative the small samples (typically 100–500 mg) used for analysis are. The "nugget effect" that plagues HSE analyses for terrestrial rocks is even more of a problem with the small, extremely HSE-poor lunar basalt samples. The nugget effect is presumably maximal for small lithic fragments. On the other hand, analyses of pyroclastic spherule glasses (*Chou et al.,* 1975; *Wasson et al.,* 1976; *Morgan and Wandless,* 1984) may be relatively immune to the nugget effect. Contamination is also not a great concern with these spherule analyses, because sieving and leaching procedures (which, significantly, were undertaken after the samples were irradiated) made it possible to determine the compositions of the spherule interiors. The leaching might even conceivably have led to spuriously low HSE data.

Another difficulty associated with the large dataset of *Wolf et al.* (1979) is that these authors determined no compatible, nonvolatile element to give context to their HSE data (albeit they determined numerous trace volatile elements and one nonvolatile incompatible element: U). For this chapter, we compiled complementary data from a wide range of literature sources (Table 3). We believe that by weeding out unreliable older data, and giving context to nearly all the *Wolf et al.* (1979) data, we have assembled a database for HSE in lunar basalts that is considerably improved over any previously used for the purpose of constraining the HSE composition of the lunar mantle (Table 3). Unfortunately, even RNAA is not sufficiently sensitive to determine most of the HSE in lunar basalts. Older Chicago and UCLA analyses typically determined only Ir and Au. In later years, Re and Os were added, but the database is still far more extensive for Ir and Au than for any other HSE element (and Au is notoriously "irregular" in its geochemical behavior). Also, we include in our survey results for several highland rocks that are compositionally "pristine" (i.e., not adulterated by meteoritic-impact mixing) and either basalts or cumulates of mafic to marginally ultramafic composition (in contrast to the anorthositic compositions typical of highland cumulates). These samples are 14434, 15382, 15386, 61224,11, and 72275 (*Arai and Warren,* 1997; *Warren and Wasson,* 1978, 1980; *Wolf et al.,* 1979). Because trends can be well established for Ni that suggest the possibility of analogous trends for Ir and other HSE, we also cover in this section an element that is not quite so thoroughly siderophile: Ni. The lunar database for Ni is more comprehensive than for any of the HSE. In addition to the sources mentioned above, in connection with HSE, important sources of Ni data include *Annell and Helz* (1970), *Taylor et al.* (1971), *Wänke et al.* (1971, 1976), *Brunfelt et al.* (1972), *Helmke et al.* (1972), *Ma et al.* (1976, 1980), and *Rhodes et al.* (1977).

### 4.3. Correlations with Lithophile Elements

Within a planetary mantle-crust system, after core-mantle separation, some siderophile elements (e.g., Ni, Co) may behave during igneous differentiation much like compatible lithophile elements (e.g., Sc). Positive correlations vs. MgO among extensive sets of terrestrial and lunar igneous rock analyses (Figs. 4a,b) imply such compatible behavior (*Naldrett and Barnes,* 1986; *Brügmann et al.,* 1987). The partitioning behavior of NiO as a compatible transition element is also well documented (e.g., *Beattie et al.,* 1991; *Jones,* 1995). As noted above, however, the partitioning of HSE among silicate phases is not so well known. *Naldrett and Barnes* (1986) inferred that the *bulk* solid/melt D for Ir is ~2 for ultramafic melts with olivine as the sole liquidus phase, and higher (of order 5–10) for basaltic melts with

both olivine and orthopyroxene as liquidus phases. The bulk D is, of course, much higher still if melt-sulfide segregation is involved. This compatible behavior results in analogous correlations between siderophile elements and Cr (Fig. 5). Lunar mare basalts manifest an analogous correlation between V and Ni. However, because olivine/melt partitioning of V is strongly anticorrelated with $f_{O_2}$ (*Canil,* 1997), no comparable Ni-V correlation exists among terrestrial basalts.

As discussed above, the HSE may not actually be partitioning into the mafic silicates *per se*, but even if they are instead concentrating into tiny (spinel?) inclusions, they evidently do so in a highly systematic way. For example, the MgO vs. Ir trend for lunar mare basalts (Fig. 4b) is not a simple linear correlation, but rather resembles a rounded "Γ." Nonetheless, this peculiar trend closely mimics the analogous MgO vs. Ni trend (Fig. 4a). The Ni/Ir ratio implied by the two trends remains constant within a factor of ~60% as MgO increases from 9 and 18 wt%, even as [Ir] increases by a factor of 6.

These correlations can be useful for estimating bulk-mantle HSE composition. Bulk-mantle concentrations of the major species MgO, and compatible-lithophile, nonvolatile elements like Cr and V, are relatively well constrained, and, at least for MgO, similar in the Moon and the Earth. Assuming mantle-crust differentiation processes have been analogous, displacements between the lunar and terrestrial trends on plots such as Figs. 4 and 5 may indicate a significant difference between the lunar and terrestrial mantle siderophile concentrations.

The differentiations that engendered the lunar mantle source regions, and the mechanisms of melting, may have been somewhat different. Another noteworthy complication is that some of the samples plotted in Figs. 4 and 5 are mild cumulates, the main cumulus phase being olivine (e.g., *Rhodes et al.,* 1977; *Neal et al.,* 1994). Terrestrial trends (*Naldrett and Barnes,* 1986; *Brügmann et al.,* 1987) indicate that concentrations of HSE are enhanced (vs. the parent melt) in cumulus mafic silicates, in much the same way as MgO and Cr are enhanced. Unless the concentrations of MgO (or Cr) and a given siderophile element in the cumulate rock are enhanced to the exact same extent, the melt-to-rock compositional path may deviate from the trend manifested by the overall database. However, since the more extremely MgO-rich terrestrial samples also typically contain cumulus olivine, the presence of cumulus olivine in some of the most MgO-rich lunar basalts should not be a major concern for purposes of interplanetary comparison. If these general rules were not being obeyed (i.e., if the siderophile element were incompatible and/or if it were being controlled by another phase such as sulfides), the data would not distribute along coherent, positive correlations like those in Figs. 4 and 5. Indeed, if the two elements are controlled by similar compatible partitioning behavior, the trend should pass close to the bulk mantle composition.

Adopting MgO as the prime lithophile cohort of the siderophile elements (Fig. 4), the lunar trends for Ni and Ir

TABLE 3.  Compilation of averaged literature data for HSE, Ni, and selected lithophile elements in lunar basaltic rocks.

| | MgO wt% | Ti mg/g | V μg/g | Cr mg/g | Ni μg/g | Re ng/g | Os ng/g | Ir ng/g | Au ng/g | Data Sources for HSE and Ni | Data Sources for Lithophile Elements |
|---|---|---|---|---|---|---|---|---|---|---|---|
| *Low-Ti mare basalts* | | | | | | | | | | | |
| 12002 | 14.8 | 15.7 | 190 | 5.84 | 120 | | | <0.05 | 0.024 | [4,41,45,57] | [10,41,45] |
| 12004 | 12.1 | 17.2 | 194 | 4.34 | 80 | | | | | [45] | [43,45] |
| 12005 | 20.0 | 16.5 | | 5.17 | 90 | | | | | [34] | [34] |
| 12009 | 11.2 | 19.6 | 153 | 4.19 | 56 | 0.0075 | | 0.031 | 0.024 | [52,57] | [34,52] |
| 12014 | 13.9 | 16.1 | | 4.42 | | | | <0.08 | | [4] | [34] |
| 12017 | 7.6 | 20.2 | | | | | | <0.20 | 0.072 | [57] | [34] |
| 12018 | 14.7 | 15.7 | 140 | 4.05 | 100 | | | <0.1 | | [10,45] | [10,45] |
| 12020 | 15.3 | 17.0 | 168 | 4.28 | | | | <0.03 | 0.036 | [57] | [15,43,45] |
| 12021 | 7.4 | 21.1 | 160 | 2.59 | 6.1 | 0.005 | | 0.0055 | 0.017 | [52,57] | [10,52] |
| 12022 | 10.4 | 29.4 | 165 | 3.59 | 43 | 0.0065 | | 0.035 | 0.052 | [41,52,57] | [41,52] |
| 12024,15 | 11.7 | 28.4 | 181 | 3.91 | 51 | <0.04 | 0.046 | 0.023 | 0.041 | [54] | [54] |
| 12036 | 16.7 | 19.2 | | 4.90 | 60 | | | | | [34] | [34] |
| 12038 | 6.8 | 19.4 | 112 | 2.13 | 5 | 0.056 | | 0.021 | 0.006 | [1,41,52] | [10,15,41,52] |
| 12040 | 16.3 | 15.2 | 153 | 4.45 | | | | <0.17 | 0.012 | [57] | [15,22] |
| 12046 | 7.3 | 27.6 | 138 | 2.01 | 11.3 | 0.0004 | | 0.002 | 0.013 | [57] | [32] |
| 12051 | 6.7 | 28.5 | 137 | 2.11 | | | | <0.09 | 0.008 | [57] | [10,15,43] |
| 12052 | 8.4 | 19.4 | 156 | 3.42 | 28 | 0.002 | | 0.004 | 0.014 | [41,45,57] | [41,45] |
| 12053 | 8.1 | 22.0 | 171 | 3.48 | 28 | | | | | [45] | [45] |
| 12063 | 9.1 | 29.0 | 132 | 2.65 | 32 | | | <0.04 | | [4,41,45] | [41,43,45] |
| 12064 | 6.8 | 23.9 | | 2.43 | 1.4 | 0.003 | | 0.008 | 0.005 | [52] | [22,45,52] |
| 12065 | 8.4 | 20.0 | 150 | 3.36 | 13 | 0.015 | | 0.058 | 0.012 | [52,57] | [22,45,52] |
| 12075 | 13.9 | 16.7 | 190 | 4.33 | 68 | 0.008 | | 0.021 | 0.032 | [52] | [43,52] |
| 14053 | 8.6 | 16.3 | 135 | 2.90 | 14 | 0.007 | | 0.017 | 0.110 | [18,57] | [18,56] |
| 14072 | 10.4 | 15.4 | | 3.29 | 31 | | | | | [18] | [18,20] |
| 15016 | 11.2 | 13.4 | 250 | 5.29 | 76 | 0.003 | | 0.018 | 0.025 | [5,41,57] | [39] |
| 15058 | 9.0 | 10.6 | | 3.30 | 40 | 0.0006 | 0.020 | 0.006 | 0.081 | [19,57] | [39] |
| 15065 | 9.7 | 10.2 | | 3.41 | 63 | 0.0055 | | 0.075 | 0.018 | [5,48,57] | [39] |
| 15085 | 7.8 | 9.4 | 156 | 2.94 | 34 | 0.002 | | 0.007 | 0.012 | [19,48,57] | [39] |
| 15256 | 9.1 | 15.1 | | | | 0.005 | | 0.022 | 0.019 | [57] | [39] |
| 15385 | 18.2 | 10.2 | 187 | 5.03 | 138 | | | | 0.025 | [24] | [39] |
| 15459,28 | 17.9 | 12.7 | | 6.27 | 136 | | | | 0.081 | [57] | [39] |
| 15475 | 8.5 | 10.8 | | 3.63 | 35 | 0.003 | | 0.015 | 0.009 | [57] | [39] |
| 15476 | 8.3 | 11.4 | 130 | 3.28 | 20 | | | | | [11] | [39] |
| 15495 | 9.0 | 11.5 | 191 | 3.34 | 37 | 0.006 | 0.032 | 0.021 | 0.083 | [47,53] | [39,53] |
| 15499 | 9.2 | 10.9 | | 3.56 | 51 | 0.0007 | | 0.004 | 0.013 | [57] | [39] |
| 15535 | 10.8 | 15.0 | | 4.02 | 75 | | | | | [5,19] | [39] |
| 15536 | 11.6 | 15.0 | | 4.12 | 57 | <0.36 | 0.027 | 0.023 | 0.037 | [53] | [39,53] |
| 15545 | 10.0 | 14.5 | | 3.70 | 51 | 0.0009 | | 0.015 | 0.005 | [57] | [39] |
| 15555 | 11.2 | 12.5 | 207 | 4.45 | 90 | 0.0012 | 0.0131 | 0.006 | 0.139 | [6,11,57] | [39] |
| 15556 | 7.7 | 14.4 | | 4.72 | | 0.004 | | 0.039 | 0.026 | [11,57] | [39] |
| 15557 | 9.4 | 10.9 | | 3.46 | 56 | | | | | [5] | [39] |
| 15597 | 8.7 | 10.9 | | 3.46 | 30 | 0.0081 | | 0.0072 | 0.045 | [19,57] | [39] |

TABLE 3. (continued).

| | MgO wt% | Ti mg/g | V μg/g | Cr mg/g | Ni μg/g | Re ng/g | Os ng/g | Ir ng/g | Au ng/g | Data Sources for HSE and Ni | Data Sources for Lithophile Elements |
|---|---|---|---|---|---|---|---|---|---|---|---|
| **Small mare basalt lithic fragments** | | | | | | | | | | | |
| 14181,6 | 10.1 | 11.7 | | 3.08 | 4 | <0.003 | 0.011 | 0.002 | 0.147 | [52] | [52] |
| 14321,184,1B | 9.0 | 10.8 | | 2.94 | 39 | 0.005 | | 0.044 | 0.30 | [57] | [46] |
| 15265,6/13 | 9.3 | 14.0 | | 3.87 | 39 | 0.0065 | | 0.018 | 0.046 | [53,57] | [53] |
| 15299,196c | 9.3 | 13.3 | | 3.94 | 16 | 0.024 | | 0.126 | 0.009 | [53] | [53] |
| 15299,201c | 17.9 | 8.5 | | 11.3 | 150 | 0.024 | | 0.113 | 0.36 | [53] | [53] |
| 24067,3202 | 8.8 | 4.2 | 164 | 1.96 | <18 | | | 0.062 | 0.036 | [16] | [26] |
| 24067,3802 | 7.0 | 4.8 | 157 | 1.81 | <19 | | | 0.015 | | [16] | [26] |
| 60639,5 | 6.5 | 44.1 | 78 | 2.05 | <6 | 0.006 | | 0.048 | 0.040 | [57] | [31] |
| 70007,299 | 21.0 | 4.8 | 166 | 5.17 | 130 | | | | | [27] | [27] |
| **High-Ti mare basalts** | | | | | | | | | | | |
| 10003,x | 7.0 | 66.3 | | 1.69 | 2.7 | | | | | [2] | [2,35] |
| 10017,33 | 7.8 | 73.2 | 57 | 2.45 | | | | 0.020 | <0.72 | [1] | [42,44] |
| 10020,25 | 7.0 | 62.1 | | 2.04 | | | | 0.027 | 0.07 | [17] | [35] |
| 10044,32 | 6.8 | 57.3 | 39 | 1.31 | 4 | | | | | [9] | [35,42,44] |
| 10045 | 8.4 | 69.2 | | 2.68 | 4 | | | | | [9] | [35,42] |
| 10047,65 | 5.9 | 59.7 | 65 | 1.45 | | | | 0.005 | 0.029 | [1] | [35] |
| 10050,26 | 7.7 | 71.6 | 98 | 2.38 | | | | 0.007 | 0.031 | [17] | [35,36,42] |
| 10057,40 | 7.0 | 65.0 | 62 | 2.16 | 6.1 | | | 0.016 | 0.015 | [1,2] | [2,44] |
| 10069,x | 8.0 | 65.3 | 65 | 2.06 | 6.7 | | | | | [2] | [25] |
| 10071,x | 7.3 | 65.3 | 75 | 2.37 | 7 | | | | | [2] | [25,35] |
| 10072,23 | 7.5 | 71.9 | | 2.28 | 6 | | | 0.022 | 0.12 | [2] | [25,35] |
| 70017,21 | 9.7 | 80.2 | | 3.46 | | 0.00064 | 0.0055 | | | [1,2] | [14] |
| 70215,57 | 8.0 | 78.7 | | 2.73 | 1 | 0.002 | | 0.003 | 0.026 | [6] | [12,33,37] |
| 71055,31 | 8.9 | 83.2 | | 2.80 | 3.4 | | | | | [29,47] | [12,33,37,47] |
| 72155,30 | 8.7 | 73.2 | 103 | 3.05 | 1.5 | | | | | [8] | [12,37] |
| 75035,35 | 6.5 | 57.3 | 31 | 1.62 | 1 | | | | | [8] | [23,33,47] |
| 75055,37 | 7.1 | 65.1 | | 1.99 | 1.5 | 0.0007 | | 0.028 | 0.021 | [29,47] | [12,23,38,47] |
| 79155,34 | 9.2 | 74.6 | | 3.42 | 3.6 | 0.003 | | | 0.011 | [8] | [33,38,47] |
| **Mare pyroclastic glasses** | | | | | | | | | | | |
| 15421 green VLT | 18.1 | (low) | 116 | 4.00 | 171 | | | 0.26 | 0.05 | [13] | [13] |
| 15426 brown | 12.5 | 22.2 | | 3.78 | 85 | 0.017 | | 0.14 | | [28,30] | [28] |
| 15426 green VLT | 17.5 | 2.3 | 165 | 3.66 | 154 | 0.0058 | 0.125 | 0.117 | 0.10 | [30] | [28] |
| 74220 orange | 14.8 | 52.1 | 120 | 4.84 | 70 | | | 0.21 | 0.27 | [21,55] | [21] |

TABLE 3. (continued).

| | MgO wt% | Ti mg/g | V μg/g | Cr mg/g | Ni μg/g | Re ng/g | Os ng/g | Ir ng/g | Au ng/g | Data Sources for HSE and Ni | Data Sources for Lithophile Elements |
|---|---|---|---|---|---|---|---|---|---|---|---|
| ***Mafic highland samples*** | | | | | | | | | | | |
| 14434 | 8.8 | 5.4 | 57 | 1.50 | 12.9 | 0.0014 | 0.0076 | 0.0084 | 0.054 | [3] | [3] |
| 15382 | 8.6 | 13.1 | 60 | 2.12 | 18 | 0.009 | 0.018 | 0.013 | 0.003 | [57] | [39] |
| 15386 | 9.9 | 12.4 | 62 | 2.27 | 9.5 | 0.016 | | 0.033 | 0.116 | [49] | [39] |
| 61224,11 | 12.8 | 2.4 | | 1.99 | 8.3 | 0.013 | | 0.148 | 0.079 | [51] | [51] |
| 67667 | 26.4 | 6.2 | 96 | 2.59 | 27 | 0.011 | 0.171 | 0.070 | 0.184 | [16,50] | [50] |
| 72275 | 8.4 | 6.6 | 120 | 2.92 | 43 | 0.007 | | 0.023 | 0.045 | [57] | [7,40] |

References (meant to be fully comprehensive for HSE, but in many cases only selected sources are shown for the more often-determined lithophile elements): [1] *Anders et al.* (1971)*; [2] *Annell and Helz* (1970); [3] *Arai and Warren* (1997)**; [4] *Baedecker et al.* (1971)*; [5] *Baedecker et al.* (1973)*; [6] *Birck and Allègre* (1994); [7] *Blanchard et al.* (1975); [8] *Boynton et al.* (1975)*; [9] *Brown et al.* (1970); [10] *Brunfelt et al.* (1971); [11] *Brunfelt et al.* (1972)*; [12] *Brunfelt et al.* (1974); [13] *Chou et al.* (1975); [14] *Compston et al.* (1970); [15] *Compston et al.* (1971); [16] *Ebihara et al.* (1992); [17] *Ganapathy et al.* (1970); [18] *Helmke et al.* (1972); [19] *Helmke et al.* (1973); [20] *Hubbard et al.* (1972); [21] *Hughes et al.* (1989); [22] *Kushiro and Haramura* (1971); [23] *Laul et al.* (1974); [24] *Ma et al.* (1976); [25] *Ma et al.* (1976); [26] *Ma and Schmitt* (1978); [27] *Ma et al.* (1980); [28] *Ma et al.* (1981); [29] *Morgan et al.* (1974); [30] *Morgan and Wandless* (1984); [31] *Murali et al.* (1976); [32] *Neal et al.* (1994); [33] *Rhodes et al.* (1976); [34] *Rhodes et al.* (1977); [35] *Rhodes and Blanchard* (1980); [36] *Rose et al.* (1970); [37] *Rose et al.* (1974); [38] *Rose et al.* (1975); [39] *Ryder* (1985); [40] *Salpas et al.* (1987); [41] *Taylor et al.* (1971); [42] *Wakita et al.* (1970); [43] *Wakita et al.* (1971); [44] *Wänke et al.* (1970); [45] *Wänke et al.* (1971)*; [46] *Wänke et al.* (1972); [47] *Wänke et al.* (1975)*; [48] *Wänke et al.* (1976)*; [49] *Warren et al.* (1978); [50] *Warren and Wasson* (1979); [51] *Warren and Wasson* (1980); [52] *Warren et al.* (1986); [53] *Warren et al.* (1987); [54] *Warren and Jerde* (1990)**; [55] *Wasson et al.* (1976); [56] *Wiesmann and Hubbard* (1975); [57] *Wolf et al.* (1979).

* Data for HSE appear to suffer from contamination (but Ni may be negligibly affected).

** Plus unpublished UCLA neutron activation analysis results.

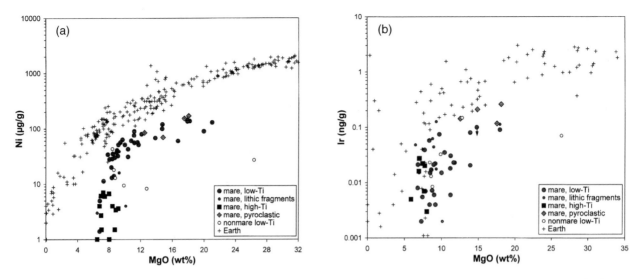

**Fig. 4.** MgO vs. **(a)** Ni and **(b)** Ir for lunar mare basalts and mare pyroclastic materials and for a wide range of terrestrial igneous rocks. For both Ni and Ir, the lunar trend is similar to its terrestrial counterpart, except displaced to lower siderophile concentration by a factor of about 4. Main data sources for mare samples are listed in Table 3.

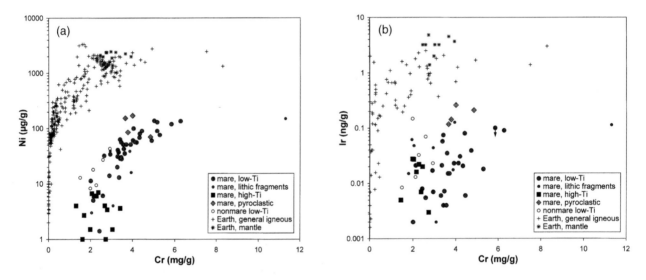

**Fig. 5.** Cr vs. **(a)** Ni and **(b)** Ir for the same dataset as in Fig. 4. Again, the lunar and terrestrial trends appear qualitatively similar, but the lunar trends are displaced to far lower siderophile concentration (or higher Cr; see text).

are clearly displaced to lower concentrations vs. their terrestrial counterparts. Depending upon the choice of [MgO] used for reference, the Moon/Earth mantle depletion factor (MDF$_{M/E}$) appears to be about 5 for Ni and roughly the same for Ir. In the case of the relatively tight, well-defined Ni trends, MDF$_{M/E}$ appears lower (about 2–4) at ~8 wt% MgO than at ~20 wt% MgO, where it appears to be roughly 5. Greatest weight should be given to the more primitive (MgO-rich) ends of the trends. It should also be borne in mind that by most estimates the Moon is slightly depleted in MgO (i.e., the lunar trends may need to be "corrected" by a rightward shift) relative to the Earth. In relation to the established estimates of ~2000 μg/g Ni and 3 ng/g Ir for

Earth's upper mantle (e.g., *McDonough and Sun*, 1995), a MDF$_{M/E}$ of 5 for both elements implies that the lunar mantle contains ~400 μg/g Ni and 0.6 ng/g Ir. If instead these concentrations are independently estimated for the lunar mantle, based on extended extrapolation of the trends in Fig. 4 to the approximate bulk Moon [MgO] (say 35 wt%), the results are similar: roughly 200–600 μg/g Ni and 0.2–0.7 ng/g Ir.

Employing Cr as another compatible lithophile that correlates with the siderophile elements (Fig. 5), the contrast between the lunar and terrestrial trends is stark. The lunar trend, at the low end of its range (Cr ~ 2 mg/g), is depleted in Ni relative to the terrestrial trend by a factor of roughly

500. This gap narrows with increasing [Cr], but the distance between the lunar and terrestrial trends remains about a factor of 30, even at Cr = 5 mg/g. In part, this huge gap is probably a result of stabilization of mantle Cr-spinel, except at high degrees of melt removal, by Earth's higher $f_{O_2}$ and pressures. The higher $f_{O_2}$ in the terrestrial mantle will tend to favor the stability of $Cr^{3+}$, in contrast with the lower $f_{O_2}$ in the lunar mantle, which instead favor $Cr^{2+}$ (*Hanson and Jones,* 1998). Also, Cr is only weakly compatible with olivine [D ~0.7 (*Jones,* 1995)], and its compatibility with low-Ca pyroxene diminishes (D falling from 7 ± 3 to 2 ± 1) as $f_{O_2}$ shifts from typical terrestrial values to typical lunar ones (*Irving,* 1978). Thus, the $MDF_{M/E}$ implications of the displacements between the lunar and terrestrial Cr-Ni and Cr-Ir trends are highly model-dependent. Even so, the correlation between Cr and Ir (Fig. 5b) reinforces the significance of the correlation between MgO and Ir (Fig. 4b). The low values of the intercepts of the Cr-Ni and Cr-Ir trends with the bulk Moon [Cr] (roughly 3–4 mg/g, intercepted at 10–200 µg/g Ni and 0.01–0.3 ng/g Ir) tend to confirm the low lunar mantle [Ni] and [Ir] inferred from the MgO correlations.

From the perspective of our refined and extended lunar database, it is clear that Ni is significantly more depleted in the lunar mantle than in Earth's mantle. Figure 4 implies that for Ni the $MDF_{M/E}$ is about 5. And the lunar mantle closely resembles the terrestrial mantle in terms of Ni/Ir (Fig. 6). Assuming this ratio of trace elements follows a lognormal distribution, the mean for 39 lunar mare basalts and pyroclastics (logµ = 3.09 ± 0.59) is virtually identical to that for 94 terrestrial mafic igneous rocks (logµ = 2.95 ± 0.59), and distinct from the highest-Ni/Ir chondrite type (EH, logµ = 1.49) by 2.7 σ.

In contrast to Ni/Ir, the few available data from simultaneous Re and Os determinations in lunar basalts (Fig. 7a) scatter about evenly across both sides of the CI-chondritic Re/Os ratio. Among lunar rocks, presumably due to the Moon's extremely low $f_{O_2}$, Re shows only faint signs of the incompatible behavior it displays in terrestrial igneous processes. Instead, Re seems to behave much like Os and Ir; e.g., the scatter in Re/Os (Fig. 7a) does not appear appreciably greater than the scatter in Os/Ir (Fig. 7b). Based on a very small dataset, the average Os/Ir for lunar basalts might be slightly enhanced vs. the uniform ratio (1.09 ± 0.03) of chondritic meteorites (*Wasson and Kallemeyn,* 1988). A more impressive case for a nonchondritic ratio between HSE is manifested by the lunar basalt data for Ir vs. Re (Fig. 7c). Even ignoring one exceptionally high-Re sample (12038, *Warren et al.,* 1986; a petrologically distinctive "aluminous mare basalt," but note that the result plotted is an average from two chips that differ in apparent [Re] by a factor of 25!), the basalt Re/Ir ratios appear consistently enhanced vs. the nearly uniform Re/Ir (0.085 ± 0.006) of chondrites. Assuming this ratio follows a log-normal distribution, the mean for 37 lunar mare basalts and pyroclastics (logµ = –0.69 ± 0.31) is distinct from the highest-Re/Ir chondrite type (EH, logµ = –1.04) by 1.1 σ. While not so dramatic as in the case of Ni/Ir, this result still implies, with a statistical significance of well over 99%, that Re/Ir is superchondritic in the sample population of basalts and pyroclastics (also noted by *Drake,* 1987).

However, these basalt compositions should not be assumed to imply a comparably superchondritic Re/Ir for the lunar mantle. Although the incompatible tendency of Re is subdued among lunar materials, it is not altogether suppressed. A weak but statistically significant (r = 0.38, n =

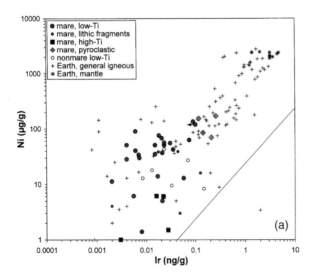
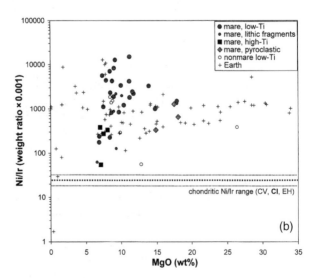

**Fig. 6.** **(a)** Ir vs. Ni and **(b)** Ni/Ir vs. MgO for the same dataset as in Fig. 4. Lunar and terrestrial basalts show remarkably similar enhancements in Ni/Ir relative to the CI-chondritic ratio (diagonal line), again showing similarity in Ni/Ir between lunar and terrestrial basalts. Note that terrestrial basalts have, on average, nearly the same Ni/Ir as terrestrial mantle xenoliths (e.g., *Jagoutz et al.,* 1979).

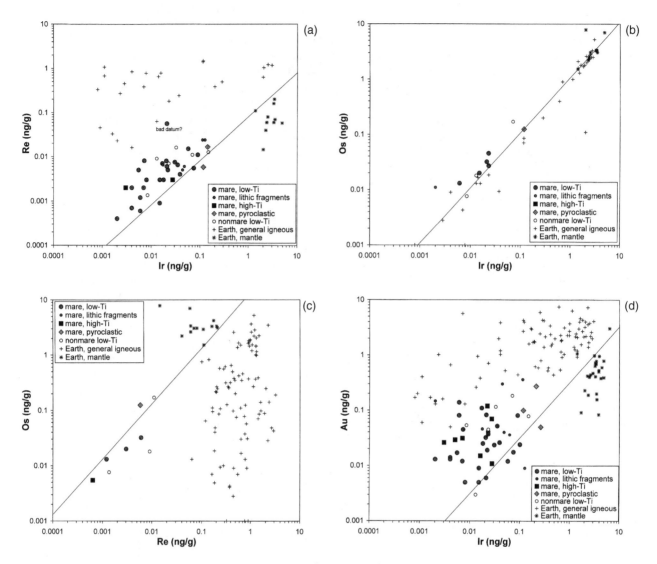

**Fig. 7.** (a) Re vs. Ir, (b) Os vs. Ir, (c) Re vs. Os, and (d) Au vs. Ir for the same dataset as in Fig. 4. The diagonal lines show CI-chondritic ratios. Among lunar basalts, both Re/Ir and Au/Ir are generally enhanced over the chondritic ratios. However, the mantle ratios are probably more nearly chondritic. The basalt compositions reflect the mildly incompatible behavior of Re and possibly the labile, volatile, and possibly incompatible behavior of Au.

38) correlation exists between Re/Ir and the heavy rare earth element Yb (Fig. 8). As all the basalts are probably enriched in Yb compared to the bulk mantle, this correlation implies they also have higher Re/Ir. The Re/Ir of the primitive green variety of pyroclastic glass (15426; *Morgan and Wandless,* 1984) is subchondritic.

Gold lives up to its terrestrial reputation for capriciousness. Even by lunar basalt HSE standards, the Au/Ir ratio shows a high degree of scatter (Fig. 7d). As in the case of Re, most samples have higher than chondritic Au/Ir ratio. Some authors have speculated that Au (and Au/Ir) may be enhanced among geochemically evolved types of lunar material, i.e., the Moon's version of pegmatite (*Hughes et al.,* 1973; *Warren et al.,* 1989). Our database of mare basalts (and a few KREEP basalts) shows no correlation between Au (or Au/Ir) and incompatible elements. However, Au is

more volatile and "labile" than other HSE, so it may be unrealistic to expect a correlation among samples typically only 100–300 mg in mass. It should also be noted that avoidance of anthropogenic contamination of such HSE-poor samples is even more difficult for Au than for HSE in general. Although the data of Fig. 7d suggest that Au/Ir may be enhanced to roughly 4× chondritic in the lunar mantle, we recommend viewing this indication more as an upper limit than as a direct measure of the lunar mantle composition. The average Au/Ir of two primitive green varieties of pyroclastic glass (15421; *Chou et al.,* 1975; 15426; *Morgan and Wandless,* 1984) is almost precisely chondritic.

Data for Ru, Rh, Pd, and Pt in mare basalts were recently published (in abstract form) for a set of seven Apollo 12 samples, mostly small lithic fragments, by a new ICP-MS team (*Snyder et al.,* 1997a; *Neal et al.,* 1999). Earlier at-

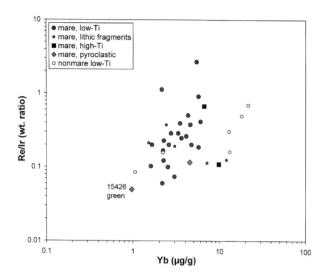

**Fig. 8.** Yb vs. Re/Ir for the same dataset as in Fig. 4. The positive correlation suggests that Re is weakly incompatible, even in the low-$f_{O_2}$ environment of the Moon.

tempts at determining these elements in mare basalts were reviewed by *Haskin and Warren* (1991). The seven ICP-MS samples appear extraordinarily HSE-rich. Apparently (Table 1 of *Snyder et al.*, 1997b; Fig. 1 of *Neal et al.*, 1999), none have <0.19 ng/g of Ir. For comparison, in our database (e.g., Fig. 4b), all but a tiny minority of extraordinarily MgO- and Cr-rich samples have <0.1 ng/g Ir. Yet most of the seven ICP-MS samples have moderate (≤3.6 mg/g) Cr (*Snyder et al.*, 1997b) (their MgO contents are mostly unknown). The apparently anomalous HSE compositions of the seven ICP-MS samples complicate interpretation of the Ru, Rh, Pd, and Pt data. However, the new analyses consistently show that Ru/Ir is nearly chondritic, with Rh and Pd about 10× less depleted (vs. chondrites) than Ir and Ru, and Pt still less depleted, by a further factor of ~3. The Pt/Pd ratio is thus ~3× chondritic, and Pt/Ir is roughly 30× chondritic. *Neal et al.* (1999) note that the Pt/Pd, Pt/Rh, and Pt/Ir ratios are all higher than typically found in terrestrial basalts. *Neal et al.* (1999) interpret this disparity as a result of the Moon acquiring its HSE composition through selective sampling of an earlier body (the proto-Earth) in which HSE had differentiated within the core. It should be noted, however, that a large province of Australian basalts analyzed by *Vogel and Keays* (1997) features similarly high Pt/Pd and Pt/Ir (Rh was not determined).

In summary, the concentrations of the HSE in the lunar mantle must be derived indirectly through correlations with other elements. Relationships vs. compatible elements (Cr, Ni, and especially MgO) suggest that the Moon's mantle is modestly depleted in at least one HSE (Ir) relative to Earth's upper mantle. The Ni/Ir ratio is remarkably similar in the two mantles. The bulk lunar mantle abundances of other HSE are harder to estimate, but Re and Os probably occur at close (within a factor of 2, but perhaps much closer) to

chondritic proportion vs. Ir. Very few reliable data are available for Ru, Rh, Pd, and Pt in mare basalts or "pristine" highland samples, so bulk lunar mantle abundances for these elements are especially uncertain.

## 5. RHENIUM-OSMIUM AND PLATINUM-OSMIUM ISOTOPIC CONSTRAINTS ON THE ORIGIN OF HIGHLY SIDEROPHILE ELEMENTS IN THE MANTLES OF THE EARTH AND MOON

The $^{187}$Re-$^{187}$Os and $^{190}$Pt-$^{186}$Os isotopic systems provide important constraints on the origin of HSE in the mantles of the Earth and Moon, and ultimately other planetary bodies. The three elements that compose these two radiogenic isotope systems are HSE, and consequently, the determination of Os-isotopic compositions can be used to monitor the long-term elemental ratios of Re/Os and Pt/Os in materials derived from planetary mantles. Rhenium-187 decays to $^{187}$Os via the $^{-}\beta$ transition, with a decay constant of $1.666 \times 10^{-11}$a$^{-1}$. In contrast, $^{190}$Pt decays to $^{186}$Os via the $\alpha$ transition, with a decay constant of $1.477 \times 10^{-12}$a$^{-1}$. The Re-Os system is the more diagnostic system for studies of long-term relative abundances of HSE in mantles because $^{187}$Re is a major isotope of Re (62.60%) and has a relatively large decay constant. The Pt-Os system is less precise in monitoring long-term Pt/Os because $^{190}$Pt is a very minor isotope of Pt (0.0130%), and its decay constant is relatively small. Nonetheless, with recent advances in high-precision measurement techniques of $^{186}$Os (*Walker et al.*, 1997b), the system has significant utility in placing constraints on the long-term Pt/Os of the Earth's upper mantle.

### 5.1. Uncertainties in Estimating Rhenium/Osmium and Platinum/Osmium from Bulk Rock Analysis

Although core formation almost certainly segregated the major proportion of HSE from the silicate portion of the Earth, the estimated relative abundances of these elements in the upper mantle, largely based on the examination of mantle xenoliths and orogenic lherzolites (e.g., *Morgan*, 1986; *Pattou et al.*, 1996), has been accepted to be generally chondritic (e.g., *Newsom*, 1990). As noted above, estimations of the absolute abundances of HSE and their relative abundances in the bulk silicate Earth are as yet imprecisely determined. This results from analytical constraints, but even more significantly from uncertainties in the selection of materials that are representative of the upper mantle. As has been shown with a variety of major and trace elements, and also long-lived radiogenic isotope systems, the upper mantle is variably melt-depleted. For example, all materials derived from the depleted mid-ocean ridge mantle (DMM) have large positive $\varepsilon_{Nd}$ values reflecting a long-term enrichment in Sm/Nd ratio of this reservoir relative to chondrites (*DePaolo and Wasserburg*, 1976). The deviance of the Sm-Nd isotopic system from chondritic results from the segregation of the light rare earth elements into the continental crust. Consequently, no one sample of the upper mantle can be representative of a true primitive (undepleted) upper mantle (PUM) composition in all chemical and isotopic characteristics. Viable estimates of the composition of the PUM must be made via extrapolations from nonprimitive materials. In addition to these problems, it should also be noted that the concentrations of HSE in the lower mantle are much more poorly constrained than the upper mantle. It is conceivable, for example, that a late accretionary veneer added HSE to the upper mantle subsequent to core formation, but that these HSE were either not mixed into the lower mantle (i.e., a chemical boundary layer exists between upper and lower mantle), or if they were, the mixing process required a significant portion of Earth history. Because of this, we believe that it is more prudent to use the term PUM, to describe the undifferentiated compositions of the HSE in the upper mantle, than to use the term bulk silicate earth (BSE), as is conventionally used for many lithophile trace elements.

Given the problems inherent in constraining the absolute abundances and relative ratios of the HSE in the present upper mantle and the PUM, determining Re/Os and Pt/Os ratios from the isotopic compositions of Os of upper mantle materials is much more straightforward and more precisely achieved. This is because the $^{187}Os/^{188}Os$ and $^{186}Os/^{188}Os$ ratios of both modern mantle materials (e.g., xenoliths) and mantle-derived materials (e.g. basalts, picrites) reflect the Re/Os and Pt/Os ratios of their mantle sources integrated over Earth history.

## 5.2. Rhenium-Osmium and Osmium-187/Osmium-188 Ratios in the Primitive Mantle

Before we consider the implications of the Os-isotopic compositions of the upper mantle, we must first consider how this composition can be constrained. Indeed, the $^{187}Os/^{188}Os$ ratio of both the modern upper mantle and the PUM can be independently estimated via the examination of several different types of materials. For example, *Allègre and Luck* (1980) initially examined the Os-isotopic compositions of Os-Ir alloys from various ultramafic rocks of variable, presumed ages. They argued that the Os-isotopic compositions of these materials were representative of the upper mantle and reported the first evidence that the Os-isotopic evolution of the mantle was very similar to the evolution trend predicted for chondritic meteorites, as estimated from modern chondritic $^{187}Os/^{188}Os$ and Re/Os ratios. The main problem with this approach is that some uncertainties exist regarding the age and provenance of the Os-Ir alloys. It should be noted that the Os-Ir alloys were chosen for analysis in this early study primarily because they contain large quantities of Os and could be easily analyzed using the techniques available at that time. Subsequent studies have emphasized the examination of materials of more well-constrained upper mantle provenance, including mid-ocean ridge basalts (MORB), abyssal peridotites, upper mantle peridotite xenoliths, orogenic lherzolites, and the ultrama-

fic sections of ophiolites. Each of these classes of materials has positive and negative aspects for determining the isotopic composition of the upper mantle and the PUM.

*5.2.1. Mid-ocean ridge basalts.* The analysis of MORB is important because it is undisputedly derived from the DMM. Most reported MORB isotopic compositions range from approximately chondritic to very much more radiogenic than chondrites (*Martin*, 1991; *Roy-Barman and Allègre*, 1994; *Schiano et al.*, 1997; *Roy-Barman et al.*, 1998). Unfortunately, several aspects of the Re-Os systematics of MORB make them less than ideal candidates for studying the Os-isotopic composition of the DMM. First, Os abundances are normally very low in MORB, and the Os contained within them is subject to contamination from much more radiogenic seawater either once erupted, or via interactions between the melts or source rocks and seawater prior to eruption ($^{187}Os/^{188}Os$ of seawater is ~1 whereas the upper mantle ratio is ~0.13; *Sharma et al.*, 1997). Second, the Re/Os ratios of most MORB are very high (because of the low Os), so that once formed, basalts grow to radiogenic Os-isotopic compositions very rapidly. Consequently, the age of a MORB sample must be well-constrained in order to examine the isotopic composition of its source. Finally, preferential melting of basaltic or eclogitic domains within the upper mantle can generate melts with low Os abundance, but with suprachondritic Os-isotopic compositions. Combined, these factors argue that the least-radiogenic MORB samples analyzed are probably the most representative of the bulk composition of the DMM (Fig. 9). The $^{187}Os/^{188}Os$ of the DMM as suggested from MORB is therefore in the range of 0.127–0.130 (*Martin*, 1991; *Roy-Barman and Allègre*, 1994; *Schiano et al.*, 1997; *Roy-Barman et al.*, 1998).

*5.2.2. Abyssal peridotites.* Most abyssal peridotites are believed to be residues of upper mantle melting and have much higher concentrations of Os than MORB. However, the Os-isotopic compositions of abyssal peridotites are moderately variable and may reflect some open system behavior as a consequence of the nearly ubiquitous alteration that is present in these materials (*Martin*, 1991; *Roy-Barman and Allègre*, 1994; *Snow and Reisberg*, 1995). Nonetheless, a limited suite of carefully chosen samples suggests that the Os-isotopic composition of abyssal peridotites and their presumed upper mantle source is chondritic to slightly less radiogenic than chondrites, with an average $^{187}Os/^{188}Os$ of 0.1246 (*Snow and Reisberg*, 1995) (Figs. 9 and 10). The variability of Os-isotopic compositions in abyssal peridotites may reflect ancient, variable depletions of Re from a chondritic source (*Brandon et al.*, 1998a). This result is consistent with the MORB data discussed above. The range of compositions could also reflect the variable addition of radiogenic $^{187}Os$ to a long-term Re-depleted source, either by melt-rock interactions or seawater interactions.

*5.2.3. Ophiolites.* *Walker et al.* (1998a) reported Os-isotopic compositions for chromites separated from the ultramafic sections of a large number of ophiolites. This type of study adds a new dimension to our understanding of the

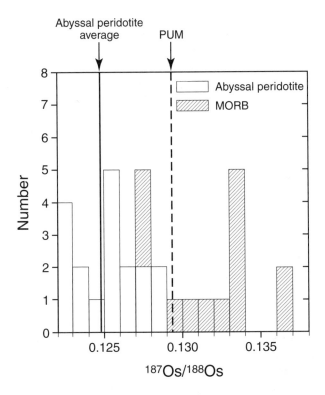

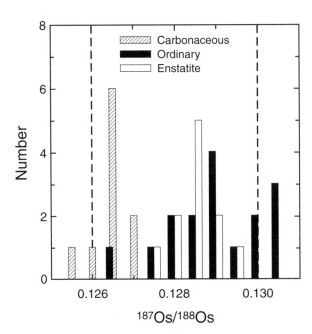

**Fig. 9.** Modal histogram of $^{187}Os/^{188}Os$ data for abyssal peridotites and MORB. Heavy vertical lines show the abyssal peridotite average as defined by *Snow and Reisberg* (1995) and the primitive upper mantle (PUM) estimate of *Meisel et al.* (1996). Note that the upper end of the abyssal peridotite and lower end of the MORB arrays are very similar to the PUM estimate. Data are from *Snow and Reisberg* (1995), *Roy-Barman and Allègre* (1995), *Martin* (1991), and *Schiano et al.* (1997).

**Fig. 10.** Modal histogram of $^{187}Os/^{188}Os$ whole-rock data for ordinary, enstatite, and carbonaceous chondrites. Note that there is little overlap between the compositions of carbonaceous chondrites relative to enstatite and ordinary chondrites. Heavy vertical lines mark the modern-day composition of Os in reservoirs whose $^{187}Re/^{188}Os$ ratio diverged from that of 0.4143 (average for this array) ± 6%. Data are from *Meisel et al.* (1996) and *Walker et al.* (1999b).

evolution of Os isotopes in the upper mantle because ophiolites of different age permit the accessing of the Os-isotopic composition of the upper mantle at those times. Such studies focus on chromites because they normally have very low Re/Os ratios (little age correction necessary) and are very resistant to the ubiquitous alteration processes that occur during ophiolite obduction. *Walker et al.* (1998a) examined 15 ophiolites whose Os-isotopic compositions delineated an evolution trajectory that is indistinguishable from that of the MORB estimate for the upper mantle, projecting to a modern $^{187}Os/^{188}Os$ ratio of 0.129 ± 1.

*5.2.4. Upper mantle lherzolites.* Examination of upper mantle xenoliths and orogenic lherzolites, as noted above, has been used to estimate absolute and relative abundances of HSE in the PUM and upper mantle. As with those estimations, the Os-isotopic composition of a xenolith (or orogenic lherzolite) is likely to have been modified from PUM evolution at some time in its past as a result of partial melting or enrichment processes. Such processes are likely to modify the Re/Os ratio of mantle peridotite, because Re is an incompatible trace element, and subsequently leads to a modification in the Os-isotopic evolution of that sample. Some metasomatic processes may also enrich mantle peri-

dotite in $^{187}Os$ (*Brandon et al.*, 1996). To circumvent these problems, *Meisel et al.* (1996) used the same extrapolation techniques for xenolith suites employed by *Morgan* (1986) to constrain hypothetical undepleted compositions. By extrapolating correlations between $^{187}Os/^{188}Os$ ratios and Lu concentration (an indicator of degree of melt depletion) for several suites of subcontinental lithospheric mantle (SCLM) xenoliths and orogenic lherzolites, they obtained a consistent $^{187}Os/^{188}Os$ of 0.1290 ± 9. Because of the different melt extraction ages in each of the systems examined, this composition is probably more representative of the PUM than of the modern upper mantle. This was considered a minimum value because the upper mantle from which SCLM is ultimately separated may have experienced long-term Re loss, via incorporation of Re into the continental crust or removal to the lower mantle via subduction of oceanic basalts. Hence, the composition of the source regions of the various suites may have been modified from the PUM prior to isolation in the SCLM.

The estimates for the $^{187}Os/^{188}Os$ of PUM and the results for MORB and ophiolites are very similar. These compositions are somewhat more radiogenic than the abyssal peridotite average, although some abyssal peridotites have compositions that overlap with the PUM estimate. It is currently unknown whether the Os-isotopic composition of the convecting upper mantle is better constrained by the abyssal peridotite average or by the ophiolite and least-radiogenic MORB compositions. If the abyssal peridotite average is a more accurate reflection of the true composition of the convecting upper mantle, then it can be argued that the DMM has experienced long-term Re depletion, and thus that there exists an additional Re-rich component in the mantle to balance this depletion (e.g., *Hauri and Hart*, 1997). The necessary additional component is not continental crust or subcontinental lithosphere, and is likely to be recycled oceanic crust for several reasons: (1) the limited amount of Re evidently contained in the continental crust (*Esser and Turekian*, 1993); (2) the fact that the Re contained in the continental crust was probably largely removed from what are now SCLM keels underlying the continents, which can remain isolated from the convecting upper mantle for billions of years (*Walker et al.*, 1989); and (3) most Re extracted from the mantle via basalt production is eventually recycled back into the upper and lower mantles. If the ophiolite and least-radiogenic MORB Os compositions are more representative of the convecting upper mantle, however, then it follows that the upper mantle has experienced little long-term Re depletion. It appears that additional work on abyssal peridotites is necessary to resolve this controversy.

### 5.3. Platinum-Osmium and Osmium-186/Osmium-188 Ratios in the Primitive Mantle

Constraints are much more limited for the $^{186}Os/^{188}Os$ of the upper mantle. The first study of $^{186}Os$ in the upper mantle, as with the first study of $^{187}Os$, reported data for Os-Ir alloys that sample presumed upper-mantle-derived ultramafic systems (*Walker et al.*, 1997a). Within uncertainty, these materials have the same $^{186}Os/^{188}Os$ ratios as chondritic meteorites. Subsequent studies of ophiolite-derived chromites and abyssal peridotites support the supposition that the $^{186}Os/^{188}Os$ of the upper mantle is chondritic (e.g., *Brandon et al.*, 1998a,b). The present database indicates an upper mantle $^{186}Os/^{188}Os$ of 0.119834 ± 2. Although few data exist for chondrites, there is not likely to be a detectable range in $^{186}Os/^{188}Os$ ratios for different types of chondrites, as is observed for $^{187}Os/^{188}Os$. This is because the limited range in Pt/Os (~±6%) in whole-rock samples of chondrites (*Horan et al.*, 1999) will not lead to detectable differences in $^{186}Os/^{188}Os$.

### 5.4. Implications

The chondritic $^{187}Os/^{188}Os$ and $^{186}Os/^{188}Os$ ratios in the modern upper mantle require that the time-integrated Re/Os and Pt/Os ratios of the upper mantle were also generally chondritic. It is instructive to consider how much perturbation in Re/Os and Pt/Os can be tolerated and still retain Os-isotopic compositions that are within the ranges of chon-

drites. First, we'll consider the Re-Os system constraints. Although the database is still limited, the total range in $^{187}Os/^{188}Os$ ratios reported for modern whole-rock carbonaceous, enstatite, and ordinary chondrites using high-precision techniques range from 0.1254 to 0.1304 (*Meisel et al.*, 1996; *Chen et al.*, 1999). Most chondrites, however, range only from 0.126 to 0.130 (*Meisel et al.*, 1996; *Walker et al.*, 1999b). We will presume the latter range reflects the Re/Os of the chondrites that are the most likely representatives of possible chondritic building blocks of planets. Since the major issue involved in this volume concerns the effects of early solar system processes, we will consider what effect perturbations from chondritic would be assuming they occurred when core formation occurred. For that age we will assume 4.5 Ga, although the absolute age within even a few hundred million years is not important for these calculations. In brief, any deviation outside of ±6% from the chondritic average Re/Os of 0.083 ($^{187}Re/^{188}Os$ = 0.4142) would lead to a modern $^{187}Os/^{188}Os$ outside the range specified (Fig. 10). This is a very strong constraint, and indicates that the long-term Re/Os of the upper mantle has been essentially identical to that of chondrites.

The constraint on upper mantle Pt/Os is less robust. If the $^{186}Os/^{188}Os$ of 0.119834 ± 2 is assumed for the upper mantle and for chondrites (*Walker et al.*, 1997a; *Brandon et al.*, 1998b), variations in the Pt/Os of the upper mantle could not vary beyond ±25% of the chondritic average. As with Re/Os, this result requires that the long-term Pt/Os was very similar, although not necessarily identical to that of chondritic meteorites.

### 5.5. Rhenium-Osmium Systematics of the Lunar Mantle

Not only do lunar basalts typically have very low concentrations of the HSE, including Re and Os (*Wolf et al.*, 1979; *Birck and Allègre*, 1994), the determination of absolute Re and Os abundances, as with the other HSE, in the lunar mantle are virtually impossible to precisely estimate from their basaltic abundances, as noted above. Nonetheless, the determination of the initial Os-isotopic compositions of younger lunar materials, such as mare basalts and orange and green volcanic glasses, *can* potentially provide the time-integrated Re/Os of lunar mantle reservoirs from the time of formation (~4.5 Ga) to the time of volcanism (~3.1–3.8 Ga). In the roughly 1 b.y. of growth, reservoirs within the lunar mantle could have grown to significantly more or less radiogenic compositions than chondrites.

The only Re-Os data that as yet exist for lunar volcanic materials is an analysis of Apollo 17 orange glasses rich soil 74220 (*Walker et al.*, 1998b). They concluded that the initial isotopic composition of two size fractions of this glass were "chondritic," although with relatively large uncertainty resulting from the analytical uncertainties associated with the Re measurement and extrapolation of its isotopic composition to 3.6 b.y. In that sense, the source region for the orange glasses appears to mimic the terrestrial upper mantle.

If additional data for other lunar volcanic materials support this observation, many of the conclusions regarding the origin of the HSE in the terrestrial mantle will also have to be applied to the lunar mantle.

## 6. HIGHLY SIDEROPHILE ELEMENTS AND TERRESTRIAL AND LUNAR MANTLE EVOLUTION: MULTIPLE WORKING HYPOTHESES

Accounting for current Os-isotopic and HSE concentrations of the upper mantle depends on an understanding of (1) how these elements were fractionated during core formation and metal-silicate segregation in the early Earth and (2) how mantle reservoirs evolved with time. As discussed above, the Os-isotopic composition of the primitive upper mantle has been well characterized in a variety of samples and approaches. It should be noted, however, that there are portions of the mantle that have significantly higher (radiogenic) Os-isotopic compositions (both $^{187}$Os and $^{186}$Os) than the PUM. Reconciling these compositions with models for the evolution of mantle reservoirs has proven to be a challenge, and we discuss the different types of models that are currently being considered.

### 6.1. Near-Chondritic Relative Highly Siderophile Element Ratios and Osmium-Isotopic Ratios in Earth's Primitive Upper Mantle

As demonstrated in the previous section, whatever process generated the Re/Os and Pt/Os ratios of the primitive upper mantle, it did not fractionate Re and Pt from Os by more than a maximum of ~±6% and ±25%, respectively, relative to likely chondritic precursors. Similarly, though less precisely, the HSE data for upper mantle materials indicates that the relative abundances of the other HSE are nearly chondritic (although not necessarily completely overlapping with chondritic ratios). These are important constraints that must be satisfied in any successful model for the origin of the HSE in the upper mantle. Core-mantle disequilibrium and heterogeneous accretion models (late veneer), which mix small amounts of chondritic material into the upper mantle after core formation, have been most successful in accounting for both moderately siderophile elements (MSE) and HSE in the primitive upper mantle (e.g., *Chou*, 1978; *Morgan*, 1986; *Jones and Drake*, 1986; *Newsom*, 1990; *O'Neill*, 1991b). Given the apparently ample opportunity for fractionation of these elements from chondritic relative ratios (see discussion in previous section), such a scenario is compelling. It should be noted, however, that the concept of a late veneer is one that has been applied to many geochemical subdisciplines and seems to have specific, and not necessarily compatible, meanings in different fields. For instance, the need for an oxidized veneer led some to propose (or imply) that the veneer must have been carbonaceous chondrite (e.g., *Chou et al.*, 1983). Osmium isotopes and noble gas isotopic measurements on chondritic and

mantle materials, however, permit a role for a reduced, or enstatite chondrite, veneer (*Meisel et al.*, 1996; *Pepin*, 1991). Additional high-precision Os-isotopic and HSE concentration data for terrestrial mantle materials and chondrites should help to further elucidate the similarities and differences between the terrestrial PUM and different chondrite groups in the future.

New high-pressure and high-temperature experimental results suggest the possibility of an equally simple scenario. Primitive upper mantle abundances of many of the moderately siderophile elements and Re may be reconciled with a scenario in which metal and peridotite magma equilibrated at the base of a deep magma ocean (e.g., *Li and Agee*, 1996; *Righter et al.*, 1997). This model for siderophile elements in the early Earth remains to be tested for the other HSE, but if metal segregation occurred at the relatively high temperatures and pressures that prevailed during such a magma ocean scenario, then the Os-isotopic constraints require that the D (Os) (LM/LS) must be within ±6% of D (Re) (LM/LS) in order to retain a chondritic Re/Os ratio in the mantle. A figure of ±25% must be assumed for the bulk distribution coefficients between Pt and Os. Whether these can be achieved with a high P-T metal-silicate equilibrium model will become clear when more experimental data are available.

### 6.2. Highly Siderophile Elements in Lunar Basalts and Mantle

Interpretation of the HSE contents of lunar material is model dependent, as there is still no consensus as to whether the Moon formed from material from the Earth's mantle or from the impactor, or whether the Moon was affected by the late veneer (addition of 0.008 Earth masses of chondritic material after metal had segregated from the upper mantle) that may have affected the Earth's upper mantle. Any interpretation of the lunar HSE data also requires some understanding of the phases that can fractionate HSE — metal, sulfide, and chromite — and thus the size and composition of the lunar core, whether there is a sulfide phase in the lunar mantle, and how chromite is distributed in the lunar interior. Given these uncertainties, we highlight a few more salient aspects of the HSE data to date.

*6.2.1. Nickel/iridium in the Earth and Moon.* Our inference that the most extremely siderophile elements (Ir, and with greater uncertainty Re, Os, and Au) in the Moon's mantle are depleted relative to Earth's mantle by only a modest factor, ~5, places an important constraint on the history of accretion and core/mantle differentiation of these two bodies. The Ni/Ir ratio of ~30× CI chondritic in the mantles of both Earth and the Moon is another important constraint. This ratio can be affected by a number of the processes identified above, such as metal-silicate equilibrium during core formation, or addition of a late veneer of chondritic material. For example, recent Lunar Prospector results (*Konopliv et al.*, 1998; *Hood and Zuber*, 2000) have confirmed that the Moon possesses a small core of its own, corresponding to about 1–3 wt% of the Moon's mass (depending on Fe-metal/FeS ratio). The depletion factor $\Delta$ implied by equilibrium between the silicate portion of the planet and its core can be calculated from the simple mass balance

$$\Delta = 1/(Dc + [1 - c])$$

where c is the weight fraction of the core and D is the metal/silicate distribution coefficient (bulk D for a liquid or solid mantle). For illustrative purposes, consider the case of a molten mantle in which the low pressure metal/silicate D for Ir is of the order $5 \times 10^5$ (*Jones, 1995*) and perhaps much higher (*Borisov and Palme, 1995*). Iridium should have been depleted in the lunar mantle to about 0.0002× the bulk-Moon concentration. For Ni, the metal/silicate D is roughly 25× lower (*Jones, 1995*), so core-mantle equilibration should have resulted in $\Delta \approx 0.005$ and Ni/Ir ≈ 25× the bulk-Moon ratio. Recent experimental work has shown that Ni becomes less siderophile at the high pressure and temperature conditions peculiar to the deep interior of the Earth (e.g., *Li and Agee, 1996*; *Walter et al., 2000*). Possibly pressure may be able to fractionate Ni from Ir in different core formation scenarios. However, according to very recent reports (*Fortenfant et al., 1998*; *Tschauner et al., 1998*), the high pressure of Earth's deep interior also moderates the metal/silicate D for many other siderophile elements, including Re, Os, and Ir. The origin of Ni/Ir fractionation awaits further study of Ir at higher pressures and temperatures. In any case, unless the bulk-Moon started with a much lower Ni/Ir than the Earth's mantle, the fractionation of Ni/Ir from a low-pressure core formation scenario is not observed.

*6.2.2. Evidence for a late veneer in the lunar mantle.* If the Moon formed almost entirely from material recycled from Earth's mantle, as advocated by *Ringwood* (1992) and implied by some simple forms of the giant impact hypothesis, the low-pressure (certainly <5 GPa) differentiation of the lunar core might be expected to deplete the silicate Moon by more than just a factor of 5 in a HSE such as Ir. Even supposing that the bulk-Moon HSE concentrations are independent of the terrestrial mantle, but perhaps in proportion to the metal + sulfide content (in analogy with ordinary chondrites of roughly similar MgO/FeO ratio), the bulk-Moon [Ir] would be no higher than about 23 ng/g. Equilibration with a 1 wt% core would result in a further $\Delta$ of 0.0002, so the Moon's mantle would be left with [Ir] = 0.005 ng/g, or about 0.001× [Ir] in Earth's mantle. Furthermore, large differences between metal-silicate D for Re and Ir would suggest that Re and Ir would be fractionated from each other during a metal segregation event, yet lunar Re/Ir ratios are close to chondritic (see *Drake, 1987,* and Fig. 7c). It seems difficult to avoid appealing to a "veneer" style accretion of primitive material to the lunar mantle, after the time when chemical exchange between the Moon's core and the main mass of its mantle had slowed to the point where mantle and core failed to equilibrate. Assuming a chondritic composition for the "veneer," our inference that mantle [Ir] is roughly 0.9 ng/g implies the veneer's mass

may constitute 0.2 wt% of the Moon. Even though mantle HSE concentrations appear to be 5× higher in the case of the Earth, it is far more difficult to assess whether a substantial veneer is present.

Any geochemical suggestions requiring a late veneer must of course be consistent with dynamical and physical arguments. Orbital integration calculations for lunar ejecta (*Gladman et al., 1995*) suggest that much of the material ejected from the Moon during an impact will attain velocities greater than the Earth's escape velocity. Such a situation may lead to an overall loss (or erosion) of material from the Moon. Even if material did accrete late to the Moon, it then must be mixed thoroughly into the lunar mantle (as also for the Earth; *Jones and Palme, 2000*) such that later basaltic magmatism can tap this material. Given that mantle convection in the Moon is likely to be inefficient, due to its rapid initial cooling (*Pritchard and Stevenson, 2000*), the most efficient way to mix this material into the lunar mantle may be while it is still molten (but after the core has separated), by turbulent convection of the magma ocean. If this were the case then the timing of the late veneer would have to be before 4.4 Ga, as this is the age of solidification of the lunar magma ocean (*Snyder et al., 2000*), but after 4.5 Ga, the differentiation age of the Moon as determined by Hf-W-isotopic studies (e.g., *Lee et al., 1997*). It should be borne in mind that in reality both a late veneer and high-P moderation of siderophile D could have played major roles in determining the final HSE composition of Earth's mantle, but we would like to emphasize that successful geochemical models will be those that also satisfy the physical and dynamical constraints highlighted above.

*6.2.3. Sulfide and chromite in the lunar mantle.* Although late-veneer scenarios provide a simple explanation for the elevated and chondritic relative proportions of some HSE in lunar basalts, there are gaps in our understanding of the behavior of critical HSE-bearing phases in the lunar mantle and during magma genesis. For instance, on the Earth, the nonchondritic-relative HSE patterns of mantle-derived basalt can be reconciled with derivation from a mantle with chondritic-relative HSE patterns. Some HSE ratios in lunar basalt are near-chondritic (e.g., Os/Ir, Re/Ir) while others (Au/Ir) are distinctly nonchondritic. Highly-siderophile-element concentrations of the lunar mantle must be calculated with a full understanding of the role of HSE-bearing phases in magma genesis: whether there is a sulfide phase in the lunar mantle, and how chromite is distributed in the lunar interior. The role of metal segregation in the Moon is clearly important, as discussed above, but of nearly equal importance is the role of chromite and sulfides. Chromite is likely to have been a liquidus phase in the lunar magma ocean and could have captured a fraction of HSE early in the Moon's history. And the most important question is whether sulfide is present in the lunar mantle.

The bulk S content of the Moon is not well constrained, but conservative estimates place it close to 100 ± 50 ppm (S. R. Taylor, personal communication), as compared to

150–300 ppm S in the primitive terrestrial mantle (*Lorand*, 1990). Sulfur contents of high $TiO_2$ basalts are ~2.5× higher than low $TiO_2$ basalt (1810 ± 280 µg/g compared to 690 ± 130 µg/g, based on data from *Gibson et al.*, 1977). High $TiO_2$ basalts from Apollos 11 and 17 were originally interpreted to be sulfide-saturated (*Gibson et al.*, 1976, 1977) on the basis of available experimental data (*Haughton et al.*, 1974). However, subsequent work by *Danckwerth et al.* (1979) showed that high $TiO_2$ basaltic melts will have a greater capacity for S (also borne out by steel-making studies; e.g., *Kärsrud*, 1984) and thus concluded that the high $TiO_2$ basalts are also S-undersaturated. Similarly, low $TiO_2$ basalts from Apollos 12 and 15, including some of the most MgO-rich, are sulfide-undersaturated (*Gibson et al.*, 1977). Because many lunar basalts are sulfide-undersaturated, they are likely to have high HSE concentrations.

## 6.3. Radiogenic Osmium and Fractionation of Rhenium, Platinum, and Osmium in the Modern Mantle

In the past 30 years, 5 geochemically unique mantle reservoirs have been identified based on major- and trace-element studies (*Hofmann*, 1997; *Carlson*, 1995) and lithophile isotope (Sr, Nd, Pb, and Hf) measurements: depleted MORB mantle (DMM), enriched mantle 1 (EM1), enriched mantle 2 (EM2), old recycled oceanic crust (HIMU), and focal zone (FOZO). Although the processes that have been hypothesized to have created these reservoirs are sometimes consistent with the enrichments in $^{187}Os$ and $^{186}Os$ observed (*Hauri and Hart*, 1993), the Os-isotopic characteristics of certain other systems, including some OIB and komatiite sources, are difficult to reconcile with these processes (cf. *Widom*, 1997; *Walker et al.*, 1999a; *Brandon et al.*, 1999).

Several mechanisms have been proposed to explain the $^{187}Os$ and $^{186}Os$ isotopic enrichments documented in some presumed plume-derived materials. These mechanisms include (1) incorporation of ancient recycled oceanic crust into a mantle source (*Hauri and Hart*, 1993), (2) the addition of small amounts of outer core metal (with presumably long-term enriched Re/Os) to an upwelling plume (*Walker et al.*, 1995, 1997a; *Brandon et al.*, 1998a), (3) addition of lower mantle materials (again with presumably long-term enriched Re/Os) to an upwelling mantle plume (*Walker et al.*, 1997b), and (4) existence of ancient mantle heterogeneities composed of garnetite (e.g., *Ringwood*, 1991) or trapped residual silicate melt, originating from a magma ocean (e.g., *Agee and Walker*, 1993). Mechanisms 1 and 2 have been extensively discussed in the literature, and largely involve processes that occurred relatively late in Earth history, so are not discussed here (e.g., *Hauri and Hart*, 1993; *Widom and Shirey*, 1996; *Bennett et al.*, 1996). Mechanisms 3 and 4, however, could reflect early Earth processes and are considered as follows.

Although the last decade of geophysical research abounds with evidence for whole mantle convection, there is still a strong possibility for some isolation of the upper

and lower mantles (e.g., *van der Hilst and Karason*, 1999), especially during the first 2 b.y. of Earth history. Petrologists have long sought convincing evidence for a compositionally layered mantle, yet even the most fundamental constraints, such as lower mantle *mg* ratio (*Arculus et al.*, 1990) or Fe/Si ratio (*Anderson*, 1989) remain uncertain. It is also not clear that the current 670-km boundary between the upper and lower mantle has been fixed over geological time. The nature of deep mantle-core equilibrium is only just beginning to be studied with cosmochemical balance and perspective (e.g., *Zerr et al.*, 1998; *Tschauner et al.*, 1998); initial work has revealed that if the lower mantle equilibrated with FeNi metal, it should contain 1 wt% Ni (*Tschauner et al.*, 1998). Primitive peridotite samples contain only 2000 µg/g (0.20 wt%) worldwide, so such Ni-rich lower mantle material has apparently not been tapped. If the upper and lower mantles have, at least in part, remained chemically distinct, this raises the question of whether the lower mantle has a distinctly different HSE composition, and consequently, different $^{187}Os$ and $^{186}Os$ characteristics relative to the generally well-characterized upper mantle.

A vertically stratified mantle could have been produced by at least two processes. One scenario that could lead to a mantle that is stratified primarily with respect to HSE is the late-veneer model noted above. The formation of the core could have stripped the entire mantle of the vast majority of HSE during a relatively short duration event. Subsequent accretion of chondritic materials, with their relatively high abundances of HSE, would potentially have established much higher concentrations of the HSE in the upper mantle than the lower, especially if there was little mixing between upper and lower mantles. Consequently, the lower mantle may have HSE concentrations and relative abundances that reflect metal-silicate partitioning. In such a scenario, Re (and Pt) may be much more compatible with silicate than Os, so a lower mantle developing very radiogenic $^{187}Os$ and $^{186}Os$ can be envisioned. Derivation of some portion of a mantle plume from such a source could potentially produce the $^{187}Os$ and $^{186}Os$ enrichments observed in some systems.

An alternate scenario that could generate both major-element and HSE compositional stratification between the upper and lower mantles is if the early Earth went through a stage with a deep (700–1000 km) magma ocean. The convective style would have initially been layered, given the distinct rheological behavior of a solid lower mantle and liquid upper mantle. The lower mantle would likely have developed a major-element composition that was slightly different from the upper mantle. The concentration of the HSE, and their relative abundances in a lower mantle generated in this manner, would reflect their partitioning behavior between deep mantle phases such as ferroan periclase, $MgSiO_3$ perovskite, majorite-rich garnet, and clinopyroxene. At present, this information is completely lacking. It is also interesting to consider the possibility that both scenarios could have occurred in unison.

Trapped silicate melt from an early magma ocean could also generate mantle reservoirs with distinct HSE characteristics. For example, an early magma ocean may convect turbulently in the interval between 0% and 70% crystallization, allowing crystals to be entrained in the convective movement. However, as the percentage of crystals reaches a critical point, turbulent convection ceases, gravitational settling can become an important force of differentiation, and crystals will tend to settle and liquids segregate to the surface (see discussion in *Solomatov,* 2000). Because silicate melts are more compressible than mantle silicates (such as olivine, garnet, pyroxenes, and feldspar), they can become denser than the surrounding mantle and be trapped at depth. In early Earth scenarios, such deep melts (perhaps komatiitic liquid with Re/Os ~1 and Pt/Os ~6; *Walker et al.,* 1988; *Rehkämper et al.,* 1999) may have been trapped within the mantle, and with further cooling, such melt pockets could have been "frozen" in the mantle as significant heterogeneities. Such trapped material would be easy to confuse with an early basaltic crust, but nonetheless would have a different origin and evolution.

## 7.  SUMMARY AND CONCLUSIONS

It is clear that analytical advances in both bulk sample and isotopic measurements have enhanced our understanding of the distribution of HSE in terrestrial and lunar samples. And experimental studies coupled with microbeam analytical techniques will lead to a better understanding of HSE in a variety of processes. Osmium-isotopic measurements on terrestrial mantle samples have constrained the PUM Re/Os and Pt/Os ratios to within ±6% and ±25%, respectively, of chondritic values. These are tight constraints that must be satisfied in any model for the origin of the Earth. Hopefully, lunar Re/Os and Pt/Os ratios will be constrained this well after much more lunar material has been analyzed. Several aspects of the Earth-Moon system remain poorly defined, including (1) the HSE composition of the lunar mantle, especially those other than Ir; (2) whether terrestrial HSE have been fractionated between the outer core and inner core, or between the lower mantle and upper mantle; and (3) the timing, composition, and areal extent of the late veneer on both the Earth and Moon. Future experimental and analytical efforts must focus on collecting data that can be used to resolve these issues, and we offer some suggestions.

First, any knowledge of lunar mantle HSE contents relies upon a full understanding of HSE behavior during magma genesis, because basaltic material is our probe into the lunar interior. Phases that control the distribution of HSE in magma genesis must be better defined in both terrestrial and lunar applications. Although metal, troilite, pyrrhotite, and sulfide liquid have all been identified as important host phases for HSE, oxides such as ilmenite, chromite, and magnetite have also been recognized as important. Yet a systematic understanding of HSE in oxides is not currently possible. Future work in this area will lead to a better understanding and appreciation of HSE correlations with Cr and V, two elements with affinities for cubic oxides. Test-

ing geochemical models on terrestrial igneous rock suites will also be important when applying such models to lunar suites.

Second, the behavior of HSE in magmatic settings is only beginning to be understood, but our knowledge of HSE in deep mantle phases is even worse. Testing geochemical and isotopic models for mantle evolution requires knowledge of the distribution of HSE between the upper and lower mantles and the inner and outer core. Partitioning studies involving deep mantle phases such as perovskites, garnets, pyroxenes, and ferroan periclases must be undertaken in order to examine possible HSE distributions within the mantle. Partitioning of HSE between solid and liquid metal must be studied across a range of pressures and metallic liquid compositions (S-, C-, O-, H-, and Si-bearing) so that the effect of light elements can be quantified (see *Hillgren et al.,* 2000). Such data will also allow more realistic modeling of core metal additions to the mantle over the history of Earth evolution.

Third, if a late veneer significantly affected the HSE abundances in the terrestrial mantle, there would most likely be evidence for such an event in lunar samples. Consequently, Os-isotopic studies of carefully selected and well-dated lunar samples, particularly lunar breccias, may yield information critical to evaluating this possibility.

*Acknowledgments.*   Reviews and comments of E. Hauri, J. Jones, J. W. Morgan, and M. J. Drake improved the clarity of this manuscript. We would like to acknowledge the support of NASA grants 57634 (RJW), NAG5-4215 (PHW), and NAG5-4084 (PI M. J. Drake), and NSF grants EAR9711454 (RJW) and EAR9706024 (M. J. Drake and KR).

## REFERENCES

Agee C. B. and Walker D. (1993) Olivine flotation in mantle melt. *Earth Planet. Sci. Lett., 114,* 315–324.

Agiorgitis G. and Wolf R. (1978) Aspects of osmium, ruthenium and iridium contents in some Greek chromites. *Chem. Geol., 23,* 267–272.

Allègre C. J. and Luck J.-M. (1980) Osmium isotopes as petrogenetic and geological tracers. *Earth Planet. Sci. Lett., 48,* 148–154.

Anders E., Ganapathy R., Keays R. R., Laul J. C., and Morgan J. W. (1971) Volatile and siderophile elements in lunar rocks: Comparison with terrestrial and meteoritic basalts. *Proc. Lunar Sci. Conf. 2nd,* pp. 1021–1036.

Anderson D. L. (1989) Composition of the Earth. *Science, 243,* 367–370.

Annell C. S. and Helz A. W. (1970) Emission spectrographic determination of trace elements in lunar samples from Apollo 11. *Proc. Apollo 11 Lunar Sci. Conf.,* pp. 991–994.

Arai T. and Warren P. H. (1997) "Large" (1.7 g) compositionally pristine diabase 14434: A lithology transitional between Mg-gabbronorite and high-Al mare basalt (abstract). *Meteoritics & Planet. Sci., 32,* A7–A8.

Arculus R. J., Holmes R. D., Powell R., and Righter K. (1990) Metal-silicate equilibrium and core formation. In *The Origin of the Earth* (H. E. Newsom and J. H. Jones, eds.), pp. 251–

271. Oxford, New York.

Auge T. (1988) Platinum-group minerals in the Tiebaghi and Vourinos ophilitic complexes: Genetic implications. *Can. Mineral., 26,* 177–192.

Baedecker P. A., Schaudy R., Elzie J. L., Kimberlin J., and Wasson J. T. (1971) Trace element studies of rocks and soils from Oceanus Procellarum and mare Tranquillitatis. *Proc. Apollo 11 Lunar Sci. Conf.,* pp. 1037–1061.

Baedecker P. A., Chou C.-L., Grudewicz E. B., and Wasson J. T. (1973) Volatile and siderophilic trace elements in Apollo 15 samples: Geochemical implications and characterization of the long-lived and short-lived extralunar materials. *Proc. Lunar Sci. Conf. 4th,* pp. 1177–1196.

Barnes S.-J. and Picard C. (1993) The behaviour of platinum-group elements during partial melting, crystal fractionation and sulphide segregation: An example from the Cape Smith Fold Belt, Northern Quebec. *Geochim. Cosmochim. Acta, 57,* 79–88.

Barnes S.-J., Naldrett A. J., and Gorton M. P. (1985) The origin of the fractionation of platinum group elements in terrestrial magmas. *Chem. Geol., 53,* 303–323.

Beattie P., Ford C., and Russell D. (1991) Partition coefficients for olivine-melt and orthopyroxene-melt systems. *Contrib. Mineral. Petrol., 109,* 212–224.

Bennett V. C., Esat T. M., and Norman M. D. (1996) Two mantle-plume components in Hawaiian picrites inferred from correlated Os-Pb isotopes. *Nature, 381,* 221–225.

Bernard A., Symonds R. B., and Rose W. I. (1990) Volatile transport and deposition of Mo, W and Re in high temperature magmatic fluids. *Appl. Geochem., 5,* 317–326.

Birck J. L. and Allègre C. J. (1994) Contrasting Re/Os magmatic fractionation in planetary basalts. *Earth Planet. Sci. Lett., 124,* 139–151.

Blanchard D. P., Haskin L. A., Jacobs J. W., Brannon J. C., and Korotev R. (1975) Major and trace element chemistry of Boulder 1 at Station 2, Apollo 17. *Moon, 14,* 359–371.

Borisov A. and Palme H. (1995) The solubility of iridium in silicate melts: New data from experiments with $Ir_{10}Pt_{90}$ alloys. *Geochim. Cosmochim. Acta, 59,* 481–485.

Borisov A. and Palme H. (1996). Experimental determination of the solubility of Au in silicate melts. *Mineral. Petrol., 56,* 297–312.

Borisov A. and Palme H. (1997) Experimental determination of the solubility of platinum in silicate melts. *Geochim. Cosmochim. Acta, 61,* 4349–4357.

Boynton W. V., Baedecker P. A., Chou C.-L., Robinson K. L., and Wasson J. T. (1975) Mixing and transport of lunar surface materials: Evidence obtained by the determination of lithophile, siderophile and volatile elements. *Proc. Lunar Sci. Conf. 6th,* pp. 2241–2259.

Brandon A. D., Creaser R. A., Shirey S. B., and Carlson R. W. (1996) Osmium recycling in subduction zones. *Science, 272,* 861–864.

Brandon A., Walker R. J., Morgan J. W., Norman M. D., and Prichard H. M. (1998a) Coupled [186]Os and [187]Os evidence for core-mantle interaction. *Science, 280,* 1570–1573.

Brandon A., Snow J., Walker R. J., and Morgan J. W. (1998b) [186]Os systematics of abyssal peridotites and Pt-Os evolution of the upper mantle. *Eos Trans. AGU, 79,* F1012.

Brandon A. D., Norman M. D., Walker R. J., and Morgan J. W. (1999) [186]Os-[187]Os systematics of Hawaiian picrites. *Earth Planet. Sci. Lett., 172,* 25–42.

Brown G. M., Emeleus C. H., Holland J. G., and Phillips R. (1970) Mineralogical, chemical and petrological features of Apollo 11 rocks and their relationship to igneous processes. *Proc. Apollo 11 Lunar Sci. Conf.,* pp. 195–219.

Brueckner H. K., Elhaddad M. A., Hamelin B., Hemming S., Kroner A., Reisberg L., and Seyler M. (1995) A Pan African origin and uplift for the gneisses and peridotites of Zabargad Island, Red Sea: A Nd, Sr, Pb and Os isotope study. *J. Geophys. Res., 100,* 22283–22297.

Brügmann G. E., Arndt N. T., Hofmann A. W., and Tobschall H. J. (1987) Noble metal abundances in komatiite suites from Alexo, Ontario, and Gorgona Island, Colombia. *Geochim. Cosmochim. Acta, 51,* 2159–2169.

Brunfelt A. O., Heier K. S., and Steinnes E. (1971) Determination of 40 elements in Apollo 12 materials by neutron activation analysis. *Proc. Lunar Sci. Conf. 2nd,* pp. 1281–1290.

Brunfelt A. O., Heier K. S., Nilssen B., Steinnes E., and Sundvoll B. (1972) Elemental composition of Apollo 15 samples (abstract). In *The Apollo 15 Lunar Samples* (J. W. Chamberlain and C. Watkins, eds.), pp. 195–197. The Lunar Science Institute, Houston.

Brunfelt A. O., Heier K. S., Nilssen B., Steinnes E., and Sundvoll B. (1974) Elemental composition of Apollo 17 fines and rocks. *Proc. Lunar Sci. Conf. 5th,* pp. 981–990.

Bulanova G. P., Griffin W. L., Ryan C. G., Shestakova O. Y., and Barnes S.-J. (1996) Trace elements in sulfide inclusions from Yakutian diamonds. *Contrib. Mineral. Petrol., 124,* 111–125.

Canil D. (1997) Vanadium partitioning and the oxidation state of Archaean komatiitic magmas. *Nature, 389,* 842–845.

Capobianco C. J. (1993). On the thermal decomposition of $MgRh_2O_4$ spinel and the solid solution $Mg(Rh, Al)_2O_4$. *Thermochim. Acta, 220,* 7–16.

Capobianco C. J. (1998) Ruthenium solubility in hematite. *Am. Mineral., 83,* 1152–1160.

Capobianco C. J. and Drake M. J. (1990) Partitioning of ruthenium, rhodium, and palladium between spinel and silicate melt and implications for platinum group element fractionation trends. *Geochim. Cosmochim. Acta, 54,* 869–874.

Capobianco C. J., Drake M. J., and Rogers P. S. Z. (1991) Crystal/melt partitioning of Ru, Rh, and Pd for silicate and oxide basaltic liquidus phases (abstract). In *Lunar and Planetary Science XXII,* pp. 179–180.

Capobianco C. J., Hervig R. L., and Drake M. J. (1994) Experiments on crystal/liquid partitioning of Ru, Rh and Pd for magnetite and hematite solid solutions crystallized from silicate melt. *Chem. Geol., 113,* 23–43.

Carlson R. W. (1995) Isotopic inferences on the chemical structure of the mantle. *J. Geodynamics, 20,* 365–386.

Chen J. H. and Wasserburg G. J. (1996) Live [107]Pd in the early solar system and implications for planetary evolution. In *Earth Processes: Reading the Isotopic Code* (A. Basu and S. Hart, eds.), pp. 1–20. AGU, Washington, DC.

Chen J. H., Papanastassiou D. A., and Wasserburg G. J. (1999) Re-Os systematics in chondrites and the fractionation of the platinum group elements in the early solar system. *Geochim. Cosmochim. Acta, 62,* 3379–3391.

Chesley J. T. and Ruiz J. (1998) Crust-mantle interaction in large igneous provinces: Implications from the Re-Os isotope systematics of the Columbia River flood basalts. *Earth Planet. Sci. Lett., 154,* 1–11.

Choi B.-G., Ouyang X., and Wasson J. T. (1995) Classification and origin of IAB and IIICD iron meteorites. *Geochim. Cosmochim. Acta, 59,* 593–612.

Chou C. L. (1978) Fractionation of siderophile elements in the Earth's upper mantle. *Proc. Lunar Planet. Sci. Conf. 9th*, pp. 219–230.

Chou C.-L., Boynton W. V., Sundberg L. L., and Wasson J. T. (1975) Volatiles on the surface of Apollo 15 green glass and trace-element distributions among Apollo 15 soils. *Proc. Lunar Sci. Conf. 6th*, pp. 1701–1727.

Chou C.-L., Shaw D. M., and Crocket J. H. (1983) Siderophile trace elements in the Earth's oceanic crust and upper mantle. *Proc. Lunar Planet. Sci. Conf. 13th*, in *J. Geophys. Res., 88*, A507–A518.

Cobb J. C. (1967) A trace-element study of iron meteorites. *J. Geophys. Res., 72*, 1329–1341.

Compston W., Chappell B. W., Arriens P. A., and Vernon M. J. (1970) The chemistry and age of Apollo 11 lunar material. *Proc. Apollo 11 Lunar Sci. Conf.*, pp. 1007–1027.

Compston W., Berry H., Vernon M. J., Chappell B. W., and Kaye M. J. (1971) Rubidium-strontium chronology and chemistry of lunar material from the Ocean of Storms. *Proc. Lunar Sci. Conf. 2nd*, pp. 1471–1485.

Creaser R. A., Papanastassiou D. A., and Wasserburg G. J. (1991) Negative thermal ion mass spectrometry of osmium, rhenium and iridium. *Geochim. Cosmochim. Acta, 55*, 397–401.

Crocket J. H. (1972) Some aspects of the geochemistry of Ru, Os, Ir and Pt in iron meteorites. *Geochim. Cosmochim. Acta, 36*, 517–535.

Crocket J. H. (1979) Platinum-group elements in mafic and ultra-mafic rocks: A survey. *Can. Mineral., 17*, 391–402.

Crocket J. H. and MacRae W. E. (1986) Platinum group element distribution in komatiitic and tholeiitic volcanic rocks from Munro Township, Ontario. *Econ. Geol., 81*, 1242–1251.

Crocket J. H., Fleet M. E., and Stone W. E. (1992) Experimental partitioning of osmium, iridium and gold between basalt melt and sulphide liquid at 1300°C. *Austral. J. Earth Sci., 39*, 427–432.

Crocket J. H., Fleet M. E., and Stone W. E. (1997) Implications of composition for experimental partitioning of platinum-group elements and gold between sulfide liquid and basalt melt: The significance of nickel content. *Geochim. Cosmochim. Acta, 61*, 4139–4149.

Daltry V. D. C. and Wilson A. H. (1997) Review of platinum-group mineralogy: compositions and elemental associations of the PG-minerals and unidentified PGE-phases. *Mineral. Petrol., 60*, 185–229.

Danckwerth P. A., Hess P. C., Rutherford M. J. (1979) The solubility of sulfur in high-TiO$_2$ mare basalts. *Proc. Lunar Planet. Sci Conf. 10th*, pp. 517–530.

DePaolo D. J. and Wasserburg G. J. (1976) Nd isotopic variations and petrogenetic models. *Geophys. Res. Lett., 3*, 249–252.

Drake M. J. (1987) Siderophile elements in planetary mantles and the origin of the Moon. *Proc. Lunar Planet. Sci Conf. 17th*, in *J. Geophys. Res., 92*, E377–E386

Drake M. J., Newsom H. E., and Capobianco C. J. (1989) V, Cr and Mn in the Earth, Moon, EPB, and SPB and the origin of the Moon. *Geochim. Cosmochim. Acta, 53*, 2101–2111.

Dulac J. (1969) Composes spinelles formes entre l'oxyde de ruthenium RuO$_2$ et les oxydes de certains metaux de transition. *Bull. Soc. fr. Mineral. Cristallogr., 92*, 487–488.

Ebihara M., Wolf R., Warren P. H., and Anders E. (1992) Trace elements in 59 mostly highlands Moon rocks. *Proc. Lunar Planet. Sci., Vol. 22*, pp. 417–426.

Echeverría L. M. (1982) Komatiites from Gorgona Island, Colum-

bia. In *Komatiites* (N. T. Arndt and E. G. Nisbet, eds.), pp. 199–209. Allen and Unwin.

Esser B. K. and Turekian K. K. (1993) The osmium isotopic composition of the continental crust. *Geochim. Cosmochim. Acta, 57*, 3093–3104.

Finnegan D. L., Miller T. L., and Zoller W. H. (1990) Iridium and other trace metal enrichments from Hawaiian volcanoes. *GSA Spec. Paper 247*, pp. 111–116.

Fleet M. E., Stone W. E., and Crockett J. H. (1991a) Partitioning of palladium, iridium, and platinum between sulfide liquid and basalt melt: Effects of melt composition, concentration, and oxygen fugacity. *Geochim. Cosmochim. Acta, 55*, 2545–2554.

Fleet M. E., Tronnes R. G., and Stone W. E. (1991b) Partitioning of platinum group elements in the Fe-O-S system to 11 GPa and their fractionation in the mantle and meteorites. *J. Geophys. Res., 96*, 21949–21958.

Fleet M. E. and Stone W. E. (1991) Partitioning of platinum-group elements in the Fe-Ni-S system and their fractionation in nature. *Geochim. Cosmochim. Acta, 55*, 245–253.

Fleet M. E., Chryssoulis S. L., Stone W. E., and Weisener C. G. (1993). Partitioning of platinum-group elements and Au in the Fe-Ni-Cu-S system: experiments on the fractional crystallization of sulfide melt. *Contrib. Mineral. Petrol., 115*, 36–44.

Fleet M. E., Crocket J. H., and Stone W. E. (1996) Partitioning of platinum-group elements (Os, Ir, Ru, Pt, Pd) and gold between sulfide liquid and basalt melt. *Geochim. Cosmochim. Acta, 60*, 2397–2412.

Fortenant S., Dingwell D. B., Rubie D. C., Hofmann A., Schiano P., Birck J. L., Gessmann C., Tubrett M., and Jenner G. (1998) Experimental investigation of osmium partitioning and implications for core formation (abstract). In *Origin of the Earth and Moon*, p. 8. LPI Contribution No. 957, Lunar and Planetary Institute, Houston.

Foster J. G., Lambert D. D., Frick L. R., and Maas R. (1996). Re-Os isotopic evidence for genesis of Archaean nickel ores from uncontaminated komatiites. *Nature, 382*, 703–706.

Freydier C., Ruiz J., Chelsey J., McCandless T., Munizaga F., and Rogers P. S. Z. (1997) Re-Os isotope systematics of sulfides from felsic igneous rocks: Application to base metal porphyry mineralization in Chile. *Geology, 25*, 775–778.

Fryer B. J. and Greenough J. D. (1992) Evidence for mantle heterogeneity from platinum group element abundances in Indian Ocean basalts. *Can. J. Earth Sci., 29*, 2329–2340.

Ganapathy R., Keays R. R., Laul J. C., and Anders E. (1970) Trace elements in Apollo 11 lunar rocks: Implications for meteorite influx and origin of Moon. *Proc. Apollo 11 Lunar Sci. Conf.*, pp. 1117–1142.

Garuti G., Oddone M., and Torres-Ruiz J. (1997) Platinum group element distribution in subcontinental mantle: evidence from the Ivrea Zone (Italy) and the Betic-Rifean cordillera (Spain and Morocco). *Can. J. Earth Sci., 34*, 444–465.

Gibson E. K. Jr., Usselman T. M., and Morris R. V. (1976) Sulfur in the Apollo 17 basalts and their source regions. *Proc. Lunar Sci. Conf. 7th*, pp. 1491–1505.

Gibson E. K. Jr., Brett R., and Andrawes F. (1977) Sulfur in lunar mare basalts as a function of bulk composition. *Proc. Lunar Sci. Conf. 8th*, pp. 1417–1428.

Gijbels R. H., Millard H. T. Jr., Desborough G. A., and Bartel A. J. (1974) Osmium, ruthenium, iridium and uranium in silicates and chromite from the eastern Bushveld Complex, South Africa. *Geochim. Cosmochim. Acta, 38*, 319–337.

Gladman B. J., Burns J. A., Duncan M. J., and Levison H. F.

(1995) The dynamical evolution of lunar impact ejecta. *Icarus, 118,* 302–321.

Gueddari K., Piboule M., and Amosse J. (1996) Differentiation of platinum group elements (PGE) and of gold during partial melting of peridotites in the lherzolitic massifs of the Betico-Rifean range (Ronda and Beni Bousera). *Chem. Geol., 134,* 181–197.

Halliday A. N., Lee D.-C., Christensen J. N., Rehkämper M., Yi W., Luo X., Hall C. M., Ballentine C. J., Pettke T., and Stirling C. (1998) Applications of multiple collector-ICP-MS to cosmochemistry, geochemistry, and paleoceanography. *Geochim. Cosmochim. Acta, 62,* 919–940.

Handler M. R., Bennett V. C., and Esat T. M. (1997) The persistence of off-cratonic lithospheric mantle: Os isotopic systematics of variably metasomatised southeast Australian xenoliths. *Earth Planet. Sci. Lett., 151,* 61–75.

Hanson B. Z. and Jones J. H. (1998) The systeamtics of $Cr^{3+}$ and $Cr^{2+}$ partitioning between olivine and liquid in the presence of spinel. *Am. Mineral., 83,* 669–684.

Hart S. R. and Kinloch E. D. (1989) Osmium isotope systematics in Witwatersrand and Bushveld ore deposits. *Econ. Geol., 84,* 1651–1655.

Hart S. R. and Ravizza G. E. (1996) Os partitioning between phases in lherzolite and basalt. In *Earth Processes: Reading the Isotopic Code* (A. Basu and S. Hart, eds.), pp. 123–134. AGU, Washington, DC.

Haskin L. A. and Warren P. H. (1991) Chemistry. In *Lunar Sourcebook: A User's Guide to the Moon* (G. H. Heiken, D. T. Vaniman, and B. M. French, eds.), pp. 357–474. Cambridge Univ., New York.

Hattori K. and Hart S.R. (1991) Osmium isotope ratios of platinum group minerals associated with ultramafic intrusions: Os isotopic evolution of the mantle. *Earth Planet. Sci. Lett., 107,* 499–514.

Haughton D. R., Roeder P. L., and Skinner B. F. (1974) Solubility of sulfur in mafic magmas. *Econ. Geol., 69,* 451–462.

Hauri E. H. and Hart S. R. (1993) Re-Os isotope systematics of HIMU and EMII oceanic island basalts from the south Pacific Ocean. *Earth Planet. Sci. Lett., 114,* 353–371.

Hauri E. H. and Hart S. R. (1997) Rhenium abundances and systematics in oceanic basalts. *Chem. Geol., 139,* 185–205.

Helmke P. A., Haskin L. A., Korotev R. L., and Ziege K. (1972) Rare earths and other trace elements in Apollo 14 samples. *Proc. Lunar Sci. Conf. 6th,* pp. 1275–1292.

Helmke P. A., Blanchard D. P., Haskin L. A., Telander K., Weiss C., and Jacobs J. W. (1973) Major and trace elements in igneous rocks from Apollo 15. *Moon, 8,* 129–148.

Hertogen J., Janssens M.-J., and Palme H. (1980) Trace elements in ocean ridge basalt glasses: implications for fractionations during mantle evolution and petrogenesis. *Geochim. Cosmochim. Acta, 44,* 2125–2143.

Hillgren V. J., Gessmann C. K., and Li J. (2000) An experimental perspective on the light element in Earth's core. In *Origin of the Earth and Moon* (R. M. Canup and K. Righter, eds.), this volume. Univ. of Arizona, Tucson.

Hood L. L. and Zuber M. (2000) Recent refinements in geophysical constraints on lunar origin and evolution. In *Origin of the Earth and Moon* (R. M. Canup and K. Righter, eds.), this volume. Univ. of Arizona, Tucson.

Hofmann A. W. (1997) Mantle geochemistry: the message from oceanic volcanism. *Nature, 385,* 219–229.

Horan M. F., Walker R. J., and Morgan J. W. (1999) High precision measurements of Pt and Os in chondrites. In *Lunar and Planetary Science XXX,* Abstract #1412. Lunar and Planetary Institute, Houston (CD-ROM).

Hubbard N. J., Gast P. W., Rhodes J. M., Bansal B. M., Wiesmann H., and Church S. E. (1972) Nonmare basalts: Part II. *Proc. Lunar Sci. Conf. 3rd,* pp. 1161–1179.

Hughes S. S., Delano J. W., and Schmitt R. A. (1989) Petrogenetic modeling of 74220 high-Ti orange volcanic glasses and the Apollo 11 and 17 high-Ti mare basalts. *Proc. Lunar Planet. Sci. Conf. 19th,* pp. 175–188.

Hughes T. C., Keays R. R., and Lovering J. F. (1973) Siderophile and volatile trace elements in Apollo 14, 15 and 16 rocks and fines: Evidence for extralunar component and Ti-, Au-, and Ag-enriched rocks in the ancient lunar crust (abstract). In *Lunar Science IV,* pp. 400–402.

Irving A. J. (1978) A review of experimental studies of crystal/liquid trace element partitioning. *Geochim. Cosmochim. Acta, 42,* 743–770.

Jagoutz E., Palme H., Baddenhausen H., Blum K., Cendales M., Dreibus G., Spettel B., Lorenz V., and Wänke H. (1979) The abundances of major, minor and trace elements in the Earth's mantle as derived from primitive ultramafic nodules. *Proc. Lunar Planet. Sci. Conf. 10th,* pp. 2031–2050.

Jones J. H. (1995) Experimental trace element partitioning. In *Rock Physics and Phase Relations, A Handbook of Physical Constants* (T. J. Ahrens, ed.), pp. 73–104. AGU, Washington, DC.

Jones J. H. and Drake M. J. (1986) Geochemical constraints on core formation in the Earth, *Nature, 322,* 221–228.

Jones J. H. and Palme H. (2000) Geochemical constraints on the origin of the Earth and Moon. In *Origin of the Earth and Moon* (R. M. Canup and K. Righter, eds.), this volume. Univ. of Arizona, Tucson.

Kärsrud K. (1984) Sulphide capacities in $TiO_2$ containing slags I. Sulphide capacities in $FeO$-$TiO_2$ melts at 1500°C. *Scand. J. Metallur., 13,* 173–175.

Kimura K., Lewis R. S., and Anders E. (1974) Distribution of gold and rhenium between nickel-iron and silicate melts: implications for the abundance of siderophile elements on the Earth and Moon. *Geochim. Cosmochim. Acta, 38,* 683–701.

Konopliv A. S., Binder A. B., Hood L. L., Kucinskas A. B., Sjogren W. L., and Williams J. G. (1998) Improved gravity field of the Moon from Lunar Prospector. *Science, 281,* 1476–1480.

Korzhinsky M. A., Tkachenko S. I., Shmulovich K. I., Tarau Y. A., and Steinberg G. S. (1994) Discovery of a pure rhenium mineral at Kudriavy volcano. *Nature, 369,* 51–52.

Kushiro I. and Haramura H. (1971) Major element variation and possible source materials of some Apollo 12 crystalline rocks. *Science, 171,* 1235–1237.

Krutzsch B. and Kemmler-Sack S. (1983) Sauerstoff — Spinelle mit ruthenium und iridium. *Materials Res. Bull., 18,* 647–652.

Lambert D. D., Walker R. J., Morgan J. W., Shirey S. B., and Carlson R. W. (1989) Rhenium-osmium and samarium-neodymium isotope systematics of the Stillwater Complex. *Science, 244,* 1169–1174.

Lauer H. V. Jr. and Jones J. H. (1998) Partitioning of Pt and Os between solid and liquid metal in the iron-nickel-sulfur system (abstract). In *Lunar and Planetary Science XXIX,* Abstract #1796. Lunar and Planetary Institute, Houston (CD-ROM).

Laul J. C., Hill D. W., and Schmitt R. A. (1974) Chemical studies of Apollo 16 and 17 samples. *Proc. Lunar Sci. Conf. 5th,* pp. 1047–1066.

Lee D-C., Halliday A. N., Snyder G. A., and Taylor L. A. (1997)

The age and origin of the Moon. *Science, 278,* 1098–1103.

Leeman W. P. and Lindstrom D. J. (1978) Partitioning of Ni⁻ between basaltic and synthetic melts and olivines — an experimental study. *Geochim. Cosmochim. Acta, 42,* 801–816.

Li J. and Agee C. B. (1996) Geochemistry of mantle-core formation at high pressure. *Nature, 381,* 686–689.

Li C., Naldrett A. J., Rucklidge J. C., and Kilius L. R. (1993) Concentrations of platinum-group elements and gold in sulfides from the Strathcona Deposit, Sudbury, Ontario. *Can. Mineral., 30,* 523–531.

Lorand J. P. (1990) Are spinel lherzolite xenoliths representative of the abundance of sulfur in the upper mantle? *Geochim. Cosmochim. Acta, 54,* 1487–1493.

Ma M.-S. and Schmitt R. A. (1978) Chemistry of 10085 fragments and Apollo 11 basalts and breccias (abstract). In *Lunar and Planetary Science IX,* pp. 678–680. Lunar and Planetary Institute, Houston.

Ma M.-S. and Schmitt R. A. (1980) Luna 24 VLT microgabbro and recrystallized basalt — New chemical data (abstract). In *Lunar and Planetary Science XI,* pp. 646–648. Lunar and Planetary Institute, Houston.

Ma M.-S., Murali A. V., and Schmitt R. A. (1976) Chemical constraints for mare basalt genesis. *Proc. Lunar Sci. Conf. 7th,* pp. 1673–1696.

Ma M.-S., Schmitt R. A., Wentworth S., Taylor G. J., Warner R. D., and Keil K. (1980) VLT mare basalt and aluminous mare basalt from Apollo 17 drill core (abstract). In *Lunar and Planetary Science XI,* pp. 655–657. Lunar and Planetary Institute, Houston.

Ma M.-S., Liu Y.-G., and Schmitt R. A. (1981) A chemical study of individual green glasses and brown glasses from 15426: Implications for their petrogenesis. *Proc. Lunar Planet. Sci. 12B,* pp. 915–933.

Malvin D. J., Drake M. J., Benjamin T. M., Duffy C. J., and Hollander M. (1986) Experimental partitioning studies of siderophile elements amongst lithophile phases: preliminary results using PIXE microprobe analysis (abstract). In *Lunar and Planetary Science XVII,* pp. 514–515. Lunar and Planetary Institute, Houston.

Mao H.-K. and Hemley R. J. (1998) New windows on the Earth's deep interior. In *Ultra High Pressure Mineralogy: Physics and Chemistry of the Earth's Deep Interior* (R. J. Hemley, ed.), pp. 1–30. *Rev. Mineral., 37.*

Marcantonio F., Zindler A., Reisberg L., and Mathez E. A. (1993) Re-Os isotopic systematics in chromites from the Stillwater Complex, Montana, USA. *Geochim. Cosmochim. Acta, 57,* 4029–4037.

Markey R. J., Stein H. J., and Morgan J. W. (1998) Highly precise Re-Os age for molybdenite using alkaline fusion and NTIMS. *Talanta, 45,* 935–946.

Martin C. E. (1991) Osmium isotopic characteristics of mantle derived rocks. *Geochim. Cosmochim. Acta, 55,* 1421–1434.

Martin C. E., Carlson R. W., Shirey S. B., Frey F. A., and Chen C.-Y. (1994) Os isotopic variations in basalts from Haleakala Volcano, Maui, Hawaii: A record of magmatic processes in oceanic mantle and crust. *Earth Planet. Sci Lett., 128,* 287.

Mathez E. A. and Peach C. L. (1990) Geochemistry of platinum group elements in mafic and ultramafic rocks. In *Ore Deposition Associated with Magmas* (J. A. Whitney and A. J. Naldrett, eds.), pp. 33–41. *Rev. Econ. Geol., 4.*

McCandless T. E., Ruiz J., and Campbell A. R. (1993) Rhenium behavior in molybdenite in hypogene and near-surface environments: Implications for Re-Os geochronometry. *Geochim.*

*Cosmochim. Acta, 57,* 889–905.

McCandless T. E., Ruiz J., Adair B. I., and Freydier C. (1999) Re-Os and Pd/Ru variations in chromitites from the Critical Zone, Bushveld Complex, South Africa. *Geochim. Cosmochim. Acta, 63,* 911–923.

McDonough W. F. and Sun S.-s. (1995) The composition of the Earth. *Chem. Geol., 120,* 223–253.

Meisel T., Walker R. J., and Morgan J. W. (1996) The osmium isotopic composition of the Earth's primitive upper mantle. *Nature, 383,* 517–520.

Mitchell R. H. and Keays R. R. (1981) Abundance and distribution of gold, palladium and iridium in some spinel and garnet lherzolites: implications for the nature and origin of precious metal-rich intergranular components in the upper mantle. *Geochim. Cosmochim. Acta, 45,* 2425–2442.

Morgan J. W. (1986) Ultramafic xenoliths: clues to Earth's late accretionary history. *J. Geophys. Res., 91,* 12375–12387.

Morgan J. W. and Baedecker P. A. (1983) Elemental composition of sulfide particles from an ultramafic xenolith and the siderophile element content of the upper mantle (abstract). In *Lunar and Planetary Science XV,* pp. 513–514. Lunar and Planetary Institute, Houston.

Morgan J. W. and Wandless G. A. (1984) Surface correlated trace elements 15426 lunar glasses (abstract). In *Lunar and Planetary Science XV,* pp. 562–563. Lunar and Planetary Institute, Houston.

Morgan J. W., Wandless G. A., Petrie R. K., and Irving A. J. (1981) Composition of the Earth's upper mantle — I. Siderophile trace elements in ultramafic nodules. *Tectonophysics, 75,* 47–67.

Morgan J. W., Ganapathy R., Higuchi H., Krähenbühl U., and Anders E. (1974) Lunar basins: Tentative characterization of projectiles, from meteoritic elements in Apollo 17 boulders. *Proc. Lunar Sci. Conf. 5th,* pp. 1703–1736.

Morgan J. W., Ganapathy R., Higuchi H., Krähenbühl U. (1976) Volatile and siderophile elements in anorthositic rocks from Fiskenasset, West Greenland: comparison with lunar and meteoritic analogues. *Geochim. Cosmochim. Acta, 40,* 861–887.

Morgan J. W., Horan M. F., Walker R. J., and Grossman J. N. (1995) Rhenium-osmium concentration and isotope systematics in group IIAB iron meteorites. *Geochim. Cosmochim. Acta, 59,* 2331–2344.

Muller O. and Roy R. (1965) Synthesis and crystal structure of $Mg_2PtO_4$ and $Zn_2PtO_4$. *Material. Res. Bull., 4,* 39–43.

Murali A. V., Ma M.-S., and Schmitt R. A. (1976) Mare basalt 60639, another eastern mare basalt (abstract). In *Lunar Science VII,* pp. 583–584. The Lunar Science Institute, Houston.

Naldrett A. J. and Barnes S.-J. (1986) The behavior of platinum group elements during fractional crystallization and partial melting. *Fortschr. Mineral., 64,* 113–133.

Naldrett A. J., Hoffman E. L., Green A. H., Chou C.-L., and Naldrett S. R. (1979) The composition of Ni-Sulfide ores, with particular reference to their content of PGE and Au. *Can. Mineral., 17,* 403–415.

Neal C. R., Hacker M. D., Snyder G. A., Taylor L. A., Liu Y.-G., and Schmitt R. A. (1994) Basalt generation at the Apollo 12 site, Part 1: New data, classification, and re-evaluation. *Meteoritics, 29,* 334–348.

Neal C. R., Jain J. C., Snyder G. A., and Taylor L. A. (1999) Platinum group elements from the ocean of storms: evidence of two cores forming? (abstract). In *Lunar and Planetary Science XXX.* Lunar and Planetary Institute, Houston (CD-ROM).

Nell J. and O'Neill H. S. C. (1997a) Gibbs free energy of formation and heat capacity of PdO: A new calibration of the Pd-

PdO buffer to high temperatures and pressures. *Geochim. Cosmochim. Acta, 60,* 2487–2493.

Nell J. and O'Neill H. S. C. (1997b) The Gibbs free energy of formation and heat capacity of β-$Rh_2O_3$ and $MgRh_2O_4$, the MgO-Rh-O phase diagram, and constraints on the stability of $Mg_2Rh^{4+}O_4$. *Geochim. Cosmochim. Acta, 61,* 4159–4171.

Newsom H. (1990) Accretion and core formation in the Earth: Evidence from siderophile elements. In *The Origin of the Earth* (H. E. Newsom and J. H. Jones, eds.), pp. 273–288. Oxford Univ., New York.

Newsom H. (1995) Composition of the solar system, planets, meteorites, and major terrestrial reservoirs. In *Rock Physics and Phase Relations, A Handbook of Physical Constants* (T. J. Ahrens, ed.), pp. 159–189. AGU, Washington, DC.

Nichiporuk W. and Brown H. (1965) The distribution of platinum and palladium metals in iron meteorites and in the metal phase of ordinary chondrites. *J. Geophys. Res., 70,* 459–470.

Nyquist L. E. and Shih C.-Y. (1992) The isotopic record of lunar basaltic volcanism. *Geochim. Cosmochim. Acta, 56,* 2213–2234.

O'Neill H. S. C. (1991) The origin of the Moon and the early history of the Earth — a chemical model. Part 1: The Moon. *Geochim. Cosmochim. Acta, 55,* 1135–1157.

O'Neill H. St. C. (1991) The origin of the Moon and the early history of the Earth — A chemical model. Part 2: The Earth. *Geochim. Cosmochim. Acta 55,* 1159–72.

O'Neill, H. S. C. and Nell J. (1997) Gibbs free energies of formation of $RuO_2$, $IrO_2$, and $OsO_2$: A high-temperature electrochemical and calorimetric study. *Geochim. Cosmochim. Acta, 61,* 5279–5293.

Page N. J. (1984) Palladium, platinum, rhodium, ruthenium and iridium in peridotites and chromitites from ophiolite complexes in Newfoundland. *Can. Mineral., 22,* 137–149.

Parry S. J. (1984) Abundance and distribution of palladium, platinum, iridium and gold in some oxide minerals. *Chem. Geol., 43,* 115–125.

Pattou L., Lorand J., and Gros M. (1996) Non-chondritic platinum group element ratios in the Earth's mantle. *Nature, 379,* 712–715.

Peach C. L., Mathez E. A., and Keays R. R. (1990) Sulfide melt–silicate melt distribution coefficients for noble metals and other chalcophile elements as deduced from MORB: Implications for partial melting. *Geochim. Cosmochim. Acta, 54,* 3379–3389.

Peach C. L. and Mathez E. A. (1996) Constraints on the formation of platinum-group element deposits in igneous rocks. *Econ. Geol., 91,* 439–450.

Peach C. L., Mathez E. A., Keays R. R., and Reeves S. J. (1994) Experimentally determined sulfide-melt silicate melt partition coefficients for iridium and palladium. *Chem. Geol., 117,* 361–377.

Pearson D. G., Shirey S. B., Harris J. W., and Carlson R. W. (1998) Sulphide inclusions in diamonds form the Koffiefontein kimberlite, S. Africa: constraints on diamond ages and mantle Re-Os systematics. *Earth Planet. Sci. Lett., 160,* 311–326.

Pegram W. J. and Allègre C.-J. (1992) Osmium isotopic compositions from oceanic basalts. *Earth Planet. Sci. Lett., 111,* 59–68.

Pepin R. O. (1991) On the origin and early evolution of terrestrial planet atmospheres and meteoritic volatiles. *Icarus, 92,* 2–79.

Pernicka A. E. and Wasson J. T. (1987) Palladium, rhenium, osmium, platinum and gold in iron meteorites. *Geochim. Cosmochim. Acta, 51,* 1717–1726.

Peucker-Ehrenbrink B. and Blum J. D. (1998) Re-Os isotope systematics and weathering of Precambrian crustal rocks: implications for the marine isotope record. *Geochim. Cosmochim. Acta, 62,* 3193–3203.

Prichard H. M. and Tarkian M. (1988) Platinum and palladium minerals from two PGE-rich localities in the Shetland ophiolite complex. *Can. Mineral., 26,* 979–990.

Pritchard M. E. and Stevenson D. (2000) Thermal aspects of a lunar origin by giant impact. In *Origin of the Earth and Moon* (R. M. Canup and K. Righter, eds.), this volume. Univ. of Arizona, Tucson.

Ravizza G. and Pyle D. (1997) PGE and Os isotopic analyses of single sample aliquots with NiS fire assay preconcentration. *Chem. Geol., 141,* 251–268.

Razin L. V. and Khomenko G. A. (1969) Accumulation of osmium, ruthenium, and the other platinum-group metals in chrome spinel in platinum-bearing dunites. *Geokhimiya, 6,* 659–672.

Rehkämper, M., Halliday A. N., Barfod D., Fitton J. G., and Dawson J. B. (1997) Platinum-group element abundance patterns in different mantle environments. *Science, 278,* 1595–1598.

Rehkämper M., Halliday A. N., Fitton J. G., Lee D.-C., Wieneke M., and Arndt N. T. (1999) Ir, Ru, Pt and Pd in basalts and komatiites: New constraints for the geochemical behavior of the platinum-group elements in the mantle. *Geochim. Cosmochim. Acta,* in press.

Reisberg L., Zindler A., Marcantonio F., White W., Wyman D., and Weaverm B. (1993) Os isotope systematics in ocean island basalts. *Earth Planet. Sci. Lett., 120,* 149–167.

Rhodes J. M. and Blanchard D. P. (1980) Chemistry of Apollo 11 low-K mare basalts. *Proc. Lunar Planet. Sci. Conf. 11th,* pp. 49–66.

Rhodes J. M., Hubbard N. J., Wiesmann H., Rodgers K. V., Brannon J. C., and Bansal B. M. (1976) Chemistry, classification, and petrogenesis of Apollo 17 mare basalts. *Proc. Lunar Sci. Conf. 7th,* pp. 1467–1490.

Rhodes J. M., Blanchard D. P., Dungan M. A., Brannon J. C., and Rodgers K. V. (1977) Chemistry of Apollo 12 mare basalts: Magma types and fractionation processes. *Proc. Lunar Sci. Conf. 8th,* pp. 1305–1338.

Righter K. and Hauri E. H. (1998) Compatibility of rhenium in garnet during mantle melting and magma genesis. *Science, 280,* 1737–1741.

Righter K., Capobianco C. J., and Drake M. J. (1995) Experimental constraints on the partitioning of Re between augite, olivine, melilite and silicate liquid at high oxygen fugacities (>NNO). *Eos Trans. AGU, 76,* F698.

Righter K., Drake M. J., and Yaxley G. (1997) Prediction of siderophile element metal-silicate partition coefficients to 20 GPa and 2800°C: the effect of pressure, temperature, $f_{O_2}$ and silicate and metallic melt composition. *Phys. Earth Planet. Inter., 100,* 115–134.

Righter K., Chesley J. T., Geist D., and Ruiz J. (1998) Behavior of Re during magma fractionation: an example from Volcan Alcedo, Galapagos Archipelago. *J. Petrol., 39,* 785–795.

Ringwood A. E. (1991) Phase transformations and their bearing on the constitution and dynamics of the mantle. *Geochim. Cosmochim. Acta, 55,* 2083–2110.

Ringwood A. E. (1992) Volatile and siderophile element geochemistry of the Moon: a reappraisal. *Earth Planet. Sci. Lett., 111,* 537–555.

Rose H. J. Jr., Cuttitta F., Dwornik E. J., Carron M. K., Christian R. P., Lindsay J. R., Ligon D. T., and Larson R. R. (1970) Semimicro x-ray fluorescence analysis of lunar samples. *Proc.*

*Apollo 11 Lunar Sci. Conf.*, pp. 1493–1497.

Rose H. J., Cuttitta F., Berman S., Brown F. W., Carron M. K., Christian R. P., Dwornik E. J., and Greenland L. P. (1974) Chemical composition of rocks and soils at Taurus-Littrow. *Proc. Lunar Sci. Conf. 5th*, pp. 1119–1133.

Rose H. J. Jr., Baedecker P. A., Berman S., Christian R. P., Dwornik E. J., Finkelman R. B., and Schnepfe M. M. (1975) Chemical composition of rocks and soils returned by the Apollo 15, 16 and 17 missions. *Proc. Lunar Sci. Conf. 6th*, pp. 1363–1373.

Roy-Barman M. and Allègre C. J. (1994) $^{187}Os/^{186}Os$ rations of mid-ocean ridge basalts and abyssal peridotites. *Geochim. Cosmochim. Acta, 58*, 5043–5054.

Roy-Barman M. and Allègre C. J. (1995) $^{187}Os/^{186}Os$ in oceanic island basalts: tracing oceanic crust recycling in the mantle. *Earth Planet. Sci. Lett., 129*, 145–161.

Roy-Barman M., Luck J.-M., and Allègre C. J. (1996) Os isotopes in orogenic lherzolite massifs and mantle heterogeneites. *Chem. Geol., 130*, 55–64.

Roy-Barman M., Wasserburg G. J., Papanastassiou D. A., and Chaussidon M. (1998) Osmium isotopic compositions and Re-Os concentrations in sulfide globules from basaltic glasses. *Earth Planet. Sci. Lett., 154*, 331–347.

Ryder G. (1985) *Catalog of Apollo 15 Rocks*. Curatorial Facility Publ. No. 20787, NASA Johnson Space Center, Houston. 1296 pp.

Salpas P. A., Taylor L. A., and Lindstrom M. M. (1987) Apollo 17 KREEPy basalts: evidence for nonuniformity of KREEP. *Proc. Lunar Planet. Sci. Conf. 17th*, in *J. Geophys. Res., 92*, E340–E348.

Schiano P., Birck J.-L., and Allègre C. J. (1997) Osmium-strontium-neodymium-lead isotopic covariations in mid-ocean ridge basalt glasses and the heterogeneity of the upper mantle. *Earth Planet. Sci. Lett., 150*, 363–379.

Scott E. R. D. and Wasson J. T. (1976) Chemical classification of iron meteorites — VII. Groups IC, IIE, IIIF and 97 other irons. *Geochim. Cosmochim. Acta, 40*, 103–115.

Seifert S. and Ringwood A. E. (1988) The lunar geochemistry of chromium and vanadium. *Earth Moon Planets, 40*, 45–70.

Sharma M., Papanastassiou D. A., and Wasserburg G. J. (1997) The concentration and isotopic composition of osmium in the oceans. *Geochim. Cosmochim. Acta, 61*, 3287–3299.

Shen J. J., Papanastassiou D. A., and Wasserburg G. J. (1996) Precise Re-Os determinations and systematic of iron meteorites. *Geochim. Cosmochim. Acta, 60*, 2887–2900.

Shirey S. B. and Walker R. J. (1998) The Re-Os isotope system in cosmochemistry and high-temperature geochemistry. *Annu. Rev. Earth Planet. Sci., 26*, 423–500.

Smales A. A., Mapper D., and Fouché K. F. (1967) The distribution of some trace elements in iron meteorites, as determined by neutron activation. *Geochim. Cosmochim. Acta, 31*, 673–720.

Smoliar M. I., Walker R. J., and Morgan J. W. (1996) Re-Os ages of Group IIA, IIIA, IVA, and IVB iron meteorites. *Science, 271*, 1099–1102.

Snow J. E. and Reisberg L. (1995) Os isotope systematics of the MORB mantle: results from altered abyssal peridotites. *Earth Planet. Sci. Lett., 133*, 411–421.

Snow J. E. and Schmidt G. (1998) Constraints on Earth accretion deduced from noble metals in the oceanic mantle. *Nature, 391*, 166–169.

Snyder G. A., Neal C. R., Jain J., and Taylor L. A. (1997a) A

"stormy" sortie for pristine rocks in lunar soils: 1. Trace-element compositions of basalts and impact melts from Apollo 12 (abstract). In *Lunar and Planetary Science XXVIII*, pp. 1349–1350. Lunar and Planetary Institute, Houston.

Snyder G. A., Neal C. R., O'Neill J. A., Jain J., and Taylor L. A. (1997b) Platinum-group elements (PGEs) and gold (Au) in the lunar regolith: Routine analysis by ultrasonic nebulization-inductively coupled plasma-mass spectrometry (USN-ICP-MS) (abstract). In *Lunar and Planetary Science XXVIII*, pp. 1353–1354. Lunar and Planetary Institute, Houston.

Snyder G. A., Borg L. E., Nyquist L., and Taylor L. A. (2000) Chronology and isotopic constraints on lunar evolution. In *Origin of the Earth and Moon* (R. M. Canup and K. Righter, eds.), this volume. Univ. of Arizona, Tucson.

Solomatov V. (2000) Fluid dynamics of magma oceans. In *Origin of the Earth and Moon* (R. M. Canup and K. Righter, eds.), this volume. Univ. of Arizona, Tucson.

Stanton R. L. (1993) *Ore Elements in Arc Lavas*. Clarendon, Oxford. 391 pp.

Stein H. J., Sundblad K., Markey R., Morgan J. W., and Motuza G. (1998a) Re-Os ages for Archean molybdenite and pyrite, Kuittila, Finland and Proterozoic molybdenite, Kabeliai, Lithuania: A metamorphic and metasomatic test for the chronometer. *Mineralium Deposita, 33*, 329–345.

Stein H. J., Markey R., Morgan J. W., Du A., and Sun Y. (1998b) Highly precise and accurate Re-Os ages for molybdenite, East Quinling molybdenite belt, Shanxi Province, China. *Econ. Geol., 92*, 827–835.

Stein H. J., Morgan J. W., and Markey R. J. (1998c) A Re-Os study of the Suwalki anorthosite massif, northeast Poland. Sixth EUROBRIDGE workshop, Kiev, Ukraine. *Geophys. J., 4*, 111–114.

Stockman H. W. and Hlava P. F. (1984). Platinum-group minerals in Alpine chromitites from southwestern Oregon. *Econ. Geol., 79*, 491–508.

Stone W. E., Crocket J. H., and Fleet M. E. (1990) Partitioning of palladium, iridium, platinum, and gold between sulfide liquid and basalt melt at 1200°C. *Geochim. Cosmochim. Acta, 54*, 2341–2344.

Taylor S. R. (1982) *Planetary Science: A Lunar Perspective*. Lunar and Planetary Institute, Houston. 512 pp.

Taylor S. R., Rudowski R., Muir P., Graham A., and Kaye M. (1971) Trace element chemistry of lunar samples from the Ocean of Storms. *Proc. Lunar Sci. Conf. 2nd*, pp. 1083–1099.

Tschauner O., Zerr A., Specht S., Rocholl A., Boehler R., and Palme H. (1998) Partitioning of nickel and cobalt between metal and silicate perovskite up to 80 GPa. *Nature, 368*, 604–607.

Van der Hilst R. D. and Karason H. (1999) Compositional heterogeneity in the bottom 1000 kilometers of Earth's mantle: toward a hybrid convection model. *Science, 283*, 1885–1888.

Vogel D. and Keays R. R. (1997) The petrogenesis and platinum-group element geochemistry of the Newer Volcanic Province, Victoria, Australia. *Chem. Geol., 136*, 181–204.

Volkening J., Walczak T., and Heumann K. G. (1991) Osmium isotope ratio determinations by negative thermal ionization mass spectrometry. *Intl. J. Mass Spectrom. Ion Proc., 105*, 147–159.

Walker R. J., Shirey S. B., and Stecher O. (1988) Comparative Re-Os, Sm-Nd and Rb-Sr isotope and trace element systematics for Archean komatiite flows from Munro Township, Abitibi Belt, Ontario. *Earth Planet. Sci. Lett., 87*, 1–12.

Walker R. J., Carlson R. W., Shirey S. B., and Boyd F. R. (1989) Os, Sr, Nd, and Pb isotope systematics of southern African peridotite xenoliths: Implications for the chemical evolution of subcontinental mantle. *Geochim. Cosmochim. Acta, 53,* 1583–1595.

Walker R. J., Morgan J. W., Naldrett A. J., Li C., and Fassett J. D. (1991) Re-Os isotope systematics of Ni-Cu sulfide ores, Sudbury Igneous Complex, Ontario: evidence for a major crustal component. *Earth Planet. Sci. Lett., 105,* 416–429.

Walker R. J., Morgan J. W., and Horan M. F. (1995) Osmium-187 enrichment in some plumes: evidence for core-mantle interaction? *Science, 269,* 819–822.

Walker R. J., Hanski E. J., Vuollo J., and Liipo J. (1996) The Os isotopic composition of Proterozoic upper mantle: evidence for chondritic upper mantle from the Outokumpu ophiolite, Finland. *Earth Planet. Sci. Lett., 141,* 161–173.

Walker R. J., Morgan J. W., Beary E. S., Smoliar M. I., Czamanske G. K., and Horan M. F. (1997a) Applications of the $^{190}Pt$-$^{186}Os$ isotope system to geochemistry and cosmochemistry. *Geochim. Cosmochim. Acta, 61,* 4799–4807.

Walker R. J., Morgan J. W., Hanski E. J., and Smolkin V. F. (1997b) Re-Os systematics of Early Proterozoic ferropicrites, Pechenga Complex, northwestern Russia: Evidence for ancient $^{187}Os$-enriched plumes. *Geochim. Cosmochim. Acta, 61,* 3145–3160.

Walker R. J., Tsuru A., and Prichard H. M. (1998a) The Os isotopic composition of the upper mantle: evidence from the ophiolite perspective. *Eos Trans. AGU, 79,* S372.

Walker R. J., Morgan J. W., Shearer C. K., and Papike J. J. (1998b) Rhenium-osmium isotopic systematics of lunar orange glass (abstract). In *Lunar and Planetary Science XXIX,* Abstract #1271. Lunar and Planetary Institute, Houston (CD-ROM).

Walker R. J., Storey M., Kerr A. C., Tarney J., and Arndt N. T. (1999a) Implications of $^{187}Os$ isotopic heterogeneities in a mantle plume: evidence from Gorgona Island and Curacao. *Geochim. Cosmochim. Acta, 63,* 713–728.

Walker R. J., Becker H., and Morgan J. W. (1999b) Comparative Re-Os isotope systematics of chondrites: implications regarding early solar system processes (abstract). In *Lunar and Planetary Science XXX,* Abstract #1208. Lunar and Planetary Institute, Houston (CD-ROM).

Walter M. J., Newsom H. E., Ertel W., and Holzheid A. (2000) Experimental and physical constraints on core formation: behavior of moderately and highly siderophile elements. In *Origin of the Earth and Moon* (R. M. Canup and K. Righter, eds.), this volume. Univ. of Arizona, Tucson.

Wänke H. (1981) Constitution of the terrestrial planets. *Philos. Trans. Roy. Soc. London, A303,* 287–302.

Wänke H., Rieder R., Baddenhausen H., Spettel B., Teschke F., Quijano-Rico M., and Balacescu A. (1970) Major and trace elements in lunar material. *Proc. Apollo 11 Lunar Sci. Conf.,* pp. 1719–1727.

Wänke H., Wlotzka F., Baddenhausen H., Balacescu A., Spettel B., Teschke F., Jagoutz E., Kruse H., Quijano-Rico M., and Rieder R. (1971) Apollo 12 samples: Chemical composition and its relation to sample locations and exposure ages, the two component origin of the various soil samples and studies on lunar metallic particles. *Proc. Lunar Sci. Conf. 2nd,* pp. 1187–1208.

Wänke H., Baddenhausen H., Balacescu A., Teschke F., Spettel B., Dreibus G., Palme H., Quijano-Rico M., Kruse H., Wlotzka F., and Begemann F. (1972) Multielement analyses of lunar samples and some implications of the results. *Proc. Lunar Sci. Conf. 3rd,* pp. 1251–1268.

Wänke H., Palme H., Baddenhausen H., Dreibus G., Jagoutz E., Kruse H., Palme C., Spettel B., Teschke F., and Thacker R. (1975) New data on the chemistry of lunar samples: Primary matter in the lunar highlands and the bulk composition of the Moon. *Proc. Lunar Sci. Conf. 6th,* pp. 1313–1340.

Wänke H., Palme H., Kruse H., Baddenhausen H., Cendales M., Dreisbus G., Hofmeister H., Jagoutz E., Palme C., Spettel B., and Thacker R. (1976) Chemistry of lunar highland rocks: A refined evaluation of the composition of the primary matter. *Proc. Lunar Sci. Conf. 7th,* pp. 3479–3500.

Wakita H., Schmitt R. A., and Rey P. (1970) Elemental abundances of major, minor and trace elements in Apollo 11 lunar rocks, soil and core samples. *Proc. Apollo 11 Lunar Sci. Conf.,* pp. 1685–1717.

Wakita H., Rey P., and Schmitt R. A. (1971) Abundances of the 14 rare-earth elements and 12 other trace elements in Apollo 12 samples: Five igneous and one breccia rocks and four soils. *Proc. Lunar Sci. Conf. 2nd,* pp. 1319–1329.

Warren P. H. (1986) The bulk-Moon MgO/FeO ratio: a highlands perspective. In *Origin of the Moon* (W. K. Hartmann, R. J. Phillips, and G. J. Taylor, eds.), pp. 279–310. Lunar and Planetary Institute, Houston.

Warren P. H. and Jerde E. A. (1990) Olivine-vitrophyre 12024,15: A sample of the margin of a lunar lava flow (abstract). In *Lunar and Planetary Science XXI,* pp. 1293–1294. Lunar and Planetary Institute, Houston.

Warren P. H. and Wasson J. T. (1978) Compositional-petrographic investigation of pristine nonmare rocks. *Proc. Lunar Planet. Sci. Conf. 9th,* pp. 185–217.

Warren P. H. and Wasson J. T. (1979) The compositional-petrographic search for pristine nonmare rocks — third foray. *Proc. Lunar Planet. Sci. Conf. 10th,* pp. 583–610.

Warren P. H. and Wasson J. T. (1980) Further foraging for pristine nonmare rocks: Correlations between geochemistry and longitude. *Proc. Lunar Planet. Sci. Conf. 11th,* pp. 431–470.

Warren P. H., Afiattalab F., and Wasson J. T. (1978) Investigation of unusual KREEPy samples: Prestine rock 15386, Cone Crater soil fragments 14143, and 12023, a typical Apollo 12 soil. *Proc. Lunar Planet. Sci. Conf. 9th,* pp. 653–660.

Warren P. H., Shirley D. N., and Kallemeyn G. W. (1986) A potpourri of pristine Moon rocks, including a VHK mare basalt and a unique, augite-rich Apollo 17 anorthosite. *Proc. Lunar Planet. Sci. Conf. 16th,* in *J. Geophys. Res., 91,* D319–D330.

Warren P. H., Jerde E. A., and Kallemeyn G. W. (1987) Pristine Moon rocks: A "large" felsite and a metal-rich ferroan anorthosite. *Proc. Lunar Planet. Sci. Conf. 17th,* in *J. Geophys. Res., 92,* E303–E313.

Warren P. H., Jerde E. A., and Kallemeyn G. W. (1989) Lunar meteorites: siderophile element contents, and implications for the composition and origin of the Moon. *Earth Planet. Sci. Lett., 91,* 245–260.

Wasson J. T. and Kallemeyn G. W. (1988) Compositions of chondrites. *Philos. Trans. Roy. Soc. London, A325,* 535–544.

Wasson J. T., Boynton W. V., Kallemeyn G. W., Sundberg L. L., and Wai C. M. (1976) Volatile compounds released during lunar lava fountaining. *Proc. Lunar Sci. Conf. 7th,* pp. 1583–1596.

Watson E. B., BenOthman D., Luck J. M., and Hofmann A. W. (1987) Partitioning of U, Pb, Cs, Yb, Hf, Re, and Os between chromian diopside pyroxene and haplobasaltic liquid. *Chem. Geol., 62,* 191–208.

Widom E. (1997) Sources of oceanic basalts: a review of the Os isotopic evidence. *Phys. Acta, 244,* 484–496.

Widom E. and Shirey S. B. (1996) Os isotope systematics in the Azores: implications for mantle plume sources. *Earth Planet. Sci. Lett., 142,* 451–465.

Wiesmann H. and Hubbard N. J. (1975) *A Compilation of the Lunar Sample Data Generated by the Gast, Nyquist and Hubbard Lunar Sample PI-Ships.* NASA Johnson Space Center, Houston. 36 pp.

Wolf R., Woodrow A., and Anders E. (1979) Lunar basalts and pristine highland rocks: Comparison of siderophile and volatile elements. *Proc. Lunar Planet. Sci. Conf. 10th,* pp. 2107–2130.

Wood S. A. (1987) Thermodynamic calculations of the volatility of the platinum group elements (PGE): The PGE content of fluids at magmatic temperatures. *Geochim. Cosmochim. Acta, 51,* 3041–3050.

Zerr A., Diegeler A., and Boehler R. (1998) Solidus of Earth's deep mantle. *Science, 281,* 243–246.

# Fluid Dynamics of a Terrestrial Magma Ocean

## V. S. Solomatov

*New Mexico State University*

The scenario of crystallization of a terrestrial magma ocean that seems to be consistent with both fluid dynamical and geochemical constraints is as follows. Even the largest impact is unlikely to melt the Earth completely. After the isostatic adjustment the temperatures at the bottom of the mantle were probably near or somewhat above the solidus. In less than a thousand years the solidification front propagates toward the surface and stops at some critical pressure, leaving the mantle below undifferentiated. Differentiation occurs mainly in the remaining shallow magma ocean, the lifetime of which extends well beyond the formation period. Iron delivered by subsequent impacts accumulates at the bottom of this shallow magma ocean and segregates into the Earth's core. This boundary can correspond to the metal-silicate equilibrium pressure of 28 GPa suggested by experiments on fractionation of siderophile elements.

## 1. INTRODUCTION

A number of arguments suggest that the Earth underwent a substantial melting during the accretion period. In particular, it has been realized that some of the Earth-forming planetesimals were inevitably of the size of Mars or the Moon (*Safronov,* 1978; *Wetherill,* 1985, 1990, 1992; *Weidenschilling et al.,* 1997). Collisions with bodies of this size could melt and even partially vaporize the Earth (*Safronov,* 1978; *Benz and Cameron,* 1990; *Melosh,* 1990). Such large impactors also gave a successful explanation for the origin of the Moon, its composition, and the angular momentum of the Earth-Moon system (*Benz et al.,* 1986, 1987, 1989; *Stevenson,* 1987; *Newsom and Taylor,* 1989; *Canup and Esposito,* 1996; *Ida et al.,* 1997; *Cameron,* 1997; *Canup and Agnor,* 2000). This and other factors such as greenhouse effect, core formation, and radiogenic heating by short-lived isotopes (*Flasar and Birch,* 1973; *Safronov,* 1978; *Kaula,* 1979; *Abe and Matsui,* 1986; *Matsui and Abe,* 1986; *Zahnle et al.,* 1988) imply a significant melting of the early Earth and lead to the hypothesis of a magma ocean.

The magma ocean hypothesis provided a new basis for the explanation of the present-day composition of the Earth (*Ohtani,* 1985; *Ohtani and Sawamoto,* 1987; *Agee and Walker,* 1988; *Herzberg and Gasparik,* 1991; *Gasparik and Drake,* 1995). However, geochemical models of differentiation of a terrestrial magma ocean faced serious limitations imposed by the observed nearly chondritic abundances of minor and trace elements, which forbid any substantial differentiation of perovskite in the lower mantle (*Kato et al.,* 1988a,b; *Ringwood,* 1990; *McFarlane and Drake,* 1990; *McFarlane et al.,* 1994). On the other hand, the partition coefficients are not well constrained at realistic temperatures, pressures, and compositions of magma oceans, and the range of possible scenarios of chemical differentiation is still unclear (e.g., *Presnall et al.,* 1998).

The problem of the apparent excess of siderophile elements in the Earth's mantle also seems to have a solution based on the magma ocean hypothesis. Recent experiments suggest that metal/silicate chemical equilibrium was established around 28 GPa and 2200 K, which can be explained with the help of a core formation model involving a deep magma ocean (*Li and Agee,* 1996; *Righter et al.,* 1997; *Righter and Drake,* 1997). However, the physical nature of this particular pressure remains unclear.

Fluid dynamics provided additional constraints on crystallization of a terrestrial magma ocean (*Tonks and Melosh,* 1990; *Davies,* 1990; *Miller et al.,* 1991a,b; *Abe,* 1993, 1995, 1997; *Solomatov and Stevenson,* 1993a,b,c). It was suggested that a low-viscosity, vigorously convecting magma ocean is more similar to an atmosphere rather than to a solid mantle. Even the temperature distribution is analogous to one of a "wet" atmosphere with condensation and evaporation (equivalent to crystallization and melting).

Important for geochemical implications was the idea of nonfractional (or equilibrium) crystallization as an alternative to the commonly assumed scenario of fractional crystallization. *Tonks and Melosh* (1990) argued that convection can prevent crystal settling so that melting does not necessarily cause differentiation. Further studies by *Solomatov and Stevenson* (1993a) and *Solomatov et al.* (1993) showed that the physics of suspension and differentiation in magma oceans is more complicated. Although crystal settling can be prevented, it is not as easy as was originally thought. It seems that many theories developed for rather simple situations cannot be applied to magma oceans because of the extreme conditions and the diversity of physical processes in magma oceans.

This is particularly true for convection. Early analyses of convection in magma oceans were based on a "classical" model of thermal convection that is applicable only to relatively weak convection. However, a magma ocean is likely to be in a different convective regime, perhaps "hard" turbulence convection (*Spera, 1992*). Kinetics of crystal nucleation and growth is another example of a poorly understood process. It controls the size of crystals and therefore the rate of differentiation.

The goal of this chapter is to review the physical processes in a terrestrial magma ocean and suggest a scenario of crystallization of the molten Earth that is consistent with geochemical constraints.

## 2.  THERMODYNAMICS

Thermodynamics of a partially molten mantle is affected by phase changes and can be constructed with the help of a standard thermodynamic approach (*Ghiorso, 1997; Asimow et al., 1997*). The phase diagram of the upper mantle shown in Fig. 1 is calculated with the help of a three-component model of magma oceans described by *Solomatov and Stevenson* (1993b). We assume that at P < 10 GPa the three components olivine, clinopyroxene, and orthopyroxene form a eutectic-like system. Variation of melt fraction between liquidus and solidus for this model is given in Fig. 2 along with the experimental data from *McKenzie and Bickle* (1988). Since the Earth's mantle is not ideal, the melting temperatures, their gradients, and the fractions of the three components are adjusted to fit the experimental data shown in Figs. 1 and 2.

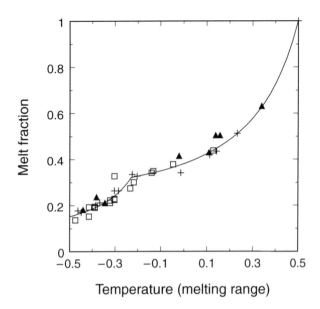

**Fig. 2.**  Variation of melt fraction between liquidus (T = –0.5) and solidus (T = 0.5) suggested by our model (solid line) fits well the experimental data collected by *McKenzie and Bickle* (1988): 0 ≤ P ≤ 0.5 GPa (crosses), 0.5 ≤ P 1.5 (squares), 1.5 < P (triangles). This is a typical three-component eutectic-like system (olivine-orthopyroxene-clinopyroxene) with a step-like melting at the solidus. Solid solutions would replace this jump by a steep gradient.

In vigorously convecting systems such as magma oceans the temperature distribution is nearly adiabatic and isentropic

$$\frac{dT}{dP} = \frac{\alpha T}{\rho c_p} \qquad (1)$$

where T is the temperature, P is the pressure, $\alpha$ is the coefficient of thermal expansion, $c_p$ is the isobaric heat capacity, and $\rho$ is the density. Adiabats starting at different potential temperatures (potential temperature is the adiabatic temperature at P = 0) are shown in Fig. 1.

The three components in the lower mantle are perovskite, $MgSiO_3$, periclase, MgO, and wüstite, FeO. Perovskite has only a small amount of Fe and is considered to be a pure $MgSiO_3$, while periclase and wüstite are assumed to form an ideal solid solution. Figures 3a and 3b show the melting temperatures of $MgSiO_3$, MgO, and FeO, liquidus, solidus, and the adiabats for the lower mantle. The melting temperatures of $MgSiO_3$, MgO, and FeO (Fig. 3a) are based on laboratory experiments and extrapolation to high pressures by *Boehler* (1992) and *Zerr and Boehler* (1993, 1994).

The liquidus and solidus curves predicted by this model for the lower mantle are significantly steeper than single-phase adiabats. The solidus is very close to one estimated by *Holland and Ahrens* (1997) and *Zerr et al.* (1998). The liquidus is high compared to previously published estimates because of the high melting temperature of perovskite and the assumption that perovskite forms eutectic-like subsystems with periclase and wüstite. *Abe* (1997), for instance,

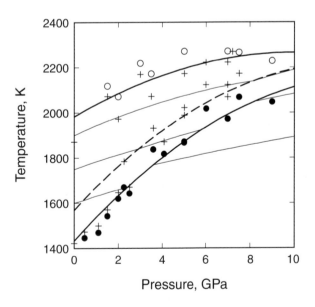

**Fig. 1.**  Three adiabats in the upper mantle are shown with thin solid lines starting at 1600, 1750, and 1900 K. Liquidus, solidus (heavy solid lines), and the beginning of crystallization of orthopyroxene (dashed line) are shown together with experimental data for peridotites from *McKenzie and Bickle* (1988), *Scarfe and Takahashi* (1986), and *Ito and Takahashi* (1987): liquid (open circles), solid (solid circles), and partial melt (crosses).

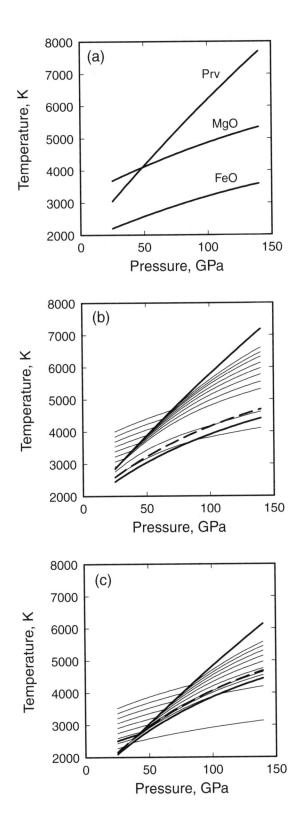

**Fig. 3.** (a) Melting curves of perovskite, MgO, and FeO according to *Boehler* (1992) and *Zerr and Boehler* (1993, 1994). (b) Adiabats in the convective magma ocean. Liquidus and solidus are shown with a heavy solid line. The beginning of magnesiowüstite crystallization is shown with a dashed line. (c) An example of calculations where magnesiowüstite is the first liquidus phase at the top of the lower mantle as observed by *Agee* (1990) and *Zhang and Herzberg* (1994).

assumed an ideal mixture of MgSiO₃ and MgO, while *Miller et al.* (1991b) and *Solomatov and Stevenson* (1993b) assumed a low melting temperature of perovskite based on the data by *Knittle and Jeanloz* (1989). For an ideal eutectic-like system, the liquidus $T_{liq}$ approximately follows the melting temperature $T_{prv}$ of pure perovskite MgSiO₃ lowered by the presence of MgO and FeO

$$T_{prv} - T_{liq} \approx \frac{k_B}{\Delta S_{prv}} n_{mw} T_{prv} \qquad (2)$$

where $\Delta S_{prv}$ is the entropy change per atom upon melting of pure perovskite, and $n_{mw}$ is the molar fraction of magnesiowüstite. For $n_{mw} \approx 0.3$, $\Delta S_{prv} \approx 5$ $k_B$ [assuming that the entropy change is approximately $k_B$ per atom (*Stishov*, 1988)] and $T_{prv} \approx 7500$ K we obtain $T_{prv} - T_{liq} \approx 450$ K.

*Agee* (1990) and *Zhang and Herzberg* (1994) found that magnesiowüstite is the liquidus phase in both chondrites and peridotites just slightly below the upper mantle/lower mantle boundary. A small fraction of Fe in the solid perovskite and the presence of other components (e.g., Ca-perovskite) in melt would decrease the liquidus temperature. The estimates of the entropy change are also not very accurate. As a result, magnesiowüstite can be the first liquidus phase at low pressures, although at higher pressures it would still be substituted by perovskite (Fig. 3c).

## 3. FORMATION OF MAGMA OCEANS BY GIANT IMPACTS

### 3.1. Mantle Temperature Before Impacts

It is quite common to assume that at the latest stages of Earth's formation the mantle temperature is close to solidus (*Abe and Matsui*, 1986; *Sasaki and Nakazawa*, 1986; *Zahnle et al.*, 1988). However, since the mantle solidus is steeper than the adiabat in the solid mantle, such temperature distribution is gravitationally unstable. The effective density contrast between the upper and lower parts of the mantle is

$$\Delta \rho \approx \frac{1}{2} \alpha \rho H \left( \frac{dT_{sol}}{dz} - \frac{dT_{ad}}{dz} \right) \approx 120 \text{ kg m}^{-3} \qquad (3)$$

where (see Table 1) $\rho = 4000$ kg m⁻³, $\alpha \approx 3 \times 10^{-5}$ K⁻¹, H ≈ $3 \times 10^6$ m is the thickness of the mantle, $dT_{sol}/dz$ is the solidus gradient, and $dT_{ad}/dz$ is the adiabatic gradient (Fig. 3). This density contrast causes gravitational instability that is somewhat similar to a classical Rayleigh-Taylor instability in a two-layer system. The timescale for the overturn can then be estimated as (*Turcotte and Schubert*, 1982)

$$t_{RT} \approx 26 \frac{\eta_s}{\Delta \rho g H} \qquad (4)$$

where $\eta_s$ is the viscosity and g is the gravity.

TABLE 1. "Typical" values of physical parameters for
a magma ocean in the early stages of crystallization
(lower mantle, the liquidus phase is perovskite).

| | |
|---|---|
| Density, $\rho$ | $4 \times 10^3$ kg m$^{-3}$ |
| Temperature, T | $4 \times 10^3$ K |
| Thermal expansion, $\alpha$ | $5 \times 10^{-5}$ K$^{-1}$ |
| Thermal capacity, $c_p$ | $10^3$ J kg$^{-1}$ K$^{-1}$ |
| Gravity, g | 10 m s$^{-2}$ |
| Crystal/melt density difference, $\Delta\rho$ | 300 kg m$^{-3}$ |
| Viscosity of melt at the liquidus, $\eta_l$ | 0.1 Pa s |
| Viscosity of solids above the solidus, $\eta_s$ | $10^{17}$ Pa s |
| Diffusion coefficient, D | $10^{-9}$ m$^2$ s$^{-1}$ |
| Enthalpy change upon melting, $\Delta H$ | $10^6$ J kg$^{-1}$ |
| Apparent surface energy, $\sigma_{app}$ | 0.02 J m$^{-2}$ |
| Pre-factor in the nucleation rate, a | $10^{20}$ m$^{-3}$ s$^{-1}$ |
| Liquidus gradient, $dT_l/dz$ | $2 \times 10^{-3}$ K m$^{-1}$ |
| Adiabatic gradient, $dT_{ad}/dz$ | $6 \times 10^{-4}$ K m$^{-1}$ |
| Variation of liquidus with $\phi$, $dT_l/d\phi$ | $10^3$ K |
| Heat flux, F | $10^6$ J m$^{-2}$ s$^{-1}$ |
| Angular velocity, $\Omega$ | $10^{-4}$ s$^{-1}$ |
| Length scale, H | $3 \times 10^6$ m |
| Convective velocity, $u_0$ | 10 m s$^{-1}$ |
| Timescale for Ostwald ripening, t | $10^6$ s |
| Effective cooling rate in plumes, $\dot{T}$ | 0.2 K s$^{-1}$ |
| Crystal size, d | $10^{-3}$ m |

This timescale depends critically on $\eta_s$. The present-day mantle viscosity is around $10^{22}$ Pa s (*King*, 1995). For the viscosity $\eta_s \approx 10^{20}$ Pa s corresponding to 200 K hotter mantle (the viscosity decreases roughly at the rate of one order of magnitude per 100 K)

$$t_{RT} \approx 2 \times 10^4 \left( \frac{\eta_s}{10^{20} \, \text{Pa s}} \right) \text{yr} \qquad (5)$$

This is very fast compared to the timescale of Earth's formation, which is of the order of $10^8$ yr as suggested by numerical simulations (*Wetherill*, 1990) and by isotopic constraints on the timing of the core formation (*Lee and Halliday*, 1995; *Halliday et al.*, 1996). In fact, a super-adiabatic temperature gradient of only 100–200 K over the entire mantle would be eliminated within $10^5$ yr or so. Therefore, the mantle had enough time to maintain a nearly adiabatic temperature during the accretion period. However, the situation can be more complicated if a sufficiently large density gradient was established due to crystal/melt differentiation or migration of liquid iron diapirs (*Elsasser*, 1963; *Stevenson*, 1981, 1990; *Karato and Murthy*, 1997a). In particular, an increase of the viscosity with depth could cause a higher concentration of liquid diapirs in high-viscosity regions and make a near-solidus mantle gravitationally stable. On the other hand, if the viscosity variations were small, liquid diapirs could stir the mantle and maintain its nearly adiabatic state. The answer depends on the poorly constrained rheological stratification of the growing Earth.

## 3.2. Impact-induced Melting

Simulations of impacts (*Melosh*, 1990; *Tonks and Melosh*, 1993; *Pierazzo et al.*, 1997) suggest that the impact completely melts a region with the radius of about half the radius of the Earth. The temperature increase in the remaining part of the Earth is due to the shock wave. It decays very rapidly with the distance from the molten region so that the opposite side of the Earth is heated very little. The gravitational energy associated with this buoyant blob of partially molten material is converted into the heat during the subsequent isostatic adjustment. This process redistributes the mass to form a gravitationally stable system. The flow of solid, dense material replaces the melt, which eventually moves on top of the solid or partially molten layers. The gravitational energy released during viscous deformation results in an additional temperature increase of 300–400 K on average (*Tonks and Melosh*, 1990). A somewhat smaller amount of energy comes from gravitational separation of Fe from the impactor (*Tonks and Melosh*, 1992). The opposite side of the Earth was probably heated by only about 500 K or so due to the net effect of the shock-wave heating, isostatic adjustment, and segregation of Fe from the impactor. If the initial thermal state was adiabatic, then a significant portion of the Earth remained solid. If, however, the initial temperature distribution was close to solidus, then the impact would produce a completely molten region and partially molten region of comparable sizes. In any case, a complete melting of the mantle is unlikely because of the high liquidus temperature.

The stresses $\sim \Delta\rho g R$ associated with the molten blob located on one side of the Earth are on the order of 10 GPa. This exceeds the ultimate strength of rocks, which is about 1–2 GPa (*Davies*, 1982). Therefore, isostatic adjustment could be an extremely fast process. Crystallization is a much slower process that starts effectively at the latest stages of the isostatic adjustment.

## 4.   VISCOSITY OF MAGMA AND "RHEOLOGICAL TRANSITION"

A large degree of melting of the mantle implies a very small viscosity of the magma ocean. Experimental and theoretical studies suggest that the viscosity of many near-liquidus ultramafic silicates at low pressures is around $\eta_l \sim$ 0.1 Pa s (*Bottinga and Weill*, 1972; *Shaw*, 1972; *Persikov et al.*, 1990; *Bottinga et al.*, 1995). Molecular dynamic simulations of $MgO$-$SiO_2$ give a value of $3 \times 10^{-3}$–$5 \times 10^{-3}$ Pa s at 5 GPa (*Wasserman et al.*, 1993a,b), which is almost independent of temperature and composition.

The viscosity of a completely depolymerized melt weakly increases with pressure (*Andrade*, 1952; *Gans*, 1972). An increase of less than one order of magnitude can be expected along the liquidus throughout the lower mantle. Thus, the viscosity of magma oceans near the liquidus is probably around $10^{-2}$–$10^{-1}$ Pa s. In the region between solidus and liquidus it is somewhat higher because of lower temperatures. However, the temperature effect is small for low-viscosity liquids (<1 Pa s), which exhibit a power-law rather than Arrhenius behavior (*Bottinga et al.*, 1995). The value of $10^{-1}$ Pa s with the uncertainty of a factor of 10 will be assumed in all subsequent estimates (Table 1).

Many authors pointed out that magma oceans can contain substantial amounts of water (*Holloway*, 1988; *Ahrens*, 1992; *Righter et al.*, 1997). Although water reduces the viscosity of magmas, in the limit of high temperatures the viscosities and diffusivities of many liquids including water are very similar (*Persikov et al.*, 1990). Therefore, the viscosity of completely depolymerized, high-temperature hydrous magma cannot be much different from that of anhydrous magma.

The viscosity of a melt/crystal mixture jumps abruptly near a critical crystal fraction as suggested by theoretical and experimental studies of concentrated suspensions (*Mooney*, 1951; *Roscoe*, 1952; *Brinkman*, 1952; *Krieger and Dougherty*, 1959; *Murray*, 1965; *Frankel and Acrivos*, 1967; *McBirney and Murase*, 1984; *Cambell and Forgacs*, 1990) and by experiments with partial melts (*Arzi*, 1978; *van der Molen and Paterson*, 1979; *Lejeune and Richet*, 1995). This can be called a "rheological transition." It depends on the crystal size distribution, the crystal shape, and other factors and is around 60%.

## 5.   CONVECTION

### 5.1.   "Classical" Model of Turbulent Convection

At small viscosities of the magma ocean, before the rheological transition, convection is extremely turbulent and is driven by cooling from the surface. The convective velocity scales as (*Priestly*, 1959; *Kraichnan*, 1962)

$$u_0 \approx 0.6 \left( \frac{\alpha g l F}{\rho c_p} \right)^{1/3} \qquad (6)$$

where F is the surface heat flux, l is the mixing length, and the coefficient in front of this equation is constrained by laboratory experiments (*Deardorff*, 1970; *Willis and Deardorff*, 1974) and atmospheric measurements (*Caughey and Palmer*, 1979). The mixing length is approximately equal to the depth of the magma ocean l ~ H. For either one-phase values of the parameters given in Table 1 or two-phase values (*Solomatov and Stevenson*, 1993b), $u_0 \approx 4$ m/s.

It is interesting that the parameter $(F/\rho)^{1/3}$ is almost the same for the magma ocean and the atmosphere. This explains why the above estimate is very similar to the observed velocities in the convective boundary layer in the atmosphere (*Caughey and Palmer*, 1979).

### 5.2.   "Hard" Turbulence

At very high Rayleigh numbers (such as those relevant to magma oceans) convection changes to a regime sometimes called "hard" turbulence convection (*Castaing et al.*, 1989; *Shraiman and Siggia*, 1990; *Grossman and Lohse*,

1992; *Siggia*, 1994). By contrast, the ordinary turbulence is called "soft" turbulence. One of the important features of "hard" turbulence convection is the existence of a large-scale circulation ("wind"). The equations suggested by *Shraiman and Siggia* (1990) can be rewritten in the form similar to equation (6)

$$u_0 \approx a \left( \frac{\alpha g l F}{\rho c_p} \right)^{1/3} \qquad (7)$$

where the coefficient a is calculated as

$$a \approx \frac{0.052}{0.22^{1/3}} x^* \qquad (8)$$

where 0.22 and 0.052 are the pre-factors in the scaling relationships for Nusselt and Reynolds numbers correspondingly, and $x^*$ is related to $u_0$ through

$$x^* = 2.5 \ln \left[ \frac{\rho u_0 l}{\eta} \frac{1}{x^*} \right] + 6 \qquad (9)$$

The solution to the above transcendental equations gives a ≈ 5.9, which varies weakly with $\rho u_0 l / \eta$.

In the "hard" turbulence regime, the velocity increases by a factor of 10. This brings the estimate of the convective velocity in the magma ocean up to 40 m/s. At even higher Rayleigh numbers, convection was expected to enter a new regime of turbulent convection (*Kraichnan*, 1962; *Siggia*, 1994). However, recent experiments did not show any evidence of such an "ultrahard" regime, and "hard" turbulence was suggested to be the truly asymptotic regime of thermal convection (*Glazier et al.*, 1999). Therefore, "hard" turbulence convection is probably applicable to the extreme conditions of magma oceans.

### 5.3.   Effect of Rotation

Despite recent attempts to study "hard" turbulence in the presence of rotation (*Julien et al.*, 1996), there is no velocity scaling in this case. The problem has not been completely solved yet even for "soft" turbulence. However, the length scale

$$l \sim \frac{u_0}{\Omega} \qquad (10)$$

explains rather well the experimentally observed reduction in the convective velocity (*Golitsyn*, 1980, 1981; *Hopfinger et al.*, 1982; *Hopfinger*, 1989; *Boubnov and Golitsyn*, 1986, 1990; *Chen et al.*, 1989; *Fernando et al.*, 1991; *Solomatov and Stevenson*, 1993a). Applying this scaling to "hard" turbulence convection we obtain

$$u_0 \approx 14 \left( \frac{\alpha g F}{\rho c_p \Omega} \right)^{1/2} \qquad (11)$$

This gives velocities around 16 m/s with the uncertainty of a factor of 3. A moderate value of 10 m/s is used in the estimates below (Table 1).

## 6.  HEAT FLUX

In the early stages of crystallization, when the surface temperature is high, the atmosphere is presumably a "silicate" one (*Thompson and Stevenson,* 1988) and the heat flux can be calculated with the help of the blackbody model

$$F_r = \sigma_B T_s^4 \qquad (12)$$

where $T_s$ is the surface temperature and $\sigma_B = 5.67 \times 10^{-8}$ J m$^{-2}$ K$^{-4}$ is the Stefan-Boltzmann constant.

This heat flux must match the heat flux transported to the surface by convection. The convective heat flux depends on whether convection is in the regime of "soft" turbulence or "hard" turbulence. In the "soft" turbulence regime

$$F_{soft} = 0.089 \frac{k(T-T_s)}{H} Ra^{1/3} \qquad (13)$$

where

$$Ra = \frac{\alpha g (T - T_s) H^3}{\kappa \nu} \qquad (14)$$

is the Rayleigh number, T is the potential temperature, k is the coefficient of thermal conductivity, $\kappa = k/\rho c_p$ is the coefficient of thermal diffusivity, and $\nu = \eta/\rho$ is the kinematic viscosity (*Kraichnan,* 1962; *Siggia,* 1994).

In the "hard" turbulence regime (*Shraiman and Siggia,* 1990; *Siggia,* 1994)

$$F_{hard} = 0.22 \frac{k(T-T_s)}{H} Ra^{2/7} Pr^{-1/7} \lambda^{-3/7} \qquad (15)$$

where $Pr = \nu/\kappa$ is the Prandtl number and $\lambda$ is the aspect ratio for the mean flow. The exact value of $\lambda$ is unknown for spherical geometry but is probably of the order of unity.

Figure 4 shows that during the initial period of crystallization, when the temperature of the magma ocean is high and convection is in the "hard" turbulence convection regime, the heat flux is close to $10^6$ W m$^{-2}$. During the latest stages of crystallization (upper mantle) the surface temperature drops below 1500 K and the heat flux decreases to $10^2$–$10^3$ W m$^{-2}$ due to a greenhouse effect (*Abe and Matsui,* 1986; *Zahnle et al.,* 1988; *Kasting,* 1988). In the "hard" turbulence regime and for $\eta_l = 0.1$ Pa s this does not happen until almost a complete crystallization of the magma ocean.

## 7.  CONDITIONS FOR EQUILIBRIUM CRYSTALLIZATION

### 7.1.  "Sufficient" Condition for Equilibrium Crystallization

In this section we are concerned with the early crystallization of the magma ocean, before the crystal fraction reaches the rheological transition.

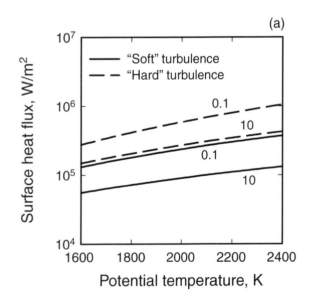

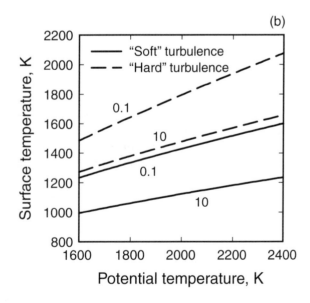

**Fig. 4.**  **(a)** Surface heat flux and **(b)** surface temperature are shown as functions of the potential temperature, the convective regime ("soft" vs. "hard" turbulence), and the viscosity of magma (0.1–10 Pa s).

The simplest scenario in which a magma ocean can crystallize without differentiation is one in which the crystal size and settling velocity are so small that the magma ocean freezes before any crystal-melt segregation occurs. What is the critical crystal size for this to happen?

The crystallization time is around

$$t_{cr} \approx \frac{(\Delta H \phi + c_p \Delta T)M}{FA} \approx 400 \, \text{yr} \quad (16)$$

where $\Delta T \sim 1000$ K is the average temperature drop upon crystallization of the magma ocean up to the crystal fraction $\phi \sim 60\%$, M is the mass of the mantle, and s is the surface area of the Earth.

The timescale for crystal settling is (*Martin and Nokes,* 1988, 1989)

$$t_s \approx H/u_s \quad (17)$$

where

$$u_s = f_\phi \frac{\Delta \rho g d^2}{18 \eta_l} \quad (18)$$

is the settling velocity and $f_\phi$ is a hindered settling function such that $f_\phi = 1$ at $\phi = 0$ (*Davis and Acrivos,* 1985). For $\phi \sim 30\%$ (on average) $f_\phi \sim 0.15$.

Requiring that crystallization is faster than settling, that is $t_{cr} < t_s$, we find that the crystal size should be smaller than

$$d_1 = \left( \frac{18 H \eta_l}{f_\phi g \Delta \rho t_{cr}} \right)^{1/2} \quad (19)$$

or

$$d_1 \approx 10^{-3} \left( \frac{\eta_l}{0.1 \, \text{Pa s}} \right)^{1/2} \left( \frac{F}{10^6 \, \text{W m}^{-2}} \right)^{1/2} \text{m} \quad (20)$$

### 7.2. "Necessary" Condition for Equilibrium Crystallization

The presence of crystals can suppress convection and slow down cooling of the magma ocean as a result of viscous heating and density stratification associated with crystal settling. This happens if the crystal size is larger than (*Solomatov and Stevenson,* 1993a)

$$d_2 = \left( \frac{18 \alpha \eta_l F}{f_\phi g c_p \Delta \rho^2 \phi} \right)^{1/2} \quad (21)$$

where $f_\phi$ is the same as in equation (18).

We obtain

$$d_2 \approx 10^{-3} \left( \frac{\eta_l}{0.1 \, \text{Pa s}} \right)^{1/2} \left( \frac{F}{10^6 \, \text{W m}^{-2}} \right)^{1/2} \text{m} \quad (22)$$

### 7.3. Critical Crystal Size for Equilibrium Crystallization

Since both "necessary" and "sufficient" conditions approximately coincide, the critical crystal size for equilibrium crystallization up to about 60% crystal fraction is

$$d_{crit} \approx 10^{-3} \left( \frac{\eta_l}{0.1 \, \text{Pa s}} \right)^{1/2} \left( \frac{F}{10^6 \, \text{W m}^{-2}} \right)^{1/2} \text{m} \quad (23)$$

The fact that these two estimates are very close to each other is not surprising; their ratio scales approximately as

$$\frac{d_1}{d_2} \sim \left( \frac{\Delta \rho / \rho}{\alpha T} \frac{c_p}{\Delta S} \right)^{1/2} \sim 1 \quad (24)$$

The interval between $d_1$ and $d_2$ is negligible compared to the uncertainties in the crystal size. Therefore, consideration of reentrainment of crystals from the bottom (*Tonks and Melosh,* 1990; *Solomatov and Stevenson,* 1993a; *Solomatov et al.,* 1993) is unnecessary for these order-of-magnitude estimates; at $d < d_1$ settling is negligible and reentrainment is unimportant, while at $d > d_2$ convection is suppressed and reentrainment cannot occur.

## 8. HOW BIG ARE THE CRYSTALS IN MAGMA OCEANS?

### 8.1. Overview

In general, if the crystal fraction is $\phi$ and the total number of crystals per unit volume is N then the crystal diameter (assuming spherical shape) is

$$d = \left( \frac{6\phi}{\pi N} \right)^{1/3} \quad (25)$$

The initial number of crystals N per unit volume is controlled by the nucleation process in the downgoing flow. If the number of crystals N does not change with time, then the crystals grow simply because the equilibrium crystal fraction $\phi$ increases with depth along the adiabat. However, dissolution of smaller crystals and growth of larger crystals decreases N (Ostwald ripening) and increases the average size of crystals according to equation (25). Below we compare two end-member cases: nucleation controlled crystal size (negligible Ostwald ripening) and Ostwald ripening controlled crystal size.

### 8.2. Crystal Size Controlled by Nucleation

When the downgoing convective flow enters the two-phase region, formation of crystals proceeds via nucleation and growth mechanism (Figs. 5 and 6). This is very similar to bulk crystallization of continuously cooling liquids. In both cases the number of nuclei formed depends on the

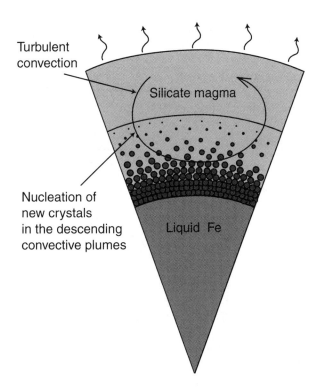

**Fig. 5.** Magma ocean in the beginning of crystallization. The size of crystals is determined by the number of nuclei produced in the descending plumes upon entering the two-phase region.

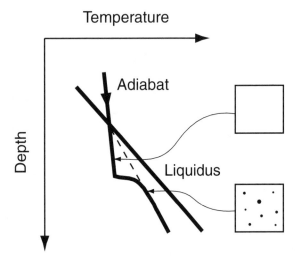

$$\dot{T} = \left( \nabla T_{liq} - \nabla T_{ad} \right) u_0$$

**Fig. 6.** Nucleation in descending flow in the magma ocean. The temperature in the superliquidus descending flow follows a one-phase adiabat (in an ideal equilibrium, the temperature would follow the two-phase adiabat shown with a dashed line). When the adiabatic temperature drops below liquidus, the temperature keeps following the one-phase adiabat. Nucleation starts after a small metastable overshoot and the temperature quickly approaches the two-phase adiabat.

cooling rate $\dot{T}$. *Solomatov and Stevenson* (1993c) solved this problem for interface kinetics controlled growth. Further analysis (*Solomatov*, 1995) showed that diffusion-controlled growth during nucleation is more consistent with various experimental data on silicates and alloys (Fig. 7), although the effect of transient nucleation (*Kashchiev*, 1969) still needs to be investigated.

The number of nuclei produced per unit volume during the nucleation period is

$$N = 0.2 \frac{\zeta^{3/2} \dot{T}^{3/2} p^{3/2}}{A \left| dT_l/d\phi \right| D^{3/2}} \qquad (26)$$

where D is the diffusion coefficient, $\zeta = \Delta T/\Delta C$ determines the relationship between the supersaturation $\Delta C$ and the supercooling $\Delta T$ and is of the order of the difference between liquidus and solidus temperatures, $\zeta \sim T_l - T_s$, A is the coefficient in the nucleation rate function

$$J(\Delta T) = a \exp\left( -\frac{A}{T'^2} \right) \qquad (27)$$

$$A = \frac{16\pi\sigma_{app}^3 T_0^2}{3k_B T \rho^2 \Delta H^2} \qquad (28)$$

where T' is the supercooling, a is a constant, $T_0$ is the melting temperature of the crystallizing phase, $k_B$ is Boltzmann's constant, $\Delta H$ is the enthalpy change per unit mass upon melting, and $\sigma_{app}$ is the apparent surface energy.

The parameter p in equation (26) is nearly constant (~30). It can be found from

$$p = \ln\left[ \frac{6.05 \left| dT_l/d\phi \right| aA^{3/2} D^{3/2}}{\zeta^{3/2} \dot{T}^{5/2} p^3} \right] \qquad (29)$$

Note that because of a large value of the logarithm, p is insensitive to the uncertainties in the parameters in the above equation.

The effective cooling rate of an adiabatically descending plume (the rate of change of the difference between the liquidus temperature $T_l$ and the actual temperature $T_{ad}$) is

$$\dot{T} = u_0 \left( \frac{dT_l}{dz} - \frac{dT_{ad}}{dz} \right) \qquad (30)$$

where $dT_l/dz$ is the liquidus gradient and $dT_{ad}/dz$ is the adiabatic gradient.

From equations (25) and (26) with $\phi \sim 60\%$ we estimate

$$d_{nuc} \approx 3 \times 10^{-4} \left( \frac{\sigma_{app}}{0.02 \text{ J m}^{-2}} \right)$$

$$\left( \frac{D}{10^{-9} \text{ m}^2 \text{ s}^{-1}} \right)^{1/2} \left( \frac{u_0}{10 \text{ m s}^{-1}} \right)^{-1/2} \text{ m} \qquad (31)$$

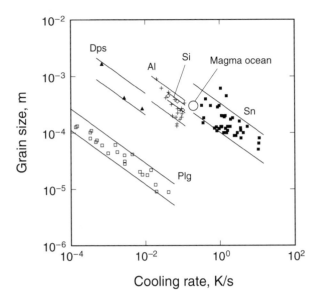

**Fig. 7.** Crystal size vs. cooling rate for five crystallizing phases (from *Solomatov*, 1995): Sn (from Sn-Pb melt, solid boxes), Al (from Al-Cu melt, pluses), Si (from Al-Si melt, diamonds), diopside (from diopside-plagioclase melt, solid triangles), and plagioclase (from diopside-plagioclase, open boxes), which is a solid solution between albite and anorthite. The data are from *Lofgren et al.* (1974), *Flemings et al.* (1976), *Grove and Walker* (1977), *Walker et al.* (1978), *Ichikawa et al.* (1985), *Grove* (1990), *Smith et al.* (1991), and *Cashman* (1993). The theoretical fit to each dataset consists of two curves showing the range of uncertainties. The curves depend only on one fitting parameter, the apparent surface energy. The location of the magma ocean is shown with a large circle (for the parameters given in Table 1).

It should be emphasized that the value of the surface energy, $\sigma_{app}$, is not well constrained. It is not the normal surface energy but rather an apparent one for nucleation. The value of 0.02 J m$^{-2}$ (Table 1) is based on the analogy with some simple cases that suggest a value of $\sigma_{app}$, which is almost 1 order of magnitude smaller compared to the usual one (*Dowty*, 1980; *Solomatov*, 1995). It is subject to large uncertainties when extrapolating to magma oceans. The presence of water reduces $\sigma_{app}$ and increases the nucleation rates (*Fenn*, 1977; *Swanson*, 1977; *Dowty*, 1980; *Davis et al.*, 1997) while only slightly increasing D (at high temperatures it is almost the same for all liquids). Therefore, the presence of water might decrease the crystal size, which is somewhat counterintuitive since one would expect an enhanced crystal growth in the presence of water.

### 8.3. Crystal Size Controlled by Ostwald Ripening

A simple estimate of the crystal size $d_{ost}$ due to Ostwald ripening can be obtained from diffusion controlled growth (if interface kinetics is the rate-limiting process then the

crystal size will be reduced further). In this case (*Lifshitz and Slyozov*, 1961; *Voorhees*, 1992)

$$d_{ost}^3 - d_0^3 = \frac{32}{9} b_\phi \alpha_0 Dt \qquad (32)$$

where $d_0$ is the initial crystal size, $\alpha_0 = \sigma c_\infty v_m / 2RT$, $\sigma$ is the surface energy, $c_\infty$ is the equilibrium concentration of the crystallizing mineral in the melt, $v_m$ is the molar volume of the crystallizing mineral, t is the characteristic residence time in the two-phase region ($\sim H/u_0$), and $b_\phi$ is a correction to the original *Lifshitz and Slyozov's* (1961) equation for large $\phi$ such that $b_\phi = 1$ at very small $\phi$ and $b_\phi \approx 3$ for $\phi \approx 30\%$. Neglecting the initial crystal size we obtain

$$d_{ost} \approx 4 \times 10^{-4} m \qquad (33)$$

This crystal size is about the same as the one controlled by nucleation (equation (31)).

When the completely molten layer disappears (temperature drops below liquidus everywhere), crystals never exit the two-phase region and the characteristic time for crystal growth is much larger. It takes about $10^9$–$10^{10}$ s for the potential temperature to drop from liquidus to the critical temperature for the rheological transition so that the crystals are at least 10 times larger during this period of crystallization

$$d_{ost}' \sim 10^{-2} m \qquad (34)$$

### 8.4. Fractional Versus Equilibrium Crystallization

The above estimates show that the crystal size during the early crystallization of the magma ocean is very close to the critical crystal size separating fractional and equilibrium crystallization of the early magma ocean (Fig. 8). This means that both equilibrium and fractional crystallization (up to 60% crystal fraction) are equally acceptable within the uncertainties of the physical parameters. The potential temperature at which fractional crystallization begins is in the ~300 K interval between the liquidus and the critical temperature for the rheological transition.

## 9. CRYSTALLIZATION BEYOND RHEOLOGICAL TRANSITION

At the crystal fraction around $\phi_{cr} \sim 60\%$, the melt/crystal mixture undergoes a very sharp rheological transition to a solid-like behavior and the deformation is controlled by the viscosity of the solid phase. Below we will explore what happens beyond this.

Suppose that when the viscosity jumps up to the viscosity of solids the material just stops convecting. A layer with the crystal fraction $\phi \sim \phi_{cr}$ would start accumulating from the bottom of the magma ocean. In about 400 yr (equation (16)), the temperature in the entire magma ocean would

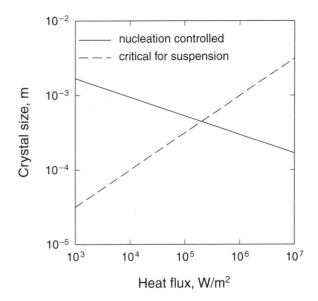

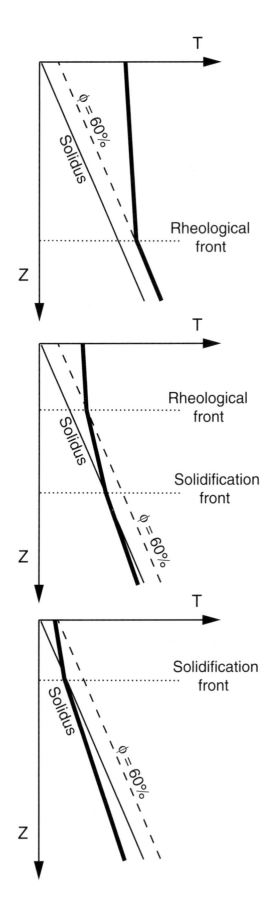

**Fig. 8.** Crystal size controlled by nucleation (solid line) and the critical crystal size for suspension (dashed line) as functions of the heat flux. The crystal size decreases with the heat flux since the convective velocities, and thus the cooling rates in the descending plumes, increase with the heat flux. On the other hand, the critical crystal size (above which crystals would settle down) increases with the heat flux. The early magma ocean is located near the intersection of these two curves.

approximately follow the curve $\phi = \phi_{cr} = $ const (the dashed line in Fig. 9).

This partially crystallized magma ocean has a huge temperature gradient that follows the curve $\phi = \phi_{cr}$ and therefore is gravitationally unstable. The timescale for Rayleigh-Taylor instability can be estimated from equation (4), in which the effective density difference across the mantle is now $\Delta\rho \approx \rho\alpha\Delta T_{superad}/2$ where $\Delta T_{superad}$ is the superadiabatic temperature contrast across the partially crystallized magma ocean (Fig. 3b).

An important parameter is the viscosity $\eta_s$ of partially molten rocks. Based on recent estimates of the viscosity in the lower mantle (*Karato and Li*, 1992; *Li et al.*, 1996; *Ita and Cohen*, 1998) the viscosity just above the solidus is expected to be around $10^{18}$ Pa s for the grain size of about $10^{-3}$ m. Small amounts of melt reduce the viscosity further. Experimental data suggest that the viscosity is likely to be 1–2 orders of magnitude smaller depending on the melt fraction (*van der Molen and Paterson*, 1979; *Cooper and Kohlstedt*, 1986; *Jin et al.*, 1994; *Rutter and Neumann*, 1995; *Hirth and Kohlstedt*, 1995a,b; *Kohlstedt and Zimmerman*, 1996). For $\eta_s \sim 10^{17}$ Pa s, equation (4) gives $t_{RT} \approx$ 20 yr.

It is clear that the magma ocean would not "wait" for 400 yr until it suddenly "decides" to resolve the instability in 20 yr time. The instability starts developing soon after the critical crystal fraction is reached at the bottom of the magma ocean and takes the form of solid-state convection,

**Fig. 9.** Propagation of the solidification and rheological fronts in the magma ocean.

which is the primary mechanism of crystallization beyond the critical crystal fraction (*Solomatov and Stevenson,* 1993b). Therefore a complete solidification front would follow the rheological front toward the surface (Fig. 9).

## 10. DIFFERENTIATION IN THE SHALLOW MAGMA OCEAN

After the rheological front reaches the surface, the driving force for this type of convection motion disappears and it stops. All important changes occur when the potential temperature decreases from liquidus to the critical temperature for the rheological transition. The crystal size becomes larger than the critical one for suspension, the heat flux drops, convection slows down, and a stable crust forms at the surface, reducing further the intensity of convection and the cooling rate of the magma ocean.

The lower bound on the depth $H_{shallow}$ of the remaining shallow magma ocean can be obtained from an adiabat starting at the critical temperature for the rheological transition. Figure 1 suggests that such an adiabat intersects solidus around 10 GPa or $H_{shallow}$ ~300 km. This estimate is rather uncertain since even a 100 K error in the estimate of the rheological transition increases $H_{shallow}$ by a factor of 2. Besides, small superadiabatic gradients can be preserved for a long time and can increase the estimate of $H_{shallow}$ substantially — note that both solidus and liquidus are quite flat between 10 and 23 GPa and even small superadiabatic temperatures can extend the depth of the shallow magma ocean all the way through the bottom of the upper mantle. The presence of water would also increase $H_{shallow}$ although the melt fraction in the temperature range between "dry" solidus and "wet" solidus is small so that "dry" solidus might still be a reasonable basis for defining the depth of the shallow magma ocean.

Moreover, when the potential temperature drops below liquidus, the completely molten layer disappears and the sequence "nucleation-growth-dissolution" changes to just "growth." In this case, the crystals had enough time to reach the critical size for the cessation of suspension (equations (23) and (34)). Although liquidus is only 300 K above the critical temperature for the rheological transition, the bottom of the shallow magma ocean can extend well into the lower mantle, probably around 40 GPa or so (Fig. 3; see also *Miller et al.,* 1991b; *Solomatov and Stevenson,* 1993b; *Abe,* 1997).

Differentiation in the shallow magma ocean takes

$$t_{diff} \sim \frac{H_{shallow}}{u_{perc}} \qquad (35)$$

where

$$u_{perc} = \frac{g\Delta\rho d^2 \phi_l^2}{150\eta_l(1-\phi_l)} \qquad (36)$$

is the percolation velocity according to the Ergun-Orning

formula (*Soo,* 1967; *Dullien,* 1979) and $\phi_l$ is the melt fraction.

For example, it takes

$$t_{diff} \sim 10^8 \left(\frac{\phi_l}{0.02}\right)^2 \left(\frac{d}{10^{-3}\,m}\right)^2 \left(\frac{\eta_l}{100\,Pa\,s}\right) yr \qquad (37)$$

to reduce the average melt fraction to about 2%. Note that here we used the value of the viscosity of a low-pressure, high-silica, polymerized magma just above the solidus (*Kushiro,* 1980, 1986). Melt/crystal density inversions (*Agee,* 1998) might result in the formation of more than one molten layer. In particular, a gravitationally stable molten layer could be formed at the bottom of the upper mantle in addition to the shallow magma ocean beneath the crust. Cooling and further crystallization of these magmatic layers would take an even longer time depending on the global thermal evolution of the mantle. It can be as fast as $10^7$–$10^8$ yr (*Davies,* 1990). However, if surface recycling was inefficient, convection beneath the solid crust would be slower (e.g., *Solomatov and Moresi,* 1996), and crystallization time would be significantly larger.

## 11. DISCUSSION AND CONCLUSION

Although the uncertainties in fluid dynamics of magma oceans are substantial, the following scenario of evolution of a terrestrial magma ocean can be suggested.

A giant impact melts a significant part of one hemisphere of the Earth. Isostatic adjustment quickly redistributes the mass to create a more stable spherically symmetric configuration. In the beginning of crystallization of the magma ocean, the temperatures in the deepest parts of the Earth were probably near the solidus and depended on the poorly constrained preimpact thermal state of the mantle. This implies that some portion of the lower mantle could retain substantial amounts of primordial volatiles.

Crystallization starts from the bottom and in less than 1000 yr propagates through the lower mantle. During this period of time, convection can prevent fractionation of crystals up to the critical crystal fraction around 60% (rheological transition from a low-viscosity suspension to a high-viscosity partially molten solid), provided the crystals are smaller than $10^{-3}$ m. Kinetics of nucleation and crystal growth suggest that this is a possibility. Solid-state convection helps further cooling and crystallization of the partially molten mantle down to the solidus and below. The product of this period of crystallization is essentially undifferentiated mantle with the remaining shallow partially molten layer. This precludes a severe fractionation of minor and trace elements (*Kato et al.,* 1988a,b; *Ringwood,* 1990; *McFarlane and Drake,* 1990; *McFarlane et al.,* 1994). Small amounts of crystals might settle down and contribute to the formation of an Fe-rich D" layer.

The shallow magma ocean is the only part of the mantle that undergoes any substantial differentiation. This can be reconciled with the observed ratios of major and minor el-

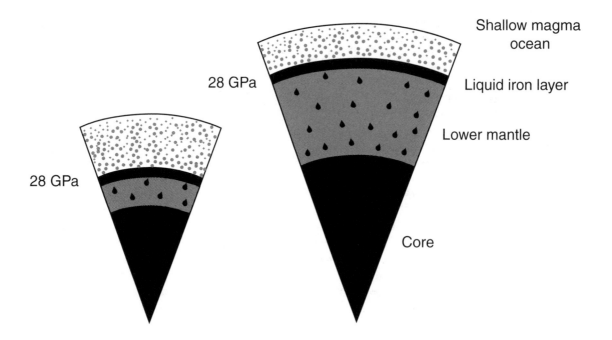

**Fig. 10.** A possible thermal structure of the growing proto-Earth: a shallow magma ocean where crystal settling/flotation and segregation of liquid Fe delivered by impacts take place; a solid lower mantle with liquid Fe diapirs and a liquid Fe core. The bottom of the shallow magma ocean is located at P = 28 GPa.

ements (*Gasparik and Drake*, 1995). Crystallization of the shallow magma ocean takes more than $10^8$ yr and is probably a rather complicated process because of the continuous supply of the new material by meteorites, remelting due to the impacts, and tidal heating (*Sears*, 1993), and a complex pattern of crystal/melt segregation due to crystal/melt density inversions (*Agee*, 1998). This part of the evolution of magma oceans merges with the subsequent thermal evolution of the Earth controlled by radiogenic heating and solid-state convection.

The two very different timescales, $10^3$ yr for crystallization of the lower mantle and more than $10^8$ yr for crystallization of the shallow magma ocean, suggest that the lower mantle healed very quickly whenever it was molten by a large impact, while the upper mantle never had enough time to crystallize completely. If so, then Fe delivered by impacts accumulated at the boundary of the shallow magma ocean. It formed a gravitationally unstable Fe layer that sank into the Earth's core as discussed by *Elsasser* (1963), *Stevenson* (1981, 1990), and *Karato and Murthy* (1997a,b). These instabilities probably developed in the form of liquid Fe diapirs that were big enough to preclude any significant chemical exchange with the lower mantle (*Stevenson*, 1990; *Karato and Murthy*, 1997a). Therefore chemical equilibrium between Fe and silicates was established at the bottom of this shallow magma ocean (Fig. 10) rather than in the lower mantle (*Tschauner et al.*, 1999).

The sudden drop in the cooling rate of the magma ocean is the factor that determines the location of the bottom of the shallow magma ocean. This happens due to two major events: (1) The potential temperature drops below liquidus so that crystals can grow to much bigger sizes (instead of going through a relatively short "nucleation-growth-dissolution" cycle) and crystal settling/flotation turns off convection; and (2) the potential temperature drops below the critical temperature for the rheological transition (about 60% crystal fraction) at which a large viscosity jump drastically reduces the vigor of convection. The difference in the potential temperature between these two events is only about 300 K, yet the estimates of the corresponding depths of the shallow magma ocean vary from 10 GPa to somewhere around 40 GPa. The experiments on metal/silicate partition coefficients for siderophile elements suggest that chemical equilibrium between liquid Fe and liquid silicate was established at P = 28 GPa (*Li and Agee*, 1996; *Righter et al.*, 1997; *Righter and Drake*, 1997). The bottom of the shallow magma ocean is controlled by pressure rather than by depth. Therefore, it is conceivable that the pressure for metal/silicate equilibrium is the pressure (perhaps an average one) that corresponds to the bottom of the shallow magma ocean.

Many aspects of fluid dynamics of magma oceans remain poorly understood. The structure and evolution of the shallow magma ocean, migration of iron toward the Earth's core, and the effect of water on all processes are among the problems that need to be investigated. It is also clear that the problems of crystallization of magma oceans cannot be separated from other aspects of Earth's formation such as impacts, core formation, and evolution of the Earth's atmosphere. A self-consistent model of Earth's formation is yet to be developed.

***Acknowledgments.*** This work was supported by NSF grant EAR-9506722, NASA grant NAG5-6897, and the Alfred P. Sloan Foundation. The author thanks K. Righter, F. J. Spera, and D. J. Stevenson for their constructive and thorough reviews.

# REFERENCES

Abe Y. (1993) Physical state of the very early Earth. *Lithos, 30,* 223–235.

Abe Y. (1995) Early evolution of the terrestrial planets. *J. Phys. Earth, 43,* 515–535.

Abe Y. (1997) Thermal and chemical evolution of the terrestrial magma ocean. *Phys. Earth Planet. Inter., 100,* 27–39.

Abe Y. and Matsui T. (1986) Early evolution of the Earth: Accretion, atmosphere formation, and thermal history. *Proc. Lunar Planet. Sci. Conf. 17th,* in *J. Geophys. Res., 91,* E291–E302.

Agee C. B. (1990) A new look at differentiation of the Earth from melting experiments on the Allende meteorite. *Nature, 346,* 834–837.

Agee C. B. (1998) Crystal-liquid density inversions in terrestrial and lunar magma oceans. *Phys. Earth Planet. Inter., 107,* 63–74.

Agee C. B. and Walker D. (1988) Mass balance and phase density constraints on early differentiation of chondritic mantle. *Earth Planet. Sci. Lett., 90,* 144–156.

Ahrens T. J. (1992) A magma ocean and the Earth's internal water budget. In *Workshop on the Physics and Chemistry of Magma Oceans from 1 bar to 4 Mbar* (C. B. Agee and J. Longhi, eds.), pp. 5–6. LPI Tech. Rpt. 92-03, Lunar and Planetary Institute, Houston.

Andrade E. N. C. (1952) Viscosity of liquids. *Proc. Roy. Soc. London, A215,* 36–43.

Arzi A. A. (1978) Critical phenomena in the rheology of partially melted rocks. *Tectonophys., 44,* 173–184.

Asimow P. D., Hirschmann M. M., and Stolper E. M. (1997) An analysis of variations in isentropic melt productivity. *Philos. Trans. R. Soc. Lond., A355,* 255–281.

Benz W. and Cameron A. G. W. (1990) Terrestrial effects of the giant impact. In *Origin of the Earth* (H. E. Newsom and J. H. Jones, eds.), pp. 61–67. Oxford Univ., New York.

Benz W., Slattery W. L., and Cameron A. G. W. (1986) The origin of the Moon and the single impact hypothesis, I. *Icarus, 66,* 515–535.

Benz W., Slattery W. L., and Cameron A. G. W. (1987) The origin of the Moon and the single impact hypothesis, II. *Icarus, 71,* 30–45.

Benz W., Cameron A. G. W., and Melosh H. J. (1989) The origin of the Moon and the single impact hypothesis, III. *Icarus, 81,* 113–131.

Boehler R. (1992) Melting of the Fe-FeO and the Fe-FeS systems at high pressure: Constraints on core temperatures. *Earth Planet. Sci. Lett., 111,* 217–227.

Bottinga Y. and Weill D. F. (1972) The viscosity of magmatic silicate liquids: A model for calculation. *Amer. J. Sci., 272,* 438–475.

Bottinga Y., Richet P., and Sipp A. (1995) Viscosity regimes of homogeneous silicate melts. *Amer. Mineral., 80,* 305–318.

Boubnov B. M. and Golitsyn G. S. (1986) Experimental study of convective structures in rotating fluids. *J. Fluid Mech., 167,* 503–531.

Boubnov B. M. and Golitsyn G. S. (1990) Temperature and velocity field regimes of convective motions in a rotating plane fluid layer. *J. Fluid Mech., 219,* 215–239.

Brinkman H. C. (1952) The viscosity of concentrated suspensions and solutions. *J. Chem. Phys., 20,* 571.

Cambell G. A. and Forgacs G. (1990) Viscosity of concentrated suspensions: An approach based on percolation theory. *Phys. Rev., 41A,* 4570–4573.

Cameron A. G. W. (1997) The origin of the Moon and the single impact hypothesis. 5. *Icarus, 126,* 126–137.

Canup R. M. and Agnor C. (2000) Title of chapter. In *Origin of the Earth and Moon* (R. M. Canup and K. Righter, eds.), this volume. Univ. of Arizona, Tucson.

Canup R. M. and Esposito L. W. (1996) Accretion of the Moon from an impact-generated disk. *Icarus, 119,* 427–446.

Cashman K. V. (1993) Relationship between plagioclase crystallization and cooling rate in basaltic melts. *Contrib. Mineral. Petrol., 113,* 126–142.

Castaing B., Gunaratne G., Heslot F., Kadanoff L., Libchaber A., Thomae S., Wu X.-Z., Zaleski S., and Zanetti G. (1989) Scaling of hard thermal turbulence in Rayleigh-Benard convection. *J. Fluid Mech., 204,* 1–30.

Caughey S. J. and Palmer S. G. (1979) Some aspects of turbulence structure through the depth of the convective layer. *Quart. J. R. Met. Soc., 105,* 811–827.

Chen R. H., Fernando J. S., and Boyer D. L. (1989) Formation of isolated vortices in a rotating convecting fluid. *J. Geophys. Res., 94,* 18445–18453.

Cooper R. F. and Kohlstedt D. L. (1986) Rheology and structure of olivine-basalt partial melts. *J. Geophys. Res., 91,* 9315–9323.

Davies G. F. (1982) Ultimate strength of solids and formation of planetary cores. *Geophys. Res. Lett., 11,* 1267–1270.

Davies G. F. (1990) Heat and mass transport in the early Earth. In *Origin of the Earth* (H. E. Newsom and J. H. Jones, eds.), pp. 175–194. Oxford Univ., New York.

Davis M. J., Ihinger P. D., and Lasaga A. C. (1997) Influence of water on nucleation kinetics in silicate melt. *J. Non-Cryst. Solids, 219,* 62–69.

Davis R. H. and Acrivos A. (1985) Sedimentation of noncolloidal particles at low Reynolds numbers. *Annu. Rev. Fluid Mech., 17,* 91–118.

Deardorff J. W. (1970) Convective velocity and temperature scales for the unstable planetary boundary layer and for Rayleigh convection. *J. Atmos. Sci., 27,* 1211–1213.

Dowty E. (1980) Crystal growth and nucleation theory and the numerical simulation of igneous crystallization. In *Physics of Magmatic Processes* (R. V. Hargraves, ed.), pp. 419–485. Princeton Univ., Princeton.

Dullien F. A. L. (1979) *Porous Media: Fluid Transport and Pore Structure.* Academic, San Diego.

Elsasser W. M. (1963) Early history of the Earth. In *Earth Science and Meteorites* (J. Geiss and E. Goldberg, eds.), pp. 1–30. North-Holland, Amsterdam.

Fenn P. M. (1977) The nucleation and growth of alkali feldspars from hydrous melts. *Can. Mineral., 15,* 135–161.

Fernando H. J. S., Chen R. R., and Boyer D. L. (1991) Effects of rotation on convective turbulence. *J. Fluid Mech., 228,* 513–547.

Flasar F. M. and Birch F. (1973) Energetics of core formation: A correction. *J. Geophys. Res., 78,* 6101–6103.

Flemings M. C., Riek R. G., and Young K. P. (1976) Rheocasting. *Mater. Sci. Eng., 25,* 103–117.

Frankel N. A. and Acrivos A. (1967) On the viscosity of a concentrated suspension of solid spheres. *Chem. Eng. Sci., 22,* 847–853.

Gans R. F. (1972) Viscosity of the Earth's core. *J. Geophys. Res., 77,* 360–366.

Gasparik T. and Drake M. J. (1995) Partitioning of elements among two silicate perovskites, superphase B, and volatile-bearing melt at 23 GPa and 1500–600°C. *Earth Planet. Sci. Lett., 134,* 307–318.

Ghiorso M. S. (1997) Thermodynamic models of igneous processes. *Annu. Rev. Earth Sci., 25,* 221–241.

Glazier J. A., Segawa T., Naert A., and Sano M. (1999) Evidence against 'ultrahard' thermal turbulence at very high Rayleigh numbers. *Nature, 398,* 307–310.

Golitsyn G. S. (1980) Geostrophic convection. *Dokl. Akad. Nauk SSSR, 251,* 1356–1360.

Golitsyn G. S. (1981) Structure of convection in rapid rotation. *Dokl. Akad. Nauk SSSR, 261,* 317–320.

Grossmann S. and Lohse D. (1992) Scaling in hard turbulent Rayleigh-Bénard flow. *Phys. Rev. A, 46,* 903–917.

Grove T. L. (1990) Cooling histories of lavas from Serocki volcano. *Proc. Ocean Drilling Prog., 106/109,* 3–8.

Grove T. L. and Walker D. (1977) Cooling histories of Apollo 15 quartz-normative basalts. *Proc. Lunar Sci. Conf. 8th,* pp. 1501–1520.

Halliday A., Rehkämper M., Lee D.-C., and Yi W. (1996) Early evolution of the Earth and Moon: New constraints from Hf-W isotope geochemistry. *Earth Planet. Sci. Lett., 142,* 75–89.

Herzberg C. and Gasparik T. (1991) Garnet and pyroxenes in the mantle: A test of the majorite hypothesis. *J. Geophys. Res., 96,* 16263–16274.

Hirth G. and Kohlstedt D. L. (1995a) Experimental constraints on the dynamics of the partially molten upper mantle: Deformation in the diffusion creep regime. *J. Geophys. Res., 100,* 1981–2001.

Hirth G. and Kohlstedt D. L. (1995b) Experimental constraints on the dynamics of the partially molten upper mantle 2. Deformation in the dislocation creep regime. *J. Geophys. Res., 100,* 15441–15449.

Holland K. G. and Ahrens T. J. (1997) Melting of $(Mg,Fe)_2SiO_4$ at the core-mantle boundary of the Earth. *Science, 275,* 1623–1625.

Holloway J. R. (1988) Planetary atmospheres during accretion: The effect of C-O-H-S equilibria (abstract). In *Lunar and Planetary Science XIX,* pp. 499–500. Lunar and Planetary Institute, Houston.

Hopfinger E. J. (1989) Turbulence and vortices in rotating fluids. In *Theoretical and Applied Mechanics* (P. Germain, M. Piau, and D. Caillerie, eds.), pp. 117–138. Elsevier, New York.

Hopfinger E. J., Browand F. K., and Gagne Y. (1982) Turbulence and waves in a rotating tank. *J. Fluid Mech., 125,* 505–534.

Ichikawa K., Kinoshita Y., and Shimamura S. (1985) Grain refinement in Al-Cu binary alloys by rheocasting. *Trans. Japan Inst. Metals, 26,* 513–522.

Ida S., Canup R. M., and Stewart G. R. (1997) Lunar accretion from an impact-generated disk. *Nature, 389,* 353–357.

Ita J. and Cohen R. E. (1998) Diffusion in MgO at high pressure: Implications for lower mantle rheology. *Geophys. Res. Lett., 25,* 1095–1098.

Ito E. and Takahashi E. (1987) Melting of peridotite at uppermost

lower-mantle conditions. *Nature, 328,* 514–517.

Jin Z.-M., Green H. W., and Zhou Y. (1994) Melt topology in partially molten mantle peridotite during ductile deformation. *Nature, 372,* 164–167.

Julien K., Legg S., McWilliams J., and Werne J. (1996) Hard turbulence in rotating Rayleigh-Bénard convection. *Phys. Rev. E, 53,* 5557–5560.

Karato S. and Li P. (1992) Diffusion creep in perovskite: Implications for the rheology of the lower mantle. *Science, 255,* 1238–1240.

Karato S.-I. and Murthy V. R. (1997a) Core formation and chemical equilibrium in the Earth — I. Physical consideration. *Phys. Earth Planet. Inter., 100,* 61–79.

Karato S.-I. and Murthy V. R. (1997b) Core formation and chemical equilibrium in the Earth — II. Chemical consequences for the mantle and the core. *Phys. Earth Planet. Inter., 100,* 81–95.

Kashchiev D. (1969) Solution of the nonsteady state problem in nucleation kinetics. *Surf. Sci., 14,* 209–220.

Kasting J. F. (1988) Runaway and moist greenhouse atmosphere and the evolution of Earth and Venus. *Icarus, 74,* 472–494.

Kato T., Ringwood A. E., and Irifune T. (1988a) Experimental determination of element portioning between silicate perovskites, garnets and liquids: Constraints on early differentiation of the mantle. *Earth Planet. Sci. Lett., 89,* 123–145.

Kato T., Ringwood A. E., and Irifune T. (1988b) Constraints on element partition coefficients between $MgSiO_3$ perovskite and liquid determined by direct measurements. *Earth Planet. Sci. Lett., 90,* 65–68.

Kaula W. M. (1979) Thermal evolution of earth and moon growing by planetesimals impacts. *J. Geophys. Res., 84,* 999–1008.

King S. D. (1995) Models of mantle viscosity. In *Mineral Physics and Crystallography: A Handbook of Physical Constants* (T. J. Ahrens, ed.), pp. 227–236. AGU, Washington, DC.

Knittle E. and Jeanloz R. (1989) Melting curve of $(Mg,Fe)SiO_3$ perovskite to 96 GPa: Evidence for a structural transition in lower mantle melts. *Geophys. Res. Lett., 16,* 421–424.

Kohlstedt D. L. and Zimmerman M. E. (1996) Rheology of partially molten mantle rocks. *Annu. Rev. Earth Planet. Sci., 24,* 41–62, 1996.

Kraichnan R. H. (1962) Turbulent thermal convection at arbitrary Prandtl number. *Phys. Fluids, 5,* 1374–1389.

Krieger I. M. and Dougherty T. J. (1959) A mechanism for non-Newtonian flow in suspensions of rigid spheres. *Trans. Soc. Rheol., 3,* 137–152.

Kushiro I. (1980) Viscosity, density, and structure of silicate melts at high pressures, and their petrological applications. In *Physics of Magmatic Processes* (R. B. Hargraves, ed.), pp. 93–120. Princeton Univ., Princeton.

Kushiro I. (1986) Viscosity of partial melts in the upper mantle. *J. Geophys. Res., 91,* 9343–9350.

Lee D.-C. and Halliday A. N. (1995) Hafnium-tungsten chronometry and the timing of terrestrial core formation. *Nature, 378,* 771–774.

Lejeune A.-M. and Richet P. (1995) Rheology of crystal-bearing silicate melts: An experimental study at high viscosities. *J. Geophys. Res., 100,* 4215–4229.

Li J. and Agee C. B. (1996) Geochemistry of mantle-core differentiation at high pressure. *Nature, 381,* 686–689.

Lifshitz I. M. and Slyozov V. V. (1961) The kinetics of precipitation from supersaturated solid solution. *J. Phys. Chem. Solids, 19,* 35–50.

Li P., Karato S., and Wang Z. (1996) High-temperature creep in fine-grained polycrystalline $CaTiO_3$, an analogue material of $(Mg,Fe)SiO_3$. *Phys. Earth Planet. Inter., 95,* 19–36.

Lofgren G., Donaldson C. H., Williams R. J., Mullins O., and Usselman T. M. (1974) Experimentally reproduced textures and mineral chemistry of Apollo 15 quartz-normative basalts. *Proc. Lunar Sci. Conf. 5th,* pp. 549–567.

Martin D. and Nokes R. (1988) Crystal settling in a vigorously convecting magma chamber. *Nature, 332,* 534–536.

Martin D. and Nokes R. (1989) A fluid dynamical study of crystal settling in convecting magmas. *J. Petrol., 30,* 1471–1500.

Matsui T. and Abe Y. (1986) Formation of a "magma ocean" on the terrestrial planets due to the blanketing effect of an impact-induced atmosphere. *Earth Moon Planets, 34,* 223–230.

McBirney A. R. and Murase T. (1984) Rheological properties of magmas. *Annu. Rev. Earth Planet. Sci., 12,* 337–357.

McFarlane E. A. and Drake M. J. (1990) Element partitioning and the early thermal history of the Earth. In *Origin of the Earth* (H. E. Newsom and J. H. Jones, eds.), pp. 135–150. Oxford Univ., New York.

McFarlane E. A., Drake M. J., and Rubie D. C. (1994) Element partitioning between Mg-perovskite, magnesiowüstite, and silicate melt at conditions of the Earth's mantle. *Geochim. Cosmochim. Acta, 58,* 5161–5172.

McKenzie D. and Bickle M. J. (1988) The volume and composition of melt generated by extension of the lithosphere. *J. Petrol., 29,* 625–679.

Melosh H. J. (1990) Giant impacts and thermal state of the early Earth. In *Origin of the Earth* (H. E. Newsom and J. H. Jones, eds.), pp. 69–83. Oxford Univ., New York.

Miller G. H., Stolper E. M., and Ahrens T. J. (1991a) The equation of state of a molten komatiite, 1, Shock wave compression to 36 GPa. *J. Geophys. Res., 96,* 11831–11848.

Miller G. H., Stolper E. M., and Ahrens T. J. (1991b) The equation of state of a molten komatiite, 2, Application to komatiite petrogenesis and the Hadean mantle. *J. Geophys. Res., 96,* 11849–11864.

Mooney M. (1951) The viscosity of a concentrated suspension of spherical particles. *J. Colloid. Sci., 6,* 162–170.

Murray J. D. (1965) On the mathematics of fluidization, Part I, Fundamental equations and wave propagation. *J. Fluid Mech., 21,* 465–493.

Newsom H. E. and Taylor S. R. (1989) Geochemical implications of the formation of the Moon by a single great impact. *Nature, 338,* 29–34.

Ohtani E. (1985) The primordial terrestrial magma ocean and its implication for stratification of the mantle. *Phys. Earth Planet. Inter., 38,* 70–80.

Ohtani E. and Sawamoto H. (1987) Melting experiment on a model chondritic mantle composition at 25 GPa. *Geophys. Res. Lett., 14,* 733–736.

Persikov E. S., Zharikov V. A., Bukhtiyarov P. G., and Polskoy S. F. (1990) The effect of volatiles on the properties of magmatic melts. *Eur. J. Mineral., 2,* 621–642.

Pierazzo E., Vickery A. M., and Melosh H. J. (1997) A reevaluation of impact melt production. *Icarus, 127,* 408–423.

Presnall D. C., Weng Y.-H., Milholland C. S., and Walter M. J. (1998) Liquidus phase relations in the system $MgO-MgSiO_3$ at pressures up to 25 GPa — constraints on crystallization of a Hadean mantle. *Phys. Earth Planet. Inter., 107,* 83–95.

Priestly C. H. B. (1959) *Turbulent Transfer in the Lower Atmosphere.* Univ. of Chicago, Chicago.

Righter K. and Drake M. J. (1997) Metal-silicate equilibrium in a homogeneously accreting Earth: New results for Re. *Earth Planet. Sci. Lett., 146,* 541–553.

Righter K., Drake M. J., and Yaxley G. (1997) Prediction of siderophile element metal-silicate partition coefficients to 20 GPa and 2800°C: The effects of pressure, temperature, oxygen fugacity, and silicate and metallic melt compositions. *Phys. Earth Planet. Inter., 100,* 115–134.

Ringwood A. E. (1990) Earliest history of the Earth-Moon system. In *Origin of the Earth* (H. E. Newsom and J. H. Jones, eds.), pp. 101–134. Oxford Univ., New York.

Roscoe R. (1952) The viscosity of suspensions of rigid spheres. *Brit. J. Appl. Phys., 3,* 267–269.

Rutter E. H. and Neumann D. H. K. (1995) Experimental deformation of partially molten Westerly granite under fluid-absent conditions, with implications for the extraction of granitic magmas. *J. Geophys. Res., 100,* 15697–15715.

Safronov V. S. (1978) The heating of the Earth during its formation. *Icarus, 33,* 3–12.

Sasaki S. and Nakazawa K. (1986) Metal-silicate fractionation in the growing Earth: Energy source for the terrestrial magma ocean. *J. Geophys. Res., 91,* 9231–9238.

Shaw H. R. (1972) Viscosities of magmatic silicate liquids: An empirical method of prediction. *Am. J. Sci., 272,* 870–893.

Scarfe C. M. and Takahashi E. (1986) Melting of garnet peridotite to 13 GPa and the early history of the upper mantle. *Nature, 322,* 354–356.

Sears W. D. (1993) Tidal dissipation and the giant impact origin for the Moon (abstract). In *Lunar and Planetary Science XXIII,* pp. 1255–1256. Lunar and Planetary Institute, Houston.

Shraiman B. I. and Siggia E. D. (1990) Heat transport in high-Rayleigh-number convection. *Phys. Rev. A, 42,* 3650–3653.

Siggia E. D. (1994) High Rayleigh number convection. *Annu. Rev. Fluid Mech., 26,* 137–168.

Smith D. M., Eady J. A., Hogan L. M., and Irwin D. W. (1991) Crystallization of a faceted primary phase in a stirred slurry. *Metall. Trans., 22A,* 575–584.

Solomatov V. S. (1995) Batch crystallization under continuous cooling: Analytical solution for diffusion limited crystal growth. *J. Crystal Growth, 148,* 421–431.

Solomatov V. S. and Moresi L.-N. (1996) Stagnant lid convection on Venus. *J. Geophys. Res., 101,* 4737–4753.

Solomatov V. S. and Stevenson D. J. (1993a) Suspension in convective layers and style of differentiation of a terrestrial magma ocean. *J. Geophys. Res., 98,* 5375–5390.

Solomatov V. S. and Stevenson D. J. (1993b) Nonfractional crystallization of a terrestrial magma ocean. *J. Geophys. Res., 98,* 5391–5406.

Solomatov V. S. and Stevenson D. J. (1993c) Kinetics of crystal growth in a terrestrial magma ocean. *J. Geophys. Res., 98,* 5407–5418.

Solomatov V. S., Olson P., and Stevenson D. J. (1993) Entrainment from a bed of particles by thermal convection. *Earth Planet. Sci. Lett., 120,* 387–393.

Soo S. L. (1967) *Fluid Dynamics of Multiphase Systems.* Blaisdell, Waltham. 524 pp.

Spera F. J. (1992) Lunar magma transport phenomena. *Geochim. Cosmochim. Acta, 56,* 2253–2265.

Stevenson D. J. (1981) Models of the Earth's core. *Science, 214,* 611–619.

Stevenson D. J. (1987) Origin of the Moon — The collision hypothesis. *Annu. Rev. Earth Planet. Sci., 15,* 271–315.

Stevenson D. J. (1990) Fluid dynamics of core formation. In *Origin of the Earth* (H. E. Newsom and J. H. Jones, eds.), pp. 231–249. Oxford Univ., New York.

Stishov S. M. (1988) Entropy, disorder, melting. *Sov. Phys. Uspekhi, 31,* 52–67.

Swanson S. E. (1977) Relation of nucleation and crystal-growth rate to the development of granitic textures. *Amer. Mineral., 62,* 966–978.

Thompson C. and Stevenson D. J. (1988) Gravitational instabilities in 2-phase disks and the origin of the Moon. *Astrophys. J., 333,* 452–481.

Tonks W. B. and Melosh H. J. (1990) The physics of crystal settling and suspension in a turbulent magma ocean. In *Origin of the Earth* (H. E. Newsom and J. H. Jones, eds.), pp. 151–174. Oxford Univ., New York.

Tonks W. B. and Melosh H. J. (1992) Core formation by giant impacts. *Icarus, 100,* 326–346.

Tonks W. B. and Melosh H. J. (1993) Magma ocean formation due to giant impacts. *J. Geophys. Res., 98,* 5319–5333.

Turcotte D. L. and Schubert G. (1982) *Geodynamics: Applications of Continuum Physics to Geological Problems.* Wiley, New York.

Tschauner O., Zerr A., Specht S., Rocholl A., Boehler R., and Palme H. (1999) Partitioning of nickel and cobalt between silicate perovskite and metal at pressures up to 80 GPa. *Nature, 398,* 604–607.

van der Molen I. and Paterson M. S. (1979) Experimental deformation of partially-melted granite. *Contrib. Mineral. Petrol., 70,* 299–318.

Voorhees P. W. (1992) Ostwald ripening of two-phase mixtures. *Annu. Rev. Mater. Sci., 22,* 197–215.

Walker D., Powell M. A., Lofgren G. E., and Hays J. F. (1978) Dynamic crystallization of a eucrite basalt. *Proc. Lunar Planet. Sci. Conf. 9th,* pp. 1369–1391.

Wasserman E. A., Yuen D. A., and Rustad J. R. (1993a) Compositional effects on the transport and thermodynamic properties of MgO-SiO$_2$ mixtures using molecular dynamics. *Phys. Earth Planet. Inter., 77,* 189–203.

Wasserman E. A., Yuen D. A., and Rustad J. R. (1993b) Molecular dynamics study of the transport properties of perovskite melts under high-temperature and pressure conditions. *Earth Planet. Sci. Lett., 114,* 373–384.

Weidenschilling S. J., Spaute D., Davis D. R., Marzari F., and Ohtsuki K. (1997) Accretional evolution of a planetesimal swarm. 2. The terrestrial zone. *Icarus, 126,* 429–455.

Wetherill G. W. (1985) Occurrence of giant impacts during the growth of the terrestrial planets. *Science, 228,* 877–879.

Wetherill G. W. (1990) Formation of the Earth. *Annu. Rev. Earth Planet. Sci., 18,* 205–256.

Willis G. E. and Deardorff J. W. (1974) A laboratory model of the unstable planetary boundary layer. *J. Atmos. Sci., 31,* 1297–1307.

Zahnle K. J., Kasting J. F., and Pollack J. B. (1988) Evolution of a steam atmosphere during Earth's accretion. *Icarus, 74,* 62–97.

Zerr A. and Boehler R. (1993) Melting of (Mg,Fe)SiO$_3$-perovskite to 625 kilobars: Indication of a high melting temperature in the lower mantle. *Science, 262,* 553–557.

Zerr A. and Boehler R. (1994) Constraints on the melting temperature of the lower mantle from high-pressure experiments on MgO and magnesiowüstite. *Nature, 371,* 506–508.

Zerr A., Diegeler A., and Boehler R. (1998) Solidus of Earth's deep mantle. *Science, 281,* 243–246.

Zhang J. and Herzberg C. T. (1994) Melting experiments on anhydrous peridotite KLB-1 from 5.0 to 22.5 GPa. *J. Geophys. Res., 99,* 17729–17742.

# Evolution of the Moon's Mantle and Crust as Reflected in Trace-Element Microbeam Studies of Lunar Magmatism

## C. K. Shearer
*University of New Mexico*

## C. Floss
*Washington University*

Ion microprobe trace-element studies of lunar cumulates [ferroan anorthosites (FAN), highlands Mg suite (HMS), and highlands alkali suite (HAS)] and volcanic glasses have provided an additional perspective in reconstructing lunar magmatism and early differentiation. Calculated melt compositions for the FANs indicate that a simple lunar magma ocean (LMO) model does not account for differences between FANs with highly magnesian mafic minerals and "typical" ferroan anorthosites. The HMS and HAS appear to have crystallized from magmas that had incompatible trace-element concentrations equal to or greater than KREEP. Partial melting of distinct, hybridized sources is consistent with these calculated melt compositions. However, the high-Mg silicates with relatively low Ni content that are observed in the HMS are suggestive of other possible processes (reduction, metal removal). The compositions of the picritic glasses indicate that they were produced by melting of hybrid cumulate sources produced by mixing of early and late LMO cumulates. The wide compositional range of near-primitive mare basalts indicates small degrees of localized melting preserved the signature of distinct mantle reservoirs. The relationship between ilmenite anomalies and $\varepsilon_{182_W}$ in the mare basalts suggests that the LMO crystallized over a short period of time.

## 1. INTRODUCTION

Magmatism is a fundamental process that occurs on all the terrestrial planets, the Moon, and some asteroids, and accounts for planetary differentiation and crustal formation. The magmatic history of a planetary body is an expression of its global dynamic and thermal processes, therefore products of magmatism provide a partial record of planetary evolution. Because small planetary bodies, such as the Moon and 4 Vesta, lose internal heat more quickly than large terrestrial planets like the Earth, they provide a more accurate and complete record of the earliest planetary magmatism in our solar system.

Although incomplete because of the imperfect and somewhat random sampling of rock types by the Apollo and Luna missions, the history of lunar magmatism has been reconstructed by numerous researchers over the past three decades. These reconstructions have illustrated its continuous nature (4.6 to ~2.0 Ga) and the influence of early differentiation on lunar mantle dynamics, magmatism, and eruptive style. Many gaps in this record exist not only because of sampling bias, but also because many products of lunar magmatism are cumulates or nonprimary basaltic magmas. Early cumulate rocks provide an incomplete record because their bulk compositions do not represent basaltic liquids, their parental melts were modified by a range of magmatic processes, and they were fragmented and modified by impact processes on the lunar surface. Most mare basalts are the end products of shallow fractional crystallization pro-

cesses and therefore have lost primary magma characteristics that provide insightful information about the lunar mantle. Trace-element microbeam analyses of lunar cumulates and primary volcanic glasses have provided an additional perspective in reconstructing aspects of lunar magmatism and planetary differentiation. The major thrust of this paper is to integrate previous and ongoing ion microprobe trace-element studies into models that have been proposed for episodes of lunar magmatism.

Using ion microprobe data of lunar materials as a further constraint in reconstructing lunar history has numerous advantages as well as some drawbacks. The ion microprobe can provide contamination-free (i.e., without coatings or inclusions) microanalyses of trace elements whose abundances are too low to analyze by other techniques (e.g., electron microprobe). For the analysis of lunar volcanic glasses, this allows determination of chemical variation within basaltic melts represented by individual volcanic glass beads and glass bead populations. Calculating parent melt compositions from the trace-element analysis of cumulate phases also benefits from a small analytical area (5–50 μm) and low concentration capabilities (*Floss et al.,* 1991, 1998; *Fowler et al.,* 1995, 1996; *Papike,* 1996; *Papike et al.,* 1994, 1996; *Shervais and McGee,* 1997, 1999; *Snyder et al.,* 1998) and eliminates the need to attempt to obtain "representative" analyses of bulk rock samples that are minute and/or only preserved in thin section. In addition, this approach ideally yields melt compositions from which the cumulates crystallized and not the composition of a bulk cumulate rock.

Obvious complications to this approach result from some of the assumptions that are made in these calculations, namely, that the magmatic signature is preserved in the individual phases and that the partition coefficients are valid. However, careful examination of the mineral compositions and selection of appropriate partition coefficients allow quanitative to semiquantitative estimates of parent melt compositions to be made. The opportunity to thus examine trace-element data within a petrographic context can provide important constraints on processes important for the evolution of the Moon.

## 2.   IGNEOUS ROCKS ON THE MOON

One of the fundamental discoveries of the Apollo missions is that the Moon consists of a variety of igneous rock types that differ widely in their mineralogy, composition, and age. The most visible evidence of these differences is the existence of two distinct terrains on the Moon: the light-colored feldspathic rocks of the highlands and the dark basalts of the maria.

The lunar crust is approximately 63 km thick on the nearside and perhaps over 100 km thick on the farside (*Taylor,* 1982). The highlands surface on the Moon has a bulk composition equivalent to anorthositic gabbro (26–28% $Al_2O_3$) and possibly represents the composition of the upper lunar crust (*Ryder and Wood,* 1977; *Spudis and Davis,* 1986). The observed enrichment in noritic components in basin ejecta with increasing basin size suggests that the bulk composition of the lower lunar crust is noritic (~20% $Al_2O_3$) (*Ryder and Wood,* 1977; *Mueller et al.,* 1988). The bulk composition of the total lunar crust may correspond to anorthositic norite (*Ryder and Wood,* 1977; *Spudis and Davis,* 1986). A similar crustal bulk composition and compositional distribution has been corroborated by recent interpretation of *Clementine* data (*Tompkins and Pieters,* 1999)

Because of intense bombardment early in the history of the Moon, the upper lunar crust is fractured and brecciated to a depth of ~20 km and is covered by thick polymict ejecta and comminuted melt sheets (≥2 km). To unravel the complex history of the lunar crust, it is essential to separate rocks that formed through endogenous igneous activity from those formed by impact-generated melting. According to *Warren* (1985, 1993), the least ambiguous means of identifying pristine rocks that were neither formed nor contaminated by impact is by analysis of siderophile elements. This is because most meteorites have high siderophile contents, and rocks formed by impact can usually be identified by elevated concentrations of these elements. *Warren* (1993) presented a list of pristine rock candidates.

Pristine highland rocks can be divided into three major groups by molar ratios of Ca/(Ca + Na) (or An content) in plagioclase vs. Mg/(Mg + Fe) (Mg') in mafic minerals (Fig. 1) (*Warner et al.,* 1976). These three major rock types, the ferroan anorthosite suite (FAN), the highlands Mg suite (HMS), and the highlands alkali suite (HAS) are cumulates from distinct parental magmas (*Warner et al,* 1976; *James,*

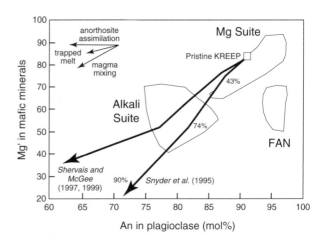

**Fig. 1.**   Plot of anorthite content (An) in plagioclase vs. Mg' of coexisting mafic silicates. Superimposed are fractionation trends calculated for Apollo 15 pristine KREEP basalt by *Snyder et al.* (1995) and *Shervais and McGee* (1997, 1999).

1980). The ferroan anorthosites have been divided into several subgroups based on compositional variations in minerals (*James et al.,* 1989). They make up at least 80% of the lunar crust. The HMS consists of dunites, troctolites, norites, and gabbronorites, whereas the HAS consists of norites, gabbros, gabbronorites, alkali anorthosites, quartz monzodiorites, and granites. The HAS is distinguished from the HMS by its evolved mineral chemistry (Fig. 1) and elevated large-ion-lithophile-element signature in the whole-rock chemistry (*Snyder et al.,* 1995). Prebasin volcanics such as KREEP basalts and high-Al basalts were erupted, perhaps contemporaneously with the emplacement of HAS and HMS plutons.

The pristine highland rocks, and the breccias and regolith derived from them, make up the lunar highlands (~83% of the lunar surface). The remaining ~17% of the lunar surface or ~1% of the lunar crust is composed of mare basalts (*Head and Wilson,* 1992). Mare basalts are enriched in FeO and $TiO_2$ and have higher $CaO/Al_2O_3$ ratios than highland rocks. These chemical differences reflect the enrichment of highland rocks in plagioclase. Mare basalts display a large variation in $TiO_2$ content and this is used in their classification (*Papike et al.,* 1976, 1998; *BVSP,* 1981; *Neal and Taylor,* 1992).

## 3.   PETROGENETIC HISTORY OF LUNAR MAGMATISM

### 3.1.   Introduction

Reconstructions of the petrogenetic history of lunar magmatism have illustrated its continuous nature (4.6 to ~2.0 Ga) and the influence of early differentiation and catastrophic bombardment on lunar mantle dynamics, magmatism, and eruptive style. In the following discussion, we grouped magmatism into stages of activity. Episodes of lunar magmatism are illustrated in Fig. 2 and their duration is summarized in Fig. 3. The reader is reminded that this grouping of lunar magmatic activity is primarily based on sampled rock types returned by manned and robotic missions to the Moon. In most cases, lunar magmatism was a continuous expression of previous dynamic and thermal global processes. The following is a brief review of models suggested for lunar differentiation and magmatism. A much more detailed discussion is presented by *Papike et al.* (1998) and *Shearer and Papike* (1999).

## 3.2.  Early Lunar Differentiation and Associated Magmatism

In many models for the early evolution of the Moon, partial melting of the Moon soon after accretion was responsible for producing an anorthositic crust and a differentiated lunar interior that was the source for subsequent periods of lunar magmatism. The extent and duration of early lunar melting and mantle processing depends upon the mechanism providing primordial heat (*Wetherill*, 1976, 1981; *Alfvén and Arrhenius*, 1976; *Sonett and Reynolds*, 1979; *Hostetler*

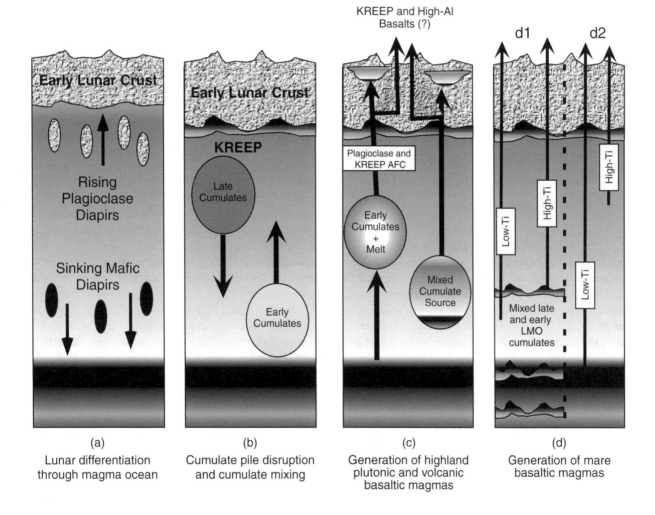

**Fig. 2.**  Summary for the evolution of the lunar mantle and magmatism. **(a)** Early lunar crust formation and lunar mantle differentiation through magma ocean formation and crystallization. **(b)** Disruption of cumulate products of magma ocean crystallization due to gravitational destabilization. **(c)** Generation of parental highland basaltic magmas through pressure release melting. The high Mg' and incompatible-element signature of the HMS rock types are a result of either assimilation of KREEP and early crust by the parental magmas followed by fractional crystallization or partial melting of hybrid cumulate sources and assimilation of KREEP. Alternatively, the high Mg' and low Ni may be a result of metal removal during the formation of the source, melting of the source, or magma evolution. The HAS and highland volcanics may be related to HMS magmas through fractional crystallization and crystal accumulation or contemporaneous magmas produced through partial melting of distinctively different sources. **(d)** Generation of mare basalts through small to moderate degrees of partial melting of mineralogically-chemically varied LMO cumulates. Melting to produce these magmas was either initiated in the deep lunar mantle (d1) or over a wide range of pressure conditions (d2).

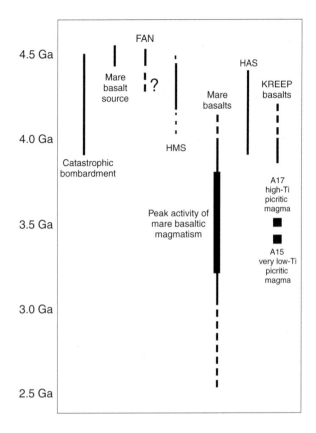

**Fig. 3.** Bar diagram of ages of major events on the Moon.

*and Drake,* 1980; *BVSP,* 1981). Potential sources for primordial heat that have been advocated are accretionary heating from large impacts (*Safronov,* 1978; *Wetherill,* 1976, 1980, 1981; *Ransford,* 1982), electromagnetic induction due to enhanced solar winds during early intense solar activity (*Herbert et al.,* 1977), short-lived radioactive species, enhanced solar luminosity, core formation, tidal dissipation, and inherited heat (*Binder,* 1978).

If heat derived by accretion is the most likely cause of lunar melting, the conditions under which it occurred are key to the existence, size, and character of the lunar magma ocean (LMO). With accretion involving relatively small projectiles (*Safronov,* 1978; *Wetherill,* 1980, 1981; *Ransford,* 1982), heat would have been lost when subsequent impacts ejected material across the lunar surface (*Wetherill,* 1976; *Warren,* 1985). Slow accretion (greater than $10^8$ yr) would have resulted in the inefficient storage of heat in the lunar mantle. Under these circumstances, the Moon would have experienced limited melting inconsistent with lunar differentiation through LMO processes. On the other hand, substantial degrees of melting would have occurred if the projectile radii during accretion were larger than 40 km (*Warren,* 1985) and if accretion occurred over a short period of time ($10^6$–$10^8$ yr). Therefore, rapid accretion is required to produce whole-Moon melting, similar to models proposed by *Wood* (1972, 1975), *Taylor and Jakeš* (1974), and *Longhi* (1977, 1980). Slightly slower accretion rates and

smaller projectiles would have produced a partially molten magmifer alluded to by *Shirley* (1983) and *Warren* (1985).

The crystallization history of the LMO was modeled by *Schnetzler and Philpotts* (1971), *Taylor and Jakeš* (1974), *Taylor* (1982), *Longhi* (1977, 1981), and *Snyder et al.* (1992). Schematic diagrams for the cumulate pile produced by LMO crystallization are presented in Fig. 4. The sequence of crystallization is highly dependent on magma ocean bulk composition and the pressure and flow regimes under which crystallization occurred and is therefore difficult to predict. In a dynamically simple LMO, the crystallization sequence advocated by many of these models is olivine ⟹ orthopyroxene ± olivine ⟹ olivine + clinopyroxene ± plagioclase ⟹ clinopyroxene + plagioclase ⟹ clinopyroxene + plagioclase + ilmenite. The extent of the olivine cumulate assemblage in the magma ocean cumulate pile has been estimated to be between 30% and 40%. The effect of the olivine-orthopyroxene field boundary line on cumulate assemblages depends on both the pressure of crystallization and the efficiency of crystal accumulation (equilibrium vs. fractional crystallization). As pressure increases, this boundary moves toward the olivine apex of the olivine-anorthite-$SiO_2$ pseudoternary. This has two effects. First, it effectively decreases the volume of the monomineralic olivine assemblage in the cumulate pile. Second, the olivine-orthopyroxene boundary becomes a cotectic surface, resulting in coprecipitation of olivine + orthopyroxene regardless of the efficiency of crystal separation. The appearance of low-Ca and high-Ca clinopyroxene follows the precipitation of orthopyroxene and olivine. The exact sequence of the appearance of these pyroxenes is compositionally dependent. The relationship among clinopyroxenes and plagioclase in the magma ocean crystallization sequence is discussed in detail by *Longhi* (1980).

The appearance of plagioclase in the crystallization sequence is extremely important to our fundamental understanding of the development and evolution of the early lunar crust. In simple dynamic models, the lunar crust is thought to have been formed by plagioclase crystallization and flotation after substantial amounts of the magma ocean had crystallized. Using the initial bulk composition suggested by *Warren* (1985) in which the $Al_2O_3$ was equal to 7%, *Snyder et al.* (1992) calculated that plagioclase would be a liquidus phase after 57% of the magma ocean crystallized. Using estimated bulk compositions with lower abundances of $Al_2O_3$ (<5%), *Snyder et al.* (1992) estimated that plagioclase would appear in the crystallization sequence after 70–80% crystallization. The ability of plagioclase to float and accumulate in "rockbergs" at the surface also depends on the Mg' of the magma ocean. Flotation of anorthositic rocks (plagioclase + minor mafic minerals) may have been impossible until the magmasphere became more Fe-rich after substantial crystallization of mafic silicates (*Warren and Wasson,* 1977; *Warren,* 1985). The separation and concentration of plagioclase in the early lunar crust produced the positive Eu anomaly associated with the primordial lunar highlands and implanted a negative Eu anomaly in the mafic

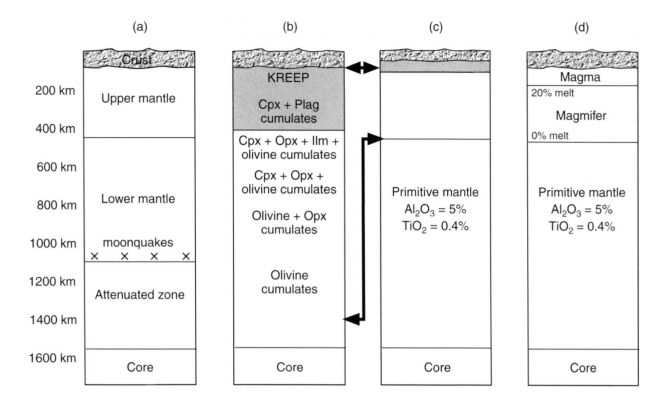

**Fig. 4.** Different types of lunar magma oceans. **(a)** Geophysical structure of the Moon. **(b)** Deep LMO that consumes all the primitive lunar mantle. **(c)** Shallow LMO that preserves the primitive lunar mantle at depth. **(d)** Magmifer processing of the primitive lunar mantle.

cumulates that sank and formed mantle sources for subsequent periods of lunar magmatism. A large portion of the cumulate pile has either no Eu anomaly or a minor negative Eu anomaly [because of orthopyroxene crystallization (*Shearer and Papike*, 1989)].

In more complex models for magma ocean dynamics (*Spera*, 1992), plagioclase is expected to be a liquidus phase during a substantial period of crystallization (*Longhi*, 1980). Because of rapid heat loss of the upper magma ocean boundary, plagioclase may have become a liquidus phase earlier in the upper boundary than in the lower boundary layer. This may have resulted in formation of a plagioclase-rich protocrust prior to 60% magma ocean crystallization. In addition, mafic components that "saw" this early plagioclase crystallization may have imposed a negative Eu anomaly on early basal cumulates (olivine + orthopyroxene) by sinking into the central zone of the magma ocean.

Ilmenite-bearing cumulates precipitated after 90% crystallization of olivine + pyroxene + plagioclase with ilmenite making up between 3% to 12% of the cumulate minerals. Incompatible trace elements excluded from the crystal structures of olivine, pyroxene, and plagioclase were concentrated in the residual melt. These residual liquids were enriched in K, REE, and P (acronym KREEP), Th, U, Zr, Hf, and Nb. The signature of this KREEP component was later incorporated into the feldspathic highland crust through

remobilization, assimilation, and mixing. Soon after crystallization of most of the LMO, the cumulate pile may have experienced gravitational overturn. This resulted in transport of late-forming cumulates into the deep lunar mantle and mixing of LMO cumulates on a variety of scales (*Ringwood and Kesson*, 1976; *Spera*, 1992; *Shearer and Papike*, 1993; *Hess and Parmentier*, 1995).

### 3.3. Early Remelting of Lunar Magma Ocean Cumulates: Post-Magma-Ocean Highland Magmatism

Even though fairly complex, quantitative models have been advanced to explain how both FAN and HMS could have come from the LMO, it appears more likely that the HMS and HAS were generated by melting events (4.5–3.9 Ga) that postdated the crystallization of most, if not all, of the magma ocean. The HMS magmas, which apparently intruded the early anorthositic crust, may have formed layered mafic complexes (*James*, 1980). Because these rocks have combined primitive (high Mg minerals) and evolved (high KREEP) geochemical signatures, identifying the sources of their parental magmas poses a considerable dilemma for petrologists. Highlands Mg suite and HAS plutonic rocks may have been generated either by decompressional melting of early magma ocean cumulates during cumulate pile overturn or by impact induced melting of shallow cumulates. In the former case, a KREEP and crustal signature must have been incorporated into these primitive basaltic magmas through assimilation near the base of the lunar crust or through melting of a hybridized mantle. The HAS could represent either the differentiation products of HMS parental magmas or a separate, but contemporaneous, episode of basaltic magmatism.

## 3.4.  Early Remelting of Lunar Magma Ocean Cumulates: Prebasin Volcanism

Although the petrologic record has been obscured by the early catastrophic impact history of the Moon, there is abundant evidence of pre-3.9-Ga nonmare basaltic volcanism. Most of this record is retained in highland soils and breccias as clasts (*Dickinson et al.,* 1985; *Shervais et al.,* 1985; *Neal et al.,* 1988; 1989a,b) or identified through remote sensing (*Schultz and Spudis,* 1979; *Bell and Hawke,* 1984; *Hawke et al.,* 1990). Numerous lines of evidence such as igneous textures, lack of iron-metal particles, low siderophile-trace-element abundances, and nonmeteoritic siderophile-element ratios indicate that clasts representing premare basalt types are volcanic and not impact derived (*James,* 1980), although an impact origin for some samples has not been excluded (*Snyder et al.,* 2000). Several studies imply that large volumes of KREEP and high-alumina basaltic magmas (cryptomaria) were erupted (*Metzger and Parker,* 1979; *Davis and Spudis,* 1985, 1987; *Head and Wilson,* 1992). *Head and Wilson* (1992) have suggested that perhaps up to one-third of the erupted basalts at the lunar surface are cryptomaria.

The KREEP basalts thus far sampled have crystallization ages of 3.85–4.10 Ga (*Nyquist and Shih,* 1992). The $\varepsilon_{Nd}$ values of KREEP basalts and HMS rock types are consistent with a cogenetic relationship (*Nyquist and Shih,* 1992). If this is true, generation, emplacement, and eruption of KREEP basaltic magmas may have extended for a longer period of time ($\approx$0.6 Ga). KREEP basaltic volcanism is thought to be primarily prebasin, premare highland volcanism. However, some interpretations of basin formation ages (*Deutsch and Stöffler,* 1987; *Stadermann et al.,* 1991) imply that later KREEP eruptive episodes may have been triggered by basin formation. *Nyquist and Shih* (1992) speculated that the melting episodes that generated both the KREEP basalts and the high-Al basalts were triggered by the early impact of a large bolide. The high-Al basalts clasts from the Apollo 14 site have ages that range from 4.0 to 4.3 Ga (*Shih et al.,* 1992; *Nyquist and Shih,* 1992), some 200 m.y. older than the oldest known KREEP basalt. However, the presence of high-alumina clasts in breccias from the Fra Mauro Formation suggests that their provenance was similar to that of the KREEP basalts. The relationship between early stages of lunar volcanism and the contemporaneous plutonism is not clear. KREEP basalts may be the volcanic equivalents of both the HMS and HAS.

## 3.5.  Late Remelting of Magma Ocean Cumulates and the Eruption of Mare Basalts

Basin-associated eruption of mare basalts occurred during and after the late stages of catastrophic bombardment. This volcanic activity may have been an extension of the thermal event that initiated prebasin volcanism. Mare basalts exhibit a wide range of compositions resulting from near-surface fractionation of a large range of primary basaltic magmas. Experiments at elevated temperatures and pressures suggest that these primary mare basalts were produced by small to moderate degrees of partial melting of either LMO cumulate assemblages over a range of pressure regimes or hybrid LMO cumulate sources in the deep lunar mantle. Alternatively, the mixed chemical signatures observed in many mare basalts may be interpreted as indicating assimilation of late-stage, evolved cumulates by melts produced deep in the LMO cumulate pile. The wide range of compositions exhibited by the mare basalts compared with earlier episodes of basaltic magmatism may reflect the thermal regime in the lunar mantle that limited the extent of partial melting and melt-source homogenization.

## 4.  FERROAN ANORTHOSITE PETROGENESIS

### 4.1.  Introduction

The concept of a LMO was born out of the discovery of anorthositic fragments in the samples returned from the Apollo 11 mission. Early workers (*Smith et al.,* 1970; *Wood et al.,* 1970) suggested that these highly evolved rocks represent remnants of the Moon's earliest crust that formed by crystal fractionation and plagioclase flotation. Additional evidence of a global differentiation event is provided by the complementary nature of this anorthositic crust and the inferred characteristics of mare basalt source regions [e.g., Al, Ca-poor with negative Eu anomalies (*Papike et al.,* 1998)], although these source regions appear to have been subsequently modified (*Ryder,* 1991). Thus, the magma ocean hypothesis has survived in one form or another to the present day (see *Taylor,* 1982, *Warren,* 1985, and *Shearer and Papike,* 1999, for reviews) and, despite alternative views over the years (e.g., *Walker,* 1983), most investigators recognize that some type of extensive differentiation is required to account for the fact that the upper part of the lunar crust consists of about 70% plagioclase (*Korotev and Haskin,* 1988).

Although a wide variety of rock types occur in the lunar highlands [see *Papike et al.* (1998) for an extensive review of lunar samples], the ferroan anorthosites, or FANs, have come to be recognized as a coherent and mineralogically and chemically distinctive group, constituting approximately half of all ancient rocks returned from the Apollo missions. They generally contain more than 90 vol% extremely calcic plagioclase, with a total range in plagioclase compositions of only $An_{94}$ to $An_{99}$ (*James,* 1980). The associated mafic minerals are primarily low-Ca pyroxene (e.g., pigeonite that has exsolved augite and inverted to orthopyroxene (*Dixon and Papike,* 1975)], although olivine and primary high-Ca pyroxene may also be present. Trace phases include Fe-Ni metal, ilmenite, silica, and chromite. Some FANs contain higher abundances of mafic minerals and are unusually rich in olivine (*McGee,* 1988; *Papike et al.,* 1998). As implied by their name, the mafic minerals in FANs are quite ferroan, with Mg' ranging between about 50 and 70. On a graph of An content in plagioclase vs. Mg' of coexisting mafic minerals (Fig. 1) the FANs plot on a steep vertical trend and are easily distinguished from other highland rock types such as the HMS and HAS.

Few ages are available for ferroan anorthosites; the nearly monomineralic nature of these rocks and their low incompatible-element abundances have created numerous analytical difficulties. Nevertheless, low initial $^{87}Sr/^{86}Sr$ ratios for FANs and Sm-Nd internal isochron ages of $4.44 \pm 0.02$ and $4.56 \pm 0.07$ Ga for 60025 and 67016 respectively (*Carlson and Lugmair,* 1988; *Alibert et al.,* 1994) indicate early formation and differentiation of the Moon. However, *Borg et al.* (1999) have obtained a young Sm-Nd isochron age of only $4.29 \pm 0.06$ Ga for ferroan anorthosite 62236, an age similar to those of younger HMS rocks.

Many attempts to understand ferroan anorthosite petrogenesis have focused on explaining the large variation in Mg' of their mafic minerals with very little change in plagioclase composition (Fig. 1). Indeed the very nature of the trend has remained unclear. While early studies noted a negative correlation between An content of plagioclase and Mg' of coexisting mafic minerals, contrary to the positive correlation expected for normal differentiation (*Wood et al.,* 1970; *Steele and Smith,* 1973), others observed either a positive correlation of these parameters (*Warren and Wasson,* 1977; *Ryder,* 1982) or no correlation (*James,* 1980; *Raedeke and McCallum,* 1980) when only pristine rocks were considered. *James et al.* (1989) suggested that the variation seen in the ferroan anorthosite trend is real and indicates the complexity of this suite of rocks. *Raedeke and McCallum* (1980) noted that rocks from the Stillwater Complex exhibit a similar steep trend and modeled the origins of both these rocks and the FANs in terms of equilibrium crystallization of trapped intercumulus liquid in a plagioclase-rich mush, in which the plagioclase composition is buffered by the abundance of plagioclase relative to trapped liquid. By varying the proportions of trapped liquid and mafic minerals relative to the amount of plagioclase, they were able to generate the entire ferroan anorthosite suite. However, a problem with this model is that neither the Stillwater anorthosites nor the FANs contain large amounts of trapped residual liquid (*Haskin et al.,* 1981).

Estimates of parent magma compositions calculated from whole-rock abundances suggest that ferroan anorthosites crystallized from parent magmas with close to chondritic refractory incompatible trace-element (ITE) ratios. Rare-earth-element patterns are approximately flat (*James et al.,* 1989; *Phinney,* 1991), with abundances ranging from about 10× chondritic to about 30× chondritic with increasing degree of fractionation. These values suggest onset of plagioclase crystallization at 75–80% solidification of the melt, assuming total melting of the silicate portion of the Moon (*Phinney,* 1991).

*James et al.* (1989) divided the FANs into four subgroups on the basis of their modes and mineral compositions: anorthositic ferroan (typical FANs), anorthositic sodic, mafic magnesian, and mafic ferroan. Although they pointed out that the significance of these subgroups is unknown, they, as well as *Palme et al.* (1984), noted complex trace-element variations in the anorthosites that are inconsistent with the derivation of all these rocks by fractionation of a common parent magma. Specifically, primitive (i.e., ITE-poor) plagioclase is associated with evolved (Fe-rich) mafic minerals, whereas anorthosites associated with more magnesian (or primitive) mafic minerals appear to have crystallized from a more ITE-rich evolved magma.

## 4.2. Trace-Element Contributions to Understanding Ferroan Anorthosite Suite Petrogenesis

With the advent of the ion microprobe it has become possible to make routine *in situ* measurements (i.e., in polished thin sections) of the trace-element compositions of individual minerals, thus allowing trace-element geochemical data to be interpreted within a petrographic context, much as major-element compositions determined by electron microprobe analysis commonly are. A number of ion microprobe investigations of lunar samples have been published over the last 5–10 years (see *Papike et al.,* 1998 for an overview), and several of these have focused on ferroan anorthosites (*Floss et al.,* 1991, 1998; *Jolliff and Hsu,* 1996; *Papike et al.,* 1997; *James et al.,* 1998; *Snyder et al.,* 1998).

Trace-element compositions (REE and other trace elements) have been measured in plagioclase from a wide variety of ferroan anorthosites, largely from the Apollo 16 site, although samples from Apollo 15 have also been studied (*Papike et al.,* 1997; *Snyder et al.,* 1998); in addition, trace-element compositions of coexisting mafic minerals (pyroxene and olivine) were determined in some of the studies (*Floss et al.,* 1998; *James et al.,* 1998; *Snyder et al.,* 1998). The concept of pristinity is important for understanding the origin of ferroan anorthosites (as well as other highland rock types). However, most FANs are highly brecciated and/or cataclasized and relict igneous textures are rarely preserved. Many of the samples analyzed in these studies are considered compositionally pristine according to the criteria of *Warren* (1993). Furthermore, *Floss et al.* (1998) focused on analyzing bimineralic or polymineralic lithic clasts exhibiting relict igneous textures in samples from the four subgroups of ferroan anorthosites identified by *James et al.* (1989).

One of the important issues addressed by all these studies is the question of redistribution of the trace elements, especially the REE, within and between mineral phases.

Many anorthosites are know to have undergone extensive recrystallization and prolonged thermal metamorphism (*James, 1972; Phinney, 1991*). Rare-earth-element compositions in plagioclase are generally uniform, both within grains and between individual grains from any given sample, suggesting widespread equilibration of this mineral, although this equilibration would only affect REE abundances and not the shapes of the patterns (*Floss et al., 1998*). Although *Jolliff and Hsu* (1996) and *Papike et al.* (1997) noted minor effects of subsolidus requilibration involving pyroxene, plagioclase REE compositions generally have not been seriously compromised in the ferroan anorthosites and may be used to obtain information about primary igneous processes that have acted on these rocks (*Papike et al., 1997; Floss et al., 1998*). Pyroxene REE compositions record a more complex history. Although most grains are not large enough to determine intermineral variation, compositions often differ from grain to grain within a given sample and low-Ca pyroxenes are generally LREE-depleted relative to expected compositions (*Floss et al., 1998*). Subsolidus equilibration with plagioclase will result in LREE-depleted pyroxene. In addition, redistribution of the REE during inversion of pigeonite to augite and orthopyroxene may also be an important factor in many samples (*James et al., 1998*).

Plagioclase REE compositions were used to estimate the compositions of equilibrium melts for the ferroan anorthosites. Melt compositions for the "typical" ferroan anorthosites [i.e., the anorthositic ferroan subgroup of *James et al.* (1989)] studied by *Floss et al.* (1998) and the larger group of *Papike et al.* (1997) have relatively flat REE patterns, some with small negative Eu anomalies, and abundances ranging from ~10 to 50× chondritic (Fig. 5). A sodic anorthositic sample has a melt composition that falls at the high end of this range (*Floss et al., 1998*). *Snyder et al.* (1992) calculated the trace-element evolution of the lunar

magma ocean using a model that assumed a bulk $Al_2O_3$ content of 5%, chondritic ratios of the REE, and overall REE abundances of 3× CI. When plagioclase becomes the liquidus phase, after ~78% equilibrium crystallization of olivine and orthopyroxene, the LMO shows a slight LREE-enrichment; as plagioclase crystallization proceeds, the residual melt develops a significant negative Eu anomaly and the final residual liquids of the magma ocean resemble those of high-K KREEP (*Warren, 1985*). The liquids in equilibrium with plagioclase from "typical" ferroan anorthosites have compositions that are consistent with those calculated by *Snyder et al.* (1992) for 78% crystallization, for the most primitive anorthosites, up to about 90% crystallization, for the most evolved (*Papike et al., 1997; Floss et al., 1998; Snyder et al., 1998*). Thus the plagioclase from this group of anorthosites has trace-element characteristics that appear to record an extended period of fractional crystallization from a lunar magma ocean. This is illustrated in Fig. 6, which shows the magnitude of the negative Eu anomaly, $(Eu-Eu^*)_n$, in the equilibrium melts as a function of chondrite-normalized Nd abundances of the melt $(Nd_n)$, which represents a fractionation index for melt evolution. Trace-element variations in the coexisting pyroxenes from some of these anorthosites are broadly consistent with the trends shown by plagioclase (*Floss et al., 1998*). However, others exhibit a decoupling of plagioclase and pyroxene compositional parameters.

Additional complexities arise when samples from the other FAN subgroups are considered. Samples with highly magnesian mafic minerals have REE abundances in both plagioclase and pyroxene that are higher than would be expected on the basis of their major-element compositions if they are cogenetic with the "typical" ferroan anorthosites discussed above. Furthermore, plagioclase from a member of the mafic ferroan subgroup has unusually high trace-element abundances, corresponding to somewhat more than 95% crystallization of the lunar magma ocean (*Snyder et al., 1992*), although its major-element composition is similar to those of other FANs (*Floss et al., 1998*).

Decoupling of the plagioclase and pyroxene compositions recorded in the FANs can probably be explained by the complexities inherently present in crystallization from a large-scale magmatic system. Entrainment of small amounts of mafic cumulate material, heterogeneous nucleation of plagioclase on available (mafic) mineral surfaces, and equilibration of early-formed mafic minerals with trapped intercumulus liquid are just some of the processes that may account for the wide variety of pyroxene compositions associated with compositionally similar plagioclase. A potentially more serious difficulty is the fact that major-element compositions in some of the anorthosites discussed above do not reflect the evolved compositions recorded by the trace elements. Although plagioclase compositions may be buffered to some extent by extended growth in the melt from which they are crystallizing, it is not clear that such a process can account for almost uniform major-element compositions while preserving trace-element evolutionary sequences that span 10–15% of the crystallization history

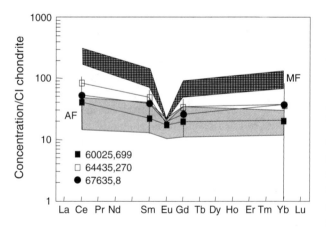

**Fig. 5.** C1 chondrite-normalized REE patterns of liquids in equilibrium with plagioclase from samples of the four FAN subgroups. Shaded area = range for samples from the anorthositic ferroan subgroup; circles = anorthositic sodic sample 67635,8; squares = mafic magnesian samples; crosshatched area = range for mafic ferroan sample 67215,6. After *Floss et al.* (1998).

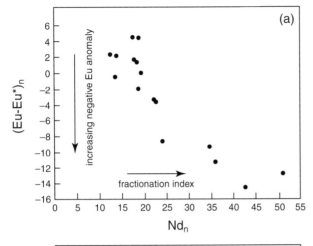

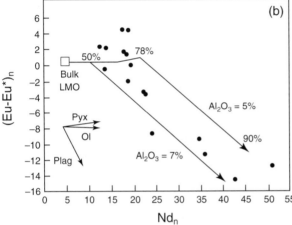

**Fig. 6.** Plot of the magnitude of the Eu anomaly $(Eu-Eu^*)_n$ vs. $Nd_n$ (chondrite-normalized Nd abundances) for melts in equilibrium with plagioclase from "typical" ferroan anorthosites. As fractionation proceeds (i.e., $Nd_n$ becomes larger), $(Eu-Eu^*)_n$ becomes more negative as a result of plagioclase crystallization. After *Papike et al.* (1997).

of the LMO (*Floss et al.,* 1998). Infiltration and/or contamination of the local magma with late-stage liquids may, however, provide a partial answer to this dilemma.

The relatively simple picture, discussed above, of early FAN petrogenesis through flotation of plagioclase to form a primordial lunar crust is put into question by the recent determination of a very young age of 4.29 ± 0.06 Ga for the ferroan anorthosite 62236 (*Borg et al.,* 1999). This young age, as well as the old ages of some Mg-suite rocks, require that multiple sources of magmas were present on the Moon very early in its history. Furthermore, all three ferroan anorthosites for which ages have been determined (see above) have positive initial $\varepsilon_{Nd}$ values, suggesting that they were derived from sources depleted in the LREE. In order to reconcile these observations with the LMO model, *Snyder et al.* (1998), *Shearer and Newsom* (1999), and *Borg et al.* (1999) note that the LMO must have existed for a very short period of time and may have had subchondritic Nd/Sm ratios. Furthermore, young ferroan anorthosites such as 62236 cannot be cumulates from the magma ocean, but must

have formed by other processes. Two-stage models such as those of *Longhi and Ashwal* (1985), in which early-formed layered intrusions are reheated to form FAN diapirs, are one possibility (*Snyder et al.,* 2000).

## 5. POST-LUNAR-MAGMA-OCEAN HIGHLANDS MAGMATISM

### 5.1. Introduction

Following the formation and consolidation of the early anorthositic lunar crust, it was intruded by the products of slightly younger episodes of magmatism, referred to as the HMS and HAS. The HMS consists primarily of troctolites and norites and lesser amounts of gabbros and dunites and contains rocks with both primitive and evolved chemical signatures. For example, very magnesian mafic minerals with Mg' > 90 are often coupled with high REE abundances and low Ni (*Warren and Wasson,* 1980; *Ryder,* 1983, 1991; *Warren,* 1985; *Papike et al.,* 1996, 1998). The parent magmas appear to have been saturated with plagioclase and at least one mafic silicate. It was recognized fairly early in studies of the lunar highland rocks that the ferroan anorthosites and the HMS defined different compositional fields in plots of An content of plagioclase vs. Mg' of mafic phases (Fig. 1).

As with the ferroan anorthosites, the petrologic history of the HMS is obscured to various degrees because of the effects of intense meteoritic bombardment and prolonged subsolidus cooling. Isotopic data indicate that HMS magmatism immediately followed the generation of the ferroan anorthositic lunar crust and extended over a period of 400 m.y. (*Carlson and Lugmair,* 1981). More recent studies (*Shih et al.,* 1993; *Borg et al.,* 1999; *Snyder et al.,* 2000) have illustrated that a more complex relationship possibly exists between FAN and HMS. Initial Sr-isotopic compositions of the HMS rocks are higher than those of ferroan anorthosites (*Carlson and Lugmair,* 1981). The HMS appears to mark the transition between LMO crystallization and serial magmatism associated with melting of a lunar mantle consisting of LMO cumulates. This transition period may have occurred as early as 10 m.y. (*Shearer and Newsom,* 2000) or as late as 200 m.y. after LMO formation (*Solomon and Longhi,* 1977). Some of the younger ages for the HMS may not be crystallization ages, but may reflect subsolidus closure (*Carlson and Lugmair,* 1981).

Early studies of the lunar highlands also identified rocks with anomalously high alkali-element contents and highly evolved mineral and chemical signatures. These alkali-rich lunar rocks are generally defined as containing greater than 0.1% $K_2O$ and 0.3% $Na_2O$ (*Warren and Wasson,* 1980). The HAS is also characterized by elevated incompatible-lithophile-element abundance. This distinctive suite of highland rocks consists of anorthosites, gabbros, monzodiorites, quartz monzodiorites, and a range of siliceous varieties referred to as granites, rhyolites, and felsites (*Warren and Wasson,* 1980; *Snyder et al.,* 1995). It appears to make up a much smaller portion of the lunar crust than the ferroan anorthosites and

HMS rocks. The period of HAS magmatism extended from ~4.4 Ga to 3.9 Ga. On the basis of ages, mineral compositions, and relative incompatible-element contents, *Warren and Wasson* (1980) demonstrated that these rock types were unrelated to the ferroan anorthosites or the mare basalts.

## 5.2.  Models for Highland Magmatism

Numerous petrogenetic models have been proposed for the generation of post-FAN, highland parental magmas. Most attempt to reconcile the contrasting primitive (e.g., high Mg') and evolved (e.g., saturated with plagioclase, high concentrations of incompatible trace elements) characteristics of the HMS. Models for the petrogenesis of the HMS include (1) impact origin, (2) products of magma ocean crystallization that were contemporaneous with the FANs, and (3) remelting of magma ocean cumulates with KREEP or anorthositic crust assimilation. Because of inconsistencies for models 1 and 2 (*Hess*, 1994; *Nyquist and Shih*, 1992; *Shih et al.*, 1993; *Raedeke and McCallum*, 1980; *Warren and Wasson*, 1980), many models have advocated a post-magma-ocean origin for both the HMS and HAS, calling upon remelting and remobilization of late-magma-ocean cumulates and/or KREEP-infiltrated lower crust (*Hess et al.*, 1978), melting of the lower portions of the cumulate pile, followed by crystallization and shallow assimilation of KREEP or anorthositic crust or both (*Warren and Wasson*, 1980; *Ryder*, 1991; *Papike et al.*, 1994, 1996), or the crystallization of basaltic magmas produced by melting of deep, hybrid cumulate sources (*Hess*, 1994).

*Hess et al.* (1978) suggested that KREEPy highland basalts that may have been parental to the HMS were generated by partial melting of magma ocean cumulates that crystallized soon after extensive ilmenite crystallization, but prior to the formation of KREEP (between 95% and 99% crystallization of the magma ocean). These magmas may have assimilated KREEP. Alternatively, *Hess* (1994) suggested that KREEP-rich magmas were a product of partial melting of a lower lunar crust that had been metasomatically altered by KREEP. Pressure release melting and remobilization of these rock types may be related to catastrophic impacts on the lunar surface (*Snyder et al.*, 1995; *Papike*, 1996; *Papike et al.*, 1994, 1996, 1997). One problem with this model is that late-stage LMO cumulates should have Mg' that are substantially less [0.78 to <0.40 (*Snyder et al.*, 1992)] than that observed in the more primitive HMS rocks (Mg' ≈ 0.90). Therefore, if these models are correct, only significant reduction and metal removal during the petrogenesis of these basaltic magmas could account for both the high Mg' and low Ni content (*Ryder*, 1983)

Models that require the HMS to represent crystallization products of high-Mg' magmas generated in the deep lunar mantle were developed to resolve some of the problems with a shallow cumulate source (*James*, 1980; *Warren and Wasson*, 1980; *Longhi*, 1981, 1992; *Longhi and Boudreau*, 1979; *Ryder*, 1991; *Smith*, 1982; *Warren*, 1985; *Hess*, 1994). *Hess* (1994) demonstrated that magmas with Mg' appropriate for

HMS parent melts could be generated by melting of early magma ocean cumulates. A magma ocean with a Mg' value equivalent to the bulk Moon (80–84) upon crystallization at high pressures would produce early cumulates of olivine with Mg' greater than 91. Subsequent melting of these cumulates would produce magmas with Mg' equivalent to that of the HMS parental melts. Because of the pressure dependence of the FeO-MgO exchange equilibrium between olivine and basaltic melt, crystallization of these high pressure melts near the lunar surface would result in liquidus olivine that is slightly more magnesian than residual olivine in the mantle source.

Partial melting of early magma ocean cumulates could have produced primitive melts with high Mg', but they would not have the same geochemical characteristics as the HMS. For example, these primitive magmas would not have high incompatible-element enrichments, fractionated Eu/Al and Na/(Na + Ca), and plagioclase as a liquidus phase until the melt Mg' was less than 0.42. Two processes have been proposed to resolve this problem: assimilation of evolved crystallization products of the magma ocean, and melting of hybrid cumulate sources.

The HMS and HAS may be either part of a continuum of crystallization products of parental basaltic magmas similar to the KREEP basalts (*Ryder*, 1976; *Warren and Wasson*, 1980; *James*, 1980; *James et al.*, 1987; *Snyder et al.*, 1995; *Shervais and McGee*, 1999) or separate episodes of basaltic magmatism (*Warren and Wasson*, 1980). Using a KREEP basalt (from the Apollo 15 site) as a starting parental magma, *Snyder et al.* (1995) demonstrated that fractional crystallization (0–99.8%) and the accumulation of mineral phases and trapped KREEPy residual liquid (2–15%) could produce the range of mineral and rock compositions observed in the HMS and HAS (Fig. 1). Alternatively, the HMS and HAS may represent contemporaneous, but separate, episodes of basaltic magmatism (*Warren and Wasson*, 1980; *James*, 1980; *James et al.*, 1987). There is some compositional evidence to suggest genetically distinct highland rock types. For example, *James et al.* (1987) subdivided many of these highland rock types into various groups on the basis of their mineral chemistry and mineral associations. Whether these subdivisions are artificial or petrologically significant is open to debate. Within this scenario, however, differences between the two suites may be attributed to the depth of initial melting prior to assimilation. For example, HMS magmatism would be a product of deep mantle melting followed by KREEP assimilation just below the lunar crust, whereas HAS magmatism would involve initial melting at shallower mantle levels followed by assimilation.

## 5.3.  Insights from Trace-Element Microbeam Analyses of Minerals

Based on the previous discussion, there are two important questions concerning the origin of the post-magma-ocean highlands magmatism. What is the magmatic relationship between the HMS, HAS, and KREEPy volcanism? What petrologic models best reconcile the contrasting primitive and evolved characteristics of the magmas parental to these highland plutonic rocks? These questions may be partially answered through ion microprobe studies of pyroxene and plagioclase from these plutonic lithologies.

To date, trace-element ion microprobe analyses have been reported for pyroxene and plagioclase for norites, troctolites, and anorthosites from the HMS (*Papike et al.,* 1994, 1996; *Shervais and McGee,* 1997) and norites and anorthosites from the HAS (*Snyder et al.,* 1994; *Shervais and McGee,* 1997, 1999). Based on these analyses and the assumption that the trace-element compositions of mineral cores could be inverted to determine melt compositions, these investigators determined the trace-element characteristics of HMS and HAS magmas. Both suites of plutonic rocks appear to have crystallized from magmas with REE and selected incompatible trace-element concentrations equal to or greater than the high K KREEP estimated by *Warren* (1985) (Fig. 7). In addition, relative to high-K KREEP, the calculated magmas for the HMS norites are depleted in Eu (*Papike et al.,* 1994, 1996), whereas those for HAS anorthosites and norites are slightly depleted in HREE and enriched in Eu (*Shervais and McGee,* 1997, 1999).

As a first approximation, *Snyder et al.* (1995) suggested that fractional crystallization of a KREEPy magma could potentially produce these two plutonic suites (Fig. 1). In this model, most of the HAS would be produced by 40–90% fractional crystallization of the magmas parental to the HMS norites. Based on the calculated parental magmas, it appears that either these two highland suites are related by processes much more complex than simple fractional crystallization or that they are contemporaneous, but petrogenetically unrelated. This is because the calculated magmas for the HMS have both higher Mg' and higher incompatible-element abundances than the calculated alkali suite magmas (Fig. 7). *Shervais and McGee* (1997, 1999) envisioned a more complex process to account for the REE characteristics of the alkali suite magmas and the difference in the albite content of the plagioclase between the calculated liquid lines of descent for KREEP magmas and alkali suite magmas. They concluded that in addition to fractional crystallization, assimilation of FANs, magma mixing, and local equilibrium crystallization were important processes relating these magmas. In particular, FAN assimilation could result in the production of abundant, sodic feldspar, lower REE content of the parental magma, and a positive Eu anomaly. Although *Shervais and McGee* (1997, 1999) demonstrated the effect of feldspar assimilation on the abundance and composition of feldspar crystallizing from a basaltic melt, reasonable amounts of HMS parental magma fractionation and FAN assimilation will not produce the trace-element patterns calculated for the alkali suite parent magmas. As shown in Fig. 8, over 80% FAN assimilation is required to produce a magma with a KREEP-normalized REE pattern with a positive Eu anomaly. In addition, there are other arguments that can be made against FAN assimilation as a model for gen

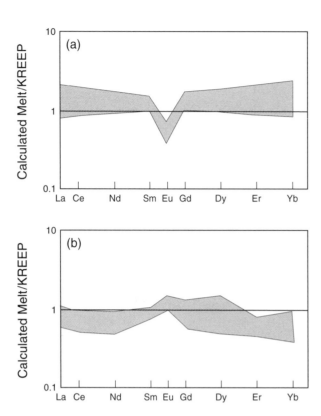

**Fig. 7.** Calculated equilibrium parent melt compositions for **(a)** Mg-suite norites (*Papike et al.,* 1997) and **(b)** alkali-suite anorthosites and norites (*Shervais and McGee,* 1997) based on ion microprobe analyses of pyroxene and plagioclase. Melt compositions are normalized to KREEP.

erating the HAS. *Hess* (1994) explored the thermal and chemical implications of anorthosite melting and plagioclase dissolution by high-Mg basaltic magmas. In his analysis of anorthosite melting and mixing as a mechanism to drive high-Mg' magmas to plagioclase saturation, he concluded that the diffusion rates for $Al_2O_3$ in basaltic melts are extremely slow and indicate that the timescales to dissolve even a small amount of plagioclase are of the same order as the characteristic solidification times of a large magma body. Therefore, based on the available ion microprobe data, the intellectually less satisfying model in which the HMS and HAS magmas result from the melting of different lunar mantle sources is preferred.

*Snyder et al.* (1995) suggested that KREEP basalts were potential parental magmas for the HMS. The KREEP basalts in the lunar collection have Mg' of ~61–66. This range in Mg' is slightly lower than would be expected for parental magmas of the HMS cumulates (mafic minerals with Mg' ≈ 90). However, it is possible that more Mg-rich KREEP basalts exist but have not been sampled. KREEPy basaltic magmas parental to the HMS would have been produced by partial melting of deep LMO cumulates with a high Mg'. However, there are two problems with this model that have

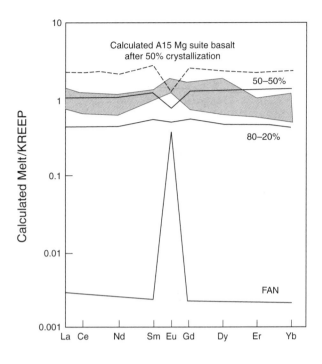

**Fig. 8.**  REE pattern for alkali-suite magmas (shaded) calculated by *Shervais and McGee* (1997), Apollo 15 Mg-suite parent magmas after 50% fractional crystallization, FAN, and mixture of FAN and fractionated melt in proportions of 50-50 and 80-20. The REE pattern for the alkali suite magmas (*Shervais and McGee,* 1997) appear not to represent products of Mg-suite magma fractionation and FAN assimilation.

not been resolved. If these high Mg' reflect melting of early LMO cumulates, why are the Ni contents of silicates so low? What mechanism produced the high KREEP signature in these magmas: incorporation of KREEP at shallow levels by assimilation, or during melting of a deep hybridized lunar mantle?

Although the Mg' of olivine ($Fo_{92-88}$) in the HMS is high and suggests a primitive magma derived from deep within the LMO cumulate pile, the Ni content of the olivine determined by microprobe (*Ryder,* 1982) and ion microprobe (*Shearer and Papike,* 2000) ranges from 10 to 320 ppm. This is fairly low for a primitive lunar basalt. In comparison, ion microprobe analyses of olivine cores ($Fo_{75-72}$) from Apollo 12 olivine basalts yield Ni concentrations of 515–450 ppm (*Papike et al.,* 1999; *Shearer and Papike,* 2000). The calculated olivine composition in equilibrium with mare basalts more primitive than the Apollo 12 olivine basalts (i.e., Apollo 15 green glass) is ~$Fo_{85}$ and Ni ≈ 1500 ppm. This poses a dilemma for interpretation of the high Mg' of the HMS and hints at the involvement of a metal phase either during melting of the source or during evolution of the basaltic magma.

Calculated melt compositions for the HMS norites have Mg' that overlap or are slightly higher than the most primitive mare basalts (i.e., picritic glasses) and a KREEP component equivalent to or slightly higher than high-K KREEP. These data pose both dilemmas and constraints for assimi-

lation as well as mantle hybridization models. As shown in assimilation and mixing models in Fig. 9, calculated melt compositions plot at incompatible-element concentrations equal to or exceeding estimates for KREEP. For simple mixing to work, the KREEP component must be more enriched in incompatible elements, and either the primitive HMS magma must contain greater than 25 wt% MgO, or the KREEP component must have higher MgO abundances because substantial fractional crystallization would accompany large amounts of assimilation, which would result in a substantial decrease in MgO (Fig. 9a). It therefore appears that the KREEP signature in the HMS cannot be accounted for exclusively by assimilation.

Partial melting of a hybridized source under certain conditions is more consistent with the calculated magma compositions. In a very simple case, a hybridized mantle consisting of ~93% early LMO cumulates (olivine and orthopyroxene, ≈40% MgO) and 7% KREEP component would, at small to moderate degrees of partial melting (5–15%), produce incompatible-element enrichments and Mg' equivalent to or approaching those calculated for the HMS norites (Fig. 9b). This model for the generation of high-Mg', KREEPy magmas is consistent with other observations. *Ryder* (1976) argued that the KREEP basalts were products of partial melting of distinct source regions and that assimilation was not a prominent process in their origin. Such a model requires the KREEP trace-element signature to be a characteristic of the source and implies a hybridized mantle. *Nyquist and Shih* (1992) proposed that the observed variations in the $I_{sr}$ and $\varepsilon_{Nd}$ were compatible with the systematic melting of a hybridized lunar mantle to produce KREEP basalts. Within this scenario, differences between the HMS and HAS may be attributed to the depth of initial melting or the characteristics of the cumulate source. Although this model accounts for the high Mg' and KREEP signature, it also has several pitfalls. In its simple form, it does not adequately explain plagioclase saturation at Mg' appropriate for HMS magmas or certain other fractionated incompatible-element signatures (e.g., fractionated Eu/Al). Less-important processes (*Warren,* 1986) or components (*Hess,* 1994) are required to explain these characteristics.

## 6.  MARE BASALT MAGMATISM

### 6.1.  Introduction

Mare basalts primarily fill multiring basins and irregular depressions on the Earth-facing hemisphere of the Moon. They are especially profuse in Nectarian- and post-Nectarian-aged basins. The thickness of these flood basalts ranges from 0.5 to 1.3 km in irregular basins to up to 4.5 km in the central portions of younger basins. The general style of volcanic activity was the eruption of large volumes of magma from relatively deep sources (not shallow crustal reservoirs) with very high effusion rates (*Head and Wilson,* 1992). Evidence of lunar fire-fountaining is preserved in the form

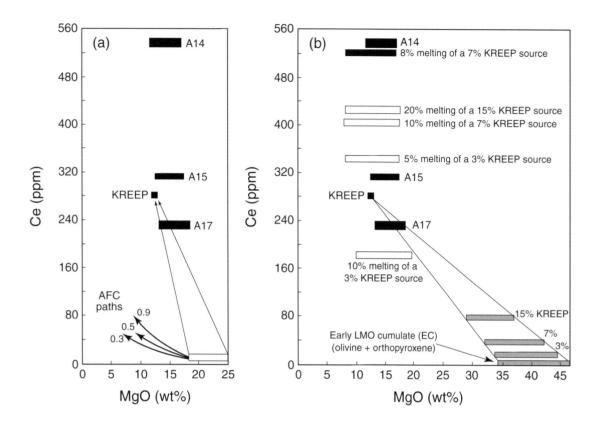

**Fig. 9.** Plot of MgO (wt%) vs. Ce (ppm) for KREEP and calculated parent magmas of Mg suite norites (filled rectangles) from *Papike et al.* (1997). **(a)** Calculated mixing and AFC (assimilation-fractional crystallization) paths for assimilation of KREEP by a proto-Mg-suite magma (open rectangle) generated by partial melting in the deep lunar mantle. **(b)** Possible hybridized source consisting of early LMO cumulates and KREEP (shaded rectangles). Also illustrated is the range of possible melts (open rectangles) produced by small degrees of partial melting of that hybridized source.

of spherical glass beads (*Delano,* 1986) and dark mantling deposits (*Head and Wilson,* 1992).

Within the context of terrestrial basalt classification, the normative mineral assemblage of mare basalts ranges from quartz to olivine. Volcanic glass beads are more Mg-rich than the crystalline mare basalts and are all olivine normative. Mare basalts exhibit numerous distinct mineralogical and chemical characteristics: (1) a spectacular range in $TiO_2$; (2) mineralogical (Fe metal) and chemical (reduced valence states for Fe, Ti, Cr) signatures reflecting extremely low O fugacities (≈IW buffer); (3) depletions in alkali, volatile, and siderophile elements; (4) absence of water and hydrous phases; (5) a negative Eu anomaly in all crystalline mare basalts and picritic glasses; (6) in general, from very low $TiO_2$ to the high $TiO_2$ mare basalts, increases in the abundances of incompatible-elements (REE, Zr, Ba, and Nb), increases in the depths of the negative Eu anomalies, and decreases in the abundances of compatible elements (e.g., Ni); and (7) radiometric ages for sampled mare basalts ranging from 4.0 to 2.93 Ga (*Nyquist and Shih,* 1992; *Burgess and Turner,* 1998). Although the Apollo sample collection implies a bimodal distribution of low-Ti and high-Ti basalts, recent interpretation of *Clementine* and *Galileo* remote sensing data indicates that mare basalts with intermediate Ti

content are more abundant than are those with high Ti contents (*Giguere et al.,* 1999).

Using major- and trace-element characteristics of picritic mare basalt glasses, *Delano* (1986) defined 25 distinct groups. These volcanic glass beads also exhibit a wide range in $TiO_2$ (0.2–17%), yet are consistently higher in Mg' than crystalline mare basalts with similar $TiO_2$. This suggests that they are the best candidates for primary mare basaltic magmas and that most crystalline mare basalts experienced fractional crystallization. The picritic glasses also imply that a range of primary melts is responsible for the diversity of mare basalts and that the sources for the basaltic magmas are fractionated (e.g., negative Eu anomaly).

### 6.2. Models for Mare Basalt Magmatism

All models for the generation of mare basalts concur that shallow, fractional crystallization was an important process in producing diverse basalt compositions. They differ in their concept of the role that early lunar differentiation played in the formation of the mantle sources for the basalts, the subsequent evolution of these sources, and the conditions under which melting occurred. Fundamental source models are (1) primitive source models, (2) LMO cumulate source models, and (3) serial magmatism source models. The source for the mare basalts is clearly differentiated on the basis of its nonchondritic composition, therefore a primitive lunar mantle as the sole source for the mare basalts is not discussed further.

To account for the differentiated signature of the mare basalts, the LMO cumulate source model was advanced relatively early during the Apollo program and matured substantially through many other studies (*Wood*, 1975; *Taylor and Jakeš*, 1974; *Solomon and Longhi*, 1977; *Warren*, 1985). In the *Taylor and Jakeš* (1974) model, early differentiation of the Moon through extensive melting resulted in the formation of a layered mantle composed of cumulates. The source of the low-Ti mare basalts consisted of relatively early-formed ultramafic cumulates (predominantly olivine and low-Ca pyroxene). Late-stage cumulates, which contain olivine, pyroxene, and ilmenite, were proposed as a source for the high-Ti basalts. Plagioclase flotation during crystallization of the LMO imposed a negative Eu anomaly on the crystallizing mafic cumulates and produced a complementary positive Eu anomaly in the feldspathic lunar crust. In addition to explaining the negative Eu anomaly in the mare basalts, it also placed into a single conceptual model the observations that the source for the mare basalts was differentiated, and that the sources for the various mare basalts were isotopically, chemically, and mineralogically distinct.

However, a model in which the source for mare basalts is analogous to a terrestrial layered intrusion has major problems. First, the multiple saturation pressures for both high-Ti (late cumulate source) and very-low-Ti (early cumulate source) basalts that experienced limited fractionation (i.e, high Mg') suggest that they were derived by melting initiated deep within the lunar mantle (*Delano*, 1986). Second, there are no significant differences in Mg' and Cr concentrations between the most primitive of the high- and low-Ti basalts. Numerous modifications to the simple layered intrusion analogy have been made in response to these objections. These models invoke either assimilation of evolved cumulates by magmas produced deep within the cumulate pile, or partial melting of hybridized mantle sources created by redistribution of the cumulate pile.

*Hubbard and Minear* (1975), *Ringwood* (1975), and *Wagner and Grove* (1995) called upon assimilation models to account for some of the discrepancies in the layered cumulate source model. In these models, low-Ti basaltic liquids that were generated in the deep (and early) LMO cumulate pile assimilated the high-Ti cumulates in the shallow portion of the cumulate pile. The wide compositional spectrum observed in mare basalts (i.e., $TiO_2$) was the result of the compositional diversity in high-Ti cumulates and the low-Ti magmas and the varying degrees of high-Ti cumulate assimilation by the low-Ti magmas.

As an alternative to assimilating high-Ti cumulates at shallow mantle levels, *Ringwood and Kesson* (1976) argued that dense, late-stage Ti- and Fe-enriched cumulates sank into deeper portions of the cumulate pile. Exchange of material, different degrees of partial melting, and different domains of partial melting of a heterogeneous source produced a wide spectrum of chemically distinct basaltic magmas (*Hughes et al.*, 1989; *Ryder*, 1991; *Shearer and Papike*, 1993; *Hess and Parmentier*, 1995; *Shearer and Newsom*, 1999). This mechanism resulted in the transport of more evolved components of the LMO cumulate pile (KREEP, high-Ti cumulates) to zones of melting, and provided a heat source (decay of incompatible, radioactive elements) to initiate melting at depth. *Spera* (1992) and *Hess and Parmentier* (1995) have modeled large-scale overturn of the cumulate pile, whereas *Snyder et al.* (1992) suggested a more localized overturn of the cumulate pile.

As an alternative to a mare basalt source that was processed by LMO formation and crystallization, *Walker* (1983) suggested that the source region for the mare basalts was produced during early lunar differentiation by serial magmatism. In this model, mafic cumulates from large layered intrusions were separated from their anorthositic cap and transported by convection into the deep lunar mantle. Here, they were mixed with undifferentiated mantle and formed the mare basalt source region. In this model, the contrasting Eu anomalies observed in the ferroan anorthosite crust and mare basalts are inconsequential. *Walker* (1983) and *Haskin* (1989) suggested that the bulk lunar crust does not exhibit the positive Eu anomaly predicted by the cumulate model. *Walker* (1983) furthermore suggested that the negative Eu anomaly exhibited by mare basalts was not characteristic of the lunar mantle. Rather, it reflected the mixing of primitive magmas with basaltic magmas that had experienced plagioclase fractionation.

As probes of planetary interiors, basalts can be used to better understand the evolution of the lunar mantle. Unfortunately, most crystalline mare basalts are products of postmelting and source extraction processes (fractional crystallization, crystal accumulation) and therefore their usefulness as mantle probes is compromised. However, picritic glass beads found at all the Apollo sampling sites represent liquid compositions and are closer approximations of primary mare basalts. Trace-element analysis of individual glass beads by ion microprobe have yielded numerous insights into the formation of mantle sources for the mare basalts, the dynamics of the lunar mantle, and its compositional variability (*Shearer and Papike*, 1993; *Shearer and Newsom*, 1999).

## 6.3. Insights from Trace-Element Microbeam Analysis of Picritic Glasses

The volcanic glasses are diverse in their trace-element characteristics, with a wide range of REE patterns and concentrations (*Shearer and Papike*, 1993; *Papike et al.*, 1998). Like the mare basalts, all have negative Eu anomalies. However, unlike the mare basalts, there is little correlation between overall REE concentration and the size of the Eu anomaly (*Shearer and Papike*, 1993). The similarities between crystalline mare basalts and picritic glasses with regard to their ranges in major- and trace-element concentrations and the nonchondritic nature of numerous trace-element ratios indicate that both were derived from a differentiated (i.e., LMO cumulate) rather than a primitive lunar mantle. The negative Eu anomalies in all near-primary, picritic glasses also argues against serial magmatism models. In an attempt to explain the negative Eu anomaly of the mare basalts within the context of a serial magmatism model, *Walker* (1983) suggested that it reflected mixing of primitive lunar magmas with basaltic magmas that had experienced plagioclase fractionation. Clearly, this is not the case as these primitive basalts also have Eu anomalies. In addition, the bulk lunar crust as estimated by *Walker* (1983) and *Haskin* (1989) includes igneous rocks produced during and following magma ocean crystallization. Therefore, even if the bulk lunar crust does not exhibit a positive Eu anomaly, the products of LMO crystallization (FAN) most certainly do.

Although the picritic glasses and crystalline mare basalts were derived by melting of LMO cumulate sources, several lines of evidence are consistent with the view that picritic magmas were derived from mantle sources that are compositionally distinct from the sources for many mare basalts. *Longhi* (1992) and *Shearer and Papike* (1993) demonstrated that most crystalline mare basalts and picritic glasses had parallel, but no common liquid lines of descent. Other trace-element comparisons (Sc vs. FeO) indicate that most mare basalts are unrelated to the picritic glasses by fractional crystallization and that these differences must be attributed to different mantle sources (Fig. 10). Differences in Pb isotopes and the mode of eruption (i.e., fire-fountaining for the picritic glasses) provide circumstantial evidence that the picritic glasses represent magmas derived from a more volatile-rich source.

What then are the parental magmas for the mare basalts? Two alternatives are that (1) they are similar to the picritic glasses and were derived from initial melting in the deep lunar mantle (>400 km) or (2) mare basalts represent the crystallization of more Fe-rich, olivine-poor primary magmas that were produced at shallower depths. We propose that the former conclusion is correct for a number of reasons. First, the crystalline mare basalts and picritic mare glasses define the same ranges in $TiO_2$ and overlap in age. This is interpreted as indicating that they were produced by melting of similar mantle assemblages under comparable conditions. Second, crystalline mare basalts have all the chemical attributes (lower Ni, Mg', higher Al) associated with substantial olivine fractionation. Although *Longhi* (1992) and *Shearer and Papike* (1993) demonstrated that the picritic mare glasses and crystalline mare basalts, in most cases, were not related by shallow fractional crystallization, calculated liquid lines of descent are parallel. Therefore the preservation of primitive magmas as picritic mare glasses is a function of magma transport and eruptive mechanism rather than dramatically difference source regions. Third, the generation of mare basalts from a shallow, subcrustal source between 3.9 and 3.0 Ga is not reconcilable with the ther-

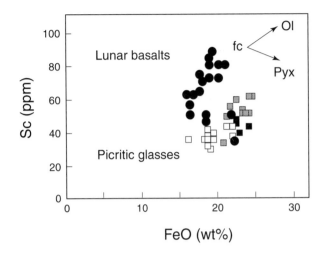

**Fig. 10.** Plot of FeO vs. Sc for picritic glasses and crystalline mare basalts. In most cases, the glasses and basalts define two separate fields that are not related by fractional crystallization. These populations probably represent distinct source regions in the lunar mantle. For the picritic glasses, the open squares = very-low-Ti glasses, black squares = low-Ti glasses, and gray squares = high-Ti glasses.

mal nature of the upper lunar mantle. The liquidus temperatures of mare basalts exceed 1200°C at depths corresponding to about 0.5 GPa (*Longhi,* 1992). There is no mechanism to reheat the shallow lunar mantle (between 200°C and 400°C) 600 m.y. after the formation of the Moon and to maintain this temperature for 700 m.y. Furthermore, the existence and preservation of mascons in multiring basins that formed at about 3.9 Ga indicate a relatively cool (1000°C) and rigid upper mantle.

The wide compositional variability observed in the picritic glasses and the limited flux of basaltic magmas erupting at the lunar surface implies a transition in the thermal evolution of the lunar mantle. Whereas high degrees of melting were asociated with the production of early crustal rocks (FAN, HMS, HAS) prior to 3.9 Ga, smaller degrees are implied for the mare basalts (after 3.9 Ga). Most models require the melting of sources consisting of olivine + orthopyroxene + ilmenite ± plagioclase ± clinopyroxene, which results in a residuum assemblage of olivine and orthopyroxene (*Walker et al.,* 1975; *Delano,* 1986; *Hughes et al.* 1989; *Snyder et al.,* 1992, *Shearer and Papike,* 1993). Melting models based strictly on the compositions of picritic glasses indicate they were generated by rather small percentages (2–15%) of partial melting (*Hughes et al.,* 1989; *Shearer and Papike,* 1993). The smaller degree of mantle melting between 3.9 and 3.0 Ga resulted in the preservation of localized mantle heterogeneities in the mare basalts (i.e., the wide range of $TiO_2$). Large-scale mantle melting would have essentially erased the signature of these small-scale mantle heterogeneities by mixing and homogenizing discrete packages of mare basalt.

The trace-element characteristics of picritic glasses also provide insights into the dynamics of the lunar mantle following crystallization of the LMO. These primitive magmas exhibit evidence for variable amounts of a KREEP component and continuous increases of ilmenite in their sources, facts that are somewhat puzzling. KREEP and ilmenite were produced during the last stages of LMO crystallization (*Taylor and Jakeš*, 1974; *Snyder et al.*, 1992; *Hess and Parmentier*, 1995), and yet the picritic basalts were produced by melting that initiated at depths greater than 400 km (*Delano*, 1986; *Longhi*, 1992; *Papike et al.*, 1998). *Shearer and Papike* (1993) pointed out that shallow assimilation of a KREEP component by magmas generated in the deep lunar mantle would result in substantial crystallization of olivine. Yet those glasses with substantial KREEP component plot on a mixing line rather than assimilation-fractional crystallization trajectories (Fig. 11). *Longhi* (1992) demonstrated that the pattern of chondrite-normalized spider diagrams reflects the presence or absence of ilmenite in the cumulate source for mare basalts. Patterns with positive anomalies of Ta, Nb, and Ti over a wide range of $TiO_2$ indicate ilmenite accumulation in these sources. The large range of $TiO_2$, Ta, and Nb in these basalts is the result of a nearly continuous variation in the ilmenite abundance of the source region (*Longhi*, 1992). Large negative anomalies of Ta, Nb, and Ti in KREEP and KREEP basalts indicate that ilmenite crystallized in the LMO prior to the formation of their cumulate sources. A more quantitative expression of the "ilmenite anomaly" is shown in a plot of Ti vs. Yb (Fig. 12). The ratio of these two elements provides a means for evaluating the "ilmenite anomaly." As shown in Fig. 12, all the primitive mare basalts that are represented by picritic glasses plot in the positive "ilmenite anomaly" field, implying that KREEP and ilmenite were not incorporated into these magmas by assimilation at shallow depths but by melting of a hybridized, cumulate source that contained a mixture of early and late magma ocean cumulates. Large-scale mixing of early and late cumulates over distances of hundreds of kilometers has been proposed and modeled by *Ringwood and Kesson* (1976), *Spera* (1992), and *Hess and Parmentier* (1995). The transport of late-stage LMO cumulates enriched in incompatible, radiogenic isotopes provides a heat source in the deep lunar mantle for initiating partial melting of early cumulates.

Another fascinating result of the trace-element studies of lunar glasses is the relationship between the "ilmenite anomaly" and radiogenic W ($\varepsilon_{182W}$). The trace-element dataset for the basalts analyzed by *Lee et al.* (1997) for radiogenic W ($\varepsilon_{182W}$) is incomplete and somewhat compromised due to the derivative nature of many of the basaltic magmas. Yet those that plot in the positive ilmenite anomaly field (Fig. 12) have nonchondritic $\varepsilon_{182W}$ values, whereas basalts that do not have positive ilmenite anomalies have chondritic $\varepsilon_{182W}$ values. This relationship suggests that both the ilmenite anomaly and the nonchondritic $\varepsilon_{182W}$ signature in the lunar mantle may be the result of the fractionation of Hf from W during ilmenite crystallization from the LMO and accumulation in LMO cumulates. *Snyder et al.* (1992) predicted that ilmenite would be a liquidus phase for the LMO after >95% of the

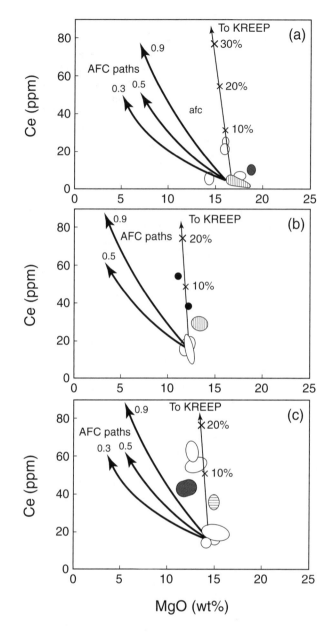

**Fig. 11.** Plots of MgO (wt%) vs. Ce (ppm) for the picritic glasses. Mixing and AFC paths are shown for **(a)** very-low-$TiO_2$ picritic magmas, **(b)** intermediate-$TiO_2$ picritic magmas, and **(c)** high-$TiO_2$ picritic magmas.

LMO crystallized and that it would be accompanied by clinopyroxene. Preliminary natural and experimental studies of Hf-W partitioning between ilmenite-melt and clinopyroxene-melt indicate ilmenite and clino-pyroxene will enrich late-stage ilmenite-clinopyroxene cumulates in Hf relative to W (*Shearer and Newsom*, 1999; *McKay and Le*, 1999; *Righter and Drake*, 1999). In ilmenite the ratio between $D^{Hf}/D^{W}$ is approximately 4 to 10 (*Shearer and Newsom*, 1999; *McKay and Le*, 1999) and in clinopyroxene the ratio between $D^{Hf}/D^{W}$ is greater than 35 (*Righter and Drake*, 1999). If Hf-W fractionation by ilmenite-clinopyroxene is

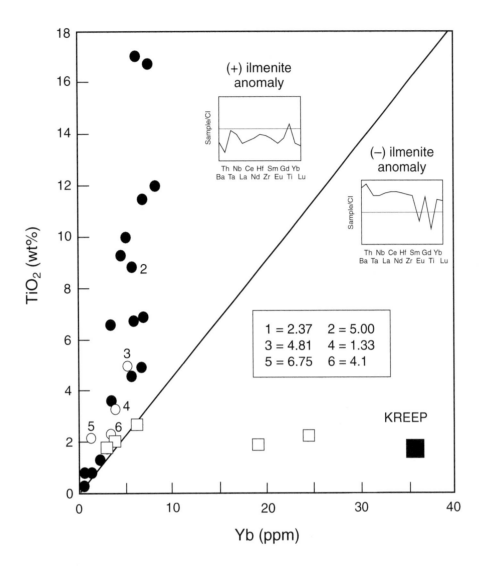

**Fig. 12.** Plot of Yb vs. TiO$_2$ for picritic glasses (filled circles), mare basalts (open circles and squares), and KREEP (filled square) illustrating the extent of the ilmenite anomaly (*Longhi,* 1992). The line represents the absence of an ilmenite anomaly. At Ti values above the line, the spider diagram has a positive ilmenite anomaly [(Ti/Yb)$_{chondrite\ normalized}$ > 1]. At Ti values below the line, the spider diagram has a negative ilmenite anomaly [(Ti/Yb)$_{chondrite\ normalized}$ < 1]. Numbers index samples with positive $\varepsilon_{182W}$ values. Mare basalts with $\varepsilon_{182W}$ of ~0 are represented by open squares. The linear relationship between (Ti/Yb)$_{chondrite\ normalized}$ and $\varepsilon_{182W}$ values should not be expected for these samples. The (Ti/Yb)$_{chondrite\ normalized}$ and $\varepsilon_{182W}$ is the result of the trace-element characteristics of the samples analyzed by *Lee et al.* (1997) being compromised to varying degrees due to the nature of the samples (near-primary basalts, highly fractionated basalts, cumulates), both ilmenite and clinopyroxene in the late-stage LMO cumulate fractionating Hf from W, and the variable degree of cumulate mixing to produce the basalt sources.

the mechanism by which cumulate lithologies obtain a radiogenic W-isotopic signature, then the duration of most of LMO crystallization may be constrained by the Hf decay, because of the late appearance of ilmenite on the LMO liquidus and the short half-life of $^{182}$Hf (9 m.y.). Depending on the extent of Hf/W fractionation by ilmenite and clinopyroxene, more than 95% of the LMO crystallized over a period of 10–40 m.y. This short time period contrasts with many other models that suggest a duration of 100 m.y. to more than 200 m.y. Rapid LMO crystallization implies that the early lunar crust (FAN) was not as thermally insulating

as some LMO models assume and that the transition from magma ocean (FAN) to serial magmatism (HMS) occurred early.

## 7. CONCLUSIONS

Combined with isotopic studies, ion microprobe investigations of trace elements in lunar material (minerals, glasses) have given us additional insights into the origin and evolution of the Moon's crust and mantle. The magma ocean event that formed the earliest lunar crust and a differentiated mantle was a short-lived event, perhaps on the order of 20–50 m.y. Subsequent periods of plutonism-volcanism reflected the thermal evolution of the lunar mantle, the changing structure of the lunar crust, and the early differentiation event. The transition from FANs to HMS and HAS (both plutonic and volcanic) to mare volcanism reflect a decrease in mantle heat resulting in a decrease in the degree of source melting and magma homogenization. Changes in the emplacement of magma reservoirs from crustal (HMS, HAS) to subcrustal (mare basalts) reflect the thermal and structural evolution of the lunar crust. Sources for post-magma-ocean episodes of magmatism were influenced by early and extensive mantle differentiation through crystallization of a magma ocean and subsequent mixing of early and late cumulates by gravitational destabilization and overturn of the LMO cumulate pile. The HMS and HAS plutonic rocks do not represent crystallization products of a single parental magma, but of several parental magmas. The HMS was probably produced by moderate degrees of melting of a hybridized mantle consisting of late (KREEP) and early (olivine + orthopyroxene) cumulates. However, the possible role of reduction and metal removal during the formation of the HMS may complicate this simple model. The mare basalts represent a continuation of this melting event. However, the dwindling melting process produced basaltic magmas that preserved local mantle heterogeneities attributed to cumulate mixing.

## REFERENCES

Alfvén H. and Arrhenius G. (1976) *Evolution of the Solar System.* NASA, Washington, DC. 599 pp.

Alibert C., Norman M. D., and McCulloch M. T. (1994) An ancient Sm-Nd age for a ferroan noritic anorthosite clast from lunar breccia 67016. *Geochim. Cosmochim. Acta, 58,* 2921–2926.

BVSP (1981) *Basaltic Volcanism on the Terrestrial Planets.* Pergamon, New York.

Bell J. F. and Hawke B. R. (1984) Lunar dark-haloed impact craters: Origin and implications for early mare volcanism. *J. Geophys. Res., 89,* 6899–6910.

Binder A. B. (1978) On fission and the devolatilization of a Moon of fission origin. *Earth Planet. Sci. Lett., 41,* 381–385.

Borg L., Norman M., Nyquist L., Bogard D., Snyder G., Taylor L., and Lindstrom M. (1999) Isotopic studies of ferroan anorthosite 62236: a young lunar crustal rock from a light rare-earth element-depleted source. *Geochim. Cosmochim. Acta, 63,* 2679–2691.

Burgess R. and Turner G. (1998) Laser argon-40–argon-39 age determinations of Luna 24 mare basalts. *Meteoritics & Planet. Sci., 33,* 921–935.

Carlson R. W. and Lugmair G. W. (1981) Time and duration of lunar highlands crust formation. *Earth Planet. Sci. Lett., 52,* 227–238.

Carlson R. W. and Lugmair G. W. (1988) The age of ferroan anorthosite 60025: oldest crust on a young moon? *Earth Planet. Sci. Lett., 90,* 119–130.

Davis P. A. and Spudis P. D. (1985) Petrologic province maps of the lunar highlands derived from orbital geochemical data.

*Proc. Lunar Planet. Sci. Conf. 16th,* in *J. Geophys. Res., 90,* D61–D74.

Davis P. A. and Spudis P. D. (1987) Global petrologic variations on the Moon: A ternary-diagram approach. *Proc. Lunar Planet. Sci. Conf. 17th,* in *J. Geophys. Res., 92,* E387–E395.

Delano J. W. (1986) Pristine lunar glasses: Criteria, data and implications. *Proc. Lunar Planet. Sci Conf. 16th,* in *J. Geophys. Res., 91,* D201–D213.

Deutsch A. and Stöffler D. (1987) Rb-Sr analyses of Apollo 16 melt rocks and a new age estimate for the Imbrium basin: Lunar basin chronology and the early bombardment of the moon. *Geochim. Cosmochim. Acta, 51,* 1951–1964.

Dickinson T., Taylor G. J., Keil K., Schmitt R. A., Hughes S. S., and Smith M. R. (1985) Apollo 14 aluminous basalts and their possible relationship to KREEP. *Proc. Lunar Planet. Sci. Conf. 15th,* in *J. Geophys. Res., 90,* C365–C374.

Dixon J. R. and Papike J. J. (1975) Petrology of anorthosites from the Descartes region of the moon: Apollo 16. *Proc. Lunar Sci. Conf. 6th,* pp. 263–291.

Floss C., James O. B., McGee J. J., and Crozaz G. (1991) Lunar ferroan anorthosites: rare earth element measurements of individual plagioclase and pyroxene grains (abstract). In *Lunar and Planetary Science XXII,* pp. 391–392. Lunar and Planetary Institute, Houston.

Floss C., James O. B., McGee J. J., and Crozaz G. (1998) Lunar ferroan anorthosite petrogenesis: clues from trace-element distributions in FAN subgroups. *Geochim. Cosmochim. Acta, 62,* 1255–1283.

Fowler G., Shearer C. K., and Papike J. J. (1995) Diogenites as asteroidal cumulates: Insights from orthopyroxene trace element chemistry. *Geochim. Cosmochim. Acta, 59,* 3071–3084.

Giguere T. A., Taylor G. J., Hawke B. R., and Lucey P. G. (1999) Distribution of the titanium contents of lunar mare basalts: not bimodal (abstract). In *Lunar and Planetary Science XXX,* Abstract #1465. Lunar and Planetary Institute, Houston (CD-ROM).

Haskin L. A. (1989) Rare earth elements in lunar materials. In *Reviews in Mineralogy, Geochemistry and Mineralogy of Rare Earth Elements* (B. R. Lipin and G. A. McKay, eds.), pp. 227–258.

Haskin L. A., Lindstrom M. M., Salpas P. A., and Lindstrom D. J. (1981) On compositional variations among lunar anorthosites. *Proc. Lunar Planet. Sci. 12B,* pp. 41–66.

Hawke B. R., Lucey P. G., Bell J. F., and Spudis P. D. (1990) Ancient mare volcanism. In *LPI-LAPST Workshop on Mare Volcanism and Basalt Petrogenesis: Astounding Fundamental Concepts (AFC) Developed Over the Last Fifteen Years,* pp. 5–6. Lunar and Planetary Institute, Houston.

Head J. W. and Wilson L. (1992) Lunar mare volcanism: Stratigraphy, eruption conditions, and the evolution of secondary crusts. *Geochim. Cosmochim. Acta, 56,* 2155–2175.

Herbert F., Drake M. J., Sonett C. P., and Wiskerchen M. J. (1977) Some constraints on the thermal history of the lunar magma ocean. *Proc. Lunar Sci. Conf. 8th,* pp. 573–582.

Hess P. C. (1994) Petrogenesis of lunar troctolites. *J. Geophys. Res., 99,* 19083–19093.

Hess P. C. and Parmentier E. M. (1995) A model for the thermal and chemical evolution of the Moon's interior: Implications for the onset of mare volcanism. *Earth Planet. Sci. Lett., 134,* 501–514.

Hess P. C., Rutherford M. J., and Campbell H. W. (1978) Ilmenite crystallization in non-mare basalt: Genesis of KREEP and high-Ti mare basalts. *Proc. Lunar Planet. Sci. Conf. 9th,* pp. 705–724.

Hostetler C. J. and Drake M. J. (1980) On the early global melting of the terrestrial planets. *Proc. Lunar Planet. Sci. Conf. 11th*, pp. 1915–1929.

Hubbard N. J. and Minear J. W. (1975) A physical and chemical model of early lunar history. *Proc. Lunar Sci. Conf. 6th*, pp. 1057–1085.

Hughes S. S., Delano J. W., and Schmitt R. A. (1989) Petrogenetic modeling of 74220 high-Ti orange volcanic glasses and the Apollo 11 and 17 high-Ti mare basalts. *Proc. Lunar Planet. Sci. Conf. 19th*, pp. 175–188.

James O. B. (1972) Lunar anorthosite 15415: texture, mineralogy and metamorphic history. *Science, 175*, 432–436.

James O. B. (1980) Rocks of the early lunar crust. *Proc. Lunar Planet. Sci. Conf. 11th*, pp. 365–393.

James O. B., Lindstrom M. M., and Flohr M. K. (1987) Petrology and geochemistry of alkali gabbronorites from lunar breccia 67975. *Proc. Lunar Planet. Sci. Conf. 17th*, in *J. Geophys. Res., 92*, E314–E330.

James O. B., Lindstrom M. M., and Flohr M. K. (1989) Ferroan anorthosite from lunar breccia 64435: implications for the origin and history of lunar ferroan anorthosites. *Proc. Lunar Planet. Sci. Conf. 19th*, pp. 219–243.

James O. B., Floss C., and McGee J. J. (1998) Rare-earth distributions in pyroxenes from a lunar mafic-magnesian ferroan anorthosite (abstract). In *Lunar and Planetary Science XXIX*, Abstract #1292. Lunar and Planetary Institute, Houston (CD-ROM).

Jolliff B. L. and Hsu W. (1996) Geochemical effects of recrystallization and exsolution of plagioclase of ferroan anorthosite (abstract). In *Lunar and Planetary Science XXVII*, pp. 611–612. Lunar and Planetary Institute, Houston.

Korotev R. L. and Haskin L. A. (1988) Europium mass balance in polymict samples and implications for plutonic rocks of the lunar crust. *Geochim. Cosmochim. Acta, 52*, 1795–1813.

Lee D.-C., Halliday A. N., Snyder G. A., and Taylor L. A. (1997) Age and origin of the Moon. *Science, 278*, 1098–1133.

Longhi J. (1977) Magma oceanography 2: chemical evolution and crustal formation. *Proc. Lunar Sci. Conf. 8th*, pp. 601–621.

Longhi J. (1980) A model of early lunar differentiation. *Proc. Lunar Planet. Sci. Conf. 11th*, pp. 289–315.

Longhi J. (1981) Preliminary modeling of high-pressure partial melting: Implications for early lunar differentiation. *Proc. Lunar Planet. Sci. 12B*, pp. 1001–1018.

Longhi J. (1992) Experimental petrology and petrogenesis of mare volcanics. *Geochim. Cosmochim. Acta, 56*, 2235–2251.

Longhi J. and Boudreau A. E. (1979) Complex igneous processes and the formation of the primitive lunar crustal rocks. *Proc. Lunar Planet. Sci. Conf. 10th*, pp. 2085–2105.

Longhi J. and Ashwal L. D. (1985) Two-stage models for lunar and terrestrial anorthosites: petrogenesis without a magma ocean. *Proc. Lunar Planet. Sci. Conf. 15th*, in *J. Geophys. Res., 90*, C571–C584.

McGee J. J. (1988) Petrology of brecciated ferroan noritic anorthosite 67215. *Proc. Lunar Planet. Sci. Conf. 18th*, pp. 21–31.

McKay G.A. and Le L. (1999) Partitioning of W and Hf between ilmenite and mare basaltic melt (abstract). In *Lunar and Planetary Science XXX*, Abstract #1996. Lunar and Planetary Institute, Houston (CD-ROM).

Metzger A. E. and Parker R. E. (1979) The distribution of titanium on the lunar surface. *Earth Planet. Sci. Lett., 45*, 155–171.

Mueller S., Taylor G. J., and Phillips R. J. (1988) Lunar composition: A geophysical and petrological synthesis. *J. Geophys. Res., 93*, 6338–6352.

Neal C. R. and Taylor L. A. (1992) Petrogenesis of mare basalts: A record of lunar volcanism. *Geochim. Cosmochim. Acta, 56*, 2177–2211.

Neal C. R., Taylor L. A., and Lindstrom M. M. (1988) Apollo 14 mare basalt petrogenesis: Assimilation of KREEP-like components by a fractionating magma. *Proc. Lunar Planet. Sci. Conf. 18th*, pp. 139–153.

Neal C. R., Taylor L. A., and Patchen A. D. (1989a) High alumina (HA) and very high potassium (VHK) basalt clasts from Apollo 14 breccias, Part 1 — Mineralogy and petrology: Evidence of crystallization from evolving magmas. *Proc. Lunar Planet. Sci. Conf. 19th*, pp. 137–145.

Neal C. R., Taylor L. A., Schmitt R. A., Hughes S. S., and Lindstrom M. M. (1989b) High alumina (HA) and very high potassium (VHK) basalt cclasts from Apollo 14 breccias, Part 2 — Whole rock geochemistry: Further evidence of combined assimilation and fractional crystallization within the lunar crust. *Proc. Lunar Planet. Sci. Conf. 19th*, pp. 147–161.

Nyquist L. E. and Shih C.-Y. (1992) The isotope record of lunar volcanism. *Geochim. Cosmochim. Acta, 56*, 2213–2234.

Palme H., Spettel B., Wänke H., Bischoff A., and Stöffler D. (1984) Early differentiation of the moon: evidence from trace elements in plagioclase. *Proc. Lunar Planet. Sci. Conf. 15th*, in *J. Geophys. Res., 89*, C3–C15.

Papike J. J. (1996) Pyroxene as a recorder of cumulate formational processes; Asteroids, Moon, Mars, Earth: Reading the record with the ion microprobe. *Am. Mineral., 81*.

Papike J. J., Hodges F. N., Bence A. E., Cameron M., and Rhodes J. M. (1976) Mare basalts: crystal chemistry, mineralogy, and petrology. *Rev. Geophys. Space Phys., 14*, 475–540.

Papike J. J., Fowler G. W., and Shearer C. K. (1994) Orthopyroxene as a recorder of lunar crust evolution: An ion microprobe investigation of Mg suite norites. *Amer. Mineral., 79*, 790–800.

Papike J. J., Fowler G. W., Shearer C. K., and Layne G. D. (1996) Ion microprobe investigation of plagioclase and orthopyroxene from lunar Mg suite norites: Implications for calculating parental melt REE concentrations and for assessing post-crystallization REE redistribution. *Geochim. Cosmochim. Acta, 60*, 3967–3978.

Papike J. J., Fowler G. W., and Shearer C. K. (1997) Evolution of the lunar crust: SIMS study of plagioclase from ferroan anorthosites. *Geochim. Cosmochim. Acta, 61*, 2343–2350.

Papike J. J., Ryder G., and Shearer C. K. (1998) Lunar samples. In *Planetary Materials* (J. J. Papike, ed.), pp. 1–234. Mineralogical Society of America, Washington, DC.

Papike J. J., Fowler G. W., Adcock C. T., and Shearer C. K. (1999) Systematics of Ni and Co in olivine from planetary melt systems: lunar mare basalts. *Amer. Mineral., 84*, 392–399.

Phinney W. C. (1991) Lunar anorthosites, their equilibrium melts and the bulk moon. *Proc. Lunar Planet. Sci., Vol. 21*, pp. 29–49.

Raedeke L. D. and McCallum I. S. (1980) A comparison of fractionation trends in the lunar crust and the Stillwater Complex. *Proc. Conf. Lunar Highlands Crust*, pp. 133–153.

Ransford G. A. (1982) The accretional heating of the terrestrial planets: a review. *Phys. Earth Planet. Inter., 29*, 209–217.

Righter K. and Drake M. J.(1999) Partitioning of W between liquid metal, solid silicates, and liquid silicates at high pressures and temperatures: Implications for the $^{182}$W isotopic anomalies in lunar and martian samples (abstract). In *Lunar and Planetary Science XXX,* Abstract #1381. Lunar and Planetary Institute, Houston (CD-ROM).

Ringwood A. E. (1975) *Composition and Petrology of the Earth's Mantle.* McGraw Hill, New York.

Ringwood A. E. and Kesson S. E. (1976) A dynamic model for mare basalt petrogenesis. *Proc. Lunar Sci. Conf. 7th,* pp. 1697–1722.

Ryder G. (1976) Lunar sample 15405: Remnant of a KREEP-granite differentiated pluton. *Earth Planet. Sci. Lett., 29,* 255–268.

Ryder G. (1982) Lunar anorthosite 60025, the petrogenesis of lunar anorthosites, and the composition of the moon. *Geochim. Cosmochim. Acta, 46,* 1591–1601.

Ryder G. (1983) Nickel in olivines and parent magmas pf lunar pristine rocks (abstract). In *Workshop on Pristine Highlands Rocks and the Early Evolution of the Moon,* pp. 66–68. LPI Tech. Rpt. 83-02, Lunar and Planetary Institute, Houston.

Ryder G. (1991) Lunar ferroan anorthosites and mare basalt sources: the mixed connection. *Geophys. Res. Lett., 18,* 2065–2068.

Ryder G. and Wood J. A. (1977) Serenitatis and Imbrium impact melts: Implications for large-scale layering in the lunar crust. *Proc. Lunar Planet. Sci. Conf. 8th,* pp. 655–668.

Safronov V. S. (1978) The heating of the Earth during its formation. *Icarus, 33,* 3–12.

Schnetzler C. C. and Philpotts J. A. (1971) Alkali, alkaline earth, and rare-earth element concentrations in some Apollo 12 soils, rocks, and separated phases. *Proc. Lunar Sci. Conf. 2nd,* pp. 1101–1122.

Schultz P. H. and Spudis P. D. (1979) Evidence for ancient mare volcanism. *Proc. Lunar Planet. Sci. Conf. 10th,* pp. 2899–2918.

Shearer C. K. and Newsom H. E. (2000) W-Hf isotope abundances and the early origin and evolution of the earth-moon system. *Geochim. Cosmochim. Acta, 64,* in press.

Shearer C. K. and Papike J. J. (1989) Is plagioclase removal responsible for the negative Eu anomaly in the source regions of mare basalts? *Geochim. Cosmochim. Acta, 53,* 3331–3336.

Shearer C. K. and Papike J. J. (1993) Basaltic magmatism on the Moon: A perspective from volcanic picritic glass beads. *Geochim. Cosmochim. Acta, 57,* 4785–4812.

Shearer C. K. and Papike J. J. (1999) Magmatic evolution of the Moon (invited review). *Am. Mineral., 84,* 1469–1494.

Shearer C. K. and Papike J. J. (2000) Compositional dichotomy of the Mg suite: Origin and implications for the thermal and compositional structure of the lunar mantle (abstract). In *Lunar and Planetary Science XXXI,* Abstract #1405. Lunar and Planetary Institute, Houston (CD-ROM).

Shervais J. W. and McGee J. J. (1997) Alkali suite anorthosites and norites: Flotation cumulates from pristine KREEP with magma mixing and the assimilation of older anorthosite (abstract). In *Lunar and Planetary Science XXIX,* Abstract #1699. Lunar and Planetary Institute, Houston (CD-ROM).

Shervais J. W. and McGee J. J. (1999) Ion and electron microprobe study of troctolites, norite, and anorthosites from Apollo 14: Evidence for urKREEP assimilation during petrogenesis of Apollo 14 Mg-suite rocks. *Geochim. Cosmochim. Acta, 62,* 3009–3024.

Shervais J. W., Taylor L. A., Laul J. C., Shih C.-Y., and Nyquist L. E. (1985) Very high potassium (VHK) basalt: Complications in lunar mare basalt petrogenesis. *Proc. Lunar Planet. Sci. Conf.*

16th, in *J. Geophys. Res., 90,* D3–D18.

Shih C.-Y., Nyquist L. E., Bansal B. M., and Wiesmann H. (1992) Rb-Sr and Sm-Nd chronology of an Apollo 17 KREEP basalt. *Earth Planet. Sci. Lett., 108,* 203–215.

Shih C.-Y., Nyquist L. E., Dasch E. J., Bogard D. D., Bansal B. M., and Wiesmann H. (1993) Ages of pristine noritic clasts from lunar breccias 15445 and 15455. *Geochim. Cosmochim. Acta, 57,* 915–931.

Shirley D. N. (1983) A partially molten magma ocean model. *Proc. Lunar Planet. Sci. Conf. 13th,* in *J. Geophys. Res., 88,* A519–A527.

Smith J. V. (1982) Heterogeneous growth of meteorites and planets, especially the Earth and Moon. *J. Geol., 90,* 1–48.

Smith J. V., Anderson A. T., Newton R. C., Olsen E. J., Wyllie P. J., Crewe A. V., Isaacson M. S., and Johnson D. (1970) Petrologic history of the moon inferred from petrography, mineralogy, and petrogenesis of Apollo 11 rocks. *Proc. Apollo 11 Lunar Sci. Conf.,* pp. 897–925.

Snyder G. A., Taylor L. A., and Neal C. R. (1992) A chemical model for generating the sources of mare basalts: combined equilibrium and fractional crystallization of the lunar magmasphere. *Geochim. Cosmochim. Acta, 56,* 3809–3823.

Snyder G. A., Taylor L. A., Jerde E. A., and Riciputi L. R. (1994) Evolved QMD-melt parentage for lunar highlands alkali suite cumulates: evidence from ion-probe rare-earth element analyses of individual minerals (abstract). In *Lunar and Planetary Science XXV,* pp. 1311–1312. Lunar and Planetary Institute, Houston.

Snyder G. A., Taylor L. A., and Halliday A. N. (1995) Chronology and petrogenesis of the lunar highlands alkali suite: Cumulates from KREEP basalt crystallization. *Geochim. Cosmochim. Acta, 59,* 1185–1203.

Snyder G. A., Floss C., and Taylor L. A. (1998) The origin of ferroan anorthosites and rethinking lunar dogma: ion probe trace-element analyses of minerals in Apollo 15 rocks (abstract). In *Lunar and Planetary Science XXIX,* Abstract #1133. Lunar and Planetary Institute, Houston (CD-ROM).

Snyder G. A., Borg L. E. Nyquist L. E., and Taylor L. A. (2000) Chronology and isotopic constraints on lunar evolution. In *Origin of the Earth and Moon* (R. M. Canup and K. Righter, eds.), this volume. Univ. of Arizona, Tucson.

Solomon S. C. and Longhi J. (1977) Magma oceanography: 1. Thermal evolution. *Proc. Lunar Sci. Conf. 8th,* pp. 583–599.

Sonett C. P. and Reynolds R. T. (1979) Primordial heating of asteroidal parent bodies. In *Asteroids* (T. Gehrels, ed.), pp. 822–848. Univ. of Arizona, Tucson.

Spera F. J. (1992) Lunar magma transport phenomena. *Geochim. Cosmochim. Acta, 56,* 2253–2266.

Spudis P. D. and Davis P. A. (1986) A chemical and petrologic model of the lunar crust and implications for lunar crustal origin. *Proc. Lunar Planet. Sci. Conf. 17th,* in *J. Geophys. Res., 91,* E84–E90.

Stadermann F. J., Heusser E., Jessberger E. K., Lingner S., and Stöffler D. (1991) The case for a younger Imbrium basin: New $^{40}$Ar-$^{39}$Ar ages of Apollo 14 rocks. *Geochim. Cosmochim. Acta, 55,* 2339–2349.

Steele I. M. and Smith J. V. (1973) Mineralogy and petrology of some Apollo 16 rocks and fines: general petrologic model of moon. *Proc. Lunar Sci. Conf. 4th,* pp. 519–536.

Taylor S. R. (1982) *Planetary Science: A Lunar Perspective.* Lunar and Planetary Institute, Houston. 481 pp.

Taylor S. R. and Jakeš P. (1974) The geochemical evolution of the Moon. *Proc. Lunar Sci. Conf. 5th,* pp. 1287–1305.

Tompkins S. and Pieters C. M. (1999) Mineralogy of the lunar crust: Results from Clementine. *Meteoritics & Planet. Sci., 34,* 25–41.

Wagner T. P. and Grove T. L. (1995) Origin of high-Ti lunar magma by erosion of ilmenite (abstract). In *Lunar and Planetary Science XXVI,* pp. 1455–1456. Lunar and Planetary Institute, Houston.

Walker D. (1983) Lunar and terrestrial crust formation. *Proc. Lunar Planet. Sci. Conf. 14th,* in *J. Geophys. Res., 88,* B17–B25.

Walker D., Longhi J., Stolper E. M., Grove T. L., and Hays J. F. (1975) Origin of titaniferous lunar basalts. *Geochim. Cosmochim. Acta, 39,* 1219–1235.

Warner J. L., Simonds C. H., and Phinney W. C. (1976) Genetic distinction between anorthosites and Mg-rich plutonic rocks: new data from 76255 (abstract). In *Lunar Science VII,* pp. 915–917. Lunar Science Institute, Houston.

Warren P. H. (1985) The magma ocean concept and lunar evolution. *Annu. Rev. Earth Planet. Sci., 13,* 201–240.

Warren P. H. (1986) Anorthosite assimilation and the origin of the Mg/Fe-related bimodality of pristine moon rocks: Support for the magmasphere hypothesis. *Proc. Lunar Planet. Sci. Conf. 16th,* in *J. Geophys. Res., 91,* D331–D343.

Warren P. H. (1993) A concise compilation of petrologic information on possibly pristine nonmare Moon rocks. *Amer. Mineral., 78,* 360–376.

Warren P. H. and Wasson J. T. (1977) Pristine nonmare rocks and the nature of the lunar crust. *Proc. Lunar Sci. Conf. 8th,* pp. 2215–2235.

Warren P. H. and Wasson J. T. (1980) Early lunar petrogenesis, oceanic and extraoceanic. *Proc. Conf. Lunar Highlands Crust,* pp. 81–99. Pergamon, New York.

Wetherill G. W. (1976) The role of large bodies in the formation of the Earth and Moon. *Proc. Lunar Sci. Conf. 7th,* pp. 3245–3257.

Wetherill G. W. (1980) Formation of the terrestrial planets. *Annu. Rev. Astron. Astrophys., 18,* 77–113.

Wetherill G. W. (1981) Nature and origin of basin-forming projectiles. In *Multi-Ring Basins, Proc. Lunar Planet. Sci. 12A* (P. H. Schultz and R. B. Merrill, eds.), pp. 1–18. Pergamon, New York.

Wood J. A. (1972) Fragments of terra rock in the Apollo 12 soil samples and a structural model of the moon. *Icarus, 15,* 462–501.

Wood J. A. (1975) Lunar petrogenesis in a well-stirred magma ocean. *Proc. Lunar Sci. Conf. 6th,* pp. 1087–1102.

Wood J. A., Dickey J. S., Marvin U. B., and Powell B. N. (1970) Lunar anorthosites. *Science, 167,* 602–604.

# Chronology and Isotopic Constraints on Lunar Evolution

**Gregory A. Snyder**
*University of Tennessee, Knoxville*

**Lars E. Borg**
*NASA Johnson Space Center*

**Laurence E. Nyquist**
*NASA Johnson Space Center*

**Lawrence A. Taylor**
*University of Tennessee, Knoxville*

---

Isotopic systematics of lunar rocks indicate three major, distinct, reservoirs in the Moon: (1) the urKREEP residuum of a global lunar magma ocean with high $^{238}U/^{204}Pb$ ($\mu$) >500, high Rb/Sr and thus elevated $^{87}Sr/^{86}Sr$, and low Sm/Nd and consequent negative $\varepsilon_{Nd}$ values; (2) a "primordial" deep mantle source with $\mu$ values more typical of Earth, low Rb/Sr and $^{87}Sr/^{86}Sr$, high Sm/Nd, and extremely positive $\varepsilon_{Nd}$ values, and positive to variable $\varepsilon_W$ values; and (3) a shallower mantle reservoir that has similar $\mu$ values to the second, intermediate $\varepsilon_{Nd}$ values, low to intermediate $^{87}Sr/^{86}Sr$, and chondritic $\varepsilon_W$ values. The vast majority of lunar samples can be modeled by mixing these three reservoirs. A possible fourth source, with $\mu$ values from 35 to 100, is represented by a few early crustal rocks, the ferroan anorthosites.

Ferroan anorthosites, ostensibly the earliest lunar crustal rocks, exhibit a range of ages from 4.56 to 4.29 Ga and initial $\varepsilon_{Nd}$ values (0.9 to 3.1). These ages are inconsistent with derivation of all these rocks from a short-lived magma ocean, as suggested by $^{182}W$ and $^{142}Nd$ anomalies in lunar highland rocks and basalts. The positive $\varepsilon_{Nd}$ values of the ferroan anorthosites indicate time-integrated LREE-depletion, which is also inconsistent with direct derivation from a progressively LREE-enriched magma ocean. Instead, the derivation of ferroan anorthosites may involve convective overturn of a magma ocean and consequent mixing of LREE-enriched, plagioclase-rich, lower crust with underlying LREE-depleted, mafic cumulate sources. Later modification of this early anorthositic crust involved serial KREEP basalt magmatism, ponding in the crust, and crystallization of highland alkali suite and magnesian suite plutons from 4.4 to 3.9 Ga. The end of this major period of crustal evolution roughly coincides in time with a fall-off in large basin-forming impacts. Argon-40–argon-39 analyses of a variety of lunar samples at the different landing sites have allowed the dating of some of the larger impacts: Nectaris (3.9 Ga), Crisium (3.895 Ga), Serenitatis (3.893 Ga), Imbrium (3.85 Ga), Autolycus (2.1 Ga), Aristillus (1.29 Ga), and Copernicus (800 Ma).

The single most salient feature of mare basalts is their extremely positive $\varepsilon_{Nd}$ values, suggesting early differentiation in the Moon and generation of depleted source regions. Isotopic studies of mare basalts lead to the following general model for evolution of the mantle of the Moon: (1) Formation of a progressively incompatible-element (IE) depleted cumulate upper mantle and IE-enriched residual liquid as a consequence of a global magma ocean. (2) The IE enriched liquid was trapped in varying proportions in the differentiated cumulates. (3) Earliest, extensive mare magmas (high-Ti mare basalts) were generated at shallow depths in the mantle from cumulate source-regions that had trapped relatively large proportions of this IE-enriched, residual, magma ocean liquid and had precipitated the dense mineral ilmenite. The trapped liquid component contained elevated abundances of heat-producing elements (K, U, and Th), increasing the fertility of associated source-regions. (4) Magmatism at a particular landing site appears to begin with more IE-enriched sources (less-positive $\varepsilon_{Nd}$ values) and proceed to less IE-ennriched sources (more positive $\varepsilon_{Nd}$ values). (5) With time, melting moved progressively deeper in the mantle to source-regions with less of the residual, heat-producing, magma ocean liquid. (6) Denser ilmenite-bearing cumulates sank into the cumulate pile carrying entrained K-U-Th enriched trapped liquids and fertilizing deeper, more Mg-rich source regions (which were then melted to form the high-Ti picritic glass beads). The major controlling factors in melting of the lunar interior are likely to be the proportion of trapped, IE-enriched, magma-ocean liquid in the cumulate source, and the sinking of ilmenite-bearing, more fertile material into the lower mantle.

# 1.  INTRODUCTION

Both short-lived and long-lived geochronometers are key to our understanding of the thermal and mechanical evolution of planets and satellites. These isotopic studies not only allow us to set time constraints on various processes (core-formation, melting, crystallization, impact), but also allow us to determine the chemical and mineralogic characteristics of plausible sources. Prior to and during the Apollo lunar program, there was a flurry of activity by a host of research groups and laboratories in studying extraterrestrial samples using radiogenic isotopic techniques. In fact, these early studies were key to the development of various new isotopic systems and to the rapid development of new chemical separation schemes and instrumentation for their analysis. Notable among these developments were the automation and online processing of data using a thermal ionization mass-spectrometer (*Wasserburg et al.,* 1969), the development of the SHRIMP ion microprobe for *in situ,* U-Pb zircon dating (*Compston et al.,* 1984), and the development of analytical methods for the analysis of Sr isotopes in small samples (*Papanastassiou and Wasserburg,* 1969), for routinely analyzing Sm-Nd isotopes (*Lugmair,* 1974), and for the analysis of the Lu-Hf isotopic system (*Patchett and Tatsumoto,* 1980a,b). In the intervening years, isotopic work on lunar rocks fell to a few groups, although this did not lessen the pace or the vigor with which these studies were performed. In the last few years, the advent of new analytical techniques, particulary the magnetic sector, multicollector inductively coupled plasma mass spectrometer (ICP-MS) (*Walder et al.,* 1993; *Halliday et al.,* 1995), has once again revolutionized the isotope geochemistry of extraterrestrial materials. These new and improving techniques have allowed the rapid and precise analysis of $^{176}$Lu-$^{176}$Hf on rather low-abundance samples (*Blichert-Toft et al.,* 1997; *Lee et al.,* 2000) and on a variety of short-lived nuclides (e.g., $^{146}$Sm-$^{142}$Nd, *Nyquist et al.,* 1995; $^{182}$Hf-$^{182}$W, *Lee and Halliday,* 1995), leading to renewed interest in the analysis of lunar rocks (*Lee et al.,* 1997; *Snyder et al.,* 1997a; *Halliday and Lee,* 1999; *Lee et al.,* 2000).

We present an overview of the chronologic and radiogenic isotopic work that has been performed on lunar rocks over the last 30 years, since the Apollo 11 landing and return. In this regard, the most recent review by *Nyquist and Shih* (1992), ostensibly for mare basalts but including a substantial amount of material on highland rocks, has proven of paramount importance. However, we do not wish only to review previous and ongoing work, but to provide models and their alternatives for formation of these samples. To this end, we will present a coherent model for lunar evolution that is consistent not only with the isotopic data, but with the trace-element chemistry of the rocks and their minerals. Finally, we will attempt to delineate areas where future research could prove most fruitful.

## 2.  ISOTOPIC SYSTEMATICS AND DATA PRESENTATION CONVENTIONS

Geochronometers can be split into two groups: those containing short-lived nuclides [$^{53}$Mn-$^{53}$Cr, $^{182}$Hf-$^{182}$W,

$^{146}$Sm-$^{142}$Nd; $t_{1/2}$ = 3.7 m.y., 9 m.y., and 103 m.y. respectively) and those containing long-lived nuclides ($^{147}$Sm-$^{143}$Nd: $t_{1/2}$ = 106 G.y.; $^{87}$Rb-$^{87}$Sr: $t_{1/2}$ = 49.4 G.y.; $^{176}$Lu-$^{176}$Hf: $t_{1/2}$ = 3.57 G.y.; $^{187}$Re-$^{187}$Os: $t_{1/2}$ = 42.3 G.y.; $^{232}$Th-$^{208}$Pb: $t_{1/2}$ = 14.01 G.y.; $^{235}$U-$^{207}$Pb: $t_{1/2}$ = 0.704 G.y.; $^{238}$U-$^{206}$Pb: $t_{1/2}$ = 4.47 G.y.; $^{40}$K-$^{40}$Ar ($^{40}$Ar-$^{39}$Ar): $t_{1/2}$ = 1.25 G.y.]. Short-lived nuclides are useful for constraining the timing of events early in the solar system, such as core formation ($^{182}$Hf-$^{182}$W) or an early silicate differentiation ($^{146}$Sm-$^{142}$Nd). These will be treated in more depth in the paper by *Halliday et al.* (2000) in this volume. Long-lived nuclides are used as tracers of silicate evolution in planets and satellites, and this will be the focus of our presentation.

By convention, Nd-, Hf-, and W-isotopic data are often presented in epsilon ($\varepsilon$) units, or the deviation relative to a chondritic uniform reservoir, CHUR (*DePaolo and Wasserburg,* 1976). As an example, for Nd-isotopic data

$$\varepsilon_{Nd} = [(^{143}Nd/^{144}Nd_{sample} - {}^{143}Nd/^{144}Nd_{CHUR})/ {}^{143}Nd/^{144}Nd_{CHUR}] \times 10^4$$

where present-day ($^{143}$Nd/$^{144}$Nd)$_{CHUR}$ = 0.512638. Similar equations can be written for Hf and W isotopes. Model ages (or single-stage evolution ages) have been calculated for both Nd and Sr isotopes and are given under as $T_{LUNI}$ and $T_{LUM}$ respectively where

$$T_{LUNI} = 1/\lambda \times ln[((^{87}Sr/^{86}Sr - 0.69903)/^{87}Rb/^{86}Sr) + 1]$$

and

$$T_{LUM} = 1/\lambda \times ln[(^{143}Nd/^{144}Nd - 0.516149)/ (^{147}Sm/^{144}Nd - 0.318) + 1]$$

The $T_{LUNI}$ model age is determined relative to a suggested lunar initial $^{87}$Sr/$^{86}$Sr = LUNI = 0.69903 at 4.55 Ga (*Nyquist et al.,* 1973). The value for $T_{LUM}$ yields a model age at which the sample was in equilibrium with a lunar upper mantle (LUM) with a $^{147}$Sm/$^{144}$Nd = 0.318 (*Snyder et al.,* 1994) and which has a present-day $^{143}$Nd/$^{144}$Nd = 0.516149. This model LUM is considered to be the *most*-depleted upper mantle and thus yields maximum ages for separation from an adcumulate lunar upper mantle (*Snyder et al.,* 1994). This lunar adcumulate source was formed late in the crystallization of a lunar magma ocean (*Snyder et al.,* 1992b).

In the presentation below, previous $^{40}$Ar/$^{39}$Ar, Rb-Sr, and Sm-Nd age determinations for the high-Ti basalts from Apollo 11 have been recalculated using the decay constants of *Steiger and Jaeger* (1977) for all systems except Rb-Sr. For Rb-Sr we have used a $\lambda_{87Rb}$ of 1.402 × 10$^{-11}$ yr$^{-1}$ as suggested by *Minster et al.* (1982) and *Nyquist et al.* (1986). This revised decay constant for Rb is based on the convergence of ages for Rb-Sr whole-rock data and U-Th-Pb data from chondritic meteorites (*Minster et al.,* 1982).

## 3.  HISTORICAL DEVELOPMENT OF CONCEPTS OF LUNAR EVOLUTION

In order to better understand the use (and abuse) of radiogenic isotopes in the study of lunar evolution, we must first review some of the historical background and the development of thought about lunar evolution. For more in-depth discussions on these concepts, see chapters presented in this book, namely those of *Righter et al.* (2000) and *Shearer and Floss* (2000).

### 3.1. The Magma Ocean Concept

With the return of the first samples from the Moon by the Apollo 11 mission, two groups had the keen insight to suggest, based upon a few calcic plagioclase crystals, that the earliest history of the Moon involved flotation of plagioclase in a global magma ocean (*Smith et al.,* 1970; *Wood et al.,* 1970). Since that time, the notion of a Moon-wide magma ocean or magmasphere has achieved much support through the petrology and chemistry of basalts and highland rocks and geophysical modeling (e.g., *Warren,* 1985). Seismic studies of the Moon have been interpreted as evidence of an upper mantle that is ~500 km in depth (*Nakamura,* 1983; *Hood and Jones,* 1987). This depth is confirmed by petrologic and chemical mass balance constraints (*Mueller et al.,* 1988). The lunar crust is known to be 60 km thick on the nearside and up to 100 km thick on the farside. Assuming that the crust was derived from the lunar upper mantle after 75–80% crystallization of a lunar magma ocean (e.g., *Snyder et al.,* 1992b), and considering a bulk composition of the upper crust of 26–28 wt% $Al_2O_3$, lower crust of 18.8 wt% $Al_2O_3$ (*Spudis and Davis,* 1986), and a minimum upper mantle $Al_2O_3$ content that is 0.4 wt%, it would take an ~500-km-deep pool of magma to generate the lunar crust by flotation of plagioclase cumulates (*Mueller et al.,* 1988).

With cooling, this lunar magma ocean (LMO) differentiated into an upper mantle of mafic cumulates and a progressively large-ion-lithophile-enriched residual liquid. Specifically, this liquid was enriched in K, REE, and P (KREEP) and may have been trapped between the cumulus minerals in the upper part of the lunar upper mantle. The upper mantle mafic cumulates are considered the sources of the mare basalts. The residual KREEPy liquid could then be easily remobilized by later magmatic events such as the melting of the upper mantle cumulates to form basalts with a KREEPy signature, the so-called KREEP basalts. Once plagioclase became a liquidus phase in the LMO, late (last 20–30 vol% of LMO crystallization; *Snyder et al.,* 1992b; *Snyder and Taylor,* 1993) in its crystallization history, it likely floated and entrained a small proportion of Fe-enriched mafic material to form an early anorthositic crust. Thus, rocks that are thought to be remnants of this early crust have been termed ferroan anorthosites or FANs. Several of these FANs have yielded ages that attest to the antiquity of the samples (see below).

## 4. EARLIEST LUNAR DIFFERENTIATION AND FERROAN ANORTHOSITES

Much of the extant work on FANs has assumed their derivation as flotation cumulates from a global LMO. Previous workers (*Papike et al.,* 1997; *Floss et al.,* 1998; and *Snyder et al.,* 1998) have shown that FANs from the Apollo 15 and 16 landing sites were precipitated from liquids whose REE compositions are similar to residual liquids generated after 70–95% crystallization of a model LMO (as per *Snyder et al.,* 1992b; *Snyder and Taylor,* 1993). However, the heavy rare-earth-elemenet (HREE) pattern of the calculated liquids does not match the modeled LMO liquids, although this may be attributed to HREE redistribution in the samples (*Papike et al.,* 1996; *Trieman,* 1996) or flaws in the LMO model. *Floss et al.* (1998) and *Shearer and Floss* (2000) have indicated complexities in a simple LMO crystallization model. They have shown that mafic minerals in FANs may have reequilibrated with trapped intercumulus liquid and that some FANs may have been autometasomatized by KREEPy liquids. Furthermore, major-element compositions for FAN plagioclase are not consistent with the histories recorded by their trace-element compositions (*Floss et al.,* 1998; *Snyder et al.,* 1998).

Further complications are provided by recent Sm-Nd isotopic data of Apollo 16 FANs. Three samples yield ages of 4.56 ± 0.07 Ga (67016), 4.44 ± 0.02 Ga (60025), and 4.29 ± 0.03 Ga (62236) respectively (*Alibert et al.,* 1994; *Carlson and Lugmair,* 1988; *Borg et al.,* 1997, 1999). All these samples point to an old large ion lithophile element (LILE)-depleted (i.e., initial $\varepsilon_{Nd}$ is positive: 60025 = +0.9 ± 0.6; 67016 = +0.9 ± 0.2; 62236 = +3.1 ± 0.9; all values calculated assuming present-day CHUR of $^{143}Nd/^{144}Nd$ = 0.512638, $^{147}Sm/^{144}Nd$ = 0.1967) source for FANs. Such an early depletion event is consistent with a signficant positive $^{142}Nd$ anomaly observed in 62236 (*Borg et al.,* 1999). If FANs were indeed derived from residual LMO liquids (which are increasingly LILE-enriched with fractionation), then they should yield signatures of this enrichment (i.e., their initial $\varepsilon_{Nd}$ values should be negative). Instead, the three FANs indicate a source that is progressively more radiogenic, relative to a chondritic bulk composition, over time. This is suggestive of a source that was LILE-depleted early in lunar evolution.

### 4.1. Formation of the Early Lunar Crust

Two of the most significant questions associated with the early Moon are what its relationship is to the proto-Earth and what is the nature of the initial differentiation. The current view is that the Moon formed as a result of a giant impact on the primordial Earth and that the earliest lunar crust was produced soon thereafter through flotation of plagioclase cumulates in the lunar magma ocean. In the magma ocean model the modern lunar crust is composed of early crystallized plagioclase cumulates, i.e., FANs that are intruded by ancient, but generally younger rocks of the highland Mg-suite (HMS). The isotopic systematics of the FANs have been difficult to unravel because these samples have undergone extensive impact metamorphism so that their Ar-Ar and Rb-Sr ages have been reset or disturbed and yield ambiguous results. Although the Sm-Nd system tends to yield less-disturbed ages, low abundances of REE and the nearly monomineralic nature of FANs has made them extremely difficult to date. In fact, only three FANs have been dated by this method and yield ages of 4.53 to 4.29 Ga (Fig. 1; *Carlson and Lugmair,* 1988; *Alibert et al.,* 1994;

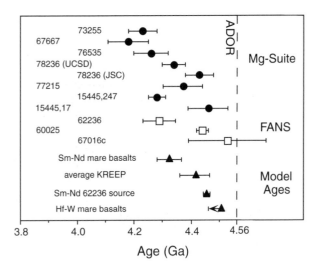

**Fig. 1.** Compilation of Sm-Nd isochron ages of the lunar crust and Sm-Nd and Hf-W model ages of lunar source differentiation. The Sm-Nd isochron dating technique is probably the least susceptible to resetting by impact metamorphism, and so best represents the ages of lunar crustal rocks. Data from *Lugmair et al.* (1976), *Nakamura et al.* (1976), *Carlson and Lugmair* (1981b), *Nyquist et al.* (1981b), *Carlson and Lugmair* (1988), and *Shih et al.* (1993a). The age of 67016c was recalculated from the data of *Alibert et al.* (1994). Model ages are from *Lee et al.* (1997), *Borg et al.* (1999), *Nyquist et al.* (1995), and *Nyquist and Shih* (1992). This figure shows that the ages of the FANs and Mg-suite rocks overlap, suggesting that by about 4.4 Ga both types of magmas were being produced on the Moon. Note that 62236 is younger than the preferred model age for KREEP from *Nyquist and Shih* (1992). Even the oldest ages of the FANs and Mg-suite rocks are within error of one another, requiring the magma ocean to exist for only a short period of time.

closure of the Sm-Nd isotopic system was prolonged past the time of crystallization as a result of slow cooling in the warm lunar crust.

The existence of some HMS rocks (such as 15445,17) that are significantly older than some FANs (such as 62236) cannot be explained in this fashion, however. Instead, these ages seem to require that both FAN and HMS magmas coexisted in the lunar mantle and crust at about 4.3–4.4 Ga. In addition, the FAN 60025 has an age of 4.44 ± 0.02 Ga (*Carlson and Lugmair*, 1988), which is indistinguishable from the average KREEP model age of 4.42 ± 0.07 Ga (*Nyquist and Shih*, 1992). This KREEP model age may be reflecting slow cooling of primordial material, remobilization of a KREEPy material formed in a short-lived LMO, or our preferred interpretation, final crystallization of the LMO. These ages are also consistent with hypothesized Sm-Nd source formation closure at 4.46 ± 0.17 Ga, based upon a rough high-Ti basalt whole-rock "isochron" (*Snyder et al.*, 1994). Hafnium-182–tungsten-182 isotopic systematics suggest an even shorter-lived silicate differentiation (and consequent isotopic closure) event of 50–60 Ma (*Lee et al.*, 1997). All this evidence indicates that the LMO existed for a relatively short period of time and solidified (at least locally) prior to the crystallization of some FANs. Thus, 60025 and especially 62236 may not be flotation cumulates of the LMO and may have formed by another process.

The second aspect of the FAN Sm-Nd isotopic data that seems inconsistent with the magma ocean model is the presence of positive initial $\varepsilon_{Nd}^{143}$ values in all analyzed FANs (Fig. 2; *Borg et al.*, 1999). Small positive $\varepsilon_{Nd}^{142}$ anomalies have also been measured on two fractions of the FAN 62236 (*Borg et al.*, 1999). Excess $^{142}$Nd (and $^{182}$W) is expressed

*Borg et al.*, 1999). In addition, all three samples have positive initial $\varepsilon_{Nd}^{143}$ values, indicative of derivation from a LREE-depleted source (Fig. 2). The young ages and positive initial $\varepsilon_{Nd}^{143}$ values are not predicted by the LMO model and may reflect disturbance of the isochrons by impact metamorphism. However, no plausible mechanism has been proposed for this disturbance, suggesting that the FAN Sm-Nd results should be taken at face value for now (*Borg et al.*, 1999).

There are two aspects of the FAN data that are not consistent with the lunar magma ocean model. First, from a comparison of Sm-Nd ages of lunar crustal rocks (Fig. 1), it is apparent that there is significant overlap between the ages of FANs and HMS. Contemporaneous crystallization of FAN and HMS magmas is not predicted by the standard LMO model, which requires FAN formation to precede HMS magmatism. It is important to note that this overlap includes relatively old HMS rocks as well as relatively young FANs. Several explanations can account for apparent overlap between closely spaced crystallization events. One explanation is that radiometric dating techniques are not precise enough to distinguish the events. Another is that

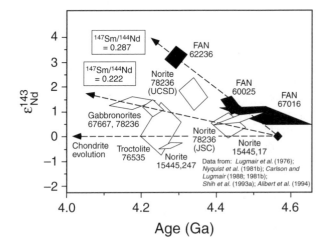

**Fig. 2.** T-$\varepsilon_{Nd}^{143}$ diagram illustrating the evolution of the magma sources of 62236. T-$\varepsilon_{Nd}^{143}$ diagram of FANs and Mg-suite rocks (sources of data same as Fig. 1). Note that all FANs have positive initial $\varepsilon_{Nd}^{143}$ values, consistent with derivation from light-REE-depleted sources. The $^{147}$Sm/$^{144}$Nd ratio estimated for the source of 62236, assuming the source formed at 4.56 Ga, is 0.287. This value is within error of the value calculated for 60025 and 67016c.

in ε units, which represent deviation from chrondritic values in parts per $10^4$. These excesses are produced by the decay of $^{146}Sm$ and $^{182}Hf$ early in the history of the solar system (prior to the extinction of the parent nuclides) in differentiated reservoirs with nonchondritic Sm/Nd and Hf/W ratios. The whole-rock 62236 fraction has a $\varepsilon_{Nd}^{142}$ value of 0.25 ± 0.12. Although this anomaly is not large, the whole-rock fraction that was measured has a $^{147}Sm/^{144}Nd$ ratio that is substantially below the CHUR value and should yield a negative $\varepsilon_{Nd}^{142}$ anomaly of about −0.1 if 62236 was derived from a chondritic source. As a result, 62236 has a positive $\varepsilon_{Nd}^{142}$ anomaly that is significantly elevated above the expected value, but is consistent with its positive initial $\varepsilon_{Nd}^{143}$ value.

These positive initial $\varepsilon_{Nd}^{143}$ values measured on 60025, 62236, and 67016c suggest that they are derived from sources that were depleted in LREE. Single-stage models presented in Fig. 2 demonstrate that all three of these FANs could be derived from a single source with a $^{147}Sm/^{144}Nd$ ratio of ~0.29. Such a strongly depleted source is typical of some of the most depleted mare basalt sources (e.g., *Snyder et al., 1992b, 1994; Misawa et al., 1993*). It is difficult to envision a mechanism to generate flotation cumulates with such isotopic systematics in the LMO. This stems from the fact that plagioclase-rich crust should form from a LREE-enriched magma produced after much of the magma ocean had crystallized, and as a result should have chrondritic or slightly below chondritic $\varepsilon_{Nd}^{143}$ values.

If the FAN Sm-Nd data has not been disturbed by some, as yet undetermined, metamorphic event, then it suggests that we may have not yet dated flotation cumulates from the magma ocean. This suggestion does not invalidate the concept that many, perhaps most, FANs form as flotation cumulates in a magma ocean, but instead suggests that 60025, 62236, and 67016c may have been derived by another process. FAN 60025 is the only one of the three that approximates in trace-element composition what would be expected of most pristine FANs analyzed from the Moon, whereas the other two samples have distinct trace-element compositions. It may also be that our sample requirements for Sm-Nd analysis of FANs, i.e., that they contain a significant (>10 modal%) mafic component, have biased our interpretations of FANs. The vast majority of FANs in the Apollo collections contain vanishingly small proportions (<1 modal%) of this requisite mafic component.

## 4.2. Temporal Constraints on Initial Silicate Differentiation

Isotopic analysis of lunar and terrestrial samples place constraints on the timing of lunar differentiation processes. However, the inability to sample ancient pristine terrestrial and lunar rocks, combined with relatively large errors in most long-lived radionuclide chronometers compared to the time resolution required to discriminate between closely associated events, has led to the use of short-lived radionuclide chronometers to constrain the earliest events in the

Earth-Moon system. Short-lived chronometers, such as $^{53}Mn$-$^{53}Cr$ ($t_{1/2}$ = 3.7 m.y.), $^{182}Hf$-$^{182}W$ ($t_{1/2}$ = 9 m.y.), and $^{146}Sm$-$^{142}Nd$ ($t_{1/2}$ = 103 m.y.), only yield relative ages. Therefore, all ages derived from these chronometers are discussed here relative to the Pb-Pb ages of angrites of 4.558 Ga (*Lugmair and Galer,* 1992). Another feature of these short-lived chronometers is that they are generally used to date the time of parent-daughter fractionation in the source region and not crystallization, and are in fact dependent on estimates of the degree of parent-daughter fractionation. The Sm-Nd system is particularly powerful in this regard because the long-lived $^{147}Sm$-$^{143}Nd$ chronometer ($t_{1/2}$ = 106 G.y.) provides an independent estimate of the degree of fractionation so that highly constrained Sm-Nd model ages result when both long- and short-lived Sm-Nd chronometers are used in conjunction.

Positive $\varepsilon_{Nd}^{142}$ values have been measured in one FAN by *Borg et al.* (1999) and in several mare basalts by *Nyquist et al.* (1995). In addition, positive $\varepsilon_{W}^{182}$ values have been observed in several mare basalts, KREEP basalts, Apollo 17 orange glasses, and FANs by *Lee et al.* (1997). However, whereas the Hf-W isotopic system may be influenced by terrestrial and lunar core formation, the composition and size of the giant impactor, accretion of a siderophile-element-rich late veneer, as well as lunar differentiation processes (*Lee et al.,* 1997; *Shearer and Newsom,* 1998), the Sm-Nd system is expected to reflect anomalies inherited from lunar mantle sources with variable Sm/Nd ratios produced during early global differentiation. Thus $^{142}Nd$ may provide the clearest constraints on the timing of crust-mantle differentiation processes for the Moon.

The excess $^{142}Nd$ observed in 62236 can be used in conjunction with the initial $^{143}Nd/^{144}Nd$ ratio and crystallization age of 62236 to constrain the time of source formation. *Borg et al.* (1999) modeled the $^{143}Nd/^{144}Nd$ and $^{142}Nd/^{144}Nd$ ratios of the 62236 source and the whole-rock fraction assuming a two-stage evolution. In the first stage the source has a chondritic $^{147}Sm/^{144}Nd$ ratio until it is depleted at the beginning of the second stage. The second stage ends with the extraction of the 62236 magmas from the source. This model constrains both the $^{147}Sm/^{144}Nd$ ratio of the 62236 source and the time of source formation (depletion). These two variables are dependent and can vary by only a small amount and still reproduce the Sm-Nd isotopic systematics of 62236. A source with a $^{147}Sm/^{144}Nd$ ratio of 0.29–0.31 is required to reproduce the Sm-Nd isotopic data. Significantly lower $^{147}Sm/^{144}Nd$ ratios will not satisfy the $^{147}Sm$-$^{143}Nd$ isotopic data, whereas significantly higher $^{147}Sm/^{144}Nd$ ratios are above the ratios postulated for the most depleted lunar sources (e.g., *Misawa et al.,* 1993; *Snyder et al.,* 1994). These models indicate that the 62236 source formed at 4.45 ± 0.01 Ga, placing a minimum age on lunar differentiation.

The Sm-Nd isotopic system has also been used to constrain the age of formation of the lunar mare source region. $\varepsilon_{Nd}^{142}$ measurements on mare basalts, KREEP basalts, and the lunar meteorite Asuka 881757 range from +0.29 ± 0.11 to

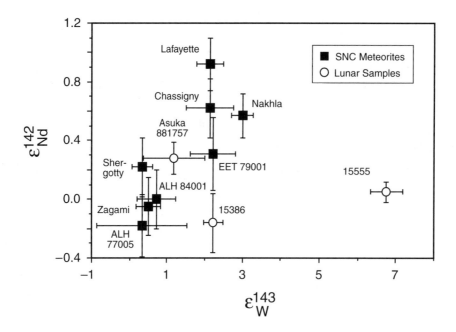

**Fig. 3.** Comparison of $\varepsilon_{Nd}^{142}$ and $\varepsilon_{W}^{182}$ values in lunar samples and SNC meteorites. Tungsten data are from *Halliday et al.* (1996); Nd data from *Harper et al.* (1995) and *Nyquist et al.* (1995); ?? (ALH). Clear correlation between $\varepsilon_{Nd}^{142}$ and $\varepsilon_{W}^{182}$ values in SNC meteorites suggest that core, mantle, and crust differentiation occurred essentially contemporaneously. Decoupling of the $\varepsilon_{Nd}^{142}$ and $\varepsilon_{W}^{182}$ values in lunar samples is indicative of a more complicated differentiation process on the Moon.

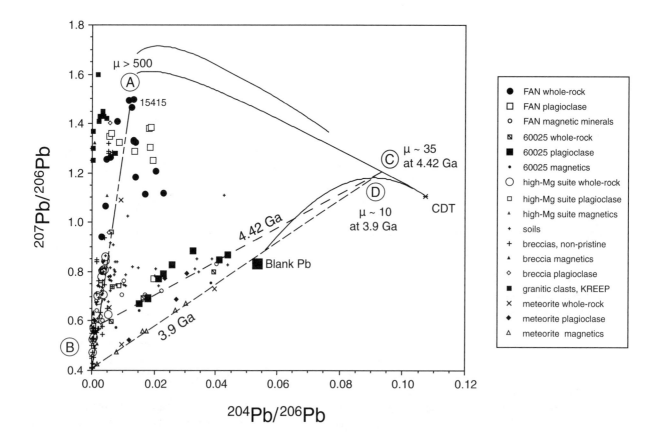

**Fig. 4.** Tera-Wasserburg Pb-Pb plot of all lunar highland Pb data. The data show essentially three major, isotopically distinct reservoirs (see text). From *Premo et al.* (1999).

$-0.16 \pm 0.20$ (*Nyquist et al.*, 1995). The observed $\varepsilon_{Nd}^{142}$ values correlate with $^{147}Sm/^{144}Nd$ ratios of the basalt source regions estimated using their initial $^{143}Nd/^{144}Nd$ ratios and crystallization ages. This correlation suggests that the source regions were depleted in LREE at approximately the same time and can be used to estimate their average formation interval. The slope of this isochron corresponds to an age of $238_{-40}^{+58}$ Ma, indicating that various mare source regions formed about 4.32 Ga (*Nyquist et al.*, 1995). This mare basalt formation age falls within the lower limits suggested by the average KREEP basalt model ages (>4.35 Ga) (*Nyquist and Shih*, 1992) and the high-Ti basalt whole-rock isochron (>4.29 Ga) (*Snyder et al.*, 1994).

The Sm-Nd age of 4.32 Ga for mare source formation is inconsistent with the presence of large $\varepsilon_W^{182}$ anomalies observed in several mare basalts, which seem to require differentiation of mare sources to have occurred prior to 4.46 Ga (*Lee et al.*, 1997; *Shearer and Newsom*, 1998; Fig. 1). Furthermore, there appears to be a lack of correlation between $\varepsilon_{Nd}^{142}$ and $\varepsilon_W^{182}$ in three lunar basalts analyzed for both isotopic systems (Fig. 3; *Nyquist et al.*, 1995; *Lee et al.*, 1997). This observation contrasts with martian meteorite samples, which show a clear correlation between $\varepsilon_W^{182}$ and $\varepsilon_{Nd}^{142}$ (Fig. 3), reflecting nearly contemporaneous segregation of the martian crust, mantle, and core (*Lee and Halliday*, 1997). If the lack of correlation between $\varepsilon_W^{182}$ and $\varepsilon_{Nd}^{142}$ in lunar samples is born out with additional analyses, it suggests that Sm/Nd and Hf/W ratios were fractionated at different times; i.e., core, mantle, and crust formation were not contemporaneous on the Moon.

Although the interpretation of U-Th-Pb isotopes in lunar samples has proven complex, due to the extreme volatility and mobility of Pb, experience has allowed useful information to be gleaned (e.g., *Premo and Tatsumoto*, 1992; *Premo et al.*, 1999). The Pb-isotopic systematics of over 90 samples of highland rocks have led to the delineation of three isotopically distinct signatures (*Premo et al.*, 1999). Over 20 years ago, a very distinctive Pb-isotopic signature was determined for many FANs (15415, 62237), most primitive HMS rocks (e.g., 76535, 78235), all granitic clasts, and many soils and breccias (*Tatsumoto et al.*, 1972; *Tera et al.*, 1972). Most lunar rocks fall on or near an array in Pb-isotopic space (Tera-Wasserburg Pb-Pb plot), which is most easily explained by mixing of two distinct Pb reservoirs, primary radiogenic Pb ($^{207}Pb/^{204}Pb = 0.4$) and radiogenic initial Pb with very high $^{207}Pb/^{206}Pb$ values (~1.45) (Fig. 4). These elevated $^{207}Pb/^{206}Pb$ values require very high $\mu$ ($^{238}U/^{204}Pb$) values (>500) in the source (*Premo et al.*, 1999). This source is thought to be related to the U- and Th-rich LMO residuum, urKREEP. The oldest Pb-isotopic signature is derived solely from FAN 60025. Plagioclase separates from 60025 (*Premo and Tatsumoto*, 1993, 2000), along with a few other samples, yield a $^{207}Pb/^{206}Pb$ age of $4.421 \pm 0.074$ (*Premo et al.*, 1999). The source for this rock is characterized by source $\mu$ values of 35–100 at the suggested formation age of 4.42–4.44 Ga (*Carlson and Lugmair*, 1988; *Premo et al.*, 1999). Due to the biased sampling of the Apollo and Luna missions, these very high (compared

to the Earth) $\mu$ values may not be representative of the bulk Moon and may be particular to the lunar nearside. The third Pb-isotopic signature is found only in lunar meteorites (*Misawa et al.*, 1993; *Torigoye-Kita et al.*, 1995) and green and orange glasses (*Tatsumoto et al.*, 1973, 1987; *Tera and Wasserburg*, 1976). The lunar meteorites are thought to have been launched from farside lunar crust. Their presumed source $\mu$ values (10–50) are only slightly elevated over terrestrial source $\mu$ values (*Premo et al.*, 1999). Thus, it appears that the primordial $\mu$ value for the Moon may not have been nearly as elevated relative to that of the Earth as originally thought. Lead-isotopic systematics point to two major, distinct reservoirs with suggestion of a minor third reservoir. The two prominent reservoirs are represented by the urKREEP residuum from a LMO with very high $\mu$ (>500), and a low-$\mu$ Earth-like mantle. This high-$\mu$ component is pervasive in lunar rocks from all landing sites.

## 5. MODIFICATION OF THE EARLY LUNAR CRUST

The evolution of the lunar highland crust, as determined by returned samples from the Moon, is commonly thought to have involved two distinct stages. The first stage was the formation of anorthositic (FANs) to leuconoritic crust as flotation cumulates from an incipient lunar magma ocean ~4.4–4.6 Ga ago. The second stage involved the modification of this early crust from 4.4 to 3.9 Ga through the crystallization of mafic and ultramafic plutonic rocks (and occasional, minor felsic rocks); these are the so-called highlands magnesian suite (HMS) and highlands alkali suite (HAS) rocks. The petrogenesis of these later mafic and ultramafic plutonic rocks has been the subject of intense debate. However, most of this debate has centered around trace-element analyses of cumulus orthopyroxene and plagioclase (*Steele et al.*, 1980; *Papike et al.*, 1994, 1996; *Shervais*, 1994; *Shervais and Stuart*, 1995; *Papike*, 1996; *Shervais and McGee*, 1998b; *Snyder et al.*, 1994; *Shervais and McGee*, 1997, 1998a) in an attempt to determine plausible parental liquids for these minerals. Very little isotopic work has been performed since the last major review of these rocks was published in 1992 (*Nyquist and Shih*, 1992). We will summarize that work below.

By definition, HAS rocks are characterized by elevated LILE (La = 30–1000×, Ba = 70–400×, and Rb = 2–10× C1 chondrites, compared to 1–10×, 2–10×, and 0.001–0.1× C1 chondrites for FANs respectively) along with evolved mineral chemistry including plagioclase and pyroxene compositions of $An_{86-76}$ and $En_{70-40}$ respectively. These HAS characteristics are also in contrast to HMS rocks, which may have similar LILE contents, but relatively Ca-rich plagioclase ($An_{97-90}$) and Mg-rich pyroxene ($En_{90-65}$).

### 5.1. Highlands Magnesian Suite (HMS)

The ages and isotopic characteristics of HMS rocks strongly suggest the interaction of a KREEP component in their evolution. *Nyquist and Shih* (1992) compiled age in-

TABLE 1.   Lunar highlands magnesian suite (HMS) chronology (in Ma).

| | $^{40}Ar/^{39}Ar$ | Rb-Sr | Sm-Nd | U-Pb Zircon | U-Pb | References |
|---|---|---|---|---|---|---|
| *Norites* | | | | | | |
| 14305,91 | | | | 4211 ± 5 | | [1] |
| 15445,17 | | | 4460 ± 70 | | | [2] |
| 15445,247 | | | 4280 ± 30 | | | [2] |
| 15455,228 | 3830 | 4590 ± 130 | 4500 ± 300 | | | [2] |
| 72255 | 3940 ± 30 | 4170 ± 50 | | | | [3,4] |
| 73215,46,25 | 4190 ± 10 | | | | | [5] |
| 77215 | 3980 ± 30 | 4420 ± 40 | 4370 ± 70 | | | [6,7] |
| 78155 | 4170 ± 30 | | | | 4170 ± 20 | [8] |
| | 4150 ± 40 | | | | | [9] |
| 78235 | | | | | 4426 ± 65 | [10] |
| 78236 | 4390 | 4380 ± 20 | 4430 ± 50 | | | [11] |
| | 4110 ± 20 | | 4340 ± 40 | | | [12,13] |
| | | | | | | |
| *Gabbronorites* | | | | | | |
| 14306,60 | | | | | 4200 ± 30 | [14] |
| 67667 | | | 4180 ± 70 | | | [15] |
| 73255,27,45 | | | 4230 ± 50 | | | [13] |
| | | | | | | |
| *Troctolites* | | | | | | |
| 14306,150 | | | | | 4245 ± 75 | [1] |
| 76535 | | 4485 ± 75 | 4330 ± 64 | | 4236 ± 15 | [16] |
| | 4190 ± 20 | | | | | [17,18] |
| | | | | | | [19,20] |
| | 4160 ± 40 | | | | | [21] |
| | 4270 ± 80 | | | | | [22] |
| | | | | | | |
| *Dunite* | | | | | | |
| 72417 | | 4550 ± 100 | | | | [23] |

References:   [1] *Meyer et al.* (1989); [2] *Shih et al.* (1993a); [3] *Leich et al.* (1975); [4] *Compston et al.* (1975); [5] *Jessberger et al.* (1977); [6] *Stettler et al.* (1974); [7] *Nakamura et al.* (1976); [8] *Oberli et al.* (1979); [9] *Turner and Cadogan* (1975); [10] *Premo and Tatsumoto* (1991); [11] *Nyquist et al.* (1981b); [12] *Aeschlimann et al.* (1982); [13] *Carlson and Lugmair* (1981a); [14] *Compston et al.* (1984); [15] *Carlson and Lugmair* (1981b); [16] *Premo and Tatsumoto* (1992); [17] *Husain and Schaeffer* (1975); [18] *Papanastassiou and Wasserburg* (1976); [19] *Lugmair et al.* (1976); [20] *Hinthorne et al.* (1975); [21] *Huneke and Wasserburg* (1975); [22] *Bogard et al.* (1975); [23] *Papanastassiou and Wasserburg* (1975).

formation from the lunar HMS and found that these ages range from 4.61 ± 0.07 to 4.17 ± 0.02 Ga (Table 1). However, ages (with the exception of one Rb-Sr determination) for the most plagioclase-rich clasts (gabbronorites and troctolites) fall in a range from 4.27 to 4.16 Ga. A compilation of U-Pb zircon ages for Apollo 14 HMS rocks (14066,47: gabbronorite 4141 ± 5 Ma; 14305,91: norite 4211 ± 5 Ma; 14306,150: troctolite 4245 ± 75 Ma; 14306,60: gabbronorite 4200 ± 30 Ma; *Meyer et al.,* 1989) yields a possible range of 4136 to 4320 Ma and a weighted average age of 4191 ± 4 Ma. *Shih et al.* (1993a) determined precise Sm-Nd ages for two HMS norites (15445,17: 4.46 ± 0.07 Ga; 15445,247: 4.28 ± 0.03 Ga) and an Rb-Sr age on one other (15455,228: 4.55 ± 0.13 Ga) from the Apollo 15 landing site that indicate that at least some HMS rocks are in excess of 4.4 b.y. old. Current isotopic data indicates that HMS rocks were produced by a variety of parental magmas over

an extended period of early lunar history, although the bulk of HMS rocks are younger than the ferroan anorthosites.

Highlands magnesian suite anorthosites have remained an enigmatic and much-debated lunar rock type. They contain mineral-chemical compositions consistent with their derivation from a primitive source, have REE contents consistent with an evolved parentage, but are relatively depleted in the alkali metals (e.g., K and Rb). Rare-earth-element contents vary widely, but in all cases, the REE are at least an order of magnitude higher than those in FANs (*Lindstrom et al.,* 1984). *Shih et al.* (1992) and *Snyder et al.* (1995a,b) have shown that many highland plutonic rocks exhibit enriched Nd-isotopic signatures (i.e., $\varepsilon_{Nd}$ is negative) that obviate a derivation from depleted mantle cumulates alone (which would have exhibited positive $\varepsilon_{Nd}$), and suggest either the possible presence of some KREEP-like contaminant or derivation from a KREEP basalt (Fig. 5). Further-

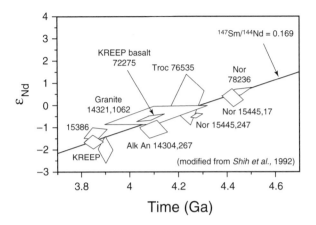

**Fig. 5.** Initial $\varepsilon_{Nd}$ values vs. ages for KREEP basalts and related highland plutonic rocks. The related plutonic rocks include a granite, a troctolite, three norites, and a noritic anorthosite. KREEP basalts and these highland plutonic rocks form a linear array corresponding to a LREE-enriched source with $^{147}Sm/^{144}Nd = 0.169$. Modified from *Shih et al.* (1992).

more, Hf-isotopic data indicate that HMS rocks lie along an array that extends from more-radiogenic FANs to the least-radiogenic KREEP basalts (*Lee et al.*, 2000).

It is considered possible that HMS magmatism was initiated at different times in different regions of the Moon. *Snyder et al.* (1995b) have suggested that HMS magmatism on the nearside occurred first in the northeast and then swept slowly to the southwest over a period of 300–400 m.y. *Snyder et al.* (1995b) speculated that this regional progression could be due to the progressive freezing of the final residual liquids of the lunar magma ocean. This final liquid layer spread beneath the surface on the lunar nearside and was deepest where the largest concentration of urKREEP has been found — at the Apollo 14 landing site. As this final liquid layer crystallized, it did so from northeast to southwest, and culminated in the production of the most evolved KREEPy liquids beneath the Apollo 14 landing site. Radiogenic heat from the decay of isotopes of U, Th, and K then began building and could have expedited melting earliest in the northeast and progressively to the southwest on the lunar nearside. These KREEPy trapped liquids were then "melted out" by ascending primitive melts from the deep lunar interior that subsequently fractionated and precipitated cumulates, first of the HMS and then of the highlands alkali suite (HAS).

## 5.2. Highlands Alkali Suite (HAS)

More evolved lunar highland rocks have been known to exist on the Moon since the earliest days of the Apollo missions (e.g., *Hubbard et al.*, 1971; *Brown et al.*, 1972). However, it was not until later that the importance of such alkali-rich highland rocks for understanding lunar petrogenesis was placed into proper context (*Warren and Wasson*,

1980). Since that time, there has been a concerted effort to find more of these rocks through labor-intensive breccia pull-apart studies of clasts (e.g., *Warren and Wasson*, 1980; *Hunter and Taylor*, 1983) and analyses of coarse-fine fractions of soils (e.g., *Jolliff et al.*, 1991; *Snyder et al.*, 1992a). It has been recognized that these clasts are not related to mare basalts or FANs (*Warren and Wasson*, 1980; *Warren et al.*, 1981). Instead, *Warren et al.* (1981) concluded that the HAS rocks were either evolved differentiates of the HMS or represent a separate, unique group. *Hunter and Taylor* (1983) also suggested, based on plagioclase and mafic mineral compositions, that these rocks represent cumulates that crystallized from evolved magmas, possibly related to the HMS. These evolved magmas could represent either the "dregs" of the early lunar magma ocean, or the crystallization products of alkalic magmatism that occurred later in the Moon's history.

In a study of zircons from various lunar rock types, *Meyer et al.* (1989) achieved precise ages for two Apollo 14 rocks of the HAS: Anorthosite 14321,16 and gabbronorite 14066,47 yielded ages of 4028 ± 6 Ma and 4141 ± 5 Ma respectively. These two ages bracket a Sm-Nd age obtained for noritic anorthosite 14304,267 (4108 ± 53 Ma) (*Snyder et al.*, 1995a) and suggest ~100 m.y. of alkalic magmatism on the western limb of the Moon (Table 2). Therefore, the parental liquid(s) of the HAS must be a later melt of an evolved pluton possibly generated from the crystallized products of a residual KREEPy magma ocean. Evidence that these rocks are indeed melts of a KREEP protolith (original LMO residuum, or urKREEP) is found in the initial $^{143}Nd/^{144}Nd$ of the three HAS anorthosites analyzed by *Snyder et al.* (1995a). The initial $\varepsilon_{Nd}$ of the mineral isochron for noritic anorthosite 14304,267 is –1.0, indistinguishable from KREEPy rocks at this time (*Shih et al.*, 1992). In this respect, it is interesting to note that a single zircon age on an alkali gabbronorite from the Apollo 16 landing site (eastern limb of the Moon) yielded an age of 4339 ± 5 Ma, pointing to the chronologically continuous and Moonwide nature of alkalic magmatism. However, HAS rocks are rare and volumetrically minor on the eastern nearside (based on limited sampling).

*5.2.1. Quartz monzodiorites (QMDs).* Other, even more evolved, HAS plutonic rocks include granites, rhyolites, felsites, and quartz monzodiorites. Quartz monzodiorites (QMDs) are the least evolved mineralogically and chemically and appear to be intermediate between granitic rocks and alkali- and ferro-gabbros (*Marvin et al.*, 1991). As such, QMDs have been implicated in the petrogenesis of HAS cumulates (*Warren et al.*, 1990; *Snyder et al.*, 1992a). The designation of a 1-cm clast from breccia 15405 as a "quartz monzodiorite" by *Ryder* (1976), and the discovery of other such clasts at the Apollo 15 landing site (*Lindstrom et al.*, 1988; *Martinez and Ryder*, 1989), led early workers to conclude that QMD was only of regional importance. However, discovery of other QMD clasts from the Apollo 14 collection (*Jolliff*, 1991) allow for the widespread existence of QMDs in the Moon.

TABLE 2.    Lunar highland alkali suite (HAS) chronology (in Ma).

| | $^{40}$Ar/$^{39}$Ar | K-Ca | Rb-Sr | Sm-Nd | U-Pb Zircon | U-Pb | References |
|---|---|---|---|---|---|---|---|
| *Anorthosite* | | | | | | | |
| 14066,47 | | | | | 4141 ± 5 | | [1] |
| 14304,267 | | | 4336 ± 81 | 4108 ± 53 | | | [2] |
| 14321,16 | | | | | 4028 ± 6 | | [1] |
| | | | | | | | |
| *Gabbronorite* | | | | | | | |
| 67975,131 | | | | | 4339 ± 5 | | [1] |
| | | | | | | | |
| *Quartz Monzodiorite* | | | | | | | |
| 15405,57 | | | | | 4370 ± 30 | | [3] |
| 15405,145 | | | | | 4309 +120/–85 | | [3] |
| 15405,88 | | | | | | 4000 ± 100 | [4] |
| | | | | | | | |
| *Granites* | | | | | | | |
| 12013,141 | 4010 ± 70 | 3760 ± 720 | 3990 ± 50 | | | | [5,6] |
| | | | 4010 ± 90 | | | | [6] |
| 12033,507 | 796 ± 10 | 3620 ± 110 | 2200 ± 650 | | 3883 ± 3 | | [3,7,8] |
| 12034,106 | | | | | 3916 ± 17 – 3986 ± 18 | | [3] |
| 14082,49 | | | | | 4216 ± 7 | | [3] |
| 14303,204 | 3830 ± 30 | 4040 ± 640 | 3950 ± 380 | | 4325 ± 12 | | [3,7,8] |
| | | | | | 4308 ± 3 | | [3] |
| 14311,90 | | | | | 4250 ± 2 | | [3] |
| 14321 B1 | | | | | 4010 ± 2 | | [3] |
| 14321,1027 | 3880 ± 30 | 4060 ± 70 | 4090 ± 110 | 4110 ± 200 | 3965 +20/–30 | | [3,8,9] |
| 73215,43 | 3870 ± 10 | | 3900 ± 50 | | | | [10,11] |
| 73217 | | | | | 4360 ± 20 | | [12] |
| 73235,60A | | | | | 4218 ± 4 | | [3] |
| 73235,63 | | | | | 4320 ± 2 | | [3] |
| 73235,73 | | | | | ≥4156 ± 3 | | [3] |
| 73255,27,3 | 3890 ± 30 | | | | | | [13] |

References:    [1] *Meyer et al.* (1989); [2] *Snyder et al.* (1995a); [3] *Meyer et al.* (1996); [4] *Tatsumoto and Unruh* (1976); [5] *Turner* (1971); [6] *Lunatic Asylum* (1970); [7] *Bogard et al.* (1994); [8] *Shih et al.* (1993b); [9] *Shih et al.* (1985); [10] *Jessberger et al.* (1977); [11] *Compston et al.* (1977); [12] *Compston et al.* (1984); [13] *Staudacher et al.* (1979).

*Meyer et al.* (1996) analyzed zircons in two QMD clasts from lunar breccia 15405 (Table 2). The zircons in these QMDs yielded a composite upper-intercept concordia age of 4294 ± 26 and a lower intercept of 1320 ± 280 (*Meyer et al.,* 1996). The significance of the lower intercept age is confirmed by evidence of a single large thermal (impact?) event that disturbed the $^{40}$Ar-$^{39}$Ar data at 1290 ± 40 Ma (*Bernatowitz et al.,* 1978). Isotope dilution U-Pb analyses of whole-rocks and mineral separates from a similar clast in 15405 yielded a rather imprecise age of 4.0 ± 0.1 Ga (*Tatsumoto and Unruh,* 1976).

*5.2.2. Lunar granites.*    Granitic samples are thought to be produced on the Moon by silicate liquid immiscibility (e.g., *Neal and Taylor,* 1989, and references therein) of a late-stage KREEPy (K, REE, and P-rich) liquid. Uranium-lead geochronology of lunar zircons has provided a backdrop for further understanding of granite genesis. A total of 18 samples of zircons have been analyzed for U-Pb isotopes, mostly taken directly from granitic ("granophyre")

materials (*Meyer et al.,* 1996). These zircons indicate a range in ages for lunar granites spanning nearly 500 m.y., from 4.36 to 3.88 Ga (Table 2). By comparing these ages with other chemical and mineralogic characteristics of the granites, *Meyer et al.* (1996) were able to distinguish between young granites and old granites. Old granites are pyroxene-bearing and have REE patterns similar to KREEP and could have formed by late-stage differentiation of a global lunar magma. Young granites have bow-shaped, concave-upward REE patterns indicative of formation by silicate liquid immiscibility (*Taylor et al.,* 1980).

Several other granitic clasts from the Moon, the largest of which is 1.8 g in mass (*Warren et al.,* 1983), have been analyzed in an attempt to determine their ages (Table 1). A precise K-Ca age (4.06 ± 0.07 Ga) was determined for the largest of these granitic clasts, 14321,1062 (,1027) (*Shih et al.,* 1993b). This age is in perfect agreement with less precise and previously determined Sm-Nd and Rb-Sr ages (4.11 ± 0.20 Ga and 4.09 ± 0.11 Ga respectively) (*Shih et al.*, 1985). However, this age is demonstrably higher than that determined by U-Pb zircon geochronology (3965 +20/–30) (Table 2; *Meyer et al.*, 1996).

Based on mineral chemistry and whole-rock major-element chemical compositions, the HAS granites are the most evolved rocks on the Moon. However, the QMDs exhibit the highest REE abundances of any lunar samples. This apparent paradox has been explained in two different scenarios. Both involve the fractional crystallization of a KREEP basalt liquid that eventually precipitates evolved cumulates such as the QMDs (*Ryder,* 1976; *Shih,* 1977; *Hess,* 1989; *Jolliff,* 1991; *Jolliff et al.,* 1993). In the first scenario, crystallization of phosphates (specifically whitlockite) and zircon deplete the liquid in REE, Zr, and Hf. This produces cumulates depleted in these elements but enriched in K, Rb, Cs, and Ba (*Marvin et al.,* 1991). In the second scenario, the liquid continues to evolve until it becomes enriched enough in Fe and Si that it will separate into two immiscible liquids, one Si-rich and relatively REE-poor and the other Fe-, Ti-, V-, and REE-rich (*Ryder,* 1976; *Taylor et al.,* 1980; *Warren et al.,* 1983; *Shih et al.,* 1985). Lunar granites are the crystallization products of this Si-rich liquid. *Marvin et al.* (1991) support the whitlockite-zircon crystallization scenario over the silicate-liquid immiscibility scenario for Apollo 15 granites. However, there is abundant evidence that certain felsites and Si-rich glasses were produced by silicate liquid immiscibility (e.g., *Hess et al.,* 1975; *Quick et al.,* 1977; *Blanchard and Budahn,* 1979; *Taylor et al.,* 1980; *Warren et al.,* 1983; *Snyder et al.,* 1993). *Jolliff* (1991) and *Jolliff et al.* (1993) have argued that whitlockite crystallization did occur prior to silicate liquid immiscibility and that both processes played important roles in the formation of lunar granite. Whitlockite crystallization thus contributes to the trace-element characteristics of lunar granite prior to its ultimate derivation via immiscibility (*Jolliff et al.,* 1993).

### 5.3. Serial KREEP Basalt Magmatism Model

Although several previous workers (e.g., *Warren,* 1988) had speculated on a relationship between KREEP and highland rocks, *Snyder et al.* (1995a,b) were the first to quantify it. They suggest that early basaltic melts of the lunar interior would have been susceptible to contamination by trapped interstitial liquid in upper mantle cumulates, so-called urKREEP. These KREEP basalts could have crystallized within the crust to form cumulate gabbros, norites, anorthosites, monzodiorites, and possibly granites, with the proportion of trapped KREEPy residual liquid determining the LILE enrichment of the rock. The earliest-formed cumulates represent the HMS, and later cumulates comprise the HAS.

Apollo 14 KREEP basalts typically have relatively high $Al_2O_3$ contents (~19 wt%), suggesting assimilation of anorthosite. This observation, combined with relatively high siderophile-element contents, indicates that Apollo 14

KREEP basalts are not pristine and only Apollo 15 KREEP basalts were used to calculate a primitive precursor magma. Even such a primitive, parental, KREEP basaltic magma has a relatively low Mg# (= $Mg^{2+}/(Mg^{2+} + Fe^{2+})$) of 64, suggesting some fractionation either during ascent or at the surface. It is speculated that these basaltic melts are the products of deep melting in the lunar interior and subsequent assimilation of primitive KREEPy residual liquids during passage of the basalt to the lunar surface. These assimilated KREEPy liquids were residual from the original LMO and could have been trapped within upper level mantle cumulates. Those basalts that did not reach the surface could have crystallized within the upper lunar crust.

*Snyder et al.* (1995a,b) modeled the progressive fractional crystallization assemblages of a primitive KREEP basalt. These were then used to model trace elements using the simple Rayleigh fractionation equation and associated cumulates were calculated using the simple, equilibrium partition-coefficient relationship. The first cumulates to crystallize from a KREEP basalt magma are enriched in Mg (for mafic minerals) and Ca (for plagioclase) and thus quite primitive. In fact, the first 55% of crystallization yields cumulates that are similar in mineralogy and chemistry to HMS rocks (*Snyder et al.,* 1995b). The next 45% of crystallization precipitates minerals with compositions similar to the HAS rocks. Furthermore, *Snyder et al.* (1995a,b) showed that much of the variation in mineral and bulk chemistry in the HMS and HAS rocks can be modeled by addition of residual KREEP basalt liquid to the crystallizing mineral assemblage. Thus, through simple fractional crystallization and addition of trapped residual liquid, the pyroxene, olivine, and plagioclase compositions of most HMS and HAS cumulates can be reproduced.

Based on similarities in mineral chemistry and modeling of trace-element ratios, the interpretation of a kinship of the HAS cumulates with KREEP basalt seems plausible. The ages of KREEP basalts from the Apollo 14 and 15 sample collections, which do not exceed 3.98 b.y. in age, are much younger than most HAS rocks (*Shih et al.,* 1992). However, *Shih et al.* (1992) determined a much older age of 4.08 ± 0.07 Ga for an Apollo 17 KREEP basalt, which led them to postulate that KREEP basalt volcanism occurred over an extended period of at least 200 m.y. This earlier age is within analytical uncertainty of QMDs and several other HAS rocks. *Shih et al.* (1992) also observed that "crystallization ages and $\varepsilon_{Nd}$ values for KREEP basalts, granites, troctolites, and norites form a linear array, suggesting that these rocks could be genetically related" (Fig. 5), although a scenario for this relationship was not presented. Importantly, they pointed out that although a genetic relationship is likely, these rocks are not coeval and yield crystallization ages that span ~700 m.y. of the Moon's history. This suggests that if KREEP basalt magmas were the parents for HAS cumulates, a similar process of KREEP basalt production and crystallization must have occurred in a serial fashion throughout the early development of the lunar crust. In fact, KREEP basalt magmatism and differentiation may have played a very important role in the later modification of the lunar crust.

## 6.  AGES AND ISOTOPIC SYSTEMATICS OF IMPACT EVENTS

### 6.1.  Argon-40–Argon-39 Studies and Basin Ages

Although the impact history of the Moon is covered by *Ryder et al.* (2000) and *Hartmann et al.* (2000) in this volume, we present a figure (Fig. 6) that yields the "absolute" chronology of the basins formed by several of these impacts. The chronology of impact events on the Moon has been based on telescopic observations, crater counting, and disturbed isotopic systematics. These disturbed isotope systems have included Rb-Sr and Sm-Nd for a variety of highland rocks, but, most importantly, precise $^{40}$Ar-$^{39}$Ar age spectra of impact melt rocks. The vast majority of impact melt rocks from the Moon yield $^{40}$Ar-$^{39}$Ar plateau ages that are seldom in excess of 3.85 Ga (*Ryder,* 1990). This has led to the speculation of a major "lunar cataclysm" at this time (*Tera et al.,* 1974), or, at least, to the cessation of major, intense, lunar bombardment at this time. Figure 6 presents a compilation of $^{40}$Ar-$^{39}$Ar determinations for various lunar samples, in order of decreasing age, that have been interpreted as representing the ages of specific impact craters on the Moon.

### 6.2.  Rubidium-Strontium and Samarium-Neodymium Isotopic Systematics of Impact Melt Breccias

A total of 26 impact-rock samples have been analyzed for Sr-isotopic composition (*Nyquist et al.,* 1973, 1974; *Nyquist,* 1977; *Reimold et al.,* 1985; *Deutsch and Stöffler,* 1987; *Mitchell et al.,* 1999), yet only 8 of these samples were also analyzed for Nd-isotopic composition (*Lugmair and Carlson,* 1978; *Reimold et al.,* 1985; *Mitchell et al.,* 1999). The Rb-Sr-isotopic compositions of these samples can be subdivided into four main groups based upon their bulk geochemical compositions (*Korotev,* 1994). There is a consistent increase in $^{87}$Rb/$^{86}$Sr from group 4 to group 1. These groups yield the following weighted average "ages" and initial Sr-isotopic ratios: group 1 = 4.45 ± 1.60 Ga, $(^{87}Sr/^{86}Sr)_i = 0.69910 ± 37$; group 2 = 3.93 ± 0.19 Ga, $(^{87}Sr/^{86}Sr)_i = 0.69958 ± 25$; group 3 = 4.19 ± 1.34 Ga, $(^{87}Sr/^{86}Sr)_i = 0.69910 ± 60$; group 4 = 3.89 ± 1.92 Ga, $(^{87}Sr/^{86}Sr)_i = 0.69921 ± 32$ (*Mitchell et al.,* 1999). The relatively small error in the weighted average "age" of group 2 samples is due, in large part, to the extensive studies carried out on these breccias from Apollo 16 (*Reimold and Reimold,* 1984; *Reimold et al.,* 1985; *Deutsch and Stöffler,* 1987). These studies determined distinct ages for certain samples and for different lithologies within the breccias, which are masked somewhat by averaging the samples. However, because many of the other samples are composed of various heterogeneous crystal fragments, clasts, and impact melt glasses, these breccias were likely derived from more than one source. Thus, the "ages" do not indicate a single event, but likely represent an average age of the crust. It is interesting to note that group 2 breccias yield a rather tight age of 3.93 ± 0.19 Ga, which is indistinguishable from an average $^{40}$Ar/$^{39}$Ar age for group 2 breccias (*McKinley et al.,* 1984) and corresponds to the known age of the Nectaris impact (3.92 Ga). Furthermore, the $(^{87}Sr/^{86}Sr)i$ ratios of the four breccia groups are indistinguishable at the level of the errors given.

The isotopic compositions of impact-melt breccias are readily explained by local derivation and mixing of crustal materials. Most impact-melt breccias have $\varepsilon_{Nd}$ values at 3.9 Ga that are intermediate between FAN and HMS samples (*Mitchell et al.,* 1999). This contrasts with the wide range in $(^{87}Sr/^{86}Sr)_i$ ratios for these breccias; the most elevated $(^{87}Sr/^{86}Sr)_i$ ratios of the group are only matched by similar values from the HAS. *Mitchell et al.* (1999) indicated that the isotopic and trace-element composition of all these breccias can be defined by three-component mixing of FAN (60025,26; *Nyquist et al.,* 1977), HMS troctolitic anorthosite (15445,17; *Shih et al.,* 1993a), and HAS norite (14304,272; *Snyder et al.,* 1995a). The Fe-Ni metal in these samples also implies up to 7% meteoritic component (*Koro-*

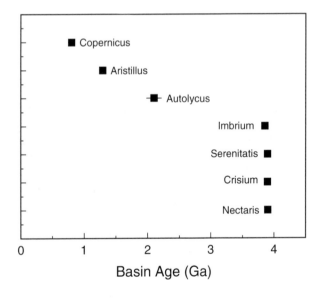

**Fig. 6.**  Histogram of age of impact basins as determined by $^{40}$Ar-$^{39}$Ar dating. Data sources are Nectaris: *Schaeffer and Husain* (1974), *Schaeffer and Schaeffer* (1977), *Maurer et al.* (1978), *Reimold et al.* (1985), *James* (1981). Crisium: *Podosek et al.* (1973), *Cadogan and Turner* (1977), *Swindle et al.* (1991). Serenitatis: *Cadogan and Turner* (1976), *Jessberger et al.* (1978), *Dalrymple and Ryder* (1996). Imbrium: *Stadermann et al.* (1991). Autolycus: *Ryder et al.* (1991). Aristillus: *Bernatowicz et al.* (1978), *Ryder et al.* (1991). Copernicus: *Eberhardt et al.* (1973), *Bogard et al.* (1994).

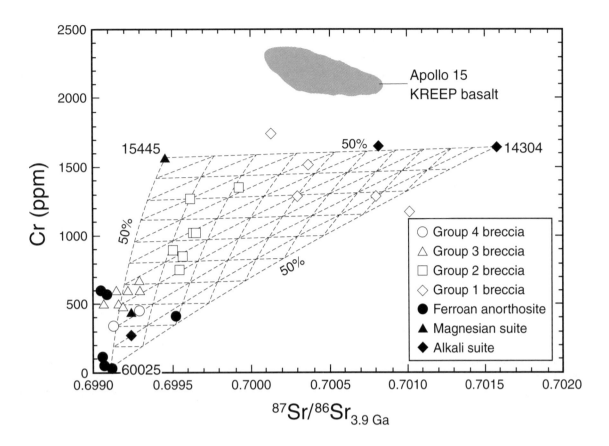

**Fig. 7.** Strontium-isotopic mixing diagram with analyses of impact-melt breccias and other lithologic units. The mixing net was calculated by mixing a Cr-poor, low-$^{87}$Sr/$^{86}$Sr$_{3.9-Ga}$ ferroan anorthosite (60025,26; *Nyquist et al.,* 1979); a Cr-rich, low-$^{87}$Sr/$^{86}$Sr$_{3.9-Ga}$ HMS troctolitic anorthosite (15445,17; *Shih et al.,* 1993a); and a Cr-rich, high-$^{87}$Sr/$^{86}$Sr$_{3.9-Ga}$ HAS norite (14304,272; *Snyder et al.,* 1995a). From *Mitchell et al.* (1999).

*tev,* 1994) (Fig. 7). The feldspathic breccias (groups 3 and 4) can be derived simply from melting and mixing of locally derived FAN (*Mitchell et al.,* 1999), consistent with trace-element and mineral-chemical modeling of components (*Korotev,* 1997). The isotopic composition of group 1 impact-melt breccias can be explained by mixing of relatively FAN-poor crustal materials: <20% FAN, 0–60% HMS, and 35–80% HAS. Group 2 impact-melt breccias are explained by mixing of 10–20% HAS, 15–55% FAN, and 30–70% HMS (Fig. 7). Groups 1 and 2 are thought to represent material transported to the Apollo 16 landing site by the Imbrium impact. Deep excavation of the Moon would lead to the uncovering of more mafic crustal material. Because the isotopic and trace-element characteristics of impact-melt breccias are indicative of melting and mixing of locally derived crustal material, they are not particularly useful in determining primary lunar differentiation and evolution, and will not be discussed further.

## 7.  MARE VOLCANISM

Although there is a growing body of evidence to the contrary (*Ringwood,* 1989; *Warren,* 1992; *Ruzicka et al.,* 1999), the currently favored model for the origin of the

Earth-Moon system involves the so-called "collision hypothesis" (*Hartmann and Davis,* 1975; *Melosh,* 1990). Thus, studies of the composition of the lunar interior would be important, not only in their own right, but also in order to achieve a better understanding of the chemistry of the Earth's early mantle. An appropriate window through which to view the lunar interior is the chemistry of basalts extruded onto the surface of the Moon.

The origin of mare basalts has been the subject of considerable study and controversy (*Neal and Taylor,* 1992, and references therein). Three principal models have been discussed. The first is the primitive source model, which proposes that mare basalts formed by varying degrees of partial melting of primitive material of bulk Moon composition (*Ringwood and Essene,* 1970; *Green et al.,* 1971). This model fails to explain many of the features (e.g., negative Eu anomalies) later realized to be ubiquitous in mare basalts (see the discussion in *Taylor,* 1982) and has long since been abandoned. The most damning criticism of such a scenario is the paucity of isotopic data suggestive of a primitive, undifferentiated source in the Moon (see discussion of picritic glasses in section 7.4). Such a source may, indeed, exist, but is not prevalent in analyzed samples returned by either the Apollo missions or lunar meteorites (see Fig. 10).

A second model suggests that the source of the mare basalt liquids is, indeed, the primitive lunar interior, but that the ascending magmas are hybridized through assimilation of the magma-ocean residuum (i.e., urKREEP as per *Warren and Wasson,* 1979; *Hubbard and Minear,* 1976; *Ringwood and Kesson,* 1976). This model predicts a positive correlation between incompatible elements in mare basalts and those in highland rocks, whereas a negative correlation is actually observed (*Taylor,* 1982). The third model places the origin of the mare basalt liquids in the mafic cumulates formed through crystallization of the lunar magma ocean. This model was proposed by *Philpotts and Schnetzler* (1970) and *Smith et al.* (1970), with further development and modification by *Taylor and Jakeš* (1974), *Walker et al.* (1975), *Drake and Consolmagno* (1976), *Shih and Schonfeld* (1976), and *Nyquist et al.* (1976). *Binder* (1982) revised the magma ocean cumulate model further by considering the effects of a convecting magma system and a trapped liquid component on the cumulate pile. This third model, albeit in various forms and consisting of differing components, successfully explains the complementary nature of the highlands and mare chemistry (*Binder,* 1982; *Snyder et al.,* 1992b; *Snyder and Taylor,* 1993). Another major advantage is that it can explain the two-stage evolution indicated by radiogenic isotopes (*Nyquist and Shih,* 1992, and references therein).

Although Rb-Sr and Sm-Nd isotopic systematics have shown that the sources for mare basalts can be quite varied (e.g., *Nyquist,* 1977; *Nyquist and Shih,* 1992; see below), Nd-Hf isotopic correlations have indicated clearly that the lunar mantle can be subdivided into two major and distinct reservoirs (Fig. 8): (1) a low-Ti mare basalt source ($\varepsilon_{Nd} \sim 4\times \varepsilon_{Hf}$) that is unlike any source recognized from the Earth, and (2) a high-Ti source ($\varepsilon_{Nd} \sim \varepsilon_{Hf}$) (*Unruh et al.,* 1984; *Beard et al.,* 1998). Furthermore, these two source arrays appear to intersect at KREEP basalt, possibly indicating mixing with the urKREEP reservoir in the Moon (Fig. 8). *Beard et al.* (1998) have shown that the differences between these two sources are a result of real, time-integrated differences in the two sources. Unlike the high-Ti mare basalt source, the low-Ti source does not contain ilmenite but subequal proportions of olivine and orthopyroxene with minor clinopyroxene. The requisite mineral chemistry for these source minerals (*Beard et al.,* 1998) leads to the interpretation that this source was created early in the evolution of the lunar magma ocean (LMO) and was thus precipitated deep in the lunar upper mantle (*Snyder et al.,* 1992b).

Several mare basalts also exhibit positive anomalies in $^{142}Nd$ (*Nyquist et al.,* 1995) generated by decay of the short-lived nuclide $^{146}Sm$ ($t_{1/2}$ = 103 m.y.). Whereas most high-Ti mare basalts indicate significantly positive $^{142}Nd$ anoma-

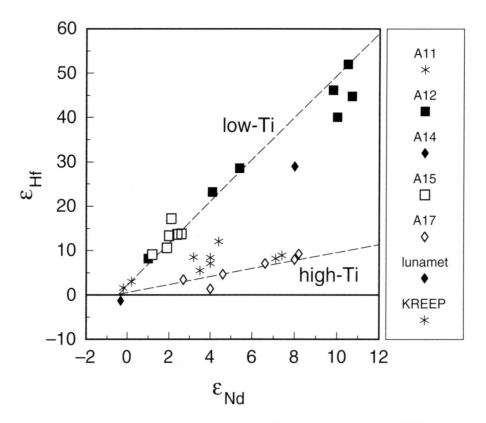

**Fig. 8.** A plot of $\varepsilon_{Nd}$ vs. $\varepsilon_{Hf}$ for lunar basalts. All samples plot into two distinct regions (low-Ti and high-Ti) with one minor subregion (high-Ti, high-K). Data sources are *Lugmair and Carlson* (1978), *Carlson and Lugmair* (1979), *Unruh et al.* (1984), *Paces et al.* (1991), *Misawa et al.* (1993), *Snyder et al.* (1994, 1997b), *Beard et al.* (1998), *Lee et al.* (2000, and unpublished data).

lies, only two low-Ti basalts, 12039 and lunar meteorite Asuka 881757, exhibit anomalies. These include the very-low-Ti (VLT) lunar meteorite Asuka 881757 ($\varepsilon_{142Nd}$ = 0.28 ± 0.11; 2σ); Apollo 17 high-Ti basalts 74255 ($\varepsilon_{142Nd}$ = 0.17 ± 0.08), 70135 ($\varepsilon_{142Nd}$ = 0.25 ± 0.15), and 75075 ($\varepsilon_{142Nd}$ = 0.29 ± 0.11); Apollo 12 ilmenite basalt 12056 ($\varepsilon_{142Nd}$ = 0.19 ± 0.20); and low-Ti basalt 12039 ($\varepsilon_{142Nd}$ = 0.25 ± 0.12). The remaining low-Ti samples have $\varepsilon_{142Nd}$ values indistinguishable from CHUR. If this difference is not due to a sampling bias, then the obvious conclusion is that high-Ti sources were either formed earlier than most low-Ti sources (in direct contrast to LMO modeling), or that low-Ti sources have been reprocessed by addition of younger chondritic material or remained open to isotopic diffusion. These slight, albeit signficantly positive $\varepsilon_{142Nd}$ values for mare basalts indicate that their sources were formed within 240 m.y. after accretion of the Moon at 4.56 Ga (*Nyquist et al.*, 1995). This corresponds to a mantle source, formed by 4.32 Ga, with a $^{147}Sm/^{144}Nd$ ratio of 0.28 (*Nyquist et al.*, 1995), similar to that determined by *Snyder et al.* (1992b) based on trace-element modeling of an initially chondritic lunar magma ocean. This 240-m.y. formation interval for the lunar mantle is consistent with Sm-Nd $T_{CHUR}$ model ages for KREEP basalts of 4.36 ± 0.06 Ga (*Lugmair and Carlson,* 1978; *Carlson and Lugmair,* 1979). Three KREEP basalts yielded an average $\varepsilon_{142Nd}$ anomaly that is slightly, but signficantly, negative ($\varepsilon_{142Nd}$ = –0.05 ± 0.04; 2σ). These slightly *negative* $\varepsilon_{142Nd}$ values for KREEP basalts indicate that these were also generated during this time interval from a reservoir with $^{147}Sm/^{144}Nd$ that was slightly less than chondritic. This reservoir corresponds to the residual, LREE-enriched (low Sm/Nd) magma ocean after crystallization of a mafic cumulate mantle (*Snyder et al.,* 1992b).

## 7.1. Earliest Mare Volcanism: Cryptomaria and High-Aluminum Basalts

The age and character of the earliest lunar basalts are important for several reasons. As products of internal melting, the basalts serve as indicators of the thermal regime within the Moon, the depth of regolith and its effects on thermal insulation of the lunar interior, the causes of melting and formation of mare magmas, and the chemical and petrologic evolution of mantle sources. Therefore, many workers have attempted to find, study, and date old basaltic rocks from both the mare and highland regions of the Moon.

The oldest mare basalts were once thought to occur at the Apollo 11 landing site, where high-Ti basalts as old as 3.88 ± 0.06 Ga were found (Sm-Nd; *Papanastassiou et al.,* 1977). *Snyder et al.* (1996) and *Shih et al.* (1999) have analyzed fragments of what are thought to be the oldest high-Ti basalts (group D), but these Sm-Nd ages essentially overlap and confirm the oldest age of *Papanastassiou and Wasserburg* (1971a). Furthermore, *Misawa et al.* (1993) determined a Sm-Nd age of 3.87 ± 0.06 Ga on the predominantly basaltic lunar meteorite Asuka 881757, which is similar in composition to low-Ti and very-low-Ti (VLT)

basalts in the Apollo collections. *Taylor et al.* (1983) found a basaltic clast (,122) in Apollo 14 breccia 14305 that yielded a Rb-Sr internal isochron age of 4.19 ± 0.05 Ga. *Dasch et al.* (1987) analyzed several high-Al basalt clasts from Apollo 14 breccia 14321 that extended apparent mare volcanism back to 4.29 ± 0.13 Ga (Rb-Sr). The isotopic systematics of the high-Al basalt clasts led *Dasch et al.* (1987) to speculate a relationship between these rocks and HMS rocks. *Neal et al.* (1988b, 1989) and *Neal and Taylor* (1990) have shown that trace-element and Sr-isotopic systematics of high-Al basalts can be explained by serial magmatism of high-Al basalts at 4.3, 4.1, and 3.95 Ga. These magmatic events were similar in that they included KREEP assimilation by a fractionating high-Al basalt magma.

Although the geochemical and petrologic evolution of high-Al basalts has been successfully modeled, the ultimate source of a parental high-Al basalt has never been satisfactorily determined. The high-Al nature has lead many to speculate that plagioclase resided in the source region of these rocks. However, *Ridley* (1975) used trace-element and mineralogic arguments to argue that plagioclase was not required and that a slight increase in silica activity in an olivine + low-Ca pyroxene + high-Ca pyroxene source would result in stronger partitioning of Al into the melt. He also stated the possibility that high-Al basalts could be residual liquids after fractionation of a more primitive mare basalt. Based upon major-element similarities, *Dickinson et al.* (1985) speculated that the source for the high-Al basalts could be similar to that for a low-Ti Apollo 12 basalt (12038): 87% olivine, 8% plagioclase, and 5% clinopyroxene. They calculated that 0.3–3% equilibrium melting of a such a source could generate the REE patterns seen in many of the high-Al basalts. However, *Neal et al.* (1988b) have since determined that all high-Al REE patterns (as well as other elements and their ratios) can be modeled through combined assimilation of KREEP and fractional crystallization of a primitive high-Al parent at the lunar surface. Thus, the ultimate source of the high-Al basalts is still in question.

These older rocks seem to fit nicely with the suggestions from clast characteristics and ages that mare emplacement began prior to 3.9 Ga (*Ryder and Taylor;* 1976; *Ryder and Spudis,* 1980). Subsequent remote sensing and photogeologic studies also indicated that mare volcanism may have predated the so-called "late terminal bombardment" of the Moon (*Hartmann et al.,* 1981). Specifically, dark halo deposits in the highlands, which are covered by basin and crater ejecta, appear to have mare affinities (*Hawke and Spudis,* 1980; *Bell and Hawke,* 1984). These dark halo deposits have been dubbed "cryptomaria" (*Head and Wilson,* 1992).

Although it is intriguing to suggest that we may have sampled some of this cryptomaria material in the form of the high-Al basalts, *Snyder et al.* (1999a) have suggested that high-Al basalts may not be true basalts. Instead, they could be impact-melt rocks, with the old ages determined for some of these rocks simply reflecting a weighted average age of the target materials (*Snyder et al.,* 1999a). The old Rb-Sr age (4.19 Ga) for the nonaluminous clast from 14305 (,122) is likely real, and could be reflecting an age of the coarse-grained portion of a mare lava flow. Thus, although mare volcanism probably occurred early in lunar crustal evolution, impact metamorphism likely proceeded to such an extent that few, if any, pristine basalts remain. Voluminous mare volcanism initiated by about 3.9 Ga was high-Ti in tenor on the lunar nearside, but may have included low-Ti and VLT volcanism on the lunar farside.

## 7.2.  High-Titanium Basalt Volcanism

It appears that the Moon is unique among planetary bodies in having a relative abundance of high-Ti (>9 wt% $TiO_2$) basalt. Among the basalts collected at the Apollo 11 and 17 landing sites, high-Ti basalts predominate. Four principal groups (A, B1-B3, B2, and D) have been delineated at the Apollo 11 landing site (*Beaty and Albee,* 1978; *Beaty et al.,* 1979) and four other groups (Types A, B1, B2, and C) have been determined from the Apollo 17 collection (*Rhodes et al.,* 1976; *Neal et al.,* 1990).

*Apollo 11.*     Group A consists of high-K basalts, enriched in incompatible trace elements. *Jerde et al.* (1994) have speculated that the group A basalts are fractionates from the Apollo 17 orange glasses, the source of which they suggested extended beneath the Apollo 11 landing site. Group B mare basalts are divided into subgroups 1, 2, and 3. Groups B1 and B3 form a continuum of compositions (*Jerde et al.,* 1994), and share many of the same textural features (*Beaty and Albee,* 1978). Group B3 basalts display the most primitive (highest Mg#, lowest incompatible-element abundances) compositions among the Apollo 11 samples. Groups B1, B2, and B3 basalts are all "low-K", but those of B2 are otherwise enriched in incompatible trace elements relative to B1 and B3 basalts. The least-represented basalts are from group D, of which only four examples have been described (*Beaty et al.,* 1979; *Snyder et al.,* 1996; *Shih et al.,* 1999). These tiny samples (two are <50 mg, the two others <200 mg) are rich in incompatible trace elements such as the REEs, comparable to the group A basalts, but are low in K like the B3 and B2 samples. Trace-element studies of the high-Ti basalts have yielded information on the projected chemistry of sources and on near-surface processes such as fractional crystallization, but cannot give information on the timing of melting and crystallization events. For such purposes, radiogenic isotope studies have proven fruitful.

Although internal mineral isochrons are most useful for dating individual rocks and flows, whole rocks may also place important constraints on isotopic closure in specific sources, diffusion of chemical species, and temperatures of the lunar interior over time. Rubidium-strontium isotopic data for all Apollo 11 high-Ti basalts (*Papanastassiou and Wasserburg,* 1971a; *Papanastassiou et al.,* 1977; *Snyder et al.,* 1994) define a line that yields an age of 3.74 ± 0.07 Ga (*Snyder et al.,* 1994). However, this line is essentially a two-point "isochron" with the two points being the high-K

(group A) and low-K (groups B1, B2, and B3) high-Ti basalts. Thus, this "isochron" is likely a mixing array between younger group A basalts (3.59 Ga) and older low-K basalts (ranging in age from 3.67 to 3.85 Ga) and has no real age significance. The low-K basalt whole rocks alone are defined by reference lines of 3.85 Ga and 4.10 Ga (*Snyder et al.,* 1994). This array converges at the y-intercept, yielding a fairly well-defined $^{87}Sr/^{86}Sr$ "initial ratio" of 0.69916 ± 8. The "age" range of this array is similar to a Rb-Sr age obtained by *Paces et al.* (1991) on a selected group of Apollo 17 basalts (4.02 ± 0.05 Ga).

A regression of all Sm-Nd whole-rock samples from Apollo 11 (*Papanastassiou et al.,* 1977; *Snyder et al.,* 1994), excluding the rare group D basalts (*Snyder et al.,* 1996; *Shih et al.,* 1999), yields a linear array that is equivalent to the age of the Moon, 4.55 ± 0.30 Ga (MSWD = 70) (*Snyder et al.,* 1994). The relatively large error in the line, as well as the large MSWD, indicate that the scatter cannot be due to analytical error alone and may indicate isotopically distinct sources for some samples. However, even if some of the scatter is due to statistical mixing of sources, the indication is that these sources were formed early in the Moon's history. A regression of all high-Ti basalts from the Moon (with the exception of type B2 basalts from Apollo 17; type B2 basalts likely were contaminated during surface processes, see *Paces et al.,* 1991) gives a line that yields an age of 4.46 ± 0.17 Ga (MSWD = 45) (*Snyder et al.,* 1994). This whole-rock isochron can be taken as evidence that the source from which high-Ti basalts were derived had a relatively homogeneous $^{143}Nd/^{144}Nd$, yet variable Sm/Nd, at 4.4 Ga. Apollo 11 group B1 basalts plot slightly above the line and indicate a relatively long-lived depleted-mantle component. The presence of a second component, which is LILE-enriched, is reflected in both the low $^{147}Sm/^{144}Nd$ values and high $^{87}Rb/^{86}Sr$ and $K_2O$ in group A samples from Apollo 11 (*Snyder et al.,* 1994).

Initial Sr isotopic ratios and $\varepsilon_{Nd}$ values for high-Ti basalts are shown in Figs. 9 and 10. Combined Sr and Nd isotopic data indicate distinct fields for the different basalt types, although group A basalts exhibit a relatively wide range in Sr initial ratios. These data alone suggest three plausible, competing interpretations: (1) four different sources for the Apollo 11 basalts, commensurate with their previously determined classification; (2) extreme heterogeneity in the lunar mantle leading to a multitude of sources [a variation on (1)]; or (3) real, albeit small, differences in the crystallization ages of samples from the same group.

*Apollo 17.*     Although these basalts are similar in terms of their major-element chemistry, trace-element chemical compositions have indicated that each basalt group (A, B1, B2, and C) requires a separate parent magma (*Shih et al.,* 1975; *Rhodes et al.,* 1976; *Neal et al.,* 1990). Early isotopic work on these basalts focused mainly on their Sr-isotopic character and was not able to resolve a difference in either ages or initial isotopic compositions between the four groups (*Nyquist et al.,* 1976). Prior to the comprehensive Nd-Sr isotopic study of *Paces et al.* (1991), the Apollo 17 high-Ti mare basalts had only been treated as part of larger studies on Moon-wide basaltic magmatism (e.g., *Nyquist et al.*, 1979; *Unruh et al.*, 1984). Neodymium and Sr isotopic compositions of the Apollo 17 high-Ti basalts are shown in Figs. 9 and 10. A single high-Ti basalt, 70017,505, has been analyzed for its $^{187}$Re-$^{187}$Os-isotopic composition (*Birck and Allegre*, 1994). This sample contains very low abundances of both Re and Os (0.67 and 3.94 ppt respectively) and a chondritic $^{187}$Os/$^{186}$Os-isotopic composition. *Birck and Allegre* (1994) suggested that these characteristics are representative of lunar basalts as well as eucrite meteorites and may be typical of environments with very low $f_{O_2}$.

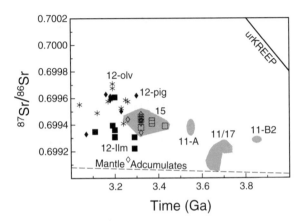

**Fig. 9.** Plot of initial $^{87}$Sr/$^{86}$Sr for mare basalts over time. Also indicated are arrays for evolution of the most depleted lunar mantle ("mantle adcumulates") and urKREEP.

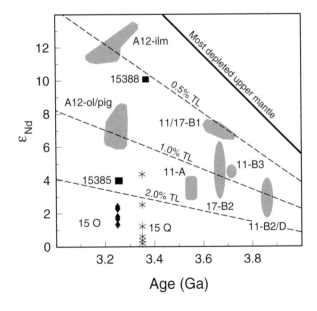

**Fig. 10.** Plot of initial $\varepsilon_{Nd}$ for mare basalts over time. Also shown in the array for the most depleted upper mantle (LUM).

*Paces et al.* (1991) were able to resolve age differences between the type A and type B1–B2 basalts from Taurus-Littrow. Four samples of type A basalt yielded a weighted average age of 3.75 ± 0.02 Ga and three samples of type B1–B2 basalt gave a weighted average age of 3.69 ± 0.02 Ga. The paucity of reliable type C Nd-Sr isotopic and geochronologic data did not allow for a clear indication of the age of this style of volcanism. However, *Paces et al.* (1991) did suggest intermediate ages (3.72 ± 0.07 and 3.73 ± 0.32 Ga) for two type C basalts from weighted averages of previously determined Rb-Sr dates (*Murthy and Coscio*, 1976; *Nyquist et al.*, 1976). Obviously, the rather large errors do not allow for confident determination of these ages. *Paces et al.* (1991) went on to speculate that high-Ti mare volcanism at the Apollo 17 landing site was relatively short-lived, lasting at least 50 m.y., but most probably much less than 200 m.y.

The initial (calculated for the age of the eruption) isotopic compositions of these basalts exhibit some interesting features. Initial $^{87}$Sr/$^{86}$Sr is quite uniform for the type A (0.699235–0.699240) and type C (0.699274–0.699301) basalts, whereas type B1–B2 basalts indicate lower ratios and a much broader range (0.699141–0.699239) in $^{87}$Sr/$^{86}$Sr. Likewise, type A, B1, and C basalts show a very restricted range in initial $\varepsilon_{Nd}$ values from 6.3 to 7.2 (see Fig. 10). Although type B2 exhibit generally lower $\varepsilon_{Nd}$ values, they also exhibit a much broader range of variation (2.7 to 6.0) (*Paces et al.*, 1991). These variable Sr and Nd initial ratios for type B2 basalts are suggestive of mixing of several distinct components during magma evolution. Furthermore, *Paces et al.* (1991) determined that the elevated $^{87}$Sr/$^{86}$Sr in type A and C basalts is likely a result of metasomatic Rb enrichment in their sources.

Isotopic studies indicate strong depletion of LILE in the upper mantle of the Moon within 200 m.y. of its formation. High-titanium mare basalts can be modeled as melts of depleted clinopyroxene-pigeonite-olivine-ilmenite adcumulates from a LMO that contain <2% intercumulus trapped liquid. Ilmenite, which is required in the source, does not become a liquidus phase in the LMO until it is more than 90% crystallized, at which point the residual LMO liquid has become extremely LILE-enriched (KREEPy) with low Sm/Nd and high Rb/Sr (*Snyder et al.*, 1992b). This KREEPy trapped liquid, though a volumetrically minor component of the cumulate pile, controls the LILE abundances in the source. With time, the mafic accumulate rapidly diverges from a chondritic reference reservoir (i.e., CHUR) to elevated, positive $\varepsilon_{Nd}$ values, whereas the trapped liquid also diverges from CHUR, but to more negative $\varepsilon_{Nd}$ values. Even small proportions of this trapped liquid will greatly effect the "initial" isotopic ratios of basalts derived from the upper mantle of the Moon.

### 7.3. Low-Titanium Basalt Volcanism

Low-potassium, high-Ti mare basalts from Apollo 11 and 17 form whole-rock "isochrons" that indicate isotopic closure of their source(s) just 100–200 m.y. after formation of the Moon and 300–500 m.y. prior to melting and subsequent eruption (*Snyder et al., 1994*). In contrast, low-Ti mare basalts lie along whole-rock arrays that define ages similar to their eruption ages. Thus, it would appear that isotopic closure was either much later for the sources of the low-Ti basalts, or that the magmas came from heterogeneous/mixed source regions.

*7.3.1. Apollo 12.* Argon-isotopic analyses have been performed on most Apollo 12 samples and have yielded ages of 3.1–3.3 Ga (e.g., *Turner, 1971; Stettler et al., 1973; Alexander and Davis, 1974; Horn et al., 1975*). Rubidium-strontium-isotopic analyses of these samples confirmed the [40]Ar-[39]Ar ages (e.g., *Papanastassiou and Wasserburg, 1970, 1971; Compston et al., 1971*). However, *Nyquist et al.* (1977, 1979, 1981a) were the first to systematically study all Apollo 12 basalt groups using the Rb-Sr and Sm-Nd isotopic methods. They found that there were systematic differences between the ilmenite basalts and the olivine-pigeonite basalts that could only be explained by derivation from separate sources. Initial Nd and Sr isotopic compositions of Apollo 12 basalts are shown in Figs. 9 and 10.

*Snyder et al.* (1994) speculated that the source for the ilmenite basalts at Apollo 12 may have been the same, or a similar source, to that which yielded the older high-Ti volcanics at the Apollo 11 and 17 landing sites. Thus, the mantle source for these rocks would have to be widespread within the lunar interior and would have had to persist from 3.85 Ga (earliest high-Ti volcanism; groups B2 and D basalts; *Snyder et al., 1996*) to 3.2 Ga, a period of nearly 700 m.y. *Snyder et al.* (1994) postulated that the source of the high-Ti basalts was depleted in ilmenite and trapped liquid by continued melting, yielding a source that was less Ti-enriched and more LREE-depleted. Thus, magmas generated at 3.2 Ga would reflect this changing source, yielding the characteristic high $\varepsilon_{Nd}$ ilmenite basalts. However, recent Lu-Hf isotopic studies tend to refute this argument. *Beard et al.* (1994, 1996, 1998) point out that the ilmenite basalts are similar in their Lu-Hf isotopic compositions to all other low-Ti basalts, including Apollo 12 and 15 basalts, and differ markedly from the high-Ti mare basalts.

*Snyder et al.* (1994, 1996) assumed that the Sm/Nd ratios of high-Ti basalts were similar to their source regions. However, *Nyquist et al.* (1977, 1979) have shown that, whereas this is roughly true for high-Ti mare basalts, it is not true for low-Ti mare basalts from the Apollo 12 landing site. In fact, they suggest that the [147]Sm/[144]Nd ratios of the ilmenite and olivine-pigeonite basalts may have decreased by 13–18%, respectively, as a consequence of small degrees of partial melting and smaller proportions of high-LREE phases in the residue (i.e., clinopyroxene and plagioclase). Thus, the relatively low $T_{LUM}$ model ages (4.13–4.15 Ga for olivine-pigeonite basalts and 3.82–3.87 Ga for ilmenite basalts; *Snyder et al., 1997b*) for the low-Ti basalts are evidence of Sm/Nd fractionation during melting, with ilmenite basalts showing greater fractionation than olivine-pigeonite basalts, consistent with the calculations of *Nyquist*

*et al.* (1979). Anomalies in the abundances of the short-lived radionuclide [142]Nd for several low-Ti mare basalts is convincing evidence that their source regions were created early in lunar evolution, reached isotopic closure at ~4.32 Ga, and were not significantly modified until melting to form the basalts (*Nyquist et al., 1995*).

The upper mantle source for the ilmenite basalts was likely formed earlier in the crystallization of the LMO, prior to formation of the source for the high-Ti basalts. Furthermore, ilmenite is not required in the residue for the ilmenite basalts (*Neal et al., 1994; Shearer et al., 1996*). This might suggest that the ilmenite-bearing source only existed in the region beneath Mare Serenitatis and Mare Tranquillitatis, in the eastern nearside. However, analyses of picritic glasses from all landing sites (*Delano, 1986*) have shown the proclivity for these samples to be high-Ti in tenor. Even picritic glasses retrieved from notably low-Ti mare sites, such as Apollo 12 and Apollo 15, have high-Ti abundances. Furthermore, Apollo remote sensing data, in concert with telescopic observations of the volcanic stratigraphy in Oceanus Procellarum, show clearly that the oldest units (Repsold Formation) contain high-Ti flows that are possibly correlative with those in Mare Serenitatis and Mare Tranquillitatis (*Whitford-Stark and Head, 1980*). Thus, it is likely that high-Ti basalt volcanism was one of the earliest, widespread, volcanic events on the lunar nearside.

*7.3.2. Apollo 14.* We have already discussed the chronology and isotopic character of high-Al basalts at the Apollo 14 landing site and whether they, indeed, represent the earliest expression of mare volcanism. However, we have not discussed the chronology and isotopic systematics of a subset of the high-Al basalts, the very-high-K (VHK) basalts. The ages of the VHK basalts are much younger than their other high-Al counterparts, ranging from 3.83 ± 0.08 Ga to 3.82 ± 0.12 Ga (Rb-Sr; *Shih et al., 1986*). These basalts are distinguished by elevated K (0.3 wt% $K_2O$), Rb, and Ba (*Warner et al., 1980; Shervais et al., 1985*), and the highest measured [87]Rb/[86]Sr and [87]Sr/[86]Sr recorded in purported mare basalts (*Shih et al., 1986*). However, the low initial [87]Sr/[86]Sr ratios in these basalts (0.6995 ± 5 and 0.6997 ± 4) mitigate against long-term alkali enrichment in the source (*Shih et al., 1986*). Although the Sm-Nd systematics of the two analyzed samples (14305,304,371 and 14168,39) have been somewhat disturbed, the [147]Sm/[144]Nd ratios (0.2007 and 0.1984 respectively) and intial $\varepsilon_{Nd}$ values (–0.1 ± 0.6 and +0.4 ± 0.5 respectively) suggest a chondritic source (*Shih et al., 1986*). Thus, the obvious alkali enrichment of these samples is in contrast to the chondritic Sm/Nd.

Models for the petrogenesis of the VHK basalts have assumed a high-Al basalt parent and concentrated on the modification of these magmas during ascent and extrusion. Due to their similar $K_2O$/La ratios (~1000), *Shih et al.* (1986) suggested that the VHK basalts could have resulted from contamination by lunar granite. The isotopic data support assimilation of up to 5% granite by a high-Al basalt (*Shervais et al., 1985; Shih et al., 1986*). *Neal et al.* (1988b) showed that the trace-element and isotopic systematics of the high-Al basalts were best explained by serial magmatism, fractional crystallization, and assimilation of a KREEP component (with a mass assimilated/mass crystallized ratio, r, of 0.22). They extended this model to include the VHK basalts by periodic assimilation of lunar granite (where r is much higher, i.e., 0.5) (*Neal et al.,* 1988a).

*7.3.3. Apollo 15.* Hafnium-tungsten-isotopic analyses show that some Apollo 15 olivine normative basalts (ONBs) show positive W anomalies and others (quartz-normative basalts or QNBs) have chondritic W-isotopic compositions (*Lee et al.,* 1997) (Fig. 11) . Olivine normative basalt 15016 is also one of two Apollo 15 basalts (both ONBs) that exhibit large $^{182}$W-isotopic anomalies. Such large $^{182}$W-isotopic anomalies are thought to represent a heterogeneous Moon with deeper portions of the mantle remaining more radiogenic, whereas shallower mantle source regions have lost this early signature, possibly due to infall of late chondritic meteorite material (see below).

Much of the variation in W-isotopic ratios in lunar basalts can be attributed to relative depths of melting of sources (e.g., *Snyder et al.,* 1997b; *Lee et al.,* 1997) for ONBs vs.

QNBs (Fig. 12). In a critical study of previous experimental petrologic work on lunar mare basalts, *Longhi* (1992b) indicated that only four of the Apollo 15 basaltic samples yielded reliable pressure-temperature information. The Apollo 15 green glass gave multiple-mineral saturation at 1450°C and 17–18 kbar pressure, corresponding to minimum depths of melting of 350–400 km. However, the mare basalts that have been studied are fractionated. Thus, the information derived from these rocks must be considered to be the *minimum* depth of segregation of the melt from the source. Regardless, the pressures determined on these samples provide "important constraints on the depth of melting" (*Longhi,* 1992b). Providing that these rocks have undergone roughly similar degrees of fractionation from their sources, relative depths of melting can be determined. A single QNB, 15065, yielded multiple-mineral saturation at a relatively low and indeterminate pressure (<5 kbar at >1270°C). Two ONBs, 15016 and 15555, yielded multiple-mineral saturation at 1310°–1350°C and 11–12 kbar pressure and 1300°C and 8.5 kbar pressure respectively. These pressures correspond to minimum depths of melting in the Moon of <100 km for the QNBs, and 170 km (15555) and 250–300 km (15016) for the ONBs (pressure-depth relationships as per *Warren,* 1985).

This knowledge, combined with the supposition that ONBs are younger (3.25 Ga) than QNBs (3.35 Ga) (*Snyder et al.,* 1998, 1999b), leads to the following scenario. Trapped liquid-enriched sources were melted first at the shallowest depths to form the QNBs. Later, deeper melting of more refractory sources led to magmas parental to the ONBs. The LILE-enriched nature of the QNBs relative to the ONBs can be explained in one of two ways: (1) smaller degrees of partial melting of a common source for QNBs, followed by higher degrees of partial melting to generate the ONBs; or (2) melting of distinct sources. The distinctly lower $\varepsilon_{Nd}$ values, lower Sm/Nd, and higher Rb/Sr for the older QNBs are suggestive of a source that was more LILE-enriched than that for the ONBs. These relationships suggest that portions of this common source that had a higher proportion of LILE and low-temperature-of-melting components (such as trapped residual liquid from a magma ocean; *Snyder et al.,* 1994, 1997b), were melted first (Fig. 12). More refractory portions of the source were melted later and generated parental magmas for the ONBs.

However, Re-Os systematics of a single Apollo 15 low-Ti basalt, 15555,854, have shown that, although its $^{187}$Os/$^{186}$Os-isotopic composition is chondritic, it contains very low abundances of Re and Os (1.35 and 13.94 ppt respectively) (*Birck and Allegre,* 1994). Such low abundances are inconsistent with later meteoritic infall, which would contain orders of magnitude more Re and Os. *Birck and Allegre* (1994) suggest that the Re/Os fractionation seen in planetary and lunar basalts is a reflection of the O fugacity and consequent relative partitioning behavior of Re and Os during basalt genesis. This must be kept in mind when attempting to reconcile the Hf-W data with a depth-related model of meteoritic infall.

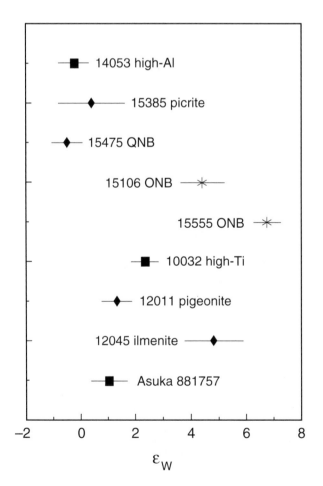

**Fig. 11.** Plot of $\varepsilon_W$ for mare basalts (data from *Lee et al.,* 1997). Only ilmenite-rich basalts 12045 and 10032 and olivine-normative basalts 15106 and 15555 exhibit significant $\varepsilon_W$ anomalies.

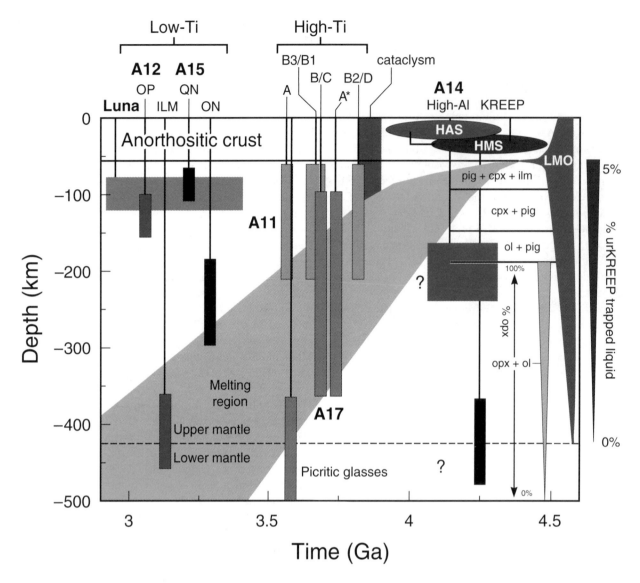

**Fig. 12.** Schematic representation of lunar accretion, magma ocean formation, and magmatism with postulated depths of melting (from multisaturation experiments, as compiled in *Longhi,* 1992b). Key: A = Apollo 11 group A high-Ti basalt; A* = Apollo 17, type A high-Ti basalt; B/C = Apollo 17 types B and C basalts; B2/D = Apollo 11 groups B2 and D basalts; B3/B1 = Apollo 11 groups B3 and B1 basalts; HAS = highland alkali suite; HMS = highland magnesian suite; ILM = Apollo 12 ilmenite basalts; LMO = lunar magma ocean; Luna = Luna 16 and 24 basalts; ON = Apollo 15 olivine-normative basalts; OP = Apollo 12 olivine/pigeonite basalts; QN = Apollo 15 quartz-normative basalts. Modified from *Nyquist and Shih* (1992).

*7.3.4. Lunas 16 and 24.   Papanastassiou and Wasserburg* (1972) analyzed the Rb-Sr isotopic compositions of mineral and total-rock separates from the largest fragment (ophitic basalt B-1) from the Luna 16 soil. This sample was unique in composition among lunar basalts in having intermediate $TiO_2$ (4.04 wt%) and only one pyroxene (pigeonite) (*Albee et al.,* 1972). Four mineral separates and a total-rock split yielded a poorly defined "age" of 3.39 ± 0.18 Ga (1σ) and a rather low initial $^{87}Sr/^{86}Sr$ of 0.69906 ± 4 (1σ). This age was in agreement with a more precise K-Ar age on the same sample (3.45 ± 0.04 Ga) (*Huneke et al.,* 1972). If this age and $^{87}Sr/^{86}Sr$ initial isotopic ratio are representative of mare volcanism in Mare Fecunditatis (*Papanastassiou and Wasserburg,* 1972), the source region for volcanism must have been severely depleted in Rb at its formation.

Based upon mineral chemistry, zonation in pyroxene, and trace-element concentrations, *Wasserburg et al.* (1978) determined that fragments in sample 24170 from the Luna 24 landing site in Mare Crisium were from a single basalt ("microgabbro"). This sample was determined to be a fine-grained low-Ti ferrobasalt similar to other low-Ti basalt fragments from this landing site. Due to the extremely low Rb/Sr ratios of individual minerals and the bulk clast ($^{87}Rb/^{86}Sr$ <0.013 for all) an age was not well-defined by this system (3.7 ± 0.6 Ga). The supposed initial $^{87}Sr/^{86}Sr$ was, curiously, the same as the Luna 16 basalt, B-1 (see above). However, *Wasserburg et al.* (1978) did determine a two-point (pyroxene, plagioclase) Sm-Nd crystallization age of 3.30 ± 0.05 Ga and an initial $\varepsilon_{Nd}$ of 1.9 ± 0.6 on this sample. This age was consistent with two $^{40}Ar$-$^{39}Ar$ determinations from separate fragments of 24170 (both 3.30 ± 0.04 Ga) (*Wasserburg et al.,* 1978). They interpreted this sample as a product of melting a source that was only slightly fractionated in Sm/Nd (6%) from chondritic, but with an extremely low Rb/Sr ratio (0.002). Based upon trace-element and Sr-isotopic systematics of a variety of small fragments from the Luna 24 soil, *Nyquist et al.* (1978) modeled the "low-Mg" basalts by 3–10% partial melting of an LMO cumulate source containing equal proportions of olivine and orthopyroxene, 7–14% clinopyroxene, and small amounts of plagioclase (0–7%). The small amount of plagioclase was thought to be needed to explain the extremely low $^{87}Rb/^{86}Sr$ (0.0041) and initial $^{87}Sr/^{86}Sr$ (0.69914 ± 6) of the "low-Mg" basalt (*Nyquist et al.,* 1978).

### 7.4. Picritic Glasses

Picritic glass beads are ubiquitous at the Apollo and Luna landing sites and are thought to represent some of the most primitive magmatism on the Moon (e.g., *Wentworth and McKay,* 1988; *Shearer et al.,* 1990). The Apollo 17 orange glass has yielded among the most radiogenic $\varepsilon_W$ value (5.0) from the Moon (*Lee et al.,* 1997). Furthermore, many of these glasses may come from great depths in the Moon, some even in excess of 700 km (*Hess,* 1991). *Longhi* (1992a) has suggested that these picritic magmas may have formed by polybaric fractional melting, with initiation of melting below 1000 km depth. Thus, the geochemistry of the picritic glasses could afford a direct connection to mare sources and an unadulterated look into the lunar mantle. However, the relationship of these glass beads to mare basalts remains enigmatic (*Longhi,* 1987). Although picritic glasses undoubtedly represent primitive magmatism, their source may be much deeper than many (most?) mare sources.

Samples of picritic glasses have been studied extensively, but little isotopic work (other than $^{40}Ar/^{39}Ar$) has been forthcoming. Due to the fine-grained nature of these samples and lack of distinct minerals, internal mineral-mineral isochrons have not been determined for these samples. Although over 25 groups of pristine glasses have been identified (*Delano,* 1986), many of the samples are also rather heterogeneous and difficult to separate in quantities acceptable for isotopic analyses. Only the homogeneous Apollo 15 green glasses and the Apollo 17 orange glasses are found in the abundances required for isotopic analyses. Furthermore, these primitive magmas appear to have been derived from the deep interior Moon much later than many of the more evolved mare basalts.

Various groups have determined $^{40}Ar$-$^{39}Ar$ ages for the Apollo 17 orange glasses that show a range from 3.50 ± 0.05 to 3.66 ± 0.06 Ga (*Huneke et al.,* 1973; *Husain and Schaeffer,* 1973; *Huneke,* 1978; *Alexander et al.,* 1980). Samarium-neodymium-isotopic data for orange-glass-rich soil sample 74220,65 is indicative of a depleted source ($^{147}Sm/^{144}Nd = 0.2249$), although much less depleted than any Apollo 17 high-Ti mare basalt (with $^{147}Sm/^{144}Nd$ from 0.24141 to 0.26234). *Jerde et al.* (1994) have suggested that the orange glasses are parental to the high-K, Apollo 11 group A basalts. The orange glass sample 74220,65 is simi-

lar to the Apollo 11 group A basalts in having lower $^{147}Sm/^{144}Nd$ and unrealistically old $T_{CHUR}$ ages, although its initial $\varepsilon_{Nd}$ value (6.6) is elevated relative to the group A basalts ($\varepsilon_{Nd} = 3.1$) (Snyder et al., unpublished data). Regardless, both the Apollo 11 group A basalts and Apollo 17 orange glass must involve a two-stage process of depletion, or ultradepletion followed by later enrichment. Such a two-stage process is not required to explain the Sm-Nd systematics of other high-Ti basalts from Apollo 11 and Apollo 17.

Apollo 15 green glasses have been analyzed both for their U-Pb (*Tatsumoto et al.,* 1987) and Sm-Nd (*Lugmair and Marti,* 1978) isotopic composition. Lead derived from the interior of the glass beads gave a $^{207}Pb/^{206}Pb$ age of 3.41 ± 0.33 Ga and defined a primitive source (Fig. 4) with a relatively low μ (= 19–55) (*Tatsumoto et al,* 1987). Argon-40–argon-39 analyses give ages that range from 3.25 ± 0.06 Ga (15086) to 3.35 ± 0.18 Ga (15426; *Podosek et al.,* 1973; *Spangler et al.,* 1984) and are indistinguishable from the $^{207}Pb/^{206}Pb$ age. Green glass separated from breccia 15426,48 yielded a $^{147}Sm/^{144}Nd$ (0.2011–0.2016 for leached and unleached samples respectively) slightly elevated above chondritic and a Nd model age of 3.8 ± 0.4 Ga (*Lugmair and Marti,* 1978). These values determine an initial $\varepsilon_{Nd}$ value that is indistinguishable from chondrites at this time. Thus, U-Pb and Sm-Nd systematics are consistent with a relatively undifferentiated reservoir for the green glasses.

## 8. MODELS OF EVOLUTION OF LUNAR CRUST AND MANTLE

Age constraints on earliest lunar differentiation discussed above include (1) Hf-W model ages of lunar basalts of ~4.51 Ga, (2) Sm-Nd model ages of 62236 of ~4.45 Ga, (3) Sm-Nd model age of lunar basalts of ~4.32 Ga, (4) the high-Ti mare basalt whole-rock isochron of 4.46 ± 0.17 Ga, and (5) whole-rock and mineral separate Pb ages of 4.42 Ga (Fig. 4). In addition, any scenario for early lunar differentiation must provide a mechanism to decouple the short-lived Hf-W and Sm-Nd isotopic systems so that they yield different model ages. Central to this discussion is the geologic interpretation of the parent-daughter fractionation events recorded in these different rock types by these two isotopic systems. Mechanisms to generate the various lunar sources are discussed below.

### 8.1. Primary Lunar Differentiation

From Figs. 3 and 8 it is apparent that lunar basalts must be derived from at least three distinct sources and that these sources are characterized by varying $^{182}W$ excesses and excess, normal, and deficient $^{142}Nd$ abundances. These sources can be inherited from differentiated bodies that accreted to form the Moon (heterogeneous accretion model) or reflect early differentiation of the Moon. The problem with a heterogeneous accretion model is preserving differentiated geochemical reservoirs throughout the accretion process given the immense energetics of accretion and the formation of a putative magma ocean. On the other hand, the $\varepsilon_W^{182}$ and $\varepsilon_{Nd}^{142}$ anomalies in the mare source regions could reflect early lunar differentiation. In this case differentiation must fractionate Hf from W without fractionating Sm from Nd and fractionate Sm from Nd without significantly fractionating Hf from W. Fractionation of Hf and W has been attributed to the core formation process and may account for the absence of negative $^{182}$W anomalies in any lunar sample analyzed so far. However, it seems difficult to produce variable $^{182}$W anomalies in mare sources solely by core formation. It is considered more likely that a particular phase, such as ilmenite and/or clinopyroxene, is present in the source region that can fractionate Hf/W (*McKay and Le*, 1999; *Righter and Drake*, 1999), but have essentially no effect on Sm/Nd. Magma ocean modeling suggests that late-stage cumulate sources that are rich in ilmenite are also rich in clinopyroxene (*Snyder et al.*, 1992b). Several high-Ti basalts have positive $\varepsilon_W^{182}$ anomalies supporting this contention (*Shearer and Newsom*, 1998; 1999).

Discrepancies between the Sm-Nd and Hf-W model ages of mare source formation should not result from simple differentiation in an undisturbed system. We speculate that the decoupling of the Sm-Nd and Hf-W isotopic systems could reflect a partial rehomogenization of the magma ocean cumulates by a process such as convective overturn. Convective overturn of the lunar mantle has been postulated by a number of authors, beginning with the ilmenite-sinking models of *Ringwood and Kesson* (1976). *Hess and Parmentier* (1995), for example, appeal to convective overturn as the cause for melting mantle basalt source regions. In this scenario, mantle source regions with variable $\varepsilon_W^{182}$ anomalies are produced by Hf-W fractionation during initial differentiation in the lunar mantle. Decoupling of the Sm-Nd and Hf-W isotopic systems might occur if convective overturn occurred after ~4.46 Ga, when $^{182}$Hf was extinct, but prior to ~4.1 Ga when $^{146}$Sm was still live and if rehomogenization of Sm-Nd, but not W, was associated with this process. Partial rehomogenization of the cumulate pile requires W to be confined to a component that did not interact with other components during overturn. If this scenario is correct, then the Hf-W model ages of ~4.51 Ga, as well as the high-Ti mare basalt whole-rock isochron of 4.46 ± 0.17 Ga, reflect the time of initial crystallization of the magma ocean, whereas the Sm-Nd model age of 4.32 Ga derived from individual mare basalts represents the time of mare source reconstitution and reequilibration after convective overturn of the cumulate pile. In this regard, it is interesting that *Spera* (1992) calculated a characteristic lunar mantle mixing time of 200 m.y.

The ~4.45-Ga age of the FAN 62236 source region is intermediate between the proposed time of magma ocean crystallization at 4.51–4.46 Ga and possible convective overturn of the magma ocean cumulates at 4.32 Ga. If the 62236 source was depleted as a result of crystallization of the magma ocean it implies that the magma ocean may have existed for at least 60 m.y. This age is consistent with the average KREEP model age of 4.42 ± 0.07 Ga (*Nyquist and Shih*, 1992), which likely dates solidification of the last dregs of the magma ocean. It is also near the middle of the

age range of 2500 yr to 200 m.y. calculated for the crystallization of the magma ocean by *Solomon and Longhi* (1977), *Herbert et al.* (1978), *Minear* (1980), *Shirley* (1983), and *Tonks and Melosh* (1990) using thermal models. Formation of a depleted lunar source region at 4.45 Ga is relatively late in comparison to the formation of depleted martian source regions estimated at ~4.53 Ga from the analysis of SNC meteorites (*Harper et al.*, 1995; *Borg et al.*, 1997). This result is not predicted by scenarios in which both bodies formed contemporaneously and differentiated according to cooling rates based on their relative sizes. Instead, it suggests that the Moon either cooled more slowly than Mars or that it formed significantly later.

Evidence of derivation from a LREE-depleted source combined with relatively young crystallization ages of FANs suggest that these samples could form by a secondary melting process in the lunar mantle or crust. *Longhi and Ashwal* (1985) have proposed such a model for the genesis of some anorthosites that could account for the Sm-Nd isotopic systematics of these FANs. *Ryder* (1991) has postulated that convective overturn of the cumulate pile could account for the generation of the HMS magmas and explain how mare basalt sources with large Eu anomalies could be present at depths below the plagioclase stability field. The convective overturn of the mantle around 4.32 Ga may also be able to account for the young ages, high initial $\varepsilon_{Nd}^{143}$, and positive $\varepsilon_{Nd}^{142}$ anomalies observed in 62236, and possibly 60025 and 67016c. Before overturn, the Moon is postulated to be stratified with plagioclase-rich cumulates on the top of Fe-rich late-stage mafic cumulates that overlie early formed Mg-rich cumulates. Overturn of the cumulate pile is thought to be driven by sinking of Fe-rich cumulates (e.g., *Spera*, 1992). Plagioclase-rich ferroan magmas could be produced if a mixture of plagioclase and Fe-rich mafic cumulates sank past the depth of plagioclase stability and was melted. *Solomon and Longhi* (1977) have suggested a similar scenario of "subduction" of "mini-plates" of gravitationally unstable quenched crust early in LMO evolution for the production of HMS plutonic rocks. This would probably require intermingling of small proportions of plagioclase and large proportions of mafic cumulates at a relatively small scale to produce a cumulate with a high density. It also requires KREEP-rich cumulates to be absent. Melting at depth could produce magmas that were strongly enriched in plagioclase components because of the shrinkage of the plagioclase stability field with depth (*Lindsley*, 1968). Reequilibration of the cumulate pile during partial melting could produce magmas with Sm-Nd isotopic systematics indicative of derivation from a LREE-depleted source.

## 8.2. Partial Melting of the Lunar Interior and Mare Basalt Formation

By combining telescopic observations and remote sensing data with petrologic, elemental, and isotopic data from the Apollo collections, a clearer picture of the evolution and partial melting of the lunar interior is starting to emerge (Fig. 12). Relatively rare "cryptomare" are probably the first evidence of melting of the lunar interior and subsequent extrusion of melts (*Head and Wilson, 1992*). However, we appear to have no samples of this volcanic event in the Apollo collections and little idea on its character from telescopic observations. KREEP (K, REE, and P-enriched) basalts are among the oldest, pristine, volcanic rocks dated from the Apollo collections and vary in age from 3.8 to 4.1 Ga (*Shih et al., 1992*). However, although the KREEP signature is nearly ubiquitous at the western landing sites, pristine KREEP basalts are rare and there is, as yet, no evidence that they formed extensive flows or maria. The high-Al basalts are also among the oldest volcanic rocks in the Apollo collections. Based upon trace-element and siderophile-element considerations, *Ridley (1975)* contended that these rocks are not hybrids of crust and a basaltic magma, but are true melts of the interior formed at about 200 km depth (*Walker et al., 1972*). In contrast, we have suggested that these rocks are not true mare basalts, but are probably impact melts of mixed derivation, the older ages reflecting the antiquity of the highland component (*Snyder et al., 1999a*). Therefore, group D and B2 high-Ti basalts from the Apollo 11 landing site represent not only the earliest volcanic event (~3.85 Ga; *Papanastassiou et al., 1977; Snyder et al., 1994, 1996*) in the Apollo collections, but among the earliest evidence of mare volcanism. That earliest mare volcanism was high-Ti in tenor is supported by telescopic observations (*Whitford-Stark and Head, 1980; Head and Wilson, 1992*).

*8.2.1. The importance of trapped liquid.* If KREEP basalts and group D and B2 high-Ti basalts are truly representative of early lunar volcanism (~3.85 Ga), then the earliest melts of the lunar interior were enriched in the incompatible elements. Some incompatible elements, namely K, U, and Th, have radioactive isotopes that are capable of producing the heat required for partial melting. Late-stage, residual liquids from crystallization of the LMO would also be enriched in incompatible elements such as the REE, K, U, and Th. It is considered likely that a small proportion (roughly <1–5%) of the residual, LMO liquid was trapped in upper-mantle cumulates as they crystallized (*Snyder et al., 1992b*) and is reflected in the changes in $^{147}Sm/^{144}Nd$ ratios of derivative, ilmenite-bearing, lunar basalts over time (*Snyder et al., 1994, 1996, 1997b*) (Fig. 13). This trapped liquid model is similar in terms of its affects on trace-element ratios and isotopic systematics as the urKREEP assimilation model of *Binder (1985)*. However, in the trapped liquid scenario the KREEPy material is part of the source and not added to the magma during ascent. Enrichment of these incompatible elements in the residual LMO liquid would increase with crystallization, thus late-forming LMO cumulates would trap progressively more incompatible-element enriched liquid. Furthermore, this residual LMO liquid would have been lighter than surrounding cumulates and would have likely migrated upward in the mantle over time. The uppermost portions of the lunar mantle should have had the highest proportion of trapped residual liquid with the most elevated incompatible-element abundances. Lunar-magma-ocean phase modeling also

shows that the uppermost (>95% crystallization of the LMO) cumulate mantle should contain ilmenite (*Snyder et al., 1992b*). Thus, the most fertile site for partial melting is also the most Ti-enriched in the Moon. Radiogenic isotopic dating of Apollo 11 and Apollo 17 low-K, high-Ti mare volcanics indicates that this volcanic episode continued until about 3.67 Ga, i.e., for about 200 m.y. (*Snyder et al., 1994*).

The fertility of cumulate sources in the lunar upper mantle could have been controlled largely by the proportion of trapped residual liquid. The earliest volcanics, KREEP basalts, have proven difficult to interpret (e.g., *Ryder, 1988; Shih et al., 1992*), but probably began their journey to the surface as melts of the deep lunar interior (for one model, see *Ryder, 1994*). It may be that these melts are akin to the komatiitic plumes in the early Earth and may be formed by perturbations deep in the Moon. However, there is no doubt that these picritic basalts have somehow obtained a KREEP signature, possibly during transit to the surface through assimilation of KREEPy, residual, LMO liquid trapped in upper-mantle cumulates. Those sources with 1.5–2% trapped, residual, LMO liquid were melted later to form basalts represented by the Apollo 11 group D and B2 high-Ti mare basalts (Figs. 12 and 13). Even later melting (at 3.67–3.71 Ga) included sources with 0.8–1.5% trapped, residual, LMO liquid and generated basalts represented by the Apollo 11 groups B1–B3 and Apollo 17 group C basalts (Figs. 12–13). Based upon limited experimental pressure-temperature estimates, the sources of these

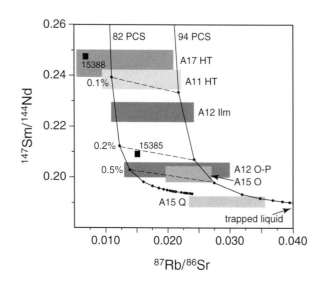

**Fig. 13.** Parent/daughter ratio plot of $^{87}Rb/^{86}Sr$ vs. $^{147}Sm/^{144}Nd$ indicating a mixing curve for the model most-depleted adcumulate ($^{147}Sm/^{144}Nd = 0.318$, $^{87}Rb/^{86}Sr = 0.005$) and a KREEPy trapped liquid (represented by 15382; $^{147}Sm/^{144}Nd = 0.168$, $^{87}Rb/^{86}Sr = 0.235$) (*Snyder et al., 1994, 1997b*). Star marks along the curve indicate percentages of KREEPy trapped liquid. Fields for particular mare basalts also shown. The displacement of these fields to the right of this curve is due to fractional crystallization within the flow unit or in transit to the surface.

low-K, high-Ti basalts were located at depths of <100 km (Apollo 11 basalts), although possibly as deep as 250 km (one Apollo 17 basalt) (*Longhi*, 1992b). Considering that the average thickness of the Moon's crust is 60 km on the nearside (*Mueller et al.*, 1988), the sources for these high-Ti mare basalts were likely very shallow in the mantle, consistent with LMO modeling (*Snyder et al.*, 1992b).

Later volcanic activity (~3.59 Ga) at the Apollo 11 and 17 landing sites consists of high-K, group A, high-Ti mare basalts at Apollo 11 and coeval high-K, high-Ti pyroclastic volcanism (picritic orange glass, represented by 74220) at both the Apollo 11 and Apollo 17 landing sites. Indeed, high-Ti picritic (orange glass) magmas are thought to be parental to the Apollo 11 group A, high-Ti mare basalts (*Jerde et al.*, 1994). The evolution of these high-K, high-Ti magmas is distinct from the other low-K, high-Ti basalts in that extensive KREEP assimilation is strongly suggested (*Jerde et al.*, 1994). Furthermore, the high-Ti orange glasses were probably derived from great depth (~500 km; *Walker et al.*, 1975; *Longhi*, 1992b) (Fig. 12).

Low-titanium mare basalts were extruded at the Apollo 15 landing site at 3.4 to 3.3 G.y. ago (Fig. 12). These basalts also were derived from sources at relatively shallow depths, from <100 to 300 km (*Longhi*, 1992a). Finally, low-Ti mare basalts at the Apollo 12 landing site were extruded between 3.28 and 3.18 Ga and were derived from upper-mantle, cumulate sources that had even less trapped, residual, LMO liquid (0.15–0.5%) than previous sources (Figs. 12 and 13). However, the sources for these basalts lie at two distinct depths (Fig. 12). The olivine and pigeonite basalts were melted from sources located at similarly shallow depths (100–200 km; *Longhi*, 1992b) as the high-Ti mare basalts from Apollos 11 and 17. In contrast, ilmenite basalts were melted from a source that may have been as deep as 350–400 km. The ilmenite-basalt source is the deepest of the known mare basalt sources, but comparable to other deep sources suggested by studies of high-Ti picritic glass beads (400 to >500 km) (*Longhi*, 1992b). The key feature of these rocks is the presence of signficant amounts of modal ilmenite.

*8.2.2. Ilmenite sinking and source fertilization.* Over time, it is clear that ilmenite-rich mafic magmas (picrites and mare basalts), which would require ilmenite to have been originally in the source, can be generated at a variety of depths. However, early on, the ilmenite-bearing melts were solely basaltic and significantly enriched in incompatible elements. This attests to the shallow, more evolved sources for these rocks. With time, these ilmenite-rich melts became less incompatible-element enriched and less evolved, reflecting slightly deeper sources with less trapped KREEPy liquid. Later melting to form ilmenite-bearing basalts and picrites, after a possible hiatus (?) of ~300 m.y., was at much greater depths (Fig. 12). Yet trace-element, major-element, and phase modeling indicates that ilmenite was one of the last phases to crystallize from the incipient LMO and thus should be found only in the uppermost mantle. Where did these deeper ilmenite-bearing sources come from?

*Spera* (1992) stated that, due to density contrasts in lunar magma ocean cumulates, ilmenite-bearing layers, formed late in the LMO and precipitated in the uppermost portion of the upper mantle, will sink relative to other cumulates. This was first suggested by *Ringwood and Kesson* (1976). However, *Hess and Parmentier* (1993) further projected that most of the ilmenite would continue to sink, eventually forming the lunar core. They also considered it likely that some of this sinking ilmenite would mix with the lunar mantle in transit, thus creating fertile, Ti-enriched source regions throughout the mantle. Combined pressure-temperature information on mare basalts and high-Ti picritic glasses are consistent with such a model. In fact, in may be possible to track the descent of some of these sinking blobs of ilmenite-bearing material by looking at the ages and depths of melting of ilmenite-rich basalts and picrites.

The earliest ilmenite-rich basalts are those found at Apollos 11 and 17 and indicate melting of shallow sources. Sparse age data from high-Ti picritic magmas (as evidenced by picritic glass beads) from the these landing sites seem to give younger ages than the mare basalts, in some cases (i.e., Apollo 17) much younger (possibly up to 200 m.y. younger). These high-Ti picritic melts were probably derived from very deep sources (400 to >500 km; *Longhi*, 1992b). The ilmenite basalts from the Apollo 12 landing site are extruded much later and also come from a very deep source (350–400 km). Extant data suggest that fertile, ilmenite-bearing sources were melted at greater depths over time, consistent with progressive sinking of the ilmenite-bearing, late LMO, cumulate source. Thus, the two major controlling factors for mare basalt source melting may prove to be the proportion of trapped, residual, incompatible-element-enriched LMO liquid and the sinking of fertile, ilmenite-bearing material into the lower mantle.

## 9. SUMMARY

Isotopic systematics of lunar rocks indicate three major, distinct, sources. One source has a high $^{238}U/^{204}Pb$ ($\mu$) > 500, high Rb/Sr and thus elevated $^{87}Sr/^{86}Sr$, and low Sm/Nd and consequent negative $\varepsilon_{Nd}$ values. This source is represented by the urKREEP residuum from the lunar magma ocean. A second source has $\mu$ values more typical of Earth, low Rb/Sr and $^{87}Sr/^{86}Sr$, high Sm/Nd and extremely positive $\varepsilon_{Nd}$ values, and positive to variable $\varepsilon_W$ values. This source is represented by a primordial lunar mantle and may have melted to form the low-Ti basalts. A third source has similar $\mu$ values to the second, intermediate $\varepsilon_{Nd}$ values, low to intermediate $^{87}Sr/^{86}Sr$, and chondritic $\varepsilon_W$ values. This source is likely represented by shallow regions of the lunar mantle that were involved in early processing and remelting in a lunar magma ocean and may have been contaminated by late meteoritic infall. This source also gave rise to the high-Ti mare basalts. Development of these reservoirs and their subsequent evolution is clearly outlined in Fig. 14 (*Beard et al.*, 1998). A possible fourth source is suggested by $\mu$ values from a few ferroan anorthosites, which, although

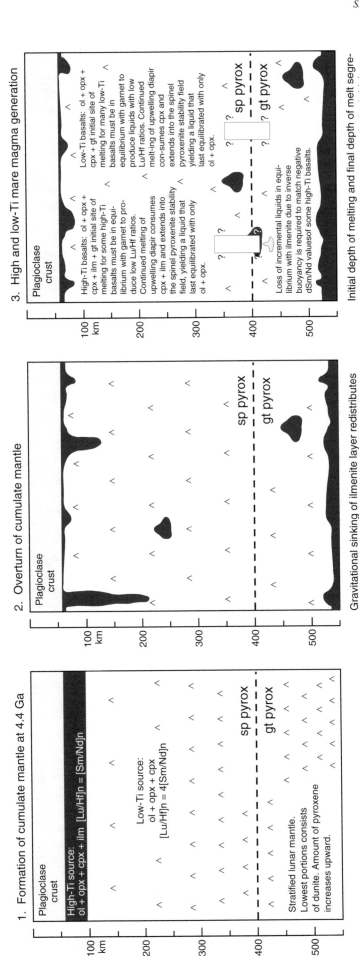

**Fig. 14.** Schematic representation of mare basalt cumulate source formation and melting (from *Beard et al.*, 1998). Source formation and melting progressed in three distinct stages. Stage 1: The formation of the lunar mantle by crystallization of a lunar magma ocean (LMO) by 4.4 Ga. This LMO cumulate was composed of an anorthositic flotation crust, a high-Ti, ilmenite-rich source just beneath, and a very thick olivine + orthopyroxene ± clinopyroxene "layer." The silicate layer is composed of olivine at the base with increasing proportions of orthopyroxene in shallower regions (see Fig. 12). In reality, a variety of silicate layers varying widely in their proportions of minerals were formed (*Snyder et al.*, 1992b; *Snyder and Taylor*, 1993), but a three-layer model is a simplification that is consistent with Hf-Nd isotopic systematics (*Beard et al.*, 1998). Stage 2: Sinking of dense, ilmenite-rich cumulates into the lower silicate mantle. These dense cumulates also likely contained entrained KREEPy material that would have been instrumental in "fluxing" deeper, less fertile source regions. As the ilmenite-rich material sank below about 400 km, aluminous pyroxenes were converted to garnet. This ilmenite-rich material could have continued to sink to form the lunar core (*Hess and Parmentier*, 1995), or reached chemical and density equilibrium somewhere in the lower mantle. Stage 3: Production of mare basalt magmas by decompression melting. Initial melting of low-Ti mare basalts is believed to have occurred in the garnet stability field and to have continued during diapiric rise into the spinel stability field (*Beard et al.*, 1998). Final incremental liquids were in equilibrium with only olivine and orthopyroxene (*Longhi*, 1992a). See *Beard et al.* (1998) for further discussion.

containing elevated μ values relative to Earth, are not nearly as elevated as the urKREEP reservoir.

The oldest rocks on the Moon, dated at between 4.56 and 4.29 Ga, are the ferroan anorthosites (FANs), which are thought to be flotation cumulates from a large ion lithophile element (LILE)-enriched, global, lunar magma ocean. However, initial $\varepsilon_{Nd}$ values for the three FANs that have been analyzed are all positive, suggesting derivation from a LILE-depleted source. This source could have formed by mixing of early-formed crustal cumulates with mafic cumulates in the lunar upper mantle. This early crust was modified by serial intrusion of Mg-rich magmas with a distinct KREEPy signature (most likely KREEP basalts that did not reach the surface) (Fig. 12). These magmas differentiated and crystallized to form the highlands magnesian suite (HMS) and later to form the highlands alkali suite (HAS). The HMS overlaps the FANs in age, ranging from 4.50 to 4.17 Ga (Fig. 12). The HAS signficantly overlaps the HMS in age, although many of the samples tend to be distinctly younger (4.37–3.88 Ga). This period of major crustal evolution was terminated by and/or reset by several intense impact events that formed the Nectaris (3.9 Ga), Crisium (3.895), Serenitatis (3.893), and Imbrium (3.85) Basins.

The sources for the mare basalts were formed in a three-stage process that is most readily defined by Hf-Nf isotopic systematics (Fig. 14). First, a global lunar magma ocean differentiated into a plagioclase-rich flotation crust, a thin high-Ti source (just below the crust), and a thick low-Ti source consisting mostly of variable proportions of olivine and orthopyroxene with minor clinopyroxene (Fig. 14.1). Olivine predominates near the base of this "layer" and orthopyroxene near the top (Fig. 12). Second, the denser ilmenite-rich layer sank into the lower silicate layer over time (Fig. 14.2). As the ilmenite-rich layer sinks (likely carrying lower-temperature melting components such as KREEP with it) below 400 km, the aluminous pyroxenes will break down to garnet. Sinking of the ilmenite-rich component will also serve to homogenize the layered silicate portion of the LMO cumulate (*Beard et al.,* 1998). The ilmenite-rich material could sink as far as the Moon's core (*Hess and Parmentier,* 1995). Third, mare magmas are produced from these sources by decompression melting (Fig. 14.3). Earliest melts were possibly formed prior to extensive ilmenite-sinking at relatively shallow depths, with later high-Ti basalts being generated at greater depths (Fig. 12).

Mare volcanism may have been operating since the formation of the earliest crust as evidenced by scattered cryptomaria. KREEP basalts formed relatively early, but may have been extruded only as localized flows and may not have formed more extensive mare. High-aluminum basalts are the oldest (4.19–4.29 Ga) dated volcanic material from the Moon and may either represent a truly aluminous magmatic event or mixing of highland materials (Fig. 12). The oldest unequivocal basalts in the present collections are group B2 and D high-Ti basalts from the Apollo 11 landing site (3.88–3.83 Ga). These basalts initiated an intense phase of active global volcanism on the Moon. Volcanism of high-Ti tenor

extended to 3.55 Ga at the Apollo 11 and 17 landing sites and likely at other locales on the nearside (i.e., Oceanus Procellarum). Telescopic observations have indicated that another pulse of high-Ti volcanism, not sampled by the Apollo collections, was also widespread at 2.5 ± 0.5 Ga. High-titanium basaltic volcanism was initiated by melting of sources rich in trapped liquid from the primordial lunar magma ocean. This trapped liquid was rich in LILE, including such heat-producing elements as K, U, and Th, and would have allowed melting of relatively shallow source regions containing both ilmenite and clinopyroxene. Thus the earliest high-Ti volcanics had relatively low $\varepsilon_{Nd}$ values and elevated $^{87}Sr/^{86}Sr$. As melting proceeded and sources with less trapped liquid were melted, the $\varepsilon_{Nd}$ values of the volcanics also became more positive. Sinking of dense, ilmenite-rich material with entrained LILE-rich liquid may have allowed later melting of both high-Ti and low-Ti sources at even greater depths.

Low-titanium basalts are prevalent at the Apollo 12, Apollo 15, Luna 16, and Luna 24 landing sites. They are younger than the high-Ti basalts in the lunar collections, ranging in age from 3.5 to 2.9 Ga. These basalts have the highest $\varepsilon_{Hf}$ values among lunar rocks and are well-correlated with $\varepsilon_{Nd}$ values ($\varepsilon_{Hf} \sim 4\times \varepsilon_{Nd}$). The most LILE-depleted basalts in the lunar collections, which also exhibit the highest initial $\varepsilon_{Nd}$ values and among the lowest $^{87}Sr/^{86}Sr$, are the ilmenite basalts from Apollo 12.

The short-lived nuclide $^{182}W$ is heterogeneously distributed in lunar samples. Several lunar basalts exhibit $^{182}W$ anomalies, mostly high-Ti or ilmenite-rich basalts, and low-Ti basalts thought to come from the deep lunar interior. High-aluminum basalts from Apollo 14 and quartz-normative basalts from Apollo 15 do not exhibit $^{182}W$ anomalies, and could be derived from shallow mantle sources that were affected by late infall of chondritic material. Conversely, the $^{182}W$ anomalies in lunar basalts could be the result of the fractionation of W from Hf afforded by such phases as ilmenite and clinopyroxene.

There is telescopic, remote sensing, and stratigraphic evidence that suggests that magmatism persisted on the Moon until 1 G.y. ago. However, samples in our collections do not extend beyond about 2.9 G.y. ago.

## 10.   FUTURE RESEARCH

Only a return to the Earth's Moon and return of samples from previously unsampled regions will allow us to answer many of the outstanding questions in lunar petrology and geochemistry. However, much can still be gleaned from the lunar samples currently in our possession and those that have been spalled off the lunar surface by impacts, occasionally wandering into the Earth's gravitational field. The important recent strides made in analytical geochemistry have allowed "new" isotopic systems to be exploited. The burgeoning field of short-lived nuclides, especially those of $^{182}Hf$-$^{182}W$ and $^{146}Sm$-$^{142}Nd$, has particular promise for understanding the earliest silicate and core-mantle differentiation in the Moon. In particular, the fractionation of Hf/ W afforded by the important lunar mantle mineral ilmenite may allow important constraints to be placed on early lunar differentiation and the assessment of its role in the lunar mantle. The long-lived $^{176}$Lu-$^{176}$Hf system has also been, and will continue to be, applied to those previously intractable low-abundance samples, such as the ferroan anorthosites and other highland rocks (e.g., *Lee et al.,* 2000, and unpublished data). Because of the extreme fractionation of Re/Os during melting processes and crust formation, it should prove to be a very useful isotopic system in the future, once analytical difficulties have been surmounted. To date only a few basalts (*Birck and Allegre,* 1994) and basaltic glasses (*Walker et al.,* 1998) and no highland rocks have been analyzed by the $^{187}$Re-$^{187}$Os method.

Microsampling techniques (such as MC-ICP-MS) could allow isotopic determinations to be made on small clasts in breccias. This would allow a large increase in the sampling of lunar materials, as many of the samples that represent particular rock types, such as the HAS and high-Al basalts, are found only as clasts in breccias and small soil fragments. Pristine KREEP basalts have been underanalyzed, due to both their paucity in the Apollo collections and analytical problems. Because the pristine KREEP basalts represent one of the three major reservoirs in the Moon, it is crucial to obtain a full complement of isotopic data on these important samples. Picritic glassses represent some of the most primitive magmas on the Moon, but have not been adequately studied due to their paucity in lunar soils and their relatively fine grain-sizes and lack of discrete minerals.

One poorly understood process in lunar petrogenesis is impact metamorphism (e.g., *Nyquist et al.,* 1987, 1988). In most terrestrial metamorphic rocks, Rb-Sr ages are almost always lower than, or the same as, Sm-Nd ages. However, impact metamorphosed lunar rocks often exhibit Rb-Sr ages that are older than Sm-Nd ages (e.g., *Snyder et al.,* 1995a). Is the Rb-Sr system more robust to impact metamorphism than the Sm-Nd system? Possibly, but we think it more likely that the answer will be found in studies of preferential melting of mineral phases during impacts. For instance, plagioclase, a major component in lunar highland rocks, contains the bulk of the Sr in a sample, but virtually none of the Rb. Furthermore, plagioclase is thought to be one of the first samples to melt and/or "open" to Rb-Sr diffusion during a meteorite impact. But, due to the extremely low Rb/Sr ratio, radiogenic Sr is not rapidly generated in plagioclase-rich samples. Thus, later impact events would not effectively reset the age of the sample, even if the plagioclase is completely remelted. This is an area that should receive further attention, both for lunar and meteorite studies.

Although the number of lunar samples is currently limited, and probably will be for decades to come, there is no paucity of new venues for lunar isotopic geochemistry. Breakthroughs in analyzing smaller samples and lower concentrations of transition metals, lanthanides, and actinides, along with their isotopes, will be key to continuing studies on lunar samples.

***Acknowledgments and Dedication.*** On the eve of the thirtieth anniversary of the first lunar landing, July 1969, this paper is dedicated to those lunar explorers and scientists who have gone before us. First, we especially would like to dedicate this paper to Mitsunobu Tatsumoto, isotope geochemist without parallel, who passed away earlier this year (January 3, 1999). His pioneering efforts and the sheer volume and quality of his work in the isotopic study of extraterrestrial materials were an inspiration to us all. Second, we would like to dedicate this paper to those intrepid explorers of the Apollo program who have since passed on to greater adventures: Pete Conrad (Apollo 12), Jack Swigert (Apollo 13), Alan Shepard and Stu Roosa (Apollo 14), Jim Irwin (Apollo 15), and Ron Evans (Apollo 17). Reviews by Chip Shearer and Kevin Righter helped to clarify the presentation and are greatly appreciated. This research was supported by NASA grants NAG 5-8154 (LAT) and 344-31-30-21 (LEN).

## REFERENCES

Aeschlimann U., Eberhardt J., Geiss J., Grogler N., Kurtz J., and Marti K. (1982) On the age of cumulate norite 78236: An $^{39}$Ar-$^{40}$Ar study (abstract). In *Lunar and Planetary Science XIII,* pp. 1–2. Lunar and Planetary Institute, Houston.

Albee A., Chodos A. A., Gancarz A. J., Haines E. L., Papanastassiou D. A., Ray L., Tera F., Wasserburg G. J., and Wen T. (1972) Mineralogy, petrology, and chemistry of a Luna 16 basaltic fragment, sample B-1. *Earth Planet. Sci. Lett., 13,* 353–367.

Alexander E. C. Jr. and Davis P. K. (1974) $^{40}$Ar-$^{39}$Ar ages and trace element contents of Apollo 14 breccias: an interlaboratory cross-calibration of $^{40}$Ar-$^{39}$Ar standards. *Geochim. Cosmochim. Acta, 38,* 911–928.

Alexander E. C. Jr., Coscio M. R. Jr., Dragon J. C., Pepin R. O., and Saito K. (1980) K/Ar dating of lunar soils IV: Orange glass from 74220 and agglutinates from 14259 and 14163. *Proc. Lunar Planet. Sci. Conf. 11th,* pp. 1663–1677.

Alibert C., Norman M. D., and McCulloch M. T. (1994) An ancient age for a ferroan anorthosite clast from lunar breccia 67016. *Geochim. Cosmochim. Acta, 58,* 2921–2926.

Beard B. L., Snyder G. A., and Taylor L. A. (1994) Deep melting and residual garnet in the sources of lunar basalts: Lu-Hf isotopic systematics (abstract). In *Lunar and Planetary Science XXV,* pp. 73–74. Lunar and Planetary Institute, Houston.

Beard B. L., Taylor L. A., Scherer E. E., Johnson C. M., and Snyder G. A. (1996) The source mineralogy of high- and low-Ti basalts based on their Hf isotopic composition (abstract). In *Lunar and Planetary Science XXVII,* pp. 81–82. Lunar and Planetary Institute, Houston.

Beard B. L., Taylor L. A., Scherer E. E., Johnson C. M., and Snyder G. A. (1998) The source region and melting mineralogy of high-titanium and low-titanium lunar basalts deduced from Lu-Hf isotope data. *Geochim. Cosmochim. Acta, 62,* 525–544.

Beaty D. W. and Albee A. L. (1978) Comparative petrology and possible genetic relations among the Apollo 11 basalts. *Proc. Lunar Planet. Sci. Conf. 9th,* pp. 359–463.

Beaty D. W., Hill S. M. R., Albee A. L., Ma M.-S., and Schmitt R. A. (1979) The petrology and chemistry of basaltic fragments from the Apollo 11 soil, Part I. *Proc. Lunar Planet. Sci. Conf. 10th,* pp. 41–75.

Bell J. F. and Hawke B. R. (1984) Lunar dark-haloed impact craters: Origin and implications for early mare volcanism. *J. Geo-*

*phys. Res., 89,* 6899–6910.

Bernatowicz T. J., Hohenberg C. M., Hudson B., Kennedy B. M., and Podosek F. (1978) Argon ages for lunar breccias 14064 and 15405. *Proc. Lunar Planet. Sci. Conf. 9th,* pp. 905–919.

Binder A. B. (1982) The mare basalt magma source region and mare basalt magma genesis. *Proc. Lunar Planet. Sci. Conf. 13th,* in *J. Geophys. Res., 87,* A37–A53.

Binder A. B. (1985) Mare basalt genesis: Modeling trace elements and isotopic ratios. *Proc. Lunar Planet. Sci. Conf. 16th,* in *JGR, 90,* C396–C404.

Birck J. L. and Allegre C. J. (1994) Contrasting Re/Os magmatic fractionation in planetary basalts. *Earth Planet. Sci. Lett., 124,* 139–148.

Blichert-Toft J., Chauvel C., and Albarede F. (1997) Separation of Hf and Lu for high-precision isotope analysis of rock samples by magnetic sector-multiple collector ICP-MS. *Contrib. Mineral. Petrol., 127,* 248–260.

Blanchard D. P. and Budahn J. R. (1979) Remnants from the ancient lunar crust: Clasts from consortium breccia 73255. *Proc. Lunar Planet. Sci. Conf. 10th,* pp. 803–816.

Bogard D. D., Nyquist L. E., Bansal B. M., Wiesmann H., and Shih C.-Y. (1975) 76535: An old lunar rock. *Earth Planet. Sci. Lett., 26,* 69–80.

Bogard D. D., Garrison D. H., Shih C.-Y., and Nyquist L. E. (1994) [39]Ar-[40]Ar dating of two lunar granites: The age of Copernicus. *Geochim. Cosmochim. Acta, 58,* 3093–3100.

Borg L. E., Norman M. D., Nyquist L. E., Snyder G. A., Taylor L. A., Lindstrom M. M., and Wiesmann H. (1997) A relatively young samarium-neodymium age of 4.36 Ga for ferroan anorthosite 62236 (abstract). In *Meteoritics, 32,* A18.

Borg L. E., Norman M. D., Nyquist L. E., Bogard D. D., Snyder G. A., Taylor L. A., and Lindstrom M. M. (1999) Isotopic studies of ferroan anorthosite 62236: A young lunar crustal rock from a light rare-earth-element-depleted source. *Geochim. Cosmochim. Acta, 63,* in press.

Brown G. M., Emeleus C. H., Holland J. G., Peckett A., and Phillips R. (1972) Mineral-chemical variations in Apollo 14 and Apollo 15 basalts and granitic fractions. *Proc. Lunar Sci. Conf. 3rd,* pp. 141–157.

Cadogan P. H. and Turner G. (1976) The chronology of Apollo 17 Station 6 boulder. *Proc. Lunar Sci. Conf. 7th,* pp. 2267–2285.

Cadogan P. H. and Turner G. (1977) [40]Ar-[39]Ar dating of Luna 16 and Luna 20 samples. *Philos. Trans. Roy. Soc. London, A284,* 167–177.

Carlson R.W. and Lugmair G. W. (1979) Sm-Nd constraints on early lunar differentiation and the evolution of KREEP. *Earth Planet. Sci. Lett., 45,* 123–132.

Carlson R. W. and Lugmair G. W. (1981a) Time and duration of of lunar highlands crust formation. *Earth Planet. Sci. Lett., 52,* 227–238.

Carlson R. W. and Lugmair G. W. (1981b) Sm-Nd age of lherzolite 67667: Implications for the processes involved in lunar crustal formation. *Earth Planet. Sci. Lett., 52,* 227–238.

Carlson R. W. and Lugmair G. W. (1988) The age of ferroan anorthosite 60025: Oldest crust on a young Moon? *Earth Planet. Sci. Lett., 90,* 119–130.

Compston W., Berry H., Vernon M. J., Chappell B. W., and Kaye M. J. (1971) Rubidium-strontium chronology and chemistry of lunar material from the Ocean of Storms. *Proc. Lunar Sci. Conf. 2nd,* pp. 1471–1485.

Compston W., Foster J. J., and Gray C. M. (1975) Rb-Sr ages of clasts from within Boulder 1, Station 1, Apollo 17. *Moon, 14,* 445–462.

Compston W., Foster J. J., and Gray C. M. (1977) Rb-Sr systematics in clasts and aphanites from consortium breccia 73215. *Proc. Lunar Sci. Conf. 8th,* pp. 2525–2549.

Compston W., Williams I. S., and Meyer C. (1984) U-Pb geochronology of zircons from breccia 73217 using a sensitive high mass-resolution ion microprobe. *Proc. Lunar Planet. Sci. Conf. 14th,* in *J. Geophys. Res., 89,* B525–B534.

Dalrymple G. B. and Ryder G. (1996) [40]Ar-[39]Ar age spectra of Apollo 17 highlands breccia samples by laser step-heating and the age of the Serenitatis basin. *J. Geophys. Res., 101,* 26069–26084.

Dasch E. J., Shih C.-Y., Bansal B. M., Wiesmann H., and Nyquist L. E. (1987) Isotopic analysis of basaltic fragments from lunar breccia 14321: Chronology and petrogenesis of pre-Imbrium mare volcanism. *Geochim. Cosmochim. Acta, 51,* 3241–3254.

Delano J. W. (1986) Pristine lunar glasses: Criteria, data, and implications. *Proc. Lunar Planet. Sci. Conf. 16th,* in *J. Geophys. Res., 91,* D201–D213.

DePaolo D. J. and Wasserburg G. J. (1976) Nd isotopic variations and petrogenetic models. *Geophys. Res. Lett., 3,* 249–252.

Deutsch A. and Stöffler D. (1987) Rb-Sr analyses of Apollo 16 melt rocks and a new estimate for the Imbrium basin: Lunar basin chronology and the early heavy bombardment of the Moon. *Geochim. Cosmochim. Acta, 51,* 1951–1964.

Dickinson T., Taylor G. J., Keil K., Schmitt R. A., Hughes S. S., and Smith M. R. (1985) Apollo 14 aluminous mare basalts and their possible relationship to KREEP. *Proc. Lunar Planet. Sci. Conf. 15th,* in *J. Geophys. Res., 90,* C365–C374.

Drake M. J. and Consolmagno G. J. (1976) Critical review of models for the evolution of high-Ti mare basalts. *Proc. Lunar Sci. Conf. 7th,* pp. 1633–1657.

Eberhardt P., Geiss J., Grogler, and Stettler A. (1973) How old is the crater Copernicus? *Moon, 8,* 104–114.

Floss C., James O. B., and Crozaz G. (1998) Lunar ferroan anorthosite petrogenesis: Clues from race element distributions in FAN subgroups. *Geochim. Cosmochim. Acta, 62,* 1255–1283.

Green D. H., Ringwood A. E., Ware N. G., Hibberson W. O., Major A., and Kiss E. (1971) Experimental petrology and petrogenesis of Apollo 12 basalts. *Proc. Lunar Sci. Conf. 2nd,* pp. 601–615.

Halliday A. N. and Lee D. C. (1999) Tungsten isotopes and the early development of the Earth and Moon. *Chem. Geol.,* in press.

Halliday A. N., Lee D.-C., Christensen J. N., Walder A. J., Freedman P. A., Jones C. E., Hall C. M., Yi W., and Teagle D. (1995) Recent developments in inductively coupled plasma magnetic sector multiple collector mass spectrometry. *Intl. J. Mass Spec. Ion Proc., 146/147,* 21–34.

Halliday A. N., Rehkamper M., Lee D.-C., and Yi W. (1996) Early evolution of the Earth and Moon: new constraints from Hf-W isotope geochemistry. *Earth Planet. Sci. Lett., 142,* 75–89.

Halliday A. N., Lee D.-C., and Jacobsen S. B. (2000) Tungsten isotopes, the timing of metal-silicate fractionation, and the origin of the Earth and Moon. In *Origin of the Earth and Moon* (R. M. Canup and K. Righter, eds.), this volume. Univ. of Arizona, Tucson.

Harper C. L., Nyquist, L. E., Bansal, B., Wiesmann, H., and Shih C. -Y. (1995) Rapid accretion and early differentiation of Mars indicated by [142]Ns/[144]Nd in SNC meteorites. *Science, 267,* 213–217.

Hartmann W. K. and Davis D. R. (1975) Satellite-sized planetesimals and lunar origin. *Icarus, 24,* 504–515.

Hartmann W. K., Strom R. G., Weidenschilling S. J., Blasius K. R., Woronow A., Dence M. R., Grieve R. A. F., Diaz J., Chapman C. R., Shoemaker E. M., and Jones K. L. (1981) Chronology of planetary volcanism by comparative studies of planetary cratering. In *Basaltic Volcanism on the Terrestrial Planets*, pp. 1049–1127. Pergamon, New York.

Hartmann W. K., Ryder G., Dones L., and Grinspoon D. (2000) The time-dependent intense bombardment of the primordial Earth/Moon system. In *Origin of the Earth and Moon* (R. M. Canup and K. Righter, eds.), this volume. Univ. of Arizona, Tucson.

Hawke B. R. and Spudis P. D. (1980) Geochemical anomalies on the eastern limb and farside of the Moon. *Geochim. Cosmochim. Acta, Suppl. 12,* 467–481.

Head J. W. and Wilson L. (1992) Lunar mare volcanism: Stratigraphy, eruption conditions, and the evolution of secondary crusts. *Geochim. Cosmochim. Acta, 56,* 2155–2175.

Herbert F., Drake M. J., and Sonett C. P. (1978) Geophysical and geochemical evolution of the lunar magma ocean. *Proc. Lunar Planet. Sci. Conf. 9th,* pp. 249–262.

Hess P. C. (1989) Highly evolved liquids from the fractionation of mare and nonmare basalts. In *Workshop on the Moon in Transition* (G. J. Taylor and P. H. Warren, eds.), pp. 46–52. LPI Tech. Rpt. 89-03, Lunar and Planetary Institute, Houston.

Hess P. C. (1991) Diapirism and the origin of high $TiO_2$ mare glasses. *Geophys. Res. Lett., 18,* 2069–2072.

Hess P. C. and Parmentier E. M. (1993) Overturn of magma ocean ilmenite cumulate layer: implications for lunar magmatic evolution and formation of a lunar core (abstract). In *Lunar and Planetary Science XXIV,* pp. 651–652. Lunar and Planetary Institute, Houston.

Hess P. C. and Parmentier E. M. (1995) A model for thermal and chemical evolution of the Moon's interior: implications for the onset of mare volcanism. *Earth Planet. Sci. Lett., 134,* 501–514.

Hess P. C., Rutherford M. J., Guillemette R. N., Ryerson F. J., and Tuchfeld H. A. (1975) Residual products of fractional crystallization of lunar magmas: an experimental study. *Proc. Lunar Sci. Conf. 6th,* pp. 895–910.

Hinthorne J. R., Conrad R., and Andersen C. A. (1975) Lead-lead age and trace element abundances in lunar troctolite 76535 (abstract). In *Lunar Science VI,* pp. 373–375. Lunar Science Institute, Houston.

Hood L. L. and Jones J. H. (1987) Geophysical constraints on lunar bulk composition and structure: a reassessment. *Proc. Lunar Planet. Sci. Conf. 17th,* in *J. Geophys. Res., 92,* E396–E410.

Horn P., Jessberger E. K., Kirsten T., and Richter H. (1975) [39]Ar-[40]Ar dating of lunar rocks: Effects of grain size and neutron irradiation. *Proc. Lunar Sci. Conf. 6th,* pp. 1563–1591.

Hubbard N. J. and Minear J. W. (1976) Petrogenesis in a moderately endowed moon. *Proc. Lunar Sci. Conf. 7th,* pp. 3421–3437.

Hubbard N. J., Gast P. W., Meyer C., Nyquist L. E., Shih C., and Wiesmann H. (1971) Chemical composition of lunar anorthosites and their parent liquids. *Earth Planet. Sci. Lett., 13,* 71–75.

Huneke J. C. (1978) [40]Ar-[39]Ar microanalysis of single 74220 glass balls and 72435 breccia clasts. *Proc. Lunar Planet. Sci. Conf. 9th,* pp. 2345–2362.

Huneke J. C. and Wasserburg G. J. (1975) Trapped [40]Ar in troctolite 76535 and evidence for enhanced [40]Ar-[39]Ar age plateaus (abstract). In *Lunar Science VI,* pp. 417–419. Lunar Science

Institute, Houston.

Huneke J. C., Podosek F. A., and Wasserburg G. J. (1972) Gas retention and cosmic-ray exposure ages of a basalt fragment from Mare Fecunditatis. *Earth Planet. Sci. Lett., 13,* 375–383.

Huneke J. C., Jessberger E. K., Podosek F. A., and Wasserburg G. J. (1973) [40]Ar/[39]Ar measurements in Apollo 16 and 17 samples and the chronology of metamorphic and volcanic activity in the Taurus-Littrow region. *Proc. Lunar Sci.Conf. 4th,* pp. 1725–1756.

Hunter R. H. and Taylor L. A. (1983) The magma ocean from the Fra Mauro shoreline: An overview of the Apollo 14 crust. *Proc. Lunar Planet. Sci. Conf. 13th,* in *J. Geophys. Res., 88,* A591–A602.

Husain L. and Schaeffer O. A. (1973) Lunar volcanism: Age of the glass in Apollo 17 orange soil. *Science, 180,* 1358–1360.

Husain L. and Schaeffer O. A. (1975) Lunar evolution: The first 600 million years. *Geophys. Res. Lett., 2,* 29–32.

James O. B. (1981) Petrologic and age relations of the Apollo 16 rocks: Implications for subsurface geology and the age of the Nectaris basin. *Proc. Lunar Planet. Sci. 12B,* pp. 209–233.

Jerde E. A., Snyder G. A., Taylor L. A., Liu Y.-G., and Schmitt R. A. (1994) The origin and evolution of lunar high-Ti basalts: Periodic melting of a single source at Mare Tranquillitatis. *Geochim. Cosmochim. Acta, 58,* 515–527.

Jessberger E. K., Kirsten T., and Staudacher T. (1977) One rock and many ages — further K-Ar data on consortium breccia 73215. *Proc. Lunar Sci. Conf. 8th,* pp. 2567–2580.

Jessberger E. K., Staudacher T., Dominik B., and Kirsten T. (1978) Argon-argon ages of aphanite samples from consortium breccia 73255. *Proc. Lunar Planet. Sci. Conf. 9th,* pp. 841–854.

Jolliff B. L. (1991) Fragments of quartz monzodiorite and felsite in Apollo 14 soil particles. *Proc. Lunar Planet. Sci., Vol. 21,* pp. 101–118.

Jolliff B. K., Korotev R. L., and Haskin L. A. (1991) Geochemistry of 2–4 mm particles from Apollo 14 soil (14161) and implications regarding igneous components and soil-forming processes. *Proc. Lunar Planet. Sci., Vol. 21,* pp. 193–220.

Jolliff B. L., Haskin L. A., Colson R. O., and Wadhwa M. (1993) Partitioning in REE-saturating minerals: Theory, experiment, and modeling of whitlockite, apatite, and evolution of lunar residual magmas. *Geochim. Cosmochim. Acta, 57,* 4069–4094.

Korotev R. L. (1994) Compositional variation in Apollo 16 impact-melt breccias and inferences for the geology and bombardment history of the Central Highlands of the Moon. *Geochim. Cosmochim. Acta, 58,* 3931–3969.

Korotev R. L. (1997) Some things we can infer about the composition of the Apollo 16 regolith. *Meteoritics & Planet. Sci., 32,* 447–478.

Lee D.-C. and Halliday A. N. (1995) Hafnium-tungsten chronometry and the timing of terrestrial core formation. *Nature, 378,* 771–774.

Lee D.-C. and Halliday A. N. (1997) Core formation on Mars and differentiated asteroids. *Nature, 388,* 854–857.

Lee D.-C., Halliday A. N., Snyder G. A., and Taylor L. A. (1997) Age and origin of the Moon. *Science, 278,* 1098–1103.

Lee D.-C., Halliday A. N., Snyder G. A., and Taylor L. A. (2000) Lu-Hf isotopic constraints on lunar evolution. In preparation.

Leich D. A., Kahl S. B., Kirschbaum A. R., Niemeyer S., and Phinney D. (1975) Rare gas constraints on the history of Boulder 1, Station 2, Apollo 17. *Moon, 14,* 407–444.

Lindsley D. H. (1968) Melting relations of plagioclase at high pressures. In *Origin of Anorthosite and Related Rocks* (Y. W. Isachsen, ed.), pp. 39–46. New York State Museum and Science Service, Memoir 18, Albany, New York.

Lindstrom M. M., Knapp S. A., Shervais J. W., and Taylor L. A. (1984) Magnesian anorthosites and associated troctolites and dunite in Apollo 14 breccias. *Proc. Lunar Planet. Sci. Conf. 15th*, in *J. Geophys. Res., 89*, C41–C49.

Lindstrom M. M., Marvin U. B., Vetter S. K., and Shervais J. W. (1988) Apennine Front revisited: diversity of Apollo 15 highland rock types. *Proc. Lunar Planet. Sci. Conf. 18th*, pp. 169–185.

Longhi J. (1987) On the connection between mare basalts and picritic volcanic glasses. *Proc. Lunar Planet. Sci. Conf. 17th*, in *J. Geophys. Res., 92*, E349–E360.

Longhi J. (1992a) Origin of picritic green glass magmas by polybaric fractional fusion. *Proc. Lunar Planet. Sci., Vol. 22*, pp. 343–353.

Longhi J. (1992b) Experimental petrology and petrogenesis of mare volcanics. *Geochim. Cosmochim. Acta, 56*, 2235–2251.

Longhi J. and Ashwal L. D. (1985) Two-stage models for lunar and terrestrial anorthosites: Petrogenesis without a magma ocean. *Proc. Lunar Planet. Sci. Conf. 15th*, in *J. Geophys. Res., 90*, C571–C584.

Lugmair G. W. (1974) Sm-Nd ages: A new dating method. *Meteoritics, 9*, 369.

Lugmair G. W. and Carlson R. W. (1978) The Sm-Nd history of KREEP. *Proc. Lunar Planet. Sci. Conf. 9th*, pp. 689–704.

Lugmair G. W. and Galer S. J. G. (1992) Age and isotopic relationships among angrites Lewis Cliffs 86010 and Angra dos Reis, *Geochim. Cosmochim. Acta, 56*, 1673–1694.

Lugmair G. W. and Marti K. (1978) Lunar initial $^{143}$Nd/$^{144}$Nd: Differential evolution of the lunar crust and mantle. *Earth Planet. Sci. Lett., 39*, 349–357.

Lugmair G. W., Marti K., Kurtz J. P., and Scheinin N. B. (1976) History and genesis of lunar troctolite 76535. *Proc. Lunar Sci. Conf. 7th*, pp. 2009–2033.

Lunatic Asylum (1970) Mineralogic and isotopic investigations on lunar rock 12013. *Earth Planet. Sci. Lett., 9*, 137–163.

Martinez R. and Ryder G. (1989) A granite fragment from the Apennine Front — brother of QMD? (abstract). In *Lunar and Planetary Science XX*, pp. 620–621. Lunar and Planetary Institute, Houston.

Marvin U. B., Lindstrom M. M., Holmberg B. B., and Martinez R. R. (1991) New observations on the quartz monzodiorite-granite suite. *Proc. Lunar Planet. Sci., Vol. 21*, pp. 119–135.

Maurer P., Eberhardt P., Geiss J., Grogler N., Stettler A., Brown G. M., Peckett A., and Krahenbuhl U. (1978) Pre-Imbrian craters and basins: ages, compositions, and excavation depths of Apollo 16 breccias. *Geochim. Cosmochim. Acta, 42*, 1687–1720.

McKay G. A. and Le L. (1999) Partitioning of tungsten and hafnium between ilmenite and mare basaltic melt (abstract). In *Lunar and Planetary Science XXX*, Abstract #1996. Lunar and Planetary Institute, Houston (CD-ROM).

McKinley J. P., Taylor G. J., Keil K., Ma M.-S., and Schmitt R. A. (1984) Apollo 16: Impact melt sheets, contrasting nature of the Cayley Plains and Descartes Mountains, and geologic history. *Proc. Lunar Planet. Sci. Conf. 14th*, in *J. Geophys. Res., 89*, B513–B524.

Melosh H. J. (1990) Giant impacts and the thermal state of the early Earth. In *Origin of the Earth* (H. E. Newsom and J. H. Jones, eds.), pp. 69–83. Oxford Univ., New York.

Meyer C., Williams I. S., and Compston W. (1989) $^{207}$Pb/$^{206}$Pb ages of zircon-containing rock fragments indicate continuous magmatism in the lunar crust from 4350 to 3900 million years (abstract). In *Lunar and Planetary Science XX*, pp. 691–692. Lunar and Planetary Institute, Houston.

Meyer C., Williams I. S., and Compston W. (1996) Uranium-lead ages for lunar zircons: Evidence for a prolonged period of granophyre formation from 4.32 to 3.88 Ga. *Meteoritics & Planet. Sci., 31*, 370–387.

Minear J. W. (1980) The lunar magma ocean: A transient lunar phenomenon? *Proc. Lunar Planet. Sci. Conf. 11th*, pp. 1941–1955.

Minster J.-F., Birck J.-L., and Allegre C. J. (1982) Absolute age of formation of chondrites studied by the $^{87}$Rb-$^{87}$Sr method. *Nature, 300*, 414–419.

Misawa K., Tatsumoto M., Dalrymple G. B., and Yanai K. (1993) An extremely low U/Pb source in the Moon: U-Th-Pb, Sm-Nd, Rb-Sr, and $^{40}$Ar/$^{39}$Ar isotopic systematics and age of lunar meteorite Asuka 881757. *Geochim. Cosmochim. Acta, 57*, 4687–4702.

Mitchell J. N., Snyder G. A., and Taylor L. A. (1999) Mineral-chemical and isotopic variations in Apollo 16 impact-melt breccias. In *Planetary Petrology and Geochemistry: The Lawrence A. Taylor 60th Birthday Volume* (G. A. Snyder, C. R. Neal, and W. G. Ernst, eds.), pp. 173–193. Bellwether Publishing, GSA Intl. Geoscience Series, Vol. 2, Columbia, Maryland.

Mueller S., Taylor G. J., and Phillips R. J. (1988) Lunar composition: A geophysical and petrological synthesis. *J. Geophys. Res., 93*, 6338–6352.

Murthy V. R. and Coscio M. R. Jr. (1976) Rb-Sr ages and isotopic systematics of some Serenitatis mare basalts. *Proc. Lunar Sci. Conf. 7th*, pp. 1529–1544.

Nakamura N., Tatsumoto M., Nunes P., Unruh D. M., Schwab A. P., and Wildeman T. R. (1976) 4.4 by old clast in Boulder 7, Apollo 17: A comprehensive chronological study by U-Pb, Rb-Sr, and Sm-Nd methods. *Proc. Lunar Sci. Conf. 7th*, pp. 2309–2333.

Nakamura Y. (1983) Seismic velocity structure of the lunar mantle. *J. Geophys. Res., 88*, 677–686.

Neal C. R. and Taylor L. A. (1989) The nature of barium partitioning between immiscible melts: a comparison of experimental and natural systems with reference to lunar granite petrogenesis. *Proc. Lunar Planet. Sci. Conf. 19th*, pp. 209–218.

Neal C. R. and Taylor L. A. (1990) Modeling of lunar basalt petrogenesis: Sr isotope evidence from Apollo 14 high-alumina basalts. *Proc. Lunar Planet. Sci. Conf. 20th*, pp. 101–108.

Neal C. R. and Taylor L. A. (1992) Petrogenesis of mare basalts: A record of lunar volcanism. *Geochim. Cosmochim. Acta, 56*, 2177–2211.

Neal C. R., Taylor L. A., and Lindstrom M. M. (1988a) Importance of lunar granite and KREEP in very high potassium (VHK) basalt petrogenesis. *Proc. Lunar Planet. Sci. Conf. 18th*, pp. 121–137.

Neal C. R., Taylor L. A., and Lindstrom M. M. (1988b) Apollo 14 mare basalt petrogenesis: assimilation of KREEP-like components by a fractionating magma. *Proc. Lunar Planet. Sci. Conf. 18th*, pp. 139–153.

Neal C. R., Taylor L. A., Schmitt R. A., Hughes S. S., and Lindstrom M. M. (1989) High alumina (HA) and very high potassium (VHK) basalt clasts from Apollo 14 breccias, Part 2 — whole rock geochemistry: Further evidence for combined assimilation and fractional crystallization within the lunar crust. *Proc. Lunar Planet. Sci. Conf. 19th*, pp. 147–161.

Neal C. R., Taylor L. A., Patchen A. D., Hughes S. S., and Schmitt

R. A. (1990) The signficance of fractional crystallization in the petrogenesis of Apollo 17 Type A and B high-Ti basalts. *Geochim. Cosmochim. Acta, 54,* 1817–1833.

Neal C. R., Hacker M. D., Snyder G. A., Taylor L. A., Liu Y.-G., and Schmitt R. A. (1994) Basalt generation at the Apollo 12 site, Part 2: Source heterogeneity, multiple melts, and crustal contamination. *Meteoritics, 29,* 349–361.

Nyquist L. E. (1977) Lunar Rb-Sr chronology. *Phys. Chem. Earth, 10,* 103–142.

Nyquist L. E. and Shih C.-Y. (1992) On the chronology of and isotopic record of lunar basaltic volcanism. *Geochim. Cosmochim. Acta, 56,* 2213–2234.

Nyquist L. E., Hubbard N. J., Gast P. W., Bansal B. M., Wiesmann H., and Jahn B.-M. (1973) Rb-Sr systematics for chemically defined Apollo 15 and 16 materials. *Proc. Lunar Sci. Conf. 4th,* pp. 1823–1846.

Nyquist L. E., Bansal B. M., Wiesmann H., and Jahn B.-M. (1974) Taurus-Littrow chronology: Some constraints on early lunar crustal development. *Proc. Lunar Sci. Conf. 5th,* pp. 1515–1540.

Nyquist L. E., Bansal B. M., and Wiesmann H. (1976) Sr isotopic constraints on the petrogenesis of Apollo 17 mare basalts. *Proc. Lunar Sci. Conf. 7th,* pp. 1507–1528.

Nyquist L. E., Bansal B. M., Wooden J. L., and Wiesmann H. (1977) Sr-isotopic constraints on the petrogenesis of Apollo 12 mare basalts. *Proc. Lunar Sci. Conf. 8th,* pp. 1383–1415.

Nyquist L. E., Wiesmann H., Bansal B., Wooden J., and McKay G. (1978) Chemical and Sr-isotopic characteristics of the Luna 24 samples. In *Mare Crisium: The View from Luna 24* (R. B. Merrill and J. J. Papike, eds.), pp. 631–656. Pergamon, New York.

Nyquist L. E., Shih C.-Y., Wooden J. L., Bansal B. M., and Wiesmann H. (1979) The Sr and Nd isotopic record of Apollo 12 basalts: Implications for lunar geochemical evolution. *Proc. Lunar Planet. Sci. Conf. 10th,* pp. 77–114.

Nyquist L. E., Wooden J. L., Shih C.-Y., Wiesmann H., and Bansal B. M. (1981a) Isotopic and REE studies of lunar basalt 12038: implications for petrogenesis of aluminous mare basalts. *Earth Planet. Sci. Lett., 55,* 335–355.

Nyquist L. E., Reimold W. U., Bogard D. D., Wooden J. L., Bansal B. M., Weismann H., and Shih C.-Y. (1981b) A comprehensive Rb-Sr, Sm-Nd, and K-Ar study of shocked norite 78236: Evidence of slow cooling in the lunar crust? *Proc Lunar Planet. Sci. 12B,* pp. 67–97.

Nyquist L. E., Takeda H., Bansal B. M., Shih C.-Y., Wiesmann H., and Wooden J. L. (1986) Rb-Sr and Sm-Nd internal isochron ages of subophitic basalt clast and a matrix sample from the YU75011 eucrite. *J. Geophys. Res., 91,* 8137–8150.

Nyquist L. E., Horz F., Wiesmann H., Shih C.-Y., and Bansal B. (1987) Isotopic studies of shergottite chronology. I. Effect of shock metamorphism on the Rb-Sr system. In *Lunar and Planetary Science XVIII,* pp. 732–733. Lunar and Planetary Institute, Houston.

Nyquist L. E., Bansal B., Wiesmann H., Shih C.-Y., and Hörz F. (1988) Isotopic studies of shock metamorphism: II. Sm-Nd (abstract). In *Lunar and Planetary Science XIX,* pp. 875–876.

Nyquist L. E., Wiesmann H., Bansal B., Shih C.-Y., Keith J. E., and Harper C. L. (1995) [146]Sm-[142]Nd formation interval for the lunar mantle. *Geochim. Cosmochim. Acta, 59,* 2817–2837.

Oberli F., Huneke J. C., and Wasserburg G. J. (1979) U-Pb and K-Ar sytematics of cataclysm and precataclysm lunar impactites (abstract). In *Lunar and Planetary Science X,* pp. 940–942.

Lunar and Planetary Institute, Houston.

Paces J. B., Nakai S., Neal C. R., Taylor L. A., Halliday A. N., and Lee D. C. (1991) A strontium and neodymium isotopic study of Apollo 17 high-Ti mare basalts: Resolution of ages, evolution of magmas, and origins of source heterogeneities. *Geochim. Cosmochim. Acta, 55,* 2025–2043.

Papanastassiou D. A. and Wasserburg G. J. (1969) Initial strontium isotopic abundances and the resolution of small time differences in the formation of planetary objects. *Earth Planet. Sci. Lett., 5,* 361–370.

Papanastassiou D. A. and Wasserburg G. J. (1970) Rb-Sr ages from the Ocean of Storms. *Earth Planet. Sci. Lett., 8,* 269–278.

Papanastassiou D. A. and Wasserburg G. J. (1971) Lunar chronology and evolution from Rb-Sr studies of Apollo 11 and 12 samples. *Earth Planet. Sci. Lett., 11,* 37–62.

Papanastassiou D. A. and Wasserburg G. J. (1972) Rb-Sr age of a Luna 16 basalt and the model age of lunar soils. *Earth Planet. Sci. Lett., 13,* 368–374.

Papanastassiou D. A. and Wasserburg G. J. (1975) Rb-Sr study of a lunar dunite and evidence for early lunar differentiates. *Proc. Lunar Sci. Conf. 6th,* pp. 1467–1489.

Papanastassiou D. A. and Wasserburg G. J. (1976) Rb-Sr age of troctolite 76535. *Proc. Lunar Sci. Conf. 7th,* pp. 2035–2054.

Papanastassiou D. A., DePaolo D. J., and Wasserburg G. J. (1977) Rb-Sr and Sm-Nd chronology and genealogy of mare basalts from the Sea of Tranquillity. *Proc. Lunar Sci. Conf. 8th,* pp. 1639–1672.

Papike J. J. (1996) Pyroxene as a recorder of cumulate formational processes in asteroids, Moon, Mars, Earth: Reading the record with the ion microprobe. *Amer. Mineral., 81,* 525–544.

Papike J. J., Fowler G. W., and Shearer C. K. (1994) Orthopyroxene as a recorder of lunar crust evolution: An ion microprobe investigation of Mg-suite norites. *Am. Mineral., 79,* 796–800.

Papike J. J., Fowler G. W., Shearer C. K., and Layne G. D. (1996) Ion microprobe investigation of plagioclase and orthopyroxene from lunar Mg-suite norites: Implications for calculating parental REE concentrations and for assessing postcrystallization REE distribution. *Geochim. Cosmochim. Acta, 60,* 3967–3978.

Papike J. J., Fowler G. W., and Shearer C. K. (1997) Evolution of the lunar crust. *Geochim. Cosmochim. Acta, 61,* 2343–2350.

Patchett P. J. and Tatsumoto M. (1980a) Lu-Hf total-rock isochron for the eucrite meteorites. *Nature, 288,* 571–574.

Patchett P. J. and Tatsumoto M. (1980b) A routine high-precision method for Lu-Hf isotope geochemistry and chronology. *Contrib. Mineral. Petrol., 75,* 263–267.

Philpotts J. A. and Schnetzler C. C. (1970) Apollo 11 lunar samples: K, Rb, Sr, Ba, and rare-earth concentrations in some rocks and separated phases. *Proc. Apollo 11 Lunar Sci. Conf.,* pp. 1471–1486.

Podosek F. A., Huneke J. C., Gancarz A. J., and Wasserburg G. J. (1973) The age and petrography of two Luna 20 fragments and inferences for widespread lunar metamorphism. *Geochim. Cosmochim. Acta, 37,* 887–904.

Premo W. R. and Tatsumoto M. (1991) U-Th-Pb isotopic systematics of lunar norite 78235. *Proc. Lunar Planet. Sci., Vol. 21,* pp. 89–100.

Premo W. R. and Tatsumoto M. (1992) U-Th-Pb, Rb-Sr, and Sm-Nd isotopic systematics of lunar troctolitic cumulate 76535: Implications on the age and origin of the early lunar, deep-seated cumulate. *Proc. Lunar Planet. Sci., Vol. 22,* 381–397.

Premo W. R. and Tatsumoto M. (1993) U-Pb isotopic systematics of ferroan anorthosite 60025 (abstract). In *Lunar and Planetary Science XXIV,* pp. 1173–1174. Lunar and Planetary Institute, Houston.

Premo W. R. and Tatsumoto M. (2000) Contrasting U-Th-Pb, Rb-Sr, and Sm-Nd isotopic systematics of lunar ferroan anorthosites 60025 and 62237: Implications on the age and origin of the Moon. In preparation.

Premo W. R., Tatsumoto M., Misawa K., Nakamura N., and Kita N. I. (1999) Pb-isotopic systematics of lunar highland rocks (>3.9 Ga): Constraints on early lunar evolution. In *Planetary Petrology and Geochemistry: The Lawrence A. Taylor 60th Birthday Volume* (G. A. Snyder, C. R. Neal, and W. G. Ernst, eds.), pp. 207–241. Bellwether Publishing, GSA Intl. Geoscience Series, Vol. 2, Columbia, Maryland.

Quick J. E., Albee A. L., Ma M.-S., Murali A. V., and Schmitt R. A. (1977) Chemical compositions and possible immiscibility of two silicate melts in 12013. *Proc. Lunar Sci. Conf. 8th,* pp. 2153–2189.

Reimold W. U. and Reimold J. N. (1984) The mineralogical, chemical, and chronological characteristics of the crystalline Apollo 16 impact melt rocks. *Fortschr. Mineral., 62,* 269–301.

Reimold W. U., Nyquist L. E., Bansal B. M., Wooden J. L., Shih C.-Y., Wiesmann H., and Mackinnon I. D. R. (1985) Isotope analysis of crystalline impact-melt rocks from Apollo 16 stations 11 and 13, North Ray Crater. *Proc. Lunar Planet. Sci. Conf. 15th,* in *J. Geophys. Res., 90,* C597–C612.

Rhodes J. M., Hubbard N. J., Wiesmann H., Rodgers K. V., Brannon J. C., and Bansal B. M. (1976) Chemistry, classification, and petrogenesis of Apollo 17 mare basalts. *Proc. Lunar Sci. Conf. 7th,* pp. 1467–1489.

Ridley W. I. (1975) On high-alumina mare basalts. *Proc. Lunar Sci. Conf. 6th,* pp. 131–145.

Righter K. and Drake M. J. (1999) Partitioning of W between liquid metal, solid silicates, and liquid silicates at high pressures and temperatures: Implications for the $^{182}$W isotope anomalies in lunar and martian samples (abstract). In *Lunar and Planetary Science XXX,* Abstract #1381. Lunar and Planetary Institute, Houston (CD-ROM).

Righter K., Walker R. J., and Warren P. H. (2000) Significance of highly siderophile elements and osmium isotopes in the lunar and terrestrial mantles. In *Origin of the Earth and Moon* (R. M. Canup and K. Righter, eds.), this volume. Univ. of Arizona, Tucson.

Ringwood A. E. (1989) Flaws in the giant impact hypothesis of lunar origin. *Earth Planet. Sci. Lett., 95,* 208–214.

Ringwood A. E. and Essene E. (1970) Petrogenesis of Apollo 11 basalts, internal constitution and origin of the moon. *Proc. Apollo 11 Lunar Sci. Conf.,* pp. 769–799.

Ringwood A. E. and Kesson S. E. (1976) A dynamic model for mare basalt petrogenesis. *Proc. Lunar Sci. Conf. 7th,* pp. 1697–1722.

Ruzicka A., Snyder G. A., and Taylor L. A. (1999) Giant impact and fission hypotheses for the origin of the Moon: A critical review of some geochemical evidence. In *Planetary Petrology and Geochemistry: The Lawrence A. Taylor 60th Birthday Volume* (G. A. Snyder, C. R. Neal, and W. G. Ernst, eds.), pp. 121–134. Bellwether Publishing, GSA Intl. Geoscience Series, Vol. 2, Columbia, Maryland.

Ryder G. (1976) Lunar sample 15405: Remnant of a KREEP basalt-granite differentiated pluton. *Earth Planet. Sci. Lett., 29,* 255–268.

Ryder G. (1988) Quenching and disruption of lunar KREEP lava

flows by impacts. *Nature, 336,* 751–754.

Ryder G. (1990) Lunar samples, lunar accretion and the early bombardment of the Moon. *Eos Trans. AGU, 71,* 313, 322–323.

Ryder G. (1991) Lunar ferroan anorthosites and mare basalt sources: the mixed connection. *Geophys. Res. Lett., 18,* 2065–2068.

Ryder G. (1994) Coincidence in time of the Imbrium basin impact and Apollo 15 KREEP volcanic flows: The case for impact-induced melting. In *Large Meteorite Impacts and Planetary Evolution* (B. O. Dressler, R. A. F. Grieve, and V. L. Sharpton, eds.), pp. 11–18. GSA Spec. Paper 293, Boulder, Colorado.

Ryder G. and Spudis P. D. (1980) Volcanic rocks in the lunar highlands. *Geochim. Cosmochim. Acta, Suppl. 12,* 353–375.

Ryder G. and Taylor G. J. (1976) Did mare-type volcanism commence early in lunar history? *Proc. Lunar Sci. Conf. 7th,* pp. 1741–1755.

Ryder G., Bogard D., and Garrison D. (1991) Probably age of Autolycus and calibration of lunar stratigraphy. *Geology, 19,* 143–146.

Ryder G., Koeberl C., and Mojzsis S. J. (2000) Evidence of early bombardment on the Earth and Moon. In *Origin of the Earth and Moon* (R. M. Canup and K. Righter, eds.), this volume. Univ. of Arizona, Tucson.

Schaeffer G. A. and Schaeffer O. A. (1977) $^{40}$Ar-$^{39}$Ar ages of lunar rocks. *Proc. Lunar Sci. Conf. 8th,* pp. 2253–2300.

Schaeffer O. A. and Husain L. (1974) Chronology of lunar basin formation. *Proc. Lunar Sci. Conf. 5th,* pp. 1541–1555.

Shearer C. K. and Floss C. (2000) Evolution of the Moon's mantle and crust as reflected in trace-element microbeam studies of lunar magmatism. In *Origin of the Earth and Moon* (R. M. Canup and K. Righter, eds.), this volume. Univ. of Arizona. Tucson.

Shearer C. H. and Newsom H. E. (1998) The origin of mantle reservoirs for mare basalts and implications for the thermal and chemical evolution of the lunar magma ocean (abstract). In *Origin of the Earth and Moon,* pp. 39–40. LPI Contribution No. 957, Lunar and Planetary Institute, Houston.

Shearer C. K., Papike J. J., Galbreath K. C., Wentworth S. J., and Shimizu N. (1990) A SIMS study of lunar "komatiitic glasses": Trace element characteristics and possible origin. *Geochim. Cosmochim. Acta, 54,* 1851–1857.

Shearer C. K., Papike J. J., and Layne G. D. (1996) The role of ilmenite in the source region for mare basalts: Evidence from niobium, zirconium, and cerium in picritic glasses. *Geochim. Cosmochim. Acta, 60,* 3521–3530.

Shervais J. W. (1994) Ion microprobe studies of lunar highland cumulate rocks: Preliminary results (abstract). In *Lunar and Planetary Science XXV,* pp. 1265–1266. Lunar and Planetary Institute, Houston.

Shervais J. W. and McGee J. J. (1997) KREEP in the western lunar highlands: An ion microprobe study of alkali and Mg suite cumulates from the Apollo 12 and 14 sites. In *Lunar and Planetary Science XXVIII,* pp. 1301–1302. Lunar and Planetary Institute, Houston.

Shervais J. W. and McGee J. J. (1998a) KREEP in the western lunar highlands: ion and electron microprobe study of alkali suite anorthosites and norites from Apollo 12 and 14. *Am. Mineral.,* in press.

Shervais J. W. and McGee J. J. (1998b) Ion and electron microprobe study of troctolites, norite, and anorthosites from Apollo 14: Evidence of urKREEP assimilation during petrogenesis of

Apollo 14 Mg-suite rocks. *Geochim. Cosmochim. Acta,* in press.

Shervais J. W. and Stuart J. B. (1995) Ion microprobe studies of lunar highland cumulate rocks: New results (abstract). In *Lunar and Planetary Science XXVI,* pp. 1285–1286. Lunar and Planetary Institute, Houston.

Shervais J. W., Taylor L. A., Laul J. C., Shih C.-Y., and Nyquist L. E. (1985) Very high potassium (VHK) basalt: Complications in in lunar mare basalt petrogenesis. *Proc. Lunar Planet. Sci. Conf. 16th,* in *J. Geophys. Res., 90,* D3–D18.

Shih C.-Y. (1977) Origins of KREEP basalts. *Proc. Lunar Sci. Conf. 8th,* pp. 2375–2401.

Shih C.-Y. and Schonfeld E. (1976) Mare basalt genesis: a cumulate-remelting model. *Proc. Lunar Sci. Conf. 7th,* pp. 1757–1792.

Shih C.-Y., Haskin L. A., Wiesmann H., Bansal B. M., and Brannon J. C. (1975) On the origin of high-Ti mare basalts. *Proc. Lunar Sci. Conf. 6th,* pp. 1255–1285.

Shih C.-Y., Nyquist L. E., Bogard D. D., Wooden J. L., Bansal B. M., and Wiesmann H. (1985) Chronology and petrogenesis of a 1.8 g lunar granitic clast: 14321,1062. *Geochim. Cosmochim. Acta, 49,* 411–426.

Shih C.-Y., Nyquist L. E., Bogard D. D., Bansal B. M., Wiesmann H., Johnson P., Shervais J. W., and Taylor L. A. (1986) Geochronology and petrogenesis of Apollo 14 very high potassium mare basalts. *Proc. Lunar Planet. Sci. Conf. 16th,* in *J. Geophys. Res., 91,* D214–D228.

Shih C.-Y., Nyquist L. E., Bansal B. M., and Wiesmann H. (1992) Rb-Sr and Sm-Nd chronology of an Apollo 17 KREEP basalt. *Earth Planet. Sci. Lett., 108,* 203–215.

Shih C.-Y., Nyquist L. E., Dasch E. J., Bogard D. D., Bansal B. M., and Wiesmann H. (1993a) Ages of pristine noritic clasts from lunar breccias 15445 and15455. *Geochim. Cosmochim. Acta, 57,* 915–931.

Shih C.-Y., Nyquist L. E., and Wiesmann H. (1993b) K-Ca chronology of lunar granites. *Geochim. Cosmochim. Acta, 57,* 4827–4841.

Shih C.-Y., Nyquist L. E., Bogard D. D., Reese Y., Wiesmann H., and Garrison D. (1999) Rb-Sr, Sm-Nd and $^{40}$Ar-$^{39}$Ar isotopic studies of an Apollo 11 group D basalt. In *Lunar and Planetary Science XXX,* Abstract #1787. Lunar and Planetary Institute, Houston (CD-ROM).

Shirley D. N. (1983) A partially molten magma ocean model. *Proc. Lunar Planet. Sci. Conf. 13th,* in *J. Geophys. Res., 88,* A519–A527.

Smith J. V., Anderson A. T., Newton R. C., Olsen E. J., Wyllie P. J., Crewe A. V., Isaacson M. S., and Johnson D. (1970) Petrologic history of the moon inferred from petrography, mineralogy, and petrogenesis of Apollo 11 rocks. *Proc. Apollo 11 Lunar Sci. Conf.,* pp. 897–925.

Snyder G. A. and Taylor L. A. (1993) Constraints on the genesis and evolution of the Moon's magma ocean and derivative cumulate sources as supported by lunar meteorites. *Proc. NIPR Symp. Antarct. Meteorites, 6,* 246–267.

Snyder G. A., Taylor L. A., Liu Y.-G., and Schmitt R. A. (1992a) Petrogenesis of the western highlands of the Moon: Evidence from a diverse group of whitlockite-rich rocks from the Fra Mauro Formation. *Proc. Lunar Planet. Sci., Vol. 22,* pp. 399–416.

Snyder G. A., Taylor L. A., and Neal C. R. (1992b) A chemical model for generating the sources of mare basalts: Combined equilibrium and fractional crystallization of the lunar magma-sphere. *Geochim. Cosmochim. Acta, 56,* 3809–3823.

Snyder G. A., Taylor L. A., and Crozaz G. (1993) Rare earth element selenochemistry of immiscible liquids and zircon at Apollo 14: An ion probe study of evolved rocks on the Moon. *Geochim. Cosmochim. Acta, 57,* 1143–1149.

Snyder G. A., Lee D-C., Taylor L. A., Halliday A. N., and Jerde E. A. (1994) Evolution of the upper mantle of the Earth's Moon: Neodymium and strontium isotopic constraints from high-Ti mare basalts. *Geochim. Cosmochim. Acta, 58,* 4795–4808.

Snyder G. A., Taylor L. A., and Halliday A. N. (1995a) Chronology and petrogenesis of the lunar highlands alkali suite: cumulates from KREEP basalt crystallization. *Geochim. Cosmochim. Acta, 59,* 1185–1203.

Snyder G. A., Neal C. R., Taylor L. A., and Halliday A. N. (1995b) Processes involved in the formation of magnesian-suite plutonic rocks from the highlands of the Earth's moon. *J. Geophys. Res., 100,* 9365–9388.

Snyder G. A., Hall C. M., Lee D.-C., Taylor L. A., and Halliday A. N. (1996) Earliest high-Ti volcanism on the Moon: $^{40}$Ar-$^{39}$Ar, Sm-Nd, and Rb-Sr isotopic studies of Group D basalts from the Apollo 11 landing site. *Meteoritics & Planet. Sci., 31,* 328–334.

Snyder G. A., Borg L. E., Lee D. C., Taylor L. A., Nyquist L. E., and Halliday A. N. (1997a) Nd-Sr-Hf isotopic and geochronologic studies of Apollo 15 basalts (abstract). In *Lunar and Planetary Science XXVIII,* pp. 1347–1348. Lunar and Planetary Institute, Houston.

Snyder G. A., Neal C. R., Taylor L. A., and Halliday A. N. (1997b) Anatexis of lunar cumulate mantle in time and space: Clues from trace-element, strontium, and neodymium isotopic chemistry of parental Apollo 12 basalts. *Geochim. Cosmochim. Acta, 61,* 2731–2747.

Snyder G. A., Borg L. E., Taylor L. A., Nyquist L. E., and Halliday A. N. (1998) Volcanism in the Hadley-Apennine region of the Moon: Geochronology, Nd-Sr isotopic systematics, and depths of melting (abstract). In *Lunar and Planetary Science XXIX.* Lunar and Planetary Institute, Houston.

Snyder G. A., Lee D.-C., Taylor L. A., and Halliday A. N. (1999a) Earliest lunar volcanism: An alternative interpretation of the Apollo 14 high-Al basalts from Nd-Sr-Hf isotopic studies. *Meteoritics & Planet. Sci.,* in press.

Snyder G. A., Borg L. E., Lee D. C., Nyquist L. E., Taylor L. A., and Halliday A. N. (1999b) Volcanism in the Hadley-Apennine region of the Moon: Chronology, Nd-Sr-Hf isotopic systematics, and petrogenesis of Apollo 15 mare basalts. *Geochim. Cosmochim. Acta,* submitted.

Solomon S. C. and Longhi J. (1977) Magma oceanography: 1. Thermal evolution. *Proc. Lunar Planet. Sci. Conf. 8th,* pp. 583–599.

Spangler R. R., Warasila R., and Delano J. W. (1984) $^{39}$Ar-$^{40}$Ar ages for the Apollo 15 green and yellow volcanic glasses. *Proc. Lunar Planet. Sci. Conf. 14th,* in *J. Geophys. Res., 89,* B487–B497.

Spera F. J. (1992) Lunar magma transport phenomena. *Geochim. Cosmochim. Acta, 56,* 2253–2265.

Spudis P. D. and Davis P. A. (1986) A chemical and petrological model of the lunar crust and implications for lunar crustal origin. *Proc. Lunar Planet. Sci. Conf. 17th,* in *J. Geophys. Res., 91,* E84–E90.

Stadermann F. J., Heusser E., Jessberger E. K., Lingner S., and Stoffler D. (1991) The case for a younger Imbrium basin: New

[40]Ar-[39]Ar ages of Apollo 14 rocks. *Geochim. Cosmochim. Acta,* *55,* 2339–2349.

Staudacher T., Jessberger E. K., Flohs I., and Kirsten T. (1979) [40]Ar-[39]Ar age systematics of consortium breccia 73255. *Proc. Lunar Planet. Sci. Conf. 10th,* pp. 745–762.

Steele I. M., Hutcheon I. D., and Smith J. V. (1980) Ion microprobe analysis and petrogenetic interpretations of Li, Mg, Ti, K, Sr, and Ba in lunar plagioclase. *Proc. Lunar Planet. Sci. Conf. 11th,* pp. 571–590.

Steiger R. H. and Jaeger E. (1977) Subcommission on geochronology: Convention on the use of decay constants in geo- and cosmochronology. *Earth Planet. Sci. Lett., 36,* 359–362.

Stettler A., Eberhardt P., Geiss J., Grögler N., and Maurer P. (1973) [39]Ar-[40]Ar ages and [37]Ar-[38]Ar exposure ages of lunar rocks. *Proc. Lunar Sci. Conf. 4th,* pp. 1865–1888.

Stettler A., Eberhardt P., Geiss J., and Grögler N. (1974) [39]Ar-[40]Ar ages of samples from Apollo 17 station 7 Boulder and implications for its formation. *Earth Planet. Sci. Lett., 23,* 453–461.

Swindle T. D., Spudis P. D., Taylor G. J., Korotev R. L., Nichols R. H. Jr., and Olinger C. T. (1991) Searching for Crisium basin ejecta: Chemistry and ages of Luna 20 impact melts. *Proc. Lunar Planet. Sci., Vol. 21,* pp. 167–181.

Tatsumoto M. and Unruh D. M. (1976) KREEP basalt age: Grain by grain U-Th-Pb systematics study of the quartz monzodiorite clast 15405,88. *Proc. Lunar Sci. Conf. 7th,* pp. 2107–2129.

Tatsumoto M., Hedge C. E., Knight R. J., Unruh D. M., and Doe B. R. (1972) U-Th-Pb, Rb-Sr, and K measurements on some Apollo 15 and Apollo 16 samples. In *The Apollo 15 Lunar Samples* (J. W. Chamberlain and C. Watkins, eds.), pp. 391–395. Lunar Science Institute, Houston.

Tatsumoto M., Nunes P. D., Knight R. J., Hedge C. E., and Unruh D. M. (1973) U-Th-Pb, Rb-Sr, and K measurements of two Apollo 17 samples. *Eos Trans. AGU, 54,* 614–615.

Tatsumoto M., Premo W. R., and Unruh D. M. (1987) Origin of lead from green glass of Apollo 15426: A search for primitive lunar lead. *Proc. Lunar Planet. Sci. Conf. 17th,* in *J. Geophys. Res., 92,* E361–E371.

Taylor G. J., Warner R. D., Keil K., Ma M.-S., and Schmitt R. A. (1980) Silicate liquid immiscibility, evolved lunar rocks and the formation of KREEP. In *Proc. Conf. Lunar Highlands Crust* (J. J. Papike and R. B. Merrill, eds.), pp. 339–352. Pergamon, New York.

Taylor L. A., Shervais J. W., Hunter R. H., Shih C.-Y., Bansal B. M., Wooden J., Nyquist L. E., and Laul J. C. (1983) Pre-4.2 AE mare-basalt volcanism in the lunar highlands. *Earth Planet. Sci. Lett., 66,* 33–47.

Taylor S. R. (1982) *Planetary Science: A Lunar Perspective.* Lunar and Planetary Institute, Houston. 481 pp.

Taylor S. R. and Jakeš P. (1974) The geochemical evolution of the Moon. *Proc. Lunar Sci. Conf. 5th,* pp. 1287–1305.

Tera F. and Wasserburg G. J. (1976) Lunar ball games and other sports (abstract). In *Lunar Science VII,* pp. 858–860. Lunar Science Institute, Houston.

Tera F., Ray L. A., and Wasserburg G. J. (1972) Distribution of Pb-U-Th in lunar anorthosite 15415 and inferences about its age. In *The Apollo 15 Lunar Samples* (J. W. Chamberlain and C. Watkins, eds.), pp. 396–401. Lunar Science Institute, Houston.

Tera F., Papanastassiou D., and Wasserburg G. J. (1974) Isotopic evidence for a terminal lunar cataclysm. *Earth Planet. Sci. Lett., 22,* 1.

Tonks W. B. and Melosh H. J. (1990) The physics of crystal settling and suspension in a turbulent magma ocean. In *Origin of*

*the Earth* (H. Newsom and J. H. Jones, eds.), pp. 151–174. Oxford Univ., New York.

Torigoye-Kita N., Misawa K., Dalrymple G. R., and Tatsumoto M. (1995) Further evidence for a low U/Pb source in the Moon: U-Th-Pb, Sm-Nd, and Ar-Ar isotopic systematics of lunar meteorite Yamato-793169. *Geochim. Cosmochim. Acta, 59,* 2621–2632.

Treiman A. H. (1996) The perils of partition: Difficulties in retrieving magma compositions from chemically equilibrated basaltic meteorites. *Geochim. Cosmochim. Acta, 60,* 147–155.

Turner G. (1971) [40]Ar-[39]Ar ages from the lunar maria. *Earth Planet. Sci. Lett., 11,* 169–191.

Turner G. and Cadogan P. H. (1975) The history of lunar bombardment inferred from [40]Ar-[39]Ar dating of highland rocks. *Proc. Lunar Sci. Conf. 6th,* pp. 1509–1538.

Unruh D. M., Stille P., Patchett P. J., and Tatsumoto M. (1984) Lu-Hf and Sm-Nd evolution in lunar mare basalts. *Proc. Lunar Planet. Sci. Conf. 14th,* in *J. Geophys. Res., 89,* B459–B477.

Walder A. J., Koller D., Reed N. M., Hutton R. C., and Freedman P. A. (1993) Isotope ratio measurement by inductively coupled plasma multiple collector mass spectrometry incorporating a high efficiency nebulization. *J. Analyt. Atomic Spectrometry, 8,* 1037–1041.

Walker D., Longhi J., and Hays J. F. (1972) Experimental petrology and origin of Fra Mauro rocks and soil. *Proc. Lunar Sci. Conf. 3rd,* pp. 797–817.

Walker D., Longhi J., Stolper E. M., Grove T., and Hays J. F. (1975) Origin of titaniferous lunar basalts. *Geochim. Cosmochim. Acta, 39,* 1219–1235.

Walker R. J., Morgan J. W., Shearer C. K., and Papike J. J. (1998) Rhenium-osmium isotopic systematics of lunar orange glass (abstract). In *Lunar and Planetary Science XXIX,* Abstract #1271. Lunar and Planetary Institute, Houston (CD-ROM).

Warner R. D., Taylor G. J., Keil K., Ma M.-S., and Schmitt R. A. (1980) Aluminous mare basalts: New data from Apollo 14 coarse-fines. *Proc. Lunar Planet. Sci. Conf. 11th,* pp. 87–104.

Warren P. H. (1985) The magma ocean concept and lunar evolution. *Annu. Rev. Earth Planet. Sci., 13,* 201–240.

Warren P. H. (1988) The origin of pristine KREEP: Effects of mixing between urKREEP and the magmas parental to the Mg-rich cumulates. *Proc. Lunar Planet. Sci. Conf. 18th,* pp. 233–241.

Warren P. H. (1992) Inheritance of silicate differentiation during lunar origin by giant impact. *Earth Planet. Sci. Lett., 112,* 101–116.

Warren P. H. and Wasson J. T. (1979) The origin of KREEP. *Rev. Geophys. Space Phys., 17,* 73–88.

Warren P. H. and Wasson J. T. (1980) Further foraging of pristine nonmare rocks: Correlations between geochemistry and longitude. *Proc. Lunar Planet Sci. Conf. 11th,* pp. 431–470.

Warren P. H., Taylor G. J., Keil K., Marshall C., and Wasson J. T. (1981) Foraging westward for pristine nonmare rocks: Complications for petrogenetic models. *Proc. Lunar Planet. Sci. 12B,* pp. 21–40.

Warren P. H., Taylor G. J., Keil K., Shirley D. N., and Wasson J. T. (1983) Petrology and chemistry of two "large" granite clasts from the Moon. *Earth Planet. Sci. Lett., 64,* 175–185.

Wasserburg G. J., Papanastassiou D. A., Nenow E. V., and Bauman C. A. (1969) A programmable magnetic field mass spectometer with on-line data processing. *Rev. Sci. Instrum., 40,* 288–291.

Wasserburg G. J., Radicati di Brozolo F., Papanastassiou D. A., McCulloch M. T., Huneke J. C., Dymek R. F., DePaolo D. J.,

Chodos A. A., and Albee A. L. (1978) Petrology, chemistry, age and irradiation history of Luna 24 samples. In *Mare Crisium: The View from Luna 24* (R. B. Merrill and J. J. Papike, eds.), pp. 657–678. Pergamon, New York.

Wentworth S. J. and McKay D. S. (1988) Glasses in ancient and young Apollo 16 regolith breccia: Populations and ultra-Mg' glass. *Proc. Lunar Planet. Sci. Conf. 18th,* pp. 67–77.

Whitford-Stark J. L. and Head J. W. (1980) Stratigraphy of Oceanus Procellarum basalts: Sources and styles of emplacement. *J. Geophys. Res., 85,* 6579–6609.

Wood J. A., Dickey J. S., Marvin U. B., and Powell B. N. (1970) Lunar anorthosites and a geophysical model of the Moon. *Proc. Apollo 11 Lunar Sci. Conf.,* pp. 965–988.

# Recent Refinements in Geophysical Constraints on Lunar Origin and Evolution

## Lon L. Hood
*University of Arizona*

## Maria T. Zuber
*Massachussetts Institute of Technology*

Recent, more complete geophysical mapping of the Moon by the Clementine and Lunar Prospector missions, as well as continued analysis and interpretation of the Apollo lunar seismic dataset, have significantly improved existing geophysical constraints on lunar origin and evolution. Gravity and topography data indicate large lateral variations in crustal thickness (~20–110 km) and a range of compensation states that imply a more complex near-surface thermal history than had previously been considered. A recent P-wave mantle velocity model together with earlier quantitative modeling strongly suggests that initial melting and differentiation of the lunar interior was limited to depths <500 km. Recent analyses of Lunar Prospector magnetometer, gravity, and groundbased laser-ranging data strengthen earlier conclusions that the Moon most probably possesses a metallic core with a mass of 1–3% of the lunar mass.

## 1. INTRODUCTION

Geophysical observations impose quantitative constraints on the internal structure, thermal state, thermal history, and bulk composition of the Moon. Gravity and topography data can, with reasonable assumptions, be applied to estimate crustal thickness and its lateral variability on a global scale. These data can also be applied to assess the rigidity of the lithosphere as a function of time during lunar evolution. Seismic data impose quantitative constraints on crustal structure beneath the Apollo landing sites and on mantle structure, thermal state, and composition as a function of depth. Gravity (moment of inertia), electromagnetic sounding, and laser ranging data together impose significant constraints on the existence and size of a possible metallic core. These results can, in turn, be applied to address several fundamental issues that are relevant to lunar origin and evolution, including (1) the early thermal state and lateral heterogeneity of the lunar lithosphere, (2) the bulk composition of the crust and mantle, (3) the depth of initial melting and differentiation, and (4) the bulk lunar metallic Fe content.

Previous reviews by *Hood* (1986) (hereinafter *H86*) and *Solomon* (1986) have summarized geophysical constraints derived from Apollo-era measurements on lunar origin and early evolution. Since that time, a series of developments has led to significant refinements of these constraints. First, several comprehensive assessments and syntheses of available geophysical and petrological constraints on lunar bulk composition have been reported (*Hood and Jones,* 1987; *Mueller et al.,* 1988). These analyses were stimulated, in part, by a more accurate seismic velocity model of the lunar mantle (*Nakamura,* 1983), based on the complete five-

year Apollo dataset. In addition, *Khan et al.* (2000) have recently derived an improved P-wave mantle velocity model using more advanced computational resources than were available during the immediate post-Apollo period. Second, more complete orbital geophysical measurements have recently been acquired by instruments onboard the Clementine and Lunar Prospector spacecraft (*Nozette et al.,* 1994; *Binder et al.,* 1998). These new analyses and measurements have led to notable refinements of geophysical constraints on lunar origin and evolution. In this chapter, these refined constraints are reviewed. In section 2, improved analyses of topography and gravity data with implications for near-surface structure and evolution are summarized. In section 3, the interpretation of mantle seismic velocity models is reviewed with emphasis on implications for bulk composition and depth of initial melting and differentiation. In section 4, a brief review is presented of geophysical constraints on the existence and size of a metallic core, including several recent results based on Lunar Prospector data. In section 5, a summary of those results that are most relevant to lunar origin and evolution models is given.

## 2. NEAR-SURFACE STRUCTURE AND THERMAL STATE

Observations from topography and gravity have provided useful information regarding lunar internal structure. When the gravitational attraction of surface topography is removed from the gravity field, the remaining (Bouguer) gravity represents the distribution of density anomalies in the interior that are indicative of lateral variations in composition and/or thermal state. However, gravity/topography inversions are inherently nonunique and benefit from independent geophysical (mainly seismic) and geochemical constraints. Topography and gravity can be used to address the rigidity of the lunar lithosphere at the time of loading by surface topography or subsurface density variations and provide evidence for early rigidity of the lunar lithosphere. Certain aspects of the long-wavelength lunar shape may also preserve the record of rigidity in the outer parts of the Moon early in its evolution.

The first high-quality lunar topography was obtained from laser altimetry on Apollos 15, 16, and 17 (e.g., *Kaula et al.,* 1974). A total of 7080 range measurements was collected from the three missions and these measurements, when converted to lunar radii by correcting for the orbit of the spacecraft, provided the first information on the shape of the Moon in a geodetic, center-of-mass reference frame. The along-track sampling of the measurements was 30–43 km, and the vertical resolution was ~12 m. The absolute radial accuracy of the measurements, which was controlled by radial orbit knowledge of the Apollo Command and Service Modules, was ~400 m. In addition, a radar sounder experiment on Apollo 17 acquired essentially continuous data along the orbit track but had lower vertical resolution (*Brown et al.,* 1974). The greatest limitation of the Apollo altimeter data was the coverage, which was restricted to within 26° of the equator. However, these data provided a measure of the Moon's equatorial radius and in addition led to the identification of the ~2.55-km offset between the Moon's center of figure and center of mass (*Kaula et al.,* 1974).

Prior to the Clementine Mission (sponsored by the Ballistic Missile Defense Organization and NASA; *Nozette et al.,* 1994), the best global-scale representation of lunar topography was a 12th degree and order spherical harmonic model by *Bills and Ferrari* (1977) that incorporated Apollo laser altimeter data, orbital photogrammetry and landmark measurements, and limb profiles. The estimated error for individual data types in that model ranged from 300 to 900 m; however, vast regions on the lunar farside lacked any measured topography. Near-global topography obtained from the Clementine lidar (*Zuber et al.,* 1994; *Smith et al.,* 1997) precipitated a significant advance in characterizing the structure of the lunar interior. The best current representation of the shape of the Moon is shown in Fig. 1a. The field is Goddard Lunar Topography Model-2 (GLTM-2), a 72nd degree and order spherical harmonic expansion of lunar radii derived from the ~73,000 valid Clementine lidar range measurements. The GLTM-2 model has an absolute vertical accuracy of ~100 m and a spatial resolution of 2.5° (76 km at the equator). The model shows that the Moon can be represented as a sphere with maximum positive and negative deviations of ~8 km, both occurring on the farside, in the areas of the Korolev and South Pole-Aitken Basins. The amplitude spectrum of the topography field exhibits significantly more power at longer wavelengths as compared to previous models due to more complete sampling of the lunar surface, particularly on the farside in areas of large topographic variance. The mechanism of support for this topography has implications for the Moon's internal structure and geodynamical evolution.

Gravity models of the Moon have been developed from S- and X-band Doppler tracking of orbiting spacecraft, principally the Lunar Orbiters, Apollo Command modules and subsatellites, and the Clementine and Lunar Prospector spacecraft. The Lunar Orbiters were typically in elliptical orbits with periapses of 50–100 km above the lunar surface, while the Apollo spacecraft occupied near-circular orbits at low inclinations with a mean altitude of 100 km, although via the subsatellites on Apollos 15 and 16 some tracking was acquired from altitudes as low as 10–20 km. Clementine occupied a polar elliptical orbit with a periapsis altitude of 400 km. Lunar Prospector spent its first year in an ~100-km altitude circular polar orbit, but the orbit altitude was lowered to a mean of ~30 km for the final seven months of the mission.

Earliest lunar gravity analyses utilized differentiated Doppler residuals, which correspond to the signal that remains after numerical integration of the equations of motion and correction for all forces on the spacecraft, including the gravitational attraction of Earth, the Sun, and the other planets, radiation pressure, and relativistic effects. During the Apollo era only the Moon's central mass term was accounted for, so the residuals contained the total resolvable information on the lunar gravity field (*Phillips et al.,* 1978). Doppler-derived accelerations are in the line of sight between spacecraft and radio tracking stations on Earth and thus necessarily sample the vertical component of gravity in a nonuniform manner over the surface. Analysis of line-of-sight gravity led to the discovery of the lunar mass concentrations (mascons) over the nearside maria (*Muller and Sjogren,* 1968). Subsequent spherical harmonic solutions based on Doppler observations from multiple satellites (*Bills and Ferrari,* 1980) combined pre-Clementine spacecraft tracking data with Earth-based lunar laser ranging observations to produce a 16th degree and order spherical harmonic solution (340-km half-wavelength resolution). *Konopliv et al.* (1993) produced the first high-resolution spherical harmonic model from the tracking of the Lunar Orbiters and Apollo subsatellites. This 60th degree and order expansion exhibited excellent performance in terms of orbit prediction but contained short wavelength noise that made it nonoptimal for regional scale geophysical studies. Clementine gravity models (*Zuber et al.,* 1994; *Lemoine et al.,* 1997) with a maximum resolution of 70th degree and order were characterized by improved short-wavelength control to enable basin modeling, and in addition provided improved knowledge of the low-degree harmonics and sectorial terms of the lunar field. The combination of tracking observations from Lunar Prospector with historical observations yielded a 75th degree and order field (*Konopliv et al.,* 1998), shown in Fig. 1b, as well as a very recent 100th degree and order field (A. Konopliv, personal communication, 1999) with a half-wavelength resolution of 55 km where the data permit. Compared to Clementine models, the Lunar Prospector gravity fields are characterized by better long-wavelength control and improved short-wavelength

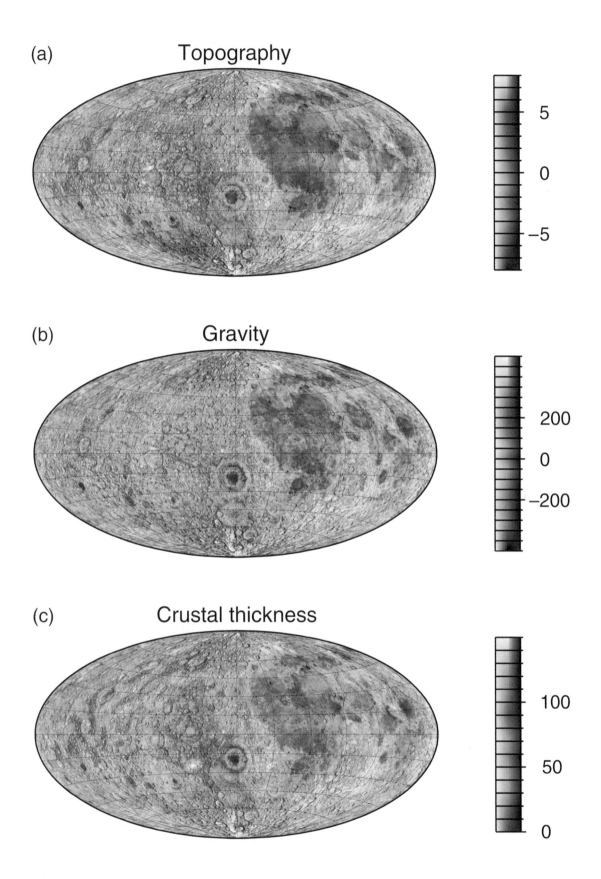

**Fig. 1.** Lunar topography, gravity, and crustal thickness. **(a)** Topography model GLGM-2 (*Smith et al., 1997*). **(b)** Gravity model LP75G (*Konopliv et al., 1998*). **(c)** Single-layer Airy compensation crustal thickness model updated from *Wieczorek and Phillips* (1998) using topography model GLTM-2 and gravity model LP75D.

resolution, the latter of which is facilitating more detailed regional modeling of nearside basins.

Despite the global nature of spherical harmonic solutions, all the gravitational models produced thus far are plagued by a lack of direct tracking on the lunar farside. Because the spacecraft is sensitive to perturbations of a range of wavelengths, some information on the spatial distribution of lunar farside gravity is obtained from spacecraft orbital changes as spacecraft enter and exit occultation, even when the spacecraft is not directly above features of interest. However, the magnitudes of anomalies are unreliable and thus cannot be used for geophysical modeling.

The thickness of the lunar crust provides an indication of the extent of melting of the lunar exterior in its early history. Crustal thickness has traditionally been determined by modeling the lunar gravity and topography fields. Given simplifying assumptions (cf. *Solomon,* 1978; *Thurber and Solomon,* 1978) including the densities and possible lateral variations of density in the crust and mantle, as well as the mode of compensation, Bouguer gravity can be converted to crustal thickness. Published crustal thickness models differ in detail, but all are "anchored" by the seismically determined crustal structure beneath the Apollo 12 and 14 sites. An early inversion (*Bratt et al.,* 1985) captured essential features such as the nearside-farside crustal thickness difference and the thinning of the crust beneath major impact basins, but that analysis was limited by the absence of reliable, near-global topography.

A common feature of the Clementine and Lunar Prospector gravity models is that highland regions are characterized by low-amplitude anomalies that reflect a state of isostatic compensation of surface topography. In principle, such compensation may be achieved either by an Airy mechanism (thickness variations of a constant-density crust) or by a Pratt mechanism (lateral density variations within a constant-thickness crust or in the upper mantle), or by some combination of the two. Evidence against a significant component of Pratt compensation has been presented, for example, by *Wieczorek and Phillips* (1997, their Fig. 8) based on a near lack of correlation of surface Fe abundance with elevation for the nearside highlands. Crustal thickness models based on near-global Clementine topography combined with the gravity models (*Zuber et al.,* 1994; *Neumann et al.,* 1996; *Solomon and Simons,* 1996; *Wieczorek and Phillips,* 1998) have therefore most commonly assumed compensation of topography by an Airy mechanism that assumes a strengthless lunar lithosphere in which pressures in vertical columns are balanced by relief along a single subsurface interface. Such models show a large range of global crustal thickness (~20–110 km) that would require major spatial variations in melting of the lunar exterior and/or significant impact-related redistribution of lunar crust (*Zuber et al.,* 1994). The ~61-km average crustal thickness, constrained by a depth-to-Moho measured during the Apollo 12 and 14 missions, is preferentially distributed toward the farside, accounting for much of the offset in

center-of-figure from the center-of-mass (*Kaula et al.,* 1974). While the average farside crustal thickness is greater than that on the nearside, the distribution is nonuniform. Most significantly, the South Pole-Aitken Basin on the lunar farside is an extensive region of crust that is much thinner than the nearside average. In Airy models the anorthositic lunar crust comprises ~10% of the volume of the Moon. Global multispectral observations show the interior of the South Pole-Aitken Basin to be more mafic than its surroundings (*Lucey et al.,* 1994, 1998), which would suggest internal compositional layering. The nature of such layering has been explored in crustal thickness inversions by *Wieczorek and Phillips* (1997, 1998), who demonstrated the viability of an internal structure with two crustal layers. Future attempts to refine shallow compositional structure from crustal thickness models will require incorporation of geochemical information, particularly in areas surrounding impact basins that have excavated deeply into the crust and reveal the shallow subsurface structure.

On a regional scale, the crustal structure in the vicinity of major impact structures has been recently improved. Gravity data show that nearside and some farside basins are characterized by strong positive mascon anomalies often surrounded by negative rings that represent subsurface mass deficiencies (*Muller and Sjogren,* 1968; *Zuber et al.,* 1974). Recent geophysical prospecting has revealed that these negative rings also characterize the Chicxulub basin on Earth and major impact basins on Mars (*Smith et al.,* 1999) and are thus fundamental features of impact basins. Modeling studies that incorporate recent estimates of mare fill thickness (*Williams and Zuber,* 1998; *Budney and Lucey,* 1998) show that the mascons are a consequence of a combination of mare filling and uplifted crust associated with rebound that occurred as part of the impact process (*Wise and Yates,* 1970; *Phillips and Dvorak,* 1981; *Melosh,* 1989). It follows that a mascon can, in principle, be produced without any mare fill (*Phillips and Dvorak,* 1981) and several such structures have recently been resolved on the farside from Lunar Prospector gravity fields (*Konopliv et al.,* 1998). Figure 2 illustrates the inferred subsurface structure via an inversion for crustal structure beneath the Orientale Basin. Surrounding the thinned central region is an apparent ring of thickened crust. While brecciation or comminution of crustal material associated with the impact process has been invoked to explain these features (*von Frese et al.,* 1996), such effects cannot in themselves plausibly account for the magnitudes of the observed mass deficiencies.

On the basis of elastic plate models constrained by observed tectonic features, it has been realized for some time that the lunar lithosphere exhibited demonstrable rigidity at the time of mare loading (e.g., *Comer et al.,* 1979; *Solomon and Head,* 1980). Clementine and Lunar Prospector observations now indicate that the lithosphere was strong enough to support loads even earlier in lunar history. In regional models of the subsurface structure of some mare basins, local isostatic models cannot fit the gravity data and it appears that relief at the surface and along the Moho was

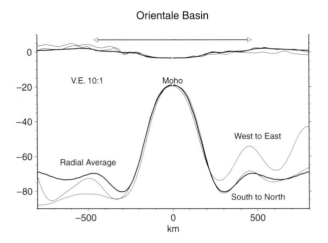

**Fig. 2.** West to east, south to north, radial average profiles of surface topography and Moho depth from an inversion using the method of *Neumann et al.* (1996) using GLTM-2 and LP75G. The model assumed a 1.7-km-thick mare load and an average depth to Moho as determined by *Wieczorek and Phillips* (1997).

supported by the strength of the lithosphere prior to mare filling (*Zuber et al.,* 1994). For example, the data for Orientale (Fig. 2) indicates that if this basin was isostatic before mare emplacement, then more than 5 km of mare fill would be required to account for the observed gravity field. Such a value is implausible since in this basin the maximum mare fill thickness is well-constrained to be less than 2 km (*Solomon and Head,* 1980; *Williams and Zuber,* 1998). Both the central uplift beneath the basin and the surrounding depression of the Moho are nonisostatic effects indicative of early lithospheric rigidity. *Neumann et al.* (1996) have identified a negative correlation between Moho relief and age, which indicates that basins formed at later times maintained a greater amount of the uplift resulting from the impact process. This observation provides qualitative evidence for cooling and thickening of the outer shell of the early Moon.

Additional evidence for early lunar rigidity comes from the Moon's long-wavelength shape. The present topographic shape of the Moon is flattened by 2.2 km, while its gravity field is flattened only ~0.5 km (*Smith and Zuber,* 1998). The hydrostatic component of the flattening, i.e., that which arises from the Moon's present-day rotation, is only 7 m. This difference between the topographic and gravitational equipotential shape has previously been explained as the "memory" of an earlier Moon that was rotating faster and had a correspondingly larger hydrostatic flattening (*Jeffreys,* 1970; *Lambeck and Pullan,* 1980; *Willemann and Turcotte,* 1981). Recent analysis (*Smith and Zuber,* 1998) has shown that other degree-2 terms of the lunar topographic field also exceed significantly their counterparts in the gravity field (*Konopliv et al.,* 1998) and are mutually consistent with a lunar rotation rate about 15× greater than at present. As-

suming synchronous rotation, the degree-2 topography would require that the Moon "froze in" its long-wavelength shape at a distance of 13 to 16 Earth radii. Such a scenario is problematic given dynamical studies (e.g., *Goldreich,* 1966; *Touma,* 2000) of the evolution of the Earth-Moon system that suggest that the Moon, which had a mostly molten exterior early in its history (*Wood et al.,* 1970), receded from Earth rapidly, before significant freezing out of the magma ocean likely occurred. Dynamical mechanisms that could slow the retreat of the Moon from Earth would cause significant heating of the lunar interior (*Touma and Wisdom,* 1994), and would exacerbate the problem of how to cool the Moon quickly enough to explain the degree-2 shape in the context of an earlier, faster rotation. Previous analysis of the compensation of long-wavelength loads (*Willemann and Turcotte,* 1981), and a more recent treatment of time-dependent load relaxation (*Zhong and Zuber,* 1999), show that compensation is strongly dependent on planetary radius, such that significant long wavelength loads can be supported over lunar history by a thin elastic lithosphere. However, current thermal models (e.g., *Pritchard and Stevenson,* 2000), which admittedly have significant uncertainties, favor hot early conditions. Joint consideration of dynamical and thermal models constrained by long wavelength topography and gravity will ultimately be required to resolve the inconsistency.

## 3. MANTLE COMPOSITION AND THERMAL STATE

The lunar mantle comprises about five-sixths of the lunar volume and therefore makes the main contribution to the Moon's bulk composition. To some extent, the composition of the upper mantle (60–500 km depth) can be inferred from surface mare basalt samples that experienced partial melting at depth and subsequent upward migration (e.g., *Ringwood and Essene,* 1970; *Delano,* 1986; *Taylor,* 1987). For example, a Mg number (Mg#) [molar MgO/(MgO + FeO) × 100] of the upper mantle in the range of 75–80 and a bulk $Al_2O_3$ content of <1 wt% may be estimated on this basis. However, few (if any) surface samples are believed to have originated in the middle and lower mantle (>500 km depth), representing more than one-third of the lunar volume. This region is therefore essentially inaccessible to direct or indirect chemical sampling and can only be characterized geophysically.

Geophysical constraints on the composition of the mantle come almost entirely from interpretations of seismic velocity models, supplemented by measurements of the lunar mean density and moment-of-inertia. As reviewed by *H86*, the small number (4) and areal distribution of the Apollo seismic stations combined with a low signal-to-noise ratio for lunar seismic signals has hindered accurate determinations of the seismic velocity structure of the mantle. A number of estimations of the velocity structure in the mantle were published based on early analyses of the Apollo seismic dataset (*Nakamura et al.,* 1976; *Goins,*

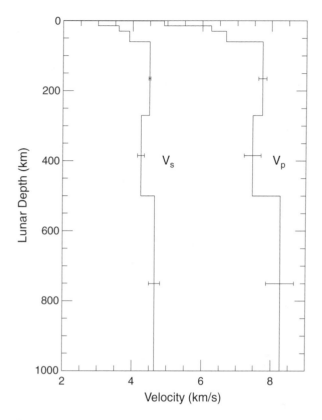

**Fig. 3.** P- and S-wave velocity model for the lunar mantle derived by *Nakamura* (1983). One-standard-deviation error limits resulting mainly from the variance in arrival time readings are shown.

1978; *Goins et al.*, 1981; *Nakamura*, 1983). Of these, the model of *Nakamura* (1983) (Fig. 3) is probably most accurate because it was the first to be based on the complete five-year Apollo record and analyzed signals from more deep moonquake sources (41) than had been considered previously. The Nakamura model is characterized by an apparent velocity decrease with depth in the upper mantle followed by a substantial velocity increase at ~500 km depth. As noted originally by *Nakamura* (1983), the 500-km velocity increase would imply either an increase in Mg# below this depth or a phase change, presumably involving the transition from the spinel to the garnet stability field for Al-bearing silicates.

Recently, *Khan* (1998) and *Khan et al.* (2000) have reported a new inversion of the complete Apollo seismic dataset using more modern, computationally intensive, inverse Monte Carlo methods. Only P-wave arrivals were analyzed, with results for the P-wave velocity structure shown in Fig. 4. Unlike the Nakamura model, the upper mantle velocity in the Khan model (8.1 ± 1.5 km/s) is essentially constant with increasing depth. Like the Nakamura model, the Khan model is characterized by a substantial velocity increase (to 9.0 ± 1.5 km/s) near 500 km. Moreover, as seen in Fig. 4, the velocity structure in the middle and lower mantle appears to be more inhomogeneous than in the up-

per mantle. In the deep moonquake source region, P-wave velocities are even larger (11.0 ± 1.2 km/s). Because of the more rigorous statistical technique, the Khan model represents an improvement over the Nakamura model. However, significant statistical uncertainties characterize both models and systematic errors (relating, for example, to lateral heterogeneities beneath the sparse Apollo seismic network) could well be present. In general, the middle and lower mantle P-wave velocities of the Khan model (9–11 km/s) are larger than those of the Nakamura model (8.26 ± 0.40 km/s) and the uncertainties are somewhat larger. In

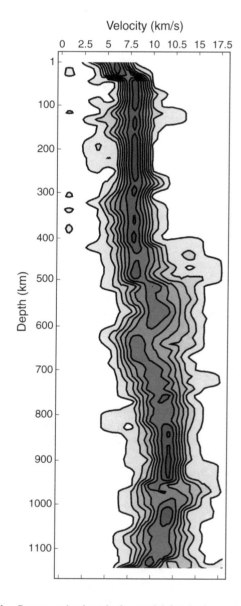

**Fig. 4.** P-wave seismic velocity model for the lunar mantle derived from the Apollo lunar seismic dataset by *Khan et al.* (2000) using an inverse Monte Carlo method. Contour lines define eight equal-spaced probability density intervals for the velocity distribution. The velocity increase near 500 km depth marks the transition from the upper mantle to the lower mantle. The deep moonquake source region is in the 850–960-km depth range. From *Khan et al.* (2000).

particular, it is possible that the more inhomogeneous middle mantle velocities may reflect, at least in part, a "leverage effect" due to data errors and/or model insensitivity at these depths.

Because of the significant temperature dependences of mineral elastic constants (see, e.g., *Hood and Jones,* 1987), quantitative interpretation of lunar seismic velocity models requires a prior consideration of bounds on the lunar mantle temperature profile. As reviewed, for example, by *H86*, constraints on the thermal structure of the lunar mantle come from surface heat flow data, seismic data, gravity data, and electromagnetic sounding data. At the surface, the mean temperature is measurable as ~0°C. At 300 km depth, a constraint on upper mantle temperatures of <800°C has been inferred from the apparent maintenance of mascon anisostasy over 3–4 eons (*Pullan and Lambeck,* 1980). This upper limit is consistent with the high mean seismic inverse dissipation (Q) values (4000–7000; *Nakamura and Koyama,*1982) in this depth range. In the middle mantle, Q-values are reduced to >1000, implying somewhat higher temperatures. However, the locations of most deep moonquake foci in the depth range from 850 to 960 km (*Khan et al.,* 2000) still requires temperatures well below the solidus at these depths to allow the implied accumulation of tidal stresses. At 850 km depth, the lunar solidus temperature may be estimated as 1400° ± 100°C, increasing to 1500° ± 100°C by 1000 km depth (*Ringwood and Essene,* 1970; *Hodges and Kushiro,* 1974). At depths greater than ~1000 km, S-wave Q-values decrease to ~100, implying greater dissipation and temperatures probably approaching the solidus. From the above constraints, a temperature profile that starts at 0°C at 0 km depth, increases to ~750°C at 300 km depth, to ~1200°C at 800 km depth, and to ~1400°C at 1100 km depth may be suggested. Such a profile is approximately consistent with some thermal history models (e.g., *Toksöz et al.,* 1978). The latter model assumed melting and differentiation to a depth of ~500 km and an average U concentration of 35 ppb. The model indicated that convective heat transport was important during early lunar history but is presently weak and confined mainly to the lower mantle.

As reviewed by *H86*, surface heat flow measurements were obtained at the Apollo 15 and 17 landing sites (21 and 16 mW/m$^2$ respectively) and have been applied to constrain both the thermal state and radioactive elemental composition of the lunar interior. By extrapolating these isolated measurements using orbital data on surface Th abundances and inferred crustal thicknesses, *Langseth et al.* (1976) originally estimated a global mean heat flow rate of 18 mW/m$^2$. Such a high mean heat flow value would imply bulk lunar U abundances ranging from 35 to 46 ppb, depending on whether secular cooling is important or whether a simple steady-state balance exists between heat production and loss respectively. These values are much larger than that for the bulk Earth (18 ppb) and would therefore support the view that the Moon is enriched in refractory elements (e.g., *Taylor,* 1987). However, *Rasmussen and Warren* (1985) later

argued that the Apollo 15 and 17 landing sites were near the edges of maria where megaregolith thickness may be unusually shallow, leading to anomalously high heat flow rates. On this basis, they proposed an alternate global mean heat flow rate of 11 mW/m$^2$. Such a global mean value would imply bulk lunar U abundances ranging from 20 to 27 ppb. The lower limit of this range is not significantly different from the terrestrial value. They therefore argued that the Apollo heat flow data alone do not definitively require a lunar excess of refractory elements. It may be concluded from these results that additional direct lunar heat flow measurements in a wider variety of geologic settings are probably needed before a firm constraint on either lunar bulk composition or thermal state will be possible.

Several of the Apollo-era seismic models have been analyzed to infer constraints on the structure and composition of the lunar mantle (*Buck and Toksöz,* 1980; *Basaltic Volcanism Study Project,* 1981; *Hood and Jones,* 1987; *Mueller et al.,* 1988). Of most interest here are the studies by *Hood and Jones* (1987) and *Mueller et al.* (1988), who analyzed the later and more accurate *Nakamura* (1983) model. In the study by *Hood and Jones* (1987), forward calculations were performed of seismic velocity profiles for several possible lunar bulk compositions, differentiation schemes, and thermal gradients. Comparisons of the calculated profiles to the Nakamura model favored models that assumed differentiation of the upper mantle only, that were more aluminous in bulk composition than the terrestrial mantle, and that were characterized by an increase in Mg# in the middle and lower mantle relative to the upper mantle. Although seismic velocities tended to decrease in the upper mantle because of the large thermal gradient in this zone, the calculated decrease was generally too small to match that of the Nakamura model. (Note, however, that the Khan model has no significant velocity decrease in the upper mantle.) In the study by *Mueller et al.* (1988), an attempt was made to "invert" the Nakamura model to find the range of lunar compositional and evolutionary models consistent with the observational seismic velocities. These authors also inferred that an increase in aluminous phases and an increase in Mg# was probably required in the middle and lower mantle to explain the P-wave velocity increase at 500 km depth. They concluded that the 500-km discontinuity probably marks the lowest extent of lunar differentiation and may represent a transition to a less fractionated, possibly primordial middle and lower mantle.

Although the *Khan et al.* (2000) seismic model has not yet been quantitatively interpreted, the authors have noted several probable implications of the model for lunar structure and evolution. In particular, the nearly homogeneous upper mantle velocity structure and the more complex middle mantle structure were interpreted to imply early melting and differentiation to 500 km depth. Beneath this transition, the middle and lower mantle were suggested to consist of relatively "pristine" and inhomogeneous material. (Note again, however, that increased data errors and/or model insensitivity in the middle and lower mantle may contribute to the more complex velocity structure at those depths.)

The above interpretations of the inferred seismic velocity structure of the lunar mantle have several interesting derivative implications for lunar origin and evolution models. In general, a preferred interpretation of the seismic velocity increase at 500 km depth is a change in composition to more aluminous, more MgO-rich mafic silicates. This inferred composition change, combined with the evidence for a homogeneous upper mantle and an inhomogeneous middle and lower mantle, is most easily understood if the Moon was initially melted and differentiated only to a depth of 500 km. Initial differentiation to 500 km means that all the Al now in the lunar crust must have been extracted from the upper mantle. This inference, combined with the increased alumina abundance in the middle and lower mantle, as well as petrologic evidence for a reduced Mg# in the lunar upper mantle, increases the likelihood that the Moon's bulk composition differs significantly from that of the terrestrial mantle. As noted by *Mueller et al.* (1988), this compositional dissimilarity is sufficient to dismiss classical fission models even without considering dynamical objections.

Of the remaining models for lunar origin (capture, binary accretion, impact-fission), the impact-fission model is currently considered to be the most dynamically plausible (e.g., *Cameron,* 1997; *Canup and Esposito,* 1996). As pointed out by *Mueller et al.* (1988), a primitive, undifferentiated middle and lower mantle consisting of more aluminous and MgO-rich phases could be consistent with such a model. Because the impactor mass and composition are poorly constrained, the initial composition of the impact-produced vapor cloud is not precisely known. However, MgO is more refractory than FeO under vapor conditions (*Grossman and Larimer,* 1974) and alumina is a very refractory component of a silicate vapor. It may therefore be suggested that fractional condensation of the impact vapor would favor a more aluminous and MgO-rich composition for the earliest condensates. As also noted by *Mueller et al.* (1988), these condensates may have accreted to form a small proto-Moon that could have remained unmelted because of smaller accretional energy deposition, larger surface area to volume ratio, and a relatively high melting temperature. Later addition of less refractory material at higher impact velocities may have melted only the outer 500 km of the final proto-Moon, producing the upper mantle and crust.

## 4.  CORE

As reviewed in more detail by *H86*, independent geophysical constraints on the existence and size of a lunar metallic core come from gravity data, seismic data, electromagnetic sounding data, laser ranging data, and paleomagnetic data. Recent measurements by instruments on the Lunar Prospector spacecraft as well as continued ground-based laser ranging measurements have led to significant refinements of these constraints. Although none is individu-

ally definitive, as reviewed below, together they provide strong evidence for the existence of a small lunar metallic core representing 1–3% of the lunar mass.

One indirect approach toward investigating the existence of a dense metallic core is to apply the moment of inertia constraint together with thermal and density models of the lunar crust and mantle (*Hood and Jones,* 1987; *Mueller et al.,* 1988). During the post-Apollo period, the best available determination of the normalized polar moment of inertia was 0.3905 ± 0.0023 (*Ferrari et al.,* 1980). Adopting an upper bound of 0.3928, both *Hood and Jones* (1987) and *Mueller et al.* (1988) found that mantle density increases alone were insufficient to match this adopted limit for plausible thermal, compositional, and evolutionary models of the lunar interior. A small dense core was therefore suggested. Considering independent limits on the maximum core size from electromagnetic sounding data, *Hood and Jones* (1987) concluded that a core representing 1–4% of the lunar mass was most probable. Corresponding core radius limits are in the range of 200–450 km for an Fe composition. Recently, *Konopliv et al.* (1998) have reported a significantly improved determination of the normalized polar moment of 0.3932 ± 0.0002 using Lunar Prospector Doppler gravity data. The corresponding upper bound of 0.3934 is only marginally larger than that adopted previously by *Hood and Jones* (1987) and by *Mueller et al.* (1988). Their conclusions with respect to the existence of a small core remain unchanged. However, the more accurate and larger value of the normalized moment increases confidence that no more than a small core is present.

Ideally, by analogy with the terrestrial case, seismic arrival time data could be used to determine directly the existence and size of a metallic core since seismic velocities are significantly reduced in a metallic medium. However, the low signal-to-noise ratio of lunar seismic signals limited the efficacy of this approach in the case of the Apollo seismic dataset (*H86*). Future lunar seismic data may be acquired by the planned Japanese LUNAR-A penetrator mission (*Mizutani,* 1995). Three penetrators containing ultrasensitive seismometers and heat-flow probes will be emplaced at locations diametrically opposite to one another and to the known locations of deep moonquake sources. The mission is designed to exploit knowledge gained from the Apollo network in order to better investigate the existence and size of a metallic core.

A third approach toward investigating the existence and size of a metallic core takes advantage of the core's high electrical conductivity relative to that of the mantle. In addition to time-dependent electromagnetic induction studies that normally employ simultaneous data from at least two magnetometers (*H86*), an alternate approach toward core detection requires data from a single magnetometer in low-altitude lunar orbit (Fig. 5). In this approach, time intervals are identified when the Moon is exposed for prolonged periods to a nearly spatially uniform magnetic field in a near-vacuum environment. Conditions that approximate this idealized situation occur occasionally each month when the Moon passes through a lobe of the geomagnetic tail. Ex-

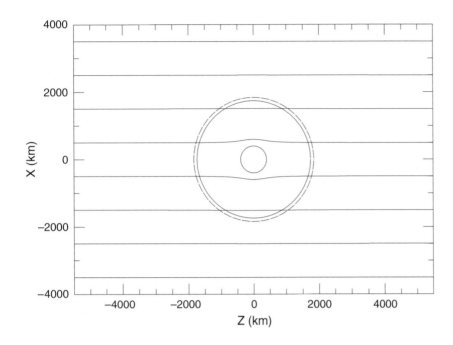

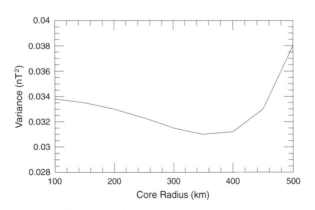

**Fig. 5.** Perturbation of spatially uniform magnetic field lines by an induced magnetic field caused by currents at the surface of a highly electrically conducting core of radius 400 km (inner circle). The outer solid circle represents the projection of the lunar surface and the outermost dashed circle indicates a nominal Lunar Prospector orbit at 100 km altitude. The orbit plane is parallel to the applied field orientation. From *Hood et al.* (1999).

posure of the Moon to such a steady field induces electrical currents in the lunar interior with an associated negative induced magnetic dipole moment oriented opposite to the applied field. The sum of the steady field and the induced field produces a perturbation as shown in Fig. 5. After currents in the lunar mantle have decayed (~5 h or less), any residual steady induced moment is expected to be due to currents near the surface of a possible highly electrically conducting core. The amplitude of the residual moment is then relatable to the core radius.

Measurements of the lunar induced moment in the geomagnetic tail were originally obtained using data from magnetometers on the Apollo 15 and 16 subsatellites (*Russell et al.*, 1981). Careful editing to eliminate periods when significant plasma densities were present resulted in a final estimated induced moment of $-4.23 \pm 0.64 \times 10^{22}$ Gauss-cm$^3$ per Gauss of applied field. Assuming that this moment was entirely caused by a metallic core, the core radius was estimated as $439 \pm 22$ km. As discussed by *H86*, possible error sources include paramagnetic effects (mean permeability >1) and incomplete decay of mantle-induced fields resulting from short-term changes in the mean field amplitude.

Since the launch of Lunar Prospector in January 1998, an effort has been made to obtain independent measurements of the lunar induced moment in the geomagnetic tail (*Hood et al.*, 1999). Because Lunar Prospector is in a near-polar orbit, measurements of the induced moment are geometrically most feasible when the orbit plane is nearly parallel to the Sun-Moon line (parallel to the Z axis in Fig. 5). This occurs approximately at six-month intervals. The first opportunity for optimal measurements, i.e., the first period when the orbit plane was optimally aligned, occurred in April 1998. During this month, the Moon fortuitously entered the north tail lobe and remained there for an un-

usually long period (>2 days) during which the tail lobe field was relatively strong and steady. Careful editing and averaging of individual orbit segments increased the signal-to-noise ratio and allowed a measurement of the induced moment as $-2.4 \pm 1.6 \times 10^{22}$ Gauss-cm$^3$ per Gauss of applied field. Figure 6 plots the variance between the mean observed field perturbation and that expected theoretically as a function of core size. It is seen that the data strongly exclude cores with radii larger than 450 km, but cores smaller than 300 km radius are less strongly excluded. The preferred moment amplitude is slightly smaller than that obtained by *Russell et al.* (1981), but the two estimates are marginally consistent within the 1σ uncertainties of the measurements. Assuming that the induced field is caused

**Fig. 6.** Variances of model field perturbations from observed mean field perturbations as a function of induced moment amplitude. Equivalent core radii are also indicated on the lower scale. From *Hood et al.* (1999).

entirely by electrical currents near the surface of a highly electrically conducting metallic core, the preferred core radius is 340 ± 90 km, representing 1–3% of the lunar mass. However, when uncertainties related to the possible presence of residual mantle-induced fields are considered, this measurement should be regarded as an upper limit on the core radius (430 km at the 1σ level).

One additional property of a metallic core that may be detectable geophysically is its liquid state. Thermal history models (e.g., *Konrad and Spohn,* 1998) indicate that Fe cores containing a small amount of S can remain liquid through the entire 4.6 b.y. of lunar evolution. A fluid metallic core enhances friction at the core-mantle boundary, causing a larger dissipation of rotational energy in the Moon than would be expected without a fluid core (*Yoder,* 1981, 1984; *Ferrari et al.,* 1980). Lunar laser ranging measurements of physical libration parameters have shown that the lunar rotation axis is advanced by about 0.2 arcsec from the Cassini alignment, indicating significant internal dissipation. In addition to solid-state friction in the lunar mantle, friction at the boundary between a fluid core and a solid silicate mantle appears to be necessary to explain the magnitude of this advance. On the basis of a turbulent coupling model, *Yoder* (1981) originally inferred a probable core radius of 330 km. More recently, additional laser ranging data and an improved gravity field from Lunar Prospector Doppler tracking have allowed a reevaluation of this constraint on the existence and size of a fluid lunar core. Taking into account all uncertainties, including possible topography on the core-mantle boundary as well as theoretical uncertainties, a 1σ upper limit of 352 km on the core radius for an assumed Fe composition has been reported (*Williams et al.,* 1999). For an Fe-FeS eutectic composition, the upper limit is increased to 374 km. It should be noted that an additional solid inner core is neither excluded nor required by the available ranging data.

Finally, one additional geophysical indication of the possible existence of a lunar metallic core is the observed pervasive paleomagnetism of lunar crustal materials (for reviews, see *Fuller,* 1974; *Fuller and Cisowski,* 1987; *Hood,* 1995). Paleointensity estimates for returned lunar samples suggest relatively large (~1 Gauss) surface magnetizing fields between about 3.6 and 3.8 eons b.p. (*Cisowski et al.,* 1983; *Cisowski and Fuller,* 1986). This would be suggestive of a former core dynamo. On the other hand, orbital measurements indicate an important role for impact processes in producing the strongest concentrations of crustal magnetic anomalies. First, low-altitude passes across the lunar nearside by the Apollo 16 subsatellite magnetometer showed that fields are relatively weak across the maria but are larger over exposures of the Fra Mauro and Cayley Formations (primary and secondary basin ejecta respectively) (*Hood et al.,* 1979). It had previously been shown from sample studies that mare basalts contain relatively little microscopic Fe remanence carriers, while impact-produced breccias and soils contain much more of these carriers (*Strangway et al.,* 1973). Thus the absence of strong fields over the maria as compared to the highlands and impact basin ejecta could be understood since impact processes played an important role in producing larger concentrations of remanence carriers. Second, the four largest concentrations of magnetic anomalies within the region mapped by the Apollo subsatellites occurred near the antipodes of the four youngest and largest impact basins: Imbrium, Orientale, Serenitatis, and Crisium (*Lin et al.,* 1988). More recent mapping using improved data coverage by the Lunar Prospector magnetometer and electron reflectometer has confirmed this correlation (*Lin et al.,* 1998). A model for the origin of magnetic anomalies antipodal to lunar impact basins involving expansion of a massive, partially ionized impact vapor cloud around the Moon has been partially developed (*Hood and Huang,* 1991). Nevertheless, the origin of the ambient magnetic field remains unclear. A core dynamo origin cannot be excluded by the orbital data alone since impact processes have greatly complicated the distribution and orientation of crustal anomaly sources. On the other hand, a core dynamo is also not required to be present according to some impact field models (e.g., *Hood and Huang,* 1991).

In summary, gravity, electromagnetic sounding, and laser ranging data appear to be converging in indicating the existence of a small dense, highly electrically conducting, and probably fluid core within the Moon. The most probable core radius is in the range of 300–400 km, representing ~1–3% of the lunar mass. However, when all uncertainties are considered, both the laser ranging data and the electromagnetic sounding data are most conservatively interpreted as implying only an upper limit on the core radius, 352 km in the case of the ranging data and 430 km in the case of the induced moment measurements. Likewise, uncertainties in mantle thermal, composition, and evolutionary models as constrained by seismic velocity models are sufficiently large that the existence of a dense core is only marginally indicated (*Mueller et al.,* 1988). Consequently, although the existence of a metallic core is the most probable interpretation of the available data, additional measurements (possibly those planned for the Japanese LUNAR-A mission) are needed to more definitively resolve this fundamental issue.

## 5.  SUMMARY

As discussed in section 2, Clementine and Lunar Prospector gravity and topography data demonstrate a wide range of global crustal thickness variations (~20–110 km), implying either impact-related redistribution of crustal materials or lateral heterogeneities in early melting and differentiation, or both. The mean crustal thickness is ~61 km with larger mean values on the farside except beneath the South Pole-Aitken Basin. Recent modeling of mascon gravity anomalies supports the view that they are caused by a combination of postimpact rebound of the crust followed by mare filling. Support for this view comes from observations of mascons associated with several farside basins that appear to lack mare fill. Modeling of regional and long-wavelength Clementine and Lunar Prospector data provide evidence for a rigid lunar lithosphere well before the mare emplacement epoch. The observed topographic flattening of the Moon and the ellipticity of the lunar equator appear to imply that the Moon was unexpectedly close (13–16 $R_E$) to the Earth when its present shape was "frozen in." Such rapid early cooling to produce the significant degree-2 power in the lunar shape is contrary to expectations from dynamical models of the evolution of the Earth-Moon system, as well as thermal models of the early lunar thermal state. Resolution of the inconsistency will require coupled dynamical and thermal models constrained by topography and gravity.

As reviewed in section 3, quantitative compositional modeling of the *Nakamura* (1983) seismic velocity model strongly suggests that initial melting and differentiation of the lunar interior was confined to the upper mantle (<500 km depth). An increase in alumina abundance and in MgO content is also probably needed in the middle and lower mantle. If differentiation was limited to 500 km depth, it becomes difficult to explain the alumina abundance in the crust unless the initial alumina content in the upper mantle was enriched relative to that of the terrestrial mantle. The recent *Khan et al.* (2000) P-wave velocity model is most consistent with a compositionally homogeneous upper mantle and a much more heterogeneous middle mantle. These results also independently suggest initial melting and differentiation of the upper mantle only. As also discussed in section 3, *Mueller et al.* (1988) specifically argue that the middle and lower mantle may represent a primordial, unmelted, refractory-rich basement on which less refractory material accreted and subsequently melted. Such an interpretation could, in principle, be consistent with fractional condensation of a vapor cloud produced according to the giant impact model, for example. A primitive, unmelted deep interior would require that any metallic core have formed via heterogeneous accretion or incremental separation of more mobile metallic phases accompanying partial melting events (without melting and homogenizing the entire proto-Moon). As reviewed by *Pritchard and Stevenson* (2000), the giant impact model does not yet accurately predict the precise thermal state of the proto-Moon because of a poor understanding of the dynamical continuum disk phase that immediately follows the impact event.

As reviewed in section 4, quantitative assessments of all available geophysical constraints, including the moment of inertia, suggest the presence of a small dense core with a mass exceeding ~1% of the lunar mass. Recent analyses of Lunar Prospector magnetometer data, gravity data, and groundbased laser ranging data independently suggest the presence of a metallic core with a radius in the range of 300–400 km. However, in the most conservative interpretation, only an upper bound on the core radius is implied by both the magnetometer data and the laser ranging data. Thus, future measurements [such as those to be obtained during the planned LUNAR-A mission (*Mizutani,* 1995)]

may be needed to more definitively establish the existence of a metallic core.

***Acknowledgments.*** Thanks are due to A. Konopliv, G. Neumann, D. Smith, and M. Wieczorek for contributing data and models presented herein. A. Khan kindly provided a copy of Fig. 4 in advance of publication. Finally, we thank Kevin Righter for the initial invitation to write this chapter and Roger Phillips for a constructive, critical review.

***Note added in proof:*** Further analysis of Apollo seismic data, including S-wave data as well as P-wave data, yields a revised estimate for the maximum depth of the upper mantle as 550 ± 20 km (A. Khan personal communication, 1999).

## REFERENCES

Basaltic Volcanism Study Project (1981) *Basaltic Volcanism on the Terrestrial Planets.* Pergamon, New York. 1286 pp.

Bills B. G. and Ferrari A. J. (1977) A harmonic analysis of lunar topography. *Icarus, 31,* 244–259.

Bills B. G. and Ferrari A. J. (1980) A harmonic analysis of lunar gravity. *J. Geophys. Res., 85,* 1013–1025.

Binder A. B., Feldman W. C., Hubbard G. S., Konopliv A., Lin R. P., Acuna M. A., and Hood L. L. (1998) Lunar Prospector searches for polar ice, a metallic core, gas release events, and the Moon's origin. *Eos Trans. AGU, 79,* 97–108.

Bratt S. R., Solomon S. C., Head J. W., and Thurber C. H. (1985) The deep structure of lunar basins: Implications for basin formation and modification. *J. Geophys. Res., 90,* 3049–3064.

Brown W. E. Jr., Adams G. F., Eggleton R. E., Jackson P., Jordan R., Kobrick M., Peeples W. J., Phillips R. J., Porcello L. J., Schaber G., Sill W. R., Thompson T. W., Ward S. H., and Zelenka J. S. (1974) Elevation profiles of the Moon. *Proc. Lunar Sci. Conf. 5th,* pp. 3037–3048.

Buck W. R. and Toksöz M. N. (1980) The bulk composition of the Moon based on geophysical constraints. *Proc. Lunar Planet. Sci. Conf. 11th,* pp. 2043–2058.

Budney C. J. and Lucey P. G. (1998) Basalt thickness in Mare Humorum: The crater excavation method. *J. Geophys. Res., 103,* 16855–16870.

Cameron A. G. W. (1997) The origin of the Moon and the single impact hypothesis V. *Icarus, 126,* 126–137.

Canup R. M. and Esposito L. W. (1996) Accretion of the Moon from an impact-generated disk. *Icarus, 119,* 427–446.

Cisowski S. M. and Fuller M. (1986) Lunar paleointensities via the IRMs normalization method and the early magnetic history of the Moon. In *Origin of the Moon* (W. K. Hartmann, R. J. Phillips, and G. J. Taylor, eds.), pp. 411–424. Lunar and Planetary Institute, Houston.

Cisowski S. M., Collinson D. W., Runcorn S. K., Stephenson A., and Fuller M. (1983) A review of lunar paleointensity data and implications for the origin of lunar magnetism. *Proc. Lunar Planet. Sci. Conf. 13th,* in *J. Geophys. Res., 88,* A691–A704.

Comer R. P., Solomon S. C., and Head J. W. (1979) Elastic lithosphere thickness on the Moon from mare tectonic features: A formal inversion. *Proc. Lunar Planet. Sci. Conf. 10th,* pp. 2441–2463.

Delano J. W. (1986) Abundances of cobalt, nickel, and volatiles in the silicate portion of the Moon. In *Origin of the Moon* (W. K. Hartmann, R. J. Phillips, G. J. Taylor, eds.), pp. 231–247. Lunar and Planetary Institute, Houston.

Ferrari A. J., Sinclair W. S., Sjogren W. L., Williams J. G., and Yoder C. F. (1980) Geophysical parameters of the Earth-Moon system. *J. Geophys. Res., 85,* 3939–3951.

Fuller M. (1974) Lunar magnetism. *Rev. Geophys. Space Phys., 12,* 23–70.

Fuller M. and Cisowski S. (1987) Lunar paleomagnetism. In *Geomagnetism* (J. A. Jacobs, ed.), pp. 307–456.

Goins N. R. (1978) The internal structure of the Moon. Ph.D. thesis, Massachusetts Institute of Technology, Cambridge. 666 pp.

Goins N. R., Dainty A. M., and Toksöz M. N. (1981) Lunar seismology: The internal structure of the Moon. *J. Geophys. Res., 86,* 5061–5074.

Goldreich P. (1966) History of the lunar orbit. *Rev. Geophys., 4,* 411–439.

Grossman L. and Larimer J. W. (1974) Early chemical history of the solar system. *Rev. Geophys., 12,* 71–101.

Hodges F. N. and Kushiro I. (1974) Apollo 17 petrology and experimental determination of differentiation sequences in model Moon compositions. *Proc. Lunar Sci. Conf. 5th,* pp. 505–520.

Hood L. L. (1986) Geophysical constraints on the lunar interior. In *Origin of the Moon* (W. K. Hartmann, R. J. Phillips, and G. J. Taylor, eds.), pp. 361–410. Lunar and Planetary Institute, Houston.

Hood L. L. (1995) Frozen fields. *Earth, Moon, and Planets, 67,* 131–142.

Hood L. L. and Huang Z. (1991) Formation of magnetic anomalies antipodal to lunar impact basins: Two-dimensional model calculations. *J. Geophys. Res., 96,* 9837–9846.

Hood L. L. and Jones J. (1987) Geophysical constraints on lunar bulk composition and structure: A reassessment. *Proc. Lunar Planet. Sci. 17th,* in *J. Geophys. Res., 92,* E396–E410.

Hood L. L., Coleman P. J. Jr., and Wilhelms D. E. (1979) The Moon: Sources of the crustal magnetic anomalies. *Science, 204,* 53–57.

Hood L. L., Lin R. P., Mitchell D. L., Acuna M. H., and Binder A. B. (1999) Initial measurements of the lunar induced magnetic moment in the geomagnetic tail using Lunar Prospector data (abstract). In *Lunar and Planetary Science XXX,* Lunar and Planetary Institute, Houston.

Jeffreys H. (1970) *The Earth.* Cambridge Univ., New York. 525 pp.

Kaula W. L., Schubert G., Lingenfelter R. E., Sjogren W. L., and Wollenhaupt W. R. (1974) Apollo laser altimetry and inferences as to lunar surface structure. *Proc. Lunar Sci. Conf. 5th,* pp. 3049–3058.

Khan A. (1998) A selenological enquiry. M.S. thesis, Odense University. 99 pp.

Khan A., Mosegaard K., and Rasmussen K. L. (2000) Seismic evidence for a lunar magma ocean extending to 560 km depth. *Geophys. Res. Lett.,* submitted.

Konopliv A. S., Sjogren W. L., Wimberly R. N., Cook R. A., and Vijayaraghavan A. (1993) A high-resolution lunar gravity field and predicted orbit behavior. AAS Paper 93-622, AAS/AIAA Astrodynamics Specialist Conference, Victoria, British Columbia.

Konopliv A. S., Binder A., Hood L., Kucinskas A., Sjogren W. L., and Williams J. G. (1998) Gravity field of the Moon from Lunar Prospector. *Science, 281,* 1476–1480.

Konrad W. and Spohn T. (1998) The influence of lunar mantle convection on partial melting and the cooling of a small core (abstract). In *Origin of the Earth and Moon,* pp. 20–21. LPI Contrib. No. 957, Lunar and Planetary Institute, Houston.

Lambeck K. and Pullan S. (1980) The lunar fossil bulge hypothesis revisited. *Phys. Earth Planet. Inter., 22,* 29–35.

Langseth M. G., Keihm S. J., and Peters K. (1976) Revised lunar heat flow values. *Proc. Lunar Sci. Conf. 7th,* pp. 3143–3171.

Lemoine F. G., Smith D. E., Zuber M. T., Neumann G. A., and Rowlands D. D. (1997) A 70th degree lunar gravity model from Clementine and other tracking data. *J. Geophys. Res., 102,* 16339–16359.

Lin R. P., Anderson K. A., and Hood L. L. (1988) Lunar surface magnetic field concentrations antipodal to young large impact basins. *Icarus, 74,* 529–541.

Lin R. P., Mitchell D. L., Curtis D. W., Anderson K. A., Carlson C. W., McFadden J., Acuña M., Hood L., and Binder A. (1998) Lunar surface magnetic fields and their interaction with the solar wind: Results from Lunar Prospector. *Science, 281,* 1480–1484.

Lucey P. G., Spudis P. D., Zuber M., Smith D., and Malaret E. (1994) Topographic-compositional units on the Moon and the early composition of the lunar crust. *Science, 266,* 1855–1858.

Lucey P. G., Taylor G., Hawke B. R., and Spudis P. D. (1998) FeO and $TiO_2$ concentrations in the South Pole-Aitken basin: Implications for mantle composition and basin formation. *J. Geophys. Res., 103,* 3701–3708.

Melosh H. J. (1989) *Impact Cratering: A Geologic Process.* Oxford Univ., New York, 245 pp.

Mizutani H. (1995) Lunar interior exploration by Japanese lunar penetrator mission, LUNAR-A. *J. Phys. Earth, 43,* 657–670.

Mueller S., Taylor G. J., and Phillips R. J. (1988) Lunar composition: A geophysical and petrological synthesis. *J. Geophys. Res., 93,* 6338–6352.

Muller P. M. and Sjogren W. L. (1968) Mascons: Lunar mass concentrations. *Science, 161,* 680–684.

Nakamura Y. (1983) Seismic velocity structure of the lunar upper mantle. *J. Geophys. Res., 88,* 677–686.

Nakamura Y. and Koyama J. (1982) Seismic Q of the lunar upper mantle. *J. Geophys. Res., 87,* 4855–4861.

Nakamura Y., Duennebier F., Latham G., and Dorman H. (1976) Structure of the lunar mantle. *J. Geophys. Res., 81,* 4818–4824.

Neumann G. A., Zuber M. T., Smith D. E., and Lemoine F. G. (1996) The lunar crust: Global signature and structure of major basins. *J. Geophys. Res., 101,* 16,841–16,863.

Nozette S., Rustan P. L., Plesance L. P., Horan D. M., Regeon P., Shoemaker E. M., Spudis P. D., Acton C., Baker D. N., Blamont J. E., Buratti B. J., Corson M. P., Davies M. E., Duxbury T. C., Eliason E. M., Jakosky B. M., Kordas J. F., Lewis I. T., Lichtenberg C. L., Lucey P. G., Malaret E., Massie M. A., Resnick J. H., Rollins C. J., Park H. S., McEwen A. S., Priest R. E., Pieters C. M., Reisse R. A., Robinson M. S., Simpson R. A., Smith D. E., Sorenson T. C., Vorder Bruegge R. W., and Zuber M. T. (1994) The Clementine mission to the Moon: Scientific overview. *Science, 266,* 1835–1839.

Phillips R. J. and Dvorak J. (1981) The origin of lunar mascon basins: Analysis of the Bouguer gravity associated with Grimaldi. In *Multi-Ring Basins, Proc. Lunar Planet. Sci. 12A* (P. H. Schultz and R. B. Merrill, eds.), pp. 91–104.

Phillips R. J., Sjogren W. L., Abbott E. A., and Zisk S. H. (1978) Simulation gravity modeling to spacecraft tracking data: Analysis and application. *J. Geophys. Res., 83,* 5455–5464.

Pritchard M. E. and Stevenson D. J. (2000) Thermal aspects of a lunar origin by giant impact. In *Origin of the Earth and Moon* (R. M. Canup and K. Righter, eds.), this volume. Univ. of Arizona, Tucson.

Pullan S. and Lambeck K. (1980) On constraining lunar mantle temperatures from gravity data *Proc. Lunar Planet. Sci. Conf. 11th*, pp. 2031–2041.

Rasmussen K. L. and Warren P. H. (1985) Megaregolith thickness, heat flow, and the bulk composition of the Moon. *Nature, 313*, 121–124.

Ringwood A. E. and Essene E. (1970) Petrogenesis of Apollo 11 basalts, internal constitution and origin of the Moon. *Proc. Apollo 11 Lunar Sci. Conf.*, pp. 769–799.

Russell C. T., Coleman P. J. Jr., and Goldstein B. E. (1981) Measurements of the lunar induced magnetic moment in the geomagnetic tail: Evidence for a lunar core. *Proc. Lunar Planet. Sci. 12B*, pp. 831–836.

Smith D. E. and Zuber M. T. (1998) Inferences about the early Moon from gravity and topography (abstract). In *Origin of the Earth and Moon*, pp. 56–57. LPI Contribution No. 957, Lunar and Planetary Institute, Houston.

Smith D. E., Zuber M. T., Neumann G. A., and Lemoine F. G. (1997) Topography of the Moon from the Clementine LIDAR. *J. Geophys. Res., 102*, 1591–1611.

Smith D. E., Tyler G. L., Balmino G., and Sjogren S. L. (1999) The gravity field of Mars from Mars Global Surveyor. *Science*, submitted.

Solomon S. C. (1978) The nature of isostasy on the Moon: How big a Pratt-fall for Airy models? *Proc. Lunar Planet. Sci. Conf. 9th*, pp. 3499–3511.

Solomon S. C. (1986) On the early thermal state of the Moon. In *Origin of the Moon* (W. K. Hartmann, R. J. Phillips, and G. J. Taylor, eds.), pp. 435–452. Lunar and Planetary Institute, Houston.

Solomon S. C. and Head J. W. (1980) Lunar mascon basins: Lava filling, tectonics and evolution of the lithosphere. *Rev. Geophys., 18*, 107–141.

Solomon S. C. and Simons M. (1996) The isostatic state of the lunar highlands from spatio-spectral localization of gravity, topography, and surface chemistry (abstract). In *Lunar and Planetary Science XXVII*, pp. 1245–1246.

Strangway D. W., Sharpe H., Gose W., and Pearce G. (1973) Magnetism and the history of the Moon. In *Magnetism and Magnetic Materials — 1972* (C. D. Graham Jr. and J. J. Rhyne, eds.), pp. 1178–1187. American Institute of Physics, New York.

Taylor S. R. (1987) The unique lunar composition and its bearing on the origin of the Moon. *Geochim. Cosmochim. Acta, 51*, 1297–1309.

Thurber C. H. and Solomon S. C. (1978) An assessment of crustal thickness variations on the lunar nearside: Models, uncertainties, and implications for crustal differentiation. *Proc. Lunar Planet. Sci. Conf. 9th*, 3481–3497.

Toksöz M. N., Hsui A. T., and Johnston D. H. (1978) Thermal evolutions of the terrestrial planets. *Moon and Planets, 18*, 281–320.

Touma J. (2000) The phase space adventure of Earth and Moon. In *Origin of the Earth and Moon* (R. M. Canup and K. Righter, eds.), this volume. Univ. of Arizona, Tucson.

Touma J. and Wisdom J. (1994) Evolution of the Earth-Moon system. *Astron. J., 108*, 1943–1961.

von Frese R. R. B., Tan L., Potts L. V., Merry C. J., and Bossler J. D. (1996) Lunar crustal analysis of Mare Orientale from Clementine satellite observations. *J. Geophys. Res., 102*, 25657–25676.

Wieczorek M. A. and Phillips R. J. (1997) The structure and compensation of the lunar highland crust. *J. Geophys. Res., 102*, 10933–10943.

Wieczorek M. A. and Phillips R. J. (1998) Potential anomalies on a sphere: Applications to the thickness of the lunar crust. *J. Geophys. Res., 103*, 1715–1724.

Willemann R. J. and Turcotte D. L. (1981) Support of topographic and other loads on the Moon and on the terrestrial planets. *Proc. Lunar Planet. Sci. 12B*, pp. 837–851.

Williams J. G., Boggs D. H., Ratcliff J. T., and Dickey J. O. (1999) The Moon's molten core and tidal Q (abstract). In *Lunar and Planetary Science XXX*, Lunar and Planetary Institute, Houston.

Williams K. K. and Zuber M. T. (1998) Measurement and analysis of major lunar basin depths from Clementine altimetry. *Icarus, 131*, 107–122.

Wise D. U. and Yates M. Y. (1970) Lunar anorthosites and a geophysical model for the Moon. (1970) *Proc. Apollo 11 Lunar Sci. Conf.*, pp. 958–988.

Wood J. A., Dickey J. S. Jr., Marvin U. B., Powell B. N., and Yates M. Y. (1970) Mascons as subsurface relief on a lunar "Moho". *J. Geophys. Res., 75*, 261–268.

Yoder C. F. (1981) The free librations of a dissipative Moon. *Philos. Trans. Roy. Soc. London, A303*, 327–338.

Yoder C. F. (1984) The size of the lunar core (abstract). In *Papers Presented to the Conference on the Origin of the Moon*, p. 6. Lunar and Planetary Institute, Houston.

Zhong S. and Zuber M. T. (1999) Long-wavelength topographic relaxation for self-gravitating planets and implications for the time-dependent compensation of surface topography. *J. Geophys. Res.*, submitted.

Zuber M. T., Smith D. E., Lemoine F. G., and Neumann G. A. (1994) The shape and internal structure of the Moon from the Clementine mission. *Science, 266*, 1839–1843.

*Part V:*

*Origin of Terrestrial Volatiles*

# Water in the Early Earth

## Yutaka Abe
*University of Tokyo*

## Eiji Ohtani
*Tohoku University*

## Takuo Okuchi
*Nagoya University*

## Kevin Righter and Michael Drake
*University of Arizona*

In this chapter we discuss the behavior of water in the early Earth. Earth likely formed through accretion of water-bearing planetesimals. An $H_2O$-rich proto-atmosphere should have formed during accretion by degassing from planetesimals and/or gravitational attraction of solar-nebula gas, and it may be a direct ancestor of Earth's present atmosphere-hydrosphere. A hydrous magma ocean can form in response to the thermal blanketing effect of an early proto-atmosphere. The presence of water in a magma ocean lowers the liquidus and solidus temperatures, affects phase relations, modifies mineral-melt and Fe-melt elemental partitioning, and alters its physical and thermochemical properties. In addition, because of the presence of water, a large amount of H may be partitioned into metallic Fe under high pressure and delivered to the core. Under plausible water concentrations in source materials, H can reconcile a major fraction of the observed density deficit in the present core. Many hydrous magnesian silicates are stable at pressures equivalent to those in Earth's transition zone, and the mantle may have long-term water storage capacity. Evidence for a hydrous Archean mantle is provided by experimental work demonstrating the ability to generate komatiitic magma by melting hydrous peridotite.

## 1. INTRODUCTION

The distribution of water in the inner solar system is poorly understood, but may be roughly in scale with the mass of the bodies. Among those bodies from which we are confident we have samples, Earth contains the most water, followed by Mars (*Carr and Wänke,* 1992), and the Moon and the eucrite parent body (asteroid 4 Vesta) are dry. To our knowledge, we have no direct samples of Venus or Mercury. Both direct observation of water in the venusian atmosphere and the high D/H ratio of the venusian atmosphere (*Hunten et al.,* 1989) point to Venus being accreted as a hydrous planet. In contrast, there is no evidence of indigenous water on Mercury, although there is a tantalizing hint of the possibility of ice in permanently shadowed craters at the mercurian poles. By analogy with a similar discovery on the Moon, this ice may be the result of trapped cometary ices. Why this range of water contents in inner solar system bodies exists is unknown, but it must be explained in any successful model of the formation of the Earth-Moon system.

What was the source of the terrestrial water? Until recently, there have been competing views. A view prevalent in the astronomical community held that Earth accreted as an anhydrous body and water was subsequently added through cometary impact. This view was driven by several separate ideas. One was that hydrous minerals would not form in the solar nebula at 1 AU from the Sun, so hydrated material was not available to be accreted to Earth. Another was that organic material would not survive the high temperatures associated with accretion, and would have to be delivered after accretion through cometary and perhaps hydrated asteroidal impacts. So why not bring in the water that way as well? Yet another idea was the notion, prevalent in European and American thought, that the terrestrial planets accreted after the solar nebula dissipated.

There was a competing view that examined accretion of water-bearing planetesimals. This view may appeal to the geological community, as it is well known that deep mantle minerals and magmas are capable of storing large quantities of water. While this view evolved largely in ignorance of dynamical studies of accretion, it is an environment that increasingly seems to accommodate our understanding of early Earth. The critical evidence that allows discrimination between these viewpoints has been the relatively recent determinations of the D/H ratios of comets (see *Owen and Bar-Nun,* 2000). The D/H ratios in Comets P-Halley, P-Hyakutake, and P-Hale-Bopp fall within the range $3.16 \pm 0.34 \times 10^{-4}$. The bulk Earth's D/H ratio is about $1.5 \times 10^{-4}$. If these three comets are broadly representative of all comets, a plausible but unproved assumption, then the post-accretion influx of water to Earth from comets is limited to less than ~20%. Therefore, Earth must have accreted at least in part from hydrated materials. Earth was born "wet."

There was another view, known as the "Kyoto School," that examined accretion in the presence of the H gas of the solar nebula. This view evolved from the theory of star formation (see *Hayashi et al.,* 1985, for a review of the Kyoto model). Since H is converted to water through reaction with silicate, Earth accreted wet in this scenario, too.

In this chapter, we explore the behavior of water in the early Earth. The recognition that Earth must have accreted water has profound consequences for early Earth, including the environment during accretion, the nature of an early magma ocean, the nature of the light element in Earth's core, and the subsequent stability of hydrous high-pressure minerals and their behavior as reservoirs for meteoric water. We will also examine the implications of a hydrous early Earth for other volatiles.

TABLE 1.   Water contents of various planetary materials.

| Material | $H_2O$ (wt%) | References |
|---|---|---|
| CO | 5.9–7.29 | [1,5] |
| CM | 6.54–~9.0 | [1,5] |
| CV | 0.617 | [1] |
| Semarkona matrix | 0.5–4.0 | [2] |
| Renazzo matrix | 2.0–8.2 | [2] |
| Enstatite | 0.21–1.37 | [3] |
| Mars mantle | 0.0036 | [1] |
| Earth mantle | 0.048 | [1] |
| Comet ice | ~80 | [4] |

References:   [1] *Dreibus and Wanke* (1989); [2] *Deloule and Robert* (1995); [3] *Mason* (1966); [4] *Jessberger et al.* (1989); [5] *Kerridge* (1985).

## 2.  HYDROUS EARLY ATMOSPHERE

### 2.1.  Accretion Model and Atmosphere

The behavior of water in the early Earth strongly depends on accretion conditions and the source of volatile materials. Candidates for the source of volatile materials include comets, the solar-nebula, and planetesimals. Cometary impact after the end of accretion is one possibility of how volatile materials were delivered to Earth, but the D/H ratios of comets are not the same as terrestrial ocean water (*Meier et al.,* 1997). In addition, recent study of the dynamics of comets indicates that the impact probability of comets is too low to supply all of the terrestrial water (*Zahnle and Dones,* 1998). Thus, most terrestrial water should originate from other sources. The solar-nebula cannot be the main source of major volatile components ($H_2O$, $CO_2$, $N_2$, etc.), because the relative concentration of rare gases to other gas species is much larger than that of the present atmosphere of Earth (e.g., *Porcelli and Pepin,* 1999). Since preferential loss of rare gases relative to other major volatiles is highly unlikely, major volatile components of the present atmosphere must have originated from another source (*Brown,* 1949; *Suess,* 1949). (These arguments do not rule out the possibility that some of the rare gases contained in the present atmosphere originated from gravitationally trapped solar nebula.) Thus, a plausible source of volatile materials is planetesimals (see Table 1). This view is partly supported by the existence of water-bearing meteorites such as carbonaceous meteorites and the similarity of the relative abundance of volatile elements between meteorites and Earth's atmosphere (*Brown,* 1949; *Suess,* 1949).

A major uncertainty is whether water-bearing planetesimals are formed at Earth's orbit or only at distant parts of the solar nebula, say at the asteroid belt. If water-bearing planetesimals formed at Earth's orbit, Earth should have accreted entirely from water-bearing planetesimals. On the other hand, if only anhydrous planetesimals are formed at

Earth's orbit, Earth-forming planetesimals should have changed from dry to wet during accretion. If this is the case, the behavior of water is strongly affected by the timing of transition from dry to wet.

Depending on the availability of volatile material during accretion, there are three possibilities for the environment of the accreting Earth. If no volatile-bearing planetesimals were available during accretion, no atmosphere would be formed during accretion. In that model, the surface temperature is always kept under the melting temperature of mantle silicate and only a subsurface magma ocean is formed, core formation proceeds under dry conditions, and volatile elements are not partitioned into metallic Fe. If Earth formed in a nebular gas, it would have attracted surrounding gas to form a distended solar-composition ($H_2$-He) atmosphere. Although this atmosphere could not be a direct ancestor of the present atmosphere, as discussed above, it would affect the distribution of volatile materials within Earth through its thermal effect. In this scenario, the surface is kept above the liquidus temperature of mantle silicate and a surface magma ocean is formed, due to the very strong blanketing effect of the atmosphere. Some volatile elements dissolve into the surface magma ocean.

If Earth formed from volatile-bearing planetesimals, its atmosphere would have formed through impact degassing; this is often called the "secondary atmosphere." However, this name is misleading, because impact degassing may start earlier than the formation of a solar-type atmosphere, which is often called the "primordial atmosphere." Hence, we refrain from using the words "primordial" or "secondary" atmosphere. A degassed atmosphere is sometimes called a "steam atmosphere," because the major volatile contained in planetesimals would be water. However, if we consider reaction with metallic Fe, it is likely to be rich in H, as noted later. This atmosphere also has a blanketing effect and would produce a surface magma ocean under certain conditions. Some volatile materials would then dissolve into the magma ocean and react with metallic Fe. The degassed atmosphere can be a direct ancestor of the present atmosphere-hydrosphere. So, this atmosphere is important in the context of the origin of the atmosphere-hydrosphere.

In the following subsections we discuss various atmospheres during accretion; solar-type and "steam" atmospheres should be regarded as end members. Because impact degassing may start before the dissipation of the solar-type atmosphere, we discuss the solar-type and steam atmospheres first, then discuss their mixture briefly.

## 2.2.  Solar-type Atmosphere

When Earth accreted in the solar nebula, a solar-type atmosphere could have formed through gravitational attraction of the surrounding nebula gas. The mass of this atmosphere becomes substantial when the proto-Earth becomes greater than about 0.1 $M_\oplus$ (1 $M_\oplus$, or Earth mass, = 5.97 × $10^{24}$ kg) (*Hayashi et al.*, 1979; *Nakazawa et al.*, 1985; *Sasaki*, 1990). Whether or not Earth had a solar-type atmosphere depends on the timing of the dissipation of the nebula gas. The mechanism and timing of gas dissipation are not well constrained at this time. Previous studies often considered strong UV radiation from the young Sun as a dissipation agent, because it was well expected for T-Tauri stars. However, recent observations of T-Tauri stars pose questions about the presence of strong UV radiation on the early Earth. Strong UV radiation is observed for young stars that have gas-dust disks accreting onto the stars. This radiation is believed to be emitted from the boundary layers between the slowly rotating stars and the rapidly rotating accretion disks (e.g., *Basri and Bertout*, 1993) and should fade once disks stop accretion.

Though the solar-type atmosphere cannot be a direct ancestor of the present atmosphere, it affects the behavior of water in Earth through its thermal blanketing effects on the accreting Earth. The blanketing effects have been estimated by calculating its structure under the assumption of spherical symmetry and hydrostatic equilibrium. *Hayashi et al.* (1979) showed that the temperature at the bottom of the atmosphere, namely at the surface of the proto-Earth, is about 1200°C for M = 0.25 $M_\oplus$ due to the blanketing effect of the opaque atmosphere. This implies that the surface temperature of the proto-Earth was high enough to melt most of the materials. Generally speaking, the surface temperature depends on the accretion rate very weakly but is sensitive to the density of the nebula and the abundance of the dust grains (*Nakazawa et al.*, 1985).

Very high-surface temperature would promote chemical reaction between the atmosphere and magma ocean. *Sasaki* (1990) took into account reduction of Fe-bearing silicate melt by atmospheric $H_2$. This reaction should raise the ratio $pH_2O/pH_2$ in the atmosphere. If oxygen fugacity is buffered by such a reaction, it would be close to that of the iron-wüstite buffer, and $pH_2O/pH_2$ would be higher than 0.1. Enhancement of $H_2O$ abundance increases the surface temperature through a rise in opacity and the mean molecular weight of the gas. The surface temperature of the proto-Earth can be as high as 4500°C when its mass is 1 $M_\oplus$. Such

high temperature should also promote extensive vaporization of silicate material into the atmosphere. Also, very high surface temperatures would result in a deep magma ocean, and would affect partitioning of elements between core and mantle materials.

Some amount of $H_2O$ should dissolve into the magma ocean. Dissolution reaction of $H_2O$ into the magma ocean is controlled by atmospheric partial pressure of $H_2O$, $pH_2O$, and is highly dependent on the $pH_2O/pH_2$ ratio (*Sasaki*, 1990). If the solar $pH_2O/pH_2$ ratio is assumed, $pH_2O$ is less than $10^4$ Pa at the surface of the magma ocean. However, if $pH_2O/pH_2$ ratio is buffered by the reaction at the surface of the magma ocean, $pH_2O$ is as large as $5 \times 10^6$ Pa. This difference is very large if we consider the behavior of $H_2O$ contained in planetesimals. If a few percent of the Earth-forming planetesimals have carbonaceous-chondrite like composition, which contains up to 9% of water by weight (Table 1; *Boato*, 1954) (including water formed by oxidization of hydrocarbons), the average water content of the source materials may range from 0.1 to a few percentages by weight. When $pH_2O$ is <$10^4$ Pa, the amount of dissolved water can be smaller than that contained in planetesimals and degassing will occur. On the other hand, if $pH_2O$ is as high as $5 \times 10^6$ Pa, most of $H_2O$ contained in planetesimals dissolves into the magma ocean.

## 2.3.  "Steam" Atmosphere

Volatile materials in the present Earth's atmosphere were probably contained in solids in Earth-forming planetesimals (*Brown*, 1949; *Suess*, 1949). If impacts of volatile-bearing planetesimals occur during Earth formation, the formation of an atmosphere due to impact degassing is expected.

Experimental and theoretical studies of impact degassing (e.g., *Lange and Ahrens*, 1982a,b; *Tyburczy et al.*, 1986) have shown that incipient and complete dehydration occur when the impact velocity exceeds about 2 km/s and 4 km/s respectively. Moreover, a substantial fraction of the impactor should vaporize once the impact velocity exceeds about 10 km/s (*Melosh and Vickery*, 1989). Because the impact velocities of Earth-forming planetesimals are greater than the escape velocity of the proto-Earth, efficient impact dehydration should occur and a proto-atmosphere should have formed during Earth's formation. The formation of the steam atmosphere starts when the mass of the proto-Earth reaches about 0.01 $M_\oplus$.

The composition and mass of the degassed atmosphere are controlled by the following factors: (1) the compositions of source materials, (2) the partitioning of volatile materials between the proto-atmosphere and planetary interior, and (3) loss from the protoplanet. As noted before, average water content of the source materials may range from 0.1 to a few percentages by weight.

The partitioning of volatile materials between the degassed atmosphere and proto-mantle depends on the temperature and pressure at the surface of the proto-Earth. Since $H_2O$ gas has strong absorption bands for infrared radiation, the degassed atmosphere plays an important role in preventing the impact energy released at the surface of the growing planet from escaping into interplanetary space. The surface of the growing Earth could be heated to the melting temperature (*Abe and Matsui*, 1985, 1986, 1988; *Matsui and Abe,* 1986a,b; *Zahnle et al.,* 1988). Although the gas nebula should have existed at the earliest stage of the Earth formation, they assumed Earth formation in gas-free space for simplicity.

The surface temperature depends on composition and mass of the atmosphere, and the average heat flux from the planet, which corresponds to the sum of the net solar flux and gravitational energy flux released by accretion. Figure 1a shows the relation among the heat flux, atmospheric $H_2O$ abundance, and surface temperature on an Earth-sized planet. Here, an $H_2$-$H_2O$ atmosphere with the $pH_2O/pH_2$ ratio fixed by iron-wüstite buffer is considered, and the mole fraction of $H_2O$ is ~10%. Therefore, this is mainly a $H_2$ atmosphere rather than a steam atmosphere. Attached numerals indicate the atmospheric mass by the surface pressure when all hydrogen in the atmosphere is converted to $H_2O$. A mass of $2.7 \times 10^7$ Pa corresponds to 1 Earth-ocean mass $(1.4 \times 10^{21}$ kg) equivalent of total H. Note that this pressure does not correspond to either the total pressure or the partial pressure of $H_2O$. If energy flux is a few hundred $W/m^2$ and the mass of the atmosphere is $1 \times 10^7$ Pa equivalent of $H_2O$, the surface temperature exceeds the melting temperature of mantle silicate. However, regardless of the atmospheric mass, very large heat flux is required for keeping the surface temperature higher than ~1700°C. Also, regardless of the atmospheric mass, surface temperature is lower than a few hundred degrees Celsius, if the heat flux is smaller than ~250 $W/m^2$. The former phenomenon is due to transparency of water vapor for short wavelength radiation. The latter phenomenon is due to condensation and collapse of a water-rich atmosphere. A flux of 250 $W/m^2$ corresponds to the critical flux required for runaway greenhouse effect (*Nakajima et al.,* 1992) for an $H_2$-$H_2O$ atmosphere. If we take into account other gases such as $CO_2$, CO, and $CH_4$, a higher surface temperature results from a smaller heat flux. The average energy flux released by accretion is given by

$$F = \frac{1}{4\pi R^2} \frac{GM}{R} \dot{M}$$

where F, R, M, and $\dot{M}$ are energy flux, radius, mass of the protoplanet, and mass-accretion rate respectively. In the

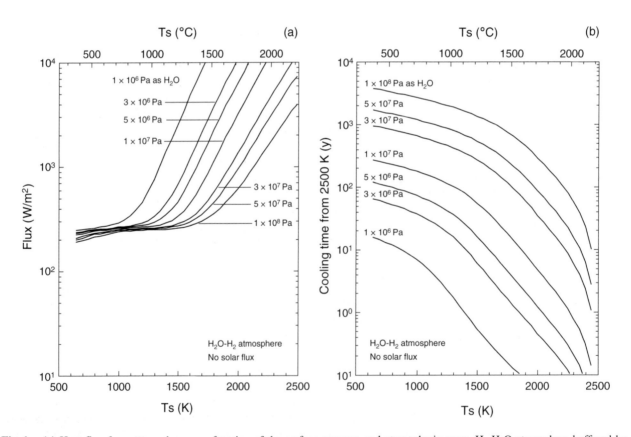

**Fig. 1.**   (a) Heat flux from atmosphere as a function of the surface pressure and atmospheric mass. $H_2$-$H_2O$ atmosphere buffered by the iron-wüstite O buffer is considered. Atmospheric mass is indicated by the surface pressure when all H in the atmosphere is converted to $H_2O$. $2.7 \times 10^7$ Pa corresponds to 1 Earth-ocean mass $(1.4 \times 10^{21}$ kg) equivalent of total H. Note that these values are different from neither the total pressure nor the partial pressure of $H_2O$. The total pressure is about one-tenth and $H_2O$ partial pressure is about one-hundredth of the value indicated by numerals, respectively. **(b)** Cooling time from 2227°C (2500 K) for the atmosphere shown in **(a)**.

runaway growth stage, Mars-sized protoplanets form in $10^6$ yr (see *Kortenkamp et al., 2000*), $\dot{M}$ is $\sim 2 \times 10^{10}$ kg/s, and the energy flux is on the order of $10^3$ W/m$^2$. This is large enough to create a molten surface on an accreting Earth with a moderate atmosphere. However, since the actual energy release occurs intermittently, the atmosphere may cool before the next planetesimal impact if the mean impact interval of planetesimals is longer than the cooling time of a hot atmosphere (*Abe, 1988; Rintoul and Stevenson, 1988*). Therefore, it is important to compare impact interval and cooling time of the atmosphere. Figure 1b shows the cooling time of atmospheres from 2227°C (2500 K) to given temperatures. In the runaway growth stage, if the planetesimal size is $\sim 10^{19}$ kg, the average impact interval is $\sim 60$ yr. This interval is shorter than the cooling time from 2227°C (2500 K) to 1227°C (1500 K) for the atmosphere with $10^7$ Pa $H_2O$ equivalent. Since heating by the solar radiation is ignored in this calculation of cooling time, a real cooling time is longer than this estimate. Moreover, much longer cooling time is likely, because greenhouse gases such as CO or $CO_2$ are likely in the H-steam atmosphere. Generally speaking, the cooling time is controlled by the opacity at the most transparent wavelengths of $H_2O$ absorption bands, which can be called "window" wavelengths. Opacity of the window wavelength is easily affected by the minor addition of greenhouse gases that have absorption bands at window wavelengths. This results in a significant increase of the cooling time. For example, the cooling time from 2227°C (2500 K) to 1227°C (1500 K) increases from about 60 yr to 1000 yr by the addition of 16% of CO-$CO_2$ mixture gas to the $10^7$ Pa $H_2O$ equivalent atmosphere. If the atmosphere is formed by degassing from chondritic sources, such an amount of CO-$CO_2$ is likely in the atmosphere equilibrated with metallic Fe. Thus, if a sufficiently large amount of $H_2O$ and other gases are released during the runaway accretion stage, a surface magma ocean can be sustained during this stage.

Core formation (iron-silicate segregation) also releases gravitational energy. The energy flux released by core formation is $\sim 5$–10% of the average energy flux released by accretion. Since convection regulates heat transport in the mantle (magma ocean), the temporal variation of the energy flux due to core formation would be small. If the accretion rate is sufficiently large, energy flux due to core formation would exceed the critical flux for the runaway greenhouse. In that case, a surface magma ocean could be sustained regardless of the cooling time of the atmosphere or impact interval.

In the later stage of accretion, behavior of the atmosphere depends on accretion models. If the accreting planetesimals remain small, a hot atmosphere and magma ocean will be sustained. On the other hand, if large planetesimals (or protoplanets) impact intermittently, the water ocean stage and the hot atmosphere with a magma ocean stage will repeat. At this stage, we have to consider the loss of atmosphere at the time of giant impacts. However, the loss efficiency is not well understood at this time.

Once a surface magma ocean is formed, the mass of the steam atmosphere and the surface temperature will not vary much during accretion (*Abe and Matsui, 1986, 1988; Matsui and Abe, 1986a,b; Zahnle et al., 1988*). This is because a large amount of water will dissolve into the surface magma ocean. Under $10^7$ Pa steam atmosphere, the concentration of water in silicate melt is $\sim 1\%$ by weight (e.g., *Fricker and Reynolds, 1968; Holland, 1984*). Since the concentration of $H_2O$ is comparable to or larger than that in the source materials, almost all $H_2O$ supplied by accreting planetesimals will be dissolved into the surface magma ocean. If the abundance of $H_2O$ in the atmosphere decreases, the efficiency of the blanketing effect is lowered and the surface temperature and the melt fraction decrease. Decrease of the melt fraction in turn results in a decrease of the average concentration of $H_2O$ in the magma ocean and a decrease in the loss of $H_2O$ from the atmosphere by dissolution. By such negative feedback, the abundance of $H_2O$ in the atmosphere is kept just above the critical abundance required for sustaining a surface magma ocean. The resulting abundance of water vapor in the atmosphere is expected to be close to the mass of the present ocean ($\sim 10^{21}$ kg, or $\sim 3 \times 10^7$ Pa) at the end of accretion (*Abe and Matsui, 1986, 1988; Matsui and Abe, 1986a,b; Zahnle et al., 1988*).

The degassed atmosphere might be lost either by impact erosion (e.g., *Melosh and Vickery, 1989*) or by the hydrodynamic escape of the atmosphere. The former depends critically on the impact velocity. During the accretion stage, the estimated impact velocity is close to the escape velocity. This is not sufficient to invoke a large amount of impact erosion. The rate of mass loss by the hydrodynamic escape depends on availability of far and extreme UV at the upper atmosphere. As noted before, estimates of UV flux on early Earth are uncertain. At the later stage of accretion, giant impacts may play an important role as a loss mechanism. But this process has not been well investigated yet.

The abundance of water vapor and C-bearing gases might not be significantly affected by the escape of the degassed atmosphere during accretion, if their abundances are buffered by the surface magma ocean (*Abe and Matsui, 1986; Zahnle et al., 1988*).

## 2.4. Mixture of Solar-type and "Steam" Atmospheres

All the studies on the degassed atmosphere assumed planetary formation in gas-free space. However, a gas nebula should have existed at least at the earliest stages of planetary formation. Hence, in reality, a mixed proto-atmosphere of solar-type and degassed components might have formed. Although properties of such mixed atmospheres have not been investigated thus far, we speculate on their properties briefly.

It should be noted that the solar-type atmosphere might have survived the nebula dissipation, because the atmosphere is tightly bound by the Earth's gravity field. Therefore, we should consider two cases: mixed atmosphere embedded in the nebula gas and mixed atmosphere detached from the nebula gas. The mixed atmosphere embedded in the nebula gas is basically similar to that of the solar-type atmosphere, with enhanced $H_2O$ abundance near the bottom. Thus, a very high surface temperature is expected. On the other hand, a mixed atmosphere detached from a nebular gas should have a structure similar to a degassed atmosphere diluted by large amount of H.

As long as the gas nebula is lost before completion of planetary formation, the chemistry for major components ($H_2O$, $CO_2$, etc.) would be similar to that of the degassed atmosphere, because these components are mainly derived through impact degassing. However, D/H ratios (*Owen and Bar-Nun*, 2000) and elemental and isotopic abundance of rare gases (*Porcelli and Pepin*, 2000) would have been significantly affected by the existence of the mixed proto-atmosphere of solar-type and degassed components. It seems inevitable that the early Earth was hot and substantial water was dissolved in a magma ocean.

## 3.  CHARACTERISTICS AND PROPERTIES OF A HYDROUS MAGMA OCEAN

Modeling and calculations show that a thick blanketing water-rich atmosphere early in Earth's history (see previous section), would foster surface temperatures as high as 1500°–1800°C, enough to melt the outer portion of the planet. The higher pressure of water in the atmosphere would allow a significant amount of water to dissolve into the silicate liquid in the magma. Because moderately siderophile element concentrations in the primitive upper mantle can be explained by metal-silicate equilibrium at high P-T conditions just below the dry peridotite solidus, *Righter et al.* (1997) also proposed that the magma ocean was hydrous (up to 2–3 wt% $H_2O$), so that the peridotite liquidus and solidus were depressed to allow a magma ocean. Although the general characteristics of a water-rich magma ocean can be appreciated, we lack detailed knowledge of several important aspects of such a system: the water capacity of a magma ocean, the effect of water on the partitioning of elements between metal, solid silicates, and liquid silicate, and the effect of dissolved water on thermochemical properties of silicate liquid. The latter will be crucial for effective modeling of the crystallization of such a system.

### 3.1.  Water Capacity of a Magma Ocean and Phase Relations in Hydrous Basic Systems

It has been known for nearly 35 years that water has a large solubility in magma, with as much as 15 wt% water entering a basaltic magma in a system saturated with a pure $H_2O$ fluid at 1.5 GPa (*Hamilton et al.*, 1964). Water solubility as a function of pressure and temperature is well known now for a large range of natural liquids — much research has focused on systems relevant to terrestrial magmatic systems (basalt, andesite, dacite, rhyolite), both in

water-saturated and mixed speciation (e.g., $CO_2$-$H_2O$) conditions. Not until recently, however, has a general model for the solubility of water as a function of P, T, and melt composition been proposed and that up to the pressure of only 0.3 GPa (*Moore et al.*, 1995a, 1998). They showed that the solubility of water in magma has a strong dependence upon melt composition, most notably $Na_2O$ causing a large increase. Using this model as a guide, we can estimate the potential for water capacity of a peridotitic magma at low pressures. A liquid of the composition of the upper mantle and saturated with a fluid of $XH_2O = 1$, will have ~7.6 wt% water dissolved at 1700°C and 0.3 GPa. At these same conditions but with a fluid with $XH_2O = 0.2$, a peridotite magma will have 2.8 wt% $H_2O$. The latter example may be more appropriate for the less-oxidizing conditions of an early terrestrial magma ocean where fluids are likely to be CO, $CO_2$, $H_2$, $CH_4$, and $H_2O$ (*Holloway*, 1988). The solubility at higher pressures will be even greater, but cannot be calculated here because this is outside of the range of calibration of the *Moore et al.* (1998) model.

The effect of water on phase relations of magmatic systems can be dramatic. The olivine phase field is expanded in basaltic systems by the addition of even a small amount of water (e.g., *Kushiro*, 1969), and plagioclase is generally suppressed by dissolution of water in basic magmas (*Sisson and Grove*, 1993). Large changes have been documented in more basic systems as well. The liquidus in the enstatite system is 600°C lower in the water-saturated system (*Inoue*, 1994) than it is in the anhydrous system (*Gasparik*, 1993; see Fig. 2a). Similarly, the forsterite liquidus has a huge depression at water contents of 20 wt% compared to dry conditions. Experiments on hydrous peridotitic materials have not only shown enormous reduction in liquidus and solidus temperatures (*Inoue and Sawamoto*, 1992; *Kawamoto and Holloway*, 1997; Fig. 2b), but have also documented a change in the liquidus phases from olivine in a dry system to orthopyroxene and garnet in a hydrous system (*Inoue and Sawamoto*, 1992). So far, these experiments have been carried out at pressures <11 GPa, and thus it is not clear how dramatic the effect of water will be at the base of the upper mantle.

### 3.2.  Effect of Water on Element Partitioning: Mineral-hydrous Melt and Solid Metal-hydrous Melt

Silicate melts can be thought of as solutions containing cations (tetrahedral = Si, Al; octahedral = Mg, Ca, Na, etc.) with two types of surrounding oxygens — bridging (between tetrahedral cations) and nonbridging (between octahedral cations). Water can have a profound effect on the structure of silicate melts, generally causing depolymerization in the form of breaking bridging oxygens between tetrahedra. In the last several decades melt structural studies have been performed on simple and specific compositions such as aluminosilicate or albitic melts (e.g., *McMillan and Wolf*, 1995; *Stebbins*, 1995). There are still many uncertain aspects of silicate melt structure. For instance, there is evidence for excess nonbridging oxygens in tectosilicate melts

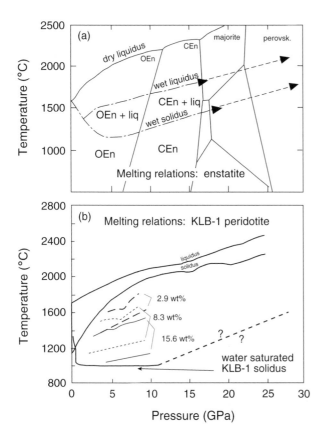

**Fig. 2.** (a) Comparison of hydrous and anhydrous phase relations in the enstatite system. Dry phase boundaries were reported by *Gasparik* (1993). Hydrous phase relations were determined by *Inoue* (1994). **(b)** Comparison of melting relations of KLB-1 for hydrous and anhydrous conditions. Highest liquidus and solidus temperatures are for the dry system, determined by *Zhang and Herzberg* (1994) (solid lines). Experiments from 4.0 to 10.0 GPa at 2.9, 8.3, and 15.6 wt% $H_2O$ were reported by *Inoue and Sawamoto* (1992). Water-saturated experiments to 11 GPa were performed by *Kawamoto and Holloway* (1997).

in which the oxygens are expected to be entirely bridging (*Stebbins and Xu,* 1997). And even though conventional wisdom holds that dissolved water will cause melt depolymerization, more recent structural studies of hydrous and anhydrous aluminosilicate melts have shown that there is no change in the ratio of nonbridging to bridging oxygens (*Kohn et al.,* 1998). Implications for more basic systems such as basaltic, komatiitic or peridotitic are unclear. Given the potential for such a large effect on silicate melt structure, it is natural to wonder how element partitioning between minerals or solid metal and silicate melt may be different in wet and dry systems. There has been some work in this area, but there is not yet a comprehensive understanding.

Although it is anticipated that dissolved water may cause oxidation of FeO to $Fe_2O_3$, such a reaction has not been observed in rhyolitic, andesitic or basaltic systems (*Moore et al.,* 1995b; *Righter and Delaney,* 1997). Nor has any difference in olivine-liquid FeO-MgO Kd been detected in

comparisons of hydrous and anhydrous systems (*Gaetani et al.,* 1995). These initial studies suggest that while dissolved water in a magma may cause changes in the phase relations of certain compositional systems (see discussion in previous section), the composition of ferromagnesian minerals precipitating from a hydrous magma will not be affected by the presence of water. So far this has been demonstrated for a small pressure range in basaltic systems and must be verified for more ultrabasic and ultramafic systems at the higher pressures relevant to deep magma ocean scenarios.

Due to the large amount of water involved in arc magma genesis, the effect of water on trace-element partitioning has been considered in some hydrous basaltic and peridotitic systems. *Gaetani et al.* (1997) found only minor differences in clinopyroxene-melt trace-element partitioning between hydrous and anhydrous systems, even where the silicate melt and clinopyroxene were of variable compositions. On the other hand, *Nielsen et al.* (1997) found substantial differences between magnetite-melt partitioning in hydrous versus anhydrous systems, but this is most likely due to changes in the magnetite composition precipitating from hydrous and anhydrous melts. There is clearly more work to be done in this area, both at low pressures (for basaltic minerals such as olivine, pyroxenes, and garnet) and ultra-high pressures (for majoritic garnet, ringwoodite, MgSi-perovskite).

Because silicate-melt composition has been shown to affect metal-silicate partitioning of siderophile trace elements, it might be expected that dissolved water will be important. Almost all previous studies of metal-silicate partitioning were done under dry conditions. Several studies have attempted to isolate the effect of dissolved water on the partitioning of siderophile elements between metal and silicate liquid (*Righter and Drake,* 1999; *Jana and Walker,* 1999). The *Righter and Drake* (1999) experiments show that solid metal-liquid silicate partition coefficients for Ni, Co, Mo, and W remain unchanged under hydrous conditions, whereas those for P remain unchanged only up to approximately 1.5 wt% dissolved water, above which they increase. *Jana and Walker* (1999) studied systems with H-C-O fluids present, and also see no changes in partitioning behavior of Ni, Co, Mo, W, P, and Ge outside of already known changes due to oxygen fugacity (hydrous experiments tend to be more oxidized than dry ones, and thus the partition coefficients are smaller). Both studies recognized that Fe would be oxidized to FeO by reaction with water, but not entirely. Iron in reduced chondritic planetary building blocks such as enstatite chondrites may be oxidized under hydrous conditions, but may still leave some $H_2O$ in an FeO-bearing mantle.

### 3.3. Thermochemical and Physical Properties of a Hydrous Magma Ocean

Although rapid decompression of hydrous magma can lead to explosive volcanic eruptions, our knowledge of the thermochemical properties of hydrous silicate melts remains primitive. Thermodynamics-based modeling of melts and crystals in anhydrous silicate melts (e.g., *Ghiorso and Sack,* 1994; *Lesher and Walker,* 1986) have had successful application to many terrestrial problems. However, they cannot currently be extended to hydrous systems for most P-T conditions. Despite this status, there has been notable progress in several areas that bear heavily on the issue of water in magmas.

The addition of water to a silicate melt has a large effect on its volumetric properties. For instance, *Ochs and Lange* (1997) have shown that a hydrous albitic melt is both more compressible and expansive than in a dry system, due to the large isothermal compressibility and thermal expansion coefficient for the $H_2O$ component. There is spectroscopic evidence for large compressibility changes in the $H_2O$ component in more basic systems as well — *Clossman and Williams* (1995) report H-O bond length changes at higher pressures. The lower density of hydrous melts will have an impact on the physics of crystal settling in a hydrous magma ocean. For instance, because garnet and orthopyroxene are stable liquidus phases in hydrous peridotite at pressures between 5.0 and 10.0 GPa (see phase relation discussion above), they will be more likely to settle out of a relatively less dense hydrous melt.

Very few thermochemical data exist for hydrous systems. Enthalpies of mixing in the albite-water system were measured by *Clemens and Navrotsky* (1987). These values can be reproduced in thermodynamic modeling based on an associated solution model that accounts for the formation of three membered rings in albite melt and the hydrolysis of Al-O-Al linkages (*Zeng and Nekvasil,* 1996). Thermodynamic modeling of hydrous magmatic systems has been extended across a range of liquids from rhyolitic to basanitic using a sub-regular solution model for silicate melts (*Ghiorso et al.,* 1983; *Ghiorso and Sack,* 1994). These models are based on high-quality solid-hydrous liquid experiments. Because such data are scarce for peridotite and ultrabasic systems, it has not yet been possible to extend this type of modeling.

## 4. HYDROGEN IN THE CORE

### 4.1. The Problem of Hydrogen

The Earth's outer core is about 10% less dense than molten Fe at the relevant pressure and temperature conditions (*Birch,* 1964; *Fukai,* 1992). This density deficit has been attributed to alloying of light elements in the Fe-rich core, such as S, C, H, and O (*Jeanloz,* 1990; *Wood,* 1997; *Hillgren et al.,* 2000). Hydrogen has been proposed as an important light element in the core in the last two decades, because the iron-water reaction at high pressure yields iron hydride (FeHx) due to a negative volume change in this reaction (*Stevenson,* 1977; *Fukai,* 1984). However, experiments on the Fe-H system are by far the most difficult among the binary system of Fe and light elements, due to rapid decomposition of FeHx during decompression after its synthesis at high pressure (*Fukai and Akimoto,* 1983).

The molar H concentration of FeHx was determined at high pressure for H/Fe = 0.13 at 3 GPa to H/Fe ~ 1 at 62 GPa, by its quenching into liquid N temperature (*Antonov et al.,* 1980), or by X-ray diffraction measurements at relevant (*in situ*) conditions (*Fukai et al.,* 1982; *Badding et al.,* 1991; *Yamakata et al.,* 1992). The latter technique was also applied to solid FeHx coexisting with silicate minerals and $H_2O$ fluid, and a H concentration up to H/Fe = 0.4 was measured at pressures to 4.9 GPa (*Yagi and Hishinuma,* 1995). In spite of a partial pressure of H lower than total pressure in this system (*Okuchi and Takahashi,* 1998), a large amount of H also dissolves in Fe as a consequence of the iron-water reaction.

Molten FeHx may be an important constituent of the molten outer core. However, the *in situ* X-ray diffraction technique cannot be applied to such experimental materials. In previous high-pressure melting experiments on FeHx (*Suzuki et al.,* 1984; *Hishinuma et al.,* 1994), many Fe droplets were recovered at temperatures ~500 K lower than the melting curve of pure Fe. The dissolution of H was inferred from this observation, but H was never detected from the quenched Fe droplets. Therefore, the concentration of H in the Earth's outer core has been uncertain until very recently.

### 4.2. Hydrogen Partitioning Experiments

Because the iron-water reaction may have proceeded extensively in a magma ocean on the primordial Earth (*Fukai and Suzuki,* 1986; *Stevenson,* 1990), the essential experiments for the problem of H in the core are those for H partitioning between molten FeHx and hydrous silicate melt at high pressure. These were recently carried out using the instant decompression technique (Table 2; *Okuchi,* 1997). A number of newly developed experimental techniques to conduct the H-partitioning experiments were previously reported in *Okuchi and Takahashi* (1998) in detail. It was found that H partitioning into molten FeHx increases with increasing temperature and increasing $H_2O$ concentration in the silicate melt (Fig. 3). To provide a comprehensive dataset for H partitioning in a primordial magma ocean, a systematic thermochemical analysis was made on the results.

At 7.5 GPa, H in molten FeHx increases from ~0.4 at 1200°C to ~0.7 at 1500°C (Fig. 3). At 1200°C, solid and molten FeHx coexist, and the latter is enriched in H by ~20 mol%. This is consistent with thermodynamic calculation, which predicts a slightly higher H concentration in the melt than in the solid (*Fukai,* 1991). The H-distribution coefficient between molten and solid FeHx, which is an analog of that between the outer core and the inner core, is therefore

$$D_{M/S}^H = \frac{(H/Fe)_{melt}}{(H/Fe)_{solid}} = 1.2 \quad (1)$$

The silicate melt composition in Table 2 was obtained by mass-balance calculation using the quenched mineral com-

TABLE 2. Run condition and results of the H-partitioning experiments at
high pressure. All experiments were conducted at 7.5 GPa.

| Run No. | P (GPa) | T (°C) | Time (s) | Starting Composition $H_2O/MgSiO_3$ (mole) | Silicate Melt Composition | | | Melt Fraction | FeHx Solid | (Fig. X) Melt | ln K (±1σ) |
|---|---|---|---|---|---|---|---|---|---|---|---|
| | | | | | Mg/Si | Fe/Si | $H_2O/Si$ | | | | |
| 499 | 7.5 | 1200 | 300 | 1.40 | 0.83 | 0.87 | 1.03 | ~20% | × | • | −3.6 ± 0.1 |
| 504 | 7.5 | 1200 | 1200 | 1.60 | 0.96 | 0.74 | 1.03 | 90% | — | • | −3.4 |
| 500 | 7.5 | 1300 | 300 | 1.60 | 1.00 | 0.77 | 0.83 | 99% | — | • | −2.4 ± 0.6 |
| 508 | 7.5 | 1400 | 60 | 1.55 | 1.00 | 0.77 | 0.78 | 100% | — | • | −2.0 |
| 603 | 7.5 | 1500 | 10 | 1.45 | 1.00 | 0.72 | 0.73 | 100% | — | • | −1.4 ± 0.2 |
| 570 | 7.5 | 1500 | 15 | 1.25 | 1.00 | 0.70 | 0.55 | 100% | — | • | −1.5 ± 0.3 |
| 514 | 7.5 | 1500 | 60 | 1.00 | 1.00 | 0.77 | 0.23 | 100% | — | • | −1.5 ± 0.2 |

position, as described in *Okuchi* (1997). From the composition of the coexisting FeHx and silicate melt at 7.5 GPa, the equilibrium constant of H-partitioning reaction in the $MgO$-$SiO_2$-Fe-O-H system was determined as a function of temperature. Hydrogen is a univalent cation in silicate melt, while it is neutral in FeHx. Therefore, during the FeHx formation in hydrous silicate melt, the H cation accepts one electron from the Fe atom. Thus the reactions of H reduction and Fe oxidation are coupled

$$\underset{\text{metal}}{Fe} + \underset{\text{silicate melt}}{H_2O} \leftrightarrow \underset{\text{silicate melt}}{Fe^{2+}} + \underset{\text{silicate melt}}{O^{2-}} + \underset{\text{metal}}{2H} \quad (2)$$

The equilibrium constant, K, is determined from the activities, $a_i$, of the products divided by the reactants

$$K = \frac{a_{Fe^{2+}} \times a_{O^{2-}} \times a_H^2}{a_{Fe} \times a_{H_2O}} \quad (3)$$

The natural log of K has a linear dependence on the reciprocal absolute temperature:

$$\ln K = -\frac{\Delta G°}{RT} = -\frac{\Delta H°}{RT} + \frac{\Delta S°}{R} \quad (4)$$

Here, $\Delta G°$, $\Delta H°$, and $\Delta S°$ are Gibbs free energy, enthalpy, and entropy change of the reaction, and R is gas constant.

The activity of each component in equation (3) was defined as a simple mole fraction of the component to all available sites in the host phase (*Okuchi*, 1997). Chemical equilibrium was confirmed because ln K is independent of starting condition and run duration at each temperature (Table 2 and Fig. 4). From the slope of the regression line, the enthalpy change ($\Delta H°$) of reaction (2) was obtained as 140 kJ/mol. This positive $\Delta H°$ indicates that H partitioning into molten FeHx increases with increasing temperature. The linear correlation of K against reciprocal absolute temperature enables its extrapolation to the slightly higher temperature relevant to H partitioning during Earth's core formation.

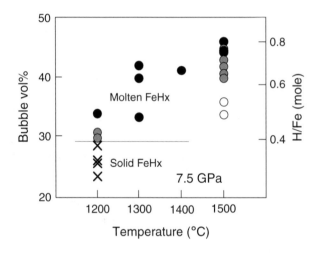

**Fig. 3.** Bubble fraction (vol%; left ordinate) of recovered Fe grains and H concentration in FeHx (right ordinate) as a function of temperature (abscissa). Symbols correspond to the state of FeHx and the $H_2O$ concentration in the starting material (see Table 2).

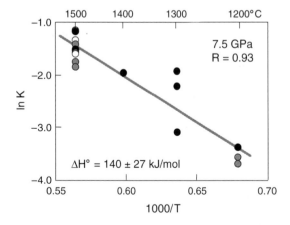

**Fig. 4.** Equilibrium constant of the H partitioning reaction (2) plotted against the reciprocal absolute temperature (×1000). The equilibrium constant increases with temperature so that the reaction is endothermic. Enthalpy change ($\Delta H°$) of the reaction is estimated from the slope of the regression line (see equation (4)). Symbols correspond to Table 2.

### 4.3.  Depth of the Earth's Magma Ocean

In planetary core formation processes, the depth of the magma ocean is an important parameter because any element partitioning between metal and silicate is dependent upon pressure and temperature (e.g., *Walter et al.,* 2000). As demonstrated in the previous section, this is also likely for the partitioning of H and/or water. As a protoplanet grows, its surface is covered by an impact-induced steam atmosphere that fosters formation of a magma ocean (see section 2). The metallic component in the falling planetesimals may sink through a magma ocean and accumulate at the bottom to form "iron ponds" (*Sasaki and Nakazawa,* 1986; *Stevenson,* 1990). The accumulation rate of Fe onto these ponds is comparable to the growth rate of the protoplanet on the order of 100 mm per year (*Hayashi et al.,* 1985; *Safronov and Ruzmaikina,* 1985). It is slower than H diffusion in Fe (*Fukai and Sugimoto,* 1985), so that the magma ocean and the ponded molten Fe may be in chemical equilibrium at least for H partitioning. A Rayleigh-Taylor instability may occur as the "ponds" grow at the top of the solid protomantle. Metallic liquid can then be transported to the center of the protoplanet within a relatively short timescale (*Stevenson,* 1981; *Sasaki and Nakazawa,* 1986), possibly so short that the molten Fe will not equilibrate with silicate. Therefore, the depth of the magma ocean is a critical parameter for understanding metal-silicate partitioning of H.

Here the H partitioning at the bottom of the magma ocean is discussed using the $\Delta G°$ determined from Fig. 4. To estimate the effect of pressure, we assume the volume change ($\Delta V°$) of the reaction (2) and the dissociation fraction of $H_2O$ as shown in Table 3. As discussed in section 4.3, neither experimental nor theoretical work define these values at present, so the following calculations are not necessarily appropriate for all pressure ranges, but are at least relevant for pressures around 7.5 GPa.

The steam protoatmosphere, magma ocean, and iron "ponds" form through differentiation of the accreting planetesimals. The atmospheric $H_2O$ mass is negligible compared with that in the magma ocean. Convection may homogenize the hydrous magma ocean, and H diffusion may homogenize the iron "ponds." Accordingly, the bulk composition of the system relevant to metal-silicate equilibrium during core formation can be approximated by that of the planetesimals. As discussed in the earlier section, large uncertainties exist for the $H_2O$ concentration of Earth's source material, so that of 0.5, 1.0, or 2.0 wt% was assumed for calculation. The hydrous magma ocean is a mixture of anhydrous minerals and hydrous silicate melt at pressures <14 GPa (*Inoue,* 1994), and the $H_2O$ concentration in the silicate melt becomes a function of the averaged melt fraction of the magma ocean from the surface to the bottom. This is numerically determined from an isentropic P-T path in the solid-melt mixture with the code developed by *Iwamori et al.* (1995). The temperature at the bottom of the

**TABLE 3.**  Assumptions for H-partitioning calculations at various pressures.

| Parameters | Pressure (GPa) | Formulation | References |
|---|---|---|---|
| $\Delta V°$ (cm³) | 3.0–10.0 | 10-P | [1–3] |
| $n_{H_2O}/X_{H_2O}$ | 3.0–7.5 | P/7.5 | [4] |
| | 7.5–10.0 | 1 | [5] |
| $2n_{OH}/X_{H_2O}$ | 3.0–7.5 | 1-P/7.5 | [3] |
| | 7.5–10.0 | 0 | [6] |

$\Delta V°$ = volume change of reaction (2). $n_{H_2O}/X_{H_2O}$ = molar fraction of $H_2O$ species to total $H_2O$ concentration in the silicate melt; $2n_{OH}/X_{H_2O}$ = molar fraction of OH species to total $H_2O$ concentration in the silicate melt. References: [1] *Boehler et al.* (1990); [2] *Saxena et al.* (1993); [3] *Dixon et al.* (1995); [4] *Stolper* (1982); [5] *Okuchi* (1997); [6] *Nowak and Behrens* (1995).

magma ocean is dependent upon the melting behavior of Earth's mantle peridotite at high pressures (*Takahashi et al.,* 1993). The temperature is not strictly at the chemical solidus of the peridotite, but at the critical melt fraction at which the solid silicate minerals are connected to form a three-dimensional network in the melt (e.g., *Taylor,* 1992).

The results of these calculations show that if the magma ocean at the final stage of Earth's accretion was a few gigapascals deep, most H accreted as $H_2O$ was partitioned into the core (Fig. 5). It is very possible that such a shallow magma ocean existed due to the atmospheric blanketing effect of accreted $H_2O$ itself, as discussed in section 2. Accordingly, most $H_2O$ in the planetesimals should be transported into the core, rather than into the atmosphere or the mantle. In other words, the mantle must contain much less H than the core after the equilibrium proceeds in the magma ocean

$$\underset{\text{metal}}{Fe} + \underset{\text{silicate melt}}{H_2O} \leftrightarrow \underset{\text{silicate melt}}{Fe^{2+}} + \underset{\text{silicate melt}}{O^{2-}} + \underset{\text{metal}}{2H}$$

The extent to which this reaction proceeds as a function of pressure and temperature must be further quantified before more quantitative conclusions may be drawn about the degree to which the silicate mantle loses water upon equilibration with core-forming metal.

### 4.4.  Hydrogen Abundance in the Present Core

Hydrogen abundance in the core has been discussed by many authors during the past two decades (*Stevenson,* 1977; *Fukai and Akimoto,* 1983; *Fukai,* 1984; *Suzuki et al.,* 1984; *Fukai and Suzuki,* 1986; *Badding et al.,* 1991; *Fukai,* 1992; *Yagi and Hishinuma,* 1995). If the $H_2O$ concentration in Earth's source material averaged 2% as proposed by *Ringwood* (1977), the density of molten Fe in the outer core can be reduced by about 6% by available H. This much exceeds the contribution of S and C (*Fukai,* 1992). The expected

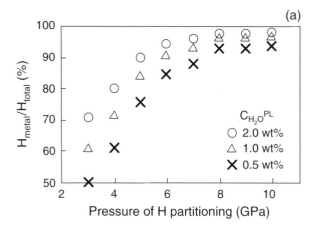 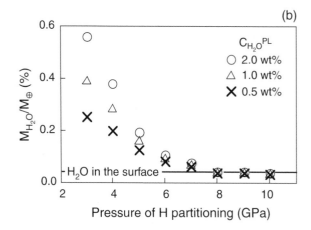

**Fig. 5.** (a) Molar percentage of H partitioned into molten Fe at high pressures at the bottom of the magma ocean. Only at a few gigapascal of pressures, H is partitioned into molten Fe rather than silicate melt. Each symbol corresponds to each assumed $H_2O$ concentration in the planetesimals ($C_{H_2O}^{PL}$). The silicate melt fraction at the bottom of the magma ocean was assumed as 30%. (b) The $H_2O$ mass left in the magma ocean after the metal-silicate partitioning. The left $H_2O$ mass ($M_{H_2O}$) is shown by weight percent of the present Earth mass ($M_{\oplus}$). If the magma ocean extends to a few gigapascal of pressures, it should become almost dry, keeping only a few times the amount of $H_2O$ existing in the present Earth's surface (ocean + crust). Symbols are the same in (a).

density reduction of ~9% by these three elements agrees well with the observed outer-core density deficit (*Okuchi, 1997*), so that further density reduction by dissolution of O, which requires higher pressure for the metal-silicate partitioning (*Hillgren et al., 2000*), is not necessary. Hydrogen may be the primary light element in the core. This is true not only for the molten outer core but also for the solid inner core, because equation (1) indicates that H may reconcile about two-thirds of the observed inner-core density deficit (*Okuchi, 1997*). Hydrogen in the inner core is also important for the estimation of core temperature, because about 600°C lower core geotherm is expected if its dissolution is taken into account (*Okuchi, 1998*).

## 5. WATER IN THE ARCHEAN AND PRESENT MANTLE

A surface magma ocean disappears when the average heat flux decreases below the critical energy flux described in section 2. At this stage, the interior of the proto-Earth, the upper mantle region, is still in a partially molten magma ocean state (*Abe, 1993*). Since the surface is already solidified, some water left after core formation may be concentrated in this subsurface magma ocean. Evolution of the Hadean and Archean mantle should be understood as the cooling and differentiation process of this magma ocean. However, this process is not yet well understood. If water is concentrated in the magma ocean, the differentiation process will be affected by the existence of water. Thus, evidence of this process may be left in Archean rock samples. Here, we discuss the wetness of the Archean mantle from the formation environment of komatiites. We then discuss

the water storage capacity of the present mantle from the stability condition of high-pressure minerals.

### 5.1. Origin of Komatiites and Hydrous Melting in the Mantle

*5.1.1. Chemistry of komatiites.* Komatiites are ultramafic magmas that erupted mainly in the Archean and also the Proterozoic and are characterized by a high MgO/SiO₂ ratio. The main groups of komatiite magmas can be divided into two types: the aluminum-depleted type (ADK), which is mainly observed in the early Archean, and the alumina-undepleted type (AUK), which is mainly observed in the late Archean. The ADK magma is characterized by superchondritic Ca/Al, Ti/Al, and Yb/La ratios, which have been interpreted to have been caused by residual garnet in the source mantle (e.g., *Green, 1975*; *Ohtani, 1984*). On the other hand, AUK magmas have near-chondritic Ca/Al, Ti/Al, and Yb/La ratios. The near-chondritic ratios of these elements in AUK implies that these magmas were formed by a relatively high degree of melting, and the magmas did not coexist with garnet in the source.

*5.1.2. The phase equilibria of the dry peridotite mantle.* Melting of peridotite under dry conditions has been studied intensively to clarify the genesis of komatiites and to discuss the processes in the early magma ocean stage (e.g., *Takahashi, 1990*; *Herzberg, 1992*; *Ohtani et al., 1995*; *Walter, 1998*). The most important change of the melting relation of the dry peridotite is expansion of the liquidus field of garnet relative to olivine with increasing pressure, and majorite garnet becomes the liquidus phase at pressures corresponding to the transition zone. This change of melting relation of dry peridotite is also consistent with the elemental fractionations in ADK and AUK magmas discussed above: ADK magmas show a chemical signature of garnet in the source and can form at greater depths than AUK magmas, which do not exhibit the garnet fractionation signature. On the basis of these arguments, it has been suggested that the secular change of the chemical signature of komatiite from ADK to AUK can be explained by the secular cooling of the potential temperature of the mantle (e.g., *Nisbet et al.,* 1993). However, this simple interpretation is challenged by recent experimental and field observations suggesting the importance of water in generation of komatiite magmas, as will be discussed in the following sections.

*5.1.3. Melting of the hydrous peridotite mantle and origin of komatiite.* A possibility of hydrous melting of the komatiite magmas was previously suggested by *Allegre* (1982). Recent high-pressure experiments revealed that the solidus and liquidus temperatures decrease drastically and the partial melt is ultramafic under the hydrous conditions (e.g., *Inoue and Sawamoto,* 1992; *Ohtani et al.,* 1996; *Asahara et al.,* 1998). This suggests that the komatiite magmas were formed by melting of the hydrous mantle. At pressures around 6.5–7.7 GPa the stability field of orthopyroxene expands with increasing water content (Fig. 6), and orthopyroxene is the liquidus phase under the hydrous conditions with the water content greater than 8 wt%. The liquidus temperature decreases by 100°–150°C, and the amount of melt increases from 1 wt% to more than 60 wt% within the temperature interval of ~50°C under the conditions of 1–2 wt% water.

Because melts formed under hydrous conditions are different from those formed in dry melting (Fig. 7), the AUK magmas could be generated by a high degree of partial melting of hydrous peridotite with olivine as the residue at the pressures around 4–6.5 GPa. This is discussed in detail in *Asahara et al.* (1998) and *Ohtani and Asahara* (1999).

A high degree of partial melting could be achieved in the peridotite mantle containing about 1–2 wt% water within a temperature interval <50° at around 1700°C, which is about 100°–150°C lower than the dry melting. Although the compositional spectrum of ADK magmas could be generated by ~40 wt% partial melting under the dry condition at 6.5 GPa, and subsequent addition or subtraction of olivine (Fig. 7), the magma can also be generated by hydrous melting of the deeper hydrous mantle. According to *Inoue and Sawamoto* (1992), the stability fields of garnet and orthopyroxene ex-

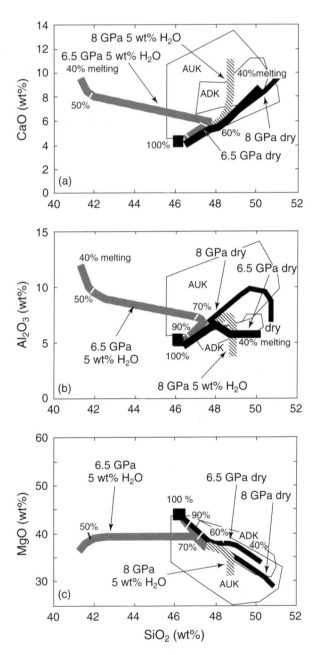

**Fig. 7.** The compositions of the komatiites and the melts formed by partial melting under the dry and wet (5 wt% $H_2O$) at 6.5 GPa and 8 GPa.

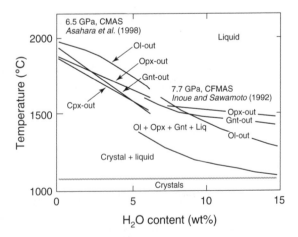

**Fig. 6.** The change of the melting relation of the model peridotite with varying water contents at high pressure.

pand relative to that of olivine with increasing water content at higher pressures ~7.7 GPa. The recent melting experiments of the hydrous peridotite made at 8 GPa (*Ohtani and Asahara*, 1999) indicates that the ADK magmas may also be generated by partial melting of a hydrous peridotite at ~8 GPa.

*5.1.4. Field evidence for hydrous melting petrogenesis of komatiites.* There are several observations that suggest the existence of water in the source mantle of komatiites. Komatiites have a characteristic texture called spinifex, which is believed to be formed by the rapid quenching of several hundreds degrees per minute due to eruption at the surface or the seafloor. Such a high cooling rate can only be reached near the surface (~3 cm depth from the surface) of the komatiite flows. However, in natural komatiites, the spinifex texture was observed in the flow units about 7 m from the surface, where the cooling rate is as low as about 0.5°C/hr (*Grove et al.*, 1996). On the basis of these observations, *Grove et al.* (1996) argued that formation of the spinifex texture of komatiites in such a slow cooling rate in the flow units can be accounted for by crystallization under hydrous conditions. Dissolved water in komatiite magmas decreases the nucleation rate whereas it enhances the growth rate, resulting in formation of the spinifex textures even at slow cooling rates.

*Stone et al.* (1997) found igneous amphibole in the komatiites from Abitibi formation and suggested that the primary komatiite magma might have contained about 2 wt% water. *Parman et al.* (1997) suggested that the initial melt of the Barberton komatiites contained about 4–6 wt% water on the basis of the composition of clinopyroxene in komatiites, which shows evidence for crystallization under the hydrous conditions on the basis of their experiments on crystallization of clinopyroxene from the komatiite melt. *Shimizu et al.* (1998) also observed a few weight percent water in a glass inclusion in a chromite grain in komatiite, suggesting that the komatiite magma contained about 0.6 wt% water. The above observations provide strong evidence for generation of komatiite magmas by hydrous melting of the peridotite mantle.

## 5.2. Hydrous Phases in Peridotite Mantle

The abundance of water in the present upper mantle has been estimated by many authors (*Michael*, 1988; *Dixon et al.*, 1997) based on the water content in minerals from the mantle xenolith estimates and water contents in the basaltic glasses. These estimations suggest that the water content in the MORB source mantle is around 140–350 ppm (*Michael*, 1988), whereas that of the OIB source is slightly wetter than the MORB source, i.e., 450–525 ppm (*Wallace*, 1998; *Dixon et al.*, 1997). Since the maximum solubility of water in olivine increases with pressure (*Kohlstedt et al.*, 1996), we may be able to assume that most of the water is accommodated in olivine in the normal upper mantle conditions. On the other hand, the water can be transported into the deep mantle by slab subduction, of which temperature is a few hundred degrees lower than the normal geotherm of the mantle. Thus the upper mantle in the subduction zone could contain more water compared to the MORB source mantle. On the basis of these estimations of the water content in the mantle, it seems clear that there is significant heterogeneity in the water content in the mantle, including high water content of the upper mantle in the cold subducting regions and relatively low water content in the MORB source mantle in the hot spreading ridge regions. The deep source mantle of OIB may have relatively higher water content.

Many hydrous minerals are expected to be stable in water-bearing peridotite. The major hydrous minerals are listed in Table 4. Serpentine, chlorite, talc, and amphibole are stable under upper mantle conditions. Serpentine and chlorite are the major hydrous phases and the dominant water reservoirs in hydrous peridotite to depths of ~150 km. Although there is no direct information on the amount of water in the harzburgite layer of slabs, the layer may contain some amount of water due to serpentinization or chloritization. Serpentine certainly forms at transform faults and fracture zones. *Schmidt and Poli* (1998) estimated that the upper 5 km of the harzburgite layer receive ~10% serpentinization, whereas *Kesson and Ringwood* (1989) suggested more extensive serpentinization.

There is a choke point for stabilizing water in the upper mantle, i.e., ~7 GPa and 600°C (*Kawamoto and Holloway*, 1997). When the temperature of the slab at 7 GPa is lower, the stability field of serpentine crosses that of phase A, which is one of the most important DHMS (dense hydrous magnesium silicate) phases stable in the deep upper mantle (*Ulmer and Trommsdorff*, 1995; *Luth*, 1995).

Recent high-pressure experiments on hydrous minerals in the deep mantle revealed that the major constituent phases in the transition zone, wadsleyite (β-phase) and ringwoodite (γ-phase) can accommodate up to about 2 wt% water (e.g., *Smyth*, 1987; *Inoue et al.*, 1995; *Kohlstedt et al.*, 1996). Thus, the transition zone is a potential water reservoir in the Earth. Hydrous wadsleyite and hydrous ringwoodite are stable at high temperatures exceeding 1600°C. These phases can coexist with silicate melt, and the mineral-liquid partition coefficient for $H_2O$ between hydrous wadsleyite and magma is estimated to be ~0.1 (*Kawamoto and Holloway*, 1997), which is significantly greater than that of olivine under the shallow upper mantle conditions. The ilmenite phase, which is stable in depleted harzburgite mantle, can also accommodate water up to 0.2 wt% at about 1600°–1750°C near the ilmenite-perovskite phase boundary (*Bolfan-Casanova et al.*, 1997).

Hydrous wadsleyite and hydrous ringwoodite are stable at relatively high-temperature conditions, whereas a number of DHMS are stable at lower temperatures typical of descending slabs (Table 4). Although there are many hydrous phases at high pressures and temperatures, some phases may not be stable in mantle compositions. The

TABLE 4.   Major hydrous minerals in the peridotite mantle, and MORB, sediment, and peridotite layers of the slab.

| Minerals | Formula | $H_2O$, wt% | Paragenesis | References |
|---|---|---|---|---|
| Chlorite | $Mg_5Al_2Si_3O_{10}(OH)_8$ | 13% | MORB, Sed, Per | |
| Amphibole group | $AX_2Y_5Z_8O_{22}(OH,F)_2$ where | 1–3% | MORB, Per | |
| | $A = Na,K; X = Na,Ca,Fe^{2+}, Mg$ | | | |
| | $Y = Mg,Fe^{2+},Fe^{3+}Al; Z = Si,Al$ | | | |
| Talc | $Mg_3Si_4O_{10}(OH)_2$ | 5% | MORB, Sed, Per | |
| Zoicite-clinozoicite | $Ca_2Al_3Si_3O_{12}(OH)$ | 2% | MORB, Sed | |
| Lausonite | $CaAl_2(Si_2O_7)(OH)_2H_2O$ | 11% | MORB | |
| Staurolite | $(Mg,Fe^{2+})_4(Al,Fe^{3+})_{18}Si_8O_{46}(OH)_2$ | 2% | MORB, Sed | |
| Chloritoid | $MgAl_2SiO_5(OH)_2$ | 8% | MORB, Sed | |
| Phengite | $K_2Al_2(Mg,Fe)_2Si_8O_{20}(OH)_4$ | 4% | Sed | |
| Phlogopite | $KMg_3AlSi_3O_{10}(OH)_2$ | 12.50% | MORB, Sed, Per | |
| Pumpellyite | $MgAl_5Si_6O_{21}(OH)_7$ | 7.10% | MORB, Sed | |
| Topaz-OH | $Al_2SiO_4(OH)_2$ | 10.70% | Sed | [1] |
| Egg-phase | $AlSiO_3(OH)$ | 7.50% | Sed | [2] |
| Serpentine | $Mg_3Si_2O_5(OH)_4$ | 28% | Per | |
| Chondrodite* | $Mg_5Si_2O_8(OH)_2$ | 5.30% | Per | [3] |
| Clinohumite* | $Mg_9Si_4O_{16}(OH)_2$ | 3% | Per | [4] |
| 10Å Phase* | $Mg_3Si_4O_{14}H_6$ | 13% | Per | [5] |
| Phase A* | $Mg_7Si_2O_8(OH)_6$ | 12% | Per | [6] |
| Phase E* | $Mg_{2.3}Si_{1.25}H_{2.4}O_6$ | 11.40% | Per | [7] |
| Superhydrous phase B* | $Mg_{10}Si_3O_{14}(OH)_4$ | 5.80% | Per | [8] |
| Phase G (phase D)* | $Mg_{1.14}Si_{1.73}H_{2.81}O_6$ | 14.5–18% | Per | [9] |
| Hydrous wadslyite | $Mg_{1.75}SiH_{0.5}O_4$ | <3% | Per | [10] |
| Hydrous ringwoodite | $Mg_{1.75}SiH_{0.5}O_4$ | <2.2% | Per | [11] |

MORB = basaltic crust layer of the slab; Sed = sediment; Per = mantle and peridotite layer in the slab. References: [1] *Schmidt and Poli* (1998); [2] *Schmidt et al.* (1998); [3] *Yamamoto and Akimoto* (1977); [4] *Yamamoto and Akimoto* (1977), the same phase as phase D by *Yamamoto and Akimoto* (1974); [5] *Sclar et al.* (1965); [6] *Ulmer and Trommsdorff* (1995) and *Luth* (1995); [7] *Kanzaki* (1991); [8] *Gasparik* (1993), same as phase C reported by *Ringwood and Major* (1967); [9] *Ohtani et al.* (1997) and *Kudoh et al.*(1997), same as phase D reported by *Yang et al.* (1997) and *Frost and Fei* (1998), *Inoue et al.* (1995); [10] *Kohlstedt et al.* (1996). *Dense hydrous magnesium silicate (DHMS) phases.

phases that were confirmed experimentally in hydrous peridotite are phase E, superhydrous phase B, phase G (which is the same phase as phase D by *Yang et al.* (1997)). These three phases are stable at transition zone conditions at temperatures below 1100°C, and the latter two phases are stable also in the upper part of the lower mantle.

### 5.3. Hydrous Phases Observed in Hydrous Mid-Ocean Ridge Basalts and Hydrous Sediment

*5.3.1. Hydrous mid-ocean ridge basalts.*   In the hydrous MORB compositions, several hydrous minerals are stable depending on pressure and temperature conditions and bulk compositions. In the descending slab conditions, chlorite and amphibole are stable up to about 2–3 GPa, and zoisite and chloritoid are stable at the pressures above the stability field of amphibole up to about 3–4 GPa (*Poli and Schmidt*, 1995; *Schmidt and Poli*, 1998). Lawsonite is stable from 3 GPa to about 8–9 GPa and below 800°C. The decomposition products of lawsonite are the assemblages composed of clinopyroxene, garnet, stishovite, kyanite (and phengite in the K-bearing system). Phengite decomposes at pressures

above 10 GPa to a mixture of clinopyroxene, garnet, and stishovite. In the hydrous MORB composition, there are no hydrous phases at pressures above the decomposition of lawsonite and/or phengite. However, a recent study by *Pawley et al.* (1993) showed stishovite contains up to 0.08 wt% of water when it contains 1.51 wt% $Al_2O_3$ at 10 GPa and 1200°C, although the maximum content of water in stishovite at high pressure and temperature is not yet clarified. Therefore, stishovite could be an important reservoir of water in the MORB component in the descending slabs at the depths greater than 300 km.

*5.3.2. Sediment.*   Approximately 10% of the upper oceanic crust is composed of sediments, and the sediment component of the plate also descends into the mantle associated with slab subduction. Major hydrous phases observed in hydrated sediments are also listed in Table 4. Phase relations of metasediments relevant to the conditions in subduction zones were studied by some authors (e.g., *Domanik and Holloway*, 1996). The assemblages of chlorite + kyanite and staurolite + quartz are stable to pressures of 2 GPa, and the assemblages composed of talc + chloritoid + phengite are stable at higher pressures. Talc breaks down to form the assemblage containing Mg-Al pumpellyite at around 4–5 GPa. The water carriers in sediments exceeding 200 km depths (P >6 GPa) are Mg-Al pumpellyite, phengite, and lawsonite. These hydrous minerals decompose by the pressure of 10 GPa (e.g., *Fockenberg, 1998*). At higher pressures, topaz-OH is stable up to about 12 GPa; it is then replaced by a denser hydrous phase, phase Egg, which is stable up to the pressures greater than 15 GPa. Phase Egg has all silicon ions in its octahedral sites, and thus expected to be stable at higher pressures (*Schmidt et al., 1998*). Although the sediment components contribute only a small portion to the water budget in the slab, there is the hydrous phase Egg which is stable even at the depths greater than the transition zone and a potential water carrier into the lower mantle (*Ono, 1998; Schmidt and Poli, 1998*).

### 5.4.  Water-Storage Capacity of the Mantle

The hydrous phases existing at high pressure and high temperature are summarized above. Many hydrous minerals including DHMS phases are stable only at the low temperature relevant to the slab conditions, and they are generally not stable under the normal mantle temperatures. The main water reservoir phases in the subducting slab conditions are schematically shown in Fig. 8. In the upper mantle, olivine can accommodate only a limited amount of water, although the amount of water stored in olivine increases with pressure (*Kohlstedt et al., 1996*). A small amount of water, however, has a great effect on its rheology, i.e., hydrolytic weakening (e.g., *Karato, 1986*).

The major constituent minerals in the transition zone, wadsleyite and ringwoodite, can accommodate a large amount of water up to ~2 wt% at 1000°–1200°C, and at least up to about 0.1 wt% of water at 1600°C. The maximum water storage potential of wadsleyite and ringwoodite is estimated

to be about 0.1–1 wt% along the normal geothermal gradient (e.g., *Inoue and Sawamoto, 1992; Kawamoto and Holloway, 1997*).

The lower mantle minerals such as Mg-perovskite do not contain a significant amount of water; preliminary data on the water content in Mg-perovskite synthesized at 1600°C under the hydrothermal conditions indicate only ~100 ppm water (*Meade et al., 1994*). Superhydrous phase B and phase G in the lower mantle decompose at temperatures around 1400°C (*Shieh et al., 1998*). Thus, the water storage potential is very small under lower mantle conditions, although there is a possibility that Al-bearing Mg-perovskite might have a greater water storage capacity as is observed in Al-bearing stishovite (*Pawley et al., 1993*).

There is a possible layered structure of the mantle for its water storage capacity; i.e., the upper and lower mantle have relatively small water storage capacities, whereas the transition zone has larger capacity. Thus, the water content in the transition zone might be large because the primordial water trapped in the lower mantle and the recycled water circulated due to the slab subduction have been stored in the transition zone during the geological time.

## 6.  CONCLUSIONS

It seems inescapable that most of the water currently in Earth arrived as part of the primary accretion process and was not delivered subsequent to accretion through impacts of comets and hydrous asteroids. Therefore, the behavior of water during accretion and primary differentiation into core, mantle, crust, hydrosphere, and atmosphere must be reconsidered. Accretion of hydrous materials leads to formation of an opaque atmosphere, which, in turn, serves as a thermal blanket and allows for the long-term (>10³ yr) persistence of a terrestrial surface magma ocean. That hydrous

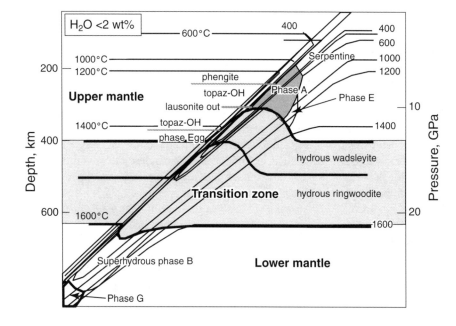

**Fig. 8.**  The schematic figure showing the major water reservoir minerals in the subducting slab conditions. The temperature distributions in the slab are those estimated for dip angle θ = 45°C, temperature gradient of 0.7°C/km, and subduction rate of 7 cm/yr (*Rubie and Ross, 1989*).

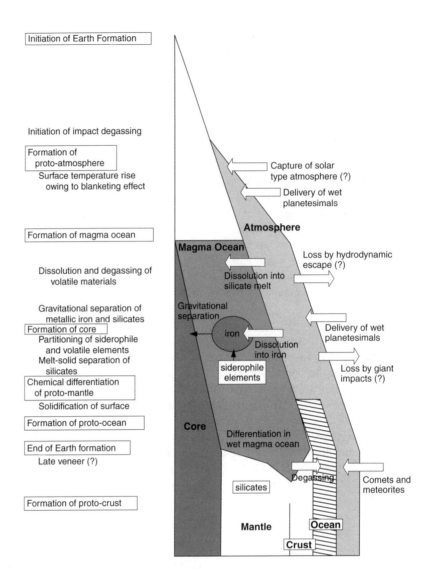

**Fig. 9.** A schematic diagram of the behavior of water during the early evolution of Earth. White broad arrows indicate the behavior of water.

magma ocean contained substantial dissolved water, which served to lower the liquidus and solidus and, upon crystallization, provided a source for water in the hydrosphere and atmosphere and for a mantle reservoir of stable hydrous high-pressure silicate phases. The hydrous magma ocean of depth corresponding to ~27 GPa is consistent with equilibration of metal that sank to the base of an ocean of silicate magma. The abundances of moderately siderophile elements and the highly siderophile element Re are explained in such a scenario. Hydrogen is known to be soluble in metal at elevated pressures, and solution of H in the metal sinking to make the core could account for some or most of the density deficit of the molten outer core. Behavior of water during accretion of Earth is schematically shown in Fig. 9.

This picture of the early Earth is satisfying, but leaves many phenomena unexplained. The issue of whether Earth accreted homogeneously or heterogeneously has long been debated. To some extent Earth must have accreted heterogeneously, as we have chondritic meteorites with different O-isotopic compositions from Earth accreting to Earth today. However, if Earth underwent a magma ocean stage, the

question becomes moot for most of the mass of the Earth, as any record of changing composition of accreting material with time would be erased in the magma ocean. A possible exception might be the highly siderophile elements. The abundances of these elements, which are less than 1% chondritic although their ratios are within a few percent of chondritic, might be established by metal-silicate equilibrium in the magma ocean. However, it is also possible that these elements were introduced as a "late veneer" constituting <1% of Earth's mass, which arrived after metal was segregated to the core and were very effectively mixed into the molten magma such that they are effectively homogeneously distributed globally at the hand specimen scale in the mantle.

Also unexplained is the reason why different planetary bodies have different water contents. For those planetary bodies for which we have samples, water content is quantized. Earth and Mars are wet; the Moon and Vesta are dry. The Earth and Moon clearly formed in the same region of the solar system, as indicated by their identical O-isotopic composition within measurement error, so one cannot appeal to formation of the bulk of the bodies in different parts of the solar system. One possibility is that the accretion of water is stochastic, with some planetesimals becoming hydrous, perhaps through the accretion of ices ejected into the inner solar system by gravitational interactions with the giant planets, while others remained anhydrous because they did not accrete any ices. In this scenario, the body which impacted Earth to make the Moon must have been anhydrous, and most of the Moon must be derived from this body, consistent with current theory (see *Cameron*, 2000). Another possibility is that the Moon-forming material lost its water during the impact event, due to the low pressures in the lunar accretion disk and hydrodynamic loss. For instance, if the Moon formed from a postimpact, circumterrestrial accretion disk, the low pressures in the disk would not have allowed dissolution of any water in the hot silicate melt. Because the accretion of the Moon from an impact-generated protolunar disk is so rapid, the Moon-forming materials are easily heated and vaporized (*Abe et al.*, 2000). Volatile elements may have been lost from the Moon through hydrodynamic escape, but still remained in the Earth's gravity field, because the original disk was gravitationally bounded by the Earth. This process may also produce a dry Moon and a wet Earth.

Several areas of future work may help to better define the role of water in the early Earth, especially as relevant to the following issues. First, the question of whether metallic Fe is fully oxidized in the presence of water remains unanswered, yet is clearly important to large-scale differentiation. The experiments of *Okuchi* (1997) demonstrated that water should be partitioned mainly into liquid metal as H if the magma ocean exceeds a threshold depth of a few gigapascals, but the results of *Righter and Drake* (1999) indicate that metal and hydrous magma can coexist. These two studies were done at different conditions. Future efforts should try to utilize S-, C-, and H-bearing metallic liquids and experimental water contents in order to yield results most applicable to the early Earth. Also, efforts should be made to define redox equilibria involving water across a wide range of pressure and temperature. Second, the effect of water on phase relations, melting properties, and thermodynamic properties of ultramafic systems must be explored in detail in order to have a more fundamental understanding of these systems as applicable to early Earth conditions. Third, the relation between the proto-atmosphere and the present atmosphere-hydrosphere must be clarified. The proto-atmosphere may likely be a mixed atmosphere of the solar-type and "steam" atmospheres. The properties of this mixed atmosphere are not yet fully understood. Though most of water in the present hydrosphere should have been delivered in planetesimals, some fraction may be lost by giant impacts and hydrodynamic escape. The effect of giant impacts and hydrodynamic escape on the proto-atmosphere must be addressed in the future.

*Acknowledgments.*    The authors wish to express their thanks to T. Owen and S. Sasaki for their constructive reviews.

# REFERENCES

Abe Y. (1988) Conditions required for sustaining a surface magma ocean. In *Proc. ISAS Lunar Planet. Symp. 21st*, pp. 225–231. Institute of Space and Astronautical Science, Sagamihara.

Abe Y. (1993) Physical state of very early Earth. *Lithos, 30*, 223–235.

Abe Y. and Matsui T. (1985) The formation of an impact-generated $H_2O$ atmosphere and its implications for the thermal history of the Earth. *Proc. Lunar Planet. Sci. Conf. 15th*, in *J. Geophys. Res., 90*, C545–C559.

Abe Y. and Matsui T. (1986) Early evolution of the Earth: accretion, atmosphere formation, and thermal history. *Proc. Lunar Planet. Sci. Conf. 17th*, in *J. Geophys. Res., 91*, E291–E302.

Abe Y. and Matsui T. (1988) Evolution of an impact-generated $H_2O$-$CO_2$ atmosphere and formation of a hot proto-ocean on Earth. *J. Atmos. Sci., 45*, 3081–3101.

Abe Y., Zahnle K., and Hashimoto A. (1998) Elemental fractionation during rapid accretion of the Moon triggered by a giant impact. In *Origin of the Earth and Moon*, p.1. LPI Contribution No. 957, Lunar and Planetary Institute, Houston.

Abe Y., Zahnle K., and Hashimoto A. (2000) Elemental fractionation during rapid accretion of the Moon. In preparation.

Allegre C. J. (1982) Genesis of Archean komatiites in a wet ultramafic subducted plate. In *Komatiites* (N. T. Arndt and E. G. Nisbet, eds.), pp. 495–500. Allen and Unwin, London.

Antonov V. E., Belash I. T., Degtyareva V. F., Ponyatovskii E. G., and Shiryaev V. I. (1980) Obtaining iron hydride under high hydrogen pressure. *Sov. Phys. Dokl., 25*, 490–492.

Asahara Y., Ohtani E., and Suzuki A. (1998) Melting relations of hydrous and dry mantle compositions and the genesis of komatiites. *Geophys. Res. Lett., 25*, 2201–2204.

Badding J. V., Hemley R. J., and Mao H. K. (1991) High-pressure chemistry of hydrogen in metals: in situ study of iron hydride. *Science, 253*, 421–424.

Basri G. and Bertout C. (1993) T Tauri stars and their accretion disks. In *Protostars and Planets III* (E. H. Levy and J. I. Lunine, eds.), pp. 543–567. Univ. of Arizona, Tucson.

Birch F. (1964) Density and composition of the mantle and core. *J. Geophys. Res., 69*, 4377–4388.

Boato G. (1954) The isotopic composition of hydrogen and carbon in the carbonaceous chondrites. *Geochim. Cosmochim. Acta, 6*, 209–220.

Boehler R., von Bargen N., and Chopelas A. (1990) Melting, thermal expansion, and phase transitions of iron at high pressures. *J. Geophys. Res., 95*, 21731–21736.

Bolfan-Casanova N., Keppler H., and Rubie D. C. (1997) Distribution of water between lower mantle phases: $MgSiO_3$-ilmenite, $MgSiO_3$-perovskite and stishovite. *Eos Trans. AGU, 78*, F736.

Brown H. (1949) Rare gases and the formation of the Earth's atmosphere. In *The Atmosphere of the Earth and Planets* (G. P. Kuiper, ed.), pp. 258–266. Univ. of Chicago, Chicago.

Cameron A. G. W. (2000) Higher-resolution simulations of the giant impact. In *Origin of the Earth and Moon* (R. M. Canup and K. Righter, eds.), this volume. Univ. of Arizona, Tucson.

Carr M. H. and Wanke H. (1992) Earth and Mars: water inventories as clues to accretional histories. *Icarus, 98*, 61–71.

Clemens J. D. and Navrotsky A. (1987) Mixing properties of $NaAlSi_3O_8$ melt-$H_2O$: new calorimetric data and some geological implications. *J. Geol., 95*, 173–186.

Clossman C. and Williams Q. (1995) In situ spectroscopic investigations of high pressure hydrated (Mg,Fe)$SiO_3$ glasses: OH vibrations as a probe of glass structure. *Am. Mineral., 80,* 201–212.

Deloule E. and Robert F. (1995) Interstellar water in meteorites? *Geochim. Cosmochim. Acta, 59,* 4695–4706.

Dixon J. E., Clague D. A., Wallace P., and Poreda R. J. (1997) Volatiles in alkalic basalts from the North Arch volcanic field, Hawaii: Extensive degassing of deep submarine-erupted alkalic series lavas. *J. Petrol., 38,* 911–939.

Domanik K. J. and Holloway J. R. (1996) The stability and composition of phenigitic muscovite and associated phases from 5.5 to 11 GPa: implications for deeply subducted sediments. *Geochim. Cosmochim. Acta, 60,* 4133–4151.

Dreibus G. and Wanke H. (1989) Supply and loss of volatile constituents during the accretion of terrestrial planets. In *Origin and Evolution of Planetary and Satellite Atmospheres* (S. K. Atreya, J. B. Pollack, and M. S. Matthews, eds.), pp. 268–289. Univ. of Arizona, Tucson.

Fockenberg T. (1998) An experimental study of the pressure-temperature stability of MgMgAl-pumpellyite in the system $MgO$-$Al_2O_3$-$SiO_2$-$H_2O$. *Am. Mineral, 83,* 220–227.

Fricker P. E. and Reynolds R. T. (1968) Development of the atmosphere of Venus. *Icarus, 9,* 221–230.

Frost D. J. and Fei Y. (1998) Stability of phase D at high pressure and high temperature. *J. Geophys. Res., 103,* 7463–7474.

Fukai Y. (1984) The iron-water reaction and the evolution of the Earth. *Nature, 308,* 174–175.

Fukai Y. (1991) From metal hydrides to the metal-hydrogen system. *J. Less-Common Metals, 8,* 172–174.

Fukai Y. (1992) Some properties of the Fe-H system at high pressures and temperatures, and their implications for the Earth's core. In *High-Pressure Research: Application to Earth and Planetary Sciences* (Y. Syono and M. H. Manghnani, eds.), pp. 373–385. TERRAPUB/American Geophysical Union, Tokyo/Washington, DC.

Fukai Y., Fukizawa A., Watanabe K., and Amano M. (1982) Hydrogen in iron: its enhanced dissolution under pressure and stabilization of the γ phase. *Jpn. J. Appl. Phys., 21,* L318–L320.

Fukai Y. and Akimoto S. (1983) Hydrogen in the Earth's core. *Proc. Jpn. Acad., 59B,* 158–162.

Fukai Y. and Sugimoto H. (1985) Diffusion of hydrogen in metals. *Adv. Phys., 34,* 263–326.

Fukai Y. and Suzuki T. (1986) Iron-water reaction under high pressure and its implication in the evolution of the Earth. *J. Geophys. Res., 91,* 9222–9230.

Gaetani G. A., Kent A. J. R., Grove T. L., Hutcheon I. D., and Stolper E. M. (1997) Experimental determination of the partitioning of trace elements between peridotite minerals and silicate melts with variable water contents. *Eos Trans. AGU, 78,* F839.

Gaetani G. A., Grove T. L., and Bryan W. B. (1995) The influence of water on the petrogenesis of subduction-related igneous rocks. *Nature, 365,* 332–336.

Gasparik T. (1993) The role of volatiles in the transition zone. *J. Geophys. Res., 98,* 4287–4300.

Ghiorso M. S., Carmichael I. S. E., Rivers M. L., and Sack R. O. (1983) The Gibbs free energy of mixing of natural silicate liquids: an expanded regular solution approximation for the calculation of magmatic intensive variables. *Contrib. Mineral. Petrol., 84,* 107–145.

Ghiorso M. S. and Sack R. O. (1994) Chemical mass transfer in magmatic processes IV: a revised an internally consistent thermodynamic model for the interpolation and extrapolation of liquid-solid equilibria in magmatic systems at elevated temperatures and pressures. *Contrib. Mineral. Petrol., 119,* 197–212.

Green D. H. (1975) Genesis of Archaean peridotitic magmas and constraints on Archaean geothermal gradients and tectonics. *Geology, 3,* 15–18.

Grove T. L., Gaetani G. A., and Parman S. W. (1996) Origin of olivine spinifex textures in 3.49 Ga komatiite magmas from Barberton Mountainland, South Africa. *Eos Trans. AGU, 77,* F281.

Hamilton D. L., Burnham C. W., and Osborn E. F. (1964) The solubility of water and effects of oxygen fugacity and water content on crystallization in mafic magmas. *J. Petrol., 5,* 21–39.

Hayashi C., Nakazawa K., and Mizuno H. (1979) Earth's melting due to the blanketing effect of the primordial dense atmosphere. *Earth Planet. Sci. Lett., 43,* 22–28.

Hayashi C., Nakazawa K., and Nakagawa Y. (1985) Formation of the solar system. In *Protostars and Planets II* (D. C. Black and M. S. Matthews, eds.), pp. 1100–1153. Univ. of Arizona, Tucson.

Herzberg C.T. (1992) Depth and degree of melting of komatiite. *J. Geophys. Res., 97,* 4521–4540.

Hillgren V. J., Gessmann C., and Li J. (2000) An experimental perspective on the light element in Earth's core. In *Origin of the Earth and Moon* (R. M. Canup and K. Righter, eds.), this volume. Univ. of Arizona, Tucson.

Hishinuma T., Yagi T., and Uchida T. (1994) Surface tension of iron hydride formed by the reaction of iron-silicate-water under pressure. *Proc. Jpn. Acad., 70B,* 71–76.

Holland H. D. (1984) *The Chemical Evolution of the Atmosphere and Oceans.* Princeton Univ., Princeton, New Jersey. 582 pp.

Holloway J. R. (1988) Planetary atmospheres during accretion: the effect of C-O-H-S equilibria (abstract). In *Lunar and Planetary Science XIX,* pp. 499–500. Lunar and Planetary Institute, Houston.

Hunten D. M., Donohue T. M., Walker J. C. G., and Kasting J. F. (1989) Escape of atmospheres and loss of water. In *Origin and Evolution of Planetary and Satellite Atmospheres* (S. K. Atreya, J. B. Pollack, and M. S. Matthews, eds.), pp. 386–422. Univ. of Arizona, Tucson.

Inoue T. (1994) Effect of water on melting phase relations and melt composition in the system $Mg_2SiO_4$-$MgSiO_3$-$H_2O$ up to 15 GPa. *Phys. Earth Planet. Inter., 85,* 237–263.

Inoue T. and Sawamoto H. (1992) High pressure melting of pyrolite under hydrous conditions and its geophysical implications. In *High-Pressure Research: Application to Earth and Planetary Sciences* (Y. Syono and M. H. Manghnani, eds.), pp. 323–331. TERRAPUB/American Geophysical Union, Tokyo/Washington, DC.

Inoue T., Yurimoto H., and Kudoh Y. (1995) Hydrous modified spinel, $Mg_{1.75}SiH_{0.5}O_4$: A new water reservoir in the mantle transition region. *Geophys. Res., Lett., 160,* 117–120.

Iwamori H., McKenzie D., and Takahashi E. (1995) Melt generation by isentropic mantle upwelling. *Earth Planet. Sci. Lett., 134,* 253–266.

Jana D. and Walker D. (1999) Core formation in the presence of various C-H-O volatiles. *Geochim. Cosmochim. Acta, 63,* 2299–2310.

Jeanloz R. (1990) The nature of the Earth's core. *Annu. Rev. Earth Planet. Sci., 18,* 357–386.

Jessberger E. K., Kissel J., and Rahe J. (1989) The composition of comets. In *Origin and Evolution of Planetary and Satellite Atmospheres* (S. K. Atreya, J. B. Pollack, and M. S. Matthews, eds.), pp. 167–191. Univ. of Arizona, Tucson.

Kanzaki M. (1991) Stability of hydrous magnesium silicates in the mantle transition zone. *Phys. Earth Planet. Inter., 66,* 307–312.

Karato S. (1986) Does partial melting reduces the creep strength of the upper mantle? *Nature, 319,* 309–310.

Kawamoto T. and Holloway J. R. (1997) Melting temperature and partial melt chemistry of $H_2O$-saturated mantle peridotite to 11 gigapascals. *Science, 276,* 240–243.

Kerridge J. F. (1985) Carbon, hydrogen and nitrogen in carbonaceous chondrites: abundances and isotopic compositions in bulk samples. *Geochim. Cosmochim. Acta, 49,* 1707–1714.

Kesson S. E. and Ringwood A. E. (1989) Slab-mantle interactions: 1. Sheared and refertilised garnet peridotite xenoliths-samples of Wadati-Benioff zones? *Chem. Geol., 78,* 83–96.

Kohlstedt D. L., Keppler H., and Rubie D. C. (1996) Solubility of water in the alpha, beta, and gamma phases of $(Mg,Fe)_2SiO_4$, *Contrib. Mineral. Petrol., 123,* 345–357.

Kohn S. C., Smith M. E., Dirken P. J., van Eck E. R. H., Kentgens A. P. M., and Dupree R. (1998) Sodium environments in dry and hydrous albite glasses: improved 23Na solid state NMR data and their implications for water dissolution mechanisms. *Geochim. Cosmochim. Acta, 62,* 79–87.

Kortenkamp S. J., Kokubo E., and Weidenschilling S. J. (2000) Formation of planetary embryos. In *Origin of the Earth and Moon* (R. M. Canup and K. Righter, eds.), this volume. Univ. of Arizona, Tucson.

Kudoh Y., Nagase T., Mizobata H., Ohtani E., Sasaki S., and Tanaka M. (1997) Structure and crystal chemistry of phase G, a new hydrous magnesium silicate synthesized at 22 GPa and 1050°C. *Geophys. Res. Lett., 24,* 1051–1054.

Kushiro I. (1969) The system forsterite-diopside-silica with and without water at high pressures. *Am. J. Sci., 267A,* 269–294.

Lange M. A. and Ahrens T. J. (1982a) The evolution of the impact generated atmosphere. *Icarus, 51,* 96–120.

Lange M. A. and Ahrens T. J. (1982b) Impact induced dehydration of serpentine and the evolution of planetary atmospheres. *Proc. Lunar Planet. Sci. Conf. 13th,* in *J. Geophys. Res., 87,* A451–A456.

Lesher C. E. and Walker D. (1986) Solution properties of silicate liquids from thermal diffusion experiments. *Geochim. Cosmochim. Acta, 50,* 1397–1411.

Luth R. W. (1995) Is phase A relevant to the Earth's mantle? *Geochim. Cosmochim. Acta, 59,* 679–682.

Mason B. (1966) The enstatite chondrites. *Geochim. Cosmochim. Acta, 30,* 23–39.

Matsui T. and Abe Y. (1986a) Evolution of an impact-induced atmosphere and magma ocean on the accreting Earth. *Nature, 319,* 303–305.

Matsui T. and Abe Y. (1986b) Impact-induced atmosphere and oceans on Earth and Venus. *Nature, 322,* 526–528.

McMillan P. F. and Wolf G. H. (1995) Vibrational spectroscopy of silicate liquids. In *Structure, Dynamics and Properties of Silicate Melts* (J. F. Stebbins, P. F. McMillan, and D. B. Dingwell, eds.), pp. 121–144. Mineralogical Society of America, Washington, DC.

Meade C., Reffner J. A., and Ito E. (1994) Synchrotron infrared absorbance measurements of hydrogen in $MgSiO_3$ perovskite. *Science, 264,* 1558–1560.

Meier R., Owen T. C., Matthews H. E., Jewitt D.C., Bocklee-

Morvan D., Biver N., Crovisier J., and Gautier D. (1997) A determination of the $DHO/H_2O$ ratio in comet C/1995 O1 (Hale-Bopp). *Science, 279,* 842–844.

Melosh H. J. and Vickery A. M. (1989) Impact erosion of the primordial atmosphere of Mars. *Nature, 338,* 487–489.

Michael P. J. (1988) The concentration, behavior and storage of $H_2O$ in the suboceanic upper mantle: implication for mantle metasomatism. *Geochim. Cosmochim. Acta, 52,* 555–566.

Moore G., Vennemann T., and Carmichael I. S. E. (1995a) Solubility of water in magmas to 2 kbar. *Geology, 23,* 1099–1102.

Moore G., Righter K., and Carmichael I. S. E. (1995b) The effect of dissolved water on the oxidation state of iron in natural silicate liquids. *Contrib. Mineral. Petrol., 120,* 170–179.

Moore G., Vennemann T., and Carmichael I. S. E. (1998) An empirical model for the solubility of water in magmas to 3 kilobars. *Am. Mineral., 83,* 36–45.

Nakajima S., Hayashi Y.-Y., and Abe Y. (1992) A study on the "runaway greenhouse effect" with a one dimensional radiative convective equilibrium model. *J. Atmos. Sci., 49,* 2256–2266.

Nakazawa K., Mizuno H., Sekiya M., and Hayashi C. (1985) Structure of the primordial atmosphere surrounding the early Earth. *J. Geomag. Geoelectr., 37,* 781–799.

Nielsen R. L., Beard B. S., and Hilyard M. L. (1997) Temperature and compositional control on the mineral-melt partitioning of high field strength and rare earth elements for amphibole and magnetite. In *Seventh Annual V. M. Goldschmidt Conference,* pp. 150–151. LPI Contribution No. 921, Lunar and Planetary Institute, Houston.

Nisbet E. G., Cheadle M. J., Arndt N. T., and Bickle M. J. (1993) Constraining the potential temperature of the Archean mantle: A review of the evidence from komatiites. *Lithos, 30,* 291–307.

Ochs F. A. III and Lange R. A. (1997) The partial molar volume, thermal expansivity and compressibility of $H_2O$ in $NaAlSi_3O_8$ liquid: new measurements and an internally consistent model. *Contrib. Mineral. Petrol., 129,* 155–165.

Ohtani E. (1984) Generation of komatiite magma and gravitational differentiation in the deep upper mantle. *Earth Planet. Sci. Lett., 67,* 261–272.

Ohtani E. and Asahara Y. (1999) Role of water in formation of komatiite and cratonic peridotites in the early Earth. *Eos Trans. AGU, 80,* S371.

Ohtani E., Nagata Y., Suzuki A., and Kato T. (1995) Melting relations of peridotite and the density crossover in planetary mantles. *Chem. Geol., 120,* 207–221.

Ohtani E., Mibe K., and Kato T. (1996) Origin of cratonic peridotite and komatiite: evidence for melting in the wet Archean mantle. *Proc. Jpn. Acad., Ser. B, 72,* 113–117.

Ohtani E., Mizobata H., Kudoh Y., Nagase T., Arashi H., Yurimoto H., and Miyagi I. (1997) A new hydrous silicate, a water reservoir, in the upper part of the lower mantle. *Geophys. Res. Lett., 24,* 1047–1050.

Okuchi T. (1997) Hydrogen partitioning into molten iron at high pressure: implications for Earth's core. *Science, 278,* 1781–1784.

Okuchi T. (1998) Melting temperature of iron hydride at high pressures and its implications for temperature of the Earth's core. *J. Phys. Condensed Matter, 10,* 11595–11598.

Okuchi T. and Takahashi E. (1998) Hydrogen in molten iron at high pressure: the first measurement. In *Properties of Earth and Planetary Materials at High Pressure and Temperature* (M. H. Manghnani and T. Yagi, eds.), pp. 249–260. American

Geophysical Union, Washington, DC.

Ono S. (1998) Stability limits of hydrous minerals in sediment and mid-ocean ridge basalt compositons: implications for water transport in subduction zones. *J. Geophys. Res., 103*, 18253–18267.

Owen T. C. and Bar-Nun A. (2000) Volatile contributions from icy planetesimals. In *Origin of the Earth and Moon* (R. M. Canup and K. Righter, eds.), this volume. Univ. of Arizona, Tucson.

Parman S. W., Dann J. C., Grove T. L., and de Wit M. J. (1997) Emplacement conditions of komatiite magmas from the 3.49 Ga Komati Formation, Barberton Greenstone Belt, South Africa. *Earth Planet. Sci. Lett., 150*, 303–323.

Pawley A. R., McMillan P. F., and Holloway J. R. (1993) Hydrogen in stishovite, with implications for mantle water content. *Science, 261*, 1024–1026.

Poli S. and Schmidt M. W. (1995) $H_2O$ transport and release in subduction zone: experimental constraints on basaltic and andesitic systems. *J. Geophys. Res., 100*, 22299–22314.

Porcelli D. and Pepin R. O. (2000) Rare gas constraints on early Earth history. In *Origin of the Earth and Moon* (R. M. Canup and K. Righter, eds.), this volume. Univ. of Arizona, Tucson.

Righter K., Drake M. J., and Yaxley G. (1997) Prediction of siderophile element metal-silicate partition coefficients to 120 kb and 2800°C: the effect of pressure, temperature, $f_{O_2}$ and silicate and metallic melt composition. *Phys. Earth Planet. Inter., 100*, 115–134.

Righter K. and Delaney J. S. (1997) Effect of dissolved water on the ferric/ferrous ratio of natural basic magma at 10 kb, 1300°C. *GSA Abstr. with Progr., 29*, A191.

Righter K. and Drake M. J. (1999) Effect of water on metal-silicate partitioning of siderophile elements: a high pressure and temperature terrestrial magma ocean and core formation. *Earth Planet. Sci. Lett., 171*, 383–399.

Ringwood A. E. (1977) Composition of the core and implications for origin of the Earth. *Geochem. J., 11*, 111–135.

Ringwood A. E. and Major A. (1967) High pressure reconnaissance investigations in the system $Mg_2SiO_4$-$MgO$-$H_2O$. *Earth Planet. Sci. Lett., 2*, 130–133.

Rintoul D. and Stevenson D. J. (1988) The role of large infrequent impacts in the thermal state of the primordial earth. In *Papers Presented to the Conference on the Origin of the Earth*, pp. 75–76. Lunar and Planetary Institute, Houston.

Safronov V. S. and Ruzmaikina T. V. (1985) Formation of the solar nebula and the planets. In *Protostars and Planets II* (D. C. Black and M. S. Matthews, eds.), pp. 959–980. Univ. of Arizona, Tucson.

Sasaki S. (1990) The primary solar-type atmosphere surrounding the accreting Earth: $H_2O$-induced high surface temperature. In *Origin of the Earth* (H. E. Newsom and J. H. Jones, eds.), pp. 195–209. Oxford Univ., New York.

Sasaki S. and Nakazawa K. (1986) Metal-silicate fractionation in the growing Earth: energy source for the terrestrial magma ocean. *J. Geophys. Res., 91*, 9231–9238.

Saxena S. K., Chatterjee N., Fei Y., and Shen G. (1993) *Thermodynamic Data on Oxides and Silicates*. Springer-Verlag, New York. 428 pp.

Schmidt M. W., Finger L. W., Angel R. J., and Dinnebier R. E. (1998) Synthesis, crystal structure, and phase relations of $AlSiO_3OH$, a high-pressure hydrous phase. *Am. Mineral., 83*, 881–888.

Schmidt M. W. and Poli S. (1998) Experimental base water bud-

gets for dehydrating slabs an consequences for arc magma generation. *Earth Planet. Sci. Lett., 163*, 361–379.

Sclar C. B., Carrison L. C., and Schwartz C. M. (1965) High pressure synthesis and stability of a new hydronium bearing layer silicate in the system $MgO$-$SiO_2$-$H_2O$. *Eos Trans. AGU, 46*, 184.

Shimizu K., Komiya T., Maruyama S., and Hirose K. (1998) The water content in the melt inclusion of chromian spinel in komatiite from Jimbabue et al. erupted in 2.7 Ga. *Japan Earth Planet Science Society Meeting, Tokyo*, Ac-015, p. 15.

Shieh S. R., Mao H. K., and Ming J. C. (1998) Decomposition of phase D in the lower mantle and the fate of dense hyrous silicates in subducting slabs. *Earth Planet. Sci. Lett., 159*, 13–23.

Sisson T. W. and Grove T. L. (1993) Experimental investigations of the role of $H_2O$ in calc-alkaline differentiation and subduction zone magmatism. *Contrib. Mineral. Petrol., 113*, 143–166.

Smyth J. R. (1987) β-$Mg_2SiO_4$: a potential host of water in the mantle? *Am. Mineral., 72*, 1051–1055.

Stebbins J. F. (1995) Dynamics and structure of silicate and oxide melts: nuclear magnetic resonance studies. In *Structure, Dynamics and Properties of Silicate Melts* (J. F. Stebbins, P. F. McMillan, and D. B. Dingwell, eds.), *Rev. Mineral., 32*, 145–246.

Stebbins J. F. and Xu Z. (1997) NMR evidence for excess nonbridging oxygen in an aluminosilicate glass. *Nature, 390*, 60–63.

Stevenson D. J. (1977) Hydrogen in the Earth's core. *Nature, 268*, 130–131.

Stevenson D. J. (1981) Models of the Earth's core. *Science, 214*, 611–619.

Stevenson D. J. (1990) Fluid dynamics of core formation. In *Origin of the Earth* (H. E. Newsom and J. H. Jones, eds.), pp. 231–249. Oxford Univ., New York.

Stolper E. (1982) The speciation of water in silicate melts. *Geochim. Cosmochim. Acta, 46*, 2609–2620.

Stone W. E., Deloule E., Larson M. S., and Lesher C. M. (1997) Evidence for hydrous high-MgO melts in the Precambrian. *Geology, 25*, 143–146.

Suess H. E. (1949) Die häufigkeit der edelgase auf der erde und im kosmos. *J. Geol., 57*, 600–607.

Suzuki T., Akimoto S., and Fukai Y. (1984) The system iron-enstatite-water at high pressures and temperatures: formation of iron hydride and some geophysical implications. *Phys. Earth Planet. Inter., 36*, 135–144.

Takahashi E. (1990) Speculations on the Archean mantle: missing link between komatiite and depleted garnet peridotite. *J. Geophys. Res., 95*, 15941–15964.

Takahashi E., Shimazaki T., Tsuzaki Y., and Yoshida H. (1993) Melting study of a peridotite KLB-1 to 6.5 GPa, and the origin of basaltic magmas. *Philos. Trans. R. Soc. Lond., 342*, 105–120.

Taylor G. J. (1992) Core formation in asteroids. *J. Geophys. Res., 97*, 14717–14726.

Tyburczy J. A., Frisch B., and Ahrens T. J. (1986) Shock-induced volatile loss from a carbonaceous chondrite: implications for planetary accretion. *Earth Planet. Sci. Lett., 80*, 201–207.

Ulmer P. and Trommsdorff V. (1995) Serpentine stability to mantle depths and subduction related magmatism. *Science, 268*, 858–861.

Wallace P. J. (1998) Water and partial melting in mantle plumes inferences from the dissolved $H_2O$ concentraitions of Hawaiian basaltic magmas. *Geophys. Res. Lett., 25*, 3639–3642.

Walter M. J. (1998) Melting of garnet peridotite and the origin of komatiites and depleted lithosphere. *J. Petrol., 39,* 29–60.

Walter M. J., Newsom H. E., Ertel W., and Holzheid A. (2000) Siderophile elements in the Earth and Moon: Metal/silicate partitioning and implications for core formation. In *Origin of the Earth and Moon* (R. M. Canup and K. Righter, eds.), this volume. Univ. of Arizona, Tucson.

Wood B. J. (1997) Hydrogen: An important constituent of the core? *Science, 278,* 1727.

Yagi T. and Hishinuma T. (1995) Iron hydride formed by the reaction of iron, silicate, and water: implications for the light element of the Earth's core. *Geophys. Res. Lett., 22,* 1933–1936.

Yamakata M., Yagi T., Utsumi W., and Fukai Y. (1992) In situ x-ray observation of iron hydride under high pressure and high temperature. *Proc. Jpn. Acad., 68B,* 172–176.

Yamamoto K. and Akimoto S. (1974) High pressure and high temperature investigations in the system $MgO$-$SiO_2$-$H_2O$. *J. Solid State Chem., 9,* 187–195.

Yamamoto K. and Akimoto S. (1977) The system $MgO$-$SiO_2$-$H_2O$ at high pressures and high temperatures-stability field for hydroxyl-clinohumite and 10A phase. *Am. J. Sci., 277,* 288–312.

Yang H., Prewitt C. T., and Frost D. J. (1997) Crystal structure of the dense hydrous magnesium silicate, phase D. *Am. Mineral., 82,* 651–654.

Zahnle K., Kasting J. F., and Pollack J. B. (1988) Evolution of a steam atmosphere during Earth's accretion. *Icarus, 74,* 62–97.

Zahnle K. and Dones L. (1998) Source of terrestrial volatiles. In *Origin of the Earth and Moon,* pp. 55–56. LPI Contribution No. 957, Lunar and Planetary Institute, Houston.

Zeng Q. and Nekvasil H. (1996) An associated solution model for albite-water melts. *Geochim. Cosmochim. Acta, 60,* 59–73.

Zhang J. and Herzberg C. (1994) Melting experiments on anhydrous peridotite KLB-1 from 5.0 to 22.5 GPa. *J. Geophys. Res., 99,* 17729–17742.

# Rare Gas Constraints on Early Earth History

## D. Porcelli
*California Institute of Technology*

## R. O. Pepin
*University of Minnesota*

In addition to the rare gases in the atmosphere, substantial inventories appear trapped within Earth. Evidence suggests that the latter are distinct in having solar nonradiogenic isotopic compositions. Various mechanisms have been proposed for the origin of terrestrial gases. Plausible models include trapping of gravitationally concentrated solar gases within the growing Earth either by adsorption and occlusion, or by dissolution in molten silicates. These require that much of planetary accretion occurred prior to nebula dispersal. Along with later subduction of atmospheric gases, these processes can account for the rare gases presently in the mantle. Atmospheric rare gases differ greatly from solar gases isotopically and elementally, and can be generated by hydrodynamic escape of a hydrogen-rich solar atmosphere, leaving a residual atmosphere preferentially depleted in lighter constituents. Radiogenic Xe isotopes indicate that substantial losses occurred late from both the interior and the atmosphere at a time consistent with the Moon's formation.

## 1.  INTRODUCTION

The origins and evolutionary histories of terrestrial volatiles are not well understood. Various models have been advanced for deriving the composition of the atmosphere from reservoirs of primordial solar system volatiles through processes operating before, during, or shortly after planetary accretion. Some are promising, but none are without problems. More limited work has been done in accounting for the acquisition of the volatiles presently stored within Earth. Signatures of the origin and physical processing of the terrestrial volatiles are expected to survive most clearly in the chemically inert noble gases. Elemental and isotopic compositions of terrestrial noble gases provide clues to the nature of the source reservoirs, the processes responsible for their evolution from primordial to present compositions, and the environments in which the processing occurred. However, these clues are not amenable to unambiguous interpretation. Models of volatile evolution based on these clues also must involve mechanisms that are consistent with what is known or plausibly inferred about the origin and evolution of the solar system as a whole.

The elemental and isotopic record of atmospheric and mantle rare gases on Earth is reviewed here, and this record then serves as a basis for a subsequent evaluation of different processes that may have controlled the acquisition, modification, and planetary losses of volatiles during early Earth history. Discussion of the origin of rare gases on other terrestrial planets is given by *Pepin* (1991, 1992, 1994, 1997), *Jakosky et al.* (1994), and *Hutchins and Jakosky* (1996).

## 2.  THE ATMOSPHERE AND SOLAR SYSTEM RESERVOIRS

### 2.1.  Rare Gas Abundances

Compared to the gases in the solar nebula, those in the atmospheres of the terrestrial planets are sharply depleted relative to the other elements and show strong elemental and isotopic fractionations. Rare gas abundances in these planetary atmospheres and CI chondrites, relative to the solar composition (which represents that of the primordial solar nebula), are shown in Fig. 1. Helium is lost from the planetary atmospheres, and is not included here. The depletions are generally less for increasing mass and are consistent between the reservoirs, although the Xe/Kr ratio in the planets is lower than in CI chondrites. This feature was initially thought to be due to the sequestration of Xe in other terrestrial reservoirs; however, such reservoirs on Earth of this "missing Xe" have not been found (see *Bernatowicz et al.*, 1985), and the lower Xe/Kr ratio has since been considered a feature of planetary atmospheres. The reproduction of the planetary elemental abundance pattern from solar nebula gases is a primary goal for models of the origin of the rare gases in the terrestrial atmosphere.

### 2.2.  Isotopic Compositions

Rare gas isotopic compositions for the atmosphere and solar system reservoirs are summarized in *Pepin* (1991). The value of $^{20}Ne/^{22}Ne = 13.8 \pm 0.1$ (*Benkert et al.*, 1993; *Pepin et al.*, 1995) derived for the solar wind is assumed to rep-

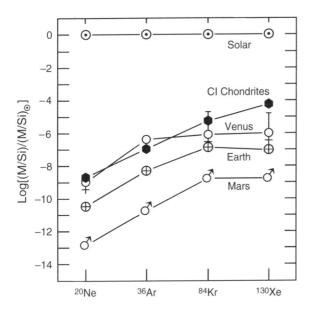

**Fig. 1.** Noble gas abundances in planetary atmospheres and CI chondrites, plotted as the atom concentration relative to Si divided by the corresponding solar ratio. Data are from a compilation by *Pepin* (1991). Note that ranges of Kr and Xe values are shown for Venus.

resent the initial solar nebula composition. Meteorites have a variety of components, including those with extrasolar compositions. Ne isotope ratios in bulk CI chondrites scatter around an average $^{20}Ne/^{22}Ne$ of 8.9 ± 1.3. The meteoritic Q component (*Wieler*, 1994), sited in the surfaces of carbonaceous phases in primitive meteorites and possibly derived by fractionating processes operating on parent bodies (*Pepin*, 1991) or in the solar nebula itself (*Ozima et al.*, 1998), and thought to be relatively free of "exotic" components, has $^{20}Ne/^{22}Ne = 10.70 ± 0.15$. The atmospheric ratio of 9.8 can be derived either from mixing meteorite and solar components or by fractionation of solar Ne. The difference between the solar (0.033) and atmospheric (0.029) $^{21}Ne/$ $^{22}Ne$ ratios is also roughly consistent with fractionation of solar Ne. Note that $^{21}Ne$ from nucleogenic reactions (see discussion in section 3), estimated from the ratio of nucleogenic $^{21}Ne$ to radiogenic $^{40}Ar$ in the bulk Earth (see *Porcelli and Wasserburg*, 1995b) and the atmospheric $^{40}Ar$ abundance, is only <3% of the atmospheric $^{21}Ne$.

The initial $^{40}Ar/^{36}Ar$ ratio in the solar system is $\sim10^{-4}$– $10^{-3}$ (*Begemann et al.*, 1976). The atmosphere has $^{40}Ar/$ $^{36}Ar = 296$, so that essentially all the $^{40}Ar$ has been produced by $^{40}K$ decay. The atmospheric $^{38}Ar/^{36}Ar$ ratio of 0.188 is similar to that found in CI chondrites of 0.189 ± 0.002 (*Mazor et al.*, 1970) but substantially higher than the current estimate of 0.173 for the solar wind (*Pepin et al.*, 1999).

Solar, bulk meteorite, and terrestrial Kr-isotopic compositions seem to be generally related to one another by mass fractionation (see *Eugster et al.*, 1967; *Pepin*, 1991), with

the atmosphere depleted in light isotopes by ~0.8%/amu relative to the solar composition (*Pepin et al.*, 1995).

The isotopic compositions of Xe components in the solar system have been more difficult to unravel. The light isotopes of chondritic Xe are approximately related to atmospheric Xe by a fractionation of ~4.2%/amu (*Krummenacher et al.*, 1962). However, it has been difficult to identify a solar system component that can be precisely related to the composition of the atmosphere directly or by simple processes (such as mass fractionation) and addition of expected radiogenic components. Multidimensional isotopic correlations of chondrite data have been used to define a composition, U-Xe, that when mass fractionated yields the light-isotope ratios of terrestrial Xe and points strongly to the presence in the atmosphere of a heavy isotope component with the composition of $^{244}Pu$ fission Xe (*Pepin and Phinney*, 1978; *Pepin*, 2000; see also *Igarashi*, 1986, 1988). Note that the substantially different U-Xe and fission Xe compositions obtained in a later analysis by *Igarashi* (1993) appear to reflect the use of a meteoritic database that does not preserve mass balance (see discussion in *Pepin*, 1999). Originally a hypothetical construct, U-Xe-like compositions have since been found in achondritic meteorites (*Michel and Eugster*, 1994; *Eugster et al.*, 1994). Solar-wind Xe (SW-Xe) appears to be isotopically identical to U-Xe except for addition of a nucleogenetic component at the two heaviest isotopes (*Pepin et al.*, 1995), as seen in Fig. 2a where both compositions are plotted relative to terrestrial Xe. The presence of this heavy-isotope component in the Sun but not in early Earth is puzzling, since solar wind Xe, presumably reflecting the Xe composition in the accretion disk, would arguably be the more plausible contributor to primordial planetary inventories. Nevertheless, as shown in Fig. 2b, solar Xe is ruled out as a parental composition from which nonradiogenic terrestrial Xe can be generated by mass fractionation. Here both U-Xe and SW-Xe have been fractionated into agreement with the nonradiogenic light-isotope ratios of atmospheric Xe, and it is clear that this processing of SW-Xe produces more $^{136}Xe$, relative to $^{130}Xe$, than the atmosphere actually contains. If air contains a later addition of a fission component, the discrepancy with the nonradiogenic air component is greater; this is also the case if chondritic Xe, rather than SW-Xe, is considered as an alternative source of primordial Xe on Earth (*Pepin*, 2000). We are left with U-Xe, and therefore with the fractionated U-Xe composition in Fig. 2b, where $(^{129}Xe/^{130}Xe)_O = 6.053$ and $(^{136}Xe/^{130}Xe)_O = 2.075$, as the present best estimate of the isotopic composition of nonradiogenic terrestrial Xe.

### 2.3. Radionuclide Abundances

The inventory of rare gases in the atmosphere is, of course, well known. The total abundance of nonradiogenic rare gases in Earth cannot be further constrained without assumptions regarding the fraction retained within Earth. However, the fraction of a radiogenic isotope within Earth can be calculated only if the concentration of the parent is

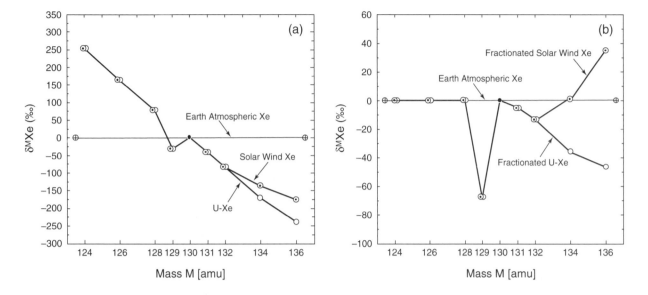

**Fig. 2.** **(a)** Relationships of unfractionated SW-Xe and U-Xe to terrestrial atmospheric Xe, plotted as permil (‰) differences from the atmospheric composition. The δ representation is defined as $\delta^M Xe + 1000[(R_M/R_{atm}) - 1]$, where $R_M = {}^M Xe/{}^{130}Xe$ is the plotted composition and $R_{atm}$ is the corresponding isotope ratio in atmospheric Xe (see *Pepin*, 1999, for the data compilation). **(b)** Same as in **(a)**, but now after hydrodynamic escape fractionation of SW-Xe and U-Xe to the extent required to match their ${}^{124-128}Xe/{}^{130}Xe$ ratios to the corresponding atmospheric values. The fractionated SW-${}^{136}Xe/{}^{130}Xe$ ratio is elevated above the air value by ~10× its 1σ uncertainty (see *Pepin*, 1999, for the data compilation).

well-constrained and the abundance of the radiogenic product in the atmosphere can be clearly resolved. Attention has focused on ${}^{40}Ar$. The K concentration of the bulk silicate Earth has been calculated using a U concentration of 21 ppb obtained from consideration of terrestrial refractory element concentrations (e.g., *O'Nions et al.,* 1981) and a K/U ratio of $1.27 \times 10^4$ obtained for the MORB source (*Jochum et al.,* 1983) and applied to the entire bulk silicate Earth; then ~40% of the ${}^{40}Ar$ produced over 4.5 G.y. is currently in the atmosphere (*Allègre et al.,* 1986, 1996; *Turcotte and Schubert,* 1988). However, the assumption that the K/U ratio of the MORB source is precisely that of the bulk silicate Earth is debatable (*Albarède,* 1998), so that the fraction of the mantle that has been degassed remains a model-dependent parameter.

Information regarding retention of rare gases on Earth can be obtained from radiogenic Xe isotopes. With the assumption that fractionated U-Xe (Fig. 2b) correctly describes the composition of nonradiogenic terrestrial Xe, about 6.8% of atmospheric ${}^{129}Xe$ (${}^{129*}Xe_{atm}$) and 4.65% of atmospheric ${}^{136}Xe$ (${}^{136*}Xe_{atm}$) are radiogenic, due respectively to β-decay of ${}^{129}I$ (${}^{129}t_{1/2}$ = 16 m.y.) and spontaneous fission of ${}^{244}Pu$ (${}^{244}t_{1/2}$ = 82 m.y.) that were incorporated into early Earth. Contributions from fission of ${}^{238}U$ (${}^{238}t_{1/2}$ = 4.5 G.y.) are minor (*Pepin,* 1999). The atmospheric abundance of each daughter nuclide can be compared to the expected initial terrestrial abundance of the parent nuclides. The terrestrial I abundance (${}^{127}I_{BSE}$) is not tightly constrained. Estimates of crustal and upper mantle I by *Mura-*

*matsu and Wedepohl* (1998) and *Déruelle et al.* (1992) yield values for the bulk silicate Earth of 3–24 ppb, partly depending upon the volume of mantle that was depleted to form the crust. *Wänke and Dreibus* (1988) estimated a bulk silicate Earth concentration of 13 ppb I based on cosmochemical arguments, and this serves as an intermediate estimate. At 4.6 G.y. ago, $({}^{129}I/{}^{127}I)_O = 1.1 \times 10^{-4}$ (see *Swindle and Podosek,* 1988), so that the bulk silicate Earth (or precursor materials) had $2.7 \times 10^{37}$ atoms ${}^{129}I$. The atmospheric ${}^{129*}Xe$ inventory is $1.7 \times 10^{35}$ atoms, so that ${}^{129*}Xe_{atm}/{}^{129}I_{BSE} \approx 10^{-2}$, and therefore is $10^{-2}\times$ the abundance expected in Earth from decay of ${}^{129}I$. The "missing ${}^{129*}Xe$" has not been found in any crustal reservoirs, and such a low value cannot be accounted for by incomplete degassing of the mantle (see below). Therefore substantial loss of ${}^{129*}Xe$ from Earth has occurred. *Wetherill* (1975) interpreted this as a closure age before which the ${}^{129}Xe$ produced during an initial "formation interval" was lost to space, followed by complete retention of ${}^{129*}Xe$, and calculated this age to be ~$10^8$ yr (~7 × ${}^{129}t_{1/2}$).

The atmospheric ${}^{136*}Xe$ inventory is $3.81 \times 10^{34}$ atoms. The initial terrestrial ${}^{244}Pu$ abundance, obtained from a present bulk silicate Earth value of 21 ppb U and $({}^{244}Pu/{}^{238}U)_O = 6.8 \times 10^{-3}$ (*Hudson et al.,* 1989), produced $2.1 \times 10^{35}$ atoms (with a fission yield of ${}^{136}Y_{244} = 7.05 \times 10^{-5}$ atoms ${}^{136}Xe$ per atom of ${}^{244}Pu$). Then ${}^{136*}Xe_{atm}/({}^{244}Pu_{BSE} {}^{136}Y_{244}) \approx 0.2$; this also indicates that there is "missing Xe." However, this calculation is less sensitive to initial losses than that of the ${}^{129}I$-${}^{129}Xe$ budget due to the longer half-life of $^{244}$Pu, and a substantial portion of the "missing $^{136*}$Xe" may be within Earth. A further constraint on Xe losses can be obtained by assuming that atmospheric $^{129*}$Xe and $^{136*}$Xe have been extracted from a single solid Earth reservoir, a model age can be calculated by combining the two systems according to the equation (*Pepin and Phinney, 1976*)

$$ t = \frac{1}{\lambda_{244} - \lambda_{129}} \ln \left[ \frac{\left( \dfrac{^{129*}Xe}{^{136*}Xe} \right) \left( \dfrac{^{244}Pu}{^{238}U} \right)_O {}^{136}Y_{244}}{\left( \dfrac{I}{U} \right)_O \left( \dfrac{^{129}I}{^{127}U} \right)_O} \right] \quad (1) $$

where t is the time since 4.6 Ga after which complete loss of radiogenic Xe no longer occurs (i.e., the system becomes closed). Note that the contribution to $^{136*}$Xe from decay of $^{238}$U compared to that of $^{244}$Pu is <10% for t <120 m.y. and can be neglected for this calculation. From the radiogenic Xe in the atmosphere and the parent nuclide concentrations quoted above, t ≈ 100 m.y., indicating that substantial and late losses of Xe have occurred from the reservoir of atmospheric gases. Although considerable uncertainty might be associated with some of the parameters used in these calculations, it seems unavoidable that there has been such substantial losses of radiogenic Xe. Note that losses may have occurred early from material forming Earth or directly from the atmosphere. *Zhang* (1998) extended this calculation by adding the $^{131*}$Xe, $^{132*}$Xe, and $^{134*}$Xe balance equations, but since the nonradiogenic atmospheric Xe composition used to obtain the amount of each radiogenic contribution is related to the present composition by the addition of $^{244}$Pu-derived Xe (*Pepin and Phinney, 1978; Pepin, 2000*), these other isotopes do not provide additional independent constraints. The calculation of equation (1) is independent of the atmospheric abundance of Xe; however, the similarity between the $^{129}$I-$^{129}$Xe age (based on the Xe atmospheric abundance) and the equation (1) $^{129}$Xe-$^{136}$Xe age (based on the Xe atmospheric isotopic composition) indicates that there has not been great alteration in the atmospheric Xe abundance (i.e., Xe loss) since the final Xe-isotopic composition was established by complete decay of the parent nuclides. In particular, the deficiency in Xe in the atmosphere of a factor of ~20 relative to the "planetary" (i.e., meteoritic) abundance pattern (see Fig. 1) cannot be explained by Xe sequestration within Earth once the atmospheric Xe has reached its present isotopic composition. Note that if there was continuous partial loss over some time rather than abrupt closure, these losses would extend over a period longer than the time calculated above.

## 3. EARTH'S MANTLE

Data for mid-ocean ridge basalts (MORB) can be used to characterize the average convecting upper mantle extending at least to the depth of the seismic discontinuity at 670 km. While ocean island basalts (OIB) sample separate and distinct mantle domains with a wide range of possible origins, some are supplied by at least one reservoir with rare gas characteristics that are distinct from both those in MORB and in the atmosphere, and this reservoir appears to require long-term isolation from the surface. As discussed below, there is evidence for nonradiogenic rare gases within the MORB and OIB sources that do not have atmospheric compositions, but rather have characteristics of solar rare gases. This provides important constraints on terrestrial rare gas evolutionary history. Isotopic variations due to radioactive decay processes are the basis for models of fractionations between parent element nuclides and rare gases, e.g., mantle degassing to the atmosphere. Available rare gas data are reviewed in detail elsewhere (e.g., *Ozima*, 1994; *Farley and Neroda*, 1998); the summary that follows highlights data that constrain the overall characteristics of rare gases in Earth. Note that measured Kr-isotopic ratios are indistinguishable from atmospheric values, and will not be discussed further.

### 3.1. Helium Isotopes

Measured mantle $^3$He/$^4$He ratios first indicated the presence of primordial rare gas isotopes within Earth, and reflect mixing between He that was initially incorporated and radiogenic $^4$He produced subsequently by U and Th decay. The ratio of $^3$He/$^4$He = 1.66 × 10$^{-4}$ for the Jupiter atmosphere (*Mahaffy et al.*, 1998) provides the best estimate for the solar nebula ratio and so an initial value for Earth; the higher present solar wind value of ~4.6 × 10$^{-4}$ is due to D burning in the Sun after accretion (*Geiss*, 1993). It has been suggested that post-D-burning solar He was incorporated in Earth (*Ozima et al.*, 1998). This is incompatible with scenarios in which terrestrial volatiles were captured directly from the solar nebula, although it might be reconciled with derivation of terrestrial He from the early solar wind. These models are considered further in section 5.

Mid-ocean ridge basalt $^3$He/$^4$He ratios fall in a narrow range (Fig. 3) around 8 × R$_A$ (where R$_A$ is the atmospheric ratio of $^3$He/$^4$He = 1.39 × 10$^{-6}$; *Lupton and Craig*, 1975, *Kurz and Jenkins*, 1981), indicating that the MORB source region has a relatively uniform rare gas isotopic signature. The $^3$He/$^4$He ratios of OIB are more variable; OIB with 6–8 × R$_A$ may include recycled components (*Kurz et al.*, 1982) or crustal contamination (*Hilton et al.*, 1995) that have lowered the ratio from that of MORB. Therefore, these OIB He compositions are not considered to constitute a significant fraction of the mantle. However, evidence for a significant long-term rare gas reservoir distinct from that of MORB comes from those having $^3$He/$^4$He ratios >~10 R$_A$. These OIB require at least one component with a time-integrated $^3$He/(U + Th) ratio greater than that of the upper mantle (*Kurz et al.*, 1982; *Allègre et al.*, 1983), and a value at least equal to the highest ratios of $^3$He/$^4$He = 4.5 × 10$^{-5}$ (32–38 × R$_A$) measured in Loihi Seamount (*Kurz et al.*, 1982; *Rison and Craig*, 1983; *Honda et al.*, 1993) and Iceland (*Hilton et al.*, 1998, 1999) samples. Ocean island basalt $^3$He/$^4$He ratios between those of MORB and Loihi are generally as-

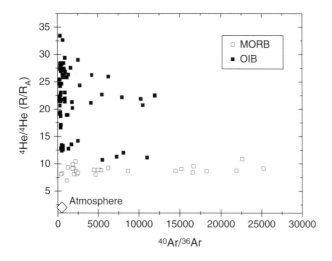

**Fig. 3.** Helium- and Ar-isotopic compositions for MORB and OIB samples (with $^3$He/$^4$He >10× $R_A$) obtained by step-heating (*Valbracht et al.*, 1996, 1997, and data compilation in *Porcelli and Wasserburg*, 1995b). The range in $^{40}$Ar/$^{36}$Ar in MORB samples is attributed to variable contamination by air Ar to mantle-derived Ar, with the highest $^{40}$Ar/$^{36}$Ar ratio therefore representing the minimum upper mantle MORB source value. Argon in OIB that have atmospheric $^{40}$Ar/$^{36}$Ar ratios have been interpreted as being dominantly contaminant air Ar (*Patterson et al.*, 1990; *Farley and Craig*, 1994), so that the $^{40}$Ar/$^{36}$Ar ratio of the OIB source is not well constrained.

sumed to be from mixing of these two components. Unfortunately, it has not been possible to define unambiguously the other geochemical attributes of the high $^3$He/$^4$He component (see *Farley and Neruda*, 1998; *Hilton et al.*, 1999).

The only present-day rare gas flux from the mantle to the atmosphere that can be constrained is that of He. Due to the low He abundance in the atmosphere (and so in seawater), the input from MORB into the oceans can be measured; the current best estimate of 1000 ± 250 mol $^3$He/yr is based upon measurements of excess $^3$He in seawater and ocean advection rates, and represents an average over the last $10^3$ yr (*Craig et al.*, 1975; *Farley et al.*, 1995).

### 3.2. Neon Isotopes

The clearest evidence for the presence of nonradiogenic mantle rare gases that are not atmospheric in composition is for Ne. Mid-ocean ridge basalt $^{20}$Ne/$^{22}$Ne and $^{21}$Ne/$^{22}$Ne ratios generally are greater than those of the atmosphere (Fig. 4a) and are correlated (*Sarda et al.*, 1988); this likely reflects variable mixing between a mantle isotopic composition represented by the highest values and atmospheric Ne contamination. The only plausible origin for the MORB mantle composition is trapped solar Ne (see section 2), with the addition of $^{21}$Ne produced largely by the reaction $^{18}$O($\alpha$,n)$^{21}$Ne within the mantle (*Sarda et al.*, 1988; *Honda et al.*, 1993). Measured ratios for OIB with high $^3$He/$^4$He

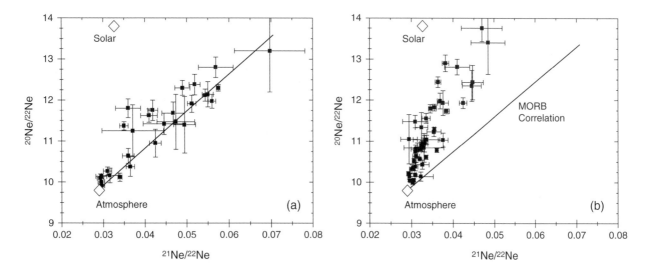

**Fig. 4.** (a) Neon-isotopic compositions for MORB samples (for references, see data compilation in *Porcelli and Wasserburg*, 1995b). Data indistinguishable from air are not shown. Also plotted is the solar Ne composition (*Benkert et al.*, 1993). The MORB correlation line is from *Sarda et al.* (1988). The $^{20}$Ne/$^{22}$Ne–$^{21}$Ne/$^{22}$Ne correlation has been interpreted as reflecting mixing of variable proportions of contaminant air Ne with a MORB component that is composed of solar Ne and radiogenic $^{21}$Ne (*Honda et al.*, 1993a; *Farley and Poreda*, 1993). (b) Neon-isotopic compositions for ocean island samples that have $^3$He/$^4$He ratios greater than that of MORB (*Valbracht et al.*, 1996, 1997; *Harrison et al.*, 1999; and data compilation in *Porcelli and Wasserburg*, 1995b). The trend in OIB data has been interpreted as reflecting mixing of variable proportions of contaminant air Ne with hotspot mantle Ne composed of solar Ne and a smaller proportion of radiogenic $^{21}$Ne than is contained in the MORB component (*Honda et al.*, 1993). The samples have $^3$He/$^4$He ratios of 11–32× $R_A$, possibly reflecting mixing of hotspot rare gases and MORB source rare gases, so that Ne from the MORB source may also be present.

ratios (Fig. 4b) span a similar range in $^{20}Ne/^{22}Ne$ ratios, but with lower corresponding $^{21}Ne/^{22}Ne$ ratios (*Sarda et al.,* 1988; *Honda et al.,* 1991, 1993), and so involve a mantle component with a similarly high $^{20}Ne/^{22}Ne$ ratio as MORB but a lower $^{21}Ne/^{22}Ne$ ratio. It should be noted that *Ozima* (1999) argued that, since precise measured $^{20}Ne/^{22}Ne$ ratios in mantle samples are generally below ~12.5, the data are more consistent with a mantle composition that has a $^{20}Ne/^{22}Ne$ ratio lower than the solar composition due to fractionation of solar gases prior to incorporation in Earth. However, an Iceland sample recently measured by *Harrison et al.* (1999) yielded a $^{20}Ne/^{22}Ne$ ratio of 13.7 ± 0.3, which is indistinguishable from an unfractionated solar wind value. The presence of other components with $^{20}Ne/^{22}Ne$ ratios between those of the atmosphere and solar cannot be discerned from the presently available data.

The ratio of the isotopic shift in He to that of Ne is similar for both the MORB and OIB component; since the production ratio of $^{21*}Ne/^{4}He$ is fixed, it appears that the two sources have similar $^{3}He/^{22}Ne$ ratios (*Honda et al.,* 1993). Calculation of the $^{21*}Ne/^{4}He$ production ratio is somewhat complex, involving various $(\alpha,n)$ and $(n,\alpha)$ reactions, with significant uncertainties in the total reaction rate of each. A range of values of $^{21*}Ne/^{4}He = (2.3–10) \times 10^{-8}$ have been estimated (*Rison,* 1980; *Eikenberg et al.,* 1993); the most recent theoretical calculations provide a value of $4.5 \times 10^{-8}$ (*Yatsevich and Honda,* 1997). Using a production ratio of $9 \times 10^{-8}$, *Porcelli and Wasserburg* (1995b) calculated a mantle ratio of $^{3}He/^{22}Ne = 4.4$. Using a production ratio of $4.5 \times 10^{-8}$ results in a value of $^{3}He/^{22}Ne = 8.8$. *Honda and McDougall* (1998) used this production ratio to calculate an average mantle value of $^{3}He/^{22}Ne = 7.7 \pm 2.6$ from a broad compilation of available data. These values have been compared to a solar wind value of ~4.1, which, however, is for the present solar wind. Using a solar nebula value of $^{3}He/^{4}He = 1.66 \times 10^{-4}$ (see above), an updated solar wind value of $^{4}He/^{20}Ne = 820$ (*Pepin et al.,* 1999), and $^{20}Ne/^{22}Ne = 13.8$, a value of $^{3}He/^{22}Ne = 1.9$ is obtained that may provide a more appropriate point of comparison. Therefore, the mantle $^{3}He/^{22}Ne$ ratio may be 2–4× greater than the solar nebula; such a difference provides a constraint on how rare gases are incorporated within Earth. *Honda and McDougall* (1998) argued that a higher mantle value is due to fractionation of mantle rare gases according to differences in solubility in a silicate melt, and could reflect either fractionation during acquisition or subsequent mantle degassing. The Henry's constants for the solubility of He in silicate melt is ~2× that of Ne (*Lux,* 1987), so that a mantle value of $^{3}He/^{22}Ne \approx 4$ (but not as high as 8.8) can be related to the solar nebula value in this way. Clearly, confirmation of the $^{21*}Ne/^{4}He$ production ratio is required to resolve this.

### 3.3.  Argon Isotopes

A large range in $^{40}Ar/^{36}Ar$ ratios measured in MORB (Fig. 3) is likely due to mixing of variable proportions of atmospheric Ar (with $^{40}Ar/^{36}Ar = 296$) with a single, more radiogenic mantle composition. The minimum value for this

mantle composition is then represented by the highest measured values of $2.8 \times 10^{4}$ (*Staudacher et al.,* 1989) to $4 \times 10^{4}$ (*Burnard et al.,* 1997). Upper mantle Ar is thus much more radiogenic than atmospheric Ar.

Measurements of $^{40}Ar/^{36}Ar$ in OIB with $^{3}He/^{4}He > 10 \times R_A$ are shown in Fig. 3. Early values for Loihi glasses of $^{40}Ar/^{36}Ar < 10^{3}$ appear to reflect substantial contamination with atmospheric Ar (*Fisher,* 1985; *Patterson et al.,* 1990) rather than a mantle composition with a $^{40}Ar/^{36}Ar$ ratio similar to that of the atmosphere (*Allègre et al.,* 1983; *Kaneoka et al.,* 1986; *Staudacher et al.,* 1986). A study of basalts from Juan Fernandez (*Farley and Craig,* 1994) found contamination contained within phenocrysts and so introduced into the magma chamber, providing an explanation for the prevalence of atmospheric Ar contamination of OIB. Recent measurements of Loihi samples found $^{40}Ar/^{36}Ar = 2600–2800$ associated with high $^{3}He/^{4}He$ ratios (*Hiyagon et al.,* 1992; *Valbracht et al.,* 1997), and so presumably containing less atmospheric contamination. *Poreda and Farley* (1992) found values of $^{40}Ar/^{36}Ar \leq 1.2 \times 10^{4}$ in Samoan xenoliths with intermediate $^{3}He/^{4}He$ ratios $(9–20) \times R_A$. It now appears that $^{40}Ar/^{36}Ar$ ratios in the high $^{3}He/^{4}He$ OIB source are likely to be >3000 but may be lower than that of MORB (see also *Matsuda and Marty,* 1995).

Measurements of MORB and OIB $^{38}Ar/^{36}Ar$ ratios typically are atmospheric within error, but have been of low precision due to the low abundance of these isotopes. Two recent analyses of MORB and OIB samples have found $^{38}Ar/^{36}Ar$ ratios lower than that of the atmosphere (*Valbracht et al.,* 1997; *Niedermann et al.,* 1997). *Pepin* (1998) argued that these values reflect a mixture of a solar upper mantle composition with atmospheric Ar contamination, and demonstrated that such mixtures could be used both to constrain the Ne/Ar ratio of the solar rare gases in the mantle, and to infer a lower-mantle $^{40}Ar/^{36}Ar$ ratio of ~5000. However, new high-precision data for the "popping rock" MORB by *Kunz* (1999) failed to find $^{38}Ar/^{36}Ar$ ratios that deviate from that of air. The unambiguous identification of solar Ar in the mantle clearly has important implications for the origin of terrestrial rare gases, and additional analyses are required to firmly establish whether there are mantle samples that contain nonatmospheric $^{38}Ar/^{36}Ar$ ratios not masked by air Ar contamination.

### 3.4.  Xenon Isotopes

Mid-ocean ridge basalt $^{129}Xe/^{130}Xe$ and $^{136}Xe/^{130}Xe$ ratios (Fig. 5) lie on a correlation extending from atmospheric ratios to higher values (*Staudacher and Allègre,* 1982), and likely reflect mixing of variable proportions of air contaminant Xe with an upper mantle Xe component having more radiogenic $^{129}Xe/^{130}Xe$ and $^{136}Xe/^{130}Xe$ ratios. The highest measured values thus provide a lower bound for the MORB source region. Xenon with atmospheric isotopic ratios in high $^{3}He/^{4}He$ OIB samples (e.g., *Allègre et al.,* 1983) appears to be dominated by air contamination (*Patterson et al.,* 1990) rather than represents mantle Xe with an air composition. Although Samoan samples with intermediate (9–

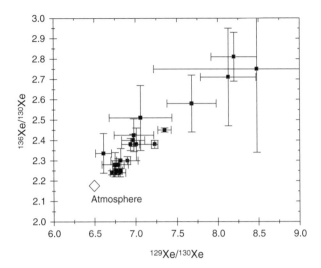

**Fig. 5.** Measured $^{136}Xe/^{130}Xe$ and $^{129}Xe/^{130}Xe$ ratios for MORB (see data compilation in *Porcelli and Wasserburg, 1995b*). Analyses within error of the atmospheric composition are not shown. The observed correlation (*Staudacher and Allègre, 1982*) is generally interpreted as reflecting different degrees of air contamination to a MORB Xe composition, so that the highest precise MORB values (*Staudacher et al., 1989*) represent the minimum values for the upper mantle.

$20 \times R_A$) He-isotopic ratios have been found with Xe-isotopic ratios distinct from those of the atmosphere (*Poreda and Farley, 1992*), the Xe in these samples may have been derived largely from the upper mantle. It is now generally accepted that the composition of mantle Xe in the OIB source is unknown.

Contributions to $^{136}Xe$ enrichments in MORB by decay of $^{238}U$ or $^{244}Pu$ in theory can be distinguished based on the spectrum of contributions to other Xe isotopes, although analyses have typically not been sufficiently precise to identify the parent nuclide. Precise measurements of Xe in $CO_2$ well gases that have $^{129}Xe$ and $^{136}Xe$ enrichments similar to those found in MORB and so are considered to be derived from the upper mantle (*Staudacher, 1987*) indicate that $^{244}Pu$ has contributed <10–20% of the $^{136}Xe$ that is in excess of the atmospheric composition (*Phinney et al., 1978; Caffee et al., 1999*). An error-weighted best fit to recent precise MORB data (*Kunz et al., 1998*) yielded a value of $32 \pm 10\%$ for the fraction of $^{136}Xe$ excesses relative to atmosphere that are $^{244}Pu$-derived, although due to the considerable scatter in the data, further work on this is warranted. It should be noted that these $^{136}Xe$ excesses have been calculated relative to the present composition of the atmosphere, which in turn contains appreciable radiogenic contributions to the heavy Xe isotopes from $^{244}Pu$ decay.

The measured ratios of the nonradiogenic isotopes in MORB are indistinguishable from those of the atmosphere. However, $CO_2$ well gases with presumably mantle-derived Xe have been found to have higher $^{124-128}Xe/^{130}Xe$ ratios

(*Phinney et al., 1978; Caffee et al., 1999*) that can be explained either by a mixture of ~10% solar Xe and ~90% atmospheric Xe, or a mantle component that has not been fractionated relative to solar Xe to the same extent as atmospheric Xe. Due to the lower precision of MORB data and the pervasive presence of atmospheric contamination in MORB samples, it has not been possible to confirm that this feature is present throughout the upper mantle.

### 3.5. Abundance Patterns

The measured rare gas abundance patterns of MORB and OIB scatter greatly. This is due to alteration as well as the fractionation effects of partitioning between the basaltic melts and a vapor phase that may be then preferentially gained or lost by a basalt. However, a common pattern appears to be that MORB Ne/Ar and Xe/Ar ratios are greater than in air. In particular, this pattern was found in a gas-rich MORB sample with high Ar- and Xe-isotopic compositions (and so less contaminated) and $^4He/^{40}Ar$ ratios that are near the expected production ratio of the upper mantle (and so not fractionated) (*Staudacher et al., 1989*). A similar pattern was identified by *Moreira et al.* (1998) in a MORB sample based on correlations between isotopic compositions and elemental ratios of repeated analyses that were due to mixing between mantle and contaminant components. Although confirmation of specific elemental ratios requires further data, the general pattern appears to be consistent.

It is often assumed that rare gases are highly incompatible during basalt genesis and are extracted from the mantle at mid-ocean ridges both efficiently and without elemental fractionation between basaltic melt and the upper mantle source region. Experimental data of partitioning between basaltic melts and olivine are consistent with this for He but not for the heavier rare gases, which have been found in higher concentrations in olivine than expected (*Hiyagon and Ozima, 1986; Musselwhite et al., 1991; Broadhurst et al., 1992*). However, these results may be due to experimental difficulties (see *Farley and Neroda, 1998*). Recent data by *Chamorro-Pérez et al.* (1998) indicate that the solubilities of rare gases in silicate melts decrease dramatically at 2 GPa (for Xe) to 10 GPa (for He), although how this affects partitioning of rare gases into melts at very low concentrations is unclear. Also, while this may provide a mechanism for preferentially retaining Xe in the upper mantle relative to other rare gases during melting at depths greater than ~60 km beneath ridges, the magnitude of the total fractionation generated during MORB genesis depends upon the fraction of material that is not melted further at shallower depths (where the rare gases are not fractionated by this mechanism) due to the geometry of upwelling.

### 3.6. Total Abundances

The upper mantle is highly depleted in rare gases relative to the bulk silicate Earth (which has an average concentration of at least the atmospheric inventory divided by the mass of the silicate Earth). As discussed above, concentrations in mantle-derived MORB are subject to modification and so are difficult to relate to the source, but the flux of $^3$He into the oceans (see above) can be divided into the mass of material added to the ocean crust of 21 km/yr (*Parsons,* 1981) that is responsible for carrying this $^3$He from the mantle; $1.5 \times 10^{15}$ atoms $^4$He/g is then obtained for undegassed MORB, similar to that measured for a gas-rich MORB sample (*Staudacher et al.,* 1989). Assuming that MORB is the result of 10% partial melting of the upper mantle and that He behaves incompatibly, then the mantle has $1.5 \times 10^{14}$ atoms $^4$He/g. Since the mantle concentration of He cannot be usefully compared with the He inventory of the atmosphere (from where He is lost), it is useful to calculate the mantle concentration of Ar. Using a production ratio of $^{4*}$He/$^{40*}$Ar $= 3$ and an upper mantle value of $^{40}$Ar/$^{36}$Ar $= 3 \times 10^4$, then the upper mantle has $2 \times 10^9$ atoms $^{36}$Ar/g. This is highly depleted compared to the value of $3 \times 10^{12}$ atoms $^{36}$Ar/g obtained by dividing the atmospheric inventory by the mass of the upper mantle. Calculation of rare gas concentrations in the high $^3$He/$^4$He OIB source requires assumptions regarding the evolution of the source region, and are discussed below.

### 3.7.  Summary

A centrally important question for the source, incorporation mechanisms, and transport of noble gases in Earth, and for the evolutionary history of the atmosphere, is whether there are rare gas compositions present in the mantle different from those of the atmosphere and characteristic of another solar system reservoir. While the evidence for the presence of solar Ne and He is strong, Ar- and Xe-isotopic compositions generally have been found to be indistinguishable or only slightly offset from those in air. However, the tendency of nonradiogenic Ar, Kr, and Xe in most mantle samples to display approximately atmospheric compositions may be due to the domination over any deep mantle signature by air-derived rare gases added through postcollection adsorption (*Burnard et al.,* 1997), seawater-magma interaction (*Patterson et al.,* 1990), or subduction into the mantle (*Porcelli and Wasserburg,* 1995a,b). Therefore the possibility that MORB Ar and Xe may be different from that of the atmosphere and have solar characteristics must be confirmed by further research. The alternative, that Ar, Kr, and Xe with nonradiogenic atmospheric compositions were initially incorporated into Earth, cannot be discounted by the available data.

The isotopic variations in the mantle due to radioactive decay processes provide the basis for mantle models of the present-day distribution and transport of mantle rare gases. For the upper mantle, the $^3$He/$^4$He and $^{21}$Ne/$^{22}$Ne ratios are narrowly constrained, while minimum values for $^{40}$Ar/$^{36}$Ar, $^{129}$Xe/$^{130}$Xe, and $^{136}$Xe/$^{130}$Xe have been established. Another significant mantle reservoir has also been identified with higher $^3$He/$^4$He and lower $^{21}$Ne/$^{22}$Ne ratios, but with unknown Ar- and Xe-isotopic compositions. The implications for these results are discussed in the next section.

## 4.  DISTRIBUTIONS AND FLUXES WITHIN EARTH

It is clear that the radiogenic nuclides in the mantle are from production within Earth. Three possible sources for the nonradiogenic mantle rare gas isotopes have been described, each with implications for the initial acquisition of deep gases. These are (1) release from the core, especially of He and Ne; (2) subduction of meteoritic material containing high concentrations of solar He and Ne or of crustal material carrying atmospheric Ar and Xe; and (3) long-term storage in the mantle, with a reservoir in the deep mantle that has relatively high concentrations of $^3$He (and therefore high $^3$He/$^4$He ratios) as well as other rare gases, and so is isolated from near-surface degassing processes.

The first and second scenarios generally have been advanced for He and Ne in the context of geochemical and geophysical models that require the processing of the entire mantle and therefore have found it difficult to preserve over long time periods within the mantle itself the distinctive $^3$He/$^4$He signatures seen in hotspot volcanics (see, e.g., *Davies,* 1990; *van Keken and Ballentine,* 1998). However, these ideas (discussed further below) have not yet been incorporated into complete models of rare gas distributions and have not been expanded to include the heavy rare gases.

### 4.1.  Rare Gases in the Core

Possible processes responsible for initially sequestering rare gases into the core and then releasing them have not been fully explored. The initial step of incorporating rare gases depends upon the specific processes of core formation. Models for segregation of the core include (1) segregation of metals into cores on protoplanetary bodies (e.g., *Taylor and Norman,* 1990) under low-pressure conditions and rapid coalescence of these cores during planetary accretion; (2) separation and percolation of metal through the solid lower mantle (*Stevenson,* 1990); and (3) ponding of separated metal at the base of a magma ocean, followed by diapiric downflow that may not involve reequilibration with the deeper mantle (e.g., *Stevenson,* 1990; *Righter et al.,* 1997). Therefore rare gases may partition into liquid metal from either solid or liquid silicates, and under either high or low pressures. The only relevant partitioning data available is for liquid silicate and metal by *Matsuda et al.* (1993), who found metal/silicate partition coefficients of $10^{-1}$–$10^{-2}$ for rare gases at 0.5 GPa, and lower values at increased pressure. *Stevenson* (1985) calculated the solubilities of rare gases in liquid metal at core pressures and indicated that Xe may be soluble. However, *Caldwell et al.* (1997) have shown experimentally that Xe does not alloy with iron at core-mantle boundary pressures. Alternatively, *Jephcoat* (1998) suggested that Xe may form dense solid inclusions in the deep mantle and segregate toward the core, thereby generating a reservoir of Xe complementary to that of the atmosphere. However, it is not clear if such a separate phase of an element of such low abundance can form in the mantle.

Rare gases will be subsequently released from the core by partitioning into the overlying silicates that are either solid or melted (*Williams and Garnero,* 1996). A constraint on the processes is provided by the present approximately solar He/Ne ratio of the mantle. If gases originally incorporated into Earth were of solar composition, then the net result of the processes of incorporation in the core and release must not result in large fractionation. It is unlikely that each step will be nonfractionating; it is only if the partitioning during incorporation and release occur with the same partition coefficients (probably requiring the same pressure, i.e., the core/mantle boundary, and between the same phases, i.e., solid or liquid silicates) that the net result will be unfractionated. The alternative for expelling rare gases from the core is if rare gas solubilities are exceeded during crystallization of a core initially near saturation. However, it is not plausible that the saturation limits for both He and Ne have been reached; the solar ratio is $^4He/^{20}Ne \approx 820$ (*Pepin,* 1999) and the saturation limits are unlikely to be in the same proportions.

## 4.2. Subduction of Rare Gases

It is possible that atmospheric rare gases incorporated in subducting materials are carried into the mantle and mixed with primordial constituents. Holocrystalline MORB and oceanic sediments have atmospheric rare gases that are greatly enriched in the heavier rare gases (*Dymond and Hogan,* 1973; *Podosek et al.,* 1980; *Matsuda and Nagao,* 1986; *Staudacher and Allègre,* 1988), although concentrations vary by several orders of magnitude and so an average value cannot be easily obtained. Also, subduction zone processing and volcanism may return much of the rare gases to the atmosphere. However, it is possible that the total rare gas abundances reaching subduction zones is sufficiently high that subduction into the mantle of only a small fraction may have a considerable impact upon the composition of Ar and Xe in the upper mantle (*Porcelli and Wasserburg,* 1995a,b). *Staudacher and Allègre* (1988) argued that subducting Xe is almost completely lost back to the atmosphere during subduction zone volcanism, although this was based upon the model-dependent requirement for preservation of a distinctive Xe-isotopic composition within the upper mantle throughout Earth history. Thus subduction of atmospheric Ar and Xe remain possible.

*Anderson* (1993) has advocated subduction of interplanetary dust material in seafloor sediments as a source of mantle $^3He$. However, the amount of material being deposited is presently several orders of magnitude too low to account for the flux of $^3He$ released at mid-ocean ridges, and it has been argued that it has been insufficient in the past as well (*Trull,* 1994). Furthermore, laboratory measurements indicate that He is lost from dust particles by diffusion at low temperatures (*Hiyagon,* 1994) and so will be lost during subduction. *Allègre et al.* (1993) pointed out that the He/Ne ratio of meteoritic dust is much lower than that of the mantle, and argued that while the Ne in the mantle could be supplied by this mechanism, the He would require another source. However, the flux of extraterrestrial material to Earth is also too low to supply the necessary Ne (*Stuart,* 1994). Thus this mechanism is no longer considered a possible dominant source of mantle rare gases (*Farley and Neroda,* 1998).

## 4.3. Models of Mantle Rare Gas Distributions

Several models have been advanced regarding the evolution of mantle rare gases that have been formulated to match at least some of the isotopic data available. While there are difficulties with each, these models can serve as guides to the possible effects of the processes considered in them.

*4.3.1. The lower mantle.* Models for the distribution of rare gases often include the assumption that the lower mantle is a distinct reservoir for rare gases where storage over the age of Earth occurs, and provide constraints on the initial characteristics of the terrestrial rare gas budget. The lower mantle is typically identified with the volume beneath the 670-km seismic discontinuity, and so comprises ~75% of the mantle (e.g., *Allègre et al.,* 1983, 1986; *Porcelli and Wasserburg,* 1995a,b). This is based on models of layered convection that isolates the lower mantle from near-surface processes. However, it is widely debated whether the lower mantle is isolated by a convective boundary at 670 km from the more depleted upper mantle (see review by *Davies,* 1998). It has been suggested that whole-mantle convection occurs and long-term storage of rare gases takes place at greater depths than 670 km in a reservoir that has remained more isolated due to higher viscosities at such depths (e.g., *Davies,* 1984; *O'Connell,* 1995; *Manga,* 1996). Alternatively, it has been argued that a compositional boundary at ~1600 km depth may isolate deeper regions (*Kellogg et al.,* 1999; *van der Hilst and Karason,* 1999). However, comprehensive models of rare gas distributions compatible with geophysical models that involve convective patterns other than the layered mantle remain to be developed. Overall, the volume of material that has not been degassed and has escaped near-surface processing is not well constrained. It should be noted that while "less degassed" or "undegassed" mantle domains are likely to be in the deeper part of the mantle, equating these with the term "lower mantle" that identifies the entire mantle deeper than 670 km remains model-dependent. Nevertheless, it is still useful to consider what the average properties of the lower mantle are in contrast to those of the upper mantle, which is the depleted region sampled at mid-ocean ridges and extends to a depth of 670 km.

The lower mantle is often assumed to be approximately a closed system with respect to both rare gas and parent nuclides. Assigning the highest OIB $^3He/^4He$ ratios to the lower mantle, a comparison between the total production of $^4He$ and the shift in $^3He/^4He$ from the initial terrestrial value to the present value provides a $^3He$ concentration. Then using the isotopic shifts discussed above and a production ratio of $^{21*}Ne/^4He = 9 \times 10^{-8}$, *Porcelli and Wasserburg* (1995b) obtained a concentration of Ne in the lower mantle of $2 \times 10^{11}$ atoms $^{20}$Ne/g. [Using the lower production ratio of *Yatsevich and Honda* (1997) results in a 50% lower Ne concentration.] These lower mantle concentrations can be compared to the concentration of $1.8 \times 10^{12}$ atoms $^{20}$Ne/g obtained by dividing the atmospheric inventory by the mass of the upper mantle; a closed system lower mantle therefore has much lower rare gas concentrations than the solid Earth reservoir that could have supplied the atmosphere. Lower mantle nonradiogenic Ar- and Xe-isotopic concentrations, and so isotopic compositions, cannot be directly calculated without assuming either lower mantle Ar/ Ne and Xe/Ne ratios or Ar- and Xe-isotopic compositions. For example, a closed system lower mantle with $^{40}$Ar/$^{36}$Ar $\geq 3000$ (see above) and K concentration of 254 ppm has $^{36}$Ar $\leq 1.9 \times 10^{11}$ atoms/g and so $^{20}$Ne/$^{36}$Ar $\geq 1$; therefore $^{20}$Ne/$^{36}$Ar ratios that are solar or greater can be accommodated. Note that it is possible to obtain high $^{40}$Ar/$^{36}$Ar ratios in the lower mantle without degassing; i.e., Ar in the lower mantle that is more radiogenic than the atmosphere is not evidence that partial degassing of the deep mantle has occurred (see *Porcelli and Wasserburg*, 1995b). Such statements have been based on the assumption that the lower mantle originally must have had the same $^{36}$Ar concentration as the source of the atmosphere.

Note that if rare gases are partially lost from the lower mantle reservoir preferentially to U (e.g., *Allègre et al.,* 1986) the initial $^3$He concentration is likely to be greater than that calculated above for a closed system. Processes such as the introduction of subducted crust that is highly depleted in both U and He into the lower mantle will reduce the present lower mantle concentration of these elements (*Albarède*, 1998) but will not change the calculated initial He concentration.

As an alternative to assigning high OIB $^3$He/$^4$He ratios to an isolated lower mantle enriched in He, *Anderson* (1998) argued that the high $^3$He/$^4$He ratios measured in ocean islands such as Loihi represents the isotopic composition of the upper mantle at the time the oceanic lithosphere in this area was formed; the $^3$He/$^4$He ratio of He trapped in the depleted part of the lithosphere, which is highly depleted in U and Th, would not be altered subsequently and is sampled by later volcanism. However, a reasonable history of the upper mantle has not been formulated in which the $^3$He/$^4$He ratio evolves from an initial value of ~120 $R_A$ to the present 8 $R_A$ while having a value as high as 32 $R_A$ <100 m.y. ago.

*4.3.2. Residual upper mantle model.* It has often been assumed that the atmosphere formed by degassing of the solid Earth (*Brown*, 1952). Early rare gas models (Fig. 6) assumed that upper mantle rare gases are residuals after extraction of the atmosphere (*Hart et al.,* 1979; *Allègre et al.,* 1983, 1986). Starting with a mantle initially uniform in rare gas and parent isotope concentrations, the upper mantle is largely degassed to form the atmosphere. The $^{40}$Ar/$^{36}$Ar, $^{129}$Xe/$^{130}$Xe, and $^{136}$Xe/$^{130}$Xe ratios in the upper mantle are determined by the rates of degassing to the atmosphere relative to the production of $^{40}$Ar, $^{129}$Xe, and $^{136}$Xe. Such models generally suppose that degassing occurs without elemen-

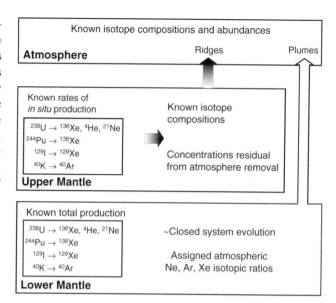

**Fig. 6.** The reservoirs and fluxes in the residual upper mantle model (*Allègre et al.,* 1986). An Earth originally uniform in U, Th, K, I, Pu, and rare gases is divided into two mantle reservoirs. The lower mantle has been essentially a closed system, and has evolved Ar- and Xe-isotopic compositions that are similar to those of the atmosphere. Rare gases in the upper mantle are residual from extraction of the atmosphere, which occurred largely early in Earth history and continued to the present. The upper mantle evolved isotopically due to this degassing as well as the addition of $^4$He, $^{21}$Ne, $^{40}$Ar, $^{129}$Xe, and $^{136}$Xe from production within the upper mantle. This reservoir does not contain any subducted gases, and is generally isolated from the lower mantle, although diffusion of He upward may occur.

tal fractionation, so that the isotopic evolution of the rare gases can be considered together. The strongly radiogenic character of measured upper mantle Ar and Xe leads to the conclusion that the atmosphere must have formed very early, prior to extinction of $^{129}$I and before most of the $^{40}$Ar was generated. It is assumed that no subduction of atmospheric rare gases back into the mantle has occurred. The lower mantle has evolved essentially as a closed system (as discussed above), and contains He with high $^3$He/$^4$He ratios as measured in OIB along with bulk silicate Earth concentrations of other rare gases. Since in this model the atmosphere plus the upper mantle has concentrations of all isotopes equal to those of the bulk Earth, the lower mantle must then have Ar and Xe with atmospheric characteristics. Mid-ocean ridge basalt $^{136}$Xe excesses are interpreted as due to the decay of $^{238}$U and the possible role of $^{244}$Pu has not been considered (*Allègre et al.,* 1983, 1986).

The important model conclusions are that (1) very radiogenic Ar and Xe in MORB required early degassing; and (2) lower mantle Ar- and Xe-isotopic compositions are like air. If OIB $^{40}$Ar/$^{36}$Ar ratios greater are than air, the model can be revised to include partial degassing of the lower mantle (*Allègre et al.,* 1986).

Several objections to the above model can be raised: (1) The difference between MORB and air $^{20}$Ne/$^{22}$Ne ratios indicates either that Ne was not initially uniformly distributed in Earth or that the Ne in the atmosphere has been modified by losses to space after degassing from the mantle. *Marty and Allé* (1994) have examined the mantle Ne isotope budgets in detail. (2) Distinctive radiogenic Xe-isotopic ratios in the upper mantle, established by degassing early in Earth history, would be obliterated by contamination from plumes rising from the gas-rich lower mantle (*Porcelli et al.*, 1986). (3) A plausible scenario for establishing an initially uniform distribution of rare gases with the characteristics of the present atmosphere has not been advanced (see further discussion below). (4) The global $^4$He flux at mid-ocean ridges is equal to production in the upper mantle, suggesting that the upper mantle $^4$He concentration is in steady state and so requires transfer of $^3$He from the lower mantle (*O'Nions and Oxburgh*, 1983). This transfer most plausibly occurs by advection of mantle material and therefore includes all the rare gases. The possibility that transfer of $^3$He from the lower mantle occurs by diffusion (*Allègre et al.*, 1986), so that the principles of the residual mantle model still can be applied to the other rare gases, has not been shown to be reasonable for maintaining the present global flux. (5) *Ozima et al.* (1985) noted that the ratio of $^{244}$Pu-derived $^{136}$Xe to radiogenic $^{129}$Xe within a residual upper mantle reservoir must be greater than that of the atmosphere extracted from that reservoir due to the greater half life of $^{244}$Pu. This would require that most of the radiogenic $^{136}$Xe in MORB is derived from $^{244}$Pu; data from MORB (*Kunz et al.*, 1998) and mantle-derived rare gases found in $CO_2$ gas fields (e.g., *Phinney et al.*, 1978; *Caffee et al.*, 1999) indicate that this may not be the case. Overall, although the conclusions of this model are often taken as firm constraints required by the rare gas data, it now appears that the model generally is not viable.

A variation of the residual upper mantle model by *Zhang and Zindler* (1989) considers degassing of MORB by partitioning of rare gases into $CO_2$ vapor for transport to the surface, with the remaining rare gases in MORB returned to the mantle. Due to different solubilities, there is fractionation between rare gases in the residual MORB and those lost to the atmosphere. The net result of this partial gas loss from MORB is fractionation during mantle degassing. However, there are difficulties with this model process. The model predicts that the $^3$He/$^{22}$Ne ratio of the upper mantle is twice that of the lower mantle; although this may be possible (*Honda and McDougall*, 1998), the data are too imprecise to make a firm prediction. The model also predicts that the upper mantle Xe/Ar ratio is less than that of air, which does not appear to be true (see above). Also, the original model was based on the assumption that lower mantle Ar- and Xe-isotopic compositions are equal to those of air, and has not been updated. A major consideration is that there is no evidence that a residual component is preserved in subducting crust that has not been degassed by hydrothermal alteration, a requirement for net fractionation between the mantle and atmosphere. The objections raised above also pertain to this variant of the residual mantle model. Nevertheless, this bubble-degassing mechanism may

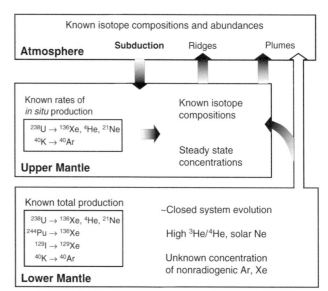

**Fig. 7.** The reservoirs and fluxes in the steady state model (after *Porcelli and Wasserburg*, 1995b). An Earth originally uniform in U, Th, K, I, and Pu (but not rare gases) is divided into two mantle reservoirs. The lower mantle has been essentially a closed system, and contains initial rare gases and radiogenic rare gas nuclides. The upper mantle is open to flows of rare gases from the lower mantle through incorporation of undegassed material rising in plumes, and from the atmosphere in subducted material. Radiogenic $^4$He, $^{21}$Ne, $^{40}$Ar, and $^{136}$Xe are added by production within the upper mantle. This reservoir is degassed in mid-ocean ridge and ocean island volcanism. Rare gas isotopic compositions of the upper mantle (constrained from MORB analyses) reflect the balance of inputs to that reservoir. For the lower mantle, Ar- and Xe-isotopic compositions are unknown and are calculated from the upper mantle balance, while the He- and Ne-isotopic compositions are obtained from OIB data.

have been important in the past during an early magma-ocean stage of terrestrial history (*Ozima and Zahnle*, 1993). At this stage it can account naturally for the preferential outgassing of Xe relative to lighter noble gases that appears to be required in models of atmospheric isotopic evolution by hydrodynamic escape (*Tolstikhin and O'Nions*, 1994; *Pepin*, 1997; see below).

*4.3.3. Steady-state upper mantle.* Other sets of models (Fig. 7) are based upon the open interaction between the upper mantle and both the lower mantle and atmosphere (*O'Nions and Oxburgh*, 1983) and have been developed most extensively by *Kellogg and Wasserburg* (1990) and *Porcelli and Wasserburg* (1995a,b,c). The upper mantle inventory is seen not as continually depleting, but rather as the result of interactions among surrounding reservoirs. In addition to the upper mantle inputs by radiogenic production and outflows by degassing at mid-ocean ridges and hotspots considered in other models, atmospheric Ar and Xe are subducted into the upper mantle, and lower mantle rare gases are transported into the upper mantle within mass fluxes of upwelling material. The upper-mantle isotope compositions are set from available MORB data and are the result of mixing of subducted and lower-mantle rare gases with radionuclides produced *in situ* at known rates from decay of U, Th, and K. The isotopic systematics of the different rare gases are linked by constraints on the relative production rates, and by the assumption that transfer of rare gases from the upper mantle to the atmosphere by volcanism, as well as the transfer of lower mantle rare gases into the upper mantle by bulk transfer, occurs without elemental fractionation. By explicitly incorporating the additional interactions between reservoirs, the objections to a residual upper mantle are overcome. It is assumed that upper mantle concentrations are in steady state, so that the inflows and outflows are equal; therefore, there are no time-dependent functions (and so additional free parameters) determining upper mantle concentrations. The effects of the time dependence of rare gas transfers into the upper mantle have been explored by *Tolstikhin and Marty* (1998) and *Kamijo et al.* (1998).

The relationship between rare gases in the atmosphere and in the lower mantle is not assumed *a priori*, nor is the concentration of lower mantle rare gases. The lower mantle is modeled as evolving as an approximately closed system (i.e., upward fluxes are small) with bulk silicate Earth parent element concentrations that are used to establish the concentrations of radiogenic nuclide concentrations. It has the high $^3He/^4He$ and $^{20}Ne/^{22}Ne$ ratios measured in ocean island basalts that are used to calculate the nonradiogenic He and Ne concentrations. Also, the isotopic shifts from the lower-mantle to upper-mantle compositions that are created by the known upper-mantle production rates of $^4He$ and $^{21}Ne$ can be used to determine the rate of He and Ne transfer from the upper mantle. An important aspect of the model is that the lower mantle is the storage reservoir of $^{129}I$-derived $^{129}Xe$ and $^{244}Pu$-derived $^{136}Xe$ ($^{238}U$-derived $^{136}Xe$ is relatively negligible). However, the lower mantle Ar and Xe isotope compositions cannot be well constrained by available data (see section 3). While the total fluxes of nonradiogenic Ar and Xe into the upper mantle can be calculated (like those of He and Ne), some undetermined proportion of these fluxes is due to subduction of atmospheric gases. The remainders are derived from the lower mantle and accompany the fixed lower mantle He flux. Therefore the Ar and Xe concentrations in the mass flux from the lower mantle, as well as the calculated Ar/He and Xe/He ratios, are dependent upon the subducted flux, with a range of values possible. The possible lower-mantle elemental abundance patterns therefore must be deduced from the plausible mechanisms for initial rare gas incorporation (see section 5).

Mechanisms operating in the model are more clearly seen by examining the systematics of Xe isotopes. This is shown in Fig. 8. The minimum upper-mantle $^{136}Xe/^{130}Xe$ and $^{129}Xe/^{130}Xe$ ratios in the upper mantle are constrained by available MORB data. The lower mantle ratios are established early in Earth history by decay of $^{129}I$ and $^{244}Pu$; however, while the $^{129}Xe/^{130}Xe$ ratio is constrained to be at least as great as that in the upper mantle, higher values are possible due to lower $^{130}Xe$ concentrations in the lower mantle. Once lower-mantle Xe is transported into the up-

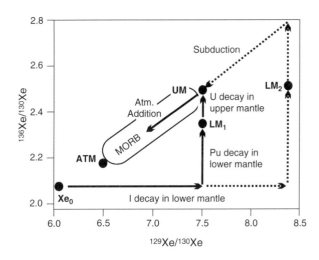

**Fig. 8.** The general pattern of mantle Xe-isotopic evolution in the steady-state model (after *Porcelli and Wasserburg,* 1995a). The present upper-mantle composition is taken as UM, the highest MORB value. Within the lower mantle, $^{129}Xe$ and $^{136}Xe$ are added to originally incorporated Xe by decay of $^{129}I$ and $^{244}Pu$, to produce the present lower-mantle composition. The magnitude of the isotopic shift is determined by the lower-mantle $^{130}Xe$ concentration; the minimum values are represented by $LM_1$, although in the absence of constraints on the lower-mantle composition higher values (e.g., $LM_2$) are possible. Xenon transferred into the upper mantle with composition $LM_1$ is mixed with $^{136}Xe$ from decay of $^{238}U$ to produce composition UM. If the lower mantle has composition $LM_2$, then composition UM is obtained by mixing of this Xe with $^{238}U$-derived $^{136}Xe$ as well as subducted atmospheric Xe. Therefore, there is a balance between the amounts of subducted Xe and the lower-mantle composition. Note that the compositions $LM_1$, $LM_2$, etc., define a narrow range of lower-mantle radiogenic $^{136}Xe$ /$^{129}Xe$ ratios.

per mantle, it is augmented by U-derived $^{136}Xe$. Subduction of atmospheric Xe is also possible, lowering the $^{136}Xe/^{130}Xe$ and $^{129}Xe/^{130}Xe$ ratios to some (unconstrained) extent, so that a range of lower-mantle compositions can be accommodated. There is therefore a trade-off between two presently unknown quantities: the amount of Xe subducted and the lower mantle $^{130}Xe$ concentration (and so $^{129}Xe/^{130}Xe$ and $^{136}Xe/^{130}Xe$ ratios). If a specific Xe/He ratio for the lower mantle is chosen (e.g., the solar ratio), then a Xe concentration can be calculated from the He concentration (which is not subducted) along with the amount of subducted Xe. Note that the possible lower mantle compositions fall on a trend of $^{136}Xe/^{130}Xe$ and $^{129}Xe/^{130}Xe$ ratios, and so with a limited range of $^{136*}Xe/^{129*}Xe$ ratios. In the model, ~50% of the radiogenic $^{136*}Xe$ in the upper mantle is derived from the lower mantle, where it was produced largely by $^{244}Pu$; this is compatible with the data of *Kunz et al.* (1998).

The important model conclusions are that (1) lower-mantle Xe has high $^{129}Xe/^{130}Xe$ and $^{136}Xe/^{130}Xe$ ratios due to decay of $^{129}I$ and $^{244}Pu$; (2) upper-mantle radiogenic Xe-isotopic enrichments are inherited from the lower mantle, with $^{136}$Xe substantially augmented in the upper mantle by production from $^{238}$U, and thus radiogenic $^{129*}$Xe is associated with a substantial fraction of $^{136*}$Xe that is $^{238}$U-derived (see Fig. 8); (3) upper-mantle nonradiogenic Xe abundances can be dominated by atmospheric gases subducted into the mantle without overwhelming $^{129}$Xe enrichments (but lowering the $^{129}$Xe/$^{130}$Xe and $^{136}$Xe/$^{130}$Xe ratios); (4) the calculated residence time of rare gases in the upper mantle is short (~1.4 G.y.) and requires that long-term storage of rare gases (including solar He and Ne) occur in the lower mantle; (5) the lower mantle has elemental ratios that are distinctive from atmospheric values and could be solar (*Burnard et al.,* 1997); and (6) upper-mantle characteristics are determined by the present fluxes, and do not constrain early mantle degassing history.

There are several issues with the model that must be resolved: (1) Correlations between He and Pb or Sr in OIB sample suites that would be expected to reflect mixing of the two mantle rare gas components suggest that the two components have comparable He concentrations. Since a closed system lower mantle is expected to have a ~$10^2$× greater concentration of He, this requires some prior degassing of this component prior to mixing to generate the sampled basalts (*Hilton et al.,* 1997; *Eiler et al.,* 1998). This must be confirmed. (2) Geophysical arguments have been advanced for greater mass exchange with the lower mantle, although a recent change in convection pattern has been suggested (see *Allègre,* 1997). (3) Other geochemical evidence has been interpreted to indicate that much of the mantle has been processed and so largely degassed (e.g., *Davies,* 1984; *Blichert-Toft and Albarède,* 1997). All these considerations generally affect the volume and the assumption of a closed-system history of the lower mantle, and may still be reconciled with an open, interacting upper mantle.

An important implication of the model is that Xe derived from $^{129}$I and $^{244}$Pu is from the deep Earth. The ratios of $^{129*}$Xe/$^{130}$Xe and $^{136*}$Xe/$^{130}$Xe in the MORB source can be estimated by subtracting the highest measured MORB $^{129}$Xe/$^{130}$Xe and $^{136}$Xe/$^{130}$Xe values (*Kunz et al.,* 1998) from those of the initial, nonradiogenic mantle Xe composition. If the atmospheric Xe composition is used as the initial composition, the ratio of $^{129*}$Xe to $^{136*}$Xe in MORB is 3.0; if the nonradiogenic atmospheric Xe composition is used (see section 2.2), the ratio is 3.3. The model places these radiogenic Xe isotopes derived from $^{244}$Pu and $^{129}$I in the deep mantle, and their relative abundances can be used to calculate a "Xe closure age" of the lower mantle using equation (1). Assuming that all the $^{136*}$Xe in MORB is derived from $^{244}$Pu in the lower mantle, then an age of ~126 Ma is obtained; if 50% of the $^{136*}$Xe in MORB is derived from $^{244}$Pu (based on the model calculations of *Porcelli and Wasserburg,* 1995b), then ~106 Ma is obtained. These results indicate that the deep Earth reservoir in which this Xe is stored suffered Xe losses early in Earth history that resulted in substantial depletion of radiogenic $^{129*}$Xe and plutogenic $^{136*}$Xe. There is no evidence for the accumulation of this radiogenic Xe elsewhere in Earth, so the losses must have been to space. The calculated time period for Xe

loss is comparable to that calculated for the atmospheric reservoir, suggesting that the mechanisms responsible affected both shallow and deep terrestrial reservoirs. Note that the residual upper mantle model assumes that the lower mantle has the same Xe-isotopic composition as the atmosphere, and therefore also requires similar Xe losses from both the lower mantle and the atmosphere.

## 5. ACQUISITION OF TERRESTRIAL RARE GASES

The abundances of rare gases both within Earth and in the atmosphere provide the lower limit to the quantities that acquisition processes must provide to the accreting planet or its source materials. These processes may account for some portion of the observed elemental and isotopic fractionations relative to solar nebula gases, or must be coupled to modification mechanisms associated with subsequent planetary losses. As a benchmark, the abundance of $^{20}$Ne in the atmosphere divided into the mass of the upper mantle (the minimum mass with which these gases were associated) is $1.8 \times 10^{12}$ atoms $^{20}$Ne/g. A substantially higher lower limit for the initial Earth inventory could be inferred from the high inventories on Venus, on the assumption that neighboring and roughly equal-mass planets initially acquired their noble gases by similar mechanisms and in comparable amounts. The atmosphere of Venus contains ~$8.7 \times 10^{12}$ atoms $^{20}$Ne/g-planet, and there may be an additional, but presently unconstrained, fraction retained within the Venus interior.

For the deep Earth reservoir associated with high $^3$He/$^4$He ratios, the $^{20}$Ne concentration calculated for the lower mantle (see section 4) of $2 \times 10^{11}$ atoms $^{20}$Ne/g provides a minimum present value. The Xe found in MORB that has been generated by $^{129}$I and $^{244}$Pu decay likely has been stored in the same deep Earth reservoir, and reflects substantial losses of radiogenic Xe over a period of ~100 m.y. (see section 4). Such losses must be accompanied by loss of the nonradiogenic Xe isotopes. *Porcelli et al.* (1998) calculated that at least 99% of the Xe must have been lost to obtain the late "formation ages" calculated above. These losses plausibly affected all the rare gases. Therefore, the present concentration of lower mantle rare gases was established after depletion by at least a factor of ~$10^2$, and so at least $2 \times 10^{13}$ atoms $^{20}$Ne/g must have been acquired initially. It is this concentration that must be provided within Earth by any acquisition mechanism considered plausible. The concentration of other rare gases depends upon the inferred deep Earth elemental abundance pattern, which presently is unconstrained (see section 3) and so is model-dependent. It may not be coincidental that the initial $^{20}$Ne inventory of Venus has also been calculated to have been ~$2 \times 10^{13}$ atoms $^{20}$Ne/g, prior to modest depletion by hydrodynamic escape (*Pepin,* 1997; see also section 6).

The possible mechanisms that have been proposed with which protoplanetary and planetary solids acquire rare gases can be conveniently related to the phases of mass accumulation in the nebula. When solids initially are predominantly small grains, rare gases can be acquired by solar wind implantation or adsorption from the surrounding gases at low ambient nebular pressures. When larger planetesimals have accumulated, gravitational capture of nebular gases can occur, and volatiles can be incorporated into the interior either by burial of surface-adsorbed species as the planetesimals grow or by dissolution into silicate melt. Bodies approaching present planetary masses can more effectively accumulate remaining nebular gases by gravitational capture and can suffer late infall of gas-rich materials from the outer portion of the nebula. These processes are discussed in more detail below, and discussion of the origin of rare gases on other planets are given in *Pepin* (1991, 1992).

## 5.1.  Solar Wind Implantation into Grains

The implantation of solar wind rare gases into dust grain planetary precursors has been proposed as a mechanism for providing atmospheric rare gases to the terrestrial planets (*Wetherill*, 1981; *McElroy and Prather*, 1981). Noble gases implanted into lunar and meteoritic dust grains by low-energy solar wind irradiation are known to display solar-like elemental abundance ratios and isotopic compositions, with typical $^{20}$Ne concentrations in lunar grains of $\leq 3 \times 10^{16}$ atoms/g (e.g., see *Pepin et al.*, 1970; *Eberhardt et al.*, 1972). Accretion of planetesimals containing at least $10^{-3}$ by mass of this solar-wind-irradiated material can account for the initial terrestrial deep Earth Ne budget. Implanted Ne (and He) can be depleted from grains preferentially by heating, or lost even at relatively low temperatures by rapid diffusion from certain mineral structures, notably plagioclase (e.g., *Frick et al.*, 1988). It may be that the Ne depletion in the atmosphere relative to Ar is due to such Ne loss; then a correspondingly greater proportion of these grains is required to provide sufficient Ne. A difficulty with this source is in having sufficient irradiation of grains by penetration of solar wind ions outward into the solar system while nebular gases were still present. However, implantation may have occurred in secondary dust created by collisions of larger bodies after dispersal of the nebular gases (*Wetherill*, 1981), or into grains located on the margins of the nebular disk (*Sasaki*, 1991); this mechanism then may be a possible source of terrestrial rare gases if sufficient material is irradiated. Subsequent processes must then generate the present terrestrial atmospheric isotopic compositions, while rare gases trapped within Earth would be isotopically approximately solar.

Accounting for deep Earth rare gases with this mechanism requires retention of rare gases within growing planetesimals during continuing accretion. There is evidence for early differentiation and core segregation in planetesimals (*Gaffey*, 1990; *Taylor and Norman*, 1990), and this likely resulted in degassing of the solid bodies. Retention of rare gases in the planetesimal atmospheres will have depended upon the masses of the bodies and the extent of blow-off during further aggregation. However, the interiors are likely to have maintained far lower rare gas concentrations than those characterizing precursor grains.

## 5.2.  Adsorption of Nebular Gases on Grains

Laboratory studies have shown that noble gases exposed to some finely divided solid materials are adsorbed on or within the surfaces of individual grains. Adsorption is most efficient for various forms of carbon, where it appears to reflect intrinsic structural properties (e.g., *Frick et al.*, 1979; *Niemeyer and Marti*, 1981; *Wacker*, 1989), but has also been experimentally demonstrated for other minerals (*e.g. Yang et al.*, 1982) and for polymineralic meteorite powder (*Fanale and Cannon*, 1972). Moreover, the process occurs in sedimentary materials (see *Bernatowicz et al.*, 1984). Adsorbed gases on these substrates generally display elemental fractionation patterns, relative to ambient gas-phase abundances, in which heavier species are enriched. For the most part the fractionations are remarkably uniform, considering the wide range of experimental and natural conditions under which they are produced. Occasional isotopic effects have been reported in natural samples (*Phinney*, 1972), but have not been observed in equilibrium adsorption on solid silicate materials in the laboratory (*Bernatowicz and Podosek*, 1986). However, *Notesco et al.* (1999) have reported isotope fractionation, most clearly seen in Ar, in a study of heavy noble gas trapping in amorphous water ice at temperatures of ~30–80 K. Also, isotopic fractionation effects are well documented in experimental environments involving ion implantation from plasmas rather than simple adsorption of neutral gases (e.g., *Frick et al.*, 1979; *Bernatowicz and Fahey*, 1986; *Bernatowicz and Hagee*, 1987; *Ponganis et al.*, 1997). The decline in solar-normalized noble gas abundance ratios from Kr to Ne in planetary atmospheres and meteorites is qualitatively similar to many of the adsorption fractionation patterns seen in the laboratory and in terrestrial sedimentary rocks. However, laboratory estimates of single-stage gas/solid partition coefficients are too low by orders of magnitude to account for planetary noble gas abundances by adsorption on free-floating dust grains in the open nebula, which therefore is not a likely mechanism for supplying the terrestrial rare gases.

## 5.3.  Gravitational Capture of Nebular Gases

Solar rare gases can be acquired by Earth by direct capture from the nebula. Gravitational capture of a substantial abundance of ambient gases with solar elemental and isotopic compositions will occur if protoplanets grow to appreciable fractions of the present Earth mass prior to dissipation of the nebular gas phase. A difficulty that presently confronts this requirement is the relative timing of nebular dissipation vs. planetary growth. Current estimates of timescales for loss of circumstellar dust and gas are on the order of ~$10^7$ yr or less (see *Podosek and Cassen*, 1994), compared to ~$10^8$ yr for planetary growth to full mass in the standard model of planetary accumulation (*Wetherill*, 1986, 1990). However, nebular lifetimes are inferred from astronomical data, and are based solely on data for the presence of fine dust. Measurements of molecular line emissions that could at least set upper limits on the longevity of the gas-phase component in "naked" T-Tauri disks do not yet exist (*Strom et al.,* 1993). There is no reason to believe that the disappearance of micrometer-sized dust from infrared detectability by accretion into larger grains would necessarily be coincident with gas loss. It is worth considering that planetary accumulation modeling predicts terrestrial planet growth to roughly 80% of final masses within $\sim 2 \times 10^7$ yr (*Wetherill,* 1986), so that if a significant remnant of gas survived in the inner solar accretion disk for only $2 \times 10^7$ yr or so after the clearing of solid material, substantial gravitational capture of nebular gases would have occurred on the proto-Earth. Another constraint comes from Hf-W data; *Lee and Halliday* (1995) found that there is no resolvable difference between the $^{182}W/^{184}W$ ratio of Earth and of chondrites, restricting the time before which Hf and W can be fractionated from one another by core formation (and W sequestration). Assuming that separation of core-forming Fe occurs concomitantly with accretion due to melting during impact, and that accretion diminishes exponentially, *Halliday et al.* (1996) calculate a minimum accretion time constant of 25 m.y. or a delay in core formation of >50 m.y. Further work is required to determine if dispersal of the nebula can extend over such time periods, or if earlier accretion can be reconciled with the W-isotopic data.

Planetary embryos that had grown to $\sim 5-10$ lunar masses in the presence of nebular gas could have captured planetary atmospheres with surface pressures approaching $\sim 10^4-10^6$ times that of the ambient nebula; calculated pressures, however, depend sensitively on atmospheric opacity, which is difficult to estimate (*Pepin,* 1991). Although the atmospheres captured by these small bodies may not provide sufficient terrestrial rare gas abundances in themselves, sufficient quantities of gases may have accumulated within the protoplanetary bodies by gas adsorption on surface materials followed by burial below the surface during continuing accretion (*Pepin,* 1991). Another consideration bearing on the probable volatile-rich nature of planetary embryos is that impact velocities of materials accreting to form these small bodies are generally too low to promote efficient degassing of the impactors themselves. Consequently their volatiles also tend to be buried within the growing embryos (*Tyburczy et al.,* 1986). If rare gases were acquired by these mechanisms, atmospheric formation would then occur by subsequent degassing and isotopic fractionation during loss to space. Rare gases trapped within Earth and forming the present deep mantle rare gases would exhibit solar isotopic compositions but elemental fractionation; for example, adsorption in laboratory experiments strongly fractionates ambient Ne and Ar (in favor of Ar) by factors of $\sim 10-50$. Evidence of such a deep Earth abundance pattern coupled with solar isotopic compositions would be a strong indicator of the role of adsorption early in Earth's accretional history (*Pepin,* 1998). Venus may have something to say about this question. Its atmospheric $^{36}Ar/^{20}Ne$ ratio is presently $\sim 4.8$, and would have been initially $\sim 3.4$ if isotopic and elemental fractionation occurred due to hydrodynamic escape (see section 6; *Pepin,* 1997); these are respectively factors of $\sim 230$ and 160 above the solar ratio of $2.1 \times 10^{-2}$.

Note, however, that although a $\sim 160 \times$ enhancement is certainly suggestive of adsorption, it is roughly a factor of 6 above the mean Ar enrichment of $\sim 25$ seen in laboratory experiments (data summarized in *Pepin,* 1991). Moreover, even with a Mars-sized terrestrial embryo and the optimistically high value of $2 \times 10^{-5}$ for the $^{20}Ne$ Henry's constant assumed in *Pepin's* (1991) treatment of adsorption, the amount of adsorbed Ne is too low, by about this same factor of 6, to account for an initial deep Earth abundance of $2 \times 10^{13}$ atoms $^{20}Ne/g$ (see above). These observations suggest that if adsorption was responsible for providing the heavier rare gases, another process, such as dissolution into surface magma, would be required to be principally responsible for augmenting the $^{20}Ne$ inventory above the level supplied by adsorption.

Another mechanism for capture of nebular gases by growing planetesimals was originally suggested by *Ozima and Nakazawa* (1980) and redeveloped and extended by *Zahnle et al.* (1990b) and *Ozima and Zahnle* (1993). Nebular gases are gravitationally segregated within the interconnected pore space of accreting bodies due to the dependence of scale height on species mass. The process terminates when growth inhibits further diffusive equilibration and loss to space. Isotopic fractionation of nebular Xe to obtain the terrestrial atmospheric composition can be achieved, although generation of the atmospheric Xe abundance requires retention of Xe within the planetesimal by adsorption as well (*Zahnle et al.,* 1990b). Fractionations of Kr and Ar isotopes in this process are given by *Ozima and Zahnle* (1993) as functions of planetesimal radius. However, while these are in the right directions, it is not clear that terrestrial isotopic compositions of all three noble gases can be generated from solar compositions, for any distribution of planetesimal masses accreted by Earth.

A consequence of this mechanism is that the atmospheric rare gas characteristics are established in accreting materials, so that rare gases presently within the deep Earth are predicted to have the same characteristics. Since this is the only model to predict isotopic ratios within Earth that are fractionated relative to solar, further constraints on mantle rare gases may determine the applicability of this mechanism. Also, this source is expected to supply isotopically similar Xe to Earth, Mars, and Venus (*Zahnle et al.,* 1990b), and since Earth and Mars are characterized by significantly different heavy Xe-isotopic ratios, additional Xe modification processes are therefore required. Analysis of Xe on Venus, for which no data presently exist, will provide a further test.

### 5.4. Gravitational Capture and Dissolution into Molten Planets

If Earth reached sufficient size in the presence of the solar nebula, a massive atmosphere of solar gases would have been gravitationally captured and supported by the luminosity provided by the growing Earth, and the underlying planet would have melted by accretional energy and the blanketing effect of the atmosphere (*Hayashi et al.,*

1979). Gases from this atmosphere would have been sequestered within the molten Earth by dissolution at the surface and downward mixing (*Mizuno et al.,* 1980). This mechanism can provide solar rare gases into the deep Earth with relative elemental abundances that have been fractionated according to differences in Henry's constants for solubility in silicates (with depletion of heavy rare gases). Initial calculations found that at least an order of magnitude more Ne than presently found in the deep mantle could be dissolved into Earth unless the atmosphere began to escape when Earth was only partially assembled (*Mizuno et al.,* 1980; *Mizuno and Wetherill,* 1984; *Sasaki and Nakazawa,* 1990; *Sasaki,* 1999). As noted above, the present abundances may be $10^{-2}\times$ that of initial concentrations, and *Porcelli et al.* (1998) and *Woolum et al.* (1999) considered the conditions required to dissolve sufficient Ne to account for the initial deep mantle inventory prior to losses at ~$10^8$ yr. If it is assumed that equilibration of the atmosphere with a thoroughly molten mantle was rapid, and uniform concentrations were maintained throughout the mantle by vigorous convection, then the initial abundances of gases retained in any mantle layer reflect surface rare gas partial pressures when that layer solidified. The depth of at which solidification occurs is determined by the surface temperature and the efficiency of convection in the molten mantle. Initial distributions of retained rare gases would then be determined by the history of surface pressure and temperature during mantle cooling and solidification, i.e., the coupled cooling of Earth and atmosphere. For typical solubility coefficients (e.g., *Jambon et al.,* 1986; *Lux,* 1987; *Broadhurst et al.,* 1992; *Carroll and Stolper,* 1993; *Shibata et al.,* 1998), a total surface pressure of the order of 100 atm under an atmosphere of solar composition is required to establish the initial deep mantle Ne concentration (*Porcelli et al.,* 1998), along with surface temperatures high enough to melt the deep mantle (~4000 K). Therefore, for this model of the origin of deep He and Ne to be viable, the conditions required to generate the necessary surface pressures to dissolve the required abundances of rare gases and the temperatures sufficient to melt to lower mantle depths must be shown to be plausible. The dense atmosphere is a balance between the gravitational attraction of the nebula-derived gases and expansion due to Earth's luminosity (energy released by accreting planetesimals and the cooling Earth). The nebular temperature and pressure provide boundary conditions, and the atmospheric opacity (which controls the rate of energy loss to space) is a critical parameter. High luminosities (or low opacities) increase the surface temperature but lower the pressure, while decreasing nebular pressures will generate lower surface pressures and temperatures. Therefore the temperature and pressure at the base of the atmosphere evolved as the energy released by accretion declined with time once planet assembly approached completion, and as the nebular pressure declined during nebula dispersal. *Woolum et al.* (1999) demonstrated that the necessary conditions were met under a range of parameter values for both convective and radiative atmospheric structures. However, the complexities in determining the effects of the composition of the lower atmosphere (which was probably strongly contaminated by evaporated terrestrial material) and the transition from optically thick to thin atmospheric conditions presently remain to be resolved.

It should be noted that not all situations facilitate the dissolution of atmospheric gases. At low nebular pressures and high initial luminosities, it is possible that the initial cooling phases occurred when the entire atmosphere was optically thin; in some such cases, rapid magma solidification occurs without the incorporation of significant concentrations of atmospheric gases. However, in the presence of a massive atmosphere which promotes gas dissolution, the mantle cooling time is greatly extended over that which would otherwise occur (*Tonks and Melosh,* 1990). Solidification times then commonly exceed 1 m.y. Overall, it appears that sufficient He and Ne (along with associated Ar, Kr, and Xe) can be incorporated into Earth over a wide range of conditions by this mechanism.

As discussed above, the plausibility of this gravitational capture mechanism depends upon whether a sufficient mass of Earth accretes prior to dispersal of nebular gases, and further work is required to determine if dispersal of the nebula can extend over such time periods.

## 5.5.  Volatile-Rich Planetesimals

Another possible source of inner planet noble gases is accretion of volatile-rich icy comets scattered inward from the outer solar system. Although noble gas distributions in comets are unknown, solar isotopic compositions would be expected. There is evidence that heavier species (Xe, Kr, and Ar) could have been trapped in ice in approximately solar elemental proportions at very low temperatures, and in appreciable concentrations (e.g., *Owen et al.,* 1992). Thus incorporation of a few percent of icy cometary matter into an accreting terrestrial planet could potentially have supplied heavy noble gases of solar composition to its primary atmosphere. Reduction of the Ne/Ar ratio relative to the solar ratio (as seen in the terrestrial atmosphere) is expected in these ices during incorporation of rare gases either by gas occlusion in amorphous ice (*Owen et al.,* 1992) or by clathration (*Lunine and Stevenson,* 1985). A limitation on the cometary contribution to the terrestrial water budget is the D/H ratio of comets, which appear to be 80% higher than that of Earth (e.g., *Bockelée et al.,* 1998), although this does not preclude comets as a major source of terrestrial rare gases (*Owen,* 1996). Infall of comets by scattering from outer solar system orbits is expected to occur after significant accretion has occurred, and so is not likely to be a direct source of deep Earth rare gases. In the cometary model (see chapter by *Owen and Bar-Nun,* 2000) no Ne is trapped in the infalling icy bodies; it is instead supplied to terrestrial planet atmospheres by degassing of another reservoir within the rocky planet. Two intrinsic constraints are therefore that the Ne/Ar ratio in the rock be high enough (e.g., with solar-like Ne/Ar) that this source did not dominate the Ar inventory as well, and that supplies of cometary Ar and rock-derived Ne to both Earth and Venus were well enough balanced between these independent sources on both planets to generate Ne/Ar ratios which differ by less than a factor of 3.

Late infall of CI chondrites has been suggested as a mechanism for supplying both volatiles and siderophiles into Earth after core formation (see, e.g., *Chou,* 1978; *Dreibus and Wänke,* 1989). Although there are similarities in the relative proportions of Ne, Ar, and Kr in the terrestrial atmosphere and chondrites (Fig. 1), neither the abundance nor isotopic composition of Xe can be explained as directly from this source. For the deep Earth, while Ne appears to be solar rather than chondritic, the relative abundance and isotopic composition of Xe is not known; it has been suggested that deep mantle rare gases are a mixture of light solar rare gases and predominantly chondritic heavy rare gases due to earlier incorporation of chondritic material (*Harper and Jacobsen,* 1996), although there is presently no evidence requiring a chondritic component. Calculations of the lower mantle in a steady state mantle model (*Porcelli and Wasserburg,* 1995b) can accommodate a chondritic Xe/Ar ratio, although with a Xe concentration 10× greater than bulk CI chondrites (cf. *Mazor et al.,* 1970). Overall, CI chondrites do not appear to be a viable source for the dominant fraction of terrestrial rare gases.

## 6. MODIFICATION AND LOSS OF TERRESTRIAL RARE GASES

The evidence available and discussed in section 3 is compatible with the incorporation of solar rare gases within Earth and with models for solar rare gases as precursors of the atmospheric gases. However, it seems very clear that modeling scenarios that appeal only to addition of rare gases from solar system sources are intrinsically unable to account for the full range of isotopic variability seen in terrestrial planet atmospheres (section 5). The processes of solar wind implantation, adsorption, and mixing of captured components do not fractionate isotopes, contrary to the strong evidence in atmospheric compositional patterns that such fractionating mechanisms have occurred. Modification of atmospheric rare gases during losses to space then must generate the present characteristics of the atmosphere. For this reason, models that postulate overabundances of solar rare gases on the planets (e.g., gravitationally captured primordial atmospheres) and subsequent dissipation to space appear fundamentally attractive. This offers the potential for isotopic fractionation during losses, and the possibility that the variable distributions of noble gas abundances and isotopes seen in present-day planetary atmospheres may be understood as reflecting different degrees of processing on the individual bodies.

As discussed in sections 2 and 4, the Xe-isotopic data provide a significant constraint upon rare gas loss from the planet, indicating that substantial (>99%) losses of rare gases occurred up to approximately $10^8$ yr after formation of the solar system from both the reservoir containing the gases presently in the atmosphere and the deep Earth source of present mantle rare gases.

### 6.1. Losses During Accretion

As growth of the protoplanet proceeds with increasing accretional energy, shock-induced devolatilization of the accreting materials occurs and volatile species are transferred into the growing atmosphere. Data summarized by *Ahrens et al.* (1989) indicate that efficient loss of $CO_2$ and $H_2O$ from accreting solids on impact occurs once ~10% of the current mass of Earth is reached. As the protoplanet increases in size, degassing will also be promoted by extensive melting due to deposition of accretional energy (*Safronov,* 1978), and further promoted by the blanketing effect of an accumulating water-rich atmosphere (see *Abe and Matsui,* 1986). However, the overall extent of this degassing depends upon the depth and duration of melting, the rate of convection, and the efficiency of degassing of material at the protoplanetary surface.

Loss of atmospheric gases to space can occur by atmospheric erosion, when a sufficient transfer of energy from accreting bodies occurs so that a substantial portion of the protoplanetary atmosphere reaches escape velocity (see *Cameron,* 1983; *Ahrens,* 1993). For smaller accreting bodies, the maximum fraction of the atmosphere that can be expelled is $~6 \times 10^{-4}$ (*Vickery and Melosh,* 1990), equivalent to the total above the plane tangent to the planetary surface at the impact location. However, atmospheric loss may be much greater for very large impacts by >lunar-sized bodies (*Chen and Ahrens,* 1997). It should be emphasized that these impact-driven losses are not expected to generate elemental or isotopic fractionations in the rare gases, and contribute only to the overall depletion of these species.

Impact of an approximately Mars-sized body to form the Moon (e.g., *Hartmann and Davis,* 1975; *Cameron and Ward,* 1976) would clearly result in catastrophic loss of volatiles from the atmosphere and may cause substantial loss of deep Earth rare gases. The W-isotopic data for lunar samples give a single-stage formation age of 53 ± 4 Ma (*Halliday et al.,* 1996; *Lee et al.,* 1997), which is consistent with earlier values obtained using long-lived nuclides of $~10^8$ yr (*Tera and Wasserburg,* 1974; *Wasserburg et al.,* 1977). These values are compatible with the Xe-isotopic "formation ages" for the terrestrial atmosphere and the deep Earth, and so the Moon-forming impact may provide an explanation for late losses of substantial fraction of the terrestrial Xe inventory. The $^{207}Pb$-$^{206}Pb$ age of Earth (*Tera and Wasserburg,* 1974; *Stacey and Kramers,* 1975; *Gancarz and Wasserburg,* 1977) also corresponds with this timescale, and suggests that Pb/U fractionation may have also been occurring during these events.

### 6.2. Hydrodynamic Escape

Hydrodynamic escape appears to be a promising mechanism for dissipation and fractionation of atmophilic species (*Zahnle and Kasting,* 1986; *Zahnle et al.,* 1990a; *Sasaki and Nakazawa,* 1988; *Pepin,* 1991, 1994, 1997; *Hunten et al.,* 1987, 1988, 1989). In this process, a H-rich primordial atmosphere is assumed to be heated at high altitudes after the nebula has dissipated, either by intense far-ultraviolet radiation from the young Sun, or by energy deposited in a large Moon-forming impact. Under these conditions, H escape fluxes can be large enough to exert upward drag forces on heavier atmospheric constituents sufficient to lift them out of the atmosphere at rates that depend on their mole fractions and masses (*Zahnle and Kasting,* 1986; *Hunten et al.,* 1987). Lighter species are entrained and lost with the outflowing H more readily than heavier ones, leading to elemental and isotopic mass fractionation of the residual atmosphere. Therefore this mechanism naturally accounts for the fact that nonradiogenic atmospheric Ne, Ar, Kr, and Xe are all isotopically heavier than their solar counterparts.

It has been shown that hydrodynamic escape of H-rich primary (i.e., gravitationally captured by the planet) atmospheres and degassed volatiles from the terrestrial planets, operating in an astrophysical environment for the early solar system inferred from observation of young star-forming regions in the galaxy, can account for most of the known details of noble gas distributions in their present-day atmospheres (*Pepin,* 1991). In this evolutionary model, a planet acquires two isotopically primordial (or "solar") volatile reservoirs during accretion, one adsorbed from the nebula into planetary embryo materials and the other coaccreted as a primary atmosphere degassed from impacting planetesimals during later planetary growth. Together, these reservoirs provide rare gases with an elemental abundance pattern necessary for obtaining the final atmospheric pattern after escape losses. Subsequent hydrodynamic losses of primary and degassed volatiles are driven by intense EUV radiation from the young evolving Sun. Hydrogen outflow fluxes strong enough to enable Xe escape from Earth, and fractionation to its present isotopic composition, require large but not unrealistic early solar EUV enhancements (by up to ~450× present levels) and atmospheric $H_2$ inventories (equivalent to water comprising a few weight percent of the planet's mass).

One restriction on the conditions required to generate the present atmospheric rare gas characteristics is the necessity of initially separating Xe from other rare gases. During the period when Xe is lost from the atmosphere and isotopically fractionated, the lighter rare gases are severely depleted and fractionated. A subsequent source of these gases, presumably from degassing of interior reservoir(s), is required to resupply the atmosphere when conditions for retention of these gases pertain. This supply must be largely unaccompanied by Xe that has not been fractionated and depleted, which if degassed after the end of fractionating loss to space would overwhelm the Xe signature previously established in the atmosphere. Since Xe is less soluble in basaltic melts than the lighter rare gases, a possible mechanism for this separation is the preferential early degassing of interior Xe into the primordial atmosphere prior to the episode of hydrodynamic loss (*Tolstikhin and O'Nions,* 1994). It should be noted that degassing of fissiogenic Xe to the atmosphere occurs after fractionation, since the iso-topic composition of fissiogenic Xe in the present atmosphere does not appear to be fractionated significantly from that of $^{244}$Pu-fission Xe (see section 2).

For hydrodynamic escape to be driven by EUV radiation, favorable conditions must be satisfied in the solar system environment. The early Sun must provide EUV energy at an intensity $10^2$–$10^3$× the present flux over a time period of 50–200 m.y., and this radiation must penetrate the midplane of the system to planetary distances, implying very low EUV opacity and thus efficient clearing of both nebular dust and gas on this same timescale. There is firm observational evidence that short-wavelength luminosities of young solar-type stars in the ~1–30 × $10^7$-yr age range are several orders of magnitude higher than that of the contemporary Sun, and that with increasing stellar age these enhanced activities decline with an exponential or power-law time dependence (see *Pepin,* 1992). Since early solar EUV radiation could not have penetrated a full gaseous nebula to planetary distances, the applicable time dependence of stellar activity is that following transition of the dense, optically thick accretion disks surrounding the classical T-Tauri stars to more tenuous disks. Infrared observations of these young stellar systems indicate that fine circumstellar dust tends to disappear from inner regions of their accretion disks within ~$10^7$ yr or less (see *Podosek and Cassen,* 1994), and provide estimates of the earliest time for nebular clearing of dust and gas. However, as noted above, while the disappearance of fine dust from IR detectability is reflected in available data, the actual timescales for gas-phase dissipation in these systems is still observationally unconstrained. Although timescales for gas dissipation as well as decay of collisionally generated dust densities to suitably low transmission opacities are very uncertain and a matter of debate (e.g., *Prinn and Fegley,* 1989), it seems likely that a very large flux of solar coronal and transition-region radiation would have penetrated the midplane of the early solar system to planetary distances within <$10^8$ yr. The observational data for T-Tauri and solar-type main-sequence stars (*Simon et al.,* 1985; *Walter et al.,* 1988) scatter considerably from a simple functional dependence of X-ray (~3–60 Å) luminosity on stellar age. *Pepin* (1991) chose an exponential decrease in total EUV radiation with a mean decay time of $9 \times 10^7$ yr, with a flux $10^3$× the present flux at $5 \times 10^7$ yr. This scale of early EUV energy deposition into a H-rich planetary atmosphere would drive an episode of hydrodynamic escape, initially intense but tapering off as the solar EUV luminosity declined with time. However, other radiation histories are possible.

Energy sources other than solar EUV absorption may have powered atmospheric escape. *Benz and Cameron* (1990) suggest that hydrodynamic loss driven by thermal energy deposited in a giant Moon-forming impact could have generated the well-known fractionation signature in terrestrial Xe. Their model of the event calls for rapid invasion of the preexisting primary atmosphere by extremely hot (~16,000 K) dissociated rock and iron vapor, emplacement of an orbiting rock-vapor disk with an inner edge at an altitude comparable to the atmospheric scale height at this temperature, and longer-term heating of the top of the atmosphere by reaccretion of dissipating disk material. *Pepin* (1997), in a variation of his 1991 model, adopted this energy source together with less-intense solar EUV radiation to drive escape of the preimpact atmosphere. This approach assumes that some fraction, probably substantial, of the atmosphere would have been ejected in the immediate aftermath of the collision by impact erosion, i.e., at escape rates for all species too high to have generated significant mass fractionation. Thereafter the system settled into a longer-term state in which dissipation of the remaining impact-deposited thermal energy drove fractionating losses of the residual atmosphere at much lower and mass-specific fluxes. This hypothesis of an early impact-driven fractionation stage on Earth sharply reduces estimates for the initial planetary H inventories needed to support hydrodynamic escape by an order of magnitude or more compared to that required by escape models in which Xe, Kr, and Ar losses from Earth are driven by EUV radiation (*Pepin,* 1991) rather than the energy of a giant impact. Solar EUV flux intensities, now required to be only high enough to enable additional postimpact Ne loss, are reduced by a similar factor. The reduction in H demand occurs because in the hydrodynamic escape formalism its required inventory is proportional to the decay rate of the energy source, and one would expect this to be more rapid for dissipation of impact-deposited energy. Here solar EUV-driven losses play no role in setting the final abundances and isotope ratios of atmospheric Xe, Kr, and Ar. However, Ne is underfractionated in the impact-driven escape episode. Replication of the present-day Ne composition therefore requires a subsequent escape episode, now powered by solar EUV radiation just intense enough for entrainment and loss of Ne but not of heavier species. The waning solar EUV flux may still have been sufficiently high ($\sim$60$\times$ present levels) to drive Ne escape at solar ages of up to $\sim$150–250 Ma (*Ayres,* 1997).

A requirement of the hydrodynamic escape mechanism is that the primordial planetary atmosphere contained a light species, presumably $H_2$, as a dominant constituent, either directly captured, generated by oxidation of metallic Fe by water in the interior, or from photolysis of water vapor. The presence of massive H-rich primordial atmospheres on the terrestrial planets at this time is conjectural. If Earth and Venus accreted to appreciable fractions of their final masses in the presence of nebular gas, such an atmosphere would be gravitationally captured from the solar nebula, although this condition may not have been met (see discussion above). Addition of water amounting to a few weight percent or less of planetary masses by accretion of icy planetesimals could also have supplied noble gases and abundant H, but much more work on modeling the probable fluxes of outer solar system objects into terrestrial planet space during the first $\sim$$10^8$ yr of solar system history is needed to judge whether this source is plausible.

It seems inevitable that some degree of hydrodynamic loss and fractionation of planetary atmospheres would nec-

essarily have occurred if the conditions noted above for energy source, H supply, and midplane transparency were even partially met.

## 7.  CONCLUSIONS

The data for rare gases both in the atmosphere and within the mantle provide important constraints regarding the history of terrestrial volatiles. There is evidence from mantle-derived samples that substantial rare gases inventories were initially trapped within Earth. The isotopic composition of mantle Ne is clearly solar-like, and there is evidence that this may be accompanied by solar Ar and Xe, although this requires further confirmation. Various mechanisms have been proposed for the origin of terrestrial gases with mixed success in accounting for the concentrations in both the atmospheric and deep Earth reservoirs. Trapping of solar gases within the growing Earth from a gravitationally concentrated nebula-derived atmosphere, either by adsorption and occlusion, or by dissolution in a partially molten mantle, may provide large quantities of solar gases. Along with later subduction of rare gases with the present atmospheric composition, these processes can account for the present mantle gases. In order to acquire sufficient atmospheric mass for such trapping to occur, planetary growth in the range of $\sim$10% (for adsorption) to possibly >50% (for dissolution) of the final Earth mass must occur prior to nebula dispersal. Atmospheric rare gases differ greatly from solar gases both isotopically and elementally, and can be generated by two-stage process of hydrodynamic escape of a H-rich isotopically solar atmosphere, which leaves a residual atmosphere preferentially depleted in lighter constituents, and planetary degassing. Radiogenic Xe isotopes indicate that substantial losses occurred late from both the interior and the atmosphere, and the timing of these losses are compatible with a hypothesized Moon-forming impact.

*Acknowledgments.*    Reviews by M. Ozima, K. Righter, and K. Zahnle are greatly appreciated. This is CIT Div. Contr. 8650. Research in Minnesota was supported by NASA grant NAG5-7094 from the Cosmochemistry Program.

## REFERENCES

Abe Y. and Matsui T. (1986) Early evolution of the Earth: accretion, atmosphere formation and thermal history. *Proc. Lunar Planet. Sci. Conf. 17th, J. Geophys. Res., 91,* E291–E302.

Ahrens T. J. (1993) Impact erosion of terrestrial planetary atmospheres. *Annu. Rev. Earth Planet. Sci., 21,* 525–555.

Ahrens T. J., O'Keefe J. D., and Lange M. A. (1989) Formation of atmospheres during accretion of the terrestrial planets. In *Origin and Evolution of Planetary and Satellite Atmospheres* (S. K. Atreya, J. B. Pollack, and M. S. Matthews, eds.), pp. 328–385. Univ. of Arizona, Tucson.

Albarède F. (1998) Time-dependent models of U-Th-He and K-Ar evolution and the layering of mantle convection. *Chem. Geol., 145,* 413–429.

Allègre C. J. (1997) Limitation on the mass exchange between the upper and lower mantle: the evolving convection regime of the

Earth. *Earth Planet. Sci. Lett., 150,* 1–6.

Allègre C. J., Staudacher T., Sarda P., and Kurz M. (1983) Constraints on evolution of Earth's mantle from rare gas systematics. *Nature, 303,* 762–766.

Allègre C. J., Staudacher T., and Sarda P. (1986) Rare gas systematics: formation of the atmosphere, evolution and structure of the Earth's mantle. *Earth Planet. Sci. Lett., 81,* 127–150.

Allègre C. J., Sarda P., and Staudacher T. (1993) Speculations about the cosmic origin of He and Ne in the interior of the Earth. *Earth Planet. Sci. Lett., 117,* 229–233.

Allègre C. J., Hofmann A., and O'Nions R. K. (1996) The argon constraints on mantle structure. *Geophys. Res. Lett., 23,* 3555–3557.

Anderson D. L. (1993) Helium-3 from the mantle: primordial signal or cosmic dust? *Science, 261,* 170–176.

Anderson D. L. (1998) The helium paradoxes. *Proc. Natl. Acad. Sci., 95,* 4822–4827.

Ayres T. R. (1997) Evolution of the solar ionizing flux. *J. Geophys. Res., 102,* 1641–1651.

Begemann F., Weber H. W., and Hintenberger H. (1976) On the primordial abundance of argon-40. *Astrophys. J. Lett., 203,* L155–L157.

Benkert J.-P., Baur H., Signer P., and Wieler R. (1993) He, Ne, and Ar from the solar wind and solar energetic particles in lunar ilmenites and pyroxenes. *J. Geophys. Res., 98,* 13147–13162.

Benz W. and Cameron A. G. W. (1990) Terrestrial effects of the giant impact. In *Origin of the Earth* (H. E. Newsom and J. H. Jones, eds.), pp. 61–67. Oxford Univ., New York.

Bernatowicz T. J. and Fahey A. J. (1986) Xe isotopic fractionation in a cathodeless glow discharge. *Geochim. Cosmochim. Acta, 50,* 445–452.

Bernatowicz T. J. and Hagee B. E. (1987) Isotopic fractionation of Kr and Xe implanted in solids at very low energies. *Geochim. Cosmochim. Acta, 51,* 1599–1611.

Bernatowicz T. J. and Podosek F. A. (1986) Adsorption and isotopic fractionation of Xe. *Geochim. Cosmochim. Acta, 50,* 1503–1507.

Bernatowicz T. J., Podosek F. A., Honda M., and Kramer F. E. (1984) The atmospheric inventory of xenon and noble gases in shales: the plastic bag experiment. *J. Geophys. Res., 89,* 4597–4611.

Bernatowicz T. J., Kennedy B. M., and Podosek F. A. (1985) Xe in glacial ice and the atmospheric inventory of noble gases. *Geochim. Cosmochim. Acta, 49,* 2561–2564.

Blichert-Toft J. and Albarède F. (1997) The Lu-Hf isotope geochemistry of chondrites and the evolution of the mantle-crust system. *Earth Planet. Sci. Lett., 148,* 243–258.

Bockelée-Morvan D., Gautier D., Lis D. C., Young K., Keene J., Phillips T., Owen T., Crovisier J., Goldsmith P. F., Bergin E. A., Despois D., and Wootten A. (1998) Deuterated water in Comet C/1996 B2 (Hyakutake) and its implications for the origin of comets. *Icarus, 133,* 147–162.

Broadhurst C. L., Drake M. J., Hagee B. E., and Bernatowicz T. J. (1992) Solubility and partitioning of neon, argon, krypton, and xenon in minerals and synthetic basaltic liquids. *Geochim. Cosmochim. Acta, 56,* 709–723.

Brown H. (1952) Rare gases and the formation of the earth's atmosphere. In *The Atmospheres of the Earth and Planets, 2nd edition* (G. P. Kuiper, ed.), pp. 258–266. Univ. of Chicago, Chicago.

Burnard P., Graham D., and Turner G. (1997) Vesicle-specific noble gas analyses of "popping rock": Implications for primordial noble gases in earth. *Science, 276,* 568–571.

Caffee M. W., Hudson G. B., Velsko C., Huss G. R., Alexander E. C. Jr., and Chivas A. R. (1999) Primordial noble gases from the Earth's mantle: identification of a primitive volatile component. *Science, 285,* 2115–2118.

Caldwell W. A., Nguyen J. H., Pfrommer B. G., Mauri F., Louie S. G., and Jeanloz R. (1997) Structure, bonding, and geochemistry of xenon at high pressures. *Science, 277,* 930–933.

Cameron A. G. W. (1983) Origin of the atmospheres of the terrestrial planets. *Icarus, 56,* 195–201.

Cameron A. G. W. and Ward W. R. (1976) The origin of the Moon (abstract). In *Lunar Science VII,* pp. 120–122. Lunar Science Institute, Houston.

Carroll M. R. and Stolper E. M. (1993) Noble gas solubilities in silicate melts and glasses: new experimental results for argon and the relationship between solubility and ionic porosity. *Geochim. Cosmochim. Acta, 57,* 5039–5051.

Chamorro-Pérez E., Gillet P., Jambon A., Badro J., and McMillan P. (1998) Low argon solubility in silicate melts at high pressure. *Nature, 393,* 352–355.

Chen G. Q. and Ahrens T. J. (1997) Erosion of terrestrial planet atmosphere by surface motion after a large impact. *Phys. Earth Planet. Inter., 100,* 21–26.

Chou C.-L. (1978) Fractionation of siderophile elements in the Earth's upper mantle. *Proc. Lunar Planet. Sci. Conf. 13th,* in *J. Geophys. Res., 88,* A507–A518.

Craig H., Clarke W. B., and Beg M. A. (1975) Excess $^3$He in deep water on the East Pacific Rise. *Earth Planet. Sci. Lett., 26,* 125–132.

Davies G. F. (1984) Geophysical and isotopic constraints on mantle convection: an interim synthesis. *J. Geophys. Res., 89,* 6017–6040.

Davies G. F. (1990) Mantle plumes, mantle stirring and hotspot chemistry. *Earth Planet. Sci. Lett., 99,* 94–109.

Davies G. F. (1998) Plates, plumes, mantle convection, and mantle evolution. In *The Earth's Mantle: Composition, Structure, and Evolution* (I. Jackson, ed.), pp. 228–258. Cambridge Univ., Cambridge.

Déruelle B., Dreibus G., and Jambon A. (1992) Iodine abundances in oceanic basalts: implications for Earth dynamics. *Earth Planet. Sci. Lett., 108,* 217–227.

Dreibus G. and Wänke H. (1989) Supply and loss of volatile constituents during the accretion of terrestrial planets. In *Origin and Evolution of Planetary and Satellite Atmospheres* (S. K. Atreya, J. B. Pollack, and M. S. Matthews, eds.), pp. 268–288. Univ. of Arizona, Tucson.

Dymond J. and Hogan J. (1973) Noble gas abundance patterns in deep-sea basalts- primordial gases from the mantle. *Earth Planet. Sci. Lett., 20,* 131–139.

Eberhardt P., Geiss J., Graf H., Grögler N., Mendia M. D., Mörgeli M., Schwaller H., Stettler A., Krähenbühl U., and von Gunten H. R. (1972) Trapped solar wind gases in Apollo 12 lunar fines 12001 and Apollo 11 breccia 10046. *Proc. Lunar Sci. Conf. 3rd,* pp. 1821–1856.

Eikenberg J., Signer P., and Wieler R. (1993) U-Xe, U-Kr, and U-Pb systematics for dating uranium minerals and investigations of the production of nucleogenic neon and argon. *Geochim. Cosmochim. Acta, 57,* 1053–1069.

Eiler J. M., Farley K. A., and Stolper E. M. (1998) Correlated helium and lead isotope variations in Hawaiian lavas. *Geochim. Cosmochim. Acta, 62,* 1977–1984.

Eugster O., Eberhardt P., and Geiss J. (1967) The isotopic composition of krypton in unequilibrated and gas rich chondrites. *Earth Planet. Sci. Lett., 2,* 385–393.

Eugster O., Weigel A., and Michel Th. (1994) Primordial Xe isotopic abundances and $^{244}$Pu-$^{136}$Xe ages of primitive differentiated meteorites. In *Noble Gas Geochemistry and Cosmochemistry* (J. Matsuda, ed.), pp. 1–9. Terra Scientific, Tokyo.

Fanale F. P. and Cannon W. A. (1972) Origin of planetary primordial rare gas: the possible role of adsorption. *Geochim. Cosmochim. Acta, 36,* 319–328.

Farley K. A. and Craig H. (1994) Atmospheric argon contamination of ocean island basalt olivine phenocrysts. *Geochim. Cosmochim. Acta, 58,* 2509–2517.

Farley K. A. and Neroda E. (1998) Noble gases in the Earth's mantle. *Annu. Rev. Earth Planet. Sci., 26,* 189–218.

Farley K. and Poreda R. J. (1993) Mantle neon and atmospheric contamination. *Earth Planet. Sci. Lett., 114,* 325–339.

Farley K. A., Maier-Reimer E., Schlosser P., and Broecker W. S. (1995) Constraints on mantle $^3$He fluxes and deep-sea circulation from an oceanic general circulation model. *J. Geophys. Res., 100,* 3829–3839.

Fisher D. E. (1985) Noble gases from oceanic island basalts do not require an undepleted mantle source. *Nature, 316,* 716–718.

Frick U., Mack R., and Chang S. (1979) Noble gas trapping and fractionation during synthesis of carbonaceous matter. *Proc. Lunar Planet. Sci. Conf. 10th,* pp.1961–1973.

Frick U., Becker R. H., and Pepin R. O. (1988) Solar wind record in the lunar regolith: nitrogen and noble gases. *Proc. Lunar Planet. Sci. Conf. 18th,* pp. 87–120.

Gaffey M. J. (1990) Thermal history of the asteroid belt: implications for accretion of the terrestrial planets. In *Origin of the Earth* (H. E. Newsom and J. H. Jones, eds.) pp. 17–28. Oxford Univ., New York.

Gancarz A. J. and Wasserburg G. J. (1977) Initial Pb of the Amitsoq gneiss, West Greenland, and implications for the age of the Earth. *Geochim. Cosmochim. Acta, 41,* 1283–1301.

Geiss J. (1993) Primordial abundances of hydrogen and helium isotopes. In *Origin and Evolution of the Elements* (N. Prantzos, E. Vangioni-Flam, and M. Cassé, eds.), pp. 89–106. Cambridge Univ., Cambridge.

Halliday A., Rehkämper M., Lee D.-C., and Yi W. (1996) Early evolution of the Earth and moon: new constraints from Hf-W isotope geochemistry. *Earth Planet. Sci. Lett., 142,* 75–89.

Harper C. L. J. and Jacobsen S. B. (1996) Noble gases and Earth's accretion. *Science, 273,* 1814–1818.

Harrison D., Burnard P., and Turner G. (1999) Noble gas behaviour and composition in the mantle: constraints from the Iceland Plume. *Earth Planet. Sci. Lett., 171/172,* 199–207.

Hart R., Dymond J., and Hogan L. (1979) Preferential formation of the atmosphere-sialic crust system from the Earth. *Nature, 278,* 156–159.

Hartmann W. K. and Davis D. R. (1975) Satellite-sized planetesimals and lunar origin. *Icarus, 24,* 504–515.

Hayashi C., Nakazawa K., and Mizuno H. (1979) Earth's melting due to the blanketing effect of the primordial dense atmosphere. *Earth Planet. Sci. Lett., 43,* 22–28.

Hilton D. R., Barling J., and Wheller G. E. (1995) Effect of shallow-level contamination on the helium isotope systematics of ocean-island lavas. *Nature, 373,* 330–333.

Hilton D. R., McMurtry G. M., and Kreulen R. (1997) Evidence for extensive degassing of the Hawaiian mantle plume from helium-carbon relationships at Kilauea volcano. *Geophys. Res. Lett., 24,* 3065–3068.

Hilton D. R., Grönvold K., Sveinbjornsdottir A. E., and Hammerschmidt K. (1998) Helium isotope evidence for off-axis degassing of the Icelandic hotspot. *Chem. Geol., 149,* 173–187.

Hilton D. R., Grönvold K., Macpherson C. G., and Castillo P. R. (1999) Extreme $^3$He/$^4$He ratios in northwest Iceland: constraining the common component in mantle plumes. *Earth Planet. Sci. Lett.,* in press.

Hiyagon H. (1994) Retention of solar helium and neon in IDPs in deep sea sediments. *Science, 263,* 1257–1259.

Hiyagon H. and Ozima M. (1986) Partitioning of noble gases between olivine and basalt melt. *Geochim. Cosmochim. Acta, 50,* 2045–2057.

Hiyagon H., Ozima M., Marty B., Zashu S., and Sakai H. (1992) Noble gases in submarine glasses from mid-oceanic ridges and Loihi Seamount: constraints on the early history of the Earth. *Geochim. Cosmochim. Acta, 56,* 1301–1316.

Honda M. and McDougall I. (1998) Primordial helium and neon in the Earth — a speculation on early degassing. *Geophys. Res. Lett., 25,* 1951–1954.

Honda M., McDougall I., Patterson D. B., Doulgeris A., and Clague D. A. (1991) Possible solar noble gas component in Hawaiian basalts. *Nature, 349,* 149–151.

Honda M., McDougall I., and Patterson D. B. (1993) Solar noble gases in the Earth: the systematics of helium-neon isotopes in mantle derived samples. *Lithos, 30,* 257–265.

Hudson G. B., Kennedy B. M., Podosek F. A., and Hohenberg C. M. (1989) The early solar system abundance of $^{244}$Pu as inferred from the St. Severin chondrite. *Proc. Lunar Planet. Sci. Conf. 19th,* pp. 547–557.

Hunten D. M., Pepin R. O., and Walker J. C. G. (1987) Mass fractionation in hydrodynamic escape. *Icarus, 69,* 532–549.

Hunten D. M., Pepin R. O., and Owen T. C. (1988) Planetary atmospheres. In *Meteorites and the Early Solar System* (J. F. Kerridge and M. S. Matthews, eds.), pp. 565–591. Univ. of Arizona, Tucson.

Hunten D. M., Donahue T. M., Walker J. C. G., and Kasting J. F. (1989) Escape of atmospheres and loss of water. In *Origin and Evolution of Planetary and Satellite Atmospheres* (S. K. Atreya, J. B. Pollack, and M. S. Matthews, eds.), pp. 167–191. Univ. of Arizona, Tucson.

Hutchins K. S. and Jakosky B. M. (1996) Evolution of Martian atmospheric argon: Implications for sources of volatiles. *J. Geophys. Res., 101,* 14933–14949.

Igarashi G. (1986) Components of xenon in carbonaceous chondrites and fission component in terrestrial atmospheric xenon. In *Abstracts for Japan–U.S. Seminar on Terrestrial Rare Gases,* pp. 20–23. Department of Physics, University of California, Berkeley.

Igarashi G. (1988) Noble gases in waters from deep-sea trenches and a volcanic lake around Japan, and investigation of Xe isotopic structure with relevance to the origin and evolution of the atmosphere. Ph.D. thesis, Geophysical Institute, Tokyo.

Igarashi G. (1993) Primitive xenon in the Earth. In *Volatiles in the Earth and Solar System* (K. A. Farley, ed.), pp. 70–80. AIP Conf. Proc. 341.

Jakosky B. M., Pepin R. O., Johnson R. E., and Fox J. L. (1994) Mars atmospheric loss and isotopic fractionation by solar-wind-induced sputtering and photochemical escape. *Icarus, 111,* 271–288.

Jambon A., Weber H., and Braun O. (1986) Solubility of He, Ne, Ar, Kr, and Xe in a basalt melt in the range 1250–1600°C. Geochemical implications. *Geochim. Cosmochim. Acta, 50,* 401–408.

Jephcoat A. P. (1998) Rare-gas solids in the Earth's deep interior. *Nature, 393,* 355–358.

Jochum K. P., Hofmann A. W., Ito E., Seufert H. M., and White

W. M. (1983) K, U, and Th in mid-oceanic ridge basalt glasses and heat production, K/U and K/Rb in the mantle. *Nature, 306,* 431–436.

Kamijo K., Hashizume K., and Matsuda J.-I. (1998) Noble gas constraints on the evolution of the atmosphere-mantle system. *Geochim. Cosmochim. Acta, 62,* 2311–2321.

Kaneoka I., Takaoka N., and Upton B. G. J. (1986) Noble gas systematics in basalts and a dunite nodule from Reunion and Grand Comore Islands, Indian Ocean. *Chem. Geol., 59,* 35–42.

Kellogg L. H. and Wasserburg G. J. (1990) The role of plumes in mantle helium fluxes. *Earth Planet. Sci. Lett., 99,* 276–289.

Kellogg L. H., Hager B. H., and van der Hilst R. D. (1999) Compositional stratification in the deep mantle. *Science, 283,* 1881–1884.

Krummenacher D., Merrihue C. M., Pepin R. O., and Reynolds J. H. (1962) Meteoritic krypton and barium versus the general isotopic anomalies in xenon. *Geochim. Cosmochim. Acta, 26,* 231–249.

Kunz J. (1999) Is there solar argon in the Earth's mantle? *Nature, 399,* 649–650.

Kunz J., Staudacher T., and Allègre C. J. (1998) Plutonium-fission xenon found in Earth's mantle. *Science, 280,* 877–880.

Kurz M. D. and Jenkins W. J. (1981) The distribution of helium in oceanic basalt glasses. *Earth Planet. Sci. Lett., 99,* 276–289.

Kurz M. D., Jenkins W. J., and Hart S. R. (1982) Helium isotopic systematics of oceanic islands and mantle heterogeneity. *Nature, 297,* 43–47.

Lee D.-C. and Halliday A. N. (1995) Hafnium-tungsten chronometry and the timing of terrestrial core formation. *Nature, 378,* 771–774.

Lee D.-C., Halliday A. N., Snyder G. A., and Taylor L. A. (1997) Age and origin of the moon. *Science, 278,* 1098–1103.

Lunine J. I. and Stevenson D. J. (1985) Thermodynamics of clathrate hydrate at low and high pressures with application to the outer solar system. *Astrophys. J. Suppl. Series, 58,* 493–531.

Lupton J. E. and Craig H. (1975) Excess $^3He$ in oceanic basalts: evidence for terrestrial primordial helium. *Earth Planet. Sci. Lett., 26,* 133–139.

Lux G. (1987) The behavior of noble gases in silicate liquids: Solution, diffusion, bubbles and surface effects, with applications to natural samples. *Geochim. Cosmochim. Acta, 51,* 1549–1560.

Mahaffy P. R., Donahue T. M., Atreya S. K., Owen T. C., and Niemann H. B. (1998) Galileo probe measurements of D/H and $^3He/^4He$ in Jupiter's atmosphere. *Space Sci. Rev., 84,* 251–263.

Manga M. (1996) Mixing of heterogeneities in the mantle: effect of viscosity differences. *Geophys. Res. Lett., 23,* 403–406.

Marty B. and Allé P. (1994) Neon and argon isotopic constraints on earth-atmosphere evolution. In *Noble Gas Geochemistry* (J. Matsuda, ed.), pp. 191–204. Terra, Tokyo.

Matsuda J. and Marty B. (1995) The $^{40}Ar/^{36}Ar$ ratio of the undepleted mantle; a reevaluation. *Geophys. Res. Lett., 22,* 1937–1940.

Matsuda J. and Nagao K. (1986) Noble gas abundances in a deep-sea sediment core from eastern equatorial Pacific. *Geochem. J., 20,* 71–80.

Matsuda J., Sudo M., Ozima M., Ito K., Ohtaka O., and Ito E. (1993) Noble gas partitioning between metal and silicate under high pressures. *Science, 259,* 788–790.

Mazor E., Heymann D., and Anders E. (1970) Noble gases in carbonaceous chondrites. *Geochim. Cosmochim. Acta, 34,* 781–824.

McElroy M. B. and Prather M. J. (1981) Noble gases in the terrestrial planets. *Nature, 293,* 535–539.

Michel T. and Eugster O. (1994) Primitive xenon in diogenites and plutonium-244 fission xenon ages of a diogenite, a howardite, and eucrites. *Meteoritics, 29,* 593–606.

Mizuno H. and Wetherill G. W. (1984) Grain abundance in the primordial atmosphere of the Earth. *Icarus, 59,* 74–86.

Mizuno H., Nakazawa K., and Hayashi C. (1980) Dissolution of the primordial rare gases into the molten Earth's material. *Earth Planet. Sci. Lett., 50,* 202–210.

Moreira M., Kunz J., and Allègre C. J. (1998) Rare gas systematics in popping rock: isotopic and elemental compositions in the upper mantle. *Science, 279,* 1178–1181.

Muramatsu Y. and Wedepohl K. H. (1998) The distribution of iodine in the earth's crust. *Chem. Geol., 147,* 201–216.

Musselwhite D. S., Drake M. J., and Swindle T. D. (1991) Early outgassing of Mars supported by differential water solubility of iodine and xenon. *Nature, 352,* 697–699.

Niedermann S., Bach W., and Erzinger J. (1997) Noble gas evidence for a lower mantle component in MORBs from the southern East Pacific Rise: Decoupling of helium and neon isotope systematics. *Geochim. Cosmochim. Acta, 61,* 2697–2715.

Niemeyer S. and Marti K. (1981) Noble gas trapping by laboratory carbon condensates. *Proc. Lunar Planet. Sci. 12B,* pp. 1177–1188.

Notesco G., Laufer D., Bar-Nun A., and Owen T. (1999) An experimental study of the isotopic enrichment in Ar, Kr and Xe when trapped in water ice. *Icarus,* in press.

O'Connell R. J. (1995) Mantle flow, viscosity structure, and geochemical reservoirs (abstract). *Eos Trans. AGU, 76,* F605.

O'Nions R. K. and Oxburgh E. R. (1983) Heat and helium in the Earth. *Nature, 306,* 429–431.

O'Nions R. K., Carter S. R., Evensen N. M., and Hamilton P. J. (1981) Upper mantle geochemistry. In *The Sea,* Vol. 7 (C. Emiliani, ed.), pp. 49–71. Wiley, New York.

Owen T. C. (1996) From planetesimals to planets: contributions of icy planetesimals to planetary atmospheres. In *From Stardust to Planetesimals* (Y. J. Pendleton and A. G. G. M. Tielens, eds.), pp. 435–450. ASP Conf. Ser. 122.

Owen T. C. and Bar-Nun A. (2000) Volatile contributions from icy planetesimals. In *Origin of the Earth and Moon* (R. M. Canup and K. Righter, eds.), this volume. Univ. of Arizona, Tucson.

Owen T., Bar-Nun A., and Kleinfeld I. (1992) Possible cometary origin of heavy noble gases in the atmospheres of Venus, Earth and Mars. *Nature, 358,* 43–46.

Ozima M. (1994) Noble gas state in the mantle. *Rev. Geophys., 32,* 405–426.

Ozima M (1999) Primordial noble gases in the Earth (abstract). In *Lunar and Planetary Science XXX,* Abstract #1096. Lunar and Planetary Institute, Houston (CD-ROM).

Ozima M. and Nakazawa K. (1980) Origin of rare gases in the Earth. *Nature, 284,* 313–316.

Ozima M. and Zahnle K. (1993) Mantle degassing and atmospheric evolution: Noble gas view. *Geochem. J., 27,* 185–200.

Ozima M., Podosek F. A., and Igarashi, G. (1985) Terrestrial xenon isotope constraints on the early history of the Earth. *Nature, 315,* 471–474.

Ozima M., Wieler R., Marty B., and Podosek F. A. (1998) Comparative studies of solar, Q-gases and terrestrial noble gases, and implications on the evolution of the solar nebula. *Geochim. Cosmochim. Acta, 62,* 301–314.

Parsons B. (1981) The rates of plate creation and consumption.

*Geophys. J. R. Astron. Soc., 67*, 437–448.

Patterson D., Honda M., and McDougall I. (1990) Atmospheric contamination: a possible source for heavy noble gases in basalts from Loihi Seamount, Hawaii. *Geophys. Res. Lett., 17*, 705–708.

Pepin R. O. (1991) On the origin and early evolution of terrestrial planet atmospheres and meteoritic volatiles. *Icarus, 92*, 2–79.

Pepin R. O. (1992) Origin of noble gases in the terrestrial planets. *Annu. Rev. Earth Planet. Sci., 20*, 389–430.

Pepin R. O. (1994). Evolution of the Martian atmosphere. *Icarus, 111*, 289–304.

Pepin R. O. (1997) Evolution of Earth's noble gases: Consequences of assuming hydrodynamic loss driven by giant impact. *Icarus, 126*, 148–156.

Pepin R. O. (1998) Isotopic evidence for a solar argon component in the Earth's mantle. *Nature, 394*, 664–667.

Pepin R. O. (1999) On the isotopic composition of primordial xenon in terrestrial planet atmospheres. In *From Dust to Terrestrial Planets* (W. Benz, R. Kallenbach, G. Lugmair, and F. Podosek, eds.), in press. ISSI Space Sciences Series 9, Kluwer, Dordrecht.

Pepin R. O. and Phinney D. (1976) The formation interval of the Earth (abstract). In *Lunar Science VII*, pp. 682–684. Lunar Science Institute, Houston.

Pepin R. O. and Phinney D. (1978) Components of xenon in the solar system. Unpublished manuscript.

Pepin R. O., Nyquist L. E., Phinney D., and Black D. C. (1970) Rare gases in Apollo 11 lunar material. *Proc. Apollo 11 Lunar Sci. Conf.*, pp. 1435–1454.

Pepin R. O., Becker R. H., and Rider P. E. (1995) Xenon and krypton in extraterrestrial regolith soils and in the solar wind. *Geochim. Cosmochim. Acta, 59*, 4997–5022.

Pepin R. O., Becker R. H., and Schlutter D. J. (1999) Irradiation records in regolith materials, I: Isotopic compositions of solar-wind neon and argon in single lunar mineral grains. *Geochim. Cosmochim. Acta, 63*, 2145–2162.

Phinney D. (1972) $^{36}$Ar, Kr, and Xe in terrestrial materials. *Earth Planet. Sci. Lett., 16*, 413–420.

Phinney D., Tennyson J., and Frick U. (1978) Xenon in $CO_2$ well gas revisited. *J. Geophys. Res., 83*, 2313–2319.

Podosek F. A. and Cassen P. (1994) Theoretical, observational and isotopic estimates of the lifetime of the solar nebula. *Meteoritics, 29*, 6–25.

Podosek F. A., Honda M., and Ozima M. (1980) Sedimentary noble gases. *Geochim. Cosmochim. Acta, 44*, 1875–1884.

Ponganis K. V., Graf T., and Marti K. (1997) Isotopic fractionation in low-energy ion implantation. *J. Geophys. Res., 102*, 19335–19343.

Porcelli D. and Wasserburg G. J. (1995a) Mass transfer of Xe through a steady-state upper mantle. *Geochim. Cosmochim. Acta, 59*, 1991–2007.

Porcelli D. and Wasserburg G. J. (1995b) Mass transfer of helium, neon, argon, and xenon through a steady-state upper mantle. *Geochim. Cosmochim. Acta, 59*, 4921–4937.

Porcelli D. and Wasserburg G. J. (1995c) A unified model for terrestrial rare gases. In *Volatiles in the Earth and Solar System* (K. A. Farley, ed.), pp. 56–69. AIP Conf. Proc. 341.

Porcelli D., Stone J. O. H., and O'Nions R. K. (1986) Rare gas reservoirs and Earth degassing (abstract). In *Lunar and Planetary Science XVII*, pp. 674–675. Lunar and Planetary Institute, Houston.

Porcelli D., Cassen P., Woolum D., and Wasserburg G. J. (1998)

acquisition and early losses of rare gases from the deep Earth (abstract). In *Origin of the Earth and Moon*, pp. 35–36. LPI Contribution No. 957, Lunar and Planetary Institute, Houston.

Poreda R. J. and Farley K. A. (1992) Rare gases in Samoan xenoliths. *Earth Planet. Sci. Lett., 113*, 129–144.

Prinn R. G. and Fegley B. Jr. (1989) Solar nebula chemistry: Origin of planetary, satellite and cometary volatiles. In *Origin and Evolution of Planetary and Satellite Atmospheres* (S. K. Atreya, J. B. Pollack, and M. S. Matthews, eds.), pp. 78–136. Univ. of Arizona, Tucson.

Righter K., Drake M. J., and Yaxley G. (1997) Prediction of siderophile element metal-silicate partition coefficients to 20 GPa and 2800 degrees C: The effects of pressure, temperature, oxygen fugacity, and silicate and metallic melt compositions. *Phys. Earth Planet. Inter., 100*, 115–134.

Rison W. (1980) Isotopic studies of rare gases in igneous rocks: Implications for the mantle and atmosphere. Ph.D. thesis, University of California.

Rison W. and Craig H. (1983) Helium isotopes and mantle volatiles in Loihi Seamount and Hawaiian Island basalts and xenoliths. *Earth Planet. Sci. Lett., 66*, 407–426.

Safronov V. S. (1978) The heating of the Earth during its formation. *Icarus, 33*, 1–12.

Sarda P., Staudacher T., and Allègre C. J. (1988) Neon isotopes in submarine basalts. *Earth Planet. Sci. Lett., 91*, 73–88.

Sasaki S. (1991) Off-disk penetration of ancient solar wind. *Icarus, 91*, 29–38.

Sasaki S. (1999) Presence of a primary solar-type atmosphere around the Earth: evidence of dissolved noble gas. *Planet. Space Sci.*, submitted.

Sasaki S. and Nakazawa K. (1988) Origin of isotopic fractionation of terrestrial Xe: hydrodynamic fractionation during escape of the primordial $H_2$-He atmosphere. *Earth Planet. Sci. Lett., 89*, 323–334.

Sasaki S. and Nakazawa K. (1990) Did a primary solar-type atmosphere exist around the proto-Earth? *Icarus, 85*, 21–42.

Shibata T., Takahashi E., and Matsuda J. (1998) Solubility of neon, argon, krypton, and xenon in binary and ternary silicate systems: A new view on noble gas solubility. *Geochim. Cosmochim. Acta, 62*, 1241–1253.

Simon T., Herbig G., and Boesgaard A. M. (1985) The evolution of chromospheric activity and the spin-down of solar-type stars. *Astrophys. J., 293*, 551–574.

Stacey J. S. and Kramers J. D. (1975) Approximation of terrestrial lead isotope evolution by a two-stage model. *Earth Planet. Sci. Lett., 26*, 207–221.

Staudacher T. (1987) Upper mantle origin for Harding County well gases. *Nature, 325*, 605–607.

Staudacher T. and Allègre C. J. (1982) Terrestrial xenology. *Earth Planet. Sci. Lett., 60*, 389–406.

Staudacher T. and Allègre C. J. (1988) Recycling of oceanic crust and sediments: the noble gas subduction barrier. *Earth Planet. Sci. Lett., 89*, 173–183.

Staudacher T., Kurz M. D., and Allègre C. J. (1986) New noble-gas data on glass samples from Loihi Seamount and Hualalai and on dunite samples from Loihi and Réunion Island. *Chem. Geol., 56*, 193–205.

Staudacher T., Sarda P., Richardson S. H., Allègre C. J., Sagna I., and Dmitriev L. M. (1989) Noble gases in basalt glasses from a mid-Atlantic ridge topographic high at 14°N: geodynamic consequences. *Earth Planet. Sci. Lett., 96*, 119–133.

Stevenson D. J. (1985) Partitioning of noble gases at extreme pressures within planets (abstract). In *Lunar and Planetary Science*

*XVI,* pp. 821–822. Lunar and Planetary Institute, Houston.

Stevenson D. J. (1990) Fluid dynamics of core formation. In *Origin of the Earth* (H. E. Newsom and J. H. Jones, eds.), pp. 231–249. Oxford Univ., New York.

Strom S. E., Edwards S., and Skrutskie M. F. (1993) Evolutionary timescales for circumstellar disks associated with intermediate and solar-type stars. In *Protostars and Planets III* (E. H. Levy, J. Lunine, and M. S. Matthews, eds.), pp. 837–866. Univ. of Arizona, Tucson.

Stuart F. M. (1994) Comment on "Speculations about the cosmic origin of He and Ne in the interior of the Earth." *Earth Planet. Sci. Lett., 122,* 245–247.

Swindle T. D. and Podosek F. A. (1988) Iodine-xenon dating. In *Meteorites and the Early Solar System* (J. F. Kerridge and M. S. Matthews, eds.), pp. 1127–1146. Univ. of Arizona, Tucson.

Taylor S. R. and Norman M. D. (1990) Accretion of differentiated planetesimals to the Earth. In *Origin of the Earth* (H. E. Newsom and J. H. Jones, eds.), pp. 29–43. Oxford Univ., New York.

Tera F. and Wasserburg G. J. (1974) U-Th-Pb systematics on lunar rocks and inferences about lunar evolution and the age of the moon. *Proc. Lunar Sci. Conf. 5th,* pp. 1571–1599.

Tolstikhin I. N. and Marty B. (1998) The evolution of terrestrial volatiles: a view from helium, neon, argon, and nitrogen isotope modeling. *Chem. Geol., 147,* 27–52.

Tolstikhin I. N. and O'Nions R. K. (1994) The Earth's missing xenon: A combination of early degassing and of rare gas loss from the atmosphere. *Chem. Geol., 115,* 1–6.

Tonks W. B. and Melosh H. J. (1990) The physics of crystal settling and suspension in a turbulent magma ocean. In *Origin of the Earth* (H. E. Newsom and J. H. Jones, eds.), pp.151–171. Oxford Univ., New York.

Trull T. (1994) Influx and age constraints on the recycled cosmic dust explanation for high $^3$He/$^4$He ratios at hotspot volcanoes. In *Noble Gas Geochemistry and Cosmochemistry* (J. Matsuda, ed.), pp. 77–88. Terra, Tokyo.

Turcotte D. L. and Schubert G. (1988) Tectonic implications of radiogenic noble gases in planetary atmospheres. *Icarus, 74,* 36–46.

Tyburczy J. A., Frisch B., and Ahrens T. J. (1986) Shock-induced volatile loss from a carbonaceous chondrite: implications for planetary accretion. *Earth Planet. Sci. Lett., 80,* 201–207.

Valbracht P. J., Honda M., Matsumoto T., Mattielli N., McDougall I., Ragettli R., and Weis D. (1996) Helium, neon, and argon isotope systematics in Kerguelen ultramafic xenoliths: implications for mantle source signatures. *Earth Planet. Sci. Lett., 138,* 29–38.

Valbracht P. J., Staudacher T., Malahoff A., and Allègre C. J. (1997) Noble gas systematics of deep rift zone glasses from Loihi Seamount, Hawaii. *Earth Planet. Sci. Lett., 150,* 399–411.

van der Hilst R. D. and Kárason H. (1999) Compositional heterogeneity in the bottom 1000 kilometers of Earth's mantle: toward a hybrid convection model. *Science, 283,* 1885–1888.

van Keken P. E. and Ballentine C. J. (1998) Whole-mantle versus layered mantle convection and the role of a high-viscosity lower mantle in terrestrial volatile evolution. *Earth Planet. Sci. Lett., 156,* 19–32.

Vickery A.M. and Melosh H. J. (1990) Atmospheric erosion and impactor retention in large impacts with application to mass extinctions. In *Global Catastrophes in Earth History* (V. L. Sharpton and P. D. Ward, eds.), pp. 289–300. Geol. Soc. Am. Spec. Paper 247.

Wacker J. F. (1989) Laboratory simulation of meteoritic noble gases. III. Sorption of neon, argon, krypton, and xenon on carbon: elemental fractionation. *Geochim. Cosmochim. Acta, 53,* 1421–1433.

Walter F. M., Brown A., Mathieu R. D., Myers P. C., and Vrba F. J. (1988) X-ray sources in regions of star formation. III. Naked T Tauri stars associated with the Taurus-Auriga complex. *Astron. J., 96,* 297–325.

Wänke H. and Dreibus G. (1988) Chemical composition and accretion history of terrestrial planets. *Philos. Trans. R. Soc. London, A325,* 545–557.

Wasserburg G. J., Papanastassiou D. A., Tera F., and Huneke J. C. (1977) I. The accumulation and bulk composition of the moon: outline of a lunar chronology. *Philos. Trans. R. Soc. London, A285,* 7–22.

Wetherill G. W. (1975) Radiometric chronology of the early solar system. *Annu. Rev. Nucl. Sci., 25,* 283–328.

Wetherill G. W. (1981) Solar wind origin of $^{36}$Ar on Venus. *Icarus, 46,* 70–80.

Wetherill G. W. (1986) Accumulation of the terrestrial planets and implications concerning lunar origin. In *Origin of the Moon* (W. K. Hartmann, R. J. Phillips, and G. J. Taylor, eds.), pp. 519–550. Lunar and Planetary Institute, Houston.

Wetherill G. W. (1990) Formation of the Earth. *Annu. Rev. Earth Planet. Sci., 18,* 205–256.

Wieler R. (1994) "Q-gases" as "local" primordial noble gas component in primitive meteorites. In *Noble Gas Geochemistry and Cosmochemistry* (J. Matsuda, ed.), pp. 31–41. Terra, Tokyo.

Williams Q. and Garnero E. J. (1996) Seismic evidence for partial melt at the base of Earth's mantle. *Science, 273,* 1528–1530.

Woolum D. S., Cassen P., Porcelli D., and Wasserburg G. J. (1999) Incorporation of solar noble gases from a nebula-derived atmosphere during magma ocean cooling (abstract). In *Lunar and Planetary Science XXX.* Lunar and Planetary Institute, Houston (CD-ROM).

Yang J., Lewis R. S., and Anders E. (1982) Sorption of noble gases by solids, with reference to meteorites. I. Magnetite and carbon. *Geochim. Cosmochim. Acta, 46,* 841–860.

Yatsevich I. and Honda M. (1997) Production of nucleogenic neon in the Earth from natural radioactive decay. *J. Geophys. Res., 102,* 10291–10298.

Zahnle K. and Kasting J. F. (1986) Mass fractionation during transonic hydrodynamic escape and implications for loss of water from Venus and Mars. *Icarus, 68,* 462–480.

Zahnle K., Kasting J. F., and Pollack J. B. (1990a) Mass fractionation of noble gases in diffusion-limited hydrodynamic hydrogen escape. *Icarus, 84,* 502–527.

Zahnle K., Pollack J. B., and Kasting J. F. (1990b) Xenon fractionation in porous planetesimals. *Geochim. Cosmochim. Acta, 54,* 2577–2586.

Zhang Y. (1998) The young age of the Earth. *Geochim. Cosmochim. Acta, 62,* 3185–3189.

Zhang Y. and Zindler A. (1989) Noble gas constraints on the evolution of Earth's atmosphere. *J. Geophys. Res., 94,* 13719–13737.

# Volatile Contributions from Icy Planetesimals

**Tobias C. Owen**
*University of Hawai'i*

**Akiva Bar-Nun**
*Tel-Aviv University*

Laboratory experiments on the trapping of gases by ice forming at low temperatures implicate comets as major carriers of the heavy noble gases to the inner planets. These icy planetesimals may also have brought the "light N" required to mix with solar $^{15}N/^{14}N$ to make atmospheric $N_2$. However, if the sample of three comets analyzed so far is typical, the Earth's oceans cannot have been produced by comets alone, they require an additional source of water with D/H < SMOW. The highly fractionated Ne in the Earth's atmosphere also indicates the importance of other carriers of volatiles, as do the noble gas abundances in meteorites from Mars. One additional carrier is probably the rocky material comprising the bulk of the mass of these planets. Venus requires a contribution from icy planetesimals formed at the low temperatures characteristic of the Kuiper Belt.

## 1. INTRODUCTION

As soon as he had successfully calculated cometary orbits, Edmund Halley (*Halley,* 1687) realized that collisions of comets with the Earth were a real possibility and might have remarkable consequences. In more modern times, *Oro* (1961) proposed that comets could have been an important source of prebiotic organic material, and other investigators have suggested that these icy planetesimals might have made major contributions to our planet's atmosphere (e.g., *Sill and Wilkening,* 1978; *Delsemme,* 1991). Recent work on this question has focused on models of the dissipation of icy planetesimals from the Uranus-Neptune region as these planets finished forming (*Fernandez and Ip,* 1997), or attempts to calculate the extent of the terrestrial cometary influx through analyses of the cratering record of the Moon (*Chyba,* 1990). All these studies suffer from the absence of evidence for a uniquely indentifiable, cometary contribution to the Earth's volatile inventory.

The difficulty in identifying the source of the Earth's volatiles is compounded by the 4.5-b.y. history of the planet, during which chemical reactions with the crust, escape of gases from the upper atmosphere, and the origin and evolution of life have completely changed the composition of the atmosphere and hydrosphere. The central problem of life's origin is intimately dependent on the early composition of the atmosphere, making our inability to define that composition all the more frustrating.

It is this dilemma that the heavy noble gases have the potential to resolve, since they are chemically inert, do not easily escape from the atmosphere, and are not involved in the activities of living organisms. The firmly entrenched idea that the atmospheric noble gas abundances (and hence other volatiles) are the result of delivery by meteorites (e.g., *Turekian and Clark,* 1975; *Anders and Owen,* 1977 ) stems from the recognition of a so-called "planetary pattern" in the noble gases found in these objects many years ago. This idea gained widespread support, despite the fact that it has never been possible to explain why the abundances of Kr and Xe in the meteorites are about the same, while Xe is much lower than Kr in the Earth's atmosphere (Fig. 1). Attempts to find the "missing" Xe buried in the sedimentary column, clathrates, or ice have failed *(Wacker and Anders,* 1984; *Bernatowicz et al.,* 1985). Furthermore, the relative abundances of the isotopes in atmospheric Xe are distinctly different from those in meteorites (or the solar wind) (Fig. 2). These difficulties with meteoritic Xe coupled with the arguments for early cometary bombardment encouraged us to pursue the idea that icy, rather than rocky, planetesimals may have delivered the heavy noble gases. If so, these icy planetesimals would have delivered many other volatiles as well.

## 2. A COMETARY MODEL FOR DELIVERY OF VOLATILES

Accordingly, we set out to determine whether or not the abundance patterns of the heavy noble gases in the atmospheres of the inner planets could be accounted for by comets. The first step was to apply the same laboratory techniques for trapping gases in amorphous ice deposited at low temperatures that had been used with $N_2$, CO, $H_2$, etc. (*Bar-Nun et al.,* 1985, 1988; *Laufer et al.,* 1987) to a mixture of heavy noble gases: Ar, Kr, and Xe (*Owen et al.,* 1991). These experiments were designed to imitate the formation of comets in the outer solar nebula. The idea is that interstellar ice grains probably sublimated as they fell toward the midplane of the nebula and recondensed on cold refractory

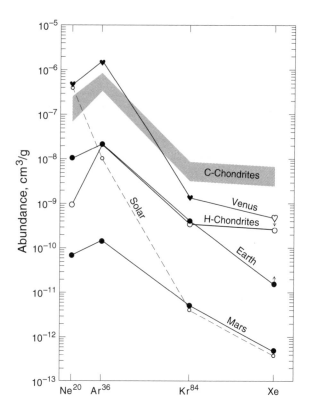

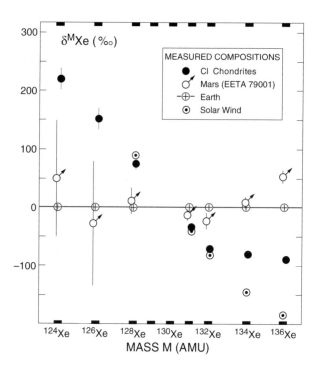

**Fig. 1.** Chondritic meteorites contain about as much Xe as Kr. The meteoritic noble gas abundances therefore do not match the abundance patterns found in inner planet atmospheres, despite the apparent agreement for Ne, Ar, and Kr. (Solar values are normalized for $^{84}Kr$ on Mars.) Note the high abundances of Ne and Ar per gram of rock and the solar type $^{36}Ar/^{84}Kr$ on Venus (*Owen and Bar-Nun,* 1995a).

**Fig. 2.** In this plot, the relative abundances of the Xe isotopes in the solar wind (as sampled in lunar surface materials), carbonaceous chondrites, and the martian atmosphere are compared with values in the Earth's atmosphere. We seem to have two "families": Mars and Earth vs. the Sun and the chondrites. But no two sources are exactly alike [(after *Pepin* (1989); solar wind data from *Ozima et al.* (1998); Mars points from *Swindle et al.* (1986), confirmed by *Garrison and Bogard* (1998) and *Mathew et al.* (1998)].

cores (*Lunine et al.,* 1991). In this recondensation process, they could trap ambient gas according to the local temperature. The laboratory results showed that temperature-dependent fractionation of the gas mixtures occurs, principally the depletion of Ar, which suggested that trapping in ice might indeed be responsible for the patterns of noble gas abundances found in the atmospheres of Mars, Earth, and Venus (*Owen et al.,* 1991). This initial analysis was inconclusive, however, since it did not seem possible to account for the variety of patterns observed on the three planets with a single model of cometary bombardment.

The SNC meteorites provided another clue. It is now widely accepted that these meteorites come from Mars (*McSween,* 1994). Yet a three-element plot of $^{36}Ar/^{132}Xe$ vs. $^{84}Kr/^{132}Xe$ shows that the abundances of these gases are very different in the different meteorites, and even in different samples of the same meteorite, with abundances from S and C meteorites forming a straight line on a log-log plot. This line passes through points corresponding to the noble gases found in the atmospheres of the Earth and Mars, but it is widely separated from the field of abundances in chondritic

meteorites or the values found in the solar wind (*Ott and Begemann,* 1985).

We have interpreted this distribution of noble gas abundances in the SNC meteorites as an effect of the mixing of two different reservoirs: one that represents gases trapped in the rocks that form the planets, the second a contribution from impacting icy planetesimals. This interpretation is based on our data for the fractionation of noble gases by trapping in amorphous ice (*Bar-Nun et al.,* 1988; *Laufer et al.,* 1987; *Owen et al.,* 1991). At a temperature of ~50 K, ice formed by condensation of water vapor traps Ar, Kr, and Xe in relative abundances that fall on an extrapolation of a mixing line drawn through the abundances from the atmospheres of Earth and Mars on the same three-element plot (Fig. 3). The analysis is complicated by the fact that the linear relation on the log-log plot (as first illustrated by *Ott and Begemann,* 1985) is very close to this mixing line (*Owen et al.,* 1992; *Ozima and Wada,* 1993; *Owen and Bar-Nun,* 1993). It may well be that more than one process is at work here. However, we have pointed out that the increase in $^{129}Xe/^{132}Xe$ with increasing $^{36}Ar/^{132}Xe$ that occurs in the

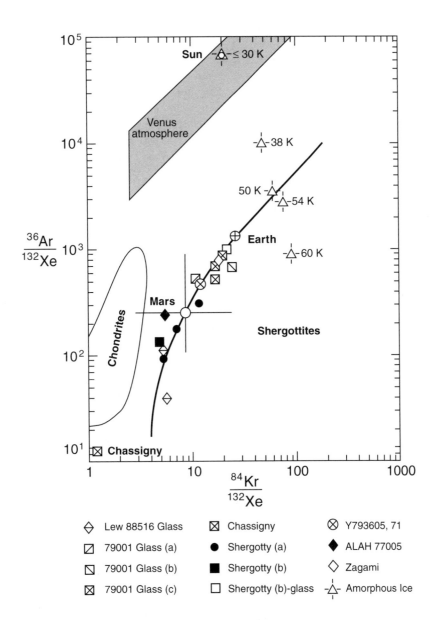

**Fig. 3.** In a three-element plot, noble gas abundances in the atmospheres of Mars (Viking) and Earth can be used to define a mixing line between internal and external volatile reservoirs. The internal reservoir lies below Mars on this plot, and consists of the rocks that formed the planet. We suggest icy planetesimals as the external reservoir, lying above the Earth mixture at the opposite end of the line. The external reservoir is represented here by the noble gases trapped in amorphous ice in laboratory experiments (the open triangles). These points correspond to solar elemental ratios multiplied by the fractionation factors found experimentally (*Owen et al., 1991*). Noble gas abundances in shergottites (meteorites from Mars) fall along this line. The gases on Venus could have been delivered by comets from the Kuiper Belt that formed at temperatures <30–35 K. The abundance of Xe on Venus is not yet known, hence the stippled trapezoid. References: Shergotty (a), Chassigny: *Ott* (1988); LEW 885516 glass: *Becker and Pepin* (1993); ALHA 77005, 79001 Glass (a): *Swindle et al.* (1986); 79001 Glass (b): *Becker and Pepin* (1984); Zagami: *Ott et al.* (1988); Shergotty (b), Y 793605, 71, 79001 glass (c): *Bogard and Garrison* (1998).

samples plotted along this same mixing line demonstrates that solubility in a melt (for example) cannot by itself explain the data (*Owen and Bar-Nun, 1993*).

The noble gases in the atmosphere of Venus do not conform to this simple picture. Xenon has not yet been measured on Venus, but the relative abundances of Ar and Kr resemble the solar ratio (Fig. 1), placing Venus somewhere

in the stippled trapezoid at the top of Fig. 3. As Fig. 1 demonstrates, Venus exhibits higher abundances of Ar and Ne per gram of rock than the most volatile-rich carbonaceous chondrites. We have suggested that the heavy noble gases on Venus were contributed by one or more icy planetesimals formed at T ≤ 30 K, i.e., objects equivalent to those populating the Kuiper Belt (*Owen et al., 1992; Owen and*

*Bar-Nun*, 1995a). A single icy planetesimal that condensed with solar elemental abundances would need a radius of ~90 km to deliver the presently observed amount of Ar. [A different approach, scaling from a now-obsolete model of Titan, gave r = 75 km (*Hunten et al.* 1988).] Obviously a cluster of smaller comets could achieve the same effect. Earth and Mars must have been struck by such planetesimals as well; the argument is simply that Venus was hit by a bigger one, or had a later series of small impacts by such objects than the other two planets. We will return to the issue of Ne in a later section.

We have further tested the applicability of the laboratory data to natural phenomena by examining the abundances of CO and $N_2$ in the comets. We pointed out that the apparently mysterious depletion of N in comets (*Geiss*, 1988; *Krankowsky*, 1991) probably results from the inability of ice to trap $N_2$ efficiently when the ice forms at T > 35 K. Our assumption is that most of the N that was present in the outer solar nebula was in the form of $N_2$, just as it is in the interstellar medium (*van Dishoeck et al.*, 1993). Hence for comets to acquire a solar ratio of N/O, the condensing ice that formed the comets would have to trap $N_2$ in the same fashion as the noble gases. Just as the abundance of Ar is sharply reduced in ice forming at 50 K, $N_2$ will be depleted as well (*Bar-Nun et al.*, 1988) . As a consequence, the atmospheric N on inner planets and on icy bodies in the outer solar system was most likely delivered in the form of $NH_3$ and other N compounds, rather than as $N_2$ . The isotopic composition of the N in the atmospheres of these relatively small bodies could therefore be different from that of N in the Sun or in the atmospheres of the giant planets, where $N_2$ should have been the dominant carrier. The isotope ratios of N compounds were presumably set by ion-molecule reactions in the presolar interstellar cloud, in which case they will exhibit systematic differences with the isotope ratio in interstellar $N_2$, as D/H measured in $H_2$ differs from D/H in H compounds (*Geiss and Bochsler*, 1982; *Owen and Bar-Nun*, 1995a). An indication that such a systematic difference indeed exists is found in the possible presence of a "light N" component in the solar system, invoked by *Geiss and Bochsler* (1982) to explain observed variations of $^{14}N/^{15}N$ in samples of lunar regolith. *Wieler et al.* (1999) have recently demonstrated that approximately 90% of the N in the lunar regolith has a nonsolar origin, opening the possibility that cometary impacts might supply the postulated "light N."

We also showed that the relative abundances of $CO^+/N_2^+$ derived from observations of comet tail spectra are consistent with the formation of the icy nuclei at temperatures near 50 K. Further work on the trapping of CO and $N_2$ in ice has substantiated our original conclusions (*Notesco and Bar-Nun*, 1996).

These consistencies between the laboratory results and the observed abundances in comets lent support to our effort to extrapolate laboratory work on ice trapping of noble gases to noble gas abundances in planetary atmospheres. Accordingly, we developed an "icy impact" model for the contribution of comets to inner planet atmospheres. An immediate consequence of this model was the realization that if comets contributed the terrestrial heavy noble gases, they would not supply enough water to fill the oceans. Impact erosion might have solved this problem by removing noble gases (plus CO, $CO_2$, and N) (*Chyba*, 1990; *Chyba et al.*, 1994), but we also suggested that some water might have been contributed from another source, such as the rocks making up most of the mass of the planet (*Owen and Bar-Nun*, 1995a).

Turning to Mars, we used the present atmospheric abundance of $^{84}Kr$ to predict that only the equivalent of 75 mbar of $CO_2$ could be accounted for. This amount of $CO_2$ would give a ratio of $C/^{84}Kr$ in the atmosphere equal to the value found in the Earth's volatile reservoir and (within existing uncertainties) in the present atmosphere of Venus. While 75 mbar of $CO_2$ is 10× the present value in the martian atmosphere, it corresponds to a layer of water only 12 m thick over the planet's surface, an insufficient amount to account for the observed fluvial erosion (*Carr*, 1986; *Greeley*, 1987). Accordingly, we invoked impact erosion, as developed for Mars by *Melosh and Vickery* (1989), as the responsible agent for diminishing the atmosphere to the 75-mbar level. We pointed out that impact erosion explains the relatively high ratio of $^{129}Xe/^{132}Xe$ in the current martian atmosphere (*Owen et al.*, 1977), well illustrated in the gases trapped in the meteorites EETA 79001 and Zagami (*Bogard and Johnson*, 1983; *Swindle et al.*, 1986; *Marti et al.*, 1995; *Garrison and Bogard*, 1998; *Mathew et al.*, 1998). Another way to achieve this same result is by massive hydrodynamic escape of the early atmosphere, as proposed by *Pepin* (1991). In each case the assumption is that the radiogenic $^{129}Xe$ or its parent $^{129}I$ is protected from loss, e.g., by differential solubility (*Musselwhite et al.*, 1991), and later outgassed to produce the observed high ratio of $^{129}Xe/^{132}Xe$ in the current martian atmosphere. Production of massive amounts of carbonates and/or adsorption in the regolith (e.g., *Pollack*, 1979), would remove $CO_2$, but it would not account for excess $^{129}Xe$ and it would leave behind large amounts of noble gases and N in the atmosphere, which are not observed.

We thus suggested a scenario in which the early atmosphere of Mars could have passed through several episodes of growth and decay, depending on the planet's bombardment history. We proposed that it is possible to see the effects of cometary bombardment in the gases trapped in the SNC meteorites, although other interpretations of these data are certainly possible (*Owen and Bar-Nun*, 1995b; *Swindle*, 1995; *Mathew et al.*, 1998).

## 3. DEUTERIUM IN COMETS: A TIE TO THE INTERSTELLAR MEDIUM

Clearly one of the corollaries of this model is that comets have probably retained a good memory of interstellar chemical conditions. One way to test this assumption is to investigate the abundances of stable isotopes of common elements, as isotope ratios should not change during the comet-formation process we have described. The recent appearance of two bright comets has allowed us to add to the isotopic data collected by *in situ* measurements of Comet Halley using groundbased techniques. Observations at radio wavelengths have been especially helpful.

The value of D/H is particularly sensitive to physical processes, since the mass ratio of these two stable isotopes is the largest in the periodic table. In Comet Halley, two independent investigations led to the result that D/H = 3.2 ± 0.1 × 10⁴ in the comet's $H_2O$ (*Balsiger et al.*, 1995; *Eberhardt et al.*, 1995). These studies were carried out *in situ*, with mass spectrometers on the Giotto spacecraft. Observations of Comet Hyakutake at radio and infrared wavelengths from Earth allowed the detection of lines of HDO and $H_2O$ that led to a value of D/H = 2.9 ± 1.0 × 10⁻⁴ (*Bockelée-Morvan et al.*, 1998). Similar investigations of Comet Hale-Bopp found (D/H) = 3.3 ± 0.8 × 10⁻⁴ in $H_2O$ (*Meier et al.*, 1998a). We thus have three Oort Cloud com-

ets exhibiting a value of D/H in $H_2O$ that is essentially twice the value of 1.56 × 10⁻⁴ in standard mean ocean water (SMOW).

In the case of Comet Hale-Bopp, it was possible to use these same remote sensing techniques to determine D/H in HCN (*Meier et al.*, 1998b). The results were quite different: (D/H)$_{HCN}$ = 23 ± 4 × 10⁻⁴. This is exactly the kind of difference between D/H in different molecular species that shows up in interstellar molecular clouds (*Millar et al.*, 1989), providing a strong case for the preservation of interstellar chemistry in comet ices. If comets had formed from solar nebula gas that was warmed and homogenized by radial mixing, the values of D/H in both molecules would be much lower and identical, as the HCN and $H_2O$ would exchange D with the huge reservoir of $H_2$ in the solar nebula, in which D/H = 0.26 ± 0.07 × 10⁻⁴ (*Mahaffy et al.*, 1998) (Fig. 4).

A simple ion-molecule reaction scheme appropriate for interstellar clouds suggests a formation temperature of 30 ± 10 K for $H_2O$ and HCN in order to produce the observed values of D/H (*Meier et al.*, 1998b; *Millar et al.*, 1989). This is noticeably higher than the canonical value of 10 K for the interiors of dark clouds, but it would be appropriate for a warm clump near a "hot core" in a cloud in which star formation is going on. This temperature is also consistent with values derived from measurements of the ortho-para ratio of H in comets (*Mumma et al.*, 1993; *Crovisier et al.*, 1997). Both the isotope ratios and the ortho-para ratio are unlikely to be changed during the simple sublimation and recondensation that takes place as icy grains settle into the disk of the solar nebula. However, the temperature at which recondensation occurs will determine how gases are trapped in the ices (*Bar-Nun et al.*, 1985; *Owen and Bar-Nun*, 1995a; *Notesco and Bar-Nun*, 1996). Hence it is not unreasonable to find solar nebula temperatures deduced from noble gas abundances to be 20°–30° higher than the original interstellar cloud (grain formation) temperatures deduced from the isotope and ortho-para ratios.

## 4. THE ORIGIN OF WATER ON EARTH AND MARS

We might also expect to see signs of cometary bombardment in the water that is present on the inner planets. Note that all three of the comets that have been studied exhibit a value of D/H in ice that is twice the value found in ocean water on Earth. All these comets came from the Oort Cloud, a spherical shell of cometary nuclei whose average radius is 50,000× the Earth's distance from the Sun. Three is certainly a tiny number compared to the 10¹¹ comets that may be present in the Oort Cloud, so one might speculate that other comets may have different values of D/H. Moreover, we have not yet sampled any comets from the Kuiper Belt, a disk of comet nuclei extending outward from Neptune. The Kuiper Belt is currently thought to be the likely source of short-period comets, which are unfortunately too faint to yield values of D/H to present groundbased techniques.

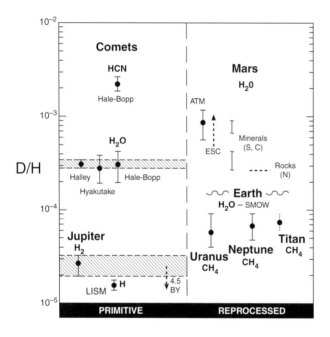

**Fig. 4.** We can divide D in the solar system into primitive (left) and reprocessed (right) reservoirs. D/H in the H gas that dominated the solar nebula can now be measured in the atmosphere of Jupiter (*Mahaffy et al.*, 1998). As expected, it is higher than the contemporary value in the local interstellar medium (LISM) (*Linsky*, 1996), which reflects the decrease in D/H over 4.5 b.y. resulting from stellar nucleosynthesis in the galaxy. The much higher values of D/H in primitive condensed matter are revealed in studies of comets (see *Meier and Owen*, 1999, for review and data sources). Exchange between H gas and volatalized condensed matter can produce a variety of D/H values, as illustrated in the right panel of the figure. The points above terrestrial SMOW are from martian rocks (nakhlites) and minerals in shergottites and Chassigny, referenced in the text. The Uranus, Neptune, and Titan points come from recent observations with the Infrared Space Observatory (*Feuchtgruber et al.*, 1999; Coustenis, personal communication, 1999).

However, these are the easiest comets for us to explore with spacecraft, because their orbits are so well known. As a result, several missions will visit short period comets in the next decades.

Meanwhile, we shall assume that the three comets that have been examined are indeed representative and see whether we can find traces of cometary water on the inner planets. Mars is the best case because the atmosphere is so thin. The total amount of water that exchanges seasonally though the martian atmosphere from pole to pole is just $2.9 \times 10^{15}$ g, equivalent to a single comet nucleus (density $\rho = 0.5$ g/cm$^3$) with a radius of only 1 km. Hence the impact of a relatively small comet could have a significant effect on the surface water supply. Indications of such an effect can be found in studies of D/H in hydrous phases of the SNC meteorites. *Watson et al.* (1994) reported a range of values of D/H in kaersutite, biotite, and apatite in Chassigny, Shergotty, and Zagami, with the highest values, 4–5.5× SMOW, occurring in Zagami apatite. Infrared spectroscopy from Earth has determined D/H in martian atmospheric water vapor to be $5.5 \pm 2\times$ terrestrial (*Owen et al.,* 1988; *Krasnopolsky et al.,* 1997a). Note that Zagami is one of the Shergottites with glassy inclusions containing the high values of $^{129}Xe/^{132}Xe$ clearly associated with trapped martian atmosphere (*Marti et al.,* 1995). The overlap between the Zagami apatite and Mars atmosphere values of D/H indicates that some mixing between atmospheric water vapor and the crustal rocks must occur. It is therefore arresting to note that the *lowest* values of D/H measured in the SNC minerals do not cluster around 1× terrestrial, but rather around ~2× terrestrial, the value measured in the three comets (Fig. 4). It thus appears possible that most of the near-surface water on Mars that accumulated after the end of the early bombardment phase was contributed primarily by cometary impact, rather than by magma from the planet or by meteoritic contributions.

This interpretation is consistent with the geochemical analysis by *Carr and Wänke* (1992), who concluded that Mars is much drier than the Earth, with roughly 35 ppm water in mantle rocks as opposed to 150 ppm for Earth. These authors suggested that one possible explanation for this difference is the lack of plate tectonics on Mars, which would prevent a volatile-rich veneer from mixing with mantle rocks. This is exactly what the D/H values in the SNC minerals appear to signify.

In whole-rock samples, D/H in water from the Shergottites is systematically higher than in Nakhla and Chassigny (*Leshin et al.,* 1996). This could be a result of the larger fraction of atmospheric H$_2$O incorporated in the shergottites by shock, a process nakhlites evidently avoided (*Bogard et al.,* 1986; *Ott,* 1988; *Drake et al.,* 1994). Furthermore, higher values of $\Delta^{17}O$ are found in water from these samples of Nakhla and Chassigny compared with the shergottites, reinforcing the idea that the shergottites sampled a different source of water. Finally, the O-isotopic ratios in water from whole-rock samples of the SNCs are also systematically different from the ratios found in silicates in these rocks

(*Karlsson et al.,* 1992), again suggesting a hydrosphere that is not strongly coupled to the lithosphere.

On Earth, the O isotopes in seawater match those in the silicates, indicating thorough mixing for at least the last 3.5 b.y. (*Robert et al.,* 1992). The Earth has also lost relatively little H to space after the postulated hydrodynamic escape (*Zahnle et al.,* 1990; *Pepin,* 1991, 1997) so the value of D/H we measure in seawater today must be close to the original value. *Yung and Dissly* (1992) estimate an increase of (D/H/(D/H)$_0$ = 1.006. If Halley, Hyakutake, and Hale-Bopp are truly representative of all comets, then we can't make the oceans out of melted comets alone. This is a very different situation from Mars. It suggests that water from the inner reservoir, the rocks making up the bulk of the Earth, must have mixed with incoming cometary water to produce our planet's oceans. On Mars, this mixing was apparently much less efficient.

Explanations for the relatively high value of O/C ($12 \pm 6$) in the terrestrial volatile inventory also suggest such mixing (*Owen and Bar-Nun,* 1995a). The solar value of O/C = 2.4, which was also the value found in Comet Halley (*Geiss,* 1988). Impact erosion on the Earth could remove CO and CO$_2$ while having less effect on water in the oceans or polar caps, thereby raising the value of O/C (*Chyba,* 1990; *Owen and Bar-Nun,* 1995a). This process would not affect the value of D/H, however. The average value of D/H in chondritic meteorites is close to that in seawater, so mixing meteoritic and cometary water would not lead to the right result. We need a contribution from a reservoir of water that is grossly deficient in C and N, with D/H $< 1.6 \times 10^{-4}$. *Lecluse and Robert* (1994) have shown that water vapor in the solar nebula at 1 AU from the Sun would have developed a value of D/H $\approx 0.8 \times 10^{-4}$ to $1.0 \times 10^{-4}$, depending on the lifetime of the nebula ($2 \times 10^5$ to $2 \times 10^6$ yr). An ocean made of roughly 35% cometary water and 65% water from the local solar nebula (trapped in planetary rocks) would satisfy the D/H constraint and would also be consistent with the observed value of O/C = $12 \pm 6$.

To accept this idea, we should be able to demonstrate that water vapor from the solar nebula was adsorbed on grains that became the rocks that formed the planets. *Robert* (1997) has reported D/H = $0.8 \times 10^{-4}$ in water from some meteorites. This is an indication that low D/H water was present in the inner solar system. However, our best hope for finding some of that original inner-nebula water appears to be on Mars, where mixing between the surface and the mantle has been so poor. The test is thus to look for water incorporated in SNC meteorites that appear to have trapped mantle gases, to see if D/H $< 1.6 \times 10^{-4}$. The best case for such a test among the rocks we have is Chassigny, which exhibits no enrichment of $^{129}Xe$ and thus appears not to have trapped any atmospheric gas. However, there is no evidence of water with low D/H in this rock (*Leshin et al.,* 1996). It may be that contamination by terrestrial water has masked the martian mantle component in Chassigny. This is a good project for a sample returned from Mars by spacecraft, where such contamination can be avoided.

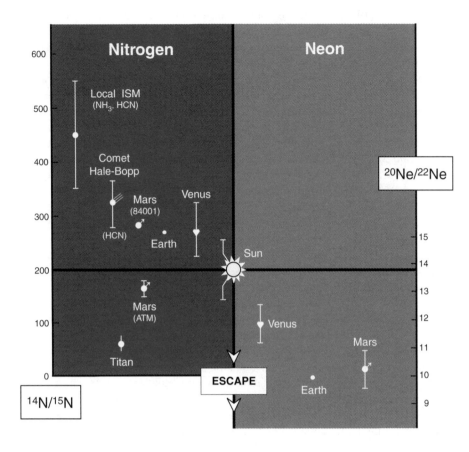

**Fig. 5.** (Left): $^{14}N/^{15}N$ is higher on Earth, Venus, and Comet Hale-Bopp and in $NH_3$ and HCN in the local interstellar medium than its value in the Sun. It is lower in the atmospheres of Mars and Titan because of escape. (Right): $^{20}Ne/^{22}Ne$ is lower on Venus, Earth, and Mars compared with the solar value, presumably as a result of early atmospheric escape. References to data points are given in text.

## 5. THE IMPORTANCE OF NEON AND NITROGEN

We have concentrated our analysis on water and the heavy noble gases: Ar, Kr, and Xe. Any model for the origin of the atmosphere must also account for Ne and N. Neon has about the same cosmic abundance as N relative to H, *viz.*, $1.2 \times 10^{-4}$ vs. $1.1 \times 10^{-4}$ (*Anders and Grevesse, 1989*). Hence we expect any atmosphere that consists of a captured remnant of the solar nebula to exhibit a ratio of $Ne/N_2 \approx 2$. On Earth, Mars, and Venus, $Ne \ll N_2$. The Ne is not only deficient in these atmospheres, the isotopes have been highly fractionated. The solar ratio of $^{20}Ne/^{22}Ne = 13.8 \pm 0.1$ (*Benkert et al., 1993*), while on Earth $^{20}Ne/^{22}Ne = 9.8$, on Venus $11.8 \pm 0.7$ (*Istomin et al., 1982*), and on Mars $10.1 \pm 0.7$ (*Wiens et al., 1986*) (Fig. 5, right).

Concentrating on Earth, we must ask how Ne can be so severely fractionated from its solar value, whereas N (mass 14 amu) is not. In atmospheric N, $^{14}N/^{15}N = 272$ (*Anders and Grevesse, 1989*) whereas in the solar wind, $^{14}N/^{15}N = 200 \pm 55$ (*Kallenbach et al., 1998*) (Fig. 5, left). Thus the difference between the atmospheric and solar wind isotope ratios for N is in the opposite sense to that for Ne: It is the heavy isotope of N that is less abundant on Earth. This cannot be achieved by an escape process like the one that fractionated the Ne isotopes. As *Kallenbach et al.* (1998) point

out, it appears to require the addition of a light N component as originally suggested by *Geiss and Bochsler* (1982) to mix with solar N. Unfortunately, this light N component has not yet been isolated. Ordinary chondrites exhibit a wide range of $^{14}N/^{15}N$ values that are uncorrelated with the N abundances in these meteorites, indicating that mass-dependent fractionation was not acting here either (*Hashizume and Sugiura, 1995*). The N in C1 and C2 chondrites is isotopically heavier than atmospheric N (*Kung and Clayton, 1978*), whereas N in the Earth's mantle is slightly depleted in $^{15}N$, but only by 5‰ (*Marty and Humbert, 1997*).

In the local interstellar medium, $^{14}N/^{15}N = 450 \pm 100$ (*Wilson and Rood, 1994; Dahmen et al., 1995*) (Fig. 5, left). This value was determined in $NH_3$ and HCN; the value in the main reservoir of N, $N_2$, remains unknown. As N compounds should be the main carriers of N in comets (*Owen and Bar-Nun, 1995a*), we thus expect cometary N to be isotopically light. However, any attempt to relate cometary and contemporary ISM N is challenged by current lack of knowledge about the sources and evolution of $^{14}N$ and $^{15}N$ in the galaxy (*Chin et al., 1999*). In any case, the best cometary measurement thus far gave $^{14}N/^{15}N = 323 \pm 46$ in the HCN of Comet Hale-Bopp (*Jewitt et al., 1997*). This measurement provides a hint that comets may indeed be the sought-for carriers of light N, but clearly this ratio needs to be determined with greater precision in more comets before it can be regarded as definitive. [*Jewitt et al.* (1997) point out that the product ($^{12}C/^{13}C$) × $^{15}N/^{14}N$) = 0.34 ± 0.06 for Comet Hale-Bopp is indistinguishable from the terrestrial value of 0.33.] If comets do carry light N, we should expect to find some evidence for that on Mars, just as in the case of the apparently cometary D/H in near-surface water. Again as for H, the problem is complicated on Mars by atmospheric escape. In this case, nonthermal escape through dissociative recombination has led to an atmospheric value of $^{14}N/^{15}N$ = 165 ± 15 (*Nier and McElroy*, 1977) (Fig. 5, left). It is therefore necessary to search for light N in a "heavy" atmospheric background. *Mathew et al.* (1998) have reported a light component in N from the martian meteorite ALH 84001 (Fig. 5, left), with $^{14}N/^{15}N \geq$ 278, which may be the first indication of the hypothesized cometary component.

Perhaps the N originally delivered to Earth was *much* lighter than the N we find in the atmosphere today. In that case, it might be possible to fractionate both Ne and N through the same escape process to achieve the present isotope ratios. Alternatively, the original N might have been just slightly lighter, like the value found on Mars by *Mathew et al.* (1998). The absence of Ne-like fractionation in such a component can again be explained in terms of cometary delivery of volatiles, albeit in a paradoxical manner. The laboratory work shows that ice forming at T = 30 K traps less than $5 \times 10^{-3}$ of the initial Ne in a starting gas mixture (*Laufer et al.*, 1987; Bar-Nun, personal communication, 1999). At T = 50 K, this fraction drops to <$10^{-5}$. We have therefore assumed that Oort Cloud comets carry a negligible amount of Ne (*Owen et al.*, 1993; *Owen and Bar-Nun*, 1995a). This assumption is supported by the apparent absence of Ne in the atmospheres of Titan, Triton, and Pluto, where the upper limits on Ne/$N_2$ are typically about 0.01. Triton and Pluto represent giant icy planetesimals that can be thought of as the largest members of the Kuiper Belt. Hence the absence of detectable Ne in their atmospheres may be taken as a good indication that ice condensing in the outer solar nebula did not trap significant amounts of this gas, and thus we do not expect to find it in comets. (Titan is a more complicated problem because it formed in Saturn's subnebula, not in the solar nebula.) This prediction is consistent with observations of Comet Hale-Bopp carried out with the Extreme Ultraviolet Explorer satellite by *Krasnopolsky et al.* (1997b). These authors established an upper limit of Ne/O < 1/200× solar in the comet.

If the comets don't carry Ne, how did this gas reach the inner planets? Once again the meteorites don't help. Even if they brought in all the Xe, the Ne they could deliver would be <10% of what we observe. Instead it seems likely that on Mars and Earth, Ne was brought in by the rocks, the internal reservoir we have discussed above. In fact, we have evidence that this was the case, because we can still find Ne whose isotope abundances approach the solar ratio in rocks derived from the mantle (*Craig and Lupton*, 1976; *Honda et al.*, 1991). Unlike the other noble gases, Ne cannot be subducted into the Earth's interior (*Hiyagon*, 1994).

Thus it is not possible to dilute the original trapped gas with highly fractionated atmospheric Ne. The record of original emplacement, perhaps by very-low-temperature planetesimals or by the early solar wind, is preserved.

If Ne, which diffuses so easily through solids, was retained by the Earth's rocks from the time of the planet's accretion, we can reasonably assume that some water was also kept in the interior, to emerge after the catastrophic formation of the Moon, mixing with incoming water from comets to form the oceans we find today. This perspective supports the idea that we may yet find evidence of this original water with D/H < SMOW in mantle-derived rocks on Mars.

Returning to the Earth, it appears that the atmospheric Ne bears a record of an early fractionating process that sharply reduced the ratio of $^{20}Ne/^{22}Ne$ from the solar value. This process must have affected all the other species in the atmosphere at the time. It may have been the massive, hydrodynamic escape of H produced by the reduction of water by contemporary crustal iron, or from the top of an early steam atmosphere (*Dreibus and Wänke*, 1989; *Zahnle et al.*, 1990; *Pepin*, 1991). The fact that we do not see evidence of such a fractionation in atmospheric N today suggests one or a combination of two possibilities, both of which implicate comets as major sources of this volatile: Either the N we breathe reached the Earth after the vigorous escape process had ended, or the N originally delivered to Earth was even lighter than the present atmospheric value of $^{14}N/^{15}N$. Cometary delivery of N (and other volatiles) *but not Ne* offers an easy means of achieving the first possibility and may satisfy the second as well. Presently available evidence suggests the second possibility is less likely (Fig. 5), but we need more measurements of $^{14}N/^{15}N$ in comets to be sure. It may well be that Ne is a kind of atmospheric fossil, a remnant of conditions that existed on the Earth before the volatiles that produced the bulk of the present atmosphere were in place.

What about Venus? The atmosphere of this planet contains 21 ± 5× the amount of Ne found on Earth (*Donahue and Pollack*, 1983). Perhaps the grains that formed Venus simply accumulated more solar wind Ne than the Earth, because of their greater proximity to the Sun (*Wetherill*, 1981). Perhaps both planets started with the same Ne content, but less Ne escaped from Venus. This would be consistent with the higher value of $^{20}Ne/^{22}Ne$ = 11.8 ± 0.7 measured on Venus by *Istomin et al.* (1982), compared with 9.8 on Earth.

On the other hand, we have argued above that the heavy noble gases were delivered to Venus by icy planetesimals that condensed at temperatures of about 30 K or less. In this case, there is the possibility that these same objects could have delivered some Ne. For example, ice forming at T = 23 ± 3 K can trap ~3% of the ambient Ne gas, while trapping all the available Ar, Kr, and Xe. Fractionation of this amount of Ne by escape could reduce the initial planetary value of Ne/Ar = 1 to the present value of 0.25 ± 0.10 while changing the isotope ratio from 13.8 to the value of 11.8 ±

0.7 reported by *Istomin et al.* (1982). Other scenarios are certainly possible, including no Ne loss and no isotope fractionation. The latter would be accommodated by the results of *Hoffman et al.* (1980), who reported a value of $^{20}Ne/^{22}Ne = 14$ (+6/–3) for this difficult measurement. It is obviously important to derive more precise values for both the abundance and the isotope ratio of Ne on Venus to determine the origin and history of this gas on the planet.

Note that even with icy planetesimal delivery of Ne to Venus, the maximum amount of N these low temperature comets could deliver would be only 30× the Ar abundance, still approximately 30× less than the value we find in the planet's present atmosphere (*Hoffman et al.,* 1980). Hence the N requires another source, which we have suggested is icy planetesimals from the Saturn-Jupiter region, as in the case of the Earth (*Owen and Bar-Nun,* 1995a). The best current value for $^{14}N/^{15}N$ on Venus is terrestrial ±20% (*Hoffman et al.,* 1979) (Fig. 5, left). As this isotope ratio depends on the mixing of at least two reservoirs with possible modifications through escape processes, it would be surprising to find identical values of $^{14}N/^{15}N$ on both Venus and Earth. Improving this measurement should be a major objective of any future atmospheric probes to Venus.

## 6.  CONCLUSIONS

How unique is the Earth? This is a perennial question in efforts to estimate the possibilities for abundant life in the universe. We have argued here that the source of our planet's atmosphere can be found in a combination of volatiles trapped in the rocks that made the planet and a late-accreting veneer of volatile-rich material delivered by icy planetesimals. The volatiles composing the atmosphere include the C, N, and water essential to life. The question then is, how common are these rocks and ices in the galaxy? At the level of detail we need for a truly satisfactory answer, we must admit that we lack the necessary data. We can draw some confidence from the often-cited similarities between the molecules found in comets and those in interstellar clouds (e.g., *Crovisier,* 1998), which may lead us to expect that icy planetesimals that form in any planetary system originating from an interstellar cloud will carry these same biogenic materials. Hence this model for the origin of our planet's atmosphere suggests that there is nothing unique about the inventory of volatiles that was delivered to the Earth. Current investigations of the martian atmosphere, aided by study of the SNC meteorites, reinforce this idea by indicating a similar inventory on that planet. The explanation we have given for the anomalous noble gas abundances on Venus suggests that the same inventory is present on Venus, with just an exotic "spike" of gases captured at low temperatures. Based on what we know today, it appears that *Oro's* (1961) original insight probably applies throughout the galaxy.

Nevertheless, we are still in the stage of finding "similarities" and "indications." We are far from a rigorous proof of the validity of the icy impact model. We have stressed the boundary conditions provided by the abundances and isotope ratios of the noble gases and determinations of cometary D/H. As we have seen, these boundary conditions are blurred by the effects of atmospheric escape and contributions from multiple molecular reservoirs. It is clear that meteoritic delivery of volatiles cannot by itself satisfy the constraints set by our present knowledge of noble gas abundances and isotope ratios. However, the cometary alternative that we have emphasized will remain conjectural until noble gases are actually measured in a comet. Although laboratory studies strongly suggest that comets can deliver the correct *elemental* abundances, we have not yet found evidence that trapping of gas in ice affects the relative abundances of Xe isotopes (*Notesco et al.,* 2000). The sensitivity of these experiments needs to be improved, and future work should include a test of multiple sublimation and deposition events. In the absence of such evidence, we are forced to assume that comets carry Xe whose isotopes resemble the distribution found in the terrestrial and martian atmospheres rather than that found in the solar wind. It is not at all obvious why this should be the case, although the need for a light N component (*Kallenbach et al.,* 1998) and the existence of several different types of Xe (e.g., *Swindle,* 1988; *Zahnle,* 1993; *Pepin,* 1994a) in the solar system allows this possibility to be taken seriously. If the cometary Xe in fact resembles solar wind Xe, it will be necessary to invoke a fractionating process that acted to produce identical results for Xe on Mars and Earth followed by subsequent, selective replacement of other volatiles, as developed and described in detail by *Pepin* (1991, 1994b, 1997).

The Xe isotopes provide the best test of a cometary connection, as the isotopes of less massive species are more easily fractionated by atmospheric processes and similarities in elemental abundances can always be attributed to coincidence. Nevertheless, it is important not to lose sight of other key volatiles. We clearly need to know whether or not comets carry the mysterious light N component that is currently missing from known inventories.

How can we move forward from this unsatisfactory situation? There are a number of possible sources of new data on the horizon:

1. *Mars.*    The Japanese Nozomi Mission launched in 1998 carries instruments that will teach us much more about possible nonthermal escape processes on Mars. This knowledge will allow a more confident reconstruction of the early mass and composition of the martian atmosphere. These parameters can then be used (again!) to test our understanding of the origin of our own atmosphere.

If present plans mature, the step forward achieved by Nozomi will soon be overshadowed by information obtained from Mars sample return missions, scheduled to begin before 2010. These missions will bring back samples of martian rocks and atmosphere for analysis on Earth, enabling far more accurate measurements of isotope ratios than we can expect from missions to the planet. These measurements will include not only the noble gases, but also isotopes of C, N, and O, the last in both $H_2O$ and $CO_2$. With the kind of precision obtainable in laboratories on Earth, great progress should be achieved in unraveling the history of the martian atmosphere from these isotope measurements, including estimates of the sizes and locations of contemporary reservoirs of $H_2O$ and $CO_2$ (*McElroy et al.,* 1977; *Owen et al.,* 1988; *Jakosky,* 1991; *Owen,* 1992; *Jakosky et al.,* 1994). Another goal of this research should be a search for low values of D/H in water from mantle-derived rocks, which should also contain Ne with $^{20}Ne/^{22}Ne$ approaching the solar value of 13.8.

2. *Comets.* Both NASA and ESA are planning missions that will rendezvous with short-period comets from the Kuiper Belt. These missions will have the capability to detect and measure the abundances of the heavy noble gases and their isotopes, and to investigate isotope ratios of N in various compounds. It is especially important to have this information from several comets, as we already know that the composition of comets can vary, both from the laboratory work on the trapping of gas in ice (Fig. 3) and from observations of variations in the abundances of C compounds in comets (*Fink,* 1992; *A'Hearn et al.,* 1995).

The ESA rendezvous mission to Comet Wirtanen is called Rosetta; it should arrive at its destination in 2012. The first NASA comet mission is called Deep Space 1 (DS-1), a technology-testing mission that uses an ion engine for continuous thrusting and is currently (2000) on track to encounter Comet Borrelly in 2001. This initial survey will be followed by the CONTOUR mission, scheduled to launch in 2002 and designed to fly past three comets starting with P/Encke in 2003. While these missions will provide a major advance over our present state of knowledge, we will ultimately need returned samples to get the level of precision we want, just as in the case of Mars. A comet sample-return mission is not yet under serious consideration.

There is hope for an inexpensive atmospheric probe to Venus within the next decade, which would tell us more about the N isotopes and the anomalous abundances of the noble gases and their isotopes on our sister planet. We await all these new results with great interest.

***Acknowledgments.*** An early version of this material appeared in *Faraday Discussion,* 109, 453–462 (1998). This research was supported in part by NASA under grant NAGW 2631. We thank Robert O. Pepin, K. J. Mathew, and Kevin Righter for helpful reviews of the manuscript, and Tom Millar for valuable references.

# REFERENCES

A'Hearn M., Millis R. L., Schleicher J. D., and Birch P. V. (1995) The ensemble properties of comets: Results from narrowband photometry of 85 comets, 1976–1992. *Icarus,* 118, 223–270.

Anders E. and Grevesse N. (1989) Abundances of the elements: Meteoritic and solar. *Geochim. Cosmochim. Acta,* 53, 197–214.

Anders E. and Owen T. ( 1977) Mars and Earth: Origin and abundances of volatiles. *Science,* 198, 453–465.

Bar-Nun A., Herman B., Laufer D., and Rappoport M. L. (1985) Trapping and release of gases by water ice and implications for icy bodies. *Icarus,* 63, 317–332.

Bar-Nun A., Kleinfeld I., and Kochavi E. (1988) Trapping of gas mixtures by amorphous water ice. *Phys. Rev.,* B38, 7749–7754.

Balsiger H., Altwegg K., and Geiss J. (1995) D/H and $^{18}O/^{16}O$ ratio in the hydronium ion and in neutral water from in situ ion measurements in Comet P/ Halley. *J. Geophys. Res.,* 100, 5827–5834.

Becker R. H. and Pepin R. O. (1984) The case for a martian origin of the Shergottites: nitrogen and noble gases in EETA 79001. *Earth Planet. Sci. Lett.,* 69, 225–242.

Becker R. H. and Pepin R. O. (1993) Nitrogen and noble gases in a glass sample from LEW 88516 (abstract). In *Lunar and Planetary Science XXIV,* pp. 77–78. Lunar and Planetary Institute, Houston.

Benkert J-P., Baur H., Signer P., and Wieler R. (1993) He, Ne, and Ar from the solar wind and solar energetic particles in lunar ilmenites and pyroxenes. *J. Geophys. Res.,* 98, 13147–13162.

Bernatowicz T. J., Kennedy B. M., and Podosek F. A. (1985) Xe in glacial ice and the atmospheric inventory of noble gases. *Geochim. Cosmochim. Acta,* 49, 2561–2564.

Bockelée-Morvan D., Gautier D., Lis D. C., Young D., Keene J., Phillips T. G., Owen T., Crovisier J., Goldsmith P. F., Bergin E. A., Despois D., and Wooten A. (1998) Deuterated water in comet C/1996 B2 (Hyakutake) and its implications for the origin of comets. *Icarus,* 133, 147–162.

Bogard D. D. and Garrison D. H. (1998) Relative abundances of argon, krypton and xenon in the Martian atmosphere as measured in Martian meteorites. *Geochim. Cosmochim. Acta,* 62, 1829–1835.

Bogard D. D. and Johnson P. H. (1983) Martian gases in an Antarctic meteorite? *Science,* 221, 651–654.

Bogard D. D., Hörz F., and Johnson P. H. (1986) Shock-implanted noble gases: An experimental study with implications for the origin of martian gases in Shergottite meteorites. *Proc. Lunar Planet. Sci. Conf. 17th,* in *J. Geophys. Res.,* 91, E99–E114.

Carr M. H. (1986) Mars: A water-rich planet? *Icarus,* 56, 187–216.

Carr M. H. and Wänke U. (1992) Earth and Mars: Water inventories as clues to accretional histories. *Icarus,* 98, 61–71.

Chin Y.-N., Henkel C., Langer N., and Mauersberger R. (1999) The detection of extragalactic $^{15}N$: Consequences for nitrogen nucleosynthesis and chemical evolution. *Astrophys. J. Lett.,* 512, L143–L146.

Chyba A. C. (1990) Impact delivery and erosion of planetary oceans in the inner solar system. *Nature,* 343, 129–133.

Chyba C., Owen T., and Ip W.-H. (1994) Impact delivery of volatiles and organic molecules to Earth. In *Hazards Due to Comets and Asteroids* (T. Gehrels, ed.), pp. 9–58. Univ. of Arizona, Tucson.

Craig H. and Lupton J. E. (1976) Primordial Ne, helium, and hydrogen in oceanic basalts. *Earth Planet. Sci. Lett.,* 31, 369–385.

Crovisier J. (1998) Physics and chemistry of comets: Recent results from comets Hyakutake and Hale-Bopp. Answers to old questions and new engimas. *Faraday Discussions,* 109, 437–452.

Crovisier J. D., Leech K., Bockelée-Morvan D., Brooke T. Y., Hanner M. S., Altieri B., Keller H. U., and Lellouch E. (1997) The spectrum of comet Hale-Bopp (C/1995 01) observed with the Infrared Space Observatory at 2.9 astronomical units from the sun. *Science,* 275, 1904–197.

Dahmen G., Wilson T. L., and Matteucci F. (1995) The nitrogen isotope abundance in the galaxy. *Astron. Astrophys.,* 295, 194–198.

Delsemme A. (1991) Nature and history of the organic compounds in comets: An astrophysical new. In *Comets in Post-Halley Era* (R. L. Newburn Jr. M. Neugebauer, and J. Rahe, eds.), pp. 337–428. Kluwer, Dordrecht.

Donahue T. M. and Pollack J. B. (1983) Origin and evolution of the atmosphere of Venus. In *Venus* (D. M. Hunten, L. Colin, T. M. Donahue, and V. I. Moroz, eds.), pp. 1003–1036. Univ. of Arizona, Tucson.

Drake M. J., Swindle T. D., Owen T., and Musselwhite D. L. (1994) Fractionated martian atmosphere in the nakhlites? *Meteoritics, 29,* 854–859.

Dreibus G. and Wänke H. (1989) Supply and loss of volatile constituents during the accretion of terrestrial planets. In *Origin and Evolution of Planetary and Satellite Atmospheres* (S. K. Atreya, J. P. Pollack, and M. S. Matthews, eds.), pp. 268–288. Univ. of Arizona, Tucson.

Eberhardt P. M., Reber D., Krankowsky D., and Hodges R. R. (1995) The D/H and $^{18}O/^{16}O$ ratios in water from Comet P/Halley. *Astron. Astrophys., 302,* 301–316.

Fernandez J. A. and Ip W.-H. (1997) Accretion of the outer planets and its influence on the surface impact process of the terrestrial planets. In *Astronomical and Biochemical Origins and the Search for Life in the Universe* (C. B. Cosmovici, S. Bowyer, and D. Werthimer, eds.), pp. 235–244. Editirice Compositori, Bologna.

Feuchtgruber H., Lellouch E., Bézard B., Encrenaz Th., de Graauw Th., and Davis G. R. (1999) Detection of HD in the atmospheres of Uranus and Neptune: a new determination of the D/H ratios. *Astron. Astrophys., 341,* 17–21.

Fink U. (1992) Comet Yanaka (1998r): A new class of carbon poor comet. *Science, 257,* 1926–1929.

Garrison D. H. and Bogard D. S. (1998) Isotopic composition of trapped and cosmogenic noble gases in several martian meteorites. *Meteorites & Planet. Sci., 33,* 721–736.

Geiss J. (1988) Composition in Halley's comet: Clues to origin and history of cometary matter. *Rev. Modern Astron., 1,* 1–27.

Geiss J. and Bochsler P. (1982) Nitrogen isotopes in the solar system. *Geochim. Cosmochim. Acta, 46,* 529–548.

Greeley R. (1987) Release of juvenile water on Mars: Estimated amounts and timing associated with volcanism. *Science, 136,* 688–690.

Halley E. (1687) As referenced in *Comets, Popular Culture and the Birth of Modern Cosmology* by S. S. Genuth, p. 162. Princeton Univ., Princeton.

Hashizume K. and Sugiura N. (1995) Nitrogen isotopes in bulk ordinary chondrites. *Geochim. Cosmochim. Acta, 59,* 4057–4069.

Hiyagon H. (1994) Retension of solar helium and Ne in IDPs in deep sea sediment. *Science, 263,* 1257–1259.

Hoffman J. H., Hodges R. R. Jr., McElroy M. B., Donahue T. M., and Kolpin M. (1979) Composition and structure of the Venus atmosphere: Results from Pioneer Venus. *Science, 205,* 49–52.

Hoffman J. H., Oyama V. I., and van Zahn U. (1980) Measurements of the Venus lower atmosphere composition: A comparison of results. *J. Geophys. Res., 85,* 7871–7881.

Honda M., McDougall I., Patterson D., Doulgeris A., and Claugue D. A. (1991) Possible solar noble-gas component in Hawaiian basalts. *Nature, 349,* 149–151.

Hunten D. M., Pepin R. O., and Owen T. C. (1988) Planetary atmospheres. In *Meteorites* (J. F. Kerridge and M. S. Matthews, eds.), pp. 565–591. Univ. of Arizona, Tucson.

Istomin V. G., Grechnev K. V., and Kochnev V. A. (1982) Preliminary results of mass-spectrometric measurements on board the Venera 13 and Venera 14 probe. *Pisma Astron. Zh., 8,* 391–398.

Jakosky B. M. (1991) Mars volatile evolution: Evidence from stable isotopes. *Icarus, 94,* 14–31.

Jakosky B. M., Pepin R. J., Johnson R. E., and Fox J. L. (1994) Mars atmospheric loss and isotopic fractionation by solar-wind-induced sputtering and photochemical escape. *Icarus, 111,* 271–288.

Jewitt D. C., Matthews H. E., Owen T., and Meier R. (1997) Measurements of $^{12}C/^{13}C$, $^{14}N/^{15}N$ and $^{32}S/^{34}S$ ratios in comet Hale-Bopp (C/1995 01). *Science, 278,* 90–93.

Kallenbach R., Geiss J., Ipavich F. M., Gloeckler G., Bochsler P.,Gliem F., Hefti S., Hilchenbach M., and Hovestadt D. (1998) Isotopic composition of solar wind nitrogen: First *in situ* determination with CELIAS/MTOF spectrometer on board SOHO. *Astrophys. J. Lett., 507,* L185–L188.

Karlsson H. R., Clayton R. N., Gibson E. K. Jr., and Mayeda T. K. (1992) Water in SNC meteorites: Evidence for a martian hydrosphere. *Science, 255,* 1409–1411.

Krankowsky D. (1991) The composition of comets. In *Comets in Post-Halley Era* (R. L. Newburn Jr., M. Neugebauer, and J. Rahe, eds.), pp. 855–879. Kluwer, Dordrecht.

Krasnopolsky V. A., Bjoraker G. L., Mumma M. J., and Jennings D. E. (1997a) High resolution spectroscopy of Mars at 3.7 and 8 μm: A sensitive search for $H_2O_2$, $H_2CO$, HCl and $CH_4$, and detection of HDO. *J. Geophys. Res., 102,* 6524–6534.

Krasnopolsky V. A., Mumma M. J., Abbott M., Flynn B. C., Meech K. J., Yoemans D. K., Feldman P. D., and Cosmovici C. B. (1997b) Detection of soft x-rays and a sensitive search for noble gases in comet Hale-Bopp (C/1995 01). *Science, 277,* 1488.

Kung C. C. and Clayton R. N. (1978) Nitrogen abundances and isotopic compositions in stony meteorites. *Earth Planet. Sci. Lett., 38,* 421–435.

Laufer D., Kochavi E., and Bar-Nun A. (1987) Structure and dynamics of amorphous water ice. *Phys. Rev., B36,* 9219–9227.

Lécluse C. and Robert F. (1994) Hydrogen isotope exchange rates: Origin of water in the inner solar system. *Geochim. Cosmochim. Acta, 58,* 2297–2339.

Leshin L. A., Epstein S., and Stolper E. M. (1996) Hydrogen isotope geochemistry of SNC meteorites. *Geochim. Cosmochim. Acta, 60,* 2635–2650.

Linsky J. L. (1996) GHRS observations of the LISM. *Space Sci. Rev., 78,* 157–164.

Lunine J. L., Engel S., Rizk B., and Horanyi M. (1991) Sublimation and reformation of icy grains in the primitive solar nebula. *Icarus, 94,* 333–343.

Mahaffy P. R., Donahue T. M., Atreya S. K., Owen T. C., and Niemann H. B. (1998) Galileo probe measurements of D/H and $^3He/^4He$ in Jupiter's atmosphere. In *Primordial Nuclei and Their Evolution* (N. Prantzos, M. Tosi, R. von Steiger, eds.), pp. 251–264. Kluwer, Dordrecht.

Marti K., Kim J. S., Thakur A. N., McCoy T. J., and Keil K. (1995) Signatures of the martian atmosphere in glass of the Zagami meteorite. *Science, 267,* 1981–1984.

Marty B. and Humbert F. (1997) Nitrogen and argon isotopes in oceanic basalts. *Earth Planet. Sci. Lett., 152,* 101–112.

Mathew K. J., Kim J. S., and Marti K. (1998) Martian atmospheric and indigeneous components of Xe and nitrogen in the Shergotty, Nakhla and Chassigny group meteorites. *Meteoritics & Planet. Sci., 33,* 655–664.

McElroy M. B., Kong T. Y., and Yung Y. L. (1977) Photochemistry and evolution of Mars' atmosphere: A Viking perspective. *J. Geophys. Res., 82,* 4379–4388.

McSween H. Y. Jr. (1994) What have we learned about Mars from SNC meteorites? *Meteoritics, 29*, 757–779.

Melosh H. J. and Vickery A. M. (1989) Impact erosion of the primordial atmosphere of Mars. *Nature, 338*, 487–489.

Meier R. and Owen T. (1999) Deuterium in comets. In *Composition and Origin of Cometary Materials* (K. Altwegg, P. Ehrenfreund, J. Geiss, and H. Huebner, eds.), pp. 33–44. Kluwer, Dordrecht.

Meier R., Owen T. C., Matthews H. E., Jewitt D. C., Bockelée-Morvan D., Biver N., Crovisier J., and Gautier D. (1998a) A determination of the HDO/$H_2O$ ratio in Comet C/1995 01 (Hale Bopp). *Science, 279*, 842–844.

Meier R., Owen T. C., Matthews H. E., Senay M., Biver N., Bockelée-Morvan D., Crovisier J., and Gautier D. (1998b) Deuterium in Comet C/1995 01 (Hale-Bopp): Detection of DCN. *Science, 279*, 1707–1710.

Millar T. J., Bennett A., and Herbst E. (1989) Deuterium fractionation in dense interstellar clouds. *Astrophys. J., 340*, 906–920.

Mumma M. J., Weissmann P. R., and Stern S. A. (1993) Comets and the origin of the solar system: Reading the Rosetta Stone. In *Protostars and Planets III* (E. H. Levy, J. I. Lunine, and M. S. Matthews, eds.), pp. 1177–1252. Univ. of Arizona, Tucson.

Musselwhite D. M., Drake M. J., and Swindle T. O. (1991) Early outgassing of Mars supported by differential water solubility of iodine and xenon. *Nature, 352*, 697–699.

Nier A. O. and McElroy M. B. (1977) Composition and structure of Mars' upper atmosphere: Results from the neutral mass spectrometers on Viking 1 and 2. *J. Geophys. Res., 82*, 4341–4350.

Notesco G. and Bar-Nun A. (1996) Enrichment of CO over $N_2$ by their trapping in amorphous ice and implications to Comet Halley. *Icarus, 122*, 118–121.

Notesco G. et al. (2000) In preparation.

Oro J. (1961) Comets and the formation of biochemical compounds on the primitive Earth. *Nature, 190*, 389–390.

Ott U. (1988) Noble gases in SNC meteorites: Shergotty, Nakhla, Chassigny. *Geochim. Cosmochim. Acta, 52*, 1937–1948.

Ott U. and Begemann F. (1985) Are all the "martian" meteorites from Mars? *Nature, 317*, 509–512.

Ott U., Löhr H. P., and Begemann F. (1988) New noble gas data from SNC meteorites: Zagami, Lafayette and etched Nakhla. *Meteoritics, 23*, 295–296.

Owen T. (1992) The composition and early history of the atmosphere of Mars. In *Mars* (H. H. Kieffer, B. M. Jakosky, C. W. Snyder, and M. S. Matthews, eds.), pp. 818–834. Univ. of Arizona, Tucson.

Owen T. and Bar-Nun A. (1993) Noble gases in atmospheres. *Nature, 361*, 693–694.

Owen T. and Bar-Nun A. (1995a) Comets, impacts and atmospheres. *Icarus, 116*, 215–226.

Owen T. and Bar-Nun A. (1995b) Comets, impacts and atmospheres II, Isotopes and noble gases. In *Volatiles in the Earth and Solar System* (K. Farley, ed.), pp. 123–138. AIP Conf. Proc. 341.

Owen T., Bar-Nun A., and Kleinfeld I. (1991) Noble gases in terrestrial planets: Evidence for cometary impacts: In *Comets in Post Halley Era* (R. L. Newburn Jr., M. Neugebauer, J. Rahe, eds.), pp. 429–438. Kluwer, Dordrecht.

Owen T., Bar-Nun A., and Kleinfeld I. (1992) Possible cometary origin of heavy noble gases in the atmospheres of Venus, Earth and Mars. *Nature, 358*, 43–46.

Owen T., Biemann K., Rushneck D. R., Biller J. E., Howarth D. W., and Lafleur A. L. (1977) The composition of the atmosphere at the surface of Mars. *J. Geophys. Res., 82*, 4635–4639.

Owen T., Maillard J. P., de Bergh C., and Lutz B. L. (1988) Deuterium on Mars: The abundance of HDO and the value of D/H. *Science, 240*, 1767–1770.

Ozima M. and Wada N. (1993) Noble gases in atmospheres. *Nature, 361*, 693.

Ozima M., Wieler R., Marty B., and Podosek F. A. (1998) Comparative studies of solar, Q-gases and terrestrial noble gases, and implications on the evolution of the solar nebula. *Geochim. Cosmochim. Acta, 62*, 301–314.

Pepin R. O. (1989) Atmospheric compositions: Key similarites and differences. In *Origin and Evolution of Planetary and Satellite Atmospheres* (S. K. Atreya, J. B. Pollack, and M. S. Matthews, eds.), pp. 293–305. Univ. of Arizona, Tucson.

Pepin R. O. (1991) On the origin and early evolution of terrestrial planet atmospheres and meteoritic volatiles. *Icarus, 92*, 2–79.

Pepin R. O. (1994a) The hunt for U-Xenon. *Meteoritics, 29*, 568–569.

Pepin R. O. (1994b) Evolution of the martian atmosphere. *Icarus, 111*, 289–304.

Pepin R. O. (1997) Evolution of Earth's noble gases: Consequences of assuming hydrodynamic loss driven by giant impact. *Icarus, 126*, 148–156.

Pollack J. B. (1979) Climatic change on the terrestrial planets. *Icarus, 37*, 479–553.

Robert F. (1997) Paper presented at Blois Symposium: *The Solar System: The Long View* (proceedings volume in press, L. Celniker and T. Van, eds.).

Robert R., Rejon-Michel A., and Javoy M. (1992) Oxygen isotopic homogeneity of the Earth: New evidence. *Earth Planet. Sci. Lett., 108*, 1–9.

Sill G. T. and Wilkening L. (1978) Ice clathrate as a possible source of the atmospheres of the terrestrial planets. *Icarus, 33*, 13–27.

Swindle T. (1995) How many martian noble gas reservoirs have we sampled? In *Volatiles in the Earth and Solar System* (K. A. Farley, ed), pp. 175–185. AIP Conf. Proc. 341.

Swindle T. D. (1988) Trapped noble gases in meteorites. In *Meteorites* (J. F. Kerridge and M. S. Matthews, eds.), pp. 535–564. Univ. of Arizona, Tucson.

Swindle T. D., Caffee M. W., and Hohenberg C. M. (1986) Xenon and other noble gases in shergottites. *Geochim. Cosmochim. Acta, 50*, 1001–1015.

Turekian K. K. and Clark S. P. Jr. (1975) The non-homogeneous accumulation model for terrestrial planet formation and the consequences for the atmosphere of Venus. *J. Atmos. Sci., 32*, 1257–1261.

Van Dishoeck E. V., Blake G. A., Draine B. T., and Lunine J. I. (1993) In *Protostars and Planets III* (E. H. Levy and J. I. Lunine, eds.), pp. 163–244. Univ. of Arizona, Tucson.

Wacker J. F. and Anders E. (1984) Trapping of xenon in ice and implications for the origin of the Earth's noble gases. *Geochim. Cosmochim. Acta, 48*, 2372–2380.

Watson L. L., Hutcheon I. D., and Stolper E. M. (1994) Water on Mars: Clues from deuterium/hydrogen and water contents of hydrous phases in SNC meteorites. *Science, 265*, 86–90.

Wetherill G. (1981) Solar wind origin of $^{36}Ar$ on Venus. *Icarus, 46*, 70–80.

Wieler R., Humbert F., and Marty B. (1999) Evidence for a predominantly non-solar origin of nitrogen in the lunar regolith revealed by single grain analyses. *Earth Planet. Sci. Lett., 167,* 47–69.

Wiens R. C., Becker R. H., and Pepin R. O. (1986) The case for a martian origin of the shergottites. II. Trapped and indigeneous gas components in EETA 79001 glass. *Earth Planet. Sci. Lett., 77,* 149–158.

Wilson T. L. and Rood R. T. (1994) Abundances in the interstellar medium. *Annu. Rev. Astron. Astrophys., 32,* 191–226.

Yung Y. and Dissly R. W. (1992) Deuterium in the solar system. In *Isotope Effects in Gas-Phase Chemistry* (J. A. Kaye, ed.), pp. 369–389. Amer. Chem. Soc. Symp. Series 502.

Zahnle K. (1993) Planetary noble gases. In *Protostars and Planets III* (E. H. Levy and J. I. Lunine, eds.), pp. 1305–1338. Univ. of Arizona, Tucson.

Zahnle K., Kasting J. E., and Pollack J. B. (1990) Mass fractionation of noble gases in diffusion-limited hydrodynamic hydrogen escape. *Icarus, 84,* 502–527.

*Part VI:*

*Conditions on the Young Earth and Moon*

# Heavy Bombardment of the Earth at ~3.85 Ga: The Search for Petrographic and Geochemical Evidence

**Graham Ryder**
*Lunar and Planetary Institute*

**Christian Koeberl**
*University of Vienna*

**Stephen J. Mojzsis**
*University of California, Los Angeles*

The Moon experienced an interval of intense bombardment peaking at ~3.85 ± 0.05 Ga; subsequent mare plains as old as 3.7 or 3.8 Ga are preserved. It can be assumed that the early Earth must have been subjected to an even more intense impact flux resulting from its larger size and because of its proximity to the Moon. Siderophile-element analyses (e.g., Ir abundance) of the oldest sediments on Earth could be used to indicate past escalated influxes of extraterrestrial material. In addition, shocked minerals may also be present in the oldest extant rocks of sedimentary origin as detrital minerals, and remnants of impact ejecta might exist in early Archean formations. Searches for impact signatures have been initiated in the oldest sediments on the Earth, from the early Archean (>3.7 Ga) terrane of West Greenland; some of these rocks have been interpreted to be at least 3.8 Ga in age. So far, unequivocal evidence of a late heavy bombardment on the early Earth remains elusive. We conclude that either the sedimentation rate of the studied sediments was too fast and therefore too diluting to record an obvious signal, or the ancient bolide flux has been overestimated, or the bombardment declined so rapidly that the Greenland sediments, some even at ~3.85 Ga in age, do not overlap in time with it.

## 1. INTRODUCTION

Collisions between planetary bodies have been fundamental in the evolution of the solar system. Studies undertaken over the last few decades have convinced most workers that the planets formed by collision and hierarchical growth starting from small objects, i.e., from dust to planetesimals to planets (e.g., *Wetherill*, 1994; *Taylor*, 1992a,b; see also chapters in section II of this volume), and not from condensation downward. Late during the accretion of the Earth (some time after ~4.5 Ga), when it had reached about 70% of its eventual mass, it was most probably impacted by a Mars-sized or larger body (see *Cameron*, 2000). The consequences of such an impact event for the proto-Earth would have been severe and seminal, ranging from almost complete melting and formation of a magma ocean, thermal loss of preexisting atmosphere, changes in spin rate and spin-axis orientation, to accretion of material from the impactor directly, or through rapid fall-out from orbital debris below the Roche limit. Much of the material blasted off in the impact eventually reimpacted the Earth; some of the ensuing Earth-orbiting debris would have rapidly coalesced to form most of the Moon and probably some smaller moonlets. Some of this geocentrically orbiting material would have continued to impact both bodies for perhaps tens of millions of years after the lunar forming impact. The essential accretion and core formation of the

Earth appears to have been completed about 50–100 m.y. after the initial collapse of the solar nebula (*Lee and Halliday*, 1995, 1996; *Halliday et al.*, 1996; *Podosek and Ozima*, 2000), in a timescale apparently more protracted than that for smaller planetesimals and Mars. As a result of later geological activity, no record of any primary accretion bombardment history remains on the surface of the Earth.

The period on Earth between the end of accretion and the production of the oldest known crustal rocks is commonly referred to as the Hadean Eon (*Cloud*, 1976, 1988; *Harland et al.*, 1989; *Taylor*, 2000), which is a *chronostratic* division (Fig. 1). Its terminal boundary is actually not defined on the Earth; *Harland et al.* (1989) equate it with the Orientale impact on the Moon. Others do not even use the term Hadean, either distinguishing the *chronometric* divisions of Archean Eon and (older) Priscoan time (*Harland et al.*, 1989), provisionally at 4.0 Ga. Here we use the term Hadean to represent the time period between the formation of the Moon at ~4.5 Ga and the beginning of the continuous terrestrial rock record at 3.8 Ga.

In contrast with the youthful age for the crust of the Earth, the surface of the Moon displays abundant evidence of an intense bombardment at some time between its original crustal formation and the outpourings of lava that form the dark mare plains. Even prior to the Apollo missions, these plains were calculated to be about 3.6 Ga in age based on crater counts and realistic flux estimates. Hence, the

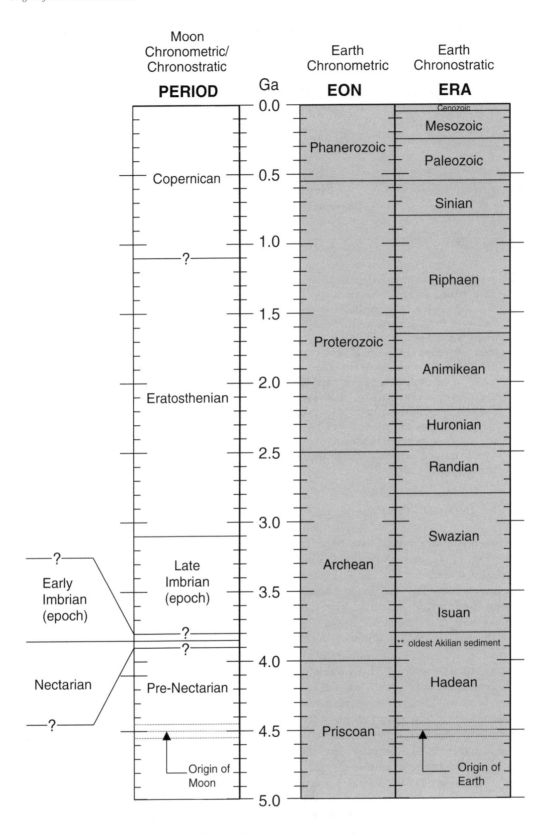

**Fig. 1.** Comparative chronostratigraphies of the Earth and Moon, based on *Harland et al.* (1989) and *Wilhelms* (1987). The times of interest in this paper are the Isuan and Hadean Eras for the Earth and the Pre-Nectarian, Nectarian, and Imbrian Periods for the Moon. The Imbrian is divided into the two Epochs of Early Imbrian and Late Imbrian, which have greatly differing styles of geological activity (rock stratigraphic units, i.e, systems, are not used in this paper). Although the chronostratic divisions into these two Epochs (the Nectarian and the pre-Nectarian) are perfectly clear, the correlation with absolute time is less established, although the age of the Fra Mauro Formation (Imbrium ejecta morphology) that defines the division of Early Imbrian and Nectarian is fairly well established at 3.84 or 3.85 Ga (e.g., *Dalrymple and Ryder,* 1993).

heavy bombardment was inferred to be ancient (*Hartmann,* 1966). Lunar highland sample data show isotopic resetting from thermal heating, for which there is abundant evidence for impact sources dominated by ages of around 3.8–3.9 Ga. The most ancient volcanic rocks from mare plains have ages of about 3.8 Ga (see, e.g., *Taylor,* 1982; *Wilhelms,* 1987). The highland ages have been interpreted to either represent a short and intense heavy bombardment period at 3.85 ± 0.05 Ga or so (e.g., *Tera et al.,* 1974; *Ryder,* 1990), or the tail end of a prolonged postaccretionary bombardment (e.g., *Baldwin,* 1974; *Hartmann,* 1975), as discussed in *Hartmann et al.* (2000). In any case, the bulk of this bombardment, which produced size-seriate scars up to multiring basins many hundreds of kilometers across, preceded 3.8 Ga. We will use the term *late heavy bombardment* to refer specifically to that bombardment of the Moon and the Earth from ~3.90 to 3.80 Ga.

In any given time-span, the Earth must have been subjected to a significantly greater bombardment than was the Moon, as it has a larger diameter and a much larger gravitational cross section, thus making it an easier target to hit (e.g., *Maher and Stevenson,* 1988; *Oberbeck and Fogleman,* 1989; *Zahnle and Sleep,* 1997). If a late heavy bombardment occurred on the Moon, the Earth was subject to a flux scaling because of the ratio of the impact cross sections (*Sleep et al.,* 1989), which may have resulted in an impact rate ≥20× greater than the lunar one, containing both more and larger impact events. The consequences for the hydrosphere, atmosphere, and even the lithosphere of Earth at that time must have been devastating (*Zahnle and Sleep,* 1997; *Grieve,* 1980; *Frey,* 1980). There is evidence that the Earth's upper mantle had already undergone some differentiation at the time of formation of the oldest igneous rocks, suggesting the prior existence of chemically evolved crust (e.g., *Harper and Jacobsen,* 1992; *McCulloch and Bennett,* 1993; *Bowring and Housh,* 1995). It has been suggested that the absence of any rocks older than about 3.9–4.0 Ga is the result of the ancient heavy bombardment, during which impact-induced mixing recycled early crustal fragments back into the upper mantle (e.g., *Grieve,* 1980; *Frey,* 1980; *Koeberl et al.,* 1998a,b). In the present contribution we outline the evidence for the character and timing of the late heavy bombardment on the Moon, and in light of this, describe petrographical and geochemical attempts to investigate if any coeval record has been preserved on the Earth.

## 2. THE BOMBARDMENT HISTORY OF THE LUNAR HIGHLANDS

### 2.1.  General

Whereas there is almost no evidence for terrestrial witnesses to the Hadean Eon, the pre-Nectarian Period, Nectarian Period, and the Early Imbrian Epoch cover this time interval on the Moon (e.g., *Harland et al.,* 1989; *Wilhelms,* 1987) (Figs. 1 and 2). The formation of a feldspathic crust was essentially complete by ~4.44 ± 0.02 Ga,

according to the recognition of lunar ferroan anorthosite of that age. The present morphology of the highlands of the Moon reflects, almost exclusively, a history of numerous subsequent impacts that occurred prior to the extrusion of the volcanic flows that form the visible mare plains (e.g., *Wilhelms,* 1984, 1987). These ancient impact structures include giant multiring basins and their debris (*Spudis,* 1993), as well as a size-seriate range of smaller craters. *Hartmann* (1965, 1966) recognized that most of this cratering occurred early in lunar history according to an estimate of the average age of mare plains of 3.6 Ga, which was calculated based on present-day cratering rates. He inferred a cratering rate averaging roughly 200× higher for the first one-seventh of lunar history than for the remainder. The general correctness of Hartmann's conclusion was demonstrated by the return of Apollo samples, and the dating of the oldest mare plains at close to 3.8 Ga (*Wilhelms,* 1987).

Geochronological studies of impact-brecciated highland samples show thermal events, most of them of impact origin, concentrated at ~3.8–3.9 Ga. These ages have been taken to represent the tail end of a heavy but declining bombardment dating back to the accretion of the Moon (e.g., *Shoemaker,* 1972, 1977; *Hartmann,* 1975, 1980; *Neukum et al.,* 1975; *Baldwin,* 1971, 1974, 1981, 1987; *Taylor,* 1982; *Wilhelms,* 1987); alternatively, they may record a sharp or cataclysmic increase in bombardment for that short interval (e.g., *Tera et al.,* 1974; *Ryder,* 1990; *Dalrymple and Ryder,* 1993, 1996). There exists a sharp drop-off in estimates for the cratering rate from the youngest highland surfaces, the Orientale and Imbrium ejecta blankets, to the oldest mare surfaces. This is according to crater counts of those surfaces, which differ by a factor of ~3–4 (e.g., *Wetherill,* 1977, 1981; *BVSP,* 1981). As a result of the difference in cratering rates, a flux at least 100× higher can be calculated for this transition period, even if those youngest highland surfaces are as much as 100 m.y. older than the oldest mare plains, which have been collecting craters for 3.8 G.y. With new studies that have expanded the age ranges for the oldest known rocks on Earth, the time span for lunar bombardment now overlaps that of these oldest rocks. Therefore, a more detailed look at the chronology and intensity of the lunar bombardment can help to understand the conditions on Earth at the time of life's emergence (*Mojzsis et al.,* 1999). The reader is referred to the paper by *Hartmann et al.* (2000) for a more complete discussion of lunar cratering history.

### 2.2.  Relative and Absolute Ages of Highland Stratigraphy

The stratigraphy of the lunar highlands has been divided on the basis of basin formation and ejecta into pre-Nectarian Period, Nectarian Period, and Early Imbrian Epoch (*Wilhelms,* 1984, 1987) (Figs. 1 and 2). These are separated by the time of production of the Nectaris Basin deposits, the Imbrium Basin deposits, and the debris blanket of the Orientale Basin respectively. Several basins were produced

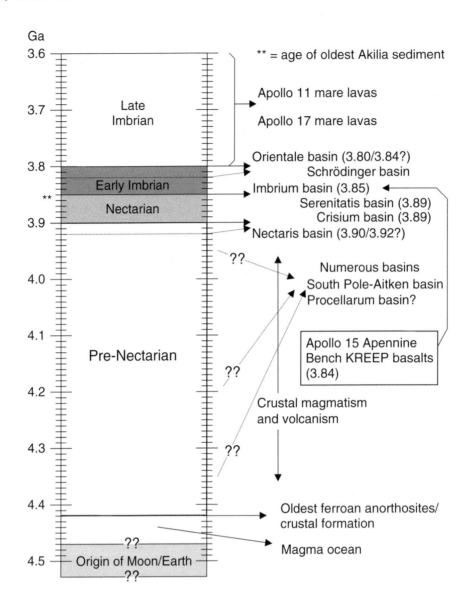

**Fig. 2.**  Stratigraphy and chronology of early lunar history, based on relative stratigraphy discussed in *Wilhelms* (1987) and absolute age inferences as discussed in this paper. The basins with underlined names define the stratigraphic column. Some other significant events or features of early lunar history are shown. While significant impacting and contraction of the geologic column is obvious at 3.8–3.9 Ga, the event/time correlations within the pre-Nectarian, and even the age of the Nectaris Basin, are much more contentious. The age of the oldest Akilia sediments, discussed in this paper, are shown (**) for comparison with lunar stratigraphy.

during the Nectarian Period, including Serenitatis and Crisium, whose ejecta regions have been sampled (Apollo 17 and Luna 20). The Schrödinger Basin is Early Imbrian, as are several large craters, including some that are almost 200 km in diameter. The oldest mare deposits were erupted in the Late Imbrian Epoch, the end of which is defined in terms of crater degradation and crater counts in the absence of any globally useful stratigraphic-datum horizons comparable to basin ejecta. The dating of these boundaries, as well as of other basins within the stratigraphic units, defines the chronology of lunar bombardment and the flux over the main period of interest here. Absolute ages quoted have been recalculated using the revised decay constants of *Steiger and*

*Jäger* (1977), and thus most are slightly younger than those given in some of the original publications.

Although these divisions for lunar time were introduced above in normal stratigraphic sequence from oldest to youngest, it is more convenient to discuss the absolute dating of the boundaries from youngest to oldest, from the simplest interpretations based on the best preserved impact craters, to the more difficult.

*2.2.1. The oldest mare surfaces.*  The Late Imbrian Epoch commenced with the formation of the Orientale Basin, the final large multiring basin to have formed on the Moon. It was followed by few large (>10 km) cratering events. The end of the Late Imbrian Epoch is arbitrarily defined, and includes the mare basalts at the Apollo 15 landing site that have been dated at ~3.25 Ga. Lavas dating to the Late Imbrian compose roughly two-thirds of the mare surfaces. Most important for the discussion here are the older mare units, including those from which mare basalt samples were collected at the Apollo 11 and Apollo 17 landing sites. The common Apollo 11 group B2 mare basalts and the rare Apollo 11 group D mare basalts are 3.80 Ga or slightly older; some Ar-Ar age determinations are as old as 3.85 Ga (*Snyder et al.,* 1994, 1996). At the Apollo 17 landing site, the oldest mare basalt so far identified formed at 3.87 ± 0.10 Ga (*Dasch et al.,* 1998), and other mare basalts from there are almost 3.80 G.y. old (see summary in *Wilhelms,* 1987). Although younger basalts were also collected from these locations, it seems likely that the presence of such old basalts close enough to the surface to be in the sample collection suggests that for all but the smallest craters (those that are a few meters across) the crater counts for these areas represent surfaces very close to 3.80 G.y. old, and perhaps slightly older. At a minimum, the crater counts for these sites represent surfaces that are at least 3.6, and probably more than 3.7, G.y. old.

*2.2.2. The age of the beginning of the Late Imbrian Epoch (the age of Orientale).* The Orientale and Schrödinger Basins are far removed from any sites sampled so far. Their ejecta have crater counts that are similar to each other and they are slightly less cratered than the Imbrium ejecta (Fra Mauro Formation, Cayley Formation). Schrödinger is older than Orientale, as it is superposed by Orientale secondaries. However, their absolute ages cannot (yet) be independently dated; they are older than the oldest affected mare plains, and thus are construed as older than ~3.80 Ga.

*2.2.3. The age of the beginning of the Early Imbrian Epoch (the age of Imbrium).* The best way to date an impact is by using the radiogenic isotopes in a clast-free or clast-poor impact-melt rock (*Ryder,* 1990; *Deutsch and Schärer,* 1994). Unfortunately, impact-melt rock that can be identified specifically as a product of the Imbrium impact is lacking, and those considered most likely (the Apollo 15 dimict breccias; *Ryder and Bower,* 1977) have disturbed Ar-Ar systems (e.g., *Bogard et al.,* 1991). Recently, *Haskin et al.* (1998) have argued that all Th-rich impact melt breccias (low-K Fra Mauro) collected on the Apollo missions are products of the Imbrium impact event. Despite the dating problems, there are ways to bracket the age of Imbrium. First, the Apennine Bench Formation is a volcanic plateau inside (hence younger than) the Imbrium Basin. Remote sensing of its morphology and chemistry allow correlation with Apollo 15 volcanic KREEP basalt samples, which have been dated. This provides a lower age limit on the Imbrium Basin of 3.84 ± 0.02 Ga (*Ryder,* 1994). Second, the Apennine Front has been little modified since the formation of the Imbrium Basin. Thus crystalline impact melt collected there should be almost entirely Imbrium impact melt, or older impact melt. *Dalrymple and Ryder* (1991, 1993) dated such melt rocks and suggested that Imbrium is

certainly no older than 3.870 ± 0.010 Ga, and probably no older than 3.836 ± 0.016 Ga. Third, similar arguments applied to the contents of the Fra Mauro Formation and the Cayley Formation suggest a similar age constraint. For example, impact-melt fragments in the white rock 14063 from Cone Crater show a range from 3.87 to 3.95 Ga; other samples that are probably not from the Fra Mauro Formation, but represent later local events (such as the 14310-group samples), are a little younger (3.82 ± 0.02 Ga). Thus, it is safe to bracket Imbrium as 3.85 ± 0.02 Ga. This is consistent with the the older Serenitatis Basin (below) having formed at about 3.89 Ga.

*2.2.4. The age of the beginning of the Nectarian Period (the age of Nectaris).* Stratigraphic and crater count data show that the Nectaris Basin is older than the Crisium Basin, but melt-rock samples from it cannot be identified with certainty. The Apollo 16 site was modified by Nectaris ejecta, and subsequently by Imbrium ejecta. Fragments within the breccias collected on the Apollo 16 mission probably include samples of melt created prior to Nectaris in several clearly recognizable large craters that underlie the site. None of the melt samples dated so far is reliably older than 3.92 Ga. The analysis of the rocks and ages by *James* (1981) strongly suggests an age for Nectaris of less than 3.92 Ga, and probably an age of ~3.90 Ga is reasonable, consistent with earlier derivations by *Turner and Cadogan* (1975) and *Maurer et al.* (1978).

The Nectarian Period also witnessed the formation of the Serenitatis and Crisium Basins, and samples were collected from their rims or ejecta. At the Apollo 17 landing site, the highland materials are dominated by coherent poikilitic melt rock, commonly in the form of boulders whose trails can be seen to run high up the massifs. These samples are most readily interpreted as melt formed in the Serenitatis Basin event. If they are not, then they are probably older, as it is inconceivable that they are ballistic ejecta from the Imbrium event. Most of these samples belong to one chemical group whose age, as determined on several samples, is now precisely established as 3.893 ± 0.009 Ga (*Dalrymple and Ryder,* 1996). This age is outside of the bracket for the Imbrium age described in the previous section. The Luna 20 sample from Crisium ejecta includes impact-melt rock samples. From these, *Swindle et al.* (1991) suggested an age of ~3.89 Ga for the Crisium Basin. These ages for Serenitatis and Crisium are consistent with an age of the older Nectaris Basin of 3.90 Ga. Several other basins, e.g., Herzsprung and Humorum, also formed after Nectaris. Thus, there was considerable bombardment of the Moon in the 60 m.y. between 3.90 Ga and ~3.84 Ga.

*2.2.5. Pre-Nectarian Period and events.* The lack of impact-melt rocks in the sample collections that are older than ~3.92 Ga cannot be due to resetting of all older ages, given the difficulties of such resetting (e.g., *Ryder,* 1990; *Deutsch and Schärer,* 1994). Most of the lunar upper crust has not been converted into impact-melt rock, which would be subject to resetting. Thus the paucity of pre-3.92-Ga impact melt can be taken as evidence that there was little impacting prior to that time, other than that expressed by the metamorphosed breccias of uncertain origin, the feldspathic granulites, that may well date back to the very earliest postaccretionary bombardment at about 4.4 Ga. Furthermore, the Pb-isotopic data of *Tera et al.* (1974) indicate events at ~3.85 Ga and events at >4.4 Ga, but not much evidence of events in between; continual resetting of Pb clocks would show up as intersects in the 4.4–3.9-Ga Pb-isotope growth curve. There is also a lack of the complement of siderophile elements that would be expected to be present in older upper crustal rocks if a heavy bombardment between 4.4 and 3.9 Ga had occurred (*Ryder,* 1999), despite claims to the contrary (e.g., *Sleep et al.,* 1989; *Chyba,* 1991). A more complete discussion of these features appears in the chapter by *Hartmann et al.* (2000).

## 2.3.  Summary of the Significance of the Lunar Cratering Record from 3.90 to 3.80 Ga

During the period from 3.90 to 3.80 Ga, a substantial amount of the extant lunar highland features, including Nectaris and many younger basins, formed on the Moon. Based on the above discussion, it is possible that this intense activity terminated at 3.85 Ga with the near-simultaneous creation of the Imbrium, Schrödinger, and Orientale Basins. Bombardment may even have finished as early as 3.87 Ga. Although the last two basins might have formed as late as 3.80 Ga, this seems unlikely given that their superposed crater populations are almost as high as those on the Imbrium ejecta, yet greater than those on the oldest mare plains, which themselves are ~3.8 G.y. old. Many lunar basins, including South Pole-Aitken, formed prior to this period, but at present there is no direct or definitive way to date their formation; South Pole-Aitken might be as young as 3.95 Ga, or as old as 4.3 Ga. In principle, future missions can obtain samples from which the chronology of ancient intense bombardment can be more reliably determined. In particular, the age of Orientale could be precisely determined, as its impact melt sheet is intact and accessible, and would constrain the younger end of bombardment. South Pole-Aitken may be datable and could provide a constraint at stratigraphically older times, because although it has been battered, remnants of the impact-melt sheet should be collectable and recognizable.

The chronology outlined above suggests a massive decline in the flux of bombardment on the Moon over a short period of time following the Nectaris event. The cratering on the Nectaris ejecta (3.90 Ga) is a factor of ~4 higher than that on Imbrium ejecta (3.85 Ga), which, in turn, is a factor of ~4× that recorded on the oldest mare plains (about 3.80 Ga, or even slightly older).

As an exercise, let us define that there are C units of craters on 3.80-G.y.-old mare plains. Then the average cratering rate from 3.80 Ga to the present is C per 3.80 units and Ga (i.e., 0.263C units/Ga). Imbrium ejecta has ~4C units of craters. Thus, the cratering rate between Imbrium formation and oldest mare plains is (4C–1C) per 0.05 units and Ga (assuming 50-m.y. age differences), which is a rate of

60C units/Ga. Therefore, the relative cratering rate of the 3.85–3.80-Ga period compared with the average since 3.80 Ga is ~228.

Furthermore, let us assume that Nectaris is 3.90 Ga and Imbrium is 3.85 Ga. There are ~16C units of craters on Nectaris ejecta. Thus the cratering rate during this period is (16C–4C) per 0.05 units and Ga, i.e., 240 units/Ga. This is 912× the average rate since 3.80 Ga.

The present rate, or the Phanerozoic rate, of cratering is probably a little lower than the average over the last 3.80 G.y., because there was a higher flux in the Late Imbrian Epoch than in the succeeding Eratosthenian and Copernican (indeed, there is evidence that suggests a higher flux in the Eratosthenian than in the Copernican; *Ryder et al.,* 1991; *Culler et al.,* 1999). In round figures the cratering rate in the period 3.90 Ga to 3.85 Ga was probably at least 1000–1500× that of the Phanerozoic, and in the period 3.85–3.80 Ga was probably at least 250–400× that of the Phanerozoic. These are higher than the rates inferred by *Hartmann* (1966), because he assumed that the observed cratering record stretched back almost to the origin of the Moon, whereas it is actually much more restricted in time. It is even possible that the later decline took place over only the first 10 or 20 m.y. after ~3.85 Ga, such that by 3.84 Ga or 3.83 Ga the flux was approaching within a few factors of that of the present day. This is the record that needs to be compared with that of the oldest rocks on Earth.

## 3.  STATE OF THE SURFACE OF THE EARTH FROM 4.5–3.8 GA

### 3.1.  Earliest Crust

Recognized extant terrestrial crustal rocks extend back to only about 89% of the history of the planet, to ~4.0 Ga; the record is improved somewhat if we include the potential information gleaned from rare detrital zircon grains that are up to 4.27 Ga in age, ~94% of Earth history. As mentioned above, is likely that the Moon-forming impact led to a large-scale melting of the Earth and the existence of an early magma ocean (e.g., papers in this volume). Mantle temperatures in the Hadean were probably much higher than today. About half of all heat produced by $^{235}$U decay to $^{207}$Pb was released during the Hadean, $^{40}$K was more abundant, as well as latent heat from accretion, all of which added several hundred degrees to the internal temperature of the Earth. Any late accretionary bodies would have added further thermal energy to the already elevated budget of heat flow in the early Earth (e.g., *Smith,* 1981; *Davies,* 1985; *Taylor,* 1993, 2000).

The nature of the earliest crust on Earth, and the amount of crust present, has been the subject of intense debate. Petrological melting concepts and comparisons with other planets suggests that the earliest crust on Earth was basaltic in composition (e.g., *Taylor,* 1989, 1992, 1993; *Arndt and Chauvel,* 1991). The existence of any substantial early feldspathic crust on the Earth seems precluded by the higher pressure at shallower depth on the Earth (in contrast to the

Moon, which does have an ancient feldspathic crust). Calcium and Al are sequestered in the deeper Earth in early high-pressure phases (particularly garnet), which delays the concentration of those elements reaching the point of plagioclase crystallization. In addition, plagioclase itself cannot crystallize at significant pressure (hence depth), so any plagioclase-bearing terrestrial crust would be thin. Neither would any plagioclase be likely to float in a water-bearing, basaltic magma ocean on the Earth, so no concentration of plagioclase toward the surface would be realized. Finally, there is no indication of any ancient reservoir of Eu or of primitive $^{87}Sr/^{86}Sr$ signatures that could have resided in an early high-Sr and low-Rb anorthositic crust (e.g., *Taylor,* 1989). Some pre-4.0-Ga differentiation of the mantle seems to have occurred, as indicated by isotopic evidence (e.g., *Harper and Jacobsen,* 1992; *Bowring and Housh,* 1995; but see also *Gruau et al.,* 1996, for cautionary remarks). Detrital zircon crystals in an Archean quartz-pebble conglomerate from the Narryer Gneiss Complex, Western Australia, are the oldest known minerals on Earth, with ages up to 4.27 Ga (e.g., *Compston and Pidgeon,* 1986). The morphological, mineralogical, and geochemical characteristics, as well as similarities with post-3.75-Ga zircons, indicate a composite granitoid source of continental provenance for these zircons (*Maas et al.,* 1992; *Mojzsis,* 1998). Thus there is evidence for at least minor amounts of felsic igneous rocks in the Hadean, which may have been present in small amounts from remelting of basaltic crust that sank back into the mantle (e.g., *Taylor,* 1989, 2000). It remains unlikely, though unproven, that significant amounts of continental crust existed on Earth during much of the Hadean Eon. The lack of initial Hf-isotopic heterogeneity and the absence of negative $\varepsilon_{Hf}$ values in early Archean rocks provides evidence against the presence of large amounts of continental crust on the Hadean Earth (*Vervoort and Blichert-Toft,* 1999). Granitic crusts require multistep derivation from the primitive mantle by recycling of subducted basaltic crust through a "wet" mantle, which will slowly lead to an increasing amount of granitic crust through time (*Taylor and McLennan,* 1995).

The lithosphere of the Hadean Earth was most probably characterized by a basaltic crust, covered by an ocean, and with little dry land and only minor amounts of felsic rocks (granitoids). Any sedimentological record, which would host information specific to surface environments such as the rate and violence of meteorite impact and the presence of life, has been almost completely lost from Hadean times, and only appears at its conclusion, near 3.90 Ga (*Mojzsis et al.,* 1996, 1999; *Mojzsis and Harrison,* 2000; *Nutman et al.,* 1996, 1997).

### 3.2. Effects of Ancient Impacting: From Basins to Dust

Individual impacts have considerable physical (morphological) and chemical effects on the target and on the atmosphere. A crater is excavated, fragmental ejecta are strewn around and into the crater, and a melt unit can be created. Some of the ejecta might be in the form of molten spherules. The projectile and some target rock are vaporized, and a fraction of the projectile vapor can be incorporated into melt and ejecta. Minerals of both the autochthonous target and the allochthonous ejecta could exhibit shock effects (e.g., planar deformation features, high-pressure polymorphs, diaplectic glasses) from the interaction of the rocks and minerals with the shock wave. If the target includes water (e.g., ocean impact) then that water gets vaporized. If the impactor contains typical chondrite-like abundances of platinum-group metals (e.g., Ir), as all meteorites other than most differentiated stony meteorites do, then these will be added to the impactite. Depending on the density of the atmosphere, there is a lower size limit below which small impactors do not penetrate the atmosphere and therefore will not form craters. At very large impactor diameters, excavation of mantle material is possible, as well as large-scale vaporization (and possible loss) of atmosphere and hydrosphere. Judging from the lunar record (see section 2), very little of the Earth's surface in the period of ~3.9–3.8 Ga should have escaped being the target of significant impacts at one time or another, and therefore escaped being covered by ejecta from craters that are at least a few kilometers in diameter. However, a more vigorous rock cycle than at present continually resurfaced the early Earth and erased (most?) evidence for such an impact environment.

Calculations that scale impact-melt production with increasing crater dimension (*Melosh,* 1989; *Cintala and Grieve,* 1994, 1998) show a breakdown of this geometric relationship for very large impact structures. As the magnitude of the impact increases, the melt volume relative to the transient crater size increases, with a larger proportion being retained inside the crater, and the depth of melting for large impact structures exceeding the depth of excavation. Therefore, the thermal effects of an impact (i.e., the large-scale melting) will actually reduce the amount of shocked rocks that are formed and preserved. In large-scale impact events, leading to the formation of craters larger than a few hundred kilometers in diameter, thermal metamorphism may be more important than shock metamorphism. However, craters smaller than a few hundred kilometers in diameter would still largely have fragmental and shocked ejecta and basement.

Most of the (considerable) speculation regarding the effects of ancient impacts on the Earth has focused on large, potentially basin-forming, events. These models attempt to understand the localization and extent of endogenic activity, such as volcanism, proto-ocean basin formation, atmospheric disturbance, continental growth and assembly, and changes in sedimentation style and topography, rather than relying on direct impact evidence. *Grieve* (1980) and *Frey* (1980) discussed the effect of impact structures with diameters exceeding 100 km on the ancient Earth, prior to about 3.8 Ga. By scaling the lunar impact record to the Earth these authors concluded that about 2500–3000 impact structures with diameters larger than about 100 km could have formed. Their simulation resulted in almost 1000 craters with diameters exceeding 200 km, and possibly about 10 structures with diameters larger than that of the Imbrium Basin on the Moon (about 1300 km diameter). This crater population would have covered about 40% of the surface of the Earth. Using the minimum estimate for the cratering frequency, *Grieve* (1980) derived a cumulative energy of about $10^{29}$ J added to the Hadean Earth from impact events, and concluded that the net effect of large impact events was to localize and accelerate a variety of endogenic geological activity.

Several studies have considered the effects of impact on the atmosphere and hydrosphere, again, particularly for very large events (*Maher and Stevenson*, 1988; *Oberbeck and Fogleman*, 1989; *Sleep et al.*, 1989; *Chyba*, 1993; *Zahnle and Sleep*, 1997). These studies have largely been expressed in the context of the early evolution of life and impact-induced sterilization. An Imbrium-scale impact onto the early Earth would have the ultimate effect of boiling off about 40 m of seawater, with a subsequent hot surface layer and annihilation of any surface ecosystems (*Zahnle and Sleep*, 1997); expected events 10× as large as this would have correspondingly larger and more devastating effects. It probably requires the impact of an asteroid several hundred kilometers in diameter to totally vaporize one present-day ocean mass of water. The scale of these events is probably too great and destructive to allow preservation of evidence. It is the probability for these vaporizing impacts on the early Earth that has led to the general impression that impact events were a negative forcing function for the development and evolution of emergent life (e.g., *Grieve*, 1998).

Along with the mega-impacts there would be numerous smaller impacts, producing more recognizable ejecta blankets, shock features, and input of siderophile elements. Simultaneously, there should be a correspondingly greater abundance of input of interplanetary particles and continuous rain of dust (that ultimately is incorporated into rocks with ongoing sedimentation) than there is at the present day. It is to these smaller-scale features that attention should be paid, to find evidence of impact in the oldest rocks.

## 4. SEARCH FOR EVIDENCE OF A LATE HEAVY BOMBARDMENT ON THE EARLY EARTH

### 4.1. Earliest Sedimentary Rocks on Earth

The critical sedimentary record of the earliest Archean is preserved in the North Atlantic province, principally in the Isua district and the Akilia association in southern West Greenland that are part of the Itsaq Gneiss complex (*Nutman et al.*, 1996). The Isua Supracrustal belt is in effect a giant version of the smaller enclaves of Akilia rocks with abundant gneisses. The Itsaq Gneiss complex of West Greenland is a 3000-km² terrane dominated by orthogneisses of granitoid compositions that intrude, in some locations, packages of associated sediments and volcanic rocks. These supracrustal rocks are composed of massive amphibolites and complex metasomatic carbonates (metamorphosed equivalents of pillow basalts and other components of early

Archean oceanic crust), ubiquitous banded iron formations (chemical sedimentary precipitate, dominated by quartz and magnetite), rare graywacke, and metapelites. The inferred environment of deposition for these volcanosedimentary successions is a sediment-poor arc or back-arc basin in relatively deep water (*Nutman et al.*, 1984). Studies of early Archean sediments from the Isua Supracrustal belt (ISB), which are ~3.80 Ga, and rocks of the Akilia association in the Godthåbsfjord region (>3.80 Ga) of southern West Greenland, suggested that they are the oldest sediments yet identified (*Nutman et al.*, 1997).

There are uncertainties concerning geological relationships on Akilia island and the nearby islets of the Godthåbsfjord archipelago that host the oldest known sediments of marine origin, and also contain evidence for life (*Mojzsis et al.*, 1996; *Nutman et al.*, 1996, 1997). These derive from reconnaissance-scale geological mapping that reveals little about the structural relationship of the banded iron formations to the polyphase, geochemically heterogeneous orthogneisses that intrude them. The geochronological relationships as they are currently inferred have been used to place a minimum age of formation for some of the sediments in excess of 3.85 Ga (*Nutman et al.*, 1997). In contrast, Moorbath and co-workers (e.g., *Moorbath et al.*, 1997; *Kamber et al.*, 1998; *Moorbath and Kamber*, 1998; *Kamber and Moorbath*, 1998; see also *Rosing*, 1999) argued that these oldest ages represent only those of zircon inherited from assimilated preexisting rocks older than the intruding granitoid orthogneisses on Akilia and the surrounding islands. However, evidence for much Pb contamination from hypothetical assimilated zirconiferous rocks is absent from the orthogneisses of the Itsaq, so the zircons are probably not inherited. Furthermore, the intruding granitoids are low in Zr, granodioritic melts are strongly undersaturated with respect to Zr, and the rocks they intrude are poor in zircon. Age estimates of 3.65 Ga derived for the intruding gneisses of southern West Greenland, which are based on whole-rock Pb-Pb, Sm-Nd, and Rb-Sr errorchrons, are susceptible to open-system REE-, Sr-, and Pb-diffusion behavior, in contrast to precise and concordant zircon geochronology (*Mojzsis and Harrison*, 2000). It is not possible to resolve the age issue here; however, this question has important implications for the search of traces of any late heavy bombardment on the Earth, as these terranes presently provide the only qualified samples to search for extraterrestrial components of a late heavy bombardment on Earth. We recognize that the evidence for a 3.85-Ga or older age for the sedimentary Akilia rocks under consideration here is stronger than that for a younger age.

### 4.2. Search Strategies and Their Rationales

We discuss three strategies used to search for evidence of a late heavy bombardment on the early Earth. First, it is possible to search for chemical evidence in sedimentary rocks that would indicate an enhanced flux of extraterrestrial materials, using different techniques and samples from both Isua and Akilia rocks from Greenland. Second, evidence of detrital shocked minerals that might have formed as a result of an incessant early bombardment may be preserved. Third, it may be possible to recognize remnants of impact ejecta (albeit strongly altered and metamorphosed) that might have been incorporated into early Archean rock formations.

## 4.3. Meteoritic Siderophile-Element Signatures

*4.3.1. Siderophile elements on the early Earth.* The Earth is a highly differentiated body, with a core, a mantle, and evolved crust. During planet formation, the highly siderophile elements (e.g., Ir, Pt, Au) partition strongly into metallic cores. The formation of the Earth's core was completed early, well before the formation of the most ancient of preserved terrestrial crustal rocks, and certainly by the time of lunar formation. Thus, Earth's earliest mantle and crustal rocks were effectively stripped of their highly siderophile element inventory early on. However, the present-day upper mantle has abundances of highly siderophile elements much higher than expected from presently known silicate-metal distribution coefficients and under the assumption of core-mantle equilibrium (Ir ~3 ppb) (*Chou*, 1978; *Chou et al.*, 1983; *Newsom*, 1990). The siderophile-element abundances show chondritic relative proportions, which plot subparallel to the CI line. The addition of ~0.75% chondritic material after termination of the core-upper mantle equilibrium under increasingly oxidizing upper mantle conditions seems necessary to explain the abundances and chondritic relative proportions of the siderophile elements in the mantle (e.g., *Chou et al.*, 1983; *Newsom*, 1990; *Holzheid and Palme*, 1998). The emplacement timing of such a veneer is not constrained by direct evidence, but is often invoked to have been as early as 4.40 Ga, or as late as 3.80 Ga.

Siderophile elements are strongly fractionated during partial melting; for example, basalts are strongly depleted in Ir (<0.05 ppb) relative to mantle peridotites (~3 ppb). In rare circumstances, siderophile elements can be concentrated in specific crustal reservoirs, e.g., platinum-rich layers in some basic intrusions; these have relative platinum-group element abundances that are strongly fractionated from chondritic values. More evolved rocks, such as pyroclastics, granites, and the sediments derived from them, contain negligible siderophile-element abundances from terrestrial sources. The source of significant abundances of siderophile elements in evolved crustal rocks, such as the melt rock at East Clearwater Lake crater (*Palme et al.*, 1979), or in the Cretaceous-Tertiary boundary clay layer (*Alvarez et al.*, 1980), can be reliably attributed to an extraterrestrial source. Thus siderophile elements in sedimentary rocks, other than in the rarest of circumstances, can be taken as an indication of an extraterrestrial flux at the time of formation of the sediments, particularly if they are in chondritic relative abundances.

Estimates of the flux of extraterrestrial material to the Earth, based on lunar stratigraphic-chronologic studies discussed above, suggest that during the peak of the late heavy bombardment this flux was, or ranged, between ~3 × 10[2]

(low estimate) and ~10[4] (high estimate) greater than at present. While this estimate is based on visible lunar craters, generally of the order of a few kilometers in diameter and larger, it is inferably true of smaller craters and of interplanetary particles and dust as well. In a geochemical sense, it does not matter whether a projectile makes a crater or burns up in the atmosphere; it will be added to the sediment as the dust settles. On the Moon the extralunar material also has high abundances of the siderophile elements (for example, the Serenitatis impactor was almost certainly an EH chondrite, *James*, 1995). All lunar impact-melt rocks from the late heavy bombardment contain Ir in the 2–20-ppb range (*Papike et al.*, 1998). On Earth, a sedimentary layer at ~3.85 Ga might show evidence for an influx of siderophile elements from an enhanced continuous background fallout, or from a specific event comparable with the Cretaceous-Tertiary boundary layer, where such events had a higher probability than at the present.

*4.3.2. Terrestrial sources of iridium in marine sediments.* Experiments have shown that ~50% of Ir in sediments is scavenged from seawater by Fe-Mn-O-OH particles (*Anbar et al.*, 1996) in oxic to suboxic environments. Anoxic environments, such as would be the case for much of the Archean hydrosphere (*Holland*, 1984), are not a major sink for Ir because of the redissolution of particulate hydroxides, except at rapid sedimentation and relatively shallow water depths. Iridium is well mixed in the oceans: The residence time for it in the hydrosphere is 2000–20,000 yr. This implies that extraterrestrial Ir could persist in seawater and be incorporated into sediments by particulate scavenging between impacts of a frequency of less than 2000 yr. It was also found by *Anbar et al.* (1996) that Ir (and Os) abundances in present-day seawater are extremely low. Thus weathering and hydrothermal alteration of ultramafic rocks, such as peridotite, which could supply Ir (and other platinum-group elements) to seawater, is insignificant in determining the abundances of these elements in present-day seawater.

Studies of REE distributions in banded iron formations demonstrate that hydrothermal activity had a strong influence on overall seawater chemistry in the Archean (*Bau and Möller*, 1993). The average concentrations of Ir in pelagic clays with sedimentation rates of ~0.001–0.003 mm a[-1] range from 0.07 to 2.0 ppb (*Barker and Anders*, 1968; *Kyte and Wasson*, 1986); in metalliferous sediments that scavenge Ir, concentrations are even higher (*Anbar et al.*, 1996). Some of these higher abundances might result from organic matter scavenging, and therefore do not reflect extraterrestrial input directly. Because these pelagic sediments are very slow to accumulate, they contain measurable Ir even at the present-day very low rates of meteoritic input.

*4.3.3. Extraterrestrial sources of iridium to the hydrosphere.* Estimates of the influx of extraterrestrial matter reaching Earth's surface during the past 100 m.y. have been the subject of numerous studies aimed at quantifying the current rate of dust accretion and the composition and source of the material. A number of methods have been used to determine this flux, using the collection of dust in the atmosphere, glacial ice, and pelagic sediments (*Love and Brownlee*, 1993). The current mass flux of extraterrestrial Ir is based on measurements of sedimentary Ir in systems with calculable sedimentation rates and calculations of the flux of infalling dusts by satellite, radar, and airborne observations. *Love and Brownlee* (1993) have estimated the amount of chondritic material raining into the Earth as dust, from measurements of abundances and sizes of microcraters developed on the Long Duration Exposure Facility (LDEF) experiment, as $40\ (\pm 20) \times 10^9$ g a$^{-1}$. Assuming chondritic relative proportions, this translates to $70 \pm 35$ mol Ir a$^{-1}$ to the whole Earth. The uncertainties reflect counting and, more importantly, the inferred encounter velocities. An assumption is that the six-year length of the LDEF experiment is adequate to be representative of the current (i.e., last few million years) flux and its possible variations. The abundances are consistent with those derived from Os isotopes in deep-sea sediments (*Esser and Turekian*, 1988) and Ir in both Antarctic ice (*Ganapathy*, 1983) and abyssal red clays (*Kyte and Wasson*, 1986). Studies by *Bonté et al.* (1987) have shown that almost all platinum-group elements present as cosmic debris occur in grains $\ll 10$ μm. These grains are quickly incorporated into sediments (*Esser and Turekian*, 1988). The finer dust grains are probably sensitive to seawater oxidation and hydrolysis after burial and have probably always contributed to a small hydrogenous component of seawater Ir.

*4.3.4. Ancient sediments and model extraterrestrial influx.* The oldest terrestrial sediments might be expected to preserve a signal of higher incident fluxes from interplanetary dust particles, micrometeorites, local impacts, airburst, cometary showers, and ablation products of such phenomena. The amount of Ir from the background that would be expected to be sampled by the water column and thus a sediment deposited or precipitated from the early Archean ocean, $[\text{Ir}]_{\text{SED}}$, can be estimated by

$$[\text{Ir}]_{\text{SED}} = (\Phi_m \times f \times [\text{Ir}]_{\text{ET}})/(\oplus_A \times \varphi_{\text{SED}} \times \rho_{\text{SED}}) \quad (1)$$

where $\Phi_m$ = estimated present extraterrestrial flux for all incoming material, f = factor increase for ancient flux, $[\text{Ir}]_{\text{ET}}$ = concentration of Ir in extraterrestrial material, $\oplus_A$ = Earth surface area, $\varphi_{\text{SED}}$ = sedimentation rate of the deposit SED, and $\rho_{\text{SED}}$ = density of SED.

A critical parameter in equation (1), other than those already discussed, is the sedimentation rate for the sample being analyzed for siderophile elements. Different sediments have deposition rates that differ by orders of magnitude. For our purposes, samples with a slow deposition rate are desirable, because they have a higher proportion of extraterrestrial material (which is why clay was investigated at the Cretaceous-Tertiary boundary; *Alvarez et al.*, 1980). The early Archean sedimentary rocks we have to work with are autochthonous precipitates, such as banded iron formations and quartzites (as metamorphosed chert). They contain negligible contributions from the weathering detritus of igneous rocks (*Dymek and Klein*, 1988), which is advantageous insofar as some of these (e.g., peridotites) *could* blur the

siderophile-element signal from hydrogenous sources of Ir (*Anbar et al.*, 1996) and other metals, although this is not a major concern. Banded iron formations from the Isua district of southern West Greenland, and also younger ones from West Australia and southern Africa, contain no significant clastic sediment components and no near-shore or evaporitic facies. Therefore, when the oldest banded iron formations formed, they must have sampled for the most part the chemistry of the water column from which they precipitated, including any extraterrestrial component.

Unfortunately, the sedimentation rate for banded-iron-formation deposition is poorly constrained. These rocks do not form in Phanerozoic environments because the pO$_2$ of the atmosphere has been too high since the Proterozoic Era, and because Fe$^{2+}$ forming by rapid oxidation to Fe$^{2+}$(Fe$^{3+}$)$_2$O(OH)$_6$ transforms to Fe$^{2+}$(Fe$^{3+}$)$_2$O$_4$ (magnetite), which has low solubility in seawater. In general, detailed sedimentological interpretations of banded-iron-formation sequences have not been available (*Klein and Beukes*, 1990). There have been considerable differences of opinion about the origin of banded iron formations and the particular environments of their deposition (*James*, 1954; *Trendall and Blockley*, 1970; *Cloud*, 1973; *Holland*, 1973; *Klein and Beukes*, 1990), with general agreement that deposition took place below wave-base. The individual bands of iron formations are considered by many workers as being equivalent to varves associated with seasonal changes in upwelling, productivity, and local O production and other factors (*Holland*, 1984, and references therein). *Trendall and Blockley* (1970) estimated the rate of deposition of the Hamersley banded iron formation (~2.5 Ga). From counting chert + magnetite ± hematite microband couplets between volcanic rocks of known age that were interbedded with the ironstones, these authors estimated a deposition rate of 0.65– 1.3 mm a$^{-1}$, which is much faster than even typical detrital sediments such as shale and siltstone. However, while banding in the Hamersley Basin may be on a scale of ~1 mm, banding elsewhere is on much coarser (centimeters) and much finer (submillimeter) scales. Indeed, banding occurs at various repetitions in any sequence, with laminae bundled into alternatively quartz-rich and magnetite-rich "beds" and higher-order packages. If banded iron formations are dominantly a reflection of hydrothermal processes and iron input (*Isley*, 1995), with or without a biogenic influence (*Cloud*, 1973; *Holm*, 1987), then the repetitions may have nothing to do with *annual* fluctuations.

More recent interpretations of accumulation rates for banded-iron-formation rates in the Hamersley Basin and Transvaal deposits are based on detailed radiogenic chronology of sequences. They suggest depositional rates orders of magnitude slower than those proposed by *Trendal and Blockley* (1970), ~0.001–0.004 mm a$^{-1}$ (*Arndt et al.*, 1991; *Barton et al.*, 1994). These rates apply to both shale and banded-iron-formation deposits; underlying dolomites may have been deposited an order of magnitude faster than those. Deposition might not have been continuous, so that individual bands might have been deposited quickly, followed by a depositional hiatus. Such hiatuses are not interpreted to reflect unconformities, and all extraterrestrial material deposited in an entire time package should be in the sequence, perhaps concentrated at grain boundaries. The essential point is that a wide range of possibilities for the overall depositional rate for banded iron formations exists, and 1 mm a$^{-1}$ is perhaps at the very high end. It is not possible to clearly establish depositional rates for the specific Isua and the Akilia banded iron formations that we analyzed (next section); we can only suggest and use a range of reasonable possibilities.

Table 1 shows the calculation Ir$_{[SED]}$ in ppb for background infall from equation (1), assuming $\Phi_m = 40 \times 10^9$ g a$^{-1}$ (*Love and Brownlee*, 1993); [Ir]$_{ET} = 480 \times 10^{-9}$ g g$^{-1}$ (= 480 ppb; chondritic) (*Anders and Grevesse*, 1989); $\oplus_A = 5.1 \times 10^{18}$ cm$^2$; $\rho_{SED} = \rho_{BIF}$ 3.3 g cm$^{-3}$, based on average mineralogy of BIF; and varied inputs of sedimentation rate from 0.100 mm a$^{-1}$ to 0.001 a$^{-1}$, and of greater extraterrestrial background flux from 300 to 10,000× the present rate.

The expected Ir abundances range from ~0.003 ppb for very rapid depositional rates and low ancient fluxes to ~11 ppb for very slow depositional rates (roughly that of Cretaceous-Tertiary boundary clay, for instance) and high ancient fluxes.

*4.3.5. Search for enhanced extraterrestrial influx of siderophile elements.* Mojzsis and co-workers studied aqueous sediments from the early Archean of southwestern Greenland for analysis for trace elements, including Ir (*Mojzsis*, 1997; *Mojzsis et al.*, 1997; *Ryder and Mojzsis*, 1998). These oldest terrestrial sediments might be expected to preserve a signal of higher flux, according to their precise age correlation with the lunar bombardment record and their depositional rate. While the methods and results will be detailed elsewhere (*Mojzsis and Ryder*, 2000), we provide a summary here.

The samples selected by Mojzsis and co-workers were early Archean banded-iron-formation enclaves from Akilia Island (the oldest currently-known sediment); banded iron formations, quartzite, and "control" granitic Amitsoq gneiss from Innersuartût Island just south of Akilia (Fig. 3); and banded iron formations from the Isua supracrustal belt (Table 2). We also analyzed the Gunflint Chert, a sample of Proterozoic banded iron formation. The samples were prepared and analyzed using neutron activation techniques

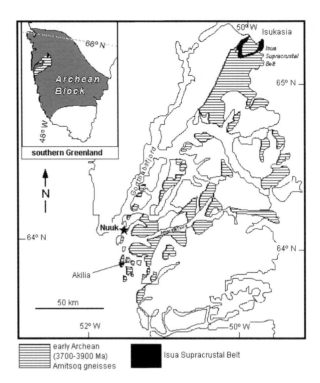

**Fig. 3.** Generalized geological map of southern West Greenland.

at the Johnson Space Center (JSC). The samples were prepared mainly as crushed, cleaned, interior, roughly whole-rock particles. Approximately 100–200 mg of particles of each sample were encapsulated in pure quartz tubes for irradiation and γ-ray counting. All samples were counted three times (~0.5 week, 1 week, and 3 weeks after irradiation); some were counted yet again a few weeks later to improve the precision (detection limit) for Ir. Data were reduced using standard procedures at the NASA Johnson Space Center laboratory (D. Mittlefehldt, personal communication). The detection limit obtained was 0.4–0.8 ppb (2σ) for Ir for all but the most Fe-rich samples, for which the detection limit was closer to 2 or 3 ppb (2σ) (Table 2).

The data showed that the samples contained little detrital material, consistent with their thin-section characteristics, with incompatible-trace-element abundances not unlike previous analyses of banded iron formations and related rocks (e.g., *Dymek and Klein*, 1988). None of the samples investigated, including the ~2.1-Ga Gunflint Chert, had Ir above its detection limit for that sample (Table 2). Clearly none of the material we analyzed was a rapid fallout similar to the Cretaceous-Tertiary boundary clay, for which we would expect several ppb Ir. However, in terms of a greater background flux at the time of even the oldest (Akilia) iron-stones, our data are open to several interpretations. If the depositional rate is truly very rapid (tenths of mm a$^{-1}$ or so), then even under the highest expected ancient meteoritic flux our data would not detect the expected Ir (<0.1 ppb, in some cases ≪0.1 ppb). However, if the depositional rate was actually more similar to that of shales or carbonates, or even somewhat faster, then our data indicate that the flux at the

TABLE 1. Calculated Ir (ppb) in sediments, from background flux.

| Sed rate, mm a$^{-1}$ | Flux times present rate | | | |
|---|---|---|---|---|
| | 300 | 1000 | 2000 | 10000 |
| 1.000 | 0.0003 | 0.0011 | 0.0023 | 0.0114 |
| 0.500 | 0.0007 | 0.0023 | 0.0046 | 0.0228 |
| 0.100 | 0.0034 | 0.0114 | 0.0228 | 0.1140 |
| 0.050 | 0.0068 | 0.0228 | 0.0456 | 0.2280 |
| 0.010 | 0.0342 | 0.1140 | 0.2280 | 1.1400 |
| 0.005 | 0.0684 | 0.2280 | 0.4560 | 2.2800 |
| 0.001 | 0.3420 | 1.1400 | 2.2800 | 11.4000 |

TABLE 2.   Neutron activation analyses of rocks from southwest Greenland.

| | FeO* (%) | Na$_2$O (%) | La (ppm) | Ir (ppb) | Cr (ppm) | Co (ppm) | Ni (ppm) |
|---|---|---|---|---|---|---|---|
| *Akilia Island banded iron formations >3.85 Ga* | | | | | | | |
| ANU-92-197/1-A | 5.8 | 0.027 | 0.52 | <.4 | 1.1 | 4.8 | 49 |
| ANU-92-197/1-B | 6.4 | 0.034 | 0.51 | <.27 | 1.5 | 4.5 | 33 |
| ANU-92-197/2-A | 7.4 | 0.051 | 1.36 | <.5 | 1.6 | 5.1 | 26 |
| ANU-92-197/2-B | 7.4 | 0.044 | 1.97 | <.3 | 1.6 | 5.0 | 33 |
| ANU-92-197/3-A1 | 7.8 | 0.030 | 0.57 | <.5 | 1.4 | 5.5 | 40 |
| ANU-92-197/3-A2 | 5.2 | 0.025 | 0.52 | <.4 | 1.1 | 3.9 | 29 |
| ANU-92-197/3-B | 9.0 | 0.030 | 0.66 | <.5 | 4.1 | 5.8 | 42 |
| ANU-92-197-X | 20.1 | 0.015 | 0.78 | | 1.6 | 8.7 | 114 |
| | | | | | | | |
| *Innersuartuut banded iron formations >3.77 Ga* | | | | | | | |
| SM/155746-A | 19.1 | 0.017 | 0.68 | <.9 | 3.9 | 8.3 | 74 |
| SM/155746-B | 23.8 | 0.018 | 0.66 | <.6 | 4.7 | 10.2 | 81 |
| SM/155746-X | 18.1 | 0.008 | 0.92 | 0.4 | 2.3 | 5.0 | 19 |
| SM/155746-C | 28.0 | 0.034 | 2.04 | <.8 | 6.7 | 9.1 | 41 |
| SM/155746-D | 36.1 | 0.030 | 2.52 | <.9 | 7.5 | 10.0 | 27 |
| SM/171770-A | 4.5 | 0.030 | 0.34 | <.4 | 1.7 | 2.7 | 21 |
| SM/171770-B | 13.0 | 0.050 | 1.49 | <.6 | 4.6 | 6.6 | 23 |
| SM/171770-X | 14.8 | 0.031 | 1.48 | | 4.4 | 8.7 | 55 |
| SM/171771-A | 70.2 | 0.129 | 1.40 | <1.5 | 87.5 | 10.1 | 56 |
| SM/171771-B | 54.3 | 0.242 | 1.70 | <.6 | 50.0 | 11.9 | 58 |
| SM/171771-X | 51.9 | 0.301 | 2.56 | | 63.9 | 11.7 | 66 |
| | | | | | | | |
| *Isua banded iron formations 3.77–3.80 Ga* | | | | | | | |
| /3446-A1 | 54.5 | 0.002 | 1.05 | <1.8 | 7.1 | 17.5 | 99 |
| /3446-A2 | 53.7 | 0.002 | 0.63 | <1.5 | 7.2 | 16.0 | 73 |
| /3446-B1 | 52.1 | 0.004 | 0.62 | <.7 | 5.8 | 16.1 | 64 |
| /3446-B2 | 51.7 | 0.002 | 0.67 | <1.1 | 6.5 | 16.1 | 76 |
| /3446-C1 | 52.9 | 0.002 | 0.58 | <.7 | 6.9 | 15.9 | 86 |
| /3446-C2 | 51.7 | 0.002 | 0.70 | <1.2 | 6.2 | 14.5 | 57 |
| /3451-A | 52.5 | 0.003 | 0.33 | <2.1 | 6.3 | 13.7 | 53 |
| /3451-B | 51.5 | 0.002 | 0.45 | <1.9 | 7.0 | 15.8 | 86 |
| | | | | | | | |
| *Isua Mt. — Isua banded iron formations* | | | | | | | |
| SM/78/248471 | 5.2 | 0.010 | 0.13 | | 0.5 | 0.6 | 0 |
| SM/GR/93/44 | 54.0 | 0.000 | 0.19 | | 4.1 | 4.1 | |
| | | | | | | | |
| *Isukasia — Isua banded iron formations* | | | | | | | |
| SM/GR/96/8 | 69.5 | 0.002 | 2.07 | | 152.9 | 38.1 | 80 |
| SM/GR/96/9 | 48.9 | 0.006 | 1.90 | | 198.7 | 20.7 | 68 |
| SM/GR/96/1 | 55.5 | 0.006 | 7.43 | | 8.9 | 27.8 | 166 |
| | | | | | | | |
| *Innersuartut Amitsoq orthogneiss >3.77 Ga* | | | | | | | |
| SM/155742 | 10.7 | 2.746 | 8.70 | | 0.0 | 4.1 | 0 |
| SM/171773-A | 2.0 | 2.872 | 9.47 | <.7 | 3.8 | 4.0 | <17 |
| SM/171773-B | 1.6 | 2.998 | 13.38 | <.8 | 3.1 | 3.0 | <18 |
| SM/171773-X | 2.8 | 2.504 | 12.89 | | 3.6 | 6.2 | 0 |
| | | | | | | | |
| *Gunflint Chert ~2.1 Ga* | | | | | | | |
| GF7-A | 4.3 | 0.011 | 0.77 | <.41 | 0.9 | 3.2 | <10 |
| GF7-B | 3.4 | 0.012 | 0.56 | <.21 | 0.5 | 2.6 | <11 |
| GF7-C | 4.2 | 0.010 | 1.01 | <.30 | 1.7 | 2.1 | <13 |

A = saw-cut free, small pieces; B = saw-cut free, larger pieces; C = saw-cut enriched; D = mafic-enriched separate; X = remainder, fines. See text for analytical information.

* Total Fe as FeO.

time of their deposition was *not* of the order of thousands of times the present flux. Clearly, at the present time we cannot provide a more definitive answer; both a better understanding of banded-iron-formation deposition rates *and* more precise methods of analysis for Ir are desirable.

More precise analyses have been made for Ir and Pt in some banded-iron-formation samples from Akilia Island (*Arnold et al.,* 1998; *Anbar et al.,* 2000). The analyses used a NiS fire assay and isotope dilution ICP-MS. Detection limits were ~0.003 ppb Ir and ~0.030 ppb Pt for the samples analyzed, which had abundances below those detection limits. This is somewhat surprising as the crustal background value is about 0.020 ppb for Ir. With such precision, even for a flux of 2000× the present and a sedimentation rate as fast as 0.5 mm a$^{-1}$, Ir should have been detected in these samples (Table 1). One can postulate even faster sedimentation rates, or nonrepresentative sampling (a nugget effect), or postdepositional loss of siderophile elements to explain these data. However, literal reading of the data would suggest that at the time of deposition, the bombardment rate was less than 2000× the present rate, probably much less.

Seventeen samples of Isua rocks, which could be up to 100 m.y. younger than the Akilia samples, were analyzed by *Koeberl et al.* (1998a,b, 2000) for their chemical composition, including siderophile-element abundances. These authors also used Ni-sulfide + Te co-precipitation fire assay and ICP-MS. The samples included metamorphic equivalents of turbidites, greywacke/felsic gneiss, conglomerate/felsic metasomatites, pelagic shale, gravity flow from the Bouma sequence, phyllite, and banded iron formations. Four of 17 samples analyzed yielded measurable amounts of Ir (ranging from 0.06 to 0.18 ppb) above the detection limit (0.03 ppb), as well as Ru and Rh. The contents of the other siderophile elements (Pt, Pd, Au) are highly varied; chondrite-normalized abundance patterns show variations by a factor of 3–4 (Fig. 4). The elevated contents were observed in a variety of different rocks: one banded-iron-formation

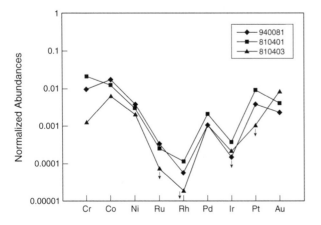

**Fig. 4.** Chondrite-normalized platinum-group element (PGE) abundance patterns in BIF (and other rocks) from Isua showing some Ir enrichment in BIF samples but a nonchondritic abundance pattern (after *Koeberl et al., 2000*).

rock, one graywacke, one gravity flow sample, and one pelagic shale. Such variation could be the result of a terrigenous detrital component. The elevated Ir content in the banded-iron-formation and pelagic shale samples may indicate a remnant meteoritic phase, but it is more likely they result from mafic contamination given the nonchondritic ratios of the other elements. If the Ir were demonstrably extraterrestrial, it would indicate a flux of ~10$^4$× present for a deposition rate of 0.05 mm a$^{-1}$, or a flux of ~10$^3$× the present with more typical sedimentation rates. However, these samples are at least 50 m.y. younger than the Imbrium event, and therefore most likely postdate the main episode of late heavy bombardment.

The chemical search for an enhanced amount of extraterrestrial matter in these Greenland samples was not deemed successful. This could indicate that either the rocks investigated were deposited very rapidly, that they do not overlap in time with the late heavy bombardment, or that the late heavy bombardment flux to the Earth was less intense than commonly predicted.

### 4.4. Search for Shocked Minerals

In normal terrestrial impact crater studies, the presence of shocked minerals is taken as confirming evidence for the impact origin of a purported astrobleme. The first petrographic search for shock features in rocks from Isua was reported by *Koeberl and Sharpton* (1988). Their study concentrated on the search for shocked quartz; however, none was found. This is understandable given the multiple upper amphibolite-grade metamorphism that these rocks underwent after their formation; such metamorphism would have repeatedly annealed the quartz. On the other hand, a variety of shocked minerals has been preserved in 2-Ga rocks from the Vredefort impact structure in South Africa. More recently, *Koeberl et al.* (1998a,b) reported on a new search for shocked minerals in Isua rocks, this time using a mineral that is more resilient in the face of recrystallization than quartz.

One of the best suited minerals for this purpose is zircon, which has been demonstrated to record a range of shock-induced features at the optical and electron microscope level (e.g., *Bohor et al.,* 1993). Furthermore, zircon is very resistant to erosion and other forms of alteration, including high-grade metamorphism. While planar deformation features in quartz may have long been annealed away, those in zircon have a good chance to survive for several billion years, as is indicated by the preservation of shocked zircons in rocks from the ~2-Ga Vredefort and Sudbury impact structures. However, the identification of suitable early Archean rocks for such a study is difficult. Whereas sedimentary rocks containing detrital shocked grains would be best for this purpose, there is some controversy as to whether actual terrigenous clastic sediments occur at Isua (e.g., *Rosing et al.,* 1996). *Koeberl et al.* (1998a,b) therefore focused on some of the samples that have not been positively proven to be plutonic, and that have mixed zircon populations either because they represent multiphase intrusives, had an extended metamorphic history, or are eroded from a mixed source.

Several samples studied by *Koeberl et al.* (1998a,b) yielded no zircons. Zircons were successfully separated from felsic schists, whose origin may be sedimentary. Grain mounts of hundreds of zircon crystals were studied; it was found that many grains are strongly fractured, but most fractures are of irregular shape or even of curved appearance. None of the crystals studied by *Koeberl et al.* (1998a,b) showed any evidence of optically visible shock deformation.

### 4.5.  Search for Impact Debris

Recent progress has been made in examining the rock record for old cosmic spherules that would be a particularly enriched carrier for extraterrestrial signatures in sediments. *Deutsch et al.* (1998) found 18 magnetic spheres in a 5-kg sample of 1.40-Ga red-bed sandstone from Finland. They assumed that all the spheres were extraterrestrial, and limited their search to the 60–125-μm size fraction. *Taylor and Brownlee* (1991) discovered numerous micrometeorites in a Jurassic (190 Ma) hardground, and *Taylor et al.* (1996) analyzed a magnetic fraction from 2 kg of Oligocene sediments and found about 250 cosmic spherules preserved. Given multiple estimates of higher impact fluxes in the early Archean and the obvious preservation of cosmic spherules even in very old rocks, the search for extraterrestrial signals in the oldest sediments is of renewed interest for estimating past fluxes.

The oldest known terrestrial impact structures are the Proterozoic Vredefort and Sudbury structures, $2023 \pm 4$ and $1850 \pm 3$ Ma respectively (cf. *Reimold and Gibson,* 1996), which represent the complete documented pre-1.85-Ga terrestrial impact record. Other evidence for early Archean impact events is less demonstrative. Some enigmatic spherule horizons at and near the contact between the ca. 3.5-Ga Fig Tree and Onverwacht Groups in the Barberton Mountain Land, South Africa, have been reported as possible impact ejecta horizons. These arguments are based on textural features, enrichments in the platinum-group elements, near-chondritic platinum-group-element patterns, and Ni-rich spinels. In the case of the Barberton spherule layers, *Koeberl and Reimold* (1995) argued that the platinum-group enrichment is not a primary feature of the spherulitic horizons, but rather is the product of secondary mineralization. None of these spherule layer samples contained any evidence for impact-characteristic shock-metamorphic deformation, such as planar deformation features in silicate minerals. Thus the impact origin of these spherule layers is debatable. On the other hand, a Late Archean (ca. 2.5 Ga) spherule layer from the Griqualand West Basin, South Africa, shows clear evidence of a primary meteoritical component (*Koeberl et al.,* 1999; *Simonson et al.,* 2000). However, no similar spherule layers have yet been reported from any of the earliest Archean rocks.

## 5.  IMPLICATIONS AND OUTLOOK

A search for any evidence on the Earth for traces of the late heavy bombardment is currently centered on petrographical and geochemical studies of the world's oldest supracrustal rocks in Greenland. Petrographic studies of zircons extracted from these rocks have so far failed to show evidence for shock metamorphism. Nor have any deposits with ejecta-like characteristics, such as spherule beds, been reported so far from field investigations. At the time of this writing, results of the chemical search for a meteoritic component in these earliest sedimentary rocks remain uncertain. Of the samples analyzed by us so far, only four samples of different composition from localities in Isua yielded Ir abundances that are somewhat above the present-day background levels for crustal rocks. However, in chondrite-normalized abundance diagrams of the platinum-group elements, even these samples show nonchondritic patterns and probably represent contamination from mafic phases. In the absence of any sign of shock metamorphism or ejecta deposits, and with ambiguous geochemical signals, no direct and unequivocal evidence of a late heavy bombardment on Earth can as yet be confirmed.

The possible reasons for the failure to obtain such direct evidence are manifold. First, the number of samples studied so far has been small. This is certainly the case for the search for shocked minerals, but the geochemical studies should have fared better. Second, the samples chosen might not have been ideal for such a search. However, given the limitations of the early rock record, the samples studied were among the best available. A petrographic search for shocked zircons should be extended to the detrital zircons of known age (3.8–4.0 Ga), and a statistically significant number of samples from different locations should be scanned. However, *Cintala and Grieve* (1998) suggested that very-large-scale impacts may yield higher relative amounts of melt and thus may not preserve much shocked material, in which case the absence of shocked zircons may not mean much. A heavy bombardment should include abundant smaller craters, quite capable of producing shocked materials, including zircon, if the target is zircon-bearing.

Third, it is possible that the formation time of the rocks studied so far do not overlap with the period of late heavy bombardment of the Moon. While the zircons from the granitoid rocks that cross-cut the Akilia samples have been interpreted to be ~3.85 Ga in age (*Nutman et al.,* 1996, 1997), it has been argued by some workers that these zircons are inherited from a preexisting rock, and that the host gneisses in the Archean of West Greenland are themselves only about 3.65 Ga in age (*Kamber and Moorbath,* 1998), but this argument has been disputed on several counts (*Mojzsis and Harrison,* 1999, 2000). The metasedimentary rocks from Isua are most likely less than 3.80 Ga in age. In this case, the late heavy bombardment would have ceased, and no direct evidence for an extraterrestrial component could be obtained. Fourth, it may be that these Akilia samples are indeed younger than the late heavy bombardment, but only slightly so, such that the uncertainties in the ages of the two overlap. It could be that Imbrium and Orientale, and the basins and craters stratigraphically between them, are all very close to 3.86 Ga, and the Akilia samples are ~3.84 Ga in age, and immediately postdate bombardment. If so, the lack of a heavy bombardment signature in the latter could result from a very rapid decline in the heavy bombardment in the 3.86–3.85 Ga timeframe. This is certainly quite possible, and would have ramifications for the bombardment history of the inner solar system and for the origin of the population of impactors that the bombardment represents. Recent studies in celestial mechanics have led to the proposal of a mechanism that could plausibly supply a short-time spike in an otherwise steady or decreasing background flux of impactors (*Zappalà et al.,* 1998); however, such sources need to be quantified. Lastly, it is possible — but not very likely — that the Akilia rocks predate intiation of bombardment at 3.9 Ga and therefore missed all the excitement (*Anbar et al.,* 2000).

Further studies will be necessary to clarify the timing of the late heavy bombardment on the Moon and its effects on the Earth, if indeed they are preserved here at all. The precise ages of the actual rocks studied, and available to be studied, need to be resolved. Furthermore, the ambiguity of using simple elemental concentrations of siderophile elements suggests that specific isotope systems, such as Os (*Koeberl and Shirey,* 1997) or Cr (*Shukolyukov and Lugmair,* 1998), would be useful in future searches for evidence of impacts on the early Earth.

*Acknowledgments.* This research was supported by the Fonds zur Förderung der wissenschaftlichen Forschung in Austria (CK), by the Lunar and Planetary Institute (GR), and by the U.S. National Science Foundation (SJM). SJM acknowledges additional support from the NASA-supported UCLA Astrobiology Center, and the Danish Scientific Research Council through the Isua Multidisciplinary Research Project directed by P. W. U. Appel. We appreciate comments on the manuscript by R. A. F. Grieve, E. Pierazzo, K. Righter, and an anonymous reviewer. The Lunar and Planetary Institute operates under contract NASW-4066 with the National Aeronautics and Space Administration. This paper is LPI Contribution No. 1003.

## REFERENCES

Alvarez L. W., Alvarez W., Asaro F., and Michel H. V. (1980) Extraterrestrial cause for the Cretaceous-Tertiary extinction. *Science, 208,* 1095–1108.

Anbar A. D., Wasserburg G. J., Papanastassiou D., and Anderson P. S. (1996) Iridium in natural waters. *Science, 273,* 1524–1528.

Anbar A. D., Arnold G. L., Mojzsis S. J., and Zahnle K. J. (2000) Extraterrestrial iridium and sediment accumulation on the Hadean Earth. *J. Geophys. Res.,* submitted.

Anders E. and Grevesse N. (1989) Abundances of the elements: meteoritic and solar. *Geochim. Cosmochim. Acta, 53,* 197–214.

Arndt N. and Chauvel C. (1991) Crust of the Hadean Earth. *Bull. Geol. Soc. Denmark, 39,* 145–151.

Arndt N. T., Nelson D. R., Compston W., Trendall A. F., and Thorne A. M. (1991) The age of the Fortescue Group, Hamersley Basin, Western Australia, from ion microprobe zircon U-Pb results. *Austr. J. Earth Sci., 38,* 261–281.

Arnold G., Anbar A., and Mojzsis S. J. (1998) Iridium and platinum in early Archean metasediments: Implications for sedimentation rate and extraterrestrial flux. *Geol. Soc. Am., Abstr. with Progr., 30(7),* A82–A83.

Baldwin R. B. (1971) On the history of lunar impact cratering: The absolute time scale and the origin of planetesimals. *Icarus, 14,* 36–52.

Baldwin R. B. (1974) Was there a "Terminal Lunar Cataclysm" $3.9–4.0 \times 10^9$ years ago? *Icarus, 23,* 157–166.

Baldwin R. B. (1981) On the origin of the planetesimals that produced the multi-ring basins. In *Multi-Ring Basins, Proc. Lunar Planet. Sci 12A* (P. H. Schultz and R. B. Merrill, eds.), pp. 19–28. Pergamon, New York.

Baldwin R. B. (1987) On the relative and absolute ages of seven lunar front face basins. II. From crater counts. *Icarus, 71,* 19–29.

Barker J. L. and Anders E. (1968) Accretion rate of cosmic matter from iridium and osmium contents of deep-sea sediments. *Geochim. Cosmochim. Acta, 32,* 627–645.

Barton E. S., Altermann W., Williams I. S., and Smith C. B. (1994) U-Pb zircon age for a tuff in the Campbell Group, Griqualand West sequence, South Africa: implications for early Proterozoic rock accumulation rates. *Geology, 22,* 343–346.

Bau M. and Möller P. (1993) Rare earth element systematics of the chemically precipitated component in early Precambrian iron-formations and the evolution of the terrestrial atmosphere-hydrosphere-lithosphere system. *Geochim. Cosmochim. Acta, 57,* 2239–2249.

Bogard D. D., Garrison D. H., and Lindstrom M. M. (1991) $^{39}Ar/^{40}Ar$ age of an Apollo 15, KREEP-poor, impact melt rock (abstract). In *Lunar and Planetary Science XXII,* pp. 117–118. Lunar and Planetary Institute, Houston.

Bohor B. F., Betterton W. J., and Krogh T. E. (1993) Impact-shocked zircons: discovery of shock-induced textures reflecting increasing degrees of shock metamorphism. *Earth Planet. Sci. Lett., 119,* 419–424.

Bonté P., Jéhanno C., Maurette M., and Brownlee D. E. (1987) Platinum metals and microstructure in magnetic deep-sea cosmic spherules. *Proc. Lunar Planet. Sci. Conf. 17th,* in *J. Geophys. Res., 92,* E641–E648.

Bowring S. A. and Housh T. (1995) The Earth's early evolution. *Science, 269,* 1535–1540.

BVSP (1981) Chronology of planetary volcanism by comparative studies of planetary cratering. In *Basaltic Volcanism on the Terrestrial Planets,* pp. 1050–1127. Pergamon, New York.

Cameron A. G. W. (2000) Higher-resolution simulations of the giant impact. In *Origin of the Earth and Moon* (R. M. Canup and K. Righter, eds.), this volume. Univ. of Arizona, Tucson.

Chou C.-L. (1978) Fractionation of siderophile elements in the Earth's upper mantle. *Proc. Lunar Planet. Sci. Conf. 9th,* pp. 219–230.

Chou C.-L., Shaw D. M., and Crockett J. H. (1983) Siderophile trace elements in the Earth's oceanic crust and upper mantle. *Proc. Lunar Planet. Sci. Conf. 13th,* in *J. Geophys. Res., 88,* A507–A518.

Chyba C. F. (1991) Terrestrial mantle siderophiles and the lunar impact record. *Icarus, 92,* 217–233.

Chyba C. F. (1993) The violent environment for the origins of life: progress and uncertainties. *Geochim. Cosmochim. Acta, 57,* 3351–3358.

Cintala M. J. and Grieve R. A. F. (1994) The effects of differential scaling of impact melt and crater dimensions on lunar and terrestrial crater: Some brief examples. In *Large Meteorite Impacts and Planetary Evolution* (B. O. Dressler, R. A. F. Grieve, and V. L. Sharpton, eds.), pp. 51–59. GSA Spec. Paper 293.

Cintala M. J. and Grieve R. A. F. (1998) Scaling impact melting and crater dimensions: Implications for the lunar cratering record. *Meteoritics & Planet. Sci., 33,* 889–912.

Cloud P. E. (1973) Paleoecological significance of the banded iron-formations. *Econ. Geol., 68,* 1135–1143.

Cloud P. E. (1976) Major features of crustal evolution. A. L. du Toit Memorial Lecture Series 14. *Trans. Geol. Soc. South Africa 79.* 33 pp.

Cloud P. E. (1988) *Oasis in Space. Earth History from the Beginning.* Norton and Company, New York. 508 pp.

Compston W. and Pidgeon R. T. (1986) Jack Hills, evidence of more very old zircons in Western Australia. *Nature, 291,* 193–196.

Culler T. S., Muller R. A., Renne P. R., Becker T., and Deino A. (1999) Laser-heating $^{40}Ar/^{39}Ar$ dating of lunar impact melt spherules from Apollo 14: New constraints on the cratering history of the Moon (abstract). *Eos Trans. AGU, 80,* S208.

Dalrymple G. B. and Ryder G. (1991) $^{40}Ar/^{39}Ar$ ages of six Apollo 15 impact melt rocks by laser step heating. *Geophys. Res. Lett., 18,* 1163–1166.

Dalrymple G. B. and Ryder G. (1993) $^{40}Ar/^{39}Ar$ age spectra of Apollo 15 impact melt rocks by laser step-heating and their bearing on the history of lunar basin formation. *J. Geophys. Res., 98,* 13085–13095.

Dalrymple G. B. and Ryder G. (1996) $^{40}Ar/^{39}Ar$ age spectra of Apollo 17 highlands breccia samples by laser step-heating and the age of the Serenitatis basin. *J. Geophys. Res., 101,* 26069–26084.

Dasch J., Ryder G., Reese Y., Wiesmann H., Chih C.-Y., and Nyquist L. (1998) Old age of formation for a distinct variety of A17 high-titanium mare basalt (abstract). In *Lunar and Planetary Science XXIX,* Abstract #1750. Lunar and Planetary Institute, Houston (CD-ROM).

Davies G. F. (1985) Heat deposition and retention in a solid planet growing by impacts. *Icarus, 63,* 45–68.

Deutsch A. and Schärer U. (1994) Dating terrestrial impact events. *Meteoritics, 29,* 301–322.

Deutsch A., Greshake A., Pesonen L. J., and Pihlaja P. (1998) Unaltered cosmic spherules in an 1.4-Ga-old sandstone from Finland. *Nature, 395,* 146–148.

Dymek R. F. and Klein C. (1988) Chemistry, petrology, and origin of banded iron-formation lithologies from the 3800 Ma Isua supracrustal belt. *Precambrian Res., 39,* 247–302.

Esser B. K. and Turekian K. K. (1988) Accretion rate of extraterrestrial particles determined from osmium isotope systematics of Pacific pelagic clay and manganese nodules. *Geochim. Cosmochim. Acta, 52,* 1383–1388.

Frey H. (1980) Crustal evolution of the early Earth: The role of major impacts. *Precambrian Res., 10,* 195–216.

Ganapathy H. (1983) The Tunguska explosion of 1908: Discovery of meteoritic debris near the explosion site and at the South Pole. *Science, 220,* 1158–1161.

Grieve R. A. F. (1980) Impact bombardment and its role in proto-continental growth on the early Earth. *Precambrian Res., 10,* 217–247.

Grieve R. A. F. (1998) Extraterrestrial impacts on Earth: The evidence and its consequences. In *Meteorites: Flux with Time and Impact Effects* (M. M. Grady, R. Hutchinson, G. J. H. McCall, and D. A. Rothery, eds.), pp. 105–131. Geol. Soc. London Spec. Publ. 140.

Gruau G., Rosing M., Bridgwater D., and Gill R. C. O. (1996) Resetting of Sm-Nd systematics during metamorphism of >3.7-Ga rocks: implications for isotopic models of early Earth differentiation. *Chem. Geol., 133,* 225–240.

Halliday A., Rehkämper M., Le D.-C., and Yi W. (1996) Early evolution of the Earth and Moon: New constraints on Hf-W geochemistry. *Earth Planet. Sci. Lett., 142,* 75–89.

Harland W. B., Armstrong R. L., Cox A. V., Craig L. E., Smith A. G., and Smith D. G. (1989) *A Geologic Time Scale 1989.* Cambridge Univ., Cambridge. 263 pp.

Harper C. L. and Jacobsen S. B. (1992) Evidence from coupled $^{147}Sm$-$^{143}Nd$ and $^{146}Sm$-$^{142}Nd$ systematics for very early (4.5-Ga) differentiation of the Earth's mantle. *Nature, 360,* 728–732.

Hartmann W. K. (1965) Secular changes in meteoritic flux through the history of the solar system. *Icarus, 4,* 207–213.

Hartmann W. K. (1966) Early lunar cratering. *Icarus, 5,* 406–418.

Hartmann W. K. (1975) Lunar "cataclysm": A misconception? *Icarus, 24,* 181–187.

Hartmann W. K. (1980) Dropping stones in magma oceans: Effects of early lunar cratering. In *Proc. Conf. Lunar Highlands Crust,* pp. 155–171. Pergamon, New York.

Hartmann W. K., Grinspoon D. H., Ryder G., and Dones L. (2000) The time-dependent intense bombardment of the primordial Earth-Moon system. In *Origin of the Earth and Moon* (R. M. Canup and K. Righter, eds.), this volume. Univ. of Arizona, Tucson.

Haskin L. A., Korotev R. L., Rockow K. M., and Jolliff B. L. (1998) The case for an Imbrium origin of the Apollo thorium-rich impact-melt breccias. *Meteoritics & Planet. Sci., 33,* 959–975.

Holland H. D. (1973) The oceans: a possible source of iron in iron-formations. *Econ. Geol., 68,* 1169–1172.

Holland H. D. (1984) *The Chemical Evolution of the Atmosphere and Oceans.* Princeton Univ., Princeton, New Jersey. 583 pp.

Holm N. G. (1987) Possible biological origin of banded iron-formations from hydrothermal solutions. *Orig. Life, 17,* 229–250.

Holzheid A. and Palme H. (1998) Early history of the Earth: Insights from siderophile elements in the Earth's mantle. In *Origin of the Earth and Moon,* pp. 13–14. LPI Contribution No. 957, Lunar and Planetary Institute, Houston.

Isley A. E. (1995) Hydrothermal plumes and the delivery of iron to banded iron-formations. *J. Geol., 103,* 169–185.

James H. L. (1954) Sedimentary facies of iron-formation. *Econ. Geol., 49,* 235–293.

James O. B. (1981) Petrologic and age relations of the Apollo 16 rocks: Implications for the subsurface geology and age of the Nectaris basin. *Proc. Lunar Planet. Sci. 12B,* pp. 209–233.

James O. B. (1995) Siderophile elements in lunar impact melts: Nature of the impactors (abstract). In *Lunar and Planetary Science XXVI,* pp. 671–672. Lunar and Planetary Institute, Houston.

Kamber B. S. and Moorbath S. (1998) Initial Pb of the Amîtsoq gneiss revisited: implication for the timing of the early Archaean crustal evolution in West Greenland. *Chem. Geol., 150,* 19–41.

Kamber B. S., Moorbath S., and Whitehouse M. T. (1998) Extreme Nd-isotope heterogeneity in the early Archean — fact or fiction? Case histories from northern Canada and West Greenland — Reply. *Chem. Geol., 148,* 219–224.

Klein C. and Beukes N. J. (1990) Geochemistry and sedimentology of facies transition from limestones to iron-formation deposition in the early Proterozoic Transvaal Supergroup, South Africa. *Econ. Geol., 84*, 1733–1774.

Koeberl C. and Reimold W. U. (1995) Early Archean spherule beds in the Barberton Mountain Land, South Africa: No evidence for impact origin. *Precambrian Res., 74*, 1–33.

Koeberl C. and Sharpton V. L. (1988) Giant impacts and their influence on the early Earth. In *Papers Presented to the Conference on Origin of the Earth*, pp. 47–48. Lunar and Planetary Institute, Houston.

Koeberl C. and Shirey S. B. (1997) Re-Os systematics as a diagnostic tool for the study of impact craters and distal ejecta. *Palaeogeogr. Palaeoclimat. Palaeoecol., 132*, 25–46.

Koeberl C., Reimold W. U., McDonald I., and Rosing M. (1998a) The late heavy bombardment on the earth? A shock petrographic and geochemical survey of some of the world's oldest rocks (abstract). In *Origin of the Earth and Moon*, pp. 19–20. LPI Contribution No. 957, Lunar and Planetary Institute, Houston.

Koeberl C., Reimold W. U., McDonald I., and Rosing M. (1998b) Search for petrographical and geochemical evidence for the late heavy bombardment on Earth (abstract). In *ESF Workshop on Impacts and the Early Earth*, pp. 15–16. Cambridge Univ., Cambridge.

Koeberl C., Simonson B. M., and Reimold W. U. (1999) Geochemistry and petrography of a Late Archean spherule layer in the Griqualand West Basin, South Africa (abstract). In *Lunar and Planetary Science XXX*, Abstract #1755. Lunar and Planetary Institute, Houston (CD-ROM).

Koeberl C., Reimold W. U., McDonald I., and Rosing M. (2000) Search for petrographical and chemical evidence for the late heavy bombardment on Earth in Early Archean rocks from Isua, Greenland. In *Impacts and the Early Earth* (I. Gilmour and C. Koeberl, eds.), pp. 73–96. Springer-Verlag, Heidelberg.

Kyte F. T. and Wasson J. T. (1986) Accretion rate of extraterrestrial matter: iridium deposited 33 to 67 million years ago. *Science, 232*, 1225–1229.

Lee D.-C. and Halliday A. N. (1995) Hafnium-tungsten chronometry and the timing of terrestrial core formation. *Nature, 378*, 771–774.

Lee D.-C. and Halliday A. N. (1996) Hf-W isotopic evidence fro rapid accretion and differentiation in the solar system. *Science, 274*, 1876–1879.

Love S. G. and Brownlee D. E. (1993) A direct measurement of the terrestrial mass accretion rate of cosmic dust. *Science, 262*, 550–553.

Maas R., Kinny P. D., Williams I. S., Froude D. O., and Compston W. (1992) The Earth's oldest known crust: A geochronological and geochemical study of 3900–4200 Ma old zircons from Mt. Narryer and Jack Hills, Western Australia. *Geochim. Cosmochim. Acta, 56*, 1281–1300.

Maher K. A. and Stevenson D. J. (1988) Impact frustration of the origin of life. *Nature, 331*, 612–614.

Maurer P., Eberhardt P., Geiss J., Grögler N., Stettler A., Brown G. M., Peckett A., and Krähenbühl U. (1978) Pre-Imbrian craters and basins: ages, compositions and excavation depths of Apollo 16 breccias. *Geochim. Cosmochim. Acta, 42*, 1687–1720.

McCulloch M. T. and Bennett V. C. (1993) Evolution of the early Earth: Constraints from $^{143}$Nd-$^{142}$Nd isotopic systematics. *Lithos, 30*, 237–255.

Melosh H. J. (1989) *Impact Cratering. A Geologic Process.* Ox-

ford Univ., New York. 245 pp.

Mojzsis S. J. (1997) Ancient sediments of Earth and Mars. Ph.D. thesis, Scripps Institution of Oceanography, University of California, San Diego. 300 pp.

Mojzsis S. J. (1998) Clues to the Hadean environment in the chemistry of ancient zircons. In *Origin of the Earth and Moon*, pp. 25–26. LPI Contribution No. 957, Lunar and Planetary Institute, Houston.

Mojzsis S. J. and Harrison T. M. (1999) Geochronological studies of the oldest known marine sediments. In *Ninth V. M. Goldschmidt Conference*. Lunar and Planetary Institute, Houston.

Mojzsis S. J. and Harrison T. M. (2000) Vestiges of a beginning: Clues to the emergent biosphere recorded in the oldest sedimentary rocks. *GSA Today, 10*, 1–6.

Mojzsis S. J. and Ryder G. (2000) Extraterrestrial accretion to the Earth and Moon ca. 3.85 Ga. In *Accretion of Extraterrestrial Matter Throughout Earth History* (B. Peuckner-Ehrinbrink and B. Schmitz, eds.). Kluwer, Dordrecht, in press.

Mojzsis S. J., Arrhenius G., McKeegan K. D., Harrison T. M., Nutman A. P., and Friend C. R. L. (1996) Evidence for life on Earth before 3,800 million years ago. *Nature, 385*, 55–59.

Mojzsis S. J., Ryder G., and Righter K. (1997) Crustal contamination by meteorites in the early Archean. *Eos Trans. AGU, 78*, 399–400.

Mojzsis S. J., Krishnamurthy R., and Arrhenius G. (1999) Before RNA and after — geophysical and geochemical constraints on molecular evolution. In *RNA World 2nd edition* (R. Gesteland, T. Cech, and J. Atkins, eds.), pp. 1–47. Cold Spring Harbor, New York.

Moorbath S. and Kamber B. S. (1998) A reassessment of the timing of early Archaean crustal evolution in West Greenland. *Geol. Greenland Surv. Bull., 180*, 88–93.

Moorbath S., Whitehouse M. J., and Kamber B. S. (1997) Extreme Nd-isotope heterogeneity in the early Archaean — fact or fiction? Case histories from northern Canada and West Greenland. *Chem. Geol., 135*, 213–231.

Neukum G., Konig B., Fechtig H., and Storzer D. (1975) Cratering in the Earth-Moon system — Consequences for age determination by crater counting. *Proc. Lunar Sci. Conf. 6th*, pp. 2597–2620.

Newsom H. E. (1990) Accretion and core formation in the Earth: Evidence from siderophile elements. In *Origin of the Earth* (H. E. Newsom and J. H. Jones, eds.), pp. 273–288. Oxford Univ., New York.

Nutman A. P., Allaart J. H., Bridgwater D., Dimroth E., and Rosing M. (1984) Stratigraphic and geochemical evidence for the depositional environment of the early Archean Isua supracrustal belt, southern West Greenland. *Precambrian Res., 25*, 365–396.

Nutman A. P., McGregor V. R., Friend C. R. L., Bennett V. C., and Kinny P. D. (1996) The Itsaq Gneiss Complex of southern West Greenland: the world's most extensive record of early crustal evolution (3,900–3,600 Ma). *Precambrian Res., 78*, 1–39.

Nutman A. P., Mojzsis S. J., and Friend C. R. L. (1997) Recognition of >3850 Ma water-lain sediments in West Greenland and their significance for the early Archean Earth. *Geochim. Cosmochim. Acta, 61*, 2475–2484.

Oberbeck V. and Fogleman G. (1989) Impacts and the origin of life. *Nature, 339*, 434.

Palme H., Göbel E., and Grieve R. A. F. (1979) The distribution of volatile and siderophile elements in the impact melt of East Clearwater (Quebec). *Proc. Lunar Planet. Sci. Conf. 10th*, pp. 2465–2492.

Papike J. J., Ryder G., and Shearer C. K. (1998) Lunar samples.

In *Planetary Materials* (J. J. Papike, ed.), pp. 5-1 to 5-234. *Rev. Mineral., 36.*

Reimold W. U. and Gibson R. L. (1996) Geology and evolution of the Vredefort impact structure, South Africa. *J. African Earth Sci., 23,* 125–162.

Rosing M. T. (1999) ¹³C-depleted carbon microparticles in >3700-Ma sea-floor sedimentary rocks from western Greenland. *Science, 283,* 674–676.

Rosing M. T., Rose N. M., Bridgwater D., and Thomsen H. S. (1996) Earliest part of Earth's stratigraphic record: a reappraisal of the >3.7 Ga Isua (Greenland) supracrustal sequence. *Geology, 24,* 43–46.

Ryder G. (1990) Lunar samples, lunar accretion, and the early bombardment history of the Moon. *Eos Trans. AGU, 71,* 313–323.

Ryder G. (1994) Coincidence in time of the Imbrium basin impact and Apollo 15 volcanic flows: The case for impact-induced melting. In *Large Meteorite Impacts and Planetary Evolution* (B. O. Dressler, R. A. F. Grieve, and V. L. Sharpton, eds.), pp. 11–18. GSA Spec. Paper 293.

Ryder G. (1999) Meteoritic abundances in the ancient lunar crust (abstract). In *Lunar and Planetary Science XXX,* Abstract #1848. Lunar and Planetary Institute, Houston (CD-ROM).

Ryder G. and Bower J. F. (1977) Petrology of Apollo 15 black-and-white rocks 15445 and 15455 — fragments of the Imbrium impact melt sheet? *Proc. Lunar Planet. Sci. Conf. 8th,* pp. 1895–1923.

Ryder G. and Mojzsis S. J. (1998) Accretion to the Earth and Moon around 3.85 Ga: What is the evidence? (abstract). *Eos Trans. AGU, 79,* F48.

Ryder G., Bogard D., and Garrison D. (1991) Probable age of Autolycus and calibration of lunar stratigraphy. *Geology, 19,* 143–146.

Shoemaker E. M. (1972) Cratering history and early evolution of the Moon (abstract). In *Lunar Science III,* pp. 696–698. Lunar Science Institute, Houston.

Shoemaker E. M. (1977) Why study impact craters? In *Impact and Explosion Cratering* (D. J. Roddy, R. O. Pepin, and R. B. Merrill, eds.), pp. 1–10. Pergamon, New York.

Shukolyukov A. and Lugmair G. W. (1998) Isotopic evidence for the Cretaceous-Tertiary impactor and its type. *Science, 282,* 927–929.

Simonson B. M., Koeberl C., McDonald I., and Reimold W. U. (2000) Geochemical evidence for an impact origin of a Late Archean spherule layer, Transvaal Supergroup, South Africa. *Geology,* submitted.

Sleep N. H., Zahnle K. J., Kasting J. F., and Morowitz H. J. (1989) Annihilation of ecosystems by large asteroid impacts on the early Earth. *Nature, 342,* 139–142.

Smith J. V. (1981) The first 800 million years of Earth's history. *Philos. Trans. Roy. Soc. London, A301,* 401–422.

Snyder G., Lee D.-C., Taylor L. A., Halliday A. N., and Jerde E. A. (1994) Evolution of the upper mantle of the Earth's Moon: Neodymium and strontium isotopic constraints from high-Ti mare basalts. *Geochim. Cosmochim. Acta, 58,* 4795–4808.

Snyder G., Hall C. M., Lee D.-C., Taylor L. A., and Halliday A. N. (1996) Earliest high-Ti volcanism on the Moon: ⁴⁰Ar-³⁹Ar, Sm-Nd, and Rb-Sr isotopic studies of Group D basalts from the Apollo 11 landing site. *Meteoritics & Planet. Sci., 31,* 328–334.

Spudis P. D. (1993) *The Geology of Multi-Ring Impact Basins.* Cambridge Univ., New York. 263 pp.

Steiger R. H. and Jäger E. (1977) Subcommission on geochronol-
ogy: Convention on the use of decay constants in geo- and cosmochronology. *Earth Planet. Sci. Lett., 36,* 359–362.

Swindle T., Spudis P. D., Taylor G. J., Korotev R., Nichols R. H., and Olinger C. T. (1991) Searching for Crisium basin ejecta: Chemistry and ages of Luna 20 impact melts. *Proc. Lunar Planet Sci., Vol. 21,* pp. 167–184.

Taylor P. L., Nusbaum R. L., Fronabarger A. K., Katuna M. P., and Summer N. (1996) Magnetic spherules in coastal plain sediments, Sullivan's Island, South Carolina, USA. *Meteoritics & Planet. Sci., 31,* 77–80.

Taylor S. R. (1982) *Planetary Science: A Lunar Perspective.* Lunar and Planetary Institute, Houston. 481 pp.

Taylor S. R. (1989) Growth of planetary crusts. *Tectonophysics, 161,* 147–156.

Taylor S. R. (1992a) The origin of the Earth. In *Understanding the Earth* (G. Brown, C. Hawkesworth, and C. Wilson, eds.), pp. 25–43. Cambridge Univ., Cambridge.

Taylor S. R. (1992b) *Solar System Evolution. A New Perspective.* Cambridge Univ., New York. 307 pp.

Taylor S. R. (1993) Early accretion history of the Earth and the Moon-forming event. *Lithos, 30,* 207–221.

Taylor S. R. (2000) Hadean Eon. In *McGraw-Hill Encyclopedia of Science and Technology.* McGraw-Hill, New York, in press.

Taylor S. R. and McLennan S. M. (1995) The geochemical evolution of the continental crust. *Rev. Geophys., 33,* 241–265.

Taylor S. and Brownlee D. E. (1991) Cosmic spherules in the geologic record. *Meteoritics, 26,* 203–211.

Tera F., Papanastassiou D. A., Wasserburg G. J. (1974) Isotopic evidence for a terminal lunar cataclysm. *Earth Planet. Sci. Lett., 22,* 1–21.

Trendall A. F. and Blockley J. G. (1970) The iron-formations of the Precambrian Hamersley Group, Western Australia, with special reference to the associated crocidolite. *Geol. Surv. W. Austral. Bull., 119,* 366 pp.

Turner G. and Cadogan P. H. (1975) The history of lunar bombardment inferred from ⁴⁰Ar/³⁹Ar dating of highland rocks. *Proc. Lunar Sci. Conf. 6th,* pp. 1509–1538.

Vervoort J. D. and Blichert-Toft J. (1999) Evolution of the depleted mantle: Hf isotope evidence from juvenile rocks through time. *Geochim. Cosmochim. Acta, 63,* 533–556.

Wetherill G. W. (1975) Late heavy bombardment of the Moon and terrestrial planets. *Proc. Lunar Sci. Conf. 6th,* pp. 1539–1561.

Wetherill G. W. (1977) Pre-mare cratering and early solar system history. In *The Soviet-American Conference on Cosmochemistry of the Moon and Planets,* pp. 553–567. NASA SP-370.

Wetherill G. W. (1981) Nature and origin of basin-forming projectiles. In *Multi-Ring Basins, Proc. Lunar Planet. Sci. 12A* (P. H. Schultz and R. B. Merrill, eds.), pp. 1–18. Pergamon, New York.

Wetherill G. W. (1994) Provenance of the terrestrial planets. *Geochim. Cosmochim. Acta, 58,* 4513–4520.

Wilhelms D. E. (1984) Moon. In *The Geology of the Terrestrial Planets* (M. H. Carr, R. S. Saunders, R. G. Strom, and D. E. Wilhelms, eds.), pp. 107–205. NASA SP-469.

Wilhelms D. E. (1987) *The Geologic History of the Moon.* U.S. Geol. Surv. Prof. Paper 1348. 302 pp.

Zahnle K. J. and Sleep N. H. (1997) Impacts and the early evolution of life. In *Comets and the Origin and Evolution of Life* (P. J. Thomas, C. F. Chyba, and C. P. McKay, eds.), pp. 175–208. Springer-Verlag, New York.

Zappalà V., Cellino A., Gladman B. J., Manley S., and Migliorini F. (1998) Asteroid showers on Earth after family breakup events. *Icarus, 134,* 176–179.

# The Time-Dependent Intense Bombardment of the Primordial Earth/Moon System

## W. K. Hartmann
*Planetary Science Institute*

## Graham Ryder
*Lunar and Planetary Institute*

## Luke Dones
*Southwest Research Institute*

## David Grinspoon
*Southwest Research Institute*

The accretion of planets involved a high flux of planetesimal impactors in the first $10^7$–$10^8$ yr, a flux on the order of $10^9\times$ the present flux. Dynamical models suggest that the early planetesimals would be used up by impacting planets or the Sun, or by being scattered out of the inner solar system, and by this process be depleted with a slowly increasing half-life of the order of 10–30 m.y.; "spikes" or transient episodes of enhanced cratering are possible in these models. Some interpretations of lunar data suggest intense bombardment prior to 3.9 Ga, phasing into the present rate; other interpretations suggest low bombardment prior to 3.9 Ga with an extraordinary or unique spike at 3.9–3.8 Ga. We review the controversial and differing models of the cratering history from 4.5 to 3.6 Ga, and the question of whether a unique cataclysm occurred at 3.9–3.8 Ga. This question has important consequences for planetary history.

## 1. GOALS

The goal of our work is to compare quantitative models of the mass flux of impactors onto the primordial Earth/Moon, and its time dependence during the first hundred million years of Earth's history, with empirical evidence from the Moon. Part of our approach is to examine the half-lives of orbits that have been calculated for planetesimals left over after Earth's formation. The assumed scenario is that each planet accumulated its mass during a formation interval, shown by meteoritic isotopic evidence to have been no more than a few tens of millions of years (*Pepin and Phinney*, 1975). By the end of that time, planets had essentially their present masses, but various planetesimals were left in nearby orbits, with many having already been scattered by near-misses during the accretion process. We would expect a rapid sweep-up of most of these bodies over a period of a few hundred million years. Initially the half-life (time to sweep up half the bodies, a parameter that may be time-dependent) was short, on the order of 10–30 m.y. (*Wetherill*, 1975), but slowly lengthened as the short-half-life bodies impacted, and as others were scattered onto increasingly remote orbits.

The plan of our paper is as follows. In section 2, we discuss the pre-Apollo and Apollo-era literature on the history of the lunar impact rate, and describe the arguments for and against a temporary, large increase in impact rate, or "cataclysm," after 3.9 Ga. In section 3, we describe a model without a cataclysm that is broadly consistent with

the lunar cratering record. In sections 4 and 5, we show that impact spikes due to a sudden influx of heliocentric impactors are possible. Section 6 describes the dynamics of different classes of small bodies in the past and present solar system, and evaluates whether these small bodies are capable of producing a cataclysm. Section 7 describes physical evidence that argues that the lunar record is more consistent with a cataclysm than with an impact rate that continues to increase monotonically into the past prior to 3.9 Ga. Section 8 points out that the issue of whether or not a cataclysm occurred is critical to a number of major issues in planetary science. Section 9 gives our conclusions.

*This paper requires the use of absolute ages for lunar samples (and inferred events) acquired from radioisotopic determinations. Many determinations for lunar samples were made prior to 1977, when revised decay constants were recommended (Steiger and Jäger, 1977) and subsequently used by most workers. For consistency, all determinations referred to in this paper are calculated using the revised constants, and thus will differ slightly (~60–80 m.y. younger) from those originally given in those papers prior to 1977.*

## 2. EARLY LITERATURE ON PRIMORDIAL INTENSE BOMBARDMENT

The understanding of the history of meteoritic bombardment of the Earth (and the Earth-Moon system) has had an interesting history, affected by changing paradigms and interpretations of Apollo and Luna samples of lunar rocks.

Even before the Apollo missions, there was direct evidence of a high bombardment rate in the earliest solar system. *Hartmann* (1965, 1966) used crater counts on the Canadian shield to estimate the impact rate and then combined that with lunar crater counts to predict that lunar maria averaged around 3.6 Ga in age; he therefore noted that the Moon must have undergone an "early intense bombardment" during its first few hundred million years in order to account for the 32× higher crater density on the uplands than on the maria. Hartmann concluded that this early intense bombardment *averaged* at least 32× (3.6/1.0), ~100× the postmare cratering rate, and possibly much more if the uplands were saturated with craters. A schematic sketch of the cratering time dependence, published in 1970, is shown in Fig. 1.

In the last few years before Apollo, Shoemaker cited classified Defense Department records of pressure waves in the atmosphere, attributed to meteoritic infall. Converting the pressure data to meteorite mass (a highly uncertain procedure), he concluded that the present-day impact rate is much higher than had been found by earlier authors, and concluded erroneously that the lunar maria were only a few hundred million years old. This seemed to remove the need for an early intense bombardment, and would have been consistent with a constant impact rate since planet formation, and an unsaturated upland cratering.

When the first lunar samples were returned, the 3.6-Ga ages for maria were confirmed. Wasserberg and the Caltech radiochronometric researchers (*Tera et al.*, 1974) were struck by the fact that the so-called "genesis rocks," primordial rocks older than ~4.0 Ga that astronauts had been instructed to find, were virtually missing from the lunar record. To explain this, and an observed clustering of highland rock ages around 3.9–3.8 Ga, they hypothesized a "ter-

minal lunar cataclysm," a disastrous, suddenly commencing, short-lived cratering episode about 3.9 Ga that was supposed to have destroyed the "genesis rocks." In some seminar presentations, Wasserberg emphasized the proposed uniqueness of the cataclysm with a cartoon Gothic cathedral inserted on his impact flux diagrams at 3.9–3.8 Ga. This idea of a cataclysm at 3.9 Ga, thought of as the end of the planet-forming period, became sufficiently pervasive that most workers adopted the phrase "terminal bombardment" or "late heavy bombardment," instead of "early heavy bombardment," even though the event was supposed to have happened in the planets' first 0.6 Ga. This terminology itself reinforced the widespread perception that the lunar rocks had proved the existence of a cataclysm. This has produced some semantic and conceptual confusion, because the terms "late heavy bombardment," "early intense bombardment," "terminal cataclysm" and other variants are widely used to refer to the same general phenomenon that the cratering rate averaged much higher in the first 0.6 Ga than after 3.9 Ga.

*Hartung* (1974), *Hartmann* (1975, 1980), and *Grinspoon* (1989) suggested that the paucity of earlier rocks might not prove that a cataclysm had occurred. They pointed out that a steadily declining flux of bodies, with lengthening half-life as planetary near-misses scattered early populations, might explain the data. The extremely intense flux in the first few hundred million years could physically destroy crustal rocks formed before 4.0 Ga as they were pulverized and as regolith formed. Indeed, *Hartmann* (1973) noted that the uplands might be saturated with craters several kilometers deep and coined the term "megaregolith" to describe the concept that the uplands had been pulverized and churned to depths as great as tens of kilometers, thus explaining the difficulty of finding more than tiny chips of "genesis rocks." Rocks exposed near the surface before 4.0 Ga would have had a high probability of being destroyed, not to mention having Ar ages reset by shock effects, but the impact rate declined so fast that rocks produced after, say, 3.8 Ga would have survived. In addition, at least some types of radioisotopic ages, such as K-Ar ages, would be frequently reset during the intense impact period.

*Grinspoon* (1989) produced quantitative models of the age distributions of surviving lunar rocks under the assumption of an exponentially declining flux, and showed that these closely matched the actual age distributions of collected lunar samples. This work demonstrated that, due to the threshold effects associated with exponential functions, a sharp peak in ages is *expected*, even if there is no peak in cratering rate. The starting point for Grinspoon's model was an extremely simplified model for age resetting by impact. He assumed that when an impact occurs, the rocks in an area of the surface equal to the area of the crater are completely reset to age zero at the time of the impact. In reality, some rocks outside the crater, in the ejecta blanket, would be reset and some rocks on the crater floor might survive resetting. The probability of the rocks at any given

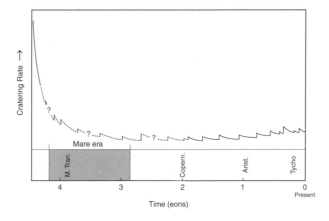

**Fig. 1.** Schematic illustration of cratering record from an early publication on "early intense bombardment." This diagram showed the cratering rate at a fixed crater diameter, such as 1 km, and predicted spikes produced by breakups of planetesimals (shown here with arbitrary timing). A modern question is whether the hypothetical "cataclysm" at 3.9 Ga could be interpreted as an unusually large spike of this type (from *Hartmann*, 1970).

point on the surface having a particular age is then the probability of that point being cratered at that given time, multiplied by the probability of that point *not* being cratered at all subsequent times. Assuming a rate of production of craters of diameter D equal to $r(D,t) = k\,D^{-2.8}\exp(t - t_0)/\tau$ (*Neukum and Ivanov*, 1994; *Baldwin*, 1971; *Hartmann et al.*, 1981), where k is the flux at time $t_0$ and $\tau$ is the decay e-folding time of the flux, then the total area of craters made, per unit surface area, in any time increment is

$$A(t) = \int_0^{D_{big}} r(D,t)\frac{\pi D^2}{4}\,dD = \left(\frac{\pi}{4}\right)\left(\frac{k}{0.2}\right)D_{big}^{0.2}\exp\frac{-(t-t_0)}{\tau} \quad (1)$$

where $D_{big}$ is the diameter of the largest craters being produced. Note that the value of A(t) is only weakly dependent on the size chosen for $D_{big}$. Letting $D_{big} = 1000$ km, $t_0 = 3.8$ Ga, and $k = 10^2\times$ the current estimated lunar flux, we find that $A(t) \approx 2 \times 10^{-9}$ e $-(t - t_0)/\tau$ per year. If production of new rocks of age t is proportional to A(t), then replacing $(t - t_0)$ with t and letting K be $A(t_0)$, the probability P(t) of survival of rocks of age t is given by

$$P(t) = k\exp\left(\frac{-t}{\tau} - K\tau\,\exp\left(\frac{-t}{\tau}\right)\right)$$

The peak of P(t) occurs when $t = \tau\ln(\tau K)$. Figure 2 is a plot of equation (1) for several plausible values of K and $\tau$. As can be seen from Fig. 2, the specific shape and peak age of this function is sensitive to the e-folding time (or to the half-life of the declining flux) or to the flux normalization chosen, but all reasonable choices result in a sharply peaked distribution of rock ages. Clearly, it is dangerous to cite an observed peak or convergence of rock ages alone to justify claims for a maximum in the cratering flux. A peak in the rock age distribution is a natural consequence of an exponentially declining flux. A more sophisticated treatment of age resetting by impact would produce results that were different in detail, but would preserve the main features shown in Fig. 2.

Meanwhile, *Wetherill* (1975) investigated dynamical evolution of planetesimals at the tail end of planet formation to account for the sudden proposed onset of cataclysmic cratering around 3.9 Ga. First, he investigated a gradual sweep-up of the planetesimals left over after planet formation, with any given initial orbit having a certain half-life against collision, but without an explicit, suddenly commencing cataclysm. So strong was the belief that such an event had been proven that Wetherill repeatedly and explicitly rejected his own evidence that a noncataclysmic solution might fit the available cratering and rock age data. For example (p. 1551), he concluded that the half-life of residual populations of accretional debris in the inner solar system "is probably only about 30 My," but finding that this would not account for a late cataclysm, he remarked, "However, in order to make the model work as well as possible, a

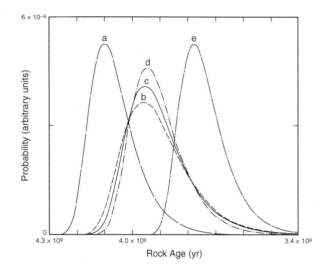

**Fig. 2.** Theoretical predictions of rock age distribution based on the work of *Grinspoon* (1989, and section 2). No "cataclysm" or spike in cratering is assumed. Rather, rocks are assumed to have ages reset (or to be destroyed) by impacts. Earliest rock ages thus are removed by the early intense bombardment. Specifically, the probability distribution of highland rock ages is given by equation (1). Curves b, c, and d are for $k = 2 \times 10^{-9}$ and $\tau = 80$ m.y., 90 m.y., and 100 m.y. respectively. Curves a and e are both for $\tau = 70$ m.y. and $k = 2 \times 10^{-8}$ and $2 \times 10^{-9}$ respectively. See text for the meaning of the variables.

longer and less probable half-life of 70 My has been used." Like the earlier authors, Wetherill spoke of impacts "resetting" ages, but did not critically examine the resetting mechanism. Applying that model, he found that his predicted histogram of upland rock ages peaked at 3.9 Ga, and, like *Hartung's* (1974), showed "considerable similarity" to the histogram of ancient rock ages. However, he argued that "the agreement is only apparent" because of scatter introduced by "interlaboratory differences." Wetherill then went on to describe three other models, all of which required storage in long-lived orbits followed by breakup of planetesimals during close encounters within Roche limits of planets, in order to throw showers of fragments into the inner solar system and provide the cataclysm. He concluded "it is believed that this model of the late heavy bombardment should be taken seriously, rather than dismissed as ad hoc..." He thus concluded that a terminal cataclysm was possible, but prudently stated that it remained to be shown "whether or not it actually occurred."

In the lunar geology community, there is a related belief that many of the largest, concentric-ring-basin-forming impacts have been convincingly dated by Apollo and Luna rock and soil samples, and that these basins were all created in a narrow interval of about 4.0–3.8 Ga, thus empirically confirming the putative "terminal cataclysm." However, serious questions have been raised as to whether rock materials collected at various landing sites have really dated a variety of basins such as Serenitatis, Nectaris, Crisium, Imbrium, and Orientale. *Haskin* (1998), for example, has argued that the Imbrium ejecta materials are much more pervasive than others have assumed, and that most of the dates assigned to the other basins, in the narrow "cataclysm" age range, actually refer to Imbrium. In other words, have we really proven that a cluster of >3 of the Moon's largest recorded impacts occurred in the 4.0–3.8-Ga interval, or only the last one or two?

In summary, it is important to revisit the whole subject of cratering, crustal evolution, basin ages, and environmental conditions in the Earth-Moon system during its first 0.6 Ga.

## 3.  A SIMPLE MODEL WITH DECLINING FLUX

Studies of crater densities at Apollo landing sites of different ages confirm that the lunar flux was several hundred times higher than it is today at the time the oldest visible surface units were emplaced, around 3.9 Ga (*Hartmann,* 1969, 1970, 1980; *Neukum et al.,* 1975; *Neukum,* 1983; *Grieve and Shoemaker,* 1994). Such studies give detailed flux data back only to about 3.9–3.8 Ga, the age of the oldest landing site. *Hartmann* (1980) estimated the mass flux at 4.55 Ga by invoking the fact that the mass of Earth aggregated within roughly 55 m.y., an interval measured by radioisotopic studies (*Pepin and Phinney,* 1975; *Podosek and Ozima,* 2000). This flux, one $M_\oplus$ in 55 m.y., he estimated at $2 \times 10^9 \times$ the present flux.

Such data led to a schematic hypothetical model of a declining flux, which we divide into four arbitrary segments in order to discuss the issues (Fig. 3). Point A represents

solar formation. Segment AB is a steeply declining initial population with a half-life only 10 m.y., starting with a flux $2 \times 10^9 \times$ today's. This flux dominates for the first 70 m.y., until point B is reached at 4.49 Ga. It represents the flux of local planetesimals from which the primordial Earth accreted, and represents the immediate tail end of planet formation. A 10-m.y. half-life is plausible during this period for planetesimals in Earth-crossing orbits, still retaining the initial low eccentricities and inclinations characteristic of planetesimals accreting in the solar nebula. The diamond, toward the end of AB, represents lunar formation, perhaps triggered by the impact of the largest of these planetesimals onto the primordial Earth (*Hartmann and Davis,* 1975; *Cameron and Ward,* 1976; *Weidenschilling et al.,* 1986). The Moon itself, in this view, is thus a consequence of the very process we discuss — the most violent event of the early intense bombardment.

Segment BC represents a population with a half-life averaging 22 m.y., dominant from 4.5 until 4.2 Ga (point C). This population corresponds to the bodies *Wetherill* (1975) posits as left over at the end of Earth's formation — planetesimals with a modest spread of semimajor axes, eccentricities, and inclinations, somewhat scattered by near encounters with Earth during its growth.

Segment CD, with a half-life 80 m.y., represents a transition period consistent with analyses by Wetherill and by *Safronov* (1972), where the remaining planetesimals are being scattered more widely, increasing the half-life of the remaining flux. Some of these bodies may even include Jupiter-scattered planetesimals from the outer solar system or asteroid belt. The proposed half-life is within the range considered by Wetherill in his modeling of the same period.

Prior to point C, megaregolith production is explosive (*Hartmann,* 1980). To show this, assume that the regolith production rate is proportional to the impact rate. If we have had 10 m of regolith production in the lunar maria at the current impact rate over 3.6 G.y., i.e., $3(10^{-9})$ m/yr. If the impact rate at some point before C was $10^6\times$ the present rate, that corresponds to ~3 km of regolith every million years (not to mention producing heating effects that may have maintained a magma ocean). Even an impact rate only $10^3\times$ the present rate grinds ~300 m of regolith every 100 m.y. The point is that physical survival of surface rock layers and surface features was inhibited before point C, and began some time later, close to 4 Ga. Finally, after point D, the flux levels out to within a factor 2 of the "modern" average flux observed after about 3.4 Ga.

Thus, the declining early bombardment rate must be examined with great care to understand its effects on the lunar geologic record.

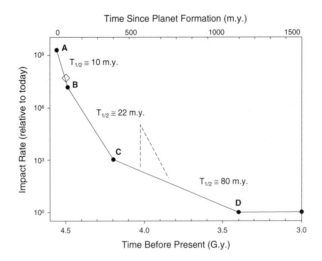

**Fig. 3.** "Empirical" history of lunar cratering flux, combining data from crater densities and rock ages at Apollo sites (after 4.0 Ga) with an initial point based on accumulation of Earth's mass in 10–100 m.y. Instantaneous half-lives for decline of the planetesimal population are shown for four arbitrarily shown segments of the curve. See text for further discussion.

## 4.  "SPIKES" IN THE CRATERING RECORD

It is now clear that "spikes" of some magnitude are likely to appear in the record of cratering rate vs. time, and if they do, they will affect the rock age distribution, in addition to the effects discussed in sections 2 and 3. For example, if a

50-km asteroid is shattered near a resonance, it could produce thousands of fragments 1 km and smaller, and the resonance could then scatter a burst of fragments into the inner solar system on million-year timescales and they would be swept up by one or more worlds with half-lives around 10 m.y. (*Gladman et al.*, 1997; *Migliorini et al.*, 1998; *Morbidelli and Gladman*, 1998; *Zappalà et al.*, 1998). A breakup of an Earth-crosser by an aphelion collision in the belt would produce a similar spike in the impact record. Furthermore, *Wetherill* (1975) and the experience of Comet Shoemaker-Levy 9 show that near encounters of weak planetesimals with planets can produce showers of impactors on short timescales. Wetherill's mechanism could also fling such fragments from the outer to the inner solar system. Since 3–4 Ga, the half-life has lengthened so much that the impact flux over the last 3 G.y. has averaged roughly constant (*Neukum and Ivanov*, 1994, pp. 389–390; *Grieve and Shoemaker*, 1994) or perhaps declined within a factor of 2 (*Culler et al.*, 2000). For these reasons, events that produce bodies swept up by the Earth-Moon system on timescales of 1–10 m.y. would appear as spikes and could have produced hundreds of multikilometer craters in 10–100 m.y. (see also section 6). Even in early discussions of this subject, *Hartmann* (1970), for example, illustrated such spikes schematically in diagrams of cratering rate vs. time, as seen in Fig. 1. One way to phrase the issue of a cataclysm is merely to ask the size of the largest spike that affected the Moon.

Note that the cratering effects of such spikes would be size dependent. For example, if the fragmented body was large enough to produce many fragments in a power law size distribution at sizes smaller than 1 km, with a largest fragment size of 10 km, then a statistically detectable spike would occur in cratering smaller than about 10 km, but no spike would be seen among craters larger than 50–100 km. Therefore, in order to produce a spike of large craters or basins in the size range >200 km, very large planetesimals must be fragmented. The view that all the dozen or more large lunar multiring impact basins were produced by a cataclysm or spike lasting only 100–200 m.y. is particularly problematic, therefore, because the projectiles producing Imbrium or Orientale were roughly 100–150 km across (*Baldwin*, 1963; *Zahnle and Sleep*, 1997). If one invokes a fragmentation event to produce the cluster of impactors, then what kind of breakup event could produce a dozen or so such large bodies, not to mention the attendant power-law size distribution of smaller bodies that would be expected in such a fragmentation?

## 5.  GEOCENTRIC VS. HELIOCENTRIC CRATERING

In principle, another possible source of a "spike" in the lunar cratering record could be collisions with geocentric material. *Ryder* (1990) pointed out that the lunar mass accumulation history and earliest cratering flux were different from that of Earth, according to the current paradigm

of lunar formation by geocentric accretion after a giant impact. This is because the accretion rate of a lunar mass of debris in geocentric orbit is of the order of $10^8 \times$ faster than accretion of the same mass in the solar nebula. However, by this very token, we conclude that lunar cratering due to circumterrestrial debris declined with an extremely short half-life, well within the first million years of lunar formation, which was about 4.5 G.y. ago. Tidal forces drove the Moon outward from its presumed initial position near the Roche limit to most of its present distance very rapidly. The precise tidal dissipation factors are not known, but *Baldwin* (1963, p. 434) showed that the average distance-time relation for tidal evolution is such that if we assume the Moon takes 4.5 G.y. to reach its present distance, it would have reached $10\,R_\oplus$ in only 40,000 yr, and one-third its present distance in 8 m.y. Thus, any impact flux of geocentric impactors is likely to have dropped below the heliocentric cratering within about 10 m.y., and is unlikely to be responsible for a hypothetical "cataclysm" or spike of cratering 500 m.y. later, at 4.0 Ga. If there was a cataclysmic spike, its origin must be sought elsewhere.

## 6.  TIME-DEPENDENCE OF THE HELIOCENTRIC FLUX

Much has been learned about the long-term dynamics of small interplanetary bodies in the solar system within the last decade. They can be subdivided by type, to clarify the discussion.

### 6.1.  Present-day Impactors

*6.1.1. Main-belt asteroids.* The main belt contains some 200 asteroids with diameters larger than 100 km, the approximate size of the impactor that created the Imbrium Basin. The main belt contains a number of strong mean-motion resonances with Jupiter, such as the 3:1 at 2.5 AU, and secular resonances like the $\nu_6$ that cut through much of the belt. These resonances can excite large eccentricities for objects whose orbits lie within the resonance. After a collision throws a fragment into a resonance, the fragment's orbit becomes unstable, in some cases changing radically in less than 1 m.y. The usual fate of the asteroid is then collision with the Sun for asteroids in the inner belt (i.e., for objects with semimajor axes <2.5 AU), and ejection from the solar system by Jupiter for asteroids in the outer belt (i.e., objects with semimajor axes >2.5 AU) (*Gladman et al.*, 1997). However, some asteroids are "decoupled" from the belt by gravitational encounters with the terrestrial planets, resonant interactions, or, for volatile-rich objects, nongravitational forces, such that their semimajor axes no longer lie in the main belt.

*6.1.2. Near-Earth asteroids.* Over 700 near-Earth asteroids, or NEAs, are known. Most NEAs are decoupled from the main belt, and it had been thought they generally remained so until they impacted a planet. *Gladman et al.* (1997), however, find that "the majority of bodies that enter the NEA population do so transiently; most come below $a = 2$ AU only for a brief residence of ~1 M[a] and then return to $a > 2$ AU to be pushed into the Sun by the $v_6$ or 3:1 resonance." *Gladman et al.* (1997, 2000) found a median lifetime of 10 m.y. for NEAs; this lifetime is at the low end of previous estimates.

*6.1.3. Trojans.* The Trojan asteroids librate around the L4 and L5 points in Jupiter's orbit. *Shoemaker et al.* (1989) estimated that the total population of Trojans larger than 15 km is about half of the main-belt population to 15 km. Like main-belt asteroids, most Trojans are stable for the age of the solar system, but the orbits of some Trojans are destabilized by gravitational perturbations from the planets or by collisions, and evolve onto orbits similar to those of ecliptic comets (*Levison et al.,* 1997).

*6.1.4. Comets.* Comets have traditionally been grouped into "long-period" and "short-period" categories. Dynamical modeling suggests a more natural division into "nearly isotropic" comets, which have a wide range of orbital inclinations, and low-inclination "ecliptic" comets (*Levison,* 1996). Nearly isotropic comets, which include Oort cloud and Halley-type comets, likely began on near-circular orbits in the region of the giant planets, and reached their current orbits through gravitational scattering by the giant planets and gravitational torques from the disk of our galaxy. Ecliptic comets (*Levison,* 1996; *Duncan and Levison,* 1997) include, in order of increasing distance from the Sun, Encke-type comets, Jupiter-family comets (JFCs), Centaurs, and scattered-disk objects (*Duncan and Levison,* 1997). Centaurs are objects such as Chiron and Pholus with planet-crossing orbits in the Jupiter-Neptune region, while scattered-disk objects have perihelia slightly outside Neptune's orbit (30 AU) and semimajor axes of order 100 AU. Ecliptic comets are generally believed to originate from the Kuiper Belt, although other sources such as Jupiter's Trojan asteroids may also contribute. Once an ecliptic comet leaves the Kuiper Belt, it typically spends ~4 m.y. as a Centaur, 0.3 m.y. as a JFC, and ~40 m.y. as a scattered-disk object.

## 6.2.  Impactor Populations at 3.9 Ga

We now attempt to estimate the populations of small bodies that were present at the time of formation of the last lunar basins (Imbrium, Orientale, and perhaps others). For each population we also assess the plausibility of a large cratering spike, as considered in various discussions of a lunar cataclysm (*Wetherill,* 1975; *Weissman,* 1989; *Chyba,* 1991).

Present-day populations of small bodies are too sparse to cause the late bombardment, so one must find a way to "store" a larger early population for some 600 m.y. The primary current reservoirs for small bodies are the main asteroid belt, Jupiter's Trojan points, and the region beyond Neptune (the Kuiper Belt, scattered disk, and Oort cloud). We assume that the Imbrium impactor had a mass of $2 \times 10^{21}$ g (*Zahnle and Sleep,* 1997), and that the total mass of

the impactors that created the late basins, $I_{\mathbb{C}}$, was three times greater, or $I_{\mathbb{C}} = 6 \times 10^{21}$ g. Depending upon the size distribution of the impactors, $I_{\oplus} = (0.8–30) \times 10^{23}$ g likely would have impacted the Earth during the same period. A question still to be discussed is whether this mass arrived on the Moon in a short-lived "spike" of <100 m.y. or was spread out over 100 m.y. or more.

[Note: The amount of mass that impacts Earth will exceed that on the Moon for two reasons. First, the Earth's cross-sectional area is larger than the Moon's by a factor, which we call x, of 13.4; for present-day near-Earth asteroids, gravitational focusing by the Earth raises the ratio of cross sections to x ~ 23 (*Zahnle and Sleep,* 1997). Second, the mass distribution of the early impactors may have been top heavy; a cumulative size distribution with index b ~ 1.5 is often used (*Maher and Stevenson,* 1988; *Chyba,* 1991), although there is some evidence for b ~ 2.7. *Hartmann* (1969) discussed a variety of natural instances of rock fragmentation (both terrestrial and extraterrestrial) and found b ranging from 1.6 to 3.5 with most values close to 2.3. The popular *Dohnanyi* (1969) collisional cascade has b = 2.5. Because Earth's cross section exceeds the Moon's by more than an order of magnitude, small number statistics imply that the Earth will likely suffer ~10 impacts by objects more massive than any that strike the Moon. When this effect is taken into account, the typical ratio R of mass impacting the Earth to that on the Moon can be shown to be $R = x^{3/b}$ for b < 3; e.g., for asteroidal impactors, $R = x^2 \sim 500$ for b = 1.5 and $R = x^{1.1} \sim 30$ for b = 2.7. For b > 3, R = x; i.e., the Earth has no advantage over the Moon beyond its larger cross section, since most of the impacting mass is in a large number of small objects.]

From the discussion above, the mass, M, of the small-body population that produced the last basins must have been at least

$$M(3.9 \text{ Ga}) = M(4.5 \text{ Ga}) \, f(4.5 \to 3.9 \text{ Ga}) = I_{\mathbb{C}}/\eta_{\mathbb{C}} = I_{\oplus}/\eta_{\oplus} \sim 10^{23}\eta_{\oplus}^{-1}\text{g} \tag{2}$$

where M(3.9 Ga) represents the mass of a given small-body reservoir at 3.9 Ga; M(4.5 Ga) is the mass of the same reservoir at 4.5 Ga; f(4.5 Ga $\to$ 3.9 Ga) = M(3.9 Ga)/M(4.5 Ga) is the fraction of the mass of the reservoir that remains at 3.9 Ga; and $\eta_{\mathbb{C}}$ and $\eta_{\oplus}$ are, respectively, the probabilities that a small body will strike the Moon or Earth during its dynamical lifetime. For inner solar system populations such as accretional remains, $\eta_{\oplus}$ can be large; e.g., $\eta_{\oplus} = \frac{1}{2}$ for a hypothetical asteroid that has equal probability of hitting Earth or Venus, and no chance of impacting the Sun or being ejected from the solar system by Jupiter. For outer solar system bodies, $\eta_{\oplus} \ll 1$ because these objects typically spend a tiny fraction of their lives in Earth-crossing orbit (see Fig. 4).

In Fig. 4, we plot the probability of Earth impact, $\eta_{\oplus}$, of a number of types of small bodies as a function of semimajor axis, a, for the current solar system. The probability of impact onto the Moon, $\eta_{\mathbb{C}}$, is lower than $\eta_{\oplus}$ by the fac-

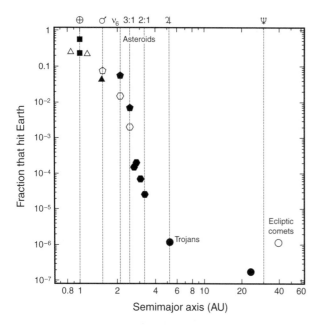

**Fig. 4.** The probability of Earth impact, $\eta_\oplus$, of a number of types of small bodies as a function of semimajor axis. Impact probabilities are for leftovers of Earth's accretion [*Dones et al.*, 1999b (open triangles)]; lunar impact ejecta [*Gladman et al.*, 1995 (solid squares)]; martian impact ejecta [*Gladman*, 1997 (open pentagon)]; near-Earth asteroids [*Gladman et al.*, 2000 (solid triangle)]; asteroids injected into the $\nu_6$ and 3:1 resonances [*Gladman et al.*, 1997 (solid pentagons); *Zappalà et al.*, 1998 (open hexagons)]; asteroids injected into resonances in the outer half of the main belt [*Zappalà et al.*, 1998 (solid hexagons)]; Trojan asteroids of Jupiter and Uranus-Neptune planetesimals [*Levison et al.*, 2000a (solid circles)]; and ecliptic comets, assumed to originate from the Kuiper Belt [*Levison et al.*, 2000b (open circle)]. For all the calculations except those of *Levison et al.* (2000a), the planets are assumed to follow their current orbits.

tor R (see note above). These results should apply to the time of the late bombardment, at least to order of magnitude, if the planetary masses and orbits at 3.9 Ga were similar to their present values. The plot can be interpreted as the likelihood that a small body ultimately impacts the Earth, once it leaves its source region due to gravitational perturbations or collisions. The giant planets, and Jupiter in particular, pose considerable obstacles to the delivery of outer solar system material to the inner solar system.

The asteroid belt shows a steep decline in the probability of Earth impact with increasing distance from the Sun. This trend can be understood by noting that the middle of the asteroid belt lies halfway between the Sun and Jupiter's orbit, and if an asteroid enters a strong main-belt resonance, its usual fate is to have its eccentricity pumped to a value near unity. In the inner half of the belt, this means that an asteroid will become Sun-grazing before it becomes Jupiter-crossing; in the outer half of the belt, the opposite is true. A typical Earth-impact probability for Earth-crossing aster-

oids from the inner belt is $10^{-2}$; $\eta_\oplus \sim t_{res}/t$ where the timescale for resonant eccentricity pumping, $t_{res}$, is ~1 m.y. and the timescale for Earth impact, t ~0.1 $(a/R_r)^2 \times$ the asteroid's orbital period, is ~100 m.y. Near-Earth asteroids have more highly evolved orbits, resulting in ~10-m.y. dynamical lifetimes and $\eta_\oplus \sim 0.1$.

In general, possible impactors from the inner solar system have higher impact probability on the Earth and Moon, but few of them are likely to have survived to 3.9 Ga. By contrast, outer solar system sources have low impact probability, but the available reservoirs have masses orders of magnitude greater. It is not clear *a priori* which should dominate at a given time.

We will now describe specific models for the bombardment, with most detail given to objects that might remain in near-Earth orbit after the Moon-forming impact. In most cases, past work on these models used Monte Carlo Öpik-Arnold calculations, or "Öpik codes," to follow the orbital evolution of small bodies. Such codes assume that orbital evolution is dominated by close encounters with planets (*Dones et al.*, 1999a). Here we describe new work using direct orbital integrations, which correctly account for all gravitational interactions at the cost of being vastly more time-consuming.

*6.2.2. Leftovers of accretion (Wetherill, 1975, Model 1; Strom and Neukum, 1988).* In this model we consider survival of small bodies in the region of the terrestrial planets. *Wetherill* (1975) assumed that this population would decay exponentially with a half-life of 70 m.y. or less (see section 2), implying that, at most, one object in 400 remained after 600 m.y. If this population produced the late lunar basins, this implies that the initial mass was of order $10^{26}$ g or more, i.e., about a lunar mass. Wetherill initially felt this model was untenable because the initial mass was implausibly large. Subsequently, *Wetherill* (1977) used an Öpik code, including an approximate treatment of the effects of the $\nu_6$ resonance, to study this problem, and found that a long-lived tail of survivors formed. The survivors appeared to be suitable impactors for producing an extended period of bombardment, though not necessarily a sudden cataclysmic spike. In this model, the long-lived survivors decouple from Earth by a three-step process: Their semimajor axes are pumped to >2 AU by close encounters with Earth and Venus; their eccentricities are then lowered by the $\nu_6$ resonance until their perihelion distances are near Mars' orbit (1.5 AU); finally, the body is removed from the resonance by an encounter with Mars. In some cases, an object can remain stored in such an orbit for hundreds of millions of years.

Wetherill found that the "half-life" of the survivors increased almost linearly with time t; his plot of the impact rate dN/dt on Earth can be fit by a power law, dN/dt ~ $t^{-2.5}$. Equivalently, the number of survivors N ~ $t^{-1.5}$. Power-law decay is equivalent to marginal stability; that is, the dynamical lifetime is always comparable to the time that an object has survived (*Laskar*, 1996). Numerical integrations of a variety of small bodies on initially unstable orbits show that at "late" times, i.e., at times when less than half the initial population remains, the number of survivors declines with time t roughly as a power law

$$N(t) \sim t^{-a} \qquad (3)$$

or logarithmically (*Holman and Wisdom*, 1993; *Evans and Tabachnik*, 1999):

$$N(t) \sim A - B \log t \qquad (4)$$

For power law decay, typically a ~1 (*Gladman et al.*, 1996, 1997; *Dones et al.*, 1996, 1999a; *Holman*, 1997; *Malyshkin and Tremaine*, 1999). Decay is not exponential because objects can (but need not; see below) evolve into "niches" in phase space that are longer-lived than their initial orbits. Logarithmic decay has the peculiar property that the population goes to zero at finite time, so at late times logarithmic decay is faster than exponential decay.

How much mass is likely to remain after the formation of the Earth and Venus is largely complete? In Monte Carlo simulations, *Wetherill* (1990) found that the Earth grew to 0.5 $M_\oplus$ in 25 m.y. and 0.99 $M_\oplus$ in 100 m.y. The Moon is generally thought to have formed as the result of the impact of a $10^{27}$-g (0.1–0.2 $M_\oplus$) body with the Earth some 50–100 m.y. after the Earth formed. However, the Moon-forming impactor was probably the largest of the other Earth-zone planetesimals; the masses of the largest Earth-crossers remaining after the Moon formed might have been orders of magnitude smaller (e.g., *Hartmann and Davis*, 1975), say $10^{24}$–$10^{25}$ g. How many of these objects would remain in orbit at 600–700 m.y.? Survival of 1% or so is needed to provide a plausible source for the bombardment that formed the last basin. Bodies in this mass range have diameters of the order 800–2000 km, and the breakup of even the single, last one might have produced a "spike" of basin-forming bodies.

Several recent orbital calculations bear on these issues. First, *Evans and Tabachnik* (1999) performed orbital integrations for an ambitious 100-m.y. survey of the stability of 2475 test particles on initially circular, zero-inclination orbits with original semimajor axes, a, in the range 0.1–2.2 AU. In four regions ("vulcanoids" with 0.09 < a < 0.21; Mercury-Venus objects with 0.58 < a < 0.68; Venus-Earth objects with 0.78 < a < 0.93; and Earth-Mars objects with 1.08 < a < 1.28), they find that many or most small bodies were stable for 100 m.y. [Note: In Evans and Tabachnik's calculations, a "stable" particle is defined as a particle that does not (1) enter the sphere of influence of a planet or the Sun, or (2) suffer ejection from the solar system. Our criterion for stability is the same, except that we consider actual collision with a planet or the Sun.] In particular, more than half the objects in the Earth-Mars zone survive for 100 m.y. They also find that logarithmic decay (equation (4)) is a good approximation at times longer than 10 m.y. Extrapolation of their results to 600 m.y. predicts

that ~17% and 38%, respectively, of the Venus-Earth and Earth-Mars objects will survive. For the Earth-Mars zone, these results predict that some objects survive up to 25 G.y.!

To test these results, we have performed, in collaboration with Hal Levison and Kevin Zahnle, a 100-m.y. integration of 200 particles with initial semimajor axes, a, between 0.7 and 1.3 AU. In Fig. 5, we plot the fraction, f, of objects with 0.7 < a < 1.0 AU (between Venus and Earth; triangles) and with 1 < a < 1.3 AU (between Earth and Mars; circles) that survive for a given time, as a function of time. In the top panel, we use a logarithmic axis for f and a linear axis for time, so that exponential decay would appear as a straight line in the plot. In the bottom panel, we use a linear axis for f and a logarithmic axis for time, so that logarithmic decay (equation (4)) appears as a straight line. As in Evans and Tabachnik's calculation, particles initially have eccentricities and inclinations of zero. We find that half the objects in the 0.7–1-AU region remain at 34 m.y. For the 1–1.3-AU region, half remain at 71 m.y. We too find that logarithmic decay is a reasonable fit to the data at late times; when we fit the data from 10 to 100 m.y. and extrapolate, we predict that 9% and 8% respectively of the Venus-Earth and Earth-Mars objects will survive for 600 m.y. The extrapolation is somewhat uncertain, since it is sensitive to which of the data are used in the fit (equation (4) is singular at t = 0, so we cannot use it to fit all the data.) We estimate that survival for objects in the 0.7–1.3-AU region at 600 m.y. could be as low as 2% or as high as 30%. The number of survivors goes to zero around 1 G.y. in our fits. We find somewhat fewer objects surviving than do Evans and Tabachnik, even though we apply a more stringent criterion for removal (see note above). Nonetheless, our overall conclusion is consistent with their findings; if accretional leftovers had initially low-inclination, near-circular orbits, the survival of ~10% of them to the time of the late heavy bombardment appears plausible.

In Fig. 5, we have also plotted results from a 60-m.y. integration of 118 near-Earth asteroids (squares) by *Gladman et al.* (2000). This sample has median orbital eccentricity and inclination of 0.5° and 10° respectively. Presumably, the Earth's early retinue of small bodies had orbital properties intermediate between the zero-eccentricity, zero-inclination test particles we assumed above and the dynamically "hot" NEAs. The NEAs have a median lifetime of 8 m.y. Again, logarithmic decay is a good fit to the data at late times, but the population tends to zero at 130 m.y. Most NEAs end their lives by hitting the Sun; only about 10–20% hit a terrestrial planet, usually the Earth or Venus.

The results in Fig. 5 are very preliminary; it is possible that survival for 600 m.y. is larger or smaller than our extrapolations would suggest. In addition, effects we ignore in our integrations, such as interparticle collisions, could reduce eccentricities and help to stabilize orbits, or could destroy small bodies. However, we emphasize that orbits in the inner solar system do not inevitably evolve into more long-lived niches, so that the "half-life" always increases

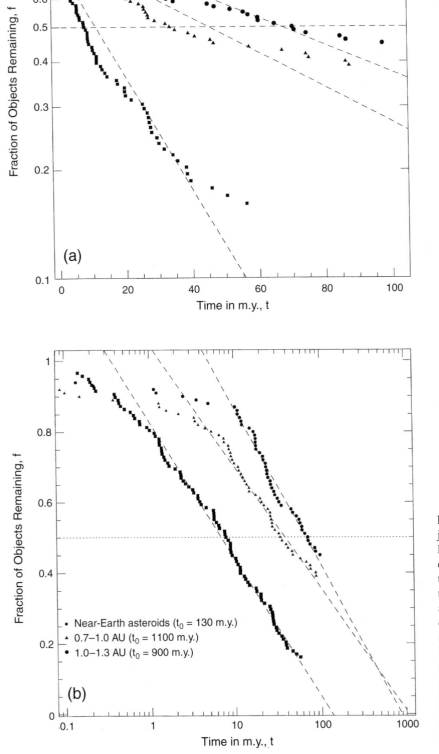

**Fig. 5.** The fraction, f, of Venus-Earth objects (i.e., 0.7 < a < 1.0 AU; triangles), Earth-Mars objects (i.e., 1 < a < 1.3 AU; circles), and near-Earth asteroids (squares) that survive for a given time, as a function of time. In **(a)**, we use a logarithmic axis for f and a linear axis for time, so that exponential decay would appear as a straight line in the plot. In **(b)**, we use a linear axis for f and a logarithmic axis for time, so that logarithmic decay (equation (4)) appears as a straight line. If the logarithmic decay law is valid at later times, the population should go to zero at a finite time, which we call $t_0$. The $t_0$ values are listed on the plot.

with time. Instead, an orbit can evolve from a stable niche into an unstable resonance that results in the rapid loss of the body, most often through collision with the Sun. It has only been appreciated in the last few years how full of these resonances the inner solar system is (*Michel, 1997*). Thus, depending upon the exact nature of the orbits of a population of small bodies, their number can decline exponentially, slower than exponentially (power law decay at late times), or faster than exponentially (logarithmic decay at late times). The survival of small bodies in the inner solar system for 600 m.y. appears plausible, but is unproven at this time.

*6.2.3. Asteroids (Turner et al., 1973; Wetherill, 1975; Chapman and Davis, 1975; Shoemaker, 1984; Zappalà et al., 1998; Morbidelli and Nesvorný, 1999).* An "asteroid shower" on the Earth and Moon following the break-up of a large main-belt asteroid was a popular Apollo-era idea for producing a pulse of cratering. *Zappalà et al.* (1998) show that "family forming event[s]" near various main-belt resonances can produce showers lasting from 5 to 80 m.y. From our equation (1) and Fig. 4, the mass required in the belt, assuming that all the collisional fragments are injected into a resonance, is ~3 × 10$^{24}$ g for a source in the innermost belt, 5 × 10$^{25}$ g for a source near the 3:1 resonance, and ~10$^{27}$ g for an outer-belt source. The inner belt mass required is comparable to the mass of the entire present-day belt, but is almost 1000× larger than the mass of the largest asteroids today with a <2.3 AU (18 Melpomene and 8 Flora). The main problem with this model is that collisional disruption of the required Ceres-sized asteroid is very improbable, unless the mass of the belt at 3.9 Ga was still of order 1 M$_\oplus$. *Davis et al.* (1985) argued that in order to preserve the observed basaltic crust of asteroid 4 Vesta (age ~4.4 Ga, based on eucrites), the belt mass when the basalts formed was lower than this, not much more than it is today. Models involving the catastrophic disruption of a Jupiter Trojan or Kuiper Belt object could also produce a pulse lasting tens of millions of years, but require even more mass.

*Morbidelli and Nesvorný* (1999) have recently found that many orbits in the inner part of the main asteroid belt evolve chaotically due to mean-motion resonances with Mars and three-body resonances involving both Mars and Jupiter. This "chaotic diffusion" allows many asteroids to eventually attain Mars-crossing orbits. Morbidelli and Nesvorný speculate that chaotic diffusion could have dynamically eroded the inner belt and caused the late heavy bombardment. This appears to be a more robust mechanism for producing the late heavy bombardment than an asteroid shower, since it does not require the unlikely disruption of a large asteroid. However, the details of this model have yet to be worked out.

*6.2.4. Formation of Uranus and Neptune and sweeping of resonances (Wetherill, 1975, Model IV; Shoemaker and Wolfe, 1984; Chyba, 1991; Levison et al., 2000a).* In this model, the late heavy bombardment is produced solely or in part by the gravitational scattering of a large mass of bodies from the vicinity of Neptune and Uranus. These plan-

ets are believed to be responsible for placing most (*Duncan et al.,* 1987), or at least many (*Hahn and Malhotra,* 1999; *Levison et al.,* 1999), of the comets in the Oort cloud. Planet formation in the Uranus-Neptune region appears to be inefficient; of the solid bodies originally between 15 and 35 AU, *Shoemaker and Wolfe* (1984) estimate that most were probably ejected from the solar system, ~20% reached the Oort cloud, and ~5% were incorporated into Neptune and Uranus. Taken at face value, this calculation implies that the Uranus-Neptune region once contained 600 M$_\oplus$ in planetesimals. The present mass of the Oort cloud, estimated to be 40 M$_\oplus$ by *Weissman* (1996), gives ~200 M$_\oplus$ for the initial mass in solids in the Uranus-Neptune. *Levison et al.* (2000a) have recently performed a calculation in which Uranus and Neptune are assumed to form at 3.9 Ga. [Note: The formation of Uranus and Neptune is an outstanding problem in planetary science (*Lissauer et al.,* 1995). The time at which these planets reached their present masses is unknown. An alternative theory that they formed earlier, some 4.5 G.y. ago, interior to 10 AU and then migrated outward, has been proposed by *Thommes et al.* (1999).] They find that if the Uranus-Neptune zone contained >100 M$_\oplus$ in solids, the amount of mass scattered into Earth-crossing orbits is large enough to produce the late lunar basins. (The impactors arrive within 10–20 m.y. of the formation of Uranus and Neptune, which is assumed to occur instantaneously.)

Gravitational scattering of planetesimals by the giant planets would have another interesting consequence. Due to an asymmetry in the fates of inward-evolving and outward-evolving planetesimals, Neptune, Uranus and, to a lesser extent, Saturn, would migrate outward, while Jupiter would move slightly inward over a period of tens of millions of years (*Fernández and Ip,* 1984; *Malhotra,* 1993; *Hahn and Malhotra,* 1999). This orbital evolution would result in the sweeping of resonances, notably the ν$_6$ secular resonance, through the asteroid belt, thereby exciting large eccentricities and destabilizing the orbits (cf. *Gomes,* 1997; *Liou and Malhotra,* 1997). Resonant sweeping can also occur due to changes in the mass of a small-body belt such as the asteroid belt. It is thus conceivable that the late heavy bombardment could be dominated by objects from the asteroid belt, even though the formation of Uranus and Neptune was the trigger. *Hartmann* (1987, 1990) cites evidence for an intense flux of scattered carbonaceous bodies near the close of planet formation, possibly fitting this scenario.

*6.2.5. Other models.* A cratering spike lasting 2–3 m.y. could be caused by a close stellar passage through the Oort cloud that results in a "comet shower" (*Hills,* 1981; *Davis et al.,* 1984; *Whitmire and Jackson,* 1984; *Hut et al.,* 1987). *Farley et al.* (1998) have found evidence for a comet shower in the late Eocene, some 35 m.y. ago. The Sun may have occupied a denser stellar environment such as an open cluster in its early days, though it probably would have moved on by the time of the late bombardment (*Gaidos,* 1995). Comet showers would have been correspondingly more frequent.

# 7. PHYSICAL EVIDENCE FOR A CATACLYSM AND AGAINST AN INTENSE IMPACT RATE BEFORE 4.0 Ga

In the view of many lunar researchers, not only did a cataclysmic spike occur, but also the petrologic and geochemical evidence for earlier intense cratering is scanty. The lunar impact history must ultimately be discerned from its record on the Moon itself, and appropriate dynamical models must make predictions consistent with that record. Impacts on the Moon after accretion and the production of a reasonably stable solid crust (prior to ~4.42 Ga, according to radiogenic isotope dating of igneous ferroan anorthosites) must have had several effects. One is the heating of the target (such that part is converted into impact melt) and the accompanying redistribution of radioactive and radiogenic isotopes that reset, or merely disturb, radiogenic clocks. A second is the introduction of meteoritic material, notably signposted by the meteoritic siderophile elements, into crustal materials. A third is the physical disruption of that solid crust. All these effects are observed, and it is their intensity that must define the lunar history.

The model outlined in the previous sections, which is partly based on the lunar record and partly on reasonable inferences and constraints about the dynamics of accretion, has a heavy but exponentially declining postaccretion impact rate, with an increasing half-life. The model does not predict any cataclysmic temporary increase in the impact rate at ~3.9 Ga, unless possibly in conjunction with a late formation of Uranus and Neptune or breakup of a very large Earth-crosser. In this section, we look more closely at the record of the Moon, updating a discussion by *Ryder* (1990).

The previous sections discussed whether the peak in a histogram of rock ages for lunar highlands (such as appears in many publications, e.g., *Horn and Kirsten,* 1977; *Wetherill,* 1981) could have been produced without a peak in impact flux ~3.9 Ga. Such a histogram of sample crystallization ages is not necessarily an accurate description of a global lunar record; published examples include multiple determinations on single rocks and multiple determinations of multiple rocks from single rock units. They also fail to discriminate among ages with varied uncertainties, and do not even distinguish rock types and their formational processes. However, such histograms effectively show that some major thermal processing at the surface of the Moon took place in the 3.9–3.8-Ga period, and the essential data are not in dispute. However, the evidence for a cataclysm is not solely such a peak in a rock-age histogram, but also the stratigraphic record of basin ages, and the record of the addition of meteoritic materials to a well-preserved crust. Furthermore, the idea that a peak in all radiogenic reset ages would occur during an exponentially declining flux (e.g., *Hartung,* 1974; *Hartmann,* 1975) appears based on an erroneous concept of how isotopic systems are reset or disturbed during cratering events. In addition, the apparent size distribution of the bombardment-era impactors around 3.9–

4.0 Ga may also suggest that a model based purely on accretional sweep-up is not complete, and that a different source may be involved.

## 7.1. Chronological Evidence for a Cataclysm

The time prior to the preservation of the oldest preserved mare plains has been stratigraphically divided, using superposition relationships, into the Early Imbrian Epoch, and the Nectarian and pre-Nectarian Periods (e.g., *Wilhelms,* 1987) (Fig. 6). The first and youngest of these ended when the dust settled on the Orientale Basin ejecta; the second ended when the projectile that made the Imbrium Basin made contact; the third and oldest ended when the Nectaris projectile made contact. The Crisium Basin and the Serenitatis Basin also assume important roles in the dating record because of the collection of samples in their vicinity. It is the dating of this stratigraphy that is highly suggestive of a cataclysm, and, as we have noted, it is critical to confirm the dates of basin-forming impacts. In order to interpret rock ages, it is important to tie the dates to the rock origins; there are no impact melt rocks known with reliable ages older than 3.95 Ga. Here we discuss two issues: (1) the resetting of isotopic systems and a general look at age determinations, regardless of the details of stratigraphy; and (2) dating the stratigraphic column and inferred fluxes.

*7.1.1. The resetting of isotopic systems.* The difficulties of resetting radiogenic clocks by impacts were reviewed by *Deutsch and Schärer* (1994); all the complexities cannot be repeated here. Suffice it to say, resetting radiogenic isotopic systems requires high temperatures over some period of time, equilibrating the existing isotopes over all phases in a rock (Rb-Sr, Nd-Sm, U, Th-Pb methods), or driving off radiogenically produced gases (K-Ar and related systems). Shock itself does not produce these effects. Very little of the material that is not actually melted degasses or has its isotopes reequilibrated, and even small clasts embedded in melt may not lose their Ar or reequilibrate other isotopes, as shown by various studies, ranging from theoretical treatments to analyses of experimentally shocked materials to inspection of natural craters (e.g., *Deutsch and Schärer,* 1994). In general, the reason is that the material is not held at high enough temperatures for long enough times; resetting is not an instantaneous event, but requires diffusion in the solid state. When a melt is produced, not only is diffusion faster but mechanical mixing homogenizes the material. Some unmelted rocks may be disturbed (partially reset) and thus show no definitive age.

Thus, the most reliable way to directly obtain the age of an impact by radiogenic methods is to analyze an impact melt rock from it. Most lunar impact melt rocks so far sampled are fine-grained and not amenable to mineral isochron dating methods such as Rb-Sr, and neither are there zircons in them large enough for U-Pb determinations. Thus, most age determinations are from the use of Ar-Ar methods. The fine grain size introduces some problems; the rapid

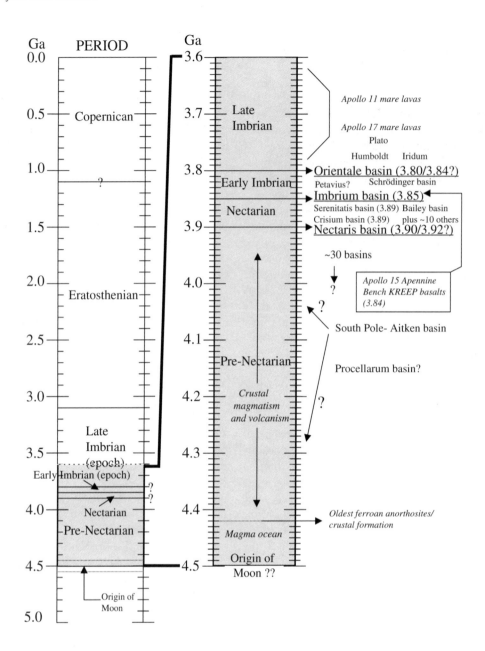

**Fig. 6.** Chronostratigraphic columns for lunar history, with relative stratigraphy based on *Wilhelms* (1984, 1987) and chronology as discussed in this paper. Left: complete column; right: expansion of the period from the origin of the Moon to ~3.6 Ga. In the righthand column, igneous events are shown in italics, and impact events in regular type. The major divisions among pre-Nectarian Period, Nectarian Period, and Early Imbrian Epoch are defined by the formation of the Nectaris Basin, the Imbrium Basin, and the Orientale Basin. The age of the Imbrium Basin, though debatable (e.g., *Deutsch and Stöffler*, 1987), is more confidently known and consensual (e.g., *Ryder,* 1994; *Dalrymple and Ryder,* 1993; *Haskin et al.,* 1998) than that of the Nectaris or Orientale Basins. While plutonic and volcanic crustal rocks, as well as some feldspathic granulitic breccias, are known with ages between ~4.45 Ga and 3.95 Ga, impact melt rocks from this period are rare to absent, though common from ~3.9 to ~3.8 Ga. The possible reasons for the absence of older melt rocks underlies the debate about linking the lunar record to models of ancient lunar (and inner solar system) impact history.

cooling suggests that some melt samples may not have degassed completely while molten, abundant small clasts may retain old ages or disturbed old ages, and the recoil artifact during analysis is common. Some have been disturbed by even later events. Nonetheless, many samples do show interpretable release structure, including good plateaus and hence reliable ages.

The formation of an impact crater has a considerable effect on the morphology of a terrain. Much of the energy of an impact event is used in moving material, and most is moved and deposited at low temperatures (e.g., *Chao,* 1977). Only very close to the projectile contact site itself is material strongly shocked and heated. Of the total displaced (allochthonous) material, about 1% is heated (and superheated) above the melting point to form an impact melt in a lunar crater about 10 km in diameter, and about 4% in a crater about 100 km in diameter, assuming velocities of 15–20 km (e.g, *Melosh,* 1989; *Cintala and Grieve,* 1994, 1998). This proportion is greater for larger craters, but still less than 10% for craters less than a few hundred kilometers in diameter (e.g., Schrödinger). The proportion of melted material to the entire affected target, allochthonous and autochthonous, is of course rather smaller.

Because of the cool nature of most ejecta and autochthonous target, virtually all the rocks collected from that crater and its ejecta will show ages of the target rocks, not of the crater-producing impact (this is what happens in terrestrial craters). Although in principle the "exponential threshold" effect (e.g., *Grinspoon,* 1989, and section 2) is correct, it is not correct to assume that the entire crater target is reset. In the case of *Grinspoon* (1989), the model is that all rocks exposed within the crater are reset; *Hartung* (1974) and *Hartmann* (1975, 1980) also appeal to physically pulverizing surface rocks into fine debris by multiple impacts during intense cratering so that earlier rock samples are difficult to find. The key to their argument is that surfaces older than 4.0 Ga are oversaturated with craters, so that individual volumes of regolith have been impacted many times in short intervals prior to 4.0 G.y. ago. However, in the view of *Ryder* (1990), virtually all the lunar crust sampled would need to have been converted to impact melt to be reset, or pulverized to fine powder to be inaccessible, and this is patently not the case. In that view, impact melt rock dates should simply appear in proportion to the impact flux, so that in a declining flux model, we would expect more older melt rock samples than younger ones in an endmember model where so little gets reset as to be negligible. (In reality, destruction of *some* older rocks by their being among the target rocks for younger melting is inevitable.) In this model, the simple peak in a histogram of ages (assuming that these are dominantly impact-related) is indeed evidence of a cataclysmic bombardment at ~3.80–3.90 Ga. However, in the competing model, the high early flux from 4.5 to 4.1 Ga pulverized the early impact melt rocks to a point where surviving samples are rare, and the age histogram does not mirror impact flux. Indeed, if we go back far enough, it is generally assumed that the high initial impact flux was capable of melting the surface layer and producing a magma ocean, obviously causing resetting! Therefore, there must have been a transition period of high heating capable of some age resetting as described here, until the impact flux became low enough that impacts were "one at a time" events. Clearly, a better understanding of impact fluxes, mechanics, and resetting is needed.

Further evidence that the abundance of ages of ~3.80–3.90 Ga is not a product of resetting comes from the nature of the rocks themselves. The younger highland surfaces, those of the Early Imbrian represented by Orientale ejecta (e.g., Hevelius Formation) and Imbrium ejecta (e.g., Fra Mauro Formation), have 2–4× the crater abundance of the oldest mare surfaces (*Hartmann et al.,* 1981), regardless of

their absolute ages. These formations contain rare igneous clasts, but most melt rock fragments within them are impact melt rocks and melt breccias. However, postmare impact-melted rocks (other than small-scale glass) and well-annealed breccias are notably rare in the mare plains, even the oldest; thus the flux over the last 3.8 G.y. has caused negligible resetting of ages on mare surfaces. Increasing this integrated flux by a factor of only 2–4 would not result in the impact melting demonstrated by highland breccias of the Fra Mauro and Cayley Formations. Thus, the ages of these melt rocks have been inferred to predate the formations from which they were collected (e.g., *Wetherill,* 1977, 1981; *Horn and Kirsten,* 1977). Indeed, there is ample evidence among the rocks from these formations for a variety of compositions, a real range in ages, and a lack of equilibration among clasts and host rocks. This is confirmed by the presence within the Fra Mauro Formation of volcanic mare basalt samples that date from more ancient times; clearly the lunar surface did not undergo general isotopic clock resetting. The survival of preimpact materials is against the recurrent idea (*Haskin et al.,* 1998) that ages have been almost entirely reset by the Imbrium basin impact.

*7.1.2. Late basin impact ages and inferred impact rates.* The flux with time needs to be derived from the dating of surfaces and the impact flux recorded on those surfaces. According to the above argument, it is erroneous to date the age of the Fra Mauro Formation (or any other) from the average age of melt rocks within it; as described by *Wetherill* (1977, 1981) and others, these rocks predate those formations. Here we outline the evidence for dating the relevant parts of the stratigraphic column.

The oldest of the classic, dark mare plains postdate the Orientale event, according to crater counts. The oldest mare rocks from a mare plain are close to 3.80 Ga (Apollo 11 and 17 high-Ti mare basalts) and one unique group appears a little older (*Dasch et al.,* 1998), and this can be taken as a reasonable age for the oldest dark mare surfaces. (Some uplands cryptomare areas, covered by a veneer of light ejecta ray material, may be older; cf. *Hartmann and Wood,* 1971; *Hawke et al.,* 1983, 1994.)

The age of the Imbrium event cannot be derived directly from impact-melt-rock samples because we do not know which samples really are derived from that melt, and those considered most likely have disturbed systems that interfere with a clear age assessment. However, the Imbrium Basin and its ejecta appear to be overlain by the Apennine Bench Formation, which is correlatable with the volcanic KREEP basalts collected at the Apollo 15 landing site. *Ryder* (1994) interprets these to give a reliable younger limit to the Imbrium event of 3.84 ± 0.02 Ga. The Apennine Front, also sampled on the Apollo 15 mission, has seen little activity since the Imbrium event formed it, and samples from it presumably represent Imbrium ejecta. Thus impact-melt-rock samples from it are either of Imbrium origin or predate it. Such samples were dated by *Dalrymple and Ryder* (1991, 1993), and suggest an age for Imbrium no older than

3.870 ± 0.010 Ga, and probably no older than 3.836 ± 0.016 Ga. (None of the samples showed any plateau age older than 3.879 ± 0.016 Ga.) A similar constraint is derived from samples of impact melt rock from the Fra Mauro Formation (e.g., *Stadermann et al.,* 1991). Impact-melt-rock fragments in the white rock 14063 from Cone Crater show a range from 3.87 to 3.95 Ga; other samples that are probably not from the Fra Mauro Formation but represent later local events (such as the 14310-group samples) are a little younger (3.82 ± 0.02 Ga). Thus, Imbrium can be fairly safely bracketed as 3.85 ± 0.02 Ga, a much tighter range than the 3.85–4.1 Ga suggested in the Basaltic Volcanism Study Project (henceforth *BVSP,* 1981). Imbrium was postdated by two multiring basins, Orientale and Schrödinger, which then must be between 3.85 and 3.80 Ga, and probably close to 3.85 Ga according to crater counts.

The Serenitatis Basin is relatively older than Imbrium, but how much older is not evident. In early studies it was believed to be one of the oldest basins because of its degradation, but later was accepted to have been greatly modified by the nearby Imbrium event. Coupled with tenuous stratigraphic information that it was younger than Crisium and hence younger than Nectaris, it was reclassified in the stratigraphic column to be one of the younger (Nectarian) basins (*Wilhelms,* 1976, 1987). This is, of course, a permissive argument, not a demonstrative one, and can be disputed. The Apollo 17 mission sampled many boulders of poikilitic impact melt rock from both North and South Massifs. These have generally been inferred to be from melt created in the formation of the Serenitatis Basin, and thus to date that event. Recent data show an age for these samples of 3.893 ± 0.009 Ga, consistent with but more precise than previous studies (*Dalrymple and Ryder,* 1996). However, the effects of Imbrium preclude using crater counts on Serenitatis ejecta to calibrate this age with cratering flux. Further, it has been suggested that these poikilitic boulders represent ballistically emplaced Imbrium ejecta (*Haskin et al.,* 1998), although for physical reasons that is unlikely. Samples collected from the Crisium ejecta blanket are sparse, but ages of melt rocks there too suggest ages of 3.85–3.90 Ga with Crisium at the upper end (*Swindle et al.,* 1991), such that Nectarian materials in general are in this range.

The age of Nectaris itself is critical. Stratigraphic and crater count data show that it is older than Crisium, but melt rock samples from it cannot be identified with certainty, and for all we know do not exist in the sample collection. The Apollo 16 site was modified by Nectaris ejecta, and subsequently by Imbrium ejecta. Fragments within the breccias collected on the Apollo 16 mission must predate these events, and probably include samples of melt rock created prior to Nectaris in several large craters that underlie the site. None of the melt rock samples reliably dated so far is older than 3.92 Ga; current work is underway to establish more clearly the ages of particular groups of melt rock composition identified by *Korotev* (1994; *Dalrymple and Ryder,* in progress). The analysis of the rocks and ages by *James* (1981) strongly suggests an age for Nectaris of less than 3.92 Ga, and probably an age of ~3.90 Ga is reasonable, consistent with earlier derivations by *Turner* (1977) and *Maurer et al.* (1978).

These data suggest a stratigraphy in which Nectaris was excavated at ~3.90 Ga, and Orientale no later than ~3.80 Ga. During this time the Crisium (3.89 Ga), Serenitatis (3.89 Ga), and several other Nectarian basins, including Humorum, Imbrium (3.85 Ga), and Schrödinger (3.80 Ga?), were excavated. From Nectaris to Imbrium, cratering declined by a factor of ~4, and by a slightly smaller factor from Imbrium to the oldest mare times (data in *BVSP*) over perhaps a similar period of absolute time.

This dating of the stratigraphic column produces a much steeper decline on a plot of flux vs. time than can be accommodated by a flux smoothly declining over 600 m.y.; in other words, if these basin ages are correct, they necessitate a cataclysmic spike. *Grinspoon* (1989) derived a flux at 4.0 Ga enhanced over the present day by a factor of 300, and at 3.8 Ga by a factor of 50, comparable to estimates by *BVSP* and *Wilhelms* (1984, 1987), who gave 1000 and 100 respectively. The basin ages cited above suggest an even steeper decline. Clearly, as noted earlier, precise dating of the lunar basin-forming impacts is crucial to understanding the early flux history that affected all planets in the inner solar system. It may require extensive *in situ* geological investigation.

## 7.2.  Siderophiles in the Ancient Highlands

The amount of meteoritic material added to the Moon following formation of a reasonably permanent crust at about 4.4 Ga is an important constraint in understanding the bombardment history. Several estimates have been made using varied lines of reasoning, but some of the models of flux vs. time are little more than schematic and they probably do not offer enough precision to make meaningful predictions about the amount of meteoritic material that hits the Moon. However, assuming that bombarding material in general stays on the Moon and assuming it is of heliocentric origin, hence on average probably chondritic material, it would add siderophile elements (e.g., Ir) to the crustal materials. The siderophiles actually measured in lunar samples (*Ryder,* 1990, 1999) can thus be used to place limits on the cumulative impacting mass.

Some rocks do contain high abundances of siderophile elements. Few rocks contain as much as 20 ppb Ir, and the average highland breccia contains less than 2 ppb Ir, corresponding to only 125 m equivalent of meteoritic material even if 2 ppb Ir represented the entire crust. Some impact melt rocks contain 5–15 (rarely higher) ppb Ir. However, they do not represent the entire crust; they form less than 10% of subregolith materials and are concentrated near the surface. A second group of rocks, the feldspathic granulites, in some cases also contain Ir abundances of 10 ppb or more, though most do not. They constitute less than 10% of highland samples, and there is no evidence that their siderophile abundances are representative of even the upper crust.

There are two ways to use such data to estimate or constrain the cumulative amount of meteoritic material added to the crust: from soil compositions, and from subregolith breccias (*Ryder,* 1990, 1999).

Highland soils are chemically averages of near-surface rocks. However, like mare soils, they must have had a micrometeorite (chondritic) component added to the upper few meters since the end of the heavy bombardment at about 3.80 Ga. This amount needs to be subtracted to find out what the meteoritic component at that time was. This is not necessarily a simple exercise, as the mixing mechanics might be somewhat different for a highland and a mare soil (this is not likely to be a major difference). The history of each local soil needs to be taken into account; for instance, soils freshened by downslope movement (e.g., on the Apennine Front) will not have their full complement of late siderophiles. The reliable soils (e.g., Apollo 16 plains) when "corrected" for a meteoritic component equivalent to that added to the Apollo 11 mare plain since 3.6 Ga (hence if anything an undercorrection) indicate older meteoritic abundances of less than 1%. Similarly, soils that might be expected to be almost completely lacking in micrometeorite component, such as the Apollo 17 South Massif landslide, do not need a "correction" at all and similarly show old meteoritic abundances of less than 1%, presumed to represent the uppermost crust at a little over 3.8 Ga.

The Apollo 16 feldspathic breccias, particularly those excavated at North Ray Crater, are commonly interpreted as subregolith breccias. They are probably ejecta from Nectaris, and match the regional chemistry prior to the addition of KREEP (as impact melt rocks, mainly) at about 3.85 Ga. These feldspathic breccias contain about 0.3% of a meteoritic component, most of which resides in impact melt rock fragments less than 3.9 Ga. The Apollo 14 white breccias (14063 and 14064) contain similarly low meteoritic abundances. Thus the evidence from these soils and breccias is that the mixed *uppermost* lunar crust contained somewhat less than 1%, and probably about 0.3%, of a meteoritic component at about 3.80–3.85 Ga.

Some estimates of the amount of meteoritic material in the crust have assumed a 2% chondritic component and a uniform depth of mixing of about 35 km (e.g., *Sleep et al.,* 1989). However, even the uppermost crust did not contain 2% meteoritic component, but probably less than 0.5% (above), and it is surely the case that the meteoritic component must decrease downward. The crust has retained a structure and it is not well-mixed (sections below). A megaregolith 35 km deep similar to the surface composition is untenable based on Clementine observations, and the surface meteoritic abundance must represent a maximum, not an average. Based on an appreciation of crustal structure and the characteristics of basin melt formation and excavation, it is clear that meteoritic addition in the 4.4–3.9-Ga period was dominantly in the upper 10–20 km. The feldspathic breccias at Apollo 16, as probable Nectaris ejecta, would represent perhaps the upper 15 km of the crust. Thus the total amount of meteoritic material added to the Moon

in the period between 4.4 Ga and the Nectaris event was roughly $0.003 \times 15$ km, or less than 50 m thickness if deposited as a surficial rock layer. This corresponds with roughly $5 \times 10^{21}$ g, which is strikingly close to the $6 \times 10^{21}$ g figure estimated in section 6 for the late basin-forming impacts. Not much more was added during the period 3.9–3.8 Ga, given that this later material is superficial and a very small proportion of the total upper crust. The value found here is much less than the amounts as high as $10^{23}$ g suggested by some models of a very early intense bombardment (e.g., *Baldwin,* 1981; *Morgan et al.,* 1977), and thus the amount of crustal siderophiles may be less than proposed for a very early intense cratering.

Much of this discussion of siderophiles is based on the assumption that early impactors matched the composition of current meteorites (asteroid belt fragments), reaching Earth's surface. Possible modification would come if early impactors included a much higher percentage of other material, such as icy/carbonaceous cometary impactors from the outer solar system.

## 7.3. Preservation of Ancient Crust

The intense early bombardment models, with a high flux declining in the first few hundred million years, generally predict a severe mixing and gardening of the crust in areas not affected by simultaneous or subsequent magmatic intrusion or extrusion. *Hartmann* (1973, 1980), for instance, shows a megaregolith equivalent to the entire crust; *Sleep et al.* (1989) assume mixing of half the crust into a uniform mass. However, the crust is not well-mixed but demonstrates a significant primary structure, as indicated even in early multidisciplinary studies (*Ryder and Wood,* 1977). Evidence from Earth-based remote sensing (e.g., *Hawke et al.,* 1983, 1994) and from the Clementine mission show considerable horizontal expanses of distinct terrain, and the preservation of vertical stratigraphies in basins (e.g., *Bussey and Spudis,* 1997). *Hartmann and Gaskell* (1997) showed that only localized parts of the lunar uplands, mainly on the farside, fully preserve crater-saturated crust. Similarly, *Hartmann and Wood* (1971), *Hawke et al.* (1983), and others have pointed out evidence of patches of premare volcanism in the highlands, apparently late in the early bombardment history. Some of these are apparently mare-like basalts, covered by thin veneers of ejected debris. Even a study of comparatively small craters, 40–180 km in diameter, shows that the crust is compositionally diverse, with preservation of the topmost anorthosites common (*Tompkins and Pieters,* 1999).

The larger sampled basins excavated pristine materials. The Serenitatis Basin melt rocks, for instance, have a clast population that is dominantly igneous plutonic. The "black and white" rocks from the Apollo 15 landing site, which might represent the Imbrium impact melt unit, show a similarly igneous set of clasts. A small proportion of fragments are feldspathic granulites of upper crustal composition, but fragments of impact melt rocks or other near-surface and brecciated material are lacking. Furthermore, the relative siderophile-element abundances in the Serenitatis melt breccias demonstrate that the impactor was an EH chondrite. It is possible that the target was contaminated with EH chondrite (alone), and the impactor was either an EH chondrite or an achondrite; but the simplest interpretation, consistent with the clast population, is that the target was noncontaminated igneous material and the impactor was an EH chondrite. The gravity data from Clementine suggest that the Serenitatis melt and excavation are from the upper and middle parts of the crust, and thus most of the deeper part of the crust lacks meteoritic contamination. A well-preserved nature of all but the uppermost crust would be inconsistent with models of continuous heavy bombardment, implying that impacting prior to 3.9 Ga must have been comparatively insignificant. Thus, further studies of the vertical lunar crustal structure, possibly *in situ*, offer important tests of the early bombardment environment.

### 7.4. Size Distribution of the Basin Impactors

The existence of late basins has long been seen as a problem for continuous declining bombardment models, because the declining bombardment should use up most impactors early, whereas the proposed number of late basins suggest not only late survival but possibly an anomalous size distribution. This problem was recognized by *Wetherill* (1975, 1977, 1981), where the size distribution of late impactors, with more large bodies than expected, was explained by breakup of bodies within the Roche limits of terrestrial planets. Regardless of the actual mechanism, accepting the existence of these late large impactors (two are proposed to be even later than the Imbrium impactor) suggests that an Earth-accretionary heliocentric population was not the entire story, and that some other process was in operation.

## 8. IMPLICATIONS FOR THE EARLY EVOLUTION OF PLANETARY SURFACES AND CLIMATES

The issue of intense bombardment in the first 600 m.y., and possibly cataclysmic spikes, has many ramifications for all terrestrial planets and the asteroid belt. To emphasize this, we provide a "shopping list" of effects of the early intense bombardment — many of which have been discussed elsewhere. Key to these effects is not only the fact that the average flux (or the flux during a cataclysm) was greater than today, but also the fact that, due to the size distribution, this produced some basin-forming impacts much larger than have occurred in recent geologic history.

- The largest terrestrial impact that occurred involved a body of roughly Mars size that ejected material from which the Moon was made (*Cameron and Ward*, 1976; see also reviews in *Weidenschilling et al.*, 1986).
- The largest impact on Mercury may have stripped off the mantle and that on Mars may have created the observed hemispheric asymmetry (*Benz et al.*, 1988).

- Large impacts may have blown off primordial atmospheres of Mars, Earth, or other planets (*Vickery and Melosh*, 1990). This effect may have influenced early martian atmospheric evolution especially.
- Early ice-rich planetesimals scattered from the outer solar system may have created a primordial, thick $H_2O$ atmosphere with attendant consequences on the climate and thermal history of the primordial Earth (*Abe and Matsui*, 1985; see also chapter by *Abe et al.*, 2000).
- A bombardment by ice-rich planetesimals might have increased the water content of Earth or other planets after they formed (*Chyba*, 1991).
- The growth of giant planets and early intense flux of bodies scattered from the outer solar system may have disrupted many bodies in the asteroid belt (*Kaula and Bigeleisen*, 1975), destabilized orbits through resonances (*Gomes*, 1997; *Liou and Malhotra*, 1997; *Levison et al.*, 2000a), and may even explain the incidence of carbonaceous inclusions in many meteorite breccias as well as captured low-albedo satellites (*Hartmann*, 1987, 1990).
- The early intense bombardment created a long-lasting or intermittent, optically thick, steady-state dusty atmosphere similar to that hypothesized after the K/T impact, throughout the first few hundred million years (*Grinspoon and Sagan*, 1987).
- Early basin-forming large impacts (of the scale that made the Imbrium and Orientale Basins on the Moon, and larger) made the Earth's crust inhomogeneous, exposing the mantle in some areas and piling up thicker crustal blocks of ejecta in others, creating protocontinental blocks on Earth and crustal asymmetry on Mars, and perhaps helping to initiate convective plate tectonic processes (*Frey*, 1977; *Hartmann*, 1980, 1988).
- The early intense bombardment may have frustrated the origins of life for several hundred million years (*Maher and Stevenson*, 1988; *Oberbeck and Fogelman*, 1989; *Sleep and Zahnle*, 1998), and on Mars by narrowing the habitable interval before liquid water disappeared from the surface (*Oberbeck and Fogelman*, 1989).

## 9. CONCLUSIONS

Because of several uncertainties and controversies about the conditions in the first few hundred million years, this review leads more toward continuing research problems than final solutions. On the one hand, direct cratering evidence indicates that the average cratering rate was higher at 4.0 Ga than it is today, and dynamical models suggest that this rate was still much higher before that. Planet accretion models suggest that the accretionary flux should have generally decreased rapidly in the first few hundred million years, with occasional short-lived spikes with a timescale on the order of 30 m.y. The half-life of planetesimal sweep-up is expected to have lengthened after planet formation at 4.55 Ga, possibly dovetailing with the declining rate picked up in the cratering record starting at about 4.0 or 3.9 Ga. On the other hand, several characteristics of the lunar petrologic record suggest a cataclysmic spike in cratering at 3.9–3.8 Ga, and even suggest, through lack of older impact melts, that the impact flux at 4.1–4.4 Ga was low instead of high. Many authors accept that four or more giant impact basins (requiring $10^2$-km-scale impactors) formed within about 200 m.y. on the Moon; if this is accepted, it leads to interesting problems about the origin of the bolides and the kind of fragmentation event and/or parent bodies that could have produced such large impactors. Controversies still exist over whether the rock age distribution on the Moon can be explained by competition between destruction and production processes, or whether they require a cratering cataclysm at 3.9–3.8 Ga to destroy older rocks and create most of the known basins at that time. Solutions for these problems would have wide ramifications for the understanding of planetary accretion, lunar formation, and the environment in which crustal structure and biology originated on Earth.

***Acknowledgments.*** The authors thank Robin Canup, Kevin Righter, Kevin Zahnle, Steve Mojzsis, and Hal Levison, who provided helpful critiques and support; Brett Gladman, who provided data prior to publication for Fig. 5; and Robin Canup for her patience and encouragement. We also thank various colleagues at our institutions. This work was supported in part by various NASA grants. This is PSI Contribution Number 353 and LPI Contribution No. 1004.

# REFERENCES

Abe Y. and Matsui T. (1985) The formation of an impact-generated H$_2$O atmosphere and its implications for the early thermal history of the Earth. *Proc. Lunar Planet. Sci. Conf. 15th,* in *J. Geophys. Res., 90,* C545–C559.

Abe Y., Ohtani E., Okuchi T., Righter K., and Drake M. (2000) Water in the early Earth. In *Origin of the Earth and Moon* (R. M. Canup and K. Righter, eds.), this volume. Univ. of Arizona, Tucson.

Baldwin R. B. (1963) *The Measure of the Moon.* Univ. of Chicago, Chicago, Illinois. 488 pp.

Baldwin R. B. (1971) On the history of lunar impact cratering: The absolute time scale and the origin of planetesimals. *Icarus, 14,* 36–52.

Baldwin R. B. (1981) On the origin of the planetesimals that produced the multi-ring basins. In *Multi-Ring Basins, Proc. Lunar Planet. Sci. 12A* (P. H. Schultz and R. B. Merrill, eds.), pp. 19–28. Pergamon, New York.

Benz W., Slattery W. L., and Cameron A. G. W. (1988) Collisional stripping of Mercury's mantle. *Icarus, 74,* 516–528.

Bussey D. B. J. and Spudis P. D. (1997) Compositional analysis of the Orientale basin using full resolution Clementine data: Some preliminary results. *Geophys. Res. Lett., 24,* 445–448.

BVSP (Basaltic Volcanism Study Project) (1981) *Basaltic Volcanism on the Terrestrial Planets.* Pergamon, New York. 1286 pp.

Cameron A. G. W. and Ward W. R. (1976) The origin of the Moon (abstract). In *Lunar Science VII,* pp. 120–122. The Lunar Science Institute, Houston.

Chao E. C. T. (1977) Basis for interpretation regarding the ages of the Serenitatis, Imbrium, and Orientale events. *Philos. Trans. Roy. Soc. London, A285,* 115–126.

Chapman C. R. and Davis D. R. (1975) Asteroid collisional evolution — Evidence for a much larger early population. *Science, 190,* 553–556.

Chyba C. F. (1991) Terrestrial mantle siderophiles and the lunar impact record. *Icarus, 92,* 217–233.

Cintala M. J. and Grieve R. A. F. (1994) The effects of differential scaling of impact melt and crater dimensions on lunar and terrestrial craters: Some brief examples. In *Large Meteorite Impacts and Planetary Evolution* (B. O. Dressler, R. A. F. Grieve, and V. L. Sharpton, eds.), pp. 51–59. GSA Spec. Paper 293, Boulder, Colorado.

Cintala M. J. and Grieve R. A. F. (1998) Scaling impact-melt and crater dimensions: Implications for the lunar cratering record. *Meteoritics & Planet. Sci., 33,* 889–912.

Culler T. S., Becker T. A., Muller R. A., and Renne P. R. (2000) Lunar impact history from $^{40}$Ar/$^{39}$Ar dating of glass spherules. *Science, 287,* in press.

Dalrymple G. B. and Ryder G. (1991) $^{40}$Ar/$^{39}$Ar ages of six Apollo 15 impact melt rocks by laser step heating. *Geophys. Res. Lett., 18,* 1163–1166.

Dalrymple G. B. and Ryder G. (1993) $^{40}$Ar/$^{39}$Ar age spectra of Apollo 15 impact melt rocks by laser step-heating and their bearing on the history of lunar basin formation. *J. Geophys. Res., 98,* 13085–13095.

Dalrymple G. B. and Ryder G. (1996) Chronology of impact melts at the Apollo 17 landing site. *Eos Trans. AGU, 74,* 197.

Dasch J., Ryder G., Reese Y., Weismann H., Shih C.-Y., and Nyquist L. E. (1998) Old age of formation for a distinct variety of A17 high-titanium mare basalt. In *Lunar and Planetary Science XXIX,* Abstract #1750, Lunar and Planetary Institute, Houston (CD-ROM; also available on line at http://cass.jsc.nasa.gov/meetings/LPSC98/pdf/1750.pdf).

Davis D. R., Chapman C., Weidenschilling S., and Greenberg R. (1985) Collisional history of asteroids: Evidence from Vesta and the Hirayama families. *Icarus, 62,* 30–53.

Davis M., Hut P., and Muller R. A. (1984) Extinction of species by periodic comet showers. *Nature, 308,* 715–717.

Deutsch A. and Schärer U. (1994) Dating terrestrial impact events. *Meteoritics, 29,* 301–322.

Deutsch A. and Stöffler D. (1987) Rb-Sr analyses of Apollo 16 melt rocks and a new age estimate for the Imbrium basin: Lunar basin chronology and the early heavy bombardment of the Moon. *Geochim. Cosmochim. Acta, 51,* 1951–1964.

Dohnanyi J. W. (1969) Collisional model of asteroids and their debris. *J. Geophys. Res., 74,* 2531–2554.

Dones L., Levison H., and Duncan M. (1996) On the dynamical lifetimes of planet-crossing objects. In *Completing the Inventory of the Solar System* (T. W. Rettig and J. Hahn, eds.), pp. 233–244. ASP Conference Series, Vol. 107, San Francisco, California.

Dones L., Gladman B., Melosh H. J., Tonks W. B., Levison H. F., and Duncan M. J. (1999a) Dynamical lifetimes and final fates of small bodies: Orbit integrations vs. Öpik calculations. *Icarus,* in press.

Dones L., Levison H. F., and Zahnle K. J. (1999b) Tossing out the leftovers: The fate of the remains of Earth's accretion. *Bull. Am. Astron. Soc., 31,* 1078–1079.

Duncan M. J. and Levison H. F. (1997) A scattered comet disk and the origin of Jupiter family comets. *Science, 276,* 1670–1672.

Duncan M., Quinn T., and Tremaine S. (1987) The formation and extent of the solar system comet cloud. *Astron. J., 94,* 1330–1338.

Evans N. W. and Tabachnik S. (1999) Possible long-lived belts in the inner solar system. *Nature, 399,* 41–43.

Farley K. A., Montanari A., Shoemaker E. M., and Shoemaker C. S. (1998) Geochemical evidence for a comet shower in the late Eocene. *Science, 280,* 1250–1253.

Fernández J. A. and Ip W.-H. (1984) Some dynamical aspects of the accretion of Uranus and Neptune — The exchange of orbital angular momentum with planetesimals. *Icarus, 58,* 109–120.

Frey H. (1977) Origin of the Earth's ocean basins. *Icarus, 32,* 235–250.

Gaidos E. (1995) Paleodynamics: Solar system formation and the early environment of the Sun. *Icarus, 114,* 258–268.

Gladman B. J. (1997) Destination: Earth. Martian meteorite delivery. *Icarus, 130,* 228–246.

Gladman B. J., Burns J. A., Duncan M. J. , and Levison H. (1995) The dynamical evolution of lunar impact ejecta. *Icarus, 118,* 302–321.

Gladman B., Burns J. A., Duncan M., Lee P., and Levison H. (1996) The exchange of impact ejecta between terrestrial planets. *Science, 271,* 1387–1392.

Gladman B., Migliorini F., Morbidelli A., Zappalà V., Michel P., Cellino A., Froeschlé Ch., Levison H. F., Bailey M., and Duncan M. (1997) Dynamical lifetimes of objects injected into asteroid belt resonances. *Science, 277,* 197–201.

Gladman B., Michel P., and Froeschlé Ch. (2000) The near-Earth object population. *Icarus,* in press.

Gomes R. S. (1997) Dynamical effects of planetary migration on the primordial asteroid belt. *Astron. J., 114,* 396–401.

Grieve R. A., and Shoemaker E. (1994) The record of past impacts on Earth. In *Hazards Due to Comets & Asteroids* (T. Gehrels, ed.), pp. 417–462. Univ. of Arizona, Tucson.

Grinspoon D. H. (1989) Large impact events and atmospheric evolution on the terrestrial planets. Ph.D. thesis, Univ. of Arizona, Tucson. 209 pp.

Grinspoon D. H. and Sagan C. (1987) *Proc. Penn State Univ. Workshop on Long-Term Stability of the Earth System.*

Hahn J. M. and Malhotra R. (1999) Orbital evolution of planets embedded in a planetesimal disk. *Astron. J., 117,* 3041–3053.

Hartmann W. K. (1965) Terrestrial and lunar flux of large meteorites in the last two billion years. *Icarus, 4,* 157–165.

Hartmann W. K. (1966) Early lunar cratering. *Icarus, 5,* 406–418.

Hartmann W. K. (1969) Lunar and interplanetary rock fragmentation. *Icarus, 10,* 201–213.

Hartmann W. K. (1970) Lunar cratering chronology. *Icarus, 13,* 299–301.

Hartmann W. K. (1973) Ancient lunar mega-regolith and subsurface structure. *Icarus, 18,* 634–636.

Hartmann W. K. (1975) Lunar "cataclysm": A misconception? *Icarus, 24,* 181–187.

Hartmann W. K. (1980) Dropping stones into magma oceans: Effects of early lunar cratering. In *Proc. Conf. Lunar Highlands Crust* (J. J. Papike and R. B. Merrill, eds.), pp. 155–171. Pergamon, New York.

Hartmann W. K. (1987) A satellite-asteroid mystery and a possible early flux of scattered C-class asteroids. *Icarus, 78,* 57–68.

Hartmann W. K. (1988) Impact strengths and energy partitioning in impacts into finite solid targets (abstract). In *Lunar and Planetary Science XVIV,* pp. 451–452. Lunar and Planetary Institute, Houston.

Hartmann W. K. (1990) Additional evidence about an early intense flux of C asteroids and the origin of Phobos. *Icarus, 87,* 236–240.

Hartmann W. K. and Davis D. (1975) Satellite-sized planetesimals and lunar origin. *Icarus, 24,* 504–515.

Hartmann W. K. and Gaskell R. W. (1997) Planetary cratering 2: Studies of saturation equilibrium. *Meteoritics & Planet. Sci., 32,* 109–121.

Hartmann W. K. and Wood C. A. (1971) Moon: Origin and evolution of multi-ring basins. *Moon, 3,* 3–78.

Hartmann W. K., Strom R. G., Weidenschilling S. J., Blasius K. R., Woronow A., Dence M. R., Grieve R. A. F., Diaz J., Chapman C. R., Shoemaker E. M., and Jones K. L. (1981) Chronology of planetary volcanism by comparative studies of planetary cratering. In *Basaltic Volcanism on the Terrestrial Planets,* pp. 1050–1129. Pergamon, New York (full text available on line at http://adsbit.harvard.edu/books/bvtp/).

Hartung J. B. (1974) Can random impacts cause the observed $^{39}$Ar/$^{40}$Ar age distribution for lunar highland rocks? *Meteoritics, 9,* 349.

Haskin L. A. (1998) The Imbrium impact event and thorium distribution at the lunar highlands surface. *J. Geophys. Res., 103,* 1679–1689.

Haskin L. A., Korotev R. L., Rockow K. M., and Jolliff B. L. (1998) The case for an Imbrium origin of the Apollo thorium-rich impact melt breccias. *Meteoritics & Planet. Sci., 33,* 959–975.

Hawke B., Bell J., and Clark P. (1983) Very ancient lunar volcanism: Implications for crustal composition and evolution (abstract). In *IAU Colloquium 77,* p. 36.

Hawke B. R., Blewett D. T., and Campbell B. A. (1994) Spectral and radar studies of the Schiller-Schickard region of the Moon (abstract). In *Lunar and Planetary Science XXV,* pp. 515–517. Lunar and Planetary Institute, Houston.

Hills J. G. (1981) Comet showers and the steady-state infall of comets from the Oort cloud. *Astron. J., 86,* 1730–1740.

Holman M. J. (1997) A possible long-lived belt of objects between Uranus and Neptune. *Nature, 387,* 785–788.

Holman M. J. and Wisdom J. (1993) Dynamical stability in the outer solar system and the delivery of short period comets. *Astron. J., 105,* 1987–1999.

Horn P. and Kirsten T. (1977) Lunar highland stratigraphy and radiometric dating. *Philos. Trans. Roy. Soc. London, A285,* 145–150.

Hut P., Alvarez W., Elder W. P., Hansen T., Kauffman E. G., Keller G., Shoemaker E. M., and Weissman P. R. (1987) Comet showers as a cause of mass extinctions. *Nature, 329,* 118–126.

James O. B. (1981) Petrologic and age relations of the Apollo 16 rocks: Implications for subsurface geology and the age of the Nectaris Basin. *Proc. Lunar Planet. Sci. 12B,* pp. 209–233.

Kaula W. and Bigeleisen P. (1975) Early scattering by Jupiter and its collision effects on the terrestrial zone. *Icarus, 25,* 18–33.

Korotev R. L. (1994) Compositional variation in Apollo 16 impact-melt breccias and inferences for the geology and bombardment history of the Central Highlands of the Moon. *Geochim. Cosmochim. Acta, 58,* 3931–3969.

Laskar J. (1996) Large scale chaos and marginal stability in the solar system. *Cel. Mech. Dyn. Astron., 64,* 115–162.

Levison H. F. (1996) Comet taxonomy. In *Completing the Inventory of the Solar System* (T. W. Rettig and J. M. Hahn, eds.), pp. 173–191. ASP Conference Proceedings, Vol. 107, San Francisco, California.

Levison H. F. and Duncan M. J. (1997) From the Kuiper Belt to Jupiter-family comets: The spatial distribution of ecliptic comets. *Icarus, 127,* 13–32.

Levison H., Dones L., Duncan M., and Weissman P. (1999) The

formation of the Oort cloud. *Bull. Amer. Astron. Soc., 31,* 1079.

Levison H. F., Dones L., Chapman C. R., Stern S. A., Duncan M. J., and Zahnle K. (2000a) Could the lunar "late heavy bombardment" have been triggered by the formation of Uranus and Neptune? *Icarus,* submitted.

Levison H. F., Duncan M. J., Zahnle K., Holman M., and Dones L. (2000b) Note: Planetary impact rates from ecliptic comets. *Icarus,* in press.

Liou J. C. and Malhotra R. (1997) Depletion of the outer asteroid belt. *Science, 275,* 375–377.

Lissauer J. J., Pollack J. B., Wetherill G. W., and Stevenson D. J. (1995) Formation of the Neptune system. In *Neptune* (D.P. Cruikshank, ed.), pp. 37–108. Univ. of Arizona, Tucson.

Maher K. A. and Stevenson D. J. (1988) Impact frustration of the origin of life. *Nature, 331,* 612–614.

Malhotra R. (1993) The origin of Pluto's peculiar orbit. *Nature, 365,* 819–821.

Malyshkin L. and Tremaine S. (1999) The Keplerian map for the planar restricted three-body problem as a model of comet evolution. *Icarus, 141,* 341–353.

Maurer P., Eberhardt P., Geiss J., Grogler N., Stettler A., Brown G., Peckett A., and Krähenbühl U. (1978) Pre-Imbrian craters and basins: Ages, compositions and excavation depths of Apollo 16 breccias. *Geochim. Cosmochim. Acta, 42,* 1687–1720.

Melosh H. J. (1989) *Impact Cratering: A Geologic Process.* Oxford Univ., New York.

Michel P. (1997) Effects of linear secular resonances in the region of semimajor axes smaller than 2 AU. *Icarus, 129,* 348–366.

Migliorini F., Michel P., Morbidelli A., Nesvorný D., and Zappalà V. (1998) Origin of multikilometer Earth- and Mars-crossing asteroids: A quantitative simulation. *Science, 281,* 2022–2024.

Morbidelli A. and Gladman B. (1998) Orbital and temporal distributions of meteorites originating in the asteroid belt. *Meteoritics & Planet. Sci., 33,* 999–1016.

Morbidelli A. and Nesvorný D. (1999) Numerous weak resonances drive asteroids towards terrestrial planets orbits. *Icarus, 139,* 295–308.

Morgan J. W., Ganapathy R., Higuchi H., and Anders E. (1977) Meteoritic material on the Moon. In *The Soviet-American Conference on the Cosmochemistry of the Moon and Planets, Part 2* (J. H. Pomeroy and N. J. Hubbard, eds.), pp. 659–689. NASA SP-370.

Neukum G. (1983) *Meteoritenbombardement und Datierung planetarer Oberflachen.* Habilitation Dissertation for Faculty Membership, Ludwig-Maximillians-University of Munich.

Neukum G. and Ivanov B. (1994) Crater size distributions and impact probabilities on Earth from lunar, terrestrial-planet, and asteroid cratering data. In *Hazards Due to Comets and Asteroids* (T. Gehrels, ed.), pp. 359–416. Univ. of Arizona, Tucson.

Neukum G., Koenig B., Fechtig H., and Storzer D. (1975). Cratering in the Earth-Moon system. *Proc. Lunar Sci. Conf. 6th,* pp. 3417–3432.

Oberbeck V. R. and Fogleman G. (1989) Estimates of the maximum time required to originate life. *Origins of Life, 19,* 549–560.

Pepin R. O. and Phinney D. (1975) The formation interval of the Earth (abstract). In *Lunar Science VII,* pp. 682–684. The Lunar Science Institute, Houston.

Podosek F. A. and Ozima M. (2000) The xenon age of the Earth. In *Origin of the Earth and Moon* (R. M. Canup and K. Righter, eds.), this volume. Univ. of Arizona, Tucson.

Ryder G. (1990) Lunar samples, lunar accretion, and the early

bombardment of the Moon. *Eos Trans. AGU, 71,* 313.

Ryder G. (1994) Coincidence in time of the Imbrium basin impact and Apollo 15 volcanic flows: The case for impact-induced melting. In *Large Meteorite Impacts and Planetary Evolution* (B. O. Dressler, R. A. F. Grieve, and V. L. Sharpton, eds.), pp. 11–18. GSA Spec. Paper 293.

Ryder G. (1999) Meteoritic abundances in the ancient lunar crust. In *Lunar and Planetary Science XXX,* Abstract #1848. Lunar and Planetary Institute, Houston (CD-ROM; also available on line at http://www.lpi.usra.edu/meetings/LPSC99/pdf/1848.pdf).

Ryder G. and Wood J. A. (1977) Serenitatis and Imbrium impact melts: Implications for large-scale layering in the lunar crust. *Proc. Lunar Planet. Sci. Conf. 8th,* pp. 655–668.

Safronov V. S. (1972) *Evolution of the Protoplanetary Cloud and Formation of the Earth and the Planets.* Israel Program for Scientific Translations, Jerusalem.

Shoemaker E. M. (1984) Large body impacts through geologic time. In *Patterns of Change in Earth Evolution* (H. D. Holland and A. F. Trendall, eds.), pp. 15–40. Dahlem Konferenzen 1984, Springer-Verlag, New York.

Shoemaker E. M. and Wolfe R. F. (1984) Evolution of the Uranus-Neptune planetesimal swarm (abstract). In *Lunar and Planetary Science XV,* pp. 780–781. Lunar and Planetary Institute, Houston.

Shoemaker E. M., Shoemaker C. S., and Wolfe R. F. (1989) Trojan asteroids — Populations, dynamical structure and origin of the L4 and L5 swarms. In *Asteroids II* (R. P. Binzel, T. Gehrels, and M. S. Matthews, eds), pp. 487–523. Univ. of Arizona, Tucson.

Sleep N. H. and Zahnle K. (1998) Refugia from asteroid impacts on early Mars and the early Earth. *J. Geophys. Res., 103,* 28529–28544.

Sleep N. H., Zahnle K. J., Kasting J. F., and Morowitz H. J. (1989) Annihilation of ecosystems by large asteroid impacts on the early Earth. *Nature, 342,* 139–142.

Stadermann F. J., Hastier E., Jessberger E. K., Lingner S., and Stöffler D. (1991) The case for a younger Imbrium basin: New $^{40}$Ar/$^{39}$Ar ages of Apollo 14 rocks. *Geochim. Cosmochim. Acta, 55,* 2339–2349.

Steiger R. H. and Jäger E. (1977) Subcommission on geochronology: Convention and the use of decay constants in geo- and cosmochronology. *Earth Planet. Sci. Lett., 36,* 359–362.

Strom R. G. and Neukum G. (1988) The cratering record on Mercury and the origin of impacting objects. In *Mercury* (F. Vilas, C. R. Chapman, and M. S. Matthews, eds.), pp. 336–373. Univ. of Arizona, Tucson.

Swindle T. D., Spudis P. D., Taylor G. J., Korotev R. L., Nichols R. H., and Olinger C. T. (1991) Searching for Crisium basin ejecta: Chemistry and ages of Luna 20 impact melts. *Proc. Lunar Planet. Sci., Vol. 21,* pp. 167–181.

Tera F., Papanastassiou D., and Wasserburg G. (1974) Isotopic evidence for a terminal lunar cataclysm. *Earth Planet. Sci. Lett., 22,* 1–21.

Thommes E. W., Duncan M. J., and Levison H. F. (1999) The formation of Uranus and Neptune in the Jupiter-Saturn region of the solar system. *Nature,* in press.

Tompkins S. and Pieters C. M. (1999) Mineralogy of the lunar crust; Results from Clementine. *Meteoritics & Planet. Sci., 34,* 25–41.

Turner G. (1977) The early chronology of the Moon: evidence for the early collisional history of the solar system. *Philos. Trans. R. Soc. London, A285,* 97–103.

Turner G., Cadogan P. H., and Yonge C. J. (1973) Argon selenochronology. *Geochim. Cosmochim. Acta, 2,* 1889–1914.

Vickery A. M. and Melosh H. J. (1990) Atmospheric erosion and impactor retention in large impacts, with application to mass extinctions. In *Global Catastrophes in Earth History* (V. Sharpton and P. Ward, eds.), pp. 289–300. GSA Spec. Paper 247, Boulder, Colorado.

Weidenschilling S. J., Greenberg R., Chapman C. R., Herbert F., Davis D. R., Drake M. J., Jones J., and Hartmann W. K. (1986) Origin of the Moon from a circumterrestrial disk. In *Origin of the Moon* (W. K. Hartmann, R. J. Phillips, and G. J. Taylor, eds.), pp. 731–762. Lunar and Planetary Institute, Houston (full text available on line at http://adsbit.harvard.edu/books/ormo/).

Weissman P. R. (1989) The impact history of the solar system — Implications for the origins of atmospheres. In *Origin and Evolution of Planetary and Satellite Atmospheres* (S. K. Atreya, J. B. Pollack, and M. S. Matthews, eds.), pp. 230–267. Univ. of Arizona, Tucson.

Weissman P. R. (1996) The Oort Cloud. In *Completing the Inventory of the Solar System* (T. W. Rettig and J. M. Hahn, eds.), pp. 265–288. ASP Conference Proceedings, Vol. 107, San Francisco, California.

Wetherill G. W. (1975) Late heavy bombardment of the Moon and terrestrial planets. *Proc. Lunar Sci. Conf. 6th,* pp. 1539–1561.

Wetherill G. W. (1977) Evolution of the earth's planetesimal swarm subsequent to the formation of the earth and Moon. *Proc. Lunar Sci. Conf. 8th,* pp. 1–16.

Wetherill G. W. (1981) Nature and origin of basin-forming projectiles. In *Multi-Ring Basins, Proc. Lunar Planet. Sci. 12A* (P. H. Schultz and R. B. Merrill, eds.), pp. 1–18. Pergamon, New York.

Wetherill G. W. (1990) Formation of the Earth. *Annu. Rev. Earth. Planet. Sci., 18,* 205–256.

Whitmire D. P. and Jackson A. A. (1984) Are periodic mass extinctions driven by a distant solar companion? *Nature, 308,* 713–715.

Wilhelms D. E. (1976) Secondary impact craters of lunar basins. *Proc. Lunar Sci. Conf. 7th,* pp. 2883–2901.

Wilhelms D. E. (1984) Moon. In *The Geology of the Terrestrial Planets* (M. H. Carr, R. S. Saunders, R. G. Strom, and D. E. Wilhelms, eds.), pp. 107–205. NASA SP-469.

Wilhelms D. E. (1987) *The Geologic History of the Moon.* U.S. Geol. Surv. Prof. Paper 1348. 302 pp.

Zahnle K. J. and Sleep N. H. (1997) Impacts and the early evolution of life. In *Comets and the Origin and Evolution of Life* (P. J. Thomas, C. F. Chyba, and C. P. McKay, eds.), pp. 175–208. Springer-Verlag, New York.

Zappalà V., Cellino A., Gladman B. J., Manley S., and Migliorini F. (1998) Asteroid showers on Earth after family break-up events. *Icarus, 134,* 176–179.

# Earth-Moon Interactions: Implications for Terrestrial Climate and Life

**D. M. Williams**
*Penn State Erie, The Behrend College*

**D. Pollard**
*The Pennsylvania State University*

The Moon is the only satellite in the solar system capable of influencing planetary climate. Its large mass and close proximity to Earth accelerates Earth's precession, and stabilizes the obliquity of the terrestrial spin axis. Included in this chapter are a rudimentary discussion and derivation of the equations of precession, and a discussion of spin-orbit coupling and evolution of Earth's obliquity, both with and without the Moon. Results from two- and three-dimensional climate models are used to examine climates of Earth and Earth-like planets at high obliquity. Finally, high obliquity and climate friction are offered as possible devices for explaining apparent low-latitude glaciation on the early Earth.

## 1. INTRODUCTION

It is remarkable to first consider that the Earth's climate can be affected by the Moon. Indeed, the evolution of at least some forms of life (e.g., humans) on Earth may have been enabled by the Moon's existence. How can this be? Fundamentally, an important driver of Earth's climate is the orientation (or obliquity) of its spin axis relative to its orbital plane, which is variable as a consequence of the joint precession of Earth's spin axis and of its orbit. As will be explained further below in section 2, the spin-axis precession results from gravitational torques exerted upon the Earth's equatorial bulge by the Sun and the Moon (but primarily by the Moon) and the orbit precession results from the gravitational influences of the other planets. Thus, the Moon, as well as the Sun and the other planets of the solar system, plays an important role in determining how Earth's obliquity evolves with time.

The evolution of obliquity is important to the long-term habitability of a planet because obliquity sets the zonation of global climate as well as the amplitudes of seasonal temperature oscillations. When the obliquity is small to moderate, as Earth's is today, the poles are the coldest part of the planet on average and the seasonal temperature swings over an orbit are globally very mild. Conversely, when the obliquity is high (>54°, which is the obliquity at which the global, annual-average insolation is uniform), the tropics are, on average, the coldest region of the planet, while temperatures in the middle and upper latitudes may be highly variable and seasonally extreme. The rapidity of changing temperatures and accompanying extremes would make Earth's climate at high obliquity dramatically different from the present, and possibly hostile to many life forms that inhabit our planet today. The specific changes to Earth's climate that life would have to cope with at high obliquity will be examined in section 3.

Whether the obliquity of early Earth were ever much different from its present value of 23.5° is an open-ended question. Earth's primordial obliquity is unconstrained by the giant impacts that occurred on Earth near the end of the main accretion era of its history (4.6–4.4 G.y. ago), and which are thought to have resulted in the origin of the Moon. A typical outcome of such energetic collisions is for the remaining planet to have a highly inclined spin axis (*Canup and Agnor*, 2000). Thus, Earth could have started out with a much higher obliquity than it exhibits today and later had it reduced to its present value. The problem is that such a reduction appears to have been dynamically unlikely or impossible with the Moon present. The Moon, as will be discussed below, acts to stabilize Earth's spin axis (that is, prevents any large secular drifts) for obliquities less than ~60°. Thus, if Earth's obliquity were ever significantly higher than it is today, but <60°, it should have remained there.

However, there is growing geologic evidence that Earth's spin axis may have been much more inclined during the Precambrian Era (3.8–0.6 G.y. ago) of its history than it is today. At least two land masses (including what is now China and Australia) appear to have been situated within 10° of Earth's equator (based on the alignment of magnetic domains within ancient magnetic minerals, and on the assumption that Earth's magnetic axis was aligned during the late Precambrian the same as today), and covered by ice between ~800 and 600 m.y. ago (*Frakes*, 1979; *Embleton and Williams*, 1986; *Zhang and Zhang*, 1985).

This implies either that the Earth was frozen pole to pole (i.e., the snowball Earth hypothesis) or that obliquity was >54° at the time, so that the tropics would have been less insolated over a seasonal cycle than the poles. The former scenario has received considerable attention recently (*Meert and Van der Voo*, 1994; *G. E. Williams et al.*, 1995; *Hoffman et al.*, 1998) and cannot yet be ruled out as a legitimate explanation for the low-latitude paleoclimatic enigma. Here we suggest an alternative hypothesis: that low-latitude glaciation during the Precambrian was a result of Earth having a high obliquity. We demonstrate that Earth's obliquity may have been subsequently lowered, as is needed to explain Earth's present obliquity, by climate friction, or obliquity-oblateness feedback, which is a resonance interaction between Earth's luni-solar precession and glacial-interglacial oscillation. The resulting obliquity reduction is enabled by Earth's rapid precession, which is driven primarily by the Moon. Thus, the stabilizing influence of the Moon on Earth's obliquity may be overridden in instances where obliquity and climate fall into an obliquity-oblateness resonance. The physics of climate friction is outlined in section 4.

It is important to note that this work has attracted the attention of several scientists with the last name Williams (e.g., D. M. Williams, this work; G. E. Williams; and G. P. Williams). Effort will be made to identify more precisely which Williams is being referred to by employing first and middle initials where appropriate.

## 2.  PRECESSION AND OBLIQUITY

### 2.1.  Precession of the Spin Axis

Fundamentally, planets precess because they are made oblate by rotation. External torques acting on a planet's asymmetric figure change the planetary momentum by causing a secular drift in the orientation of its spin axis (see Fig. 1). All objects in the solar system contribute a torque

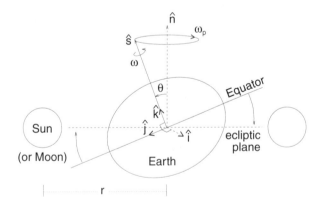

**Fig 1.** Earth-Sun (or Earth-Moon) configuration of maximum precessional torque. The torque results from Earth's asymmetric gravitational potential (i.e., its equatorial bulge), and is directed perpendicular to the spin axis, ŝ, and into the page. The sign of the torque is the same for both halves of the orbit, which causes the spin axis to precess about the ecliptic normal, n̂, with a frequency $\omega_p$. Maximum torque is applied when the Sun or the Moon is farthest from Earth's equatorial plane. Thus the solar contribution is zero at an equinox and greatest at a solstice. Adapted from *Stacey* (1992).

that acts on Earth's equatorial bulge, but almost all the torque is supplied by the Sun and the Moon. (By comparison, the combined torque contributed by the other planets is less than 0.01% the luni-solar torque.) The sum of the torques acting on Earth's equatorial bulge is equal to the time rate of change in momentum. Thus

$$\sum_{n=1}^{N} T_n = \frac{d\bar{L}}{dt} = L\frac{d\hat{s}}{dt} = C\omega\frac{d\hat{s}}{dt} \tag{1}$$

where $T_n$ is the external torque, N is the number of perturbers, C is the principle inertia moment around Earth's spin axis, $\omega$ is Earth's angular rotation rate, and $d\hat{s}/dt$ is the spin-axis precession rate. Both C and $\omega$ (and hence L) are approximately constant over the 10-k.y. timescale associated with Earth's spin-axis precession. Summing the torques from the Sun and the Moon, averaging over a seasonal cycle, and dividing by $C\omega$ yields

$$\frac{d\hat{s}}{dt} = \frac{3(C-A)(\hat{s}\cdot\hat{n})(\hat{s}\times\hat{n})}{2C\omega}$$
$$\left[\frac{GM_{sun}}{r^3(1-e^2)^{3/2}} + \frac{GM_{moon}}{r_m^3(1-e_m^2)^{3/2}}\right] \tag{2}$$

for the angular rate of precession of Earth's spin axis. Here, e and $e_m$ are the orbital eccentricity of the Earth about the Sun and of the Moon about the Earth respectively. This equation is of fundamental importance because it illustrates the linear dependence of the precession rate on dynamical flattening [(C – A)/C], obliquity [$\cos^{-1}(\hat{s}\cdot\hat{n})$], and mass of the perturbing object, as well as the $r^{-3}$ dependence on distance to the perturbing object. Equation (2) is often written in shortened form (cf. *Ward*, 1974; *Bills*, 1994) as

$$\frac{d\hat{s}}{dt} = \alpha(\hat{s}\cdot\hat{n})(\hat{s}\times\hat{n}) \tag{3}$$

where $\alpha$ is known as the precession constant and contains all the information in equation (2) excluding the vector products. It is easily demonstrated by inserting numbers into equation (2) that the contribution of the lunar torque to the precession rate, 34.8" yr⁻¹, is more than twice that of the solar torque, 15.6" yr⁻¹. The combined spin precession rate, $\alpha(\hat{s}\cdot\hat{n}) = 50.4"$ yr⁻¹, which exceeds by a large margin the rates of precession of all the other planets in the solar system. In this sense, Earth is dynamically quite peculiar.

### 2.2.  Precession of the Orbit

By itself, the luni-solar precession is not able to significantly alter Earth's obliquity. It is the concurrent motion of the spin axis and orbit normal that causes Earth's obliquity to fluctuate with time. Earth's orbit undergoes continuous dynamical evolution as a result of mutual gravitational interactions with the other planets in the solar system. These interactions allow the planets to exchange momentum, which tends to make the inclinations and eccentricities of

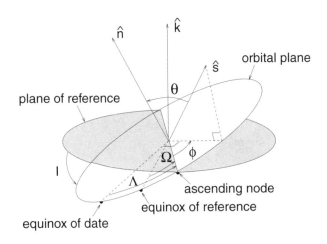

**Fig. 2.** Planes and variables of precession. The plane of reference is defined arbitrarily, and is often either the invariable plane of the solar system (*Ward*, 1974) or the J2000 ecliptic plane (*Laskar et al.*, 1993b).

the orbits of the smallest system members highly variable. These same planet-planet interactions cause the orbits of the smallest planets to precess about the invariable plane in the solar system, which is the average-momentum plane in the solar system lying between the orbital planes of Jupiter and Saturn. The invariable plane is the plane of reference in Fig. 2 having a normal vector $\hat{k}$. As the spin axis, $\hat{s}$, tries to precess about the orbit normal, $\hat{n}$, the orbit normal carries out its own precession about the invariable plane normal, $\hat{k}$. The ability of the spin axis to keep up with orbit normal and thus to maintain a fixed obliquity depends on the relative rates of precession. In general, planetary spin precession cannot keep up perfectly with the precession of its orbit, and the obliquity is caused to vary.

### 2.3. Equations of Precession

To monitor the time evolution of planetary obliquity, it is necessary to write the equations of motion for the spin vector as viewed from a reference frame co-moving with the orbital plane. In the new refererence frame, there are now two equations of motion: one to follow the precession, and one to follow the obliquity. The equations of precession, as written by *Ward* (1974), are

$$\frac{d\theta}{dt} = -\sin I \cos\phi \frac{d\Omega}{dt} + \sin\phi \frac{dI}{dt} \quad (4)$$

$$\frac{d\phi}{dt} = -\alpha \cos\theta - (\cos I - \sin I \cot\theta \sin\phi)\frac{d\Omega}{dt} + \cot\theta \cos\phi \frac{dI}{dt} \quad (5)$$

where $\phi$ is the precession angle between the projection of the spin axis on the orbital plane and the ascending node,

and $\theta$ is the obliquity (see Fig. 2). A thorough derivation of these equations is contained in *Ward* (1974) and *Colombo* (1966).

In the limit where the orbital inclination I is small, dI/dt = 0, and d$\Omega$/dt = constant [both are actually a sum of many time-varying sinusoids (*Ward*, 1974)], equations (4) and (5) may be approximately written

$$\frac{d\theta}{dt} \approx -\sin I \cos\phi \, \dot{\Omega} \quad (6)$$

$$\frac{d\phi}{dt} \approx -\alpha \cos\theta - \cos I \dot{\Omega} \quad (7)$$

where we have employed the raised-dot notation for representing the time derivatives. A lesser term containing $\sin\phi$ was eliminated from equation (7) because it contributes only a small correction. Also, $\cos\theta$ is reasonably approximated by $\cos\bar{\theta}$, where $\bar{\theta}$ is the mean obliquity over a precession cycle, which, for present Earth, is approximately the same as $\theta$ since the obliquity varies little over the cycle. As d$\theta$/dt $\propto$ cos $\phi$, and the righthand side of equation (7) is approximately constant, it is easy to see that, to first order, the obliquity executes a sinusoidal variation of frequency

$$\omega' = \left(-\alpha \cos\bar{\theta} - \cos I \dot{\Omega}\right) \quad (8)$$

and amplitude $(\dot{\Omega}/\omega')\sin I$. For Earth, the first-order obliquity response is simply the sum of the orbital and spin-axis precession velocities. Earth's orbit precesses in response to the other planets with a dominant frequency of –18.9" yr⁻¹ (the minus sign indicating that the direction of the orbit precession is opposite the motion of the spin axis), and the spin axis precesses at a rate of 50" yr⁻¹. Thus, $\omega'$ = 50" yr⁻¹ – 18.9" yr⁻¹ = 31.1" yr⁻¹, which yields a main period for the obliquity variation of ~41 k.y. The amplitude of the variation is ~$(\dot{\Omega}/\omega')\sin I$, where I = 1.58° is the inclination of Earth's orbit relative to the invariable plane. This yields an obliquity range of 1.92°.

### 2.4. Earth's Obliquity Cycle

To investigate the sensitivity of Earth's obliquity cycle on planetary and satellite parameters, the orbits of the planets in the solar system were followed for 17 m.y. using a mixed-variable symplectic orbital integrator (*Levison and Duncan*, 1994). The orbital solution was then Fourier analyzed to obtain a power spectrum of the main frequencies of orbit precession (Fig. 3a). The spectrum in Fig. 3a is a close match to the finely resolved spectrum of *Laskar et al.* (1993b), who used a technique of far greater sophistication. The integration method of *Laskar et al.* (1993b) was then used to solve the equations of precession to high precision, and to follow the evolution of Earth's obliquity over 1 m.y. (see Fig. 3b). The amplitude of Earth's present obliquity oscillation is small because the rates of spin and orbit pre-

Orbital precession rate ("/yr)

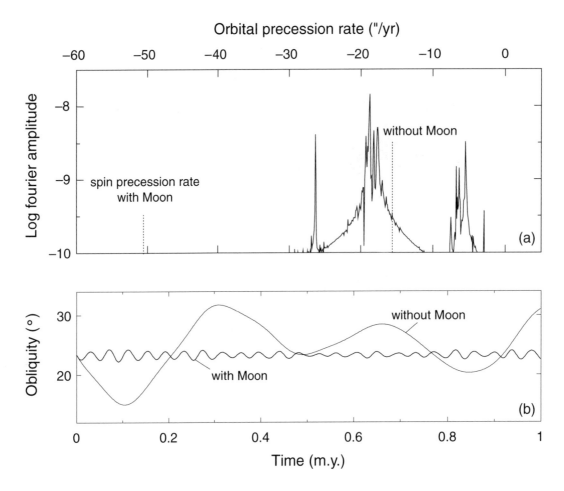

**Fig. 3.** (a) The leading orbital precession frequencies for Earth. For comparison, the spin precession rates of Earth with and without the Moon are indicated with vertical dotted lines. (b) Corresponding obliquity variations for Earth over 1 m.y.

cession are very different; however, if these two precessional motions ever happened to be in resonance, Earth's obliquity would vary considerably, and in chaotic fashion.

Clearly the condition for a stable obliquity will not be realized everywhere in nature. Other planets are unlikely to possess as large a satellite as the Moon, and therefore will tend to precess more slowly than does Earth. Mars, for example, precesses much more slowly (8" yr⁻¹), which enables its obliquity to vary between 15° and 45° with main periods of 0.12 and 12 m.y. (*Ward and Rudy*, 1991). Long-term chaotic evolution of the spin axis may enable Mars to reach obliquities as high as 45°–50°(*Touma and Wisdom*, 1993) or even 60° (*Laskar and Robutel*, 1993).

A moonless Earth would precess at a rate of ~16" yr⁻¹ (assuming a present-day spin rate), which is comparable to the leading frequencies of orbital precession (see Fig. 3a). The resulting obliquity variation (Fig. 3b) is considerably amplified and chaotic (nonperiodic). Laskar and colleagues (*Laskar et al.*, 1993a; *N'eron de Surgy and Laskar*, 1997) have demonstrated that Earth's obliquity might ultimately reach 85° under such circumstances. A similar fate would be rendered Earth with a moon if the moon were smaller than Earth's Moon or were in a slightly larger orbit, since

the spin-precession rate is ∝ M/r³ (see equation (2)). According to Fig. 3a, Earth would encounter the leading orbital precession frequencies if its precession rate, $\alpha\cos\theta$, were smaller than ~30" yr⁻¹ — the limit for spin-axis stability. *Ward* (1982) demonstrated that this hypothetical destabilizing event will be realized in 1–2 G.y. when the Moon has receded from its present distance of 60 $R_E$ to ~66 $R_E$ and Earth's obliquity has grown to 30° as a result of tidal evolution. It is interesting to note that the obliquity would vary chaotically today if the Moon were slightly less than half its present size, or 0.47 $M_m$; this serves to illuminate the fragility of our present situation.

## 3. CLIMATE AT HIGH OBLIQUITY

### 3.1. Obliquity as a Driver of Climate

The development of life on Earth has to some degree been enabled by its relatively stable obliquity and hence stable climate over most of its geologic history. Today the obliquity fluctuates by only ~±1.2° about a 23.3° mean, but it has long been realized that even these minor variations can profoundly affect Earth's climate. The last glacial maximum occurring ~21 k.y. ago, which covered upper North America and Scandinavia with nearly 4 km of ice (*Peltier*, 1994), is thought to have been triggered by the combined influences of Earth's orbital precession — which aligned the winter solstice with the time of aphelion — and obliquity, which was then near the minimum of its 41-k.y. cycle (*Berger et al.*, 1992). (As obliquity decreases, the insolation received at the poles is reduced, thereby promoting the growth of ice sheets.) Given the dramatic changes to Earth's climate that stem (at least in part) from the present 2°–3°-obliquity oscillation, it is interesting to speculate how climate might respond to much larger, and possibly chaotic, obliquity variations that could episodically tip Earth over onto its side.

### 3.2. Simulating Earth's Climate at High Obliquity

An energy-balance climate model has been used to investigate the effect of high obliquity on climates of Earth-like planets (*Williams and Kasting*, 1997). The energy-balance model (EBM) is a time-varying model of one dimension that calculates zonally averaged surface temperatures for 18 latitudinal bands over a seasonal cycle. In the model, top-of-atmosphere albedo and outgoing-infrared flux are parameterized as functions of $CO_2$ partial pressure and surface temperature (which sets the level of atmospheric water vapor) from runs performed using a one-dimensional radiative-convective model employed by Kasting to study the atmospheres of early Venus (*Kasting*, 1988), Earth (*Kasting and Ackerman*, 1986), and Mars (*Kasting*, 1991). The level of $CO_2$ in the atmosphere is adjusted by the model to balance the carbonate-silicate weathering cycle (*Walker et al.*, 1981). In this cycle, $CO_2$ is exchanged between the atmosphere and crust through chemical weathering of surface minerals, resulting in deposition of $CO_2$ on the ocean floors as $CaCO_3$, and outgassing of $CO_2$ through volcanos. The equilibrium $CO_2$ level is affected by obliquity, which controls seasonal insolation, and by the sizes and positions of continents because the rate of chemical weathering depends on the surface temperatures and expanse of exposed land areas. Interzonal heat flow is modeled in the EBM using a diffusion term of the form

$$\frac{\partial}{\partial x} D(1-x^2)\frac{\partial T}{\partial x} \qquad (9)$$

with a transport efficiency factor, D, that is proportional to the total atmospheric pressure and with x equal to the sine of latitude. In reality, heat transport within and between latitudinal bands is much more complicated, being accomplished both by winds and ocean currents with an efficiency that also depends on continental topography.

### 3.3. Model Results

The EBM was used to model the climate of Earth at both low (23.5°) and high (90°) obliquity, and the results of the calculations are shown in Figs. 4a and 4b. A comparison of

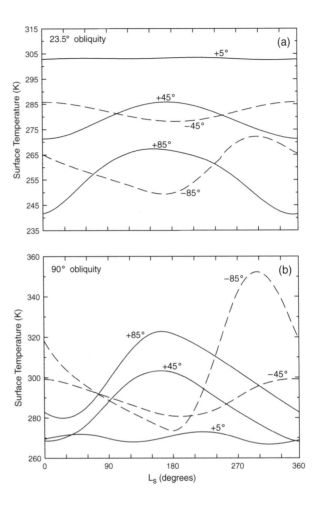

**Fig. 4.** Representative seasonal temperature cycles for Earth at (a) 23.5° obliquity and (b) 90° obliquity. $L_s$ is the orbital longitude of the planet with respect to the vernal equinox ($L_s = 0°$). Solstices occur at $L_s = 90°$ and 270°. Temperatures are shown for five latitude zones, each 10° wide and centered on the latitude labeling the curves. Northern latitudes are indicated by solid lines, and southern latitudes are indicated by dashed lines.

these figures reveals that the climatic zonation of Earth at high obliquity is reversed so that the coldest temperatures occur at low latitude This result had been predicted many years ago by *Ward* (1974), who recognized that the poles of planet receive more insolation, on average, than the equator once obliquity exceeds 54°.

By itself, a reversal in climatic zonation may not preclude habitability, but it may be key to understanding the evidence for low-latitude glacial structures during the Precambrian Era of Earth's history (*Evans et al.*, 1997; *Embleton and Williams*, 1986), which is discussed further below in section 4. What is important at high obliquity are the extraordinary swings in seasonal temperature that would be experienced by most of the planet, particularly in the middle and upper latitudes. The EBM results in Plate 9b show that temperatures over Antarctica, for example, would fluctuate seasonally between the freezing point of water and 353 K, which exceeds the survivable high-temperature limit (~330 K) for advanced life forms on Earth today. Even at 45° latitude, the seasonal cycle amplitude is at least twice what it is at 23.5° obliquity, which would allow temperatures in continental interiors at this latitude to vary between ~250 K (−10°F) and ~320 K (120°F) in only six months. Seasonal cycle amplitudes would be considerably more amplified on planets with larger continents or with smaller amounts of water because the atmosphere over the continents responds much more rapidly to changes in seasonal insolation than does the coupled atmosphere-ocean system. Also, long periods of continuous sunlight and darkness (~90 days at 45° latitude) might impose limits on photosynthetic biology. Clearly, Earth with an obliquity as high as 90° might not be able to support the types of life that inhabit the surface today.

### 3.4. Shortcomings of the Energy-Balance Climate Model

Although EBMs have been used effectively in many paleoclimate studies to map out parameter space and explore basic sensitivities to external forcing, their predictions are limited by the absence of explicit atmospheric dynamics, vertical structure, hydrologic processes, and land-surface processes. They can be tuned to yield good representations of present-day zonal mean surface temperatures, but for large perturbations away from present conditions, there is no guarantee that the omitted processes will continue to behave in the same way as today. Therefore a natural extension of our previous EBM work with high obliquities is to use an atmospheric general circulation model (GCM), coupled as comprehensively as possible with other components of the Earth-climate system: land, ice sheets, snow, and ocean. As a bonus, since a GCM simulates surface radiation, winds, and precipitation in addition to temperature, it can be used to explicitly predict the location and extent of ice sheets, which is important for investigating the mysterious low-latitude glacial climates of the Precambrian Era.

### 3.5. General Circulation Model Description

Improving upon the earlier calculations requires employing a model of greater sophistication. The model that is today being used to study high-obliquity climates is version 2.0 of the Global Environmental and Ecological Simulation of Interactive Systems (GENESIS). It was developed at the National Center for Atmospheric Research (NCAR) with emphasis on terrestrial biophysical and cryospheric processes for paleoclimatic experiments, and continues to be developed at the Earth System Science Center at Pennsylvania State University. An earlier version with coarser atmospheric GCM resolution has been described in *Thompson and Pollard* (1995a,b) and *Pollard and Thompson* (1994, 1995), and has been used extensively for both paleoclimate and future climate studies (e.g., *Bonan et al.,* 1992; *Barron et al.,* 1993; *Crowley et al.,* 1993; *Dutton and Barron,* 1996; *Fawcett et al.,* 1997; *Otto-Bliesner,* 1996; *Otto-Bliesner and Upchurch,* 1997; *Pollard and Schulz,* 1994; *Sloan and Rea,*

1995). New physics in version 2 and improvements in its present-day climate are described in *Thompson and Pollard* (1997); paleoclimatic applications are described in *Pollard and Thompson* (1997a,b), *Pollard et al.* (1998), *DeConto et al.* (1999), and *Sloan and Pollard* (1998); and interactive-vegetation experiments are described in *Foley et al.* (1998, 1999).

### 3.6. Results from GENESIS

We have recently performed a 10-year run of GENESIS version 2.0 using present-day conditions, except increasing the obliquity to 54° (from the present value of 23.45°). The value of 54° represents a high but not extreme obliquity, and happens to produce annual-mean insolation that is nearly independent of latitude.

In the run, we used the present-day continental configuration because it allows a more straightforward comparison with modern climate, while recognizing that this combination of continental distribution and obliquity may have never actually happened in Earth's history. The run was initialized from present-day conditions, and after 10 years the mixed-layer temperatures were not fully equilibrated; however, averaged over years 8–10 of the run, the temperatures are quite representative of the final state (for instance, within ~1°–3°C of local temperatures).

As expected, the seasonal amplitudes of surface temperature over land are much greater than present (Plate 9), with values rising to more than 70°C in July in northern Asia. Surprisingly, in January the northern oceans and northern coastal areas are much warmer than today, with oceanic temperatures above 10°C. In fact, no sea ice forms in either winter hemisphere. This is due to the annual-mean insolation at high latitudes being much greater than today, only slightly less than that in today's tropics. Annual-mean temperatures in the Arctic (and in most other regions except the Antarctic ice sheet) are above 20°C, and the thermal inertia of the 50-m mixed layer limits the seasonal temperature cycle over northern oceans and coastal areas.

In mid- and low-latitude continental interiors, winter temperatures are much colder than today's (for instance, 0° to −10°C in the Sahara in January). However, there are large equatorial regions where temperatures remain quite mild (between ~5° and 30°C) year round: Amazonia, equatorial Africa, and the islands of southeast Asia. The same is true for some thin coastal strips in somewhat higher latitudes of most continents that are buffered by the ocean. Most of these areas receive significant rainfall; hence terrestrial life in these refugia would presumably have little trouble surviving obliquities as high as 54°.

Global precipitation patterns (not shown) show two features that are very different from today. First, the tropical band of intense intertropical convergence zone (ITCZ) precipitation undergoes very large seasonal shifts in latitude, from ~20°S in January to ~20°N in July, as might be expected from the greater seasonal variations of insolation. This large seasonal ITCZ shift is also manifested in the vertical columns of high relative humidity and cloudiness (not shown) in zonal cross-sections. Second, over the high-latitude northern hemispheric oceans, there are large amounts of rainfall in winter (about 6 mm/day), where January temperatures do not fall much below 15°C. But in summer, poleward of ~40°N rainfall drops drastically to desert-like conditions, except over the high topography of the Rockies and Greenland. This seems to be due simply to the extremely high summer temperatures throughout the lower troposphere in the northern high latitudes, with saturation vapor pressures too high and relative humidities too low to produce significant convective or stratiform precipitation. Zonal cross-sections of relative humidity bear this out (not shown), with values of 50% or less at all levels above 50°N in July. There is also very little cloudiness in this region in July (not shown), which exacerbates the intense solar heating. Basically, the high-obliquity insolation in northern summer causes warming that outpaces evaporation, dries out the atmosphere, "burns off" the clouds, and produces a hemispheric-wide desert-like summer climate.

The large-scale circulation (not shown) is drastically different from today, and shows some correspondence with the behavior found by *G. P. Williams* (1988a,b). For instance, there are virtually no westerly jet streams in either summer hemisphere. This follows from geostrophic balance and the meridional temperature gradients, which, both at the surface (Plate 9) and throughout the troposphere (not shown), are reversed from today (increasing poleward) in both summer hemispheres.

In both winter hemispheres seasonal snow cover occurs similarly to today, except for its absence in high-latitude coastal areas that are buffered by the ocean and its occurrence much further equatorward in the colder continental interiors. All snow melts in the summer on Greenland and Antarctica, suggesting that these ice sheets would quickly melt and vanish with 54° obliquity. Conversely, unlike today there is year-round snow and net annual snow accumulation on the central Andes and Himalayas, suggesting that high obliquities favor low latitudes as sites of ice-sheet growth compared to high latitudes, at least with the present continental distribution. However, since ice-sheet topography is not well resolved by the relatively coarse GCM grid, these conclusions based on the "raw" GCM snow amounts are uncertain. One of us (D.P.) has attacked this problem using the GENESIS GCM (*Thompson and Pollard,* 1997; *Pollard and Thompson,* 1997a), and we now have the capability of driving a two-dimensional dynamic ice sheet model off line on a high-resolution (1° × 1°) grid with realistic ice-sheet topography and surface physics, driven by the GENESIS GCM climate (*Pollard and Thompson,* 1997b). We have made use of this capability to predict both the locations and the sizes of Precambrian ice sheets for the GCM simulations below.

The maximum instantaneous temperature reached in the preliminary experiment was 88°C (at the land surface in high northern latitudes in summer), not far from the maximum temperature (354 K) obtained by the two-dimension calculations using the energy-balance model at 90°. This

suggests that the earlier energy-balance climate calculations may be treated as conservative estimates for the extreme temperatures that are actually possible. Temperatures exceeding the boiling point of water can be expected to occur for obliquities greater than 54°, and some effort has been made to keep the GCM code numerically stable under such conditions.

In experiments such as these, there is the possibility of major reorganizations of the deep ocean thermohaline circulation and upper gyral system, especially with drastically different climate forcing. The mean-annual surface air temperature over oceans (not shown) for our experiment is nothing like today's, and actually decreases equatorward from 35°C at 70°N to a plateau of 20°C over most of the southern hemisphere. Our diffusive heat parameterization in the mixed-layer model (*Thompson and Pollard,* 1997) at least transports heat from hot to cold, as the upper gyral circulation would do, but whether the deep thermohaline circulation is driven by temperature or salinity gradients is an open question.

## 4. OBLIQUITY-OBLATENESS FEEDBACK

### 4.1. Low-Latitude Glaciation on the Precambrian Earth

Was Earth's own obliquity ever much different from its present value? Geologic and climatic indicators suggest that the answer to this question may be "yes," even though Earth's current obliquity is very stable. Paleomagnetic data suggest that Earth was glaciated at low latitudes during the Paleoproterozoic [~2.4–2.2 Ga (*Evans et al.,* 1997; *Williams and Schmidt,* 1999)] and during the Neoproterozoic [~820–550 Ma (*Frakes,* 1979; *Embleton and Williams,* 1986; *Zhang and Zhang,* 1985; *Schmidt and Williams,* 1995; *Park,* 1997; *Kirschvink,* 1992)]. If Earth's magnetic field was aligned more or less with its spin axis, as today, then either the polar ice caps must have extended well down into the tropics — the "snowball Earth" hypothesis (*Kirschvink,* 1992; *Hoffman et al.,* 1998) — or the present zonation of climate with respect to latitude must have been reversed. The main objection to the so-called "snowball Earth" hypothesis is that a global ice blanket would have likely eliminated all photosynthetic life on the planet, contrary to the evidence of uninterrupted life in the fossil record. The most recent work by *Hoffman et al.* (1998) cites depressed levels of $\delta^{13}C$ contained in Neoproterozoic rocks as evidence of global ice cover. The $\delta^{13}C$ isotope is a measure of carbonate burial flux and thus marine photosynthetic activity, which would be expected to decline had the Earth's oceans frozen over at that time. But it is also important to point out that the rate of burial of carbonates would have slowed even under incomplete glacial coverage, especially if the ice cover were confined to the tropics where marine life would have occupied the greatest area. Thus the $\delta^{13}C$ data does not definitely restrict the number of possible explanations of low-latitude glaciation to one.

## 4.2.  A High-Obliquity Solution: A Second Application of GENESIS

To reconcile the geologic record with the past and present evidence of life on the planet, *D. M. Williams et al.* (1998) assumed that Earth's obliquity was at one time high enough for the tropics to be colder than the poles. [Earth may have originally attained a high obliquity as a consequence of the impact thought to have formed the Moon, as argued by *G. E. Williams* (1993).] To test the strength of these hypotheses, we performed a 10-year run using GENESIS with $CO_2$ set to 50 ppm, the obliquity set to 54°, and with all the continents grouped together as a 35°-wide strip of land circumscribing the globe and centered on the equator. We inserted a small (20° × 20°) mountain having a maximum elevation of 5 km to test whether high altitude is necessary for net-annual accumulation of snow on the continent. Air temperatures decrease with height by 6.5°C/km.

Under these conditions, the equatorial supercontinent is shown in Plate 10 to be the coldest place on Earth for much of the year, as expected. Temperatures are shown to reach −10°C near sea level and −30°C over the higher elevations around the solstice when insolation is lowest. Precipitation and wind patterns (not shown) demonstrate there to be little trouble delivering moisture to the continent. The key result of this experiment is the positive net-annual accumulation of snow at high altitude over the equatorial supercontinent. Snow fraction is shown in Plate 11 to remain 100% even at the peak of the seasonal insolation cycle in April. This is significant because it implies that continents with similar elevated geographies should have had little trouble becoming glaciated at low latitude during the Precambrian.

However, even if it were possible to form ice sheets at low-latitude as a consequence of high planetary obliquity during the Precambrian, it still remains to be shown how Earth's obliquity could have decreased by several tens of degrees between ~600 Ma — the age of the youngest low-latitude glacial deposits (*Frakes,* 1979; *Embleton and Williams,* 1986; *Zhang and Zhang,* 1985) — and 430 Ma, when paleotidal data suggest that the obliquity was close to its present value (*G. E. Williams,* 1993). G. E. Williams himself suggested that core-mantle dissipation could have caused the obliquity to decrease. However, this appears to be unlikely because the viscosity of the outer core is too low (*Rochester,* 1976; *N'eron de Surgy and Laskar,* 1997). Furthermore, even if it could be shown to work, this mechanism should have operated throughout Earth's history, making it difficult to explain how the obliquity could have remained high as late as 600 Ma.

## 4.3.  Obliquity-Oblateness Feedback: "Climate Friction"

To make G. E. Williams' hypothesis work, one needs a mechanism for reducing obliquity that could have operated preferentially during the Late Proterozoic. One possibility

is for Earth's obliquity to have been reduced by obliquity-oblateness feedback (*Rubincam,* 1990), sometimes termed "climate friction." In this process, a positive or negative secular obliquity drift can occur as a result of the time delay between a planet's oscillating obliquity and cyclic variations in oblateness resulting from changes in continental ice volume and sea level. To understand the analytic origin of the obliquity drift, we will consider the case when the right-hand side of equation (7) is not constant, but variable as a result of an ice age. The growth and decay of continental glaciers, and subsequent isostatic response of Earth's upper mantle, changes the oblateness of Earth (as well as the dynamic ellipticity and the precession constant α) slightly over an obliquity cycle. Thus

$$\alpha = \alpha_0 + \Delta\alpha \qquad (10)$$

where the subscript zero denotes the value of an ice-free planet in hydrostatic equilibrium. Dividing equation (6) by (7) yields

$$\frac{d\theta}{d\phi} \approx \frac{\sin I \cos\phi \dot\Omega}{\alpha\cos\bar\theta + \cos I \dot\Omega} \qquad (11)$$

Using equations (8) and (10), this may be written

$$\frac{d\theta}{d\phi} \approx \frac{-\sin I \cos\phi \dot\Omega}{\omega'[1-(\Delta\alpha\cos\bar\theta/\omega')]} \qquad (12)$$

which after first-order Taylor expansion becomes

$$\frac{d\theta}{d\phi} \approx \frac{-\sin I \cos\phi \dot\Omega}{\omega'}\left(1+\frac{\Delta\alpha\cos\bar\theta}{\omega'}\right) \qquad (13)$$

The variable $\Delta\alpha$ depends on obliquity-induced changes to insolation through its effect on glacial mass, ocean depth, and the deformation of the solid Earth. For this work, the area of a large, pole-centered glacier is assumed to be limited by maximum diurnally-averaged insolation received at 45° latitude

$$S_{45} = \frac{q_0}{\pi}(H\sin(45°)\sin\theta + \cos(45°)\cos\theta\sin H) \qquad (14)$$

where $q_0$ is the solar constant and

$$\cos H = -\tan(45°)\tan\theta \qquad (15)$$

Milankovitch employed $S_{65}$ to model glaciers of the Pleistocene, but the glaciers modeled here are imagined to extend to much lower latitudes. Now, following *Ito et al.* (1995),

$$\Delta\alpha = \frac{3n^2\chi}{2\omega}\frac{(C-A)}{C}\frac{\Delta J_2}{J_2}$$
$$[S_{45}(\phi-\xi_i) - f(\xi_s)S_{45}(\phi-\xi_i-\xi_s)] \qquad (16)$$

where χ is the portion of equation (2) in square brackets,

$$\frac{\Delta J_2}{J_2} = \frac{(\Delta C - \Delta A)}{(C - A)} \qquad (17)$$

is the change to Earth's oblateness resulting from the ice sheets, and lagging the phase of the insolation cycle by $\xi_i$,

$$f(\xi_s) \approx 0.8 - 0.8\left(\frac{\xi_s}{90°}\right) \qquad (18)$$

is the normalized amplitude of the oblateness changes resulting from isostatic adjustment of the solid Earth, which is assumed to lag the phase of the periodic ice-volume variation by

$$\xi_s = (0.033P + 2570)\frac{360°}{P} \qquad (19)$$

(see *Ito et al.*, 1995). Here, $P = 2\pi/\omega'$ (41 k.y. at present) is the main period of the obliquity oscillation that may be estimated using equation (8). In the actual computation, $S_{45}$ was normalized so that the oblateness variation from ice loading was never greater than the largest possible value of $\Delta J_2/J_2$ for maximum glacial coverage. A useful analytic simplification is enabled by the approximation of $S_{45}(\omega t)$ in equation (16) with $\sin(\omega t)$. Inserting equation (16) into equation (13) gives

$$\frac{d\theta}{d\phi} \approx \frac{-\sin I \cos\phi \dot{\Omega}}{\omega'} -$$

$$\frac{3n^2 \chi \sin I \dot{\Omega} \cos\bar{\theta}}{2\omega\omega'^2} \frac{(C-A)}{C} \frac{\Delta J_2}{J_2} x$$

$$\left[ f(\xi_s)\left(\cos(\xi_i + \xi_s)\sin\phi\cos\phi - \right. \right. \qquad (20)$$

$$\sin(\xi_i + \xi_s)\cos^2\phi\right) - \left(\cos(\xi_i)\right)$$

$$\left. \sin\phi\cos\phi - \sin(\xi_i)\cos^2\phi\right)\right]$$

Averaging equation (20) over a precession cycle eliminates terms containing $\cos\phi$ and $\cos\phi\sin\phi$. This may be recast using $\phi = \omega t$ to give the secular drift of obliquity over time

$$\frac{\overline{d\theta}}{d\phi} \approx \frac{3n^2 \chi \sin I \dot{\Omega} \cos\bar{\theta}}{4\omega\omega'^2} \frac{(C-A)}{C} \frac{\Delta J_2}{J_2}$$

$$\left[\sin(\xi)_i - f(\xi_s)\sin(\xi_s + \xi_i)\right] \qquad (21)$$

## 4.4. Climate Friction

*Rubincam* (1993) demonstrated that such "climate friction" could account for a 10° drift in the martian obliquity over 4.5 G.y. *Bills* (1994) later calculated that a much larger drift (+60° in 100 m.y.) was possible for Earth, assuming a glacial-interglacial variation in oblateness of 1% based on O-isotopic data for the Pleistocene glaciations. This estimate

was later revised downward (see *Rubincam*, 1995) after it was pointed out that the original calculation had neglected ice-induced compression of the solid Earth (*Peltier and Jiang*, 1994). The long response time ($10^4$ yr) of the viscous upper mantle to changes in ice volume allows the solid-Earth variation to cancel only a fraction, $f(\xi_s)\Delta J_2/J_2$, of the water-ice variation. Subsequent calculations (*Peltier and Jiang*, 1994; *Mitrovica and Forte*, 1995) that included this effect lowered the estimate of the net change in oblateness to ~0.4%. Constrained by this result, subsequent authors (*Ito et al.*, 1995; *Rubincam*, 1995) have concluded that the change to Earth's obliquity cannot have been more than 10°–20° over Earth's entire (~450 m.y.) glacial history.

These calculations, however, assume that all ice ages were of comparable severity to those in the Pleistocene, which is not necessarily true. Different continental configurations, and possibly lower global-average temperatures, could have resulted in larger $\Delta J_2/J_2$ values for earlier glaciations. A lengthy, yet straightforward, calculation of the changes to Earth's inertial moments subject to continental ice loading (see *D. M. Williams*, 1998) reveals that if the continents were at one time clustered around one of the poles, as may have occurred during the Late Proterozoic (*Hecht and Scotese*, 1997), the change to oblateness from ice loading could have been as high as ~2.6%. These calculations were performed under the assumption that Earth's continents were completely covered by ice 3.5 km thick, comparable to the thickness of modern-day ice structures over Greenland and Antarctica. Under these assumptions, we calculate that the net oblateness variation (with solid-Earth response included) is only ~0.66% on average, owing to differences in insolation cycle amplitude and thus maximum ice volume over time. This is still more than 1.5× the maximum variation thought possible for the Pleistocene, which could have enabled a correspondingly high rate of secular obliquity drift for the Late Proterozoic.

The sign and magnitude of the obliquity-oblateness feedback depend on the locations and areas of the continents and on the phase lags $\xi_i$ and $\xi_s$. High-latitude ice sheets cause $J_2$ to decrease because they cancel out part of the Earth's equatorial bulge. For $25° < \xi_i < 206°$, the resulting secular change in obliquity is positive (see Fig. 5). This range can is easily obtained from equation (21), which shows the obliquity drift to be positive only if

$$\sin(\xi_i) > f(\xi_s)\sin(\xi_i + \xi_s) \qquad (22)$$

or, rewriting,

$$\sin(\xi_i) > f(\xi_s)\sin\xi_s\cos\xi_i/ \qquad (23)$$
$$\left[1 - f(\xi_s)\cos(\xi_s)\right]$$

Thus, if $\xi_s = 30°$, then the above condition is approximately satisfied if $\tan\xi_i < 0.48$, which gives the range of ice-sheet-formation phase lags written above.

Low-latitude ice sheets, which might occur at times of high obliquity, produce an obliquity drift in the same di-

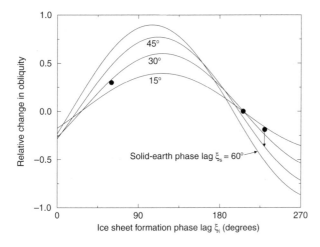

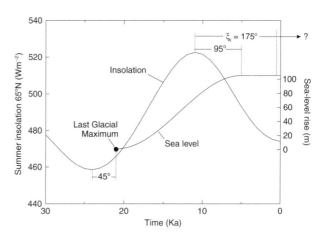

**Fig. 5.** Rate and direction of obliquity drift for different values of the phase lag between the obliquity-insolation forcing and the ice sheet variation, $\xi_i$. The four curves show the effect of varying the phase lag between the glacial extremum and the resulting isostatic response of the solid Earth, $\xi_i$. Plotted on the vertical axis is the quantity $[\sin\xi_i - f(\xi_s)\sin(\xi_i + \xi_s)]$ (see equation (21)). The filled circles mark the ice sheet formation phase lags, $\xi_i$, used for the three calculations shown in Fig. 6. The solid-Earth phase lag, $\xi_s = 28°$ for each run, but when $\xi_i = 230°$, the value $\xi_s$ grows from 28° to 45° (see *Ito et al., 1995*) over 100 m.y. (indicated by an arrow) as a result of an increase to the ice-loading frequency as the obliquity drifts downward.

**Fig. 6.** Insolation and relative sea level for the period defining the last deglaciation. The curve showing summer insolation at 65°N was obtained using the orbital solution La90 provided by *Laskar et al.* (1993b) with code to solve the equations of precession and insolation. Last glacial maximum took place at ~21 ka. The sea-level variation between 21 ka and 5 ka is a sinusoidal fit to model results of *Peltier* (1994). The slope in the sea-level variation indicates that the initial meltdown was very rapid, and that subsequent changes to the masses of the remaining (Antarctic and Greenland) ice structures have, since ~5 ka, been negligible. The value of $\xi_i$ is ambiguous, as it could be defined either as the time between minimum insolation and glacial maximum or that between maximum insolation and glacial minimum (which itself is difficult to identify precisely).

rection because both the effect on $J_2$ and the phasing with respect to the obliquity cycle are reversed. Previous studies of climate friction (*Ito et al., 1995*; *Rubincam, 1993, 1995*; *Bills, 1994*) have assumed that the feedback would be in this direction based on phase lag estimates for the Pleistocene glaciations. For example, *Imbrie et al.* (1992) favor $\xi_i \sim 80°$ (or ~9 k.y.) for the 41-k.y. obliquity cycle based on cross-spectral analysis of northern hemisphere, high-latitude, summer (NHHLS) insolation and $\delta^{18}O$ values in marine carbonates over the past 2 m.y. However, the actual ages of marine sediments are not known with great precision prior to ~30 ka, so such inferences are not very firm. A phase lag of <90° is also suggested by analogy with the seasonal cycle, in which the coldest winter temperatures and highest snow accumulation at midlatitudes occur 1–2 months after winter solstice. Ice volume need not respond in the same manner, though, to the much slower and weaker changes in insolation caused by orbital and obliquity variations. In colder climates, ice sheets might expand until or after NHHLS insolation reaches its peak, much like snow cover behaves in mountainous areas on Earth today, which would cause ice volume to be at least ~180° out of phase with the solar forcing.

Accurate age dates are available only for the last deglaciation, which is thought to have been triggered by a precessionally induced increase in NHHLS insolation. As Fig. 6

illustrates, the value of $\xi_i$ for this one event could lie anywhere above 45°, depending on how one interprets the shape of the sea level (ice volume) curve. The response of ice volume to NHHLS insolation is evidently not sinusoidal and presumably depends in a complicated manner on both climate and continental positions. With this in mind, it is quite conceivable that climate friction could act to decrease a planet's obliquity, as required to make G. E. Williams' hypothesis work.

### 4.5. Calculating the Secular Drift in Obliquity

To determine what changes to Earth's obliquity might have been possible, we integrated the equations of motion for the planets in the solar system over 100 m.y. using a symplectic orbit integrator provided by *Levison and Duncan* (1994). The equations of precession (equations (4) and (5)) were then integrated over 100 m.y. using the integrator of Laskar (*Laskar et al., 1993b*). The integrations were performed for several different values of $\xi_i$.

We assumed, somewhat arbitrarily, that Earth's obliquity was 55° at 600 Ma — the age of the youngest low-latitude glacial deposits (*Frakes, 1979*; *Embleton and Williams, 1986*; *Zhang and Zhang, 1985*). G. E. Williams (1993) suggested that the obliquity must originally have been >54° because this is the critical latitude above which the poles receive more annually averaged insolation than does the equator. However, ice sheets are more sensitive to changes in seasonal insolation extremes than to annually averaged insolation, so Earth's obliquity during the Precambrian need not actually have been this high. Figure 7 shows that under extreme oscillatory ice loading over the poles, Earth's obliquity could have been reduced to within 3° of its present value by 500 Ma if $\xi_i = 230°$. This result is in good agreement with the geologic data referenced earlier (*G. E. Williams*, 1993).

## 5. CONCLUDING REMARKS

There are at least two ways out of the low-latitude glacial paradox in Earth's history. One is to assume that Earth was at one time cold enough to have been frozen pole-to-pole. The other is to assume that Earth had a steeply inclined spin axis for the times when the glaciations occurred. Both explanations are ridden with pitfalls. The major weakness of the high-obliquity explanation advanced here is that the proposed mechanism — "climate friction" — for reducing the Earth's obliquity by ~30° in ~100 m.y. relies on a very special set of geologic and climatic circumstances to work properly. First, the global climate must have been cold to enable colossal ice sheets to form over the continents. Second, the continents affected by the glaciations must have been situated at high latitude to change Earth's ice-age oblateness by extreme amounts and to allow the obliquity to drift downward below 54°, where the coldest planetary temperatures would have occurred at the poles. Finally, the

glacial-interglacial oscillation must have operated with the same frequency as the obliquity cycle, and with a sizeable time lag between obliquity (insolation) and glacial variations to enable the obliquity of Earth to decrease with time. Whether all these events took place during the Late Precambrian is unknown. The intent here is to simply demonstrate that a large secular drift in Earth's obliquity (ultimately a consequence of Earth's luni-solar precession) is possible for a reasonable range of relevant geophysical parameters.

It is perhaps naive to conclude that Earth would be completely uninhabitable without the Moon; that is, without a stable obliquity. Whole planetary ecosystems are probably difficult to annihilate completely by even the most extreme seasonal fluctuations in temperature. Yet planets with Earth-like atmospheres, shallow oceans, or predominantly high-latitude continents whose temperatures routinely exceed 100°C will, at the very least, be biologically limiting for life dependent on liquid water. In this sense, then, perhaps Earth does owe its habitability (or at least its suitability for humans) to the Moon, which effectively maintains Earth's spin axis at low obliquity.

***Acknowledgments.*** We received valuable comments from J. Kasting and an anonymous referee. We thank H. Levison and M. Duncan for the orbital integration software and J. Laskar for software used to integrate the equations of precession.

## REFERENCES

Barron E. J., Peterson W. W., Pollard D., and Thompson S. L. (1993) Past climate and the role of ocean heat transport: model simulations for the Cretaceous. *Paleoceanography, 8,* 785–798.

Berger A. L., Loutre M. F., and Laskar J. (1992) Stability of the astronomical frequencies over the earth's history for paleoclimate studies. *Science, 255,* 560–566.

Bills B. G. (1994) Obliquity-oblateness feedback: Are climatically sensitive values of obliquity dynamically unstable? *Geophys. Res. Lett., 21,* 177–180.

Bonan G. B., Pollard D., and Thompson S. L. (1992) Effects of boreal forest vegetation on global climate. *Nature, 359,* 716–718.

Canup R. M. and Agnor C. (1999) Accretion of the terrestrial planets and the Earth-Moon system. In *Origin of the Earth and Moon* (R. M. Canup and K. Righter, eds.), this volume. Univ. of Arizona, Tucson.

Colombo G. (1966) Cassini's second and third laws. *Astron. J., 71,* 891–896.

Crowley T. J. and Baum S. K. (1993) Effect of decreased solar luminosity on Late Precambrian ice extent. *J. Geophys. Res., 98,* 16723–16732.

Crowley T. J., Baum S. K., and Kim K. Y. (1993) General circulation model experiments with pole-centered supercontinents. *J. Geophys. Res., 98,* 8793–8800.

DeConto R. M., Thompson S. L., Pollard D., Hay W. W., and Bergengren J. C. (1999) The role of terrestrial ecosystems in maintaining greenhouse paleoclimates. *J. Climate,* submitted.

Dutton J. F. and Barron E. J. (1996) GENESIS sensitivity to changes in past vegetation. *Palaeoclimates, 1,* 325–354.

Embleton B. J. and Williams G. E. (1986) Low paleolatitude of

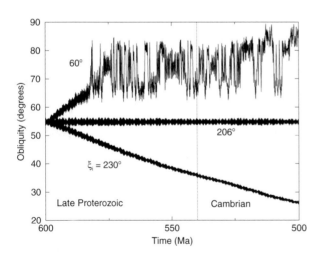

**Fig. 7.** Secular drift in obliquity over 100 m.y. spanning the Late Proterozoic-Cambrian boundary (indicated by a vertical dotted line) for three different values of the ice sheet formation phase lag, $\xi_i$. For the three calculations, $\Delta J_2/J_2 = 0.0262$ and the obliquity is started at 55°, which sets the initial period of the obliquity cycle, P ~ 58.5 k.y. and the phase lag of solid Earth deformation, $\xi_s$. The case of positive drift, with $\xi_i = 60°$, shows the spin axis entering a spin-orbit resonance at an obliquity of ~65° in under 20 m.y., where it then is able to vary chaotically upward to 90°.

deposition for late Precambrian periglacial varvites in South Australia: implications for palaeo-climatology. *Earth Planet. Sci. Lett., 79,* 419–430.

Evans D. A., Beukes N. J., and Kirschvink J. L. (1997) Low-latitude glaciation in the Palaeoproterozoic era. *Nature, 386,* 262–266.

Fawcett P. J., Agustdottir A. M., Alley R. B., and Shuman C. A. (1997) The Younger Dryas termination and North Atlantic deep water formation: insights from climate model simulations and Greenland ice cores. *Paleoceanography, 12,* 23–38.

Foley J. A., Levis S., Prentice I. C., Pollard D., and Thompson S. L. (1998) Coupling dynamic models of climate and vegetation. *Global Change Biology, 4,* 561–579.

Foley J. A., Levis S., Costa M. H., Doherty R., Kutzbach J. E., and Pollard D. (1999) Incorporating dynamic vegetation cover within global climate models. *Ecol. Applic.,* in press.

Frakes L. A. (1979) *Climates Throughout Geologic Time.* Elsevier, Amsterdam.

Hecht J. and Scotese C. R. (1997) *Ages of the Earth.* MacMillan, New York.

Hoffman P. F., Kaufman A. J., Halverson G. P., and Schrag D. P. (1998) A Neoproterozoic snowball earth. *Science, 281,* 1342–1346.

Imbrie J., Boyle E. A., Clemens S. C., Duffy A., Howard W. R., Kukla G., Kutzback J., Martinson D. G., McIntyre A., Mix A. C., Molfino B., Morley J. J., Peterson L. C., Pisias N. G., Prell W. L., Raymo M. E., Shackelton N. J., and Toggweiler J. R. (1992) On the structure and origin of major glaciation cycles: 1. Linear responses to Milankovitch forcing. *Paleoceanography, 7,* 701–738.

Ito T., Masuda K., Hamano Y., and Matsui T. (1995) Climate friction: A possible cause for secular drift of Earth's obliquity. *J. Geophys. Res., 100,* 15147–15161.

Kasting J. F. (1988) Runaway and moist greenhouse atmospheres and the evolution of Earth and Venus. *Icarus, 74,* 472–494.

Kasting J. F. (1991) $CO_2$ condensation and the climate of early Mars. *Icarus, 94,* 1–13.

Kasting J. F. and Ackerman T. P. (1986) Climatic consequences of very high carbon dioxide levels in the Earth's early atmosphere. *Science, 234,* 1383–1385.

Kirschvink J. L. (1992) Late Proterozoic low-latitude global glaciation: The snowball Earth. In *Proterozoic Biosphere: A Multidisciplinary Study* (J. W. Schopf and C. Klein, eds.), pp. 51–52. Cambridge Univ., Cambridge.

Laskar J. and Robutel P. (1993) The chaotic obliquity of the planets. *Nature, 361,* 608–614.

Laskar J., Joutel F., and Robutel P. (1993a) Stabilization of the Earth's obliquity by the Moon. *Nature, 361,* 615–617.

Laskar J., Joutel F., and Robutel P. (1993b) Orbital, precessional, and insolation quantities for the Earth from −20 Myr to +10 Myr. *Astron. Astrophys., 270,* 522–533.

Levison H. F. and Duncan M. J. (1994) The long-term dynamical behavior of short-period comets. *Icarus, 108,* 18–36.

Meert J. G. and Van der Voo R. (1994) The Neoproterozoic (1000–540 Ma) glacial intervals: No more snowball Earth? *Earth Planet. Sci. Lett., 123,* 1–13.

Mitrovica J. X. and Forte A. M. (1995) Pleistocene glaciation and the Earth's precession constant. *Geophys. J. Intl., 121,* 21–32.

N'eron de Surgy O. and Laskar J. (1997) On the long-term evolution of the spin of the Earth. *Astron. Astrophys., 318,* 975–989.

Otto-Bliesner B. L. (1996) Initiation of a continental ice sheet in a global climate model (GENESIS). *J. Geophys. Res., 101,* 16909–16920.

Otto-Bliesner B. L. and Upchurch G. R. (1997) Vegetation-induced warming of high-latitude regions during the late Cretaceous period. *Nature, 385,* 804–807.

Park J. K. (1997) Paleomagnetic evidence for low-latitude glaciation during deposition of the Neoproterozoic Rapitan Group, Mackenzie Mountains, N.W.T., Canada. *Can. J. Earth Sci., 34,* 34–49.

Peltier W. R. (1994) Ice age paleotopography. *Science, 265,* 195–201.

Peltier W. R. and Jiang X. (1994) The precession constant of the Earth: Variations through the ice-age. *Geophys. Res. Lett., 21,* 2299–2302.

Pollard D. and Schulz M. (1994) A model for the potential locations of Triassic evaporite basins driven by paleoclimatic GCM simulations. *Global Planet. Change, 9,* 233–249.

Pollard D. and Thompson S. L. (1994) Sea-ice dynamics and $CO_2$ sensitivity in a global climate model. *Atmos.-Ocean, 32,* 449–467.

Pollard D. and Thompson S. L. (1995) Use of a land-surface-transfer scheme (LSX) in a global climate model: The response to doubling stomatal resistance. *Global Planet. Change, 10,* 129–161.

Pollard D. and Thompson S. L. (1997a) Climate and ice-sheet mass balance at the last glacial maximum from the GENESIS version 2 global climate model. *Quat. Sci. Rev., 16,* 841–864.

Pollard D. and Thompson S. L. (1997b) Driving a high-resolution dynamic ice-sheet model with GCM climate: Ice-sheet initiation at 116 Kyr BP. *Ann. Glaciol., 25,* 296–304.

Pollard D., Bergengren J. C., Stillwell-Sollar L. M., Felzer B., and Thompson S. L. (1998) Climate simulations for 10000 and 6000 years BP using the GENESIS global climate model. *Palaeoclimates — Data and Modelling, 2,* 183–218.

Rochester M. G. (1976) The secular decrease of obliquity due to dissipative core-mantle coupling. *Geophys. J. Roy. Astron. Soc., 46,* 109–126.

Rubincam D. P. (1990) Mars: Change in axial tilt due to climate? *Science, 248,* 720–721.

Rubincam D. P. (1993) The obliquity of Mars and "climate friction." *J. Geophys. Res., 98,* 10827–10832.

Rubincam D. P. (1995) Has climate changed Earth's tilt? *Paleoceanography, 10,* 365–372.

Schmidt P. W. and Williams G. E. (1995) The Neoproterozoic climatic paradox: Equatorial paleolatitude for Marinoan glaciation near sea level in South Australia. *Earth Planet. Sci. Lett., 134,* 107–124.

Sloan L. C. and Pollard D. (1998) Polar stratospheric clouds: a high latitude warming mechanism in an ancient greenhouse world. *Geophys. Res. Lett., 25,* 3517–3520.

Sloan L. C. and Rea D. K. (1995) Atmospheric carbon dioxide and early Eocene climate: a general circulation modeling sensitivity study. *Palaeogeog., Palaeoclim., Palaeoecol., 119,* 275–292.

Stacey F. D. (1992) *Physics of the Earth, 3rd edition.* Brookfield, Brisbane.

Thompson S. L. and Pollard D. (1995a) A global climate model (GENESIS) with a land-surface-transfer scheme (LSX). Part 1: Present-day climate. *J. Climate, 8,* 732–761.

Thompson S. L. and Pollard D. (1995b) A global climate model (GENESIS) with a land-surface-transfer scheme (LSX). Part 2: $CO_2$ sensitivity. *J. Climate, 8,* 1104–1121.

Thompson S. L. and Pollard D. (1997) Greenland and Antarctic mass balances for present and doubled atmospheric $CO_2$ from the GENESIS version 2 global climate model. *J. Climate, 10,* 871–900.

Touma J. and Wisdom J. (1993) The chaotic obliquity of Mars. *Science, 259,* 1294–1297.

Walker J. C. G., Hays P. B, and Kasting J. F. (1981) A negative feedback mechanism for the long-term stabilization of Earth's surface temperature. *J. Geophys. Res., 86,* 9776–9782.

Ward W. R. (1974) Climatic variations on Mars: I. Astronomical theory of insolation. *J. Geophys. Res., 79,* 3375–3386.

Ward W. R. (1982) Comments on the long-term stability of the Earth's obliquity. *Icarus, 50,* 444–448.

Ward W. R. and Rudy D. J. (1991) Resonant obliquity of Mars? *Icarus, 94,* 160–164.

Williams D. M. (1998) The stability of habitable planetary environments. Ph.D. dissertation, The Pennsylvania State University.

Williams D. M. and Kasting J. F. (1997) Habitable planets with high obliquities. *Icarus, 129,* 254–268.

Williams D. M., Kasting J. F., and Frakes L. A. (1998) Low-latitude glaciation and rapid changes in the Earth's obliquity explained by obliquity-oblateness feedback. *Nature, 396,* 453–455.

Williams G. E. (1993) History of the Earth's obliquity. *Earth Sci. Rev., 34,* 1–45.

Williams G. E. and Schmidt P. W. (1999) Paleomagnetism of the Palaeoproterozoic Gowganda and Lorrain formations, Ontario: low paleolatitude for Huronian glaciation. *Earth Planet. Sci. Lett.,* in press.

Williams G. E., Schmidt P. W., Embleton B. J., Meert J. G., and Van der Voo R. (1995) The Neoproterozoic (1000–540 Ma) glacial intervals; no more snowball earth?; discussion and reply. *Earth Planet. Sci. Lett., 131,* 115–125.

Williams G. P. (1988a) The dynamical range of global circulations — I. *Climate Dynamics, 2,* 205–260.

Williams G. P. (1988b) The dynamical range of global circulations — II. *Climate Dynamics, 3,* 45–84.

Zhang H. and Zhang W. (1985) Palaeomagnetic data, late Precambrian magneto-stratigraphy and tectonic evolution of eastern China. *Precambrian Res., 29,* 65–75.

.

# The Early Earth vs. The Origin of Life

**Everett L. Shock, Jan P. Amend, and Mikhail Yu. Zolotov**
*Washington University, St. Louis*

Irrefutable evidence on how life originated does not exist. Hypotheses regarding its origin, however, are plentiful. Those that have prevailed for most of this century require an atmosphere dominated by ammonia ($NH_3$) and methane ($CH_4$), organic synthesis driven by energy sources that are external to the hydrosphere/lithosphere, and a first organism that makes its living by consuming the resulting supply of organic compounds. Diverse lines of evidence have been amassed over the last several decades that refute these particular origin of life hypotheses. For example, geologic evidence, atmospheric photochemistry, and current constraints on the formation of terrestrial planets indicate that a plausible early atmosphere was not dominated by $NH_3$ and $CH_4$, but rather by nitrogen ($N_2$) and carbon dioxide ($CO_2$). In addition, the location, duration, and quantity of external energy sources are not particularly predictable or reliable, and are not generally effective in driving the reduction reactions required to make organic compounds from $N_2$ and $CO_2$. Finally, revolutions in molecular biology have led to the observation that organisms that synthesize biomass from inorganic starting materials like $CO_2$ populate the deepest and shortest branches on the universal phylogenetic tree of life on Earth. These fundamental developments have permitted new and more geologically consistent ideas about the emergence of life. We argue that plausible hypotheses of the emergence of life on Earth call for a network of energetically favorable gradual synthesis processes in response to normal geologic forces. Inescapable chemical disequilibrium states, established and at least partially maintained in the hydrosphere at or near the dynamic surface of early Earth, can provide the energy for organic and biomolecule synthesis from inorganic source materials. Geologic conditions conducive to the formation of aqueous organic compounds, including precursors to complex biopolymers such as nucleic acids and proteins that are present in all organisms, are probably similar to those at which autotrophic organisms emerged. Hydrothermal systems are perhaps the best example of a normal geologic environment that maintains these minimum necessary conditions for life. These systems are an inevitable consequence of volcanic or impact activity in the presence of liquid water, and were likely more abundant and dynamic on early Earth than today. In further support, modern molecular phylogenies place chemosynthetic thermophilic autotrophs, organisms isolated from and probably ubiquitous in hydrothermal systems, nearest the last common ancestor in the tree of life.

## 1. WHAT CAN THE ORIGIN OF LIFE TELL US ABOUT THE ORIGIN OF THE EARTH?

In his conclusions a decade ago, *Walker* (1990) said that "the origin of life does not constrain theories of Earth origin." He reached this conclusion after reviewing the facts available and stating:

"I believe that chemical evolution, the formation of biological monomers, tells us at this stage nothing about the origin of the Earth. The key question now in understanding the origin of life is how to build biological macromolecules and, ultimately, living cells. There has been little progress on these aspects of the origin of life study. While life did originate somewhere, sometime, and probably not very far away from the Earth, we simply do not know where, when, or how, so the origin of life sets few constraints on our ideas concerning the origin of the Earth."

The doubts inherent to this position were justified at the time, and may be justified now. On the other hand, recent progress in shaping alternatives to the paradigm that prevailed a decade ago may provide a fresh perspective. Although they do not provide a direct answer to this question, these new ideas advocate a different set of conditions for the emergence of life on the Earth, and, we argue, place constraints on how and when conditions at the surface of the Earth were "normal" in the sense that they were geologically and geochemically familiar.

Familiarity is one thing grossly lacking in the paradigm for the origin of life that has prevailed for much of this century. In this model, life emerged as a heterotrophic metabolism that consumed a "prebiotic soup" of organic compounds generated by the input of (largely) external energy into the atmosphere (*Oparin,* 1924, 1936; *Haldane,* 1929; *Miller,* 1957; *Miller and Urey,* 1959). The postulated atmosphere was nothing like the Earth's, but rather a mixture of reduced gases (methane, ammonia, water, and hydrogen) whose origin was assumed to be consistent with late differentiation of the core after a relatively cool accretion of the planet (*Urey,* 1952a,b). The tremendous early success of spark discharge (*Miller,* 1953) and other experiments in producing organic compounds from reduced gas mixtures (see review by *Shock, 1990*) strengthened the case for a "prebiotic soup" into a paradigm for the origin of life.

However, many advances in the latter half of the twentieth century have challenged this paradigm. Isotopic geochemistry of mantle, meteorite, and lunar samples; planetary exploration; and dynamic simulations of nebular formation have questioned the slow differentiation model (*Safranov, 1969; Ringwood, 1979; Stevenson, 1981; O'Neill and Palme, 1998; McCulloch and Bennett, 1998*). Prevailing models have the Earth forming through violent accretionary processes that drove simultaneous melting, differentiation, and release of volatiles. Plate tectonics has emerged as a paradigm in the Earth sciences following the conclusive demonstration of the dynamic nature of the mantle. Despite the extensive recycling that has occurred, coordinated studies by geologists, paleontologists, geochronologists, and geochemists have revealed over 4 b.y. of recorded history of Earth processes and life in unprecedented detail. Simultaneously, atmospheric models have shown that, owing to photolysis, reduced gases are doomed to evanescent lifetimes in the presence of water vapor (e.g., *Levine et al., 1982; Zahnle and Walker, 1982; Kasting et al., 1983*). Finally, modern molecular phylogenies place chemosynthetic autotrophs, rather than heterotrophs or photosynthetic autotrophs, deepest in the tree of life (*Woese et al., 1990; Pace, 1997*).

These developments do not just permit but in effect require a new set of ideas about how life began, which may stand a better chance of enlightening the search for the origin of the Earth. The contrasts between these new ideas and more conventional origin of life concepts are outlined in Fig. 1. In the conventional view, abiotic organic synthesis occurred in a reduced atmosphere, and was driven by spark discharges and other energy sources that are generated by processes external to the hydrosphere and lithosphere where life occurs. A variant on this idea is that organic compounds synthesized elsewhere were brought to Earth on comets, meteorites, and interplanetary dust. Accumulation of abiotically synthesized organic compounds in the ocean is imagined to generate a "prebiotic soup" of organic compounds and nutrients, and life is thought to have arisen in the form of a heterotrophic metabolism capable of consuming the soup. In contrast, it is now thought that carbon and nitrogen in the atmosphere were not reduced, but were predominantly present as $CO_2$ and $N_2$. Furthermore, the current view holds that internal energy sources (radioactive decay, mantle convection, magmatic processes) can provide conditions for organic synthesis by inducing geochemical disequilibrium states that involve diverse oxidation/reduction potentials in the crust and ocean. In addition, these geochemical potentials, established though common geologic processes, are proposed as the driving forces for the emergence of an autotrophic metabolism capable of generating organic compounds from inorganic sources of carbon and nitrogen and making biological use of geochemical energy. In this view life is an expected consequence of normal geological processes.

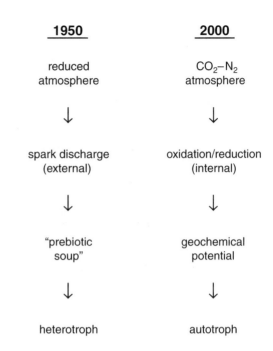

**Fig. 1.** A comparison between the mid-century paradigm for the origin of life, and more recent ideas about the emergence of life.

## 2.  WHEN DID CONDITIONS BECOME "NORMAL" ON THE EARLY EARTH?

The early Earth was a violent place that seems to have been unlivable. Accretionary energy provided enough heat, probably enhanced by short-lived radionuclides, to drive core/mantle/crust differentiation, and some models produce a magma ocean to considerable depths in the mantle (*Jones and Drake, 1986; Tonks and Melosh, 1990; O'Neill, 1991; Righter and Drake, 1997; Righter et al., 1997*). The cataclysmic nature of the giant impact hypothesis for the formation of the Moon would have vaporized silicates (*Cameron and Benz, 1988; Benz and Cameron, 1990; Melosh, 1990; Cameron, 1997*). At the final stages of accretion, an impact-generated atmosphere and proto-ocean could have formed (*Matsui and Abe, 1986; Abe and Matsui, 1988*). The lunar record of impacts extrapolated to the Earth suggests that the period of bombardment by large objects (≥100 km) continued for up to 700 m.y. after the Earth formed (*Maher and Stevenson, 1988; Oberbeck and Fogelman, 1989; Sleep et al., 1989; Sleep and Zahnle, 1998*), although it is difficult to be sure of the statistical certainty with which such events occurred. If impacts were large enough to vaporize a volume of water equivalent to the present oceans [~450 km impactor (*Sleep et al., 1989*)], they would have provided a major perturbation in the evolution of life at the Earth's surface.

It has been argued that "sterilizing" events would have required more than one origin of life, and that major impacts provided bottlenecks to evolution through which relatively few organisms could pass (*Oberbeck and Fogelman,*

1989; *Sleep and Zahnle,* 1998). Those that could survive and that found themselves at the surface would have been faced with a flux of ultraviolet radiation from the young Sun that was many orders of magnitude greater than at present and unmitigated by an ozone shield (*Levine et al.,* 1982; *Canuto et al.,* 1982, 1983; *Kasting et al.,* 1983; *Levine,* 1985), but a luminosity at visible wavelengths that was feeble by comparison (*Sagan and Mullen,* 1972; *Owen et al.,* 1979; *Gough,* 1981). Nevertheless, an early transition was made to surface or near-surface environments that were conducive to life, and, in the process, life became established as a normal process on the Earth.

The timing of events on the early Earth is open to considerable uncertainty owing to the inadequacies of the geologic record, which is often the product of chance preservation of crustal material on an extremely active planet (*Bowring and Housh,* 1995). We have a few facts that are helpful in following the transition from Earth's dramatic origin to conditions that could support life, and that would possibly have been recognizable and even quite familiar. These are assembled in Fig. 2, which shows a timeline for the first 1.2 b.y. of Earth history. The oldest preserved por-

tions of crust that are known are in the Acasta gneiss complex found in the Slave province of Canada and are ~4.03 b.y. old (Ga) (*Bowring et al.,* 1989a,b; *Stern and Bleeker,* 1998). These rocks are granites (tonalites) that are deformed, and indicate having been formed through intracrustal reprocessing of mantle-derived magma, juvenile crust, and older, hydrated mafic crust (*Bowring and Housh,* 1995), probably in a subduction-related setting. In other words, indications from the oldest known rocks are that "normal" geodynamic and geochemical processes were at work in the mantle and crust, and had been for some time.

The only older preserved terrestrial materials available are individual zircon grains with ages up to ~4.3 Ga found in younger rocks (*Froude et al.,* 1983; *Compston and Pidgeon,* 1986; *Nelson,* 1997). These zircons give indications [trace elements and mineral inclusions (*Maas et al.,* 1992)] of having formed from granitic melts, again suggesting that familiar crust-forming geologic processes including subduction and something similar to plate tectonics were underway less than 300 m.y. after the formation of the Earth. It follows that familiar crustal geochemical processes (weathering, sediment accumulation, diagenesis, hydrothermal circulation and alteration, metamorphism, etc.) were also active by this time, and have been ever since.

Evidence for life in the rock record does not yet extend back as far as the oldest surviving pieces of continental crust. Nevertheless, there are well-preserved fossils of filamentous microbes from the Apex chert of western Australia with ages of ~3.465 Ga (*Schopf,* 1993), additional filamentous fossils from the underlying Tower Formation [3.556 ± 0.032 Ga (*Awramik et al.,* 1983)] for which the exact provenance is unclear (*Awramik et al.,* 1988; *Buick,* 1988; *Schopf,* 1992), and filamentous microfossils from the Onverwacht group of South Africa of comparable age [3.540 ± 0.030 Ga (*Walsh and Lowe,* 1985)]. Additional evidence for active biological processes earlier in the rock record comes mainly from C-isotopic measurements on the metamorphosed sediments of the ~3.8-Ga Isua supracrustal rocks from western Greenland. *Schidlowski et al.* (1979) argued for the biogenicity of the stable C-isotopic signatures obtained from these rocks, and supporting evidence for such origins is provided on the scale of individual mineral grains by ion microprobe measurements conducted by *Mojzsis et al.* (1996).

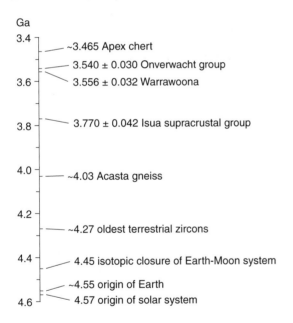

**Fig. 2.** Timeline for the first 1.2 b.y. (Ga) of Earth history. The age of the Earth based on Pb isotopes is from *Patterson* (1956), and the slightly older age of the solar system is based on studies of Ca-Al inclusions in meteorites (*Manhes et al.,* 1987; *Lugmair and Galer,* 1992; *Lugmair and Shukolyukov,* 1992). The isotopic closure of the Earth-Moon system was probably complete by 4.45 Ga (*Zhang,* 1998). Within 0.250 b.y. "normal" geodynamic and geochemical processes generated granitic rocks (as represented by surviving zircon grains), and, since about 0.4 b.y. later, pieces of continental crust (Acasta Gneiss) have survived the recycling processes engendered by mantle convection. Less than 0.25 b.y. after that in the geologic record there is isotopic evidence for biologically fractionated C isotopes (Isua, age from *Hamilton et al.,* 1978), and less than 0.25 b.y. after that easily recognized microfossils can be found in sedimentary rocks.

## 3.  DIFFICULTIES IN STRETCHING PARADIGMS TO FIT OBSERVATIONS

The transition from the violent earliest stages of Earth's history to conditions conducive to life is a major unsolved mystery. Before the current appreciation for planetary formation through accretionary processes had been attained, the view held by many was that abiotic organic synthesis processes, driven by spark discharges and other externally derived energy inputs into a reduced atmosphere, led to a "prebiotic soup" of organic compounds and nutrients, and that life arose in this soup in the guise of a heterotrophic metabolism. (In fact, the term "prebiotic" has taken on a meaning distinct from abiotic. We are using the term "prebiotic" as it is often used in the study of the origin of life where it seems to refer to conditions consistent with those assumed in the soup hypothesis, i.e., temperate water, reduced atmosphere, external energy to drive organic synthesis. Such conditions may or may not be the same as conditions that existed at any time on the Earth.) Decades of research that followed on the early success of abiotic organic synthesis experiments often assumed that there would have been a source for almost any organic compound of interest. Popular notions like the RNA-world (*Gilbert,* 1986), sparked by the discovery that some RNA molecules have catalytic as well as genetic abilities (*Kruger et al.,* 1982; *Guerrier-Takada et al.,* 1983), can be accommodated in the soup hypothesis, but not without raising concerns for the source of the RNA (see *Joyce and Orgel,* 1993). [RNA, or ribonucleic acid, is an essential component of all cells. It is usually a single-stranded genetic molecule that is used to transfer information from DNA (deoxyribonucleic acid) in the formation of proteins. Unlike RNA, DNA has not been shown to have catalytic abilities, and it was long thought that proteins were the only catalytic biomolecules in cells.] In fact, the concept of a "prebiotic" Earth popular in many textbooks is substantiated based on the success of experiments in chemistry labs rather than observations of natural systems. As a consequence, as more has been learned about the processes leading to the formation of the Earth, it is increasingly difficult to connect the popular concept of a "prebiotic" Earth with present geochemical, geophysical, and geodynamic views of the early Earth.

No evidence has been found in the geologic record to support the soup hypothesis or the "prebiotic" Earth, but the sketchy nature of that record is often invoked to counteract this criticism. Indeed, lines of evidence from other planetary bodies and objects, such as the wealth of organic compounds in carbonaceous chondrites or the smoggy haze of Titan, are produced to support the soup hypothesis despite the fact that these objects underwent histories quite distinct from that of the Earth. If anything, analysis of carbonaceous meteorites, interplanetary dust particles, presolar grains, cometary exploration, and astronomical measurements of the interstellar media and nebulae show that organic compounds are present and even abundant nearly everywhere (*Cronin et al.,* 1988; *Clemett et al.,* 1993; *Allamandola et al.,* 1989; *Messenger et al.,* 1998; *Irvine et al.,* 1999). Available evidence leads to the conclusion that organic synthesis is a normal process throughout the solar system, but that it appears to be predominantly abiotic and may or may not tell us much about the origin of life.

Another problem for conventional "prebiotic" Earth models is the instability of a reduced atmosphere at any time in Earth history. One geologically plausible abiotic source for reduced compounds in the early atmosphere is mantle-derived volcanic gases in equilibrium with metallic iron, which is unlikely given the early differentiation of the core and mantle (*Halliday et al.,* 1996; *Halliday and Lee,* 1999).

Another source might be impact-generated gases, although these tend to be disequilibrium mixtures of reduced ($H_2$, CO) and oxidized ($CO_2$, $O_2$) gases (*Mukhin et al.,* 1989), which consume each other upon reequilibration. Furthermore, photolysis of reduced gases like $NH_3$ and $CH_4$ in a water-rich atmosphere subjected to UV radiation drives their rapid oxidation on timescales of days to years (*Kuhn and Atreya,* 1979; *Kasting,* 1982; *Levine and Augustsson,* 1985). Saving the heterotroph-from-the-soup hypothesis requires methods to put a lid on the pot, after finding extrageologic ways to generate reduced gases and then protect them from some sources of energy but not others (see *Bada et al.,* 1994; *Sagan and Chyba,* 1997). On the other hand, abandoning this hypothesis altogether may be a constructive option.

## 4. THE EMERGENCE OF LIFE AS AN ORDINARY EVENT

The search for alternate pathways of organic synthesis consistent with what is known about the formation and early evolution of the Earth is a nascent field of research. Much of the search has focused on hydrothermal systems, which are an inevitable consequence of magmatic and impact activity in the presence of liquid water. Early ideas preceded (*French,* 1964, 1971; *Ingmannson and Dowler,* 1977) and followed immediately upon (*Corliss et al.,* 1981) the discovery of high-temperature submarine hydrothermal systems in the mid-1970s. These were followed by additional arguments (*Baross and Hoffman,* 1985) and theoretical models (*Shock,* 1990, 1992a) that showed the potential for organic synthesis during water-rock reactions hosted in basalt and other seafloor rocks. Hydrothermal experiments showed that abiotic organic synthesis was plausible (*French,* 1964, 1971; *Hennet et al.,* 1992; *Marshall,* 1994), although many early experiments were conducted at conditions unlike those in hydrothermal systems on the Earth (see review by *Shock,* 1992b). Extreme examples of mismatch between the laboratory and nature allowed some investigators to be highly dubious of the potential for hydrothermal organic synthesis (*Miller and Bada,* 1988). However, experiments constrained to be more like natural hydrothermal processes show great success in generating and transforming a wide variety of organic compounds (*Seewald,* 1994; *Berndt et al.,* 1996; *Huber and Wächtershäuser,* 1997, 1998; *McCollom et al.,* 1999a,b; *McCollom and Simoneit,* 1999), and recent theoretical models show an enormous potential for organic synthesis from $CO_2$ and H (*Shock and Schulte,* 1998; *Amend and Shock,* 1999) as hydrothermal fluids mix with seawater, or from $CO_2$ or CO as volcanic gases mix with crustal aqueous fluids or the atmosphere (*Zolotov and Shock,* 1999a,b). In addition, there is growing recognition of abiotic organic synthesis in various geologic materials (e.g., *Konnerup-Madsen,* 1989; *Salvi and Williams-Jones,* 1997). Indeed, the success of recent experiments and models supports the notion that abiotic organic synthesis is an ongoing process on the Earth (*Hennet et al.,* 1992). Perhaps we do not perceive its importance because it is less efficient than biological synthesis and/or geochemical transformation of biologically generated organic compounds.

The success of hydrothermal organic synthesis experiments may help to perpetuate a version of the heterotroph-before-autotroph hypothesis. Facile hydrothermal abiotic organic synthesis could provide a steady source of compounds for condensation and polymerization reactions. Many researchers have shown that construction of biomolecules can be facilitated by adsorption onto mineral surfaces (*Ferris and Ertem*, 1992; *Ertem and Ferris*, 1996, 1998; *Ferris et al.*, 1996; *Sowerby and Heckl*, 1998), some of which may be catalytic for reactions involving the reduction of carbon oxides (*Wächtershäuser*, 1988, 1992; *Huber and Wächtershäuser*, 1997, 1998; *Schoonen et al.*, 1999) or the polymerization of monomers (*Nishihama et al.*, 1997). Even the RNA-world could be accommodated in hydrothermal systems given the proclivity of nucleotides for mineral surfaces and their enhanced reactivity at higher temperatures (*Ferris*, 1998; *Ertem and Ferris*, 1998). An environment sheltered from impact roasting and UV singeing, complete with a steady supply of abiotically generated organic compounds, would even provide excellent room and board for an early heterotroph. Favorable geologic conditions for the synthesis of organic compounds might also favor biosynthetic pathways, especially if easily formulated substances provided low-tech versions of the catalytic functions now conducted by enzymes. Though possible, the heterotroph-before-autotroph hypothesis requires a reconfiguration of the deepest branches of the phylogenetic tree (see below) or considerable faith that even deeper branches will be identified — populated by obligate heterotrophs.

The geochemical drive for organic synthesis in hydrothermal systems permits a distinct departure from the soup hypothesis. Several investigators promote various scenarios in which the earliest metabolic strategy is some form of autotrophy (*Wächtershäuser*, 1988, 1990, 1992, 1998; *Shock et al.*, 1995, 1998; *Russell and Hall*, 1997; *Huber and Wächtershäuser*, 1998), and aim for a clean break from the historical paradigm by referring to these as "emergence of life" rather than "origin of life" models. These alternative approaches require no stockpiling of soup or other inventories of raw materials; they further rely on normal geological processes like hydrothermal systems to supply conducive conditions for life and envision the appearance of simple metabolic processes in response to easily established geochemical potentials. In a sense, these are habitat-based models for the emergence of life — something that has been missing from most origin of life discussions.

By emphasizing a gradual process rather than a specific event, these investigators are enlivening the realm of experimental and theoretical studies that elucidate the transformation from geochemical to biological processes. In its extreme forms, this approach takes organic synthesis as the means rather than a prerequisite, and views complex biomolecules like DNA and proteins as symptoms of life rather than its cause. Through a peculiar mix of the truly mundane with the possibly profound, these models demand that something as ubiquitous on the Earth as life developed in response to extremely normal geological processes driven by inescapable aspects of the geodynamics of an active planet. The proposal that the emergence of life is a normal process stands in contrast to the largely abandoned origin of life paradigm that seems to require special circumstances if and when applied to the Earth.

## 5. THE DIRECTION OF COOLING IS THE DIRECTION OF REDUCTION

One inescapable aspect of the transition from the accretion of the Earth to a habitable planet is that near-surface environments cooled. This transition has enormous effects on the geochemistry of volatile elements such as C, N, and S. At temperatures of magmatic processes on the Earth, these elements are oxidized as indicated by the composition of gases from Kilauea volcano given in Table 1. Carbon in volcanic gases is predominantly present as $CO_2$. At the much lower temperatures prevailing in sedimentary basins, volatile elements are preserved for enormous periods of time in reduced forms like sulfides and organic compounds. Carbon in sedimentary rocks is present as carbonate and as petroleum, coal, and natural gas rich in $CH_4$. Reasons for this difference in oxidation states of volatile elements at different temperatures are revealed by a thermodynamic analysis of the stable forms of these elements at oxidation states consistent with crustal rocks.

The following three redox reactions involving volatile forms of C, Fe-bearing minerals (fayalite, $Fe_2SiO_4$; magnetite, $Fe_3O_4$; and hematite, $Fe_2O_3$), and quartz ($SiO_2$)

$$CO_2 + 4\,H_2 = CH_4 + 2\,H_2O \tag{1}$$

$$\begin{aligned}3\text{ fayalite} + 2\,H_2O = \\ 2\text{ magnetite} + 3\text{ quartz} + 2\,H_2\end{aligned} \tag{2}$$

$$2\text{ magnetite} + H_2O = 3\text{ hematite} + H_2 \tag{3}$$

help to illustrate the interplay between temperature and oxidation state that controls C geochemistry in the crust. In

TABLE 1. Volcanic gas composition from Kilauea used to calculate energetics of autotrophic reactions during dynamic cooling from equilibrium.

| Gas | Volume % |
| --- | --- |
| $H_2O$ | 52.3 |
| $CO_2$ | 30.6 |
| $SO_2$ | 14.7 |
| $CO$ | 1.16 |
| $H_2$ | 0.786 |

Also contains traces of HCl, $S_2$, $H_2S$, $N_2$, $NH_3$, $CH_4$. This sample represents an average J probe (*Symonds et al.*, 1994). The composition is set by the oxidation state of the fayalite-magnetite-quartz buffer at 1200°C (log $f_{O_2}$ = −8.13).

each case, we can write law-of-mass-action expressions for these reactions, calculate equilibrium constants, solve for $H_2$ fugacities ($f_{H_2}$), and plot the results in a comparative fashion. [All geochemical calculations performed in this study used thermodynamic data, equations, and equation-of-state parameters from *Helgeson et al.* (1978, 1998), *Shock et al.* (1989, 1992, 1997), *Shock and Helgeson* (1990), *Schulte and Shock* (1993), *Shock and McKinnon* (1993), *Shock and Koretsky* (1993, 1995), *Shock* (1995), *Sverjensky et al.* (1997), and *Amend and Helgeson* (1997a,b). Equilibrium constants were calculated with the SUPCRT92 program (*Johnson et al.,* 1992) and reaction path calculations were conducted with the EQ3/6 software package (*Wolery,* 1992; *Wolery and Daveler,* 1992).]

As an example, the law-of-mass-action expression for equation (1) is given by

$$K_1 = \frac{f_{CH_4}(a_{H_2O})^2}{f_{CO_2}(f_{H_2})^4} \quad (4)$$

where $K_1$ represents the equilibrium constant, and a stands for activity, or in logarithmic form as

$$\log K_1 = \log(f_{CH_4}/f_{CO_2}) + 2\log a_{H_2O} - 4\log f_{H_2} \quad (5)$$

In the vast majority of aqueous geologic fluids the assumption is easily justified that $a_{H_2O} = 1$, allowing us to rearrange equation (5) to yield

$$\log f_{H_2} = 1/4[\log(f_{CH_4}/f_{CO_2}) - \log K_1] \quad (6)$$

By evaluating the equilibrium constant for reaction (1) as a function of temperature and pressure, we can plot an equilibrium $f_{H_2}$ curve for any ratio of $CH_4$ and $CO_2$ fugacities. The solid curve in Fig. 3 is for equal fugacities of $CH_4$ and $CO_2$, and shows ranges of $f_{H_2}$ and temperature over which either form of C would predominate at stable chemical equilibrium at 500 bar. Values of $f_{H_2}$ correlate with the oxidation state of the system in the same way that partial pressures of $H_2$ would; as $f_{H_2}$ increases conditions become more reduced. It is evident from this figure that, at the same value of $f_{H_2}$, $CO_2$ would predominate over $CH_4$ at higher temperatures, and $CH_4$ would predominate at lower temperatures. This seems consistent with qualitative observations of high- and low-temperature geologic systems.

Although it is relatively familiar to relate natural observations to temperature, it is less intuitive to link them to a redox parameter like $f_{H_2}$. As a framework for relating $f_{H_2}$ values to geologic processes, equilibrium values of $f_{H_2}$ consistent with reactions (2) and (3) are plotted as dashed curves in Fig. 3. It should be noted that fayalite contains only ferrous iron, magnetite contains ferrous and ferric iron, and hematite contains only ferric iron. This means that increasing oxidation leads to the replacement of fayalite with magnetite, and magnetite with hematite, consistent with steadily decreasing values of $f_{H_2}$.

It cannot be overemphasized that the cross-cutting nature of the $f_{H_2}$ curve for the C-O-H system and those for

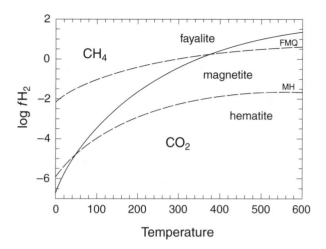

**Fig. 3.** Plot of $H_2$ fugacity against temperature at 500 bar showing the curve for equilibrium equal fugacities of $CO_2$ and $CH_4$ (solid curve), as well as the trajectories (dashed curves) of the quartz-fayalite-magnetite and magnetite-hematite buffers (in the presence of $H_2O$). The Fe-bearing minerals help to locate crustal materials on this plot, and to show why $CO_2$ predominates in high-temperature volcanic gases while $CH_4$ is abundant in low-temperature sedimentary basins.

the Fe-bearing mineral assemblages holds the key to understanding the combined effects of temperature and oxidation state on C chemistry in geochemical processes. Examination of Fig. 3 shows that at the high temperatures of igneous processes involving mafic volcanic rocks (which often contain either a fayalite component in olivine or a magnetite component in oxides, or both), $CO_2$ will predominate if stable chemical equilibrium can be reached. The fact that equilibrium often *is* reached in these systems explains the composition of many volcanic gases (*Symonds et al.,* 1994). Similarly, at the lower temperatures encountered in sedimentary basins (roughly 0°–300°C), we find $CH_4$ and organic compounds preserved in sedimentary rocks that often contain magnetite.

Methane is the volatile form of carbon that is thermodynamically stable at the temperatures and oxidation states of sedimentary basins. The slow reactions that generate $CH_4$ from organic matter drive the overall alteration process called "maturation" that transforms organic matter in sedimentary rocks into petroleum and coal (*Helgeson et al.,* 1993). These processes are accompanied by a lowering of the overall thermodynamic energy states of geochemical systems to local minima in which concentration ratios of organic compounds attain values consistent with *metastable* equilibria (*Shock,* 1988, 1989, 1994; *Helgeson et al.,* 1993; *Seewald,* 1994). Metastable equilibria, rather than stable equilibria, are attained in sedimentary basins owing to the extremely sluggish rates of mechanistically complicated reactions leading to methane and graphite at lower temperatures. The inhibition of stable equilibrium generates conditions in which numerous hydrocarbons and other organic compounds can coexist with one another and with carbonate minerals in the rocks. As demonstrated by *Shock* (1994), values of $f_{H_2}$ consistent with naturally occurring metastable equilibrium states lie just above the solid curve in Fig. 3.

Comparison of the speciation of carbon in volcanic gases and sedimentary basins with the plot in Fig. 3 not only reveals the thermodynamic explanations for easily observed natural phenomena, it also illuminates the fact that cooling from high temperature at oxidation states consistent with crustal rocks can drive reduction reactions. Therefore, reduction of $CO_2$ and/or CO is the direction for organic synthesis favored by the cooling of initially hot geologic materials (lava, intrusions, impact melts). The conditions where carbon oxides can be reduced to organic compounds fall in a band along the solid curve in Fig. 3, at temperatures lower than those at which stable equilibrium rapidly occurs (generally >500°C) but high enough so that reaction progress is likely over geologic time. As argued elsewhere (*Shock*, 1990, 1992a, 1996, 1997; *Shock and Schulte*, 1998; *Amend and Shock*, 1998, 2000; *Zolotov and Shock*, 1999, 2000), the bulk composition of geologic systems tends to generate oxidation states conducive to organic synthesis below about 300°C (see below). Before any organic synthesis can occur, disequilibrium states have to be established by geological and geophysical processes. As argued here, most of the likely disequilibrium states engendered by cooling favor reduction reactions.

## 6. DYNAMIC COOLING, QUENCHING, AND GEOCHEMICAL SUPPORT OF REDUCTIVE METABOLISMS

In eruptions of Kilauea at 1200°C, reactions between minerals, magma, and gases are so rapid that chemical equilibrium is established. Equilibrium with respect to reaction (1), together with other redox reactions involving Fe-bearing minerals, $CO_2$, CO, $CH_4$, $SO_2$, $H_2S$, $N_2$, $NH_3$, $H_2$, $O_2$, and $H_2O$ set the relative fugacities of these gases consistent with the oxidation state controlled by the bulk composition of the volcanic system. As temperature drops, the thermodynamic drive for all these reactions is to generate reduced species ($CH_4$, $H_2S$, $NH_3$, $H_2O$) at the expense of oxidized ones. We can write reactions to indicate this drive as

$$CO_2 + 4\ H_2 \Rightarrow CH_4 + 2\ H_2O \qquad (7)$$

$$CO + 3\ H_2 \Rightarrow CH_4 + H_2O \qquad (8)$$

$$SO_2 + 3\ H_2 \Rightarrow H_2S + 2\ H_2O \qquad (9)$$

$$N_2 + 3\ H_2 \Rightarrow 2\ NH_3 \qquad (10)$$

and

$$0.5\ O_2 + H_2 \Rightarrow H_2O \qquad (11)$$

and use additional calculations to quantify the thermodynamic drive that favors these reactions. As an example, the energetic drive to convert $CO_2$ into $CH_4$ is given by the over-

all Gibbs free energy of reaction ($\Delta G_r$), which is composed of two parts as revealed by the familiar expression

$$\Delta G_r = \Delta G°_r + RT \ln Q_r \qquad (12)$$

where $\Delta G°_r$ represents the standard state contribution, which is related to the equilibrium constant for the reaction, $K_r$, via

$$\Delta G°_r = -RT \ln K_r \qquad (13)$$

where R designates the gas constant and T stands for temperature, and the activity product, $Q_r$, is defined by

$$Q_r = \prod_i a_i^{\nu_{i,r}} \qquad (14)$$

where $a_i$ stands for the activity of the ith chemical species in the reaction, and $\nu_{i,r}$ represents the stoichiometric reaction coefficient for the ith species in the rth reaction, which is positive for products and negative for reactants, and can be calculated from geochemical constraints on the compounds involved in reaction (7) using

$$Q_r = \frac{f_{CH_4}(a_{H_2O})^2}{f_{CO_2}(f_{H_2})^4} \qquad (15)$$

Note that $Q_r$ has the same form as the law-of-mass-action expression for $K_r$ (equation (4)), and at equilibrium, $Q_r = K_r$, and therefore $\Delta G_r = 0$. Away from equilibrium, $Q_r \neq K_r$, and $\Delta G_r \neq 0$. As a consequence of the sign conventions implicit in thermodynamic definitions, negative values of $\Delta G_r$ indicate reactions that are favored to proceed. If these reactions are not kinetically inhibited and can proceed, the overall free energy of the system will be lowered.

As written, reactions (7)–(11) are all favored to proceed in gases cooled from volcanic temperatures. This is indicated by the negative values of $\Delta G_r$ below 1200°C shown in Fig 4a. At 1200°C, the gas is initially at thermodynamic equilibrium, and $\Delta G_r = 0$ for all reactions. If volcanic gases were trapped and cooled rapidly these gas compositions may be quenched before they are able to reequilibrate. If so, disequilibrium states with respect to reduction reactions would be established, and all these reactions could potentially yield energy if they were catalyzed. Geologic observations show and laboratory experiments confirm that all these reductive reactions are sluggish as temperature drops (see reviews by *Barnes*, 1987; *Zolotov and Shock*, 1999). As an example, the reduction of $CO_2$ is unlikely to occur in metamorphic and hydrothermal processes below ~400°C (see review by *Shock*, 1992a), and is only achieved in the laboratory at temperatures <800°C in gas mixtures by using Pt or other efficient catalysts (*Burkhard and Ulmer*, 1995; *Miyamoto and Mikouchi*, 1996; *Beckett and Mendybaev*, 1997). Note that the oxygen reduction reaction (11) has the most negative values of $\Delta G_r$, and would yield the greatest amount of energy, followed by the formation of $CH_4$ from CO (reaction (8)), which yields about exactly the same amount of energy as $H_2S$ formation from $SO_2$ (reac-

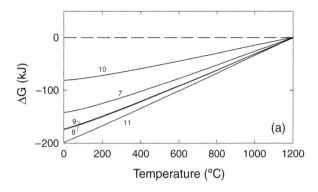

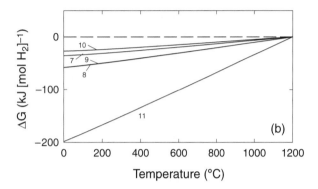

**Fig. 4.** Overall Gibbs free energies of reactions (7) through (11) in cooling volcanic gas that is quenched at its equilibrium composition at 1200°C (see Table 1). These reactions correspond to several major chemosynthetic sources of energy for autotrophic microorganisms. All these reactions are capable of releasing energy as the quenched volcanic gas cools, and therefore of supporting life. Values are given for the reactions **(a)** as written, and **(b)** per mole of $H_2$ consumed.

tion (9)), and then $CH_4$ production from $CO_2$ (reaction (7)), and $NH_3$ generation from $N_2$ (reaction (10)). The energetic ranking of these reactions stays the same when compared on the basis of energy yield per mole of $H_2$ as shown in Fig. 4b.

More compelling than geologic observations or laboratory measurements is that fact that hyperthermophilic microorganisms (both Archaea and Bacteria that live above 80°C) use at least some of these geochemical disequilibria as energy sources. For example, methanogenesis from $CO_2$ and $H_2$ (reaction (7)) fuels several hyperthermophilic methanogens (*Stetter et al.,* 1981; *Jones et al.,* 1983; *Lauerer et al.,* 1986; *Burggraf et al.,* 1990a; *Kurr et al.,* 1991; *Jeanthon et al.,* 1998), many of which have the genes for the enzymes required to reduce CO. Reduction of sulfate, thiosulfate, and/or sulfite provides energy for species of *Archeoglobus* (*Stetter,* 1988; *Burggraf et al.,* 1990b; *Stetter et al.,* 1993; *Huber et al.,* 1997), *Ferroglobus* (*Hafenbrandl et al.,* 1996), and other archaeal and bacterial genera (*Zeikus et al.,* 1983; *Huber et al.,* 1987; *Rozanova and Pivovarova,* 1988; *Fardeau et al.,* 1996; *Blöchl et al.,* 1997; *Jochimsen et al.,* 1997; *Itoh et al.,* 1998). Oxygen reduction is used as an energy-

yielding reaction by the hot-spring dwellers *Aquifex* (*Deckert et al.,* 1998; *Huber et al.,* 1992) and *Thermocrinis* (*Huber et al.,* 1998). These and other hyperthermophiles populate the deepest and shortest branches of the universal phylogenetic tree.

The advent of genetically based phylogenetic trees has revolutionized understanding about the evolutionary relations among all life on the Earth, but the major impact of these methods has been in understanding microbial evolution. One of the more useful trees for generating testable hypotheses is built on genetic information from small subunits of ribosomal RNA (16S rRNA), and a version of such a tree is shown in Fig. 5 (after *Pace,* 1997). Although the angles between branches are aesthetic, the lengths of the branches reflect the difference in 16S rRNA sequences among the various organisms, with shorter branches reflecting greater similarity to the branch point. Note that the organisms that populate the deepest and shortest branches among the Archaea and Bacteria are all thermophiles or hyperthermophiles, and include those that pursue reductive metabolisms fueled by geochemical disequilibrium established by dynamic cooling. This implies that chemosynthetic organisms have been using these geologically provided sources of geochemical energy for a very long time.

One interpretation of the 16S rRNA tree is that the course of evolution has generally moved from high temperature to low temperature. In every case of a major branch that includes mesophilic organisms together with hyperthermophilic or thermophilic organisms, the high-temperature organisms diverge more deeply (*Stetter,* 1992). Another notable feature of the 16S rRNA tree is that the deepest branching organisms do not use light energy; photosynthesis appears to be a later development than taking advantage of geochemical energy sources. Altogether, the 16S rRNA tree tells us that the simplest description of the last common ancestor of all living organisms is that it was most likely a thermophilic, nonphotosynthetic, anaerobic autotroph with a reductive metabolism (*Pace,* 1991; *Shock,* 1997). This idea is reinforced in the statement by *Baross* (1998) that "metabolic pathways are more ancient than many of the individual genes used to construct phylogenetic trees." It could be concluded that over the course of time life has emerged from propitious higher-temperature environments and infiltrated lower-temperature, more hostile environments where evolution allowed (and demanded) adaptation.

Regardless of evolutionary interpretations of the 16S rRNA tree, the fact remains that it does not necessarily demand that life began at high temperatures using reductive metabolism. The base of the tree could be an artifact of a major evolutionary bottleneck through which life passed (it could also represent a sparse dataset), as argued by various researchers (*Doolittle and Brown,* 1994; *Woese,* 1998; *Baross,* 1998). Although this proposition may give succor to ideas about life's origins that predate the exploration of the solar system, plate tectonics, and modern molecular biology, it does not explain why the phylogenetic tree is rooted in inevitable geologic processes that would have been even more pronounced on the early Earth and that are

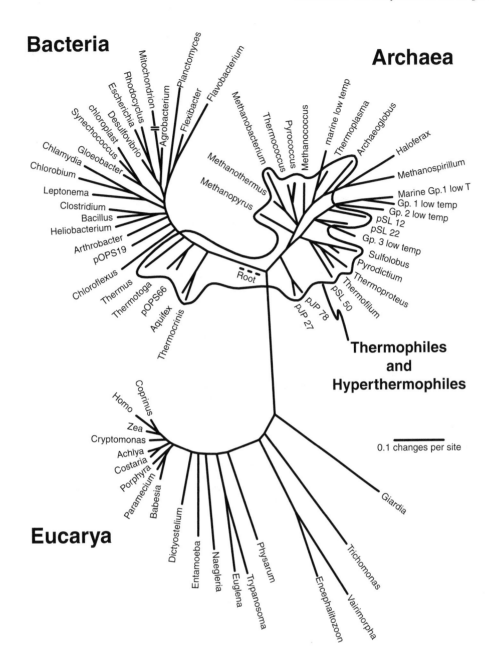

**Fig. 5.** Universal phylogenetic tree of life based on 16S rRNA data (see text). Lengths of lines separating pairs of organisms correspond to the genetic differences between them. Life is divided into three domains (Bacteria, Archaea, and Eucarya) as a result of constructing this tree. Deep, short branches indicated in the center of the tree are populated by thermophilic and hyperthermophilic organisms.

uniquely suited to supporting autotrophic chemosynthesis. There are not yet compelling reasons to assume that this consistency is a coincidence.

## 7.  HYDROTHERMAL SYSTEMS AS SOURCES OF GEOCHEMICAL DISEQUILIBRIUM AND HABITATS

Dynamic zones of hydrothermal systems support hyperthermophilic, reductive metabolisms because they generate conditions of thermodynamic disequilibrium that favor re-

ductive reactions (*McCollom and Shock,* 1997; *Shock et al.,* 1998; *Shock and Schulte,* 1998; *Amend and Shock,* 1998; *McCollom,* 2000). A major source of disequilibrium in these systems is supplied by mixing of fluids that are far from thermal or chemical equilibrium with each other. (Disequilibrium can be established in the mixture even in cases where each fluid is in chemical equilibrium with its local environment.) This is easily visualized at the surface expressions of submarine hydrothermal systems where hot reduced fluids enriched in $H_2S$, $Fe^{+2}$, $Mn^{+2}$, and $H_2$ mix with cold oxidized seawater containing $SO_4^{-2}$, and $O_2$ (see Table 2),

TABLE 2.    Endmember fluid compositions used in mixing calculations described in this study.

|  | EPR 21°N SW | Seawater |
| --- | --- | --- |
| $H_2$,aq | 0.45 | 0 |
| $O_2$,aq | 0 | 0.076 |
| $SO_4^{-2}$ | 0.6 | 27.9 |
| $H_2S$,tot | 7.5 | 0 |
| pH | 4.54 | 7.8 |
| $Na^+$ | 439 | 464 |
| $Ca^{+2}$ | 16.6 | 10.2 |
| $Mg^{+2}$ | 0 | 52.7 |
| $Ba^{+2}$ | 0.010 | 0.00014 |
| $Al^{+3}$ | 0.0047 | 0.00002 |
| Fe | 0.750 | 0.0000015 |
| $Cl^-$ | 496 | 541 |
| $\Sigma CO_2$* | 5.72 | 2.3 |
| $SiO_2$,aq | 17.3 | 0.16 |
| $K^+$ | 23.2 | 9.8 |
| $Mn^{+2}$ | 0.70 | 0 |
| $Cu^{+2}$ | 0.010 | 0.000007 |
| $Pb^{+2}$ | 0.000194 | 0 |
| $Zn^{+2}$ | 0.089 | 0.00001 |

EPR 21°N SW is taken as a typical mid-ocean-ridge hydrothermal vent fluid and seawater is an average composition (data from *Wehlan and Craig,* 1983; *Von Damm et al.,* 1985; *Von Damm,* 1990). All concentrations are millimolal.
*$\Sigma CO_2 = (CO_2,aq + HCO_3^- + CO_3^{-2})$.

resulting in rapid precipitation of minerals and highly anomalous concentrations of life compared to the rest of the seafloor or the open ocean.

As the hydrothermal fluid mixes with seawater, cooling lowers the rates of reactions and disequilibrium is generated (*McCollom and Shock,* 1997). Higher-temperature fluid mixtures (≥50°C) are dominated by the input from the hydrothermal fluids and possess disequilibrium states in which reduction reactions are thermodynamically favored. Lower-temperature fluid mixtures, dominated by seawater, develop disequilibrium states in which oxidation reactions are favored. All these reactions are relatively sluggish and can be catalyzed by enzymes. As a result, autotrophic methanogens and sulfur reducers tend to thrive at elevated temperatures, and aerobic methanotrophs and sulfide oxidizers populate the lower temperature portions of these systems (see *Karl,* 1995).

*Methanococcus jannaschii* is an example of the type of reductive autotrophic microbe supported in these systems. This autotrophic methanogen was isolated from a submarine hydrothermal system, and will grow in the lab at temperatures from 50° to 86°C (*Jones et al.,* 1983), and to 90°C at high pressures (*Miller et al.,* 1988). At its optimum growth temperature in the laboratory of 85°C the population of *M. jannaschii* will double in 26 min (*Jones et al.,* 1983). It gains its metabolic energy from dissimilatory $CO_2$ reduction as in reaction (7). The maximum amount of energy that *M. jannaschii* can obtain from this process is dic-

tated by the concentrations of $CO_2$, $H_2$, and $CH_4$ in the fluid mixtures in which it lives, as well as the standard state Gibbs free energy of reaction (7) over this range of temperatures. Calculated values of $\Delta G°_r$ are shown in Fig. 6a, together with values of $RT\ln Q_r$ calculated by mixing the two fluid compositions listed in Table 2 at all proportions necessary to obtain temperatures between 350° and 2°C (*McCollom and Shock,* 1997). The large shift in the $RT\ln Q_r$ term, occurs at the mixing ratio at which the dissolved $O_2$ in seawater is fully consumed by reaction with $H_2$ if equilibrium is reached with respect to reaction (11) (see *McCollom and Shock,* 1997). The resulting values of $\Delta G_r$ obtained with equation (12) are shown in Fig. 6b, where it can be seen that the range of temperature over which this reaction is exergonic (negative values of $\Delta G_r$, i.e., conditions where the reaction can yield energy) spans ~280°C from ~40° to ~320°C. Note that at the experimentally determined optimal growth temperature for *M. jannaschii*, ~100 kJ of energy are released for each mole of $CH_4$ generated.

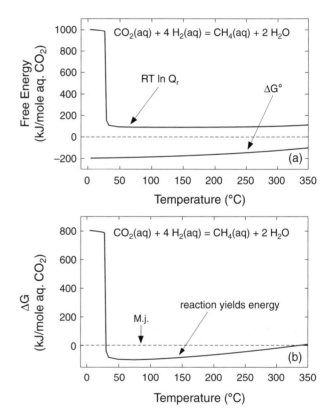

**Fig. 6.** Thermodynamics of autotrophic methanogenesis from $CO_2$ and $H_2$ at conditions prevailing in submarine hydrothermal systems. These plots are based on calculations conducted by *McCollom and Shock* (1997). **(a)** The two terms on the right side of equation (12), and **(b)** their sum, which corresponds to the overall Gibbs free energy of the reaction at the temperatures and compositions that result from mixing hot hydrothermal fluids with seawater (see Table 2). Negative values of the overall Gibbs free energy correspond to conditions at which the reaction is capable of supplying energy. The optimum temperature (85°C) for growing *Methanococcus jannaschii* in the laboratory is indicated.

Methanogenesis is by no means the only favorable $CO_2$ reduction reaction in mixtures of hydrothermal fluids and seawater. Synthesis of many ketones, alcohols, and carboxylic acids are also energy-yielding processes (*Shock and Schulte*, 1998), as is the synthesis of several amino acids (*Amend and Shock*, 1998). As an example, consider the net autotrophic synthesis of the amino acid leucine (($CH_3$)$_2$ $CHCH_2CHNH_2COOH$) consistent with

$$6\ CO_2(aq) + NH_4^+ + 15\ H_2(aq) \Rightarrow$$
$$\text{leucine(aq)} + H^+ + 10\ H_2O \qquad (16)$$

Thermodynamic analysis of this reaction (*Amend and Shock*, 1998) shows that it is exergonic at 100°C in hydrothermal systems, and it is slightly more exergonic at 85°C. This is shown in Fig. 7 where values of $\Delta G_r$ are plotted at temperatures ≤200°C for autotrophic leucine synthesis in mixtures between hydrothermal fluids and seawater. Note that autotrophic synthesis of leucine is exergonic between ~40°C and 160°C. At the optimum lab temperature for *M. jannaschii*, $\Delta G_r$ of this reaction is –129.4 kJ per mole of leucine produced, which means that energy is *released* during leucine synthesis at these conditions. Leucine is not unique; as shown in results from *Amend and Shock* (1998), 11 of the 20 net reactions for autotrophic synthesis of protein-forming amino acids are exergonic in the range of temperatures and fluid compositions where submarine hyperthermophiles live. These results stand in stark contrast to amino acid synthesis at low-temperature aerobic conditions, which always has large energetic costs. These observations may help reveal new pathways for amino acid synthesis in hyperthermophiles. At least they help to emphasize that the conditions in hydrothermal systems that seem so exotic to us are enormously more conducive to life than those we find familiar.

The fact that reductive organic synthesis can lower the energetic state of hydrothermal systems explains the occurrence of autotrophic primary production by chemosynthetic hyperthermophiles. These organisms act as catalysts for reactions that are already thermodynamically favored. As catalysts, their high degree of selectivity is afforded by the enormous amounts of energy available through dissimilatory reactions like methanogenesis, sulfate reduction, and sulfur reduction. Additionally, the existence of thermodynamic drives in the overall direction of biosynthesis provides an ingredient that is often missing in the search for how life began. The geological and geophysical processes that lead to the geochemical potentials inherent to hydrothermal systems provide something like a "habitat" for the emergence of life. Such habitats on early Earth may have differed from those present today, but geochemical disequilibrium states must have existed during the Hadean to permit the emergence of life.

## 8. CONDITIONS IN HADEAN HYDROTHERMAL SYSTEMS

As an introduction to speculations about hydrothermal systems on the early Earth that predates the rock record (the Hadean), it will help to summarize the points made so far:

1. The earliest geologic record supports the idea that familiar geological and geochemical processes were operating extremely early in Earth history.

2. Various aspects of planetary accretion, differentiation, mantle convection, atmospheric photochemistry, and molecular phylogeny make it increasingly difficult to find support for the popular scenario from midcentury that imposition of external sources of energy into a reduced atmosphere generated a "prebiotic soup" in which the first organism appeared as a heterotroph.

3. Instead, current emergence-of-life scenarios typically include internal forms of geochemical energy that establish redox environments in which autotrophic reductive metabolisms are nurtured.

4. Fluid mixing in hydrothermal systems provides energy to chemosynthetic hyperthermophiles that populate the deepest and shortest branches in the 16S rRNA tree.

5. Whether generated by radioactive decay, initial accretion, or later impacts, internal heat from the Earth flows to space through volcanic processes, which, in the presence of liquid water, establish conditions in which reductive metabolisms can catalyze thermodynamically favored but kinetically inhibited organic synthesis reactions.

In what substantive ways would conditions on the Hadean Earth affect the potential for the emergence of life in hydrothermal systems? Probably the most likely differences between then and now involve the magnitude of heat flow

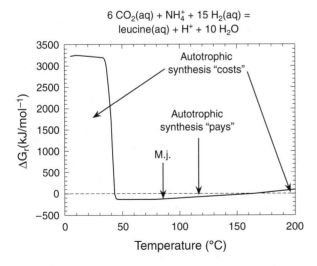

**Fig. 7.** Overall Gibbs free energy of autotrophic leucine synthesis at conditions in hydrothermal systems corresponding to those used to generate Fig. 6. This plot is based on calculations conducted by *Amend and Shock* (1998). Note that autotrophic leucine synthesis releases energy between about 40° and 160°C in hydrothermal systems, and becomes extremely costly in surface seawater. The optimum laboratory growth temperature (85°C) for *Methanococcus jannaschii* is indicated.

and the compositions of the atmosphere, seawater, upper mantle, and crustal rocks that could host hydrothermal systems. On the Hadean Earth:

1. Energy from accretion and radioactive decay of short-lived nuclides would have lead to greater heat flow from the mantle. The consequences include an increase in the amount of volcanism, and enhanced rates of subduction and seafloor spreading leading to greater recycling of crust into the mantle. Implications for the emergence of life are that there would have been more volcanic activity and thus more hydrothermal circulation of fluids. If continents were absent or less abundant (*Walker*, 1985), the extent of hydrothermal activity would have been greater than at present. It is also possible that the magnitude of hydrothermal circulation around impacts would have been comparable to that driven by volcanic activity. In addition, severe impacts into an ocean during the latest stages of accretion could have produced gas mixtures with high degrees of disequilibrium ($H_2$, $O_2$, $CO$), which could also be sources of geochemical energy for autotrophs.

2. Higher degrees of partial melting of the primordial mantle could have provided somewhat more mafic oceanic crustal material. In addition, until subduction processes oxidized a significant part of the upper mantle, the magma sources would have been more reduced than at present. The consequence for the emergence of life is that higher $H_2$ concentrations would be expected in volcanic gases and hydrothermal fluids, and those higher concentrations would provide a greater drive for reduction of carbon oxides, $SO_2$, $N_2$, and $O_2$ as temperature dropped.

3. Although not reduced, the atmosphere was likely to contain considerably less $O_2$ than at present. The implication for the emergence of life is that conditions at which reduction could occur would have extended to lower temperatures in hydrothermal systems than they do at present.

4. In some models, investigators argue that the atmosphere contained considerably higher concentrations of $CO_2$ than at present. Beyond the capacity for greenhouse warming of the surface of the Earth, the implication for the emergence of life is that the ocean may have been a greater source of dissolved inorganic carbon than at present. In active seafloor systems, hydrothermal fluids can contain more than twice as much dissolved inorganic carbon as seawater (see *McCollom and Shock*, 1997). An ocean in equilibrium with an atmosphere containing 10 bar of $CO_2$ (*Walker*, 1985) would contain about 40× as much dissolved inorganic carbon as the present oceans (see *Amend and Shock*, 2000).

Taken together, these apparent differences between the present and Hadean Earth suggest that conditions in hydrothermal systems on the early Earth were even more favorable for the emergence of life than they are at present.

## 9.  DIRECTIONS FOR FUTURE RESEARCH

Considerations of the arguments presented above lead to many directions for future research that could test the con-

nections between the early evolution of the Earth and the emergence of life. Here we list a few examples of research areas that would benefit from additional attention.

1. *Coupled models of atmospheric chemistry, ocean chemistry, water/rock reactions, and impact and igneous processes.*    Simulations of the early Earth are necessary aspects of the study of the emergence of life. At present, atmospheric models that incorporate rates of photochemical processes are conducted separately from water/rock models that simulate the consequences of hydrothermal alteration on rock and fluid compositions. Water/rock models are typically conducted separately from those that characterize igneous processes, even for applications to oceanic crust. Nevertheless, the feedback between these processes is appreciated by most who try to evaluate what dynamic processes on the early Earth were like. Additional clues about the relevance of certain processes, and their relative rates, could be obtained by fully coupling these models.

2. *Investigation of the earliest hydrothermal processes in the solar system.*    In the absence of a continuous record on the Earth, meteorites and samples from Mars provide an exclusive inventory for understanding the processes that occurred in volcanic and hydrothermal systems at the time when life emerged. Coordinated analytical and theoretical studies of the record left by these processes will provide clues about the contemporaneous processes on the Hadean Earth.

3. *Kinetic and mechanistic analysis of abiotic organic synthesis reactions at hydrothermal conditions.*    The success of hydrothermal organic synthesis and organic transformation experiments points the way to additional experiments and mechanistic studies that can reveal how mineral surfaces, potential gradients, and variations in composition affect the rates of these processes.

4. *Thermodynamic analysis of the energetic demands of biosynthetic pathways thought to be plausible for early metabolisms.*    The energetic costs of biosynthetic processes leading to proteins, nucleic acids, membranes, and other biomolecules at the actual temperatures, pressures, and compositions of hydrothermal systems are largely unknown. Without such information it is difficult to assess whether those costs are offset by sources of geochemical energy in active hydrothermal systems or their plausible Hadean counterparts.

5. *Tests of the 16S rRNA phylogenetic tree.*    Finally, identification of a low-temperature (mesophilic) organism that branches more deeply in the phylogenetic tree than closely related hyperthermophilic or thermophilic organisms would challenge many of the arguments that can be made based on phylogenetic relations.

***Acknowledgments.***    Many thanks to Gavin Chan, Andrey Plyasunov, Giles Farrant, Panjai Prapaipong, Anna-Louise Reysenbach, Tom McCollom, Mitch Schulte, Sam Bowring, Paul Hoffman, Steve Mojzsis, Kevin Zahnle, Gary Olsen, Melanie Summit, and Roger Buick for helpful, inspirational, and lively discussions; to Kevin Righter, Eric Gaidos, and Ariel Anbar for reviews; and to Barb Winston for technical assistance. This work was funded by NASA Exobiology grant NAG5-7696 and NSF LExEn grant OCE-9714288. GEOPIG Contribution #188.

# REFERENCES

Abe Y. and Matsui T. (1988) Evolution of impact-generated $H_2O$-$CO_2$ atmosphere and formation of a hot proto-ocean on Earth. *J. Atmosph. Sci., 45,* 3081–3101.

Allamandola L. J., Tielens G. G. M., and Baker J. R. (1989) Interstellar polycyclic aromatic hydrocarbons — The infrared emission bands, the excitation emission mechanism, and the astrophysical implications. *Astrophys. J., 71,* 733–775.

Amend J. P. and Helgeson H. C. (1997a) Group additivity equations of state for calculating the standard molal thermodynamic properties of aqueous organic species at elevated temperatures and pressures. *Geochim. Cosmochim. Acta, 61,* 11–46.

Amend J. P. and Helgeson H. C. (1997b) Calculation of the standard molal thermodynamic properties of aqueous biomolecules at elevated temperatures and pressures. Part I. L-α-amino acids. *J. Chem. Soc. Faraday Trans., 93,* 1927–1941.

Amend J. P. and Shock E. L. (1998) Energetics of amino acid synthesis in hydrothermal ecosystems. *Science, 281,* 1659–1662.

Amend J. P. and Shock E. L. (2000) Thermodynamics of amino acid synthesis in hydrothermal ecosystems on the early Earth. In *Perspectives in Amino Acid and Protein Geochemistry* (G. Goodfriend, ed.). Plenum, New York, in press.

Awramik S. M., Schopf J. W., and Water M. R. (1983) Filamentous fossil bacteria from the Archean of Western Australia. *Precambrian Res., 20,* 357–374.

Awramik S. M., Schopf J. W., and Water M. R. (1988) Carbonaceous filaments from North Pole, Western Australia: are they fossil bacteria in Archean stromatolites? A discussion. *Precambrian Res., 39,* 3030–3039.

Bada J. L., Bigham C., Miller S. L. (1994) Impact melting of frozen oceans on the early Earth: implications for the origin of life. *Proc. Natl. Acad. Sci. USA, 91,* 1248–1250.

Barnes H. L. (1987) In *Hydrothermal Experimental Techniques* (G. C. Ulmer and H. L. Barnes, eds.), pp. 507–514. Wiley and Sons, New York.

Baross J. A. (1998) Do the geological and geochemical records of the early Earth support the prediction from global phylogenetic models of a thermophilic cenancestor? In *Thermophiles: The Keys to Molecular Evolution and the Origin of Life?* (J. Wiegel and M. W. W. Adams, eds.), pp. 3–18. Taylor and Francis, London.

Baross J. A. and Hoffman S. E. (1985) Submarine hydrothermal vents and associated gradient environments as sites for the origin and evolution of life. *Orig. Life Evol. Biosph., 15,* 327–345.

Beckett J. R. and Mendybaev R. A. (1997) The measurement of oxygen fugacities in flowing gas mixtures at temperatures below 1200°C. *Geochim. Cosmochim. Acta, 61,* 4331–4336.

Benz W. and Cameron A. G. W. (1990) Terrestrial effects of the giant impact. In *Origin of the Earth* (H. E. Newsom and J. H. Jones, eds.), pp. 61–67. Oxford Univ., New York.

Berndt M. E., Allen D. E., and Seyfried W. E. Jr. (1996) Reduction of $CO_2$ during serpentinization of olivine at 300°C and 500 bars. *Geology, 24,* 351–354.

Blöchl E., Rachel R., Burggraff S., Hafenbradl D., Janasch H. W.,

and Stetter K. O. (1997) *Pyrolobus fumarii*, gen. nov. and sp. nov., represents a novel group of archaea, extending the upper temperature limit for life to 113°C. *Extremophiles, 1,* 14–21.

Bowring S. A. and Housh T. (1995) The Earth's early evolution. *Science, 269,* 1535–1540.

Bowring S. A., King J. E., Housh T. B., Isachsen C. E., and Podosek F. A. (1989a) Neodymium and lead isotope evidence for enriched early Archaean crust in North America. *Nature, 340,* 222–225.

Bowring S. A., Williams I. S., and Compston W. (1989b) 3.96 Ga gneisses from the Slave Province, Northwest Territories, Canada. *Geology, 17,* 971–975.

Buick R. (1988) Carbonaceous filaments from North Pole, Western Australia: are they fossil bacteria in Archean stromatolites? A reply. *Precambrian Res., 39,* 311–317.

Burggraff S., Fricke H., Neuner A., Kristjansson J., Rouvier P., Mandelco L., Woese C. R., and Stetter K. O. (1990a) *Methanococcus igneus* sp. nov., a novel hyperthermophilic methanogen from a shallow submarine hydrothermal system. *Syst. Appl. Microbiol., 13,* 263–269.

Burggraff S., Jannasch H. W., Nicolaus B., and Stetter K. O. (1990b) *Archaeglobus profundus* sp. nov., represents a new species within the sulfate-reducing archaebacteria. *Syst. Appl. Microbiol., 13,* 24–28.

Burkhard D. J. M. and Ulmer G. C. (1995) Kinetics and equilibria of redox systems at temperatures as low as 300°C. *Geochim. Cosmochim. Acta, 59,* 1699–1714.

Cameron A. G. W. (1997) The origin of the Moon and the single impact hypothesis. V. *Icarus, 126,* 126–137.

Cameron A. G. W. and Benz W. (1988) Effects of the giant impact on the Earth (abstract). In *Papers Presented to the Conference on Origin of the Earth*, pp. 11–12. Lunar and Planetary Institute, Houston.

Canuto V. M., Levine J. S., Augustsson T. R., and Imhoff C. L. (1982) UV radiation from the young Sun and oxygen and ozone levels in the prebiological palaeoatmosphere. *Nature, 296,* 816–820.

Canuto V. M., Levine J. S., Augustsson T. R., Imhoff C. L., and Giampapa M. S. (1983) The young Sun and the atmosphere and photochemistry of the early Earth. *Nature, 305,* 281–286.

Clemett S. J., Maechling C. R., Zare R. N., Swan P. D., and Walker R. M. (1993) Identification of complex aromatic molecules in individual interplanetary dust particles. *Science, 262,* 721–723.

Compston W. and Pidgeon R. T. (1986) Jack Hills, Evidence of more very old detrital zircon in Western Australia. *Nature, 321,* 766–769.

Corliss J. B., Baross J. A., and Hoffman S. E. (1981) An hypothesis concerning the relationship between submarine hot springs and the origin of life on Earth. *Ocean. Acta, No. SP,* 59–69.

Cronin J. R., Pizzarello S., and Cruikshank D. P. (1988) Organic matter in carbonaceous chondrites, planetary satellites, asteroids and comets. In *Meteorites and the Early Solar System* (J. F. Kerridge and M. S. Matthews, eds.), pp. 819–857. Univ. of Arizona, Tucson.

Deckert G., Warren P. V., Gaasterland T., Young W. G., Lenox A. L., Graham D. E., Overbeek R., Snead M. A., Keller M., Aujay M., Huber R., Feldman R. A., Short J. M., Olsen G. J., and Swanson R. V. (1998) The complete genome of the hyperthermophilic bacterium *Aquifex aeolicus*. *Nature, 392,* 353–358.

Doolittle W. F. and Brown J. R. (1994) Tempo, mode, the progenote, and the universal root. *Proc. Natl. Acad. Sci. USA, 91,* 6721–6728.

Ertem G. and Ferris J. P. (1996) Synthesis of RNA oligomers on heterogeneous templates. *Nature, 379,* 238–240.

Ertem G. and Ferris J. P. (1998) Formation of RNA oligomers on montmorillonite: Site of catalysis. *Orig. Life Evol. Biosph., 28,* 485–499.

Fardeau M.-L., Faudon C., Cayol J.-L., Magot M., Patel B. K. C., and Ollivier B. (1996) Effect of thiosulphate as electron acceptor on glucose and xylose oxidation by *Thermoanaerobacter finni* and a *Thermoanaerobacter* sp. isolated from oil field water. *Res. Microbiol., 147,* 159–165.

Ferris J. P. (1998) Catalyzed RNA synthesis for the RNA world. In *The Molecular Origins of Life: Assembling Pieces of the Puzzle* (A. Black, ed.), pp. 255–268. Cambridge Univ., Cambridge.

Ferris J. P. and Ertem G. (1992) Oligomerization of ribonucleotides on montmorillonite: reaction of the 5'-phosphorimidazolide of adenosine. *Science, 257,* 1387–1388.

Ferris J. P., Hill A. R., Liu R., and Orgel L. E. (1996) Synthesis of long prebiotic oligomers on mineral surfaces. *Nature, 381,* 59–61.

French B. M. (1964) 1. Synthesis and stability of siderite, $FeCO_3$, 2. Progressive contact metamorphism of the Biwabik Iron Formation on the Mesabi Range, Minnesota. Ph.D. thesis, The John Hopkins University, Baltimore, Maryland.

French B. M. (1971) Stability relations of siderite ($FeCO_3$), in the system Fe-C-O. *Am. J. Sci. 271,* 37–78.

Froude D. O., Ireland T. R., Kinney P. D., Williams I. S., and Compston W. (1983) Ion microprobe identification of 4,100–4,200 Myr-old terrestrial zircons. *Nature, 304,* 616–618.

Gilbert W. (1986) The RNA world. *Nature, 319,* 618.

Gough D. O. (1981) Solar interior structure and luminosity variations. *Sol. Phys., 74,* 21–34.

Guerrier-Takada C., Gardiner K., Marsh T., Pace N., and Altman S. (1983) The RNA moiety of ribonuclease P is the catalytic subunit of the enzyme. *Cell, 35,* 849–857.

Hafenbradl D., Keller M., Dirmeier R., Rachel R., Rossnagel P., Burggraf S., Huber H., and Stetter K. O. (1996) *Ferroglobus placidus* gen. nov., sp. nov., a novel hyperthermophilic archaeum that oxidizes $Fe^{2+}$ at neutral pH under anoxic conditions. *Arch. Microbiol., 166,* 308–314.

Hamilton P. J., O'Nions R. K., Evensen N. M., Bridgwater D., and Allaart J. H. (1978) Sm-Nd isotopic investigations of Isua supracrustals and implications for mantle evolution. *Nature, 272,* 41–43.

Haldane J. B. S. (1929) The origin of life. *The Rationalist Annual.*

Halliday A. and Lee D.-C. (1999) Tungsten isotopes and the early development of the Earth and Moon. *Geochim. Cosmochim. Acta, 63,* in press.

Halliday A., Rehkaemper M., Lee D.-C., and Yi W. (1996) Early evolution of the Earth and Moon: new constraints from Hf-W isotope geochemistry. *Earth Planet. Sci. Lett., 142,* 75–89.

Helgeson H. C., Delaney J. M., Nesbitt H. W., and Bird D. K. (1978) Summary and critique of the thermodynamic properties of rock-forming minerals. *Am. J. Sci., 278A,* 1–229.

Helgeson H. C., Knox A. M., Owens C. E., and Shock E. L. (1993) Petroleum, oil field watersand authigenic mineral assemblages: Are they in metastable equilibrium in hydrocarbon reservoirs? *Geochim. Cosmochim. Acta, 57,* 3295–3339.

Helgeson H. C., Owens C. E., Knox A. M., and Richard L. (1998) Calculations of the standard molar thermodynamic properties

of crystalline, liquid, and gas organic molecules at high temperatures and pressures. *Geochim. Cosmochim. Acta, 68,* 985–1081.

Hennet R. J.-C., Holm N. G., and Engel M. H. (1992) Abiotic synthesis of amino acids under hydrothermal conditions and the origin of life: a perpetual phenomenon? *Naturwissenschaften, 79,* 361–365.

Huber C. and Wächtershäuser G. (1998) Peptides by activation of amino acids with CO on (Ni,Fe)S surfaces: Implications for the origin of life. *Science, 281,* 670–674.

Huber C. and Wächtershäuser G. (1997) Activated acetic acid by carbon fixation on (Ni,Fe)S under primordial conditions. *Science, 276,* 245–246.

Huber R., Kristjansson J. K., and Stetter K. O. (1987) *Pyrobaculum* gen. nov., a new genus of neutrophilic, rod-shaped archaebacteria from continental solfataras growing optimally at 100°C. *Arch. Microbiol., 149,* 95–101.

Huber R., Wilharm T., Huber D., Trincone A., Burggraf S., König H., Rachel R., Rockinger I., Fricke H., and Stetter K. O. (1992) *Aquifex pyrophilus* gen. nov., sp. nov., represents a novel group of marine hyperthermophilic hydrogen-oxidizing bacteria. *Syst. Appl. Microbiol., 15,* 340–351.

Huber H., Jannasch H., Rachel R., Fuchs T., and Stetter K. O. (1997) *Archaeoglobus veneficus* sp. nov., a novel facultative chemolithoautotrophic hyperthermophilic sulfite reducer, isolated from abyssal black smokers. *Syst. Appl. Microbiol., 20,* 374–380.

Huber R., Eder, W., Heldwein S., Wanner G., Huber H., Rachel R., and Stetter K. O. (1998) *Thermocrinis ruber* gen. nov., sp. nov., a pink-filament-forming hyperthermophilic bacterium isolated from Yellowstone National Park. *Appl. Environ. Microbiol., 64,* 3576–3583.

Ingmannson D. E. and Dowler M. J. 1977. Chemical evolution and the evolution of the Earth's crust. *Orig. Life Evol. Biosph., 8,* 221–224.

Irvine W. M., Schloerb F. P., Crovisier J., Fegley B. Jr., and Mumma M. J. (1999) Comets: A link between interstellar and nebular chemistry. In *Protostars and Planets IV* (A. Boss et al., eds.). Univ. of Arizona, Tucson, in press.

Itoh T., Suzuki K., and Nakase T. (1998) *Thermocladium modestius* gen. nov., sp. nov., a new genus of rod-shaped, extremely thermophilic crenarchaeote. *Intl. J. Syst. Bacteriol., 48,* 879–887.

Jeanthon C., L'Haridon S., Reysenbach A. L., Vernet M., Messner P., Sleytr U. B., and Prieur D. (1998) *Methanococcus infernus* sp. nov., a novel hyperthermophilic lithotrophic methanogen isolated from a deep-sea hydrothermal vent. *Intl. J. Syst. Bacteriol., 48,* 913–919.

Jochimsen B., Penemann-Simon S., Völker H., Stüben D., Botz R., Stoffers P., Dando P. R., and Thomm M. (1997) *Stetteria hydrogenophila* gen. nov. and sp. nov., a novel mixotrophic sulfur-dependent crenarchaeote isolated from Milos, Greece. *Extremophiles, 1,* 67–73.

Johnson J. W., Oelkers E. H., and Helgeson H. C. (1992) SUPCRT92: A software package for calculating the standard molal thermodynamic properties of minerals, gases, aqueous species, and reactions from 1 to 5000 bars and 0° to 1000°. *Computers Geosci., 18,* 899–947.

Jones J. H. and Drake M. J. (1986) Geochemical constraints on core formation in the Earth. *Nature, 322,* 221–228.

Jones W. J., Leigh J. A., Mayer F., Woese C. R., and Wolfe R. S. (1983) *Methanococcus jannaschii* sp. Nov., an extremely thermophilic methanogen from a submarine hydrothermal vent. *Arch. Microbiol., 136,* 254–261.

Joyce G. F. and Orgel L. E. (1993) Prospects for understanding the origin of the RNA World. In *The RNA World* (R. F. Gesteland and J. F. Atkins, eds.), pp. 1–25. Cold Spring Harbor Laboratory.

Karl D. M. (1995) Ecology of free-living hydrothermal vent microbial communities. In *The Microbiology of Deep-Sea Hydrothermal Vents* (D. M. Karl, ed.), pp. 219–253. CRC, Boca Raton, Florida.

Kasting J. F. (1982) Stability of ammonia in the primitive terrestrial atmosphere. *J. Geophys. Res., 87,* 3091–3098.

Kasting J. F., Zahnle K. J., and Walker J. C. G. (1983) Photochemistry of methane in the Earth's early atmosphere. *Precambrian Res., 20,* 121–148.

Konnerup-Madsen J. (1989) Abiotic hydrocarbon gases associated with alkaline igneous activity. *Mem. Geol. Soc. India, 11,* 13–24.

Kruger K., Grabowski P. J., ZaugA. J., Sands J., Gottschling D. E., and Cech T. R. (1982) Self-splicing RNA: Autoexcision and autocyclization of the ribosomal RNA intervening sequence of Tetrahymena. *Cell, 31,* 147–157.

Kuhn W. R. and Atreya S. K. (1979) Ammonia photolysis and the greenhouse effect in the primordial atmosphere of the earth. *Icarus, 37,* 207–13.

Kurr M., Huber R., König H., Jannasch H. W., Fricke H., Trincone A., Kristjansson J. K., and Stetter K. O. (1991) *Methanopyrus kandleri,* gen. and sp. nov. represents a novel group of hyperthermophilic methanogens, growing at 110°C. *Arch. Microbiol., 156,* 239–247.

Lauerer G., Kristjansson J. K., Langworthy T. A., König H., and Stetter K. O. (1986) *Methanothermus sociabilis* sp. nov., a second species within the Methanothermaceae growing at 97°C. *Syst. Appl. Microbiol., 8,* 100–105.

Levine J. S. (1985) The photochemistry of the early atmosphere. In *The Photochemistry of Atmosphere* (J. S. Levine, ed.), pp. 3–38. Academic, New York.

Levine J. S. and Augustsson T. R. (1985) The photochemistry of biogenic gases in the early and present atmosphere. *Orig. Life, 15,* 299–318.

Levine J. S., Augustsson T. R., and Naturajan M. (1982) The prebiological paleoatmosphere: stability and composition. *Orig. Life, 12,* 245–259.

Lugmair G. W. and Galer S. J. G. (1992) Age and isotopic relationships among the angrites Lewis Cliff 86010 and Angra dos Reis. *Geochim. Cosmochim. Acta, 56,* 1673–1694.

Lugmair G. W. and Shukolykov A. (1992) Early solar system timescales according to $^{53}$Mn-$^{53}$Cr systematics. *Geochim. Cosmochim. Acta, 62,* 2863–2886.

Maas R., Kinny P. D., Williams I. S., Froude D. O., and Compston W. (1992) The Earth's oldest known crust: A geochronological and geochemical study of 3900–4200 Myr old detrital zircons from Mt. Narryer and Jack Hills, Western Australia. *Geochim. Cosmochim. Acta, 56,* 1281–1300.

Maher K. A. and Stevenson D. J. (1988) Impact frustration of the origin of life. *Nature, 331,* 612–614.

Manhes G., Gopel C., and Allegre C. J. (1987) High resolution chronology of the early solar system bases on lead isotopes. *Meteoritics, 22,* 453–454.

Marshall W. (1994) Hydrothermal synthesis of amino acids. *Geochim. Cosmochim. Acta, 58,* 2099–2106.

Matsui T. and Abe Y. (1986) Evolution of an impact-induced atmosphere and magma ocean on the accreting Earth. *Nature, 319,* 303–305.

McCollom T. M. (2000) Geochemical constraints on primary productivity in submarine hydrothermal vent plumes. *Deep-Sea Res. I, 47,* 85–101.

McCollom T. M. and Shock E. L. (1997) Geochemical constraints on chemolithoautotrophic metabolism by microorganisms in seafloor hydrothermal systems. *Geochim. Cosmochim. Acta, 61,* 4375–4391.

McCollom T. M. and Simoneit B. R. T. (1999) Abiotic formation of hydrocarbons and oxygenated compounds during the thermal decomposition of iron oxalate. *Orig. Life Evol. Biosph., 29,* 167–186.

McCollom T. M., Simoneit B. R. T., and Shock E. L. (1999a) Hydrous pyrolysis of polycyclic aromatic hydrocarbons and implications for the origin of PAH in hydrothermal petroleum. *Energy & Fuels, 13,* 401–410.

McCollom T. M., Ritter G., and Simoneit B. R. T. (1999b) Lipid synthesis under hydrothermal conditions by Fischer-Tropsch-type reactions. *Orig. Life Evol. Biosph., 29,* 153–166.

McCulloch M. T. and Bennett V. C. (1988) Early differentiation of the Earth: An isotopic perspective. In *The Earth's Mantle: Composition, Structure, and Evolution* (I. Jackson, ed.), p. 127. Cambridge Univ., New York.

Melosh H. J. (1990) Giant impacts and the thermal state of the early Earth. In *Origin of the Earth* (H. E. Newsom and J. H. Jones, eds.), pp. 69–83. Oxford Univ., New York.

Messenger S., Amari S., Gao, X., Walker R. M., Clemett S. J., Chiller X. D. F., Zare R. N., and Lewis R. S. (1998) Indigenous polycyclic aromatic hydrocarbons in circumstellar graphite grains from primitive meteorites. *Astrophys. J., 502,* 284–295.

Miller J. F., Shah N. N., Nelson C. M., Ludlow J. M., and Clark D. S. (1988) Pressure and temperature effects on growth and methane production of the extreme thermophile *Methanococcus jannaschii. Appl. Environ. Microbiol., 54,* 3039–3042.

Miller S. L. (1953) A production of amino acids under possible primitive Earth conditions. *Science, 117,* 528–529.

Miller S. L. (1957) The formation of organic compounds on the primitive Earth. *Ann. New York Acad. Sci., 69,* 260–275.

Miller S. L. and Bada J. L. (1988) Submarine hot springs and the origin of life. *Nature, 334,* 609–611.

Miller S. L. and Urey H. C. (1959) Organic compound synthesis on the primitive Earth. *Science, 130,* 245–251.

Miyamoto M. and Mikouchi T. (1996) Platinum catalytic effect on oxygen fugacity of $CO_2$-$H_2$ gas mixtures measured with $ZrO_2$ oxygen sensor at 105 Pa from 1300° to 700°C. *Geochim. Cosmochim. Acta, 60,* 2917–2920.

Mojzsis S. J., Arrhenius G., McKeegan K. D., Harrison T. M., Nutman A. P., and Friend C. R. L. (1996) Evidence for life on Earth before 3,800 million years ago. *Nature, 384,* 55–59.

Mukhin L. M., Gerasimov M. V., and Safonova E. N. (1989) Origin of precursors of organic molecules during evaporation of meteorites and mafic terrestrial rocks. *Nature, 340,* 46–48.

Nelson D. R. (1997) Compilation of SHRIMP U-Pb zircon geochronology data, (1996) *Geol. Surv. Western Australia Rec., 1997/2,* 189.

Nishihama S., Yamada H., and Nakazawa H. (1997) Polymerization of tetramethylcyclotetrasiloxane monomer by ion-exchanged montmorillonite catalysts. *Clay Miner., 32,* 645–651.

Oberbeck V. R. and Fogelman G. (1989) Impact constraints on

the environment for chemical evolution and the continuity of life. *Orig. Life Evol. Biosph., 20,* 181–195.

O'Neill H. St. C. (1991) The origin of the Moon and the early history of the Earth — A chemical model. Part 2: The Earth. *Geochim. Cosmochim Acta, 55,* 1159–1172.

O'Neill H. St. C. and Palme H. (1998) Composition of the silicate Earth: Implications for accretion and core formation. In *The Earth's Mantle: Composition, Structure, and Evolution* (I. Jackson, ed.), pp. 3–126. Cambridge Univ., Cambridge.

Oparin A. I. (1924) *Proiskhozhdenie zhizny.* Moskovskii Rabochii, Moscow.

Oparin A. I. (1936) *Origin of Life.* Translated from Russian by Margolis S. (1953), Dover, New York.

Owen T., Cess R. D., and Ramanathan V. (1979) Early Earth: An enhanced carbon dioxide greenhouse to compensate for reduced solar luminosity. *Nature, 277,* 640–642.

Pace N. R. (1991) Origin of life: Facing up to the physical setting. *Cell, 65,* 531–533.

Pace N. R. (1997) A molecular view of microbial diversity and the biosphere. *Science, 276,* 734–740.

Righter K. and Drake M. J. (1997) Metal-silicate equilibrium in a homogenously accreting earth: new results for Re. *Earth Planet. Sci. Lett., 146,* 541–553.

Righter K., Drake M. J., and Yaxley G. (1997) Prediction of siderophile element metal-silicate partition coefficients to 20 GPa and 2800°C: The effects of pressure, temperature, oxygen fugacity, and silicate and metallic melt compositions. *Phys. Earth Planet. Inter., 100,* 115–134.

Ringwood A. E. (1979) *Origin of Earth and Moon.* Springer-Verlag, New York. 295 pp.

Rozanova E. P. and Pivovarova T. A. (1988) Reclassification of *Desulfovibrio thermophilus* (Rozanova, Khudyakova, 1974). *Microbiologiya, 57,* 102–106.

Russell M. J. and Hall A. J. (1997) The emergence of life from iron monosulphide bubbles at a hydrothermal redox front. *J. Geol. Soc., 154,* 377–402.

Sagan C. and Chyba C. (1997) The early faint sun paradox: organic shielding of ultraviolet-labile greenhouse gases. *Science, 276,* 1217–1221.

Sagan C. and Mullen G. (1972) Earth and Mars: Evolution of atmospheres and surface temperatures. *Science, 177,* 52–56.

Salvi S. and Williams-Jones A. E. (1997) Fisher-Tropsch synthesis of hydrocarbons during sub-solidus alteration of the Strange Lake peralkaline granite, Quebec/Labrador, Canada. *Geochim. Cosmochim. Acta, 61,* 83–99.

Schidlowski M., Appel P. W. U., Eichmann R., and Junge C. E. (1979) Carbon isotope geochemistry of the $3.7 \times 10^9$-yr-old Isua sediments, West Greenland: Implications of the Archaean carbon and oxygen cycles. *Geochim. Cosmochim. Acta, 43,* 189–199.

Schoonen M. A. A., Xu Y., and Bebie J. (1999) Energetics and kinetics of the prebiotic synthesis of simple organic and amino acids with the FeS-H2S/FES2 redox couple as reductant. *Orig. Life Evol. Biosph., 29,* 5–32.

Schulte M. D. and Shock E. L. (1993) Aldehydes in hydrothermal solutions: Standard partial molal thermodynamic properties and relative stabilities at high temperatures and pressures. *Geochim. Cosmochim. Acta, 57,* 3835–3846.

Schopf J. W. (1992) Paleobiology of the Archean. In *The Proterozoic Biosphere: A Multidisciplinary Study* (J. W. Schopf and C. Klein, eds.), pp. 25–39. Cambridge Univ., New York.

Schopf J. W. (1993) Microfossils of the early Archean apex chert: New evidence of the antiquity of Life. *Science, 260,* 640–646.

Seewald J. S. (1994) Evidence for metastable equilibrium between hydrocarbons under hydrothermal conditions. *Nature, 370,* 285–287.

Shock E. L. (1988) Organic acid metastability in sedimentary basins. *Geology, 16,* 886–890.

Shock E. L. (1989) Corrections to "Organic acid metastability in sedimentary basins." *Geology, 17,* 572–573.

Shock E. L. (1990) Geochemical constraints on the origin of organic compounds in hydrothermal systems. *Orig. Life Evol. Biosph., 20,* 331–367.

Shock E. L. (1992a) Chemical environments in submarine hydrothermal systems. In *Marine Hydrothermal Systems and the Origin of Life* (N. Holm, ed.), special issue, *Orig. Life Evol. Biosph., 22,* 67–107.

Shock E. L. (1992b) Hydrothermal organic synthesis experiments. In *Marine Hydrothermal Systems and the Origin of Life* (N. Holm, ed.), special issue, *Orig. Life Evol. Biosph., 22,* 135–146.

Shock E. L. (1994) Application of thermodynamic calculations to geochemical processes involving organic acids. In *Organic Acids in Geological Processes* (E. O. Pittman and M. D. Lewan, eds.), pp. 270–318. Springer-Verlag, Berlin.

Shock E. L. (1995) Organic acids in hydrothermal solutions: Standard molal thermodynamic properties of carboxylic acids, and estimates of dissociation constants at high temperatures and pressures. *Am. J. Sci., 295,* 496–580.

Shock E. L. (1996) Hydrothermal systems as environments for the emergence of life. In *Evolution of Hydrothermal Ecosystems on Earth (and Mars?)* (G. R. Bock and J. A. Goode, eds.), pp. 40–60. Wiley, Chichester, New York.

Shock E. L. (1997) High temperature life without photosynthesis as a model for Mars. *J. Geophys. Res., 102,* 23687–23694.

Shock E. L. and Helgeson H. C. (1990) Calculation of the thermodynamic and transport properties of aqueous species at high pressures and temperatures: Standard partial molal properties of organic species. *Geochim. Cosmochim. Acta, 54,* 915–945.

Shock E. L. and Koretsky C. M. (1993) Metal-organic complexes in geochemical processes: Calculation of standard partial molal thermodynamic properties of aqueous acetate complexes at high pressures and temperatures. *Geochim. Cosmochim. Acta, 57,* 4899–4922.

Shock E. L. and Koretsky C. M. (1995) Metal-organic complexes in geochemical processes: Estimation of standard partial molal thermodynamic properties of aqueous complexes between metal cations and monovalent organic acid ligands at high pressures and temperatures. *Geochim. Cosmochim. Acta, 59,* 1497–1532.

Shock E. L. and McKinnon W. B. (1993) Hydrothermal processing of cometary volatiles — Applications to Triton. *Icarus, 106,* 464–477.

Shock E. L. and Schulte M. D. (1998) Organic synthesis during fluid mixing in hydrothermal systems. *J. Geophys. Res., 103,* 28513–28527.

Shock E. L., Helgeson H. C., and Sverjensky D. A. (1989) Calculation of the thermodynamic and transport properties of aqueous species at high pressures and temperatures: Standard partial molal properties of inorganic neutral species. *Geochim. Cosmochim. Acta, 53,* 2157–2183.

Shock E. L., Oelkers E. H., Johnson J. W., Sverjensky D. A., and Helgeson H. C. (1992) Calculation of the thermodynamic properties of aqueous species at high pressures and temperatures: Effective electrostatic radii, dissociation constants, and standard partial molal properties to 1000°C and 5 kb. *J. Chem. Soc., Faraday Trans., 88,* 803–826.

Shock E. L., McCollom T. and Schulte M. D. (1995) Geochemical constraints on chemolithoautotrophic reactions in hydrothermal systems. *Orig. Life Evol. Biosph., 25,* 141–159.

Shock E. L., Sassani D. C., Willis M., and Sverjensky D. A. (1997) Inorganic species in geologic fluids: Correlations among standard molal thermodynamic properties of aqueous ions and hydroxide complexes. *Geochim. Cosmochim. Acta, 61,* 907–950.

Shock E. L., McCollom T., and Schulte M. D. (1998) The emergence of metabolism from within hydrothermal systems. In *Thermophiles: The Keys to Molecular Evolution and the Origin of Life?* (J. Wiegel and M. W. W. Adams, eds.), pp. 59–76. Taylor and Francis, London.

Sleep N. H. and Zahnle K. (1998) Refugia from asteroid impacts on early Mars and the early Earth. *J. Geophys. Res., 103,* 28529–28544.

Sleep N. H., Zahnle K. J., Kasting J. F., and Morowitz H. J. (1989) Annihilation of ecosystems by large asteroid impacts on the early Earth. *Nature, 342,* 139–142.

Sowerby S. J. and Heckl W. M. (1998) The role of self-assembled monolayers of the purine and pyrimidine bases in the emergence of life. *Orig. Life Evol. Biosph., 28,* 283–310.

Stern R. A. and Bleeker W. (1998) Age of the world's oldest rocks refined using Canada's SHRIMP: The Acasta gneiss complex, Northwest Territories, Canada. *Geosci. Canada, 25,* 27–31.

Stetter K. O. (1988) *Archaeoglobus fulgidus* gen. nov., sp. nov.: a New taxon of extremely thermophilic archaebacteria. *Syst. Appl. Microbiol., 10,* 172–173.

Stetter K. O. (1992) Life at the upper temperature border. In *Frontiers of Life* (Trân Thahn Vân et al., eds.), pp. 195–219. Editions Frontières.

Stetter K. O., Thomm M., Winter J., Wildgruber G., Huber H., Zillig, W., Janécovic D., König H., Palm P., and Wunderl S. (1981) *Methanothermus fervidus*, sp. nov., a novel extremely thermophilic methanogen isolated from an Icelandic hot spring. *Zentralbl. Bakteriol. Mikrobiol. Hyg. I. Abt. Orig., C2,* 166–178.

Stetter K. O., Huber R., Blöchl E., Kurr M., Eden R. D., Fielder M., Cash H., and Vance I. (1993) Hyperthermophilic archaea are thriving in deep North Sea and Alaskan oil reservoirs. *Nature, 365,* 743–745.

Stevenson D. J. (1981) Models of the Earth's core. *Science, 214,* 611–619.

Symonds R. B., Rose W. I., Bluth G. J. S., and Gerlach T. M. (1994) Volcanic gas studies: methods, results, and applications. In *Volatiles in Magma* (M. R. Carrol and J. R. Holloway, eds.), pp. 1–66. Mineralogical Society of America, Washington, DC.

Sverjensky D. A., Shock E. L., and Helgeson H. C. (1997) Prediction of the thermodynamic properties of aqueous metal complexes to 1000°C and 5 kb. *Geochim. Cosmochim. Acta, 61,* 1359–1412.

Tonks W. B. and Melosh H. J. (1990) The physics of crystal settling and suspension in a turbulent magma ocean. In *Origin of the Earth* (H. E. Newsom and J. H. Jones, eds.), pp. 151–174. Oxford Univ., New York.

Urey H. C. (1952a) *The Planets: Their Origin and Development.* Yale Univ., New Haven, Connecticut. 245 pp.

Urey H. C. (1952b) On the early chemical history of the Earth and origin of life. *Proc. Natl. Acad. Sci. USA, 38,* 351–363.

Von Damm K. L. (1990) Seafloor hydrothermal activity: Black smoker chemistry and chimneys. *Annu. Rev. Earth Planet. Sci., 18,* 173–204.

Von Damm K. L., Edmond J. M., Grant B., Measures C. I., Walden B., and Weiss R. F. (1985) Chemistry of submarine hydrothermal systems at 21°N, East Pacific Rise. *Geochim. Cosmochim. Acta, 49,* 2197–2220.

Wächtershäuser G. (1988) Before enzymes and templates: Theory of surface metabolism. *Microbiol. Rev., 52,* 452–484.

Wächtershäuser G. (1990) Evolution of the first metabolic cycles. *Proc. Natl. Acad. Sci., 87,* 200–204.

Wächtershäuser G. (1992) Groundworks for an evolutionary biochemistry: The iron-sulphur world. *Prog. Biophys. Molec. Biol., 58,* 85–201.

Wächtershäuser G. (1998) Origin of life in an iron-sulfur world. In *The Molecular Origins of Life: Assembling Pieces of the Puzzle* (A. Black, ed.), pp. 206–218. Cambridge Univ., Cambridge.

Walker J. C. G. (1985) Carbon dioxide on the early Earth. *Orig. Life Evol. Biosph., 16,* 117–127.

Walker J. C. G. (1990) Origin of an Inhabited Planet. In *Origin of the Earth* (H. E. Newsom and J. H. Jones, eds.), pp. 371–375. Oxford Univ., New York.

Walsh M. M. and Lowe D. R. (1985) Filamentous microfossils from the 3,500-Myr-old Onverwacht Group, Barberton Mountain Land, South Africa. *Nature, 314,* 530–532.

Welhan J. A. and Craig H. (1983) Methane, hydrogen, and helium in hydrothermal fluids at 21°N on the East Pacific Rise. In *Hydrothermal Processes at Seafloor Spreading Centers* (P. A. Rona et al., eds.), pp. 391–409. Plenum, New York.

Woese C. (1998) The universal ancestor. *Proc. Natl. Acad. Sci. USA, 95,* 6854–6859.

Woese C. R., Kandler O., and Wheelis M. L. (1990) Towards a natural system of organisms: Proposal for the domains Archaea, Bacteria, and Eucarya. *Proc. Natl. Acad. Sci., 87,* 4576–4579.

Wolery T. J. (1992) *EQ3NR, A Computer Program for Geochemical Aqueous Speciation-Solubility Calculations: Theoretical Manual, User's Guide, and Related Documentation.* Lawrence Livermore National Laboratory, Livermore, California.

Wolery T. J. and Daveler S. A. (1992) *EQ6, A Computer Program for Reaction Path Modeling of Aqueous Geochemical Systems: Theoretical Manual, User's Guide, and Related Documents.* Lawrence Livermore National Laboratory, Livermore, California.

Zahnle K. J. and Walker J. C. G. (1982) The evolution of solar ultraviolet luminosity. *Rev. Geophys. Space Phys., 20,* 280–292.

Zeikus J. G., Dawson M. A., Thompson T. E., Ingvorsen K., and Hatchikian E. C. (1983) Microbial ecology of volcanic sulphidogenesis: Isolation and characterization of *Thermodesulfobacterium commune* gen. nov. and sp. nov. *J. Gen. Microbiol., 129,* 1159–1169.

Zhang Y. (1998) The young age of Earth. *Geochim. Cosmochim. Acta, 62,* 3185–3189.

Zolotov M. and Shock E. (1999) Abiotic synthesis of polycyclic aromatic hydrocarbons on Mars. *J. Geophys. Res., 104,* 14033–14049.

Zolotov M. Yu. and Shock E. L. (2000) A thermodynamic assessment of the potential synthesis of condensed hydrocarbons during cooling and dilution of volcanic gases. *J. Geophys. Res., 105,* 539–559.

*Color Section*

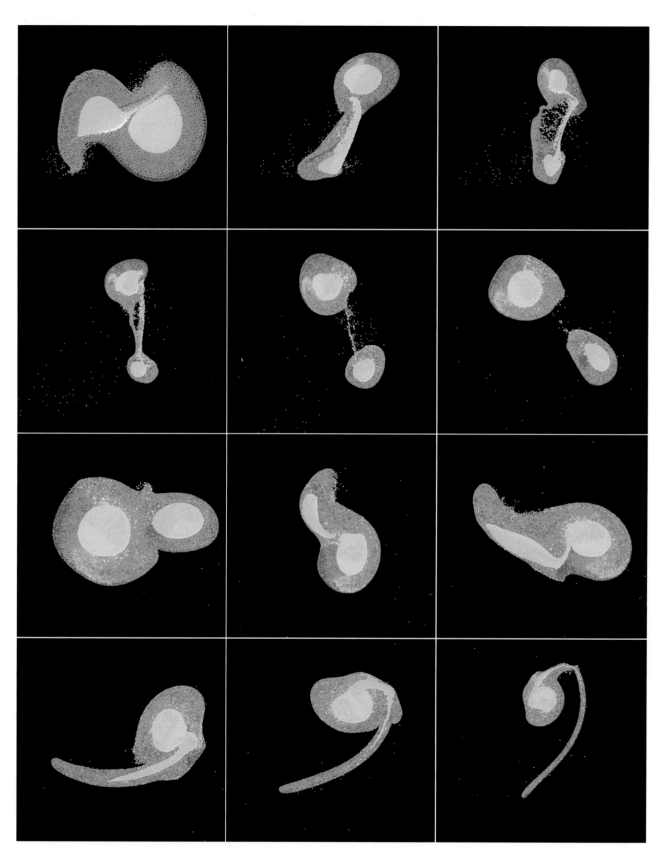

**Plate 1.** A series of 12 panels showing the dynamics of the giant impact in case AS04, for the interval between the first collision and the second collision. See the text for the explanation of the color scheme. The interval between panels is about 20 minutes.

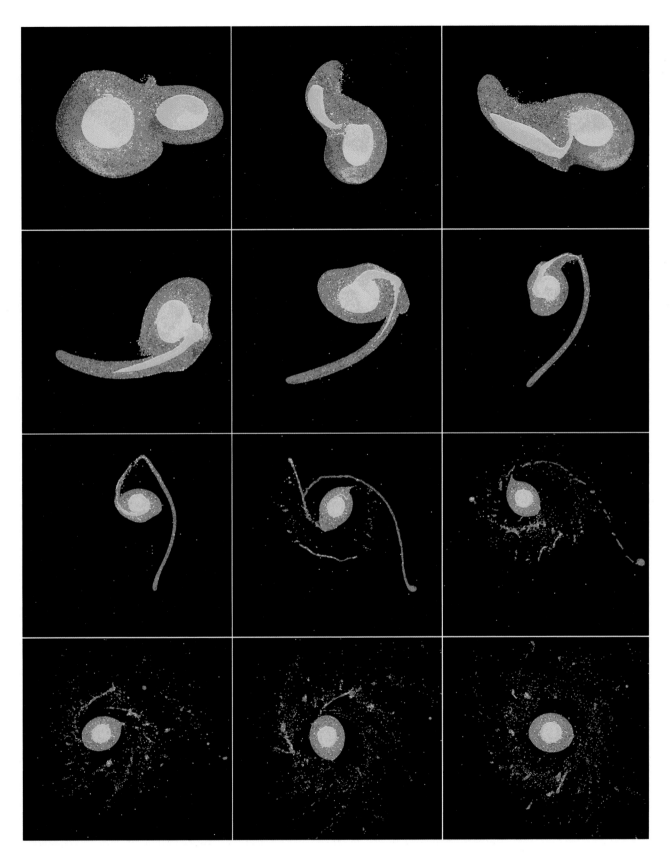

**Plate 2.** The second collision in the giant impact of case AS04. The impactor is destroyed. Its Fe drains into the proto-Earth. Its mantle forms a trailing spiral arm around the proto-Earth, which then fragments into small pieces, out of which many clusters of particles are reassembled. The intervals between the first seven panels are 40 min, and thereafter 100 min. Many curved arcs are shown that are manifestations of tidal stripping.

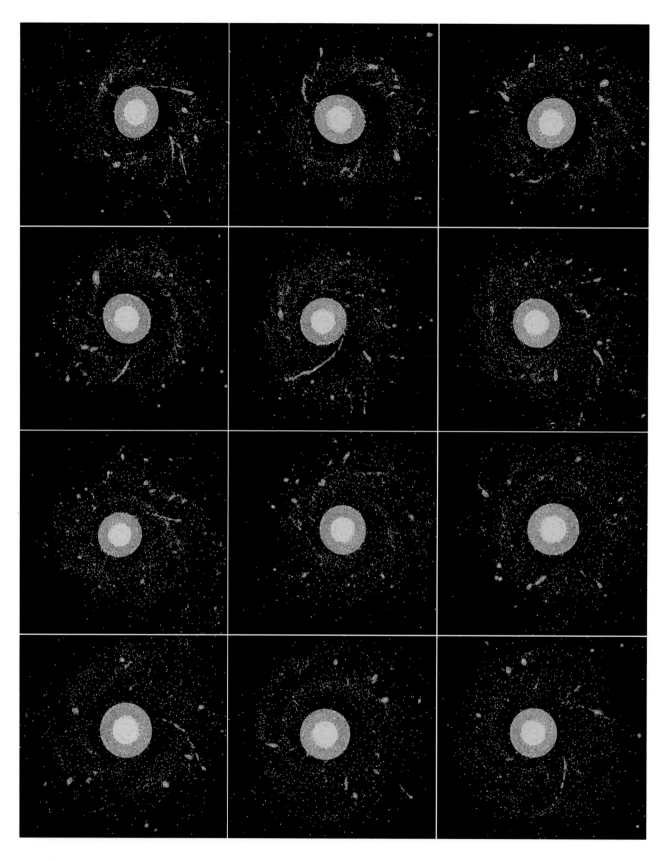

**Plate 3.** The continued evolution of the orbiting debris in the AS04 run, with intervals between the panels of 100 min.

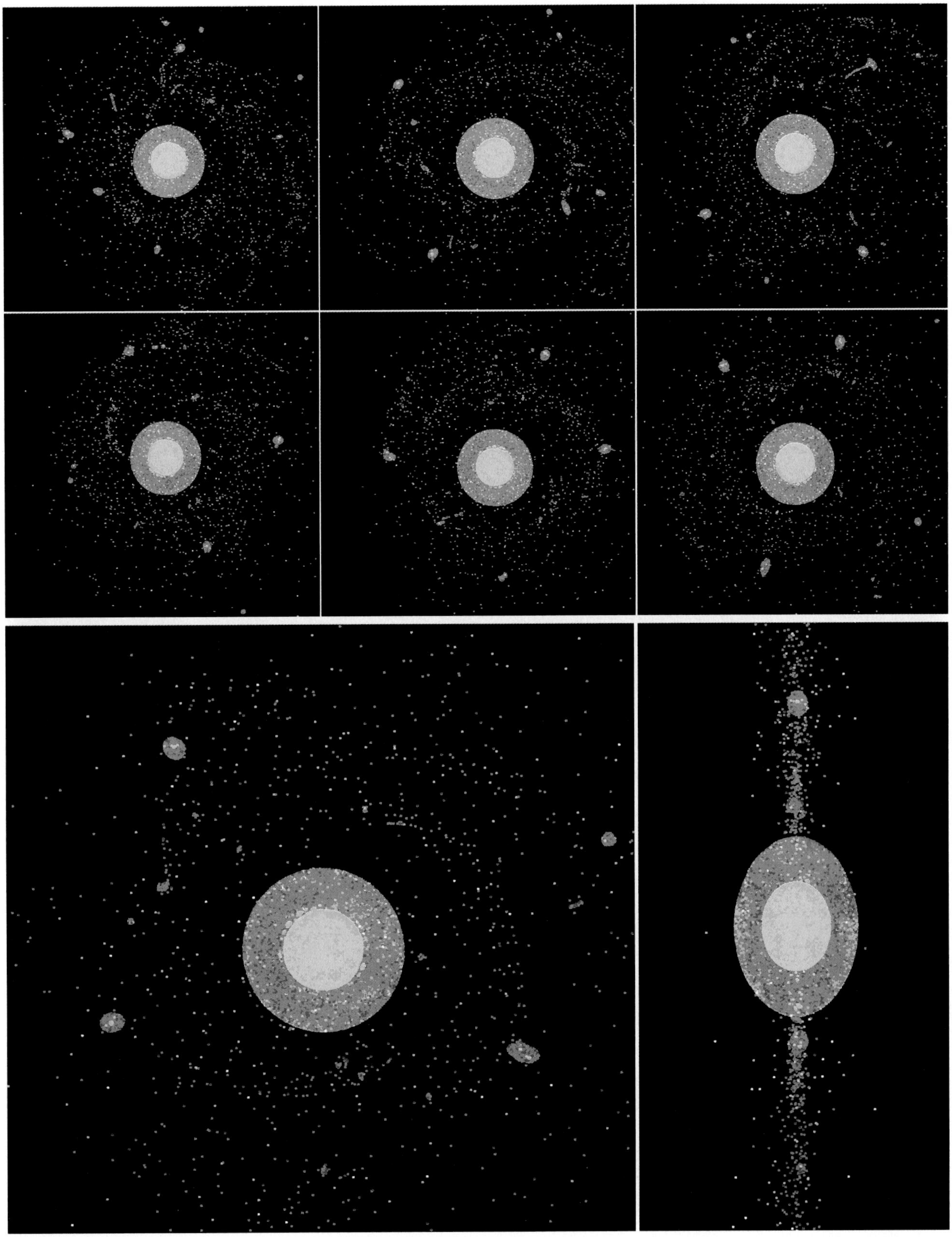

**Plate 4.** Further evolution of the AS04 run, with time intervals of 100 min. The bottom half of the sheet shows an enlargement of the image for archival dump 1500, together with an image of it rotated 90° so that it is viewed in the equatorial plane.

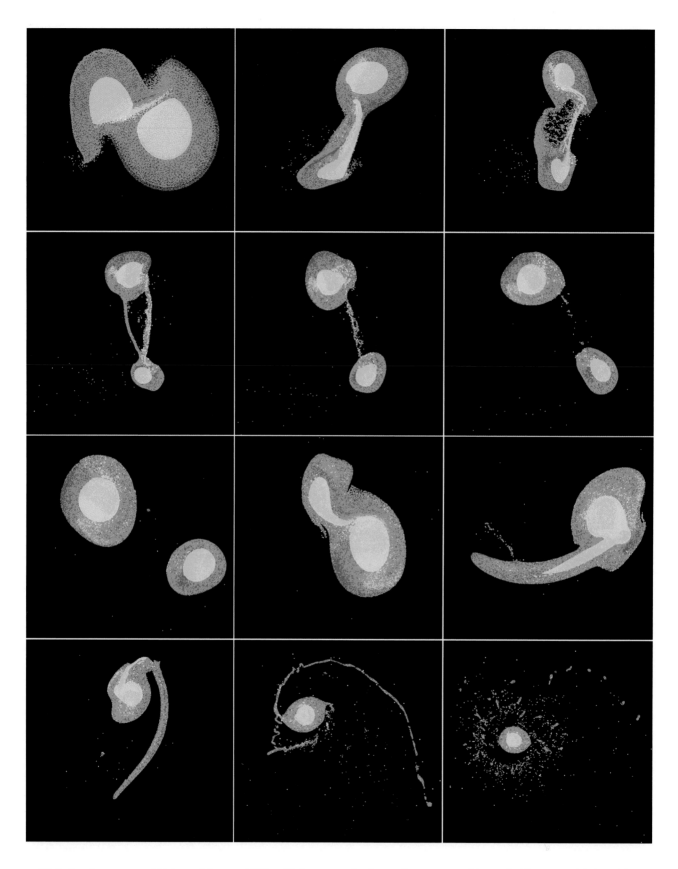

**Plate 5.** A compressed history of the run AS05, with images at the times of every second image in Plates 1 and 2 for AS04.

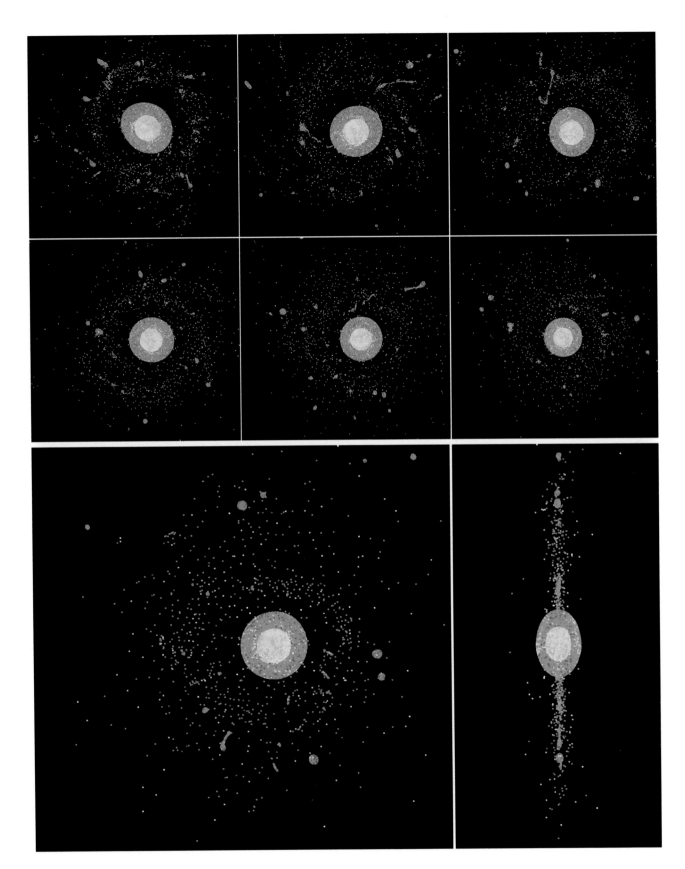

**Plate 6.** A continuation of the compressed history of run AS05, with images at the times of every third image from Plates 3 and 4 for AS04. The bottom half is an enlargement of archival dump 1500 for AS05.

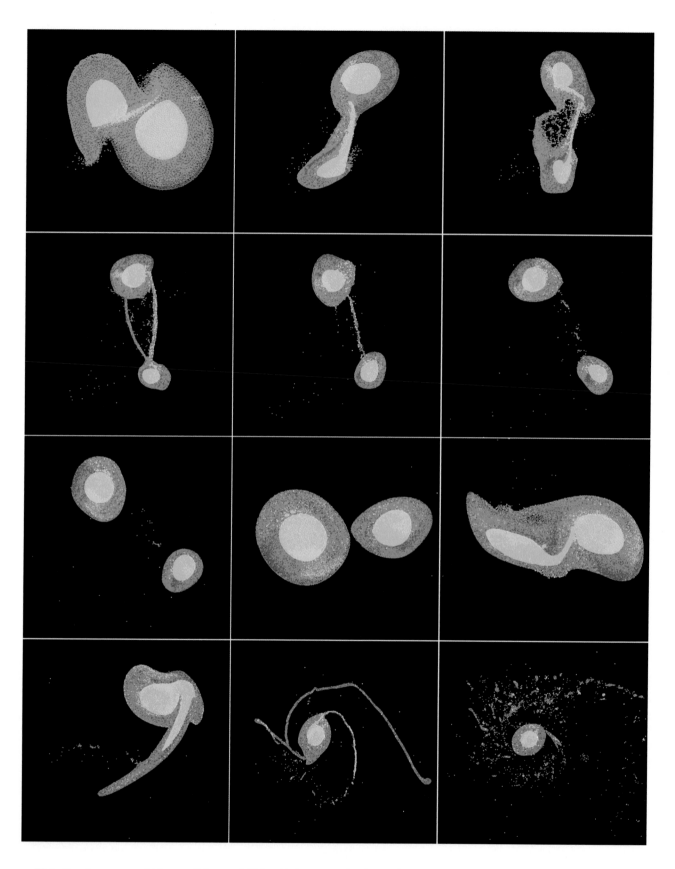

**Plate 7.** A compressed history of the run AS06, with images at the times of every second image in Plates 1 and 2 for AS04.

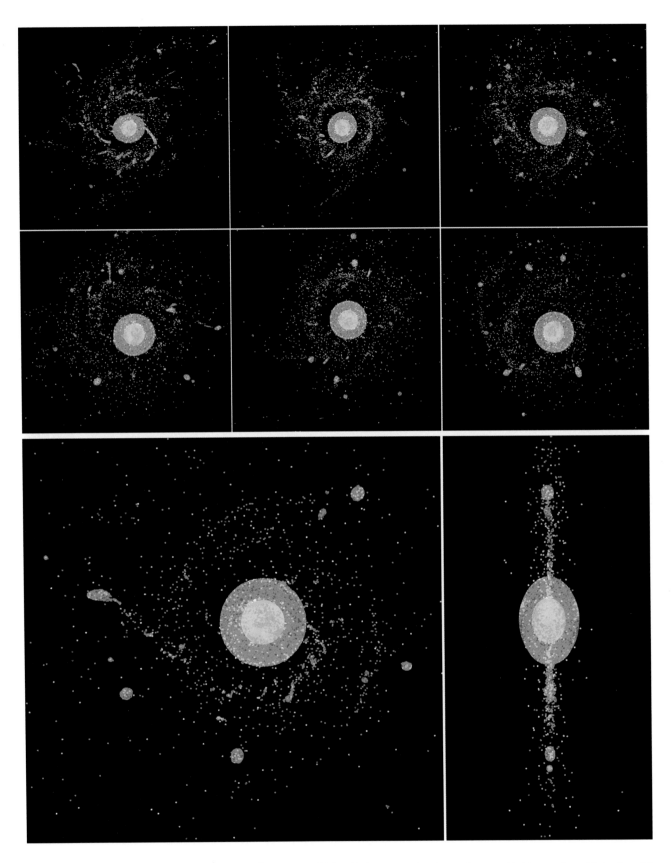

**Plate 8.** A continuation of the compressed history of run AS06, with images at the times of every third image from Plates 3 and 4 for AS04. The bottom half is an enlargement of archival dump 1500 for AS06.

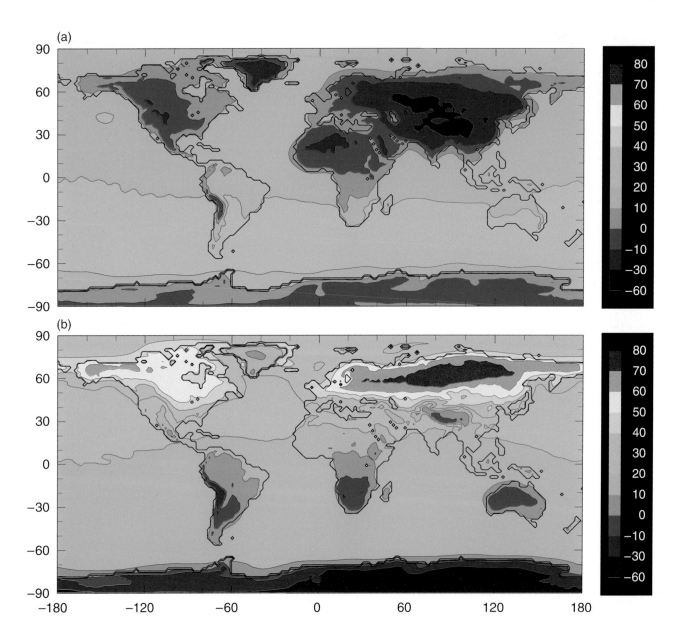

**Plate 9.** Model surface-air temperatures (°C) in (**a**) January and (**b**) July for modern conditions except with the obliquity set to 54°. Temperatures are averaged over the last three years of a 10-yr GCM simulation.

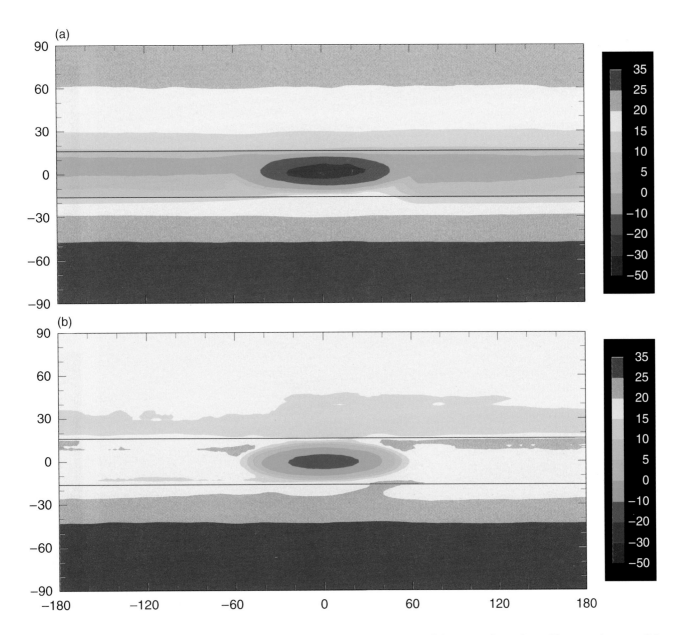

**Plate 10.**   Surface-air temperatures (°C) in (**a**) January and (**b**) April over an equatorial supercontinent situated between the two solid horizontal lines. The coldest temperatures are shown to be over a prescribed mountain located near 0° longitude.

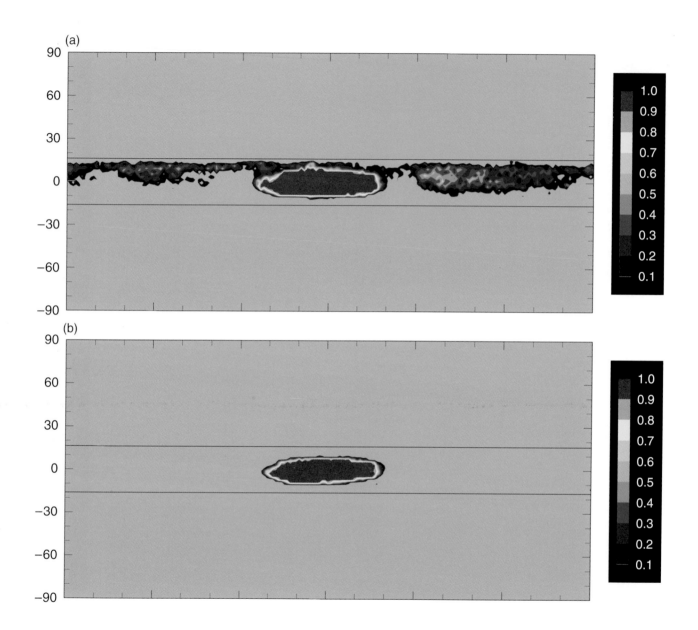

**Plate 11.**  Fraction of surface covered by snow/ice for **(a)** January and **(b)** April.

# Glossary

*abyssal peridotite* — peridotite samples recovered from the flanks of near-ridge transform faults.

*accretion* — growth of solid bodies through collisional aggregation of smaller bodies.

*achondrite* — meteorite that lacks chondrules (stony and iron meteorites).

*ANEOS* — a semi-analytic equation of state that links pressure, temperature, and density for materials of interest in impact computations.

*angrite* — achondritic meteorite consisting of dipopside-rich pyroxene, olivine, and plagioclase, with a roughly basaltic composition.

*angular momentum* — a property of rotating systems that depends upon mass and its distribution, angular velocity, and radius. The angular momentum of the Earth-Moon system is contained in the Earth's rotation and the Moon's orbital motion.

*anorthite* — the calcium-rich end member of the plagioclase feldspar mineral series; $CaAl_2Si_2O_8$.

*anorthosite* — an igneous rock made of >90% plagioclase feldspar.

*asthenosphere* — a region in the interior of the Earth that is more ductile or plastic than the overlying lithosphere.

*AU* — the mean distance between the Earth and Sun ($1.496 \times 10^{13}$ cm).

*basalt* — a fine-grained, dark-colored igneous rock comprised primarily of pyroxene, olivine, feldspar, and glass.

*breccia* — a rock composed of coarse, angular mineral or rock fragments embedded with a fine-grained matrix.

*bulk silicate Earth* — *See* primitive upper mantle.

*carbonaceous chondrites* — the most primitive stony-meteorites, in which the abundances of the nonvolatile elements are thought to approximate most closely those of the primordial solar nebula.

*CAIs (calcium aluminum-rich inclusions)* — refractory inclusions that consist of the minerals perovskite, melilite, spinel, and hibonite, and represent the oldest solid material found in the solar nebula.

*chalcophile element* — an element with a preference for sulfide minerals.

*chondrite* — the most abundant class of stony meteorites that contain chondrules.

*chondrule* — small, rounded body in meteorites (generally less than 1 mm in diameter) commonly composed of olivine and/or orthopyroxene.

*clinopyroxene* — mineral of the pyroxene group such as diopside ($CaMgSi_2O_6$) or pigeonite (($Ca,Mg,Fe)_2Si_2O_6$).

*compatible element* — a minor or trace element that partitions readily into crystalline rather than melt phases.

*core* — dense, metal- or sulfide-rich central region of a planet.

*crust* — outer, highly differentiated region of a planet.

*cumulate* — a plutonic igneous rock composed chiefly of crystals accumulated by sinking or floating in a magma.

*differentiation* — the process by which planetary bodies develop concentric zones that differ in chemical and mineralogical composition.

*diogenite* — achondrite meteorite comprised of orthopyroxene and minor amounts of olivine and chromite.

*dunite* — a peridotite comprised of >90% olivine, with accessory pyroxene, chromite, or plagioclase.

*eccentricity* — a measure of the amount by which an orbit deviates from circularity. It is the ratio of the distance from center to focus to the semimajor axis.

*eclogite* — a dense rock comprised of garnet and pyroxene, similar in composition to basalt.

*ecliptic* — the mean plane of the Earth's orbit around the Sun.

*Ekman layer* — upper boundary layer within which the amplitude of a wave changes exponentially.

*equinox* — the point where a planet's equator crosses its orbital plane.

*escape velocity* — the speed an object must attain to escape from the gravitational field of another object.

*eucrite* — a basaltic meteorite composed essentially of feldspar and clinopyroxene.

*extinct isotope* — a radioactive isotope that existed when the solar system formed, but with too short of a half-life to allow detectable amounts to remain now.

*feldspar* — a group of aluminous silicate minerals, with K-, Na-, and Ca-bearing end members.

*fractionation* — the separation of chemical elements from an initially homogeneous state into different phases or systems.

*fractional crystallization* — formation and separation of mineral phases of varying composition during crystallization of a silicate melt or magma, resulting in continuous change of composition of the magma.

*fugacity* — a thermodynamic function used instead of pressure in describing the behavior of nonideal gases.

*gabbro* — a coarse-grained, dark igneous rock made up chiefly of plagioclase (usually labradorite) and pyroxene. A coarse-grained equivalent to a basalt.

*garnet* — a group of minerals with the general formula $X_3Y_2(SiO_4)_3$, in which X = Ca, Mg, Mn, or $Fe^{2+}$, and Y = Al, $Fe^{3+}$, and $Cr^{3+}$.

*giant impact theory* — the theory that the Moon formed from material ejected into Earth orbit during an impact between a large impactor and the proto-Earth.

*gravitational instability* — condition in which slight rearrangements or concentrations of a relatively uniform distribution of mass can, by their gravitational effect, initiate substantial further contraction of mass into even more localized concentrations.

*heterogeneous accretion* — Earth formation theory in which the Earth accreted in several stages involving early reduced material and late oxidized material.

*Hill sphere* — the approximately spherical region within

which a planet, rather than the Sun, dominates the motion of particles.

*Hill's approximation* — an approximation of the equations of motion for nearly circular, co-planar bodies in a central gravitational field, which allows these equations to be linearized.

*howardite* — a type of basaltic achondrite that is brecciated.

*ilmenite* — an oxide mineral with the composition $FeTiO_3$.

*impact parameter* — term relating to the geometry of a collision between two spheres; it is the offset between the spheres' centers projected perpendicular to the line of approach.

*inclination* — the angle between the plane of a planet's orbit and the ecliptic (Earth's orbital plane), or a satellite's orbit and its planet's equator.

*incompatible element* — a minor or trace element that partitions readily into the melt phase rather than crystalline phases.

*isochron* — a line on a diagram passing through plots of samples with the same age but differing isotope ratios.

*isotopes* — atoms of a specific element that differ in number of neutrons in the nucleus; this results in different atomic weights, and slightly differing chemical properties.

*iron meteorite* — a class of meteorite composed mainly of iron or iron-nickel metal.

*Jacobi constant, or integral* — the only known integral of motion in the restricted three-body problem, where a massless particle moves under the influence of two point masses in circular orbit around one another. This constant is a pseudo-energy in a rotating coordinate system in which centrifugal force is represented by a potential field.

*Keplerian orbit* — motion of a freely orbiting body in the inverse-square-law force field of another body.

*Keplerian shear* — shearing motion of an ensemble of particles, each on a nearly circular, Keplerian orbit. Orbital velocity decreases with increasing orbital radius, yielding the shear.

*Knudsen number* — the ratio of the mean free path length of the molecules in a fluid to a characteristic length of the structure in the fluid stream.

*komatiite* — an MgO-rich rock thought to have formed by large percentages (20–30%) of melting of the Earth's mantle. Found mainly in the Archean period of Earth history.

*KREEP* — An acronym for a lunar crustal component rich in potassium (K), the rare earth elements (REE), phosphorus (P), and other incompatible elements.

*Lagrangian points* — the five equilibrium points in the restricted three-body problem. Two of the Lagrange points ($L_4$ and $L_5$) are located at the vertexes of equilateral triangles formed by the two primaries (e.g., Sun and Saturn, or Saturn and satellite) and are stable; the other three are unstable and lie on the line connecting the two primaries.

*Laplace plane* — the orientation of the orbital plane of a satellite that is perturbed by the planetary oblateness, the Sun, and nearby, massive satellites and is not fixed in inertial space. The latter perturbations cause the pole of the satellite orbit to precess about the pole of the Laplacian plane.

*late veneer* — in heterogeneous accretion theory, the late addition of material to a planet after a metallic core has formed.

*liquidus* — the line or surface in a phase diagram above which the system is completely liquid.

*lithophile element* — an element tending to concentrate in oxygen-containing compounds (silicates) as opposed to metal or sulfide.

*lithosphere* — an outer shell of a planet that has high rock strength and undergoes brittle deformation.

*Love number* — the parameter that describes the enhanced gravitational potential of a body due to the redistribution of mass in the body caused by the external perturbation; the value depends on internal structure and the nature of the perturbation.

*lunar capture theory* — the hypothesis that the Moon was gravitationally captured by the Earth.

*lunar cataclysm* — the hypothesis that the Moon underwent a heavy bombardment period at approximately 3.9 Ga.

*lunar fission theory* — the hypothesis that the Moon was formed by separation or fission from the Earth.

*lunar coaccretion theory* — the hypothesis that the Moon and Earth accreted together.

*magma ocean* — a globally extensive layer of magma on a planet or moon that consists of >50% melt.

*magnesian perovskite* — a mineral with the formula $MgSiO_3$ that is stable at high pressures (>250 kbar, as exist deep within the Earth's mantle).

*magnesiowüstite* — an oxide with the formula $(Mg,Fe)O$; likely to be stable in Earth's lower mantle.

*majorite* — a component of garnet that is stable at high pressures and temperature in which no aluminum is present, only silica. It has the general formula of $(Mg,Fe)_4Si_4O_{12}$.

*mantle* — the zone of a planet beneath its crust and above its core.

*mare basalt* — basalts that form the lunar maria, which are the dark-colored areas on the Moon.

*mascon* — regions on the Moon of excess mass concentrations per unit area identified by positive gravity anomalies and associated with mare-filled multiring basins.

*mesosiderite* — stony-iron meteorites that consist of a brecciated mixture of iron-nickel metal and silicate minerals and pieces of gabbro and basalt.

*mid ocean ridge basalt (MORB)* — basalt on the Earth that forms on the sea floor by adiabatic melting of the convecting upper mantle.

*moment of inertia* — a quantity related to the density distribution within a planet, specifically, the tendency for an increase of density with depth.

*noble gases* — the rare gases helium, neon, argon, krypton, xenon, and radon.

*noble metals* — gold (Au), rhenium(Re), and the platinum group elements: platinum (Pt), rhodium (Rh), ruthenium (Ru), iridium (Ir), osmium (Os), and palladium (Pd).

*nuclides* — atoms characterized by the number of protons (Z) and neutrons (N). The mass number (A) = Z + N; isotopes are nuclides with the same number of protons, but differing numbers of neutrons; isobars have the same mass number (A) but different numbers of protons (Z) and neutrons (N).

*oblateness* — the ratio of the difference between equatorial and polar radii to their mean value. Usually an indication of how fast a body is rotating.

*obliquity* — the tilt angle between a planet's axis of rotation and the pole of the orbit (the axis perpendicular to the orbital plane).

*ocean island basalt (OIB)* — basalt on the Earth that is found above hot spots such as Hawai'i or the Galápagos Islands.

*olivine* — a common silicate mineral within Earth's upper mantle and chondritic meteorites, with the general formula of $(Mg,Fe)_2SiO_4$.

*optical depth* — the intensity of light passing through a medium consisting of a field of particles decreases exponentially with distance through the medium. The depth of penetration into the medium (e.g., into a planetary atmosphere or ring) can be expressed in terms of optical depth, or the number of factors of *e* of diminishment.

*orbital resonances* — orbital locations where pairs of orbital frequencies are in ratios of small whole numbers.

*orthopyroxene* — an orthorhombic member of the pyroxene mineral group, with the general formula of $(Mg,Fe)_2Si_2O_6$.

*pallasite* — stony-iron meteorites comprised of roughly equal amounts of olivine and iron-nickel metal. They are thought to be pieces of the core-mantle boundary of a small diffferentiated planetesimal.

*partition coefficient* — the ratio of the concentration of a trace element in one phase to its concentration in a second phase with which it is in equilibrium. Phases can be solid or liquid, metals, or silicates.

*percolation* — the process by which a liquid settles through a solid matrix.

*periclase* — the mineral MgO.

*peridotite* — an igneous rock primarily composed of pyroxene and olivine

*picrite* — high MgO, and usually olivine-bearing, basaltic rocks thought to be derived by melting of peridotite.

*plagioclase* — the calcium and sodium-bearing mineral series within the feldspar group.

*planetesimal* — bodies from millimeter to about a kilometer in size that are believed to have formed during the early planet-forming process.

*plutonic* — a term applied to igneous rocks that have crystallized at depth, usually with coarsely crystalline texture.

*porosity* — the volume percentage of a rock (or other material) that is occupied by voids or fluid.

*ppb* — parts per billion (by weight), also ng/g.

*ppm* — parts per million (by weight), also μg/g.

*precession* — a slow, periodic conical motion of the rotation axis of a spinning body.

*primitive upper mantle* — estimated composition of Earth's mantle after core formation and before continental crust formation.

*P-wave velocity* — seismic-body wave velocity associated with particle motion (alternating compression and expansion) in the direction of wave propagation.

*radiogenic* — a term referring to an isotope having been formed from a radioactive parent.

*rare earth element (REE)* — a collective term for the elements with atomic number 57–71, or the lanthanide series.

*rare gases* — *See* noble gases.

*refractory element* — an element that vaporizes at very high temperatures, such as uranium (U), aluminum (Al), calcium (Ca), and the rare earth elements (REE).

*resonance* — selective response of any periodic system to an external stimulus of the same frequency as a natural frequency of the system.

*Reynolds number* — a dimensionless number (R = Lv/ν, where L is a typical dimension of the system, v is a measure of the velocities that prevail, and ν is the kinematic viscosity) that affects the likelihood of turbulence in the fluids.

*Roche Limit* — the critical separation between two bodies with no tensile strength at which tidal forces are so strong that the smaller body is torn apart; for the Earth and Moon, this distance is about 2.9 Earth radii.

*semimajor axis* — half the length of the major axis of an orbit (its greatest diameter).

*short-lived radionuclide* — *See* extinct nuclide.

*siderophile element* — an element that preferentially enters the metal phase ("sidero" = iron, and "phile" = loving).

*silicate* — a mineral or compound whose crystal structure contains $SiO_4$ tetrahedra.

*solar nebula* — the primitive disk-shaped cloud of dust and gas from which all bodies in the solar system originated.

*solidus* — the line or surface on a phase diagram below which the system is completely solid.

*sound speed* — in a gas, the speed of sound approximately equal to the rms speed of the molecules. By analogy, for a system of orbiting particles, sound speed is sometimes used to denote random velocity of particles.

*SPH* — smoothed particle hydrodynamics.

*spinel* — a group of oxide minerals with similar physical and chemical properties and with the general formula of $AB_2O_4$.

*synchronous orbit* — the distance from a planet of an orbiting satellite at which the satellite's period of revolution is exactly the same as the planet's diurnal rotation.

*S-wave velocity* — seismic-body wave velocity with shearing motion perpendicular to the direction of wave propagation.

*trace element* — an element found in very low (trace) amounts, 100 ppm or less.

*T-Tauri* — an early pre-main-sequence state of stellar evolution, characterized by extensive mass loss from the young star.

*turbulent convection* — mode of convection whereby convective velocity is high, viscosity is low, and entrained crystals tend to remain entrained as opposed to settling out.

*van der Waal forces* — the relatively weak attractive forces operative between neutral atoms and molecules.

*viscosity* — the resistance that a fluid system offers to flow when it is subjected to a shear stress. It is a measure of the internal friction that results when velocity gradients exist within a system.

*volatile element* — an element with a low vaporization temperature such as the alkalis (K, Na) or Pb.

*Young's modulus* — the proportionality constant between stress (force per unit area) and strain (change in length per unit length) for an elastic material. In cgs units it is expressed in dynes $cm^{-2}$ or lbs $ft^{-2}$.

# Index